Project Editor: Angelo A. Terracina

Copy Editor: Andrew Potter

Production Editors: Angelo A. Terracina
S.K. Das

Cover Design: Angelo A. Terracina
Edrees S. Saraj
CJ Media, USA

Editor-in-chief: (Mrs.) Kusum Rafiquzzaman

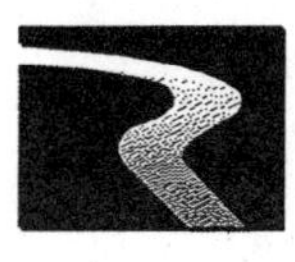

P. O. Box 5439
Diamond Bar, California 91765
USA
Telephone: (909) 593-8124
Fax: (909) 629-1034
E-mail: rafisystems@uia.net
web site: www.rafisystems.com

Library of Congress Catalog Card Number: 99-098130

Printed in the United States of America
10 9 8 7 6 5 4 3 2 1

ISBN 0-9664980-3-8

THIRD EDITION

Fundamentals of Digital Logic and Microcomputer Design

M. RAFIQUZZAMAN, Ph.D.
Professor
California State Polytechnic University
Pomona, California
and
President
Rafi Systems, Inc.

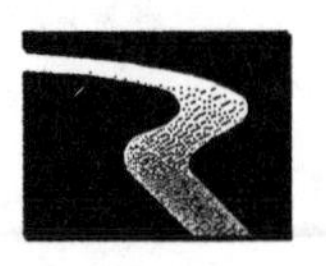

RAFI SYSTEMS, INC.

CALIFORNIA NEW YORK

PREFACE

This book covers all basic concepts of computer engineering and science from digital logic circuits to the design of a complete microcomputer system in a systematic and simplified manner. It is written to present a clear understanding of the principles and basic tools required to design typical digital systems such as microcomputers.

To accomplish this goal, the computer is first defined as consisting of three blocks: central processing unit (CPU), memory, and I/O. It has been pointed out that the CPU is analogous to the brains of a human being. Computer memory is similar to human memory. A question asked to a human being is analogous to entering a program into the computer using an input device such as the keyboard, and answering the question by the human is similar in concept to outputting the result required by the program to a computer output device such as the printer. The main difference is that human beings can think independently whereas computers can only answer questions that they are programmed for. Due to the advances in semiconductor technology, it is possible to fabricate the CPU in a single chip. The result is the microprocessor. Intel's Pentium and Motorola's Power PC are typical examples of microprocessors. Memory and I/O chips must be connected to the microprocessor chip to implement a microcomputer so that these microprocessors will be able to perform meaningful operations.

This book clearly points out that computers only understand 1's and 0's. Hence, it is important for the students to be familiar with binary numbers. Furthermore, the book focuses on the fact that computers can normally only add. Hence, all other operations such as subtraction are performed via addition. This can be accomplished via twos complement arithmetic for binary numbers. Hence, this topic is included, along with a clear explanation of signed and unsigned binary numbers.

As far as the computer programming is concerned, assembly language programming is covered in this book for typical Intel and Motorola microprocessors. An overview of C, C++, and Java high-level languages is also included. These are the only high-level languages that can perform I/O operations. This book points out the advantages and disadvantages of programming typical microprocessors in C and assembly languages.

Three design levels are covered in this book: device level, logic level, and system level. Device-level design, which designs logic gates such as AND, OR, and NOT using transistors, is included from a basic point of view. Logic-level design, on the other hand, is the design technique in which logic gates are used to design a digital component such as an adder. Finally, system-level design is covered for typical Intel and Motorola microprocessors. Microcomputers have been designed by interfacing memory and I/O chips to these microprocessors.

Digital systems at the logic level are classified into two types of circuits, combinational and sequential. Combinational circuits have no memory whereas sequential circuits contain memory. The microprocessors are designed using both combinational and sequential circuits. Therefore, these topics are covered in detail. ***The third edition of this book contains certain new topics including Motorola's AltiVec technology. This edition provides additional examples and end-of-chapter problems, and several appendices containing instrucion sets (alphabetical order) of both Intel 8086 and Motorola 68000 along with the specifications of their support chips. A brief overview of ASIC, DVD, Windows 2000, CPLD, FPGA and Verilog is also included.***

The material included in this book is divided into three sections. The first section contains Chapters 1 through 5. These chapters describe digital circuits at the gate and flip-flop levels and describe the analysis and design of combinational and sequential circuits. The second section contains Chapters 6 through 8, and Chapter 11. These chapters describe microcomputer organization/architecture, programming, design of computer instruction sets, CPU, memory, and I/O. The third section contains Chapters 9 and 10. These chapters contain typical 16-, 32-, and 64-bit microprocessors manufactured by Intel and Motorola. Future plans of Intel and Motorola for the next century are also included. Some sentences are presented in *italics* throughout the book to point out the importance of certain topics. The details of the topics covered in eleven chapters of this book follow.

- Chapter 1 presents an explanation of basic terminologies, fundamental concepts of digital integrated circuits using transistors, a comparison of LSTTL, HC, and HCT IC characteristics, the evolution of computers, the basics of the Internet, and technological forecasts. This chapter points out that HCMOS logic will dominate the future market.
- Chapter 2 provides various number systems and codes suitable for representing information in microprocessors.
- Chapter 3 covers Boolean algebra along with map simplification of Boolean functions. The basic characteristics of digital logic gates are also presented.
- Chapter 4 contains analysis and design of combinational circuits. Typical combinational circuits such as adders, decoders, encoders, multiplexers, and demultiplexers are included. Also, a summary of typical PLD programming languages such as ABEL is included.
- Chapter 5 covers various types of flip-flops. Analysis and design of sequential circuits such as counters are also provided.
- Chapter 6 presents typical microcomputer architecture, internal microprocessor organization, memory, I/O, and programming concepts.

- Chapter 7 includes the fundamentals of instruction set design. Design of registers and ALUs is introduced. Furthermore, control unit design using both hardwired and microprogrammed approaches is included. Nanomemory concepts are also covered.
- Chapter 8 explains the basics of memory and I/O design. Topics such as main memory array design, memory management concepts, and cache memory organization are included.
- Chapters 9 and 10 contain detailed descriptions of the architectures, addressing modes, instruction sets, and I/O and system design concepts of the Intel 8086 and 80386 and the Motorola MC68000 / MC68HC000 and MC68020 microprocessors. Overviews of the Intel 80486 / Pentium / Pentium Pro / Pentium II / Celeron / Pentium III, and the Motorola 68030 / 68040 / 68060 / PowerPC (32- and 64-bit) microprocessors are included. Finally, future plans by both Intel and Motorola are discussed.
- Chapter 11 provides more detailed descriptions of state machine design using ASM chart, array/ROM-based multipliers, microprogramming, cache and virtual memories, and pipeline processing. An overview of VHDL and general classification of computer architectures including SIMD is also covered.

The book can be used in a number of ways. Because the materials presented are basic and do not require any advanced mathematical background, the book can easily be adopted as a text for three quarter or two semester courses. These courses can be taught at the undergraduate level in engineering and computer science. The recommended course sequence can be digital logic design in the first course, with topics that include selected portions from Chapters 1 through 5 and Chapter 11, followed by a second course on computer architecture / organization (Chapters 6 through 8, and selected portions of Chapter 11). The third course may include Chapters 9 and 10, covering Intel and/or Motorola microprocessors.

The audience of this book can also be graduate students or practicing microprocessor system designers in the industry. Portions of Chapters 9 and 10 can be used as an introductory graduate text in electrical engineering or computer science. Practitioners of microprocessor system design in the industry will find simplified explanations along with examples and comparison considerations than are found in manufacturers manuals.

The author wishes to express his sincere appreciation to his students, Edrees S. Saraj, Louie Orbiso, Erik Esola, Husadee Pongprachakkul, Eli Castro, Fong Cheng, and Johnny Chang, and to all others for typing the manuscript. The author is grateful to his wife, Kusum for her commitment and dedication in bringing this book to publication. The author is also grateful to his good friend, US Congressman Duke Cunningham (TOPGUN, Vietnam), and to his colleague, Dr. Rajan Chandra of Cal Poly, Pomona for their inspiration. Finally, the author is especially indebted to his father and his deceased mother who were primarily responsible for the author's accomplishments. ***M. Rafiquzzaman , Pomona, California***

THE AUTHOR

M. Rafiquzzaman obtained his Ph.D. in Electrical Engineering from the University of Windsor, Canada in 1974. He worked for Esso/Exxon and Bell Northern Research for approximately 5 years. Dr. Rafiquzzaman is presently a professor of electrical and computer engineering at California State Polytechnic University, Pomona. He was Chair of the department there from 1984 to 1985. Dr. Rafiquzzaman was also an adjunct professor of electrical engineering systems at University of Southern California, Los Angeles. He consulted for ARCO, Rockwell, Los Angeles County, and Ralph M. Parsons Corporation in the areas of computer applications. He has published eight books on computers, which have been translated into Russian, Chinese, and Spanish. He authored his other books with Prentice-Hall, John Wiley, CRC Press, Harper & Row, and West/PWS.

Dr. Rafiquzzaman is the founder of Rafi Systems, Inc., a manufacturer of biomedical devices including intraocular (cataract implant) lenses using state-of-the-art CNC machines, and also a computer systems consulting firm in California. In 1984, he managed the Olympic Swimming, Diving and Synchronized Swimming teams as chairman of the Contingency commission. He was also involved in developing hardware and software for the Swiss timing, score keeping, and computer systems.

From 1984 to 1989, he was the instructor for Motorola in Southern California teaching short courses on Motorola 16-bit and 32-bit microprocessors for local industries, including Hughes Aircraft, Lockheed, Northrop, TRW, Ford Aerospace, General Dynamics, McDonnell Douglas and Rockwell. Dr. Rafiquzzaman was involved as a consultant in managing microprocessor-based Airport Remote Maintenance system for FAA (Federal Aviation Administration), Washington, D.C. Dr. Rafiquzzaman was advisor (State Minister) to the President of Bangladesh on computers from 1988 to 1990. He is currently involved in research activities in both hardware and software aspects of typical 16-bit, 32-bit, and 64-bit microprocessor-based applications.

The author strongly believes that all electrical engineers should take before graduation at least the following courses and associated laboratories in addition to all other courses in the curriculum: one programming course in a high level language such as C/C++/Java , one course in Digital Logic Design (Combinational and Sequential), two courses in microprocessor-based system design (Intel and Motorola) from chip level along with assembly language programming, one course in digital electronics (must include CMOS/HCMOS), one course in Computer architecture/organization, and one course in programming PLD/CPLD/FPGA using VHDL/Verilog. Some of these courses are usually core courses in the curriculum. In order to keep up with the rapidly changing field of microprocessors, the students should take at least one 16-bit microprocessor-based system design course (Intel or Motorola) early in the curriculum with prerequisites as high level language programming and digital logic design courses. The basics of digital electronics discussed in Chapter 1 of this book can be used to understand the microprocessor hardware. The students can then visit Intel and Motorola web sites to keep themselves current with the microprocessor technology. This will strengthen their digital background, and will allow them to cope with technological advancements in the future. Practicing electrical engineers should also be familiar with the above topics to make them more valuable in the industry.

To: The University of Windsor and the beautiful country of Canada.

The author arrived in Canada from Bangladesh approximately thirty years ago to pursue Graduate studies without any computer background at the University of Windsor. The author is indebted to the University of Windsor and the Government of Canada for providing financial support during his Graduate studies. He obtained his academic background and industrial experience in computers in Canada. The author moved to California in 1978 to enhance his experience in the rapidly changing field of microprocessors.

The author also dedicates this book to his former professor, Dr. W. C. Miller for guidance, encouragement and support in achieving his career goals.

TABLE OF CONTENTS

1

INTRODUCTION TO DIGITAL SYSTEMS

Digital systems are designed to store, process, and communicate information in digital form. They are found in a wide range of applications, including process control, communication systems, digital instruments, and consumer products. The digital computer, more commonly called the "computer," is an example of a typical digital system.

A computer manipulates information in digital, or more precisely, binary form. A binary number has only two discrete values — one or zero. Each of these discrete values is represented by the ON and OFF status of an electronic switch called a "transistor." All computers, therefore, only understand binary numbers. Any decimal number (base 10, with ten digits from 0 to 9) can be represented by a binary number (base 2, with digits 0 and 1).

The basic blocks of a computer are the central processing unit (CPU), the memory, and the input/output (I/O). The CPU of the computer is basically the same as the brains of a human being. Computer memory is conceptually similar to human memory. A question asked to a human being is analogous to entering a program into the computer using an input device such as the keyboard, and answering the question by the human is similar in concept to outputting the result required by the program to a computer output device such as the printer. The main difference is that human beings can think independently, whereas computers can only answer questions that they are programmed for. Computer hardware refers to components of a computer such as memory, CPU, transistors, nuts, bolts, and so on. Programs can perform a specific task such as addition if the computer has an electronic circuit capable of adding two numbers. Programmers cannot change these electronic circuits but can perform tasks on them using instructions.

Computer software, on the other hand, consists of a collection of programs. Programs contain instructions and data for performing a specific task. These programs, written using any programming language such as C++, must be translated into binary prior to execution by the computer. This is because the computer only understands binary numbers. Therefore, a translator for converting such a program into binary is necessary. Hence, a translator program called the *compiler* is used for translating programs written in a programming language such as C++ into binary. These programs in binary form are then stored in the computer memory for execution because computers only understand 1's and 0's. Furthermore, computers can only add. This means that all operations such as subtraction, multiplication, and division are performed by addition.

Due to advances in semiconductor technology, it is possible to fabricate the CPU in a single chip. The result is the *microprocessor*. Both metal-oxide semiconductors (MOS) and bipolar technologies were used in the fabrication process. The CPU can be placed on a single chip when MOS technology is used. However, a number of chips are required with the bipolar technology. MOS technology is used these days to fabricate the microprocessor in a single chip. Along with the microprocessor chip, appropriate memory and I/O chips can be used to design a *microcomputer*. The pins on each one of these chips are then connected to the proper lines on the system bus, which consists of address, data, and control lines. In the past, some manufacturers have designed a complete microcomputer on a single chip with limited capabilities. Both single-chip microprocessors and single-chip microcomputers are being extensively used in a wide range of industrial and home applications.

"Microcontrollers" evolved from single-chip microcomputers. The microcontrollers are typically used for dedicated applications such as automotive systems, home appliances, and home entertainment systems. Typical microcontrollers, therefore, include a microcomputer, timers, and A/D (analog to digital) and D/A (digital to analog) converters — all in a single chip. Examples of typical microcontrollers are Intel 8751 (8-bit) / 8096 (16-bit) and Motorola HC11 (8-bit) / HC16 (16-bit).

In this chapter, we first define some basic terms associated with the computers. We then describe briefly the evolution of the computers and the microprocessors. Finally, a typical practical application, basics of the Internet, and technological forecasts are included.

1.1 Explanation of Terms

Before we go on, it is necessary to understand some basic terms.

- A *bit* is the abbreviation for the term binary digit. A binary digit can have only two values, which are represented by the symbols 0 and 1, whereas a decimal digit can have 10 values, represented by the symbols 0 through 9. The bit values are easily implemented in electronic and magnetic media by two-state devices whose states portray either of the binary digits, 0 or 1. Examples of such two-state devices are a transistor that is conducting or not conducting, a capacitor that is charged or discharged, and a magnetic material that is magnetized North-to-South or South-to-North.

- The *bit size* of a computer refers to the number of bits that can be processed simultaneously by the basic arithmetic circuits of the computer. A number of bits taken as a group in this manner is called a *word.* For example, a 32-bit computer can process a 32-bit word. An 8-bit word is referred to as a byte, and a 4-bit word is known as a nibble.
- An *arithmetic logic unit* (ALU) is a digital circuit which performs arithmetic and logic operations on two *n*-bit digital words. The value of *n* can be 4, 8, 16, 32, or 64. Typical operations performed by the ALU are addition, subtraction, ANDing, ORing, and comparison of two *n*-bit digital words. The size of the ALU defines the size of the computer. For example, a 32-bit computer contains a 32-bit ALU.
- A *microprocessor* is the CPU of a microcomputer contained in a single chip and normally must be augmented with peripheral support devices in order to function. In general, the CPU contains several *registers* (memory elements), the ALU, and the control unit, which translates instructions and performs the desired task. The number of peripheral devices depends upon the particular application involved and even varies within one application. As the microprocessor industry matures, more of these functions are being integrated onto chips in order to reduce the system package count. In general, a *microcomputer* consists of a microprocessor (CPU), input and output means, and a memory in which to store programs (instructions and data). The microcomputer can be implemented on a single chip containing a CPU, program and data memory, and I/O means. *Throughout this book the terms "computer" and "CPU" will be used interchangeably with "Microcomputer" and "Microprocessor" respectively.*
- An *address* is a pattern of 0's and 1's that represents a specific location of memory or a particular I/O device. Typical 8-bit microprocessors have 16 address lines, and, these 16 lines can produce 2^{16} unique 16-bit patterns from 0000000000000000 to 1111111111111111, representing 65,536 different address combinations.
- *Read-only memory* (*ROM*) is a storage medium for the groups of bits called *words*, and its contents cannot normally be altered once programmed. A typical ROM is fabricated on a chip and can store, for example, 2048 eight-bit words, which can be individually accessed by presenting one of 2048 addresses to it. This ROM is referred to as a 2K by 8-bit ROM. 10110111 is an example of an 8-bit word that might be stored in one location in this memory. A ROM is also a nonvolatile storage device, which means that its contents are retained in the event of power failure to the ROM chip. Because of this characteristic, ROMs are used to store programs (instructions and data) that must always be available to the microprocessor.

- *Random access memory* (RAM) is also a storage medium for groups of bits or words whose contents can not only be read but also altered at specific addresses. Furthermore, a RAM normally provides volatile storage, which means that its contents are lost in the event of a power failure. RAMs are fabricated on chips and have typical densities of 4096 bits to one megabit per chip. These bits can be organized in many ways, for example, as 4096-by-1-bit words, or as 2048-by-8-bit words. RAMs are normally used for the storage of temporary data and intermediate results as well as programs that can be reloaded from a back-up nonvolatile source. RAMS are capable of providing large storage capacity in the range of Megabits.
- A *register* can then be considered as volatile storage for a number of bits. These bits may be entered into the register simultaneously (in parallel), or sequentially (serially) from right to left or from left to right, 1 bit at a time. An 8-bit register storing the bits 11110000 is represented as follows:

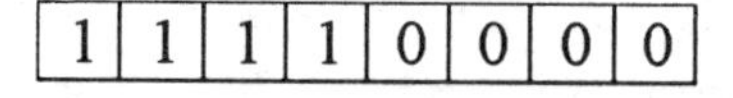

- The term *bus* refers to a number of conductors (wires) organized to provide a means of communication among different elements in a microcomputer system. The conductors in the bus can be grouped in terms of their functions. A microprocessor normally has an address bus, a data bus, and a control bus. The address bits to memory or to an external device are sent out on the address bus. Instructions from memory and data to and from memory or external devices normally travel on the data bus. Control signals for the other buses and among system elements are transmitted on the control bus. Buses are sometimes bidirectional; that is, information can be transmitted in either direction on the bus, but normally only in one direction at a time.
- The *instruction set* of a microprocessor is the list of commands that the microprocessor is designed to execute. Typical instructions are ADD, SUBTRACT, and STORE. Individual instructions are coded as unique bit patterns, which are recognized and executed by the microprocessor. If a microprocessor has 3 bits allocated to the representation of instructions, then the microprocessor will recognize a maximum of 2^3 or eight different instructions. The microprocessor will then have a maximum of eight instructions in its instruction set. It is obvious that some instructions will be more suitable to a particular application than others. For example, if a microprocessor is to be used in a calculating mode, instructions such as ADD, SUBTRACT, MULTIPLY, and DIVIDE would be desirable. In a control application, instructions inputting digitized signals into the processor and outputting digital control variables to external circuits are essential. The number of instructions necessary in an application will directly influence the amount of hardware in the chip set and the number and organization of the interconnecting bus lines.

- A microcomputer requires synchronization among its components, and this is provided by the *clock* or timing circuits. A clock is analogous to the heart beats of a human body.
- The *chip* is an integrated circuit (IC) package containing digital circuits.
- The term *gate* refers to digital circuits which perform logic operations such as AND,OR, and NOT. In an AND operation, the output of the AND gate is one if all inputs are one; the output is zero if one or more inputs are zero. The OR gate, on the other hand, provides a zero output if all inputs are zero; the output is one if one or more inputs are one. Finally, a NOT gate (also called an inverter) has one input and one output. The NOT gate produces one if the input is zero; the output is zero if the input is one.
- *Transistors* are basically electronic switching devices. There are two types of transistors. These are *bipolar junction transistors* (*BJTs*) and *metal-oxide semiconductor* (*MOS*) transistors. The operation of the BJT depends on the flow of two types of carriers: electrons (*n*-channel) and holes (*p*-channel), whereas the MOS transistor is unipolar and its operation depends on the flow of only one type of carrier, either electrons (*n*-channel) or holes (*p*-channel).
- The *speed power product* (*SPP*) is a measure of performance of a logic gate. It is expressed in picojoules (pJ). SPP is obtained by multiplying the speed (in ns) by the power dissipation (in mW) of a gate.

1.2 Design Levels

Three design levels can be defined for digital systems: systems level, logic level, and device level.

- *Systems level* is the type of design in which CPU, memory, and I/O chips are interfaced to build a computer.
- *Logic level*, on the other hand, is the design technique in which chips containing logic gates such as AND, OR, and NOT are used to design a digital component such as the ALU.
- Finally, *device level* utilizes transistors to design logic gates.

1.3 Combinational vs. Sequential Systems

Digital systems at the logic level can be classified into two types. These are *combinational and sequential.*

Combinational systems contain no memory whereas sequential systems require memory to remember the present state in order to go to the next state. A binary adder capable of providing the sum upon application of the numbers to be added is an example of a combinational system. For example, consider a 4-bit adder. The inputs to this adder will be two 4-bit numbers; the output will be the 4-bit sum. In this case, the adder will generate the 4-bit sum output upon application of the two 4-bit inputs.

Sequential systems, on the other hand, require memory. The counter is an example of a sequential system. For instance, suppose that the counter is required to count in the sequence 0, 1, 2 and then repeat the sequence. In this case, the counter must have memory to remember the present count in order to go to the next. The counter must remember that it is at count 0 in order to go to the next count, 1. In order to count to 2, the counter must remember that it is counting 1 at the present state. In order to repeat the sequence, the counter must count back to 0 based on the present count, 2, and the process continues. A chip containing sequential circuits will have a clock input pin.

In general, all computers contain both combinational and sequential circuits. However, most computers are regarded as clocked sequential systems. In these computers, almost all activities pertaining to instruction execution are synchronized with clocks.

1.4 Digital Integrated Circuits

The transistor can be considered as an electronic switch. The on and off states of a transistor are used to represent binary digits. Transistors, therefore, play an important role in the design of digital systems. This section describes the basic characteristics of digital devices and logic families. These include diodes, transistors, and a summary of digital logic families. These topics are covered from a very basic point of view. This will allow the readers with some background in digital devices to see how they are utilized in designing digital systems.

1.4.1 Diodes

A diode is an electronic switch. It is a two-terminal device. Figure 1.1 shows the symbolic representation.

The positive terminal (made with the *p*-type semiconductor material) is called the anode; the negative terminal (made with the *n*-type semiconductor material) is called a cathode. When a voltage, $V = 0.6$ volt is applied across the anode and the cathode, the switch closes and a current I flows from anode to the cathode.

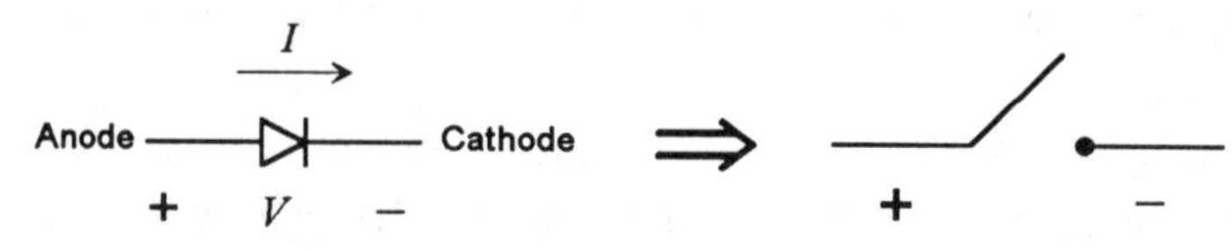

FIGURE 1.1 Symbolic representations of a diode

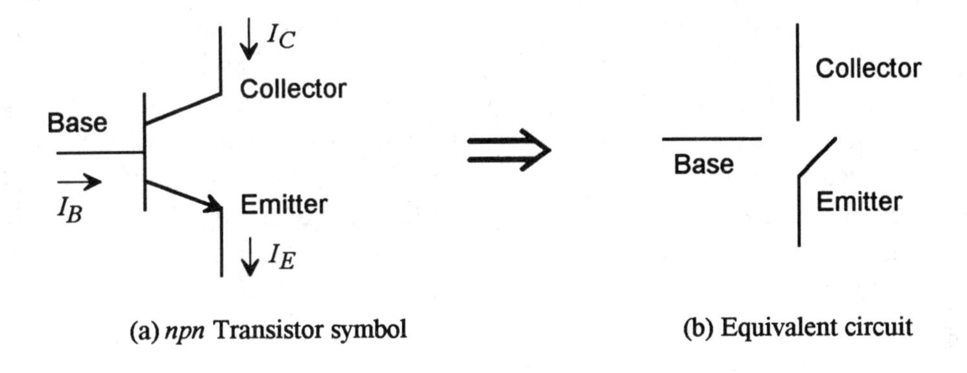

(a) *npn* Transistor symbol (b) Equivalent circuit

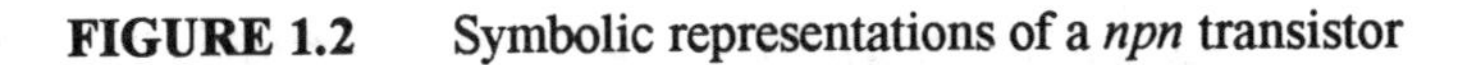

FIGURE 1.2 Symbolic representations of a *npn* transistor

1.4.2 Transistors

A bipolar junction transistor (BJT) or commonly called the *transistor is also an electronic switch* like the diode. Both electrons (*n*-channel) and holes (*p*-channel) are used for carrier flow; hence, the name "bipolar" is used. The BJT is used in transistor logic circuits that have several advantages over diode logic circuits. First of all, the *transistor acts as a logic device called an inverter*. Note that an inverter provides a LOW output for a HIGH input and a HIGH output for a LOW input. Secondly, the *transistor is a current amplifier (buffer)*. Transistors can, therefore, be used to amplify these currents to control external devices such as a light emitting diode (LED) requiring high currents. Finally, transistor logic gates operate faster than diode gates.

There are two types of transistors, namely *npn* and *pnp*. The classification depends on the fabrication process. *npn* transistors are widely used in digital circuits.

Figure 1.2 shows the symbolic representation of an *npn* transistor. The transistor is a three-terminal device. These are base, emitter, and collector. *The transistor is a current-controlled switch. This means that an adequate current at the base will close the switch allowing a current to flow from the collector to the emitter*. This current direction is identified on the *npn* transistor symbol in Figure 1.2(a) by a downward arrow on the emitter. Note that a base resistance is normally required to generate the base current.

The transistor has three modes of operation: cutoff, saturation, and active. In digital circuits, a transistor is used as a switch, which is either ON (closed) or OFF (open). When no base current flows, the emitter~collector switch is open and the transistor operates in the cutoff (OFF) mode. On the other hand, when a base current flows such that the voltage across the base and the emitter is at least 0.6 V, the switch closes. If the base current is further increased, there will be a situation in which V_{CE} (voltage across the collector and the emitter) attains a constant value of approximately 0.2 V. This is called the saturation (ON) mode of the transistor. The "active" mode is between the cutoff and saturation modes. In this mode, the base current (I_B) is amplified so that the collector current, $I_C = \beta I_B$, where β is called the gain, and is typically in the range of 10 to 100 for typical transistors. Note that when the transistor reaches saturation, increasing I_B does not drop V_{CE} below $V_{CE\,(Sat.)}$ of 0.2 V. On the other hand, V_{CE} varies from 0.8 V to 5 V in the active mode. Therefore, the cutoff (OFF) and saturation (ON) modes of the transistor are used in designing digital circuits. The active mode of the transistor in which the transistor acts as a current amplifier (also called buffer) is used in digital output circuits.

1.4.2.1 Operation of the Transistor as an Inverter

Figure 1.3 shows how to use the transistor as an inverter. When $V_{IN} = 0$, the transistor is in cutoff (OFF), and the collector-emitter switch is open. This means that no current flows from $+V_{CC}$ to ground. V_{OUT} is equal to $+V_{CC}$. Thus, V_{OUT} is high.

On the other hand, when V_{IN} is HIGH, the emitter-collector is closed. A current flows from $+V_{CC}$ to ground. The transistor operates in saturation, and $V_{OUT} = V_{CE\,(Sat)} = 0.2\text{ V} \approx 0$. Thus, V_{OUT} is basically connected to ground.

Therefore, for V_{IN} = LOW, V_{OUT} = HIGH, and for V_{IN} = HIGH, V_{OUT} = LOW. Hence, the *npn* transistor in Figure 1.3 acts as an inverter.

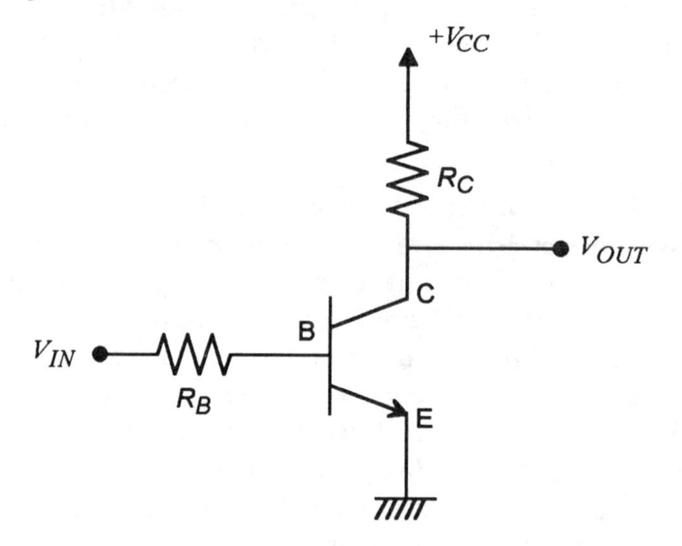

FIGURE 1.3 An inverter

Note that V_{CC} is typically +5 V DC. The input voltage levels are normally in the range of 0 to 0.8 volts for LOW and 2 volts to 5 volts for HIGH. The output voltage levels, on the other hand, are normally 0.2 volts for LOW and 3.6 volts for HIGH.

1.4.2.2 Light Emitting Diodes (LEDs) and Seven Segment Displays

LEDs are extensively used as outputs in digital systems as status indicators. An LED is typically driven by low voltage and low current. This makes the LED a very attractive device for use with digital systems. Table 1.1 provides the current and voltage requirements of red, yellow, and green LEDs.

TABLE 1.1 Current and Voltage Requirements of LEDs

LEDs	Red	Yellow	Green
Current	10 mA	10 mA	20 mA
Voltage	1.7 V	2.2V	2.4V

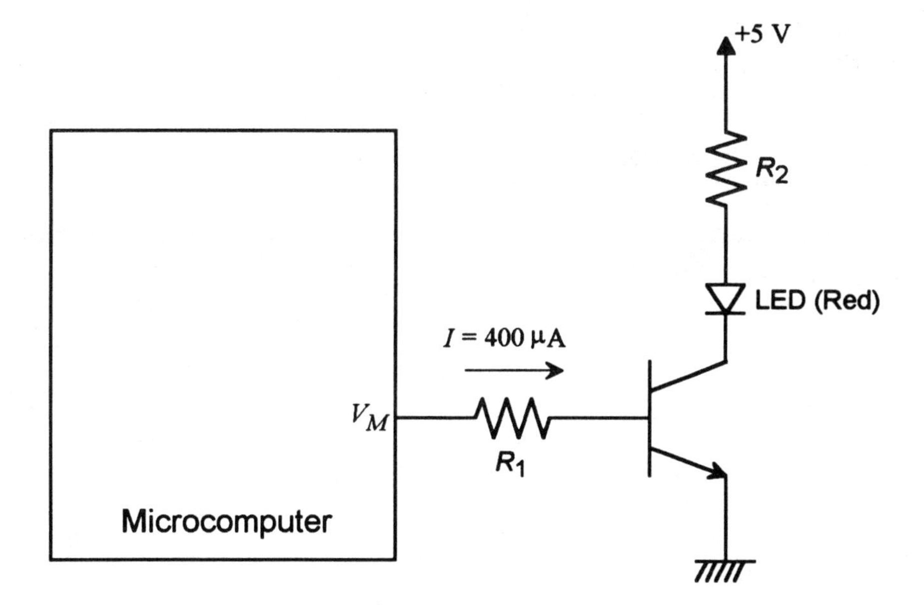

FIGURE 1.4 Microcomputer - LED interface

Basically, an LED will be ON, generating light, when its cathode is sufficiently negative with respect to its anode. A digital system such as a microcomputer can therefore light an LED either by grounding the cathode (if the anode is tied to +5 V) or by applying +5 V to the anode (if the cathode is grounded) through an appropriate resistor value. A typical hardware interface between a microcomputer and an LED is depicted in Figure 1.4.

A microcomputer normally outputs 400 μA at a minimum voltage, $V_M = 2.4$ volts for a HIGH. The red LED requires 10 mA at 1.7 volts. A buffer such as a transistor is required to turn the LED ON. Since the transistor is an inverter, a HIGH input to the transistor will turn the LED ON. We now design the interface; that is, the values of R1, R2, and the gain β for the transistor will be determined.

A HIGH at the microcomputer output will turn the transistor ON into active mode. This will allow a path of current to flow from the +5 V source through R_2 and the LED to the ground. The appropriate value of R_2 needs to be calculated to satisfy the voltage and current requirements of the LED. Also, suppose that $V_{BE} = 0.6$ V when the transistor is in active mode. This means that R_1 needs to be calculated with the specified values of $V_M = 2.4$ V and I = 400 μA. The values of R_1, R_2, and β are calculated as follows:

$$R_1 = \frac{V_M - V_{BE}}{400\ \mu\text{A}} = \frac{2.4 - 0.6}{400\ \mu\text{A}} = 4.5\ \text{K}\Omega$$

Assuming $V_{CE} \cong 0$,

$$R_2 = \frac{5 - 1.7 - V_{CE}}{10\ \text{mA}} = \frac{5 - 1.7}{10\ \text{mA}} = 330\ \Omega$$

$$\beta = \frac{I_C}{I_B} = \frac{10\ \text{mA}}{400\ \mu\text{A}} = \frac{10 \times 10^{-3}}{400 \times 10^{-6}} = 25$$

Therefore, the interface design is complete, and a transistor with a minimum β of 25, $R_1 = 4.5$ KΩ, and $R_2 = 330\ \Omega$ are required.

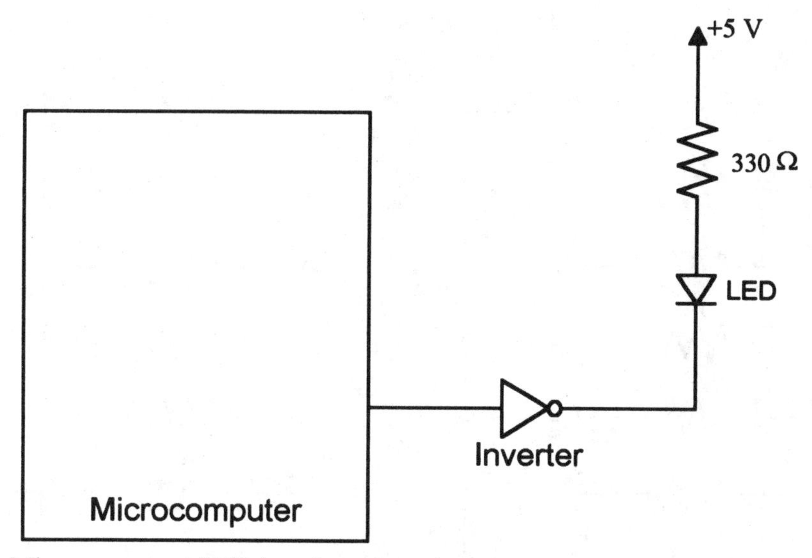

FIGURE 1.5 Microcomputer - LED interface via an inverter

An inverting buffer chip such as 74LS368 can be used in place of a transistor in Figure 1.4. A typical interface of an LED to a microcomputer via an inverter is shown in Figure 1.5. Note that the transistor base resistance is inside the inverter. Therefore, R_1 is not required to be connected to the output of the microcomputer. The symbol —▷o— is used to represent an inverter. Inverters will be discussed in more detail later. In figure 1.5, when the microcomputer outputs a HIGH, the transistor switch inside the inverter closes. A current flows from the +5 V source, through the 330-ohm resistor and the LED, into the ground inside the inverter. The LED is thus turned ON.

A seven-segment display can be used to display, for example, decimal numbers from 0 to 9. The name "seven segment" is based on the fact that there are seven LEDs — one in each segment of the display. Figure 1.6 shows a typical seven-segment display.

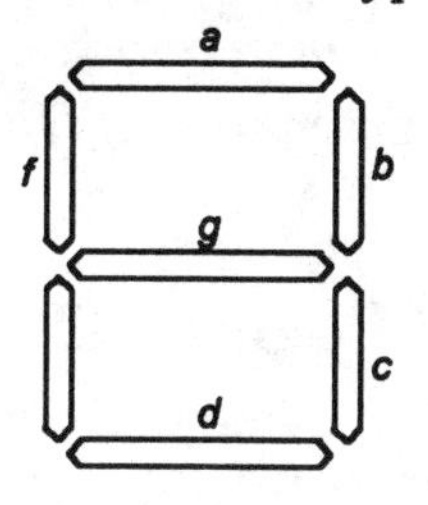

FIGURE 1.6 A seven-segment display

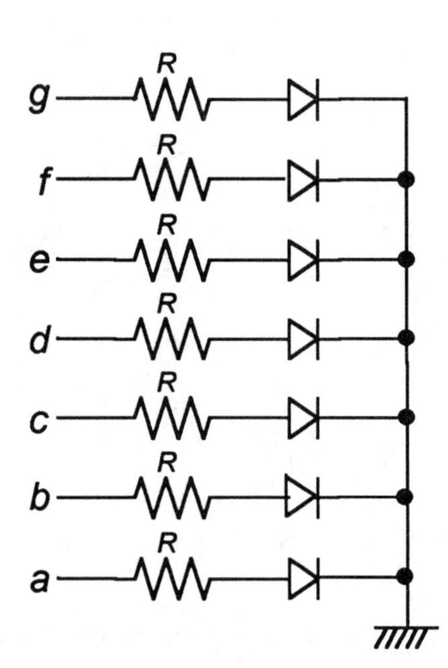

Common Cathode

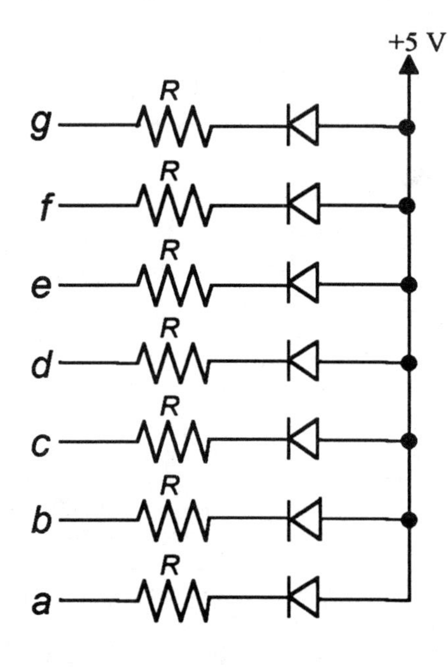

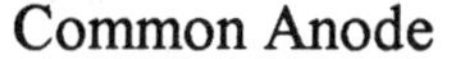

Common Anode

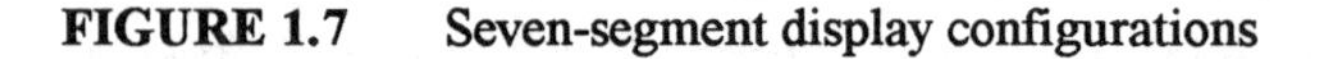

FIGURE 1.7 Seven-segment display configurations

In Figure 1.6, each segment contains an LED. All decimal numbers from 0 to 9 can be displayed by turning the appropriate segment "ON" or "OFF". For example, a zero can be displayed by turning the LED in segment g "OFF" and turning the other six LEDs in segments a through f "ON." *There are two types of seven segment displays. These are common cathode and common anode. Figure 1.7 shows these display configurations.*

In a common cathode arrangement, the microcomputer can send a HIGH to light a segment and a LOW to turn it off. In a common anode configuration, on the other hand, the microcomputer sends a LOW to light a segment and a HIGH to turn it off. In both configurations, $R = 330$ ohms can be used.

1.4.2.3 Transistor Transistor Logic (TTL) and its Variations

The transistor transistor logic (TTL) family of chips evolved from diodes and transistors. This family used to be called DTL (diode transistor logic). The diodes were then replaced by transistors, and thus the name "TTL" evolved. The power supply voltage (V_{CC}) for TTL is +5 V. The two logic levels are approximately 0 and 3.5 V.

There are several variations of the TTL family. These are based on the saturation mode (saturated logic) and active mode (nonsaturated logic) operations of the transistor. In the saturation mode, the transistor takes some time to come out of the saturation to switch to the cutoff mode. On the other hand, some TTL families define the logic levels in the active mode operation of the transistor and are called nonsaturated logic. Since the transistors do not go into saturation, these families do not have any saturation delay time for the switching operation. Therefore, the nonsaturated logic family is faster than saturated logic.

The saturated TTL family includes standard TTL (TTL), high-speed TTL (H-TTL), and low-power TTL (L-TTL). The nonsaturated TTL family includes Schottky TTL (S-TTL), low-power Schottky TTL (LS-TTL), advanced Schottky TTL (AS-TTL), and advanced low-power Schottky TTL (ALS-TTL). The development of LS-TTL made TTL, H-TTL, and L-TTL obsolete. Another technology, called emitter-coupled logic (ECL), utilizes nonsaturated logic. The ECL family provides the highest speed. ECL is used in digital systems requiring ultrahigh speed, such as supercomputers.

The important parameters of the digital logic families are fan-out, power dissipation, propagation delay, and noise margin.

Fan-out is defined as the maximum number of inputs that can be connected to the output of a gate. It is expressed as a number. The output of a gate is normally connected to the inputs of other similar gates. Typical fan-out for TTL is 10. On the other hand, fan-outs for S-TTL, LS-TTL, and ECL, are 10, 20, and 25, respectively.

Power dissipation is the power (milliwatts) required to operate the gate. This power must be supplied by the power supply and is consumed by the gate. Typical power consumed by TTL is 10 mW. On the other hand, S-TTL, LS-TTL, and ECL absorb 22 mW, 2 mW, and 25 mW respectively.

Propagation delay is the time required for a signal from input to output when the binary output changes its value. Typical propagation delay for TTL is 10 nanoseconds (ns). On the other hand, S-TTL, LS-TTL, and ECL have propagation delays of 3 ns, 10 ns, and 2 ns, respectively.

Noise margin is defined as the maximum voltage due to noise that can be added to the input of a digital circuit without causing any undesirable change in the circuit output. Typical noise margin for TTL is 0.4 V. Noise margins for S-TTL, LS-TTL, and ECL are 0.4 V, 0.4 V, and 0.2 V , respectively.

1.4.2.4 TTL Outputs

There are three types of output configurations for TTL. These are open-collector output, totem-pole output, and tristate (three-state) output.

The open-collector output means that the TTL output is a transistor with nothing connected to the collector. The collector voltage provides the output of the gate. For the open-collector output to work properly, a resistor (called the pullup resistor), with a value of typically 1 Kohm, should be connected between the open collector output and a +5 V power supply.

If the outputs of several open-collector gates are tied together with an external resistor (typically 1 Kohm) to a +5 V source, a logical AND function is performed at the connecting point. This is called wired-AND logic.

Figure 1.8 shows two open-collector outputs (*A* and *B*) are connected together to a common output point *C* via a 1 KΩ resistor and a +5 V source.

The common-output point *C* is HIGH only when both transistors are in cutoff (OFF) mode, providing *A* = HIGH and *B* = HIGH. If one or both of the two transistors is turned ON, making one (or both open-collector outputs) LOW, this will drive the common output *C* to LOW. Note that a LOW (Ground for example) signal when connected to a HIGH (+5V for example) signal generates a LOW. Thus, *C* is obtained by performing a logical AND operation of the open collector outputs *A* and *B*.

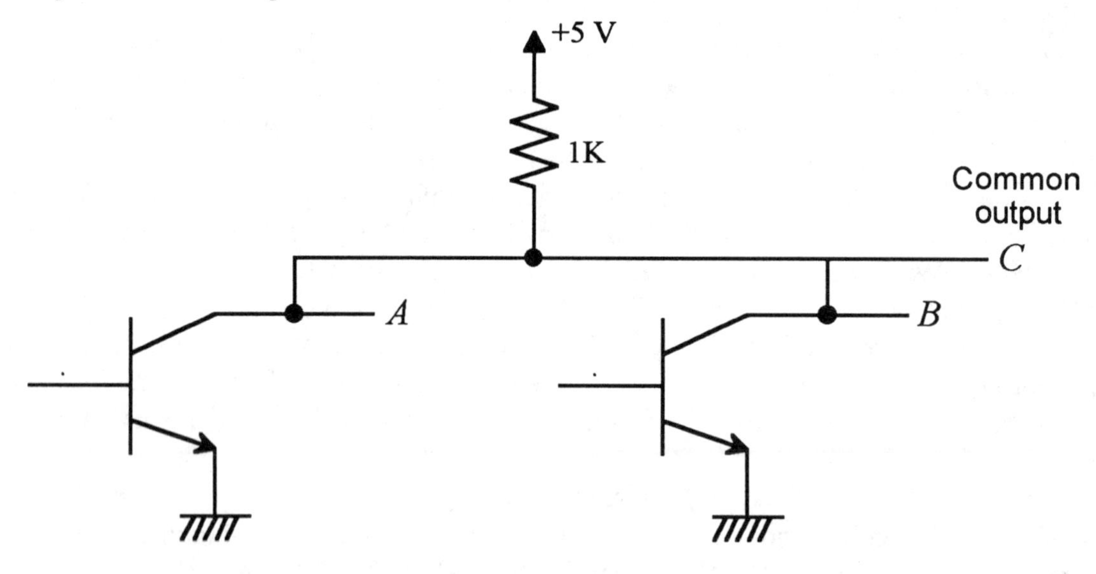

FIGURE 1.8 Two open-collector outputs A and B tied together

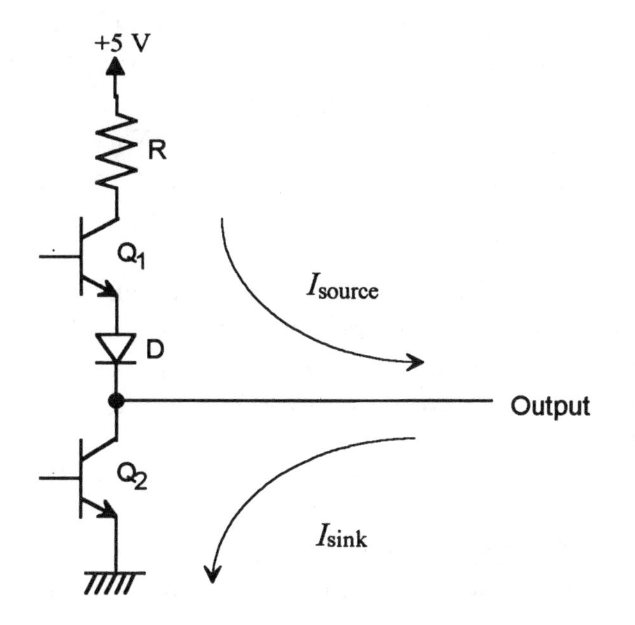

FIGURE 1.9 TTL Totem-pole output

Let us briefly review the totem-pole output circuit shown in Figure 1.9.

The circuit operates as follows:

When transistor Q_1 is ON, transistor Q_2 is OFF. When Q_1 is OFF, Q_2 is ON. This is how the totem-pole output is designed. The complete TTL gate connected to the bases of transistors Q_1 and Q_2 is not shown; only the output circuit is shown.

In the figure, Q_1 is turned ON when the logic gate circuit connected to its base sends a HIGH output. The switches in transistor Q_1 and diode D close while the switch in Q_2 is open. A current flows from the +5 V source through R, Q_1, and D to the output. This current is called I_{source} or output high current, I_{OH}. This is typically represented by a negative sign in front of the current value in the TTL data book, a notation indicating that the chip is losing current. For a low output value of the logic gate, the switches in Q_1 and D are open and the switch in Q_2 closes. A current flows from the output through Q_2 to ground. This current is called I_{sink} or Output Low current, I_{OL}. This is represented by a positive sign in front of the current value in the TTL data book, indicating that current is being added to the chip. Either I_{source} or I_{sink} can be used to drive a typical output device such as an LED. I_{source} (I_{OH}) is normally much smaller than I_{sink} (I_{OL}). I_{source} (I_{OH}) is typically –0.4 mA (or –400 μA) at a minimum voltage of 2.7 V at the output. I_{source} is normally used to drive devices that require high currents. A current amplifier (buffer) such as a transistor or an inverting buffer chip such as 74LS368 needs to be connected at the output if I_{source} is used to drive a device such as an LED requiring high current (10 mA to 20 mA). I_{sink} is normally 8 mA .

The totem-pole outputs must not be tied together. When two totem-pole outputs are connected together with the output of one gate HIGH and the output of the second gate LOW, the excessive amount of current drawn can produce enough heat to damage the transistors in the circuit.

Tristate is a special totem-pole output that allows connecting the outputs together like the open-collector outputs. When a totem-pole output TTL gate has this property, it is called a tristate (three state) output. A tristate has three output states:

1. A LOW level state when the lower transistor in the totem-pole is ON and the upper transistor is OFF.
2. A HIGH level when the upper transistor in the totem-pole is ON and the lower transistor is OFF.
3. A third state when both output transistors in the totem-pole are OFF. This third state provides an open circuit or high-impedance state which allows a direct wire connection of many outputs to a common line called the bus.

1.4.2.5 A Typical Switch Input Circuit for TTL

Figure 1.10 shows a switch circuit that can be used as a single bit into the input of a TTL gate. When the DIP switch is open, V_{IN} is HIGH. On the other hand, when the switch is closed, V_{IN} is low. V_{IN} can be used as an input bit to a TTL logic gate for performing laboratory experiments.

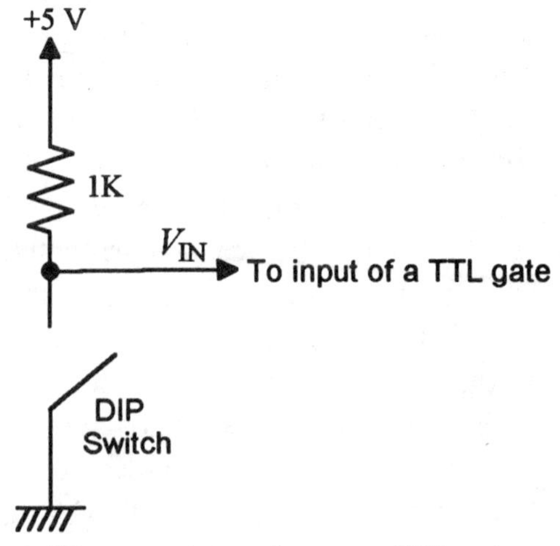

FIGURE 1.10 A typical circuit for connecting an input to a TTL gate

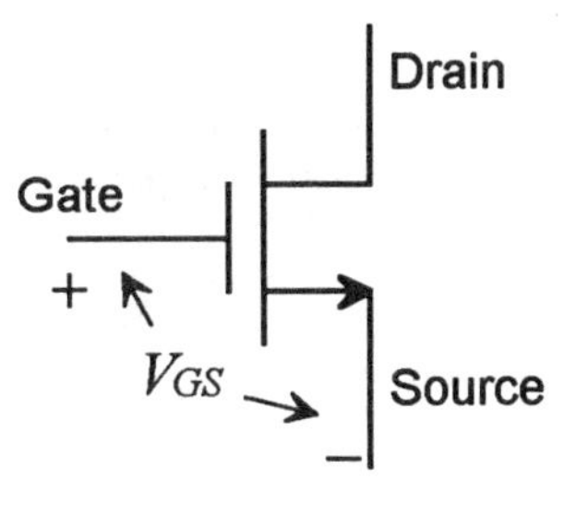

FIGURE 1.11 nMOS transistor symbol

1.4.3 MOS Transistors

Metal-Oxide Semiconductor (MOS) transistors occupy less space in the circuit and consume much less power than bipolar junction transistors. Therefore, MOS transistors are used in highly integrated circuits. The MOS transistor is unipolar. This means that one type of carrier flow, either electrons (*n*-type) or holes (*p*-type) is used. *The MOS transistor works as a voltage-controlled resistance. In digital circuits, a MOS transistor operates as a switch such that its resistance is either very high (OFF) or very low (ON).* The MOS transistor is a three-terminal device: gate, source, and drain. There are two types of MOS transistors, namely, nMOS and pMOS. The power supply (V_{CC}) for pMOS is in the range of 17 V to 24 V, while V_{CC} for nMOS is lower than pMOS and can be from 5 V to 12 V. Figure 1.11 shows the symbolic representation of an nMOS transistor.

When $V_{GS} = 0$*, the resistance from drain to source (*R_{DS}*) is in the order of megaohms (Transistor OFF state). On the other hand, as* V_{GS} *is increased,* R_{DS} *decreases to a few tens of ohms (Transistor ON state).* Note that in a MOS transistor, there is no connection between the gate and the other two terminals (source and drain). The nMOS gate voltage (V_{GS}) increases or decreases the current flow from drain to source by changing R_{DS}. Popular 8-bit microprocessors such as the Intel 8085 and the Motorola 6809 were designed using nMOS.

Figure 1.12 depicts the symbol for a pMOS transistor. The operation of the pMOS transistor is very similar to the nMOS transistor except that V_{GS} is typically zero or negative. The resistance from drain to source (R_{DS}) becomes very high (OFF) for $V_{GS} = 0$. On the other hand, R_{DS} decreases to a very low value (ON) if V_{GS} is decreased. pMOS was used in fabricating the first 4-bit microprocessors (Intel 4004/4040) and 8-bit microprocessor (Intel 8008). Basically, in a MOS transistor (nMOS or pMOS), V_{GS} creates an electric field that increases or decreases the current flow between source and drain. From the symbols of the MOS transistors, it can be seen that there is no connection between the gate and the other two terminals (source and drain). This symbolic representation is used in order to indicate that no current flows from the gate to the source, irrespective of the gate voltage.

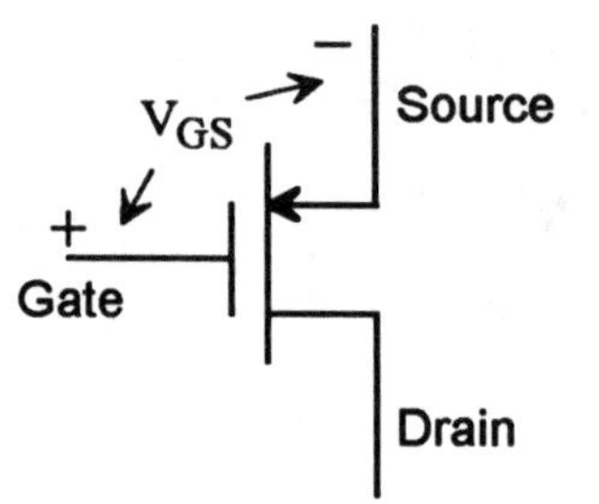

FIGURE 1.12 pMOS transistor symbol

1.4.3.1 Operation of the nMOS Transistor as an Inverter

Figure 1.13 shows an nMOS inverter. When V_{IN} = LOW, the resistance between the drain and the source (R_{DS}) is very high, and no current flows from V_{CC} to the ground. V_{OUT} is therefore high. On the other hand, when V_{IN} = high, R_{DS} is very low, a current flows from V_{CC} to the source, and V_{OUT} is LOW. Therefore, the circuit acts as an inverter.

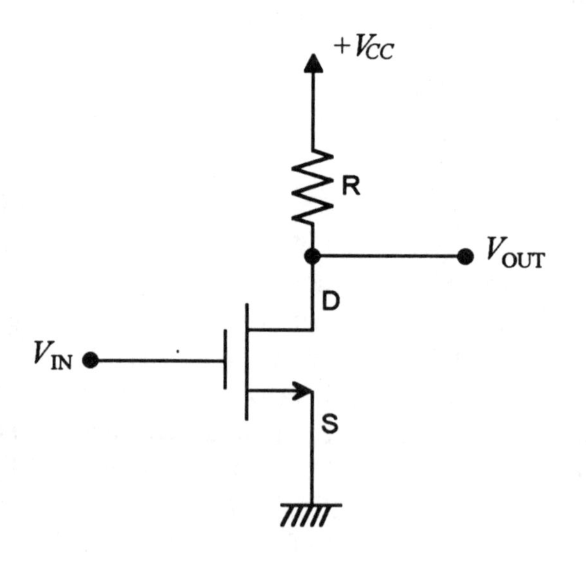

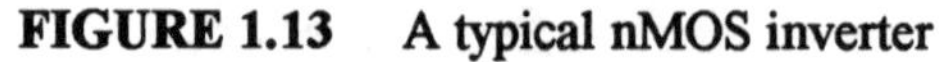

FIGURE 1.13 A typical nMOS inverter

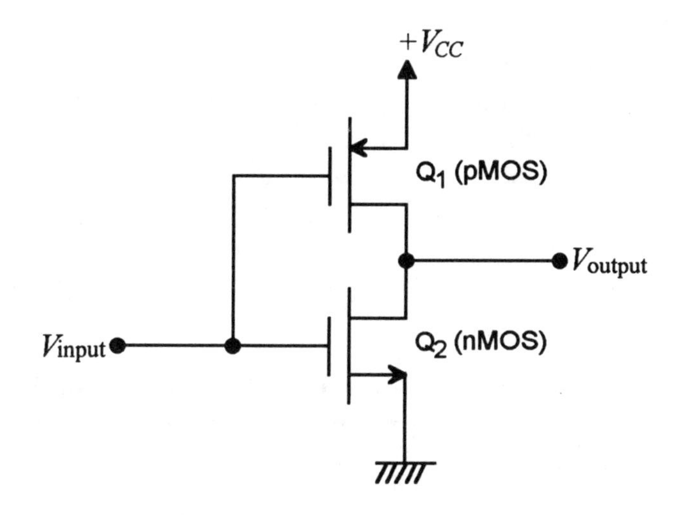

FIGURE 1.14 A CMOS inverter

1.4.3.2 Complementary MOS (CMOS)

CMOS dissipates low power and offers short propagation delays compared to TTL. CMOS is fabricated by combining nMOS and pMOS together. Figure 1.14 shows a typical CMOS inverter.

The CMOS inverter is very similar to the TTL totem-pole output circuit. That is, when Q_1 is ON (low resistance), Q_2 is OFF (high resistance), and vice versa. When V_{input} = LOW, Q_1 is ON and Q_2 is OFF. This makes V_{output} HIGH. On the other hand, when V_{input} = HIGH, Q_1 is OFF (high resistance) and Q_2 is ON (low resistance). This provides a low V_{output}. Thus, the circuit works as an inverter.

Digital circuits using CMOS consume less power than MOS and bipolar transistor circuits. In addition, *CMOS provides high circuit density. That is, more circuits can be placed in a chip using CMOS. Finally, CMOS offers high noise immunity. In CMOS, unused inputs should not be left open. Because of the very high input resistance, a floating input may change back and forth between a LOW and a HIGH, creating system problems. All unused CMOS inputs should be tied to V_{CC}, ground, or another high or low signal source appropriate to the device's function.* CMOS can operate over a large range of power supply voltages (3 V to 15 V). Two CMOS families, namely CD4000 and 54C/74C, were first introduced. CD 4000A is in the declining stage.

There are four members in the CMOS family which are very popular these days: the high-speed CMOS (HC), high-speed CMOS/TTL-input compatible (HCT), advanced CMOS (AC), and advanced CMOS/TTL-input compatible (ACT). The HCT chips have a specifically designed input circuit that is compatible with LS-TTL logic levels (2V for HIGH input and 0.8V for LOW input). LS-TTL outputs can directly drive HCT inputs while HCT outputs can directly drive HC inputs. Therefore, HCT buffers can be placed between LS-TTL and HC chips to make the LS-TTL outputs compatible with the HC inputs.

Several characteristics of 74HC and 74HCT are compared with 74LS-TTL and nMOS technologies in Table 1.2.

TABLE 1.2 Comparison of output characteristics of LS-TTL, nMOS, HC, and HCT

	V_{OH}	I_{OH}	V_{OL}	I_{OL}
LS-TTL	2.7 V	–400 μA	0.5 V	8 mA
nMOS	2.4 V	–400 μA	0.4 V	2 mA
HC	3.7 V	–4 mA	0.4 V	4 mA
HCT	3.7 V	–4 mA	0.4 V	4 mA

Note that in the table, HC and HCT have the same source (I_{OH}) and sink (I_{OL}) currents. This is because in a typical CMOS gate, the ON resistances of the pMOS and nMOS transistors are approximately the same.

The input characteristics of HC and HCT are compared in Table 1.3.

TABLE 1.3 Comparison of input characteristics of HC and HCT

	V_{IH}	I_{IH}	V_{IL}	I_{IL}	Fanout
HC	3.15 V	1μA	0.9 V	1μA	10
HCT	2.0 V	1μA	0.8 V	1μA	10

The above table shows that *LS-TTL is not guaranteed to drive an HC input.* The LS-TTL output HIGH is grater than or equal to 2.7V while an HC input needs at least 3.15V. Therefore, the HCT input requiring V_{IH} of 2.0V can be driven by the LS-TTL output, providing at least 2.7V; 74HCT244 (unidirectional) and 74HCT245 (bidirectional) buffers can be used.

1.4.3.3 MOS Outputs

Like TTL, the MOS logic offers three types of outputs. These are push-pull (totem-pole in TTL), open drain (open collector in TTL), and tristate outputs.

For example, the 74HC00 contains four independent 2-input NAND gates and includes push-pull output. The 74HC03 also contains four independent 2-input NAND gates, but has open drain outputs. The 74HC03 requires a pull-up resistor for each gate. The 74HC125 contains four independent tri-state buffers in a single chip.

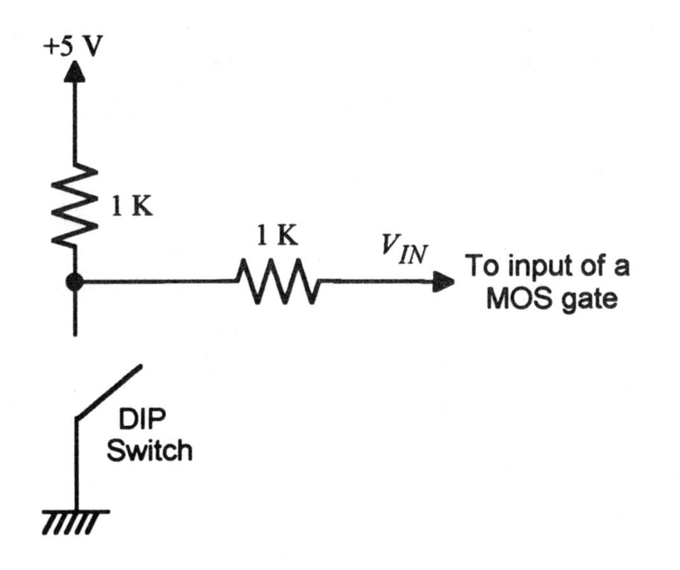

FIGURE 1.15 A typical switch for MOS input

1.4.3.4 A Typical Switch Input Circuit for MOS Chips

Figure 1.15 shows a switch circuit that can be used as a single bit into the input of a MOS gate. When the DIP switch is open, V_{IN} is HIGH. On the other hand, when the switch is closed, V_{IN} is LOW. V_{IN} can be used as an input bit for performing laboratory experiments. Note that unlike TTL, a 1K resistor is connected between the switch and the input of the MOS gate. This provides for protection against static discharge. This 1-Kohm resistor is not required if the MOS chip contains internal circuitry providing protection against damage to inputs due to static discharge.

1.5 Integrated Circuits (ICs)

Device level design utilizes transistors to design circuits called *gates*, such as AND gates and OR gates. One or more gates are fabricated on a single silicon chip by an integrated circuit (IC) manufacturer in an IC package.

An IC chip is packaged typically in a ceramic or plastic package. The commercially available ICs can be classified as small-scale integration (SSI), medium-scale integration (MSI), large-scale integration (LSI), and very large-scale integration (VLSI).

- *A single SSI IC contains a maximum of approximately 10 gates. Typical logic functions such as AND, OR, and NOT are implemented in SSI IC chips. The MSI IC, on the other hand, includes from 11 to up to 100 gates in a single chip. The MSI chips normally perform specific functions such as add.*
- *The LSI IC contains more than 100 to approximately 1000 gates. Digital systems such as 8-bit microprocessors and memory chips are typical examples of LSI ICs.*
- *The VLSI IC includes more than 1000 gates. More commonly, the VLSI ICs are identified by the number of transistors rather than the gate count in a single chip. Typical examples of VLSI IC chips include 32-bit microprocessors and one megabit memories.* For example, *the Intel Pentium is a VLSI IC containing 3.1 million transistors in a single chip.*

An IC chip is usually inserted in a printed-circuit board (PCB) that is connected to other IC chips on the board via pins or electrical terminals. In laboratory experiments or prototype systems, the IC chips are typically placed on breadboards or wire-wrap boards and connected by wires. *The breadboards normally have noise problems for frequencies over 4 MHz. Wire-wrap boards are used above 4 MHz.* The number of pins in an IC chip varies from ten to several hundred, depending on the package type. *Each IC chip must be powered and grounded via its power and ground pins. The VLSI chips such as the Pentium have several power and ground pins. This is done in order to reduce noise by distributing power in the circuitry inside the chip.*

The SSI and MSI chips normally use an IC package called *dual in-line package* (DIP). The LSI and VLSI chips, on the other hand, are typically fabricated in surface-mount or pin grid array (PGA) packages. The DIP is widely used because of its low price and ease of installation into the circuit board.

SSI chips are identified as 5400-series (these are for military applications with stringent requirements on voltage and temperature and are expensive) or 7400 series (for commercial applications). Both series have identical pin assignments on chips with the same part numbers, although the first two numeric digits of the part name are different. Typical commercial SSI ICs can be identified as follows:

74S	Schottky TTL
74LS	Low-power Schottky TTL
74AS	Advanced Schottky TTL
74F	Fast TTL (Similar to 74AS; manufactured by Fairchild)
74ALS	Advanced low-power Schottky TTL

Note that two digits appended at the end of each of these IC identifications define the type of logic operation performed, the number of pins, and the total number of gates on the chip. For example, 74S00, 74LS00, 74AS00, 74F00, and 74ALS00 perform NAND operation. All of them have 14 pins and contain four independent NAND gates in a single chip.

The gates in the ECL family are identified by the part numbers 10XXX and 100XXX, where XXX indicates three digits. The 100XXX family is faster, requires low power supply, but it consumes more power than the 10XXX. Note that 10XXX and 100XXX are also known as 10K and 100K families.

The commercially available *CMOS family is identified in the same manner as the TTL SSI ICs. For example, 74LS00 and 74HC00 (High-speed CMOS) are identical, with 14 pins and containing four independent NAND gates in a single chip.* Note that 74HCXX gates have operating speeds similar to 74LS-TTL gates. For example, the 74HC00 contains four independent two-input NAND gates. Each NAND gate has a typical propagation delay of 10 ns and a fanout of 10 LS-TTL.

Unlike TTL inputs, CMOS inputs should never be held floating. The unused input pins must be connected to V_{CC}, *ground, or an output. The TTL input contains an internal resistor that makes it HIGH when unused or floating.* The CMOS input does not have any such resistor and therefore possesses high resistance. The unused CMOS inputs must be tied to V_{CC}, ground, or other gate outputs. In some CMOS chips, inputs have internal pull-up or pull-down resistors. These inputs, when unused, should be connected to V_{CC} or ground to make the inputs high or low.

The CMOS family has become popular compared to TTL due to better performance. Some major IC manufacturers such as National Semiconductor do not make 7400 series TTL anymore. Although some others, including Fairchild and Texas Instruments still offer the 7400 TTL series, the use of the SSI TTL family (74S, 74LS, 74AS, 74F, and 74ALS) is in the declining stage, and will be obsolete in the future. On the other hand, the use of CMOS-based chips such as 74HC and 74HCT has increased significantly because of their high performance. These chips will dominate the future market.

1.6 Evolution of Computers

The first electronic computer, called ENIAC, was invented in 1946 in the Moore School of Engineering, University of Pennsylvania. ENIAC was designed using vacuum tubes and relays. This computer performed addition, subtraction, and other operations via special wiring rather than programming. The concept of executing operations by the computer via storing programs in memory became feasible later.

John Von Neumann, a student at the Moore School, designed the first conceptual architecture of a stored program computer, called the EDVAC. Soon afterward, M. V. Wilkes of Cambridge University implemented the first operational stored memory computer called the EDSAC. The Von Neumann architecture was the first computer that allowed storing of instructions and data in the same memory. This resulted in the introduction of other computers such as ILLIAC at the University of Illinois and JOHNIAC at the RAND Corporation.

The computers discussed so far were used for scientific computations. With the invention of transistors in the 1950s, the computer industry grew more rapidly. The entry of IBM (International Business Machines) into the computer industry happened in 1953 with the development of a desk calculator called the IBM 701. In 1954, IBM announced its first magnetic drum-based computer called the IBM 650. This computer allowed the use of system-oriented programs such as compilers feasible. Note that compilers are programs capable of translating high-level language programs into binary numbers that all computers understand.

With the advent of integrated circuits, IBM introduced the 360 in 1965 and the 370 in 1970. Other computer manufacturers such as Digital Equipment Corporation (DEC), RCA, NCR, and Honeywell followed IBM. For example, DEC introduced its popular real-time computer PDP 11 in the late 1960s. Note that real-time computers are loosely defined as the computers that provide fast responses to process requests. Typical real-time applications include process control such as temperature control and aircraft simulation.

Intel Corporation is generally acknowledged as the company that introduced the microprocessor successfully into the marketplace. Its first processor, the 4004, was introduced in 1971 and evolved from a development effort while making a calculator chip set. The 4004 microprocessor was the central component in the chip set, which was called the MCS-4. The other components in the set were a 4001 ROM, a 4002 RAM, and a 4003 Shift Register.

Shortly after the 4004 appeared in the commercial marketplace, three other general-purpose microprocessors were introduced. These devices were the Rockwell International 4-bit PPS-4, the Intel 8-bit 8008, and the National Semiconductor 16-bit IMP-16. Other companies such as General Electric, RCA, and Viatron had also made contributions to the development of the microprocessor prior to 1971.

The microprocessors introduced between 1971 and 1972 were the first-generation systems designed using PMOS technology. In 1973, second-generation microprocessors such as the Motorola 6800 and the Intel 8080 (8-bit microprocessors) were introduced. The second-generation microprocessors were designed using the NMOS technology. This technology resulted in a significant increase in instruction execution speed and higher chip densities compared to PMOS. Since then, microprocessors have been fabricated using a variety of technologies and designs. NMOS microprocessors such as the Intel 8085, the Zilog Z80, and the Motorola 6800/6809 were introduced based on the second-generation microprocessors. The third generation HMOS microprocessors, introduced in 1978, is typically represented by the Intel 8086 and the Motorola 68000, which are 16-bit microprocessors.

In 1980, fourth-generation HCMOS and BICMOS (combination of BIPOLAR and HCMOS) 32-bit microprocessors evolved. Intel introduced the first commercial 32-bit microprocessor, the problematic Intel 432. This processor was eventually discontinued by Intel. Since 1985, more 32-bit microprocessors have been introduced. These include Motorola's MC 68020/68030/68040/PowerPC, Intel's 80386/80486 and the Intel Pentium microprocessors.

The performance offered by the 32-bit microprocessor is more comparable to that of superminicomputers such as Digital Equipment Corporation's VAX11/750 and VAX11/780. Intel and Motorola introduced RISC (Reduced Instruction Set Computer) microprocessors, namely the Intel 80960 and Motorola MC88100/PowerPC, with simplified instruction sets. Note that the purpose of RISC microprocessors is to maximize speed by reducing clock cycles per instruction. Almost all computations can be obtained from a simple instruction set. Some manufacturers are speeding up the processors for data crunching types of applications. Compaq / Digital Equipment Corporation Alpha family includes 64-bit RISC microprocessors. These processors run at speeds in excess of 300 MHz .

The 32-bit Pentium II microprocessor is Intel's addition to the Pentium line of microprocessors, which originated from the 80X86 line. *The Pentium II can run at speeds of 333 MHz, 300 MHz, 266 MHz, and 233 MHz. Intel implemented its MMX (Matrix Math eXtensions) technology to enhance multimedia and communications operations.* To achieve this, Intel added 57 new instructions to manipulate video, audio, and graphical data more efficiently. Pentium III is also added to the Pentium family. Chapter 9 provides an overview of these processors. *Intel plans to release a new 64-bit processor called "Merced" by the year 2000.* The new processor will be a joint effort by Intel and Hewlett-Packard. The chip is expected to replace the 80X86 series in the high-performance workstation and server markets.

Motorola's PowerPC microprocessor is a product of an alliance with IBM and Apple Computer. PowerPC is a RISC microprocessor, and includes both 32-bit and 64-bit microprocessors. The newest versions of the PowerPC include: PowerPC 603e (300 MHz maximum), PowerPC 750/740 (266 MHz maximum), and PowerPC 604e (350 MHz maximum). The PowerPC 604e is intended for high-end Macintosh and Mac-compatible systems. Motorola's AltiVec technology (discussed in Chapter 10) is implemented in Apple's G4 computer. *Due to growing popularity of HCMOS technology, major manufacturers of microprocessors such as Intel and Motorola have been discontinuing HMOS-based microprocessor chips. For example, Motorola already replaced the HMOS 68000 by the HCMOS 68HC000 in 1995. Intel 8086, designed by using HMOS technology, is being replaced by*

32-bit HCMOS 80XXX microprocessors. However, it is expected that both Intel 8086 and Motorola 68000 will be available from other sources in the future.

An overview of the latest microprocessors is provided in this section. Unfortunately, this may be old news within a year. One can see, however, that both Intel and Motorola offer (and will continue to offer) quality microprocessors to satisfy demanding applications.

1.7 A Typical Microcomputer-Based Application

In order to put the microprocessor into perspective, it is important to explore a typical application. For example, consider a microprocessor-based dedicated controller in Figure 1.16. Suppose that it is necessary to maintain the temperature of the furnace to a desired level to maintain the quality of a product. Assume that the designer has decided to control this temperature by adjusting the fuel. This can be accomplished using a microcomputer along with the interfacing components as follows.

Temperature is an analog (continuous) signal. It can be measured by a temperature sensing (measuring) device such as a thermocouple. The thermocouple provides the measurement in millivolts (mV) equivalent to the temperature. Since microcomputers only understand binary numbers (1's and 0's), each analog mV signal must be converted to a binary number using an analog to digital (A/D) converter chip.

First, the millivolt signal is amplified by a mV/V amplifier to make the signal compatible for A/D conversion. A microcomputer can be programmed to solve an equation with the furnace temperature as an input. This equation compares the temperature measured with the desired temperature which can be entered into the microcomputer via the keyboard.

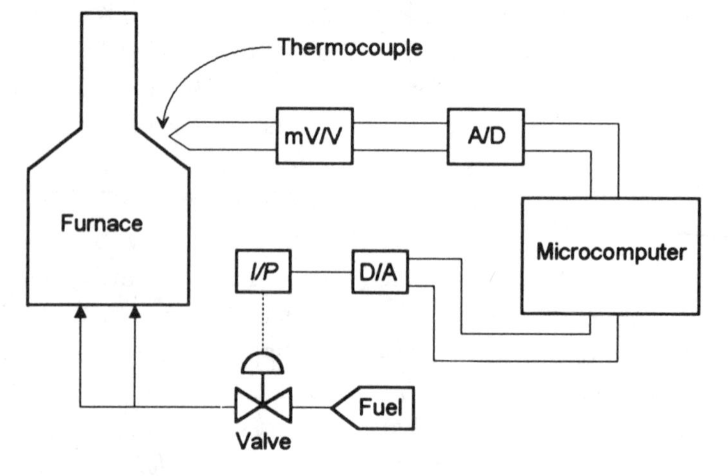

FIGURE 1.16 Furnace Temperature Control

The output of this equation will provide the appropriate opening and closing of the fuel valve to maintain the appropriate temperature. Since this output is computed by the microcomputer, it is a binary number. This binary output must be converted into an analog current or voltage signal.

The D/A (digital to analog) converter chip inputs this binary number and converts it into an analog current (I). This signal is then input into the current/pneumatic (I/P) transducer for opening or closing the fuel input valve by air pressure to adjust the fuel to the furnace. The desired temperature of the furnace can thus be achieved. Note that a transducer converts one form of energy (analog electrical current in this case) to another form (air pressure in this example).

1.8 Selected Topics in Digital Systems

1.8.1 Basics of the Internet

The Internet is like the phone system, which ties callers from around the world together via a web of data transmission lines. Figure 1.17 shows a typical internet system. An internet (with a lowercase i) consists of a number of network facilities. These include local area networks (LANs) and/or Wide Area Networks (WANs) for transmitting data through a network. The Internet (with an uppercase I, also called the Net) is a very special internet, connecting thousands of such networks. Note that the LAN in a data network connecting systems such as personal computers and servers within a relatively small area of a few miles or less. A WAN, on the other hand, is a data network connecting systems within a large geographical area.

On the Net, giant computers called "servers" are connected together via telephone lines and fiber-optic cable. Note that fiber-optic cable is a transmission medium in which data travels in the form of light. Data sent via fiber-optic cable are not degraded by noise during transmission. The binary information from the transmitting computer is converted to analog electrical signals using D/A converters and then to light signals by utilizing devices such as injection lasers. These light signals travel through fiber-optic cable. At the receiving end, these light signals are converted into electrical analog signals using phototransistors. These analog signals are then converted to binary information for the receiving computer using A/D converters.

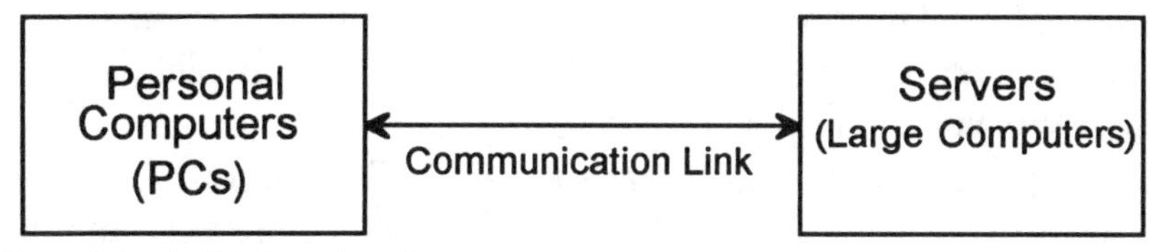

FIGURE 1.17 A typical internet system

Whether in homes, businesses, or schools, the personal computers can be linked to virtually any source of stored information via the Internet. *The Web refers to all the interconnected data sources which can be accessed by the personal computers using the Internet.*

A browser program is needed on the personal computer to see content on the World Wide Web. The browser program is a sort of remote controller that assembles photos, text, and audio onto Web pages that appear in the computer screen. Microsoft's Internet Explorer and Netscape's Navigator are two popular examples of browser programs.

Microsoft's Windows operating system is a program for managing resources such as the keyboard, disk drives, etc., on the PC and for connecting the PC to the Internet. The Windows operating system dominates the market at present.

The real work on the Internet is done on servers that process data. Web browsers type in addresses (those http:// *commands) that connect them to the servers.*

A program containing a set of rules called a "protocol" is used to govern data communication through a network. The Internet utilizes TCP/IP (Transmission Control Protocol/Internet Protocol) as its protocol. TCP contains a set of rules for reliable and accurate delivery of data; IP includes rules for routing data. TCP/IP protocol was developed by the Department of Defense (DOD) of the U.S. government in 1983. The TCP/IP program used by the Internet allows services such as the ability to copy files from one computer to another, exchange of electronic mail (E-mail) between two users, remote printing, and accessing the World Wide Web (WWW). TCP/IP is the most universally accepted networking software that made the tremendous growth of the Internet possible. The world has become a smaller place because of the Internet.

1.8.2 Trends and Perspectives in Digital Technology

This section provides a summary of technological forecasts. Topics include advancements in ICs, microprocessors, Internet, programming languages, ASIC and DVD as follows:

i) With the advent of IC technology, it is expected that it would be possible to place 750 million transistors on one chip by the year 2012. Furthermore, the replacement of aluminum wire (high resistance) on ICs by copper wire (low resistance) will reduce power consumption and improve reliability.

ii) Microprocessor designers have traditionally refined architectures by raising clock speeds and adding ALUs that can process instructions simultaneously. Many modern microprocessors can execute instructions out of order, so that one instruction waiting for data does not stall the entire processor. These microprocessors can predict in advance where branch will be taken. The drawbacks of incorporating these types of capabilities in the modern microprocessors are that the chip's circuitry is devoted to overheads.

A new microprocessor architecture called EPIC (Explicitly Parallel Instruction Computing), jointly being developed by Intel and Hewlett-Packard, minimizes these overheads. EPIC will be introduced in the year 2000 with a new Intel chip called "Merced."

Motorola, on the other hand, announced its AltiVec technology (discussed in Chapter 10) which is used as the foundation for Apple's next generation computers such as G4.

iii) Internet Explorer was part and parcel of Windows in the past. Microsoft's enhanced operating system, called "Windows 98," has been introduced. The Web browser interface and Web search capability are an integral part of Windows 98. Also, a built-in TV tuner (allows two or more monitors at a time; watching TV on one monitor while working on another), a better Dial-up network and a system shield to prevent crashes are included.

Windows 2000 is the new name for Windows NT 5.0 and is an enhanced version of Windows NT 4.0. Note that Windows NT 4.0 is a multipurpose server operating system providing applications including communication, file, and print services to advanced media and Web features. However, Users of Windows 2000 should look carefully at the Hardware Compatibility List (HCL) for Windows 2000 before installing any hardware.

iv) A high level language, called Java (developed by Sun Microsystems) has been rapidly gaining wide acceptance. Java is based on the C++ language. An overview of Java is provided in Chapter 6.

v) ASICs (Application-Specific ICs) are chips designed for a specific, limited application. ASICs normally reduce the total manufacturing cost of a product by reducing chip count, physical size, and power consumption. ASIC chips use gate arrays for rapid and low cost development of applications. Unlike chips designed from scratch, a gate array is a VLSI chip containing transistors and connections (called structures) that are pre-designed. The gate array chip is then fabricated using these structures and the connection information provided by the customers.

vi) DVD (normally stands for "Digital Video Disc" or "Digital Versatile Disc") is the next generation of optical disc technology. It is basically a larger, fast CD (Compact Disc) that can hold video as well as audio and computer information. The DVD-ROM like the CD-ROM uses a laser to read data from a disc. However, the data in DVD-ROM is stored in more compact form in more than one layer of the disc. Thus, DVD disc provides a higher capacity (typically 8.5 Megabytes --equivalent to 13 compact discs) of storage compared to CD.

DVD aims to encompass home entertainment, computers, and business information with a single digital format. It will eventually replace audio CD, videotape, laser disc, CD-ROM, and video game cartridges. There are basically three types of DVD. These are DVD-Video, DVD-ROM and DVD-RAM. DVD-Video (simply called DVD) holds information that can be played in a DVD player connected to a TV while DVD-ROM holds computer programs and can be read by DVD-ROM drive interfaced to a computer. The difference is similar to that between audio CD and CD-ROM. DVD drives can also read CD-ROMs. Therefore, DVD drives rather than CD-ROM drives are included in some Personal Computers (PCs). Most computers with DVD-ROM drives can also play DVD-Videos.

DVD-RAM can be read from and written into many times. CD-RW (CD-Rewriteable) and DVD-RAM are the read/write equivalents of CD-ROM and DVD-ROM respectively. CD-RW uses infrared laser like the CD-ROM. Both DVD-ROM and DVD-RAM, on the other hand, use a red laser, which has a shorter wavelength than infrared laser. The shorter wavelength of the red laser provides DVD with larger storage capacity compared to CD.

2

NUMBER SYSTEMS AND CODES

This chapter describes some of the fundamental concepts needed to implement and use a computer effectively. Thus the basics of number systems, codes, and error detection/correction are presented.

2.1 Number Systems

A computer, like all digital machines, utilizes two states to represent information. These two states are given the symbols 1 and 0. It is important to remember that these 1's and 0's are symbols for the two states and have no inherent numerical meanings of their own. These two digits are called binary digits (bits) and can be used to represent numbers of any magnitude. *The microcomputer carries out all the arithmetic and logic operations internally using binary numbers*. Because binary numbers are long, a more compact form using some other number system is preferable to represent them. The computer user finds it convenient to work with this compact form. Hence, it is important to understand the various number systems used with computers. These are described in the following sections.

2.1.1 General Number Representation

In general, a number N can be represented in the following form:

$$N = d_{p-1} \times b^{p-1} + d_{p-2} \times b^{p-2} + \cdots + d_0 \times b^0 + d_{-1} \times b^{-1} + \cdots + d_{-q} \times b^{-q} \qquad \textbf{2.1}$$

where b is the base or radix of the number system, the d's are the digits of the number system, p is the number of integer digits, and q is the number of fractional digits.

N can also be written as a string of digits whose integer and fractional portions are separated by the radix or decimal point (•). In this format, the number N is represented as

$$N = d_{p-1}d_{p-2}\cdots d_1 d_0 \bullet d_{-1}\cdots d_{-q} \qquad \textbf{2.2}$$

If a number has no fractional portion, (e.g., $q = 0$ in the form of Equation 2.1), then the number is called an integer number or an integer. Conversely, if the number has no integer portion (e.g., $p = 0$ in the form of Equation 2.1), the number is called a fractional number or a fraction. If both p and q are not zero, then the number is called a mixed number.

2.1.1.1 Decimal Number System

In the decimal number system (base 10), which is most familiar to us, the integer number 125_{10} can be expressed as

$$125_{10} = 1 \times 10^2 + 2 \times 10^1 + 5 \times 10^0 \qquad \textbf{2.3}$$

In this equation, the left-hand side corresponds to the form given by Equation 2.2. The right-hand side of Equation 2.3 is represented by the form of equation 2.1, where $b = 10$, $d_2 = 1$, $d_1 = 2$, $d_0 = 5$, $d_{-1} = \ldots = d_{-q} = 0$, $p = 3$, and $q = 0$.

Now, consider the fractional decimal number 0.532_{10}. This number can be expressed as

$$0.532_{10} = 5 \times 10^{-1} + 3 \times 10^{-2} + 2 \times 10^{-3} \qquad \textbf{2.4}$$

The left-hand side of Equation 2.4 corresponds to Equation 2.2. The right-hand side of Equation 2.4 is in the form of Equation 2.1, where $b = 10$, $d_{-1} = 5$, $d_{-2} = 3$, $d_{-3} = 2$, $q = 3$, $p = 0$, $d_{p-1} = \ldots = d_0 = 0$.

Finally, consider the mixed number 125.532_{10}. This number is in the form of Equation 2.2. Translating the number to the form of Equation 2.1 yields

$$125.532_{10} = 1 \times 10^2 + 2 \times 10^1 + 5 \times 10^0 + 5 \times 10^{-1} + 3 \times 10^{-2} + 2 \times 10^{-3} \qquad \textbf{2.5}$$

Comparing the right-hand side of Equation 2.5 with equation 2.1 yields $b = 10$, $p = 3$, $q = 3$, $d_2 = 1$, $d_1 = 2$, $d_0 = 5$, $d_{-1} = 5$, $d_{-2} = 3$, and $d_{-3} = 2$.

2.1.1.2 Binary Number System

In terms of Equation 2.1, the binary number system has a base or radix of 2 and has two allowable digits, 0 and 1. From Equation 2.1, a 4-bit binary number 1110_2 can be interpreted as

$$1110_2 = 1 \times 2^3 + 1 \times 2^2 + 1 \times 2^1 + 0 \times 2^0 = 14_{10}$$

This conversion from binary to decimal can be obtained by inspecting the binary number as follows

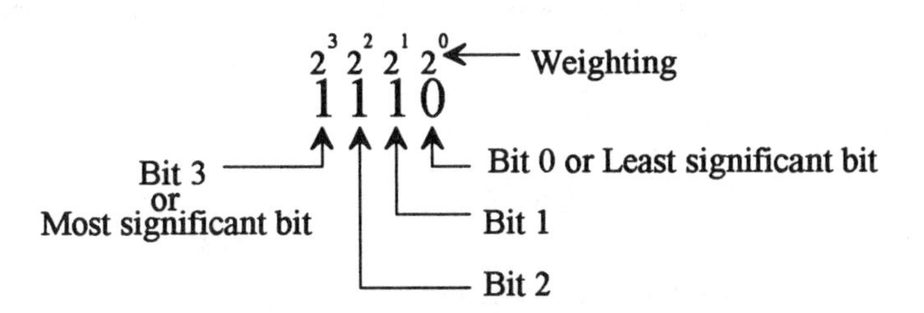

Note that bits 0, 1, 2, and 3 have corresponding weighting values of 1, 2, 4, and 8. *Because a binary number only contains 1's and 0's, adding the weighting values of only the bits of the binary number containing 1's will provide its decimal value.* The decimal value of 1110_2 is 14_{10} (2 + 4 + 8), because bits 1, 2, and 3 have binary digit 1, whereas bit 0 contains 0.

Therefore, the decimal value of any binary number can be readily obtained by just adding the weighting values for the bit positions containing 1's. Furthermore, the value of the least significant bit (bit 0) determines whether the number is odd or even. For example, if the least significant bit is 1, the number is odd; otherwise, the number is even.

Next, consider a mixed number 101.01_2 as follows:

$$101.01_2 = 1 \times 2^2 + 0 \times 2^1 + 1 \times 2^0 + 0 \times 2^{-1} + 1 \times 2^{-2} \quad \textbf{2.6}$$

The decimal or base 10 value of 101.01_2 is found from the right-hand side of Equation 2.6 as $4 + 0 + 1 + 0 + ¼ = 5.25_{10}$.

2.1.1.3 Octal Number System

The radix or base of the octal number system is 8. There are eight digits, 0 through 7, allowed in this number system.

Consider the octal number 25.32_8, which can be interpreted as:

$$2 \times 8^1 + 5 \times 8^0 + 3 \times 8^{-1} + 2 \times 8^{-2}$$

The decimal value of this number is found by completing the summation of

$$16 + 5 + 3 \times 1/8 + 2 \times 1/64 = 16 + 5 + 0.375 + 0.03125 = 21.40625_{10}$$

One converts a number from binary to octal representation easily by taking the binary digits in groups of 3 bits to an octal digit.

The octal digit is obtained by considering each group of 3 bits as a separate binary number capable of representing the octal digits 0 through 7. The radix point remains in its original position. The following example illustrates the procedure.

Suppose that it is desired to convert 1001.11_2 into octal form. First take the groups of 3 bits starting at the radix point. Where there are not enough leading or trailing bits to complete the triplet, 0's are appended. Now each group of 3 bits is converted to its corresponding octal digit.

$$\underbrace{001}_{1}\ \underbrace{001}_{1}\ .\ \underbrace{110}_{6}{}_2 = 11.6_8$$

The conversion back to binary from octal is simply the reverse of the binary-to-octal process. For example, conversion from 11.6_8 to binary is accomplished by expanding each octal digit to its equivalent binary values as shown:

$$\underbrace{1}_{001}\ \underbrace{1}_{001}\ .\ \underbrace{6}_{110}$$

2.1.1.4 Hexadecimal Number System

The hexadecimal or base-16 number system has 16 individual digits. Each of these digits, as in all number systems, must be represented by a single unique symbol. The digits in the hexadecimal number system are 0 through 9 and the letters A through F. Letters were chosen to represent the hexadecimal digits greater than 9 because a single symbol is required for each digit. Table 2.1 lists the 16 digits of the hexadecimal number system and their corresponding binary and decimal values.

TABLE 2.1 Number Systems

Hexadecimal	Decimal	Binary
0	0	0000
1	1	0001
2	2	0010
3	3	0011
4	4	0100
5	5	0101
6	6	0110
7	7	0111
8	8	1000
9	9	1001
A	10	1010
B	11	1011
C	12	1100
D	13	1101
E	14	1110
F	15	1111

2.1.2 Converting Numbers from One Base to Another

2.1.2.1 Binary-to-Decimal Conversion and Vice Versa

Consider converting 1100.01_2 to its decimal equivalent. As before,

$$
\begin{aligned}
1100.01_2 &= 1\times 2^3 + 1\times 2^2 + 0\times 2^1 + 0\times 2^0 + 0\times 2^{-1} + 1\times 2^{-2} \\
&= 8+4+0+0+0+.25 \\
&= 12.25_{10}
\end{aligned}
$$

Continuous division by 2, keeping track of the remainders, provides a simple method of converting a decimal number to its binary equivalent. As an example, to convert decimal 12_{10}

		quotient	+	remainder
$\frac{12}{2}$	=	6	+	0
$\frac{6}{2}$	=	3	+	0
$\frac{3}{2}$	=	1	+	1
$\frac{1}{2}$	=	0	+	1

$1\ 1\ 0\ 0_2$

to its binary equivalent, proceed as follows:
Thus $12_{10} = 1100_2$.

Fractions

One can convert 0.0101_2 to its decimal equivalent as follows:

$$
\begin{aligned}
0.0101_2 &= 0\times 2^{-1} + 1\times 2^{-2} + 0\times 2^{-3} + 1\times 2^{-4} \\
&= 0+0.25+0+0.0625 \\
&= 0.3125_{10}
\end{aligned}
$$

A decimal fractional number can be converted to its binary equivalent as follows:

0.8125	0.6250	0.2500	0.5000
× 2	× 2	× 2	× 2
(1).6250	(1).2500	(0).5000	(1).0000
↓	↓	↓	↓
1	1	0	1

Therefore $0.8125_{10} = 0.1101_2$.

Unfortunately, binary-to-decimal fractional conversions are not always exact. Suppose that it is desired to convert 0.3615 into its binary equivalent:

0.3615	0.7230	0.4460	0.8920	0.7840
× 2	× 2	× 2	× 2	× 2
(0).7230	(1).4460	(0).8920	(1).7840	(1).5680
↓	↓	↓	↓	↓
0	1	0	1	1

The answer is $0.01011\ldots_2$. As a check, let us convert back:

$$0.01011_2 = 0 \times 2^{-1} + 1 \times 2^{-2} + 0 \times 2^{-3} + 1 \times 2^{-4} + 1 \times 2^{-5}$$

$$= 0 + 0.25 + 0 + 0.0625 + 0.03125$$

$$= 0.34375$$

The difference is 0.3615 − 0.34375 = 0.01775. This difference is caused by the neglected remainder 0.5680. The neglected remainder (0.5680) multiplied by the smallest computed term (0.03125) gives the total error:

$$0.5680 \times 0.03125 = 0.01775$$

Mixed Numbers

Finally, convert 13.25_{10} to its binary equivalent. It is convenient to carry out separate conversions for the integer and fractional parts. Consider first the integer number 13. As before,

		quotient	+	remainder
$\frac{13}{2}$	=	6	+	1
$\frac{6}{2}$	=	3	+	0
$\frac{3}{2}$	=	1	+	1
$\frac{1}{2}$	=	0	+	1

$$13_{10} = 1\ 1\ 0\ 1_2$$

Now convert the fractional part 0.25_{10} as follows:

$$\begin{array}{cc} 0.25 & 0.50 \\ \times 2 & \times 2 \\ \hline \textcircled{0}.50 & \textcircled{1}.00 \\ \downarrow & \downarrow \\ 0 & 1 \end{array}$$

Thus $0.25_{10} = 0.01_2$. Therefore $13.25_{10} = 1101.01_2$.

Note that the same procedure applies for converting a decimal integer number to other number systems such as octal or hexadecimal; Continuous division by the appropriate base (8 or 16) and keeping track of remainders converts a decimal number from decimal to the selected number system.

2.1.2.2 Binary-to-Hexadecimal Conversion and Vice Versa

The conversions between hexadecimal and binary numbers are done in exactly the same manner as the conversions between octal and binary, except that groups of 4 are used. The following examples illustrate this:

$$1011011_2 = \underbrace{0101}_{5} \quad \underbrace{1011}_{B} = 5B_{16}$$

Note that the binary integer number is grouped in 4-bit units, starting from the least significant bit. Zeros are added with the most significant 4 bits if necessary. As with octal numbers, for fractional numbers this grouping into 4 bits is started from the radix point. Now consider converting $2AB_{16}$ into its binary equivalent as follows:

$$\begin{array}{ccccc} 2AB_{16} & = & 2 & A & B \\ & & \downarrow & \downarrow & \downarrow \\ & & 0010 & 1010 & 1011 \end{array}$$

$$= 001010101011_2$$

2.1.2.3 Hexadecimal-to-Decimal Conversion and Vice Versa

Consider converting the hexadecimal number $23A_{16}$ into its decimal equivalent and vice versa. This can be accomplished as follows:

$$23A_{16} = 2 \times 16^2 + 3 \times 16^1 + 10 \times 16^0$$

$$= 512 + 48 + 10 = 570_{10}$$

Note that in the equation, the value 10 is substituted for A. Now to convert 570_{10} back to $23A_{16}$,

	quotient	+	remainder
$\frac{570}{16}$ =	35	+	A
$\frac{35}{16}$ =	2	+	3
$\frac{2}{16}$ =	0	+	2

2 3 A

Thus, $570_{10} = 23A_{16}$.

Example 2.1

Determine by inspecting the binary equivalent of the following hexadecimal numbers whether they are odd or even. Then verify the result by their decimal equivalents.

(a) $2B_{16}$

(b) $A2_{16}$

Solution

(a)

		128	64	32	16	8	4	2	1	← Weighting
$2B_{16}$	=	0	0	1	0	1	0	1	1_2	

The number is odd, since the least significant bit is 1.

Decimal value = 32 + 8 + 2 + 1 = 43_{10}, which is odd.

(b)

		128	64	32	16	8	4	2	1	← Weighting
$A2_{16}$	=	1	0	1	0	0	0	1	0_2	

The number is even, since the least significant bit is 0.

Decimal value = 128 + 32 + 2 = 162_{10}, which is even.

2.2 Unsigned and Signed Binary Numbers

An unsigned binary number has no arithmetic sign. *Unsigned binary numbers are therefore always positive. Typical examples are your age or a memory address which are always positive numbers.* An 8-bit unsigned binary integer represents all numbers from 00_{16} through FF_{16} (0_{10} through 255_{10}).

The techniques used to represent the signed integers are:

- Sign-magnitude approach
- Ones complement approach
- Twos complement approach

Because the sign of a number can be either positive or negative, only one bit, referred to as the sign bit, is needed to represent the sign. *The widely used sign convention is that if the sign bit is zero, the number is positive; otherwise it is negative. (The rationale behind this convention is that the quantity $(-1)^s$ is positive when $s = 0$ and is negative when $s = 1$).* Also, in all three approaches, the most significant bit of the number is considered as the sign bit.

In sign-magnitude representation, the most significant bit of the given n-bit binary number holds the sign, and the remaining $n - 1$ bits directly give the magnitude of the negative number. For example, the sign-magnitude representation of +7 is 0111 and that of −4 is 1100. Table 2.2 represents all possible 4-bit patterns and their meanings in sign-magnitude form.

TABLE 2.2 All Possible 4-Bit Integers Represented in Sign-Magnitude Form

Bit Pattern	Interpretation as a Sign-Magnitude Integer
0000	+0
0001	+1
0010	+2
0011	+3
0100	+4
0101	+5
0110	+6
0111	+7
1000	−0
1001	−1
1010	−2
1011	−3
1100	−4
1101	−5
1110	−6
1111	−7

In Table 2.2, the sign-magnitude approach represents a signed number in a natural manner. With 4 bits we can only represent numbers in the range $-7 \leq x \leq +7$. In general, if there are n bits, then we can cover all numbers in the range $\pm(2^{n-1} - 1)$. Note that with $n - 1$ bits, any value from 0 to $2^{n-1} - 1$ can be represented. However, this approach leads to a confusion because there are two representations for the number zero (0000 means +0; 1000 means −0).

In complement approach, positive numbers have the same representation as they do in the sign-magnitude representation. However, in this technique negative numbers are represented in a different manner. Before we proceed, let us define the term *complement* of a number. The complement of a number A, written as $\bar{A}$ (or A') is obtained by taking bit-by-bit complement of A. In other words, each 0 in A is replaced with 1 and vice versa. For example, the complement of the number 0100_2 is 1011_2 and that of 1111_2 is 0000_2. In the ones complement approach, a negative number, $-x$, is the complement of its positive representation. For example let us find the ones complement representation of 0100_2 ($+4_{10}$). The complement of 0100 is 1011, and this denotes the negative number -4_{10}. Table 2.3 summarizes all possible 4-bit binary patterns and their interpretations as ones complement numbers.

TABLE 2.3 All Possible 4-Bit Integers Represented in Ones Complement Form

Bit Pattern	Interpretation as a Ones Complement Number
0000	+0
0001	+1
0010	+2
0011	+3
0100	+4
0101	+5
0110	+6
0111	+7
1000	−7
1001	−6
1010	−5
1011	−4
1100	−3
1101	−2
1110	−1
1111	−0

From Table 2.3, the ones complement approach does not handle negative numbers naturally. In other words, if the number is negative (when the sign bit is 1), its magnitude is not obvious from its ones complement. To determine its magnitude, one needs to take its ones complement. For example, consider the number 110110. The most significant bit indicates that this is a negative number. Because the number is negative, its magnitude cannot be obtained by directly looking at 110110. Instead, one needs to take the ones complement of 110110 to obtain 001001. The value of 001001 as a sign-magnitude number is +9. On the other hand, 110110 represents –9 in ones complement form. Like the sign-magnitude representation, the ones complement approach does not increase the range of numbers covered by a fixed number of bit patterns. For example, 4 bits cover the range –7 to +7. The same range is obtained with sign-magnitude representation. Note that the confusion of two distinct representations for zero exists in the ones complement approach. Now, let us discuss the two's complement approach.

In this method, positive integers are represented in the same manner as they are in the sign-magnitude method. In other words, if the sign bit is zero, the number is positive and its magnitude can be directly obtained by looking at the remaining $n - 1$ bits. However, a negative number $-x$ can be represented in twos complement form as follows:

- Represent $+x$ in sign magnitude form and call this result y
- Take the ones complement of y to get $\bar{y}$ (or y')
- $\bar{y} + 1$ is the twos complement representation of $-x$.

The following example illustrates this:

Example 2.2

Represent the following decimal numbers in twos complement form. Use 7 bits to represent the numbers.

(a) +39

(b) –43

Solution

(a) Because the number +39 is positive, its twos complement representation is the same as its sign-magnitude representation as shown here:

$$y = \underbrace{0}_{+}\ \underbrace{\overset{2^5}{1}\ \overset{2^4}{0}\ \overset{2^3}{0}\ \overset{2^2}{1}\ \overset{2^1}{1}\ \overset{2^0}{1}}_{39}$$

(b) In this case, the given number –43 is negative. The twos complement form of the number can be obtained as follows:

1. Step 1: Represent +43 in sign magnitude form

$$y = \underbrace{0}_{+}\ \underbrace{\overset{2^5}{1}\ \overset{2^4}{0}\ \overset{2^3}{1}\ \overset{2^2}{0}\ \overset{2^1}{1}\ \overset{2^0}{1}}_{43}$$

2. Step 2: Take the ones complement of y:

$$\bar{y} = 1\,0\,1\,0\,1\,0\,0$$

3. Step 3: Add one to $\bar{y}$ to get the final answer.

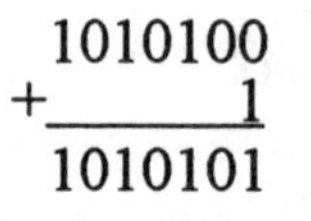

Table 2.4 lists all possible 4-bit patterns along with their twos complement forms. From Table 2.4, it can be concluded that:

- Twos complement form does not provide two representations for zero.
- Twos complement form covers up to −8 in the negative side, and this is more than can be achieved with the other two methods. In general, with n bits, and using twos complement approach, one can cover all the numbers in the range $-(2^{n-1})$ to $+(2^{n-1} - 1)$.

Computers use twos complement form for representing numbers. The computer interprets 1111_2 as +15 when representing it as an unsigned number. On the other hand, the computer interprets 1111_2 as -1_{10} (twos complement) when representing it as a signed number.

TABLE 2.4 All Possible 4-Bit Integers Represented in Twos Complement Form

Bit Pattern	Interpretation as a Twos Complement Number
0000	0
0001	+1
0010	+2
0011	+3
0100	+4
0101	+5
0110	+6
0111	+7
1000	−8
1001	−7
1010	−6
1011	−5
1100	−4
1101	−3
1110	−2
1111	−1

2.3 Codes

Codes are used extensively with computers to define alphanumeric characters and other information. Some of the codes used with computers are described in the following sections.

2.3.1 Binary-Coded-Decimal Code (8421 Code)

The 10 decimal digits 0 through 9 can be represented by their corresponding 4-bit binary numbers. The digits coded in this fashion are called binary-coded-decimal *(BCD)* digits *in* 8421 code, *or BCD digits. Two unpacked BCD numbers are usually packed into a byte to form "packed BCD."* For example, two unpacked BCD digits 2 and 5 can be combined as a packed BCD byte 25. The concept of packed and unpacked BCD digits are explained later in this section. Table 2.5 lists the BCD representation of the 10 decimal digits.

The six possible remaining 4-bit codes as shown in Table 2.5 are not used and represent invalid BCD codes if they occur.

TABLE 2.5 BCD Representation of the 10 Decimal Digits

Decimal Digits		BCD Representation
0		0000
1		0001
2		0010
3		0011
4		0100
5		0101
6		0110
7		0111
8		1000
9		1001
10	Invalid BCD Code	1010
11	Invalid BCD Code	1011
12	Invalid BCD Code	1100
13	Invalid BCD Code	1101
14	Invalid BCD Code	1110
15	Invalid BCD Code	1111

Consider converting the decimal number 356_{10} into its BCD equivalent as follows:

3	5	6
↓	↓	↓
0011	0101	0110

BCD is used with computers when it is desired to handle data that are in decimals.

2.3.2 Alphanumeric Codes

A computer must be capable of handling nonnumeric information if it is to be very useful. In other words, a computer must be able to recognize codes that represent numbers, letters, and special characters. These codes are classified as alphanumeric or character codes. A complete and adequate set of necessary characters includes these:

1. 26 lowercase letters
2. 26 uppercase letters
3. 10 numeric digits (0–9)
4. About 25 special characters, which include + / # % , and so on.

This totals 87 characters. To represent 87 characters with some type of binary code would require at least 7 bits. With 7 bits there are $2^7 = 128$ possible binary numbers; 87 of these combinations of 0 and 1 bits serve as the code groups representing the 87 different characters.

The 8-bit byte has been universally accepted as the data unit for representing character codes. The two most common alphanumeric codes are known as the American Standard Code for Information Interchange (ASCII) and the Extended Binary-Coded Decimal Interchange Code (EBCDIC). ASCII is typically used with microprocessors. IBM uses EBCDIC code. Eight bits are used to represent characters, although 7 bits suffice, because the eighth bit is frequently used to test for errors and is referred to as a parity bit. It can be set to 1 or 0, so that the number of 1 bits in the byte is always odd or even.

Table 2.6 shows a list of ASCII and EBCDIC codes. Some EBCDIC codes do not have corresponding ASCII codes. *Note that decimal digits 0 through 9 are represented by 30_{16} through 39_{16} in ASCII. On the other hand, these decimal digits are represented by $F0_{16}$ though $F9_{16}$ in EBCDIC.*

A computer program is usually written for code conversion when input/output devices of different codes are connected to the computer. For example, suppose it is desired to enter a number 5 into a computer via an ASCII keyboard and print this data on an EBCDIC printer. The ASCII keyboard will generate 35_{16} when the number 5 is pushed. The ASCII code 35_{16} for the decimal digit 5 enters into the computer and resides in the computer's memory. To print the digit 5 on the EBCDIC printer, a program must be written that will convert the ASCII code 35_{16} for 5 to its EBCDIC code $F5_{16}$. The output of this program is $F5_{16}$. This will be input to the EBCDIC printer. Because the printer only understands EBCDIC codes, it inputs the EBCDIC code $F5_{16}$ and prints the digit 5.

Let us now discuss packed and unpacked BCD codes in more detail. For example, in order to enter data 24 in decimal into a computer, the two keys (2 and 4) will be pushed on the

ASCII keyboard. This will generate 32 and 34 (32 and 34 are ASCII codes in hexadecimal for 2 and 4 respectively) inside the computer. A program can be written to convert these ASCII codes into unpacked BCD 02 and 04, and then convert to packed BCD 24 which will be represented in binary inside the computer to perform the desired operation.

TABLE 2.6 ASCII and EBCDIC Codes in Hex.

Character	ASCII	EBCDIC	Characte r	ASCII	EBCDIC	Character	ASCII	EBCDIC	Character	ASCII	EBCDIC
@	40			60		blank	20	40	NUL	00	
A	41	C1	a	61	81	!	21	5A	SOH	01	
B	42	C2	b	62	82	“	22	7F	STX	02	
C	43	C3	c	63	83	#	23	7B	ETX	03	
D	44	C4	d	64	84	$	24	5B	EOT	04	37
E	45	C5	e	65	85	%	25	6C	ENQ	05	
F	46	C6	f	66	86	&	26	50	ACK	06	
G	47	C7	g	67	87	‘	27	7D	BEL	07	
H	48	C8	h	68	88	(	28	4D	BS	08	16
I	49	C9	i	69	89	)	29	5D	HT	09	05
J	4A	D1	j	6A	91	*	2A	5C	LF	0A	25
K	4B	D2	k	6B	92	+	2B	4E	VT	0B	
L	4C	D3	l	6C	93	,	2C	6B	FF	0C	
M	4D	D4	m	6D	94	-	2D	60	CR	0D	15
N	4E	D5	n	6E	95	.	2E	4B	SO	0E	
O	4F	D6	o	6F	96	/	2F	61	SI	0F	
P	50	D7	p	70	97	0	30	F0	DLE	10	
Q	51	D8	q	71	98	1	31	F1	DC1	11	
R	52	D9	r	72	99	2	32	F2	DC2	12	
S	53	E2	s	73	A2	3	33	F3	DC3	13	
T	54	E3	t	74	A3	4	34	F4	DC4	14	
U	55	E4	u	75	A4	5	35	F5	NAK	15	
V	56	E5	v	76	A5	6	36	F6	SYN	16	
W	57	E6	w	77	A6	7	37	F7	ETB	17	
X	58	E7	x	78	A7	8	38	F8	CAN	18	
Y	59	E8	y	79	A8	9	39	F9	EM	19	
Z	5A	E9	z	7A	A9	:	3A		SUB	1A	
[	5B		{	7B		;	3B	5E	ESC	1B	
\	5C		|	7C	4F	<	3C	4C	FS	1C	
]	5D		}	7D		=	3D	7E	GS	1D	
^	5E		~	7E		>	3E	6E	RS	1E	
_	5F	6D	DEL	7F	07	?	3F	6F	US	1F	

TABLE 2.7 Excess-3 Representation of Decimal Digits

Decimal Digits	Excess-3 Representation
0	0011
1	0100
2	0101
3	0110
4	0111
5	1000
6	1001
7	1010
8	1011
9	1100

2.3.3 Excess-3 Code

The excess-3 representation of a decimal digit *d* can be obtained by adding 3 to its value. All decimal digits and their excess-3 representations are listed in Table 2.7.
The excess-3 code is an unweighted code because its value is obtained by adding three to the corresponding binary value. The excess-3 code is self-complementing. For example, decimal digit 0 in excess-3 (0011) is ones complement of 9 in excess three (1100). Similarly, decimal digit 1 is ones complement of 8, and so on. This is why some older computers used excess three code. Conversion between excess-3 and decimal numbers is illustrated below:

Decimal number	1	9	8	3
	↕	↕	↕	↕
Excess-3 Representation	0100	1100	1011	0110

2.3.4 Gray Code

Sometimes codes can also be constructed using a property called reflected symmetry. One such code is known as Gray code. The Gray code is used in Karnaugh maps for simplifying combinational logic design. This topic is covered in Chapter 4. Before we proceed, we briefly explain the concept of reflected symmetry. Consider the two bits 0 and 1, and stack these two bits. Assume that there is a plane mirror in front of this stack and produce the reflected image of the stack as shown in the following:

```
          0
          1
mirror← ─────
          1
          0
```

Appending a zero to all elements of the stack above the plane mirror and append a one to all elements of the stack that lies below the mirror will provide the following result:

```
Appended  {  0 0
 zeros    {  0 1
            ─────
Appended  {  1 1
  ones    {  1 0
```

Now, removal of the plane mirror will result in a stack of 2-bit Gray Code as follows:

0 0
0 1
1 1
1 0

Here, any two adjacent bit patterns differ only in one bit. For example, the patterns 11 and 10 differ only in the least significant bit.

Repeating the reflection operation on the stack of 2-bit binary patterns, a 3-bit Gray code can be obtained. Two adjacent binary numbers differ in only one bit. The result is shown in Figure 2.1.

Applying the reflection process to the 3-bit Gray code, 4-bit Gray Code can be obtained. This is shown in Figure 2.2.

```
0 0 0                       0 0 0
0 0 1                       0 0 1
0 1 1       Mirror          0 1 1
0 1 0  ←                    0 1 0
─────
1 1 0  ───────────────→     1 1 0
1 1 1       Result          1 1 1
1 0 1   after moving        1 0 1
1 0 0    the mirror         1 0 0
```

FIGURE 2.1 The process of obtaining 3-bit reflected binary code

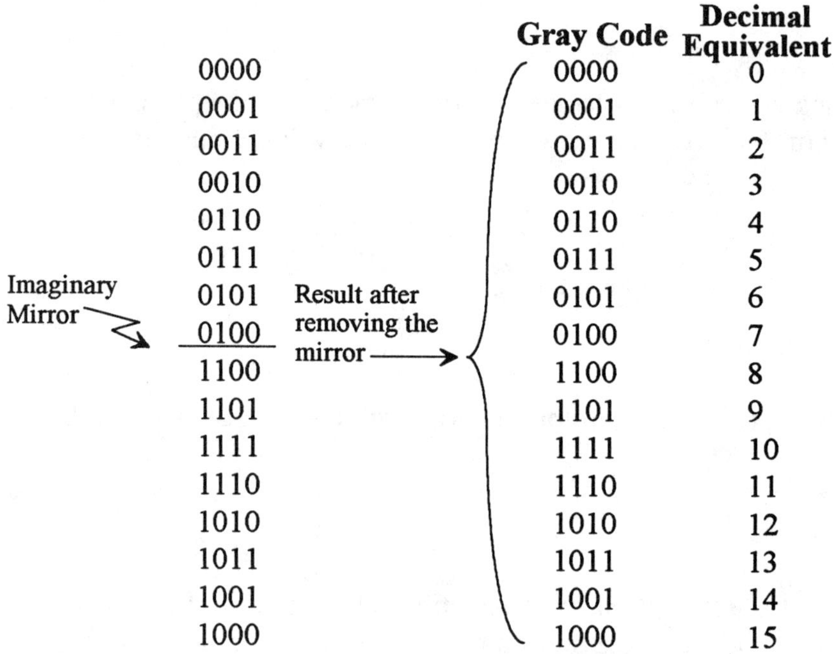

		Gray Code	Decimal Equivalent
0000		0000	0
0001		0001	1
0011		0011	2
0010		0010	3
0110		0110	4
0111		0111	5
0101	Result after	0101	6
0100	removing the	0100	7
1100	mirror →	1100	8
1101		1101	9
1111		1111	10
1110		1110	11
1010		1010	12
1011		1011	13
1001		1001	14
1000		1000	15

Imaginary Mirror (between 0100 and 1100)

FIGURE 2.2 The process of obtaining a 4-bit Gray code from a 3-bit Gray code.

The Gray code is useful in instrumentation systems to digitally represent the position of a mechanical shaft. In these applications, one bit change between characters is required. For example, suppose a shaft is divided into eight segments and each shaft is assigned a number. If binary numbers are used, an error may occur while changing segment 7 (0111_2) to segment 8 (1000_2). In this case, all 4 bits need to be changed. If the sensor representing the most significant bit takes longer to change, the result will be 0000_2, representing segment 0. This can be avoided by using Gray code, in which only one bit changes when going from one number to the next.

2.4 Fixed-Point and Floating-Point Representations

A number representation assuming a fixed location of the radix point is called *fixed-point representation*. The range of numbers that can be represented in fixed-point notation is severely limited. The following numbers are examples of fixed-point numbers:

$$0110.1100_2,\ 51.12_{10},\ DE.2A_{16}$$

In typical scientific computations, the range of numbers is very large. Floating-point representation is used to handle such ranges. A floating-point number is represented as $N \times r^p$, where N is the mantissa or significand, r is the base or radix of the number system, and p is the exponent or power to which r is raised. Some examples of numbers in floating-point notation and their fixed-point decimal equivalents are:

fixed-point decimal number	floating-point representation
0.0167_{10}	0.167×10^{-1}
1101.101_2	0.1101101×2^4
$BE.2A9_{16}$	$0.BE2A9 \times 16^2$

In converting from fixed-point to floating-point number representation, we normalize the resulting mantissas, that is, the digits of the fixed-point numbers are shifted so that the highest-order nonzero digit appears to the right of the decimal point, and consequently a 0 always appears to the left of the decimal point. This convention is normally adopted in floating-point number representation. Because all numbers will be assumed to be in normalized form, the binary point is not required to be represented in the computers.

Typical 32-bit microprocessors such as the Intel 80486/Pentium and the Motorola 68040 and PowerPC contain on-chip floating-point hardware. This means that these microprocessors can be programmed using instructions to perform operations such as addition, subtraction, multiplication, and division using floating-point numbers.

2.5 Arithmetic Operations

As mentioned before, computers can only add. Therefore, all other arithmetic operations are typically accomplished via addition. All numbers inside the computer are in binary form. These numbers are usually treated internally as integers, and any fractional arithmetic must be implemented by the programmer in the program. The *arithmetic and logic unit* (ALU) in the computer's CPU performs typical arithmetic and logic operations. The ALUs perform function such as addition, subtraction, magnitude comparison, ANDing, and ORing of two binary or packed BCD numbers. The procedures involved in executing these functions are discussed now to provide an understanding of the basic arithmetic operations performed in a typical microprocessor. The logic operations are covered in Chapter 3.

2.5.1 Binary Arithmetic

2.5.1.1 Addition

The addition of two binary numbers is carried out in the same way as the addition of decimal numbers. However, only four possible combinations can occur when adding two binary digits (bits):

augend	+	addend	=	carry	sum	decimal value
0	+	0	=	0	0	0
1	+	0	=	0	1	1
0	+	1	=	0	1	1
1	+	1	=	1	0	2

The following are some examples of binary addition. The corresponding decimal additions are also included.

```
  010  (2)              111 ← carry
+ 011  (3)              101.11    (5.75)
  101  (5)            + 011.10    (3.50)
                      1 001.01    (9.25)
                      ↑
                 final carry
```

Addition is the most important arithmetic operation in microprocessors because the operations of subtraction, multiplication, and division as they are performed in most modern digital computers use only addition as their basic operation.

2.5.1.2 Subtraction

As mentioned before, computers can usually only add binary digits; they cannot directly subtract. Therefore, the operation of subtraction in microprocessors is performed using the operation of addition using complement arithmetic.

In general, the b's complement of an m-digit number, M is defined as $b^m - M$ for $M \neq 0$ and 0 for $M = 0$. Note that for base 10, $b = 10$ and 10^m is a decimal number with a 1 followed by m 0's. For example, 10^4 is 10000; 1 followed by four 0's. On the other hand, $b = 2$ for binary and 2^m indicates 1 followed by m 0's. For example, 2^3 means 1000 in binary.

The $(b-1)$'s complement of an m-digit number, M is defined as $(b^m - 1) - M$. Therefore, the b's complement of an m-digit number, M can be obtained by adding 1 to its $(b-1)$'s complement. Next, let us illustrate the concept of complement arithmetic by means of some examples. Consider a 4-digit decimal number, 5786. In this case, $b = 10$ for base 10 and $m = 4$ since there are four digits.

$$10\text{'s complement of } 5786 = 10^4 - 5786 = 10000 - 5786 = 4214$$

Now, let us obtain 10's complement of 5786 using (10 – 1)'s or 9's complement arithmetic as follows:

$$9'\text{s complement of } 5786 = (10^4 - 1) - 5786 = 9999 - 5786 = 4213$$

Hence, 10's complement of 5786 = 9's complement of 5786 + 1 = 4213 + 1 = 4214.
Next, let us determine the 2's complement of a 3-bit binary number, 010. In this case, $b = 2$ for binary and $m = 3$ since there are three bits in the number.

$$2\text{'s complement of } 010 = 2^3 - 010 = 1000 - 010.$$

Using paper and pencil method, the result of subtraction can be obtained as follows:

$$\begin{array}{r} 1000_2 \\ -010_2 \\ \hline 110_2 \end{array}$$

Note that in the above, 110_2 is -2 in decimal when interpreted as a signed number. Therefore, 2's complement of a number negates the number being complemented. This will be explained later in this section.

The 2's complement of 010 can be obtained using its 1's complement arithmetic as follows:

$$1\text{'s complement of } 010 = (2^3 - 1) - 010 = 111 - 010 = 101$$

$$2\text{'s complement of } 101 = 101 + 1 = 110$$

From the above procedure for obtaining the 1's complement of 010, it can be concluded that the 1's complement of a binary number can be achieved by subtracting each bit of the binary number from 1. This means that when subtracting a bit (0 or 1) from 1, one can have either $1 - 0 = 1$ or $1 - 1 = 0$; that is, the 1's complement of 0 is 1 and the 1's complement of 1 is 0. *In general, the 1's complement of a binary number can be obtained by changing 0's to 1's and 1's to 0's.*

Next, let us describe the procedure of subtracting decimal numbers using addition. This process requires the use of the 10's complement form. The 10's complement of a number can be obtained by subtracting the number from 10.

Consider the decimal subtraction 7 – 4 = 3. The 10's complement of 4 is 10 – 4 = 6. The decimal subtraction can be performed using the 10's complement addition as follows:

```
                     minuend      7
10's complement of subtrahend   + 6
                                ---
                            ──> 13
ignore final carry of 1 to obtain
the subtraction result of 3.
```

When a larger number is subtracted from a smaller number, there is no carry to be discarded. Consider the decimal subtraction 4–7 = –3. The 10's complement of 7 is 10–7 = 3.

Therefore,

```
                     minuend      4
10's complement of subtrahend   + 3
                                ---
                            ──>  7
no final carry
```

When there is no final carry, the final answer is the negative of the 10's complement of 7. Therefore, the correct result of subtraction is – (10– 7) = –3.

The same procedures can be applied for performing binary subtraction. In the case of binary subtraction, the twos complement of the subtrahend is used.

As mentioned before, the twos complement of a binary number is obtained by replacing each 0 with a 1 and each 1 with a 0 and adding 1 to the resulting number. The first step generates a ones complement or simply the complement of a binary number. For example, the ones complement of 10010101 is 01101010. *Note that the ones complement of a binary number can be obtained by using inverters; eight inverters are required for generating ones complement of an 8-bit number.*

The twos complement of a binary number is formed by adding 1 to the ones complement of the number. For example, the twos complement of 10010101 is found as follows:

binary number	10010101
1's complement	01101010
add 1	+1
2's complement	01101011

Now, using the twos complement, binary subtraction can be carried out.

Consider the following subtraction using the normal (pencil and paper) procedure:

minuend	0101	(5)
subtrahend	−0011	(−3)
result	0010	2

Using the twos complement subtraction,

minuend	0101
2's complement of subtrahend	1101
	1 0010

discard final carry

The final answer is 0010 (decimal 2).

Consider another example. Using pencil and paper method:

minuend	0101	(5)
subtrahend	− 1001	(−9)
result	− 0100	(−4)

Using the twos complement,

minuend	0101
2's complement of subtrahend	0111
result	1100

no final carry

Therefore, the final answer is −(twos complement of 1100) = −0100, which is −4 in decimal.

Computers typically handle signed numbers by using the most significant bit of a number as the sign bit. If this bit is 0, then the number is positive; otherwise the number is negative. Computers use the twos complement of the number to represent negative binary numbers and obtain the sign of the result from the most significant bit. In the paper and pencil method, the sign of the result of binary subtraction using twos complement can be obtained by utilizing either the most significant bit of the result or the final carry.

For example, the number $+22_{10}$ can be represented using 8 bits as:

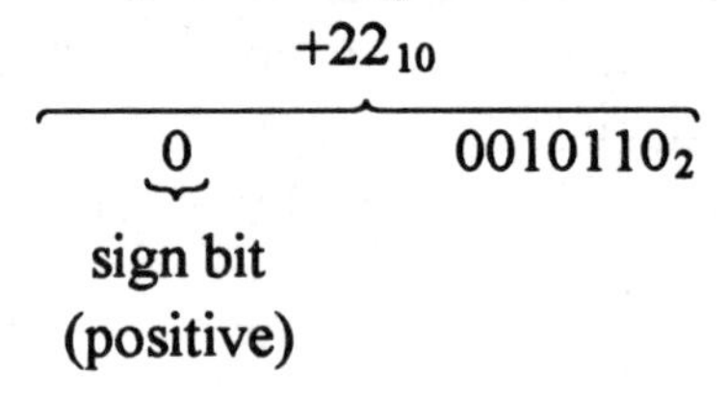

Hence,

$$-22_{10} = \overbrace{\underbrace{1}_{\substack{\text{sign bit} \\ \text{(negative)}}} \quad 1101010}^{\text{twos complement of } +22_{10}}$$

We now show the procedures for carrying out the addition and subtraction in computers using twos complement arithmetic.

Examples of arithmetic operations of the signed binary numbers are give below. Assume 5 bits to represent each number.

1. Both augend and addend are positive:

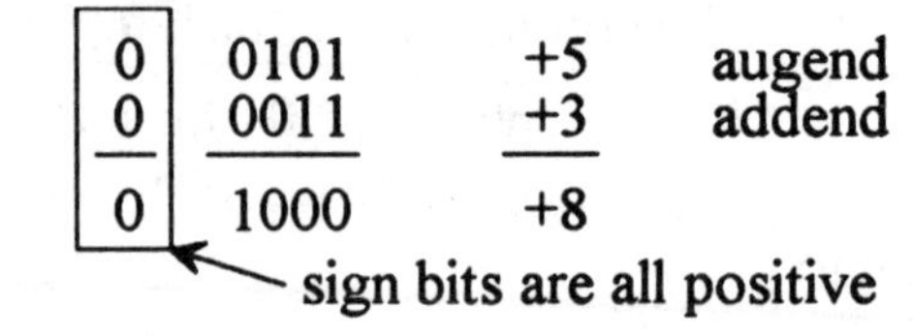

2. Augend is positive, addend is negative:

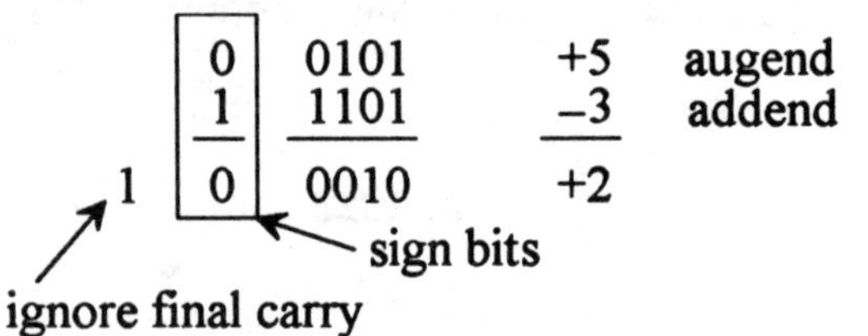

Note that the twos complement of 3 is 11101.
Consider another example:

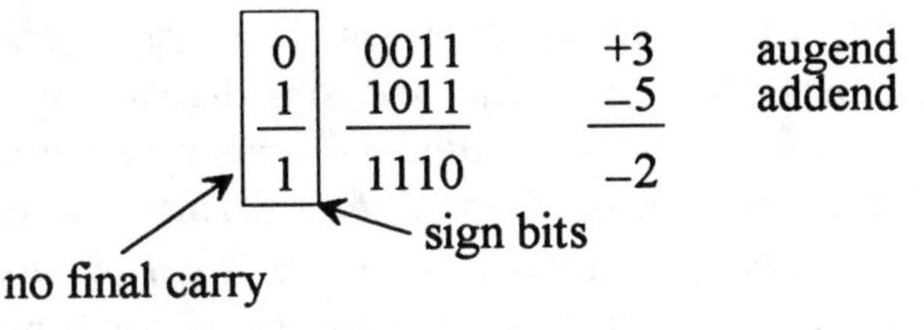

The result is the twos complement of 11110, which is 00010, and therefore, the final answer is -2_{10}.

3. Both augend and addend are negative:

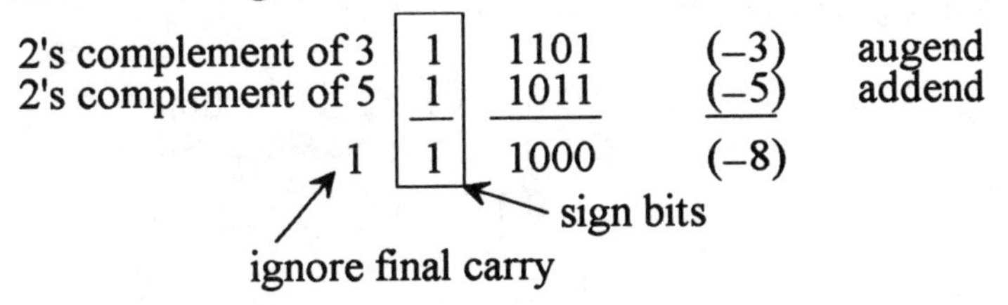

Therefore, the result in binary is 11000. Since the most significant bit is 1, the result is negative. Hence, the result in decimal will be −(twos complement of 11000), which is -8_{10}.

4. Equal augend and addend with opposite signs:

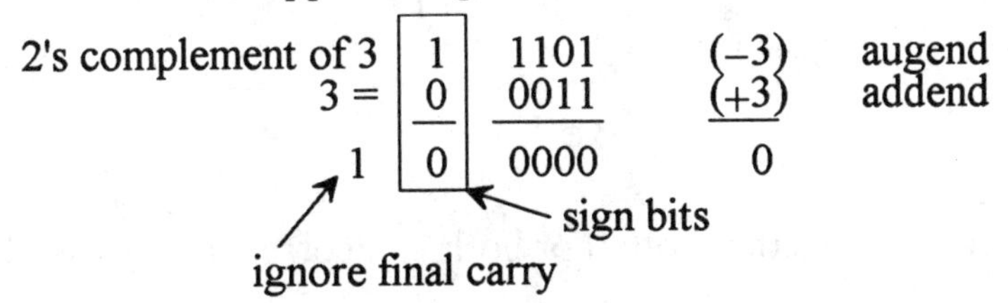

The final answer is zero.

In all these cases, the sign bit of each of the numbers is conceptually isolated from the number itself. The subtraction operation performed here is similar to twos complement subtraction. For example, when subtracting the subtrahend from the minuend using twos complement, the subtrahend is converted into its twos complement along with the sign bit. If the sign bit of the subtrahend is 1 (for negative subtrahend), its twos complement converts the sign bit from 1 to 0. To perform the subtraction, the twos complement of the subtrahend is added to the minuend. The sign bit of the result provides whether the answer is positive or negative.

However, an error (indicated by overflow in a microprocessor) may occur while performing twos complement arithmetic. *The overflow arises from the representation of the sign flag by the most significant bit of a binary number in signed binary operation. The computer automatically sets an overflow bit to 1 if the result of an arithmetic operation is too big for the computer's maximum word size; otherwise it is reset to 0.* To clearly understand the concept of overflow, consider the following examples for 8-bit numbers. Let C_7 be the carry out of the most significant bit (sign bit) and C_6 be the carry out of the previous (bit 6) data bit (seventh bit). We will show by means of numerical examples that as long as C_7 and C_6 are the same, the result is always correct. If, however, C_7 and C_6 are different, the result is incorrect and sets the overflow bit to 1. Now consider the following cases.

Case 1: C_7 and C_6 are the same.

```
              0 0 0 0 0 1 1 0        06₁₆
              0 0 0 1 0 1 0 0       +14₁₆
          → 0 0 0 0 1 1 0 1 0        1A₁₆
C7 = 0      ↑
          C6 = 0

              0 1 1 0 1 0 0 0        68₁₆
              1 1 1 1 1 0 1 0       -06₁₆
          → 1 0 1 1 0 0 0 1 0        62₁₆
C7 = 1      ↑
          C6 = 1
```

Therefore when C_7 and C_6 are either both 0 or both 1, a correct answer is obtained.

Case 2: C_7 and C_6 are different.

```
              0 1 0 1 1 0 0 1        59₁₆
              0 1 0 0 0 1 0 1       +45₁₆
          → 0 1 0 0 1 1 1 1 0       -62₁₆ ?
C7 = 0      ↑
          C6 = 1
```

$C_6 = 1$ and $C_7 = 0$ give an incorrect answer because the result shows that the addition of two positive numbers is negative.

```
                10110110        -4A16
                10000001        -7F16
             1  00110111        +3716 ?
    C7 = 1       ↑
               C6 = 0
```

$C_6 = 0$ and $C_7 = 1$ provide an incorrect answer because the result indicates that the addition of two negative numbers is positive. *Hence, the overflow bit will be set to zero if the carries C_7 and C_6 are the same, that is, if both C_7 and C_6 are either 0 or 1. On the other hand, the overflow flag will be set to 1 if the carries C_7 and C_6 are different. The answer is incorrect when the overflow bit is set to 1.* Thus,

$$\text{Overflow} = C_7 \oplus C_6.$$

Note that the symbol $\oplus$ represents exclusive-OR logic operation. Exclusive-OR means that when two inputs are the same (both one or both zero), the output is zero. On the other hand, if two inputs are different, the output is one. The overflow can be considered as the output while C_6 and C_7 are the two inputs. The exclusive-OR operation is covered in Chapter 3.

While performing signed arithmetic using pencil and paper, one must consider the overflow bit to ensure that the result is correct. *An overflow of one after a signed operation indicates that the result is too large to be accommodated in the number of bits assigned. One must increase the number of bits for the correct result.*

Example 2.3

Perform the following signed operations and comment on the results. Assume twos complement numbers.

(a) $A = 1010_2$, $B = 0100_2$. Find $A - B$.

(b) Perform $(-3_{10}) - (-2_{10})$ using twos complement and 4 bits.

Solution

(a) The most significant bit of A is 1, so A is a negative number whereas B is a positive number.

$$\begin{array}{rl} A = & 1010 \\ \text{Add 2's complement of } B = & +1100 \\ \hline & 0110 = 6 \\ C_3 = 1, & C_2 = 0 \end{array} \qquad \begin{array}{r} (-6_{10}) \\ -(+4_{10}) \\ \hline -10_{10} \end{array}$$

Because C_3 and C_2 are different, there is an overflow and the result is incorrect. Four bits are too small to hold the correct answer. If we increase the number of bits for A and B to 5, the correct result can be obtained as follows:

$$A = -6_{10} = 11010_2$$
$$B = +4_{10} = 00100_2$$

$$\begin{array}{rl} A = & 11010_2 \\ \text{Add 2's complement of } B = & +11100_2 \\ \hline & 10110_2 \\ C_4 = 1, & C_3 = 1 \end{array}$$

The result is correct because C_4 and C_3 are the same. The most significant bit of the result is 1. This means that the result is negative. Therefore, to express the result in base-10, one must take the twos complement and convert the binary number to decimal and place a negative sign in front of it. Thus, twos complement of $10110_2 = -01010_2 = -10_{10}$.

(b)

$$-3_{10} = \text{2's complement of} +3_{10}$$
$$= 1101_2$$
$$-2_{10} = \text{2's complement of } +2_{10}$$
$$= 1110_2$$

$$\begin{array}{rl} -3_{10} = & 1101_2 \\ \text{Add 2's complement of} -2_{10} = & +0010_2 \\ \hline & 1111 \\ C_3 = 0, & C_2 = 0 \end{array} \qquad \begin{array}{r} (-3_{10}) \\ -(-2_{10}) \\ \hline -1_{10} \end{array}$$

C_2 and C_3 are the same, so the result is correct. The most significant bit of the result is 1. This means that the result is negative. To find the result in decimal, one must take twos complement of the result and place a negative sign in front of it.

Twos complement of $1111_2 = -0001_2 = -1_{10}$

2.5.1.3 Multiplication of Unsigned Binary Numbers

Multiplication of two binary numbers can be carried out in the same way as is done with the decimal numbers using pencil and paper. Consider the following example:

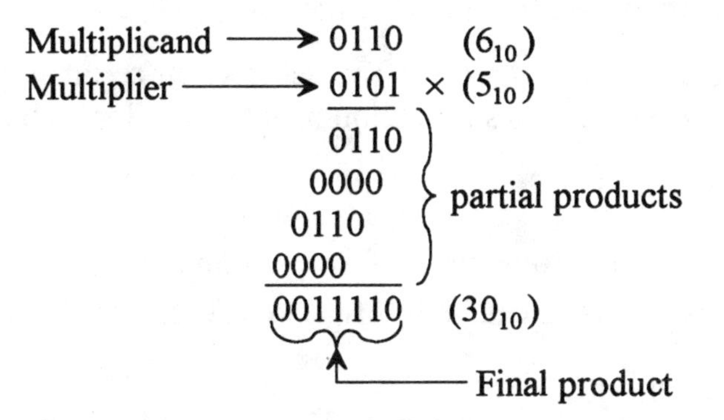

Several multiplication algorithms are available. *Multiplication of two unsigned numbers can be accomplished via repeated addition.* For example, to multiply 4_{10} by 3_{10}, the number 4_{10} can be added twice to itself to obtain the result, 12_{10}.

2.5.1.4 Division of Unsigned Binary Numbers

Binary division is carried out in the same way as the division of decimal numbers. As an example, consider the following division:

```
          110 ←— Quotient = 6₁₀
  011 )10100 ←— Dividend = 20₁₀
   ↗   011
Divisor = 3₁₀  100 ←— Partial Remainders
        011
         010
         000
          10 ←— Remainder = 2₁₀
```

```
     6 ←— quotient
  3 )20 ←— dividend
     18
      2 ←— remainder
```

Division between unsigned numbers can be accomplished via repeated subtraction. For example, consider dividing 7_{10} by 3_{10} as follows:

Dividend	Divisor	Subtraction Result	Counter
7_{10}	3_{10}	$7 - 3 = 4$	1
		$4 - 3 = 1$	$1 + 1 = 2$

Quotient = Counter value = 2
Remainder = subtraction result = 1

Here, one is added to a counter whenever the subtraction result is greater than the divisor. The result is obtained as soon as the subtraction result is smaller than the divisor.

2.5.2 BCD Arithmetic

Many computers have instructions to perform arithmetic operations using packed BCD numbers. Next, we consider some examples of packed BCD addition and subtraction.

2.5.2.1 BCD Addition

The two cases that may occur while adding two packed BCD numbers are considered next. Consider adding packed BCD numbers 25 and 33:

```
 25        0010        0101
+33        0011        0011
---   ---------------------
 58        0101        1000
```

In this example, none of the sums of the pairs of decimal digits exceeded 9; therefore, no decimal carries were produced. For these reasons, the BCD addition process is straightforward and is actually the same as binary addition.

Now, consider the addition of 8 and 4 in BCD:

```
  8        0000        1000
 +4        0000        0100
---   ---------------------
 12        0000        1100   ← invalid code group for BCD
```

The sum 1100 does not exist in BCD code. It is one of the six forbidden or invalid 4-bit code groups. This has occurred because the sum of two digits exceeds 9. Whenever this occurs, the sum has to be corrected by the addition of 6 (0110) to skip over the six invalid code groups. For example,

```
  8        0000        1000
 +4        0000        0100
---   ---------------------
 12        0000        1100        invalid sum
          +0000        0110        add 6 for correction
      ---------------------
           0001        0010        BCD for 12
            1           2
```

As another example, add packed BCD numbers 56 and 81:

56	0101	0110	BCD for 56
+81	1000	0001	BCD for 81
137	1101	0111	invalid sum in 2nd digit
	+0110		add 6 for correction
0001 (1)	0011 (3)	0111 (7)	← correct answer 137

2.5.2.2 BCD Subtraction

Subtraction of packed BCD numbers can be accomplished in a number of different ways. One method is to add the 10's complement of the subtrahend to the minuend using packed BCD addition rules, as described earlier.

One means of finding the 10's complement of a *d*-digit packed BCD number N is to take the twos complement of each digit individually, producing a number N_1. Then, ignoring any carries, add the *d*-digit factor M to N_1, where the least significant digit of M is 1010 and all remaining digits of M are 1001.

As an example, consider subtracting 26_{10} from 84_{10} using BCD subtraction. This can be accomplished as follows:

26_{10}	0010 (2)	0110 (6)

Now, the 10's complement of 26_{10} can be found according to the rules by individually determining the twos complement of 2 and 6, adding the 10's complement factor, and discarding any carries. The twos complement of 2 is 1110, and the twos complement of 6 is 1010. Therefore,

2's complement of each digit of 26_{10}		1110		1010
addition factor to find 10's complement		+1001		1010
10's complement of 26_{10}	(1)	0111 (7)	(1)	0100 (4)

ignore these carries

10's complement of 26_{10}		0111	0100
84_{10}		+1000	0100
		1111	1000
BCD correction factor		+0110	
	(1)	0101 (5)	1000 (8)

ignore carry

Therefore, the final answer is 58_{10}.

2.5.3 Multiword Binary Addition and Subtraction

In many cases, the word length of a particular microprocessor may not be large enough to represent the desired magnitude of a number. Suppose, for example, that numbers in the range from 0 to 65,535 are to be used in an 8-bit microprocessor in binary addition and subtraction operations using the twos complement number representation. This can be accomplished by storing the 16-bit numbers each in two 8-bit memory locations. Addition or subtraction of the two 16-bit numbers is implemented by adding or subtracting the lower 8 bits of each number, storing the result in 8-bit memory location or register, and then adding the two high-order parts of the number with any carry or borrow generated from the first addition or subtraction. The latter partial sum or difference will be the high-order portion of the result. Therefore, the two 8-bit operations together comprise the 16-bit result.

Here are some examples of 16-bit addition and subtraction.

16-Bit Addition

```
                        upper half of the     lower half of the
                         16-bit number         16-bit number
                          01001011              01111010
                         +00101110              00101101
intermediate
  carries ---------------> 111                  1111
                        ----------------------------------
                          01111001              10100111
                        high byte of the      low byte of the
                            answer                answer
```

The low-order 8-bit addition can be computed by using the 8-bit microprocessor's ADD *instruction and the high-order 8-bit sum can be obtained by using the* ADC *(ADD with carry) instruction in the program.*

16-Bit Subtraction

Consider $23A6_{16} - 124A_{16} = 115C_{16}$.

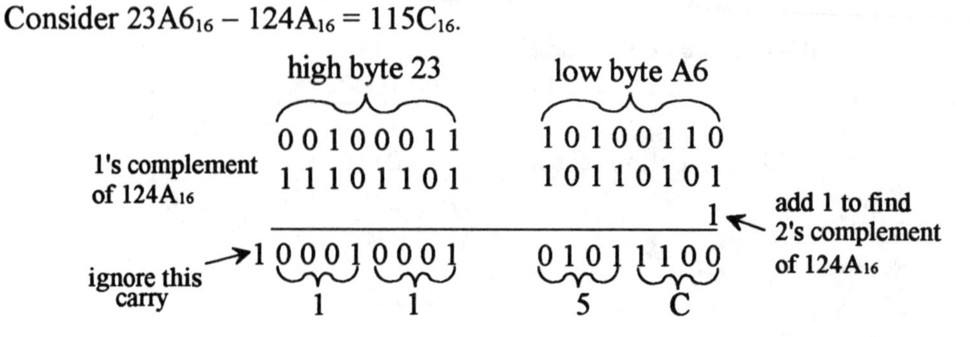

The low-order 8-bit subtraction can be obtained by using `SUB` *instruction of the 8-bit microprocessor, and the high-order 8-bit subtraction can be obtained by using* `SBB` *(SUBTRACT with borrow) instruction in the program.*

2.6 Error Correction and Detection

In digital systems, it is possible that the transmitted information is not received correctly. Note that a computer is a digital system in which information transfer can take place in many ways. For example, data may be moved from a CPU register to another device or vice versa. When the transmitted data is not received correctly at the receiving end, an error occurs. One possible cause for such errors is noise problems during transmission. To avoid these problems, error detection and correction may be necessary. In a digital system, an error occurs when a 0 is changed to a 1 and vice versa. Correction of this error means replacement of a 1 with 0 and vice versa. The reliability of digital data depends on the methods employed for error detection and correction.

The simplest way to detect the presence of an error is by adding a single bit, called the "parity" bit, to the message bits and then transmitting the message along with the parity bit. The parity bit is usually computed in two ways: even parity and odd parity. *In the even parity method, the parity bit is added in such a way that after its inclusion, the number of 1's in the message together with the parity bit is an even number. On the other hand, in an odd parity scheme, the parity bit is added in such a way that the number of 1's in the message and the parity bit is an odd number*. For example, suppose a message to be transmitted is 0110. If even parity is used by the transmitting computer, the transmitted data along with the parity will be 00110. On the other hand, if odd parity is used, the data to be transmitted will be 10110. The parity computation can be implemented in hardware by using exclusive-OR gates (to be discussed in Chapter 3). Usually for a given message, the parity bit is generated using either an even or odd parity scheme by the transmitting computer. The message is then transmitted along with the parity bit. At the receiving end, the parity is checked by the receiving computer. If there is a discrepancy, the data received will obviously be incorrect. For example, suppose that the message bits are 1101. The even parity bit for this message is 1. The transmitted data will be

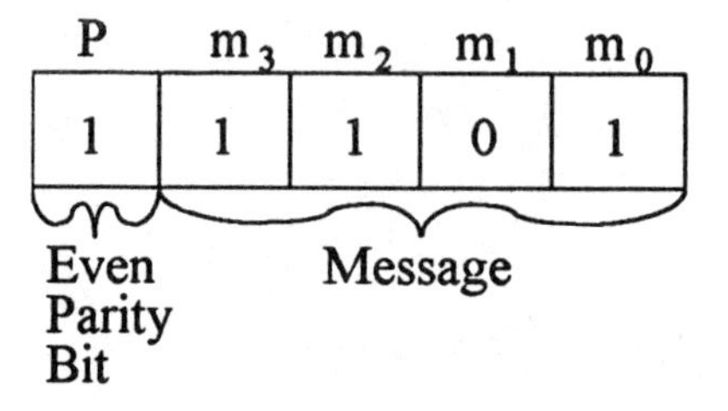

Suppose that an error occurs in the least significant bit; that is m_0 is changed from 1 to 0 during transmission. The received data will be

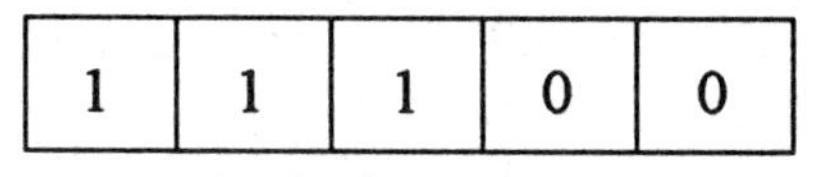

The receiving computer performs parity check on this data by counting the number of ones and finds it to be an odd number, three. Therefore, an error is detected.

With a single parity bit, an error due to a single bit change can be detected. Errors due to 2-bit changes during transmission will go undetected. In such situations, multiple parity bits are used. One such technique is the "Hamming code," which uses 3 parity bits for a 4-bit message.

QUESTIONS AND PROBLEMS

2.1 Convert the following unsigned binary numbers into their decimal equivalents: 01110101_2; 1101.101_2; 1000.111_2.

2.2 Convert the following numbers into binary:
(a) 152_{10} (b) 343_{10}

2.3 Convert the following numbers into octal:
(a) 1843_{10} (b) 1766_{10}

2.4 Convert the following numbers into hexadecimal
(a) 1987_{10} (b) 3072_{10}

2.5 Convert the following binary numbers into octal and hexadecimal numbers:
(a) 1101011100101 (b) 1100001100110000011

2.6 Using 8 bits, represent the integers –48 and 52 in
 (a) sign magnitude form
 (b) ones complement form
 (c) twos complement form

2.7 Identify the following unsigned binary numbers as odd or even without converting them to decimal: 11001100_2; 00100100_2; 01111001_2.

2.8 Convert 532.372_{10} into its binary equivalent.

2.9 Convert the following hex numbers to binary: $15FD_{16}$; $26EA_{16}$.

2.10 Represent the following BCD numbers into binary:
 (a) 11264 (b) 8192

2.11 Represent the following numbers in excess-3:
 (a) 678 (b) 32874 (c) 61440

2.12 What is the excess-3 equivalent of octal 1543?

2.13 Represent the following binary numbers in BCD:
 (a) 0001 1001 0101 0001
 (b) 0110 0001 0100 0100 0000

2.14 Express the following binary numbers into excess-3:
 (a) 0101 1001 0111
 (b) 0110 1001 0000

2.15 Perform the following unsigned binary addition. Include the answer in decimal.

$$\begin{array}{r} 1\,0\,1\,1.0\,1 \\ +\,0\,1\,1\,0.0\,1\,1 \\ \hline \end{array}$$

2.16 Perform the indicated arithmetic operations in binary. Assume that the numbers are in decimal and represented using 8 bits. Express results in decimal. Use twos complement approach for carrying out all subtractions.

(a) $\begin{array}{r} 14 \\ +17 \\ \hline \end{array}$ (c) $\begin{array}{r} 32 \\ -14 \\ \hline \end{array}$

(b) $\begin{array}{r} 34 \\ +28 \\ \hline \end{array}$ (d) $\begin{array}{r} 34 \\ -42 \\ \hline \end{array}$

2.17 Using twos complement, perform the following subtraction: $3AFA_{16} - 2F1E_{16}$. Include answer in hex.

2.18 Using 9's and 10's complement arithmetic, perform the following arithmetic operations: (a) $254_{10} - 132_{10}$ (b) $783_{10} - 807_{10}$

2.19 Perform the following arithmetic operations in binary using 6 bits. Assume all numbers are signed decimal. Use twos complement arithmetic. Indicate if there is any overflow.

(a) $\begin{array}{r} 14 \\ \underline{+8} \end{array}$ (b) $\begin{array}{r} 7 \\ \underline{+(-7)} \end{array}$ (c) $\begin{array}{r} 27 \\ \underline{+(-19)} \end{array}$

(d) $\begin{array}{r} (-24) \\ \underline{+(-19)} \end{array}$ (e) $\begin{array}{r} 19 \\ \underline{-(-12)} \end{array}$ (f) $\begin{array}{r} (-17) \\ \underline{-(-16)} \end{array}$

2.20 Perform the following unsigned multiplication in binary using a minimum number of bits required for each decimal number using pencil and paper method:

$$12 \times 52$$

2.21 Perform the following unsigned division in binary using a minimum number of bits required for each decimal number:

$$3\overline{)14}$$

2.22 Convert the following BCD numbers into binary and then perform the indicated arithmetic operations in BCD:

(a) $\begin{array}{r} 54 \\ \underline{+48} \end{array}$ (b) $\begin{array}{r} 782 \\ \underline{+219} \end{array}$ (c) $\begin{array}{r} 82 \\ \underline{-58} \end{array}$

2.23 Find the odd parity bit for the following binary message to be transmitted: 10110000.

2.24 Repeat Problem 2.20 using repeated addition.

2.25 Repeat Problem 2.21 using repeated subtraction.

2.26 If a transmitting computer sends an 8-bit binary message 11000111 using an even parity bit. Write the 9-bit data with the parity bit in the most significant bit. If the receiving computer receives the 9-bit data as 110000111, is the 8-bit message received correctly? Comment.

3

BOOLEAN ALGEBRA AND DIGITAL LOGIC GATES

This chapter describes fundamentals of logic operations, Boolean algebra, minimization techniques, and implementation of basic digital circuits.

Digital circuits contain hardware elements called "gates" that perform logic operations on binary numbers. Devices such as transistors can be used to perform the logic operations. Boolean algebra is a mathematical system that provides the basis for these logic operations. George Boole, an English mathematician, introduced this theory of digital logic. The term *Boolean variable* is used to mean the two-valued binary digit 1 or 0.

3.1 Basic Logic Operations

Boolean algebra uses three basic logic operations namely, NOT, OR, and AND. These operations are described next.

3.1.1 NOT Operation

The NOT operation inverts or provides the ones complement of a binary digit. This operation takes a single input and generates one output. The NOT operation of a binary digit provides the following result:

NOT 1 = 0
NOT 0 = 1

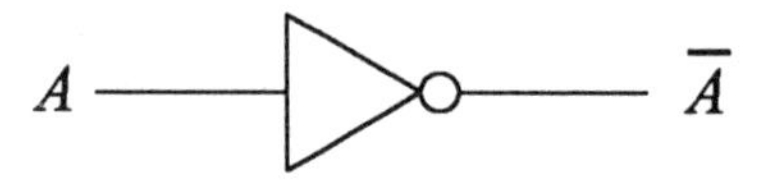

FIGURE 3.1 Symbol for a NOT gate

Therefore, NOT of a Boolean variable A, written as $\overline{A}$ (or A') is 1 if and only if A is 0. Similarly, $\overline{A}$ is 0 if and only if A is 1. This definition may also be specified in the form of a truth table:

Input	Output
A	$\overline{A}$
0	1
1	0

Note that a truth table contains the inputs and outputs of digital logic circuits. The symbolic representation of an electronic circuit that implements a NOT operation is shown in Figure 3.1.

A NOT gate is also referred to as an "inverter" because it inverts the voltage levels. As discussed in Chapter 1, *a transistor acts as an inverter.* A 0-volt at the input generates a 5-volt output; a 5-volt input provides a 0-volt output.

As an example, the 74HC04 (or 74LS04) is a hex inverter 14-pin chip containing six independent inverters in the same chip as shown in Figure 3.2.

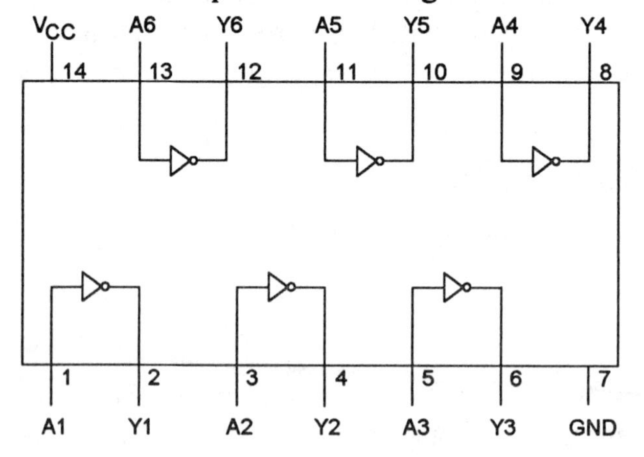

FIGURE 3.2 Pin diagram for the 74HC04 or 74LS04

Computers normally include a NOT instruction to perform the ones complement of a binary number on a bit-by-bit basis. An 8-bit computer can perform NOT operation on an 8-bit binary number. For example, the computer can execute a NOT instruction on an 8-bit binary number 01101111 to provide the result 10010000. *The computer utilizes an internal electronic circuit consisting of eight inverters to invert the 8-bit data in parallel.*

3.1.2 OR operation

The OR operation for two variables A and B generates a result of 1 if A or B, or both, are 1. However, if both A and B are zero, then the result is 0.

A plus sign + (logical sum) or ∨ symbol is normally used to represent OR. The four possible combinations of ORing two binary digits are

$$0 + 0 = 0$$
$$0 + 1 = 1$$
$$1 + 0 = 1$$
$$1 + 1 = 1$$

A truth table is usually used with logic operations to represent all possible combinations of inputs and the corresponding outputs. The truth table for the OR operation is

Inputs		
A	B	*Output* $= A + B$
0	0	0
0	1	1
1	0	1
1	1	1

Figure 3.3 shows the symbolic representation of an OR gate.

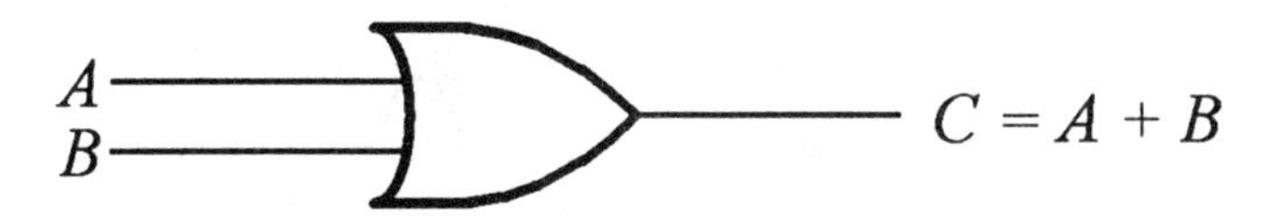

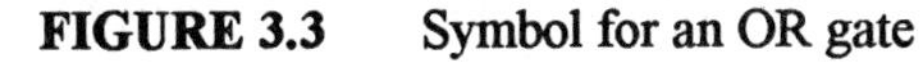
FIGURE 3.3 Symbol for an OR gate

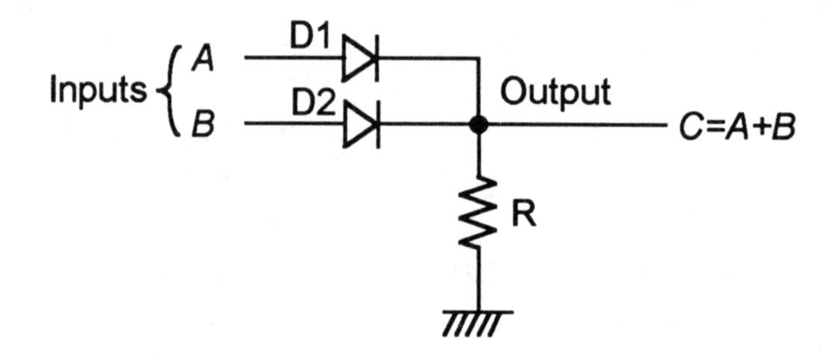

FIGURE 3.4 Diode OR gate

Logic gates using diodes provide good examples to understand how semiconductor devices are utilized in logic operations. Note that diodes are hardly used in designing logic gates. Figure 3.4 shows a two-input-diode OR gate. The diode (see Chapter 1) is a switch, and it closes when there is a voltage drop of 0.6 V between the anode and the cathode. Suppose that a voltage range of 0 to 2 V is considered as logic 0 and a voltage of 3 to 5 V is logic 1. If both *A* and *B* are at logic 0 (say 1.5 V) with a voltage drop across the diodes of 0.6 V to close the diode switches, a current flows from the inputs through *R* to ground, output *C* will be at 1.5 V - 0.6 V = 0.9 V (logic 0). On the other hand, if one or both inputs are at logic 1 (say 4.5 V) the output *C* will be at 4.5 - 0.6 V = 3.9 V (logic 1). Therefore, the circuit acts as an OR gate.

The 74HC32 (or 74LS32) is a commercially available quad 2-input 14-pin OR gate chip. This chip contains four 2-input/1-output independent OR gates as shown in Figure 3.5.

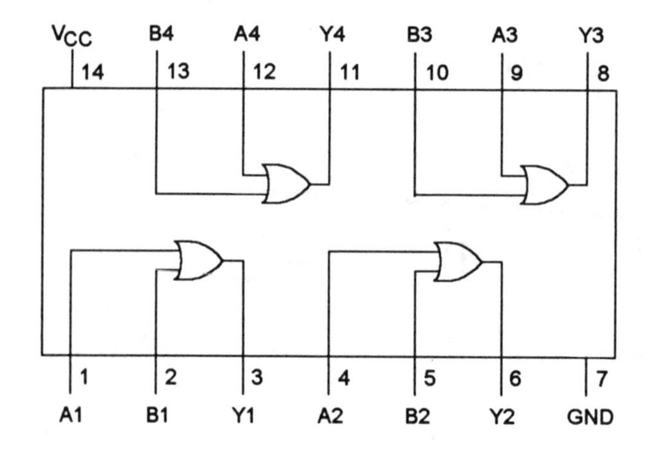

FIGURE 3.5 Pin diagram for 74HC32 or 74LS32

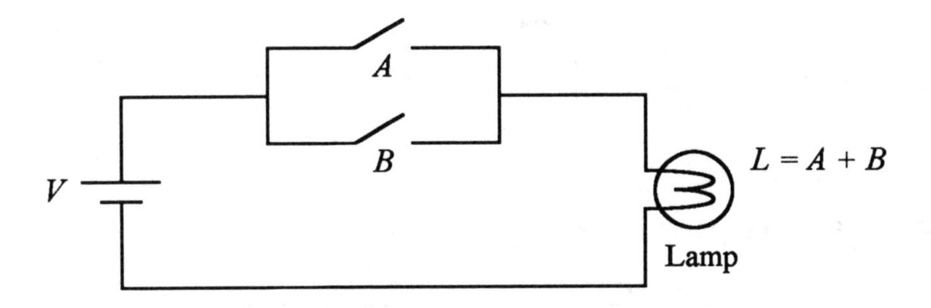

FIGURE 3.6 An example of the OR operation

To understand the logic OR operation, consider Figure 3.6. V is a voltage source, A and B are switches, and L is an electrical lamp. L will be turned ON if either switch A or B or both are closed; otherwise, the lamp will be OFF. Hence, $L = A + B$. *Computers normally contain an OR instruction to perform the OR operation between two binary numbers.* For example, the computer can execute an OR instruction to OR $3A_{16}$ with 21_{16} on a bit by bit basis:

$$\begin{aligned} 3A_{16} &= 0011\ 1010 \\ 21_{16} &= \underline{0010\ 0001} \\ & \quad \underbrace{0011}_{3}\ \underbrace{1011}_{B}\,{}_{16} \end{aligned}$$

The computer typically utilizes eight two-input OR gates to accomplish this.

3.1.3 AND operation

The AND operation for two variables A and B generates a result of 1 if both A and B are 1. However, if either A or B, or both, are zero, then the result is 0.

The dot $\cdot$ and $\wedge$ symbol are both used to represent the AND operation. The AND operation between two binary digits is

$$0 \cdot 0 = 0$$

$$0 \cdot 1 = 0$$

$$1 \cdot 0 = 0$$

$$1 \cdot 1 = 1$$

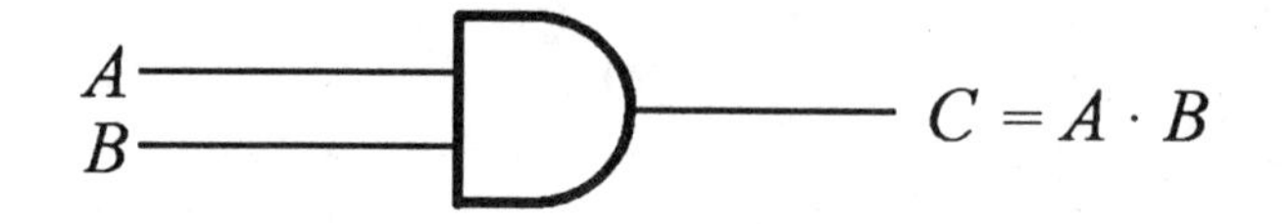

FIGURE 3.7 AND gate symbol

The truth table for the AND operation is

Inputs		
A	B	$Output = A \cdot B = AB$
0	0	0
0	1	0
1	0	0
1	1	1

Figure 3.7 shows the symbolic representation of an AND gate. Figure 3.8 shows a two-input diode AND gate.

As we did for the OR gate, let us assume that the range 0 to +2 V represents logic 0 and the range 3 to 5 V is logic 1. Now, if A and B are both HIGH (say 3.3 V) and the anode of both diodes at 3.9 V, the switches in D_1 and D_2 close. A current flows from +5 V through resistor R to +3.3 V input to ground. The output C will be HIGH (3.9 V). On the other hand, if a low voltage (say 0.5 V) is applied at A and a high voltage (3.3V) is applied at B. The value of R is selected in such a way that 1.1 V appears at the anode side of D_1; at the same time 3.9 V appears at the anode side of D_2. The switches in both diodes will close because each has a voltage drop of 0.6 V between the anode and cathode. A current flows from the +5 V input through R and the diodes to ground. Output C will be low (1.1 V) because the output will be lower of the two voltages. Thus, it can be shown that when either one or both inputs are low, the output is low, so the circuit works as an AND gate. As mentioned before, diode logic gates are easier to understand, but they are not normally used these days.

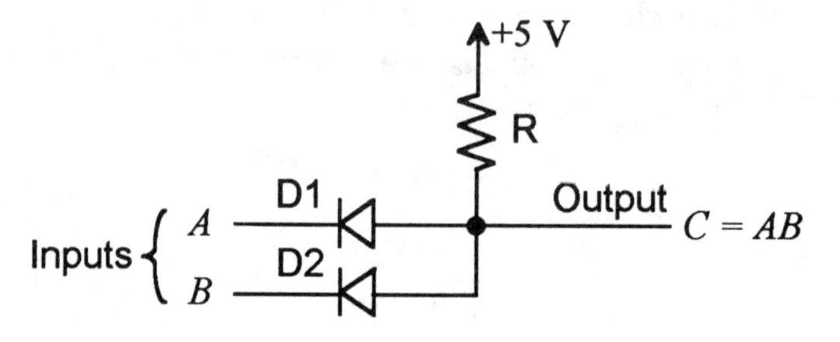

FIGURE 3.8 Diode AND gate

Transistors are utilized in designing logic gates. Diode logic gates are provided as examples in order to illustrate how semiconductor devices are utilized in designing them.

The 74HC08 (or 74LS08) is a commercially available quad 2-input 14-pin AND gate chip. This chip contains four 2-input/1-output independent AND gates as shown in Figure 3.9. To illustrate the logic AND operation consider Figure 3.10. The lamp L will be on when both switches A and B are closed; otherwise, the lamp L will be turned OFF. Hence,

$$L = A \cdot B$$

Computers normally have an instruction to perform the AND operation between two binary numbers. For example, the computer can execute an AND instruction to perform ANDing 31_{16} with $A1_{16}$ as follows:

$$\begin{array}{rl} 31_{16} & = 0011\ 0001 \\ A1_{16} & = \underline{1010\ 0001} \\ & \ \underbrace{0010}_{2}\ \underbrace{0001}_{1}\ {}_{16} \end{array}$$

The computer utilizes eight two-input AND gates to accomplish this.

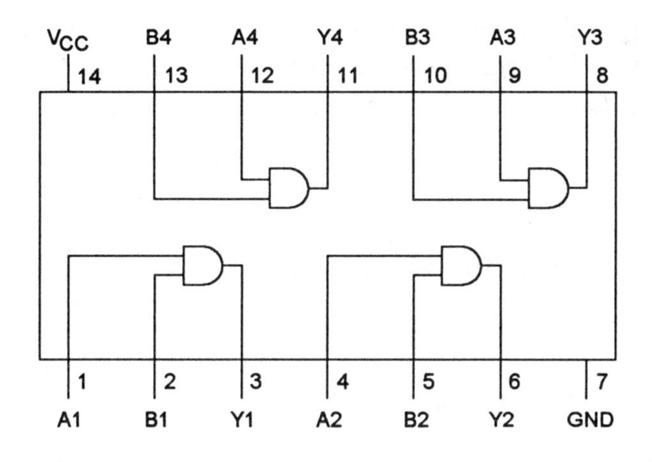

FIGURE 3.9 Pin Diagram for 74HC08 or 74LS08

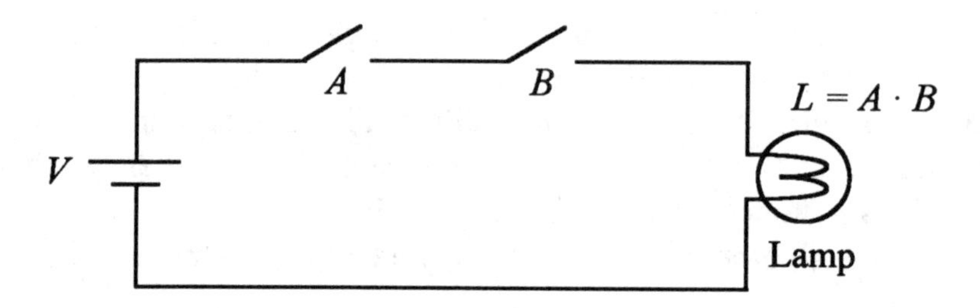

FIGURE 3.10 An example of the AND operation

3.2 Other Logic Operations

The four other important logic operations are NOR, NAND, Exclusive-OR (XOR) and Exclusive-NOR (XNOR).

3.2.1 NOR operation

The NOR output is produced by inverting the output of an OR operation. Figure 3.11 shows a NOR gate along with its truth table. Figure 3.12 shows the symbolic representation of a NOR gate. In the figure, the small circle at the output of the NOR gate is called the inversion bubble. The 74HC02 (or 74LS02) is a commercially available quad 2-input 14-pin NOR gate chip. This chip contains four 2-input/1-output independent NOR gates as shown in Figure 3.13.

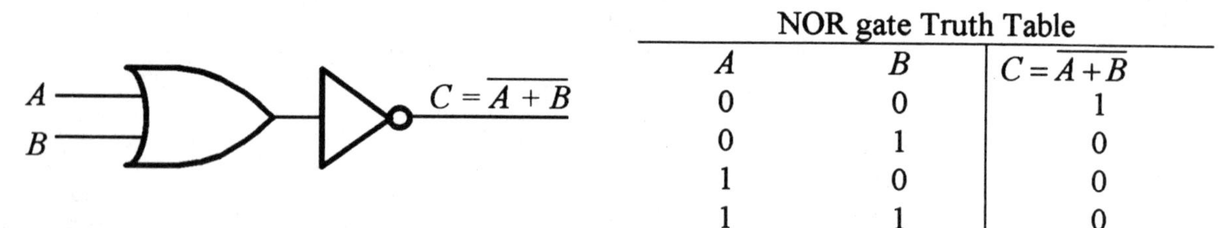

NOR gate Truth Table

A	B	$C = \overline{A+B}$
0	0	1
0	1	0
1	0	0
1	1	0

FIGURE 3.11 A NOR gate with its truth table

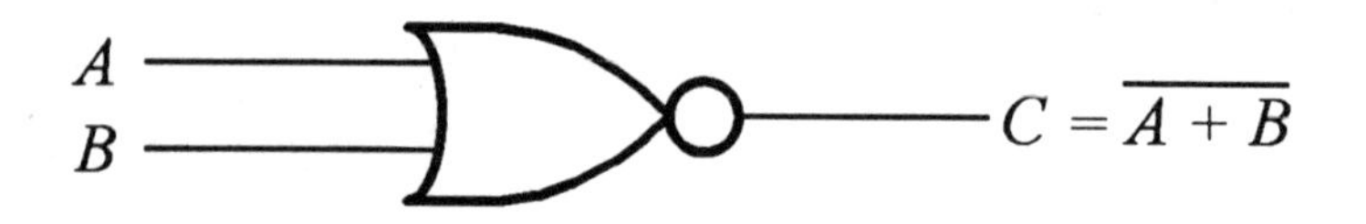

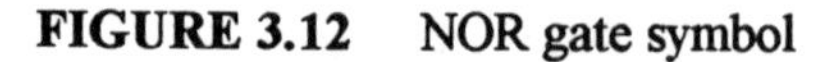

FIGURE 3.12 NOR gate symbol

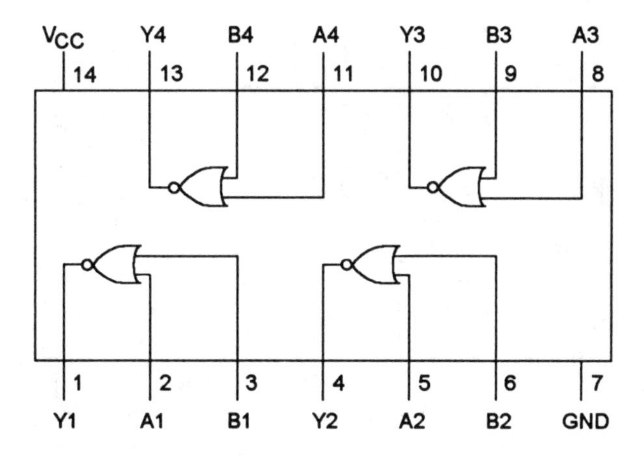

FIGURE 3.13 Pin diagram for 74HC02 or 74LS02

3.2.2 NAND operation

The NAND output is generated by inverting the output of an AND operation. Figure 3.14 shows a NAND gate and its truth table. Figure 3.15 shows the symbolic representation of a NAND gate.

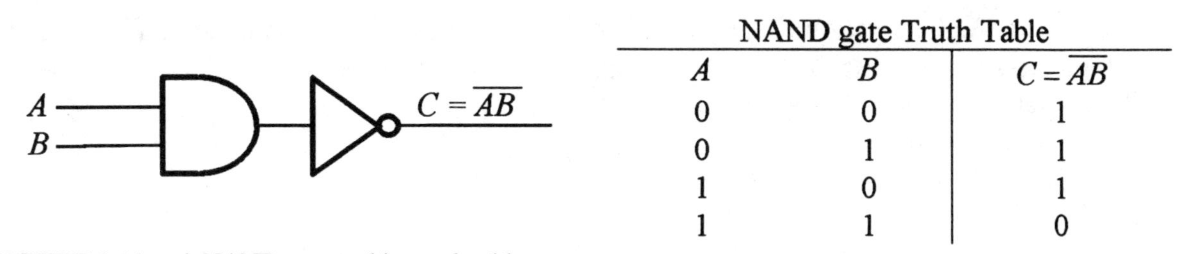

NAND gate Truth Table

A	B	$C = \overline{AB}$
0	0	1
0	1	1
1	0	1
1	1	0

FIGURE 3.14 A NAND gate and its truth table

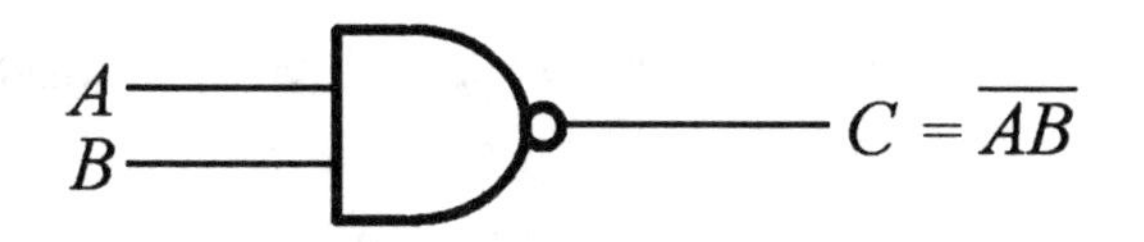

FIGURE 3.15 NAND gate symbol

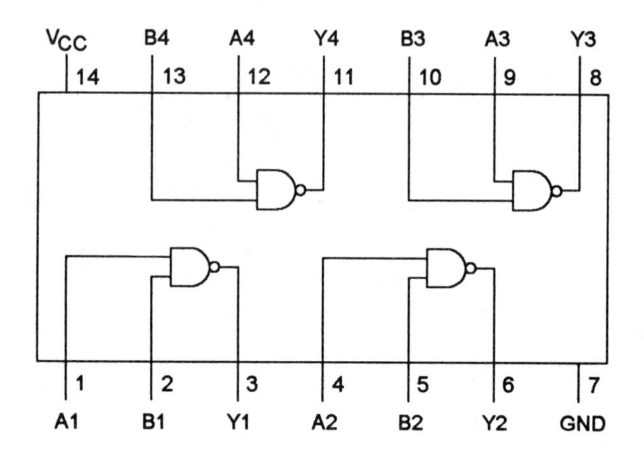

FIGURE 3.16 Pin diagram for 74HC00 or 74LS00

The 74HC00 (or 74LS00) is a commercially available quad 2-input/1-output 14-pin NAND gate chip. This chip contains four 2-input/1-output independent NAND gates as shown in Figure 3.16.

3.2.3 Exclusive-OR operation (XOR)

The Exclusive-OR operation (XOR) generates an output of 1 if the inputs are different and 0 if the inputs are the same. The ⊕ or ∀ symbol is used to represent the XOR operation. The XOR operation between binary digits is

$$0 \oplus 0 = 0$$

$$0 \oplus 1 = 1$$

$$1 \oplus 0 = 1$$

$$1 \oplus 1 = 0$$

Most computers have an instruction to perform the XOR operation. Consider XORing $3A_{16}$ with 21_{16}.

$$\begin{array}{rl} 3A_{16} = & 0011\ \ 1010 \\ 21_{16} = & \underline{0010\ \ 0001} \\ & \underbrace{0001}_{1}\ \ \underbrace{1011}_{B}{}_{16} \end{array}$$

It is interesting to note that XORing any number with another number of the same length but with all 1's will generate the ones complement of the original number. For example, consider XORing 31_{16} with FF_{16}:

$$
\begin{array}{rl}
31_{16} & 0011\ 0001 \\
\text{1's complement of } 31_{16} & \underbrace{1100}_{C}\ \underbrace{1110}_{E_{16}}
\end{array}
$$

$$
\begin{array}{rl}
31_{16} \oplus FF_{16} & 0011\ 0001 \\
 & 1111\ 1111 \\ \hline
 & \underbrace{1100}_{C}\ \underbrace{1110}_{E_{16}}
\end{array}
$$

The truth table for Exclusive-OR operation is

Inputs		*Output*
A	B	$C = A \oplus B$
0	0	0
0	1	1
1	0	1
1	1	0

From the truth table, $A \oplus B$ is 1 only when $A = 0$ and $B = 1$ or $A = 1$ and $B = 0$. Therefore,

$$C = A \oplus B = A\overline{B} + \overline{A}B$$

Figure 3.17 shows an implementation of an XOR gate using AND and OR gates. Figure 3.18 shows the symbolic representation of the Exclusive-OR gate assuming that both true and complimented values of A and B are available.

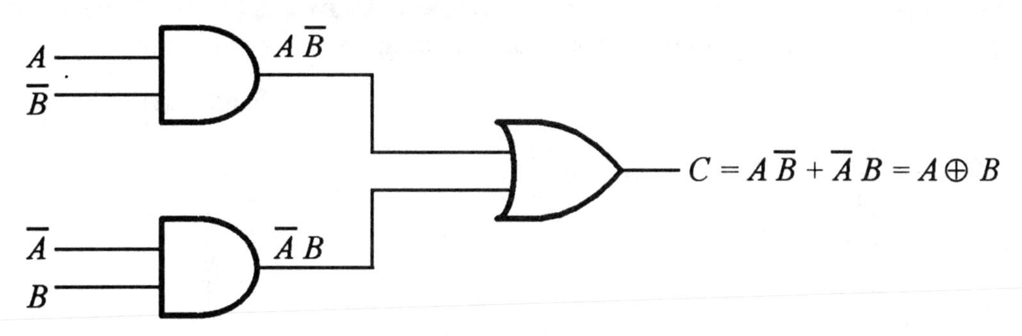

FIGURE 3.17 AND-OR Implementation of the Exclusive-OR gate

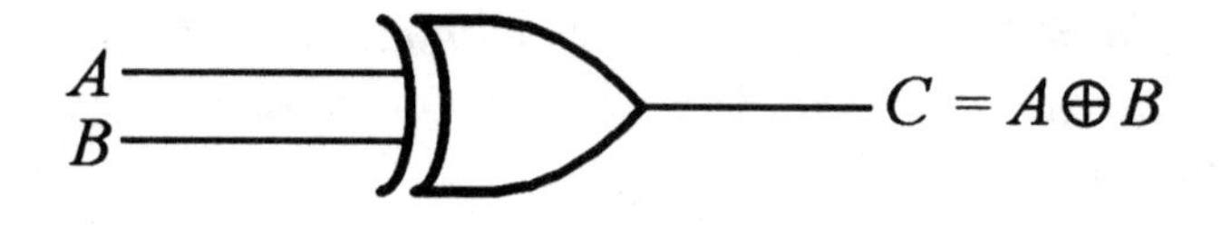

FIGURE 3.18 XOR symbol

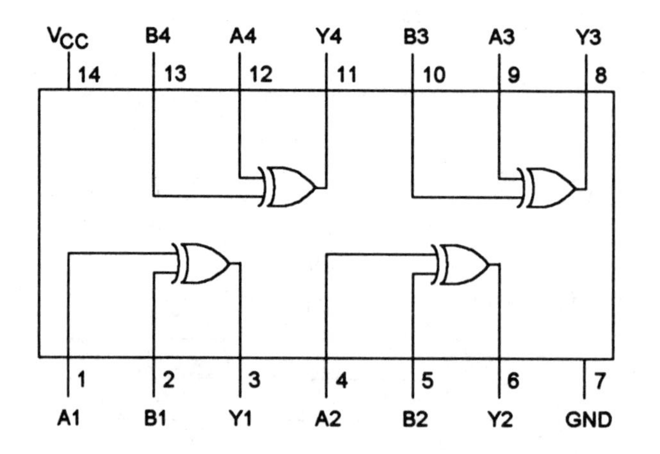

FIGURE 3.19 Pin diagram for 74HC86 or 74LS86

The 74HC86 (or 74LS86) is a commercially available quad 2-input 14-pin Exclusive-OR gate chip. This chip contains four 2-input/1-output independent exclusive-OR gates as shown in Figure 3.19.

3.2.4 Exclusive-NOR Operation (XNOR)

The one's complement of the Exclusive-OR operation is known as the Exclusive-NOR operation. Figure 3.20 shows its symbolic representation along with the truth table.

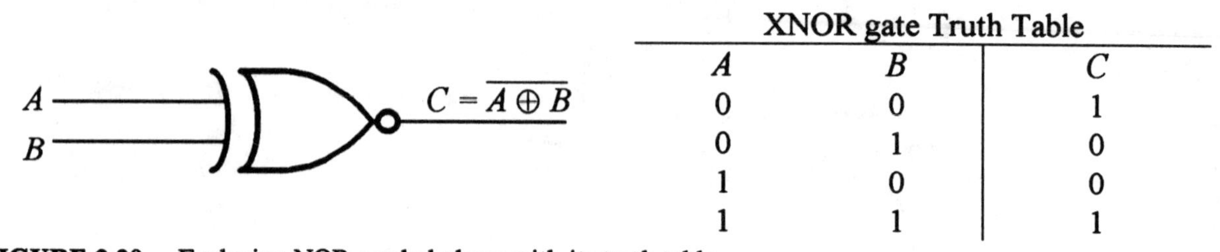

XNOR gate Truth Table

A	B	C
0	0	1
0	1	0
1	0	0
1	1	1

FIGURE 3.20 Exclusive-NOR symbol along with its truth table

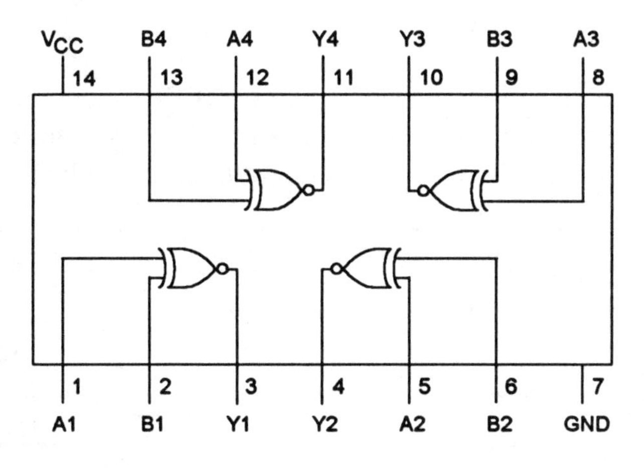

FIGURE 3.21 Pin Diagram for 74HC266 or 74LS266

The XNOR operation is represented by the symbol $\odot$. Therefore, $C = \overline{A \oplus B} = A \odot B$. *The XNOR operation is also called equivalence*. From the truth table, output C is 1 if both A and B are 0's or both A and B are 1's; otherwise, C is 0. That is, $C = 1$, for $A = 0$ and $B = 0$ or $A = 1$ and $B = 1$. Hence,

$$C = A \odot B = \overline{A}\,\overline{B} + AB$$

The 74HC266 (or 74LS266) is a quad 2-input/1-output 14-pin Exclusive-NOR gate chip. This chip contains four 2-input/1-output independent Exclusive-NOR gates shown in Figure 3.21.

Note that the symbol C is chosen arbitrarily in all the above logic operations to represent the output of each logic gate. Also, note that all logic gates (except NOT) can have at least two inputs with only one output. The NOT gate, on the other hand, has one input and one output.

3.3 IEEE Symbols for Logic Gates

The institute of Electrical and Electronics Engineers (IEEE) recommends rectangular shape symbols for logic gates: The original logic symbols have been utilized for years and will be retained in the rest of this book. IEEE symbols for gates are listed below:

Gate	Common Symbol	IEEE Symbol
AND	A, B — $f = AB$	A, B — & — $f = AB$
OR	A, B — $f = A + B$	A, B — ≥1 — $f = A + B$
NOT	A — $f = \overline{A}$	A — 1 — $f = \overline{A}$
NAND	A, B — $f = \overline{AB}$	A, B — & — $f = \overline{AB}$
NOR	A, B — $f = \overline{A + B}$	A, B — ≥1 — $f = \overline{A + B}$
Exclusive-OR	A, B — $f = A \oplus B$	A, B — =1 — $f = A \oplus B$
Exclusive-NOR	A, B — $f = \overline{A \oplus B}$	A, B — =1 — $f = \overline{A \oplus B}$

3.4 Positive and Negative Logic

The inputs and outputs of logic gates are represented by either logic 1 or logic 0. There are two ways of assigning voltage levels to the logic levels, positive logic and negative logic. The positive logic convention assigns a HIGH (H) voltage for logic 1 and LOW (L) voltage for logic 0. On the other hand, in the negative logic convention, a logic 1 = LOW (L) voltage and logic 0 = HIGH (H) voltage.

The IC data sheets typically define these levels in terms of voltage levels rather than logic levels. The designer decides on whether to use positive or negative logic. As an example, consider a gate with the following truth table:

A	*B*	*f*
L	*L*	*H*
L	*H*	*H*
H	*L*	*H*
H	*H*	*L*

Using positive logic, (H = 1 and L = 0) the following table is obtained:

A	*B*	*f*
0	0	1
0	1	1
1	0	1
1	1	0

This is the truth table for a NAND gate. However, negative logic, (H = 0 and L = 1) provides the following table:

A	*B*	*f*
1	1	0
1	0	0
0	1	0
0	0	1

This is the truth table for an NOR gate. Note that converting from positive to negative logic and vice versa for logic gates basically provides the *dual* (discussed later in this chapter) of a function. This means that changing 0's to 1's and 1's to 0's for both inputs and outputs of a logic gate, the logic gate is converted from an NOR gate to an NAND gate as shown in the example. In this book, the positive logic convention will be used.

Note that positive logic and active high logic are equivalent (HIGH = 1, LOW = 0). On the other hand, negative logic and active low logic are equivalent (HIGH = 0, LOW = 1).

A signal is "active high" if it performs the required function when HIGH (H = 1). An "active low" signal, on the other hand, performs the required function when LOW (L = 0). A signal is said to be asserted when it is active. A signal is disasserted when it is not at its active level.

Active levels may be associated with inputs and outputs of logic gates. For example, an AND gate performs a logical AND operation on two active HIGH inputs and provides an active HIGH output. This also means that if both the inputs of the AND gate are asserted, the output is asserted.

3.5 Boolean Algebra

Boolean algebra provides basis for logic operations using binary variables. Alphabetic characters are used to represent the binary variables. A binary variable can have either true or complement value. For example, the binary variable A can be either A and/or $\overline{A}$ in a Boolean function.

A Boolean function is an operation expressing logical operations between binary variables. The Boolean function can have a value of 0 or 1. As an example of a Boolean function, consider the following:

$$f = \overline{A}\,\overline{B} + C$$

Here, the Boolean function f is 1 if both $\overline{A}$ and $\overline{B}$ are 1 or C is 1; otherwise, f is 0. Note that $\overline{A}$ means that if $A = 1$, then $\overline{A} = 0$. Thus, when $B = 1$, then $\overline{B} = 0$. It can therefore be concluded that f is one when $A = 0$ and $B = 0$ or $C = 1$.

A truth table can be used to represent a Boolean function. The truth table contains a combination of 1's and 0's for the binary variables. Furthermore, the truth table provides the value of the Boolean function as 1 or 0 for each combination of the input binary variables. Table 3.1 provides the truth table for the Boolean function $f = \overline{A}\,\overline{B} + C$. In the table, if $A = 1$, $B = 1$, and $C = 0$, $f = 0.0 + 0 = 0$. Note that table 3.1 contains three input variables (A, B, C) and one output variable (f). Also, by ORing ones in the truth table, the function f contains several terms; however, the function can be simplified using the techniques to be discussed later.

TABLE 3.1 Truth Table for $f = \overline{A}\,\overline{B} + C$

A	*B*	*C*	*f*
0	0	0	1
0	0	1	1
0	1	0	0
0	1	1	1
1	0	0	0
1	0	1	1
1	1	0	0
1	1	1	1

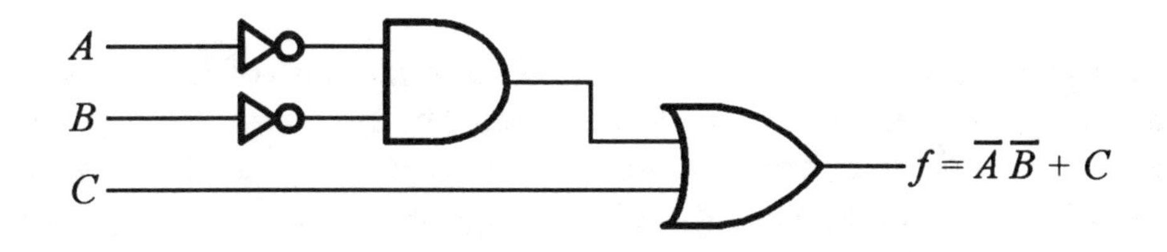

FIGURE 3.22 Logic diagram for $f = \overline{A}\,\overline{B} + C$

A Boolean function can also be represented in terms of a logic diagram. Figure 3.22 shows the logic diagram for $f = \overline{A}\,\overline{B} + C$. The Boolean expression $f = \overline{A}\,\overline{B} + C$ contains two terms, $\overline{A}\,\overline{B}$ and C, which are inputs to logic gates. Each term may include a single or multiple variables, called "literals," that may or may not be complemented. For example, $f = \overline{A}\,\overline{B} + C$ contains three literals, $\overline{A}$, $\overline{B}$, and C. *Note that a variable and its complement are both called literals. For two variables, the literals are A, B, $\overline{A}$, and $\overline{B}$.*

Boolean functions can be simplified by using the rules (identities) of Boolean algebra. This allows one to minimize the number of gates in a logic diagram, which reduces the cost of implementing a logic circuit.

3.5.1 Boolean Identities

Here is a list of Boolean identities that are useful in simplifying Boolean expressions:

1. a) $A + 0 = A$ b) $A \cdot 1 = A$
2. a) $A + 1 = 1$ b) $A \cdot 0 = 0$
3. a) $A + A = A$ b) $A \cdot A = A$
4. a) $A + \overline{A} = 1$ b) $A \cdot \overline{A} = 0$
5. a) $(\overline{\overline{A}}) = A$
6. Commutative Law:
   a) $A + B = B + A$ b) $A \cdot B = B \cdot A$
7. Associative Law:
   a) $A + (B + C) = (A + B) + C$ b) $A \cdot (B \cdot C) = (A \cdot B) \cdot C$
8. Distributive Law:
   a) $A \cdot (B + C) = A \cdot B + A \cdot C$ b) $A + B \cdot C = (A + B) \cdot (A + C)$
9. DeMorgan's Theorem:
   a) $\overline{A + B} = \overline{A} \cdot \overline{B}$ b) $\overline{A \cdot B} = \overline{A} + \overline{B}$

In the list, each identity identified by b) on the right is the dual of the corresponding identify a) on the left. *Note that the dual of a Boolean expression is obtained by changing 1's to 0's and*

0's to 1's if they appear in the equation, and AND to OR and OR to AND on both sides of the equal sign.

For example, consider identity 4. Relation 4a is the dual of relation 4b because the AND in the expression is replaced by an OR and then, 0 by 1.

The Duality Principle of Boolean algebra states that a Boolean expression is unchanged if the dual of both sides of the equal sign is taken. Consider, for example, the Boolean function,

$f = \overline{B} + \overline{A}\,\overline{B}$ Therefore, $f = \overline{B} \cdot (1 + \overline{A})$

$= \overline{B}$

The dual of f,

$$f_D = \overline{B} \cdot (\overline{A} + \overline{B})$$

$$f_D = \overline{B} \cdot \overline{A} + \overline{B} \cdot \overline{B} = \overline{B}\,\overline{A} + \overline{B}$$

$$= \overline{B}(\overline{A} + 1) = \overline{B}$$

Hence, $f = f_D$. In order to verify some of the identities, consider the following examples:

i) Identity 2a) $A + 1 = 1$
For $A = 0$, $A + 1 = 0 + 1 = 1$
For $A = 1$, $A + 1 = 1 + 1 = 1$

ii) Identity 4b) $A \cdot \overline{A} = 0$. If $A = 1$, then $\overline{A} = 0$. Hence, $A \cdot \overline{A} = 1 \cdot 0 = 0$

iii) *Identity 8b)* $A + B \cdot C = (A + B) \cdot (A + C)$ *is very useful in manipulating Boolean expressions.* This identity can be verified by means of a truth table as follows:

A	B	C	$B \cdot C$	$A + B$	$A + C$	$A + B \cdot C$	$(A + B) \cdot (A + C)$
0	0	0	0	0	0	0	0
0	0	1	0	0	1	0	0
0	1	0	0	1	0	0	0
0	1	1	1	1	1	1	1
1	0	0	0	1	1	1	1
1	0	1	0	1	1	1	1
1	1	0	0	1	1	1	1
1	1	1	1	1	1	1	1

iv) *Identities 9a) and 9b) (DeMorgan's Theorem) are useful in determining one's complement of a Boolean expression.* DeMorgan's theorem can be verified by means of a truth table as follows:

A	B	$\overline{A}$	$\overline{B}$	$\overline{A} \cdot \overline{B}$	$A + B$	$\overline{A + B}$	$A \cdot B$	$\overline{A \cdot B}$	$\overline{A} + \overline{B}$
0	0	1	1	1	0	1	0	1	1
0	1	1	0	0	1	0	0	1	1
1	0	0	1	0	1	0	0	1	1
1	1	0	0	0	1	0	1	0	0

De Morgan's Theorem can be expressed in a general form for *n* variables as follows:

$$\overline{A + B + C + D + \ldots} = \overline{A} \cdot \overline{B} \cdot \overline{C} \cdot \overline{D} \cdot \ldots$$

$$\overline{A \cdot B \cdot C \cdot D \cdot \ldots} = \overline{A} + \overline{B} + \overline{C} + \overline{D} + \ldots$$

The logic gates except for the inverter can have more than two inputs if the logic operation performed by the gate is commutative and associative (Identities 6a and 7a on Page 83). For example, the OR operation has these two properties as follows: A + B = B + A (Commutative) and (A+B) + C = A+ (B + C) = A + B + C (Associative). This means that the OR gate inputs can be interchanged. Thus, the OR gate can have more than two inputs . Similarly, using the identities 6b and 7b on Page 83, it can be shown that the AND gate can also have more than two inputs. Note that the NOR and NAND operations, on the other hand, are commutative, but not associative. Therefore, it is not possible to have NOR and NAND gates with more than two inputs. However, NOR and NAND gates with more than two inputs can be obtained by using inverted OR and inverted AND respectively. The Exclusive-OR and Exclusive-NOR operations are both commutative and associative. Thus, these gates can have more than two inputs. However, Exclusive-OR and Exclusive-NOR gates with more than two inputs are uncommon from the hardware point of view.

3.5.2 Simplification Using Boolean Identities

Although there are no defined set of rules for minimizing a Boolean expression, appropriate identities can be used to accomplish this. Consider the Boolean function

$$f = ABCD + \bar{A}BCD + \overline{BC}$$

This equation can be implemented using logic gates as shown in Figure 3.23(a). The expression can be simplified by using identities as follows:

$f = BCD(A + \bar{A}) + \overline{BC}$ By identity 4a)

$= BCD \cdot 1 + \overline{BC}$ By identity 1b)

$= BCD + \overline{BC}$

Assume $BC = E$, then $\overline{BC} = \bar{E}$ and,

$f = ED + \bar{E}$

$= (E + \bar{E})(\bar{E} + D)$ By identity 8b)

$= \bar{E} + D$ By identity 4a)

Substituting $\bar{E} = \overline{BC}$, $f = \overline{BC} + D$

The simplified form is implemented using logic gates in Figure 3.23(b). The logic diagram in Figure 3.23(b) requires only one NAND gate and an OR gate. This implementation is inexpensive compared to the circuit of Figure 3.23(a). Both logic circuits perform the same function.

The following truth table can be used to show that the outputs produced by both circuits are equivalent:

A	B	C	D	$f = ABCD + \bar{A}BCD + \overline{BC}$	$f = \overline{BC} + D$
0	0	0	0	1	1
0	0	0	1	1	1
0	0	1	0	1	1
0	0	1	1	1	1
0	1	0	0	1	1
0	1	0	1	1	1
0	1	1	0	0	0
0	1	1	1	1	1
1	0	0	0	1	1
1	0	0	1	1	1
1	0	1	0	1	1
1	0	1	1	1	1
1	1	0	0	1	1
1	1	0	1	1	1
1	1	1	0	0	0
1	1	1	1	1	1

The following are some more examples for simplifying Boolean expressions using identities:

i)

$$f = \bar{X} + \bar{Y} + \overline{XY} + \overline{XY}Z$$
$$= \overline{XY} + \overline{XY} + \overline{XY}Z$$
$$= \overline{XY} + \overline{XY}Z$$
$$= \overline{XY}(1 + Z)$$
$$= \overline{XY}$$

ii)

$$f = \overline{AB}CD + \bar{A}CD + \bar{B}CD + (1 \oplus AB)CD$$
$$= \overline{AB}CD + CD(\bar{A} + \bar{B}) + \overline{AB}CD$$
$$= \overline{AB}CD + \overline{AB}CD + \overline{AB}CD$$
$$= \overline{AB}CD$$

iii) Show that $f = \overline{(a + \bar{b})(\bar{a} + b)}$ can be implemented using one Exclusive -OR gate.

Solution: Using DeMorgan's theorem, $f = \overline{(a + \bar{b})(\bar{a} + b)}$

$$= \overline{(a + \bar{b})} + \overline{(\bar{a} + b)} = (\bar{a}.\bar{\bar{b}}) + (\bar{\bar{a}}.\bar{b}) = \bar{a}b + a\bar{b} = a \oplus b$$

iv) Show that $f = \overline{(\bar{A} + \bar{B})(\bar{E} + \bar{F})}$ can be implemented using two AND and one OR gates.

Solution: $f = \overline{\overline{(AB)}\,\overline{(EF)}} = AB + EF$ using DeMorgan's theorem.

v) Express $f = (X + \bar{X}Z)(X + Z)$ using only one two-input OR gate.

Solution: $f = (X + \bar{X})(X + Z)(X + Z)$ using distributive law. Hence, $f = X + Z$

vi) Express f for $\bar{f} = (\bar{A} + \bar{B} + \bar{C}) + \overline{ABC}$ using only one three input AND gate.

Solution: Using DeMorgan's theorem, $f = \bar{\bar{f}} = \overline{(\bar{A} + \bar{B} + \bar{C}) + \overline{ABC}}$

$$= (ABC).(ABC)$$
$$= ABC$$

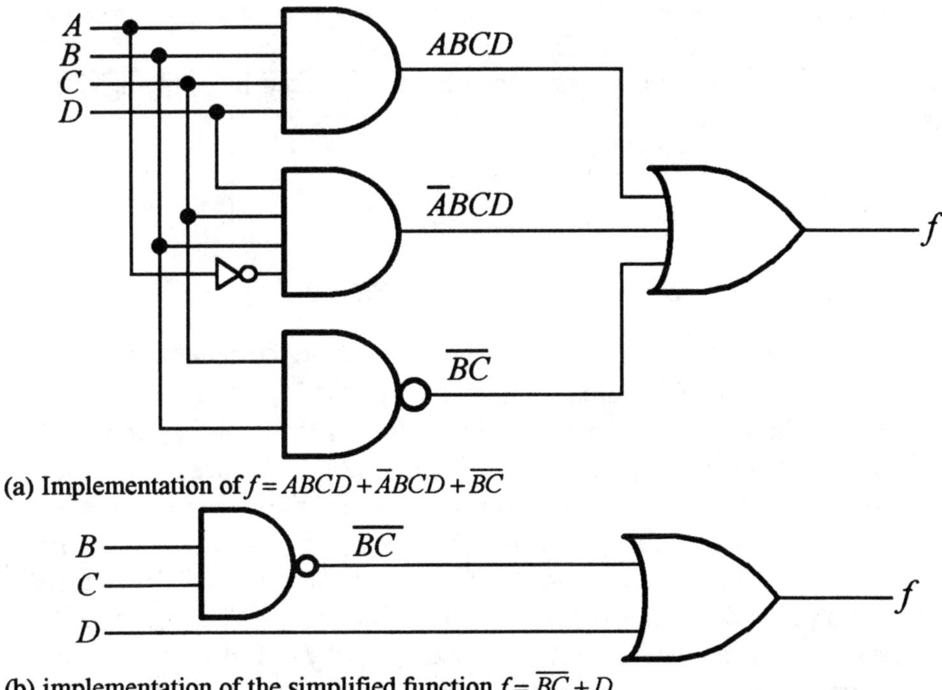

(a) Implementation of $f = ABCD + \overline{A}BCD + \overline{BC}$

(b) implementation of the simplified function $f = \overline{BC} + D$

FIGURE 3.23 Implementation of Boolean function using logic gates

3.5.3 Consensus Theorem

The Consensus Theorem is expressed as

$$AB + \overline{A}C + BC = AB + \overline{A}C$$

The theorem states that the AND term BC can be eliminated from the expression if one of the literals such as B is ANDed with the true value of another literal (A) and the other term C is ANDed with its complement ($\overline{A}$). This theorem can sometimes be applied to minimize Boolean equations.

The Consensus Theorem can be proved as follows:

$$\begin{aligned} AB + \overline{A}C + BC &= AB + \overline{A}C + BC(A + \overline{A}) \\ &= AB + \overline{A}C + ABC + \overline{A}BC \\ &= AB + ABC + \overline{A}C + \overline{A}BC \\ &= AB(1 + C) + \overline{A}C(1 + B) \\ &= AB + \overline{A}C \end{aligned}$$

The dual of the Consensus Theorem can be expressed as

$$(A+B)(\overline{A}+C)(B+C) = (A+B)(\overline{A}+C)$$

To illustrate how a Boolean expression can be manipulated by applying the Consensus Theorem, consider the following:

$$f = (B+\overline{D})(\overline{B}+C)$$

$$= B\overline{B} + BC + \overline{B}\,\overline{D} + C\overline{D}$$

$$= BC + \overline{B}\,\overline{D} + C\overline{D}, \text{ since } B\overline{B} = 0$$

Because C is ANDed with B, and $\overline{D}$ is ANDed with its complement $\overline{B}$, by using the Consensus Theorem, $C\overline{D}$ can be eliminated. Thus, $f = BC + \overline{B}\,\overline{D}$.

The Consensus Theorem can be used in logic circuits for avoiding undesirable behavior. To illustrate this, consider the logic circuits in Figure 3.24. In Figure 3.24(a), the output is one i) if B and C are 1 and $A = 0$ or ii) if B and C are 1 and $A = 1$.

Suppose that in Figure 3.24(a), $B = 1$, $C = 1$, and $A = 0$. Assume that the propagation delay time of each gate is 10 ns (nanoseconds). The circuit output f will be 1 after 30 ns (3 gate delays). Now, if input A changes from 0 to 1, the outputs of NOT gate 1 and AND gate 2 will be 0 and 1 respectively after 10 ns. This will make output $f = 1$ after 20 ns. The output of AND gate 3 will be low after 20 ns, which will not affect the output of f.

Now, assume that B and C stay at 1 while A changes from 1 to 0. The outputs of NOT gate 1 and AND gate 2 will be 1 and 0 respectively after 10 ns. Because the output of AND gate 3 is 0 from the previous case, this will change output of OR gate 4 to 0 for a brief period of time. After 10 ns, the output of AND gate 3 changes to 1, making the output of f HIGH (desired value). Note that, for $B = 1$, $C = 1$, and $A = 0$, the output f should have stayed at 1 from the equation $f = AB + \overline{A}C$. However, f changed to zero for a short period of time. This change is called a "glitch" or "hazard" and occurs from the gate delays in a circuit. Glitches can cause circuit malfunction and should be eliminated. *Application of the Consensus theorem gets rid of the glitch.* By adding the redundant term BC, the modified logic circuit for f is obtained. Figure 3.24(b) shows the logic circuit. Now, consider the case in which the glitch occurs in Figure 3.24(a) when B and C stay at 1 while A changes from 1 to 0. For the circuit in Figure 3.24(b) the glitch will disappear, because $BC = 1$ throughout any changes in values of A and $\overline{A}$. Thus, minimization of logic gates might not always be desirable; rather, a circuit without any hazards would be the main objective of the designer.

There are two types of hazards: static and dynamic. Static hazard occurs when a signal should remain at one value, but instead it oscillates a few times before settling back to its original value. Dynamic hazard occurs, when a signal should make a clean transition to a new logic value, but instead it oscillates between the two logic values before making the

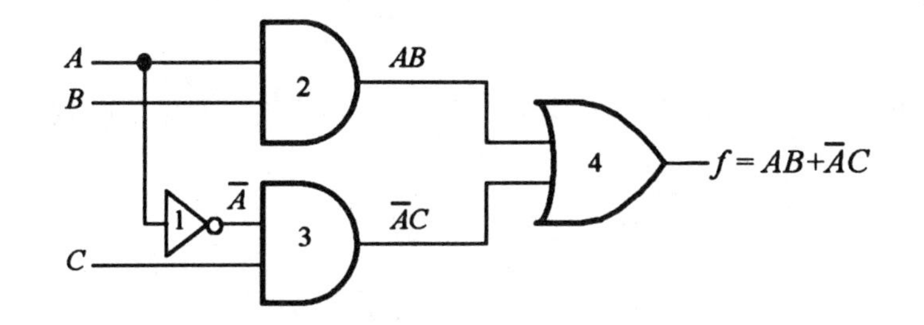

(a) Logic circuit for $f = AB + \overline{A}C$

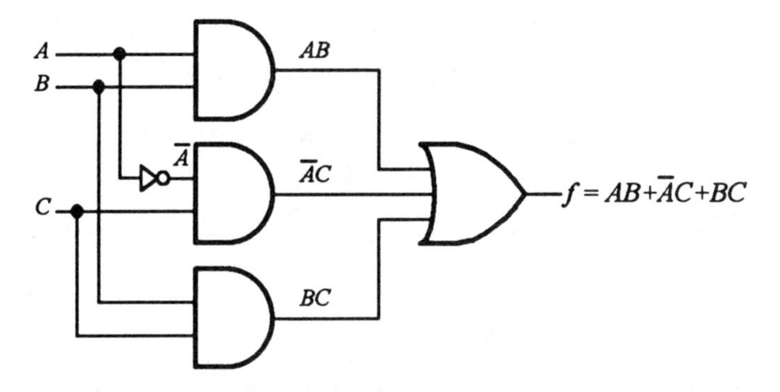

(b) Logic circuit for $f = AB + \overline{A}C + BC$

FIGURE 3.24 Logic circuit for the Consensus Theorem

transition to its final value. Both types of hazards occur because of *races* in the various paths of a circuit. A *race* is a situation in which signals traveling through two or more paths compete with each other to affect a common signal. It is, therefore, possible for the final signal value to be determined by the winner of the race. One way to eliminate races is by applying the Consensus theorem as illustrated in the preceding example.

3.5.4 Complement of a Boolean Function

The complement of a function f can be obtained algebraically by applying DeMorgan's Theorem. It follows from this theorem that the complement of a function can also be derived by taking the dual of the function and complementing each literal.

Example 3.1

Find the complement of the function $f = \overline{C}(AB + \overline{A}\,\overline{B}D + \overline{A}B\overline{D})$

i) Using DeMorgan's Theorem ii) By taking the dual and complementing each literal

Solution

i) Using DeMorgan's Theorem as many times as required, the complement of the function can be obtained:

$$\bar{f} = \overline{\bar{C}\left(AB + \bar{A}\bar{B}D + \bar{A}B\bar{D}\right)}$$

$$= \bar{\bar{C}} + \overline{\left(AB + \bar{A}\bar{B}D + \bar{A}B\bar{D}\right)}$$

$$= C + \left(\overline{AB} \cdot \overline{\bar{A}\bar{B}D} \cdot \overline{\bar{A}B\bar{D}}\right)$$

$$= C + (\bar{A} + \bar{B})(A + B + \bar{D})(A + \bar{B} + D)$$

ii) By taking the dual and complementing each literal, we have:

The dual of f: $\bar{C} + (A + B)(\bar{A} + \bar{B} + D)(\bar{A} + B + \bar{D})$

Complementing each literal: $C + (\bar{A} + \bar{B})(A + B + \bar{D})(A + \bar{B} + D) = \bar{f}$

3.6 Standard Representations

The standard representations of a Boolean function typically contain either logical product (AND) terms called "minterms" or logical sum (OR) terms called "maxterms." These standard representations make the minimization procedures easier. The standard representations are also called "Canonical forms."

A minterm is a product term of all variables in which each variable can be either complemented or uncomplemented. For example, there are four minterms for two variables, A and B. These minterms are $\bar{A}\bar{B}$, $\bar{A}B$, $A\bar{B}$, and AB. On the other hand, there are eight minterms for three variables, A, B, and C. These minterms are $\bar{A}\bar{B}\bar{C}$, $\bar{A}\bar{B}C$, $\bar{A}B\bar{C}$, $\bar{A}BC$, $A\bar{B}\bar{C}$, $A\bar{B}C$, $AB\bar{C}$, and ABC. These product terms represent numeric values from 0 through 7. In general, there are 2^n minterms for n variables.

A minterm is represented by the symbol m_j, where the subscript j is the decimal equivalent of the binary number of the minterm. For example, the decimal equivalents (j) of the binary numbers represented by the four minterms of two variables, A and B, are 0 ($\bar{A}\bar{B}$), 1 ($\bar{A}B$), 2 ($A\bar{B}$), and 3 (AB). Therefore, the symbolic representations of the four minterms of two variables are m_0, m_1, m_2, and m_3 as follows:

A	B	Minterm	Symbol
0	0	$\overline{A}\,\overline{B}$	m_0
0	1	$\overline{A}B$	m_1
1	0	$A\overline{B}$	m_2
1	1	AB	m_3

In general, the n minterms of p ($n = 2^p$) variables are: $m_0, m_1, m_2, \ldots, m_{n-1}$.

It has been shown that a Boolean function can be defined by a truth table. A Boolean function can be exressed in terms of minterms. For example, consider the following truth table:

A	B	f
0	0	1
0	1	0
1	0	1
1	1	1

One can determine the function f by logically summing (ORing) the product terms for which f is 1. Therefore,

$$f = \overline{A}\,\overline{B} + A\overline{B} + AB$$

This is called the Sum-of-Products expression. *A logic diagram of a sum-of-products expression contains several AND gates followed by a single OR gate.* In terms of minterms, f can be represented as:

$$f = \Sigma m(0, 2, 3)$$

The symbol Σ denotes the logical sum (OR) of the minterms.

A maxterm, on the other hand, can be defined as a logical sum (OR) term that contains all variables in complemented or uncomplemented form. The four maxterms of two variables are $A + B, \overline{A} + B, A + \overline{B}$, and $\overline{A} + \overline{B}$.

A maxterm is obtained from the logical sum of all the variables by complementing each variable. Each maxterm is represented by the symbol M_j, where subscript j is the decimal equivalent of the binary number of the maxterm. Therefore, the four maxterms of the two variables, A and B, can be represented as follows:

A	B	Maxterm	Symbol
0	0	$A + B$	M_0
0	1	$A + \overline{B}$	M_1
1	0	$\overline{A} + B$	M_2
1	1	$\overline{A} + \overline{B}$	M_3

In the preceding figure, consider maxterm M_2 as an example. Since $A = 1$ and $B = 0$, the maxterm M_2 is found as $\overline{A} + B$ by taking the logical sum of complement of A (since $A = 1$) and true value of B (since $B = 0$). In general, there are n maxterms (M_0, M_1, ... , M_{n-1}) for p variables, where $n = 2^p$.

The relationship between minterm and maxterm can be established by using DeMorgan's theorem. Consider, for example, minterm m_1 and maxterm M_1 for two variables:

$$m_1 = \overline{A}B, \qquad M_1 = A + \overline{B}$$

Taking the complement of m_1,

$$\overline{m_1} = \overline{\overline{A}B}$$

$$= \overline{\overline{A}} + \overline{B} \text{ by DeMorgan's Theorem}$$

$$= A + \overline{B}$$

$$= M_1$$

Therefore $m_1 = \overline{M}$, or $\overline{m_1} = M_1$. This implies that $m_j = \overline{M_j}$ or $\overline{m_j} = M_j$. *That is, a minterm is the complement of its corresponding maxterm and vice versa.*

In order to represent a Boolean function in terms of maxterms, consider the following truth table:

A	B	f	$\overline{f}$
0	0	1	0
0	1	0	1
1	0	0	1
1	1	0	1

Taking the logical sum of minterms of $\overline{f}$,

$$\overline{f} = \overline{A}B + A\overline{B} + AB$$

$$= m_1 + m_2 + m_3$$

$$= \sum m(1, 2, 3)$$

By taking complement of $\overline{f}$,

$$f = \overline{\overline{f}} = \overline{m_1 + m_2 + m_3} = \overline{m_1} \cdot \overline{m_2} \cdot \overline{m_3}$$

$$= M_1 \cdot M_2 \cdot M_3 \text{ (since } M_j = \overline{m_j}\text{)}$$

$$= (A + \overline{B})(\overline{A} + B)(\overline{A} + \overline{B})$$

This is called the *product-of-sums* expression. *The logic diagram of a product-of-sums expression contains several OR gates followed by a single AND gate*. Hence, $f = \Pi M(1, 2, 3)$ where the symbol Π represents the logical product (AND) of maxterms M_1, M_2, and M_3 in this case. Note that one can express a Boolean function in terms of maxterms by inspecting a truth table and then logically ANDing the maxterms for which the Boolean function has a value of 0.

A Boolean function that is not expressed in terms of sums of minterms or product of maxterms can be represented by a truth table. The function can then be expressed in terms of minterms or maxterms. For example, consider $f = A + B\overline{C}$. The function f is not in a sum of minterms or product of maxterms form, since each term does not include all three variables A, B, and C. The truth table for f can be determined as follows:

A	B	C	$f = A + B\overline{C}$
0	0	0	0
0	0	1	0
0	1	0	1
0	1	1	0
1	0	0	1
1	0	1	1
1	1	0	1
1	1	1	1

From the truth table, the sum of minterm form ($f = 1$) is:

$$f = \Sigma m(2, 4, 5, 6, 7) = \overline{A}B\overline{C} + A\overline{B}\,\overline{C} + A\overline{B}C + AB\overline{C} + ABC$$

From the truth table, the product of maxterm form ($f = 0$) is:

$$f = \Pi M(0, 1, 3) = (A + B + C)(A + B + \overline{C})(A + \overline{B} + \overline{C})$$

The complement of f, $\overline{f} = \Sigma m(0, 1, 3)$, is obtained by the logical sum of minterms for f=0. Also, note that a function containing all minterms is 1. This means that in the above truth table, if f=1 for all eight combinations of A, B, and C, then $f = \Sigma m(0, 1, 2, 3, 4, 5, 6, 7) = 1$.

As mentioned before, the logic diagram of a sum of minterm form contains several AND gates and a single OR gate. This is illustrated by the logic diagram for $f = \Sigma m(2, 4, 5, 6, 7) = \overline{A}B\overline{C} + A\overline{B}\,\overline{C} + A\overline{B}C + AB\overline{C} + ABC$ as shown in figure 3.25(a).

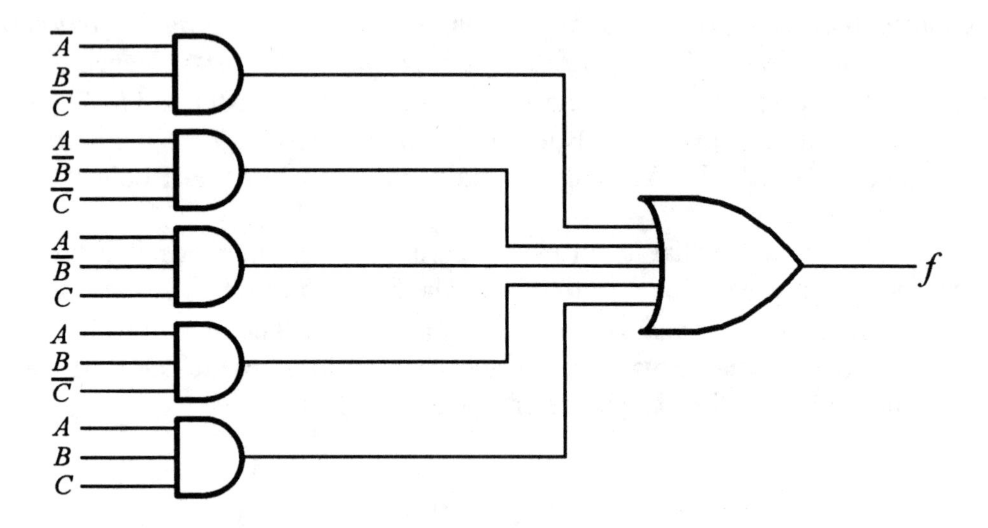

FIGURE 3.25(a) Logic diagram of a sum of minterms

Similarly, the logic diagram of a product of maxterm expression form contains several OR gates and a single AND gate. This is illustrated by the logic diagram for $f = \Pi M(0, 1, 3) = (A + B + C)(A + B + \overline{C})(A + \overline{B} + \overline{C})$ as shown in figure 3.25(b).

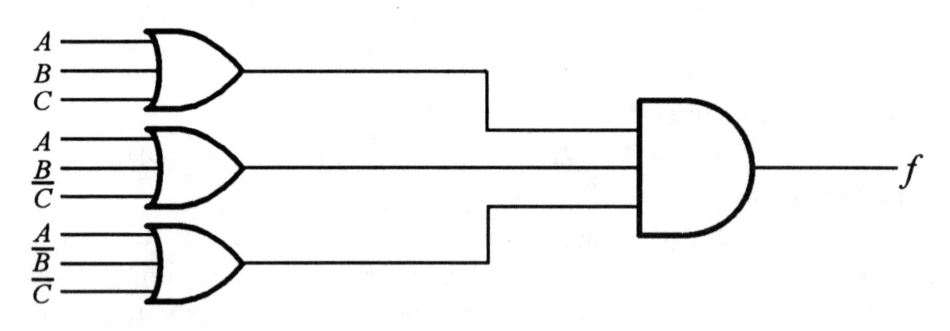

FIGURE 3.25(b) Logic diagram of a product of maxterms

Example 3.2

Using the following truth table, express the Boolean function f in terms of sum-of-products (minterms) and product-of-sums (maxterms):

A	B	C	f
0	0	0	0
0	0	1	1
0	1	0	1
0	1	1	1
1	0	0	0
1	0	1	0
1	1	0	1
1	1	1	0

Solution

From the truth table, $f = 1$ for minterms m_1, m_2, m_3, and m_6. Therefore, the Boolean function f can be expressed by taking the logical sum (OR) of these minterms as follows:

$$f = \Sigma m(1,2,3,6) = \overline{A}\,\overline{B}C + \overline{A}B\overline{C} + \overline{A}BC + AB\overline{C}$$

Now, let us express f in terms of maxterms. By inspecting the truth table, $f = 0$ for maxterms M_0, M_4, M_5, and M_7. Therefore, the function f can be obtained by logically ANDing these maxterms as follows:

$$f = \Pi M(0,4,5,7) = (A+B+C)(\overline{A}+B+C)(\overline{A}+B+\overline{C})(\overline{A}+\overline{B}+\overline{C})$$

3.7 Karnaugh Maps

A Karnaugh map or simply a K-map is a diagram showing the graphical form of a truth table. Since there is no specific set of rules for minimizing a Boolean function using identities, it is difficult to know whether the minimum expression is obtained. The K-map provides a systematic procedure for simplifying Boolean functions of typically up to five variables. K-maps for more than five variables are difficult to use. However, a computer program using a tabular method such as the Quine-McCluskey algorithm can be used to minimize Boolean functions.

The K-map is a diagram containing squares with each square representing one of the minterms of the Boolean function. For example, the K-map of two variables (A,B) contains four squares. The four minterms $\overline{A}\,\overline{B}$, $\overline{A}B$, $A\overline{B}$, and AB are represented by each square. Similarly, there are 8 squares for three variables, 16 squares for four variables, and 32 squares for five variables. Since any Boolean function can be expressed in terms of minterms, the K-map can be used to visually represent a Boolean function.

The K-map is drawn in such a way that there is only a 1-bit change from one square to the next (Gray code). Squares can be combined in groups of 2^n where n=0,1,2,3,4,5, and the Boolean function can be minimized by following certain rules. This minimum expression will reduce the total number of gates for implementation. Thus, the cost of building the logic circuit is reduced.

3.7.1 Two-Variable K-map

Figure 3.26 shows the K-map for two variables.

Since there are four minterms with two variables, four squares are required to represent them. This is depicted in the map of Figure 3.26(a). Each square represents a minterm. Figure 3.26(b) shows the K-map for two variables. Since each variable has a value of 1 or 0, in the K-map of Figure 3.26(b), the 0 and 1 shown on the left of the map corresponds to A while the 0 and 1 on the top are assigned to the variable B. *The squares containing minterms with one variable change are called "adjacent" squares. A square is adjacent of another square placed horizontally or vertically next to it.* For example, consider the minterms m_0 and m_1. Since $m_0 = \bar{A}\bar{B}$ and $m_1 = \bar{A}B$, there is a one variable change ($\bar{B}$ in m_0 and B in m_1, $\bar{A}$ is same in both squares). Therefore, m_0 and m_1 are adjacent squares. Similarly, other adjacent squares in the map include m_0 and m_2, or m_1 and m_3. $m_0(\bar{A}\bar{B})$ and $m_3(AB)$ are not adjacent squares since both variables change from 0's to 1's. The adjacent squares can be combined to eliminate one of the variables. This is based on the Boolean identities $A+\bar{A}=1$ or $B+\bar{B}=1$.

The adjacent squares can also be identified by considering the map as a book. By closing the book at the middle vertical line, m_0 and m_2 will respectively be placed on m_1 and m_3. Thus, m_0 and m_1 are adjacent; squares m_2 and m_3 are also adjacent. Similarly, by closing the map at the middle horizontal line, m_0 will fall on m_2 while m_1 will be placed on m_3. Thus, m_0 and m_2 or m_1 and m_3 are adjacent squares.

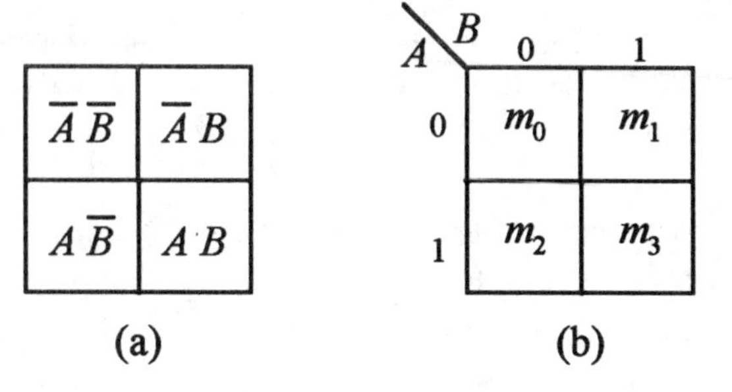

FIGURE 3.26 Two-variable K-map

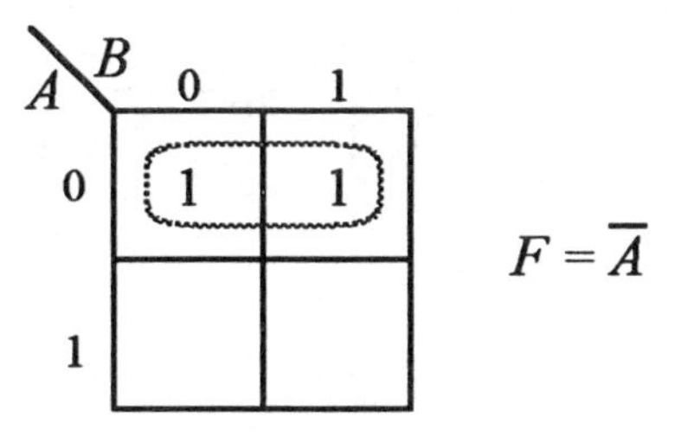

FIGURE 3.27 K-Map for $F = \Sigma m(0,1)$

Now, let us consider a Boolean function, $F = \Sigma m(0,1)$. Figure 3.27 shows that the function F containing two minterms m_0 and m_1 are identified by placing 1's in the corresponding squares of the map. In order to minimize the function F, the two squares can be combined as shown since they are adjacent. The map is then inspected for common variables looking at the squares vertically and horizontally. Since $A = 0$ is common to both squares, $F = \overline{A}$. This can be proven analytically by using Boolean identities as follows:

$$F = \sum m(0,1) = \overline{A}\,\overline{B} + \overline{A}B$$

$$= \overline{A}(\overline{B} + B) = \overline{A} \text{ (since } \overline{B} + B = 1)$$

In a two-variable K-map, adjacent squares can be combined in groups of 2 or 4.

Next, consider $F=\Sigma m(0,2,3)$. The K-map is shown in Figure 3.28. 1's are placed in the squares defined by the minterms m_0, m_2, and m_3. By combining the adjacent squares m_0 with m_2 and m_2 with m_3, the common terms can be determined to simplify the function F. For example, by inspecting m_0 and m_2 vertically and horizontally, the term $\overline{B}$ is the common term. On the other hand, by looking at m_2 and m_3 horizontally and vertically, variable A is the common term. The minimized form of the function F can be obtained by logically ORing these common terms. Therefore,

$$F = A + \overline{B}.$$

Note that the function $F = 1$ for $F = \Sigma m(0,1,2,3)$ *in which all squares in the K-map are 1.*

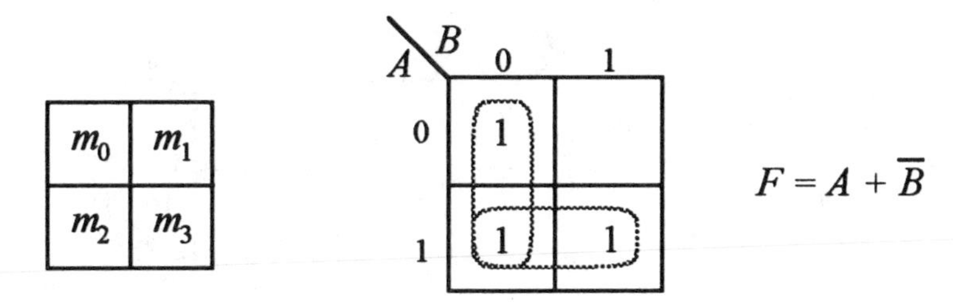

FIGURE 3.28 K-Map for $F = \Sigma m(0,2,3)$

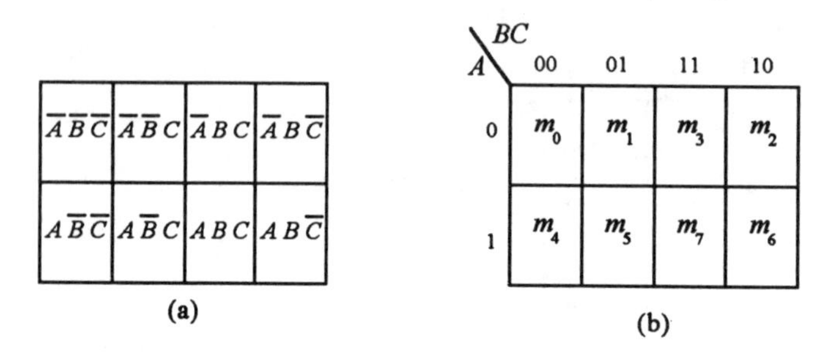

FIGURE 3.29 Three-variable K-map

3.7.2 Three-Variable K-map

Figure 3.29 shows the K-map for three variables. Figure 3.29(a) shows a map with three literals in each square. There are eight minterms (m_0, m_1, ... , m_7) for three variables. Figure 3.29(b) shows these minterms — one for each square in the K-map.

Like the two-variable K-map, a square is adjacent to the squares placed horizontally or vertically next to it. Consider the minterms m_1, m_2, m_3, and m_7. For example, m_3 is adjacent *to* m_1, m_2, and m_7; m_1 is adjacent to m_3; m_2 is adjacent to m_3; m_7 is adjacent to m_3. But, m_7 is adjacent neither to m_1 nor to m_2; m_1 is not adjacent to m_2 and vice versa.

Like the two-variable map, the K-map can be considered as a book. The adjacent squares can also be determined by closing the book at the middle horizontal and vertical lines. For example, closing the book at the middle horizontal line, the adjacent pair of squares are m_0 and m_4, m_1 and m_5, m_3 and m_7, m_2 and m_6. On the other hand, closing the book at the middle vertical line, the adjacent pair of squares are m_0 and m_2, m_1 and m_3, m_4 and m_6, m_5 and m_7.

For a three variable K-map, adjacent squares can be combined in powers of 2: 1 (2^0), 2 (2^1), 4 (2^2) and 8 (2^3). The Boolean function is 1 when all eight squares are 1. It is desirable to combine as many squares as possible. For example, grouping two (2^1) adjacent squares will provide a product term of two literals and combining four (2^2) adjacent squares will provide a product term of one literal for a three-variable K-map. The following examples illustrate this.

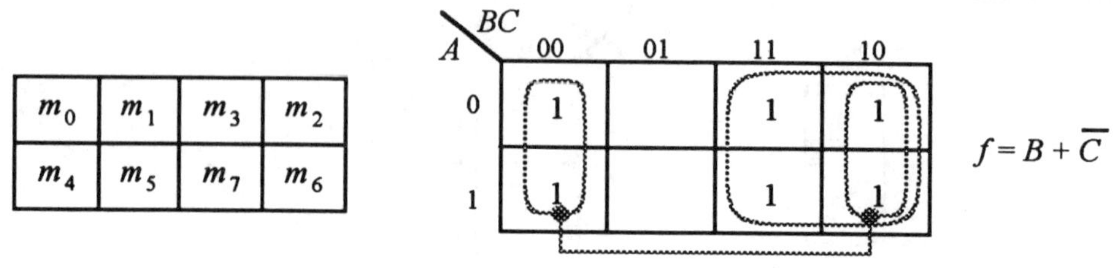

FIGURE 3.30 K-map for $f(A, B, C) = \Sigma\, m(0, 2, 3, 4, 6, 7)$

Example 3.3

Simplify the Boolean function

$$f(A, B, C) = \Sigma\, m(0, 2, 3, 4, 6, 7)$$

using a K-map.

Solution

Figure 3.30 shows the K-map along with the grouping of adjacent squares. First, a 1 is placed in the K-map for each minterm that represents the function. Next, the adjacent squares are identified by squares next to each other. Therefore, m_2, m_3, m_6, and m_7 can be combined as a group of adjacent squares. The common term for this grouping is B. Note that combining four (2^2) squares provides the result with only one literal, B. Next, by folding the K-map at the middle vertical line, adjacent squares m_0, m_2, m_4, and m_6 can be identified. Combining them together will provide the single common term $\overline{C}$. Therefore,

$$f = B + \overline{C}$$

This result can be verified analytically by using the identities as follows:

$$\begin{aligned}
f &= \sum m(0,2,3,4,6,7) \\
&= \overline{A}\,\overline{B}\,\overline{C} + \overline{A}B\overline{C} + \overline{A}BC + A\overline{B}\,\overline{C} + AB\overline{C} + ABC \\
&= \overline{B}\,\overline{C}(A + \overline{A}) + B\overline{C}(\overline{A} + A) + BC(\overline{A} + A) \\
&= \overline{B}\,\overline{C} + B\overline{C} + BC \\
&= \overline{C}(\overline{B} + B) + BC \\
&= \overline{C} + BC \\
&= (B + \overline{C})(C + \overline{C}) = B + \overline{C} \qquad \text{(using the Distributive Law)}
\end{aligned}$$

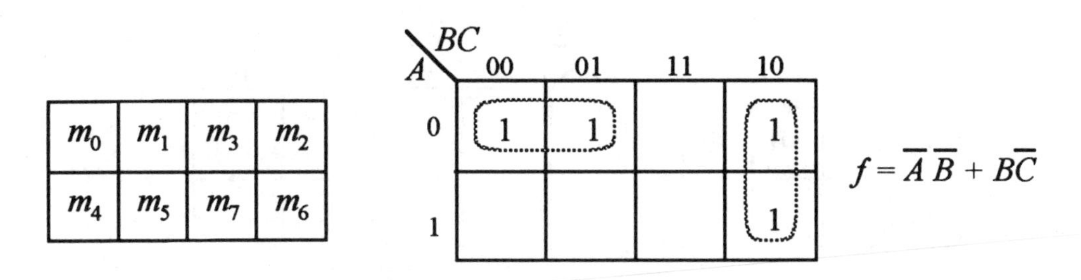

FIGURE 3.31 K-map for $f(A, B, C) = \Sigma\, m(0, 1, 2, 6)$

Example 3.4

Simplify the Boolean function

$$f(A, B, C) = \Sigma m(0, 1, 2, 6)$$

using a K-map.

Solution

Figure 3.31 shows the K-map along with the grouping of adjacent squares.

From the K-map, grouping adjacent squares and logically ORing common product terms,

$$f = \overline{A}\,\overline{B} + B\overline{C}$$

Example 3.5

Simplify the Boolean function

$$F = \overline{A}\,\overline{B}\,\overline{C} + A\overline{B}\,\overline{C} + \overline{B}C$$

using a K-map.

Solution

The function contains three variables, A, B, and C, and is not expressed in minterm form.
The first step is to express the function in terms of minterms as follows:

$$F = \overline{A}\,\overline{B}\,\overline{C} + A\overline{B}\,\overline{C} + \overline{B}C(A + \overline{A})$$

$$= \overline{A}\,\overline{B}\,\overline{C} + A\overline{B}\,\overline{C} + A\overline{B}C + \overline{A}\,\overline{B}C$$

$$= \sum m(0, 1, 4, 5)$$

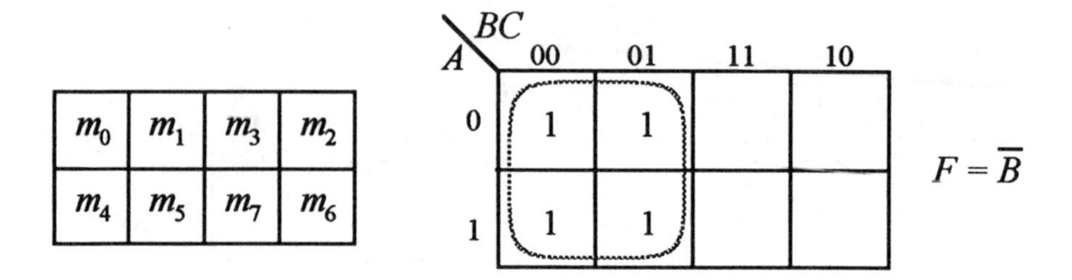

FIGURE 3.32 K-map for $F = \overline{A}\,\overline{B}\,\overline{C} + A\overline{B}\,\overline{C} + \overline{B}C$

Figure 3.32 shows the K-map. Note that the four (2^2) adjacent squares are grouped to provide a single literal $\overline{B}$ by eliminating the other literals. Therefore, $F=\overline{B}$. Although F is not expressed in minterm form, one can usually identify the squares with 1's in the K-map for the function $F=\overline{A}\,\overline{B}\,\overline{C}+A\overline{B}\,\overline{C}+\overline{B}C$ by inspection. This will avoid the lengthy process of converting such functions into minterm form.

3.7.3 Four-Variable K-map

A four-variable K-map, depicted in Figure 3.33, contains 16 squares because there are 16 minterms. Figure 3.33(a) includes four literals in each square. Figure 3.33(b) lists each minterm in its respective square. As before, a square is adjacent to the squares placed horizontally or vertically next to it. For example, m_7 is adjacent to m_3, m_5, m_6, and m_{15}. Also, by closing the K-map at the middle vertical line, the adjacent pairs of squares are m_3 and m_1, m_2 and m_0, m_4 and m_6, m_{12} and m_{14}, m_8 and m_{10}, and so on. On the other hand, closing it at the middle horizontal line will provide the following adjacent squares: m_0 and m_8, m_1 and m_9, m_3 and m_{11}, m_2 and m_{10}, and so on.

For a four-variable K-map, adjacent squares can be grouped in powers of 2: 1 (2^0), 2 (2^1), 4 (2^2), 8 (2^3), and 16 (2^4). The Boolean function is 1 when all 16 minterms are 1. Combining two adjacent squares will provide a product term of three literals; four adjacent squares will provide a product term of two literals; eight adjacent squares will yield a product term of one literal.

AB \ CD	00	01	11	10
00	$\overline{A}\,\overline{B}\,\overline{C}\,\overline{D}$	$\overline{A}\,\overline{B}\,\overline{C}D$	$\overline{A}\,\overline{B}CD$	$\overline{A}\,\overline{B}C\overline{D}$
01	$\overline{A}B\overline{C}\,\overline{D}$	$\overline{A}B\overline{C}D$	$\overline{A}BCD$	$\overline{A}BC\overline{D}$
11	$AB\overline{C}\,\overline{D}$	$AB\overline{C}D$	$ABCD$	$ABC\overline{D}$
10	$A\overline{B}\,\overline{C}\,\overline{D}$	$A\overline{B}\,\overline{C}D$	$A\overline{B}CD$	$A\overline{B}C\overline{D}$

(a)

AB \ CD	00	01	11	10
00	m_0	m_1	m_3	m_2
01	m_4	m_5	m_7	m_6
11	m_{12}	m_{13}	m_{15}	m_{14}
10	m_8	m_9	m_{11}	m_{10}

(b)

FIGURE 3.33 Four-variable K-map

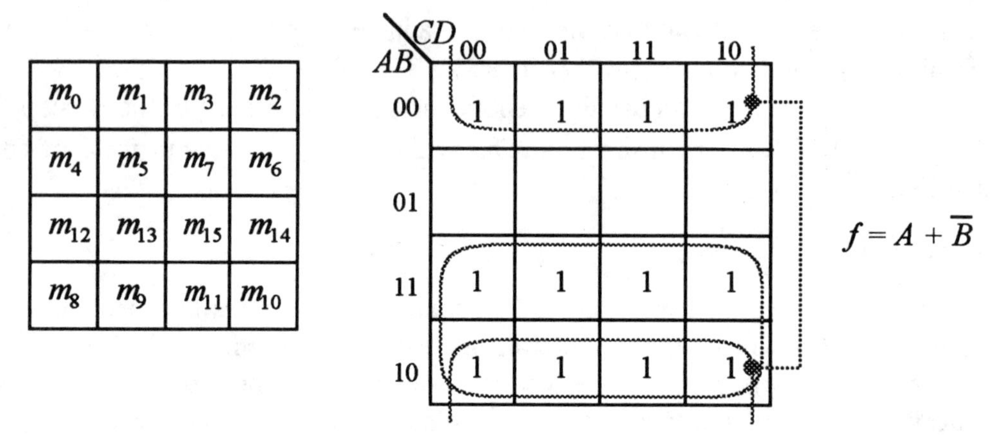

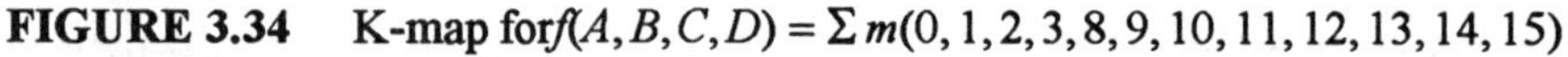

FIGURE 3.34 K-map for $f(A, B, C, D) = \Sigma m(0, 1, 2, 3, 8, 9, 10, 11, 12, 13, 14, 15)$

Example 3.6

Simplify the Boolean function

$$f(A, B, C, D) = \Sigma m(0, 1, 2, 3, 8, 9, 10, 11, 12, 13, 14, 15)$$

using a K-map.

Solution

Figure 3.34 shows the K-map. The 8 adjacent squares combined in the bottom two rows yield the common product term of one literal, A. Because the top row is adjacent to the bottom row, combining the minterms in these two rows will provide a common product term of a single literal, $\overline{B}$. Therefore, by ORing these two terms, the minimized form of the function, $F = A + \overline{B}$ is obtained.

Example 3.7

Simplify the Boolean function $F(A, B, C, D) = \Sigma m(0, 2, 4, 5, 6, 8, 10)$ using a K-map.

Solution

Figure 3.35 shows the K-map. The common product term obtained by grouping the adjacent squares m_0, m_2, m_4, and m_6 will contain $\overline{A}\,\overline{D}$. The common product term obtained by grouping the adjacent squares m_0, m_2, m_8, and m_{10} will be $\overline{B}\,\overline{D}$. Combining the adjacent squares m_4 and m_5 will provide the common term $\overline{A}B\overline{C}$. ORing these common product terms will yield the minimum function, $F(A, B, C, D) = \overline{A}\,\overline{D} + \overline{B}\,\overline{D} + \overline{A}B\overline{C}$.

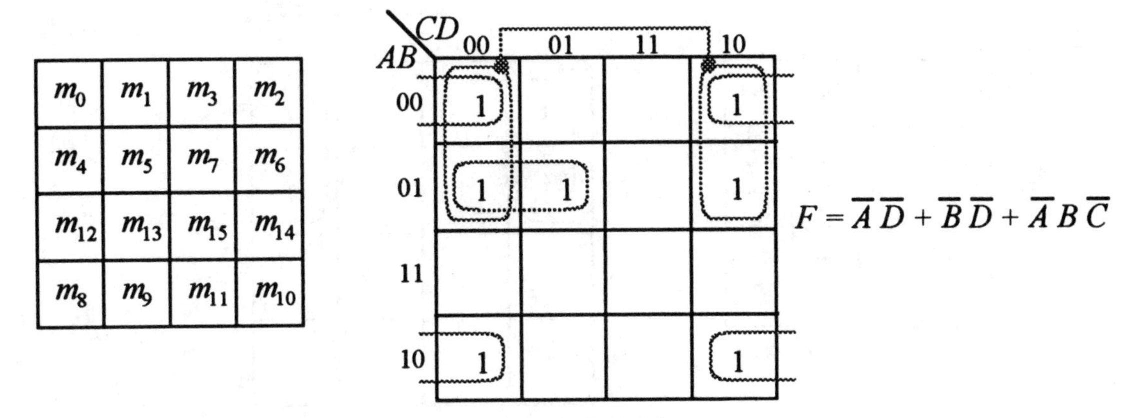

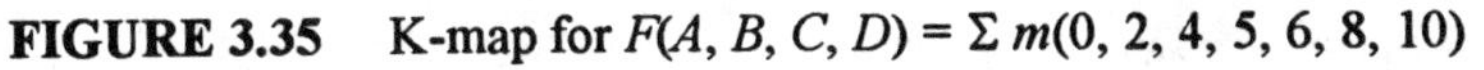

FIGURE 3.35 K-map for $F(A, B, C, D) = \Sigma\, m(0, 2, 4, 5, 6, 8, 10)$

Example 3.8

Simplify the Boolean Function, $F = \overline{A}\,\overline{B}\,\overline{C} + \overline{A}B\overline{C} + \overline{A}B\overline{D} + \overline{A}\,\overline{B}C\overline{D}$ using a K-map.

Solution

Figure 3.36 shows the K-map. In the figure, the function F can be expressed in terms of minterms as follows:

$$F = \overline{A}\,\overline{B}\,\overline{C}(D+\overline{D}) + \overline{A}B\overline{C}(D+\overline{D}) + \overline{A}B\overline{D}(C+\overline{C}) + \overline{A}\,\overline{B}C\overline{D}$$

$$= \overline{A}\,\overline{B}\,\overline{C}D + \overline{A}\,\overline{B}\,\overline{C}\,\overline{D} + \overline{A}B\overline{C}D + \overline{A}B\overline{C}\,\overline{D} + \overline{A}B\overline{D}C + \overline{A}B\overline{D}\,\overline{C} + \overline{A}\,\overline{B}C\overline{D}$$

$$= m_1 + m_0 + m_5 + m_4 + m_6 + m_4 + m_2$$

$$= m_1 + m_0 + m_5 + m_4 + m_6 + m_2$$

because $m_4 + m_4 = m_4$

Rearranging the terms:

$$F = m_0 + m_1 + m_2 + m_4 + m_5 + m_6$$

Therefore, $F = \Sigma\, m(0, 1, 2, 4, 5, 6)$

These minterms are marked as 1 in the K-map. The adjacent squares are grouped as shown. The minimum form of the function, $F = \overline{A}\,\overline{C} + \overline{A}\,\overline{D}$.

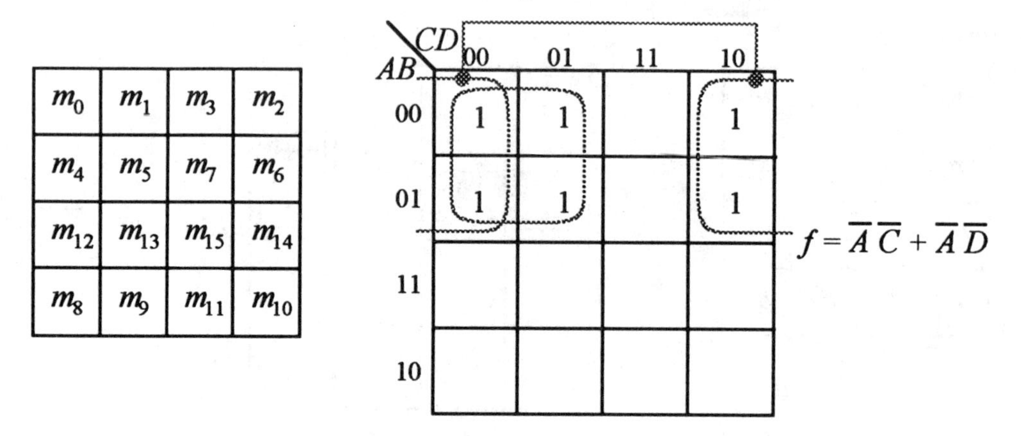

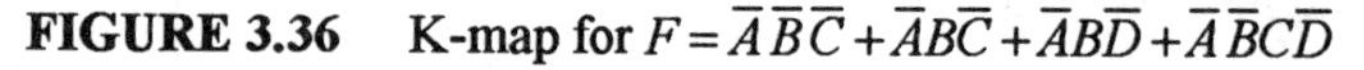

FIGURE 3.36 K-map for $F=\overline{A}\,\overline{B}\,\overline{C}+\overline{A}B\overline{C}+\overline{A}B\overline{D}+\overline{A}\,\overline{B}C\overline{D}$

3.7.4 Prime Implicants

A prime implicant is the product term obtained as a result of grouping the maximum number of allowable adjacent squares in a K-map. The prime implicant is called "essential" if it is the only term covering the minterms. A prime implicant is called "nonessential" if another prime implicant covers the same minterms. The simplified expression for a function can be determined using the K-map as follows:

i) Determine all the essential prime implicants.
ii) Express the minimum form of the function by logically ORing the essential prime implicants obtained in i) along with other prime implicants that may be required to cover any remaining minterms not covered by the essential prime implicants.

Example 3.9

Find the essential prime implicants from the K-map of Figure 3.37 and then determine the simplified expression for the function.

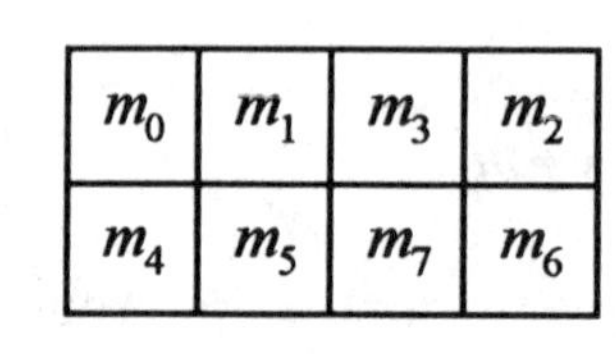

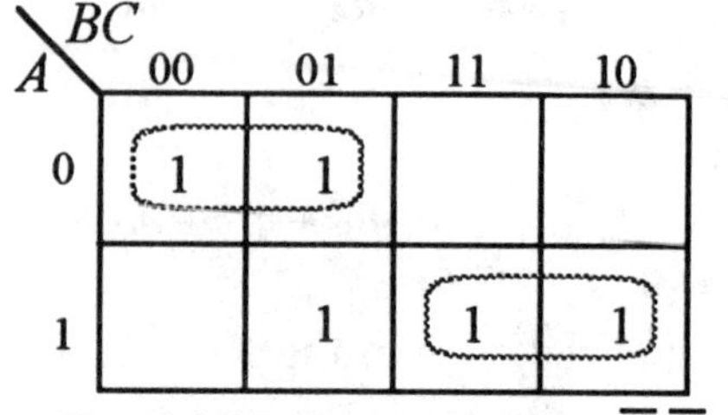

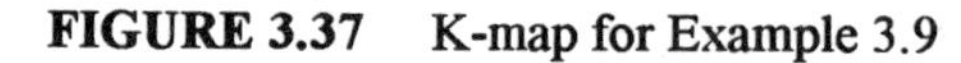

FIGURE 3.37 K-map for Example 3.9

Solution

The essential prime implicants are $AB, \overline{A}\,\overline{B}$ because minterms m_0 and m_1 can only be covered by the term $\overline{A}\,\overline{B}$ and minterms m_6 and m_7 can only be covered by the term AB.

The terms AC and $\overline{B}C$ are nonessential prime implicants because minterm m_5 can be combined with either m_1 or m_7. The term AC can be obtained by combining m_5 with m_7 whereas the term $\overline{B}C$ is obtained by combining m_5 with m_1. The function can be expressed in two simplified forms as follows:

$$f = \overline{A}\,\overline{B} + AB + AC$$

or

$$f = \overline{A}\,\overline{B} + AB + \overline{B}C$$

Example 3.10

Find the essential prime implicants from the K-map of Figure 3.38 and then find the simplified expression for the function.

Solution

The prime implicants can be obtained as follows:

1. By combining minterms m_5, m_7, m_{13}, and m_{15}, the prime implicant BD is obtained.
2. By combining minterms m_8, m_{10}, m_{12}, and m_{14}, the prime implicant $A\overline{D}$ is obtained.
3. By combining minterms m_{12}, m_{13}, m_{14}, and m_{15}, the prime implicant AB is obtained.

m_0	m_1	m_3	m_2
m_4	m_5	m_7	m_6
m_{12}	m_{13}	m_{15}	m_{14}
m_8	m_9	m_{11}	m_{10}

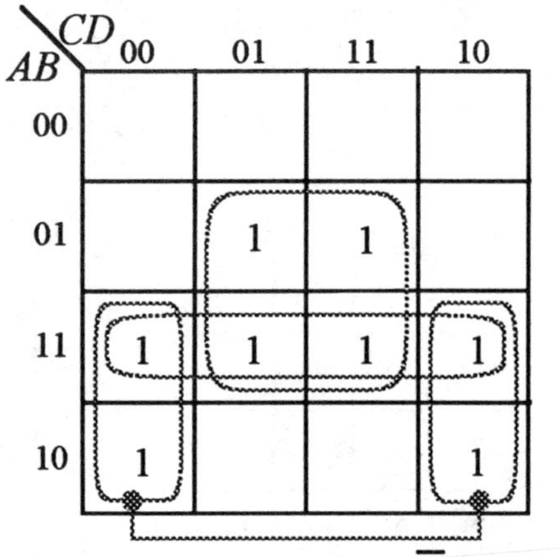

Prime Implicants BD, $A\overline{D}$, and AB

FIGURE 3.38 K-map for Example 3.10

The terms BD and $A\overline{D}$ are essential prime implicants whereas AB is a nonessential prime implicant because minterms m_5 and m_7 can only be covered by the term BD and minterms m_8 and m_{10} can only be covered by the term $A\overline{D}$. However, minterms m_{12}, m_{13}, m_{14}, and m_{15} can be covered by these two prime implicants (BD and $A\overline{D}$). Therefore, the term AB is not an essential prime implicant. Because all minterms are covered by the essential prime implicants, BD and$A\overline{D}$, the term AB is not required to simplify the function. Therefore,

$$f = BD + A\overline{D}.$$

Example 3.11

Find the essential prime implicants and then simplify the function

$$f = \Sigma m(2, 4, 5, 8, 9, 13)$$

using a K-map.

Solution

Figure 3.39 shows the K-map. The essential prime implicants are $\overline{A}\,\overline{B}C\overline{D}$, $\overline{A}B\overline{C}$, and $A\overline{B}\,\overline{C}$ because minterms m_4 and m_5 can only be covered by the term $\overline{A}B\overline{C}$,, minterms m_8 and m_9 can only be covered by the term $A\overline{B}\,\overline{C}$, and minterm m_2 can only be covered by the term $\overline{A}\,\overline{B}C\overline{D}$.

Minterm m_{13} can be combined with either m_5 or m_9. Combining m_{13} with m_5 will yield the term $B\overline{C}D$; combining m_{13} with m_9 will provide the term $A\overline{C}D$. Therefore, minterm m_{13} can be covered by either $B\overline{C}D$ or $A\overline{C}D$. Therefore, $B\overline{C}D$ and $A\overline{C}D$ are nonessential prime implicants. Hence, the function has two simplified forms:

$$f = \overline{A}\,\overline{B}C\overline{D} + \overline{A}B\overline{C} + A\overline{B}\,\overline{C} + B\overline{C}D$$

or

$$f = \overline{A}\,\overline{B}C\overline{D} + \overline{A}B\overline{C} + A\overline{B}\,\overline{C} + A\overline{C}D$$

m_0	m_1	m_3	m_2
m_4	m_5	m_7	m_6
m_{12}	m_{13}	m_{15}	m_{14}
m_8	m_9	m_{11}	m_{10}

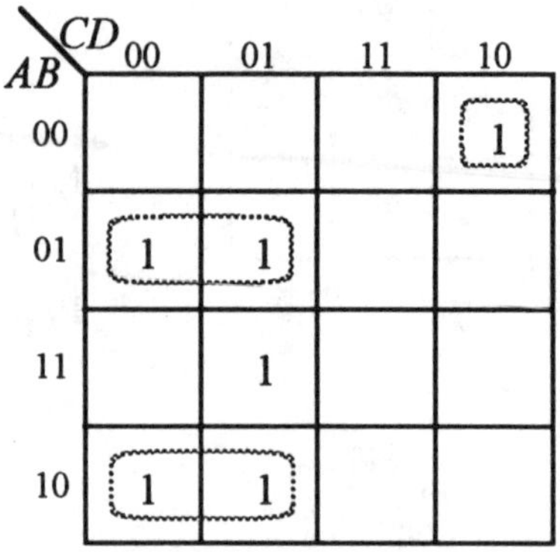

AB \ CD	00	01	11	10
00				1
01	1	1		
11		1		
10	1	1		

FIGURE 3.39 K-map for $f = \Sigma\, m(2, 4, 5, 8, 9, 13)$

3.7.5 Expressing a Function in Product-of-sums Form Using a K-Map

So far, the simplified Boolean functions derived from the K-map were expressed in sum-of-products form. This section will describe the procedure for obtaining the simplified Boolean function in product-of-sums form.

In the K-map, the minterms of a function are represented by 1's. If the empty squares in the K-map are identified as 0's, combining the appropriate adjacent squares will provide the simplified expression of the complement of the function ($\bar{f}$). By taking the complement of $\bar{f}$, the simplified expression for the function, f, can be obtained.

Example 3.12

Simplify the Boolean function
$f(A,B,C,D) = \Sigma m(0,1,4,5,6,7,8,9,14,15)$
in product-of-sums form using a K-map.

Solution

Figure 3.40 shows the K-map. Combining the 0's, a simplified expression for the complement of the function can be obtained as follows:

$$\bar{f} = \bar{B}C + AB\bar{C}$$

By DeMorgan's Theorem,

$$\bar{\bar{f}} = f = \overline{(\bar{B}C + AB\bar{C})} = \left(\overline{\bar{B}C}\right)\cdot\left(\overline{AB\bar{C}}\right) = (B+\bar{C})\cdot(\bar{A}+\bar{B}+C)$$

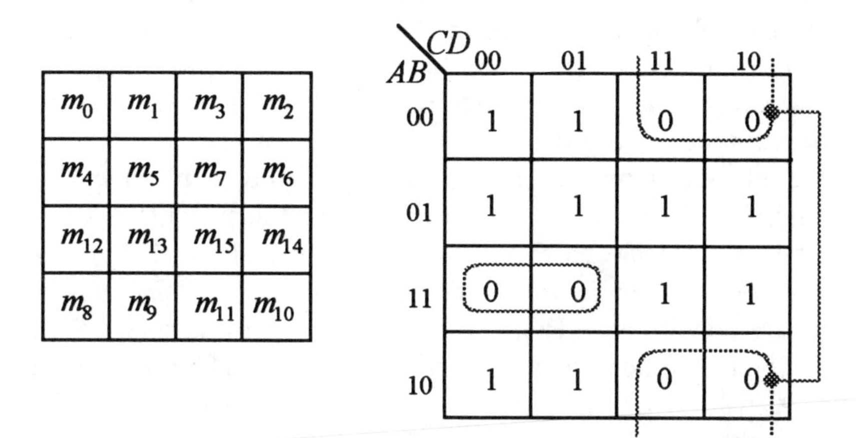

FIGURE 3.40 K-map for $f(A, B, C, D) = \Sigma m(0, 1, 4, 5, 6, 7, 8, 9, 14, 15)$

The example illustrates the procedure for simplifying a function in product-of-sums form from its expression as a sum of minterms. The procedure is similar for simplifying a function expressed in product-of-sums (maxterms).

To represent a function expressed in product-of-sums in the K-map, the complement of the function must first be taken. The squares will then be identified as 1's for the minterms of the complement of the function. For example, consider the following function expressed in maxterm form:

$$f = (\bar{A} + B + C)(A + \bar{B} + \bar{C})(A + B + C)$$

This function can be represented in the K-map by taking its complement and representing in terms of minterms as follows:

$$\bar{f} = A\bar{B}\bar{C} + \bar{A}BC + \bar{A}\bar{B}\bar{C}$$
$$= \sum m(0, 3, 4)$$

Placing 1's in the K-map for m_0, m_3, and m_4 will provide the minterms for $\bar{f}$. The simplified expression for the sum-of-products form of the function, $\bar{f}$ can be obtained by grouping 1's. Finally, the product-of-sums form of the function, f, can be obtained by complementing the function, $\bar{f}$.

3.7.6 Don't Care Conditions

The squares of a K-map are marked with 1's for the minterms of a function. The other squares are assumed to be 0's. This is not always true, because there may be situations in which the function is not defined for all combinations of the variables. Such functions having undefined outputs for certain combinations of literals are called "incompletely specified functions." One does not normally care about the value of the function for undefined minterms. Therefore, the undefined minterms of a function are called "don't care conditions." *Simply put, the don't care conditions are situations in which one or more literals in a minterm can never happen, resulting in nonoccurence of the minterm.*

As an example, BCD numbers include ten digits (0 through 9) and are defined by four bits (0000_2 through 1001_2). However, one can represent binary numbers from 0000_2 through 1111_2 using four bits. This means that the binary combinations 1010_2 through 1111_2 (10_{10} through 15_{10}) can never occur in BCD. Therefore, these six combinations (1010_2 through 1111_2) are don't care conditions in BCD. The functions for these six combinations of the four literals are unspecified. The don't care condition is represented by the symbol X. This means that the symbol X will be placed inside a square in the K-map for which the function is unspecified. The don't care minterms can be used to simplify a function. The function can be minimized by assigning 1's or 0's for X's in the K-map while determining adjacent squares. These assigned values of X's can then be grouped with 1's or 0's in the K-map, depending on the combination that provides the minimum expression. Note that a don't care condition may not be required if it does not help in minimizing the function. To help in understanding the concept of don't care conditions, the following example is provided.

Example 3.13

Simplify the function $f(A,B,C,D) = \Sigma m(0,2,5,8,10,12)$ using a K-map. Assume that the minterms m_1, m_4, m_6, m_7, and m_{15} can never occur.

Solution

The don't care conditions are

$$d(A,B,C,D) = \Sigma m(1,4,6,7,15)$$

Figure 3.41 shows the K-map. By assigning X = 1 and combining 1's as shown, f can be expressed in sum-of-products form as follows:

$$f = \overline{C}\,\overline{D} + \overline{A}B + \overline{B}\,\overline{D}$$

On the other hand, by assigning X = 0 and combining 0's as shown in Figure 3.42, $\overline{f}$ can be obtained as a product-of-sums. Thus,

$$\overline{f} = CD + AD + BC$$

m_0	m_1	m_3	m_2
m_4	m_5	m_7	m_6
m_{12}	m_{13}	m_{15}	m_{14}
m_8	m_9	m_{11}	m_{10}

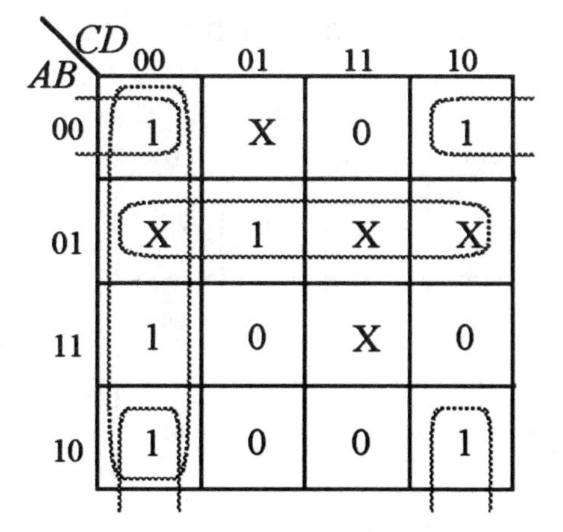

FIGURE 3.41 K-map for Example 3.13

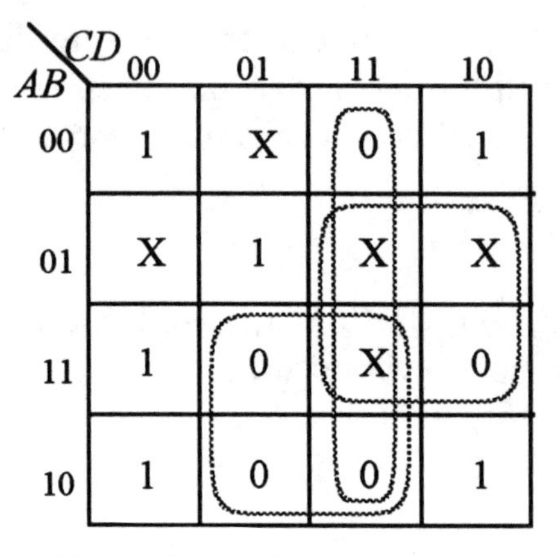

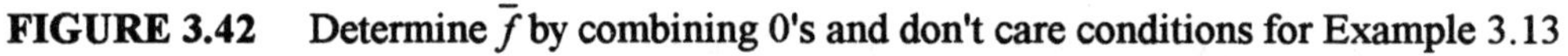
FIGURE 3.42 Determine $\bar{f}$ by combining 0's and don't care conditions for Example 3.13

$$f = \bar{\bar{f}} = \overline{CD + AD + BC}$$
$$= (\overline{CD})(\overline{AD})(\overline{BC})$$
$$= (\bar{C} + \bar{D})(\bar{A} + \bar{D})(\bar{B} + \bar{C})$$

3.7.7 Five-Variable K-map

Figure 3.43 shows a five-variable K-map. The five-variable K-map contains 32 squares. It contains two four-variable maps for *BCDE* with $A = 0$ in one of the two maps and $A = 1$ in the other. The value of a minterm in each map can be determined by the decimal value of the five literals.

$A = 0$

BC \ DE	00	01	11	10
00	m_0	m_1	m_3	m_2
01	m_4	m_5	m_7	m_6
11	m_{12}	m_{13}	m_{15}	m_{14}
10	m_8	m_9	m_{11}	m_{10}

(a)

$A = 1$

BC \ DE	00	01	11	10
00	m_{16}	m_{17}	m_{19}	m_{18}
01	m_{20}	m_{21}	m_{23}	m_{22}
11	m_{28}	m_{29}	m_{31}	m_{30}
10	m_{24}	m_{25}	m_{27}	m_{26}

(b)

FIGURE 3.43 Five-Variable K-map

For example, minterm m_{14} from Figure 3.43(a) can be expressed in terms of the five literals as $\overline{A}BCD\overline{E}$. On the other hand, minterm m_{26} can be expressed in terms of the five literals from Figure 3.43(b) as $AB\overline{C}D\overline{E}$.

When simplifying a function, each K-map can first be considered as an individual four-variable map with $A = 0$ or $A = 1$. Combining of adjacent squares will be identical to typical four-variable maps. Next, the adjacent squares between the two K-maps can be determined by placing the map in Figure 3.43(a) on top of the map in Figure 3.43(b). Two squares are adjacent when a square in Figure 3.43(a) falls on the square in Figure 3.43(b) and vice versa. For example, minterm m_0 is adjacent to minterm m_{16}, minterm m_1 is adjacent to minterm m_{17}, and so on.

Example 3.14

Simplify the function

$$f(A,B,C,D,E) = \Sigma m(3,7,10,11,14,15,19,23)$$

using a K-map.

Solution

Figure 3.44 shows the K-map.

$$f = \overline{A}BD + \overline{B}DE$$

To find the adjacent squares, the K-maps are first considered individually. From Figure 3.44(a), combining minterms m_{10}, m_{11}, m_{14}, and m_{15} will yield the product term $\overline{A}BD$.

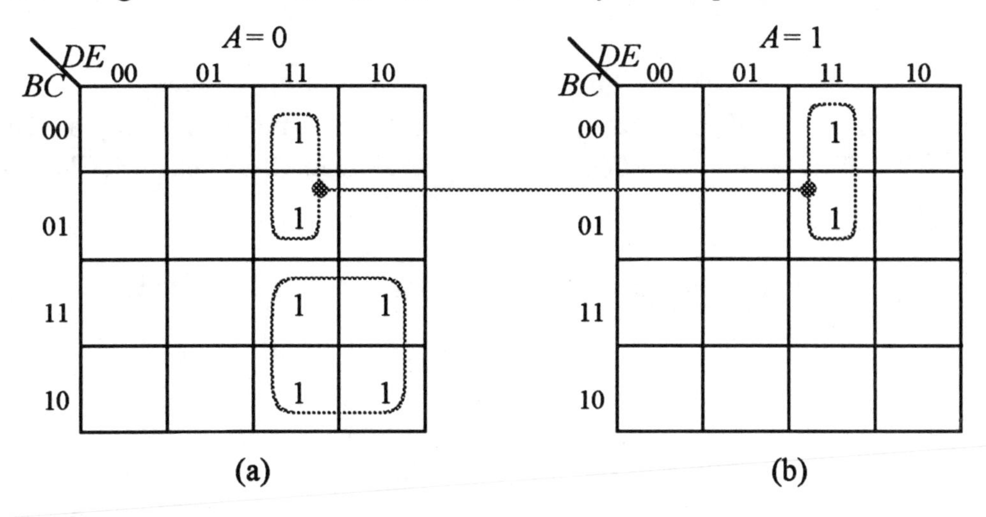

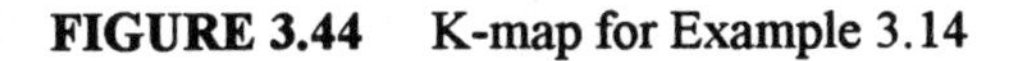

FIGURE 3.44 K-map for Example 3.14

Minterms m_{19} and m_{23} are in the K-map of Figure 3.44(b). However, they are adjacent to minterms m_3 and m_7 in Figure 3.44(a). Combining m_3, m_7, m_{19}, and m_{23} together, the product term $\overline{B}DE$ can be obtained. Literals A or $\overline{A}$ are not included here because adjacent squares belong to both $A = 0$ and $A = 1$.

Therefore, the minimum form of f is

$$f = \overline{A}BD + \overline{B}DE$$

3.8 Quine–McCluskey Method

When the number of variables in a K-map is more than five, it becomes impractical to use K-maps in order to minimize a function. A tabular method known as Quine–McCluskey can be used. A computer program is usually written for the Quine–McCluskey method. One uses this program to simplify a function with more than five variables.

Like the K-map, the Quine–McCluskey method first finds all prime implicants of the function. A minimum number of prime implicants is then selected that defines the function. In order to understand the Quine–McCluskey method, an example will be provided using tables and manual check-off procedures. Although a computer program rather than manual approach is normally used by logic designers, a simple manual example is presented here so that the method can be easily understood.

The Quine–McCluskey method first tabulates the minterms that define the function. The following example illustrates how a Boolean function is minimized using the Quine–McCluskey method.

Example 3.15

In Example 3.7, $F(A,B,C,D) = \Sigma m(0,2,4,5,6,8,10)$ is simplified using a K-map. The minimum form is $F = \overline{A}\,\overline{D} + \overline{B}\,\overline{D} + \overline{A}B\overline{C}$. Verify this result using the Quine–McCluskey method.

Solution

First arrange the binary representation of the minterms as shown in Table 3.2.

TABLE 3.2 Simplifying $F = \Sigma\, m(0, 2, 4, 5, 6, 8, 10)$ Using the Quine–McCluskey Method

(i)						(ii)						(iii)				
Minterm	*A*	*B*	*C*	*D*			*A*	*B*	*C*	*D*			*A*	*B*	*C*	*D*
0	0	0	0	0	✓	0,2	0	0	–	0	✓	0,2,4,6	0	–	–	0
2	0	0	1	0	✓	0,4	0	–	0	0	✓	0,2,8,10	–	0	–	0
4	0	1	0	0	✓	0,8	–	0	0	0	✓	0,4,2,6	0	–	–	0
8	1	0	0	0	✓	2,6	0	–	1	0	✓	0,8,2,10	–	0	–	0
5	0	1	0	1	✓	2,10	–	0	1	0	✓					
6	0	1	1	0	✓	4,5	0	1	0	–						
10	1	0	1	0	✓	4,6	0	1	–	0	✓					
						8,10	1	0	–	0	✓					

In the table, the minterms are grouped according to the number of 1's contained in their binary representations. For example, consider column (i). Because minterms m_2, m_4, and m_8 contain one 1, they are grouped together. On the other hand, minterms m_5, m_6, and m_{10} contain two 1's, so they are grouped together.

Next, consider column (ii). Any two minterms that vary by one bit in column (i) are grouped together in column (ii). Starting from the top row, proceeding to the bottom row, and comparing the binary representation of each minterm in column (i), pairs of minterms having only a one-variable change are grouped together in column (ii) with the variable bit replaced by the symbol –. For example, comparing $m_0 = 0000$ with $m_2 = 0010$, there is a one-variable change in bit position 1. This is shown in column (ii) by placing – in bit position 1 with the other three bits unchanged. Therefore, the top row of column (ii) contains 00–0. The procedure is repeated until all minterms are compared from top to bottom for one unmatched bit and are represented by replacing this bit position with – and other bits unchanged. A ✓ is placed on the right-hand side to indicate that this minterm is compared with all others and its pair with one bit change is found. If a minterm does not have another minterm with one bit change, no check mark is placed on its right. This means that the prime implicant will contain four literals and will be included in the simplified from of the function *F*. In column (i), for each minterm, which has a corresponding pair with one bit change is identified. These pairs are listed in column (ii).

Finally, consider column (iii). Each minterm pair in column (ii) is compared to the next, starting from the top, to find another pair with one bit change; for example m_0, m_2 = 00–0 and m_4, m_6 = 01–0. For this case, bit position 2 does not match. This bit position is replaced by – in the top row of column (iii). Therefore, in column (iii), the top row groups these four minterms 0, 2, 4, 6 with *ABCD* as **0 – – 0**. Similarly, all other pairs in column (ii) are compared from top to bottom for one bit change and are listed accordingly in column (iii) if an unmatched bit is found. A check mark is placed in the right of column (ii) if an unmatched bit is found between two pairs. Note that minterms 4 and 5 do not have any other pair in the list of column (ii) having one unmatched bit. Therefore, this pair is not checked on the right and must be included in the simplified form of F as a prime implicant containing three variables. The two rows of column (iii) (0,2,4,6 and 0,4,2,6) are the same and contain **0 – – 0**. Therefore, this term should be considered once. Similarly, the groups 0,2,8,10 and 0,8,2,10 containing -0-0 should be considered ones. In column (iii), there are no more groups that exist with one unmatched bit.

The comparison process stops. The prime implicants will be the unchecked terms $\bar{A}B\bar{C}$ (from column (ii)) along with, $\bar{A}\bar{D}$ and $\bar{B}\bar{D}$ [from column (iii)]. Thus, the simplified form for F is

$$F = \bar{A}\bar{D} + \bar{B}\bar{D} + \bar{A}B\bar{C}$$

This agrees with the result of Example 3.7.

3.9 Implementation of Digital Circuits with NAND, NOR, and Exclusive-OR/Exclusive-NOR Gates

This section first covers implementation of logic circuits using NAND and NOR gates. These gates are extensively used for designing digital circuits. *The NAND and NOR gates are called "universal gates" because any digital circuit can be implemented with them*. These gates are ,therefore, more commonly used than AND and OR gates. Finally, Exclusive-NOR gates are used to design parity generation and checking circuits.

3.9.1 NAND Gate Implementation

Any logic operation can be implemented by NAND gates. Figure 3.45 shows how NOT, AND, OR, and AND-invert operations can be implemented with NAND gates. A Boolean function can be implemented using NAND gates by first obtaining the simplified expression of the function in terms of AND-OR- NOT logic operations. The function can then be converted to NAND logic. A function expressed in sum-of-products form can be readily implemented using NAND gates.

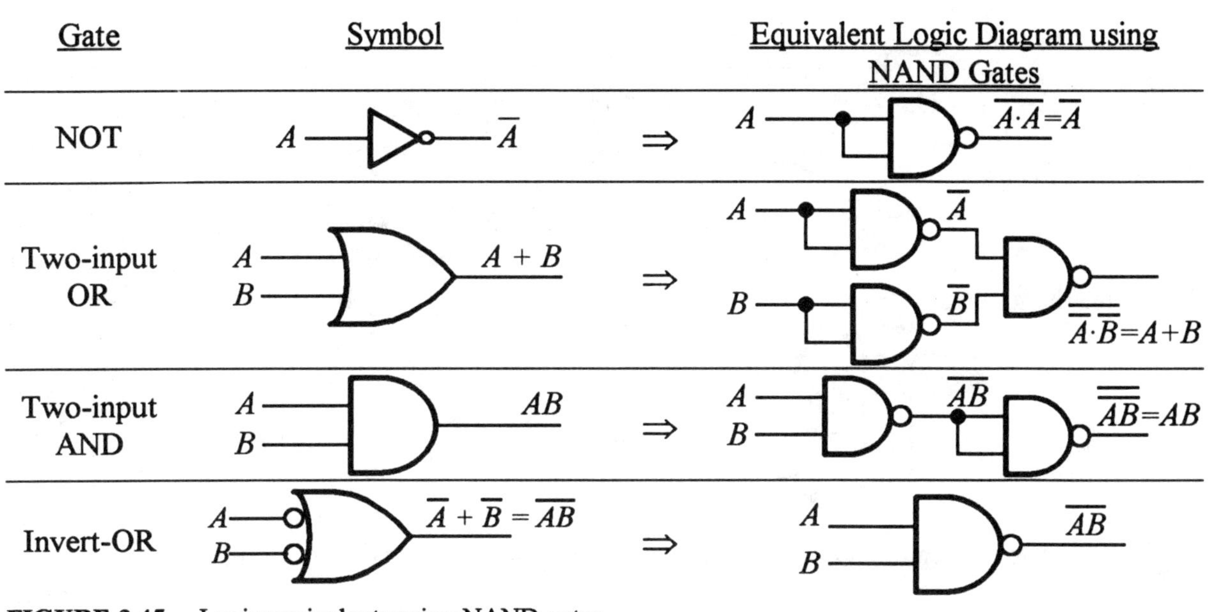

FIGURE 3.45 Logic equivalents using NAND gates

Example 3.16

Implement the simplified function $F = \overline{XY + XZ}$ using NAND gates.

Solution

First implement the function using AND, OR, and NOT gates as follows:

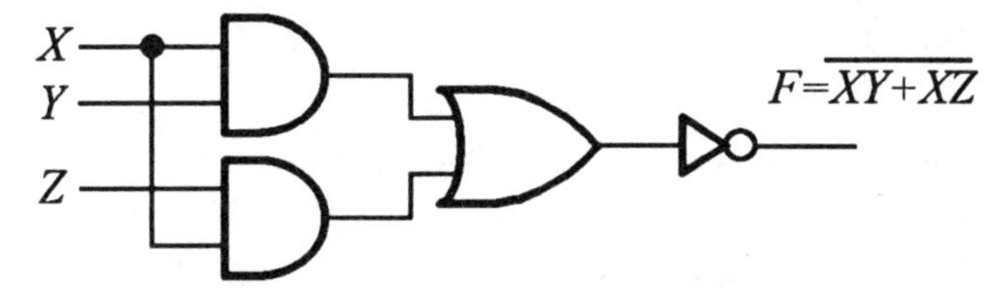

Now convert the AND, OR, and NOT gates to NAND gates as follows:

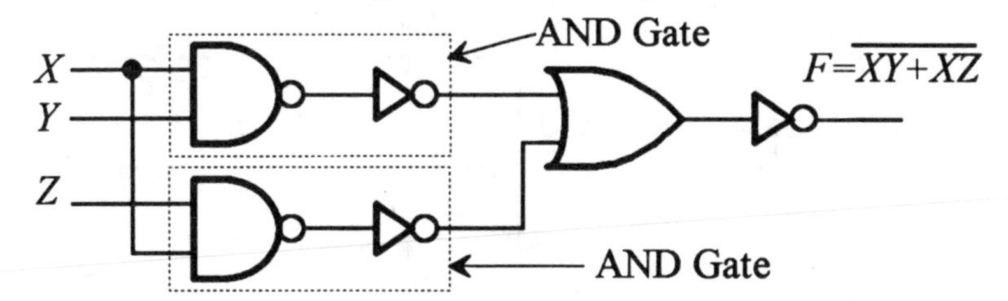

The NOT gates can be represented as bubbles at the inputs of the OR gate as follows:

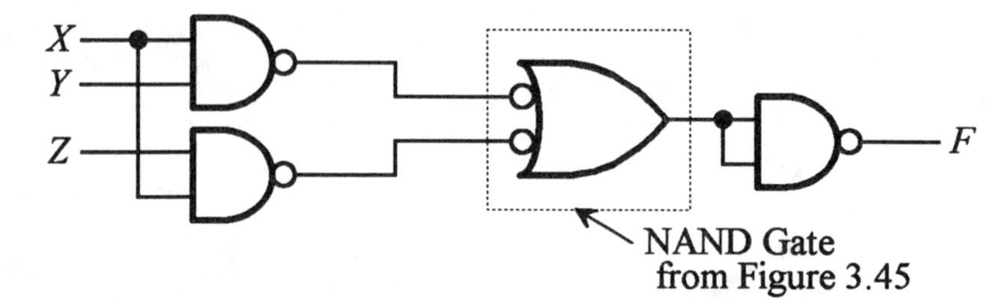

Therefore, the function $F = \overline{XY + XZ}$ can be implemented using only NAND gates as follows:

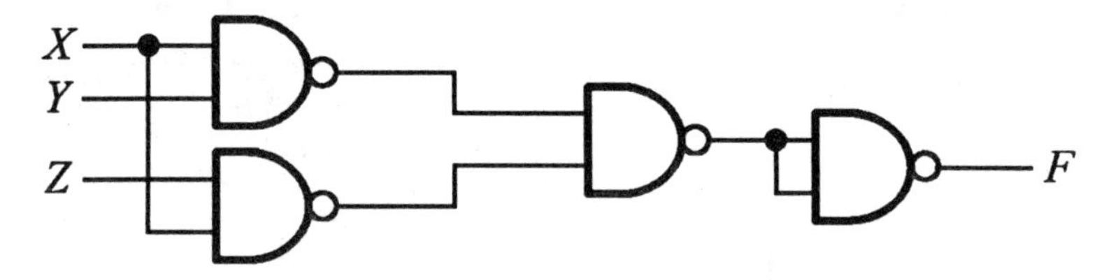

This is a three-level implementation since 3 gate delays are required to obtain the output F.

Example 3.17

Implement the following Boolean function using NAND gates:

$$f(A, B, C, D) = \Sigma(0, 3, 4, 8, 11, 12, 15)$$

Solution

From the K-map of Figure 3.46,

$$f(A, B, C, D) = \overline{C}\,\overline{D} + \overline{B}CD + ACD$$

Figure 3.47 shows the logic diagram using AND and OR gates.

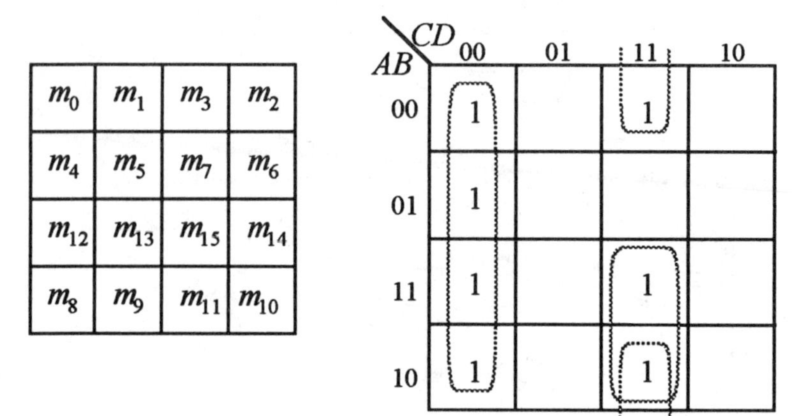

FIGURE 3.46 K-map for Example 3.17

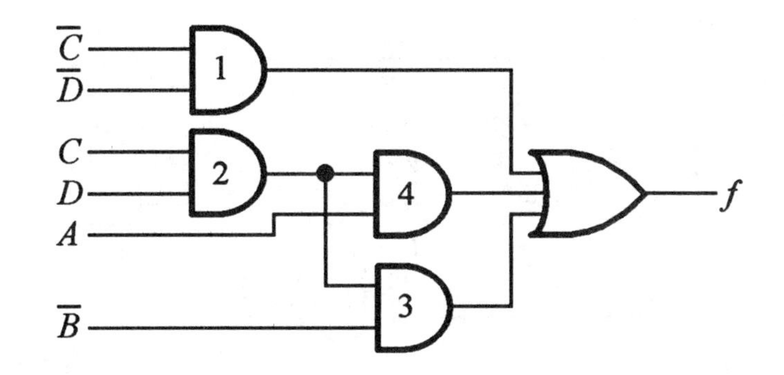

FIGURE 3.47 Logic diagram for $f = \overline{C}\,\overline{D} + \overline{B}CD + ACD$

Note that the logic circuit has three gate delays. Figure 3.48 shows the various steps for implementing this circuit using NAND gates.

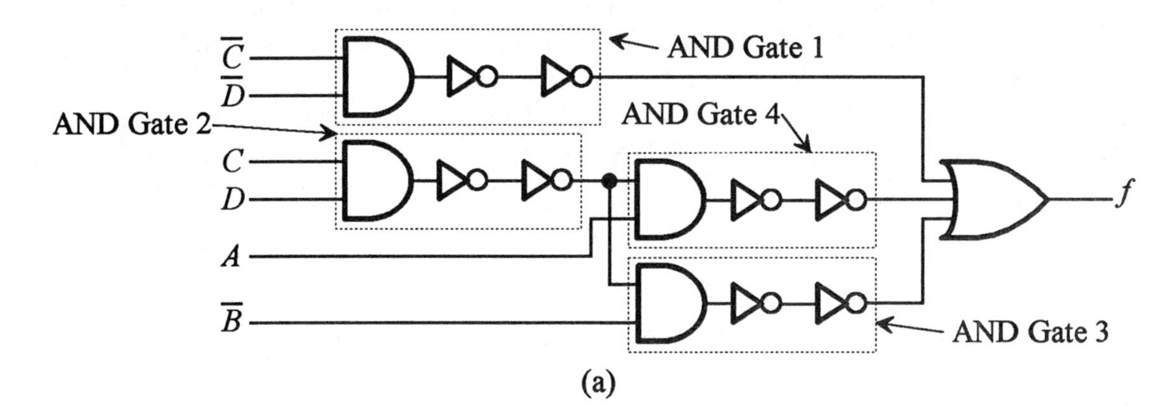

(a)

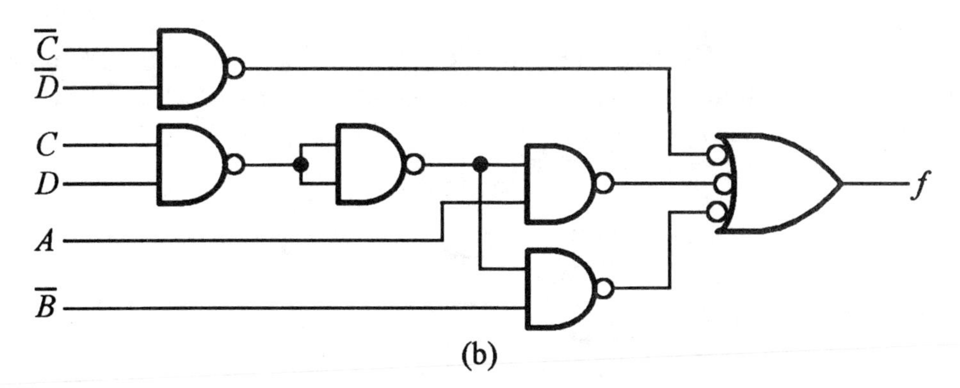

(b)

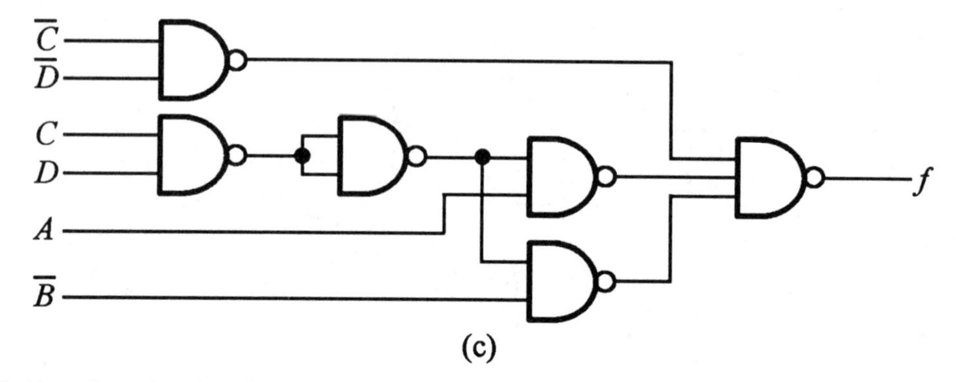

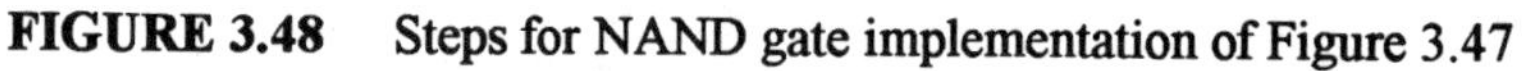

FIGURE 3.48 Steps for NAND gate implementation of Figure 3.47

In Figure 3.48(a), each AND gate of Figure 3.47 is represented by an AND gate with two inverters at the output. For example, consider AND gate 1 of Figure 3.47. The AND gate and an inverter are used to form the NAND gate shown in the top row of Figure 3.48(b) with an inverter (indicated by a bubble at the OR gate input). AND gates 3 and 4 are represented in the same way as AND gate 1 in Figure 3.48(b).

Finally, in Figure 3.48(c), the OR gate with the bubbles at the input in Figure 3.48(b) is replaced by a NAND gate. Thus, the NAND gate implementation in Figure 3.48(c) is obtained.

Example 3.18

Implement the following functions with NAND gates:

$$f = (CD + \bar{D})(AB)$$

Solution

Figure 3.49 shows the AND-OR implementation of the function. The AND-OR implementation in the figure can be converted to the NAND implementation as shown in Figure 3.50.

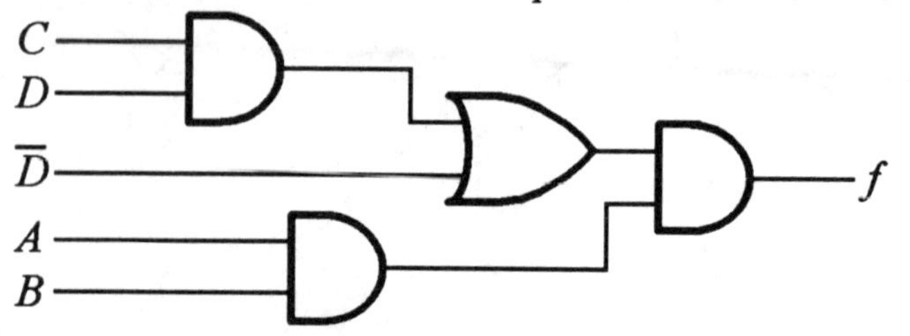

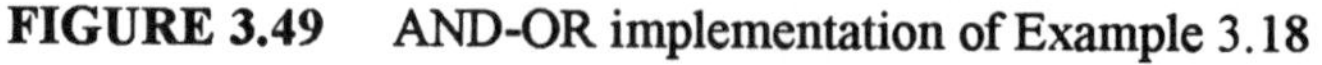

FIGURE 3.49 AND-OR implementation of Example 3.18

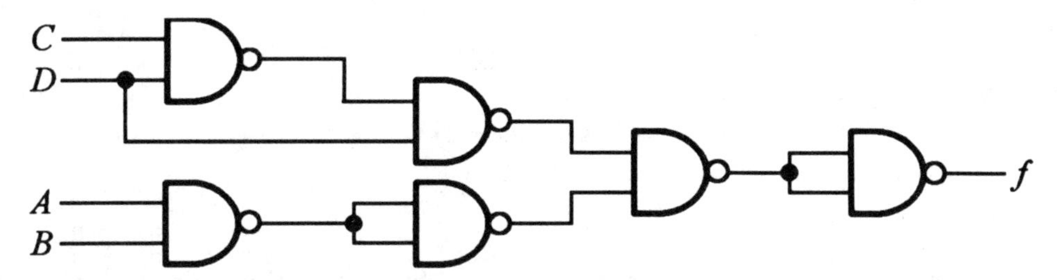

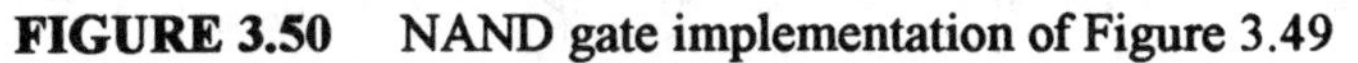
FIGURE 3.50 NAND gate implementation of Figure 3.49

3.9.2 NOR Gate Implementation

Figure 3.51 shows the NOR gate equivalent logic diagrams for NOT, OR, AND, and OR-invert logic operations.

A Boolean function can be implemented using NOR gates by first obtaining the simplified expression of the function in terms of AND and OR gates. The function can then be converted to NOR logic. A function expressed in product-of-sums can be implemented using NOR gates.

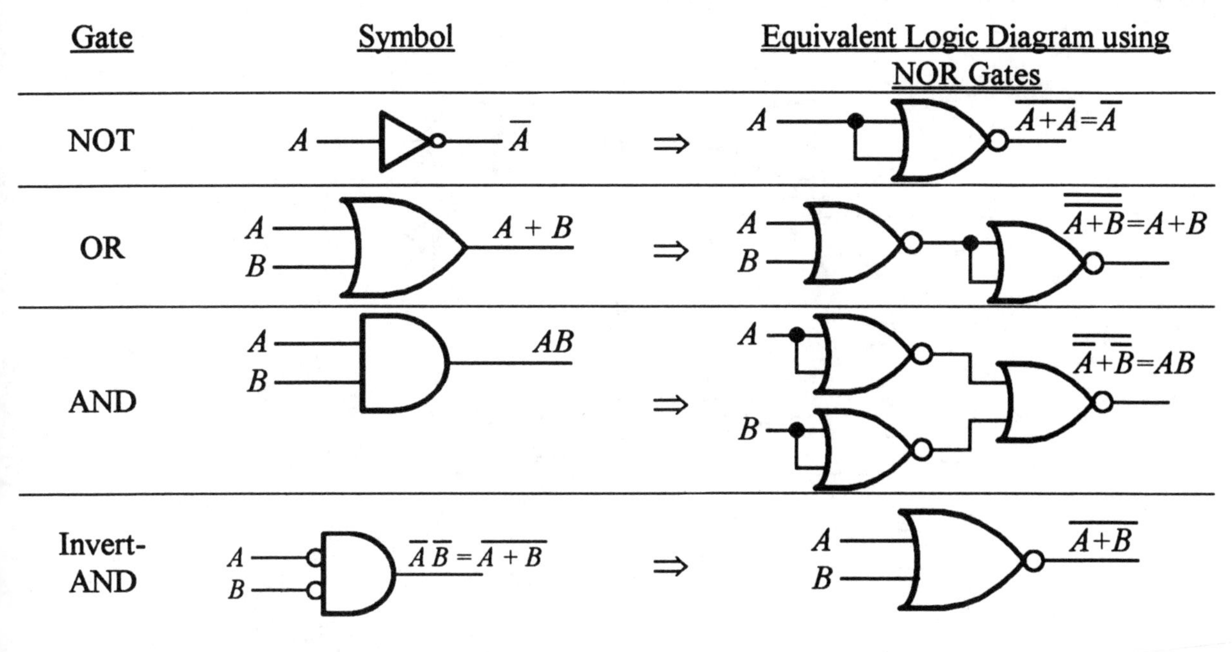

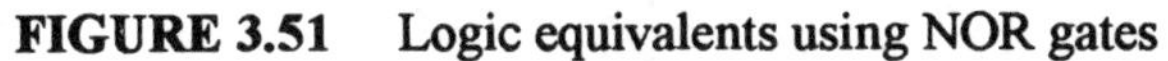
FIGURE 3.51 Logic equivalents using NOR gates

Example 3.19

Implement the following function using NOR gates:

$$f = w(x + \overline{y})(x + z)$$

Solution

Figure 3.52 shows the AND-OR implementation of the logic equation. Figure 3.53 shows the NOR implementation.

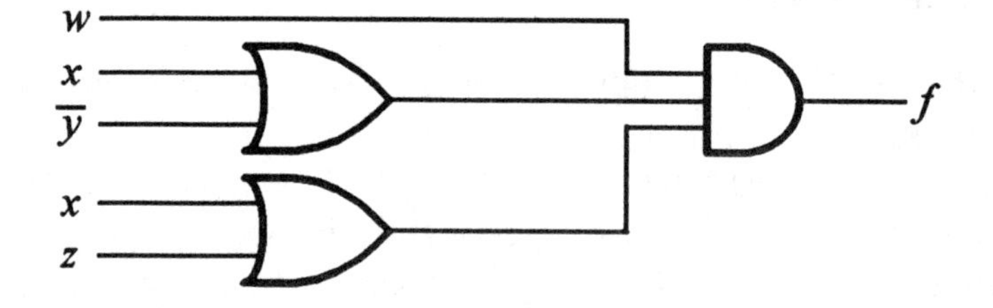

FIGURE 3.52 AND-OR implementation of Example 3.19

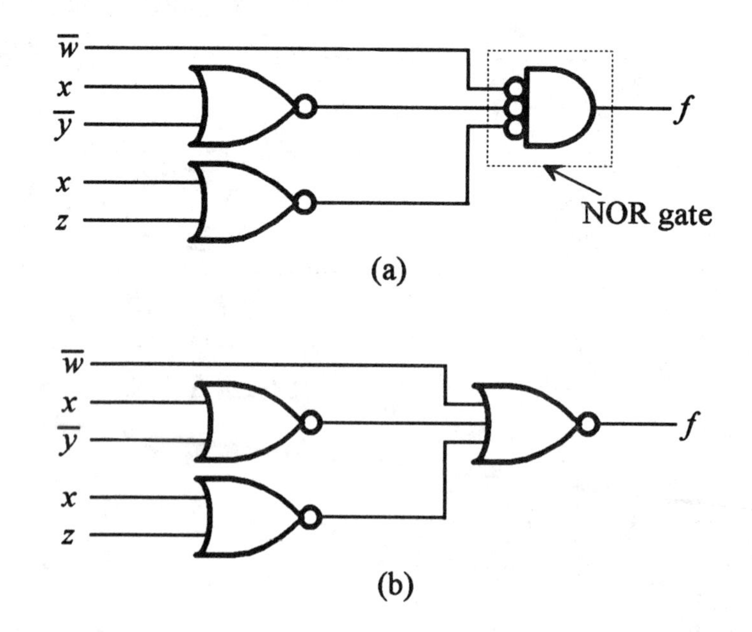

FIGURE 3.53 NOR implementation of Example 3.19

3.9.3 XOR / XNOR Implementations

As mentioned before, the Exclusive-OR operation between two variables A and B can be expressed as

$$A \oplus B = A\overline{B} + \overline{A}B.$$

The Exclusive-NOR or equivalence operation between A and B can be expressed as

$$A \odot B = \overline{A \oplus B} = AB + \overline{A}\,\overline{B}.$$

The following identities are applicable to the Exclusive-OR operation:

i) $A \oplus 0 = A \cdot 1 + \overline{A} \cdot 0 = A$

ii) $A \oplus 1 = A \cdot 0 + \overline{A} \cdot 1 = \overline{A}$

iii) $A \oplus A = A \cdot \overline{A} + \overline{A} \cdot A = 0$

iv) $A \oplus \overline{A} = A \cdot A + \overline{A} \cdot \overline{A} = A + \overline{A} = 1$

Finally, Exclusive-OR is commutative and associative:

$$A \oplus B = B \oplus A$$

$$(A \oplus B) \oplus C = A \oplus (B \oplus C)$$

$$= A \oplus B \oplus C$$

The Exclusive-NOR operation between three or more variables is called an "even function" because the Exclusive-NOR operation between three or more variables includes product terms in which each term contains an even number of 1's. For example, consider Exclusive-NORing three variables as follows:

$$f = \overline{A \oplus B \oplus C} = \overline{(A\overline{B} + \overline{A}B) \oplus C}$$

Let $D = A\overline{B} + \overline{A}B$. Then $\overline{D} = \overline{A\overline{B} + \overline{A}B} = AB + \overline{A}\,\overline{B}$.

Therefore,

$$f = \overline{D \oplus C}$$

$$= DC + \overline{D}\,\overline{C}$$

$$= (A\overline{B} + \overline{A}B)C + \overline{(A\overline{B} + \overline{A}B)}\overline{C}$$

$$= (A\overline{B} + \overline{A}\,B)C + (AB + \overline{A}\,\overline{B})\overline{C}$$

Hence, $f = A\overline{B}C + \overline{A}\,BC + AB\overline{C} + \overline{A}\,\overline{B}\,\overline{C}$

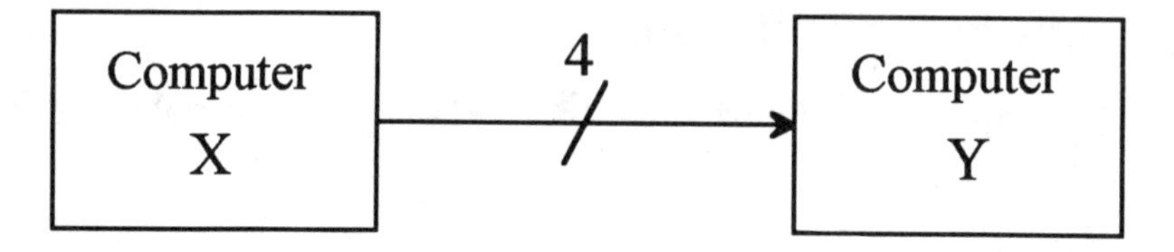

FIGURE 3.54 Parity generation and checking

Note that in this equation, $f = 1$ when one or more product terms in the equation are 1. However, by inspection, the binary equivalents of the right-hand side of the equation are 101, 011, 110, and 000. That is, the function is expressed as the logical sum (OR) of product terms containing even numbers of ones. Therefore, the function is called an even function.

Exclusive-NOR operation can be used for error detection and correction using parity during data transmission. Note that parity can be classified as either odd or even. The parity is defined by the number of 1's contained in a string of data bits. When the data contains an odd number of 1's, the data is said to have "odd parity"; On the other hand, the data has "even parity" when the number of 1's is even. To illustrate how parity is used as an error check bit during data transmission, consider Figure 3.54.

Suppose that Computer X is required to transmit a 3-bit message to Computer Y. To ensure that data is transmitted properly, an extra bit called the parity bit can be added by the transmitting Computer X before sending the data. In other words, Computer X generates the parity bit depending on whether odd or even parity is used during the transmission. Suppose that odd parity is used. The odd parity bit for the three-bit message will be as follows:

Message			Odd Parity Bit
A	B	C	P
0	0	0	1
0	0	1	0
0	1	0	0
0	1	1	1
1	0	0	0
1	0	1	1
1	1	0	1
1	1	1	0

Here $P = 1$ when the 3-bit message ABC contains an even number of 1's. Thus, the parity bit will ensure that the 3-bit message contains an odd number of 1's before transmission. $P = 1$ when the message contains an even number of 1's. Therefore, P is an even function.

Thus,

$$P = \overline{A \oplus B \oplus C}.$$

The transmitting Computer X generates this parity bit. Computer X then transmits 4-bit information (a 3-bit message along with the parity bit) to Computer Y. Computer Y receives this 4-bit information and checks to see whether each 4-bit data item contains an odd number of 1's (odd parity). If the parity is odd, Computer Y accepts the 3-bit message; otherwise the computer sends the 4-bit information back to Computer X for retransmission. Note that Computer Y checks the parity of the transmitted data using the equation

$$E = \overline{P \oplus A \oplus B \oplus C}$$

Here the error E = 1 if the four bits have an even number of ones (even parity). That is, at least one of the four bits is changed during transmission. On the other hand, the error bit, E = 0 if the 4-bit data has an odd number of ones. Figure 3.52 shows the implementation of the parity bit, $P = \overline{A \oplus B \oplus C}$, and the error bit, $E = \overline{P \oplus A \oplus B \oplus C}$.

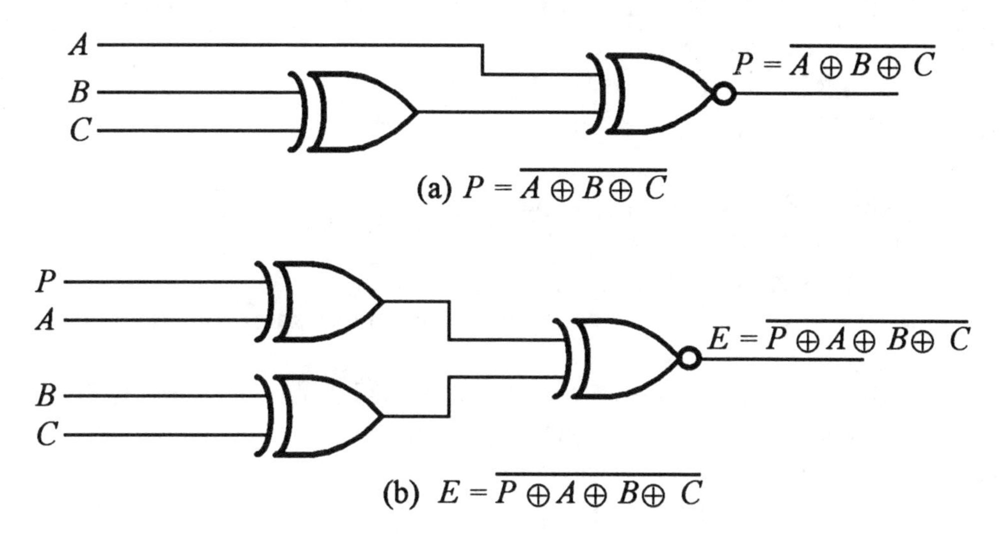

FIGURE 3.52 Implementation of parity generation and checking using XOR / XNOR gates

QUESTIONS AND PROBLEMS

3.1 Perform the following operations. Include your answers in hexadecimal.
$A6_{16}$ OR 31_{16}; $F7A_{16}$ AND $D80_{16}$; $36_{16} \oplus 2A_{16}$

3.2 Given $A = 1001_2$, $B = 1101_2$, find: A OR B; $B \wedge A$; $\bar{A}$; $A \oplus A$.

3.3 Perform the following operation: $A7_{16} \oplus FF_{16}$.
What is the relationship of the result to $A7_{16}$?

3.4 Prove the following identities algebraically and by means of truth tables:

(a) $(A+B)(\overline{\overline{A}+\overline{B}}) = 0$

(b) $A+\overline{A}B = A+B$

(c) $XY+\overline{X}\,\overline{Y}+X\overline{Y}+\overline{X}Y = 1$

(d) $\overline{(A+\overline{A}B)} = \overline{A}\,\overline{B}$

(e) $(\overline{X}+Y)(X+\overline{Y}) = \overline{X \oplus Y}$

(f) $\overline{B}\,\overline{C}+ABC+\overline{A}\,\overline{C} = \overline{C \oplus (AB)}$

3.5 Minimize the following Boolean expression using identities:

(a) $XY+(1 \oplus X)+X\overline{Z}+X\overline{Y}+XZ$

(b) $AB\overline{C}+AB\overline{CD}+AB\overline{D}$

(c) $BC+ABC\overline{D}+\overline{A}BCD+ABCD$

(d) $(\overline{X}+\overline{Y})(\overline{XY})+ZXY+X\overline{Z}Y$

3.6 Using DeMorgan's theorem, draw logic diagrams for $F = AB\overline{C}+\overline{A}\,\overline{B}+BC$

(a) Using only AND gates and inverters.

(b) Using only OR gates and inverters.

You may use two-input and three-input AND and OR gates for (a) and (b).

3.7 Using truth tables, express each one of the following functions and their complements in terms of sum of minterms and product of maxterms:

(a) $F = ABC+\overline{A}BD+\overline{A}\,\overline{B}\,\overline{C}+AC\overline{D}$

(b) $F = (W+X+Y)(W\overline{X}+Y)$

3.8 Express each of the following expressions in terms of minterms and maxterms.

(a) $F = B\overline{C}+\overline{A}B+B(A+C)$

(b) $F = (A+\overline{B}+C)(\overline{A}+B)$

3.9 Minimize each of the following functions using a K-map:

(a) $F(A, B, C) = \Sigma\, m(0, 1, 4, 5)$

(b) $F(A, B, C) = \Sigma\, m(0, 1, 2, 3, 6)$

(c) $F(X, Y, Z) = \Sigma\, m(0, 2, 4, 6)$

3.10 Minimize each of the following expressions for F using a K-map.

(a) $F(A,B,C) = \overline{B}\,\overline{C} + ABC + AB\overline{C}$

(b) $F(A,B,C) = \overline{A}B\overline{C} + BC$

(c) $F(A,B,C) = \overline{A}\,\overline{C} + A(\overline{B}\,\overline{C} + B\overline{C})$

3.11 Simplify each of the following functions for F using a K-map.

(a) $F(W,X,Y,Z) = \Sigma m(0,1,4,5,8,9)$

(b) $F(A,B,C,D) = \Sigma m(0,2,8,10,12,14)$

(c) $\overline{F}(A,B,C,D) = \Sigma m(2,4,5,6,7,10,14)$

(d) $F(W,X,Y,Z) = \Sigma m(2,3,6,7,8,9,12,13)$

(e) $\overline{F}(W,X,Y,Z) = \Sigma m(0,2,4,6,8,10,12,14)$

(f) $F(W,X,Y,Z) = \Sigma m(1,3,5,7,9,11,13,15)$

3.12 Minimize each of the following expressions for $\overline{F}$ using a K-map in sums-of-product form:

(a) $F(W,X,Y,Z) = \overline{W}\,\overline{X}YZ + WYZ$

(b) $F = \overline{A}\,\overline{B}\,\overline{C}\,\overline{D} + \overline{A}CD + ABCD$

(c) $F = (\overline{A} + \overline{B} + C + \overline{D})(\overline{A} + B + C + \overline{D})(A + \overline{B} + C + \overline{D})$

3.13 Find essential prime implicants and then minimize each of the following functions for F using a K-map:

(a) $F(A, B, C, D) = \Sigma\, m(3, 4, 5, 7, 11, 12, 15)$

(b) $F(W, X, Y, Z) = \Sigma\, m(2, 3, 6, 7, 8, 9, 12, 13, 15)$

3.14 Minimize each of the following functions for f using a K-map and don't care conditions, d.

(a) $f(A, B, C) = \Sigma\, m(1, 2, 4, 7)$
$d(A, B, C) = \Sigma\, m(5, 6)$

(b) $f(X, Y, Z) = \Sigma\, m(2, 6)$
$d(X, Y, Z) = \Sigma\, m(0, 1, 3, 4, 5, 7)$

(c) $f(A, B, C, D) = \Sigma\, m(0, 2, 3, 11)$
$d(A, B, C, D) = \Sigma\, m(1, 8, 9, 10)$

(d) $f(A, B, C, D) = \Sigma\, m(4, 5, 10, 11)$
$d(A, B, C, D) = \Sigma\, m(12, 13, 14, 15)$

3.15 Minimize the following expression using the Quine–McCluskey method. Verify the results using a K-map. Draw logic diagrams using NAND gates. Assume true and complemented inputs.

$$F(A, B, C, D) = \Sigma\, m(0, 1, 4, 5, 8, 12)$$

3.16 Minimize the following expression using a K-map:
$F = AB + \bar{A}\,\bar{B}\,\bar{C}\,\bar{D} + C\bar{D} + \bar{A}\,\bar{B}\,\bar{C}D$
and then draw schematics using:
(a) NAND gates.
(b) NOR gates.

3.17 Minimize the following function $F(A, B, C, D) = \Sigma\, m(6, 7, 8, 9)$ assuming that the condition $AB = 11_2$ can never occur. Draw schematics using:
(a) NAND gates.
(b) NOR gates.

3.18 It is desired to compare two 4-bit numbers for equality. If the two numbers are equal, the circuit will generate an output of 1. Draw a logic circuit using a minimum number of gates of your choice.

3.19 Show analytically that $A \oplus (A \oplus B) = B$.

3.20 Show that the the Boolean function, $f = A \oplus B \oplus AB$ between two variables, *A and B*, can be implemented using a single two-input OR gate.

3.21 Design a parity generation circuit for a 5-bit data (4-bit message with an even parity bit) to be transmitted by computer X. The receiving computer Y will generate an error bit, $E = 1$, if the 5-bit data received has an odd parity; otherwise, $E = 0$. Draw logic diagrams for both parity generation and checking using XOR gates.

3.22 Draw a logic diagram for an Exclusive-OR using only four two-input NAND gates.

4

COMBINATIONAL LOGIC DESIGN

4.1 Basic Concepts

Digital logic circuits can be classified into two types: combinational and sequential. A combinational circuit is designed using logic gates in which application of inputs generates the outputs at any time. An example of a combinational circuit is an adder, which produces the result of addition as output upon application of the two numbers to be added as inputs.

A sequential circuit, on the other hand, is designed using logic gates and memory elements known as "flip-flops." Note that the flip-flop is a one-bit memory. A sequential circuit generates the circuit outputs based on the present inputs and the outputs (states) of the memory elements. The sequential circuit is basically a combinational circuit with memory. Note that a combinational circuit does not require any memory (flip-flops), whereas sequential circuits require flip-flops to remember the present states. A counter is a typical example of a sequential circuit. To illustrate the sequential circuit, suppose that it is desired to count in the sequence 0, 1, 2, 3, 0, 1,... and repeat. In binary, the sequence is 00, 01, 10, 11, 00, 01, ..., and so on. This means that a two-bit memory using two flip-flops is required for storing the two bits of the counter because each flip-flop stores one bit. Let us call these flip-flops with outputs A and B. Note that initially $A = 0$ and $B = 0$. The flip-flop changes outputs upon application of a clock pulse. With appropriate inputs to the flip-flops and then applying the clock pulse, the flip-flops change the states (outputs) to $A = 0$, $B = 1$. Thus, the count to 1 can be obtained. The flip-flops store (remember) this count. Upon application of appropriate inputs along with the clock, the flip-flops will change the status to $A = 1$, $B = 0$; thus, the count to 2 is obtained. The flip-flops remember (store) this count at the outputs until a common clock pulse is applied to the flip-flops. The inputs to the flip-flops are manipulated by a combinational circuit based on A and B as inputs. For example, consider $A = 1$, $B = 0$. The inputs to the flip-flops are determined in such a way that the flip-flops change the states at the clock pulse to $A = 1$, $B = 1$; thus, the count to 3 is obtained. The process is repeated.

4.2 Analysis of a Combinational Logic Circuit

A combinational logic circuit can be analyzed by (i) first, identifying the number of inputs and outputs, (ii) expressing the output functions in terms of the inputs, and (iii) determining the truth table for the logic diagram. As an example, consider the combinational circuit in Figure 4.1 There are three inputs (X, Y, and Z) and two outputs (Z_1 and Z_2) in the circuit.

Let us now express the outputs F_1 and F_2 in terms of the inputs. The output F_1 of the AND gate #1 is $F_1 = XY$. The output F_2 of NOR gate #2 can be expressed as $F_2 = \overline{X+Y}$. The output of the XOR gate #3 is

$$F_3 = X \oplus F_1 = (X \oplus XY)$$

Because one of the inputs of the XOR gate #4 is 1, its output is inverted. Therefore,

$$Z_1 = \overline{F_2} = X + Y.$$

Finally,

$$Z_2 = X \oplus F_3 = X \oplus (X \oplus XY)$$

Therefore,

$$\begin{aligned} Z_2 &= X \oplus (X \cdot \overline{XY} + \overline{X} \cdot XY) \\ &= X \oplus (X \cdot (\overline{X} + \overline{Y})) \\ &= X \oplus (X\overline{Y}) \\ &= X(\overline{X\overline{Y}}) + \overline{X}(X\overline{Y}) \\ &= X(\overline{X} + Y) \\ &= XY \end{aligned}$$

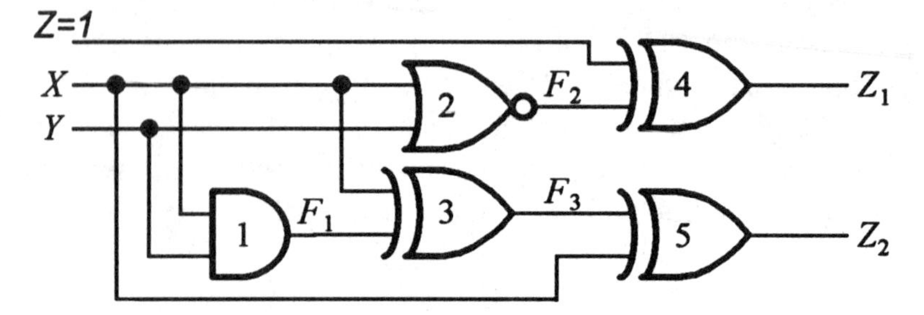

FIGURE 4.1 Analysis of a combinational logic circuit

The truth table shown in Table 4.1 can be obtained by using the logic equations for Z_1 and Z_2.

TABLE 4.1 Truth Table for Figure 4.1 with Input, $Z = 1$

Inputs		Outputs	
X	*Y*	Z_1	Z_2
0	0	0	0
0	1	1	0
1	0	1	0
1	1	1	1

4.3 Design of a Combinational Circuit

A combinational circuit can be designed using three steps as follows:

1) Determine the inputs and the outputs from problem definition and then derive the truth table.
2) Use K-maps to minimize the number of inputs (literals) in order to express the outputs. This reduces the number of gates and thus the implementation cost.
3) Draw the logic diagram

In order to illustrate the design procedure, consider the following example. Suppose it is desired to design a combinational circuit with three inputs (*A*, *B*, and *C*) and one output *F*. The output *F* is one if *A*, *B*, and *C* are not equal ($A \neq B \neq C$); $F = 0$ otherwise.

First, the number of inputs and outputs are identified. There are three inputs (*A*, *B*, and *C*) and one output, *F*. Next the truth table is obtained as shown in Table 4.2.

F in the truth table of Table 4.2 is simplified using K-maps and implemented as shown in Figure 4.2.

TABLE 4.2 Truth Table for *F*

A	*B*	*C*	*F*
0	0	0	0
0	0	1	1
0	1	0	1
0	1	1	1
1	0	0	1
1	0	1	1
1	1	0	1
1	1	1	0

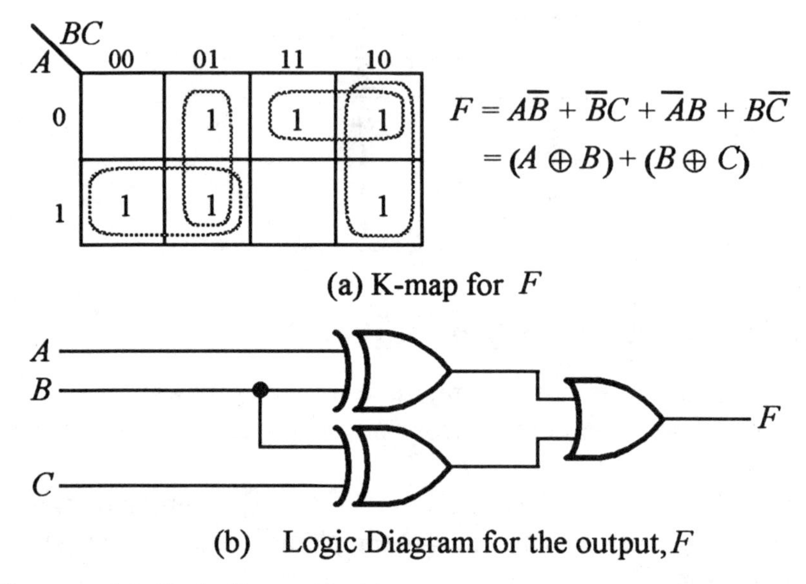

(a) K-map for F

(b) Logic Diagram for the output, F

FIGURE 4.2 K-map and the logic diagram for F

4.4 Multiple-Output Combinational Circuits

A combinational circuit may have more than one output. In such a situation, each output must be expressed as a function of the inputs. *A digital circuit called the "code converter" is an example of multiple-output circuits. A code converter transforms information from one binary code to another*. As an example, consider the BCD to seven-segment code converter shown in Figure 4.3.

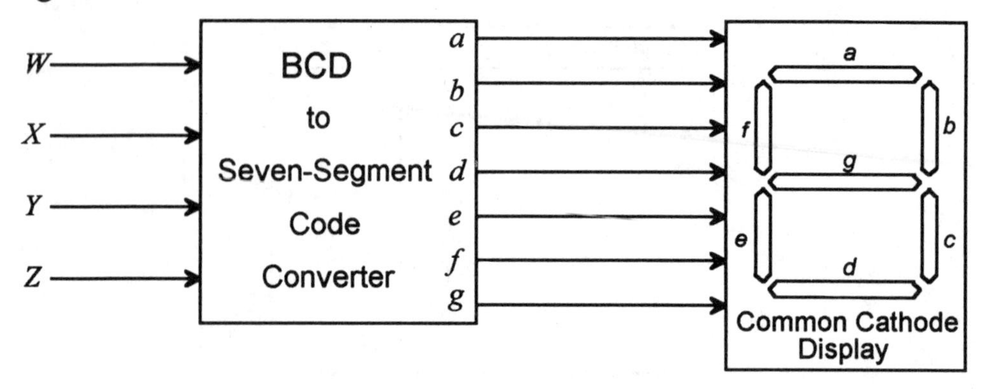

FIGURE 4.3 BCD to seven-segment code converter

The code converter in the figure can be designed to translate the BCD inputs (*W*, *X*, *Y*, and *Z*) to seven-segment code for displaying decimal digits. The inputs *W*, *X*, *Y*, and *Z* can be entered into the code converter via four switches as was discussed in Chapter 1. A combinational circuit can be designed for the code converter that will translate each digit entered using four bits into seven output bits (one bit for each segment) of the display. In this case, the code converter has four inputs and seven outputs. This code converter is commonly known as a "BCD to seven-segment decoder." *With four bits (W, X, Y, and Z), there are sixteen combinations (0000 through 1111) of 1's and 0's. BCD allows only 10 (0000 through 1001) of these 16 combinations, so the invalid numbers (1010 through 1111) will never occur for BCD and can be considered as don't cares in K-maps because it does not matter what the seven outputs (a through g) are for these invalid combinations.*
The 7447 (TTL) is a commercially available BCD to 7-segment decoder/driver chip. It is designed for driving common-anode displays. A low output is intended to light a segment. Current -limiting resistors (330 Ohms) should be used when driving a 7-segment display with this chip. For normal operation, the LT (Lamp test) and BI/RBO (Blanking Input / Ripple Blanking Input) must be open or conntected to HIGH.

To illustrate the design of a BCD to seven-segment decoder, consider designing a code converter for displaying the decimal digits 2, 4, and 9, using the diagram shown in Figure 4.3. First, it is obvious that the BCD to seven-segment decoder has four inputs and seven outputs. Table 4.3 shows the truth table.

TABLE 4.3 Truth Table for Converting Decimal Digits (Since common-cathode, a 1 will turn a segment ON and a 0 will turn it OFF)

Decimal Digit to be Displayed	BCD Input Bits				Seven-Segment Output Bits						
	W	*X*	*Y*	*Z*	*a*	*b*	*c*	*d*	*e*	*f*	*g*
2	0	0	1	0	1	1	0	1	1	0	1
4	0	1	0	0	0	1	1	0	0	1	1
9	1	0	0	1	1	1	1	0	0	1	1

For the valid BCD digits that are not displayed (0, 1, 3, 5, 6, 7, 8) in this example, the combinational circuit for the code converter will generate 0's for the seven output bits (*a* through *g*). However, these seven bits will be don't-cares in the K-map for the invalid BCD digits 10 through 15. Figure 4.4 shows the K-maps and the logic diagram. Note that non-inverting buffers, and current-limiting resistors (not shown in Figure 4.3) are required between the BCD to seven-segment code converter, and the common-cathode display.

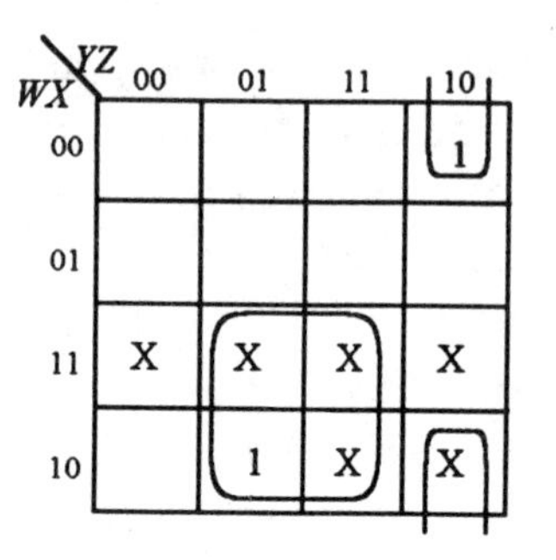

i) K-map for a: $a = WZ + \overline{X}Y\overline{Z}$

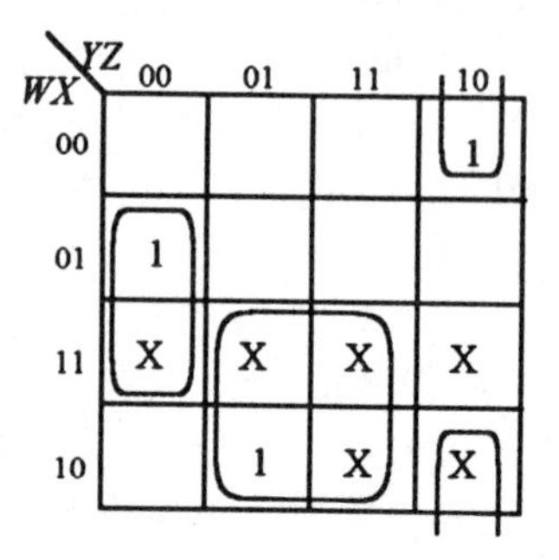

ii) K-map for b : $b = X\overline{Y}\,\overline{Z} + WZ + \overline{X}Y\overline{Z}$

$= \overline{Z}(X\overline{Y} + \overline{X}Y) + WZ$

$= \overline{Z}(X \oplus Y) + WZ$

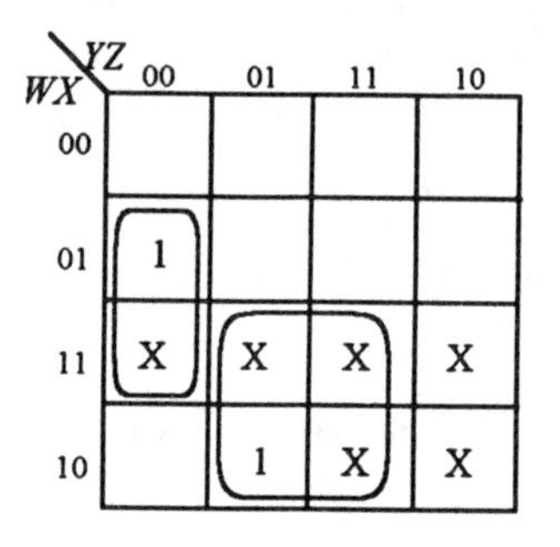

iii) K-map for c: $c = X\overline{Y}\,\overline{Z} + WZ$

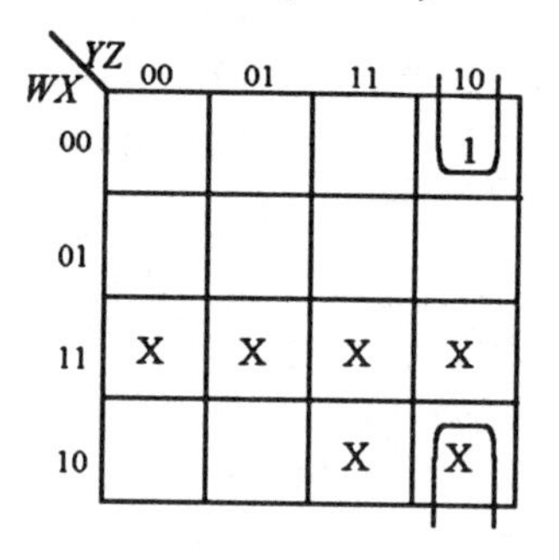

iv) K-map for d: $d = \overline{X}Y\overline{Z}$

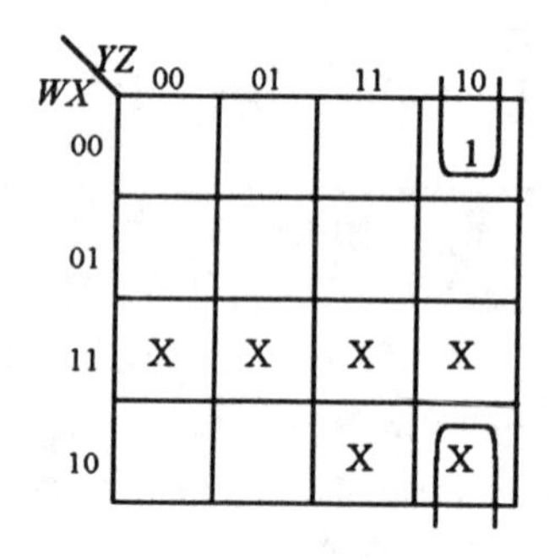

v) K-map for e: $e = \overline{X}Y\overline{Z}$

WX \ YZ	00	01	11	10
00				
01	1			
11	X	X	X	X
10		1	X	X

vi) K-map for f: $f = X\overline{Y}\,\overline{Z} + WZ$

WX \ YZ	00	01	11	10
00				1
01	1			
11	X	X	X	X
10		1	X	X

$$g = X\overline{Y}\,\overline{Z} + WZ + \overline{X}Y\overline{Z}$$
$$= \overline{Z}(X\overline{Y} + \overline{X}Y) + WZ$$
$$= \overline{Z}(X \oplus Y) + WZ$$

vii) K-map for g

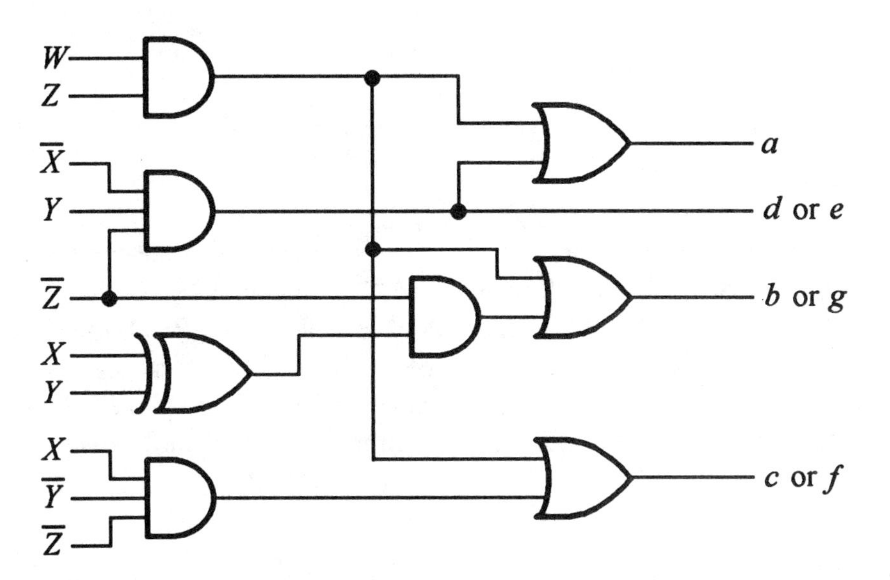

viii) Logic diagram assuming both true and complemented values of the inputs are available.

FIGURE 4.4 BCD to seven-segment decoder for decimal digits 2, 4, and 9

Example 4.1

Design a digital circuit that will convert the BCD codes for the decimal digits (0 through 9) to their Gray codes.

Solution

Because both Gray code and BCD code are represented by four bits for each decimal digit, there are four inputs and four outputs. Table 4.4 shows the truth table.

TABLE 4.4 Truth Table for Example 4.1

Decimal	Input BCD Code				Output Gray Code			
Digit	W	X	Y	Z	f_3	f_2	f_1	f_0
0	0	0	0	0	0	0	0	0
1	0	0	0	1	0	0	0	1
2	0	0	1	0	0	0	1	1
3	0	0	1	1	0	0	1	0
4	0	1	0	0	0	1	1	0
5	0	1	0	1	0	1	1	1
6	0	1	1	0	0	1	0	1
7	0	1	1	1	0	1	0	0
8	1	0	0	0	1	1	0	0
9	1	0	0	1	1	1	0	1

Note that 4-bit binary combination will provide 16 (2^4) combinations of 1's and 0's. Because only ten of these combinations (0000 through 1001) are allowed in BCD, the invalid combinations 1010 through 1111 can never occur in BCD. Therefore, these six binary inputs are considered as don't cares. This means that it does not matter what binary values are assumed by $f_3 f_2 f_1 f_0$ for $WXYZ$ = 1010 through 1111. Figure 4.5 shows the K-maps and the logic circuit.

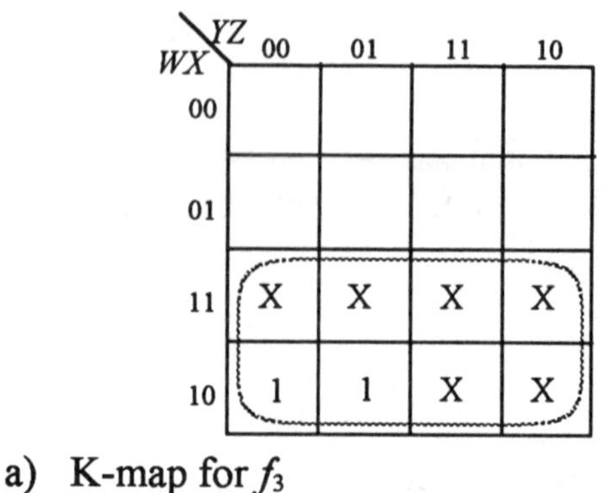

a) K-map for f_3

$f_3 = W$

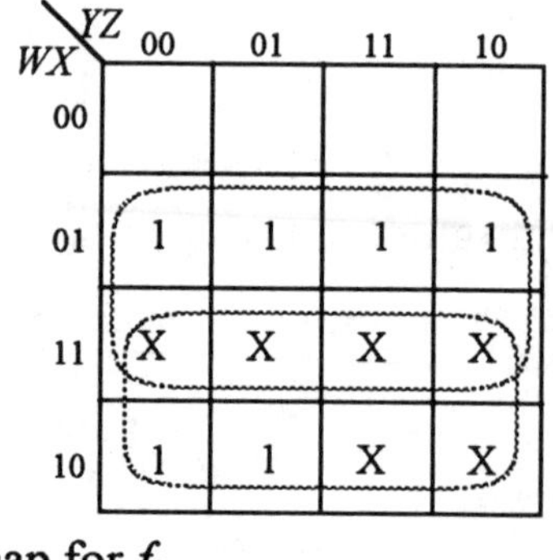

b) K-map for f_2

$f_2 = W + X$

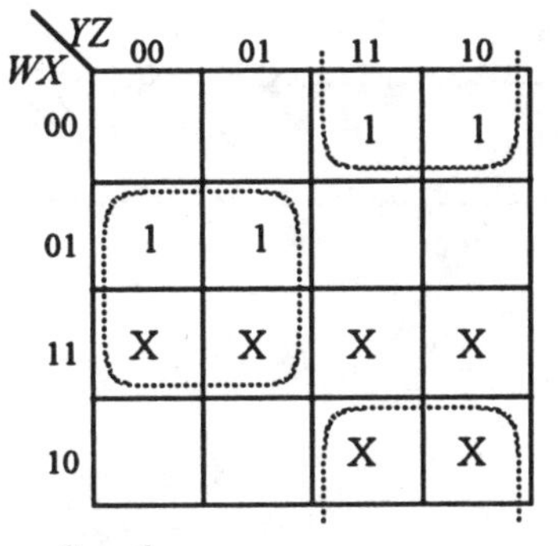

c) K-map for f_1

d) K-m

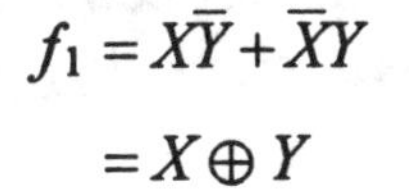

$$f_1 = X\bar{Y} + \bar{X}Y$$
$$= X \oplus Y$$

$$f_0 = \bar{Y}Z + Y\bar{Z}$$
$$= Y \oplus Z$$

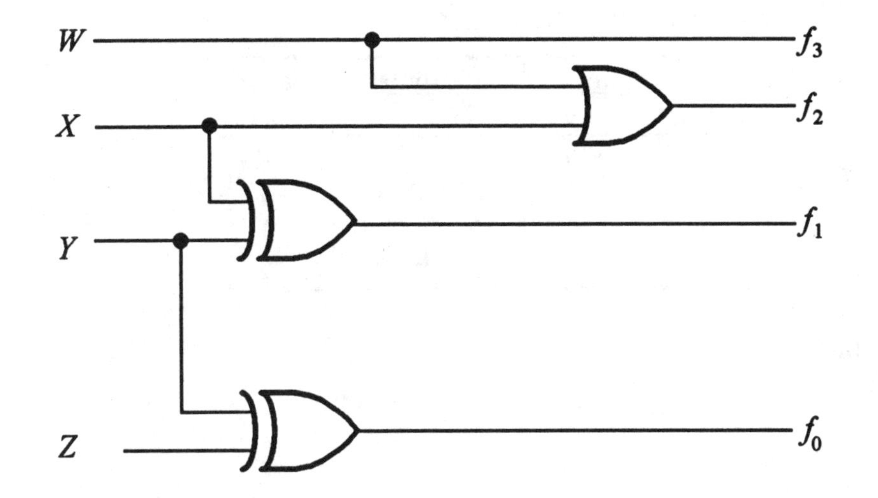

e) Logic diagram for Example 4.1

FIGURE 4.5 K-maps and Logic Circuit for Example 4.1

4.5 Typical Combinational Circuits

This section describes typical combinational circuits. Topics include binary adders, subtractors, comparators, decoders, encoders, multiplexers, and demultiplexers. These digital components are implemented in MSI chips.

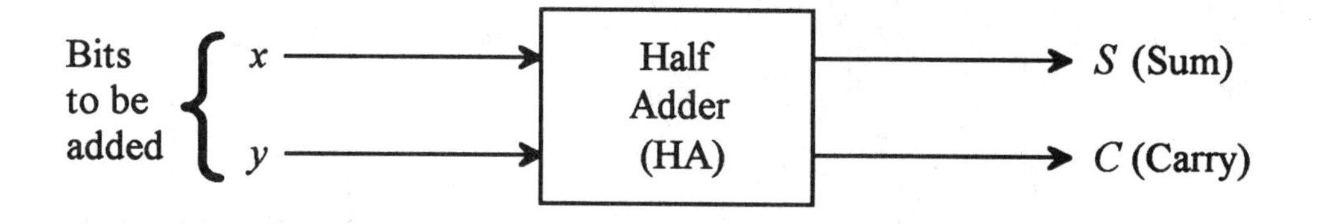

FIGURE 4.6 Block Diagram of a Half-Adder

4.5.1 Binary Adders

When two bits x and y are added, a sum and a carry are generated. *A combinational circuit that adds two bits is called a "half-adder."* Figure 4.6 shows a block diagram of the half-adder. Table 4.5 shows the truth table of the half-adder.

TABLE 4.5 Truth Table of the Half-Adder

Inputs		Outputs		Decimal
x	y	C	S	Value
0	0	0	0	0
0	1	0	1	1
1	0	0	1	1
1	1	1	0	2

From the truth table,

$$S = \bar{x}y + x\bar{y} = x \oplus y$$

$$C = xy$$

Figure 4.7 shows the logic diagram of the half-adder.

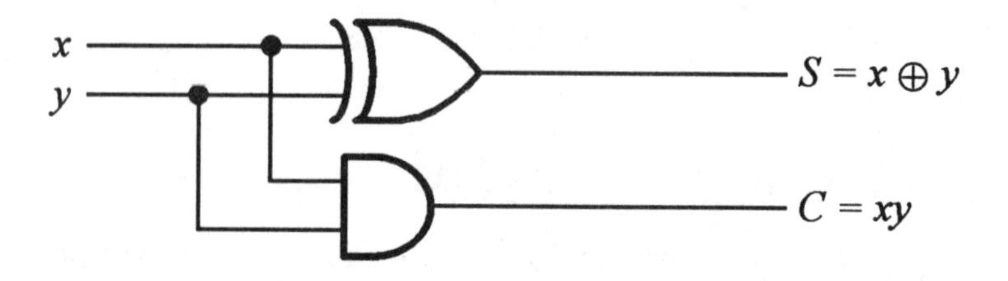

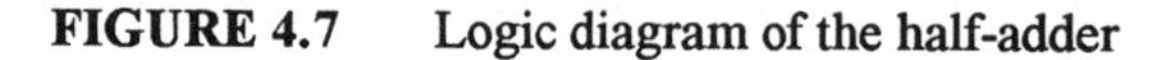
FIGURE 4.7 Logic diagram of the half-adder

Next, consider addition of two 4-bit numbers as follows:

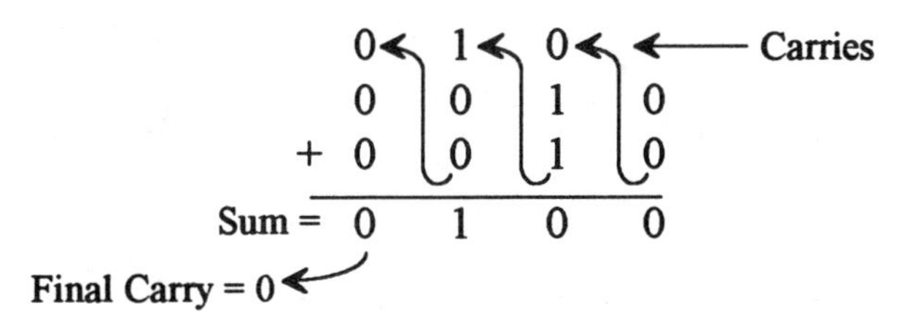

This addition of two bits will generate a sum and a carry. The carry may be 0 or 1. Also, there will be no previous carry while adding the least significant bits (bit 0) of the two numbers. This means that two bits need to be added for bit 0 of the two numbers. On the other hand, addition of three bits (two bits of the two numbers and a previous carry, which may be 0 or 1) is required for all the subsequent bits. Note that two half-adders are required to add three bits.

A combinational circuit that adds three bits, generating a sum and a carry (which may be 0 or 1), is called a "full adder." The names half-adder and full adder are based on the fact that two half-adders are required to obtain a full adder. Figure 4.8 shows the block diagram of a full adder.

The full adder adds three bits, x, y, and z, and generates a sum and a carry. Table 4.6 shows the truth table of a full adder.

TABLE 4.6 Truth Table of a Full Adder

Inputs			Outputs		Decimal
x	y	z	C	S	Value
0	0	0	0	0	0
0	0	1	0	1	1
0	1	0	0	1	1
0	1	1	1	0	2
1	0	0	0	1	1
1	0	1	1	0	2
1	1	0	1	0	2
1	1	1	1	1	3

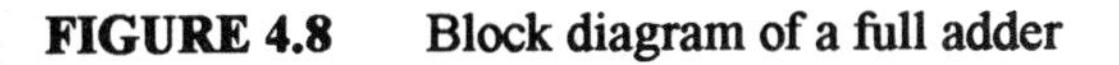

FIGURE 4.8 Block diagram of a full adder

From the truth table,

$$S = \bar{x}\bar{y}z + \bar{x}y\bar{z} + x\bar{y}\bar{z} + xyz$$
$$= (\bar{x}y + x\bar{y})\bar{z} + (xy + \bar{x}\bar{y})z$$
$$\text{Let } w = \bar{x}y + x\bar{y} = x \oplus y$$
$$\text{then } \bar{w} = \overline{x \oplus y} = xy + \bar{x}\bar{y}$$
$$\text{Therefore, } S = w\bar{z} + \bar{w}z$$
$$= w \oplus z$$
$$= x \oplus y \oplus z$$

Also, from the truth table,

$$C = \bar{x}yz + x\bar{y}z + xy\bar{z} + xyz$$
$$= (\bar{x}y + x\bar{y})z + xy(z + \bar{z})$$
$$= wz + xy \text{ where } w = \bar{x}y + x\bar{y} = x \oplus y$$
$$= (x \oplus y)z + xy$$

Figure 4.9 shows the logic diagram of a full adder.

A 4-bit parallel adder for adding two 4-bit numbers $x_3x_2x_1x_0$ and $y_3y_2y_1y_0$ can be implemented using one half-adder and three full adders as shown in Figure 4.10. A full adder adds two bits if one of its inputs $C_{in} = 0$. This means that the half-adder in Figure 4.10 can be replaced by a full adder with its C_{in} connected to ground. Figure 4.11 shows implementation of a 4-bit parallel adder using four full adders.

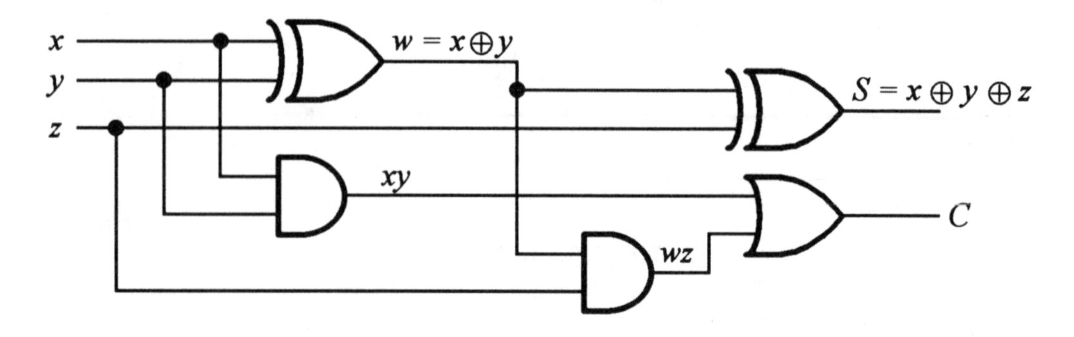

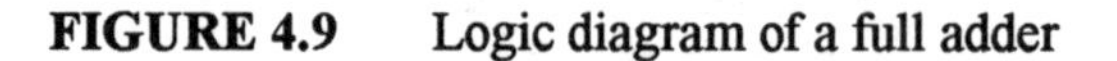
FIGURE 4.9 Logic diagram of a full adder

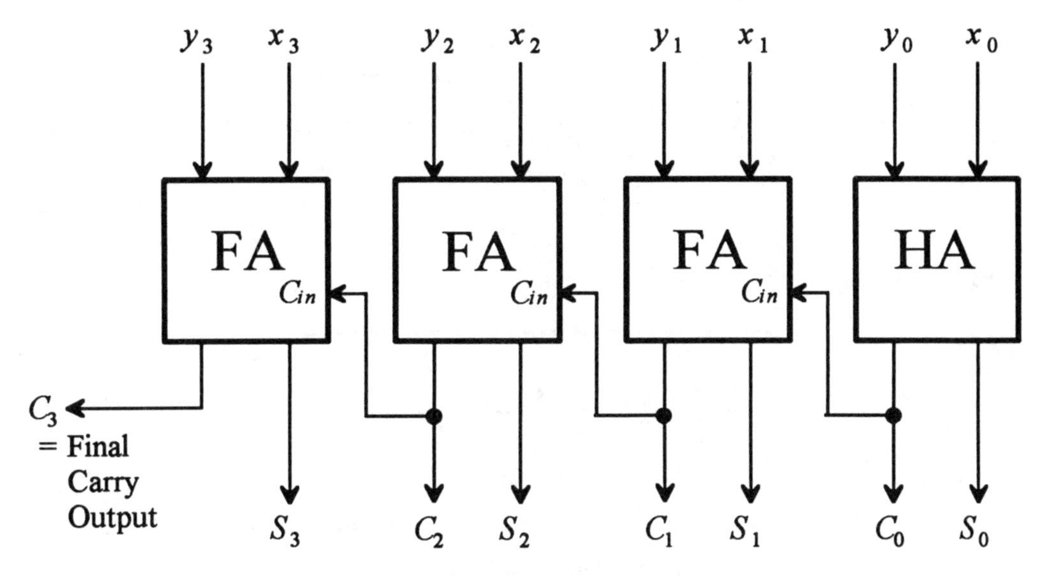

FIGURE 4.10 4-bit parallel adder using one half-adder and three full adders

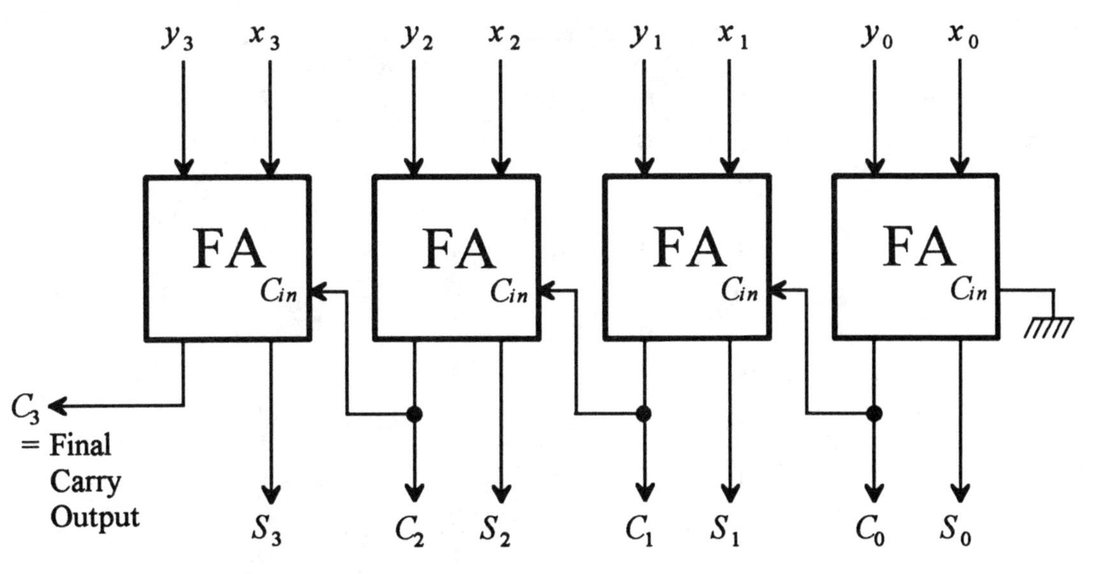

FIGURE 4.11 Four-bit parallel adder using full adders

4.5.2 Comparators

The digital comparator is a widely used combinational system. Figure 4.12 shows a 2-bit digital comparator, which provides the result of comparing two 2-bit unsigned numbers as follows:

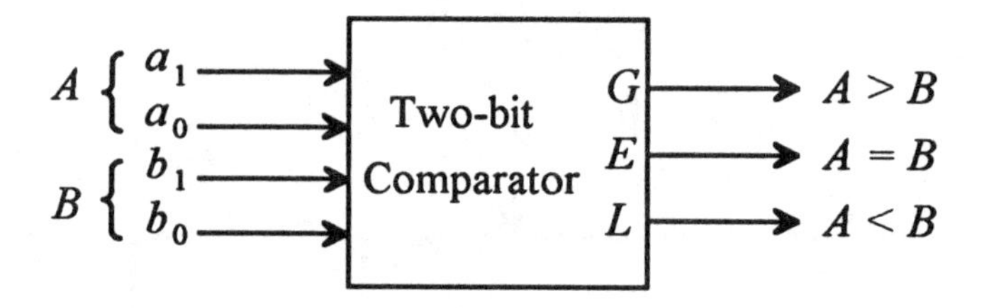

FIGURE 4.12 Block diagram of a two-bit comparator

Input Comparison	Outputs G	E	L
$A > B$	1	0	0
$A < B$	0	0	1
$A = B$	0	1	0

Table 4.7 provides the truth table for the 2-bit comparator.

TABLE 4.7 Truth Table for the 2-Bit Comparator

Inputs a_1	a_0	b_1	b_0	Outputs G	E	L
0	0	0	0	0	1	0
0	1	0	0	1	0	0
1	0	0	0	1	0	0
1	1	0	0	1	0	0
0	0	0	1	0	0	1
0	1	0	1	0	1	0
1	0	0	1	1	0	0
1	1	0	1	1	0	0
0	0	1	0	0	0	1
0	1	1	0	0	0	1
1	0	1	0	0	1	0
1	1	1	0	1	0	0
0	0	1	1	0	0	1
0	1	1	1	0	0	1
1	0	1	1	0	0	1
1	1	1	1	0	1	0

Figure 4.13 shows the K-map and the logic diagram:

K-map for G:

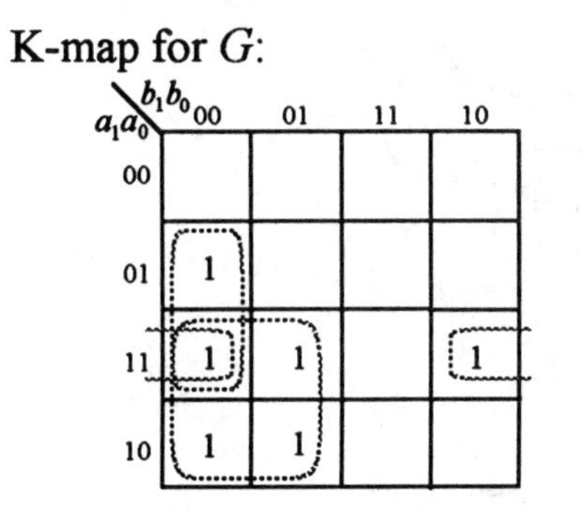

$$G = a_1\overline{b_1} + a_0\overline{b_1}\,\overline{b_0} + a_1a_0\overline{b_0}$$

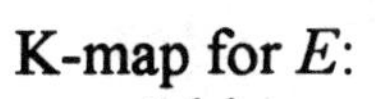

K-map for E:

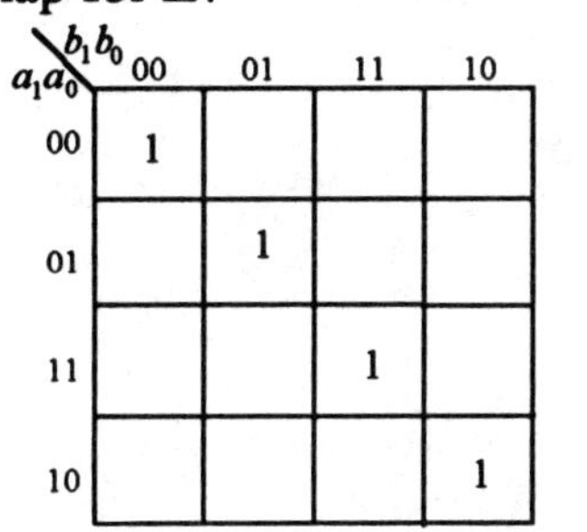

$$E = \overline{a_1}\,\overline{a_0}\,\overline{b_1}\,\overline{b_0} + \overline{a_1}a_0\overline{b_1}b_0 + a_1a_0b_1b_0 + a_1\overline{a_0}b_1\overline{b_0}$$
$$= \overline{a_1}\,\overline{b_1}(\overline{a_0}\,\overline{b_0} + a_0b_0) + a_1b_1(a_0b_0 + \overline{a_0}\,\overline{b_0})$$
$$= (a_0b_0 + \overline{a_0}\,\overline{b_0})(a_1b_1 + \overline{a_1}\,\overline{b_1})$$
$$= (a_0 \odot b_0)(a_1 \odot b_1)$$

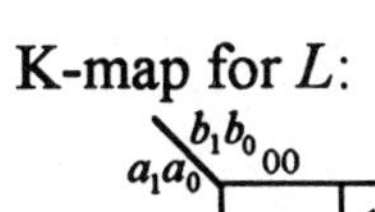

K-map for L:

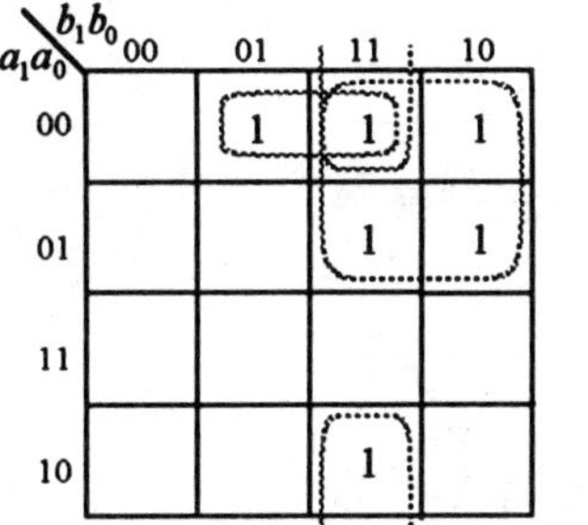

$$L = \overline{a_1}b_1 + \overline{a_0}b_1\,b_0 + \overline{a_1}\,\overline{a_0}b_0$$

a) K-maps for the 2-bit comparator

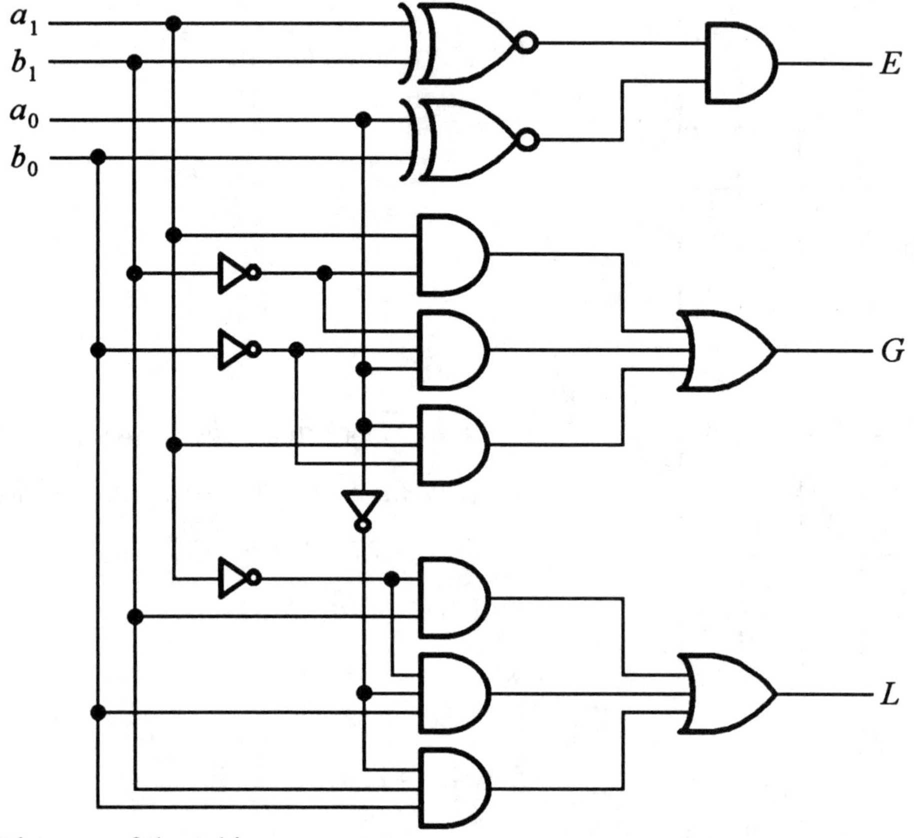

b) Logic Diagram of the 2-bit comparator

FIGURE 4.13 Design of a 2-bit comparator

4.5.3 Decoders

An n*-bit binary number provides 2^n combinations of 1's and 0's.* For example, a 2-bit binary number presents 4 (2^2) combinations of 1's and 0's: 00, 01, 10, and 11. *A decoder is a combinational circuit that transforms an* n*-bit binary number into 2^n outputs.* However, a decoder sometimes may have less than 2^n outputs. For example, the BCD to seven-segment decoder has 4 inputs and 7 outputs rather than 16 (2^4) outputs.

The block diagram of a 2-to-4 decoder is shown in Figure 4.14. Table 4.8 provides the truth table. In the truth table, the symbol X is the don't care condition, which can be 0 or 1. Also, $E = 0$ disables the decoder. On the other hand, the decoder is enabled when $E = 1$. For example, when $E = 1$, $x_1 = 0$, $x_0 = 0$, and the output d_0 is HIGH while the other outputs d_1, d_2, and d_3 are zero. Note that $d_0 = E\overline{x_1}\,\overline{x_0}$, $d_1 = E\overline{x_1}x_0$, $d_2 = Ex_1\overline{x_0}$, and $d_3 = Ex_1x_0$.

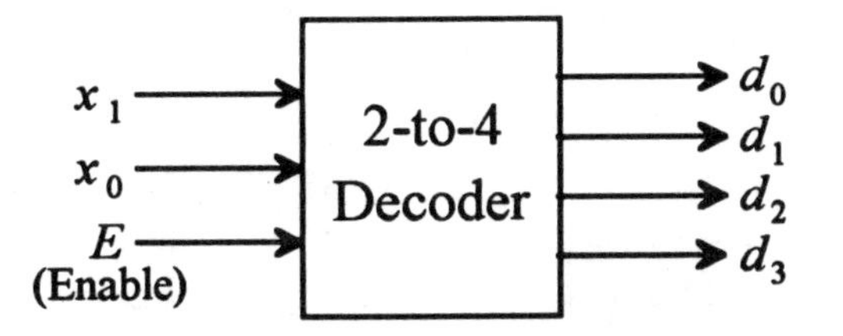

FIGURE 4.14 Block diagram of the 2-to-4 decoder

Therefore, the 2-to-4 line decoder outputs one of the four minterms of the two input variables x_1 and x_0 when $E = 1$. In general, for n inputs, the n-to 2^n decoder when enabled generates one of the 2^n minterms at the output based on the input combinations. The decoder actually provides binary to decimal conversion operation. Using the truth table of Table 4.8, a logic diagram of the 2-to-4 decoder can be obtained as shown in Figure 4.15. Large decoders can be designed using small decoders as the building blocks. For example, a 4-to-16 line decoder can be designed using five 2-to-4 decoders as shown in Figure 4.16.

TABLE 4.8 Truth Table of the 2-to-4 Decoder

Inputs			Outputs			
E	x_1	x_0	d_0	d_1	d_2	d_3
0	X	X	0	0	0	0
1	0	0	1	0	0	0
1	0	1	0	1	0	0
1	1	0	0	0	1	0
1	1	1	0	0	0	1

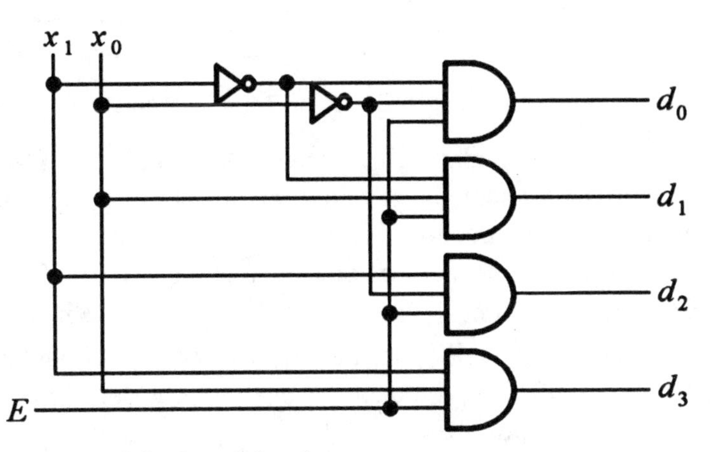

FIGURE 4.15 Logic diagram of the 2-to-4 decoder

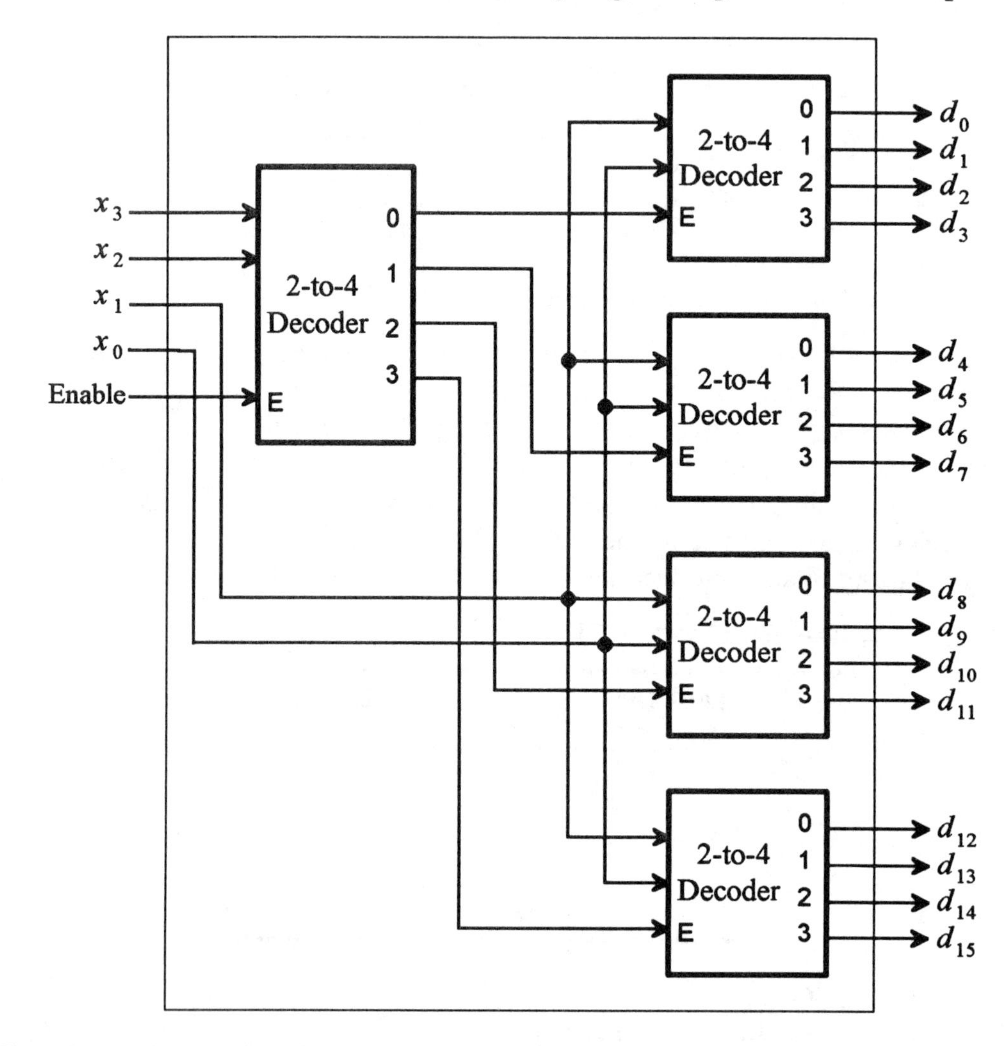

FIGURE 4.16 Implementation of a 4-to-16 Decoder Using 2-to-4 decoders

Commercially available decoders are normally built using NAND gates rather than AND gates because it is less expensive to produce the selected decoder output in its complement form. Also, most commercial decoders contain one or more enable inputs to control the circuit operation. An example of the commercial decoder is the 74HC138 or the 74LS138. This is a 3-to-8 decoder with three enable lines G_1, $\overline{G_{2A}}$, and $\overline{G_{2B}}$. When $G_1 = H$, $\overline{G_{2A}} = L$ and $\overline{G_{2B}} = L$, the decoder is enabled. The decoder has three inputs, C, B, and A, and eight outputs $Y_0, Y_1, Y_2, \ldots, Y_7$. With $CBA = 001$ and the decoder enabled, the selected output line Y_1 (line 1) goes to LOW while the other output lines stay HIGH.

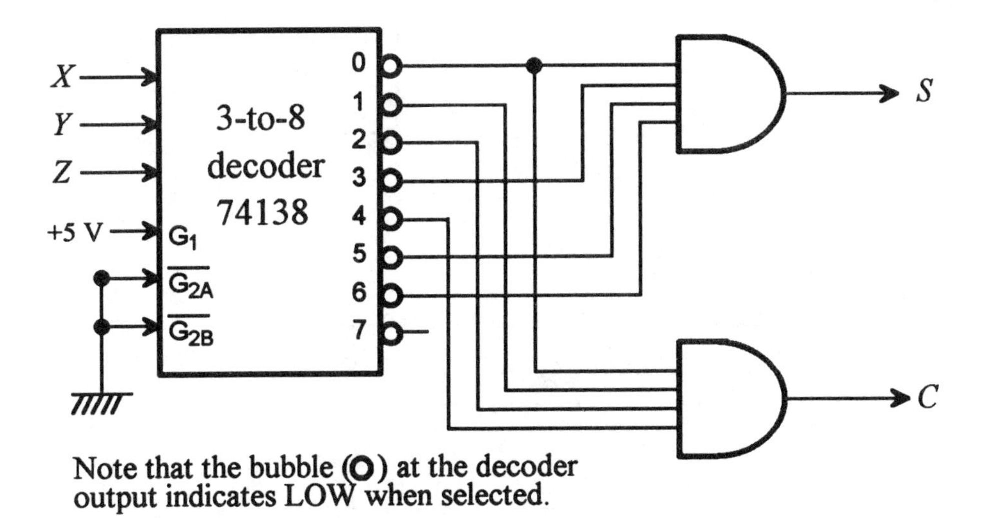

FIGURE 4.17 Implementation of a Full-adder Using a 74138 Decoder and Two 4-input AND Gates

Because any Boolean function can be expressed as a logical sum of minterms, a decoder can be used to produce the minterms. A Boolean function can then be obtained by logical operation of the appropriate minterms. However, since the 74138 generates a LOW on the selected output line, a Boolean function can be obtained by logically ANDing the appropriate minterms. For example, consider the truth table of the full adder listed in Table 4.6. The inverted sum and the inverted carry can be expressed in terms of minterms as follows:

$$\overline{S} = \sum m(0,3,5,6), \quad S = \overline{m_0} \cdot \overline{m_3} \cdot \overline{m_5} \cdot \overline{m_6}$$

$$\overline{C} = \sum m(0,1,2,4), \quad C = \overline{m_0} \cdot \overline{m_1} \cdot \overline{m_2} \cdot \overline{m_4}$$

Figure 4.17 shows the implementation of a full adder using a 74138 decoder ($C=X$, $B=Y$, $A=Z$) and two 4-input AND gates. Note that the 74138 in the Manufacturer's data book uses the symbols *C, B, A* as three inputs to the decoder with C as the most significant bit and *A* as the least significant bit.

4.5.4 Encoders

An encoder is a combinational circuit that performs the reverse operation of a decoder. An encoder has a maximum of 2^n inputs and n outputs. Figure 4.18 shows the block diagram of a 4-to-2 encoder. Table 4.9 provides the truth table of the 4-to-2 encoder.

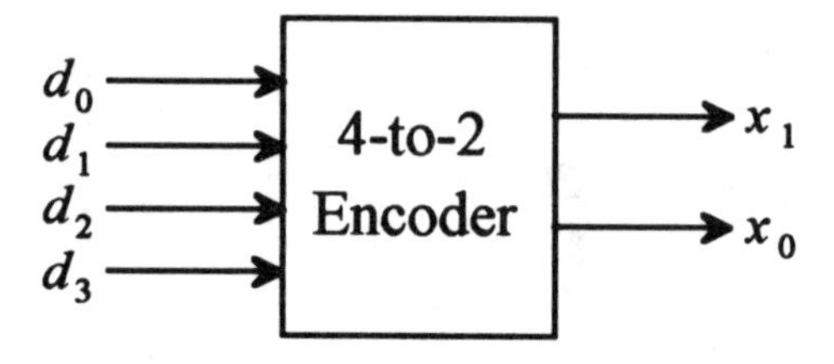

FIGURE 4.18 Block diagram of a 4-to-2 encoder

TABLE 4.9 Truth Table of the 4-to-2 Encoder

Inputs				Outputs	
d_0	d_1	d_2	d_3	x_1	x_0
1	0	0	0	0	0
0	1	0	0	0	1
0	0	1	0	1	0
0	0	0	1	1	1

From the truth table, it can be concluded that an encoder actually performs decimal-to-binary conversion. In the encoder defined by Table 4.9, it is assumed that only one of the four inputs can be HIGH at any time. If more than one input is 1 at the same time, an undefined output is generated. For example, if d_1 and d_2 are 1 at the same time, both x_0 and x_1 are 1. This represents binary 3 rather than 1 or 2. Therefore, in an encoder in which more than one input can be active simultaneously, a priority scheme must be implemented in the inputs to ensure that only one input will be encoded at the output.

A 4-to-2 priority encoder will be designed next. Suppose that it is assumed that inputs with higher subscripts have higher priorities. This means that d_3 has the highest priority and d_0 has the lowest priority. Therefore, if d_0 and d_1 become one simultaneously, the output will be 01 for d_1. Table 4.10 shows the truth table of the 4-to-2 priority encoder. Figure 4.19 shows the K-maps and the logic diagram of the 4-to-2 priority encoder.

TABLE 4.10 Truth Table of the 4-to-2 Priority Encoder

Inputs				Outputs	
d_0	d_1	d_2	d_3	x_1	x_0
1	0	0	0	0	0
X	1	0	0	0	1
X	X	1	0	1	0
X	X	X	1	1	1

X means don't care

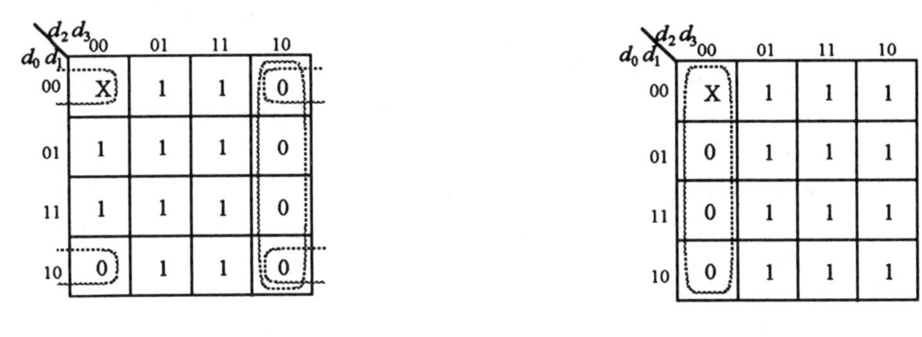

a) K-map for $\overline{x_0}$

$$\overline{x_0} = \overline{d_1}\,\overline{d_3} + d_2\overline{d_3}$$

$$x_0 = (d_1 + d_3)(\overline{d_2} + d_3)$$

b) K-map for $\overline{x_1}$

$$\overline{x_1} = \overline{d_2}\,\overline{d_3}$$

$$x_1 = d_2 + d_3$$

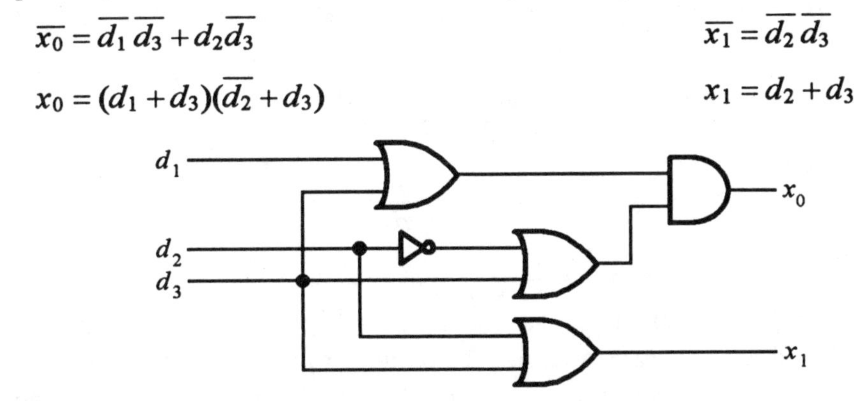

c) Logic diagram

FIGURE 4.19 K-maps and logic diagram of a 2-to-4 priority encoder

4.5.5 Multiplexers

A multiplexer (abbreviated as MUX) is a combinational circuit that selects one of n *input lines and provides it on the output.* Thus, the multiplexer has several inputs and only one output. The select lines identify or address one of several inputs and provides it on the output line. Figure 4.20 shows the block diagram of a 2-to-1 multiplexer. The two inputs can be selected by one select line, S. When $S = 0$, input line 0 (d_0) will be presented as the output. On the other hand, when $S = 1$, input line 1 (d_1) will be produced at the output.

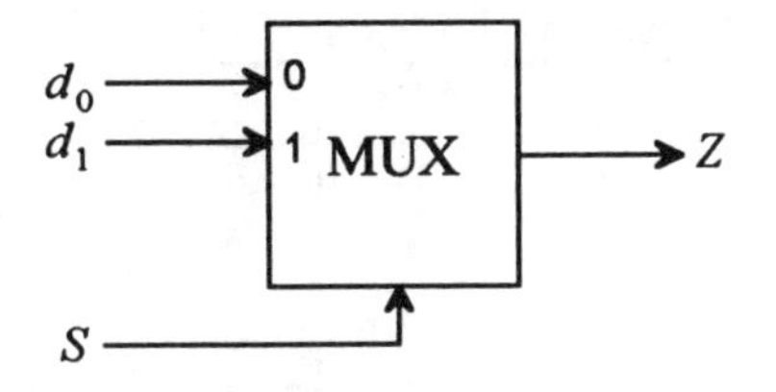

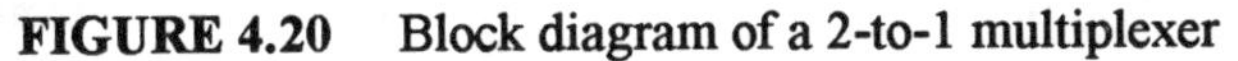

FIGURE 4.20 Block diagram of a 2-to-1 multiplexer

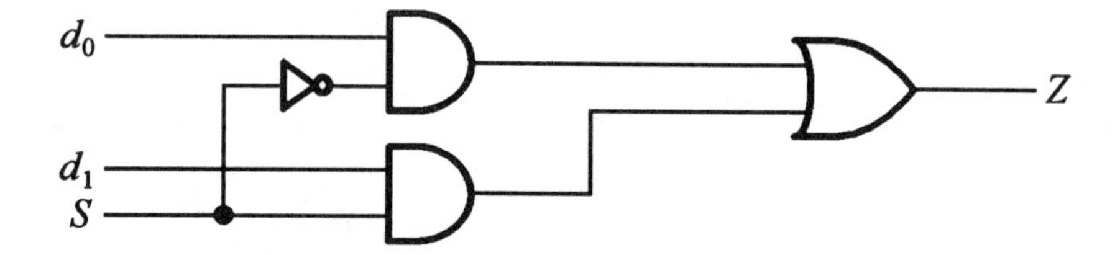

FIGURE 4.21 Logic diagram of the 2-to-1 MUX

Table 4.11 shows the truth table of the 2-to-1 multiplexer. From the truth table, $Z = \bar{S}d_0 + Sd_1$. Figure 4.21 shows the logic diagram. In general, a multiplexer with n select lines can select one of 2^n data inputs. Hence, multiplexers are sometimes referred to as "data selectors."

TABLE 4.11 Truth Table of the 2-to-1 Multiplexer

Select Input S	Output Z
0	d_0
1	d_1

A large multiplexer can be implemented using a small multiplexer as the building block. For example, consider the block diagram and the truth table of a 4-to-1 multiplexer shown in Figure 4.22 and Table 4.12 respectively. The 4-input multiplexer can be implemented using three 2-to-1 multiplexers as shown in Figure 4.23.

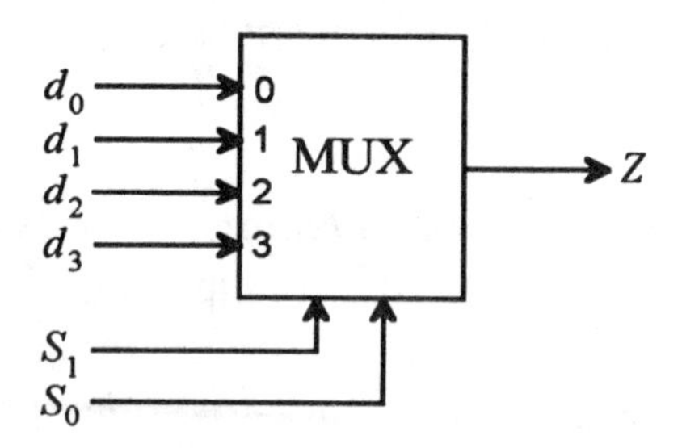

FIGURE 4.22 Block-diagram Representation of a Four-input Multiplexer

TABLE 4.12 Truth Table of the Four Input Multiplexer

S_1	S_0	Z
0	0	d_0
0	1	d_1
1	0	d_2
1	1	d_3

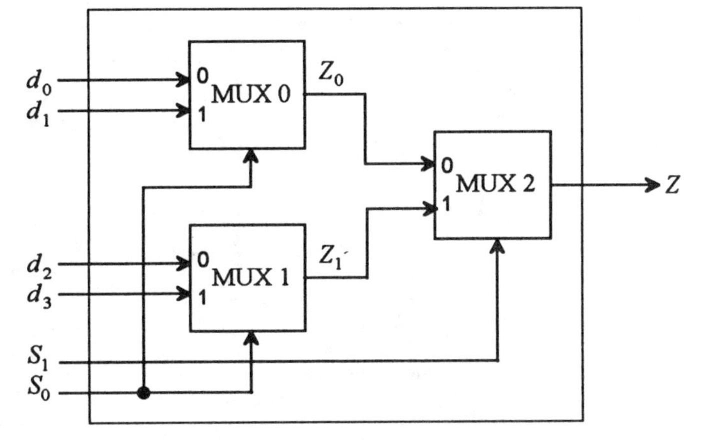

FIGURE 4.23 Implementation of a Four-Input Multiplexer Using Only Two-input Multiplexers

In Figure 4.23, the select line S_0 is applied as input to the multiplexers MUX 0 and MUX 1. This means that $Z_0 = d_0$ or d_1 and $Z_1 = d_2$ or d_3, depending on whether $S_0 = 0$ or 1. The select line S_1 is given as input to the multiplexer MUX 2. This implies that $Z = Z_0$ if $S_1 = 0$; otherwise $Z = Z_1$. In this arrangement if $S_1S_0 = 11$, then $Z = d_3$ because $S_0 = 1$ implies that $Z_0 = d_1$ and $Z_1 = d_3$ because $S_1 = 1$, the MUX 2 selects the data input Z_1, and thus $Z = d_3$. The other entries of the truth table of Table 4.12 can be verified in a similar manner.

4.5.6 Demultiplexers

The demultiplexer is a combinational circuit that performs the reverse operation of a multiplexer. The demultiplexer has only one input and several outputs. One of the outputs is selected by the combination of 1's and 0's of the select inputs. These inputs determine one of the output lines to be selected; data from the input line is then transferred to the selected output line. Figure 4.24 shows the block diagram of a 1-to-8 demultiplexer. Suppose that $i = 1$ and $S_2S_1S_0 = 010$; output line d_2 will be selected and a 1 will be output on d_2.

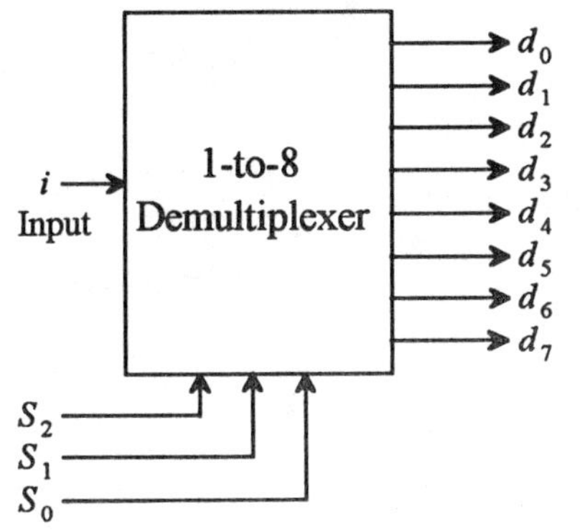

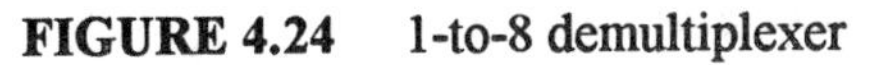

FIGURE 4.24 1-to-8 demultiplexer

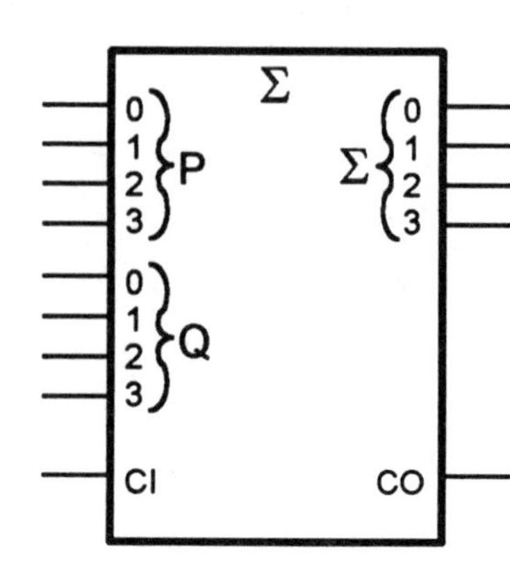

a) 4-bit Parallel Adder (74LS283 or 74HC283)

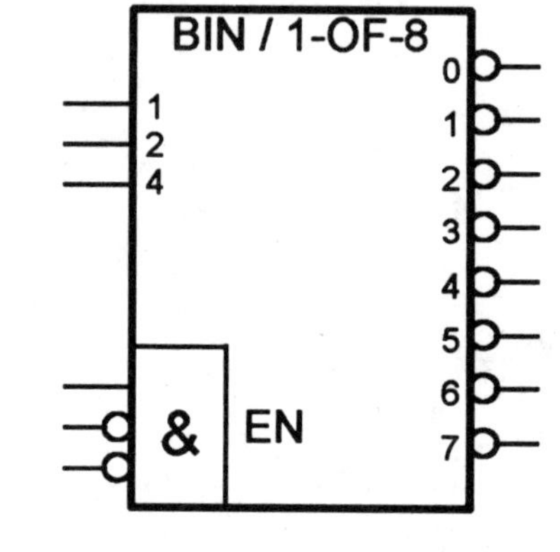

b) 3-to-8 Decoder (74LS138 or 74HC138)

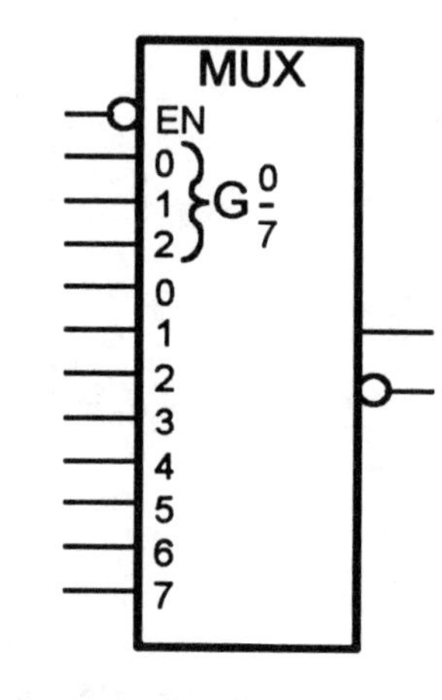

c) 8-to-1 Multiplexer (74LS151 or 74HC151) (providing both true and complemented outputs)

FIGURE 4.25 IEEE Symbols

4.6 IEEE Standard Symbols

IEEE has developed standard graphic symbols for commonly used digital components such as adders, decoders, and multiplexers. These are depicted in Figure 4.25.

Example 4.2

Design a combinational circuit using a decoder and OR gates to implement the function depicted in Figure 4.26.

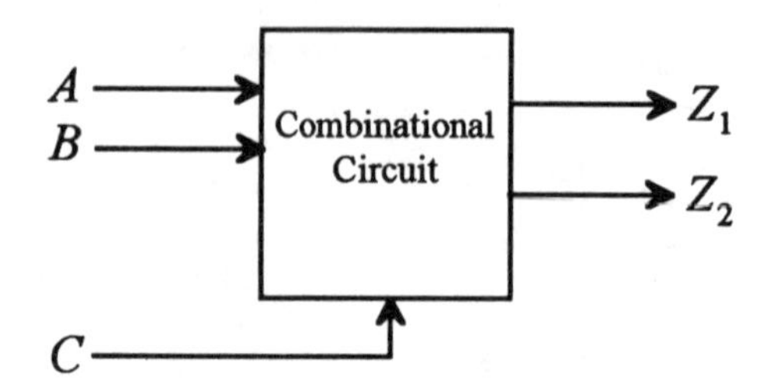

If $C = 0$, Z_1 follows B and $Z_2 = A + B$.

If $C = 1$, $Z_1 = A + B$ and $Z_2 = AB$.
Assume that the decoder output is HIGH when selected.

FIGURE 4.26 Figure for Example 4.2

Solution

The truth table is shown in Table 4.13.

TABLE 4.13 Truth Table for Example 4.2

Inputs			Outputs	
C	B	A	Z_1	Z_2
0	0	0	0	0
0	0	1	0	1
0	1	0	1	1
0	1	1	1	1
1	0	0	0	0
1	0	1	1	0
1	1	0	1	0
1	1	1	1	1

From the truth table,

$$Z_1 = \sum m(2, 3, 5, 6, 7)$$

$$Z_2 = \sum m(1, 2, 3, 7)$$

The logic diagram is shown in Figure 4.27.

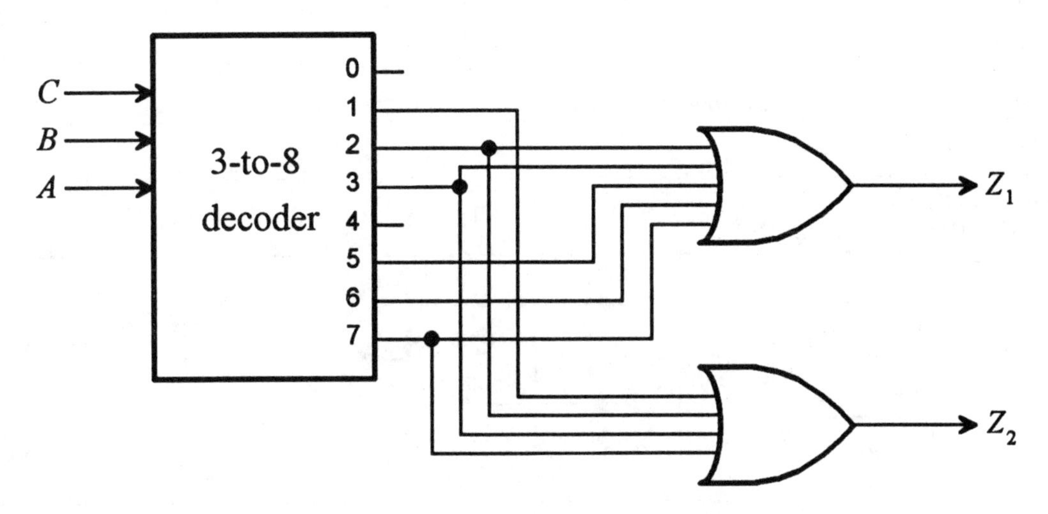

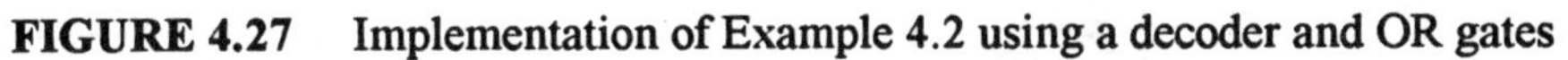

FIGURE 4.27 Implementation of Example 4.2 using a decoder and OR gates

Example 4.3

Design a combinational circuit to implement a 4-bit adder/subtractor using full adders and multiplexers.

Solution

The subtraction $x - y$ of two binary numbers can be performed using twos complement arithmetic. As discussed before, $x - y = x +$ (ones complement of y) + 1

Using this concept, parallel subtractors can be implemented. A 4-bit adder/subtractor is shown in Figure 4.28. Note that XOR gates (S and y_n as inputs) can be used in place of multiplexers.

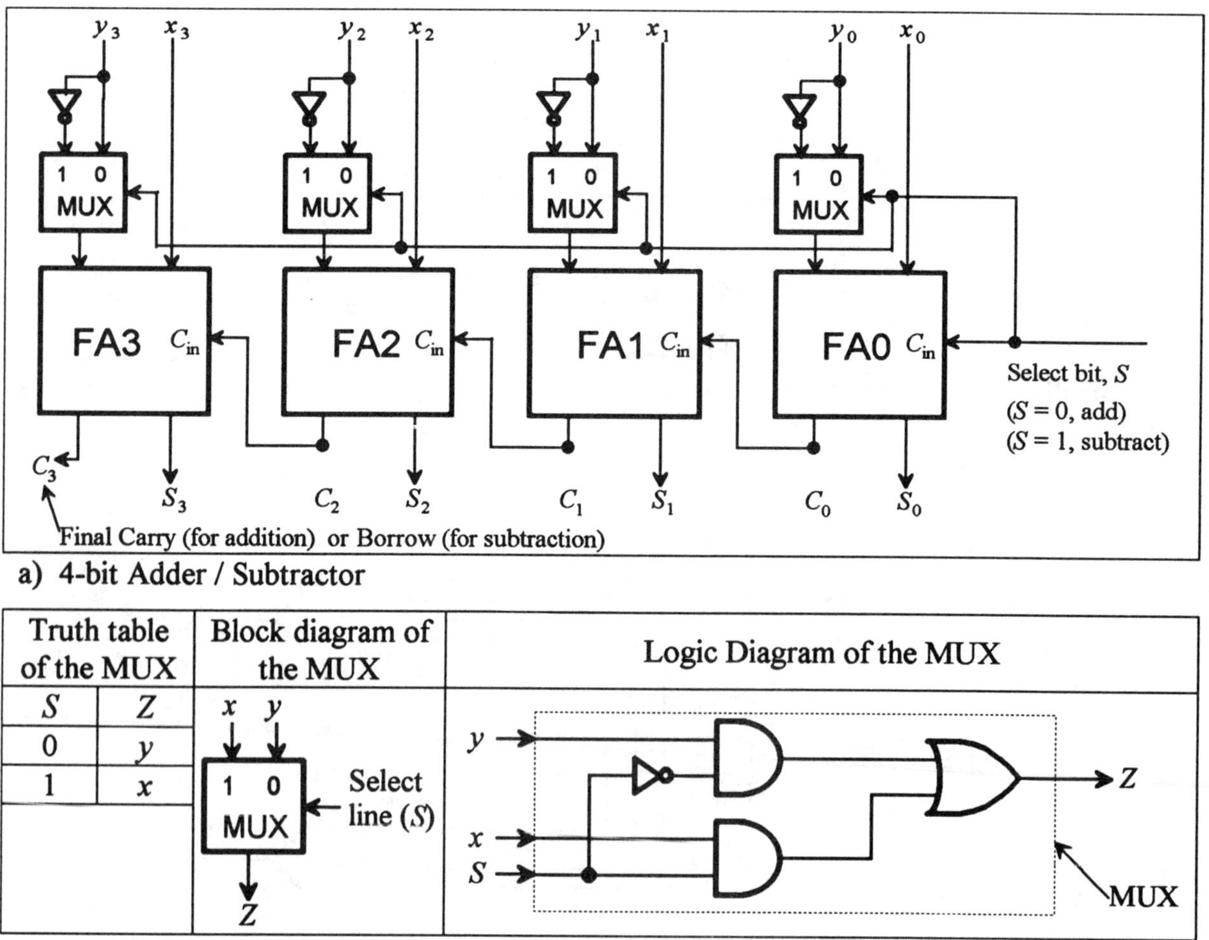

FIGURE 4.28 4-bit Adder / Subtractor

The adder/subtractor in Figure 4.28 utilizes four MUX's. Each MUX has one select line (S) and is capable of selecting one of two lines, x or y. The details of the MUX are shown in Figure 4.28(b). From the logic diagram of the MUX, when $S = 0$, the output of the MUX (Z) will be y. On the other hand, when $S = 1$, then Z will be x. In the block diagram of the MUX, x and y are respectively marked with 1 and 0, indicating that when $S = 0$, y is selected; on the other hand, x is selected for $S = 1$.

The 4-bit adder/subtractor of Figure 4.28(a) either adds two 4-bit numbers and performs $(x_3x_2x_1x_0)$ ADD $(y_3y_2y_1y_0)$ when $S = 0$ or performs the subtraction operation $(x_3x_2x_1x_0)$ MINUS $(y_3y_2y_1y_0)$ for $S = 1$. The select bit S can be implemented by a switch. When $S = 0$, each MUX outputs the true value of y_n ($n = 0$ through 3) to the corresponding input of the full adder FA_n ($n = 0$ through 3). Because $S = 0$ (C_{in} for $FA_0 = 0$), the four full adders perform the desired 4-bit addition.
When $S = 1$ (C_{in} for $FA_0 = 1$), each MUX generates the ones complement of y_n at the corresponding input of the full adder FA_n. Because $S = C_{in} = 1$, the four full adders provide the following operation:

$$(x_3x_2x_1x_0) - (y_3y_2y_1y_0) = (x_3x_2x_1x_0) + (\overline{y_3}\,\overline{y_2}\,\overline{y_1}\,\overline{y_0}) + 1$$

4.7 Read-Only Memories (ROMs)

Read-only memory, commonly called "ROM," is a nonvolatile memory (meaning that it retains information in case of power failure) that provides read-only access to the stored data. A block-diagram representation of a ROM is shown in Figure 4.29. The total capacity of this ROM is $2^n \times m$ bits. Whenever an n-bit address is placed on the address line, the m-bit information stored in this address will appear on the data lines. The m-bit output generated by the ROM is also called a "word."

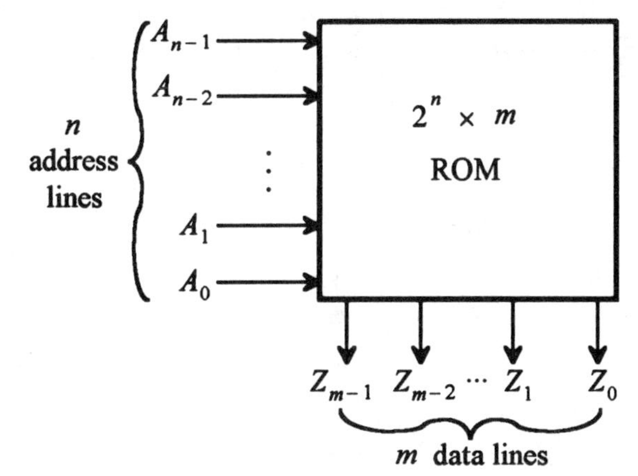

FIGURE 4.29 Block-diagram Representation of a ROM

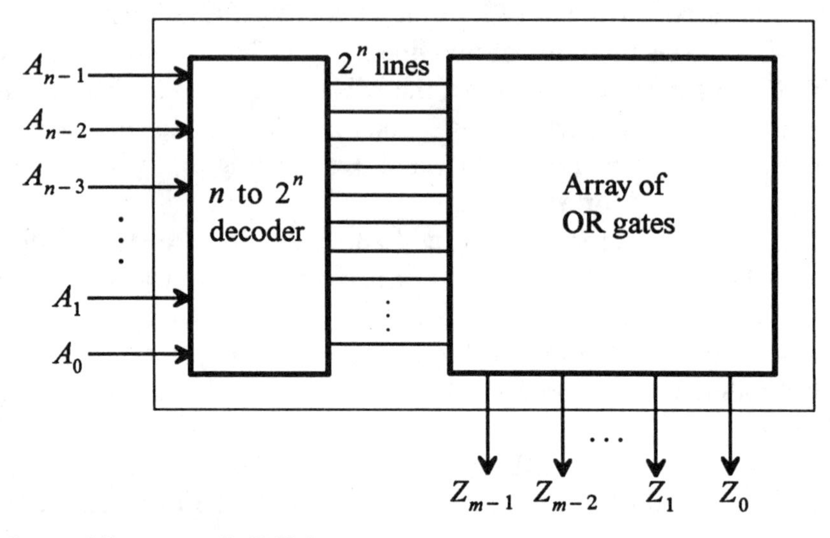

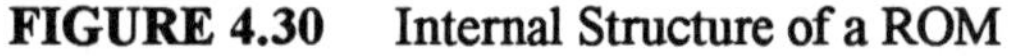
FIGURE 4.30 Internal Structure of a ROM

For example, a 1K × 8 (1024 × 8)-bit ROM chip contains 10 address pins (2^{10} = 1024 = 1K) and 8 data pins. Therefore, n = 10 and m = 8. On the other hand, an 8K × 8 (8192 × 8)-bit ROM chip includes 13 address pins (2^{13} = 8192 = 8K) and 8 data pins. Thus, n = 13 and m = 8.

A ROM is an LSI chip that can be designed using an array of semiconductor devices such as diodes, transistors, or MOS transistors. *A ROM is a combinational circuit. Internally, a ROM is composed of a decoder and OR gates; this is illustrated in Figure 4.30.*

The OR gate of the ROM may be built using diodes. A typical 3-input diode OR gate is shown in Figure 4.31. Resistor R pulls the output down to a LOW level as long as all the inputs are LOW. However, if either input is connected to a high voltage source (3 to 5 volts), the output is pulled HIGH to within one diode drop of the input. Thus, the circuit operates as an OR gate.

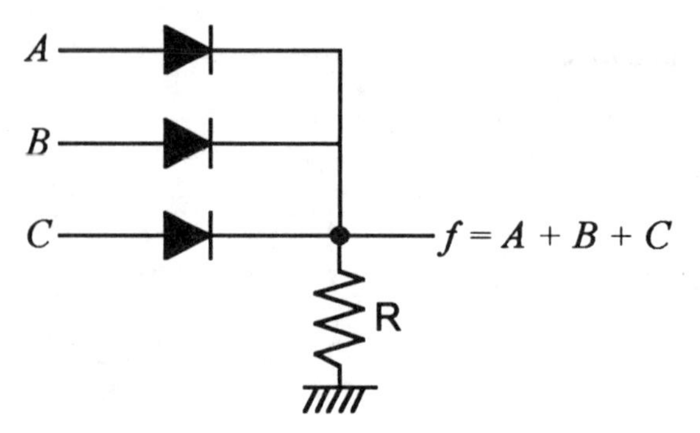

FIGURE 4.31 Diode-OR Gate

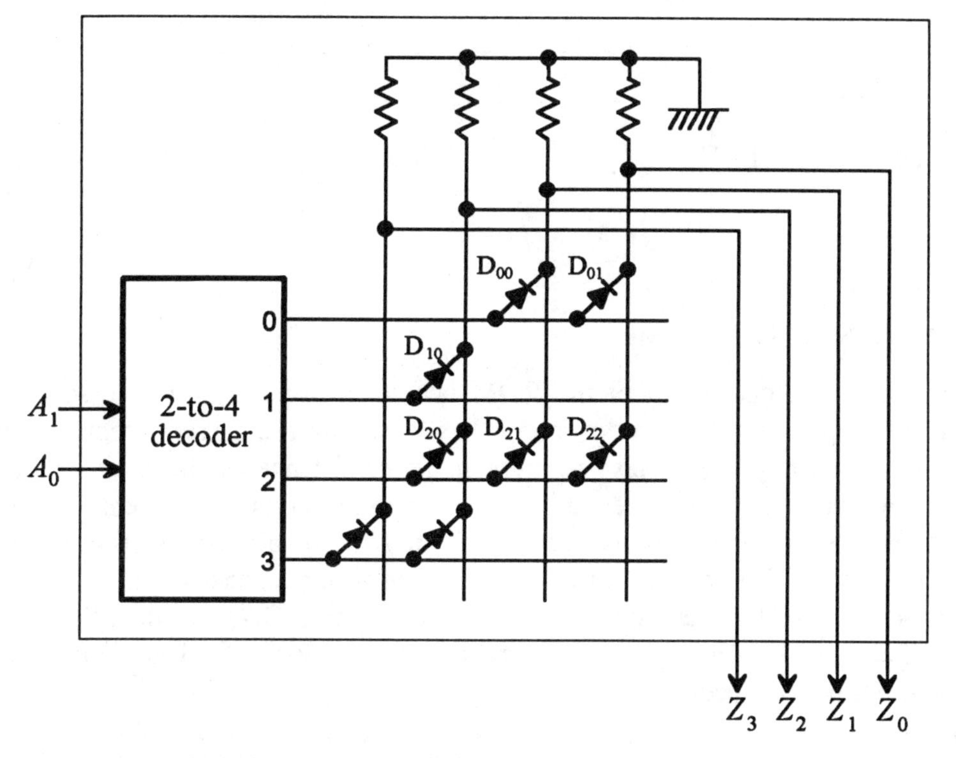

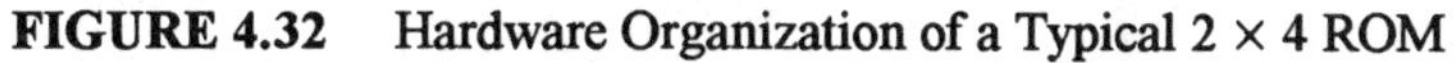

FIGURE 4.32 Hardware Organization of a Typical 2 × 4 ROM

To illustrate the operation of a ROM, consider the 2 × 4-bit ROM of Figure 4.32. In this system , when $A_1A_0 = 00$, the decoder output line 0 will be HIGH. This causes the diodes D_{00} and D_{01} to conduct, and thus the output $Z = Z_3Z_2Z_1Z_0 = 0011$. Similarly, when $A_1A_0 = 01$, the decoder output line 1 goes high, diode D_{10} conducts, and the output will be $Z = Z_3Z_2Z_1Z_0 = 0100$. Table 4.14 shows the truth table. Figure 4.33 shows the subcategories of ROMs and their associated technologies. ROM implementation offers a cost-effective solution for building circuits to perform useful tasks such as square root and transcendental function computations.

TABLE 4.14 Truth Table implemented by the ROM of Figure 4.32

A_1	A_0	Z_3	Z_2	Z_1	Z_0
0	0	0	0	1	1
0	1	0	1	0	0
1	0	0	1	1	1
1	1	1	1	0	0

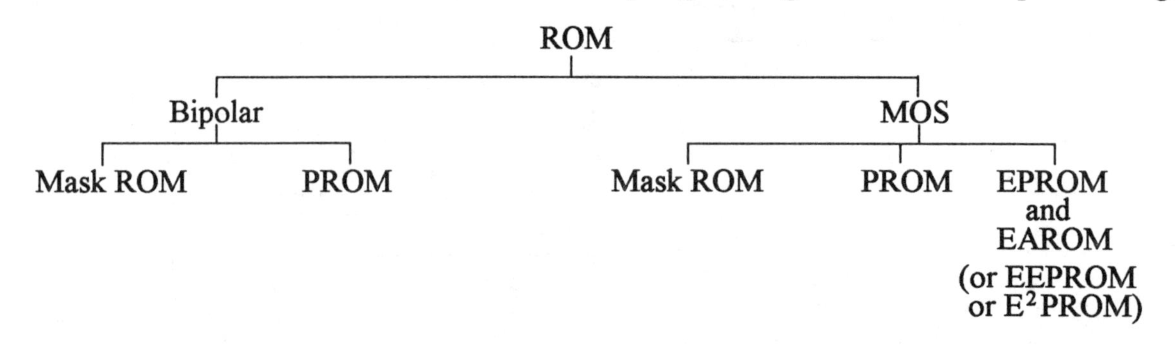

FIGURE 4.33 Subcategories of ROMs

The diode OR gate array of the ROM is also customarily referred to as the "diode matrix." In this approach, the decoder circuit has to supply the currents required by the load connected to the output lines. This requirement limits the use of the diode matrix in some applications in which the load demands substantial amounts of current. To rectify this problem, the diode matrix may be replaced with an array of bipolar junction transistors. For example, the ROM of Figure 4.32 may be redesigned using transistors; it is shown in Figure 4.34. When $A_1A_0 = 00$, the transistors T_{00} and T_{01} are turned ON, and thus the output generated is $Z_3Z_2Z_1Z_0 = 0011$. Note that the current required by the load is supplied by the ON transistor and not by the decoder.

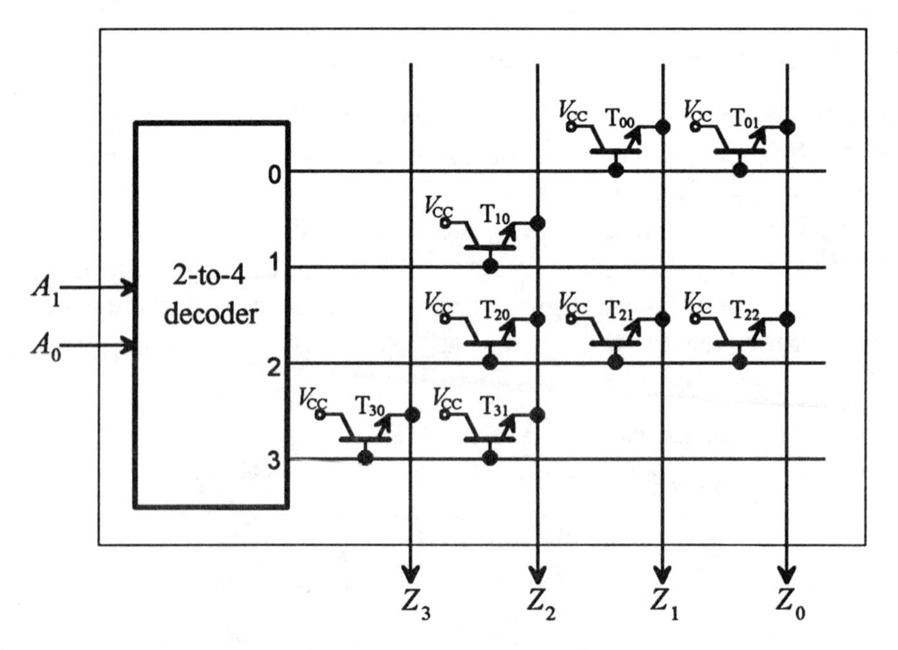

FIGURE 4.34 Alternative implementation of the ROM of Figure 4.32

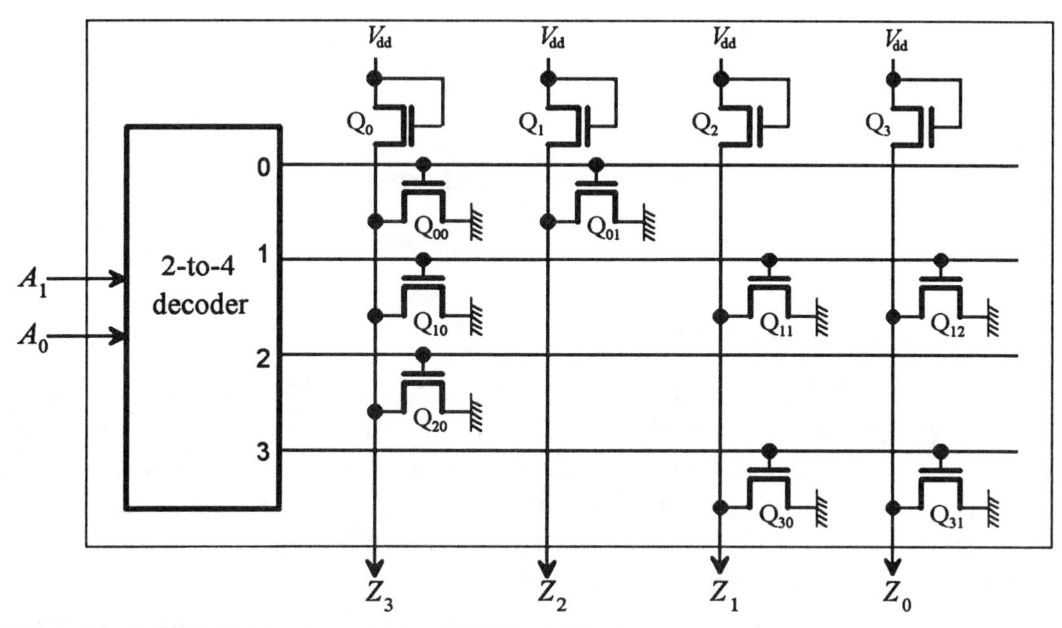

FIGURE 4.35 MOS ROM implementation of Table 4.14

The access time (read time) of a bipolar ROM typically lies in the range 16 to 50 nanoseconds. However, large ROMs cannot be constructed using bipolar devices because they occupy more space in the chip. Therefore, large ROMs are often implemented using MOS transistors. A MOS ROM implementation of Table 4.14 is shown in Figure 4.35 If $A_1A_0 = 00$, transistors Q_{00} and Q_{01} will conduct, and thus $Z_3Z_2 = 00$.Because the transistors Q_2 and Q_3 are always on, it follows that $Z_1Z_0 = 11$. Typical access time (time required for read) is 450 nanoseconds.

A ROM must be programmed before it can be used. This involves placing the switching devices at the appropriate intersection points of the row and column lines. For example, in a mask ROM the contents of the ROM are initialized by the manufacturer at the time of its production. This means that this approach is well suited for producing standard circuits (such as a bar-code generator or a 2 × 2 multiplier) that are widely used by many digital designers. Because these types of ROMs are mass-produced, their costs are also very low. However, a mask ROM cannot be reconfigured by a user. That is, a user cannot alter its contents.

Occasionally, a user may wish to develop a specific ROM-based circuit as demanded by the application area. In this case, a ROM that allows a user to initialize its contents is required. A ROM with such a flexibility is known as a PROM (programmable ROM). PROMs may be manufactured using fusible links. A typical bipolar ROM cell with a fusible link is illustrated in Figure 4.36.

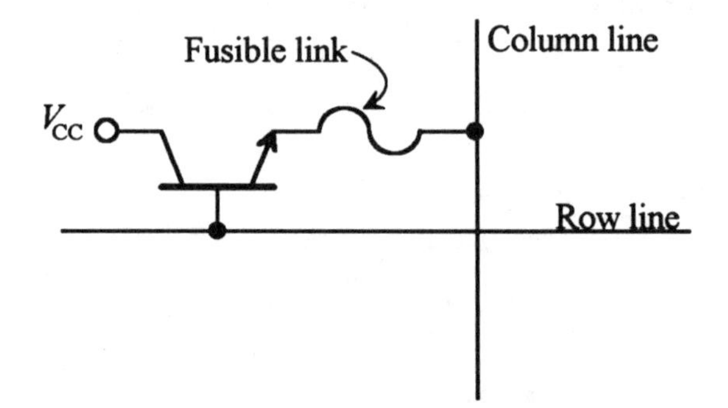

FIGURE 4.36 A ROM Cell with Fusible Link

In this device, the manufacturer places a switching element along with a fusible link at each intersection. This implies that all ROM cells are initialized with a 1. If a user desires to store a zero in a particular cell, the fuse is blown at that point. This activity is called "programming," and it may be accomplished by passing electrical impulses. It should be pointed out that in such a ROM a user can program the ROM only once. That is, it is not possible to reprogram a PROM once the fuse is blown.

When a new product is developed, it may be necessary for the designer to modify the contents of the ROM. A ROM with this capability is referred to as an EPROM (erasable programmable ROM). Usually, the contents of this memory are erased by exposing the ROM chip to ultraviolet light. Typical erase times vary between 10 and 30 minutes. After erasure the ROM may be reprogrammed by passing voltage pulses at the special inputs. The 2764 chip is a typical example of an EPROM. It is a 28-pin 8K × 8 chip contained in a dual in-line package (DIP). It has 13 address input pins and 8 data output pins. Note that the 2764 needs 13 (2^{13} = 8192) pins to address 8192 (8K) locations.

The growth in IC technology allowed the production of another type of ROM whose contents may be erased using electrical impulses. These memory devices are customarily referred to as "electrically alterable ROMs" (EAROMs) or "electrically erasable PROMs" (EEPROMs or E^2PROMs). The main advantage of an EEPROM is that its contents can be changed without removing the chip from the circuit board. Note that EPROMs and EAROMs are designed using only MOS transistors.

4.8 Programmable Logic Devices (PLDs)

A programmable logic device (PLD) is a generic name for an IC chip capable of being programmed by the user after it is manufactured. It is programmed by blowing fuses. *A PLD chip contains an array of AND gates and OR gates.* There are three types of PLDs. They are identified by the location of fuses on the AND-OR array. Figure 4.37 shows the block diagrams of these PLDs.

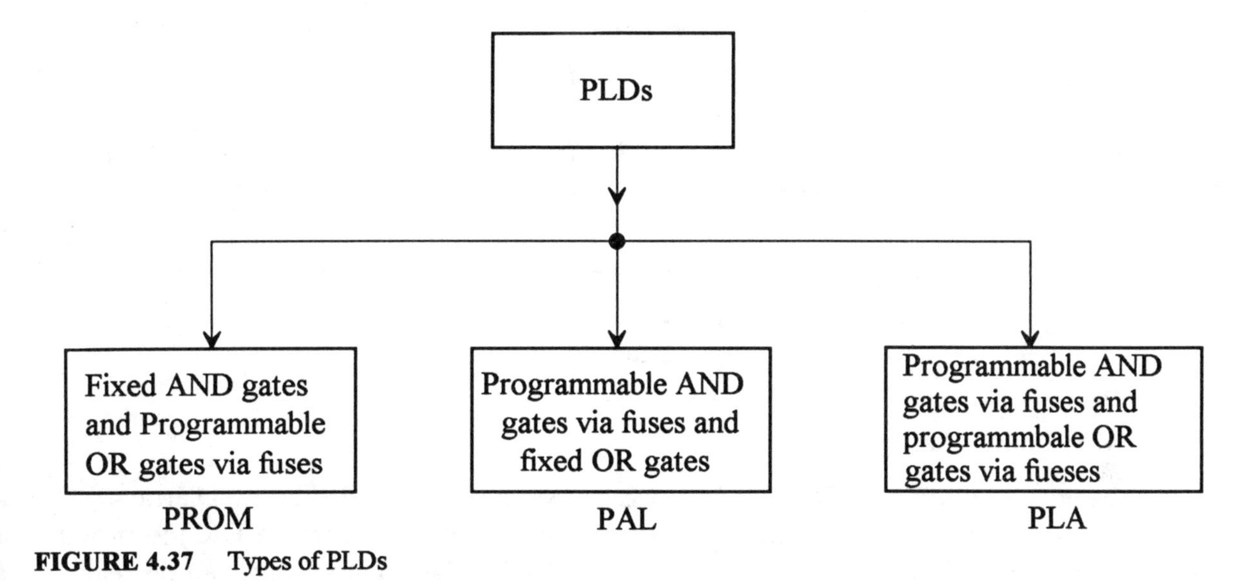

FIGURE 4.37 Types of PLDs

The PROM was discussed in the last section. A PROM contains a number of fixed AND gates and programmable OR gates. The PROM can be programmed to represent Boolean functions in sum of products (minterms) form. The PAL, on the other hand, includes programmable AND gates and fixed OR gates. The PAL can be programmed to implement Boolean functions as a logical sum (OR) of product terms. Finally, the PLA (programmable logic array) includes several AND and OR gates, both of which are programmable. The PLA is very flexible in the sense that the necessary AND terms can be logically ORed to provide the desired Boolean functions. Let us explain the basics of PLAs. In order to illustrate a PLA, a special AND gate or OR gate symbol with multiple inputs will be utilized as shown in Figure 4.38.

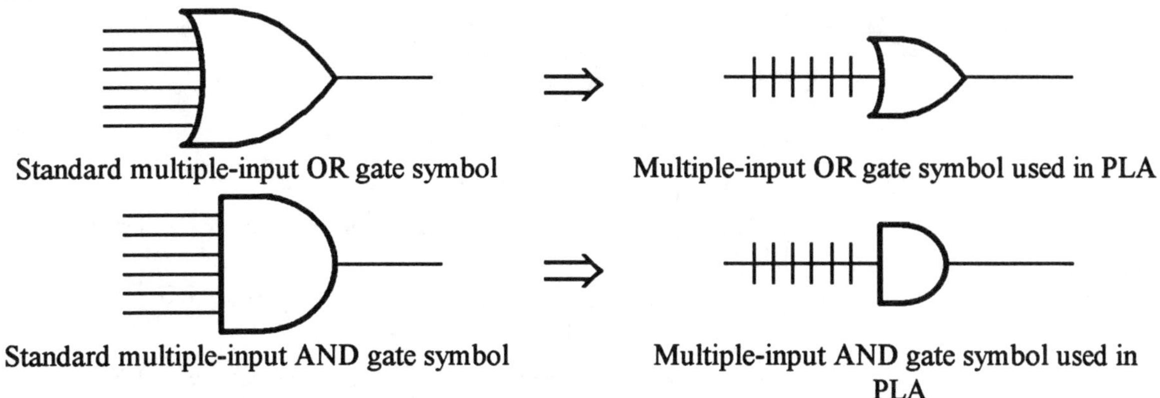

FIGURE 4.38 Multiple input AND and OR Gate Symbols for PLA

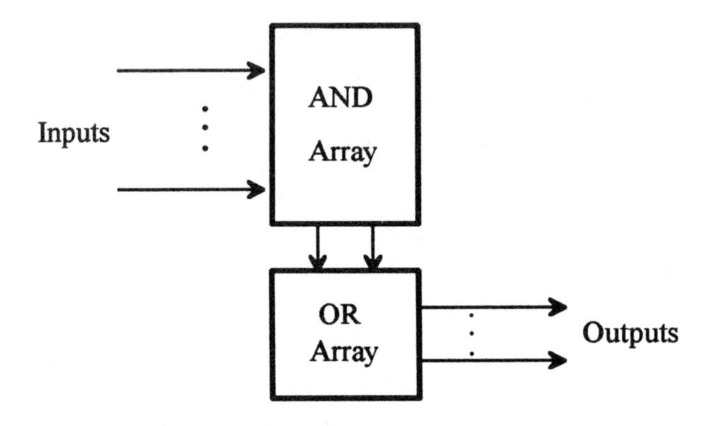

FIGURE 4.39 Internal Structure of a PLA

The internal structure of a typical PLA is shown in Figure 4.39. The AND array of this system generates the required product terms, and the OR array is used to OR the product terms generated by the array. As in the case of the ROM, these gate arrays can be realized using diodes, transistors, or MOS devices. The significance of a PLA is explained in the following example.

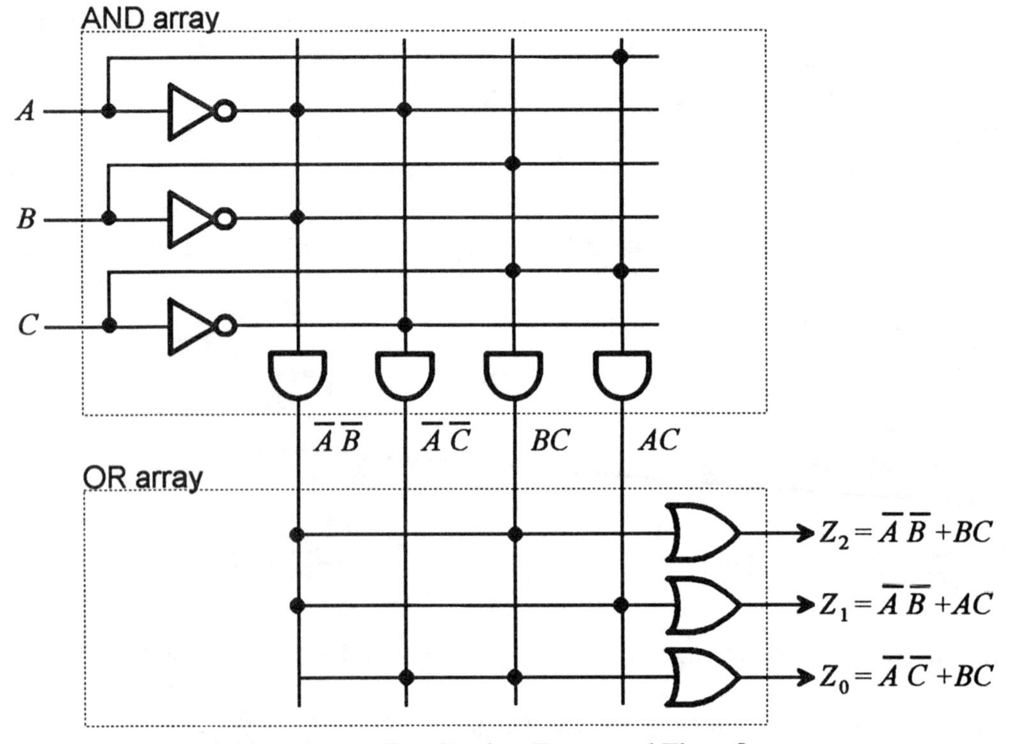

FIGURE 4.40 A PLA with Three Inputs, Four Product Terms, and Three Outputs

Consider the PLA shown in Figure 4.40. This PLA has three inputs, *A*, *B*, and *C*. The AND generates from product terms $\overline{A}\,\overline{B}$, $\overline{A}\,\overline{C}$, *BC*, and *AC*. These product terms are logically summed up in the OR array, and the outputs Z_0, Z_1, and Z_2 are generated. Note that the dot in the figure indicates the presence of a switching element such as a diode or transistor. The use of PLAs is very cost-effective when the number of inputs in a combinational circuit realized by a ROM is very high and all input combinations are not used. For example, consider the following multiple output functions:

$$W = AE + BC$$

$$X = CD + FE$$

$$Y = FG + HI$$

To implement these Boolean functions in a ROM, a 512 × 3 array is needed because there are nine inputs (*A* through *I*) ($2^9 = 512$) and three outputs (*W*, *X*, *Y*), but the same functions can be realized in a PLA using six product terms, nine inputs, and three outputs, as shown in Figure 4.41. Therefore, a considerable savings in hardware can be achieved with PLAs.

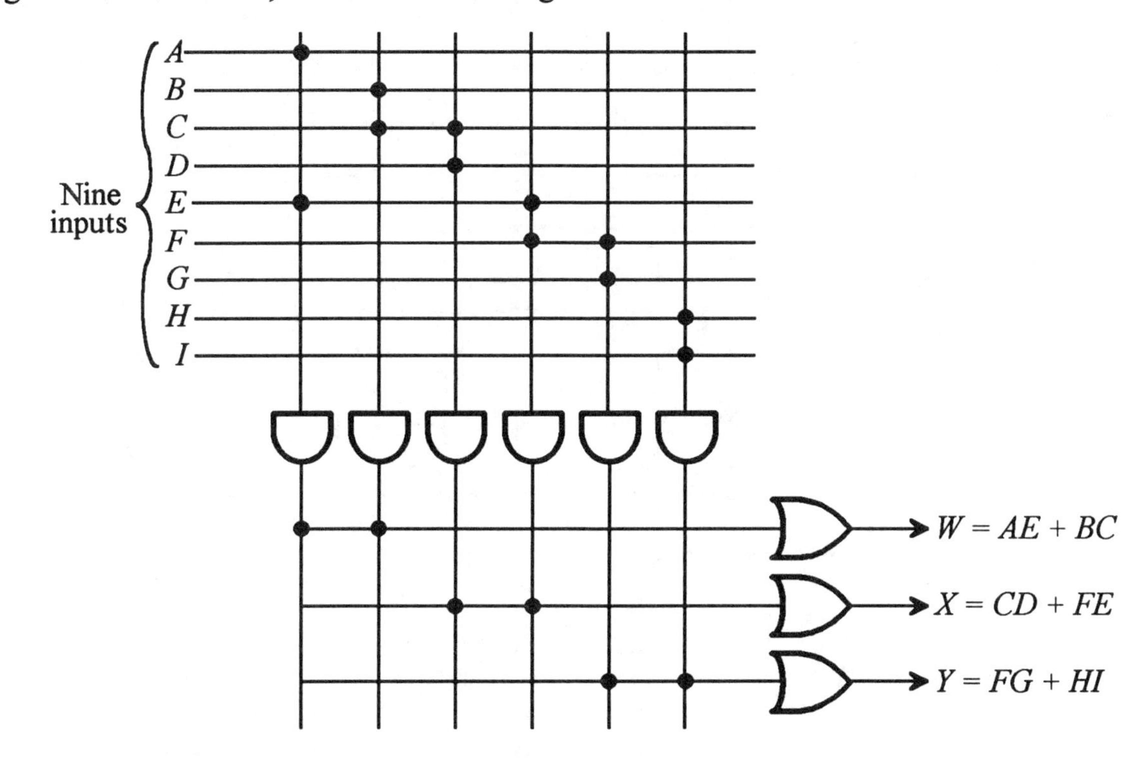

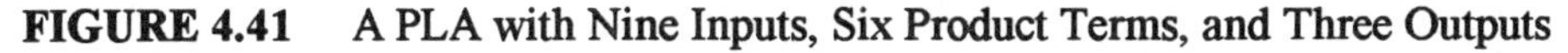
FIGURE 4.41 A PLA with Nine Inputs, Six Product Terms, and Three Outputs

Example 4.3

Implement Example 4.2 using PLAs.

Solution

From Example 4.2,

$$Z_1(A,B,C) = \sum m(2,3,5,6,7)$$
$$= \overline{C}B\overline{A} + \overline{C}BA + C\overline{B}A + CB\overline{A} + CBA$$

$$Z_2(A,B,C) = \sum m(1,2,3,7)$$
$$= \overline{C}\,\overline{B}A + \overline{C}B\overline{A} + \overline{C}BA + CBA$$

Figure 4.42 shows the PLA implementation.

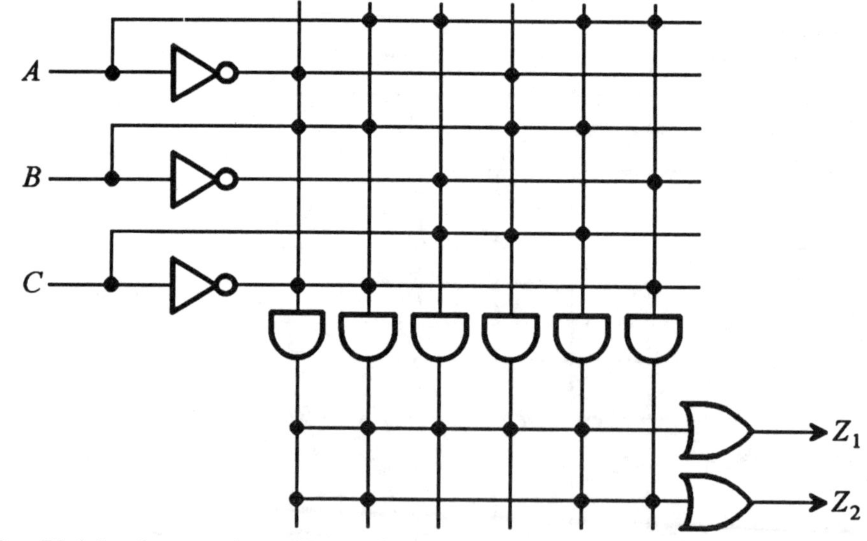

FIGURE 4.42 PLA Implementation of Example 4.3

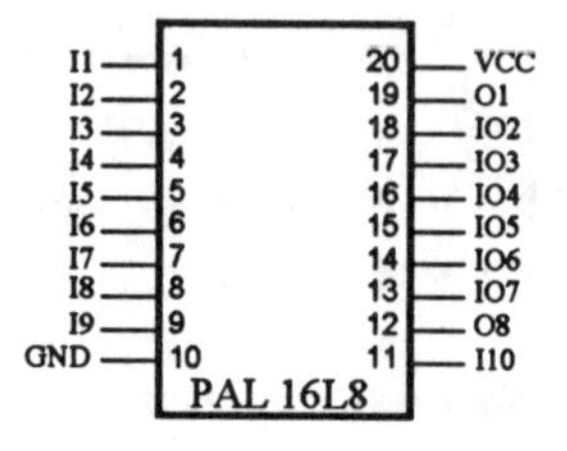

FIGURE 4.43 Pinout for PAL 16L8

4.9 Commercially Available PLDs

Both mask programmable and field programmable PLAs are available. Mask programmable PLAs are similar to mask ROMs in the sense that they are programmed at the time of manufacture. Field programmable PLAs (FPLAs) on the other hand, can be programmed by the user with a computer-aided design (CAD) program to select a minimum number of product terms to express the Boolean functions.

Among all PLDs, PALs are widely used. Note that PAL is a registered trademark of Advanced Micro Devices, Inc. (AMD). PALs were introduced by Monolithic Memories (a division of AMD) in 1970. The PAL chips are usually identified by a two-digit number followed by a letter and then one or two digits. The first two-digit number specifies the number of inputs whereas the last one or two digits define the number of outputs. The fixed number of AND gates are connected to either an OR or a NOR gate. The letter H indicates that the output gates are OR gates; the letter L is used when the outputs are NOR gates; the letter C is used when the outputs include both OR and NOR gates. Note that OR outputs generate active HIGH whereas NORs provide active LOW outputs. On the other hand, OR-NOR gates include both active HIGH and active LOW outputs.

For example, the PAL16L8 is a 20-pin chip with a maximum of 16 inputs, up to 8 outputs, one power pin, and one ground pin. The 16L8 contains 10 nonshared inputs, six inputs that are shared by six outputs, and two nonshared outputs. Figure 4.43 shows the pin diagram of the PAL16L8. Note that PEEL (Programmable Electrically Erasable Logic) devices or Erasable PLDs such as 18CV8 or 16V8 are available for instant reprogramming just like an EEPROM. These devices utilize CMOS EEPROM technology. These erasable PLDs use electronic switches rather than fuses so that they are erasable and reprogrammable like EEPROMs.

Due to advent in IC technology, larger PLDs using the basic ones are designed. The basic PLDs cannot be used for larger digital-design applications. Therefore, CPLD (complex PLD) chips are designed by the manufacturers to accomplish this. A typical CPLD contains several PLDs (each PLD containing AND and OR gates) along with all the interconnections in the same chip. Some IC manufacturers took a different approach for handling larger applications. They devised FPGA (Field Programmable Gate Array) chips. A typical FPGA chip contains several smaller individual logic blocks (each block containing any type of logic gate) along with all interconnections in a single chip. Application of either CPLD or FPGA depends on the user's choice. Both have similar advantages. Products can be developed using either one from conceptual design via prototype to production in a very short time. Hardware design languages such as VHDL or Verilog along with CAD tools, allow CPLDs and FPGAs to be programmed with millions of gates in a short time.

4.10 PLD Programming Languages

Computer-aided design (CAD) software can be used to program PLDs. Three popular PLD languages are PALASM (Advanced Micro Devices, Inc.), ABEL (Data I/O Corporation, Inc.), and VHDL (U.S. Department of Defense). VHDL is introduced in Chapter 11. An overview of ABEL will be provided here. ABEL stands for Advanced Boolean Expression Language while PAL Assembler is abbreviated as PALASM. ABEL is supported by a PLD language translator. The purpose of the translator is to provide the fuse pattern from the program written in ABEL in terms of the fuse pattern of a PLD. Note that most PLDs can be programmed using the sum of minterms form. The ABEL translator can minimize the equations in sum of minterms or in almost any other format. ABEL is basically a high-level language for hardware design similar to software design language such as Pascal or C. An ABEL program (module) consists of the following basic items: module statement, declarations, logic descriptions, and test vectors.

The `module` statement assigns a name to the program. The module name can be any valid identifier. Identifiers may be in upper, lower, or mixed (characters and digits) case and are case sensitive. This means that the identifier `START` is not the same as the identifier `Start`. Identifiers can have a maximum of 32 characters. They may include both uppercase and lowercase letters, the underscore character (_), and digits, but must not start with a digit. The **`title`** statement assigns a title to the module and must be enclosed by single quotes.

Typical declarations include pin declarations and device declarations (type of PLD). A device declaration includes a **`device`** identifier along with the PLD type. An example is `ADDBIN device 'P16L8';` in figure 4.44. Here, `ADDBIN` is the device identifier, `device` is the declaration, and `P16L8` (used for PAL16L8) is the device type. The translator uses the device ID to determine whether the PLD device is capable of performing the function according to the logic descriptions. Pin declarations include actual assignments of the device pins. Pins must be declared before they are used as inputs and outputs. The logic descriptions include equations and truth tables defining outputs in terms of the inputs.

Test vectors are optional. Digital circuits can be tested by applying appropriate combinations of inputs (test vectors) and verifying the outputs. Test vectors are like truth tables that define the inputs and expected outputs for logic simulation. In ABEL, test vectors are entered into source programs. The ABEL test vectors and ABEL truth table are identical. The test vectors are used to describe test inputs and desired outputs for simulation. The ABEL logic simulator can be used to verify the function of a design using the test vectors before programming the PLDs.

The ABEL transistor utilizes a number of directives to perform functions such as use of alternate set of Boolean operators. For example, the ABEL directive **`@ALTERNATE`** instructs the ABEL translator to use an alternate set of Boolean operator symbols as follows:

Boolean Operation	Standard ABEL Symbol Used	Alternate Symbol
NOT	!	/
AND	&	*
OR	#	+
XOR	$	:+:

Figure 4.44 shows the pin assignments for the PAL16L8 and a typical ABEL program for a 4-bit binary adder, which can be implemented in a PAL16L8.

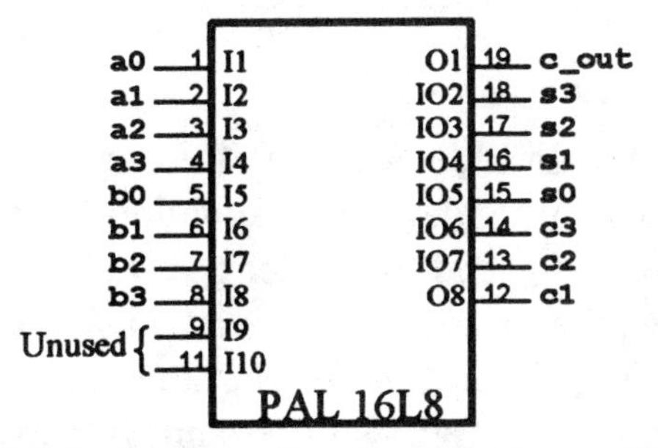

(a) Signals on the PAL16L8 for implementing the 4-bit binary adder.

```
Module adder
Title '4-bit adder using one HA and three FAS'
        ADDBIN device 'P16L8';
        @ALTERNATE
        "input pins
        a0,a1,a2,a3                 pin 1,2,3,4;
        b0,b1,b2,b3                 pin 5,6,7,8;
        "output pins
        c1,c2,c3                           pin 12,13,14;
        s0,s1,s2,s3                 pin 15,16,17,18;
        c_out                       pin 19;
Equations
        s0 = a0:+:b0;                      "half adder for bit 0
        c1 = a0*b0;
        s1 = a1:+:b1:+:c1;         "full adder for bit 1
        c2 = (a1:+:b1)*c1+a1*b1;
        s2 = a2:+:b2:+:c2;          "full adder for bit 2
        c3 = (a2:+:b2)*c2+a2*b2;
        s3 = a3:+:b3:+:c3;          "full adder for bit 3
        c_out = (a3:+:b3)*c3+a3*b3;
Test_vectors([[a3..a0], [b3..b0]]->    [[c_out,s3..s0]])
              [    2,      2      ]      ->    [      4      ];
              [    5,      4      ]      ->    [      9      ];
              [    6,      5      ]      ->    [      11     ];
              [    8,      7      ]      ->    [      15     ];
End
```

(b) An ABEL Program for a 4-bit Binary Adder

FIGURE 4.44 PAL16L8 Pin Assignments and Program for 4-Bit Adder

To program the PAL16L8 for the 4-bit binary adder, the inputs and outputs are assigned to the appropriate pins in the 16L8 chip. The 4-bit binary addition is performed as follows:

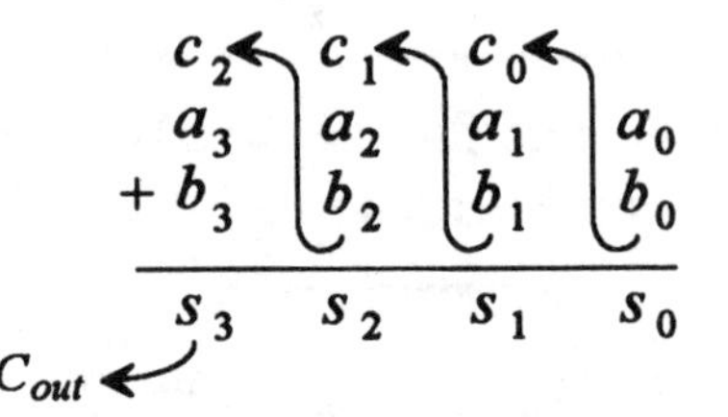

Bits a_0 and b_0 are added using one half-adder whereas the each of the three bits for the sums s_1, s_2, and s_3 are added using three full adders. The equation portion of the ABEL program utilizes the logic equations for half adder and full adder covered earlier in this chapter. For test vectors, samples of input data are taken to verify the results using the ABEL simulation. Once the results have been verified, the PAL16L8 can be programmed.

Finally, Verilog (developed by Gateway Design Automation in 1984, and later acquired by Cadence Design Systems), another hardware design language, has been becoming popular. Verilog is not an acronym, although its primary purpose is to VERIfy LOGic. Verilog (syntax based mostly on C and some Pascal) is easier to learn compared to VHDL (syntax based on Ada). Verilog provides more features than VHDL to support large project development. At present, both VHDL and Verilog have approximately equal market share.

QUESTIONS AND PROBLEMS

4.1 Find function F for the following circuit:

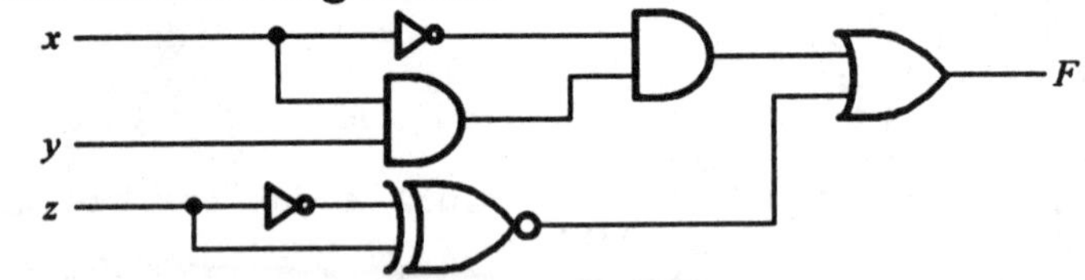

4.2 Express the following functions F_1 and F_2 in terms of the inputs A, B, and C. What is the relationship between F_1 and F_2?

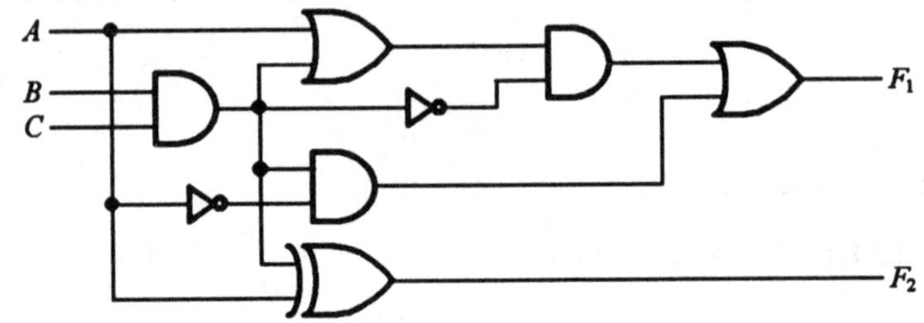

4.3 Given the following circuit:

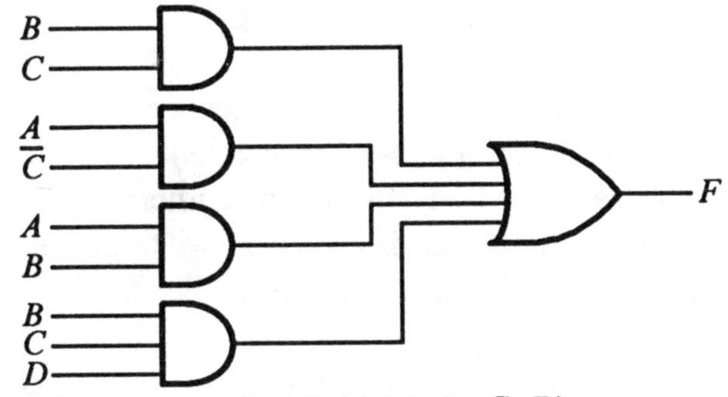

(a) Derive the Boolean expression for $F(A, B, C, D)$.
(b) Derive the truth table.
(c) Determine the simplified expression for $F(A, B, C, D)$ using a K-map.
(d) Draw the logic diagram for the simplified expression using NAND gates.

4.4 Determine the function F of the following function and then analyze the function using Boolean identities to show that $F = A + B$.

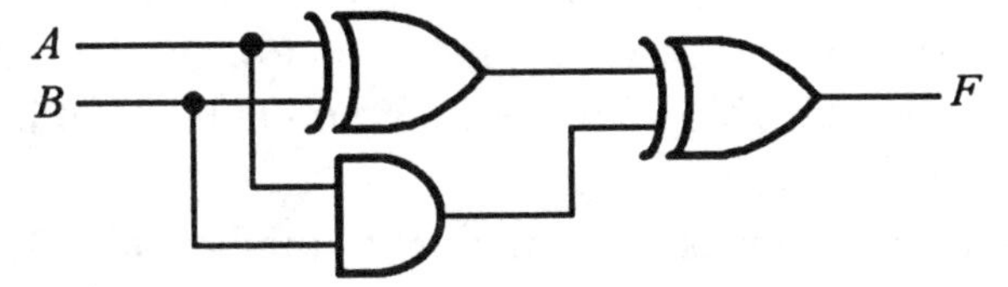

4.5 Draw a logic diagram to implement $F = ABCDE$ using only 3-input AND gates.

4.6 Draw a logic diagram using two-input AND and OR gates to implement the following function $F = P(P + Q)(P + Q + R)(P + Q + R + S)$ without any simplification; then analyze the logic circuit to verify that $F = P$.

4.7 Design a combinational circuit with three inputs (A, B, C) and one output (F). The output is 1 when $A + C = 0$ or $AC = 1$; otherwise the output is 0. Draw a logic diagram using a single logic gate.

4.8 Design a combinational circuit that accepts a 3-bit unsigned number and generates an output binary number equal to the input number plus 1. Draw a logic diagram.

4.9 Design a combinational circuit with five input bits generating a 4-bit output that is the ones complement of four of the five input bits. Draw a logic diagram. Do not use NOT, NAND, or NOR gates.

4.10 Design a combinational circuit that converts a 4-bit BCD input to its nines complement output. Draw a logic diagram.

4.11 Design a BCD to seven-segment decoder that will accept a decimal digit in BCD and generate the appropriate outputs for the segments to display a decimal digit (0–9). Use a common anode display. Turn the seven segment OFF for non-BCD digits. Draw a logic circuit. What will happen if a common cathode display is used? Comment on the interface between the the decoder and the display.

4.12 Design a combinational circuit using full adders to increment a 4-bit number by one. Draw a logic diagram using the block diagram of a full adder as the building block.

4.13 Design a combinational circuit using full adders to multiply a 4-bit unsigned number by 2. Draw a logic diagram using the block diagram of a full adder as the building block.

4.14 Design a combinational circuit that adds two 4-bit signed numbers and generates an output of 1 if the 4-bit result is zero; the output is 0 if the 4-bit result is nonzero. Assume no carry output. Draw a logic circuit using the block diagram of a 4-bit parallel adder as the building block and a minimum number of logic gates.

4.15 Design a 4×16 decoder using a minimum number of 74138 and logic gates.

4.16 Design a logic circuit using a minimum number of 74138s (3×8 decoders) to generate the minterms m_1, m_5, and m_9 based on four switch inputs *S3, S2, S1, S0.* Then display the selected minterm number (1 or 5 or 9) on a seven-segment display by generating a 4-bit input (*W, X, Y, Z)* for a BCD to seven-segment code converter. Ignore the display for all other minterms.

Note that these four inputs (*W, X, Y, Z)* can be obtained from the selected output line (1 or 5 or 9) of the decoders that is generated by the four input switches (*S3, S2, S1, S0).* Use a minimum number of logic gates. Determine the truth table, and then draw a block diagram of your implementation using the following building blocks (Figure P4.16):

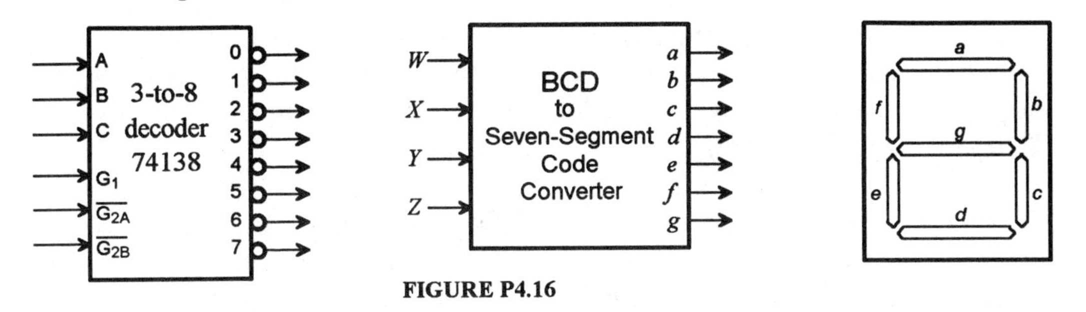

FIGURE P4.16

4.17 A combinational circuit is specified by the following equations:

$$F_0 = \overline{A}\,\overline{B}\,\overline{C} + A\overline{B}C + A\overline{B}\,\overline{C} \qquad F_2 = \overline{A}\,\overline{B}\,\overline{C} + ABC$$

$$F_1 = A\overline{B}\,\overline{C} + AB\overline{C} \qquad F_3 = \overline{A}BC + \overline{A}B\overline{C} + AB\overline{C}$$

Draw a logic diagram using a decoder and external gates. Assume that the decoder outputs a HIGH on the selected line.

4.18 Implement a combinational circuit to generate the following:

$$F_0 = \sum m(1,3,4)$$

$$F_1 = \sum m(0,2,4,7)$$

$$F_2 = \sum m(0,1,3,5,6)$$

$$F_3 = \sum m(2,6)$$

Draw a logic diagram using a 74138 decoder and external gates.

4.19 Determine the truth table for a hexadecimal-to-binary priority encoder with line 0 having the highest priority and line 15 with the lowest.

4.20 Implement a digital circuit to increment (for $C_{in} = 1$) or decrement (for $C_{in} = 0$) a 4-bit signed number by 1 generating outputs in twos complement form. Note that C_{in} is the input carry to the full adder for the least significant bit. Draw a schematic:

(a) Using only a minimum number of full adders and multiplexers.

(b) Using only a minimum number of full adders and inverters. Do not use any multiplexers.

4.21 What are the main logic elements/gates in a ROM chip?

4.22 Design a combinational circuit using a 16 X 4 ROM that will increment a 4-bit unsigned number by 1. Determine the truth table and then draw a block diagram of your implementation showing the addresses and their contents in binary along with one Output Enable (OE) input.

4.23 What are the basic differences between PROM, PLA, PAL and PEEL?

4.24 What is the technology used to fabricate EPROMs and EEPROMs?

4.25 Write an ABEL program to implement a 3×8 decoder in the PAL16L8.

4.26 Design a 4-bit adder/subtractor (Example 4.3) using only full adders and EXCLUSIVE-OR gates. Do not use any multiplexers.

4.27 Design a combinational circuit using a minimum number of full adders, and logic gates which will perform A plus B or A minus B (A and B are signed numbers), depending on a mode select input, M. If $M = 0$, addition is carried out; if $M = 1$, subtraction is carried out. Assume $A = A_4 A_3 A_2 A_1 A_0$ and $B = B_4 B_3 B_2 B_1 B_0$ (Two 5-bit numbers). The circuit should be able to carry out the subtraction even if $A < B$. Use an LED to indicate the sign of the result (LED ON for negative result and LED OFF for positive result). The result of the operation should always appear in BCD form on seven-segment displays. The Overflow bit (V) should be indicated by another LED (LED ON for V=1 and LED OFF for V=0). Do not use any multiplexers.

5

SEQUENTIAL LOGIC DESIGN

5.1 Basic Concepts

So far, we have considered the design of combinational circuits. The main characteristic of these circuits is that the outputs at a particular time *t* are determined by the inputs at the same time *t*. This means that combinational circuits require no memory. However, in practice, most digital systems contain combinational circuits along with memory. These circuits are called "sequential."

In sequential circuits, the present outputs depend on the present inputs and the previous states stored in the memory elements. These states must be fed back to the inputs in order to generate the present outputs. There are two types of sequential circuits: synchronous and asynchronous.

In a synchronous sequential circuit, a clock signal is used at discrete instants of time to ensure that all desired operations are initiated only by a train of synchronizing clock pulses. A timing device called the "clock generator" produces these clock pulses. The desired outputs of the memory elements are obtained upon application of the clock pulses and some other signal at their inputs. This type of sequential circuit is also called a "clocked sequential circuit." The memory elements used in clocked sequential circuits are called "flip-flops." The flip-flop stores only one bit. A clocked sequential circuit usually utilizes several flip-flops to store a number of bits as required. *Synchronous sequential circuits are also called "state machines." In an asynchronous sequential circuit, completion of one operation starts the operation that is next in sequence. Synchronizing clock pulses are not required. Instead, time-delay devices are used in asynchronous sequential circuits as memory elements. Logic gates are typically used as time delay devices, because the propagation delay time associated with a logic gate is adequate to provide the required delay.* A combinational circuit with feedback among logic gates can be considered as an asynchronous sequential circuit. One must be careful while designing asynchronous systems because feedback among logic gates may result in undesirable system operation. The logic designer is normally faced with many problems related to the instability of asynchronous system, so they are not commonly used. Most of the sequential circuits encountered in practice are synchronous because it is easy to design and analyze such circuits.

5.2 Flip-Flops

A flip-flop is a one-bit memory. As long as power is available, the flip-flop retains the bit. However, its output (stored bit) can be changed by the clock input. *Flip-flops are designed using basic storage circuits called "latches."* The most common latch is the SR (Set-Reset) latch. *A flip-flop is a latch with a clock input*. This convention will be used in this book.

5.2.1 SR Latch

Figure 5.1 shows a basic latch circuit using NOR gates along with its truth table. The SR latch has two inputs, S (Set) and R (Reset), and two outputs Q (true output) and $\overline{Q}$ (complement of Q). To analyze the SR latch of Figure 5.1(a), note that a NOR gate generates an output 1 when all inputs are 0; on the other hand, the output of a NOR gate is 0 if any input is 1. Now assume that $S = 1$ and $R = 0$; the $\overline{Q}$ output of NOR gate #2 will be 0. This places 0 at both inputs of NOR gate #1. Therefore, output Q of NOR gate #1 will be 1. Thus, Q stays at 1. This means that one of the inputs to NOR gate #2 will be 1, producing 0 at the $\overline{Q}$ output regardless of the value of S. Thus, when the pulse at S becomes 0, the output $\overline{Q}$ will still be 0. This will apply 0 at the input of NOR #1. Thus, Q will continue to remain at 1. This means that when the set input $S = 1$ and the reset (clear) input $R = 0$, the SR latch stores a 1 ($Q = 1$, $\overline{Q} = 0$). This means that the SR latch is set to 1.

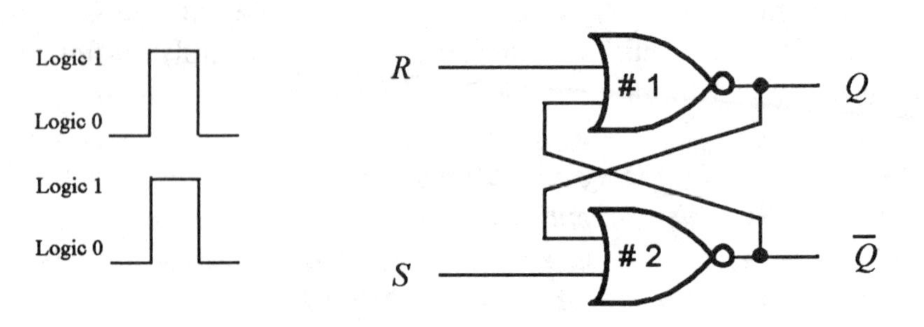

(a) NOR gate implementation

S	R	Q	$\overline{Q}$
0	0	Q	$\overline{Q}$
0	1	0	1
1	0	1	0
1	1	0	0?

(b) Truth table

FIGURE 5.1 SR Latch using NOR gates

Consider $S = 0$, $R = 1$; the Q output of NOR gate #1 will be 0. This will apply 0 at both inputs of NOR gate #2. Thus, output $\overline{Q}$ will be 1. When the reset pulse input R returns to zero, the outputs continues to remain at $Q = 0$, and $\overline{Q} = 1$. This means that with set input $S = 0$ and reset input $R = 1$, the SR latch is cleared to 0 ($Q = 0$, $\overline{Q} = 1$).

Next, consider $Q = 1$, $\overline{Q} = 0$. With $S = 0$ and $R = 0$, the NOR gate #1 will have both inputs at 0. This will generate 1 at the Q output. The output $\overline{Q}$ of NOR gate #2 will be zero. Thus, the outputs Q and $\overline{Q}$ are unchanged when $S = 0$ and $R = 0$.

When $S = 1$ and $R = 1$, both Q and $\overline{Q}$ outputs are 0. This is an invalid condition because for the SR latch Q and $\overline{Q}$ must be complements of each other. Therefore, one must ensure that the condition $S = 1$ and $R = 1$ does not occur for the SR latch. This undesirable situation is indicated by a question mark (?) in the truth table.

An SR latch can be built from NAND gates with active-low set and reset inputs. Figure 5.2 shows the NAND gate implementation of an SR latch.

The SR latch with S and R inputs will store a 1 ($Q = 1$ and $\overline{Q} = 0$) when the S input is activated by a low input (logic 0) and $R = 1$. On the other hand, the latch will be cleared or reset to 0 ($Q = 0$, $\overline{Q} = 1$) when the R input is activated by a low input (logic 0) and $S = 1$.

Note that an active low signal can be defined as a signal that performs the desired function when it is low or 0. In Figure 5.2, the SR latch stores a 1 when $S = 0 =$ active low and $R = 1$; on the other hand, the latch stores a 0 when $R = 0 =$ active low and $S = 1$.

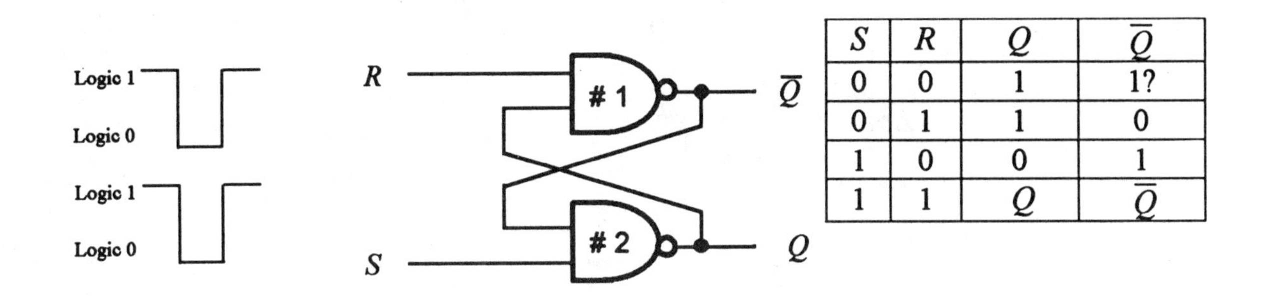

S	R	Q	$\overline{Q}$
0	0	1	1?
0	1	1	0
1	0	0	1
1	1	Q	$\overline{Q}$

(a) NAND gate implementation (b) Truth table

FIGURE 5.2 NAND implementation of an SR latch

Note that the NAND gate produces a 0 if all inputs are 1; on the other hand, the NAND gate generates a 1 if at least one input is 0. Now, suppose that $S = 0$ and $R = 1$. This implies that the output of NAND gate #2 is 1. Thus, $Q = 1$. This will apply 1 to both inputs of NAND gate #1. Thus, $\overline{Q} = 0$. Therefore, a 1 is stored in the latch. Similarly, with inputs $S = 1$ and $R = 0$, it can be shown that $Q = 0$ and $\overline{Q} = 1$. The latch stores a 0.

With $S = 1$ and $R = 1$, both outputs of the latch will remain at the previous values. There will be no change in the latch outputs. Finally, $S = 0$ and $R = 0$ will produce a invalid condition ($Q = 1$ and $\overline{Q} = 1$). This is indicated by a question mark (?) in the truth table of Figure 5.2(b).

An SR latch can be used for designing a switch debouncing circuit. Mechanical switches are typically used in digital systems for inputting binary data manually. These mechanical ON-OFF switches (e.g., the keys in a computer keyboard) vibrate or bounce several times such that instead of changing state once when activated, a key opens and closes several times before settling at its final values. These bounces last for several milliseconds before settling down.

A debouncer circuit, shown in Figure 5.3, can be used with each key to get rid of the bounces. The circuit consists of an SR latch (using NOR gates) and a pair of resistors. In the figure, a single-pole double-throw switch is connected to an SR latch. The center contact (Z) is tied to +5 V and outputs logic 1. On the other hand, contacts X or Y provide logic 0 when not connected to contact Z. The values of the resistors are selected in such a way that X is HIGH when connected to Z or Y is HIGH when connected to Z.

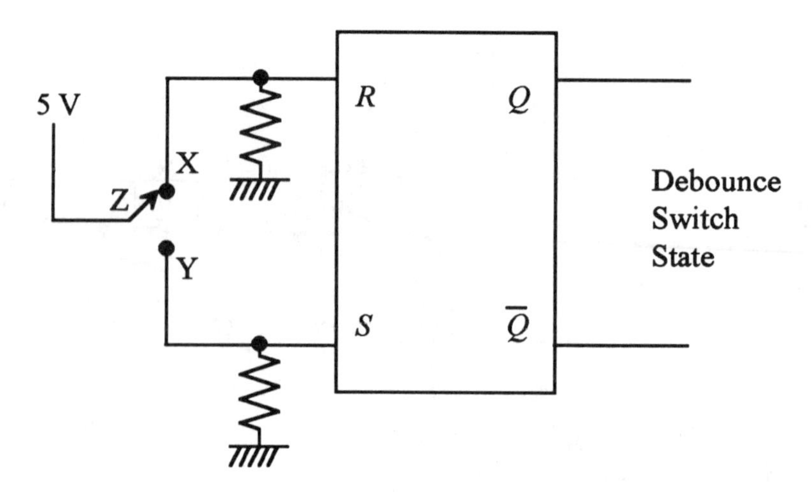

FIGURE 5.3 A debouncing circuit for a mechanical switch

When the switch is connected to X, a HIGH is applied at the R input, and $S = 0$, then $Q = 0$, $\overline{Q} = 1$. Now, suppose that the switch is moved from X to Y. The switch is disconnected from R and $R = 0$ because the ground at the R input pulls R to 0. The outputs Q and $\overline{Q}$ of the SR latch are unchanged because both R and S inputs are at 0 during the switch transition from X to Y. When the switch touches Y, the S input of the latch goes to HIGH and thus $Q = 1$ and $\overline{Q} = 0$. If the switch vibrates, temporarily breaking the connection, the S input of the SR latch becomes 0, leaving the latch outputs unchanged. If the switch bounces back connecting Z to Y, the S input becomes 1, the latch is set again, and the outputs of the SR latch do not change. Similarly, the switch transition from Y to X will get rid of switch bounces and will provide smooth transition.

5.2.2 RS Flip-Flop

An RS flip-flop is a clocked SR latch. This means that the RS flip-flop is same as the SR latch with a clock input. The SR flip-flop is an important circuit because all other flip-flops are built from it. Figure 5.4 shows an RS flip-flop.

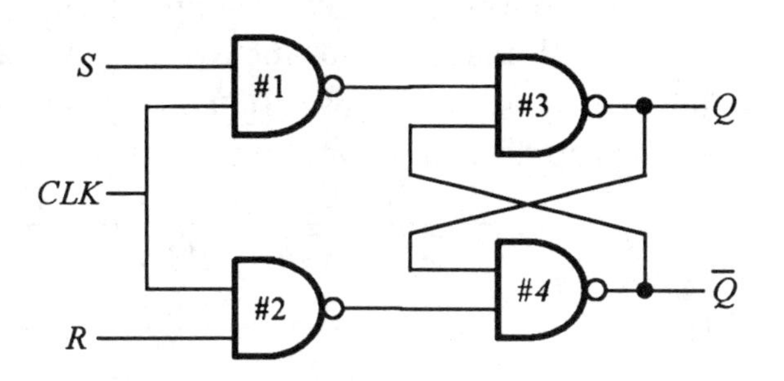

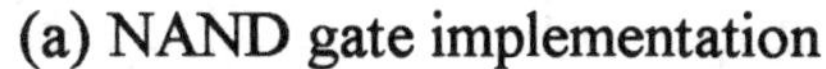

(a) NAND gate implementation

S	R	Clk	Q^+	$\overline{Q^+}$
0	0	1	Q	$\overline{Q}$
0	1	1	0	1
1	0	1	1	0
1	1	1	?	?
X	X	0	Q	$\overline{Q}$

(b) Truth Table

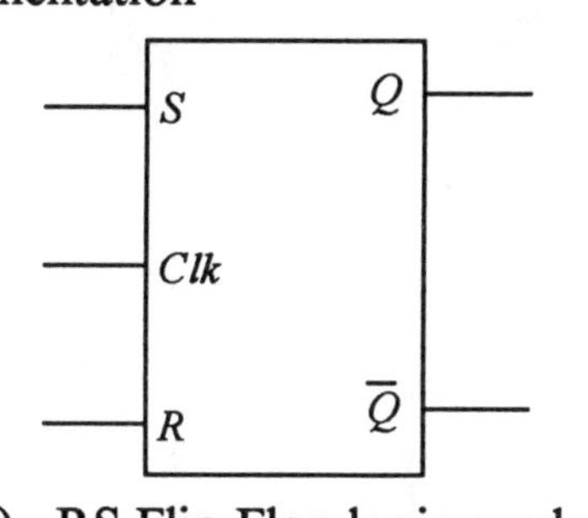

(c) RS Flip-Flop logic symbol

FIGURE 5.4 RS Flip-Flop

The RS flip-flop contains an SR latch with two more NAND gates. It has three inputs (S, CLK, R) and two outputs (Q and $\overline{Q}$). When $S = 0$ and $R = 0$ and CLK = 1, the outputs of both NAND gates #1 and #2 are 1. This means that the output of NAND gate #3 is 0 if $\overline{Q} = 1$ and is 1 if $\overline{Q} = 0$. This means that Q is unchanged as long as $S = 0$ and $R = 0$. On the other hand, the output of NAND gate #4 is 0 if $Q = 1$ and is 1 if $Q = 0$. Thus, $\overline{Q}$ is also unchanged. Suppose $S = 1$, $R = 0$, and CLK = 1. This will produce 0 and 1 at the outputs of NAND gates #1 and #2 respectively. This in turn will generate 1 and 0 at the outputs of NAND gates #3 and #4 respectively. Thus, the flip-flop is set to 1. When the clock is zero, the outputs of both NAND gates #1 and #2 are 1. This in turn will make the outputs of NAND gates #3 and #4 unchanged.

The other conditions in the function table can similarly be verified. Note that $S = 1$, $R = 1$, and CLK = 1 is combination of invalid inputs because this will make both outputs, Q and $\overline{Q}$ equal to 1. Also, Q and $\overline{Q}$ must be complements of each other in the RS flip-flop. Q^+ and $\overline{Q^+}$ are outputs of the flip-flop after the clock (CLK) is applied.

5.2.3 D Flip-Flop

Figure 5.5 shows the logic diagram and truth table of a D flip-flop (Delay flip-flop). This type of flip-flop ensures that the invalid input combinations $S = 1$ and $R = 1$ for the RS flip-flop can never occur. The D flip-flop has two inputs (D and CLK) and two outputs (Q and $\overline{Q}$). The D input is same as the S input and the complement of D is applied to the R input. Thus, R and S can never be equal to 1 simultaneously.

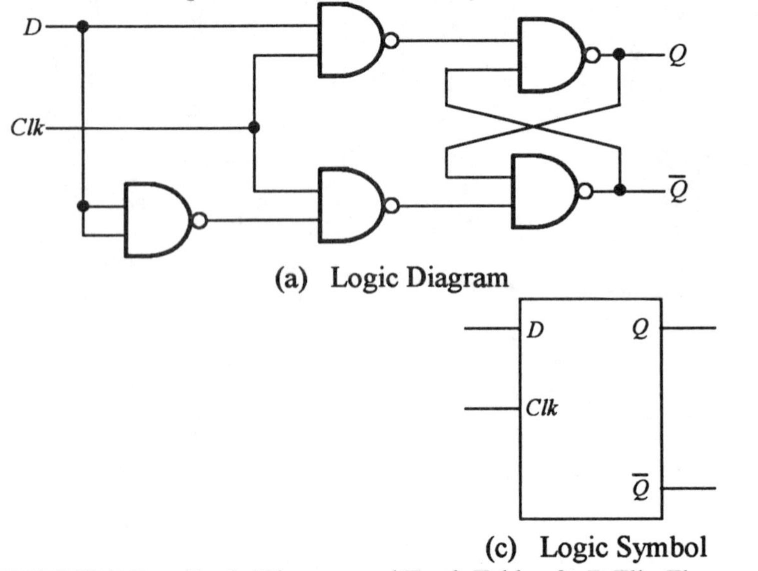

(a) Logic Diagram

D	Clk	Q^+	$\overline{Q^+}$
0	1	0	1
1	1	1	0
X	0	Q	$\overline{Q}$

(b) Truth Table

(c) Logic Symbol

FIGURE 5.5 Logic Diagram and Truth Table of a D Flip-Flop

The D flip-flop transfers the D input to output Q when CLK = 1. Note that if CLK = 0, one of the inputs to each of the last two NAND gates will be 1; thus, outputs of the D flip-flop remain unchanged regardless of the values of the D input.

The D flip-flop is also called a "transparent latch." The term "transparent" is based on the fact that the output Q follows the D input when CLK = 1. Therefore, transfer of input to outputs is transparent, as if the flip-flop were not present.

5.2.4 JK Flip-Flop

The JK flip-flop is a modified version of the RS flip-flop such that the S and R inputs of the RS flip-flop correspond to the J and K inputs of the JK flip-flop. Furthermore, the invalid inputs $S = 1$ and $R = 1$ are allowed in the JK flip-flop. *When J = 1, K = 1, and Clk = 1, the JK flip-flop complements its output*. Otherwise, the meaning of the J and K inputs is the same as that of the S and R inputs respectively. Figure 5.6 shows a logic diagram of JK flip-flop along with its truth table. This is a NAND/NOR implementation. The circuit operation of Figure 5.6(a) is discussed in the following:

i) Suppose $Q = 1$, $\overline{Q} = 0$, and CLK = 1. With $J = 0$ and $K = 0$, the outputs of inverters #2 and #5 are both 0. This means that the outputs of NOR gates #3 and #6 are 1 and 0 respectively. Therefore, the outputs of the flip-flop are unchanged

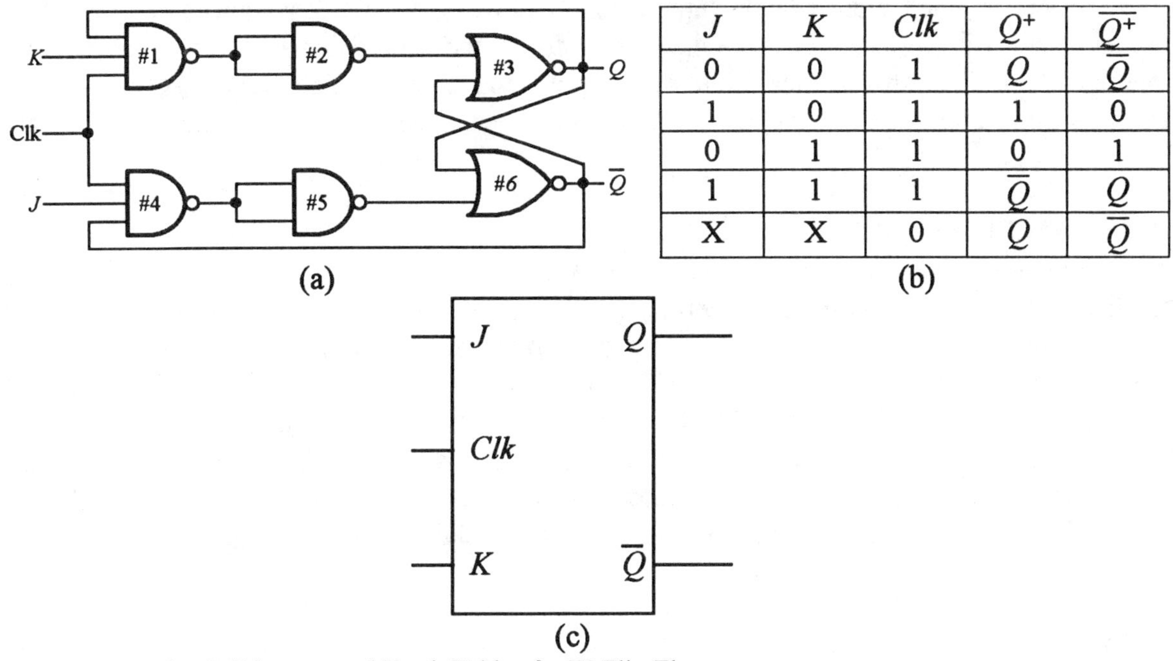

J	K	Clk	Q^+	$\overline{Q^+}$
0	0	1	Q	$\overline{Q}$
1	0	1	1	0
0	1	1	0	1
1	1	1	$\overline{Q}$	Q
X	X	0	Q	$\overline{Q}$

FIGURE 5.6 Logic Diagram and Truth Table of a JK Flip-Flop

ii) Suppose $Q = 0$, $\overline{Q} = 1$, and CLK = 1. With $J = 1$ and $K = 0$, the outputs of inverters #2 and #5 are 0 and 1 respectively. This means that a 0 is produced at the output of NOR gate #6 ($\overline{Q} = 0$). Thus, apply a 0 at one of the inputs of NOR gate #3 generating a 1 at its output ($Q = 1$). The JK flip-flop is therefore set to 1 ($Q = 1$ and $\overline{Q} = 0$).

iii) Suppose $Q = 1$, $\overline{Q} = 0$ and CLK = 1. With $J = 0$ and $K = 1$, the outputs of the inverter #2 and #5 are 1 and 0 respectively. This means that the output of NOR gate #3 is 0. This will produce a 1 at the output of NOR gate #6. Thus, the flip-flop is cleared to zero ($Q = 0$ and $\overline{Q} = 1$).

iv) Suppose $Q = 1$, $\overline{Q} = 0$, and CLK = 1. With $J = 1$ and $K = 1$, the outputs of inverters #2 and #5 are 1 and 0 respectively. This will produce a 0 at the output of NOR gate #3 ($Q =$ 0). This in turn will apply 0 at one of the inputs of NOR gate #6, making its output HIGH ($\overline{Q} = 1$). Thus, the output of the JK flip-flop is complemented. The other rows in the truth table of the JK flip-flop can similarly be verified.

5.2.5 T Flip-Flop

The T (Toggle) flip-flop complements its output when the clock input is applied with $T = 1$; the output remains unchanged when $T = 0$. The name "toggle" is based on the fact that the T flip-flop toggles or complements its output when the clock input is 1 with $T = 1$. The T flip-flop can be obtained from the JK flip-flop in two ways. In the first approach, the J and K inputs of the JK flip-flop can be tied together to provide the T input; the output is complemented when $T = 1$ at the clock while the output remains unchanged when $T = 0$ at the clock. In the second approach, the J and K inputs can be tied to high; in this case, T is the clock input.

5.3 Master-Slave Flip-Flop

As mentioned before, sequential circuits contain combinational circuits with flip-flops in the feedback loop. These flip-flops generate outputs at the clock based on the inputs from the combinational circuits. The feedback loop can create an undesirable situation if the outputs from the combinational circuits that are connected to the flip-flop inputs change values at the clock pulse simultaneously when flip-flops change outputs. This situation can be avoided if the flip-flop outputs do not change until the clock pulse goes back to 0. One way of accomplishing this is to ensure that the outputs of the flip-flops are affected by the pulse transition rather than pulse duration of the clock input.To understand this concept, consider the clock pulses shown in Figure 5.7.

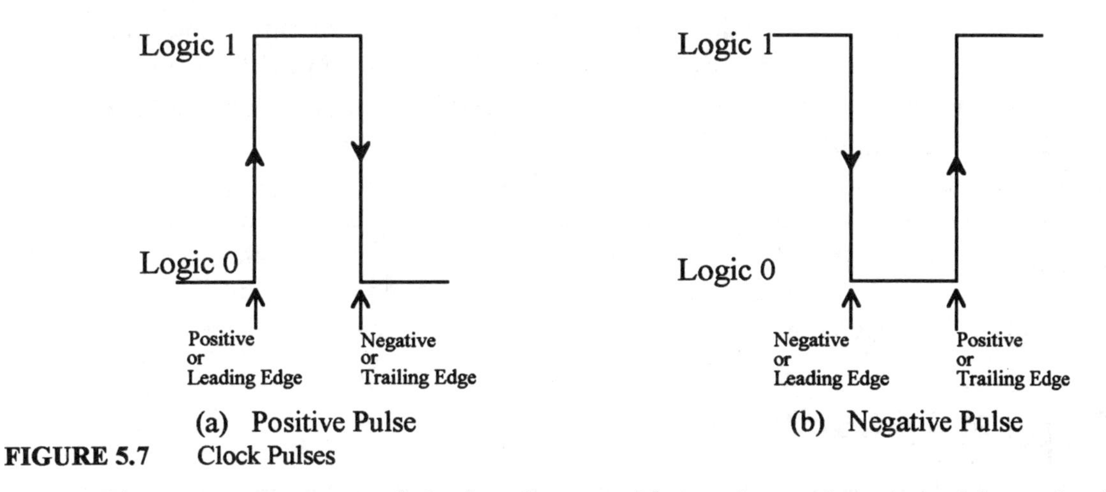

(a) Positive Pulse (b) Negative Pulse

FIGURE 5.7 Clock Pulses

There are two types of clock pulses: positive and negative. A positive pulse includes two transitions: logic 0 to logic 1 and logic 1 to logic 0. A negative pulse also goes through two transitions: logic 1 to logic 0 and logic 0 to logic 1.

Assume that a positive pulse is used as the clock input of a D flip-flop. With the *D* input = 1, the output of the flip-flop will become 1 when the clock pulse reaches logic 1. Now, suppose that the *D* input changes to zero but the clock pulse is still 1. This means that the flip-flop will have a new output, 0. In this situation, the output of one flip-flop cannot be connected to the input of another when both flip-flops are enabled simultaneously by the same clock input. This problem can be avoided if the flip-flop is clocked by either the leading or the trailing edge rather than the signal level of the pulse. A master–slave flip-flop is used to accomplish this.

Figure 5.8 shows a typical master-slave D flip-flop. A master-slave flip-flop contains two independent flip-flops. Flip-flop #1 (FF #1) works as a master flip-flop, whereas the flip-flop (FF #2) is a slave. An inverter is used to invert the clock input to the slave flip-flop.

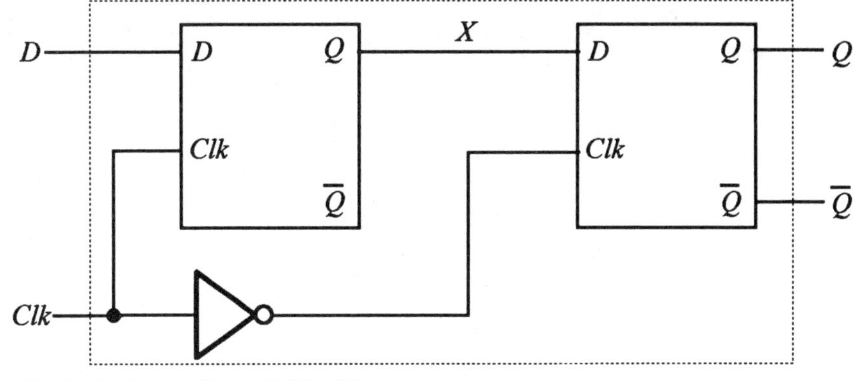

FIGURE 5.8 Typical Master-Slave D Flip-Flop

Assume that the CLK is a positive pulse. Suppose that the *D* input of the master flip-flop (FF #1) is 1 and the CLK input = 1 (leading edge). The output of the inverter will apply a 0 at the CLK input of the slave flip-flop (FF #2). Thus, FF #2 is disabled. The master flip-flop will transfer a 1 to its *Q* output. Thus, *X* will be 1.

At the trailing edge of the CLK input, the CLK input of the master flip-flop is 0. Thus, FF #1 is disabled. The inverter will apply a 1 at the CLK input of the FF #2. Thus, 1 at the *X* input (*D* input of FF #2) will be transferred to the *Q* output of FF #2. When the CLK goes back to 0, the master flip-flop is separated. This avoids any change in the other inputs to affect the master flip-flop. The slave flip-flop will have the same output as the master.

5.4 Preset and Clear Inputs

Commercially available flip-flops include separate inputs for setting the flip-flop to 1 or clearing the flip-flop to 0. These inputs are called "preset" and "clear" inputs respectively. These inputs are useful for initializing the flip-flops without the clock pulse. When the power is turned ON, the output of the flip-flop is in undefined state. The preset and clear inputs can directly set or clear the flip-flop as desired prior to its clocked operation.

Figure 5.9 shows a D flip-flop with clear inputs. The triangular symbol indicates that the flip-flop is clocked at the leading edge of the clock pulse. In Figure 5.9, a circle (inverter) is used with the triangular symbol. This means that the flip-flop is enabled at the trailing edge (inversion of the leading edge) of the clock pulse. The circle at the clear input means that clear input must be 1 for normal operation. If the clear input is tied to ground (logic 0), the flip-flop is cleared to 0 ($Q = 0$, $\overline{Q} = 1$) irrespective of the clock pulse and the *D* input. The CLR input should be connected to 1 for normal operation. Some flip-flops may have a preset input that sets *Q* to 1 and $\overline{Q}$ to 0 when the preset input is tied to ground. The preset input is connected to 1 for normal operation.

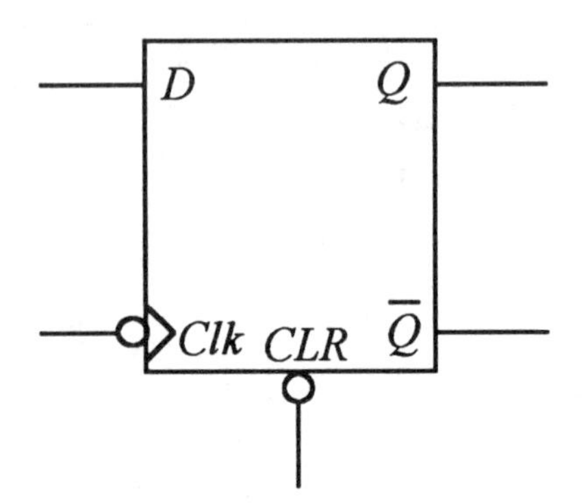

FIGURE 5.9 D Flip-Flop with Clear Input

5.5 Summary of Flip-Flops

Figures 5.10 through 5.13 summarize operations of all four flip-flops along with the symbolic representations, characteristic and excitation tables. In the figures, X represents don't care whereas $Q+$ indicates output Q after the clock pulse is applied.

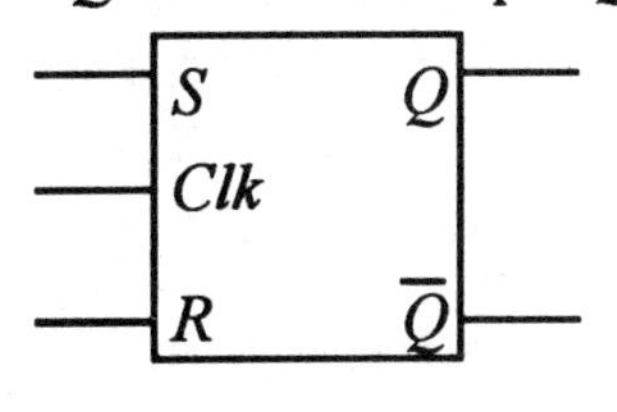

(a) Symbolic representation

S	R	$Q+$	
0	0	Q	Unchanged
0	1	0	Reset
1	0	1	Set
1	1	?	Invalid

(b) Characteristic table

Q	$Q+$	S	R
0	0	0	X
0	1	1	0
1	0	0	1
1	1	X	0

(c) Excitation table

FIGURE 5.10 RS flip-flop

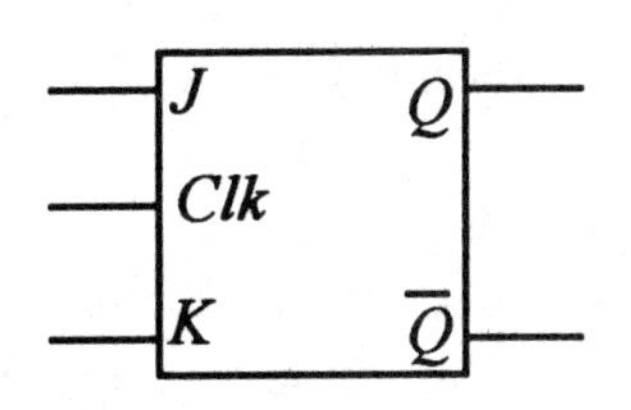

(a) Symbolic representation

J	K	$Q+$	
0	0	Q	Unchanged
0	1	0	Reset
1	0	1	Set
1	1	$\overline{Q}$	Complement

(b) Characteristic table

Q	$Q+$	J	K
0	0	0	X
0	1	1	X
1	0	X	1
1	1	X	0

(c) Excitation table

FIGURE 5.11 JK flip-flop

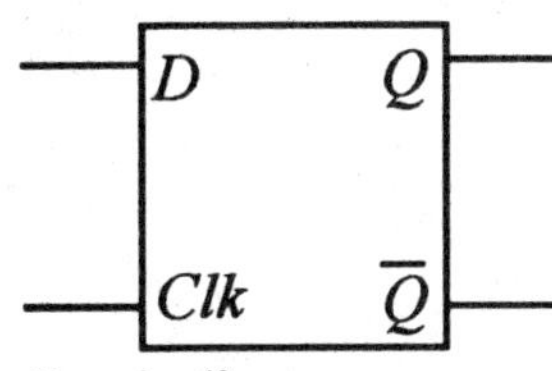

(a) Symbolic representation

D	$Q+$	
0	0	Reset
1	1	Set

(b) Characteristic table

Q	$Q+$	D
0	0	0
0	1	1
1	0	0
1	1	1

(c) Excitation table

FIGURE 5.12 D flip-flop

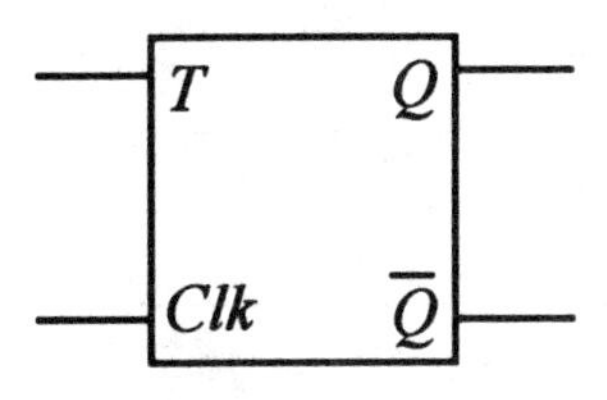

(a) Symbolic representation

T	$Q+$	
0	Q	Unchanged
1	$\overline{Q}$	Complement

(b) Characteristic table

Q	$Q+$	T
0	0	0
0	1	1
1	0	1
1	1	0

(c) Excitation table

FIGURE 5.13 T flip-flop

The characteristic table of a flip-flop is similar to its truth table. It contains the input combinations along with the output after the clock pulse. The characteristic table is useful for analyzing a flip-flop.

The present state (present output), the next state (next output) after the clock pulse, and the required inputs for the transition are included in the excitation table. This is useful for designing a sequential circuit, in which one normally knows the transition from the present to next state and wants to determine the required flip-flop inputs for the transition.

The D flip-flop is widely used in digital systems for transferring data. Several D flip-flops can be combined to form a register in the CPU of a computer. The 74HC374 is a 20-pin chip containing eight independent D flip-flops. It is designed using CMOS. The flip-flops are enabled at the leading edge of the clock. The 74LS374 is same as the 74HC374 except that it is designed using TTL.

The JK flip-flop is a universal flip-flop and is typically used for general applications. Typical commercially available flip-flops include the 74HC73 (or 74LS73A) and 74HC374 (or 74LS374). The 74HC73. is a 14-pin chip. It contains two independent JK flip-flops in the same chip, designed using CMOS. Each flip-flop is enabled at the trailing edge of the clock pulse. Each flip-flop also contains a direct clear input. The 74HC73 is cleared to zero when the clear input is LOW. The 74LS73A is same as the 74HC73 except that it is designed using TTL. The T flip-flop is normally used for designing binary counters because binary counters require complementation.The T flip-flop is not commercially available. It can be obtained from JK flip-flop by connecting the J and K inputs together as mentioned in section 5.2.5.

An example of a commercially available level-triggered flip-flop is the 74HC373 (or 74LS373). The 373 (20-pin chip) contains eight independent D latches with one enable input.

Sometimes the characteristic equation of a flip-flop is useful in analyzing the flip-flop's operation. The characteristic equations for the flip-flops can be obtained from the truth tables. Figure 5.14 through 5.16 show how these equations are obtained using K-maps for RS, JK, T, and D flip-flops.

Q	S	R	$Q+$
0	0	0	0
0	0	1	0
0	1	0	1
0	1	1	Invalid
1	0	0	1
1	0	1	0
1	1	0	1
1	1	1	Invalid

(a) Truth Table for RS-FF

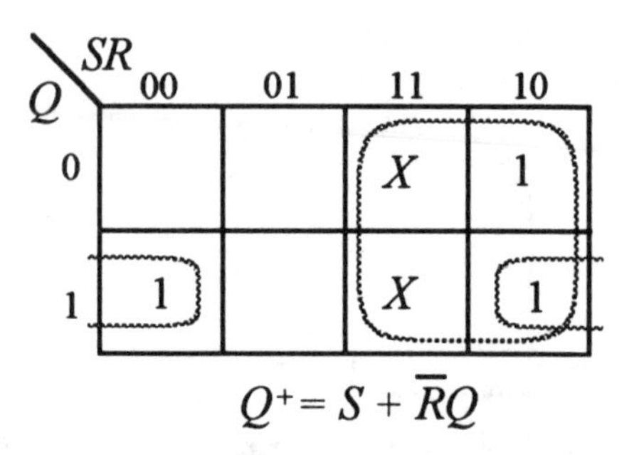

(b) K-map for characteristic equation of RS-FF

FIGURE 5.14 Truth table and K-map for the characteristic equation of RS flip-flop

Q	J	K	$Q+$
0	0	0	0
0	0	1	0
0	1	0	1
0	1	1	1
1	0	0	1
1	0	1	0
1	1	0	1
1	1	1	0

(a) Truth Table for JK-FF

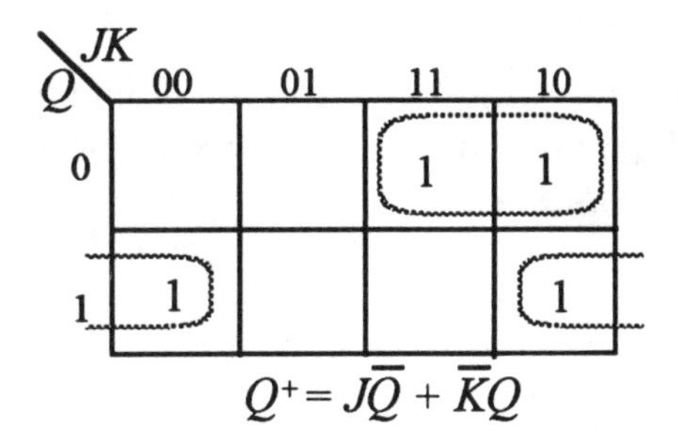

(b) K-map for characteristic equation of JK-FF

Q	T	$Q+$
0	0	0
0	1	1
1	0	1
1	1	0

(c) Truth Table for T-FF

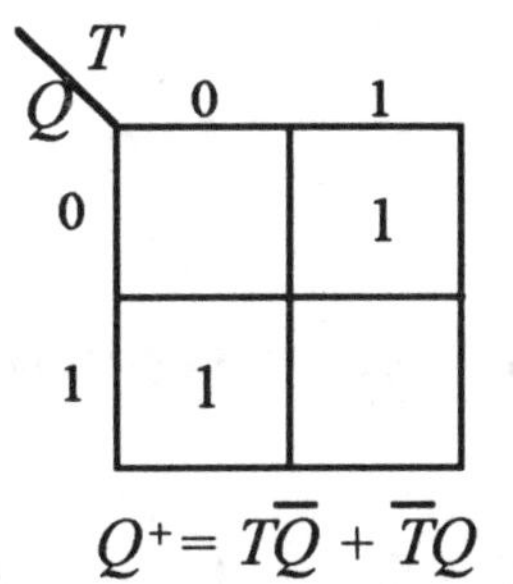

$Q^{+} = T\overline{Q} + \overline{T}Q$

(d) K-map for characteristic equation of T-FF.

FIGURE 5.15 Truth table and K-map for the characteristic equation of JK and T flip-flops

Q	D	$Q+$
0	0	0
0	1	1
1	0	0
1	1	1

(a) Truth Table for D-FF

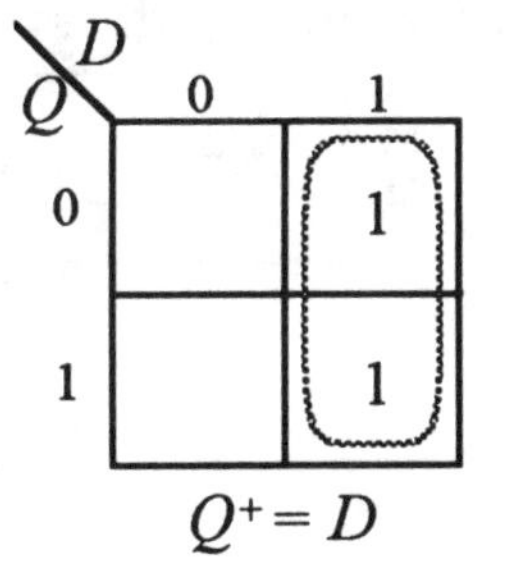

$Q^{+} = D$

(b) K-map for characteristic equation of D-FF.

FIGURE 5.16 Truth table and K-map for the characteristic equation of D flip-flop

Example 5.1

Given the following clock and the D inputs for a negative-edge-triggered D flip-flop, draw the timing diagram for the Q output for the first five cycles shown. Assume Q is preset to 1 initially. ***Solution:***

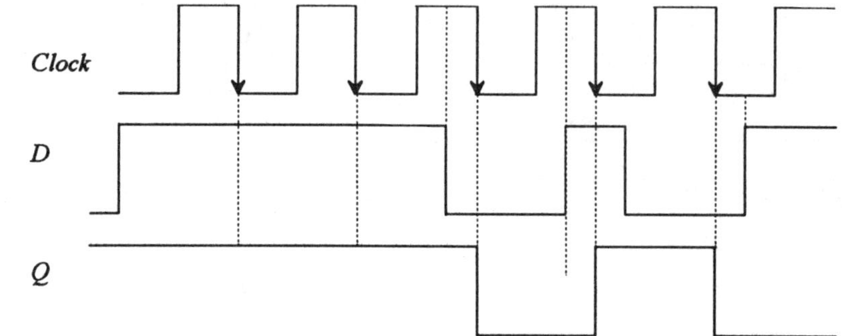

5.6 Analysis of Synchronous Sequential Circuits

A synchronous sequential circuit can be analyzed by determining the relationships between inputs, outputs, and flip-flop states. A state table or a state diagram illustrates how the inputs and the states of the flip-flops affect the circuit outputs. Boolean expressions can be obtained for the inputs of the flip-flops in terms of present states of the flip-flops and the circuit inputs. As an example consider analyzing the synchronous sequential circuit of Figure 5.17.

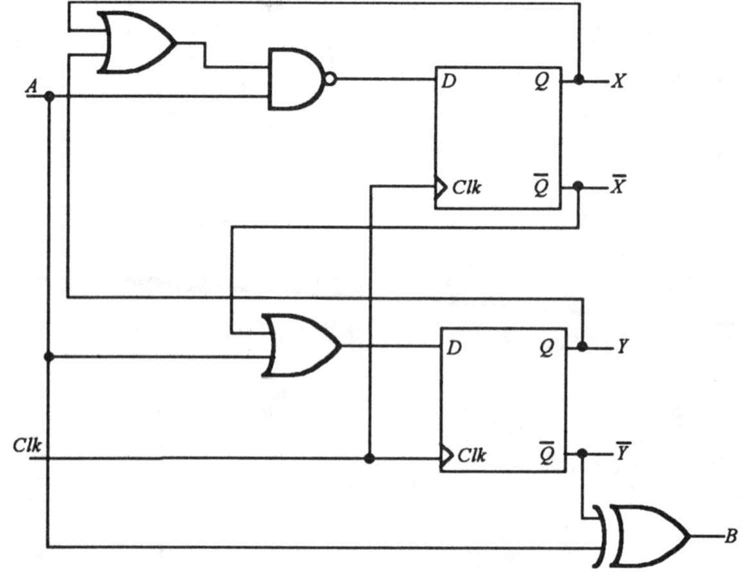

FIGURE 5.17 Analysis of a sequential circuit

The logic circuit contains two D flip-flops (outputs *X*, *Y*), one input *A* and one output *B*. The equations for the next states of the flip-flops can be written as

$$X^{+}=\overline{(X+Y)\cdot A}$$
$$Y^{+}=A+\overline{X}$$

Here *X*+ and *Y*+ represent the next states of the flip-flops after the clock pulse. The right side of each equation denotes the present states of the flip-flops (*X*, *Y*) and the input (*A*) that will produce the next state of each flip-flop. The Boolean expressions for the next state are obtained from the combinational circuit portion of the sequential circuit. The outputs of the combinational circuit are connected to the *D* inputs of the flip-flops. These *D* inputs provide the next states of the flip-flops after the clock pulse. The present state of the output *B* can be derived from the figure as follows:

$$B = A\oplus\overline{Y}$$

A state table listing the inputs, the outputs, and the states of the flip-flops along with the required flip-flop inputs can be obtained for Figure 5.17. Table 5.1 depicts a typical state table. The state table is formed by using the following equations (shown earlier):

$$X^{+}=\overline{(X+Y)\cdot A}$$
$$Y^{+}=A+\overline{X}$$

TABLE 5.1 State Table for Figure 5.16

Present State		Input	Next State		Flip Flop Inputs		Output
X	*Y*	*A*	*X*+	*Y*+	D_X	D_Y	*B*
0	0	0	1	1	1	1	1
0	0	1	1	1	1	1	0
0	1	0	1	1	1	1	0
0	1	1	0	1	0	1	1
1	0	0	1	0	1	0	1
1	0	1	0	1	0	1	0
1	1	0	1	0	1	0	0
1	1	1	0	1	0	1	1

To derive the state table, all combinations of the present states of the flip-flops and input *A* are tabulated. There are eight combinations for three variables from 000 to 111. The values for the flip-flop inputs (next states of the flip-flops) are determined using the equations. For example, consider the top row with $X = 0$, $Y = 0$, and $A = 0$. Substituting in the equations for next states.

$$X^{+} = \overline{(X+Y)\cdot A} = \overline{(0+0)\cdot 0} = 1$$

$$Y^{+} = A + \overline{X} = 0 + \overline{0} = 1$$

Now, to find the flip-flop inputs, one should consider each flip-flop separately. Two D flip-flops are used. Note that for a D flip-flop, the input at D is same as the next state. The D input is transferred to the output Q at the clock pulse. Therefore, $X+ = D_x$ and $Y+ = D_y$.

The characteristic table of a D flip-flop, discussed in Chapter 5, is used to determine the flip-flop inputs that will change present states of the flip-flops to next state. The characteristic table of D flip-flop is provided here for reference:

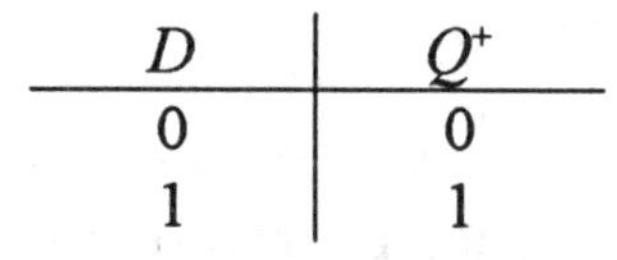

D	Q^{+}
0	0
1	1

Therefore, for D flip-flops, the next states and the flip-flop inputs will be same in the state table. By inspecting the top row of the state table, it can be concluded that $D_x = 1$ and $D_y = 1$ because the next states $X+ = 1$ and $Y+ = 1$.

Finally, the output B can be obtained from the equation,

$$B = A \oplus \overline{Y}$$

For example, consider the top row of the state table. $A = 0$ and $Y = 0$. Thus,

$$B = 0 \oplus \overline{0} = 0 \oplus 1 = 1$$

TABLE 5.2 Another Form of the State Table

Present State		Next State				Flip Flop Inputs				Outputs	
		$A=0$		$A=1$		$A=0$		$A=1$		$A=0$	$A=1$
X	Y	$X+$	$Y+$	$X+$	$Y+$	D_X	D_Y	D_X	D_Y	B	B
0	0	1	1	1	1	1	1	1	1	1	0
0	1	1	1	0	1	1	1	0	1	0	1
1	0	1	0	0	1	1	0	0	1	1	0
1	1	1	0	0	1	1	0	0	1	0	1

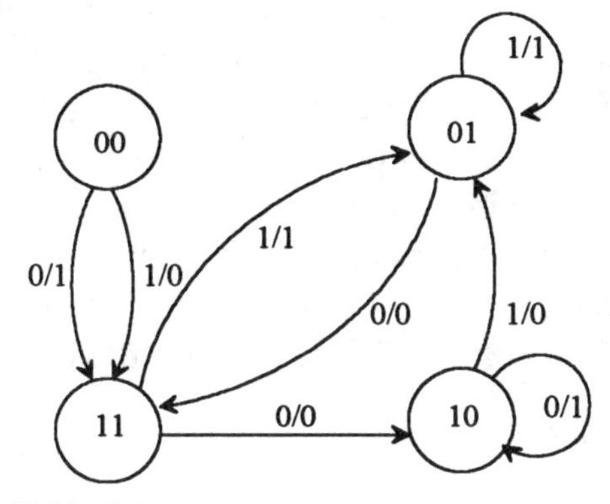

FIGURE 5.18 State diagram for Table 5.1

All other rows of the state table can similarly be verified. The state table of Table 5.1 can be shown in a slightly different manner. Table 5.2 depicts another for of the state table of Table 5.1.

A state table can be depicted in a graphical form. All information in the state table can be represented in the state diagram. A circle is used to represent a state in the state diagram. A straight line with an arrow indicator is used to show direction of transition from one state to another. Figure 5.18 shows the state diagram for Table 5.1.

Because there are two flip-flops (X, Y) in Figure 5.17, there are four states: 00, 01, 10 and 11. These are shown in the circle of the state diagram. Also, transition from one state to another is represented by a line with an arrow. Each line is assigned with a/b where a is input and b is output. From the example in Figure 5.18, with present state 10 and an input of 1, the output is 0 and the next state is 01. If the input (and/or output) is not defined in a problem, the input (and/or output) will be deleted in the state table and the state diagram.

The inputs of the flip-flops (D_x and D_y) in the state table are not necessary to derive the state diagram. In analyzing a synchronous sequential circuit, the logic diagram is given. The state equation, state table, and state diagram are obtained from the logic diagram. However, in order to design a sequential circuit, the designer has to derive the state table and the state diagram from the problem definition. The flip-flop inputs will be useful in the design. One must express the flip-flop inputs and outputs in terms of the present states of the flip-flops and the inputs. The minimum forms of these expressions can be obtained using a K-map. From these expressions, the logic diagram can be drawn.

5.7 Types of Synchronous Sequential Circuits

There are two types of Synchronous sequential circuits: the Mealy circuit and the Moore circuit. A synchronous sequential circuit typically contains inputs, outputs, and flip-flops. In the Mealy circuit, the outputs depend on both the inputs and the present states of the flip-flops. In the Moore circuit, on the other hand, the outputs are obtained from the flip-flops, and depend only on the present states of the flip-flops . Therefore, the only difference between the two types of circuits is in how the outputs are produced.

The state table of a Mealy circuit must contain an output column. . The state table of a Moore circuit may contain an output column, which is dependent only on the present states of the flip-flops. A Moore machine normally requires more states to generate identical output sequence compared to a Mealy machine. This is because the transitions are associated with the outputs in a Mealy machine.

5.8 Minimization of States

A simplified form of a synchronous sequential circuit can be obtained by minimizing the number of states. This will reduce the number of flip-flops and simplify the complexity of the circuit implementations. However, logic designers rarely use the minimization procedures. Also, there are sometimes instances in which design of a synchronous sequential circuit is simplified if the number of states is increased. The techniques for reducing the number of states presented in this section are merely for illustrative purpose.

The number of states can be reduced by using the concept of equivalent states. Two states are equivalent if both states provide the same outputs for identical inputs. One of the states can be eliminated if two states are equivalent. Thus, the number of states can be reduced.

For example, consider the state diagram of Figure 5.19. Each state is represented by a circle with transition to the next state based on either an input of 0 or 1 generating an output.

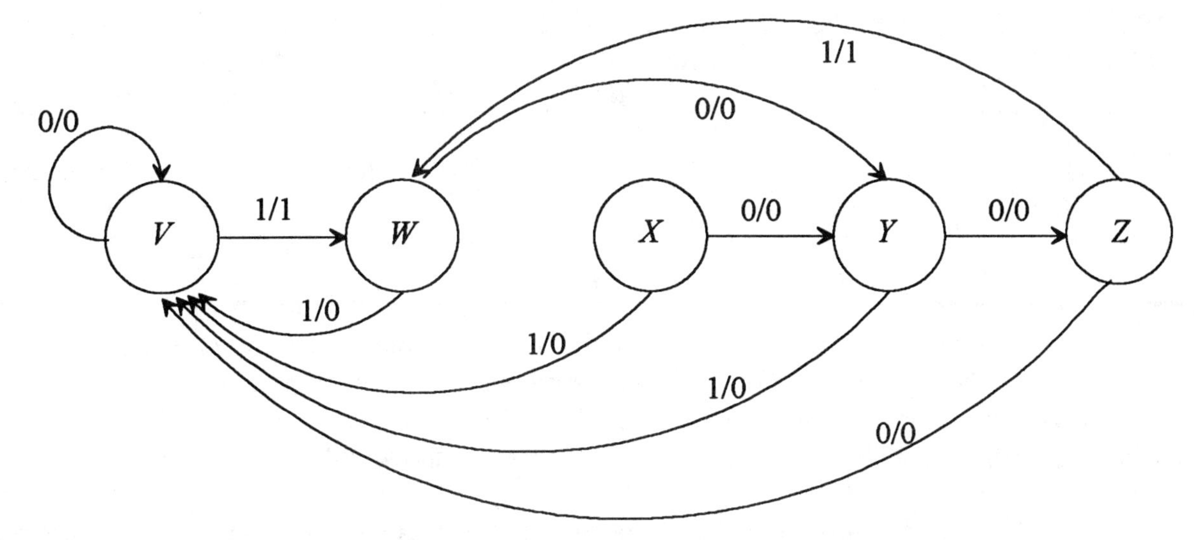

FIGURE 5.19 State diagram for minimization

Next, consider that a string of input data bits (*d*) in the sequence 0100111101 is applied at state *V* of a synchronous sequential circuit. For the given input sequence, the output and the state sequence can be obtained as follows:

State	*V*	*V*	*W*	*Y*	*Z*	*W*	*V*	*W*	*V*	*V*	*W*
Input	0	1	0	0	1	1	1	1	0	1	
Output	0	1	0	0	1	0	1	0	0	1	

With the sequential circuit in initial state *V*, a 0 input generates a 0 output and the circuit stays in state *V*, whereas in state *V*, an input of 1 produces an output 1 and the circuit will move to the next state *W*. In state *W* and input = 0, the output is 0 and the next state is *Y*. The process thus continues.

TABLE 5.3 State table for minimization of states

Present State	Next State		Output	
	d=0	*d=1*	*d=0*	*d=1*
V	*V*	*W*	0	1
W	*Y*	*V*	0	0
X	*Y*	*V*	0	0
Y	*Z*	*V*	0	0
Z	*V*	*W*	0	1

TABLE 5.4 Replacing states by their equivalents

Present State	Next State		Output	
	d=0	*d=1*	*d=0*	*d=1*
V	*V*	*W*	0	1
W	*Y*	*V*	0	0
~~*X*~~	*Y*	*V*	0	0
Y	~~*Z*~~ *V*	*V*	0	0
~~*Z*~~	*V*	*W*	0	1

The state table shown in Table 5.3 for the state diagram in Figure 5.19 can be obtained. Next, the equivalent states will be determined to reduce the number of states. *V* and *Z* are equivalent because they have same next states of *V* and *W* with identical inputs $d = 0$ and $d = 1$. Similarly, *W* and *X* are equivalent states. Table 5.4 shows the process of replacing of a state by its equivalent.

Because *V* and *Z* are equivalent, one of the states can be eliminated; *Z* is removed. Also, *W* and *X* are equivalent, so one of the states can be removed; *X* is thus eliminated in the state table. The row with present states *X* and *Z* is also eliminated. If they appear in the next state columns, they must be replaced by their equivalent states. In our case, the row for state *Y* contains *Z* in the next column. This is replaced its equivalent state *V*. By inspecting the modified state table further, no more equivalent states are found. The state table after elimination of equivalent states is shown in Table 5.5.

TABLE 5.5 State table after the elimination of equivalent states

Present State	Next State		Output	
	d=0	*d=1*	*d=0*	*d=1*
V	*V*	*W*	0	1
W	*Y*	*V*	0	0
Y	*V*	*V*	0	0

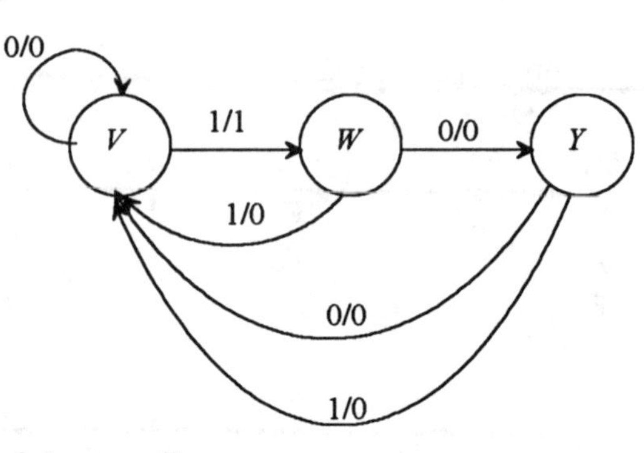

FIGURE 5.20 Reduced form of the state diagram

Note that the original state diagram in Figure 5.19 requires five states. Figure 5.20 shows the reduced form of the state diagram with only three states. Three flip-flops are required to represent five states whereas two flip-flops will represent three states. Thus, one flip-flop is eliminated and the complexity of implementation may be reduced. Note that a synchronous sequential circuit can be minimized by determining the equivalent states, provided the designer is only concerned with the output sequences due to input sequences.

5.9 Design of Synchronous Sequential Circuits

The procedure for designing a synchronous sequential circuit is a three-step process as follows:

1. Derive the state table and state diagram from the problem definition. If the state diagram is given, determine the state table.
2. Obtain the minimum form of the Boolean equations for flip-flop inputs and outputs, if any, using a K-map.
3. Draw the logic diagram. Note that a combinational circuit is designed using a truth table whereas the synchronous sequential circuit design is based on the state table.

Example 5.2

Design a synchronous sequential circuit for the state diagram of Figure 5.21 using D flip-flops.

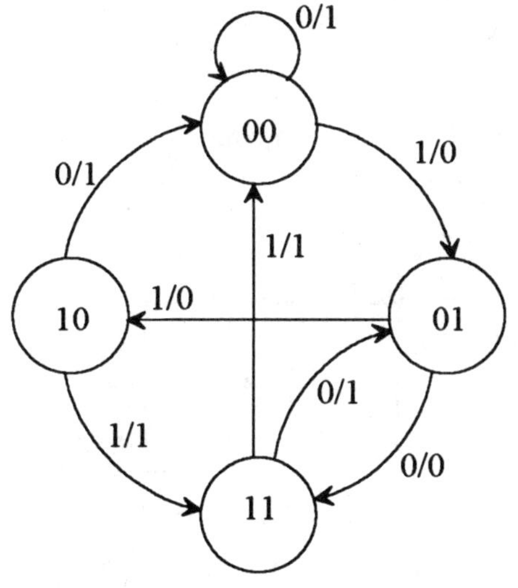

FIGURE 5.21 State diagram for Example 5.2

TABLE 5.6 State Table for Example 5.2

Present State		Input	Next State		Flip Flop Inputs		Output
X	Y	A	$X+$	$Y+$	D_X	D_Y	Z
0	0	0	0	0	0	0	1
0	0	1	0	1	0	1	0
0	1	0	1	1	1	1	0
0	1	1	1	0	1	0	0
1	0	0	0	0	0	0	1
1	0	1	1	1	1	1	1
1	1	0	0	1	0	1	1
1	1	1	0	0	0	0	1

Solution

Step 1: Derive the state table. The state table is derived from the state diagram (Figure 5.21) and the excitation table [Figure 5.12(c)] of the D flip-flop. Table 5.6 shows the state table.

The state table is obtained directly from the state diagram. In the state table, the next states are same as the flip-flop inputs because D flip-flops are used. This is evident from the excitation table of Figure 5.12(c).

Step 2: Obtain the minimum forms of the equations for the flip-flop inputs and the output. Using K-maps and the output, the equations for flip-flop inputs are simplified as shown in Figure 5.22.

Step 3: Draw the logic diagram. The logic diagram is shown in Figure 5.23.

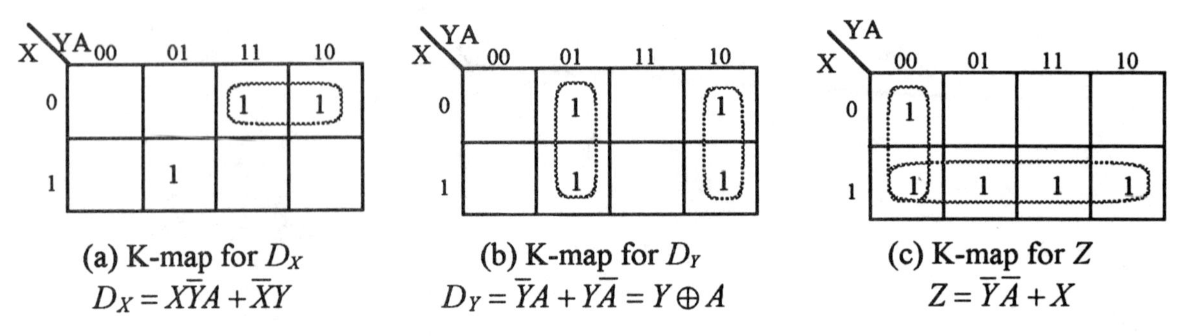

(a) K-map for D_X $D_X = X\overline{Y}A + \overline{X}Y$

(b) K-map for D_Y $D_Y = \overline{Y}A + Y\overline{A} = Y \oplus A$

(c) K-map for Z $Z = \overline{Y}\overline{A} + X$

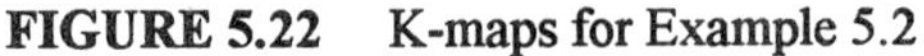

FIGURE 5.22 K-maps for Example 5.2

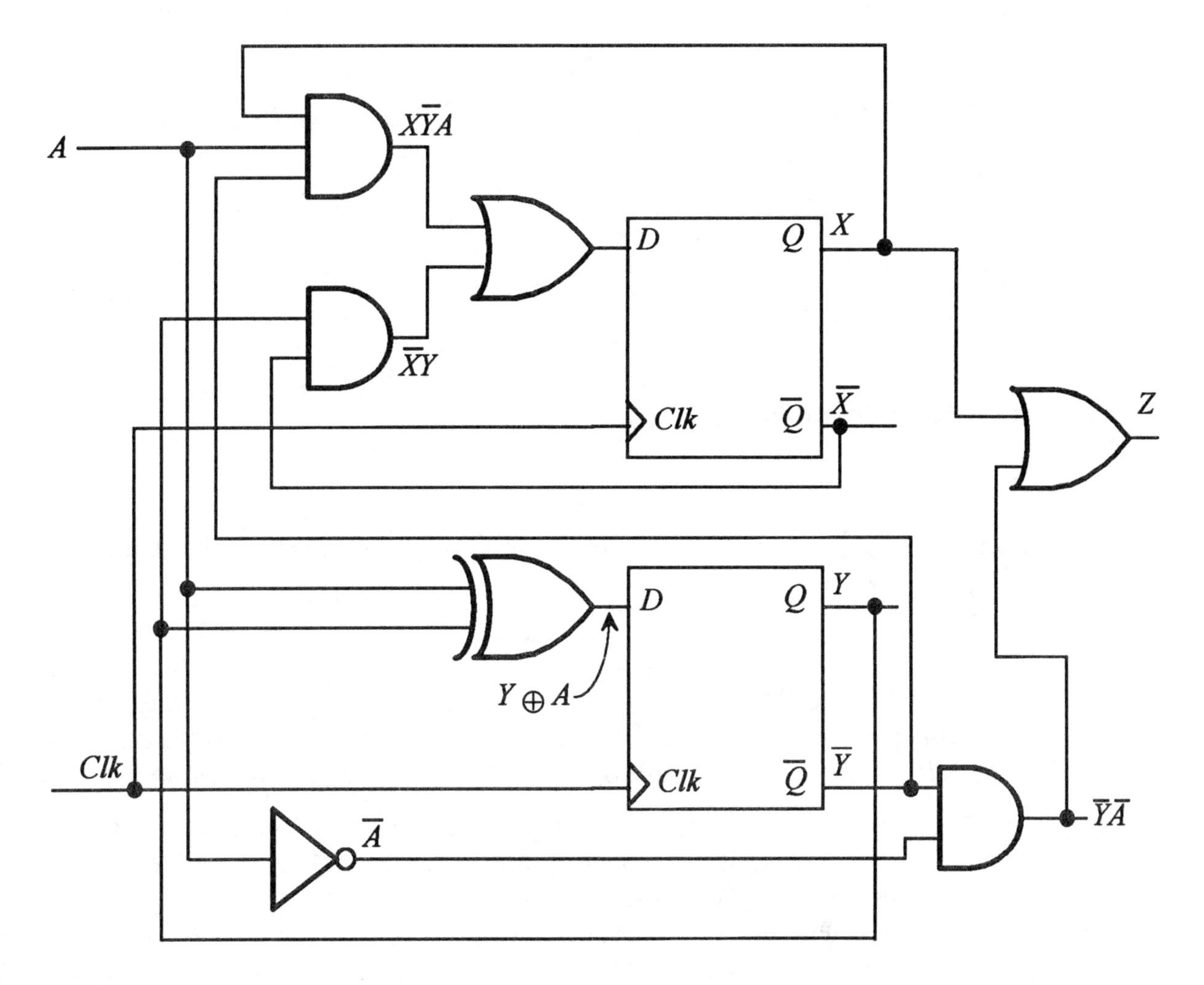

FIGURE 5.23 Logic diagram for Example 5.2

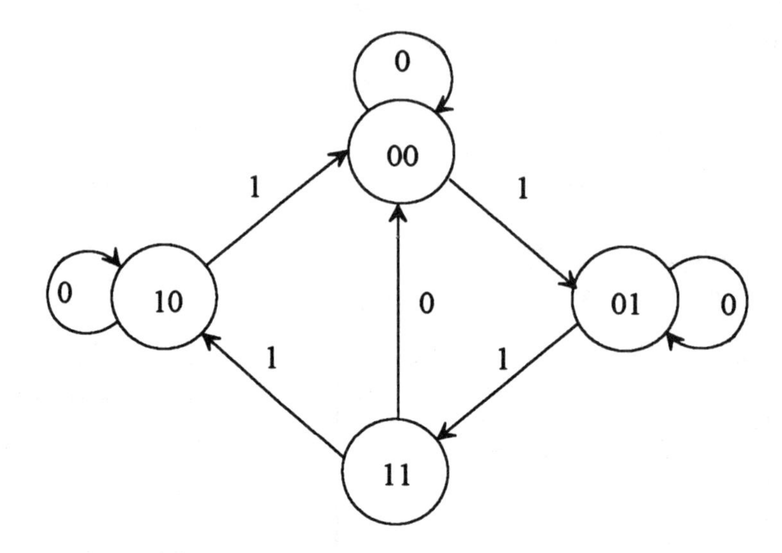

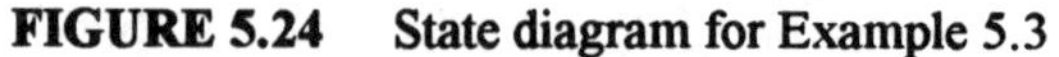

FIGURE 5.24 State diagram for Example 5.3

Example 5.3

Design a synchronous sequential circuit for the state diagram of Figure 5.24 using JK flip-flops.

Solution

Step 1: Derive the state table. The state table can be directly obtained from the state diagram (Figure 5.24) and the excitation table [Figure 5.11(c)]. Table 5.7 shows the state table. For convenience, the excitation table of JK flip-flop of Figure 5.11(c) is also included.

Let us explain how the state table is obtained. The input A is 0 or 1 at each state, so the left three columns show all eight combinations for X, Y, and A. The next state column is obtained from the state diagram. The flip-flop inputs are then obtained using the excitation table for the JK flip-flop. For example, consider the top row. From the state diagram, the present state (00) remains in the same state (00) when input A = 0 and the clock pulse is applied. The output of flip-flop X goes from 0 to 0 and the output of flip-flop Y goes from 0 to 0. From the excitation table of the JK flip-flop, $J_x = 0$, $K_x = X$, $J_x = 0$, and $K_x = X$. The other rows are obtained similarly.

TABLE 5.7 State and Excitation Tables for Example 5.3

(a) Excitation Table of JK flip-flop from Figure 5.11c

Q	*Q*+	*J*	*K*
0	0	0	X
0	1	1	X
1	0	X	1
1	1	X	0

(b) State Table for Example 5.2

Present State		Input	Next State		Flip Flop Inputs			
X	Y	A	$X+$	$Y+$	J_X	K_X	J_Y	K_Y
0	0	0	0	0	0	X	0	X
0	0	1	0	1	0	X	1	X
0	1	0	0	1	0	X	X	0
0	1	1	1	1	1	X	X	0
1	0	0	1	0	X	0	0	X
1	0	1	0	0	X	1	0	X
1	1	0	0	0	X	1	X	1
1	1	1	1	0	X	0	X	1

Step 2: Obtain the minimum forms of the equations for the flip-flop inputs. Using K-maps, the equations for flip-flop inputs are simplified as shown in Figure 5.25.

Step 3: Draw the logic diagram as shown in Figure 5.26.

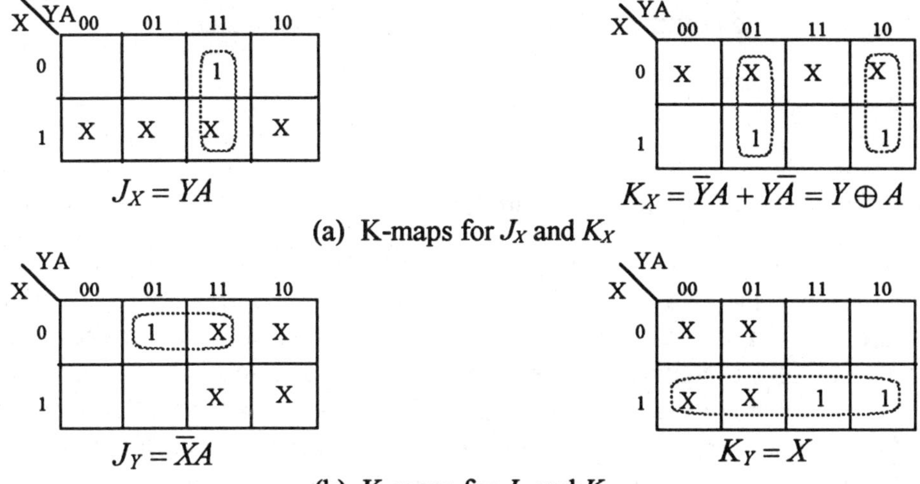

FIGURE 5.25 K-maps for Example 5.3

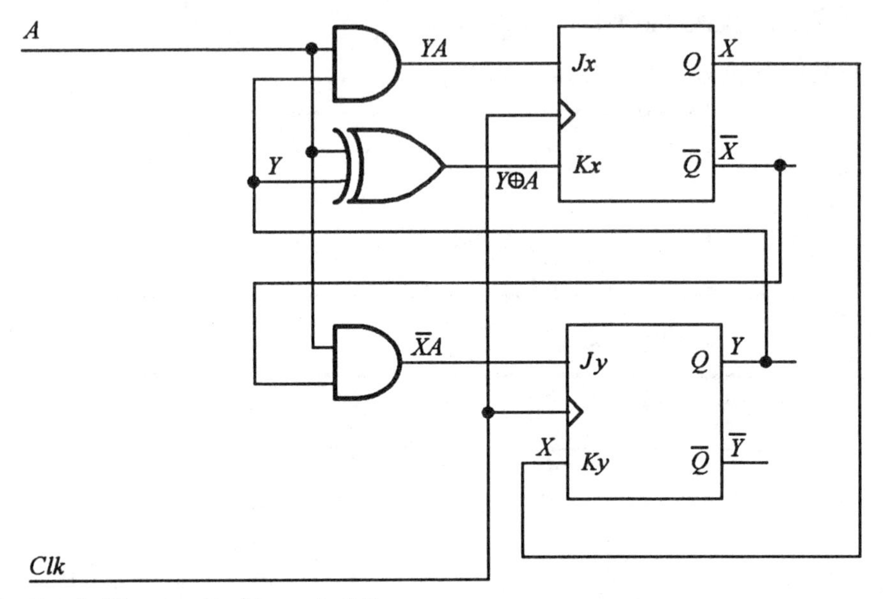

FIGURE 5.26 Logic Diagram for Example 5.3

Example 5.4

Design a synchronous sequential circuit with one input X and an output Z. The input X is a serial message and the system reads X one bit at a time. The output $Z = 1$ whenever the pattern 101 is encountered in the serial message. For example,

If input:	0 0 1 0 1 0 1 1 1 0 1 0 0 0 1 0 1
then output:	0 0 0 0 1 0 1 0 0 0 1 0 0 0 0 0 1

Use T flip-flops.

Solution

Step 1: Derive the state diagram and the state table.

Figure 5.27 shows the state diagram. In this diagram each node represents a state. The labeled arcs (lines joining two nodes) represent state transitions. For example, when the system is in state C, if it receives an input 1, it produces an output 1 and makes a transition to the state D after the clock. Similarly, when the system is in state C and receives a 0 input, it generates a 0 output and moves to state A after the clock. This type of sequential circuit is called a *Mealy machine* because the output generated depends on both the input X and the present state of the system. It should be emphasized that each state in the state diagram actually performs a bookkeeping operation; these operations are summarized as follows:

State	Interpretation
A	Looking for a new pattern
B	Received the first 1
C	Received a 1 followed by a 0
D	Recognized the pattern 101

The state diagram can be translated into a *state table,* as shown in Table 5.8. Each state can be represented by the binary assignment as follows:

Symbolic State	Binary State y_1	y_0
A	0	0
B	0	1
C	1	1
D	1	0

The state table in Table 5.8 can be modified to reflect this state assignment, as illustrated in Table 5.9. Note that the excitation table actually describes the required excitation for a particular state transition to occur. For example, with respect to a T flip-flop, for the transition $0 \rightarrow 1$ or $1 \rightarrow 0$, a 1 must be applied to the *T* input. Similarly, for transitions $0 \rightarrow 0$ or $1 \rightarrow 1$ (that is, no change of state), the *T* input must be made 0. Using this excitation table, the flip-flop input equations can be derived as illustrated in Table 5.9.

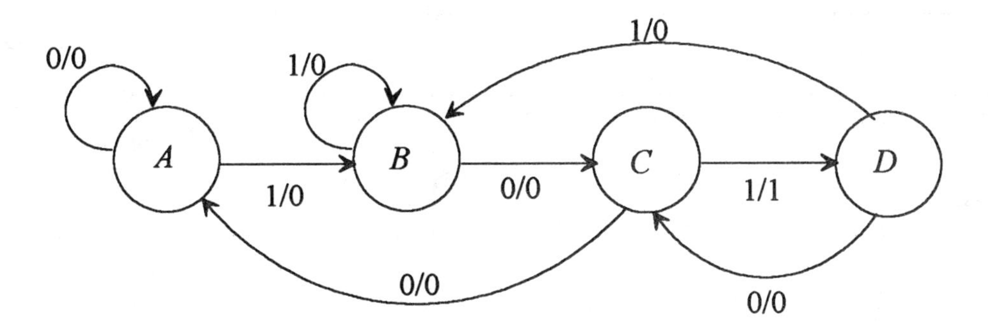

FIGURE 5.27 State Diagram for Example 5.4

TABLE 5.8 State Table for Example 5.4

Present State	Next State		Output Z	
	$X=0$	$X=1$	$X=0$	$X=1$
A	A	B	0	0
B	C	B	0	0
C	A	D	0	1
D	C	B	0	0

TABLE 5.9 Modified State Table for Example 5.4

Present State		Next State		Output Z	
y_1	y_0	$y_1^+ y_0^+$ $X=0$	$y_1^+ y_0^+$ $X=1$	Input $X=0$	Input $X=1$
0	0	0 0	0 1	0	0
0	1	1 1	0 1	0	0
1	1	0 0	1 0	0	1
1	0	1 1	0 1	0	0

In this figure, the entries corresponding to the flip-flop inputs T_{y_1} and T_{y_0} are directly derived using the T flip-flop excitation table. For example, consider the present state $y_1y_0 = 00$. When the input $X = 1$, the next state is 01. This means that flip-flop y_1 should not change its states and flip-flop y_0 must change its state to 1. It follows that $T_{y_1} = 0$ (because a $0 \rightarrow 0$ transition is required) and $T_{y_0} = 1$ (because a $0 \rightarrow 1$ transition is required). The other entries for T_{y_1} and T_{y_0} may be obtained in a similar manner.

The state table of Table 5.9 is obtained using the excitation table for T flip-flop of Figure 5.13(c) redrawn as follows:

Present State		Input	Next State		Flip Flop Inputs		Ouput
y_1	y_0	X	y_1^+	y_0^+	T_{y_1}	T_{y_0}	Z
0	0	0	0	0	0	0	0
0	0	1	0	1	0	1	0
0	1	0	1	1	1	0	0
0	1	1	0	1	0	0	0
1	0	0	1	1	0	1	0
1	0	1	0	1	1	1	0
1	1	0	0	0	1	1	0
1	1	1	1	0	0	1	1

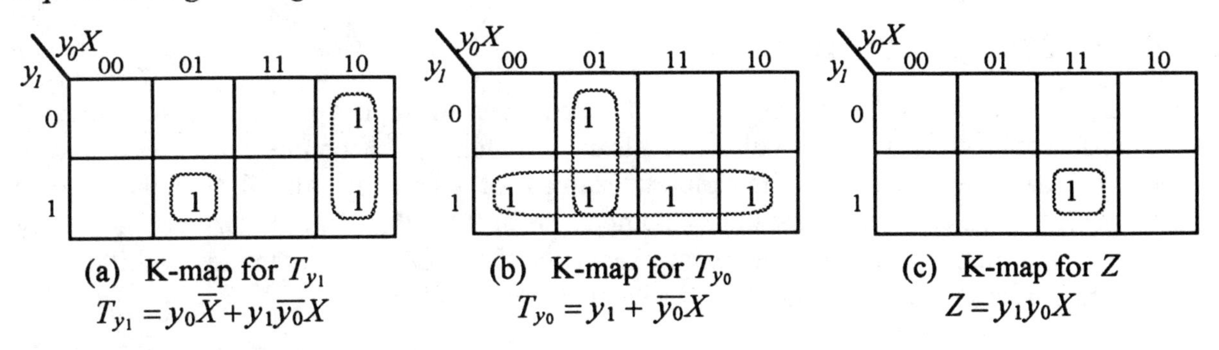

(a) K-map for T_{y_1}
$T_{y_1} = y_0\overline{X} + y_1\overline{y_0}X$

(b) K-map for T_{y_0}
$T_{y_0} = y_1 + \overline{y_0}X$

(c) K-map for Z
$Z = y_1y_0X$

FIGURE 5.28 K-maps for Example 5.4

Step 2: Derive the minimum forms of the equations for the flip-flop inputs and the output.

Using K-maps, the simplified equations for the flip-flops inputs and the output can be obtained as shown in Figure 5.28.

Step 3: Draw the logic diagram as shown in Figure 5.29.

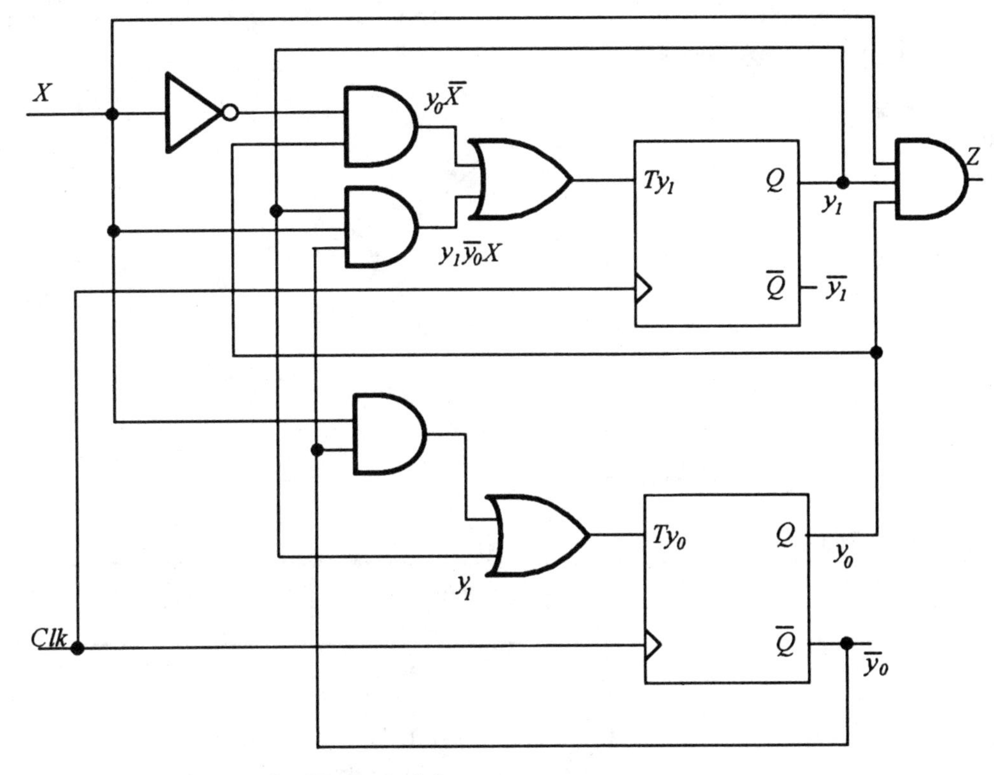

FIGURE 5.29 Logic Diagram for Example 5.4

5.10 Design of Counters

A counter is a synchronous sequential circuit that moves through a predefined sequence of states upon application of clock pulses. A binary counter, which counts binary numbers in sequence at each clock pulse, is the simplest example of a counter. An n-bit binary counter contains n flip-flops and can count binary numbers from 0 to 2^{n-1}. Other binary counters may count in an arbitrary manner in a nonbinary sequence. The following examples will illustrate the straight binary sequence and nonbinary sequence counters.

Example 5.5

Design a two-bit counter to count in the sequence 00, 01, 10, 11, and repeat. Use T flip-flops.

Solution

Step 1: Derive the state diagram and the state table.

Figure 5.30 shows the state diagram. Note that state transition occurs at the clock pulse. No state transitions occurs if there is no clock pulse. Therefore, the clock pulse does not appear as an input. Table 5.10 shows the state table.

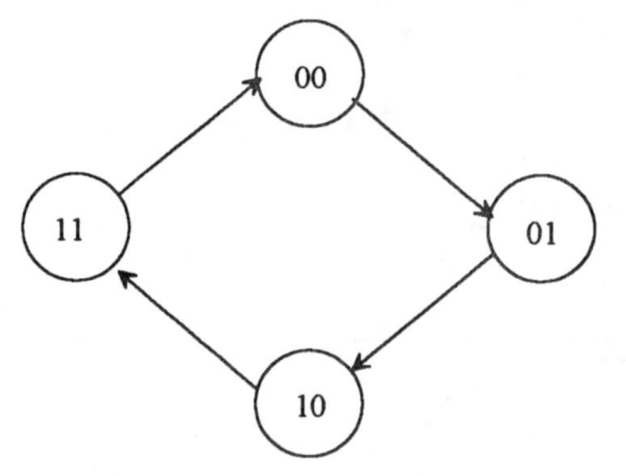

FIGURE 5.30 State Diagram for Example 5.5

TABLE 5.10 Stable table for Example 5.5

Present State		Next State		Flip Flop inputs	
a_1	a_0	a_1+	a_0+	T_{A_1}	T_{A_0}
0	0	0	1	0	1
0	1	1	0	1	1
1	0	1	1	0	1
1	1	0	0	1	1

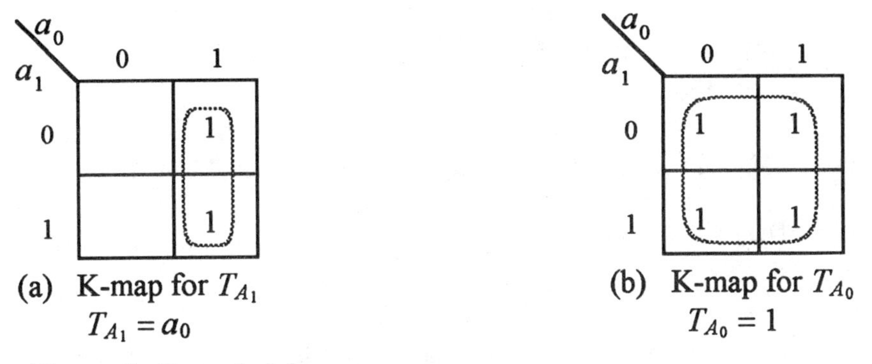

(a) K-map for T_{A_1}
$T_{A_1} = a_0$

(b) K-map for T_{A_0}
$T_{A_0} = 1$

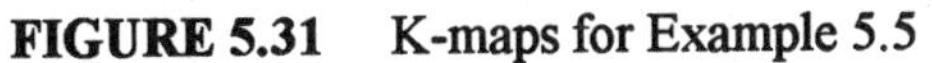

FIGURE 5.31 K-maps for Example 5.5

The excitation table of the T flip-flop is used for deriving the state table. For example, consider the top row. The state remains unchanged ($a_1 = 0$ and $a_{1+} = 0$) requiring a T input of 0 and thus $T_{A_1} = 0$. a_0 is complemented from the present state to the next state, and thus $T_{A_0} = 1$.

Step 2: Derive the minimum forms of the equations for the flip-flop inputs.

Using K-maps, the simplified equations for the flip-flop inputs can be obtained as shown in Figure 5.31.

Step 3: Draw the logic diagram as shown in Figure 5.32.

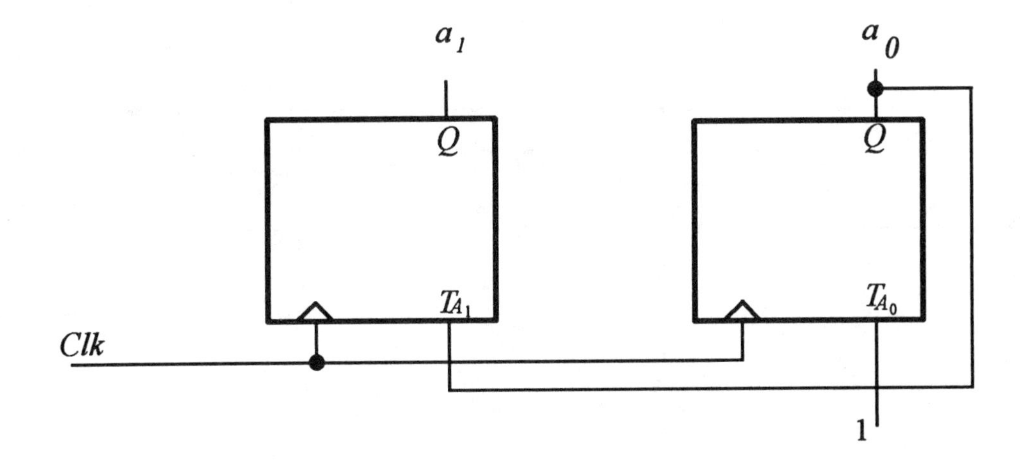

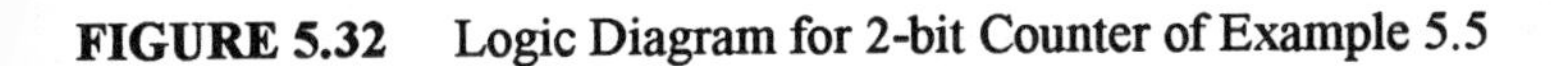

FIGURE 5.32 Logic Diagram for 2-bit Counter of Example 5.5

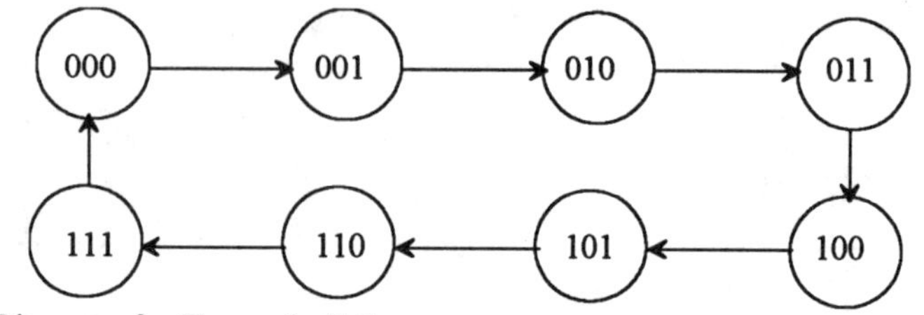

FIGURE 5.33 State Diagram for Example 5.6

Example 5.6

Design a three-bit counter to count in the sequence 000 through 111, return to 000 after 111, and then repeat the count. Use JK flip-flops.

Solution

Step 1: Derive the state diagram and the state table.

Figure 5.33 shows the state diagram. Table 5.11 shows the state table. Consider the top row. The present state of a_2 changes from 0 to 0 at the clock, a_1 changes from 0 to 0, and a_0 changes from 0 to 1. From the JK flip-flop excitation table, for these transitions, $Ja_2 = 0$, $Ka_2 = X$, $Ja_1 = 0$, $Ka_1 = X$, and $Ja_0 = 1$, $Ka_0 = X$.

TABLE 5.11 State Table for Example 5.6

(a) Excitation Table of JK Flip-flops

Q	$Q+$	J	K
0	0	0	X
0	1	1	X
1	0	X	1
1	1	X	0

(b) State Table for Example 5.6

Present State			Next State			Flip-Flop Inputs					
a_2	a_1	a_0	a_2+	a_1+	a_0+	Ja_2	Ka_2	Ja_1	Ka_1	Ja_0	Ka_0
0	0	0	0	0	1	0	X	0	X	1	X
0	0	1	0	1	0	0	X	1	X	X	1
0	1	0	0	1	1	0	X	X	0	1	X
0	1	1	1	0	0	1	X	X	1	X	1
1	0	0	1	0	1	X	0	0	X	1	X
1	0	1	1	1	0	X	0	1	X	X	1
1	1	0	1	1	1	X	0	X	0	1	X
1	1	1	0	0	0	X	1	X	1	X	1

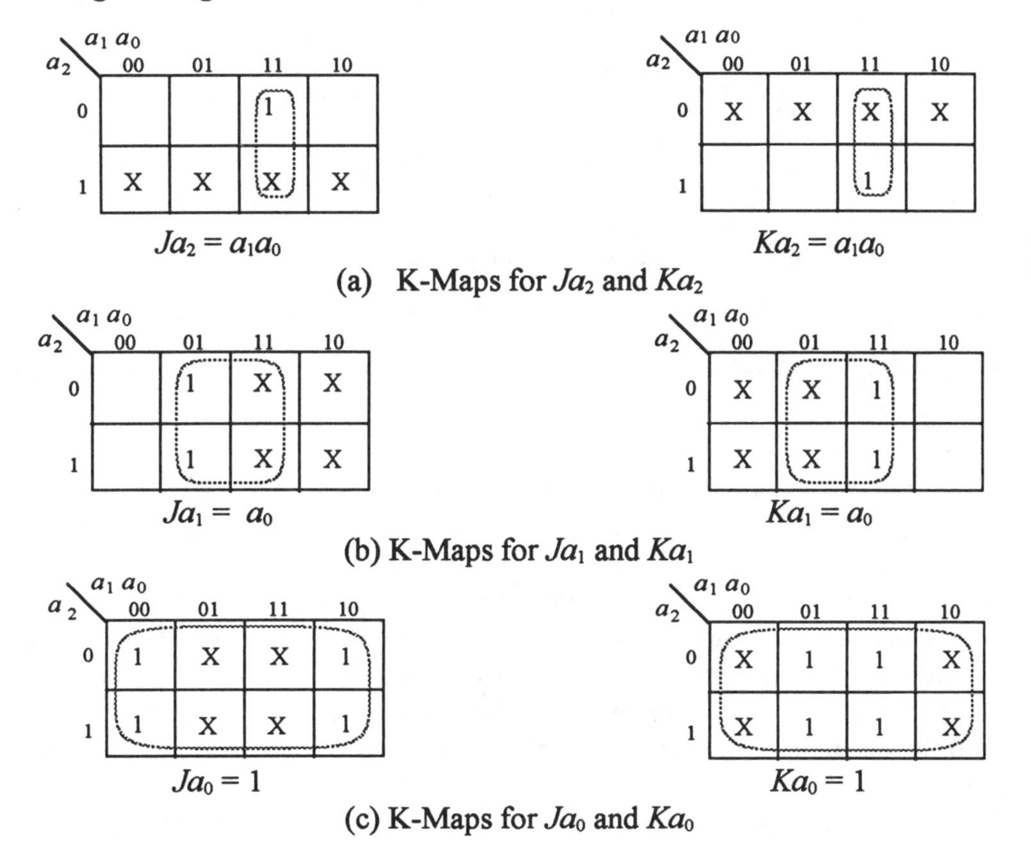

FIGURE 5.34 K-Maps for Example 5.6

Step 2: Derive the minimum forms of the equations for the flip-flop inputs. Using K-maps, the simplified equations for the flip-flop inputs can be obtained as shown in Figure 5.34.

Step 3: Draw the logic diagram as shown in Figure 5.35.

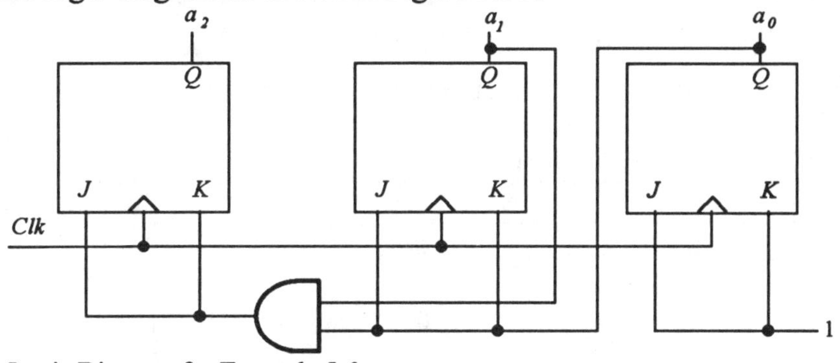

FIGURE 5.35 Logic Diagram for Example 5.6

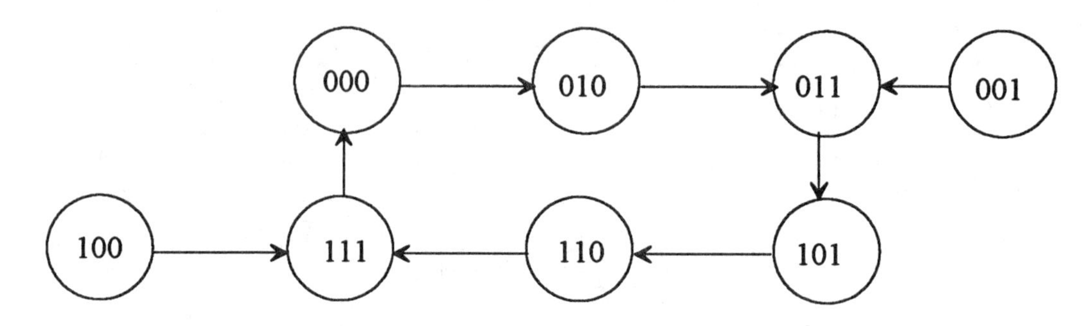

FIGURE 5.36 State Diagram for Example 5.7

Example 5.7

Design a 3-bit nonbinary counter that will count in the sequence 000, 010, 011, 101, 110, 111, and repeat the sequence. The counter has two unused states. These are 001 and 100. Implement the counter as a self-correcting such that if the counter happens to be in one of the unused states (001 or 100) upon power-up or due to error, the next clock pulse puts it in one of the valid states and the counter provides the correct count. Use T Flip-flops. Note that the initial states of the flip-flops are unpredictable when power is turned ON. Therefore, all the unused (don't care) states of the counter should be checked to ensure that the counter eventually goes into the desirable counting sequence. This is called a self-correcting counter.

Solution

Step 1: Derive the state diagram and the state table. Figure 5.36 shows the state diagram. Note that in the state diagram it is shown that if the counter goes to an invalid state such as 001 upon power-up, the counter will then go to the valid state 111 and will count correctly. Similarly, for the invalid state 100, the counter will be in state 111 and the correct count will continue. This self-correcting feature will be verified from the counter's state table using T flip-flops as shown in Table 5.12.

TABLE 5.12 State Table for Example 5.7

(a) Excitation Table for T Flip-Flop

Q	$Q+$	T
0	0	0
0	1	1
1	0	1
1	1	0

TABLE 5.12 State Table for Example 5.7 (Continued)

(b) State Table for Example 5.7

Present State			Next State			Flip Flop Inputs		
a_2	a_1	a_0	a_2+	a_1+	a_0+	Ta_2	Ta_1	Ta_0
0	0	0	0	1	0	0	1	0
0	1	0	0	1	1	0	0	1
0	1	1	1	0	1	1	1	0
1	0	1	1	1	0	0	1	1
1	1	0	1	1	1	0	0	1
1	1	1	0	0	0	1	1	1

Step 2: Derive the minimum forms of the equations for the flip-flop inputs.

Using K-maps, the simplified equations for the flip-flop inputs can be obtained, as shown in Figure 5.37. The unused states 001 and 100 are invalid and can never occur, so they are don't care conditions.

Now, let us verify the self-correcting feature of the counter. The flip-flop input equations are

$$Ta_2 = a_1 a_0$$

$$Ta_1 = \overline{a_1} + a_0$$

$$Ta_0 = a_2 + a_1\overline{a_0}$$

Suppose that the counter is in the invalid state 001 upon power-up or due to error, therefore, in this state, $a_2 = 0$, $a_1 = 0$, and $a_0 = 1$. Substituting these values in the flip-flop input equations, we get

$$Ta_2 = 0 \cdot 1 = 0$$

$$Ta_1 = \overline{0} + 1 = 1$$

$$Ta_0 = 0 + 0 \cdot \overline{1} = 0$$

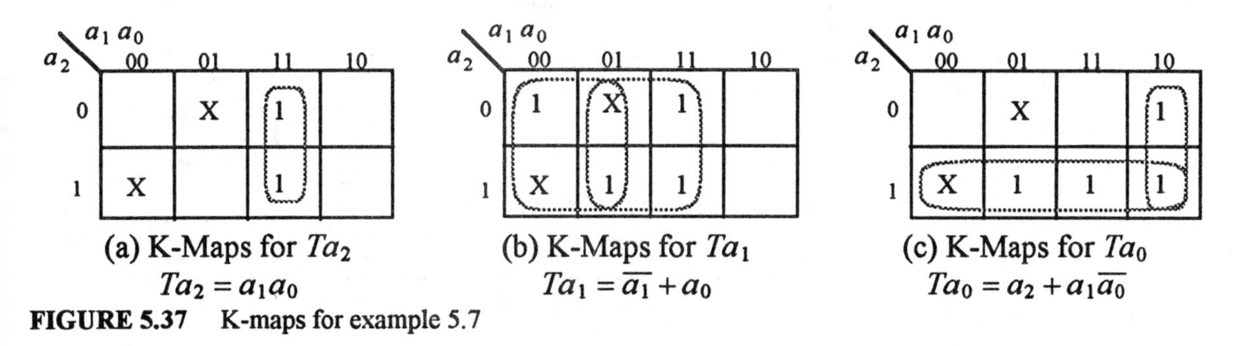

(a) K-Maps for Ta_2
$Ta_2 = a_1 a_0$

(b) K-Maps for Ta_1
$Ta_1 = \overline{a_1} + a_0$

(c) K-Maps for Ta_0
$Ta_0 = a_2 + a_1\overline{a_0}$

FIGURE 5.37 K-maps for example 5.7

Note that with $a_2a_1a_0$ = 001 and $Ta_2Ta_1Ta_0$ = 010, the state changes from 001 to 011. Therefore, the next state will be 011. The correct count will resume. Next, if the flip-flop goes to the invalid state 100 due to error or when power is turned ON. Substituting $a_2 = 1$, $a_1 = 0$, and $a_0 = 0$ gives

$$Ta_2 = 0 \cdot 0 = 0$$

$$Ta_1 = \bar{0} + 0 = 1$$

$$Ta_0 = 1 + 0 \cdot \bar{0} = 1$$

Note that with $a_2a_1a_0$ = 100 and $Ta_2Ta_1Ta_0$ = 011, the state changes from 100 to 111. Hence, the next state for the counter will be 111. The correct count will continue. Therefore, the counter is self-correcting.

Step 3: Draw the logic diagram as shown in Figure 5.38.

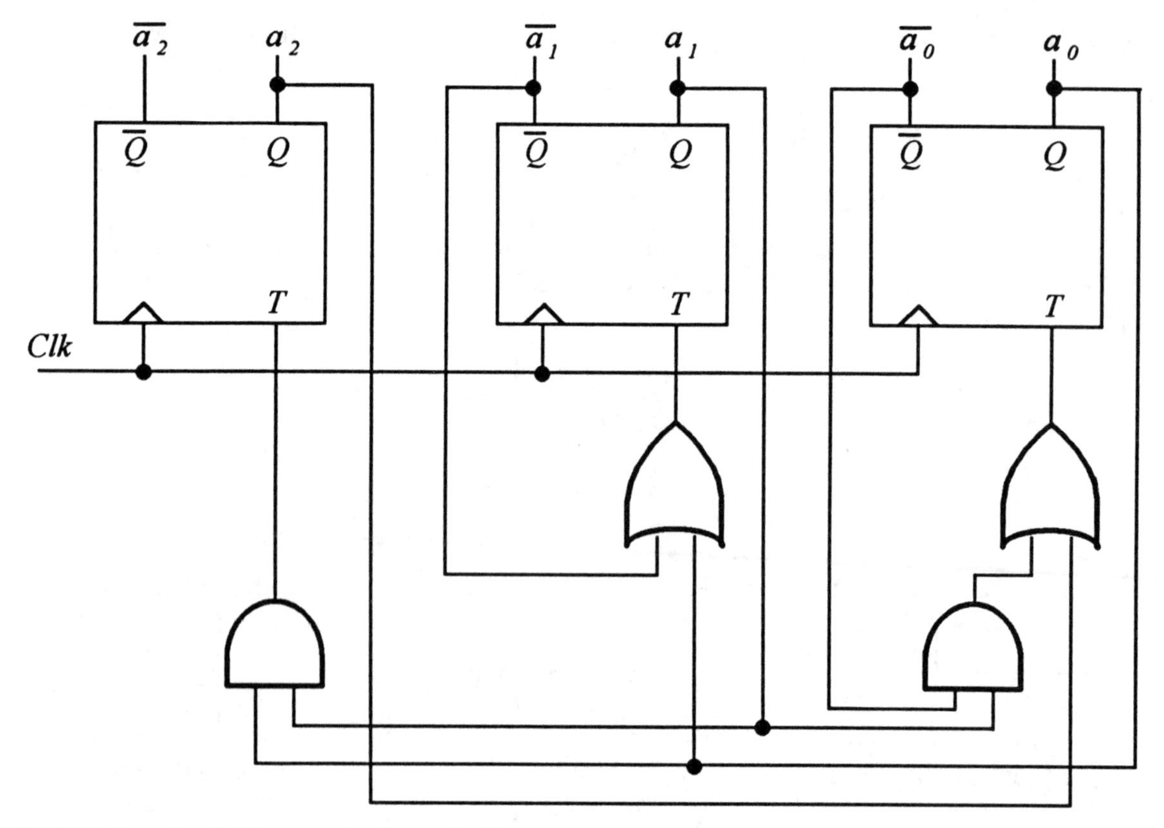

FIGURE 5.38 Logic Diagram for Example 5.7

5.11 Examples of Synchronous Sequential Circuits

Typical examples include registers, modulo-*n* counters and RAMs (Random Access Memories). They play an important role in the design of digital systems, especially computers.

5.11.1 Registers

A register contains a number of flip-flops for storing binary information in a computer. The register is an important part of any CPU. A CPU with many registers reduces the number of accesses to the main memory, therefore simplifying the programming task and shortening execution time. A general-purpose register (GPR) is designed in this section. The primary task of the GPR is to store address or data for an indefinite amount of time, then to be able to retrieve the data when needed. A GPR is also capable of manipulating the stored data by shift left or right operations. Figure 5.39 contains a summary of typical shift operations. In logical shift operation, a bit that is shifted out will be lost, and the vacant position will be filled with a 0. For example, if we have the number $(11)_{10}$, after right shift, the following occurs:

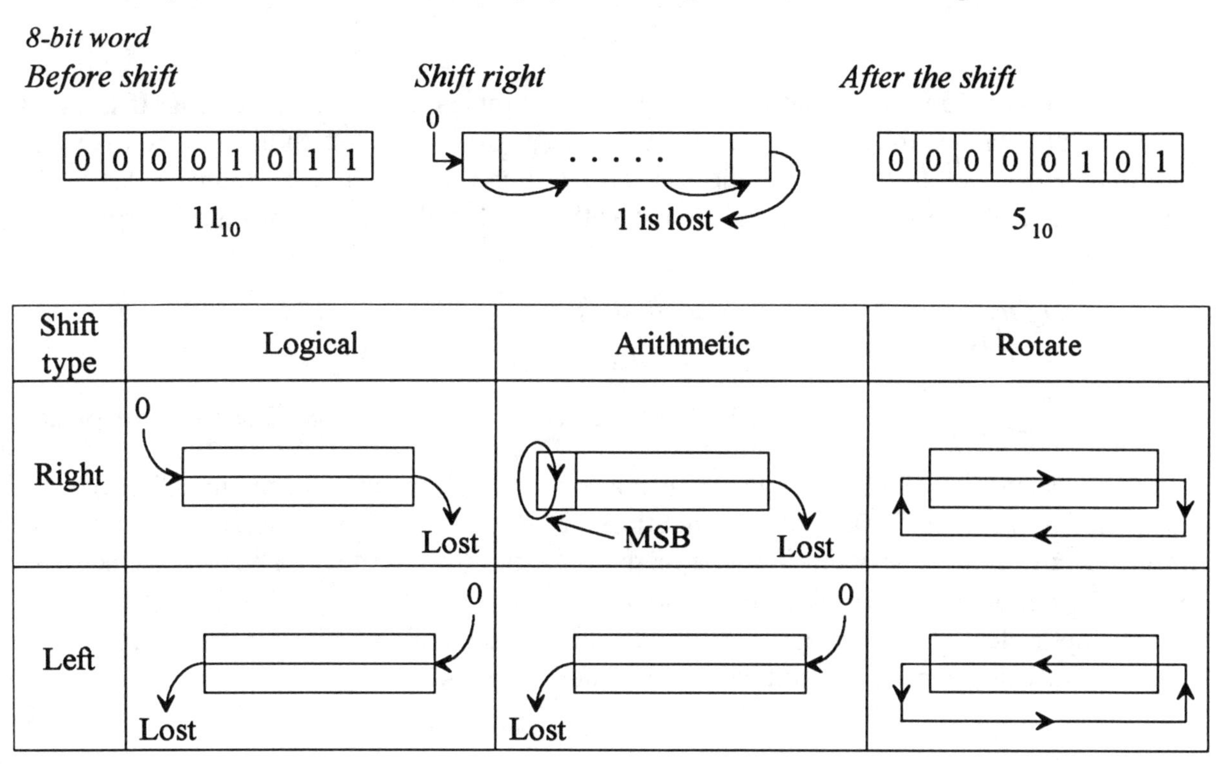

FIGURE 5.39 Summary of Typical Shift Operations

It must be emphasized that a logical left or right shift of an unsigned number by n positions implies multiplication or division of the number by 2^n, respectively, provided that a 1 is not shifted out during the operation.

In the case of true arithmetic left or right shift operations, the sign bit of the number to be shifted must be retained. However, in computers, this is true for right shift and not for left shift operation. For example, if a register is shifted right arithmetically, the most significant bit (MSB) of the register is preserved, thus ensuring that the sign of the number will remain unchanged. This is illustrated next:

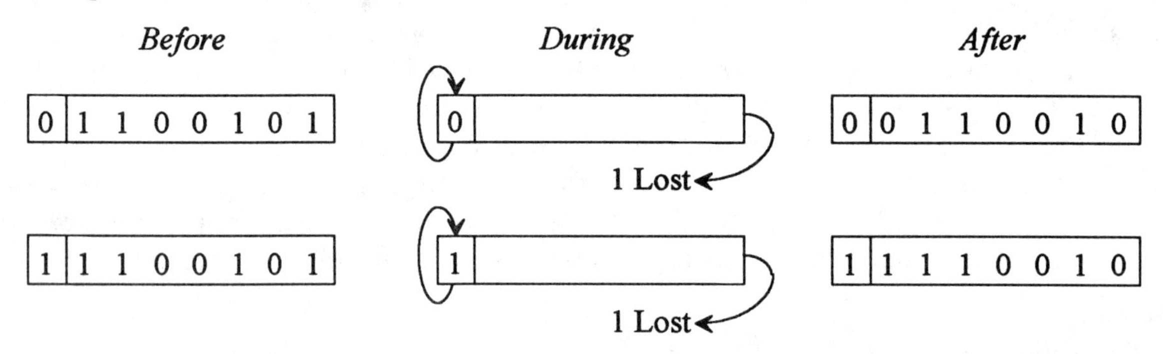

There is no difference between arithmetic and logical left shift operations. If the most significant bit changes from 0 to 1, or vice versa, in an arithmetic left shift, the result is incorrect and the computer sets the overflow flag to 1. For example, if the original value of the register is $(3)_{10}$, the results of two successive arithmetic left shift operations are interpreted as follows:

Original	*After first shift*	*After second shift*
$0011_2 = (3)_{10}$	$0110_2 = (6)_{10}$	$1100_2 = (-4)$
	$3 \times 2 = 6$, correct	$6 \times 2 = 12$ not -4, incorrect

To design a GPR, first let us propose a basic cell *S*. The internal organization of the *S* cell is shown in Figure 5.40. A 4-input multiplexer selects one of the external inputs as the D flip-flop input, and the selected input appears as the flip-flop output *Q* after the clock pulse. The $\overline{CLR}$ input is an asynchronous clear input, and whenever this input is asserted (held low), the flip-flop is cleared to zero. Using the basic cell *S* as the building block, a 4-bit GPR can be designed. Its schematic representation is shown in Figure 5.41.

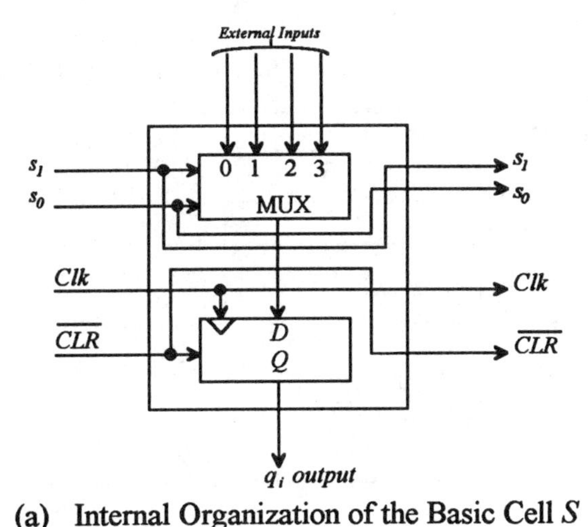

(a) Internal Organization of the Basic Cell S

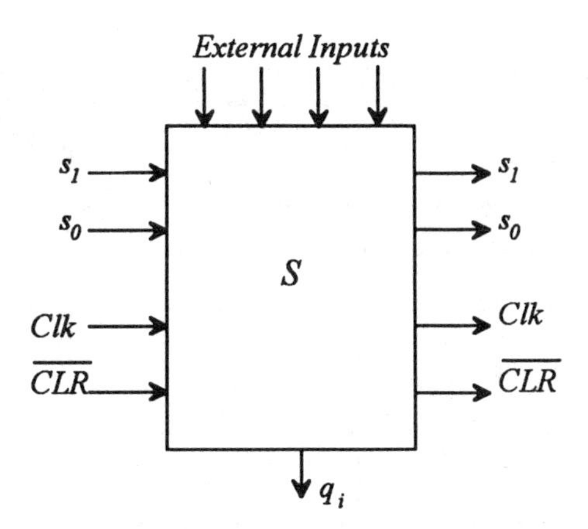

(b) Block Diagram of the Basic Cell S

FIGURE 5.40 A Basic Cell for Designing a GPR

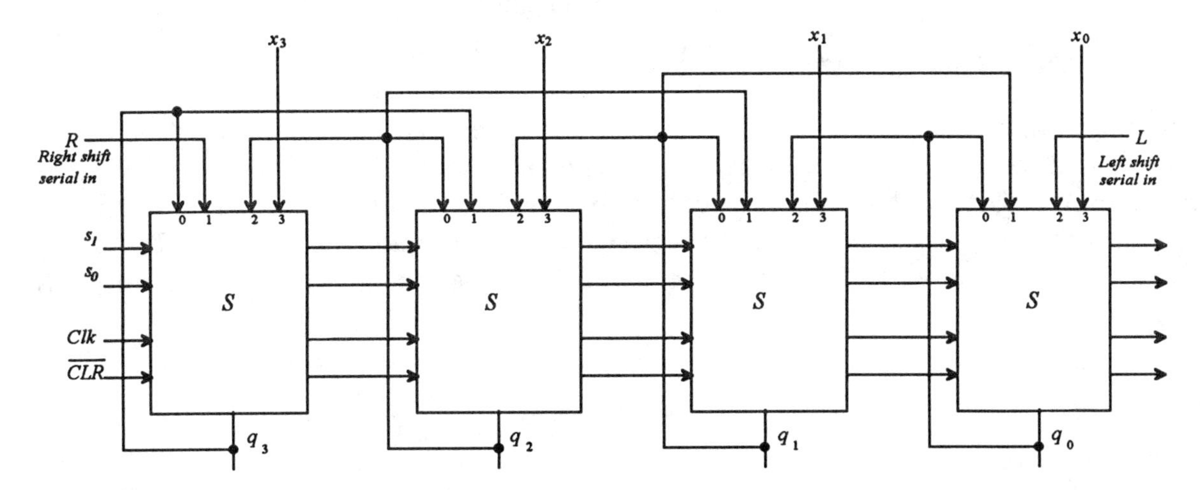

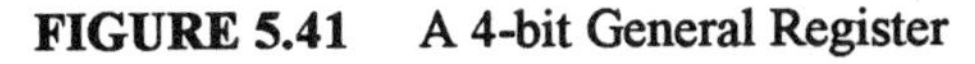
FIGURE 5.41 A 4-bit General Register

The truth table illustrating the operation of this register is shown in Table 5.13. This table shows that manipulation of the selection inputs S_1 and S_0 = 11, the external inputs x_3 through x_0 are selected as the D inputs for the flip-flop, the output q_i will follow the input x_i after the clock. By choosing the correct values for the serial shift inputs R and L, logical, arithmetic, or rotating shifts can be achieved.

This register can be loaded with any desired data in a serial fashion. For example, after four successive right shift operations, data a_3 a_2 a_1 a_0 will be loaded into the register if the register is set in the right shift mode and the required data a_3 a_2 a_1 a_0 is applied serially to input R.

TABLE 5.13 Truth Table for the General Register

Selection Input		Clock input	Clear Input	Operation
s_1	s_0	*Clk*	$\overline{CLR}$	
X	X	X	0	Clear
0	0	↑	1	No Operation
0	1	↑	1	Shift Right
1	0	↑	1	Shift Left
1	1	↑	1	Parallel Load

X means "don't care"

5.11.2 Modulo-*n* Counters

The modulo-*n* counter counts in a sequence and then repeats the count. Modulo-*n* counters can be used to generate timing signals in a computer. The control unit inside the CPU of a computer translates instructions. The control unit utilizes timing signals that determines the time sequences in which the operations required by an instruction are executed. These timing signals shown in Figure 5.42 can be generated by a special modulo-*n* counter called the ring counter.

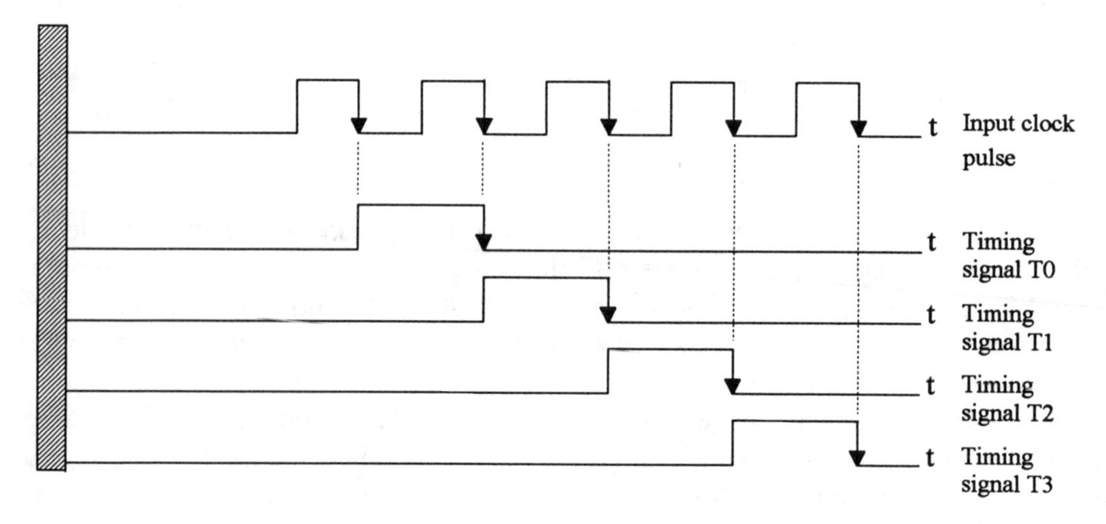

FIGURE 5.42 Timing Signals

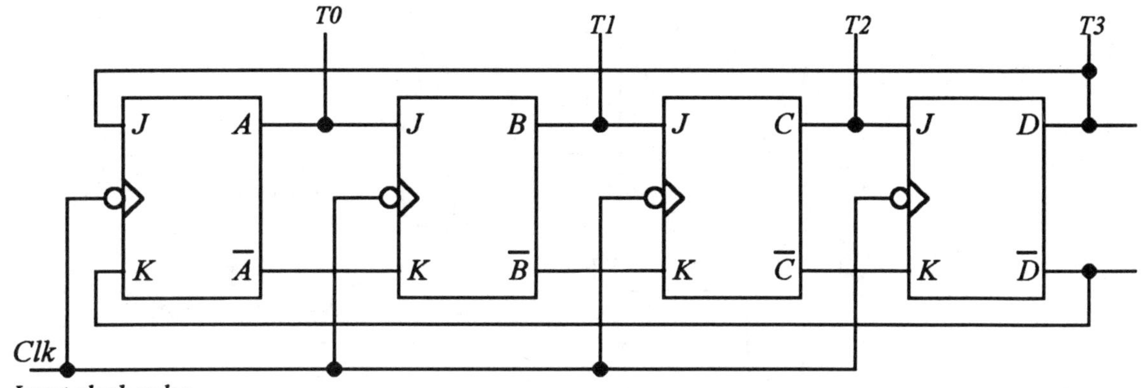

FIGURE 5.43 Ring Counter

An n-bit ring counter transfers a single bit among the flip-flops to provide n unique states. Figure 5.43 shows a 4-bit ring counter. Note that the ring counter requires no decoding but contains n flip-flops for a n-bit ring counter. This system will sequence in the following manner:

Present State				*Next State*			
A	B	C	D	A^+	B^+	C^+	D^+
1	0	0	0	0	1	0	0
0	1	0	0	0	0	1	0
0	0	1	0	0	0	0	1
0	0	0	1	1	0	0	0

This circuit is also known as a *circular shift register,* because the least significant bit shifted is not lost. The Boolean equations for each timing variable are derived by inspection as follows:

$$T_0 = A; \qquad T_1 = B; \qquad T_2 = C; \qquad T_3 = D$$

The main advantages of this circuit are design simplicity and the ability to generate timing signals without a decoder. Nevertheless, n flip-flops are required to generate n timing signals. This approach is not economically feasible for large values of n. To generate timing signals economically, a new approach is used. A modulo-2^n counter is first designed using n flip-flops. The n outputs from this counter are then connected to a n-to-2^n decoder as inputs to generate 2^n timing signals. The circuit depicted in Figure 5.44 shows how to generate four timing signals using a modulo-4 counter and a 2-to-4 decoder.

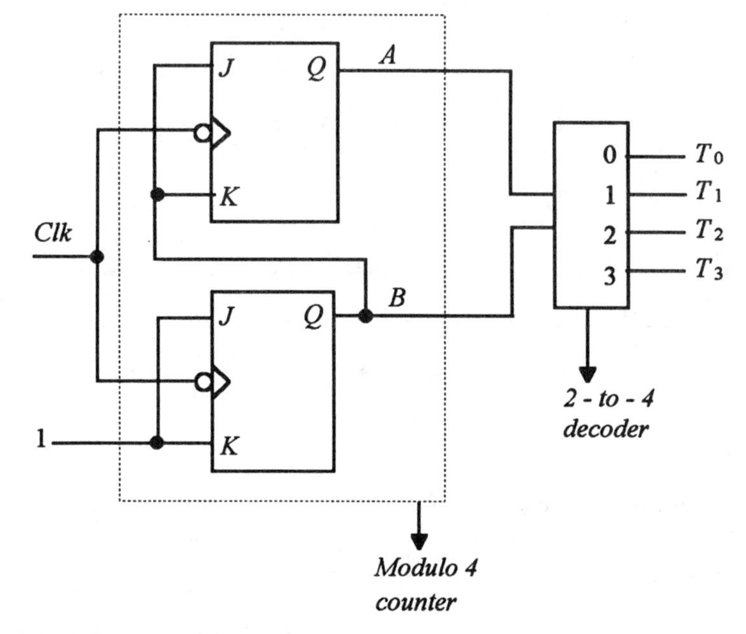

FIGURE 5.44 Modulo-4 Counter with a Decoder

In the preceding circuit, the Boolean equation for each timing signal can be derived as

$$T_0 = \overline{A}\,\overline{B}$$

$$T_1 = \overline{A}B$$

$$T_2 = A\overline{B}$$

$$T_3 = AB$$

These equations show that four 2-input AND gates are needed to derive the timing signals (assuming single-level decoding). The main advantage of this approach is that 2^n timing signals using only n flip-flops are generated. In this method, though, 2^n (n-input) AND gates are required to decode the n-bit output from the flip-flops into 2^n different timing signals. Yet the ring counter approach requires 2^n flip-flops to accomplish the same task.

Typical modulo-n counters provide trade-offs between the number of flip-flops and the amount of decoding logic needed. The binary counter uses the minimum number of flip-flops but requires a decoder. On the other hand, the ring counter uses the maximum number of flip-flops but requires no decoding logic. The Johnson counter (also called the Switch-tail counter or the Mobius counter) is very similar to a ring counter except that the $\overline{Q}$ output of the right-hand flip-flop is connected to the J input of the leftmost flip-flop. Also, the Q input of the rightmost flip-flop is connected to the K input of the leftmost flip-flop.

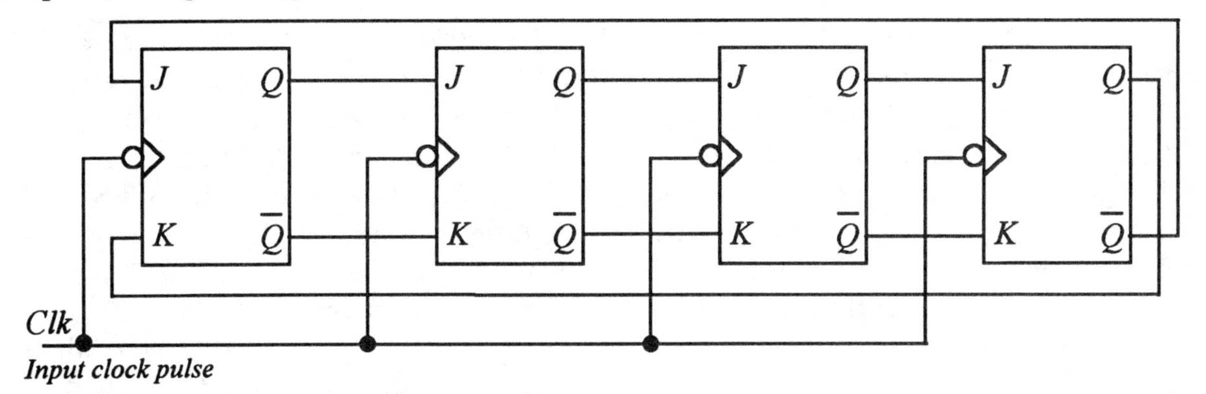

FIGURE 5.45 Four-bit Johnson Counter

Figure 5.45 shows a 4-bit Johnson counter. A Johnson counter requires the same hardware as a ring counter of the same size but can represent twice as many states. Assume that the flip-flops are initialized at 1000. The counter will count in the sequence 1000, 1100, 1110, 1111, 0111, 0011, 0001, 0000, 1000, and repeat.

5.11.3 Random-Access Memory (RAM)

As mentioned before, a RAM is read/write volatile memory. RAM can be classified into two types: static RAM (SRAM) and dynamic RAM (DRAM). A static RAM stores each bit in a flip-flop whereas the dynamic RAM stores each bit as charge in a capacitor. As long as power is available, the static RAM retains information. Because the capacitor can hold charge for a few milliseconds, the dynamic RAM must be refreshed every few milliseconds. This means that a circuit must rewrite that stored bit in a dynamic RAM every few milliseconds. Let us now discuss a typical SRAM implementation using D flip-flops. Figure 5.46 shows a typical RAM cell.

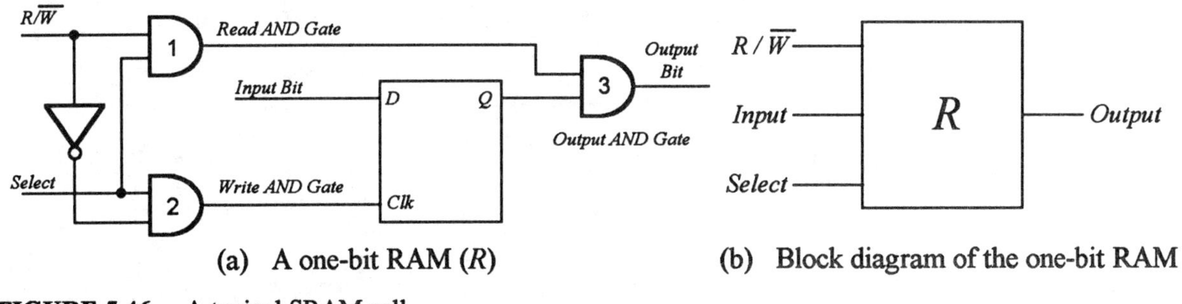

(a) A one-bit RAM (*R*)

(b) Block diagram of the one-bit RAM

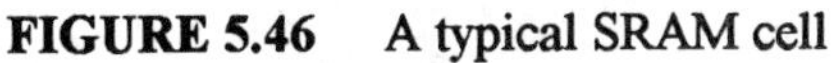
FIGURE 5.46 A typical SRAM cell

In Figure 5.46(a), $R/\overline{W}$ = 1 means READ whereas $R/\overline{W}$ = 0 indicates a WRITE operation. Select = 1 indicates that the one-bit RAM is selected. In order to read the cell, $R/\overline{W}$ is 1 and select = 1. A 1 appears at the input of AND gate 3. This will transfer Q to the output. This is a READ operation. Note that the inverted $R/\overline{W}$ to the input of AND gate 2 is 0. This will apply a 0 at the input of the CLK input of the D flip-flop. The output of the D flip-flop is unchanged. In order to write into the one-bit RAM, $R/\overline{W}$ must be zero. This will apply a 1 at the input of AND gate 2. The output of AND gate 2 (CLK input) is 1. The D input is connected to the value of the bit (1 or 0) to be written into the one-bit RAM. With CLK = 1, the input bit is transferred at the output. The one-bit RAM is, therefore, written into with the input bit. Figure 5.47 shows a 4 × 2 RAM. It includes 8 RAM cells providing 2-bit output and 4 locations.

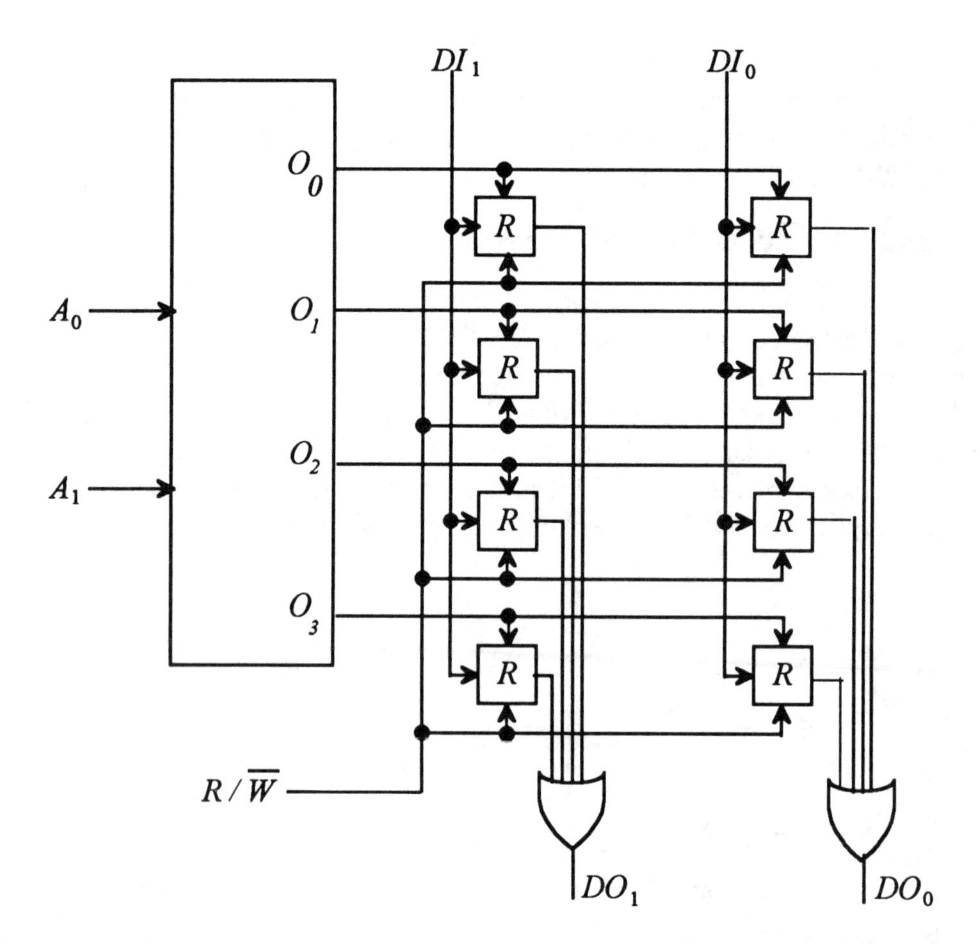

FIGURE 5.47 4 × 2 RAM

The RAM contains a 2 × 4 decoder and 8 RAM cells implemented with D flip-flops and gates. In contrast, a ROM consists of a decoder and OR gates. The four locations (00, 01, 10, 11) in the RAM are addressed by 2 bits (A_1, A_0). In order to read from location 00, the address A_1A_0 = 00 and $R/\overline{W}$ = 1. The decoder selects O_0 high. $R/\overline{W}$ = 1 will apply 0 at the clock inputs of the two RAM cells of the top row and will apply 1 at the inputs of the output AND gates, thus transferring the outputs of the two D flip-flops to the inputs of the two OR gates. The other inputs of the OR gate will be 0. Thus, the outputs of the two RAM cells of the top row will be transferred to DO_1 and DO_0, performing a READ operation. On the other hand, consider a WRITE operation: The 2-bit data to be written is presented at DI_1 DI_0. Suppose A_1A_0 = 00. The top row is selected (O_0 = 1). Input bits at DI_1 and DI_0 will respectively be applied at the inputs of the D flip-flops of the top row. Because $R/\overline{W}$ = 0, the clock inputs of both the D flip-flops of the top row are 1; thus, the *D* inputs are transferred to the outputs of the flip-flops. Therefore, data at DI_1 DI_0 will be written into the RAM.

5.12 Algorithmic State Machines (ASM) Chart

The performance of a synchronous sequential circuit (also referred to as a state machine) can be represented in a systematic way by using a flowchart called the Algorithmic State Machines (ASM) chart. This is an alternative approach to the state diagram. In the previous sections, it was shown how state diagrams could be used to design synchronous sequential circuit. An ASM chart can sometimes be used along with the state diagram for designing a synchronous sequential circuit. An ASM chart is similar to a flowchart for a computer program. The main difference is that the flowchart for a computer program is translated into software whereas an ASM chart is used to implement hardware. An ASM chart specifies the sequence of operations of the state machine along with the conditions required for their execution. Three symbols are utilized to develop the ASM chart: the state symbol, the decision symbol, and the conditional output symbol (see Figure 5.48).

The ASM chart utilizes one state symbol for each state. The state symbol includes the state name, binary code assignment, and outputs (if any) that are asserted during the specified state. The decision symbol indicates testing of an input and then going to an exit if the condition is true and to another exit if the condition is false. The entry of the conditional output symbol is connected to the exit of the decision symbol.

The ASM chart and the state diagram are very similar. Each state in a state diagram is basically similar to the state symbol. The decision symbol is similar to the binary information written on the lines connecting two states in a state diagram. Figure 5.49 shows an example of a ASM chart for a modulo-7 counter (counting the sequence 000, 001, ..., 111 and repeat) with an enable input. Q_2, Q_1, and Q_0 at the top of the ASM chart represent the three flip-flop states for the 3-bit counter.

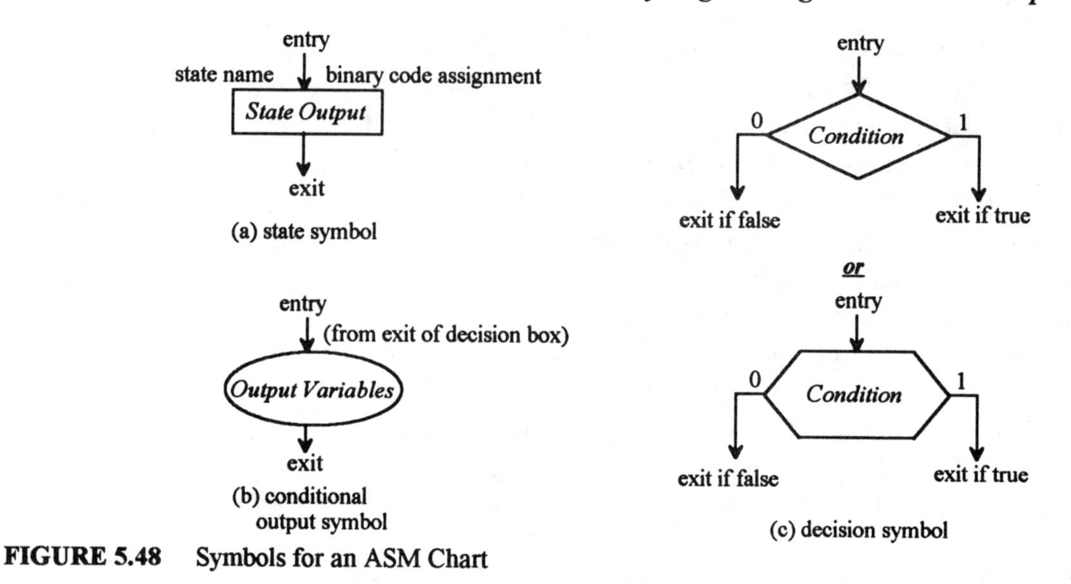

FIGURE 5.48 Symbols for an ASM Chart

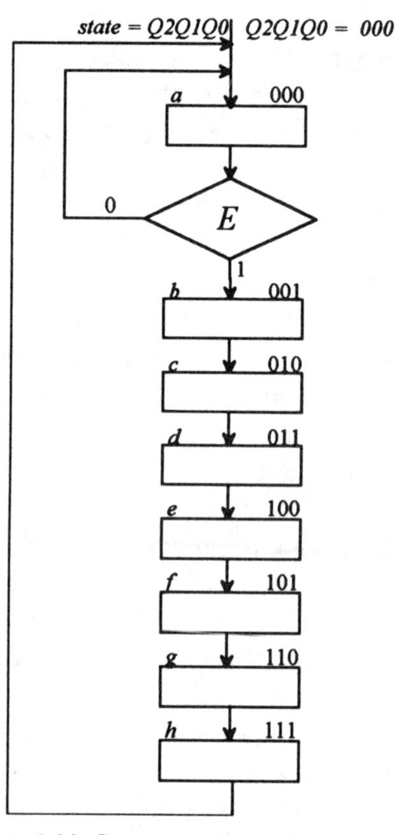

FIGURE 5.49 An ASM Chart for a 3-bit Counter with Enable Input

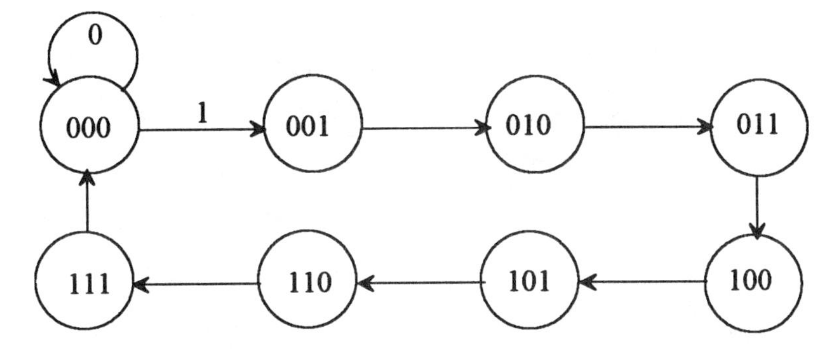

FIGURE 5.50 State Diagram for the 3-bit Counter

Each state symbol is given a symbolic name at the upper left corner along with a binary code assignment of the state at the upper right corner. For example, the state '*a*' is assigned with a binary value of 000. The enable input E can only be checked at state a, and the counter can be stopped if $E = 0$; the counter continues if $E = 1$. This is illustrated by the decision symbol. Figure 5.50 shows the equivalent state diagram of the ASM chart for the 3-bit counter.

The ASM chart describes the sequence of events and the timing relationship between the states of a synchronous sequential circuit and the operations that occur for transition from one state to the next. An arbitrary ASM chart depicted in Figure 5.51 illustrates this. The chart contains three ASM blocks. Note that an ASM block must contain one state symbol and may include any number of decisions and conditional output symbols connected to the exit. The three ASM blocks are the ASM block for T_0 surrounded by the dashed lines and the simple ASM block defined by T_1 and T_2. Figure 5.52 shows the state diagram.

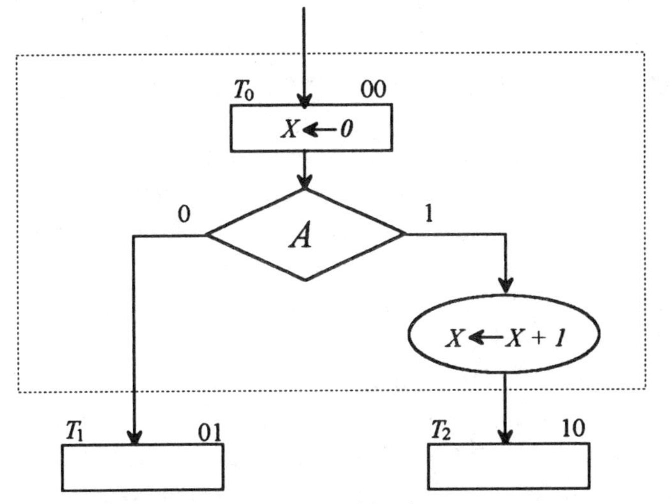

FIGURE 5.51 ASM Chart illustrating timing relationships between states

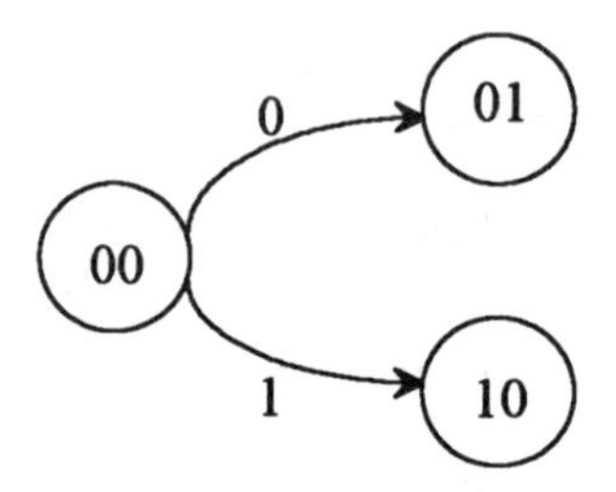

FIGURE 5.52 State Diagram for the ASM Chart of Figure 5.51

From the ASM chart of Figure 5.51, there are three states: T_0, T_1, and T_2. A ring counter can be used to generate these timing signals. During T_0, register X is cleared and flip-flop A is checked. If $A = 0$, the next state will be T_1. On the other hand, if $A = 1$, the circuit increments register X by 1 and then moves to the next state, T_2. Note that the following operations are performed by the circuit during state T_0:

1. Clear register X.
2. Check flip-flop A for 1 or 0.
3. If $A = 1$, increment X by 1.

On the other hand, state machines do not perform any operations during T_1 and T_2. Note that in contrast, state diagrams do not provide any timing relationship between states. ASM charts are utilized in designing the controller of digital systems such as the control unit of a CPU. It is sometimes useful to convert an ASM chart to a state diagram and then utilize the procedures of synchronous sequential circuits to design the control logic. *Design of state machines using ASM chart along with examples is provided in Chapter 11.*

5.13 Asynchronous Sequential Circuits

Asynchronous sequential circuits do not require any synchronizing clocks. As mentioned before, a sequential circuit basically consists of a combinational circuit with memory. In synchronous sequential circuits, memory elements are clocked flip-flops. In contrast, memory in asynchronous sequential circuits includes either unclocked flip-flop or time-delay devices. The propagation delay time of a logic gate (finite time for a signal to propagate through a gate) provides its memory capability. Note that a sequential circuit contains inputs, outputs, and states. In synchronous sequential circuits, changes in states take place due to clock pulses. On the other hand, asynchronous sequential circuits typically contain a combinational circuit with feedback. The timing problems in the feedback may cause instability. Asynchronous sequential circuits are, therefore, more difficult to design than synchronous sequential circuits.

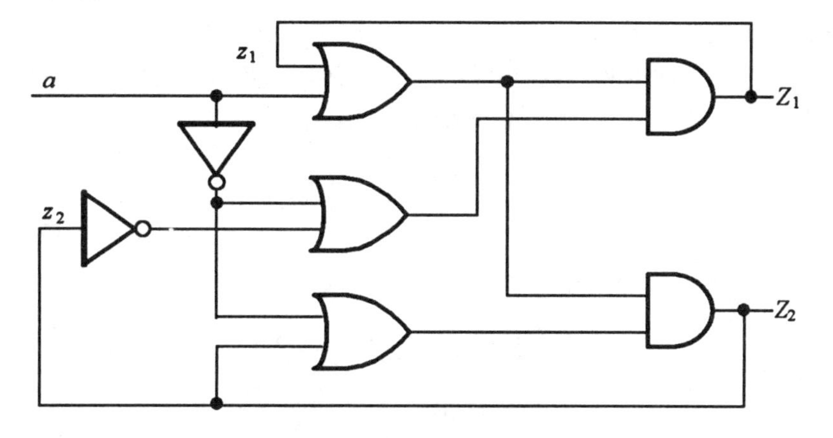

FIGURE 5.53 Asynchronous Sequential Circuit

Asynchronous sequential circuits are used in applications in which the system must take appropriate actions to input changes rather than waiting for a clock to initiate actions. For proper operation of an asynchronous sequential circuit, the inputs must change one at a time when the circuit is in a stable condition (called the "fundamental mode of operation"). The inputs to the asynchronous sequential circuits are called "primary variables" whereas outputs are called "secondary variables."

Figure 5.53 shows an asynchronous sequential circuit. In the feedback loops, the uppercase letters are used to indicate next values of the secondary variables and the lowercase letters indicate present values of the secondary variables. For example, Z_1, and Z_2 are next values whereas z_1 and z_2 are present values. The output equations can be derived as follows:

$$Z_1 = (a + z_1)(\overline{a} + \overline{z_2})$$

$$Z_2 = (a + z_1)(\overline{a} + z_2)$$

The delays in the feedback loops can be obtained from the propagation delays between z_1 and Z_1 or z_2 and Z_2. Let us now plot the functions Z_1 and Z_2 in a map, and a transition table as shown in Figure 5.54.

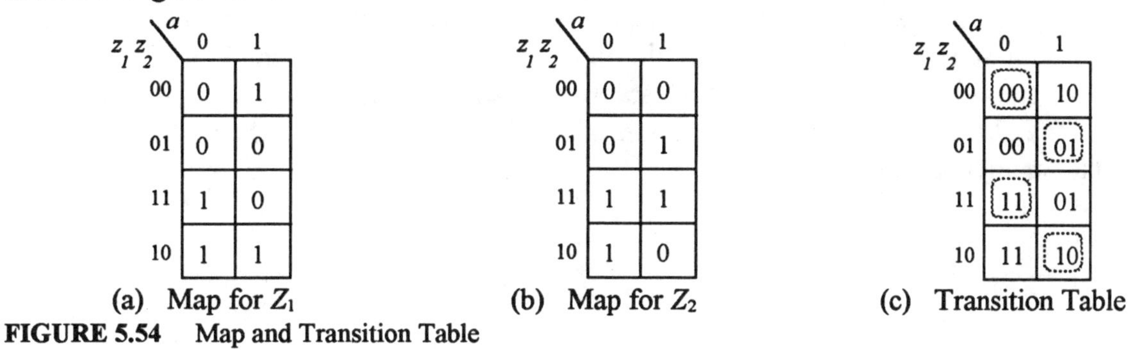

$z_1 z_2$ \ a	0	1
00	0	1
01	0	0
11	1	0
10	1	1

(a) Map for Z_1

$z_1 z_2$ \ a	0	1
00	0	0
01	0	1
11	1	1
10	1	0

(b) Map for Z_2

$z_1 z_2$ \ a	0	1
00	(00)	10
01	00	(01)
11	(11)	01
10	11	(10)

(c) Transition Table

FIGURE 5.54 Map and Transition Table

The map for Z_1 in Figure 5.51(a) is obtained by substituting the values z_1, z_2, and a for each square into the equation for Z1. For example, consider $z_1z_2 = 11$ and $a = 0$.

$$Z_1 = (a + z_1)(\bar{a} + \bar{z_2})$$

$$= (0 + 1)(\bar{0} + \bar{1})$$

$$= 1$$

$$Z_2 = (a + z_1)(\bar{a} + z_2)$$

$$= (0 + 1)(\bar{0} + 1)$$

$$= 1$$

Similarly, values for all other sequences can be obtained similarly. The transition table of Figure 5.54(c) can be obtained by combining the binary values of two squares in the same position and placing them in the corresponding square in the transition table. Thus, the variable $Z = Z_1Z_2$ is placed in each square of the transition table. For example, from the first square of Figure 5.51(a) and (b), $Z = 00$. This is shown in the first square of Figure 5.54(c). The squares in the transition table in which $z_1z_2 = Z_1Z_2$ are circled to show that they are stable. The uncircled squares are unstable states.

Let us now analyze the behavior of the circuit due to change in the input variable. Suppose $a = 0$, $z_1z_2 = 00$, then the output is 00. Thus, 00 is circled and shown in the first square of Figure 5.54(c). Z is the next value of z_1z_2 and is a stable state. Next suppose that a goes from 0 to 1 and the value of Z changes from 00 to 01. Note that this causes an interim unstable situation because Z_1Z_2 is initially equal to z_1z_2. This is because as soon as the input changes from 0 to 1, this change in input travels through the circuit to change Z_1Z_2 from 00 to 01. The feedback loop in the circuit eventually makes z_1z_2 equal to Z_1Z_2; that is, $z_1z_2 = Z_1Z_2 =$ 01. Because $z_1z_2 = Z_1Z_2$, the circuit attains a stable state. The state 01 is circled in the figure to indicate this. Similarly, it can be shown that as the input to an asynchronous sequential circuit changes, the circuit goes to a temporary unstable condition until it reaches a stable state when Z_1Z_2 = present state, z_1z_2. Therefore, as the input moves between 0 and 1, the circuit goes through the states 00, 01, 11, 10, and repeats the sequence depending on the input changes. A state table can be derived from the transition table. This is shown in Table 5.14, which is the state table for Figure 5.54(c).

TABLE 5.14 Transition Table

Present State		Next State			
		a=0		*a=1*	
0	0	0	0	1	0
0	1	0	0	0	1
1	0	1	1	1	0
1	1	1	1	0	1

A flow table obtained from the transition table is normally used in designing an asynchronous sequential circuit. A flow table resembles a transition table except that the states are represented by letters instead of binary numbers. The transition table of Figure 5.54(c) can be translated into a flow table as shown in Figure 5.55. Note that the states are represented by binary numbers as follows: w = 00, x = 01, y = 11, z = 10. The flow table in Figure 5.55 is called a "primitive flow table" because it has only one stable state in each row.

An asynchronous sequential circuit can be designed using the primitive flow table from the problem definition. The flow table is then simplified by combining squares to a minimum number of states. The transition table is then obtained by assigning binary numbers to the states. Finally, a logic diagram is obtained from the transition table. The logic diagram includes a combinational circuit with feedback.

The design of an asynchronous sequential circuit is more difficult than the synchronous sequential circuit because of the timing problems associated with the feedback loop. This topic is beyond the scope of this book.

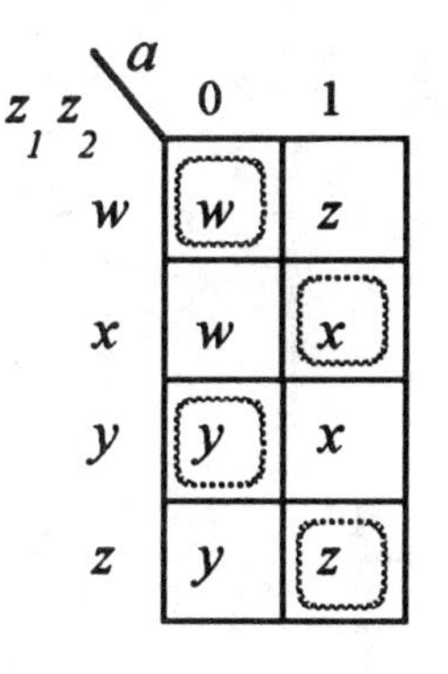

FIGURE 5.55 Flow Table

QUESTIONS AND PROBLEMS

5.1 What is the basic difference between a combinational circuit and a sequential circuit?

5.2 Identify the main characteristics of a synchronous sequential circuit and an asynchronous sequential circuit.

5.3 What is the basic difference between a latch and a flip-flop?

5.4 Draw the logic diagram of a D flip-flop using OR gates and inverters.

5.5 Assume that initially $x = 1$, $A = 0$, and $B = 1$. Determine the values of A and B after the leading edge of *Clk*.

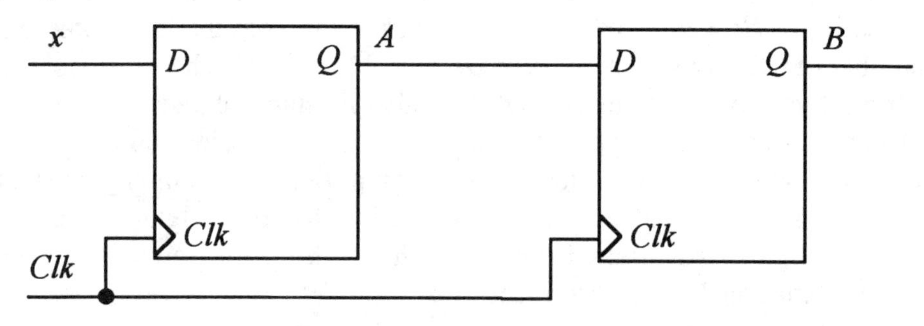

FIGURE P5.5

5.6 Draw the logic diagram of a JK flip-flop using AND gates and inverters.

5.7 Assume that initially $X = 1$, $A = 0$, and $B = 1$. Determine the values of A and B after one *Clk* pulse. Note that the flip-flops are triggered at the clock level.

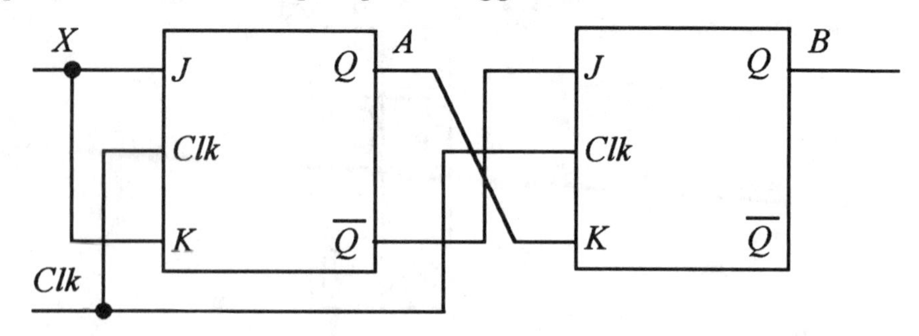

FIGURE P5.7

5.8 Given Figure P5.8, draw the timing diagram for Q and $\overline{Q}$ assuming a trailing-edge-triggered JK flip-flop. Assume Q is preset to 1 initially.

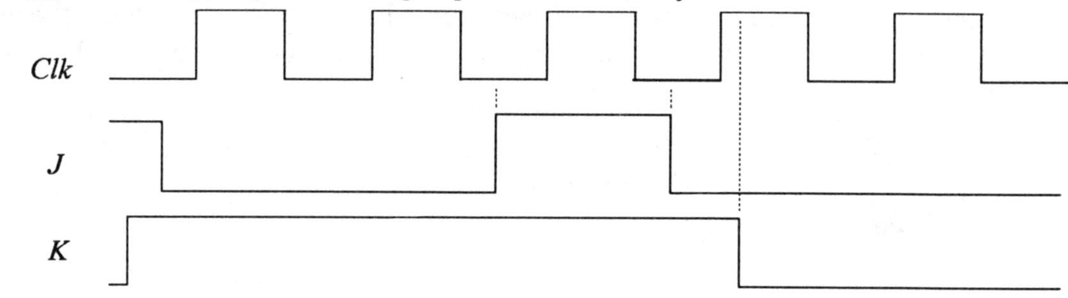

FIGURE P5.8

5.9 Given the timing diagram for a leading-edge triggered D flip-flop in Figure P5.9, draw the timing diagrams for Q and $\overline{Q}$. Assume Q is cleared to zero initially.

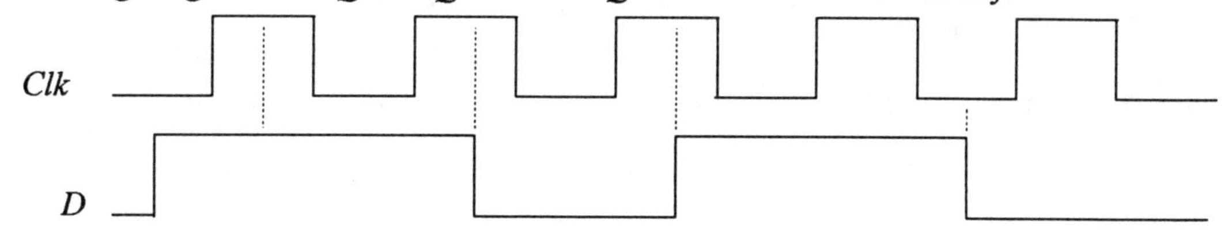

FIGURE P5.9

5.10 Given the timing diagram for a trailing-edge triggered T flip-flop in Figure P5.10, draw the timing diagram for Q. Assume Q is preset to 1 initially.

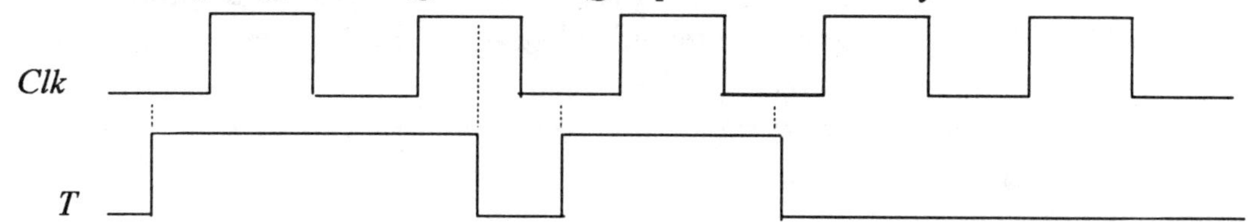

FIGURE P5.10

5.11 Why would you use an edge-triggered flip-flop rather than a level-triggered flip-flop?

5.12 What are the advantages of a master–slave flip-flop?

5.13 Draw the block diagram of a commercial T flip-flop.

5.14 Draw a logic circuit of the switch debouncer circuit using NAND gates.

5.15 Analyze the clocked synchronous circuit shown in Figure P5.15. Express the next state in terms of the present state and inputs, derive the state table, and draw the state diagram.

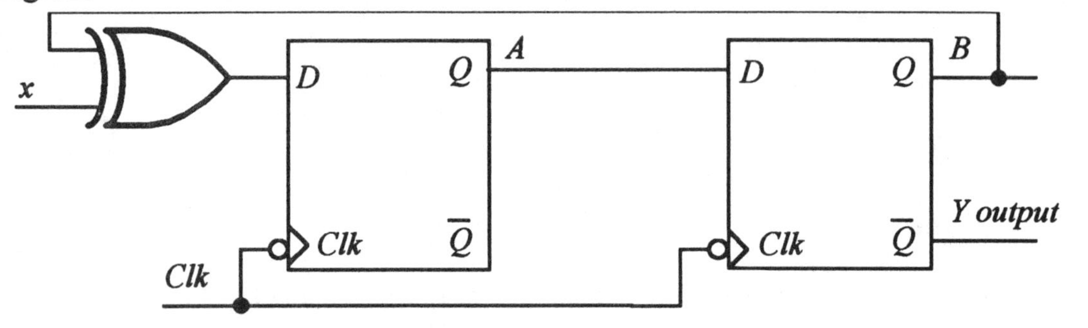

FIGURE P5.15

5.16 A synchronous sequential circuit with two D flip-flops *(a,b as outputs)*, one input *(x)*, and an output *(y)* is expressed by the following equations:

$$D_a = a\bar{b}x + \bar{a}b, \quad D_b = \bar{x}b + \bar{b}x$$

$$y = \bar{b}\,\bar{x} + a$$

(a) Derive the state table and state diagram for the circuit.
(b) Draw a logic diagram.

5.17 A synchronous sequential circuit is represented by the state diagram shown in Figure P5.17. Using JK flip-flops and undefined states as don't-cares:
(a) Derive the state table.
(b) Minimize the equation for flip-flop inputs using K-maps.
(c) Draw a logic diagram.

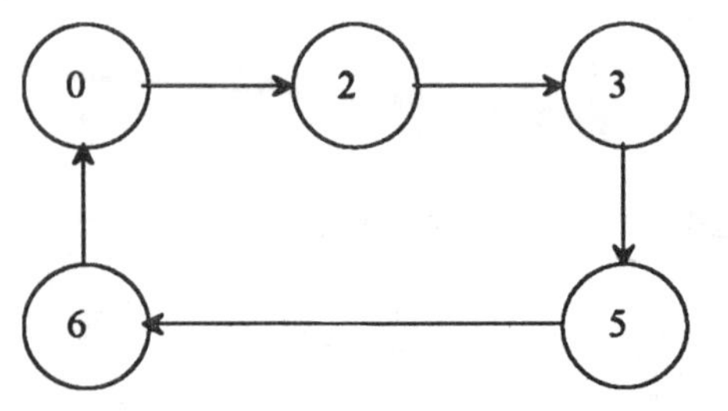

FIGURE P5.17

5.18 A sequential circuit contains two D flip-flops (A, B), one input (x), and one output (y), as shown in Figure P5.18.
Derive the state table and the state diagram of the sequential circuit.

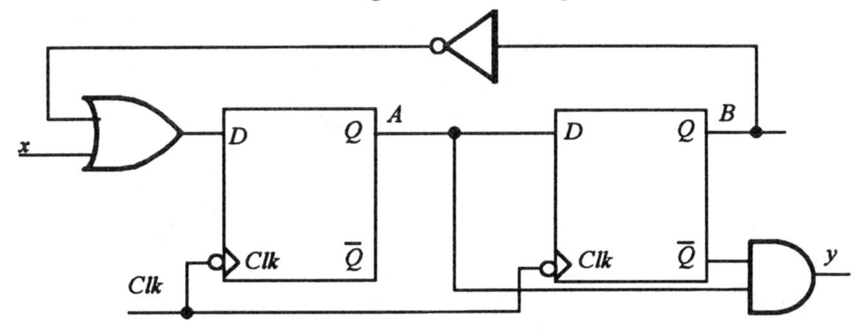

FIGURE P5.18

5.19 Design a synchronous sequential circuit using D flip-flops for the state diagram shown in Figure P5.19.

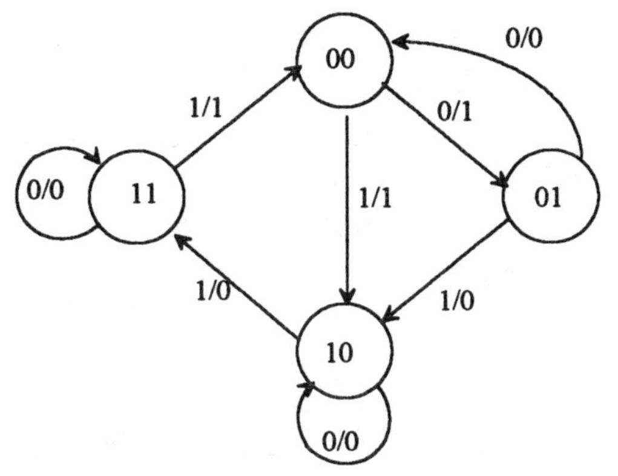

FIGURE P5.19

5.20 Design a 2-bit counter that will count in the following sequence: 00, 11, 10, 01, and repeat. Using T flip-flops:
(a) Draw a state diagram.
(b) Derive a state table.
(c) Implement the circuit.

5.21 Design a synchronous sequential circuit with one input x and one output y. The input x is a serial message, and the system reads x one bit at a time. The output y is 1 whenever the binary pattern 000 is encountered in the serial message. For example: If the input is 01000000, then the output will be 00001010. Use T flip-flops.

5.22 Analyze the circuit shown in Figure P5.22 and show that it is equivalent to a T flip-flop.

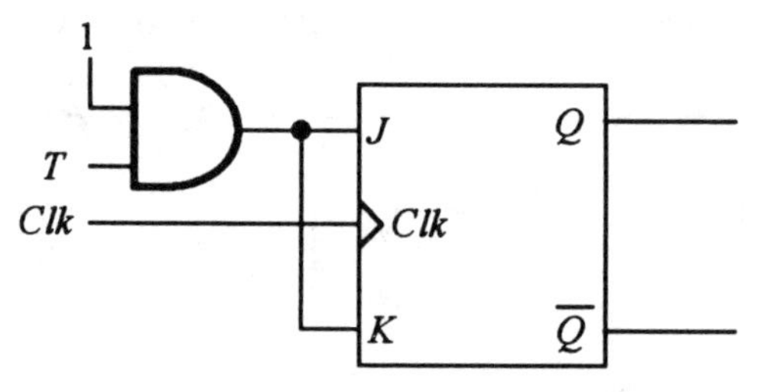

FIGURE P5.22

5.23 Design a BCD counter to count in the sequence 0000, 0001, 0010, 0011, 0100, 0101, 0110, 0111, 1000, 1001, and repeat. Use T flip-flops.

5.24 Design the following nonbinary sequence counters using the type of flip-flop specified. Assume the unused states as don't cares. Your design must produce a self-correcting counter.

(a) Design a nonbinary sequence counter with the sequence 0, 1, 3, 4, 5, 6, 7, and repeat. Use JK flip-flops.

(b) Design a nonbinary sequence counter with the sequence 0, 2, 3, 4, 6, 7, and repeat. Use D flip-flops.

(c) Design a nonbinary sequence counter with the sequence 0, 1, 2, 4, 5, 6, 7, and repeat. Use T flip-flops.

5.25 Design a 4-bit general-purpose register as follows:

S_1	S_0	Function
0	0	Load external data
0	1	Rotate left; ($A_0 \leftarrow A_3$, $A_i \leftarrow A_{i-1}$ for i = 1,2,3)
1	0	Rotate right; ($A_3 \leftarrow A_0$, $A_i \leftarrow A_{i+1}$ for i = 0,1,2)
1	1	Increment

Use Figure P5.25 as the building block:

FIGURE P5.25

5.26 Design a logic diagram that will generate 19 timing signals. Use a ring counter with JK flip-flops.

5.27 Consider the 2-bit Johnson counter shown in Figure P5.27. Derive the state diagram. Assume the D flip-flops are initialized to $A = 0$ and $B = 0$.

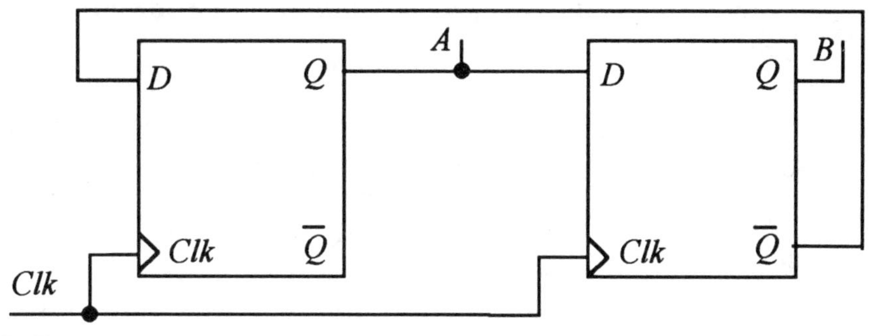

FIGURE P5.27

5.28 Assuming $AB = 10$, verify that the 2-bit counter shown in Figure P5.28 is a ring counter. Derive the state diagram.

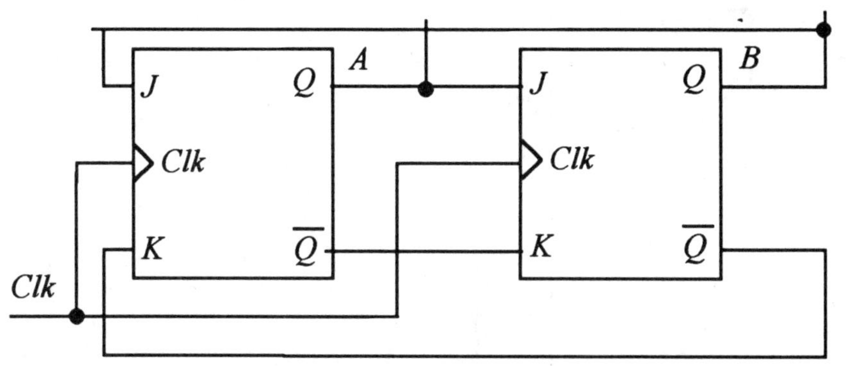

FIGURE P5.28

5.29 What is the basic difference between SRAM and DRAM?

5.30 Given a memory with a 24-bit address and 8-bit word size,
(a) How many bytes can be stored in this memory?
(b) If this memory were constructed from 1K × 1-bit RAM chips, how many memory chips would be required?

5.31 Draw an ASM chart for the following: Assume three states (a, b, c) in the system with one input x and two registers R_1 and R_2. The circuit is initially in state a. If $x = 0$, the control goes from state a to state b and, clears registers R_1 to 0 and sets R_2 to 1, and then moves to state c.

On the other hand if $x = 1$, the control goes to state c. In state c, R_1 is subtracted from R_2 and the result is stored in R_1. The control then moves back to state a and the process continues.

5.32 Derive the output equations for the asynchronous sequential circuit shown in Figure P5.32. Also, determine the state table and flow table.

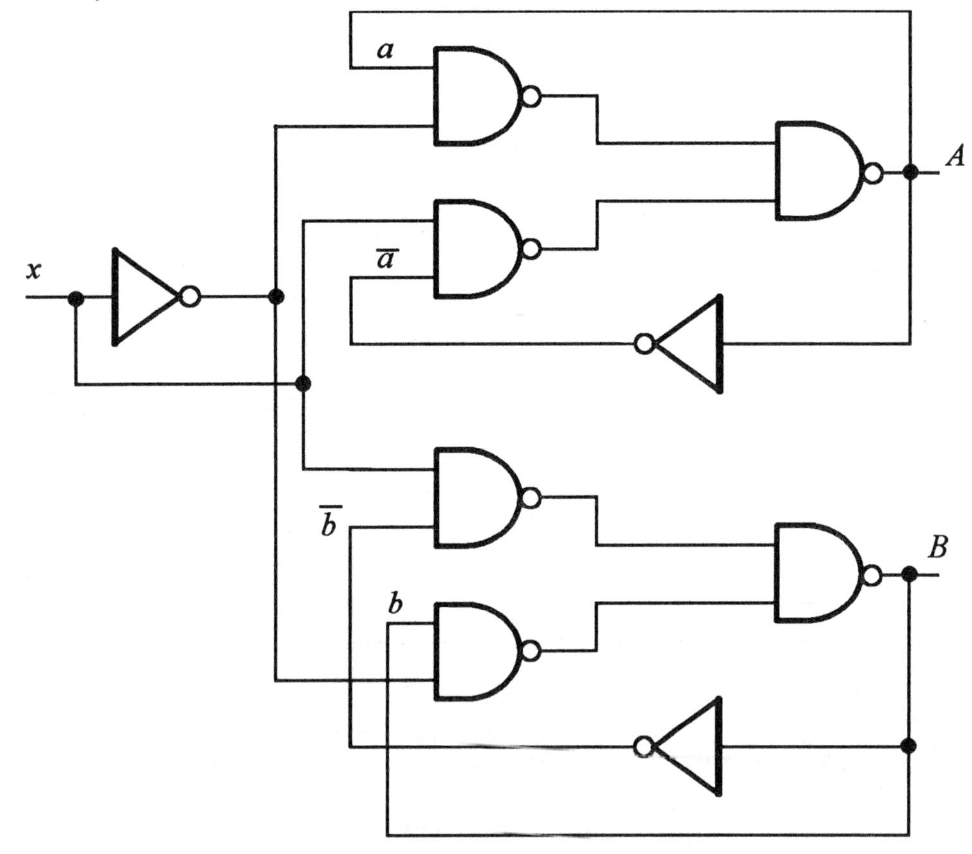

FIGURE P5.32

6

MICROCOMPUTER ARCHITECTURE, PROGRAMMING, AND SYSTEM DESIGN CONCEPTS

This chapter describes the fundamental material needed to understand the basic characteristics of microprocessors. It includes topics such as typical microcomputer architecture, timing signals, internal microprocessor structure, and status flags. The architectural features are then compared to the Intel 8086 architecture. Finally, microcomputer programming languages and system design concepts are described.

6.1 Basic Blocks of a Microcomputer

A microcomputer has three basic blocks: a central processing unit (CPU), a memory unit, and an input/output unit. The CPU *executes all the instructions and performs arithmetic and logic operations on data. The CPU of the microcomputer is called the "microprocessor."* The MOS microprocessor is typically a single VLSI (Very Large-Scale Integration) chip that contains all the registers, control unit, and arithmetic/ logic circuits of the microcomputer.

A memory unit *stores both data and instructions. The memory section typically contains ROM and RAM chips. The ROM can only be read and is nonvolatile, that is, it retains its contents when the power is turned off.* A ROM is typically used to store instructions and data that do not change. For example, it might store a table of codes for outputting data to a display external to the microcomputer for turning on a digit from 0 to 9.

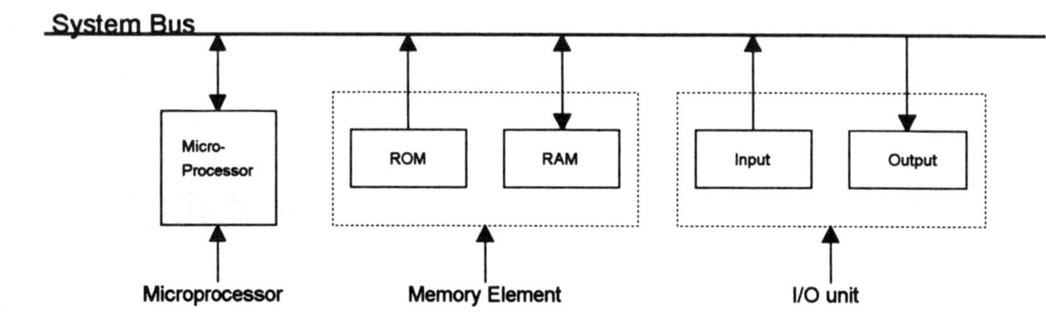

FIGURE 6.1 Basic blocks of a microcomputer

One can read from and write into a RAM. *The RAM is volatile; that is, it does not retain its contents when the power is turned off.* A RAM is used to store programs and data that are temporary and might change during the course of executing a program. An *I/O unit* transfers data between the microcomputer and the external devices. The transfer involves data, status, and control signals.

In a single-chip microcomputer, these three elements are on one chip, whereas in a single-chip microprocessor, separate chips for memory and I/O are required. *Single-chip microcomputers are also referred to as "microcontrollers." The microcontrollers are typically used for dedicated applications such as automotive systems, home appliances, and home entertainment systems. Typical microcontrollers, therefore, include on-chip timers and A/D (analog to digital) and D/A (digital to analog) converters.* Two popular microcontrollers are the Intel 8751 (8 bit)/8096 (16 bit) and the Motorola HC11 (8 bit)/HC16 (16 bit). The 16-bit microcontrollers include more on-chip ROM, RAM, and I/O than the 8-bit microcontrollers. Figure 6.1 shows the basic blocks of a microcomputer. The System bus (comprised of several wires) connects these blocks.

6.2 Typical Microcomputer Architecture

In this section, we describe the microcomputer structure in more detail. The various microcomputer architectures available today are basically the same in principle. The main variations are in the number of data and address bits and in the types of control signals they use.

To understand the basic principles of microcomputer architecture, it is necessary to investigate a typical microcomputer in detail. Once such a clear understanding is obtained, it will be easier to work with any specific microcomputer. Figure 6.2 illustrates the most simplified version of a typical microcomputer. The figure shows the basic blocks of a microcomputer system. The various buses that connect these blocks are also shown. Although this figure looks very simple, it includes all the main elements of a typical microcomputer system.

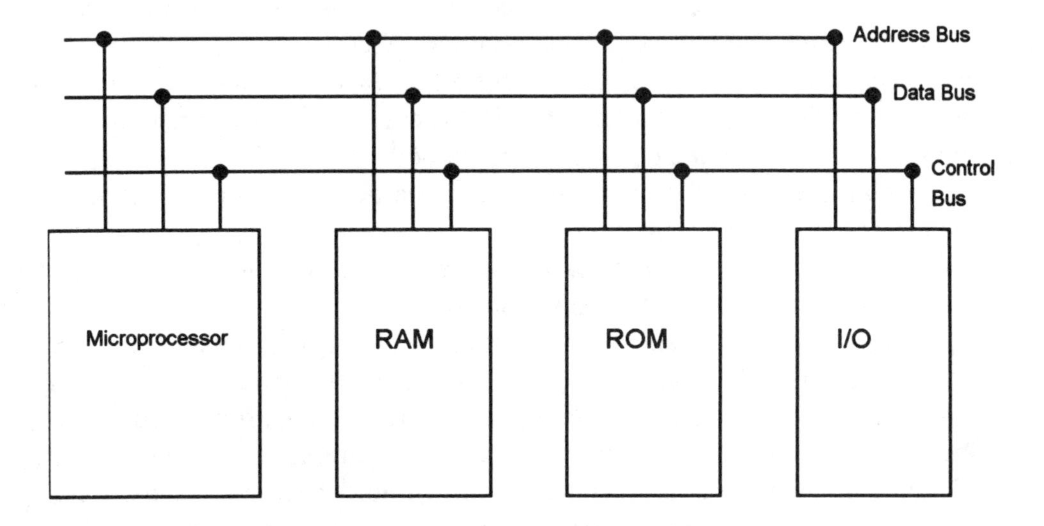

FIGURE 6.2 Simplified version of a typical microcomputer structure

6.2.1 The Microcomputer Bus

The microcomputer's system bus contains three buses, which carry all the address, data, and control information involved in program execution. These buses connect the microprocessor (CPU) to each of the ROM, RAM, and I/O elements so that information transfer between the microprocessor and any of the other elements can take place.

In the microcomputer, most information transfers are carried out with respect to the memory. When the memory is receiving data from another microcomputer element, it is called a WRITE operation, and data is written into a selected memory location. When the memory is sending data to another microcomputer element, it is called a READ operation, and data is being read from a selected memory location.

In the *address bus*, information transfer takes place only in one direction, from the microprocessor to the memory or I/O elements. Therefore, this is called a "unidirectional bus." This bus is typically 20 to 32 bits long. *The size of the address bus determines the total number of memory addresses available in which programs can be executed by the microprocessor. The address bus is specified by the total number of address pins on the microprocessor chip. This also determines the direct addressing capability or the size of the main memory of the microprocessor. The microprocessor can only execute the programs located in the main memory*. For example, a microprocessor with 20 address pins can generate 2^{20} = 1,048,576 (one megabyte) different possible addresses (combinations of 1's and 0's) on the address bus. The microprocessor includes addresses from 0 to 1,048,575 (00000_{16} through

$FFFFF_{16}$). A memory location can be represented by each one of these addresses. For example, an 8-bit data item can be stored in an address 00200_{16}.

When a microprocessor such as the 8086 wants to transfer information between itself and a certain memory location, it generates the 20-bit address from an internal register on its 20 address pins A_0–A_{19}, which then appears on the address bus. These 20 address bits are decoded to determine the desired memory location. The decoding process normally requires hardware (decoders) not shown in Figure 6.2.

In the *data bus*, data can flow in both directions, that is, to or from the microprocessor. Therefore, this is a bidirectional bus. In some microprocessors, the data pins are used to send other information such as address bits in addition to data. This means that the data pins are time-shared or multiplexed. The Intel 8086 microprocessor is an example where the 20 bits of the address are multiplexed on the 16-bit data bus and four status lines.

The *control bus* consists of a number of signals that are used to synchronize the operation of the individual microcomputer elements. The microprocessor sends some of these control signals to the other elements to indicate the type of operation being performed. Each microcomputer has a unique set of control signals. However, there are some control signals that are common to most microprocessors. We describe some of these control signals later in this section.

6.2.2 Clock Signals

The system clock signals are contained in the control bus. These signals generate the appropriate clock periods during which instruction executions are carried out by the microprocessor. The clock signals vary from one microprocessor to another. Some microprocessors have an internal clock generator circuit to generate a clock signal. These microprocessors require an external crystal or an RC network to be connected at the appropriate microprocessor pins for setting the operating frequency. For example, the Intel 80186 (16-bit microprocessor) does not require an external clock generator circuit. *However, most microprocessors do not have the internal clock generator circuit and require an external chip or circuit to generate the clock signal.* Figure 6.3 shows a typical clock signal.

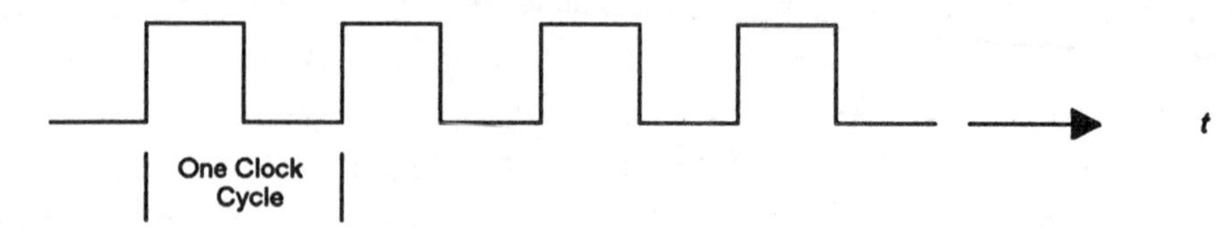

FIGURE 6.3 A typical clock signal

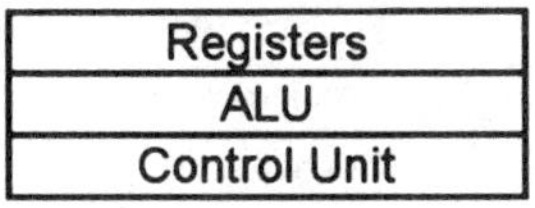

Registers
ALU
Control Unit

FIGURE 6.4 A microprocessor chip with the main functional elements

6.3 The Single-Chip Microprocessor

As mentioned before, the microprocessor is the CPU of the microcomputer. Therefore, the power of the microcomputer is determined by the capabilities of the microprocessor. Its clock frequency determines the speed of the microcomputer. The number of data and address pins on the microprocessor chip make up the microcomputer's word size and maximum memory size. The microcomputer's I/O and interfacing capabilities are determined by the control pins on the microprocessor chip.

The logic inside the microprocessor chip can be divided into three main areas: the register section, the control unit, and the arithmetic and logic unit (ALU). A microprocessor chip with these three sections is shown in Figure 6.4. We now describe these sections.

6.3.1 Register Section

The number, size, and types of registers vary from one microprocessor to another. However, the various registers in all microprocessors carry out similar operations. The register structures of microprocessors play a major role in designing the microprocessor architectures. Also, the register structures for a specific microprocessor determine how convenient and easy it is to program this microprocessor.

We first describe the most basic types of microprocessor registers, their functions, and how they are used. We then consider the other common types of registers.

6.3.1.1 Basic Microprocessor Registers

There are four basic microprocessor registers: instruction register, program counter, memory address register, and accumulator.

- **Instruction Register (IR).** *The instruction register stores instructions.* The contents of an instruction register are always decoded by the microprocessor as an instruction. After fetching an instruction code from memory, the microprocessor stores it in the instruction register. The instruction is decoded internally by the microprocessor, which then performs the required operation. The word size of the microprocessor determines the size of the instruction register. For example, a 16-bit microprocessor has a 16-bit instruction register.

- **Program Counter (PC).** The program counter contains the address of the instruction or operation code (op-code). *The program counter normally contains the address of the next instruction to be executed.* Note the following features of the program counter:
  1. *Upon activating the microprocessor's RESET input, the address of the first instruction to be executed is loaded into the program counter.*
  2. To execute an instruction, the microprocessor typically places the contents of the program counter on the address bus and reads ("fetches") the contents of this address, that is, instruction, from memory. The program counter contents are automatically incremented by the microprocessor's internal logic. The microprocessor thus executes a program sequentially, unless the program contains an instruction such as a `JUMP` instruction, which changes the sequence.
  3. The size of the program counter is determined by the size of the address bus.
  4. Many instructions, such as JUMP and conditional JUMP, change the contents of the program counter from its normal sequential address value. The program counter is loaded with the address specified in these instructions.
- **Memory Address Register (MAR).** The memory address register contains the address of data. The microprocessor uses the address, which is stored in the memory address register, as a direct pointer to memory. The contents of the address is the actual data that is being transferred.
- **Accumulator (A).** For an 8-bit microprocessor, the accumulator is typically an 8-bit register. It is used to store the result after most ALU operations. These microprocessors have instructions to shift or rotate the accumulator 1 bit to the right or left through the carry flag. The accumulator is typically used for inputting a byte into the accumulator from an external device or outputting a byte to an external device from the accumulator. Some microprocessors, such as the Motorola 6809, have more than one accumulator. In these microprocessor, the accumulator to be used by the instruction is specified in the op-code.

Depending on the register section, the microprocessor can be classified either as an accumulator-based or a general-purpose register-based machine. In an accumulator-based microprocessor such as the Intel 8085 and Motorola 6809, the data is assumed to be held in a register called the "accumulator." All arithmetic and logic operations are performed using this register as one of the data sources. The result after the operation is stored in the accumulator. *Eight-bit microprocessors are usually accumulator based.*

The general-purpose register-based microprocessor is usually popular with 16-, 32-, and 64-bit microprocessors, such as the Intel 8086/80386/80486/Pentium and the Motorola 68000/68020/68030/68040/PowerPC. The name "general-purpose" comes from the fact that these registers can hold data, memory addresses, or the results of arithmetic or logic operations. The number, size, and types of registers vary from one microprocessor to another.

Most registers are general-purpose whereas some, such as the program counter (PC), are provided for dedicated functions. *The PC normally contains the address of the next instruction to be executed. As metioned before, upon activating the microprocessor chip's RESET input pin, the PC is normally initialized with the address of the first instruction.* For example, the 80486, upon hardware reset, reads the first instruction from the 32-bit hex address FFFFFFF0. To execute the instruction, the microprocessor normally places the PC contents on the address bus and reads (fetches) the first instruction from external memory. The program counter contents are then automatically incremented by the ALU. The microcomputer thus usually executes a program sequentially unless it encounters a jump or branch instruction. As mentioned earlier, the size of the PC varies from one microprocessor to another depending on the address size. For example, the 68000 has a 24-bit PC, whereas the 68040 contains a 32-bit PC. Note that in general-purpose register-based microprocessors, the four basic registers typically include a PC, a MAR, an IR, and a data register.

6.3.1.2 Use of the Basic Microprocessor Registers

To provide a clear understanding of how the basic microprocessor registers are used, a binary addition program will be considered. The program logic will be explained by showing how each instruction changes the contents of the four registers. Assume that all numbers are in hex. Suppose that the contents of the memory location 2010 are to be added with the contents of 2012. Assume that [NNNN] represents the contents of the memory location NNNN. Now, suppose that [2010] = 0002 and [2012] = 0005. The steps involved in accomplishing this addition can be summarized as follows:

1. Load the memory address register (MAR) with the address of the first data to be added, that is, load 2010 into MAR.
2. Move the contents of this address to a data register, D0; that is, move first data into D0.
3. Increment the MAR by 2 to hold 2012, the address of the second data to be added.
4. Add the contents of this memory location to the data that was moved to the data register, D0 in step 2, and store the result in the 16-bit data register, D0. The above addition program will be written using 68000 instructions. Note that the 68000 uses 24-bit addresses; 24-bit addresses such as 002000_{16} will be represented as 2000_{16} (16-bit number) in the following.

The following instructions for the Motorola 68000 will be used to achieve this addition:

1. Load the contents of the next 16-bit memory word into the memory address register, A1. Note that register A1 can be considered as MAR in the 68000.

2. Read the 16-bit contents of the memory location addressed by MAR into data register, D0.
3. Increment MAR by 2 to hold 2012, the address of the second data to be added.
4. Add the current contents of data register, D0 to the contents of the memory location whose address is in MAR and store the 16-bit result in D0.

The following instructions for the Motorola 68000 will be used to achieve the above addition:

3279_{16}	Load the contents of the next 16-bit memory word into the memory address register, A1.
3010_{16}	Read the 16-bit contents of the memory location addressed by MAR into data register, D0.
5249_{16}	Increment MAR by 2.
$D051_{16}$	Add the current contents of data register, D0, to the contents of the memory location whose address is in MAR and store the 16-bit result in D0.

The complete program in hexadecimal, starting at location 2000_{16} (arbitrarily chosen) is given in Figure 6.5. Note that each memory address stores 16 bits. Hence, memory addresses are shown in increments of 2. Assume that the microcomputer can be instructed that the starting address of the program is 2000_{16}. This means that the program counter can be initialized to contain 2000_{16}, the address of the first instruction to be executed. Note that the contents of the other three registers are not known at this point. The microprocessor loads the contents of memory location addressed by the program counter into IR. Thus, the first instruction, 3279_{16}, stored in address 2000_{16} is transferred into IR.

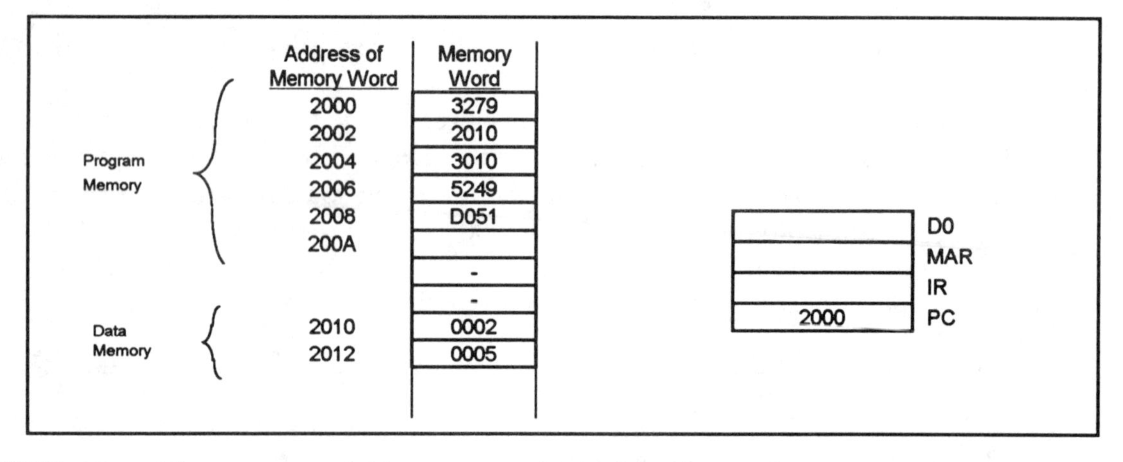

FIGURE 6.5 Microprocessor addition program with initial register and memory contents

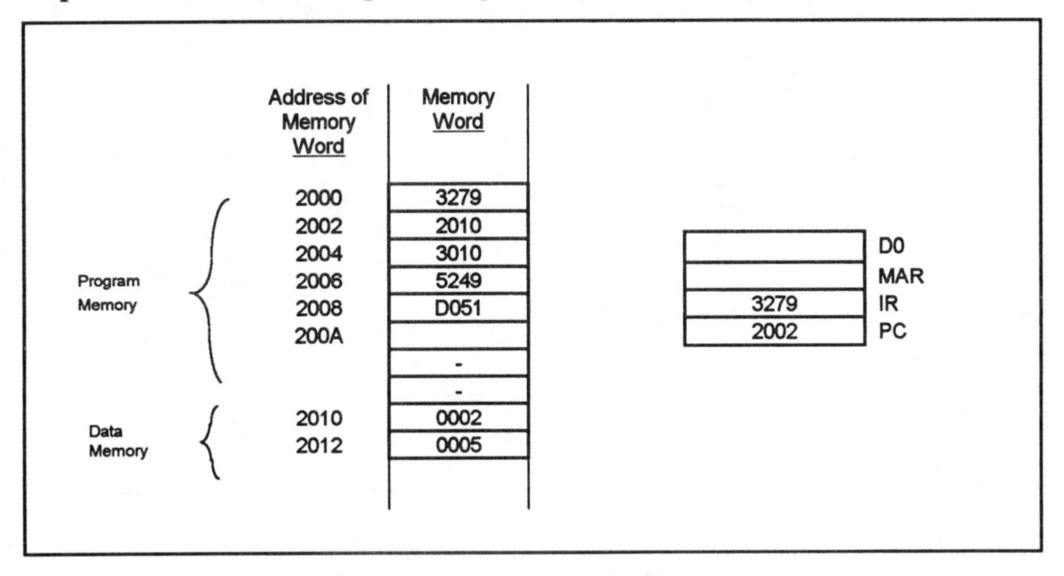

FIGURE 6.6 Microprocessor addition program (modified during execution)

The program counter contents are then incremented by 2 by the microprocessor's ALU to hold 2002_{16}. The register contents that result along with the program are shown in Figure 6.6.

The binary code 3279_{16} in the IR is executed by the microprocessor. The microprocessor then takes appropriate actions. Note that the instruction, 3279_{16}, loads the contents of the next memory location addressed by the PC into the MAR. Thus, 2010_{16} is loaded into the MAR. The contents of the PC are then incremented by 2 to hold 2004_{16}. This is shown in Figure 6.7.

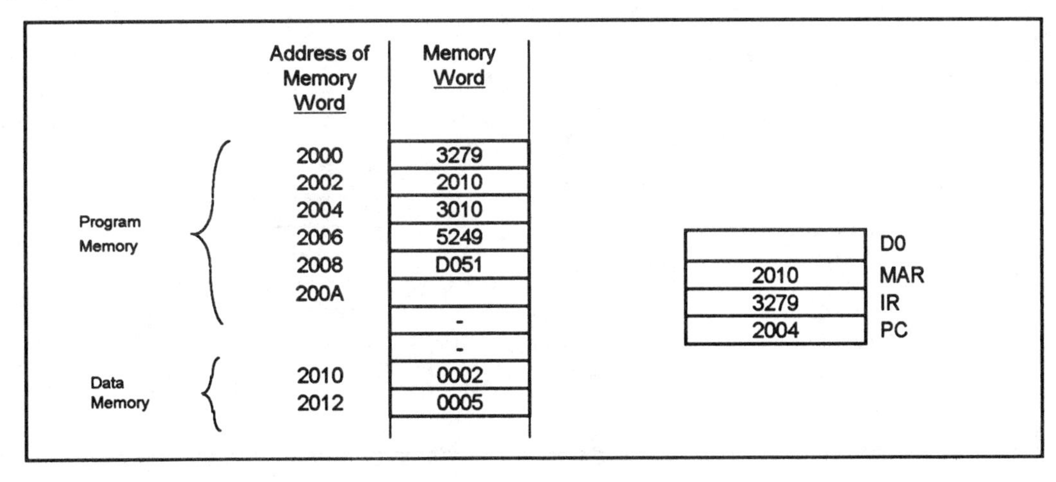

FIGURE 6.7 Microprocessor addition program (modified during execution)

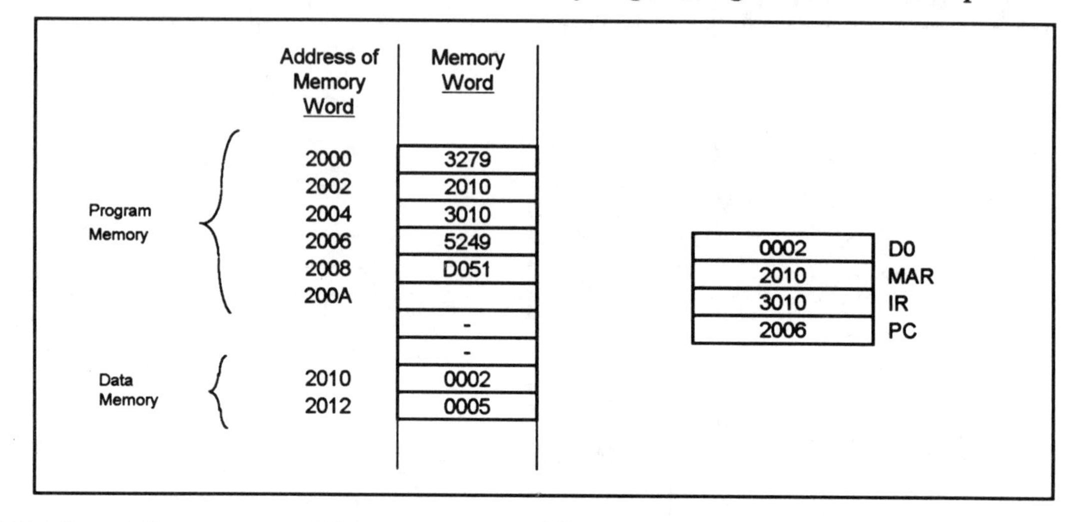

FIGURE 6.8 Microprocessor addition program (modified during execution)

Next, the microprocessor loads the contents of the memory location addressed by the PC into the IR; thus, 3010_{16} is loaded into the IR. The PC contents are then incremented by 2 to hold 2006_{16}. This is shown in Figure 6.8. In response to the instruction 3010_{16}, the contents of the memory location addressed by the MAR are loaded into the data register, D0; thus, 0002_{16} is moved to register D0. The contents of the PC are not incremented this time. This is because 0002_{16} is not immediate data. Figure 6.9 shows the details. Next the microprocessor loads 5249_{16} to IR and then increments PC to contain 2008_{16} as shown in Figure 6.10.

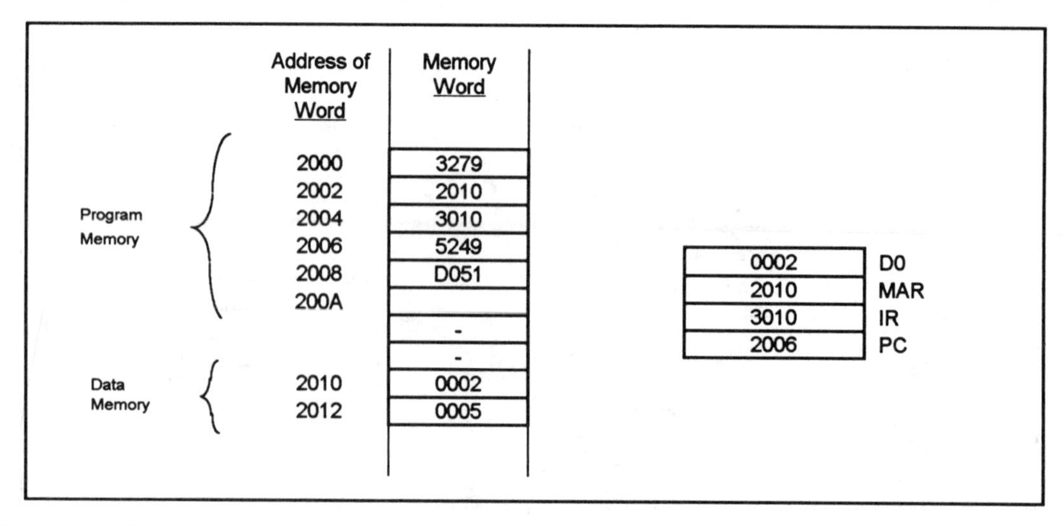

FIGURE 6.9 Microprocessor addition program (modified during execution)

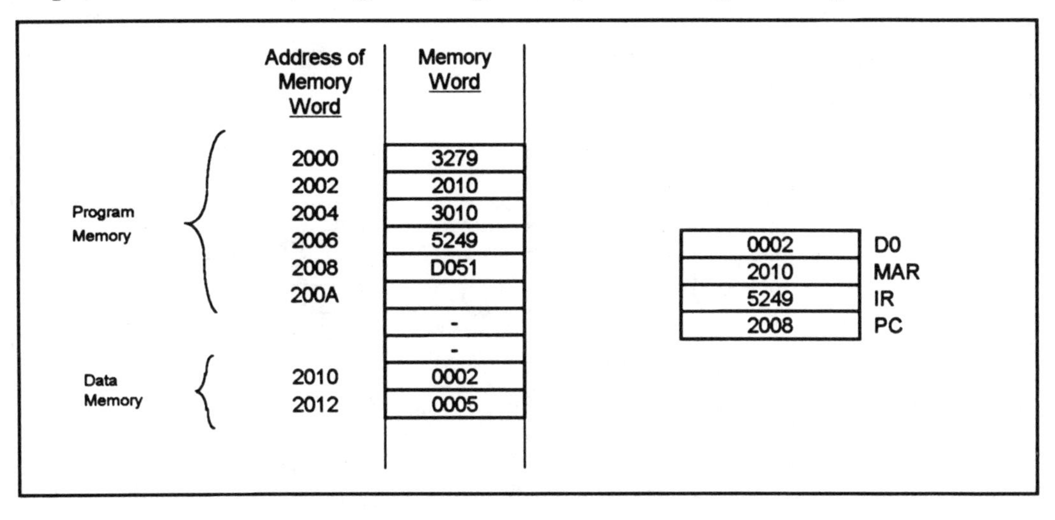

FIGURE 6.10 Microprocessor addition program (modified during execution)

In response to the instruction 5249_{16} in the IR, the microprocessor increments the MAR by 2 to contain 2012_{16} as shown in Figure 6.11. Next, the instruction $D051_{16}$ in location 2008_{16} is loaded into the IR, and the PC is then incremented by 2 to hold $200A_{16}$ as shown in Figure 6.12. Finally, in response to instruction $D051_{16}$, the microprocessor adds the contents of the memory location addressed by MAR (address 2012_{16}) with the contents of register D0 and stores the result in D0. Thus, 0002_{16} is added with 0005_{16}, and the 16-bit result 0007_{16} is stored in D0 as shown in Figure 6.13. This completes the execution of the binary addition program.

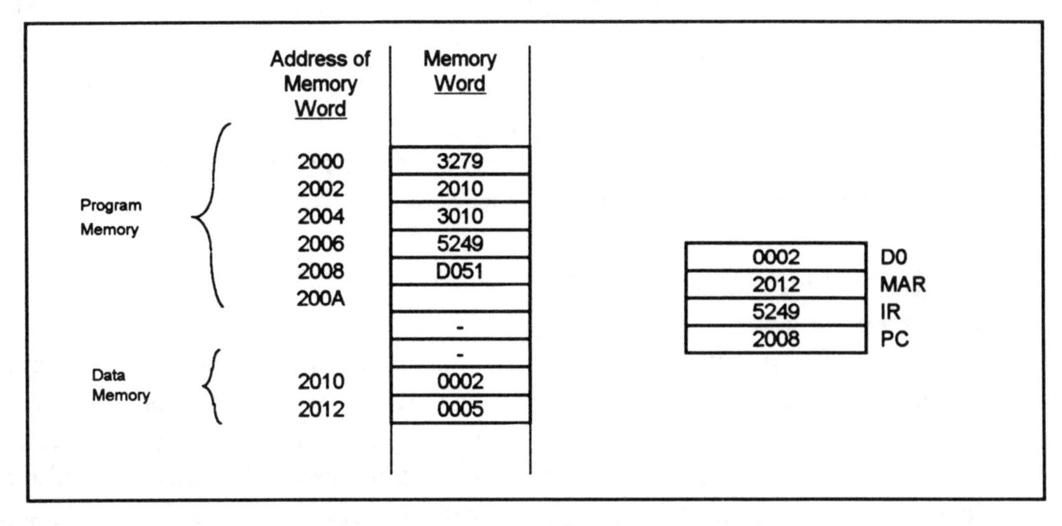

FIGURE 6.11 Microprocessor addition program (modified during execution)

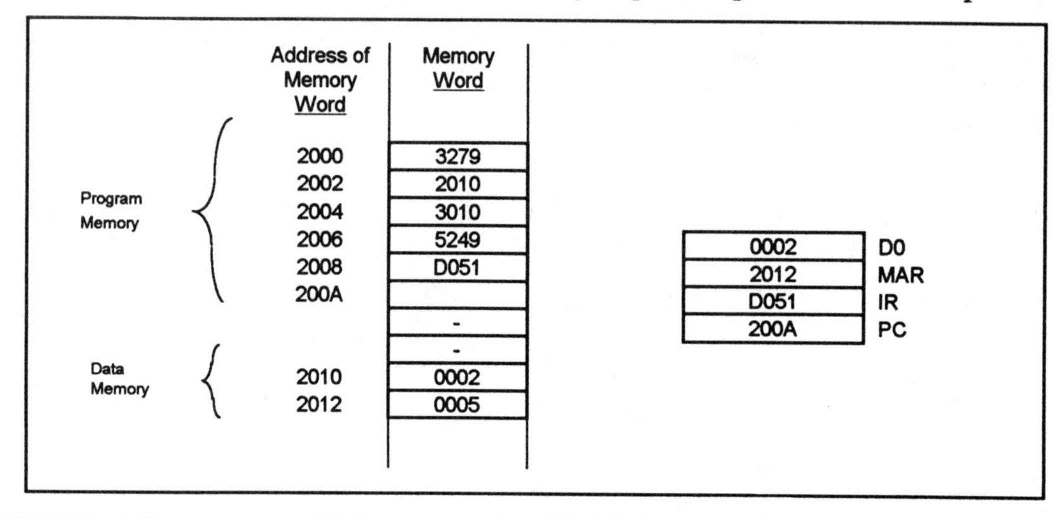

FIGURE 6.12 Microprocessor addition program (modified during execution)

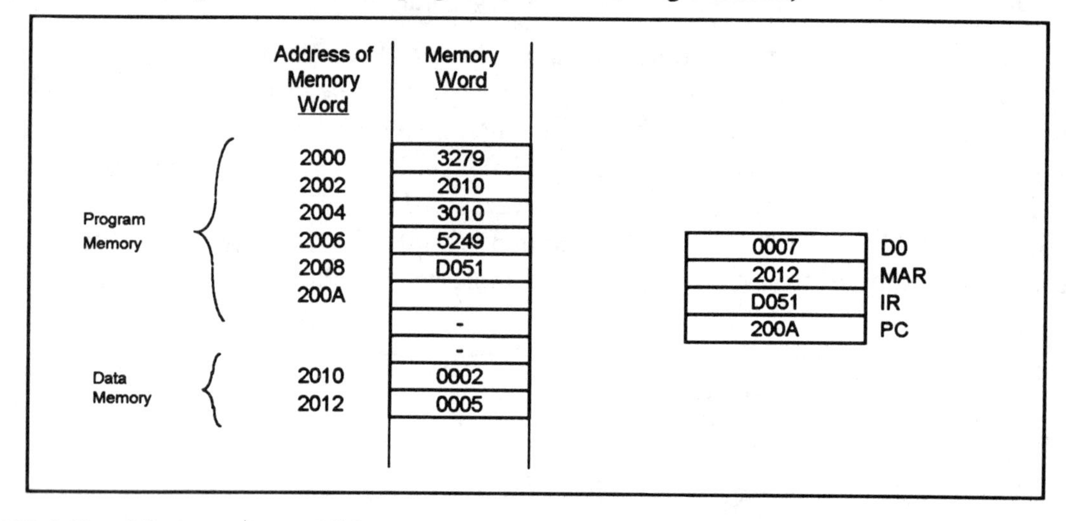

FIGURE 6.13 Microprocessor addition program (modified during execution)

6.3.1.3 Other Microprocessor Registers

- **General-Purpose Registers**
  The 16-, 32-, and 64-bit microprocessors are register oriented. They have a number of general-purpose registers for storing temporary data or for carrying out data transfers between various registers. The use of general-purpose registers speeds up the execution of a program because the microprocessor does not have to read data from external memory via the data bus if data is stored in one of its general-purpose registers. These registers are typically 16 to 32 bits. The number of general-purpose registers will vary from one

microprocessor to another. Some of the typical functions performed by instructions associated with the general-purpose registers are given here. We will use [REG] to indicate the contents of the general-purpose register and [M] to indicate the contents of a memory location.

1. Move [REG] to or from memory: [M] ← [REG] or [REG] ← [M].
2. Move the contents of one register to another: [REG1] ← [REG2].
3. Increment or decrement [REG] by 1: [REG] ← [REG] + 1 or [REG] ← [REG] - 1.
4. Load 16-bit data into a register [REG] : [REG] ← 16-bit data.

- **Index Register**
  An *index register* is typically used as a counter in address modification for an instruction, or for general storage functions. The index register is particularly useful with instructions that access tables or arrays of data. In this operation the index register is used to modify the address portion of the instruction. Thus, the appropriate data in a table can be accessed. This is called "indexed addressing." This addressing mode is normally available to the programmers of microprocessors. The effective address for an instruction using the indexed addressing mode is determined by adding the address portion of the instruction to the contents of the index register. Index registers are typically 16 or 32 bits long. In a typical 16- or 32-bit microprocessor, general-purpose registers can be used as index registers.
- **Status Register**
  The *status register,* also known as the "processor status word register" or the "condition code register," contains individual bits, with each bit having special significance. The bits in the status register are called "flags." The status of a specific microprocessor operation is indicated by each flag, which is set or reset by the microprocessor's internal logic to indicate the status of certain microprocessor operations such as arithmetic and logic operations. The status flags are also used in conditional `JUMP` instructions. We will describe some of the common flags in the following.

  The *carry flag* is used to reflect whether or not the result generated by an arithmetic operation is greater than the microprocessor's word size. As an example, the addition of two 8-bit numbers might produce a carry. This carry is generated out of the eighth position, which results in setting the carry flag. However, the carry flag will be zero if no carry is generated from the addition. As mentioned before, in multibyte arithmetic, any carry out of the low-byte addition must be added to the high-byte addition to obtain the correct result. This can illustrated by the following example:

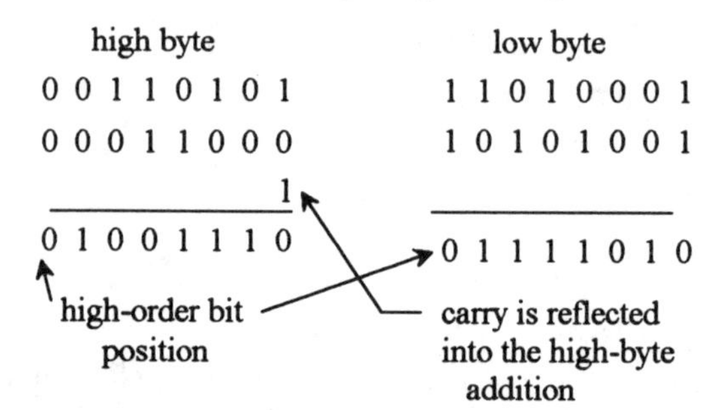

While performing BCD arithmetic with microprocessors, the carry out of the low nibble (4 bits) has a special significance. Because a BCD digit is represented by 4 bits, any carry out of the low 4 bits must be propagated into the high 4 bits for BCD arithmetic. This carry flag is known as the *auxiliary carry flag* and is set to 1 if the carry out of the low 4 bits is 1, otherwise it is 0.

A *zero flag* is used to show whether the result of an operation is zero. It is set to 1 if the result is zero, and it is reset to 0 if the result is nonzero. A *parity flag* is set to 1 to indicate whether the result of the last operation contains either an even number of 1's (even parity) or an odd number of 1's (odd parity), depending on the microprocessor. The type of parity flag used (even or odd) is determined by the microprocessor's internal structure and is not selectable. The sign flag (also sometimes called the negative flag) is used to indicate whether the result of the last operation is positive or negative. If the most significant bit of the last operation is 1, then this flag is set to 1 to indicate that the result is negative. This flag is reset to 0 if the most significant bit of the result is zero, that is, if the result is positive.

As mentioned before, the *overflow flag* arises from the representation of the sign flag by the most significant bit of a word in signed binary operation. The overflow flag is set to 1 if the result of an arithmetic operation is too big for the microprocessor's maximum word size, otherwise it is reset to 0. Let C_f be the final carry out of the most significant bit (sign bit) and C_p be the previous carry. It was shown in Chapter 2 that the overflow flag is the exclusive OR of the carries C_p and C_f.

$$\text{Overflow} = C_p \oplus C_f$$

- **Stack Pointer Register**

The *stack* consists of a number of RAM locations set aside for reading data from or writing data into these locations and is typically used by subroutines (a subroutine is a program that performs operations frequently needed by the main or calling program). The address of the stack is contained in a register called the "stack pointer." Two instructions, PUSH and POP, are usually available with the stack. The PUSH operation is defined as writing to the top or bottom of the stack, whereas the POP operation means reading from the top or bottom of the stack. Some microprocessors access the stack from the top; the others access via the bottom. When the stack is accessed from the bottom, the stack pointer is incremented after a PUSH and decremented after a POP operation. On the other hand, when the stack is accessed from the top, the stack pointer is decremented after a PUSH and incremented after a POP. Microprocessors typically use 16- or 32-bit registers for performing the PUSH or POP operations. The incrementing or decrementing of the stack pointer depends on whether the operation is PUSH or POP and also whether the stack is accessed from the top or the bottom.

We now illustrate the stack operations in more detail. We use 16-bit registers in Figures 6.14 and 6.15. In Figure 6.14, the stack pointer is incremented by 2 (since 16-bit register) to address location 20C7 after the PUSH. Now consider the POP operation of Figure 6.15. Note that after the POP, the stack pointer is decremented by 2. [20C5] and [20C6] are assumed to be empty conceptually after the POP operation. Finally, consider the PUSH operation of Figure 6.16. The stack is accessed from the top. Note that the stack pointer is decremented by 2 after a PUSH. Next, consider the POP (Figure 6.17). [20C4] and [20C5] are assumed to be empty after the POP.

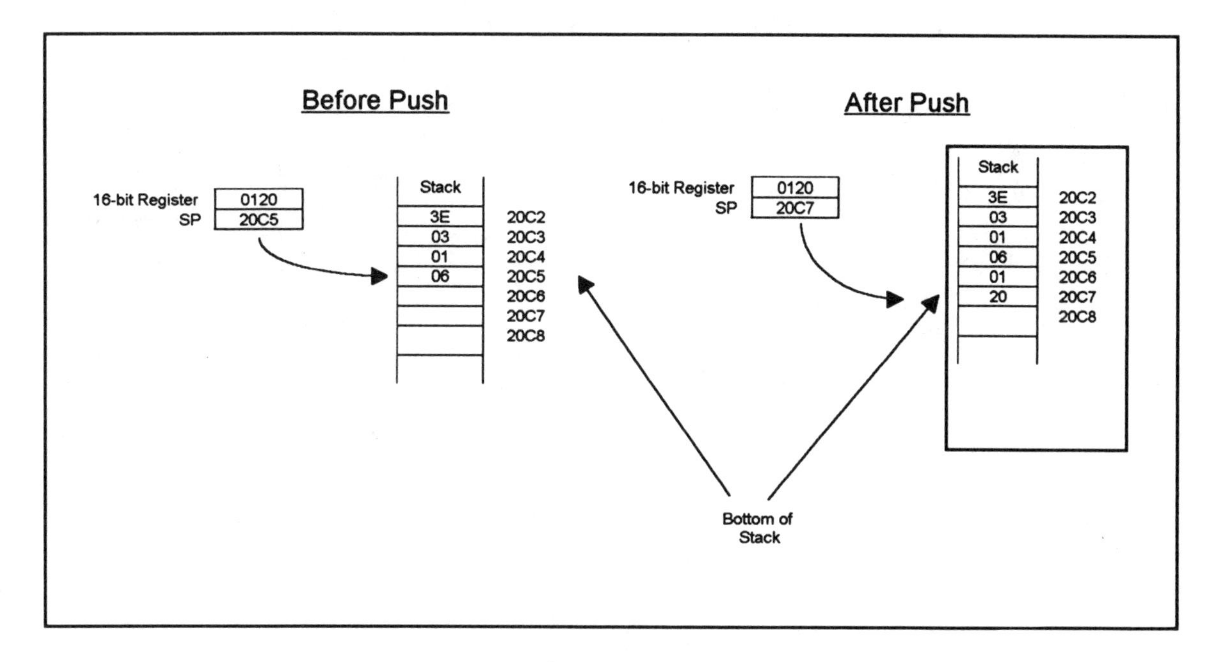

FIGURE 6.14 PUSH operation when accessing stack from bottom

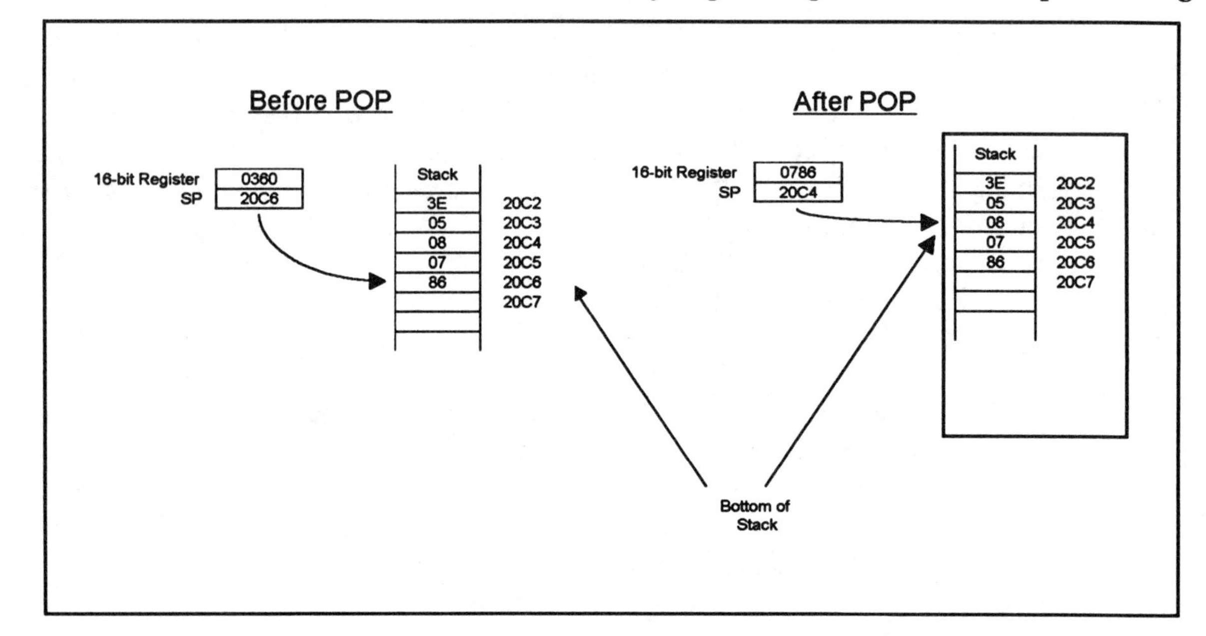

FIGURE 6.15 POP operation when accessing stack from bottom

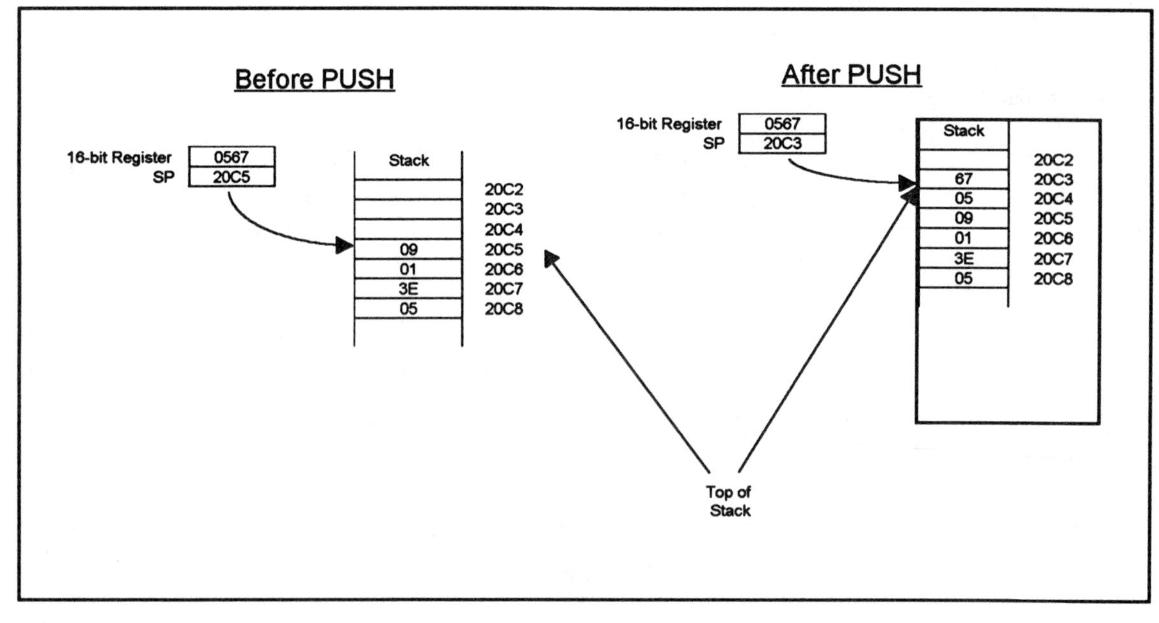

FIGURE 6.16 PUSH operation when accessing stack from top

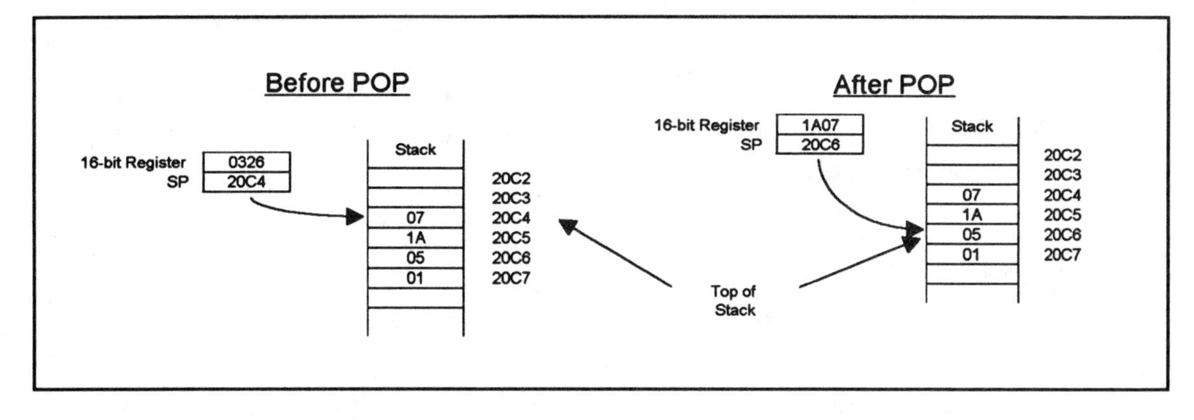

FIGURE 6.17 POP operation when accessing stack from top

Example 6.1

Determine the carry (*C*), sign (*S*), zero (*Z*), overflow (*V*), and parity (*P*) flags for the following operation: 0110_2 plus 1010_2.

Assume the parity bit = 1 for ODD parity in the result; otherwise the parity bit = 0. Also, assume that the numbers are signed. Draw a logic diagram for implementing the flags in a 5-bit register using D flip-flops; use P = bit 0, V = bit 1, Z = bit 2, S = bit 3, and C = bit 4.

Solution

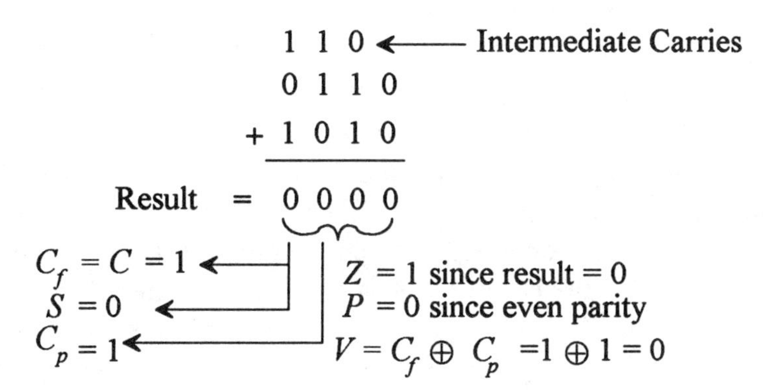

The flag register can be implemented from the 4-bit result as follows:

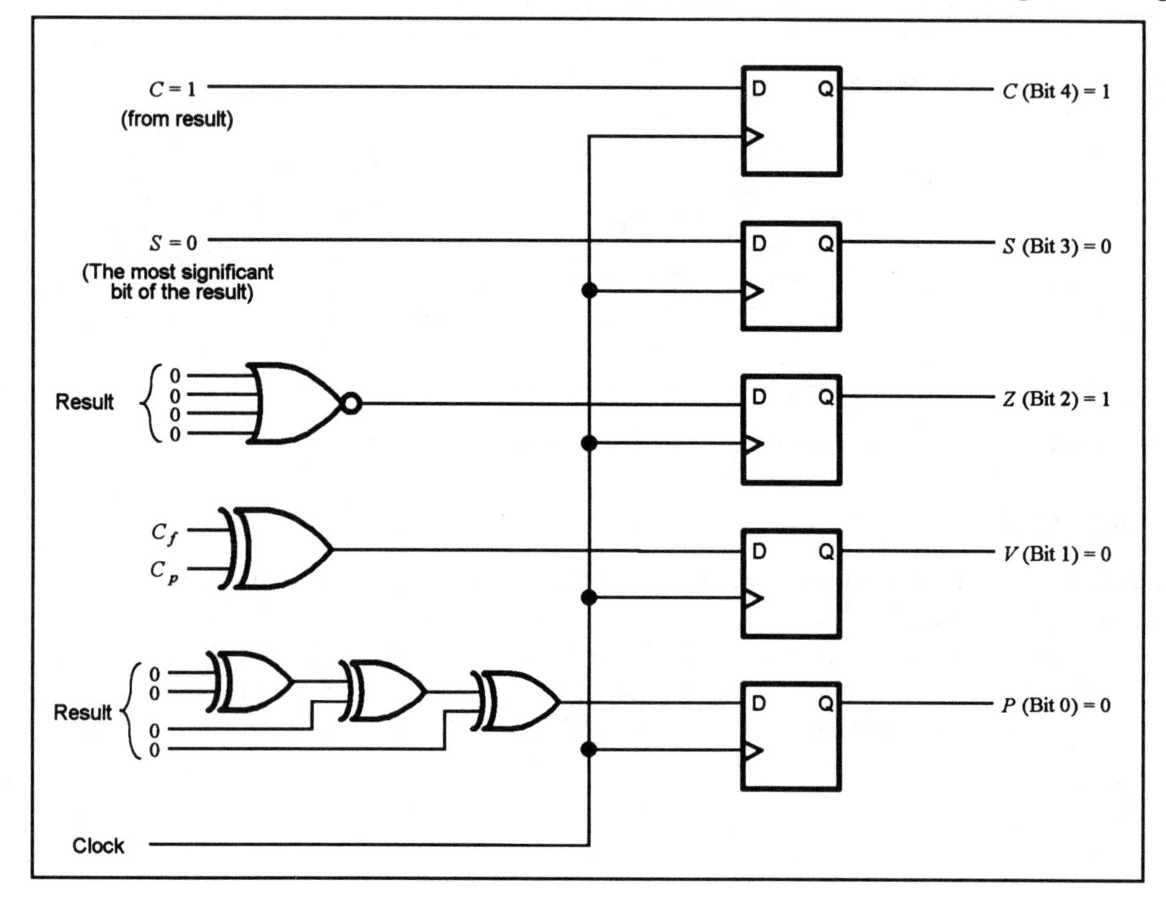

6.3.2 Control Unit

The main purpose of the control unit is to read and decode instructions from the program memory. To execute an instruction, the control unit steps through the appropriate blocks of the ALU based on the op-codes contained in the instruction register. The op-codes define the operations to be performed by the control unit in order to execute an instruction. The control unit interprets the contents of the instruction register and then responds to the instruction by generating a sequence of enable signals. These signals activate the appropriate ALU logic blocks to perform the required operation.

The control unit generates the *control signals*, which are output to the other microcomputer elements via the control bus. The control unit also takes appropriate actions in response to the control signals on the control bus provided by the other microcomputer elements.

The control signals vary from one microprocessor to another. For each specific microprocessor, these signals are described in detail in the manufacturer's manual. It is impossible to describe all the control signals for various manufacturers. However, we cover some of the common ones in the following discussion.

- **RESET.** This input is common to all microprocessors. When this input pin is driven to HIGH or LOW (depending on the microprocessor), the program counter is loaded with a predefined address specified by the manufacturer. For example, in the 80486, upon hardware reset, the program counter is loaded with $FFFFFFF0_{16}$. This means that the instruction stored at memory location $FFFFFFF0_{16}$ is executed first. In some other microprocessors, such as the Motorola 68000, the program counter is not loaded directly by activating the RESET input. In this case, the program counter is loaded indirectly from two locations (such as 000004 and 000006) predefined by the manufacturer. This means that these two locations contain the address of the first instruction to be executed.
- **READ/WRITE (R/$\overline{\text{W}}$).** This output line is common to all microprocessors. The status of this line tells the other microcomputer elements whether the microprocessor is performing a READ or a WRITE operation. A HIGH signal on this line indicates a READ operation and a LOW indicates a WRITE operation. Some microprocessors have separate READ and WRITE pins.
- **$\overline{\text{READY}}$.** This is an input to the microprocessor. Slow devices (memory and I/O) use this signal to gain extra time to transfer data to or receive data from a microprocessor. The $\overline{\text{READY}}$ signal is usually an active low signal, that is, LOW means the microprocessor is ready. Therefore, when the microprocessor selects a slow device, the device places a LOW on the $\overline{\text{READY}}$ pin. The microprocessor responds by suspending all its internal operations and enters a WAIT state. When the device is ready to send or receive data, it removes the $\overline{\text{READY}}$ signal. The microprocessor comes out of the WAIT state and performs the appropriate operation.
- **Interrupt Request (INT or IRQ).** The external I/O devices can interrupt the microprocessor via this input pin on the microprocessor chip. When this signal is activated by the external devices, the microprocessor jumps to a special program, called the "interrupt service routine." This program is normally written by the user for performing tasks that the interrupting device wants the microprocessor to do. After completing this program, the microprocessor returns to the main program it was executing when the interrupt occurred.

6.3.3 Arithmetic and Logic Unit (ALU)

The ALU performs all the data manipulations, such as arithmetic and logic operations, inside the microprocessor. *The size of the ALU conforms to the word length of the microcomputer. This means that a 32-bit microprocessor will have a 32-bit ALU.* Typically, the ALU performs the following functions:

1. Binary addition and logic operations
2. Finding the ones complement of data
3. Shifting or rotating the contents of a general-purpose register 1 bit to the left or right through carry

6.3.4 Functional Representations of a Simple and a Typical Microprocessor

Figure 6.18 shows the functional block diagram of a simple microprocessor. Note that the data bus shown is internal to the microprocessor chip and should not be confused with the system bus. The system bus is external to the microprocessor and is used to connect all the necessary chips to form a microcomputer. The buffer register in Figure 6.18 stores any data read from memory for further processing by the ALU. All other blocks of Figure 6.18 have been discussed earlier. Figure 6.19 shows the simplified block diagram of a realistic microprocessor, the Intel 8086.

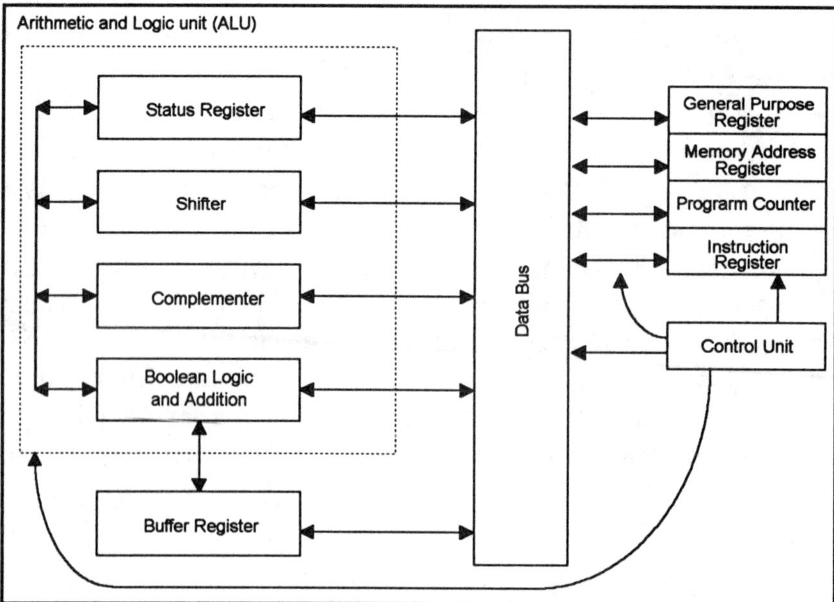

FIGURE 6.18 Functional representation of a simple microprocessor

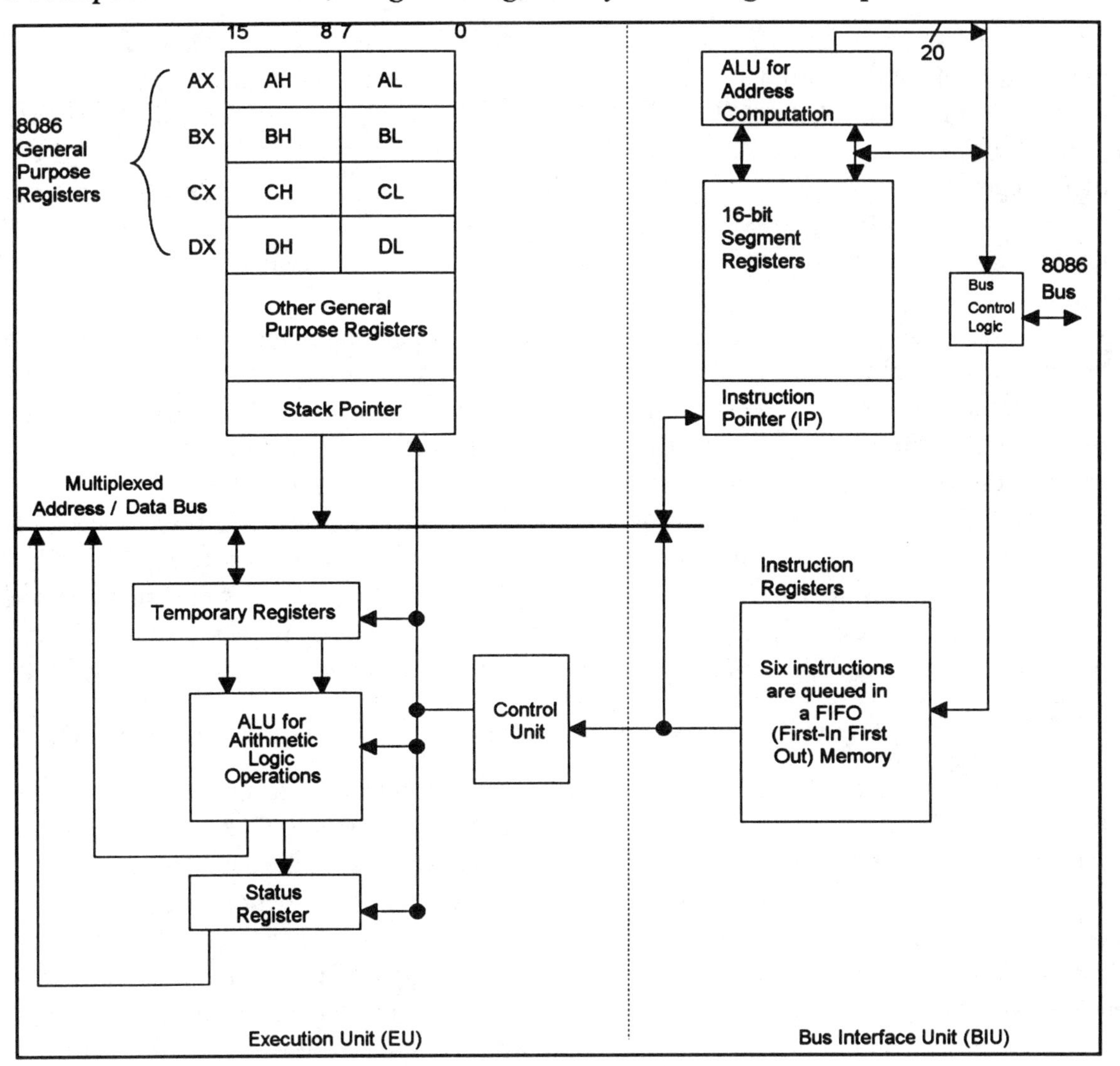

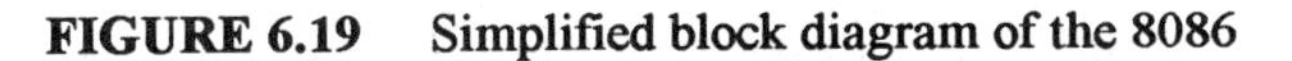
FIGURE 6.19 Simplified block diagram of the 8086

The 8086 microprocessor is internally divided into two functional units: the bus interface unit (BIU) and the execution unit (EU). The BIU interfaces the 8086 to external memory and I/O chips. The BIU and EU function independently. The BIU reads (fetches) instructions and writes or reads data to or from memory and I/O ports. The EU executes instructions that have already been fetched by the BIU. The BIU contains segment registers, the instruction pointer (IP), the instruction queue registers, and the address generation/bus control circuitry.

The 8086 uses segmented memory. This means that the 8086's 1 MB main memory is divided into 16 segments of 64 KB each. Within a particular segment, the instruction pointer (IP) works as a program counter (PC). Both the IP and the segment registers are 16 bits wide. The 20-bit address is generated in the BIU by using the contents of a 16-bit IP and a 16-bit segment register. The ALU in the BIU is used for this purpose. *Memory segmentation is useful in a time-shared system when several users share a microprocessor. Segmentation makes it easy to switch from one user program to another by changing the contents of a segment register.*

The bus control logic of the BIU generates all the bus control signals such as read and write signals for memory and I/O. The BIU's instruction register consist of a first-in–first-out (FIFO) memory in which up to six instruction bytes are preread (prefetched) from external memory ahead of time to speed up instruction execution. The control unit in the EU translates the instructions based on the contents of the instruction registers in the BIU.

The EU contains several 16-bit general-purpose registers. Some of them are AX, BX, CX, and DX. Each of these registers can be used either as an 8-bit register (AH, AL, BH, BL, CH, CL, DH, DL) or as a 16-bit register (AX, BX, CX, DX). Register BX can also be used to hold the address in a segment. The EU also contain a 16-bit status register. The ALU in the EU performs all arithmetic and logic operations. The 8086 is covered in detail in Chapter 9.

6.3.5 Microprogramming the Control Unit (A Simplified Explanation)

In this section, we discuss how the op-codes are interpreted by the microprocessor. Most microprocessors have an internal memory, called the "control memory" (ROM). This memory is used to store a number of codes, called the "microinstructions." These microinstructions are combined together to design instructions. Each instruction in the instruction register initiates execution of a set of microinstructions in the control unit to perform the operation required by the instruction. The microprocessor manufacturers define the microinstructions by programming the control memory (ROM) and thus, design the instruction set of the microprocessor. This type of programming is known as "microprogramming." Note that the control units of most 16-, 32-, and 64-bit microprocessors are microprogrammed.

For simplicity, we illustrate the concepts of microprogramming using Figure 6.18. Let us consider incrementing the contents of the register. This is basically an addition operation. The control unit will send an enable signal to execute the ALU adder logic. Incrementing the contents of a register consists of transferring the register contents to the ALU adder and then returning the result to the register. The complete incrementing process is accomplished via the five steps shown in Figures 6.20 through Figure 6.24. In all five steps, the control unit initiates execution of each microinstruction. Figure 6.20 shows the transfer of the register contents to the data bus. Figure 6.21 shows the transfer of the contents of the data bus to the adder in the

ALU in order to add 1 to it. Figure 6.22 shows the activation of the adder logic. Figure 6.23 shows the transfer of the result from the adder to the data bus. Finally, Figure 6.24 shows the transfer of the data bus contents to the register.

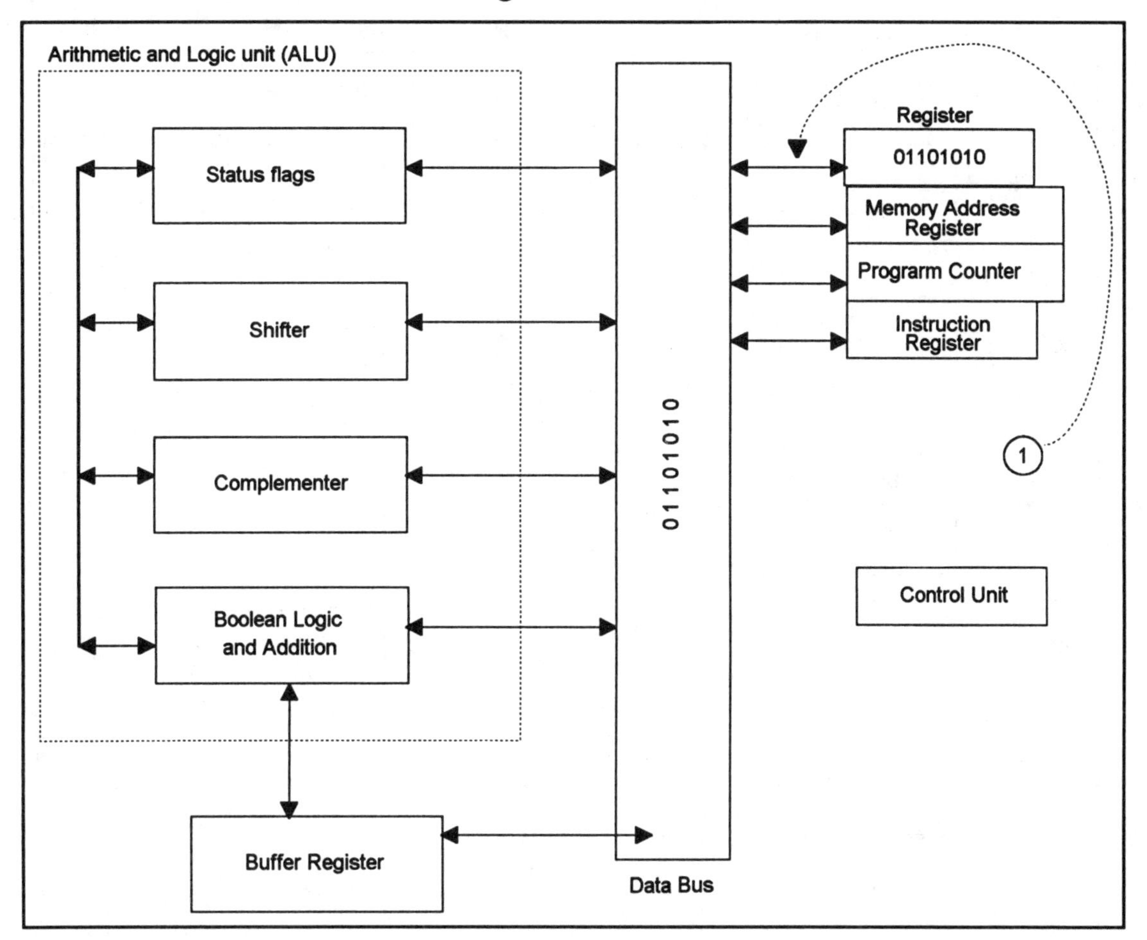

FIGURE 6.20 Transferring register contents to data bus

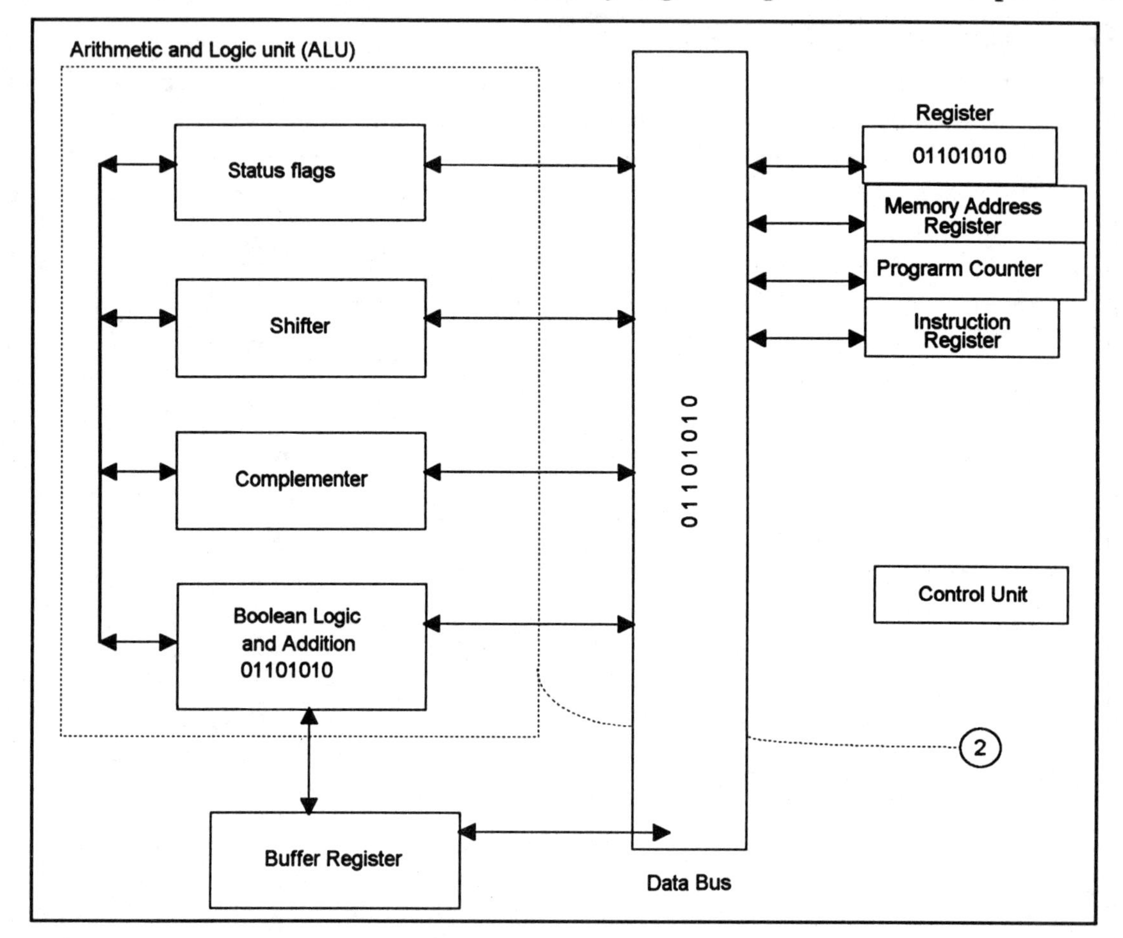

FIGURE 6.21 Transferring data bus contents to the ALU

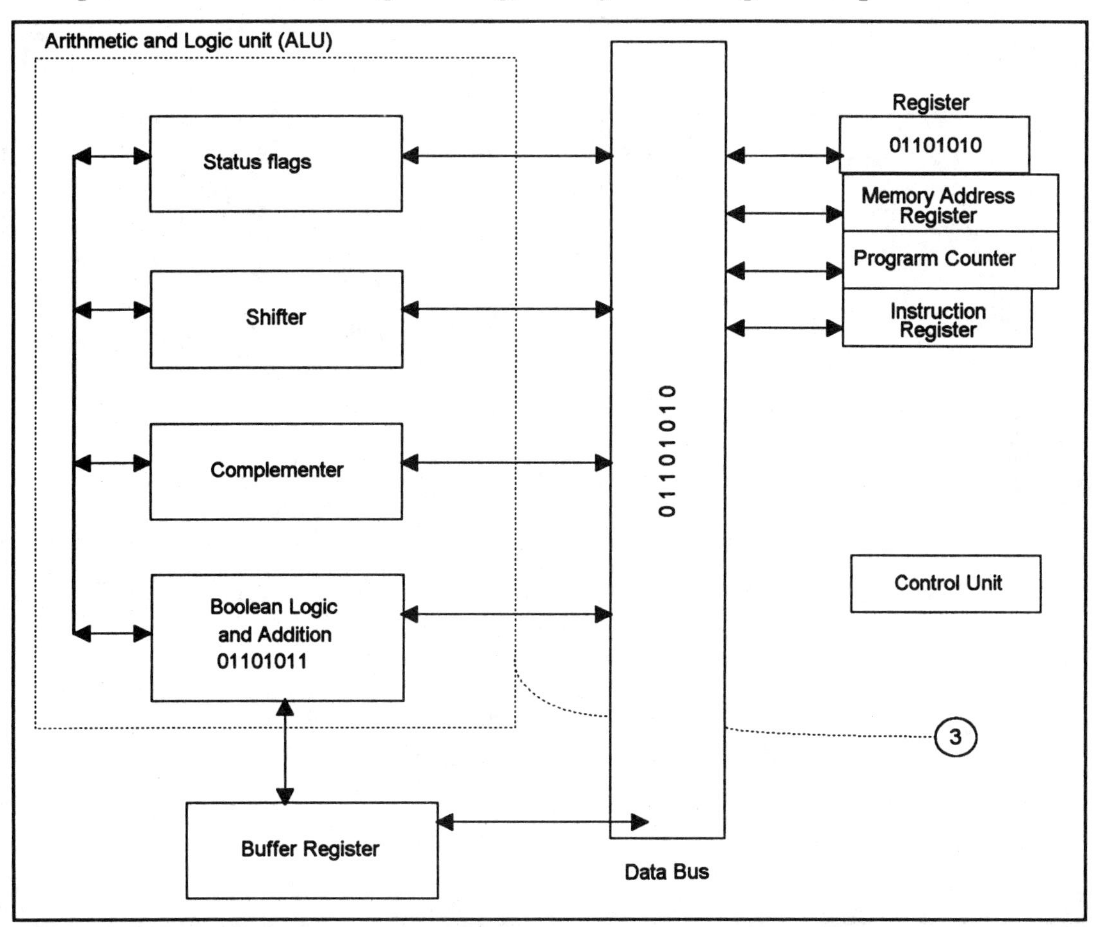

FIGURE 6.22 Activating the ALU logic

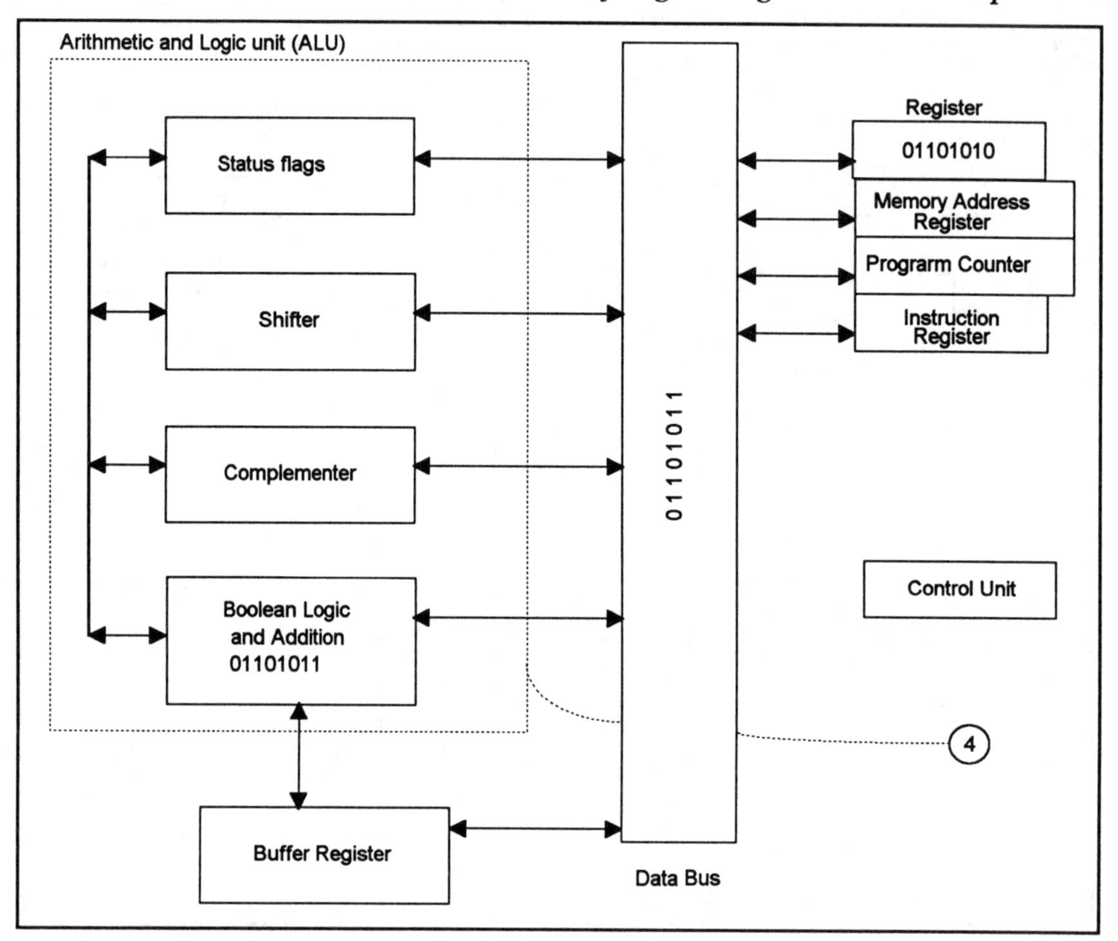

FIGURE 6.23 Transferring the ALU result to the data bus

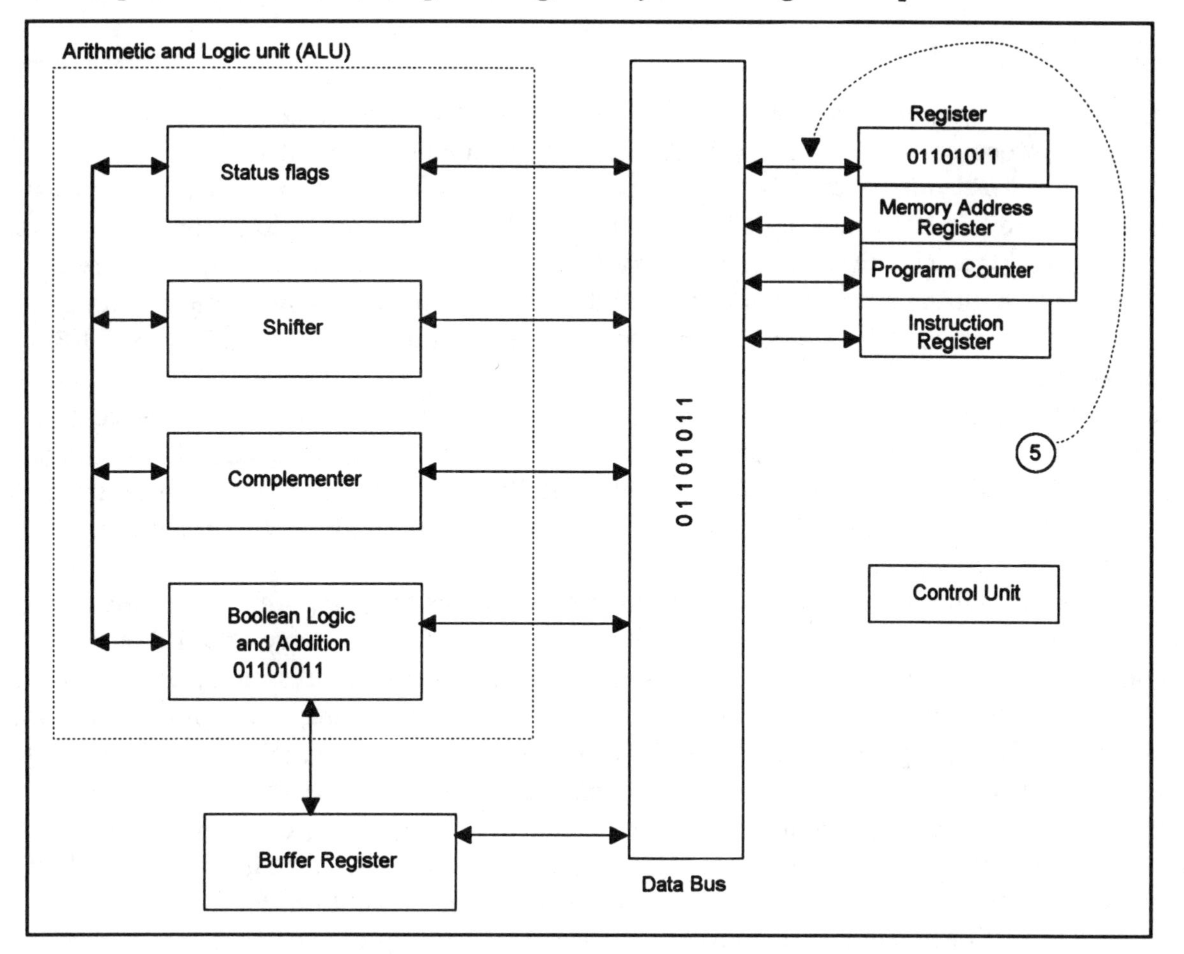

FIGURE 6.24 Transferring the data bus

Microprogramming is typically used by the microprocessor designer to program the logic performed by the control unit. On the other hand, assembly language programming is a popular programming language used by the microprocessor user for programming the microprocessor to perform a desired function. A microprogram is stored in the control unit. An assembly language program is stored in the main memory. The assembly language program is called a macroprogram*. A* macroinstruction *(or simply an instruction) initiates execution of a complete microprogram.*

A simplified explanation of microprogramming is provided in this section. *This topic will be covered in detail in Chapters 7 and 11.*

6.4 The Memory

The main or external memory (or simply the memory) stores both instructions and data. For 8-bit microprocessors, the memory is divided into a number of 8-bit units called "memory words." *An 8-bit unit of data is termed a "byte."* Therefore, for an 8-bit microprocessor, "memory word" and "memory byte" mean the same thing. For 16-bit microprocessors, a word contains two bytes (16 bits). A memory word is identified in the memory by an address. For example, the 8086 microprocessor uses 20-bit addresses for accessing memory words. This provides a maximum of 2^{20} = 1 MB of memory addresses, ranging from 00000_{16} to $FFFFF_{16}$ in hexadecimal.

As mentioned before, an important characteristic of a memory is whether it is volatile or nonvolatile. The contents of a volatile memory are lost if the power is turned off. On the other hand, a nonvolatile memory retains its contents after power is switched off. Typical examples of nonvolatile memory are ROM and magnetic memory (floppy disk). A RAM is a volatile memory unless backed up by battery.

As mentioned earlier, some microprocessors such as the Intel 8086 divide the memory into segments. For example, the 8086 divides the 1 MB main memory into 16 segments (0 through 15). Each segment contains 64 KB of memory and is addressed by 16 bits. Figure 6.25 shows a typical main memory layout of the 8086. In the figure, the high four bits of an address specify the segment number. As an example, consider address 10005_{16} of segment 1. The high four bits, 0001, of this address define the location is in segment 1 and the low 16 bits, 0005_{16}, specify the particular address in segment 1. The 68000, on the other hand, uses linear or nonsegmented memory. For example, the 68000 uses 24 address pins to directly address 2^{24}= 16 MB of memory with addresses from 000000_{16} to $FFFFFF_{16}$.

As mentioned before, memories can be categorized into two main types: read-only memory (ROM) and random-access memory (RAM). As shown in Figure 6.26, ROMs and RAMs are then divided into a number of subcategories, which are discussed next.

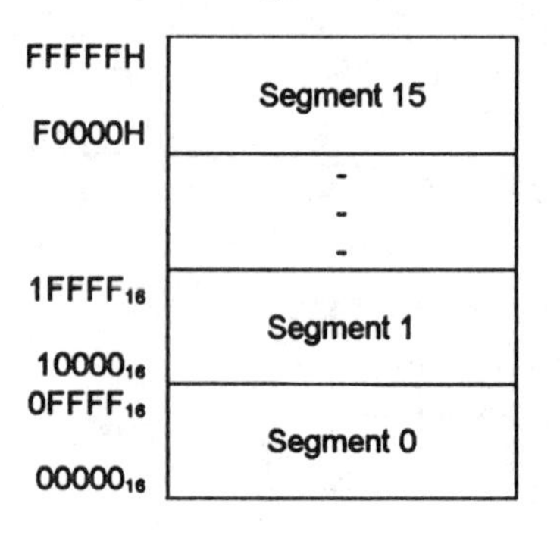

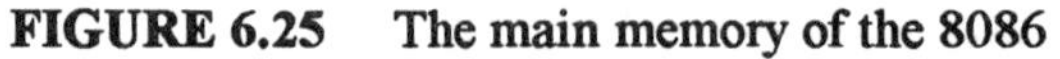

FIGURE 6.25 The main memory of the 8086

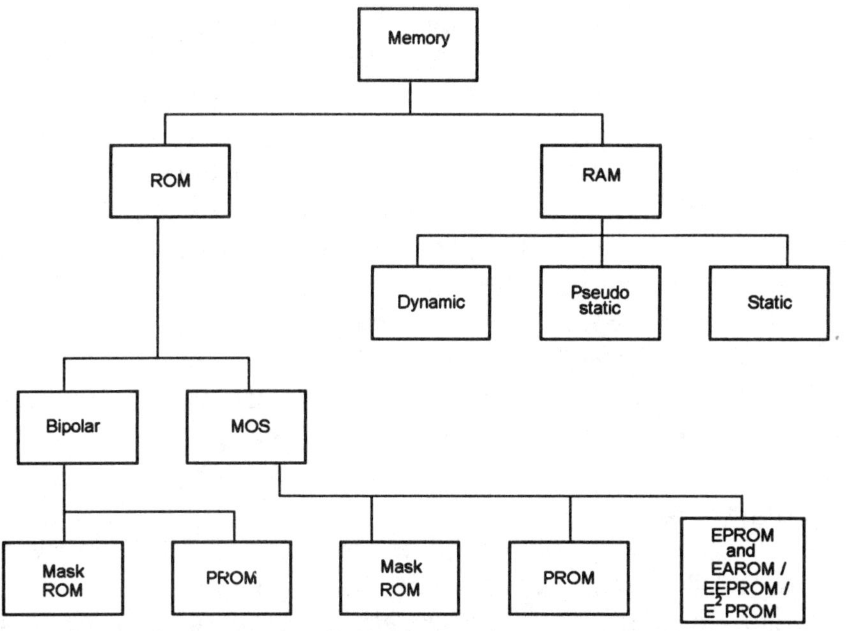

FIGURE 6.26 Summary of available semiconductor memories for microprocessor systems

6.4.1 Random-Access Memory (RAM)

There are three types of RAM: dynamic RAM, pseudo-static RAM , and static RAM. *Dynamic RAM stores data in capacitors, that is, it can hold data for a few milliseconds. Hence, dynamic RAMs are refreshed typically by using external refresh circuitry.* Pseudo-static RAMs are dynamic RAMs with internal refresh. Finally, *static RAM stores data in flip-flops. Therefore, this memory does not need to be refreshed.* RAMs are volatile unless backed up by battery. Dynamic RAMs (DRAMs) are used in applications requiring large memory. DRAMs have higher densities than Static RAMs (SRAMs). Typical examples of DRAMs are 4464 (64K × 4-bit), 44256 (256K × 4-bit), and 41000 (1M × 1-bit). DRAMs are inexpensive, occupy less space , and dissipate less power compared to SRAMs. Two enhanced versions of DRAM are EDO DRAM (Extended Data Output DRAM) and SDRAM (Synchronous DRAM). The EDO DRAM provides fast access by allowing the DRAM controller to output the next address at the same time the current data is being read. An SDRAM contains multiple DRAMs (typically 4) internally. SDRAMs utilize the multiplexed addressing of conventional DRAMs . That is, SDRAMs provide row and column addresses in two steps like DRAMs. However, the control signals and address inputs are sampled by the SDRAM at the leading edge of a common clock signal (133 MHz maximum). SDRAMs provide higher densities by further reducing the need for support circuitry and faster speeds than conventional DRAMs. The SDRAM has become popular with PC (Personal Computer) memory.

6.4.2 Read-Only Memory (ROM)

ROMs can only be read. This memory is nonvolatile. From the technology point of view, ROMs are divided into two main types, bipolar and MOS. As can be expected, bipolar ROMs are faster than MOS ROMs. Each type is further divided into two common types, mask ROM

are faster than MOS ROMs. Each type is further divided into two common types, mask ROM and programmable ROM, and MOS ROMs contain one more type, erasable PROM (EPROM such as Intel 2732 and EAROM or EEPROM or E^2PROM such as Intel 2864). Mask ROMs are programmed by a masking operation performed on the chip during the manufacturing process. The contents of mask ROMs are permanent and cannot be changed by the user. On the other hand, the programmable ROM (PROM) can be programmed by the user by means of proper equipment. However, once this type of memory is programmed, its contents cannot be changed. Erasable PROMs (EPROMs and EAROMs) can be programmed, and their contents can also be altered by using special equipment, called the PROM programmer. When designing a microcomputer for a particular application, the permanent programs are stored in ROMs. Control memories are ROMs. PROMs can be programmed by the user. PROM chips are normally designed using transistors and fuses. These transistors can be selected by addressing via the pins on the chip. In order to program this memory, the selected fuses are "blown" or "burned" by applying a voltage on the appropriate pins of the chip. This causes the memory to be permanently programmed.

Erasable PROMs (EPROMs) can be reprogrammed and erased. The chip must be removed from the microcomputer system for programming. This memory is erased by exposing the chip via a lid or window on the chip to ultraviolet light. Typical erase times vary between 10 and 30 min. The EPROM can be programmed by inserting the chip into a socket of the PROM programmer and providing proper addresses and voltage pulses at the appropriate pins of the chip. Electrically alterable ROMs (EAROMs) can be programmed without removing the memory from the ROM's sockets. These memories are also called read mostly memories (RMMs), because they have much slower write times than read times. Therefore, these memories are usually suited for operations when mostly reading rather that writing will be performed. Another type of memory called "Flash memory" (nonvolatile) invented in mid 1980s by Toshiba is designed using a combination of EPROM and EAROM technologies. Flash memory can be reprogrammed electrically while imbedded on the board. One can change multiple bytes at a time. An example of Flash memory is the Intel 28F020 (256K x 8).

6.4.3 READ and WRITE Operations

To execute an instruction, the microprocessor reads or fetches the op-code via the data bus from a memory location in the ROM/RAM external to the microprocessor. It then places the op-code (instruction) in the instruction register. Finally, the microprocessor executes the instruction. Therefore, the execution of an instruction consists of two portions, instruction fetch and instruction execution. We will consider the instruction fetch, memory READ and memory WRITE timing diagrams in the following using a single clock signal. Figure 6.27 shows a typical instruction fetch timing diagram.

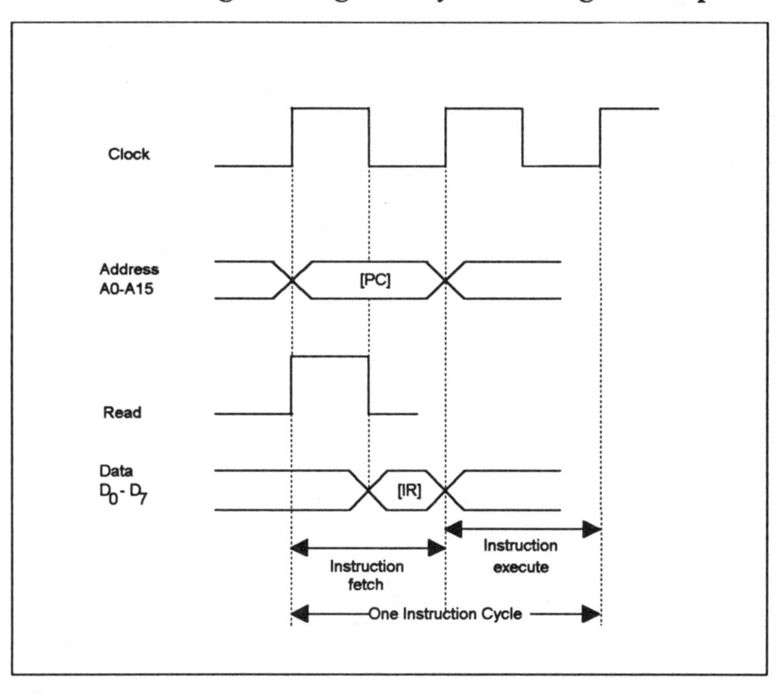

FIGURE 6.27 Typical Instruction Fetch Timing Diagram for an 8-bit Microprocessor

In Figure 6.27, to fetch an instruction, when the clock signal goes to HIGH, the microprocessor places the contents of the program counter on the address bus via the address pins A_0–A_{15} on the chip. Note that since each one of these lines A_0–A_{15} can be either HIGH or LOW, both transitions are shown for the address in Figure 6.27. The instruction fetch is basically a memory READ operation. Therefore, the microprocessor raises the signal on the READ pin to HIGH. As soon as the clock goes to LOW, the logic external to the microprocessor gets the contents of the memory location addressed by A_0–A_{15} and places them on the data bus D_0–D_7. The microprocessor then takes the data and stores it in the instruction register so that it gets interpreted as an instruction. This is called "instruction fetch." The microprocessor performs this sequence of operations for every instruction.

We now describe the READ and WRITE timing diagrams. A typical READ timing diagram is shown in Figure 6.28. Memory READ is basically loading the contents of a memory location of the main ROM/RAM into an internal register of the microprocessor. The address of the location is provided by the contents of the memory address register (MAR). Let us now explain the READ timing diagram of Figure 6.28.

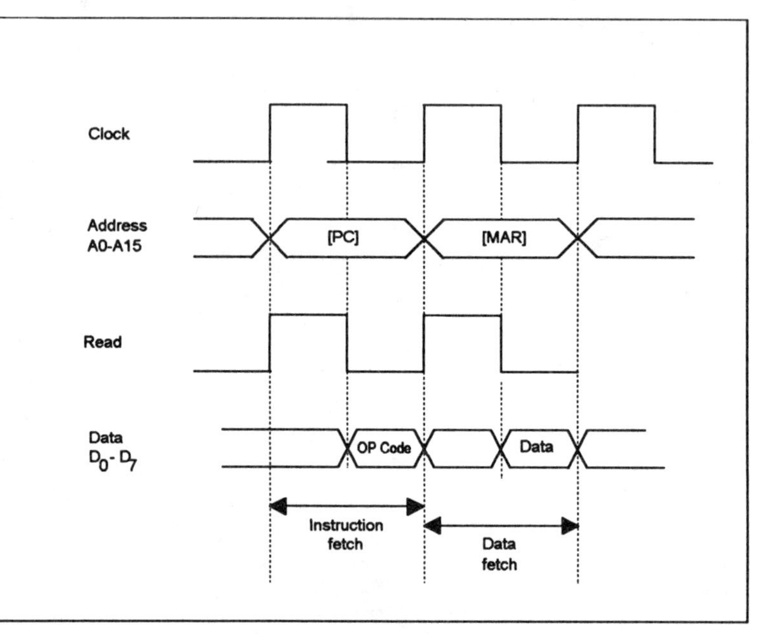

FIGURE 6.28 Typical Memory READ Timing Diagram

1. The microprocessor performs the instruction fetch cycle as before to READ the op-code.
2. The microprocessor interprets the op-code as a memory READ operation.
3. When the clock pin signal goes to HIGH, the microprocessor places the contents of the memory address register on the address pins A_0–A_{15} of the chip.
4. At the same time, the microprocessor raises the READ pin signal to HIGH.
5. The logic external to the microprocessor gets the contents of the location in the main ROM/RAM addressed by the memory address register and places them on the data from the data bus D_0–D_7.
6. Finally, the microprocessor gets this data from the data bus via its pins $D_0 - D_7$ and stores it in an internal register.

Memory WRITE is basically storing the contents of an internal register of the microprocessor into a memory location of the main RAM. The contents of the memory address register provide the address of the location where data is to be stored. Figure 6.29 shows a typical WRITE timing diagram. It can be explained in the following way:

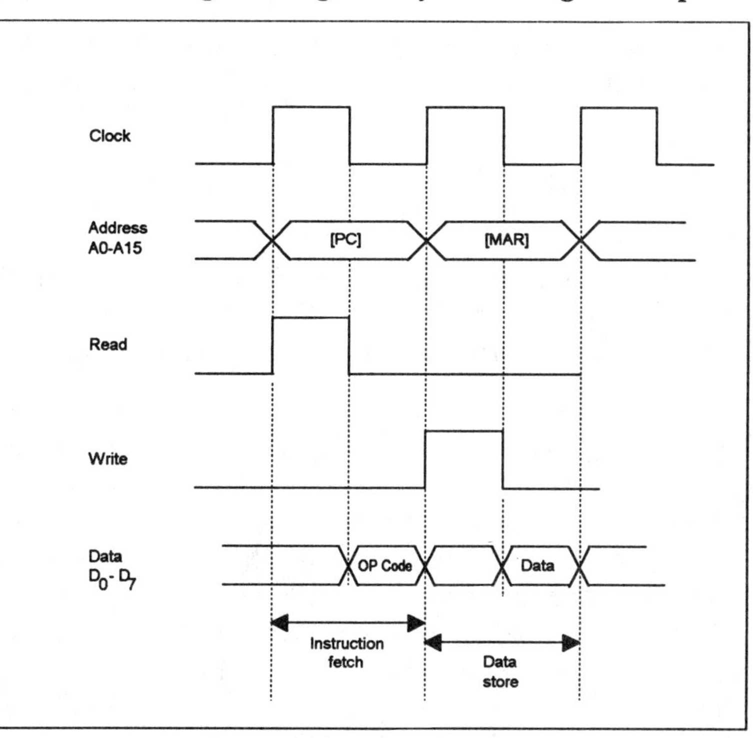

FIGURE 6.29 Typical Memory WRITE Timing Diagram

1. The microprocessor fetches the instruction code as before.
2. The microprocessor interprets the instruction code as a memory WRITE instruction and then proceeds to perform the DATA STORE cycle.
3. When the clock pin signal goes to HIGH, the microprocessor places the contents of the memory address register on the address pins A_0–A_{15} of the chip.
4. At the same time, the microprocessor raises the WRITE pin signal to HIGH.
5. The microprocessor places data to be stored from the contents of an internal register onto the data pins D_0–D_7.
6. The logic external to the microprocessor stores the data from the register into a RAM location addressed by the memory address register.

6.4.4 Memory Organization

Microcomputer memory typically consists of ROMs and RAMs. Because RAMs can be both read from and written into, the logic required to implement RAMs is more complex than that for ROMs. A microcomputer system designer is normally interested in how the microcomputer memory is organized or, in other words, how to connect the ROMS and RAMS and then determine the memory map of the microcomputer; that is, the designer would be interested in finding out what memory locations are assigned to the ROMs and RAMs. The designer can then implement the permanent programs in ROMs and the temporary programs in RAMs.

6.5 Input/Output

Input/Output (I/O) operation is defined as the transfer of data between the microcomputer system and the external world. There are typically three main ways of transferring data between the microcomputer system and the external devices: programmed I/O, interrupt I/O, and direct memory access. We now define them.

- **Programmed I/O.** Using this technique, the microprocessor executes a program to perform all data transfers between the microcomputer system and the external devices. The main characteristic of this type of I/O technique is that the external device carries out the functions as dictated by the program inside the microcomputer memory. In other words, the microprocessor completely controls all the transfers.
- **Interrupt I/O.** In this technique, an external device can force the microcomputer system to stop executing the current program temporarily so that it can execute another program, known as the "interrupt service routine." This routine satisfies the needs of the external device. After having completed this program, the microprocessor returns to the program that it was executing before the interrupt.
- **Direct Memory Access (DMA).** This is a type of I/O technique in which data can be transferred between the microcomputer memory and external devices without any microprocessor (CPU) involvement. Direct memory access is typically used to transfer blocks of data between microcomputer memory and external device. An interface chip called the DMA controller chip is used with the microprocessor for transferring data via direct memory access.

6.6 Microcomputer Programming Concepts

This section includes the fundamental concepts of microcomputer programming. Typical programming characteristics such as programming languages, microprocessor instruction sets in general, addressing modes, and instruction formats are discussed.

6.6.1 Microcomputer Programming Languages

Microcomputers are typically programmed using semi-English-language statements (assembly language). In addition to assembly languages, microcomputers use a more understandable human-oriented language called the "high-level language." No matter what type of language is used to write the programs, the microcomputers only understand binary numbers. Therefore, the programs must eventually be translated into their appropriate binary forms. The main ways of accomplishing this are discussed later.

Microcomputer programming languages can typically be divided into three main types:

1. Machine language
2. Assembly language
3. High-level language

A machine language program consists of either binary or hexadecimal op-codes. Programming a microcomputer with either one is relatively difficult, because one must deal only with numbers. The architecture and microprograms of a microprocessor determine all its instructions. These instructions are called the microprocessor's "instruction set." Programs in assembly and high-level languages are represented by instructions that use English- language-type statements. The programmer finds it relatively more convenient to write the programs in assembly or a high-level language than in machine language. However, a translator must be used to convert the assembly or high-level programs into binary machine language so that the microprocessor can execute the programs. This is shown in Figure 6.30.

An assembler translates a program written in assembly language into a machine language program. A compiler or interpreter, on the other hand, converts a high-level language program such as C or C++ into a machine language program. Assembly or high-level language programs are called "source codes." Machine language programs are known as "object codes." A translator converts source codes to object codes. Next, we discuss the three main types of programming language in more detail.

6.6.2 Machine Language

A microprocessor has a unique set of machine language instructions defined by its manufacturer. No two microprocessors by two different manufacturers have the same machine language instruction set. For example, the Intel 8086 microprocessor uses the code $01D8_{16}$ for its addition instruction whereas the Motorola 68000 uses the code $D282_{16}$. Therefore, a machine language program for one microcomputer will not usually run on another microcomputer of a different manufacturer.

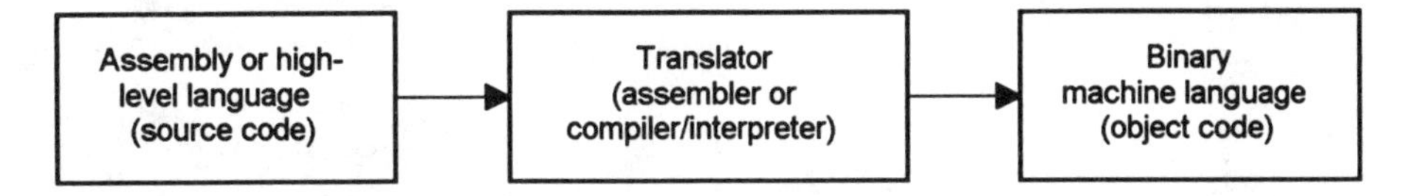

FIGURE 6.30 Translating assembly or a high-level language into binary machine language

At the most elementary level, a microprocessor program can be written using its instruction set in binary machine language. As an example, a program written for adding two numbers using the Intel 8086 machine language is

```
1011 1000 0000 0001 0000 0000
1011 1011 0000 0010 0000 0000
0000 0001 1101 1000
1111 0100
```

Obviously, the program is very difficult to understand, unless the programmer remembers all the 8086 codes, which is impractical. Because one finds it very inconvenient to work with 1's and 0's, it is almost impossible to write an error-free program at the first try. Also, it is very tiring for the programmer to enter a machine language program written in binary into the microcomputer's RAM. For example, the programmer needs a number of binary switches to enter the binary program. This is definitely subject to errors.

To increase the programmer's efficiency in writing a machine language program, hexadecimal numbers rather than binary numbers are used. The following is the same addition program in hexadecimal, using the Intel 8086 instruction set:

B80100
BB0200
01D8
F4

It is easier to detect an error in a hexadecimal program, because each byte contains only two hexadecimal digits. One would enter a hexadecimal program using a hexadecimal keyboards. A keyboard monitor program in ROM, provided by the manufacturer, controls the hexadecimal keyboard. This program converts each key actuation into binary machine language in order for the microprocessor to understand the program. However, programming in hexadecimal is not normally used.

6.6.3 Assembly Language

The next programming level is to use the assembly language. Each instruction in an assembly language is composed of four fields:

1. Label field
2. Instruction, mnemonic, or op-code field
3. Operand field
4. Comment field

As an example, a typical program for adding two 16-bit numbers written in 8086 assembly language is

Label	Mnemonic	Operand	Comment
`START`	`MOV`	`AX,1`	move 1 into AX
	`MOV`	`BX,2`	move 2 into BX
	`ADD`	`AX,BX`	add the contents of AX with BX
	`JMP`	`START`	jump to the beginning of the program

Obviously, programming in assembly language is more convenient than programming in machine language, because each mnemonic gives an idea of the type of operation it is supposed to perform. Therefore, with assembly language, the programmer does not have to find the numerical op-codes from a table of the instruction set, and programming efficiency is significantly improved.

The assembly language program is translated into binary via a program called an "assembler." The assembler program reads each assembly instruction of a program as ASCII characters and translates them into the respective binary op-codes. As an example, consider the HLT instruction for the 8086. Its binary op-code is 1111 0100. An assembler would convert HLT into 111 0100 as shown in Figure 6.31.

An advantage of the assembler is address computation. Most programs use addresses within the program as data storage or as targets for jumps or calls. When programming in machine language, these addresses must be calculated by hand. Every time the program changes size, all the addresses must be recomputed, a time-consuming task. The assembler solves this problem by allowing the programmer to assign a symbol to an address. The programmer may then reference that address elsewhere by using the symbol. The assembler computes the actual address for the programmer and fills it in automatically. ***One can obtain hands-on experience with a typical assembler for a microprocessor by downloading it from the Internet.***

Assembly Code	Binary form of ASCII Codes as Seen by Assembler	Binary OP Code Created by Assembler
H	0100 1000	
L	0100 1100	1111 0100
T	0101 0100	

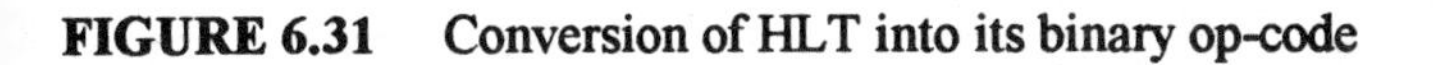

FIGURE 6.31 Conversion of HLT into its binary op-code

Most assemblers use two passes to assemble a program. This means that they read the input program text twice. The first pass is used to compute the addresses of all labels in the program. In order to find the address of a label, it is necessary to know the total length of all the binary code preceding that label. Unfortunately, however, that address may be needed in that preceding code. Therefore, the first pass computes the addresses of all labels and stores them for the next pass, which generates the actual binary code. Various types of assemblers are available today. We define some of them in the following paragraphs.

- **One-Pass Assembler.** This assembler goes through the assembly language program once and translates it into a machine language program. This assembler has the problem of defining forward references. This means that a `JUMP` instruction using an address that appears later in the program must be defined by the programmer after the program is assembled.
- **Two-Pass Assembler.** This assembler scans the assembly language program twice. In the first pass, this assembler creates a symbol table. A symbol table consists of labels with addresses assigned to them. This way labels can be used for `JUMP` statements and no address calculation has to be done by the user. On the second pass, the assembler translates the assembly language program into the machine code. The two-pass assembler is more desirable and much easier to use.
- **Macroassembler**. This type of assembler translates a program written in macrolanguage into the machine language. This assembler lets the programmer define all instruction sequences using macros. Note that, by using macros, the programmer can assign a name to an instruction sequence that appears repeatedly in a program. The programmer can thus avoid writing an instruction sequence that is required many times in a program by using macros. The macroassembler replaces a macroname with the appropriate instruction sequence each time it encounters a macroname.

  It is interesting to see the difference between a subroutine and a macroprogram. A specific subroutine occurs once in a program. A subroutine is executed by CALLing it from a main program. The program execution jumps out of the main program and then executes the subroutine. At the end of the subroutine, a RET instruction is used to resume program execution following the CALL SUBROUTINE instruction in the main program. A macro, on the other hand, does not cause the program execution to branch out of the main program. Each time a macro occurs, it is replaced with the appropriate instruction sequence in the main program. Typical advantages of using macros are shorter source programs and better program documentation. A typical disadvantage is that effects on registers and flags may not be obvious.

Conditional macroassembly is very useful in determining whether or not an instruction sequence is to be included in the assembly depending on a condition that is true or false. If two different programs are to be executed repeatedly based on a condition that can be either true or false, it is convenient to use conditional macros. Based on each condition, a particular program is assembled. Each condition and the appropriate program are typically included within IF and ENDIF pseudo-instructions.

- **Cross Assembler.** This type of assembler is typically resident in a processor and assembles programs for another for which it is written. The cross assembler program is written in a high-level language so that it can run on different types of processors that understand the same high-level language.
- **Resident Assembler.** This type of assembler assembles programs for a processor in which it is resident. The resident assembler may slow down the operation of the processor on which it runs.
- **Meta-assembler.** This type of assembler can assemble programs for many different types of processors. The programmer usually defines the particular processor being used.

As mentioned before, each line of an assembly language program consists of four fields: label, mnemonic or op-code, operand, and comment. The assembler ignores the comment field but translates the other fields. The label field must start with an uppercase alphabetic character. The assembler must know where one field starts and another ends. Most assemblers allow the programmer to use a special symbol or delimiter to indicate the beginning or end of each field. Typical delimiters used are spaces, commas, semicolons, and colons:

- Spaces are used between fields.
- Commas (,) are used between addresses in an operand field.
- A semicolon (;) is used before a comment.
- A colon (:) or no delimiter is used after a label.

To handle numbers, most assemblers consider all numbers as decimal numbers unless specified. Most assemblers will also allow binary, octal, or hexadecimal numbers. The user must define the type of number system used in some way. This is usually done by using a letter following the number. Typical letters used are

- B for binary
- Q for octal
- H for hexadecimal

Assemblers generally require hexadecimal numbers to start with a digit. A 0 is typically used if the first digit of the hexadecimal number is a letter. This is done to distinguish between numbers and labels. For example, most assemblers will require the number A5H to be represented as 0A5H.

Assemblers use pseudo-instructions or directives to make the formatting of the edited text easier. These pseudo-instructions are not directly translated into machine language instructions. They equate labels to addresses, assign the program to certain areas of memory, or insert titles, page numbers, and so on. To use the assembler directives or pseudo-instructions, the programmer puts them in the op-code field, and, if the pseudo-instructions require an address or data, the programmer places them in the label or data field. Typical pseudo-instructions are **ORIGIN** (ORG), **EQUATE** (EQU), **DEFINE BYTE** (DB), and DEFINE WORD (DW).

ORIGIN (ORG)

The pseudo-instruction **ORG** lets the programmer place the programs anywhere in memory. Internally, the assembler maintains a program-counter-type register called the "address counter." This counter maintains the address of the next instruction or data to be processed.

An ORG pseudo-instruction is similar in concept to the JUMP instruction. Recall that the JUMP instruction causes the processor to place a new address in the program counter. Similarly, the ORG pseudo-instruction causes the assembler to place a new value in the address counter.

Typical ORG statements are

```
ORG 7000H
CLC
```

The 8086 assembler will generate the following code for these statements:

```
7000 F8
```

Most assemblers assign a value of zero to starting address of a program if the programmer does not define this by means of an ORG.

Equate (EQU)

The pseudo-instruction **EQU** assigns a value in its operand field to an address in its label field. This allows the user to assign a numeric value to a symbolic name. The user can then use the symbolic name in the program instead of its numeric value. This reduces errors.

A typical example of EQU is START EQU 0200H, which assigns the value 0200 in hexadecimal to the label START. Another example is

```
PORTA     EQU  40H
          MOV  AL, 0FFH
          OUT  PORTA, AL
```

In this example, the EQU gives PORTA the value 40 hex, and FF hex is the data to be written into register AL by MOV AL, 0FFH. OUT PORTA, AL then outputs this data FF hex to port 40, which has already been equated to PORTA before.

Note that, if a label in the operand field is equated to another label in the label field, then the label in the operand field must be previously defined. For example, the EQU statement

```
BEGIN       EQU   START
```

will generate an error unless START is defined previously with a numeric value.

Define Byte (DB)

The pseudo-instruction DB is usually used to set a memory location to certain byte value. For example,

```
START       DB    45H
```

will store the data value 45 hex to the address START.

With some assemblers, the DB pseudo-instruction can be used to generate a table of data as follows:

```
            ORG   7000H
TABLE       DB    20H,30H,40H,50H
```

In this case, 20 hex is the first data of the memory location 7000; 30 hex, 40 hex, and 50 hex occupy the next three memory locations. Therefore, the data in memory will look like this:

```
7000  20
7001  30
7002  40
7003  50
```

Define Word (DW)

The pseudo-instruction DW is typically used to assign a 16-bit value to two memory locations. For example,

```
            ORG   7000H
START       DW    4AC2H
```

will assign C2 to location 7000 and 4A to location 7001. It is assumed that the assembler will assign the low byte first (C2) and then the high byte (4A).

With some assemblers, the DW pseudo-instruction can be used to generate a table of 16-bit data as follows:

```
            ORG   8000H
POINTER     DW    5000H,6000H,7000H
```

In this case, the three 16-bit values 5000H, 6000H, and 7000H are assigned to memory locations starting at the address 8000H. That is, the array would look like this:

```
8000    00
8001    50
8002    00
8003    60
8004    00
8005    70
```

Assemblers also use a number of housekeeping pseudo-instructions. Typical housekeeping pseudo-instructions are `TITLE`, `PAGE`, `END`, and `LIST`. The following are the housekeeping pseudo-instructions that control the assembler operation and its program listing.

`TITLE` prints the specified heading at the top of each page of the program listing. For example,

```
TITLE "Square Root Algorithm"
```

will print the name "Square Root Algorithm" on top of each page.

`PAGE` skips to the next line.

`END` indicates the end of the assembly language source program.

`LIST` directs the assembler to print the assembler source program.

In the following, assembly language instruction formats, instruction sets, and addressing modes available with typical microprocessors will be discussed.

Assembly Language Instruction Formats

Depending on the number of addresses specified, we have the following instruction formats:

- Three address
- Two address
- One address
- Zero address

Because all instructions are stored in the main memory, instruction formats are designed in such a way that instructions take less space and have more processing capabilities. It should be emphasized that the microprocessor architecture has considerable influence on a specific instruction format. The following are some important technical points that have to be considered while designing an instruction format:

- The size of an instruction word is chosen in such a way that it facilitates the specification of more operations by a designer. For example, with 4- and 8-bit op-code fields, we can specify 16 and 256 distinct operations respectively.

- Instructions are used to manipulate various data elements such as integers, floating-point numbers, and character strings. In particular, all programs written in a symbolic language such as C are internally stored as characters. Therefore, memory space will not be wasted if the word length of the machine is some integral multiple of the number of bits needed to represent a character. Because all characters are represented using typical 8-bit character codes such as ASCII or EBCDIC, it is desirable to have 8-, 16-, 32-, or 64-bit words for the word length.
- The size of the address field is chosen in such a way that a high resolution is guaranteed. Note that in any microprocessor, the ultimate resolution is a bit. Memory resolution is function of the instruction length, and in particular, short instructions provide less resolution. For example, in a microcomputer with 32K 16-bit memory words, at least 19 bits are required to access each bit of the word. (This is because 2^{15} = 32K and $2^4 = 16$)

The general form of a *three address instruction* is shown below:

<op-code> Addr1, Addr2, Addr3

Some typical three-address instructions are

```
MUL  A, B, C       ;     C  <- A * B
ADD  A, B, C       ;     C  <- A + B
SUB  R1, R2, R3    ;     R3 <- R1 - R2
```

In this specification, all alphabetic characters are assumed to represent memory addresses, and the string that begins with the letter R indicates a register. The third address of this type of instruction is usually referred to as the "destination address." The result of an operation is always assumed to be saved in the destination address.

Typical programs can be written using these three address instructions. For example, consider the following sequence of three address instructions

```
MUL  A, B, R1      ;     R1 <- A * B
MUL  C, D, R2      ;     R2 <- C * D
MUL  E, F, R3      ;     R3 <- E * F
ADD  R1, R2, R1    ;     R1 <- R1 + R2
SUB  R1, R3, Z     ;     Z  <- R1 - R3
```

This sequence implements the statement `Z = A * B + C * D - E * F`. *The three-address format is normally used by 32-bit microprocessors in addition to the other formats.*

If we drop the third address from the three-address format, we obtain the two-address format. Its general form is

<op-code> Addr1, Addr2

Some typical *two-address* instructions are

```
MOV    A, R1     ;    R1 <- A
ADD    C, R2     ;    R2 <- R2 + C
SUB    R1, R2    ;    R2 <- R2 - R1
```

In this format, the addresses Addr1 and Addr2 respectively represent source and destination addresses. The following sequence of two-address instructions is equivalent to the program using three-address format presented earlier:

```
MOV    A, R1     ;    R1 <- A
MUL    B, R1     ;    R1 <- R1 * B
MOV    C, R2     ;    R2 <- C
MUL    D, R2     ;    R2 <- R2 * D
MOV    E, R3     ;    R3 <- E
MUL    F, R3     ;    R3 <- R3 * F
ADD    R2, R1    ;    R1 <- R1 + R2
SUB    R3, R1    ;    R1 <- R1 - R3
MOV    R1, Z     ;    Z  <- R1
```

This format is predominant in typical general-purpose microprocessors such as the Intel 8086 and the Motorola 68000. Typical 8-bit microprocessors such as the Intel 8085 and the Motorola 6809 are accumulator based. In these microprocessors, the accumulator register is assumed to be the destination for all arithmetic and logic operations. Also, this register always holds one of the source operands. Thus, we only need to specify one address in the instruction, and therefore, this idea reduces the instruction length. The one-address format is predominant in 8-bit microprocessors. Some typical one-address instructions are

```
LDA    B    ;    Acc <- B
ADD    C    ;    Acc <- Acc + C
MUL    D    ;    Acc <- Acc * D
STA    E    ;    E <- Acc
```

The following program illustrates how one can translate the statement `Z = A * B + C * D - E * F` into a sequence of one-address instructions:

```
LDA    E     ;    Acc <- E
MUL    F     ;    Acc <- Acc * F
STA    T1    ;    T1  <- Acc
LDA    C     ;    Acc <- C
MUL    D     ;    Acc <- Acc * D
STA    T2    ;    T2 <- Acc
LDA    A     ;    Acc <- A
```

```
MUL     B    ;    Acc <- Acc * B
ADD     T2   ;    Acc <- Acc + T2
SUB     T1   ;    Acc <- Acc - T1
STA     Z    ;    Z <- Acc
```

In this program, T1 and T2 represent the addresses of memory locations used to store temporary results.

Instructions that do not require any addresses are called "zero-address instructions." All microprocessors include some zero-address instructions in the instruction set. Typical examples of zero-address instructions are `CLC` (clear carry) and `NOP`.

Typical Assembly Language Instruction Sets

An instruction set of a specific microprocessor consists of all the instructions that it can execute. The capabilities of a microprocessor are determined, to some extent, by the types of instructions it is able to perform. Each microprocessor has a unique instruction set designed by its manufacturer to do a specific task.

We discuss some of the instructions that are common to all microprocessors. We will group chunks of these instructions together which have similar functions. These instructions typically include

- **Data Processing Instructions.** These operations perform actual data manipulations. The instructions typically include arithmetic/logic operations and increment/decrement and rotate/shift operations.
- **Instructions for Controlling Microprocessor Operations.** These instructions typically include those that set the reset specific flags and halt or stop the microprocessor.
- **Data Movement Instructions.** These instructions move data from a register to memory and vice versa, between registers, and between a register and an I/O device.
- **Instructions Using Memory Addresses.** An instruction in this category typically contains a memory address, which is used to read a data word from memory into a microprocessor register or for writing data from a register into a memory location. Many instructions under data processing and movement fall in this category.
- **Conditional and Unconditional JUMPS.** These instructions typically include one of the following:
  1. Unconditional `JUMP`, which always transfers the memory address specified in the instruction into the program counter.
  2. Conditional `JUMP`, which transfers the address portion of the instruction into the program counter based on the conditions set by one of the status flags in the flag register.

Typical Assembly Language Addressing Modes

One of the tasks performed by a microprocessor during execution of an instruction is the determination of the operand and destination addresses. The manner in which a microprocessor accomplishes this task is called the "addressing mode." Now, let us present the typical microprocessor addressing modes, relating them to the instruction sets of Motorola 68000.

An instruction is said to have "implied or inherent addressing mode" if it does not have any operand. For example, consider the following instruction: RTS, which means "return from a subroutine to the main program." The RTS instruction is a no-operand instruction. The program counter is implied in the instruction because although the program counter is not included in the RTS instruction, the return address is loaded in the program counter after its execution.

Whenever an instruction/operand contains data, it is called an "immediate mode" instruction. For example, consider the following 68000 instruction:

```
ADD     #15, D0    ;     D0 <- D0 + 15
```

In this instruction, the symbol # indicates to the assembler that it is an immediate mode instruction. This instruction adds 15 to the contents of register D0 and then stores the result in D0. An instruction is said to have a register mode if it contains a register as opposed to a memory address. This means that the operand values are held in the microprocessor registers. For example, consider the following 68000 instruction:

```
ADD     D1, D0     ; D0 <- D1 + D0
```

This ADD instruction is a two-operand instruction. Both operands (source and destination) have register mode. The instruction adds the 16-bit contents of D0 to the 16-bit contents of D1 and stores the 16-bit result in D0.

An instruction is said to have an absolute or direct addressing mode if it contains a memory address in the operand field. For example, consider the 68000 instruction

```
ADD  3000, D2
```

This instruction adds the 16-bit contents of memory address 3000 to the 16-bit contents of D2 and stores the 16-bit result in D2. The source operand to this ADD instruction contains 3000 and is in absolute or direct addressing mode. When an instruction specifies a microprocessor register to hold the address, the resulting addressing mode is known as the "register indirect mode." For example, consider the 68000 instruction:

```
CLR (A0)
```

This instruction clears the 16-bit contents of a memory location whose address is in register A0 to zero. The instruction is in register indirect mode.

The conditional branch instructions are used to change the order of execution of a program based on the conditions set by the status flags. Some microprocessors use conditional branching using the absolute mode. The op-code verifies a condition set by a particular status flag. If the condition is satisfied, the program counter is changed to the value of the operand address (defined in the instruction). If the condition is not satisfied, the program counter is incremented, and the program is executed in its normal order.

Typical 16-bit microprocessors use conditional branch instructions. Some conditional branch instructions are 16 bits wide. The first byte is the op-code for checking a particular flag. The second byte is an 8-bit offset, which is added to the contents of the program counter if the condition is satisfied to determine the effective address. This offset is considered as a signed binary number with the most significant bit as the sign bit. It means that the offset can vary from -128_{10} to $+127_{10}$ (0 being positive).

Consider the following 68000 example, which uses the branch not equal (BNE) instruction:

```
BNE 8
```

Suppose that the program counter contains 2000 (address of the next instruction to be executed) while executing this BNE instruction. Now, if $Z = 0$, the microprocessor will load $2000 + 8 = 2008$ into the program counter and program execution resumes at address 2008. On the other hand, if $Z = 1$, the microprocessor continues with the next instruction.

In the last example the program jumped forward, requiring positive offset. An example for branching with negative offset is

```
BNE -14
```

Suppose that the current program counter value = 2004_{16}

= 0010 0000 0000 0100

offset = 2's complement of 14_{10} = $F2_{16}$

= [1111 1111] 1111 0010

ignore → 1 0001 1111 1111 0110

1 F F 6_{16}

reflect this 1 to the high byte
(sign extension)

Therefore, to branch backward to $1FF6_{16}$, the assembler uses an offset of F2 following the op-code for BNE.

Subroutine Calls in Assembly Language

It is sometimes desirable to execute a common task many times in a program. Consider the case when the sum of squares of numbers is required several times in a program. One could write a sequence of instructions in the main program for carrying out the sum of squares every time it is required. This is all right for short programs. For long programs, however, it is convenient for the programmer to write a small program known as a "subroutine" for performing the sum of squares, and then call this program each time it is needed in the main program.

Therefore, a subroutine can be defined as a program carrying out a particular function that can be called by another program known as the "main program." The subroutine only needs to be placed once in memory starting at a particular memory location. Each time the main program requires this subroutine, it can branch to it, typically by using a jump to subroutine (`JSR`) instruction along with its starting address. The subroutine is then executed. At the end of the subroutine, a `RETURN` instruction takes control back to the main program.

The 68000 includes two subroutine call instructions. Typical examples include `JSR 4000` and `BSR 24`. `JSR 4000` is an instruction using absolute mode. In response to the execution of `JSR`, the 68000 saves (pushes) the current program counter contents (address of the next instruction to be executed) onto the stack. The program counter is then loaded, with 4000 included in the `JSR` instruction. The starting address of the subroutine is 4000. The `RTS` (return from subroutine) at the end of the subroutine reads (pops) the return address saved into the stack before jumping to the subroutine into the program counter. The program execution thus resumes in the main program. `BSR 24` is an instruction using relative mode. This instruction works in the same way as the `JSR 4000` except that displacement `24` is added to the current program counter contents to jump to the subroutine.

6.6.4 High-Level Languages

As mentioned before, the programmer's efficiency with assembly language increases significantly compared to machine language. However, the programmer needs to be well acquainted with the microprocessor's architecture and its instruction set. Further, the programmer has to provide an op-code for each operation that the microprocessor has to carry out in order to execute a program. As an example, for adding two numbers, the programmer would instruct the microprocessor to load the first number into a register, add the second number to the register, and then store the result in memory. However, the programmer might find it tedious to write all the steps required for a large program. Also, to become a reasonably good assembly language programmer, one needs to have a lot of experience.

High-level language programs composed of English-language-type statements rectify all these deficiencies of machine and assembly language programming. The programmer does not need to be familiar with the internal microprocessor structure or its instruction set. Also, each statement in a high-level language corresponds to a number of assembly or machine language instructions. For example, consider the statement `F = A + B` written in a high-level language called FORTRAN. This single statement adds the contents of A with B and stores the result in F. This is equivalent to a number of steps in machine or assembly language, as mentioned before. It should be pointed out that the letters A, B, and F do not refer to particular registers within the microprocessor. Rather, they are memory locations.

A number of high-level languages such as C, C++, and Java are widely used these days. Typical microprocessors, namely, the Intel 8086, the Motorola 68000, and others, can be programmed using these high-level languages. A high-level language is a problem-oriented language. The programmer does not have to know the details of the architecture of the microprocessor and its instruction set. Basically, the programmer follows the rules of the particular language being used to solve the problem at hand. A second advantage is that a program written in a particular high-level language can be executed by two different microcomputers, provided they both understand that language. For example, a program written in C for an Intel 8086–based microcomputer will run on a Motorola 68000-based microcomputer because both microprocessors have a compiler to translate the C language into their particular machine language; minor modifications are required for input/output programs.

As mentioned before, like the assembly language program, a high-level language program requires a special program for converting the high-level statements into object codes. This program can be either an interpreter or a compiler. They are usually very large programs compared to assemblers.

An interpreter reads each high-level statement such as `F = A + B` and directs the microprocessor to perform the operations required to execute the statement. The interpreter converts each statement into machine language codes but does not convert the entire program into machine language codes prior to execution. Hence, it does not generate an object program. Therefore, an interpreter is a program that executes a set of machine language instructions in response to each high-level statement in order to carry out the function. A compiler, however, converts each statement into a set of machine language instructions and also produces an object program that is stored in memory. This program must then be executed by the microprocessor to perform the required task in the high-level program. In summary, an interpreter executes each statement as it proceeds, without generating an object code, whereas a compiler converts a high-level program into an object program that is stored in memory. This program is then executed. *Compilers normally provide inefficient machine codes because of the general guidelines that must be followed for designing them. C, C++, and Java are the only high-level languages that include Input/Output instructions. However, the compiled codes generate many more lines of machine code than an equivalent assembly language program. Therefore, the assembled program will take up less memory space and will execute much faster compared to the compiled C, C++, or Java codes.* I/O programs written in C are compared with assembly language programs written in 8086 and 68000 in Chapters 9 and 10. C language is a popular high-level language, the C++ language, based on C, is also very popular, and Java, developed by Sun Microsystems, is gaining wide acceptance.

Therefore, one of the main uses of assembly language is in writing programs for real-time applications. "Real-time" means that the task required by the application must be completed before any other input to the program can occur which will change its operation. Typical programs involving non-real-time applications and extensive mathematical computations may be written in C, C++, or Java. A brief description of these languages is given in the following.

C Language

The C Programming language was developed by Dennis Ritchie of Bell Labs in 1972. *C has become a very popular language for many engineers and scientists, primarily because it is portable except for I/O and however, can be used to write programs requiring I/O operations with minor modifications.* This means that a program written in C for the 8086 will run on the 68000 with some modifications related to I/O as long as C compilers for both microprocessors are available.

C is case sensitive. This means that uppercase letters are different from lowercase letters. Hence Start and start are two different variables. C is a general-purpose programming language and is found in numerous applications as follows:

- **Systems Programming.** Many operating systems, compilers, and assemblers are written in C. Note that an operating system typically is included with the personal computer when it is purchased. The operating system provides an interface between the user and the hardware by including a set of commands to select and execute the software on the system
- **Computer-Aided Design (CAD) Applications.** CAD programs are written in C. Typical tasks to be accomplished by a CAD program are logic synthesis and simulation.
- **Numerical Computation.** To solve mathematical problems such as integration and differentiation
- **Other Applications.** These include programs for printers and floppy disk controllers, and digital control algorithms using single-chip microcomputers.

A C program may be viewed as a collection of functions. Execution of a C program will always begin by a call to the function called "main." This means that all C programs should have its main program named as `main`. However, one can give any name to other functions.

A simple C program that prints "I wrote a C-program" is

```
/* First C-program */
#include <stdio.h>
main ( )
{
   printf("I wrote a C-program");
}
```

Here, `main` is a function of no arguments, indicated by ( ). The parenthesis must be present even if there are no arguments. The braces { } enclose the statements that make up the function.

The line `printf("I wrote a C-program");` is a function call that calls a function named `printf`, with the argument "`I wrote a C-program`." `printf` is a library function that prints output on the terminal. Note that `/* */` is used to enclose comments. These are not translated by the compiler.

A variation of the C program just described is

```
/* Another C program */
#include <stdio.h>
main ( )
{
   printf("I wrote");
   printf(" a C-");
   printf("program");
   printf("\n");
}
```

Here, `#include` is a preprocessor directive for the C language compiler. These directives give instructions to the compiler that are performed before the program is compiled. The directive `#include <stdio.h>` inserts additional statements in the program. These statements are contained in the file stdio.h. The file `stdio.h` is included with the standard C library. The `stdio.h` file contains information related to the input/output statement.

The `\n` in the last line of the program is C notation for the newline character. Upon printing, the cursor moves forward to the left margin on the next line. `printf` never supplies a newline automatically. Therefore, multiple `printf`'s may be used to output "I wrote a C-program" on a single line in a few steps. The escape sequence `\n` can be used to print three statements on three different lines. An illustration is given in the following:

```
#include <stdio.h>
main ( )
{
   printf("I wrote a C-Program \n");
   printf("This will be printed on a new line \n");
   printf("So also is this line \n");
}
```

All variables in C must be declared before use, normally at the start of the function before any executable statements. The compiler provides an error message if one forgets a declaration. A declaration includes a type and a list of variables that have that type. For example, the declaration `int a, b` implies that the variables `a` and `b` are integers. Next, write a program to add and subtract two integers a and b where a = 100 and b = 200. The C program is

```
#include <stdio.h>
main ( )
{
   int a = 100, b = 200;
   /*a and b are integers */
   printf("The sum is: %d \n", a + b);
   printf("The difference is: %d \n", a - b);
}
```

The `%d` in the `printf` statement represents "decimal integer." Note that `printf` is not part of the C language; there is no input or output defined in C itself. `printf` is a function that is contained in the standard library of routines that can be accessed by C programs. The values of `a` and `b` can be entered via the keyboard by using the scanf function. The scanf allows the programmer to enter data from the keyboard. A typical expression for `scanf` is

```
scanf("%d%d", &a, &b);
```

This expression indicates that the two values to be entered via the keyboard are in decimal. These two decimal numbers are to be stored in addresses `a` and `b`. Note that the symbol `&` is an address operator.

The C program for adding and subtracting two integers `a` and `b` using `scanf` is

```
/* C Program that performs basic I/O */
#include <stdio.h>
main ( )
{
   int a,b;
   printf("Input two integers: ");
   scanf("%d%d", &a, &b);
   printf("Their sum is: %d\n", a + b);
   printf("Their difference is: %d\n", a - b);
}
```

In summary, writing a working C program involves four steps as follows:

Step 1: Using a text editor, prepare a file containing the C code. This file is called the "source file."

Step 2: Preprocess the code. The preprocessor makes the code ready for compiling. The preprocessor looks through the source file for lines that start with a #. In the previous programming examples, `#include <stdio.h>` is a preprocessor. This preprocessor instruction copies the contents of the standard header file `stdio.h` into the source code. This header file `stdio.h` describes typical input/output functions such as `scanf( )` and `printf( )` functions.

Step 3: The compiler translates the preprocessed code into machine code. The output from the compiler is called object code.

Step 4: The linker combines the object file with code from the C libraries. For instance, in the examples shown here, the actual code for the library function `printf( )` is inserted from the standard library to the object code by the linker. The linker generates an executable file. Thus, the linker makes a complete program.

Before writing C programs, the programmer must make sure that the computer runs either the UNIX or MS-DOS operating system. Two essential programming tools are required. These are a text editor and a C compiler. The text editor is a program provided with a computer system to create and modify compiler files. The C compiler is also a program that translates C code into machine code.

C++

C++ is a modified version of C language. C++ was developed by Bjarne Stroustrup of Bell Labs in 1980. It includes all features of C and also supports object-oriented programming (OOP). A program can be divided into subprograms using OOP. Each subprogram is an independent object with its own instructions and data. Thus, complexity of programming is reduced. It is therefore easier for the programmer to manage larger programs.

All OOP languages including C++, have three characteristics: encapsulation, polymorphism, and inheritance. *Encapsulation* is a technique that keeps code and data together in such a way that they are protected form outside interference and misuse. A subprogram thus created is called an "object."

Code, data, or both may be private or public. Private code and/or data may be accessed by another part of the same object. On the other hand, public code and/or data may be accessed by a program resident outside the object containing them. One of the most important characteristic of C++ is the class. The class declaration is a technique for creating an object. Note that a class consists of data and functions.

Encapsulation is available with C to some extent. For example, when a library function such as `printf` is used, one uses a black box program. When `printf` is used, several internal variables are created and intialized that are not accessible to the programmer.

Polymorphism (from Greek word meaning "several forms") allows one to define a general class of actions. Within a general class, the specific action is determined by the type of data. For example, in C, the absolute value actions `abs( )` and `fabs( )` compute the absolute values of an integer and a floating point number respectively. In C++, on the other hand, one absolute value action, `abs( )` is used for both data types. The type of data is then used to call `abs( )` to determine which specific version of the function is actually used. Thus, one function name for two different data items is used.

Inheritance is the ability by which one class called subclass obtains the properties of another class called a superclass. Inheritance is convenient for code reusability. Inheritance supports hierarchy classes.

Following are some basic differences between C and C++:

1. In C, one must use `void` with the prototype for a function with no arguments. For example, in C, the prototype `int rand(void);` returns an integer that is a random number.

   In C++, the `void` is optional. Therefore, in C++, the prototype for `rand( )` can be written as `int rand( );`. Of course, `int rand(void);` is a valid prototype in C++. This means that both prototypes are allowed in C++

2. C++ can use the C type of comment mechanism. That is, a comment can start with `/*` and end with `*/`. C++ can also use a simple line comment that starts with a `//` and stops at the end of the line terminated by a carriage return. Typically, C++ uses C-like comments for multiline comments and the C++ comment mechanism for short comments.

3. In C++, local variables can be declared anywhere. In contrast, in C, local variables must be declared at the start of a block before any action statements.

4. In C++, all functions need to be prototyped. In C, prototypes are optional. Note that a function prototype allows the compiler to check that the function is called with the proper number and types of arguments. It also tells the compiler the type of value that the function is supposed to return. In C, if the function prototype is omitted, the compiler will return an integer. An example of a prototype function is `int abs(int n)` , this provides an integer that is an absolute value of `n`.

Java

Introduced in 1991 by Sun MicroSystems, Java is based on C++ and is a true object oriented language. That is, everything in a Java program is an object and everything is obtained from a single object class.

A Java program must include at least one class. A class includes data type declarations and statements. Every Java standalone program requires a main method at the beginning. Java only supports class methods and not separate functions. There is no preprocessor in Java. However, there is an `import` statement, which is similar to the `#include` preprocessor statement in C. The purpose of the `import` statement in Java is to instruct the interpreter to load the class, which exists in another compilation statement. Java uses the same comment syntax, `/* */` and `//`, as C and C++. In addition, a special comment syntax, `/** */`, that can precede declarations is used in Java.

Java does not require pointers. In C, a pointer may be substituted for the array name to access array elements. In Java, arrays are created by using the "new" operator by including the size of the array in the new expression (rather than in the declaration) as follows:

```
int array [ ] = new int[6];
```

Also, all arrays store the specified size in a variable named `length` as follows:

```
int stringsize = array.length;
```

Therefore, in Java, arrays and strings are not subject to the errors or confusion that is common to arrays and strings in C.

6.7 Monitors

A monitor consists of a number of subroutines grouped together to provide "intelligence" to a microcomputer system. This intelligence gives the microcomputer system the capabilities for debugging a user program, system design, and displays. The monitor is usually sold by the microcomputer manufacturer in a ROM, PROM, or EPROM. Note that when the microprocessor, I/O, and memory are connected together to design a microcomputer, without a monitor program a microprocessor development system is required in order to develop the programs. An example of a monitor is the Intel SDK-86 monitor, which contains debugging routines, a display routine, and many other programs.

6.8 Flowcharts

Before writing an assembly language program for a specific operation, it is convenient to represent the program in a schematic form called *flowchart.* A brief listing of the basic shapes used in a flowchart and their functions is given in Figure 6.32.

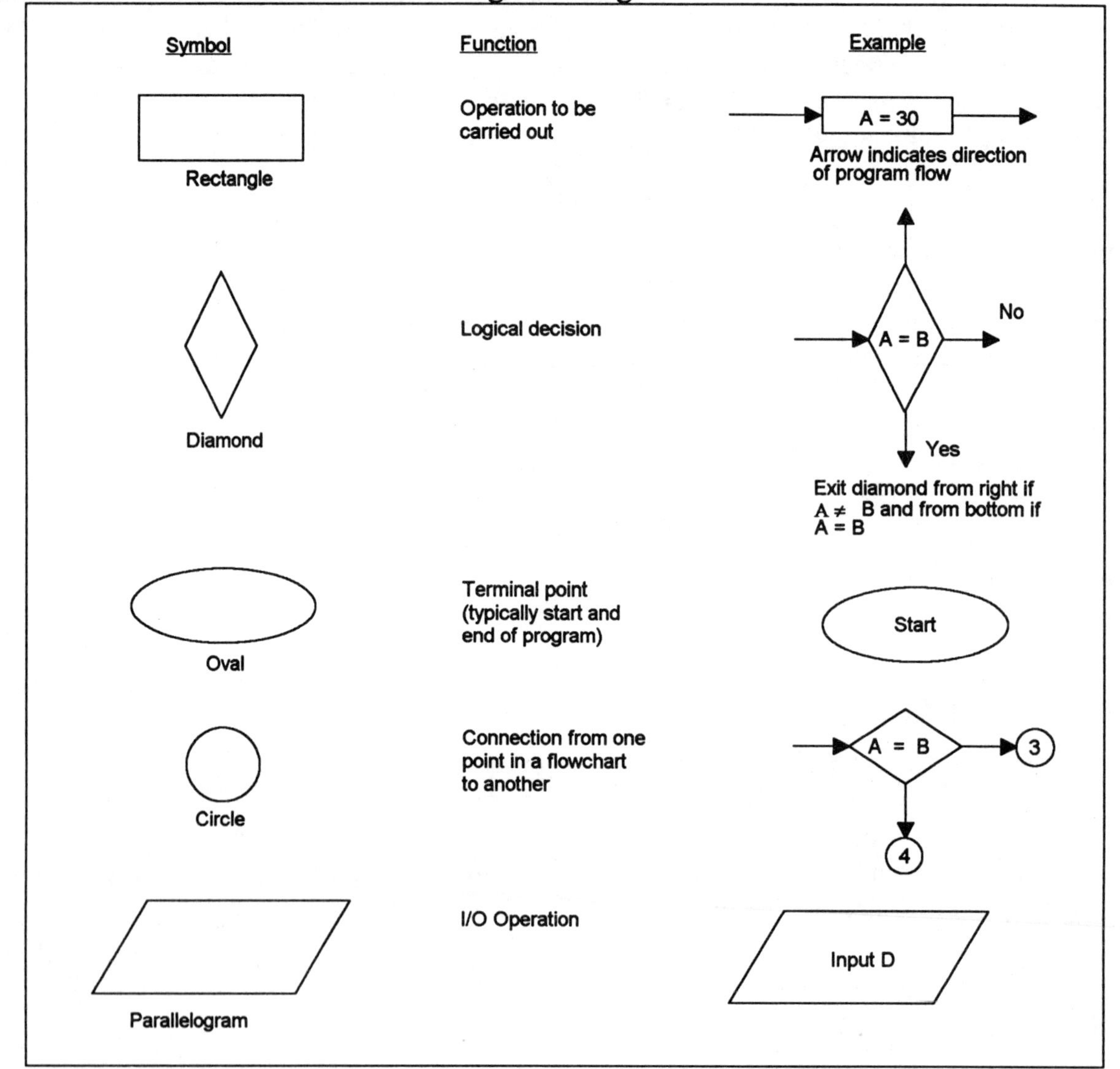

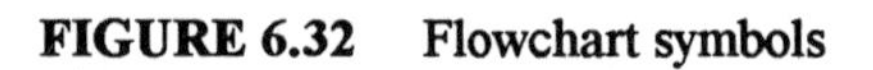
FIGURE 6.32 Flowchart symbols

6.9 Basic Features of Microcomputer Development Systems

A microcomputer development system is a tool that allows the designer to develop, debug, and integrate error-free application software in microprocessor systems.

Development systems fall into one of two categories: systems supplied by the device manufacturer (nonuniversal systems) and systems built by after-market manufacturers (universal systems). The main difference between the two categories is the range of microprocessors that a system will accommodate. Nonuniversal systems are supplied by the microprocessor manufacturer (Intel, Motorola) and are limited to use for the particular microprocessor manufactured by the supplier. In this manner, an Intel development system may not be used to develop a Motorola-based system. The universal development systems (Hewlett-Packard, Tektronix) can develop hardware and software for several microprocessors.

Within both categories of development systems, there are basically three types available: single-user systems, time-shared systems, and networked systems. A single-user system consists of one development station that can be used by one user at a time. Single-user systems are low in cost and may be sufficient for small systems development. Time-shared systems usually consist of a "dumb" type of terminal connected by data lines to a centralized microcomputer-based system that controls all operations. A networked system usually consists of a number of smart cathode ray tubes (CRTs) capable of performing most of the development work and can be connected over data lines to a central microcomputer. The central microcomputer in a network system usually is in charge of allocating disk storage space and will download some programs into the user's workstation microcomputer. A microcomputer development system is a combination of the hardware necessary for microprocessor design and the software to control the hardware. The basic components of the hardware are the central processor, the CRT terminal, mass storage device (floppy or hard disk), and usually an in-circuit emulator (ICE).

In a single-user system, the central processor executes the operating system software, handles the input/output (I/O) facilities, executes the development programs (editor, assembler, linker), and allocates storage space for the programs in execution. In a large multiuser networked system the central processor may be responsible for the I/O facilities and execution of development programs. The CRT terminal provides the interface between the user and the operating system or program under execution. The user enters commands or data via the CRT keyboard, and the program under execution displays data to the user via the CRT screen. Each program (whether system software or user program) is stored in an ordered format on disk. Each separate entry on the disk is called a *file*. The operating system software contains the routines necessary to interface between the user and the mass storage unit. When the user requests a file by a specific *file name*, the operating system finds the program stored on disk by the file name and loads it into mean memory. More advanced development systems contain *memory management* software that protects a user's files from unauthorized modification by another user. This is accomplished via a unique user identification code called USER ID. A user can only access files that have the user's unique code. The equipment listed here makes up a basic development system, but most systems have other devices such as printers and EPROM and PAL programmers attached. A printer is needed to provide the user with a hard copy record of the program under development.

After the application system software has been completely developed and debugged, it needs to be permanently stored for execution in the target hardware. The EPROM (erasable/programmable read-only memory) programmer takes the machine code and programs it into an EPROM. EPROMs are more generally used in system development because they may be erased and reprogrammed if the program changes. EPROM programmers usually interface to circuits particularly designed to program a specific EPROM.

Most development systems support one or more in-circuit emulators (ICEs). The ICE is one of the most advanced tools for microprocessor hardware development. To use an ICE, the microprocessor chip is removed from the system under development (called the target processor) and the emulator is plugged into the microprocessor socket. The ICE will functionally and electrically act identically to the target processor with the exception that the ICE is under the control of development system software. In this manner the development system may exercise the hardware that is being designed and monitor all status information available about the operation of the target processor. Using an ICE, processor register contents may be displayed on the CRT and operation of the hardware observed in a single-stepping mode. In-circuit emulators can find hardware and software bugs quickly that might take many hours to locate using conventional hardware testing methods.

Architectures for development systems can be generally divided into two categories: the master/slave configuration and the single-processor configuration. In a master/slave configuration, the master (host) processor controls the mass storage device and processes all I/O (CRT, printer). The software for development systems is written for the master processor, which is usually not the same as the slave (target) processor. The slave microprocessor is typically connected to the user prototype via a connector which links the slave processor to the master processor.

Some development systems such as the HP 64000 completely separate the system bus from the emulation bus and therefore use a separate block of memory for emulation. This separation allows passive monitoring of the software executing on the target processor without stopping the emulation process. A benefit of the separate emulation facilities allows the master processor to be used for editing, assembling, and so on while the slave processor continues the emulation. A designer may therefore start an emulation running, exit the emulator program, and at some future time return to the emulation program.

Another advantage of the separate bus architecture is that an operating system needs to be written only once for the master processor and will be used no matter what type of slave processor is being emulated. When a new slave processor is to be emulated, only the emulator probe needs to be changed.

A disadvantage of the master/slave architecture is that it is expensive. In single-processor architecture, only one processor is used for system operation and target emulation. The single processor does both jobs, executing system software as well as acting as the target processor. Because there is only one processor involved, the system software must be rewritten for each type of processor that is to be emulated. Because the system software must reside in the same memory used by the emulator, not all memory will be available to the emulation process, which may be a disadvantage when large prototypes are being developed. The single-processor systems are inexpensive.

The programs provided for microprocessor development are the operating system, editor, assembler, linker, compiler, and debugger. The operating system is responsible for executing the user's commands. The operating system handles I/O functions, memory management, and loading of programs from mass storage into RAM for execution. The editor allows the user to enter the source code (either assembly language or some high-level language) into the development system.

Almost all current microprocessor development systems use the character-oriented editor, more commonly referred to as the screen editor. The editor is called a "screen editor" because the text is dynamically displayed on the screen and the display automatically updates any edits made by the user.

The screen editor uses the pointer concept to point to the character(s) that need editing. The pointer in a screen editor is called the "cursor," and special commands allow the user to position the cursor to any location displayed on the screen. When the cursor is positioned, the user may insert characters, delete characters, or simply type over the existing characters.

Complete lines may be added or deleted using special editor commands. By placing the editor in the insert mode, any text typed will be inserted at the cursor position when the cursor is positioned between two existing lines. If the cursor is positioned on a line to be deleted, a single command will remove the entire line from the file.

Screen editors implement the editor commands in different fashions. Some editors use dedicated keys to provide some cursor movements. The cursor keys are usually marked with arrows to show the direction of the cursor movement. More advanced editors (such as the HP 64000) use soft keys. A soft key is an unmarked key located on the keyboard directly below the bottom of the CRT screen. The mode of the editor decides what functions the keys are to perform. The function of each key is displayed on the screen directly above the appropriate key. The soft key approach is valuable because it allows the editor to reassign a key to a new function when necessary.

The source code generated on the editor is stored as ASCII or text characters and cannot be executed by a microprocessor. Before the code can be executed, it must be converted to a form accessible by the microprocessor. An assembler is the program used to translate the assembly language source code generated with an editor into object code (machine code), which may be executed by a microprocessor.

The output file from most development system assemblers is an object file. The object file is usually relocatable code that may be configured to execute at any address. The function of the linker is to convert the object file to an *absolute* file, which consists of the actual machine code at the correct address for execution. The absolute files thus created are used for debugging and finally for programming EPROMs.

Debugging a microprocessor-based system may be divided into two categories: software debugging and hardware debugging. Both debugging processes are usually carried out separately because software debugging can be carried out on an out-of-circuit emulator (OCE) without having the final system hardware.
The usual software development tools provided with the development system are

- Single-step facility
- Breakpoint facility

A single stepper simply allows the user to execute the program being debugged one instruction at a time. By examining the register and memory contents during each step, the debugger can detect such program faults as incorrect jumps, incorrect addressing, erroneous op-codes, and so on. A breakpoint allows the user to execute an entire section of a program being debugged.

There are two types of breakpoints: hardware and software. The hardware breakpoint uses the hardware to monitor the system address bus and detect when the program is executing the desired breakpoint location. When the breakpoint is detected, the hardware uses the processor control lines to halt the processor for inspection or cause the processor to execute an interrupt to a breakpoint routine. Hardware breakpoints can be used to debug both ROM- and RAM-based programs. Software breakpoint routines may only operate on a system with the program in RAM because the breakpoint instruction must be inserted into the program that is to be executed.

Single-stepper and breakpoint methods complement each other. The user may insert a breakpoint at the desired point and let the program execute up to that point. When the program stops at the breakpoint the user may use a single-stepper to examine the program one instruction at a time. Thus, the user can pinpoint the error in a program.

There are two main hardware-debugging tools: the logic analyzer and the in-circuit emulator. Logic analyzers are usually used to debug hardware faults in a system. The logic analyzer is the digital version of an oscilloscope because it allows the user to view logic levels in the hardware. In-circuit emulators can be used to debug and integrate software and hardware. PC-based workstations are extensively used as development systems.

6.10 System Development Flowchart

The total development of a microprocessor-based system typically involves three phases: software design, hardware design, and program diagnostic design. A systems programmer will be assigned the task of writing the application software, a logic designer will be assigned the task of designing the hardware, and typically both designers will be assigned the task of developing diagnostics to test the system. For small systems, one engineer may do all three phases, while on large systems several engineers may be assigned to each phase. Figure 6.33 shows a flowchart for the total development of a system. Notice that software and hardware development may occur in parallel to save time.

The first step in developing the software is to take the system specifications and write a flowchart to accomplish the desired tasks that will implement the specifications. The assembly language or high-level source code may now be written from the system flowchart. The complete source code is then assembled. The assembler is the object code and a program listing. The object code will be used later by the linker. The program listing may be sent to a disk file for use in debugging, or it may be directed to the printer.

The linker can now take the object code generated by the assembler and create the final absolute code that will be executed on the target system. The emulation phase will take the absolute code and load it into the development system RAM. From here, the program may be debugged using breakpoints or single stepping.

Working from the system specifications, a block diagram of the hardware must be developed. The logic diagram and schematics may now be drawn using the block diagram as a guide, and a prototype may now be constructed an tested for wiring errors. When the prototype has been constructed it may be debugged for correct operation using standard electronic testing equipment such as oscilloscopes, meters, logic probes, and logic analyzers, all with test programs created for this purpose. After the prototype has been debugged electrically, the development system in-circuit emulator may be used to check it functionally. The ICE will verify the memory map, correct I/O operation, and so on. The next step in system development is to validate the complete system by running operational checks on the prototype with the finalized application software installed. The EPROMs and/or PALs are then programmed with the error-free programs.

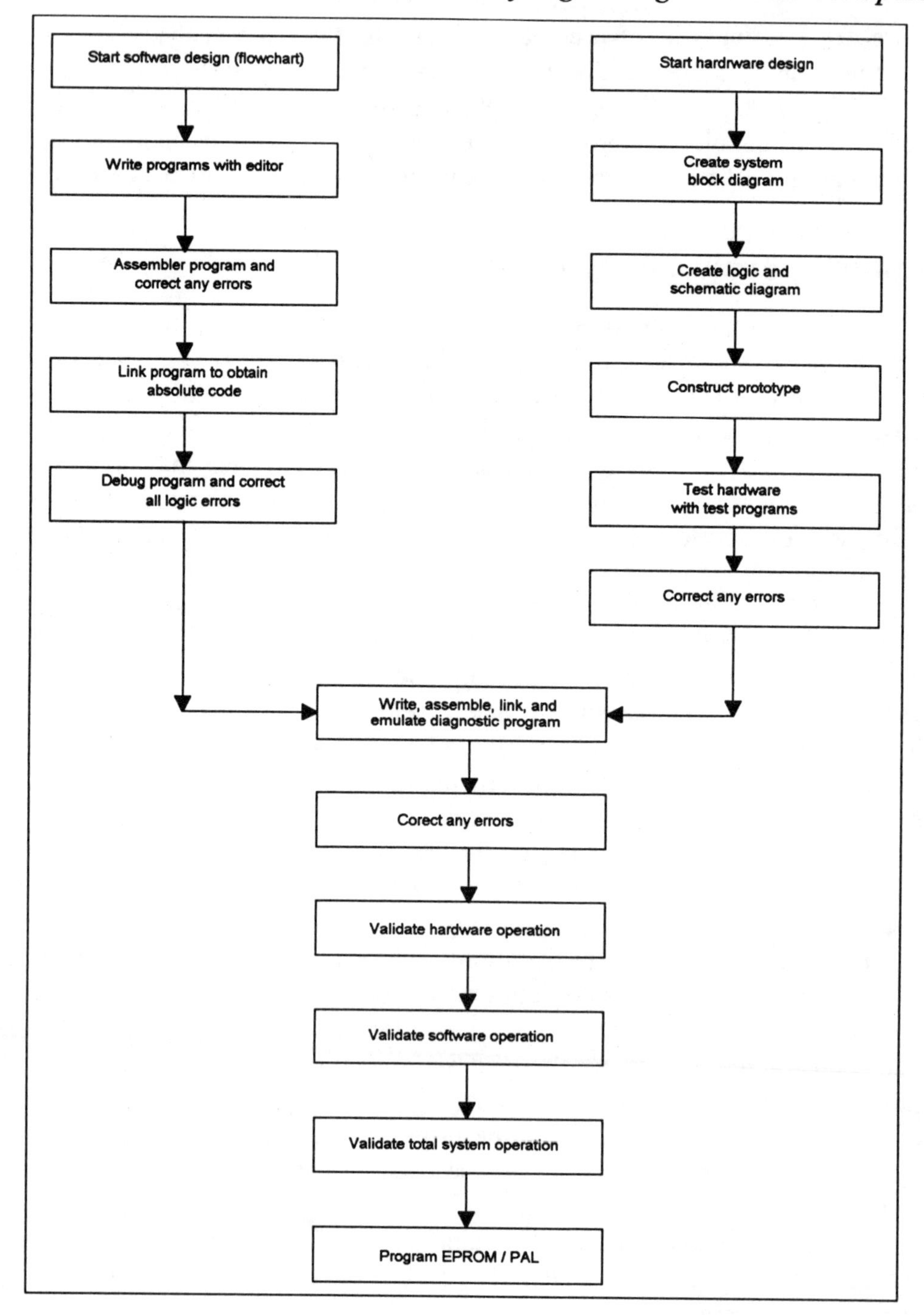

FIGURE 6.33 Microprocessor system development flowchart

6.11 Typical Features of 32-bit and 64-bit Microprocessors

This section describes the basic aspects of typical 32-bit and 64-bit microprocessors. Topics include on-chip features such as pipelining, memory management, floating-point, and cache memory implemented in typical 32-bit and 64-bit microprocessors.

Typical 16-, 32-, and 64-bit microprocessors are designed using HCMOS technology. This means that the unused inputs must not be kept floating; they must be connected to power, ground, or outputs of other chips as required by an application. Motorola's 16-bit MC68000 (68HC000), which can operate on 32-bit words internally, and the problematic 32-bit microprocessor Intel iAPX432 were introduced in 1980. Soon afterward, the concept of "mainframe on a chip" or "micromainframe" was used to indicate the capabilities of these microprocessors and to distinguish them from previous 8- and 16-bit microprocessors.

The introduction of several 32-bit microprocessors revolutionized the microprocessor world. The performance of these 32-bit microprocessors is actually more comparable to that of superminicomputers such as Digital Equipment Corporation's VAX11/750 and VAX11/780. Designers of 32-bit microprocessors have implemented many powerful features of these mainframe computers to increase the capabilities of the microprocessor chip sets. These include pipelining, on-chip cache memory, memory management, and floating-point arithmetic. *These topics are covered in more detail in Chapter 11.*

Pipelining is the technique in which instruction fetch and execute cycles are overlapped. This method allows simultaneous preparation for execution of one or more instructions while another instruction is being executed. Pipelining was used for many years in mainframe and minicomputer CPUs to speed up the instruction execution time of these machines. The 32-bit microprocessors implement the pipelining concept and simultaneously operate on several 32-bit words, which may represent different instructions or part of a single instruction.

Although pipelining greatly increases the rate of execution of nonbranching code, pipelines must be emptied and refilled each time a branch or jump instruction is in the code. This may slow down the processing rate for code with many branches or jumps. Thus, there is an optimum pipeline depth, which is strongly related to the instruction set, architecture, and gate density attainable on the processor chip. For many of the applications run on the 32-bit microprocessors, the three-stage pipeline is considered a reasonably optimal depth.

With memory management, virtual memory techniques, traditionally a feature of mainframes, are also implemented as on-chip hardware on typical 32-bit microprocessors. This allows programmers to write programs much larger than those that could fit in the main memory space available to the microprocessors; the programs are simply stored on a secondary device, such as a disk drive, and portions of the program are swapped into main memory as needed.

Segmentation circuitry has been included in many 32-bit microprocessor chips. With this technique, blocks of code called "segments," which correspond to modules of the program and have varying sizes set by the programmer or compiler, are swapped. For many applications, however, an alternative method borrowed from mainframes and superminis called "paging" is used. Basically, paging differs from segmentation in that pages are of equal sizes. Demand paging, in which the operating system automatically swaps pages as needed, can be used with all 32-bit microprocessors.

Floating-point arithmetic is yet another area in which the new chips are mimicking mainframes. With early microprocessors, floating-point arithmetic was implemented in software, largely as a subroutine. When required, execution would jump to a piece of code that would handle the tasks. This method, however, slows the execution rate considerably, so floating-point hardware, such as fast bit-slice (registers and ALU on a chip) processors and, in some cases, special-purpose chips, was developed. Other than the Intel 8087, these chips behaved more or less like peripherals. When floating-point arithmetic was required, the problems were sent to the floating-point processor and the CPU was freed to move on to other instructions while it waited for the results. The floating-point processor is implemented as on-chip hardware in typical 32-bit microprocessors, as in mainframe and minicomputer CPUs. Caching or memory-management schemes are utilized with all 32-bit microprocessors in order to minimize access time for most instructions.

A cache, used for years in minis and mainframes, is a relatively small, high-speed memory installed between a processor and its main memory. The theory behind a cache is that a significant portion of the CPU time spent running typical programs is tied up in executing loops; thus, the chances are good that if an instruction to be executed is not the next sequential instruction, it will be one of some relatively small number of instructions back, a concept known as locality of reference. Therefore, a high-speed memory large enough to contain most loops should greatly increase processing rates. Cache memory is included as on-chip hardware in typical 32-bit microprocessors.

Typical 32-bit microprocessors such as Pentium and PowerPC chips are superscalar processors. This means that they can execute more than one instruction in one clock cycle. *Also, some 32-bit microprocessors such as the PowerPC contain an on-chip real-time clock. This allows these processors to use modern multitasking operating systems that require time keeping for task switching and for keeping the calendar date.*

A few 32-bit microprocessors implement a multiple branch prediction feature. This allows these microprocessors to anticipate jumps of the instruction flow ahead of time. Also, some 32-bit microprocessors determine an optimal sequence of instruction execution by looking at decoded instructions and then determining whether to execute or hold the instructions. Typical 32-bit microprocessors use a "look ahead" approach to execute instructions.

Typical 32-bit microprocessors instruction pool for a sequence of instructions and perform a useful task rather than execute the present instruction and then go to the next.

The 64-bit microprocessors include all the features of 32-bit microprocessors. In addition, they also contain multiple on-chip integer and floating-point units, a larger address and data bus. The 64-bit microprocessors can typically execute 4 instructions per clock cycle and can run at a clock speed of more than 300 MHz.

The Pentium microprocessor is designed using a combination of mostly microprogramming (CISC--Complex Instruction Set Computer) and some hardwired control (RISC --Reduced Instruction Set Computer) whereas the PowerPC is designed using hardwired control with almost no microcode. The PowerPC is a RISC (reduced instruction set computer) microprocessor. This means that a simple instruction set is included with PowerPC. The PowerPC instruction set includes register to register, load, and store instructions. All instructions involving arithmetic operations use registers; load and store instructions are utilized to access memory. Almost all computations can be obtained from these simple instructions. *The basic concepts associated with CISC and RISC are covered in Chapters 7,10 and 11.* Finally, the 64-bit microprocessors are ideal candidates for data-crunching machines and high-performance desktop systems/workstations.

QUESTIONS AND PROBLEMS

6.1 What is the difference between a single-chip microprocessor and a single-chip microcomputer?

6.2 What is a microcontroller? Name one commercially available microcontroller.

6.3 What is the difference between:
- (a) The program counter (PC) and the memory address register (MAR)?
- (b) The accumulator (A) and the instruction register (IR)?
- (c) General-purpose register-based microprocessor and accumulator-based microprocessor. Name a commercially available microprocessor of each type.

6.4 Assuming signed numbers, find the sign, carry, zero, and overflow flags of:
- (a) $09_{16} + 17_{16}$.
- (b) $A5_{16} - A5_{16}$
- (c) $71_{16} - A9_{16}$
- (d) $6E_{16} + 3A_{16}$
- (e) $7E_{16} + 7E_{16}$

6.5 What is meant by PUSH and POP operations in the stack?

6.6 Suppose that an 8-bit microprocessor has a 16-bit stack pointer and uses a 16-bit register to access the stack from the top. Assume that initially the stack pointer and the 16-bit register contain $20C0_{16}$ and 0205_{16} respectively. After the PUSH operation:
(a) What are the contents of the stack pointer?
(b) What are the contents of memory locations $20BE_{16}$ and $20BF_{16}$?

6.7 Assuming the microprocessor architecture of Figure 6.18, write down a possible sequence of microinstructions for finding the ones complement of an 8-bit number. Assume that the number is already in the register.

6.8 What do you mean by a multiplexed address and data bus?

6.9 Name four general-purpose registers in the 8086.

6.10 Name one 8086 register that can be used to hold an address in a segment.

6.11 What is the difference between EPROM and PROM? Are both types available with bipolar and also MOS technologies?

6.12 Assuming a single clock signal and four registers (PC, MAR, Reg, and IR) for a microprocessor, draw a timing diagram for loading the memory address register. Explain the sequence of events relating them to the four registers.

6.13 Given a memory with a 14-bit address and 8-bit word size.
(a) How many bytes can be stored in this memory?
(b) If this memory were constructed from 1K × 1-bit RAMs, how many memory chips would be required?
(c) How many bits would be used for chip select?

6.14 Define the three types of I/O. Identify each one as either "microprocessor initiated" or "device initiated."

6.15 What is the basic difference between a compiler and an assembler?

6.16 Write a program equivalent to the Pascal assignment statement:

```
Z := (A + (B * C) + (D * E) - (F / G) - (H * I)
```

Use only

(a) Three-address instructions
(b) Two-address instructions

6.17 Describe the meaning of each one of the following addressing modes.

(a) Immediate
(b) Absolute
(c) Register
(d) Register indirect
(e) Relative
(f) Implied

6.18 Assume that a microprocessor has only two registers R1 and R2 and that only the following instruction is available:

```
XOR     Ri, Rj     ;     Rj <- Ri ⊕ Rj
                   ;     i,j = 1,2
```

Using this XOR instruction, find an instruction sequence in order to exchange the contents of registers R1 and R2

6.19 What are the advantages of subroutines?

6.20 Explain the use of a stack in implementing subroutine calls.

6.21 Determine the contents of address 5004_{16} after assembling the following:

(a)
```
ORG   5002H
DB    00H, 05H, 07H, 00H, 03H
```

(b)
```
ORG 5000H
DW    0702H, 123FH, 7020H, 0000H
```

6.22 What is the difference between:

(a) A cross assembler and a resident assembler
(b) A two-pass assembler and meta-assembler
(c) Single step and breakpoint

6.23 Identify some of the differences between C, C++, and Java.

6.24 How does a microprocessor obtain the address of the first instruction to be executed?

6.25 Summarize the basic features of a typical microcomputer development system.

6.26 Discuss the steps involved in designing a microprocessor-based system.

6.27 Discuss the typical features of 32-bit and 64-bit microprocessors.

7

DESIGN OF COMPUTER INSTRUCTION SET AND THE CPU

This chapter describes the design of the instruction set and the central processor unit (CPU). Topics include op-code encoding, design of typical microprocessor registers, the arithmetic logic unit (ALU), and the control unit.

7.1 Design of the Computer Instructions

A program consists of a sequence of instructions. An instruction performs operations on stored data. There are two components in an instruction: an op-code field and an address field. The op-code field defines the type of operation to be performed on data, which may be stored in a microprocessor register or in the main memory. The address field may contain one or more addresses of data. When data are read from or stored into two or more addresses by the instruction, the address field may contain more than one address. For example, consider the following instruction:

MOVE	D0,D1
Op-code field	**Address field**

Assume that this computer uses D0 as the source register and D1 as the destination register. This instruction moves the contents of the microprocessor register D0 to register D1. The number and types of instructions supported by a microprocessor vary from one microprocessor to another and primarily depend on the microprocessor architecture. The number of instructions supported by a typical microprocessor depends on the size of the op-code field. For example, an 8-bit op-code can specify a maximum of 256 unique instructions.

As mentioned before, a computer only understands 1's and 0's. This means that the computer can execute an instruction only if it is in binary. A unique binary pattern must be assigned to each op-code by a process called "op-code encoding."

The Block code method is one of the simplest techniques of designing instructions. In this approach, a fixed length of binary pattern is assigned to each op-code. For example, an n-bit binary number can represent 2^n unique op-codes. Consider for example, a hypothetical instruction set shown in Figure 7.1. In this figure, there are 8 different instructions that can be encoded using three bits i_2, i_1, i_0 as shown in Figure 7.2. A 3-to-8 decoder can be used to encode the 8 hypothetical instructions as shown in Figure 7.3.

Instruction	**Operation Performed**
MOVE reg_1, reg_2	$reg_2 \leftarrow reg_1$
CLR reg	$reg \leftarrow 0$
ADD reg_1, reg_2	$reg_2 \leftarrow reg_1 + reg_2$
SUB reg_1, reg_2	$reg_2 \leftarrow reg_2 - reg_1$
AND reg_1, reg_2	$reg_2 \leftarrow reg_1$ AND reg_2
OR reg_1, reg_2	$reg_2 \leftarrow reg_1$ OR reg_2
INC reg	$reg \leftarrow reg + 1$
JMP addr	PC $\leftarrow$ addr; Unconditionally Jump to addr

FIGURE 7.1 A hypothetical instruction set

Instruction	**3-Bit Op-Code**		
	i_2	i_1	i_0
MOVE	0	0	0
CLR	0	0	1
ADD	0	1	0
SUB	0	1	1
AND	1	0	0
OR	1	0	1
INC	1	1	0
JMP	1	1	1

FIGURE 7.2 Op-code encoding using block code

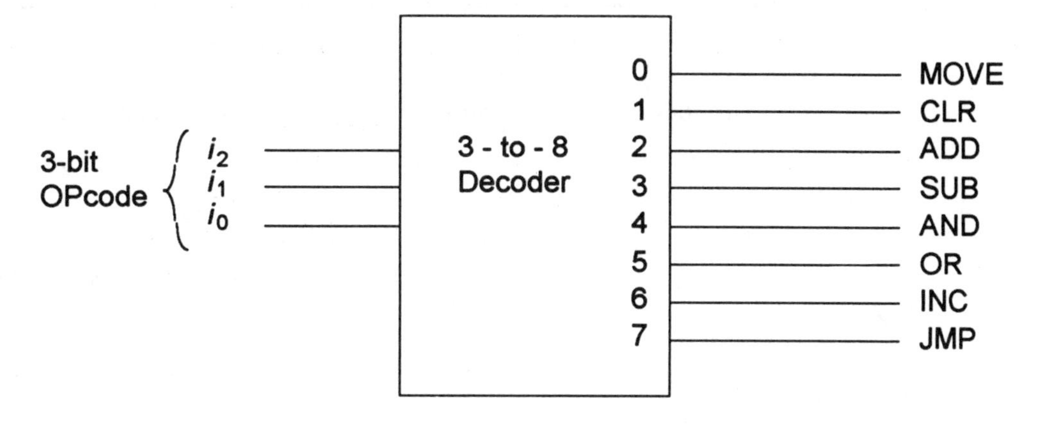

FIGURE 7.3 Instruction decoder

An n-to-2^n decoder is required for an n-bit op-code. As n increases, the cost of the decoder and decoding time will also increase. In some op-code encoding techniques such as the "expanding op-code" method, the length of the instruction is a function of the number of addresses used by the instruction. For example, consider a 16-bit instruction in which the lengths of the op-code and address fields are 5 bits and 11 bits respectively. Using such an instruction format, 32 (2^5) operations allowing access to 2048 (2^{11}) memory locations can be specified. Now, if the size of the instruction is kept at 16 bits but the address field is increased to 12 bits, the op-code length will then be decreased to 4 bits. This change will specify 16 (2^4) operations with access to 4096 (2^{12}) memory locations. Thus, the number of operations is reduced by 50% and the number of memory locations is increased by 100%. This concept is used in designing instructions using expanding op-code technique.

Consider an instruction format with 8-bit instruction length and a 2-bit op-code field. Four unique two-address (3 bits for each address) instructions can be specified. This is depicted in Figure 7.4. If three rather than four two-address instructions are used, eight one-address instructions can be specified. This is shown in Figure 7.5.

OP- Code (2-bits)	Address 1 (3-bits)	Address 2 (3-bits)
i_1 i_0		
0 0	$x_2\ x_1\ x_0$	$y_2\ y_1\ y_0$
0 1	$x_2\ x_1\ x_0$	$y_2\ y_1\ y_0$
1 0	$x_2\ x_1\ x_0$	$y_2\ y_1\ y_0$
1 1	$x_2\ x_1\ x_0$	$y_2\ y_1\ y_0$

FIGURE 7.4 Four two-address instructions

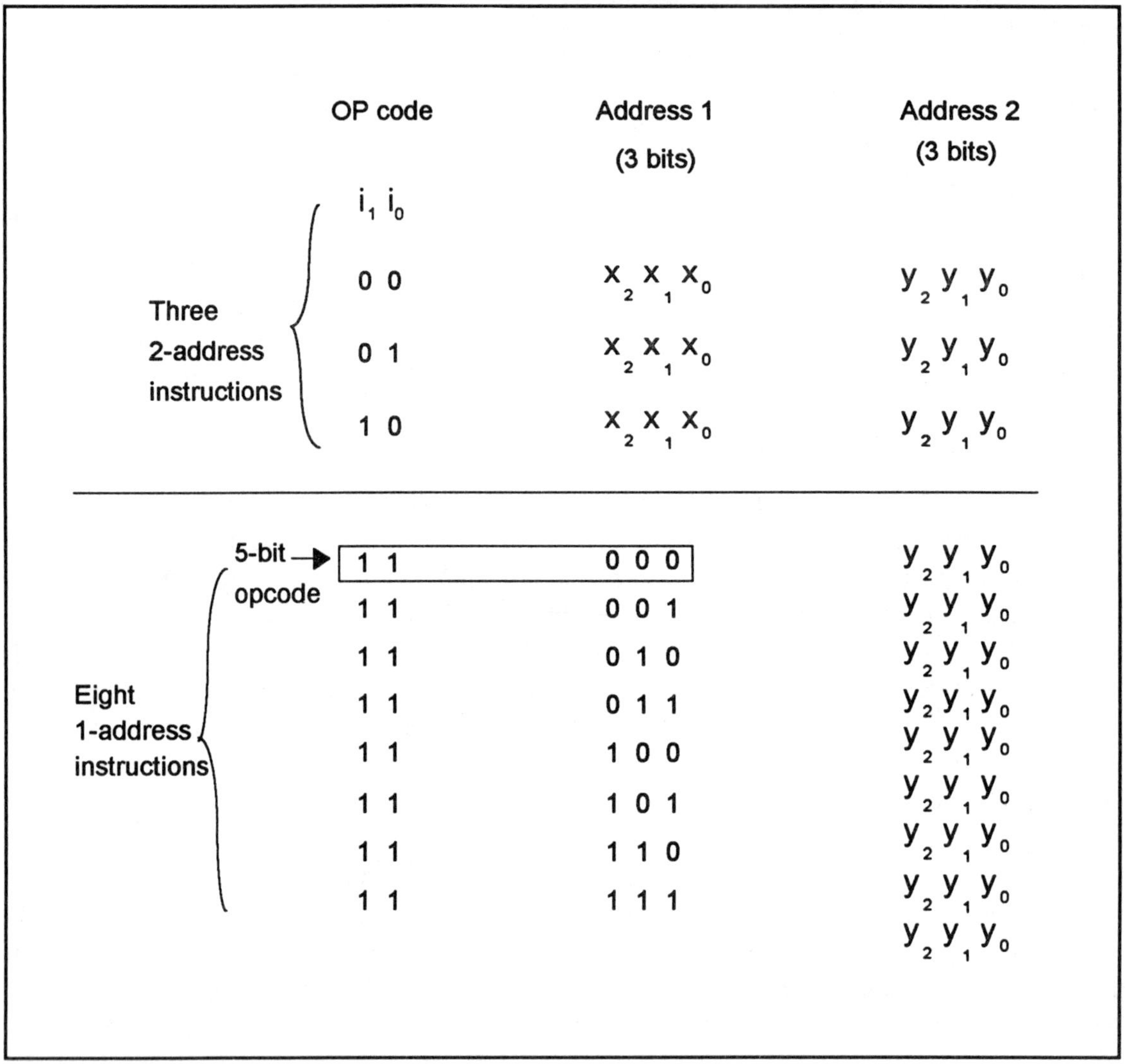

	OP code	Address 1 (3 bits)	Address 2 (3 bits)
	i_1 i_0		
Three 2-address instructions	0 0	x_2 x_1 x_0	y_2 y_1 y_0
	0 1	x_2 x_1 x_0	y_2 y_1 y_0
	1 0	x_2 x_1 x_0	y_2 y_1 y_0
5-bit opcode →	1 1	0 0 0	y_2 y_1 y_0
Eight 1-address instructions	1 1	0 0 1	y_2 y_1 y_0
	1 1	0 1 0	y_2 y_1 y_0
	1 1	0 1 1	y_2 y_1 y_0
	1 1	1 0 0	y_2 y_1 y_0
	1 1	1 0 1	y_2 y_1 y_0
	1 1	1 1 0	y_2 y_1 y_0
	1 1	1 1 1	y_2 y_1 y_0
			y_2 y_1 y_0

FIGURE 7.5 Three 2-address and eight 1-address instructions

The length of the op-code field for each one-address instruction is 5 bits. Thus, the length of the op-code field increases as the number of address field is decreased. Now, if the total number of one-address instructions is reduced from 8 to 7, then eight 0-address instructions can also be specified. This is shown in Figure 7.6.

Instructions	Opcode			
Three 2-address instructions	2-bit opcode	0 0	$x_2\, x_1\, x_0$	$y_2\, y_1\, y_0$
		0 1	$x_2\, x_1\, x_0$	$y_2\, y_1\, y_0$
		1 0	$x_2\, x_1\, x_0$	$y_2\, y_1\, y_0$
Seven 1-address instructions	5-bit opcode	1 1	0 0 0	$y_2\, y_1\, y_0$
		1 1	0 0 1	$y_2\, y_1\, y_0$
		1 1	0 1 0	$y_2\, y_1\, y_0$
		1 1	0 1 1	$y_2\, y_1\, y_0$
		1 1	1 0 0	$y_2\, y_1\, y_0$
		1 1	1 0 1	$y_2\, y_1\, y_0$
		1 1	1 1 0	$y_2\, y_1\, y_0$
Eight 0-address instructions	8-bit opcode	1 1	1 1 1	0 0 0
		1 1	1 1 1	0 0 1
		1 1	1 1 1	0 1 0
		- -	- - -	- - -
		- -	- - -	- - -
		1 1	1 1 1	1 1 1

FIGURE 7.6 3 two-address, 7 one-address, and 8 zero-address instructions

7.2 Design of the CPU

The CPU contains three elements: registers, the ALU (Arithmetic Logic Unit), and the control unit. These topics are discussed next. *The design of a microprogrammed CPU is included in Chapter 11.*

7.2.1 Register Design

The concept of general-purpose and flag registers is provided in Chapters 5 and 6. The main purpose of a general-purpose register is to store address or data for an indefinite period of time. The computer can execute an instruction to retrieve the contents of this register when needed. A computer can also execute instructions to perform shift operations on the contents of a general-purpose register. This section includes combinational shifter design and the concepts associated with barrel shifters.

A high-speed shifter can be designed using combinational circuit components such as a multiplexer. The block diagram, internal organization, and truth table of a typical combinational shifter are shown in Figure 7.7. From the truth table, the following equations can be obtained:

$$y_3 = \overline{s_1}\,\overline{s_0}i_3 + \overline{s_1}s_0i_2 + s_1\overline{s_0}i_1 + s_1s_0i_0$$

$$y_2 = \overline{s_1}\,\overline{s_0}i_2 + \overline{s_1}s_0i_1 + s_1\overline{s_0}i_0 + s_1s_0i_{-1}$$

$$y_1 = \overline{s_1}\,\overline{s_0}i_1 + \overline{s_1}s_0i_0 + s_1\overline{s_0}i_{-1} + s_1s_0i_{-2}$$

$$y_0 = \overline{s_1}\,\overline{s_0}i_0 + \overline{s_1}s_0i_{-1} + s_1\overline{s_0}i_{-2} + s_1s_0i_{-3}$$

The 4 × 4 shifter of Figure 7.7 can be expanded to obtain a system capable of rotating 16-bit data to the left by 0, 1, 2, or 3 positions, which is shown in Figure 7.8.

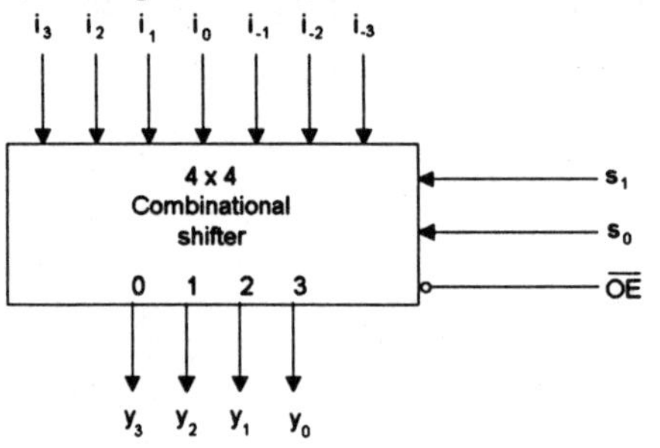

(a) Block Diagram

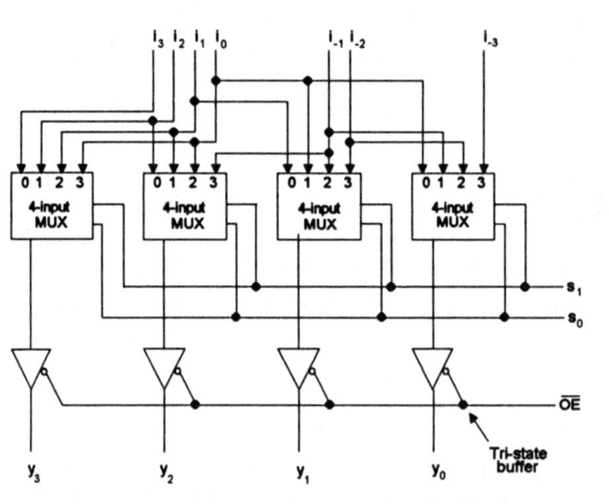

(b) Internal Schematic

	Shift Count		Output				Comment
$\overline{OE}$	s_1	s_0	y_3	y_2	y_1	y_0	
1	X	X	Z	Z	Z	Z	Outputs are tristated
0	0	0	i_3	i_2	i_1	i_0	Pass (no shift)
0	0	1	i_2	i_1	i_0	i_{-1}	Left Shift once
0	1	0	i_1	i_0	i_{-1}	i_{-2}	Left shift twice
0	1	1	i_0	i_{-1}	i_{-2}	i_{-3}	Left shift three times

(c) Truth Table (x is don't care in the above)

FIGURE 7.7 4 × 4 combinational shifter

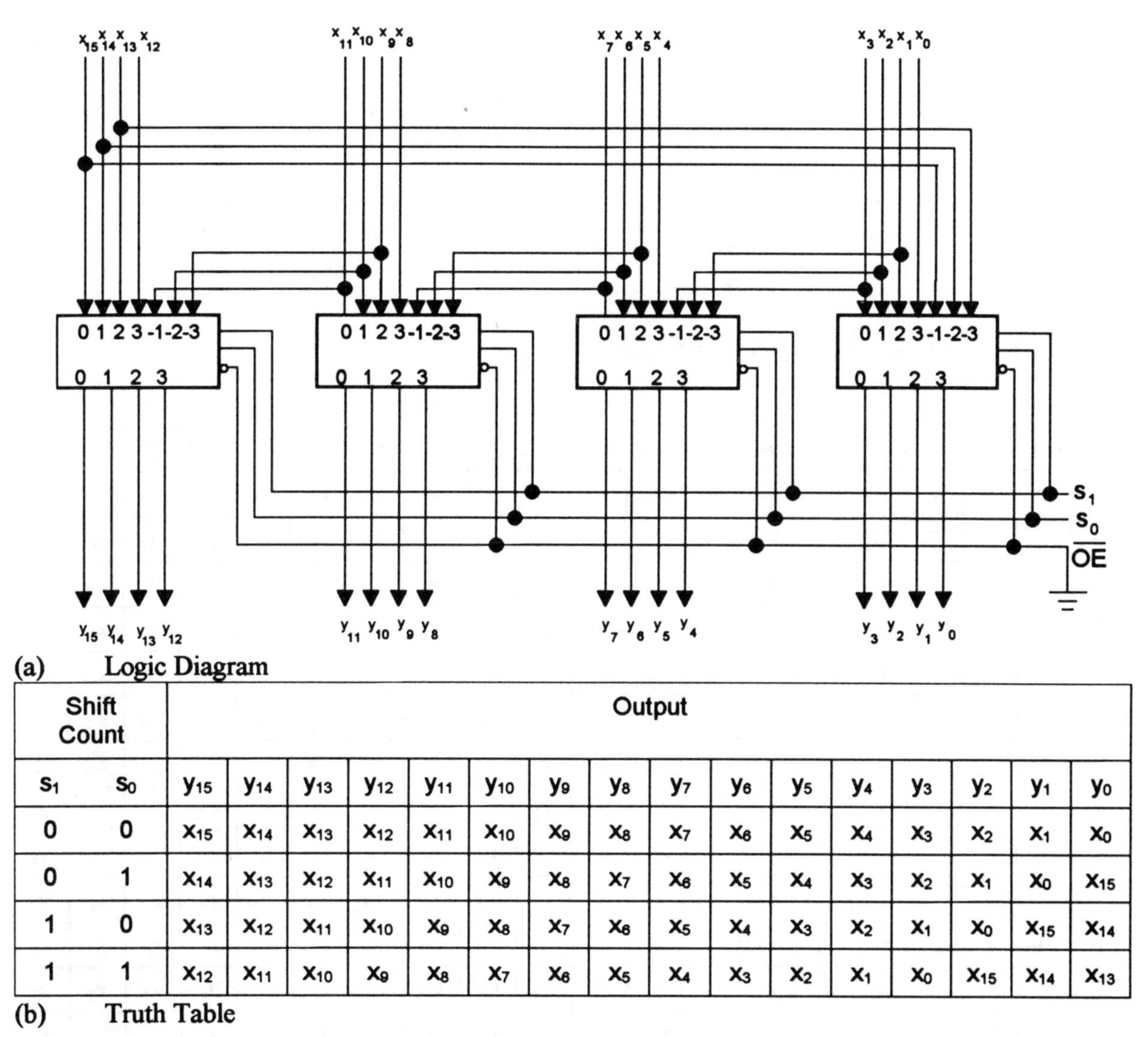

(a) Logic Diagram

Shift Count		Output															
S_1	S_0	y_{15}	y_{14}	y_{13}	y_{12}	y_{11}	y_{10}	y_9	y_8	y_7	y_6	y_5	y_4	y_3	y_2	y_1	y_0
0	0	x_{15}	x_{14}	x_{13}	x_{12}	x_{11}	x_{10}	x_9	x_8	x_7	x_6	x_5	x_4	x_3	x_2	x_1	x_0
0	1	x_{14}	x_{13}	x_{12}	x_{11}	x_{10}	x_9	x_8	x_7	x_6	x_5	x_4	x_3	x_2	x_1	x_0	x_{15}
1	0	x_{13}	x_{12}	x_{11}	x_{10}	x_9	x_8	x_7	x_6	x_5	x_4	x_3	x_2	x_1	x_0	x_{15}	x_{14}
1	1	x_{12}	x_{11}	x_{10}	x_9	x_8	x_7	x_6	x_5	x_4	x_3	x_2	x_1	x_0	x_{15}	x_{14}	x_{13}

(b) Truth Table

FIGURE 7.8 Combinational shifter capable of rotating 16-bit data to the left by 0, 1, 2, or 3 positions

This design can be extended to obtain a more powerful shifter called the *barrel shifter*. The shift is a cycle rotation, which means that the input binary information is shifted in one direction; the most significant bit is moved to the least significant position.

The block-diagram representation of a 16 × 16 barrel shifter is shown in Figure 7.9. This shifter is capable of rotating the given 16-bit data to the left by n positions, where $0 \leq n \leq 15$. Figure 7.9 shows the truth table representing the operation of the shifter. The barrel shifter is an on-chip component for typical 32-bit and 64-bit microprocessors.

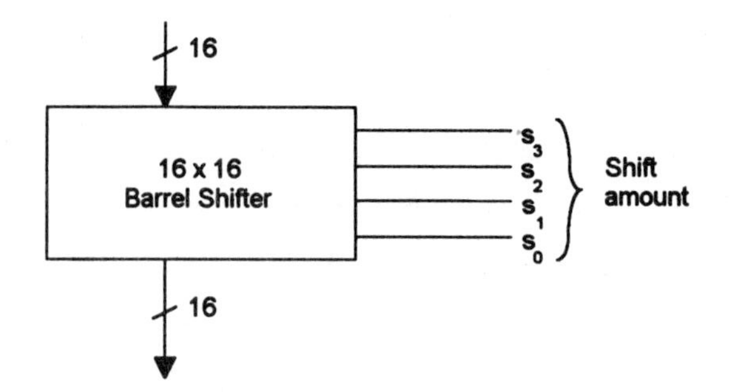

(a) Block Diagram of a 16 × 16 Barrel Shifter

Shift Count				Output															
s_3	s_2	s_1	s_0	y_{15}	y_{14}	y_{13}	y_{12}	y_{11}	y_{10}	y_9	y_8	y_7	y_6	y_5	y_4	y_3	y_2	y_1	y_0
0	0	0	0	x_{15}	x_{14}	x_{13}	x_{12}	x_{11}	x_{10}	x_9	x_8	x_7	x_6	x_5	x_4	x_3	x_2	x_1	x_0
0	0	0	1	x_{14}	x_{13}	x_{12}	x_{11}	x_{10}	x_9	x_8	x_7	x_6	x_5	x_4	x_3	x_2	x_1	x_0	x_{15}
0	0	1	0	x_{13}	x_{12}	x_{11}	x_{10}	x_9	x_8	x_7	x_6	x_5	x_4	x_3	x_2	x_1	x_0	x_{15}	x_{14}
0	0	1	1	x_{12}	x_{11}	x_{10}	x_9	x_8	x_7	x_6	x_5	x_4	x_3	x_2	x_1	x_0	x_{15}	x_{14}	x_{13}
0	1	0	0	x_{11}	x_{10}	x_9	x_8	x_7	x_6	x_5	x_4	x_3	x_2	x_1	x_0	x_{15}	x_{14}	x_{13}	x_{12}
0	1	0	1	x_{10}	x_9	x_8	x_7	x_6	x_5	x_4	x_3	x_2	x_1	x_0	x_{15}	x_{14}	x_{13}	x_{12}	x_{11}
0	1	1	0	x_9	x_8	x_7	x_6	x_5	x_4	x_3	x_2	x_1	x_0	x_{15}	x_{14}	x_{13}	x_{12}	x_{11}	x_{10}
0	1	1	1	x_8	x_7	x_6	x_5	x_4	x_3	x_2	x_1	x_0	x_{15}	x_{14}	x_{13}	x_{12}	x_{11}	x_{10}	x_9
1	0	0	0	x_7	x_6	x_5	x_4	x_3	x_2	x_1	x_0	x_{15}	x_{14}	x_{13}	x_{12}	x_{11}	x_{10}	x_9	x_8
1	0	0	1	x_6	x_5	x_4	x_3	x_2	x_1	x_0	x_{15}	x_{14}	x_{13}	x_{12}	x_{11}	x_{10}	x_9	x_8	x_7
1	0	1	0	x_5	x_4	x_3	x_2	x_1	x_0	x_{15}	x_{14}	x_{13}	x_{12}	x_{11}	x_{10}	x_9	x_8	x_7	x_6
1	0	1	1	x_4	x_3	x_2	x_1	x_0	x_{15}	x_{14}	x_{13}	x_{12}	x_{11}	x_{10}	x_9	x_8	x_7	x_6	x_5
1	1	0	0	x_3	x_2	x_1	x_0	x_{15}	x_{14}	x_{13}	x_{12}	x_{11}	x_{10}	x_9	x_8	x_7	x_6	x_5	x_4
1	1	0	1	x_2	x_1	x_0	x_{15}	x_{14}	x_{13}	x_{12}	x_{11}	x_{10}	x_9	x_8	x_7	x_6	x_5	x_4	x_3
1	1	1	0	x_1	x_0	x_{15}	x_{14}	x_{13}	x_{12}	x_{11}	x_{10}	x_9	x_8	x_7	x_6	x_5	x_4	x_3	x_2
1	1	1	1	x_0	x_{15}	x_{14}	x_{13}	x_{12}	x_{11}	x_{10}	x_9	x_8	x_7	x_6	x_5	x_4	x_3	x_2	x_1

(b) Truth Table of the 16 × 16 Barrel Shifter

FIGURE 7.9 Barrel shifter

7.2.2 ALU Design

Addition is the basic arithmetic operation performed by an ALU. Other operations such as subtraction and multiplication can be obtained via addition. Thus, the time required to add two numbers plays an important role in determining the speed of the ALU.

The basic concepts of half-adder, full adder, and parallel adder are discussed in Section 4.5.1. The following equations for the full-adder were obtained. Assume $x_i = x$, $y_i = y$, $c_i = z$, and $C_{i+1} = C$ in Table 4.6.

$$\text{Sum, } S_i = \overline{x_i}\,\overline{y_i}c_i + \overline{x_i}y_i\overline{c_i} + x_i\overline{y_i}\,\overline{c_i} + x_iy_ic_i$$

$$= x_i \oplus y_i \oplus c_i$$

From Table 4.6,

$$\text{Carry, } C_{i+1} = \overline{x_i}y_ic_i + x_i\overline{y_i}c_i + x_iy_i\overline{c_i} + x_iy_ic_i$$

$$= (\overline{x_i}y_ic_i + x_iy_ic_i) + (x_i\overline{y_i}c_i + x_iy_ic_i) + (x_iy_i\overline{c_i} + x_iy_ic_i)$$

$$= y_ic_i + x_ic_i + x_iy_i$$

The logic diagrams for implementing these equations are given in Figure 7.10.

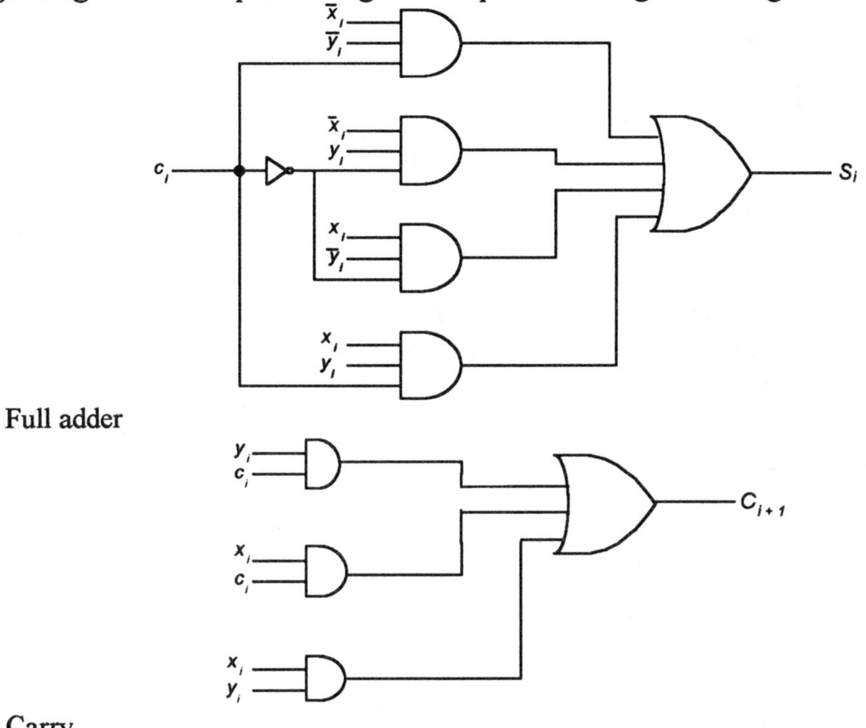

(a) Full adder

(b) Carry

FIGURE 7.10 Logic circuit of full adder

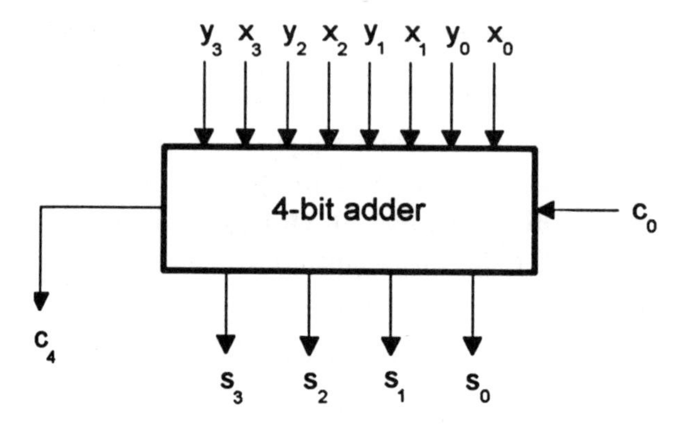

(a) Block Diagram of a 4-bit Ripple-Carry Adder

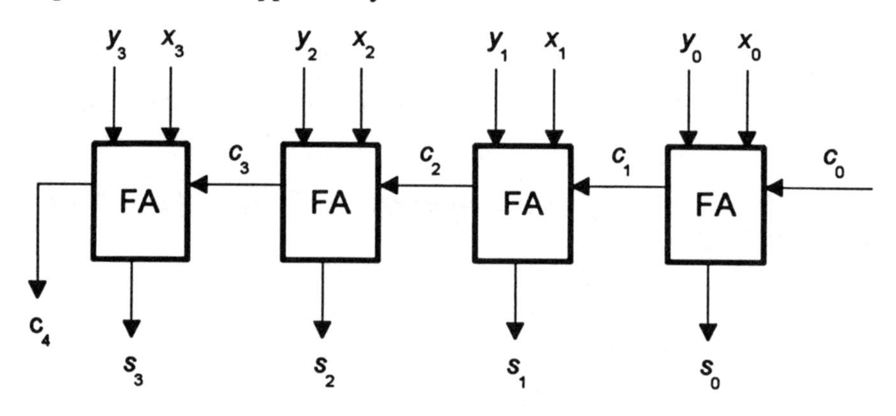

(b) Four 4-bit Full Adders are Cascaded to implement a 4-Bit Ripple-Carry Adder

FIGURE 7.11 Implementation of a 4-bit Ripple-Carry Adder

As has been made apparent by Figure 7.10, for generating C_{i+1} from c_i, two gate delays are required. To generate S_i from c_i, three gate delays are required because c_i must be inverted to obtain $\overline{c_i}$. Note that no inverters are required to get $\overline{x_i}$ or $\overline{y_i}$ from x_i or y_i, respectively, because the numbers to be added are usually stored in a register that is a collection of flip-flops. The flip-flop generates both normal and complemented outputs. For the purpose of discussion, assume that the gate delay is Δ time units, and the actual value of Δ is decided by the technology. For example, if transistor translator logic (TTL) circuits are used, the value of Δ will be 10 ns.

By cascading n full adders, an n-bit parallel adder capable of handling two n-bit operands (X and Y) can be designed. The implementation of a *4-bit ripple-carry or parallel adder* is shown in Figure 7.11. When two unsigned integers are added, the input carry, c_0, is always zero. The 4-bit adder is also called a "carry-propagate adder" (CPA), because the carry is propagated serially through each full adder. This hardware can be cascaded to obtain a 16-bit CPA, as shown in Figure 7.12; $c_0 = 0$ or 1 for multiprecision addition.

Although the design of an *n*-bit CPA is straightforward, the carry propagation time limits the speed of operation. For example, in the 16-bit CPA (see Figure 7.12), the addition operation is completed only when the sum bits s_0 through s_{15} are available.

To generate s_{15}, c_{15} must be available. The generation of c_{15} depends on the availability of c_{14}, which must wait for c_{13} to become available. In the worst case, the carry process propagates through 15 full adders. Therefore, the worst-case add-time of the 16-bit CPA can be estimated as follows:

Time taken for carry to propagate through 15 full adders (the delay involved in the path from c_0 to c_{15})	$= 15 * 2\ \Delta$
Time taken to generate s_{15} from c_{15}	$= 3\ \Delta$
Total	$= 33\ \Delta$

If Δ = 10 ns, then the worst-case add-time of a 16-bit CPA is 330 ns. This delay is prohibitive for high-speed systems, in which the expected add-time is typically less than 100 ns, which makes it necessary to devise a new technique to increase the speed of operation by a factor of 3. One such technique is known as the "carry look-ahead." In this approach the extra hardware is used to generate each carry (c_i, $i > 0$) directly from c_0. To be more practical, consider the design of a 4-bit carry look-ahead adder (CLA). Let us see how this may be used to obtain a 16-bit adder that operates at a speed higher than the 16-bit CPA.

Recall that in a full adder for adding X_i, Y_i, and C_i, the output carry C_{i+1} is related to its carry input C_i, as follows:

$$C_{i+1} = X_i Y_i + X_i C_i + Y_i C_i$$

The result can be rewritten as

$$C_{i+1} = G_i + P_i C_i$$

where $G_i = X_i Y_i$ and $P_i = X_i + Y_i$

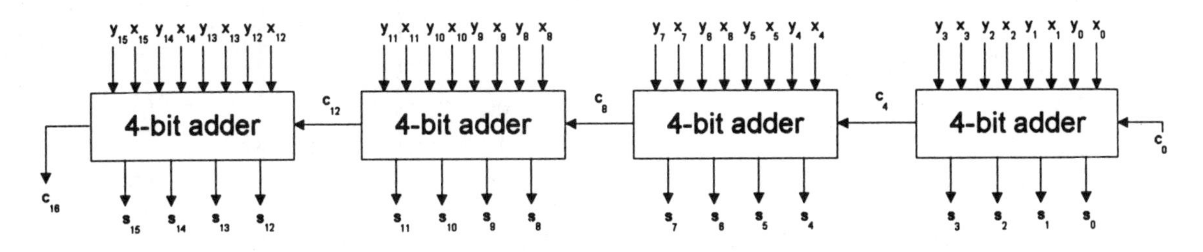

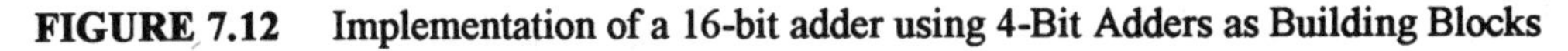

FIGURE 7.12 Implementation of a 16-bit adder using 4-Bit Adders as Building Blocks

The function G_i is called the carry-generate function, because a carry is generated when $X_i = Y_i = 1$. If X_i or Y_i is a 1, then the input carry C_i is propagated to the next stage. For this reason, the function P_i is often referred to as the "carry-propagate" function. Using G_i and P_i, C_1, C_2, C_3, and C_4 can be expressed as follows:

$$C_1 = G_0 + P_0C_0$$

$$C_2 = G_1 + P_1C_1$$

$$C_3 = G_2 + P_2C_2$$

$$C_4 = G_3 + P_3C_3$$

All high-order carries can be generated in terms of C_0 as follows:

$$C_1 = G_0 + P_0C_0$$

$$C_2 = G_1 + P_1(G_0 + P_0C_0) = G_1 + P_1G_0 + P_1P_0C_0$$

$$C_3 = G_2 + P_2C_2 = G_2 + P_2(G_1 + P_1G_0 + P_1P_0C_0)$$

$$= G_2 + P_2G_1 + P_2P_1G_0 + P_2P_1P_0C_0$$

$$C_4 = G_3 + P_3C_3 = G_3 + P_3(G_2 + P_2G_1 + P_2P_1G_0 + P_2P_1P_0C_0)$$

$$= G_3 + P_3G_2 + P_3P_2G_1 + P_3P_2P_1G_0 + P_3P_2P_1P_0C_0$$

Therefore C_1, C_2, C_3, and C_4 can generated directly from C_0. For this reason, these equations are called "carry look-ahead equations," and the hardware that implements these equations is called a "4-stage look-ahead circuit" (4-CLC).. The block diagram of such circuit is shown in Figure 7.13.

$$g_0 = G_3 + P_3G_2 + P_3P_2G_1 + P_3P_2P_1G_0$$

$$p_0 = P_3P_2P_1P_0$$

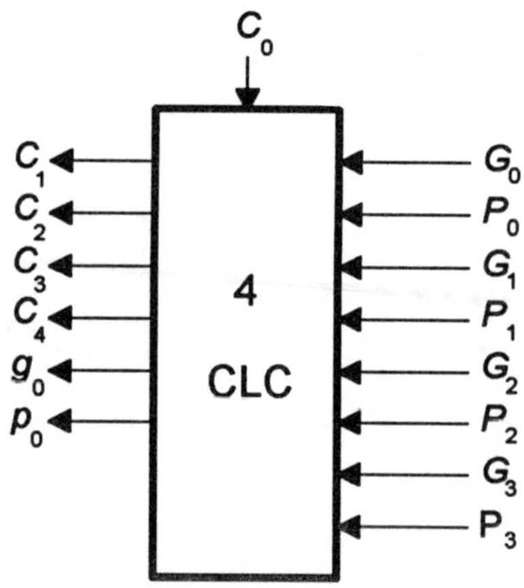

FIGURE 7.13 A Four-Stage Carry Look-ahead Circuit

The following are some important points about this system:

- A 4-CLC can be implemented as a two-level AND-OR logic circuit (The first level consists of AND gates, whereas the second level includes OR gates).
- The outputs g_0 and p_0 are useful to obtain a higher-order look-ahead system.

To construct a 4-bit CLA, assume the existence of the basic adder cell shown in Figure 7.14. Using this basic cell and 4-bit CLC, the design of a 4-bit CLA can be completed as shown in Figure 7.15. Using this cell as a building block, a 16-bit adder can be designed as shown in Figure 7.16.

The worst-case add-time of this adder can be calculated as follows:

		Delay
For P_i, G_i generation from X_i, Y_i $(0 \leq i \leq 15)$	...	Δ
To generate C_4 from C_0	...	2Δ
To generate C_8 from C_4	...	2Δ
To generate C_{12} from C_8	...	2Δ
To generate C_{15} from C_{12}	...	2Δ
To generate S_{15} from C_{15}	...	3Δ
Total delay	...	12Δ

A graphical illustration of this calculation can be shown as follows:

$$\text{Data available} \overset{\Delta}{\mapsto} G_iP_i \overset{2\Delta}{\rightarrow} C_4 \overset{2\Delta}{\rightarrow} C_8 \overset{2\Delta}{\rightarrow} C_{12} \overset{2\Delta}{\rightarrow} C_{15} \overset{3\Delta}{\rightarrow} S_{15}$$

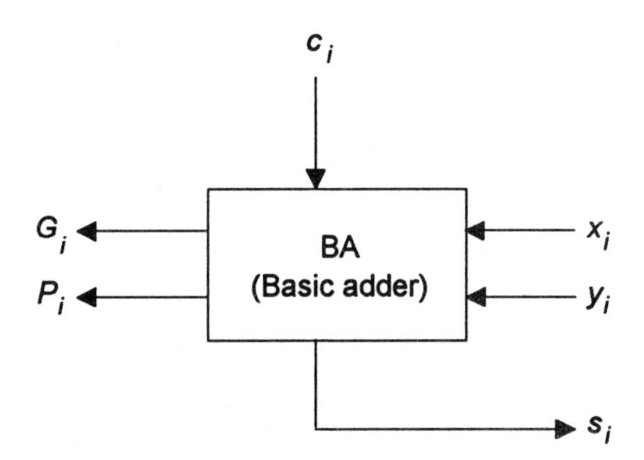

FIGURE 7.14 Basic CLA cell

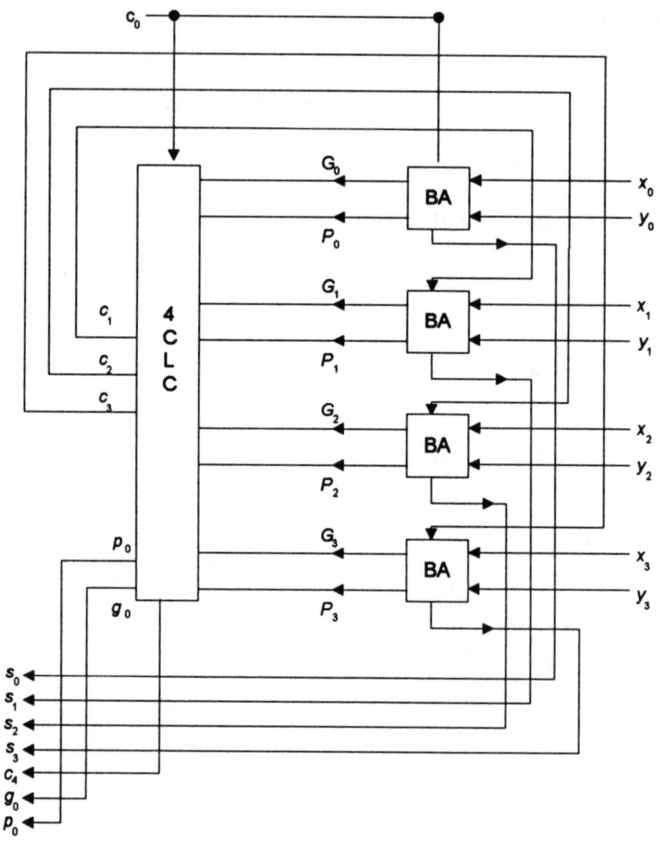

FIGURE 7.15 A 4-bit CLA

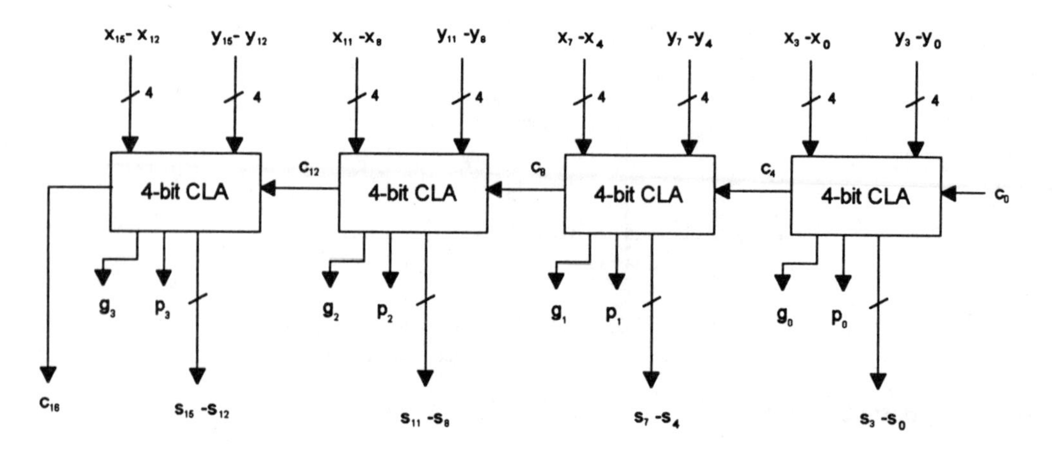

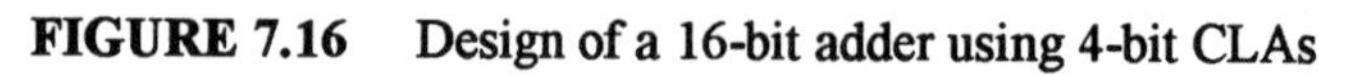
FIGURE 7.16 Design of a 16-bit adder using 4-bit CLAs

From this calculation, it is apparent that the new 16-bit adder is faster than the 16-bit CPA by a factor of 3. In fact, this system can be speeded up further by employing another 4-bit CLC and eliminating the carry propagation between the 4-bit CLA blocks. For this purpose, the g_i and p_i outputs generated by the 4-bit CLA are used. This design task is left as an exercise to the reader.

If there is a need to add more than 3 operands, a technique known as "carry-save addition" is used. To see its effectiveness, consider the following example:

```
  44
  28
  32
  79
  63←Sum vector
 12 ←Carry vector
 183←Final answer
```

In this example, four decimal numbers are added. First, the unit digits are added, producing a sum of 3 and a carry digit of 2. Similarly, the tens digits are added, producing a sum digit of 6 and a carry digit of 1. Because there is no carry propagation from the unit digit to the tenth digit, these summations can be carried out in parallel to produce a sum vector of 63 and a carry vector of 12. When all operands are exhausted, the sum and the shifted carry vector are added in the conventional manner, which produces the final answer. Note that the carry is propagated only in the last step, which generates the final answer no matter how many operands are added. The concept is also referred to as "addition by deferred carry assimilation."

Functionally, an ALU can be divided up into two segments: the arithmetic unit and the logic unit. The arithmetic unit performs typical arithmetic operations such as addition, subtraction, and increment or decrement by 1. Usually, the operands involved may be signed or unsigned integers. In some cases, however, an arithmetic unit must handle 4-bit binary-coded decimal (BCD) numbers and floating-point numbers. Therefore, this unit must include the circuitry necessary to manipulate these data types. As the name implies, the logic unit contains hardware elements that perform typical operations such as Boolean NOT and OR. In this section, the design of a simple ALU using typical combinational elements such as gates, multiplexers, and a 4-bit parallel adder is discussed. For this approach, an arithmetic unit and a logic unit are first designed separately; then they are combined to obtain an ALU.

For the first step, a two-function arithmetic unit, as shown in Figure 7.17 is designed. The key element of this system is a 4-bit parallel adder. The multiplexers select either Y or $\overline{Y}$ for the 3-input of the parallel adder. In particular, if $s_0 = 0$, then $B = Y$; otherwise $B = \overline{Y}$. Because the selection input (s_0) also controls the input carry (c_{in}), the following results:

$$\text{If } s_0 = 0 \text{ then } F = X \text{ plus } Y$$

$$\text{else } F = X \text{ plus } \overline{Y} \text{ plus } 1$$

$$= X \text{ minus } Y$$

This arithmetic unit generates addition and subtraction operations.

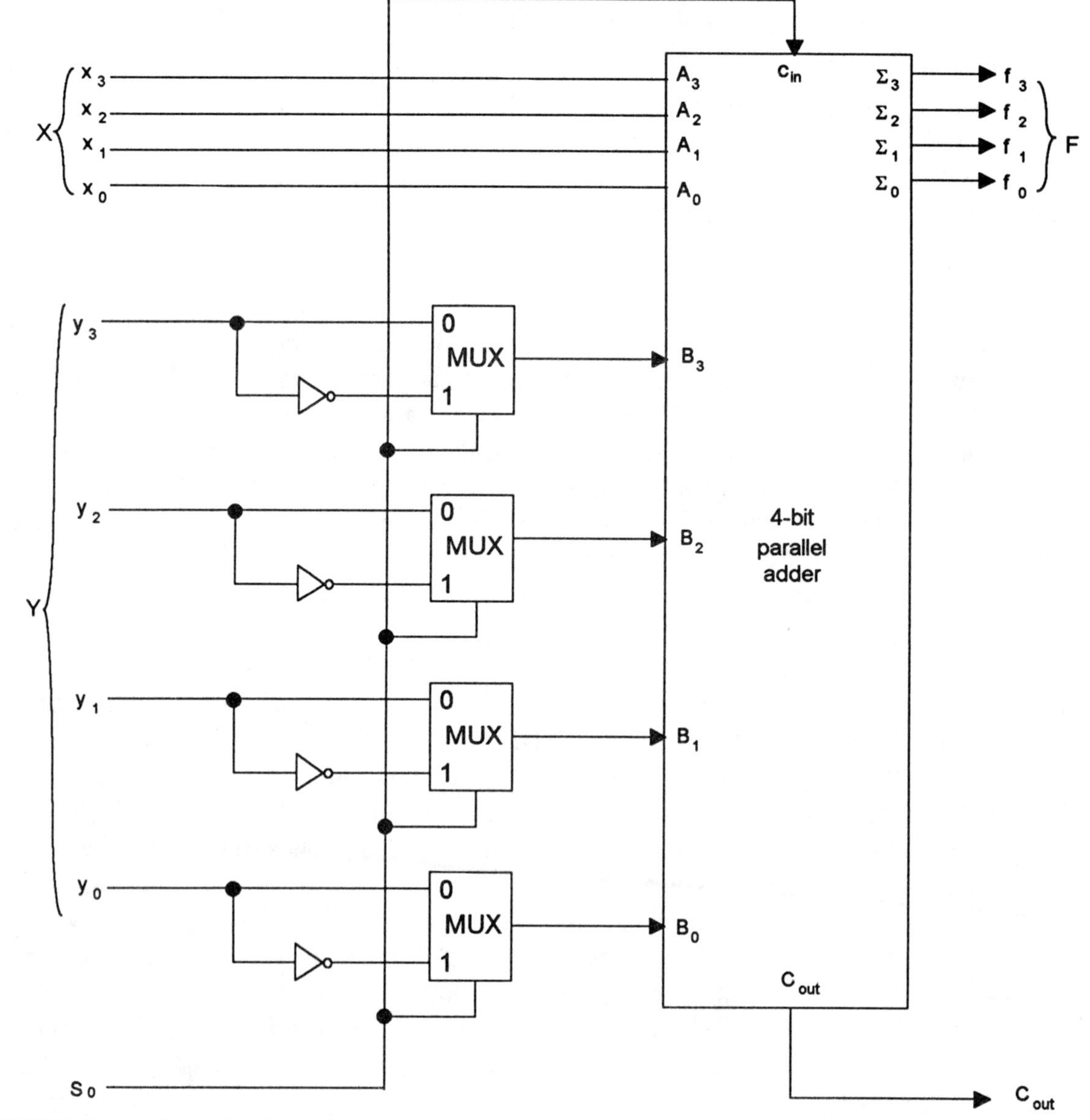

FIGURE 7.17 Organization of an arithmetic unit

For the second step, let us design a two-function logic unit; this is shown in Figure 7.18. From Figure 7.18 it can be seen that when $s_0 = 0$, the output $G = X$ AND Y; otherwise the output $G = X \oplus Y$. Note that from these two Boolean operations, other operations such as NOT and OR can be derived by the following Boolean identities:

$$1 \oplus x = \bar{x}$$

$$x \text{ OR } y = x \oplus y \oplus xy$$

Therefore, NOT and OR operations can be obtained by using additional hardware and the circuit of Figure 7.18. The outputs generated by the arithmetic and logic units can be combined by using a set of multiplexers, as shown in Figure 7.19. From this organization it can be seen that when the select line $s_1 = 1$, the multiplexers select outputs generated by the logic unit; otherwise, the outputs of the arithmetic unit are selected.

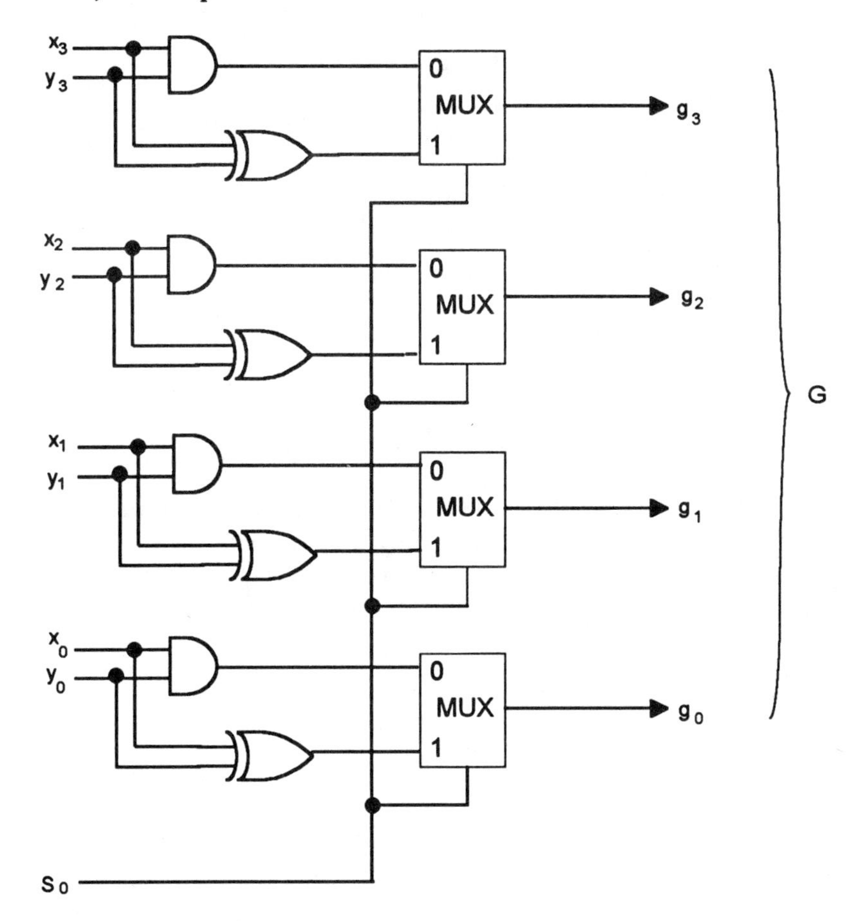

FIGURE 7.18 Organization of a 4-bit two-function logic unit

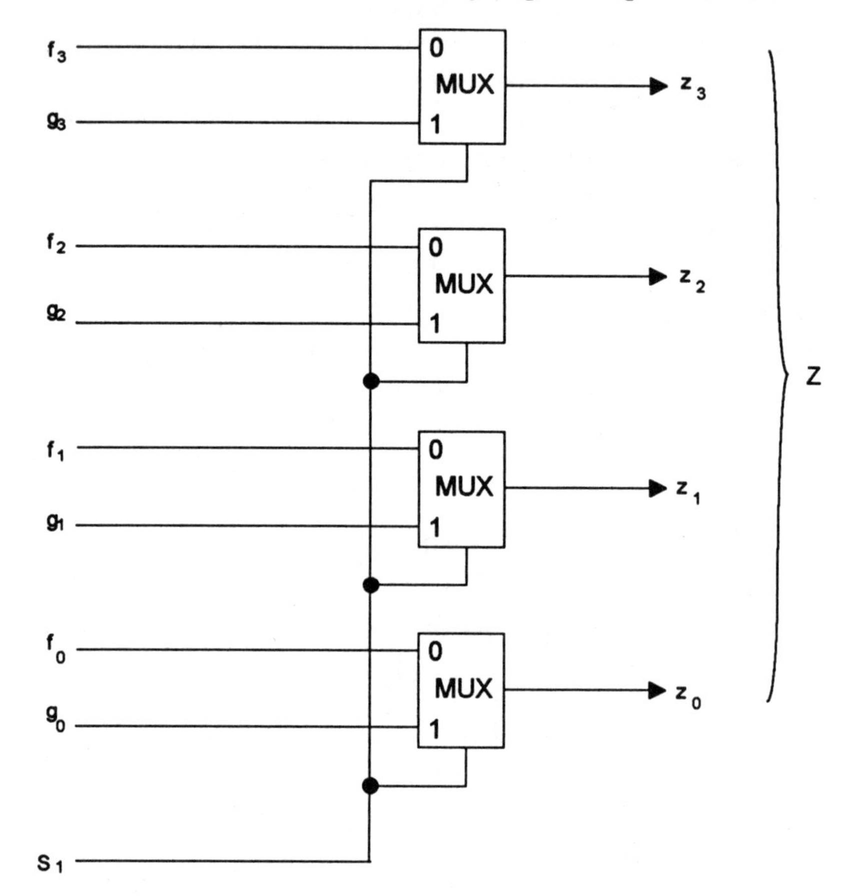

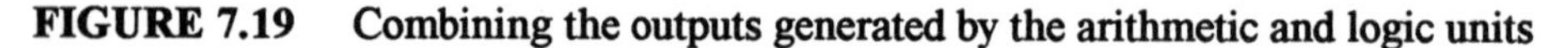

FIGURE 7.19 Combining the outputs generated by the arithmetic and logic units

More commonly, the select line, s_1, is referred to as the *mode input* because it selects the desired mode of operation (arithmetic or logic). A complete block diagram schematic of this ALU is shown in Figure 7.20. The truth table illustrating the operation of this ALU is shown in Figure 7.21. This table shows that this ALU is capable of performing 2 arithmetic and 2 logic operations on the 4-bit operands X and Y.

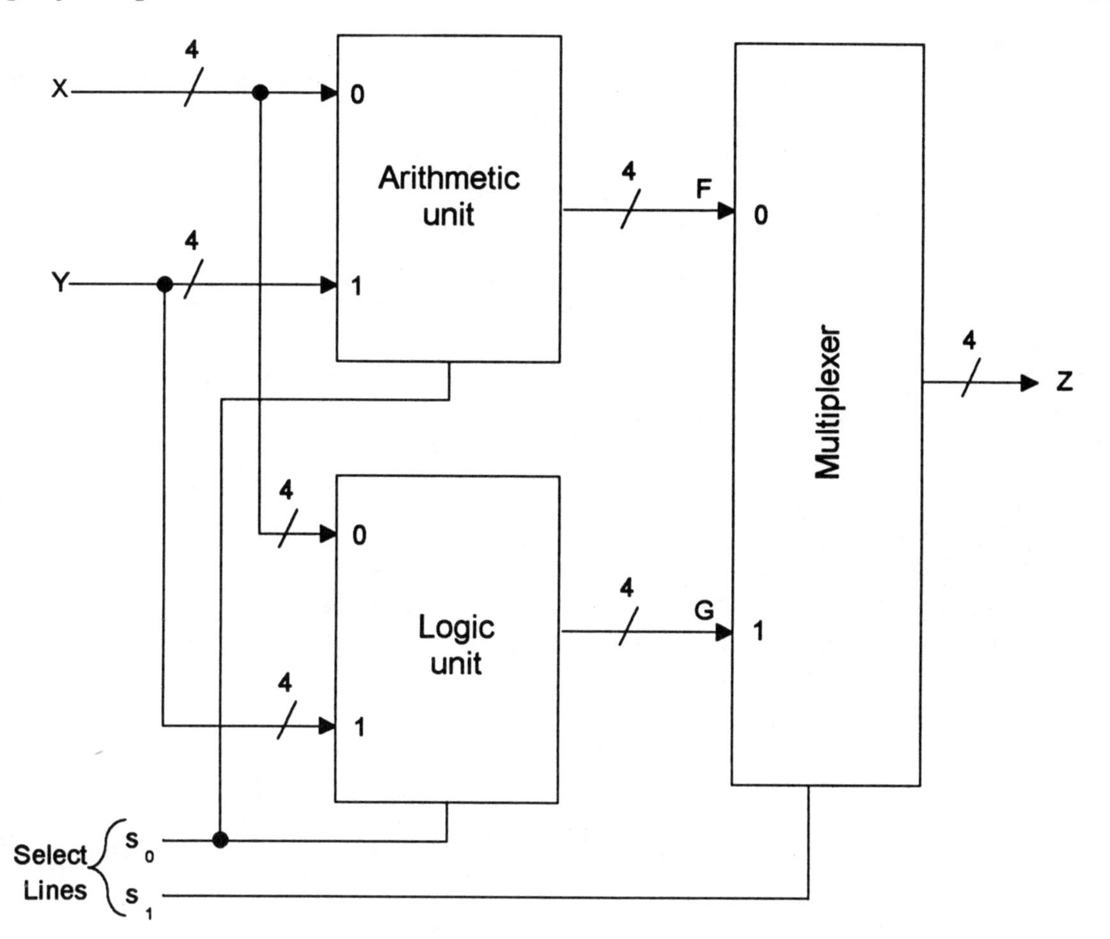

FIGURE 7.20 Schematic representation of the four functions

Select Lines		Output Z	Comment
S_1	S_0		
0	0	X plus Y	Addition
0	1	X plus $\overline{Y}$ Plus 1	2's Complement subtraction
1	0	$X \wedge Y$	Boolean AND
1	1	$X \oplus Y$	Exclusive-OR

FIGURE 7.21 Truth table controlling the operations of the ALU of Figure 7.20

The rapid growth in IC technology permitted the manufacturers to produce an ALU as an MSI block. Such systems implement many operations, and their use as a system component reduces the hardware cost, board space, debugging effort, and failure rate. Usually, each MSI ALU chip is designed as a 4-bit slice. However, a designer can easily interconnect n such chips to get a $4n$-bit ALU. Some popular 4-bit ALU chips are the 74381 and 74181. The 74381 ALU performs 3 arithmetic and 2 miscellaneous operations on 4-bit operands. The 74181 ALU performs 16 arithmetic and 16 Boolean operations on two 4-bit operands, using either active high or active low data. A complete description and operational characteristics of these devices may be found in the original-equipment manufacturer's data books.

Typical 8-bit microprocessors, such as the Intel 8085 and Motorola 6809, do not include multiplication and division instructions due to limitations in the circuit densities that can be placed on the chip. Due to advanced semiconductor technology, 16-, 32-, and 64-bit microprocessors usually include multiplication and division algorithms in a ROM inside the chip. These algorithms typically utilize an ALU to carry out the operations.

As mentioned earlier, unsigned multiplication can be carried out by repeated addition. Signed multiplication can be performed using various algorithms. A simple algorithm follows.

In the case of signed numbers, there are three possibilities:

1. M and Q are in sign-magnitude form.
2. M and Q are in ones complement form.
3. M and Q are in twos complement form.

For the first case, perform unsigned multiplication of the magnitudes without the sign bits. The sign bit of the product is determined as $M_n \oplus Q_n$, where M_n and Q_n are the most significant bits (sign bits) of the multiplicand (M) and the multiplier (Q), respectively. For the second case, proceed as follows:

Step 1: If $M_n = 1$, then compute the ones complement of M.

Step 2: If $Q_n = 1$, then compute the ones complement of Q.

Step 3: Multiply the $n - 1$ bits of the multiplier and the multiplicand.

Step 4: $S_n = M_n \oplus Q_n$

Step 5: If $S_n = 1$, then compute the ones complement of the result obtained in Step 3.

Whenever the ones complement of a negative number (sign bit = 1) is taken, the sign is reversed. Hence, with respect to the multiplier, the inputs are always a positive quantity. When the sign of the bit is negative, however ($M_n \oplus Q_n = 1$), the result must be presented in the ones complement form. This is why the ones complement of the product found by the unsigned multiplier is computed. When M and Q are in twos complement form, the same procedure is repeated, with the exception that the twos complement must be determined when $Q_n = 1$, $M_n = 1$, or $M_n \oplus Q_n = 1$. Consider M and Q as twos complement numbers. Suppose $M = 1100_2$ and $Q = 0111_2$. Because $M_n = 1$, take the twos complement of $M = 0100_2$; because Q_n

= 0, do not change Q. Multiply 0111_2 and 0100_2 using the unsigned multiplication method discussed before. The product is 00011100_2. The sign of the product $S_n = M_n \oplus Q_n = 1 \oplus 0 = 1$. Hence, take the twos complement of the product 00011100_2 to obtain 11100100_2, which is the final answer: -28_{10}.

The general equation for division is *Dividend* = *Quotient* * *Divisor* + *Remainder*. It can be used for signed division. For example, consider dividend = – 9, divisor = 2. Three possible solutions are shown below:

(a) $-9 = -4 * 2 - 1$, Quotient = – 4, Remainder = – 1.
(b) $-9 = -5 * 2 + 1$, Quotient = – 5, Remainder = +1.
(c) $-9 = -6 * 2 + 3$, Quotient = – 6, Remainder = +3.

However, the correct answer is shown in (a) in which, Quotient = – 4 and Remainder = – 1. Hence, for signed division, the sign of the remainder is the same as the sign of the dividend, unless the remainder is zero. Typical microprocessors such as Motorola 68XXX follow this convention.

7.2.3 Design of the Control Unit

The main purpose of the control unit is to translate or decode instructions and generate appropriate enable signals to accomplish the desired operation. Based on the contents of the instruction register, the control unit sends the selected data items to the appropriate processing hardware at the right time. The control unit drives the associated processing hardware by generating a set of signals that are synchronized with a master clock.

The control unit performs two basic operations: instruction interpretation and instruction sequencing. In the interpretation phase, the control unit reads (fetches) an instruction from the memory addressed by the contents of the program counter into the instruction register. The control unit inputs the contents of the instruction register. It recognizes the instruction type, obtains the necessary operands, and routes them to the appropriate functional units of the execution unit (registers and ALU). The control unit then issues necessary signals to the execution unit to perform the desired operation and routes the results to the specified destination.

In the sequencing phase, the control unit generates the address of the next instruction to be executed and loads it into the program counter. To design a control unit, one must be familiar with some basic concepts such as register transfer operations, types of bus structures inside the control unit, and generation of timing signals. These are described in the next section.

There are two methods for designing a control unit: hardwired control and microprogrammed control. In the hardwired approach, synchronous sequential circuit design procedures are used in designing the control unit. Note that a control unit is a clocked sequential circuit. The name "hardwired control" evolved from the fact that the final circuit is built by physically connecting the components such as gates and flip-flops. In the microprogrammed approach, on the other hand, all control functions are stored in a ROM inside the control unit. This memory is called the "control memory." RAMs and PALs are also used to implement the control memory. The words in this memory are called "control words," and they specify the control functions to be performed by the control unit. The control words are fetched from the control memory and the bits are routed to appropriate functional units to enable various gates. An instruction is thus executed. Design of control units using microprogramming (sometimes called *firmware* to distinguish it from hardwired control) is more expensive than using hardwired controls. To execute an instruction, the contents of the control memory in microprogrammed control must be read, which reduces the overall speed of the control unit.The most important advantage of microprogramming is its flexibility; many additions and changes are made by simply changing the microprogram in the control memory. A small change in the hardwired approach may lead to redesigning the entire system.

There are two types of microprocessor architectures: CISC (Complex Instruction Set Computer) and RISC (Reduced Instruction Set Computer). CISC microprocessors contain a large number of instructions and many addressing modes while RISC microprocessors include a simple instruction set with a few addressing modes. Almost all computations can be obtained from a few simple operations. RISC basically supports a small set of commonly used instructions which are excuted at a fast clock rate compared to CISC which contains a large instruction set (some of which are rarely used) executed at a slower clock rate. In order to implement fetch /execute cycle for supporting a large instruction set for CISC, the clock is typically slower. In CISC, most instructions can access memory while RISC contains mostly load/store instructions. The complex instruction set of CISC requires a complex control unit, thus requiring microprogrammed implementation. RISC utilizes hardwired control which is faster. CISC is more difficult to pipeline while RISC provides more efficient pipelining. An advantage of CISC over RISC is that complex programs require fewer instructions in CISC with a fewer fetch cycles while the RISC requires a large number of instructions to accomplish the same task with several fetch cycles. However, RISC can significantly improve its performance with a faster clock, more efficient pipelining and compiler optimization. PowerPC and Intel 80XXX utilize RISC and CISC architectures respectively. Intel Pentium family, on the other hand, utilizes a combination of RISC and CISC architectures for providing high performance. The Pentium uses RISC (hardwired control) to implement efficient pipelining for simple instructions. CISC (microprogrammed control) for complex instructions is utilized by the Pentium to provide upward compatibility with the Intel 8086/80X86 family.

7.2.3.1 Basic Concepts

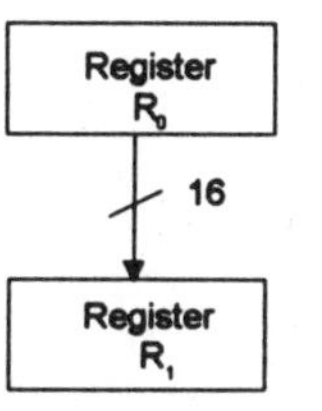

FIGURE 7.22 16-Bit register transfer from R_0 to R_1

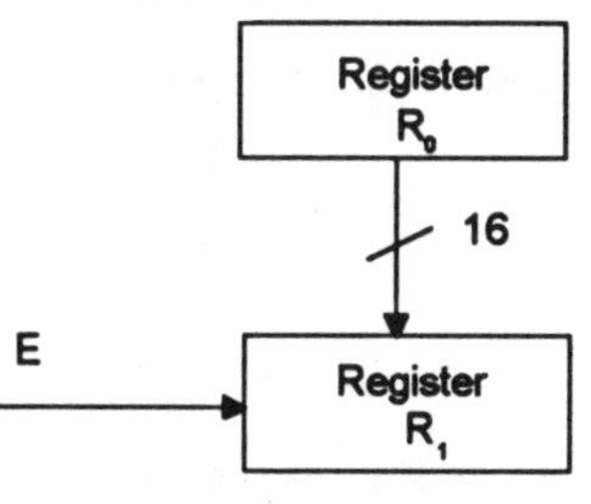

FIGURE 7.23 An enable input controlling register transfer

Register transfer notation is the fundamental concept associated with the control unit design. For example, consider the register transfer operation of Figure 7.22. The contents of 16-bit register R_0 are transferred to 16-bit register R_1 as described by the following notation:

$$R_1 \leftarrow R_0$$

The symbol $\leftarrow$ is called the transfer operator. However, this notation does not indicate the number of bits to be transferred. A declaration statement specifying the size of each register is used for the purpose:

```
Declare registers R0 [16], R1 [16]
```

The register transfer notation can also be used to move a specific bit from one register to a particular bit position in another. For example, the statement

$$R_1 [1] \leftarrow R_0 [14]$$

means that bit 14 of register R_0 is moved to bit 1 of register R_1.

An enable signal usually controls transfer of data from one register to another. For example, consider Figure 7.23. In the figure, the 16-bit contents of register R_0 are transferred to register R_1 if the enable input E is HIGH; otherwise the contents of R_0 and R_1 remain the same. Such a conditional transfer can be represented as

$$E: R_1 \leftarrow R_0$$

Figure 7.24 shows a hardware implementation of transfer of each bit of R_0 and R_1. The enable input may sometimes be a function of more than one variable. For example, consider the following statement involving three 16-bit registers: If $R_0 < R_1$ and $R_2 [1] = 1$ then $R_1 \leftarrow R_0$.

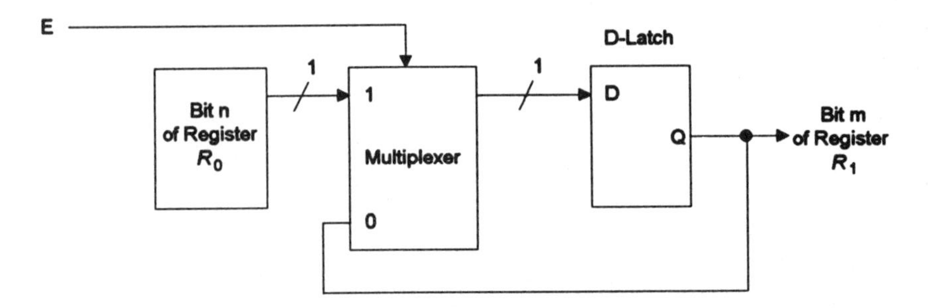

FIGURE 7.24 Hardware for each bit transfer from R_0 to R_1

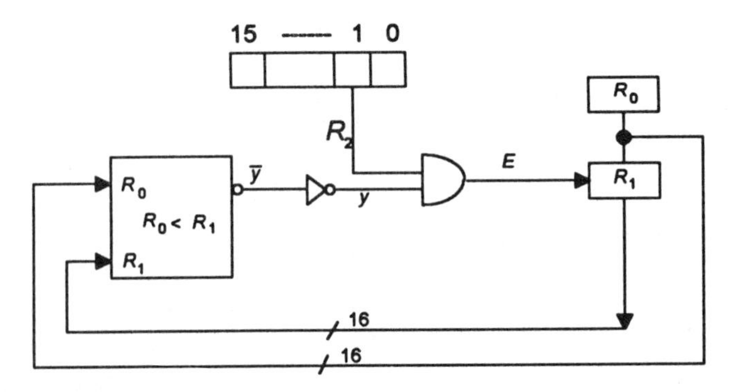

FIGURE 7.25 Hardware implementation E: $R_1 \leftarrow R_0$ where $E = y \cdot R_2$ [1]

The condition $R_0 < R_1$ can be determined by an 8-bit comparator such that the output y of the comparator goes to 0 if $R_0 < R_1$. The conditional transfer can then be expressed as follows: E: $R_1 \leftarrow R_0$ where $E = y \cdot R_2$ [1]. Figure 7.25 depicts the hardware implementation.

A number of wires called "buses" are normally used to transfer data in and out of a digital processing system. Typically, there will be a pair of buses ("inbuses" and "outbuses") inside the CPU to transfer data from the external devises into the processing section and vice versa. Like the registers, these buses are also represented using register transfer notations and declaration statements. For example, "Declare inbus [16] and outbus [16]" indicate that the digital system contains two 16-bit wide data buses (inbus and outbus). $R_0 \leftarrow$ inbus means that the data on the inbus is transferred into register R_0 when the next clock arrives. An equate (=) symbol can also be used in place of $\leftarrow$. For example, "outbus = R_1 [15:8]" means that the high-order 8 bits of the 16-bit register R_1 are made available on the outbus for one clock period. An algorithm implemented by a digital system can be described by using a set of register transfer notations and typical control structures such as if-then and go to. For example, consider the description shown in Figure 7.26 for multiplying two 8-bit unsigned numbers

```
Declare registers R[8],M[8],Q[8];
Declare buses inbus[8],outbus[8];
Start:  R ← 0, M ← inbus;              Clear register R to 0 and move multiplicand
        Q ← inbus;                     Transfer multiplier
 Loop:  R ← R + M, Q ← Q-1;            Add multiplicand
        If Q < > 0 then go to loop;    repeat if Q≠ 0
        Outbus ← R;
 Halt:  Go to Halt;
```

FIGURE 7.26 Register transfer description of 8 × 8 unsigned multiplication (Assume 8-bit result)

(Multiplication of an 8-bit unsigned multiplier by an 8-bit multiplicand) using repeated addition.

The hardware components for the preceding description include an 8-bit inbus, an 8-bit outbus, an 8-bit parallel adder, and three 8-bit registers, *R*, *M*, and *Q*. This hardware performs unsigned multiplication by repeated addition. This is equivalent to unsigned multiplication performed by assembly language instruction.

A distinguishing feature of this description is to describe concurrent operations. For example, the operations $R \leftarrow 0$ and $M \leftarrow$ inbus can be performed simultaneously. As a general rule, a comma is inserted between operations that can be executed concurrently. On the other hand, a semicolon between two transfer operations indicates that they must be performed serially. This restriction is primarily due to the data path provided in the hardware. For example, in the description, because there is only one input bus, the operations $M \leftarrow$ inbus and $Q \leftarrow$ inbus cannot be performed simultaneously. Rather, these two operations must be carried out serially. However, one of these operations may be overlapped with the operation $R \leftarrow 0$ because the operation does not use the inbus. The description also includes labels and comments to improve readability of the task description. Operations such as $R \leftarrow 0$ and M ← inbus are called "micro-operations", because they can be completed in one clock cycle. In general, a computer instruction can be expressed as a sequence of micro-operations.

The rate at which a microprocessor completes operations such as $R \leftarrow R + M$ is determined by its bus structure inside the microprocessor chip. The cost of the microprocessor increases with the complexity of the bus structure. Three types of bus structures are typically used: single-bus, two-bus, and three-bus architectures. The simplest of all bus structures is the single-bus organization shown in Figure 7.27. At any time, data may be transferred between any two registers or between a register and the ALU. If the ALU requires two operands such as in response to an ADD instruction, the operands can only be transferred one at a time. In single-bus architecture, the bus must be multiplexed among various operands. Also, the ALU must have buffer registers to hold the transferred operand.

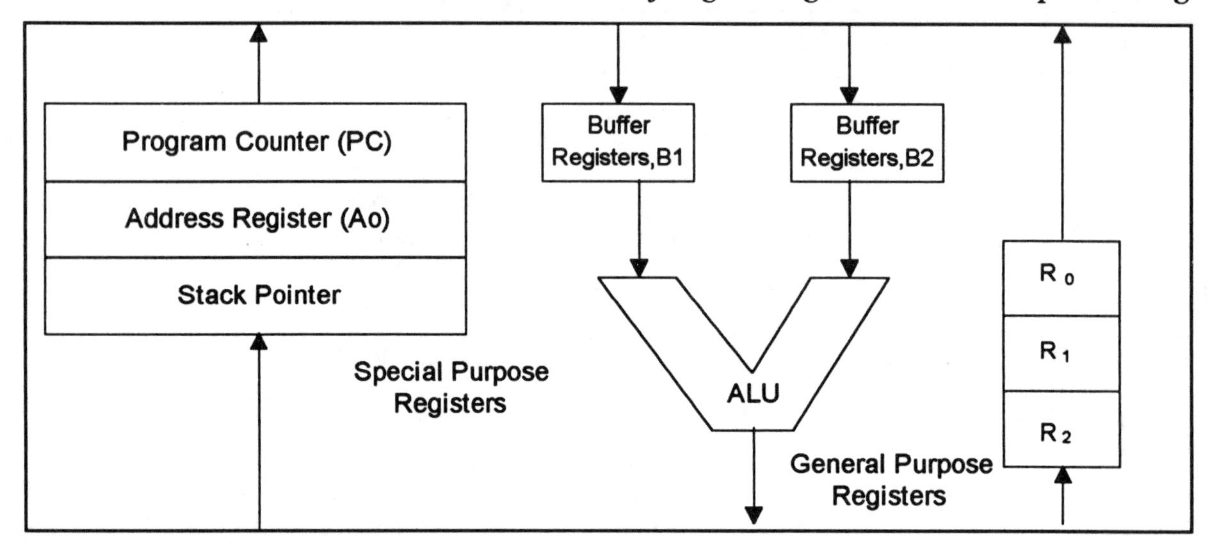

FIGURE 7.27 Single-bus architecture

In Figure 7.27, an add operation such as $R_0 \leftarrow R_1 + R_2$ is completed in three clock cycles as follows:

First clock cycle: The contents of R_1 are moved to buffer register B_1 of the ALU.

Second clock cycle: The contents of R_2 are moved to buffer register B_2 of the ALU.

Third clock cycle: The sum generated by the ALU is loaded into R_0.

A single-bus structure slows down the speed of instruction execution even though data may already be in the microprocessor registers. The instruction's execution time is longer if the operands are in memory; two clock cycles may be required to retrieve the operands into the microprocessor registers from external memory.

To execute an instruction such as ADD between two operands already in register, the control logic in a single-bus structure must follow a three-step sequence. Each step represents a control state. Therefore, a single-bus architecture requires a large number of states in the control logic, so more hardware may be needed to design the control unit. Because all data transfers take place through the same bus one at a time, the design effort to build the control logic is greatly reduced.

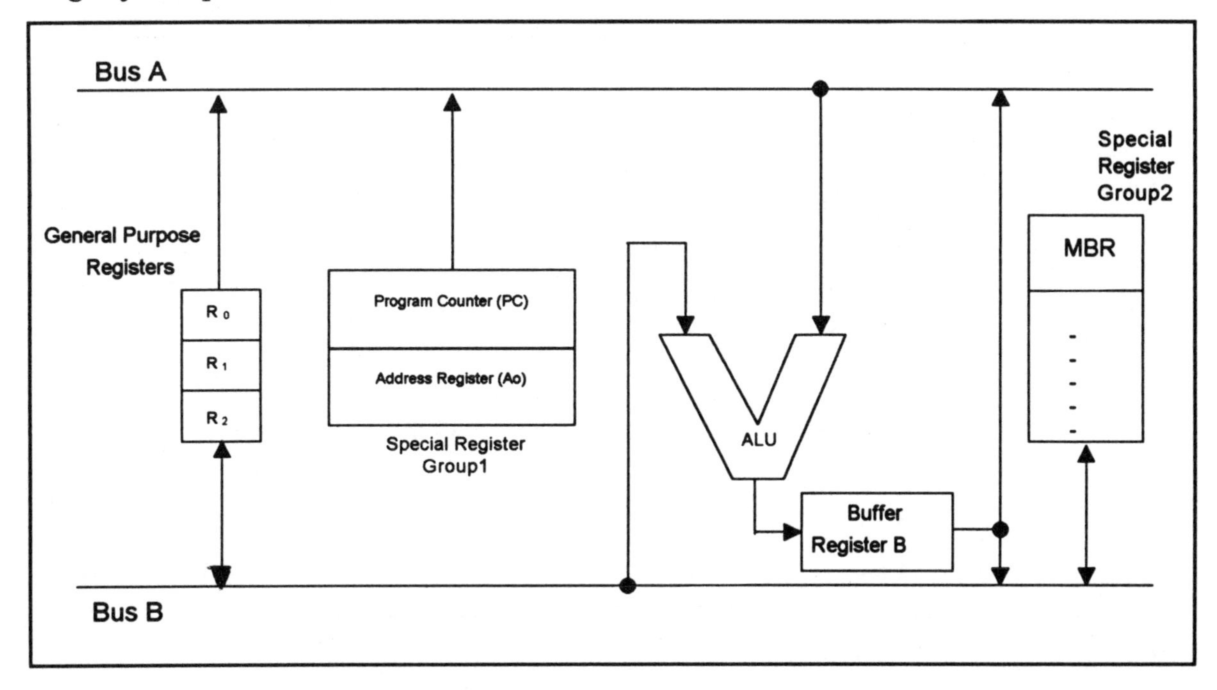

FIGURE 7.28 Two-bus architecture

Next, consider a two-bus architecture, shown in Figure 7.28. All general-purpose registers are connected to both buses (bus *A* and bus *B*) to form a two-bus architecture. The two operands required by the ALU are, therefore, routed in one clock cycle. Instruction execution is faster because the ALU does not have to wait for the second operand, unlike the single-bus architecture. The information on a bus may be from a general-purpose register or a special-purpose register. In this arrangement, special-purpose registers are often divided into two groups. Each group is connected to one of the buses. Data from two special-purpose registers of the same group cannot be transferred to the ALU at the same time.

In the two-bus architecture, the contents of the program counter are always transferred to the right input of the ALU because it is connected to bus *A*. Similarly, the contents of the special register MBR (memory buffer register, to hold up data retrieved from external memory) are always transferred to the left input of the ALU because it is connected to bus *B*.

In Figure 7.28, an add operation such as $R_0 \leftarrow R_1 + R_2$ is completed in two clock cycles as follows:

First clock cycle: The contents of R_1 and R_2 are moved to the inputs of ALU. The ALU then generates the sum in the output register.

Second clock cycle: The sum from the output register is routed to R_0.

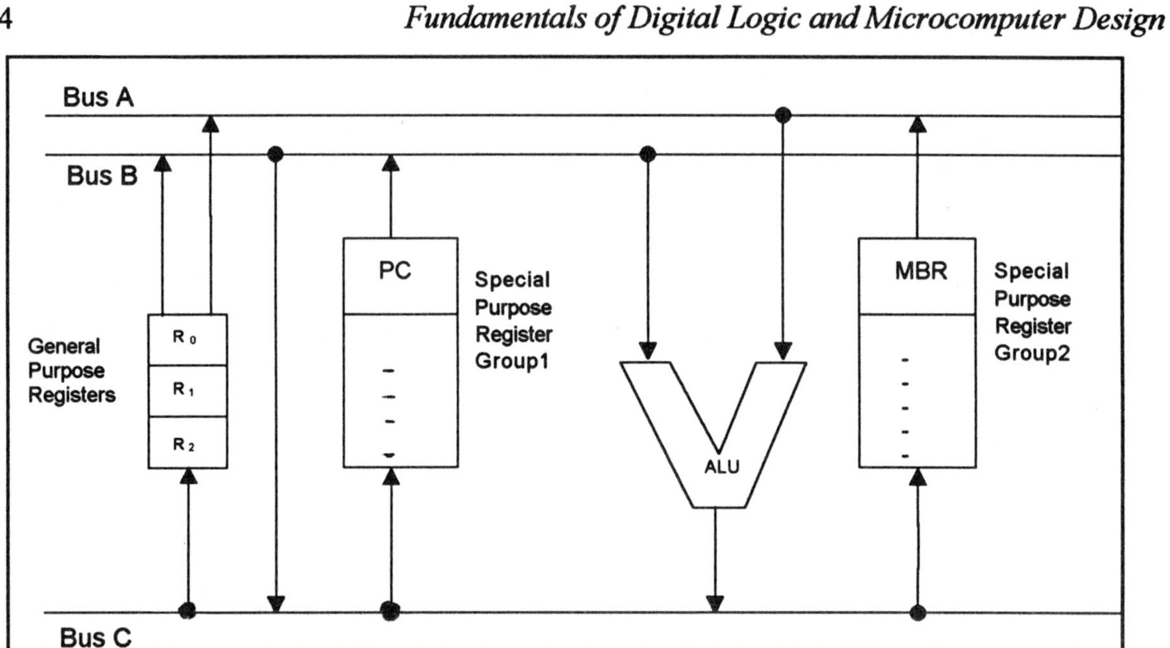

FIGURE 7.29 Three-bus architecture

The performance of a two-bus architecture can be improved by adding a third bus (bus *C*), at the output of the ALU. Figure 7.29 depicts a typical three-bus architecture. The three-bus architecture perform the addition operation $R_0 \leftarrow R_1 + R_2$ in one cycle as follows:

First cycle: The contents of R_1 and R_2 are moved to the inputs of the ALU via bus A and bus *B* respectively. The sum generated by the ALU is then transferred to R_0 via bus *C*.

The addition of the third bus will increase the system cost and also the complexity of the control unit design.

Note that the bus architectures described so far are inside the microprocessor chip. On the other hand, the system bus connecting the microprocessor, memory, and I/O are external to the microprocessor.

Another important concept required in the design of a control unit is the generation of timing signals. One of the main tasks of a control unit is to properly sequence a set of operations such as a sequence of *n* consecutive clock pulses. To carry out an operation, timing signals are generated from a master clock. Figure 7.30 shows the input clock pulse and the four timing signals T_0, T_1, T_2, and T_3. A ring counter (described in Chapter 5) can be used to generate these timing signals. To carry out an operation P_i at the *i*th clock pulse, a control unit must count the clock pulses and produce a timing signal T_i.

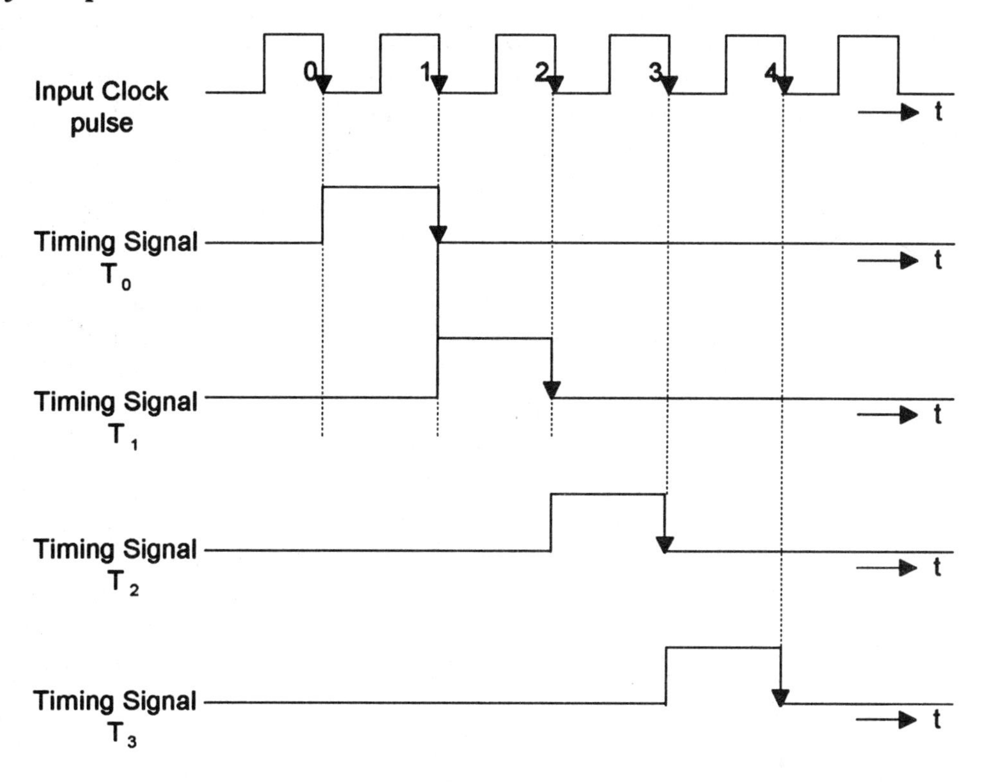

FIGURE 7.30 Timing signals

7.2.3.2 Hardwired Control Design

The steps involved in hardwired control design are summarized as follows:

1. Derive a flowchart from the problem definition and validate the algorithm by using trial data.
2. Obtain a register transfer description of the algorithm from the flowchart.
3. Specify a processing hardware along with various components.
4. Complete the design of the processing section by establishing the necessary control inputs.
5. Determine a block diagram of the controller.
6. Obtain the state diagram of the controller.
7. Specify the characteristic of the hardware for generating the required timing signals used in the controller.
8. Draw the logic circuit of the controller.

The following example is provided to illustrate the concepts associated with implementation of a typical instruction in a control unit using hardwired control. The unsigned multiplication by repeated addition discussed earlier is used for this purpose. A 4-bit by 4-bit unsigned multiplication will be considered. Assume the result of multiplication is 4 bits.

Step 1: Derive a flowchart from the problem definition and then validate the algorithm using trial data.

Figure 7.31 shows the flowchart. In the figure, M and Q are two 4-bit registers containing the unsigned multiplicand and unsigned multiplier respectively. Assume that the result of multiplication is 4-bit wide. The 4-bit result of the multiplication called the "product" will be stored in the 4-bit register, R. The contents of R are then output to the outbus.

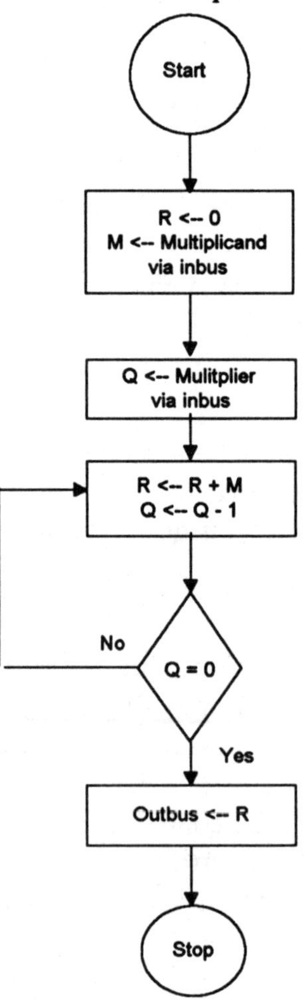

FIGURE 7.31 Flowchart for 4-bit × 4-bit multiplication

	R	M	Q
Initialization	0000	0100	0011
Iteration 1 R <-- R + M Q <-- Q - 1	0100	0100	0010
Iteration 2 R <-- R + M Q <-- Q - 1	1000	0100	0001
Iteration 3 R <-- R + M Q <-- Q - 1	1100	0100	0000

Product $=12_{10}$

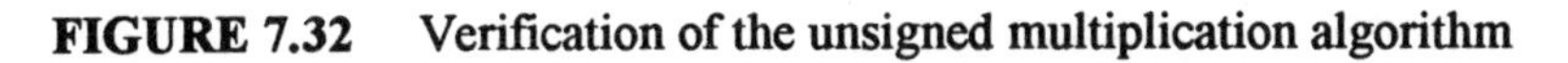

FIGURE 7.32 Verification of the unsigned multiplication algorithm

The flowchart in Figure 7.31 is similar to an ASM chart and provides a hardware description of the algorithm. The sequence of events and their timing relationships are described in the flowchart. For example, the operations, $R \leftarrow 0$ and $M \leftarrow$ multiplicand shown in the same block are executed simultaneously. Note that $M \leftarrow$ multiplicand via inbus and $Q \leftarrow$ multiplier via inbus must be performed serially because both operations use a single input bus for loading data. These operations are, therefore, shown in different blocks. Because $R \leftarrow 0$ does not use the inbus, this operation is overlapped, in our case, with initializing of M via the inbus. This simultaneous operation is indicated by placing them in the same block.

The algorithm will now be verified by means of a numerical example as shown in Figure 7.32. Suppose $M = 0100_2 = 4_{10}$ and $Q = 0011_2 = 3_{10}$; then R = product $= 1100_2 = 12_{10}$

Step 2: Obtain a register transfer description of the algorithm from the flowchart.

Figure 7.33 shows the description of the algorithm.

```
Start: R ← 0, M ← inbus;               Clear Register to 0 and move multiplicand
       Q ← inbus;                      Transfer Multiplier
 Loop: R ← R + M, Q ← Q -1;            Perform addition, decrement counter
       If Q < > 0 then goto Loop;      Repeat if Q ≠ 0
       outbus ← R;
 Halt: Go to Halt;
```

FIGURE 7.33 Register transfer description 4-bit × 4-bit unsigned multiplication

Step 3: Specify a processing hardware along with various components.

The processing section contains three main components:

- General-purpose registers
- 4-bit adder
- Tristate buffer

Figure 7.34 shows these components. The general-purpose register is a trailing edge-triggered device.

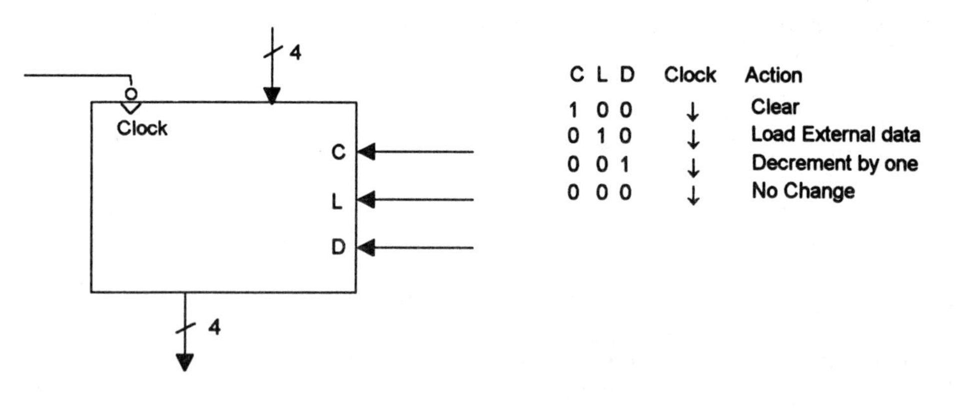

4
Clock
C
L
D
4
C L D Clock Action
1 0 0 ↓ Clear
0 1 0 ↓ Load External data
0 0 1 ↓ Decrement by one
0 0 0 ↓ No Change

(a) General Purpose Register

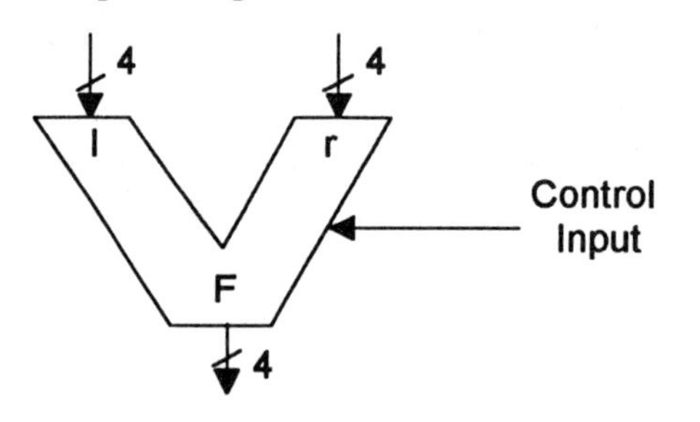

(b) 4-bit Adder

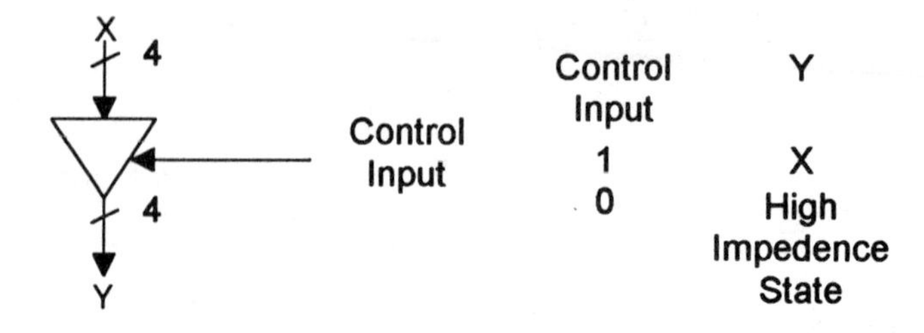

(c) Tristate Buffer

FIGURE 7.34 Components of the processing section of 4-bit by 4-bit unsigned multiplication

Three operations (clear, parallel load, and decrement) can be performed by applying the appropriate inputs at *C*, *L*, and *D*. All these operations are synchronized at the trailing (high to low) edge of the clock pulse.

The 4-bit adder can be implemented using 4-bit adder circuits. The tristate buffer is used to control data transfer to the outbus.

Step 4: Complete the design of the processing section by establishing the necessary control inputs.

Figure 7.35 shows the detailed logic diagram of the processing section, along with the control inputs.

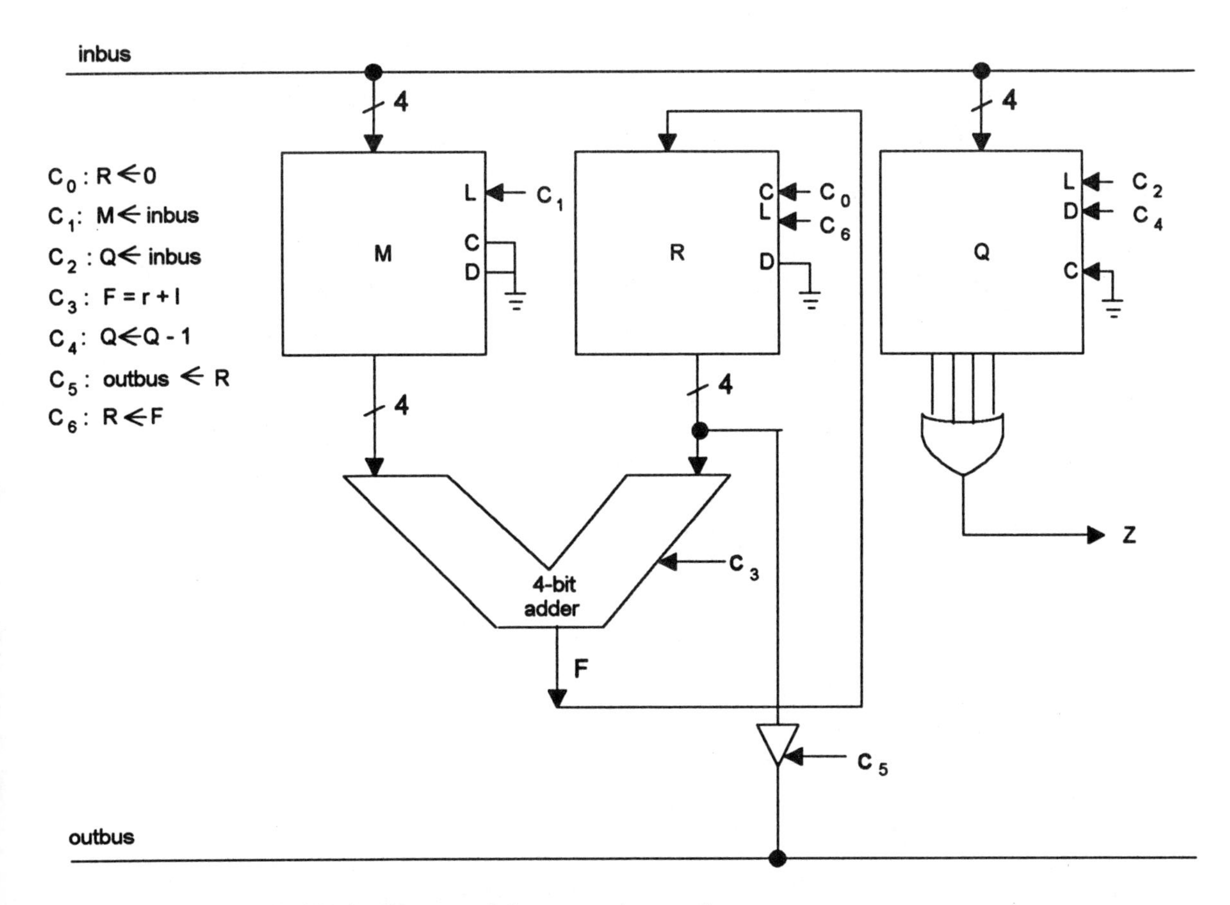

FIGURE 7.35 Detailed logic diagram of the processing section

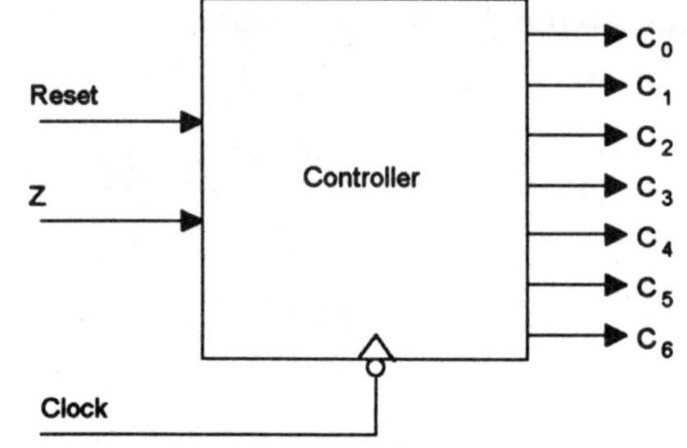

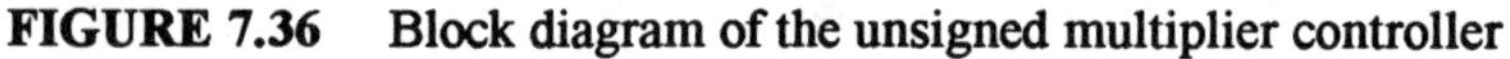

FIGURE 7.36 Block diagram of the unsigned multiplier controller

Step 5: Determine a block diagram of the controller.

Figure 7.36 shows the block diagram.

The controller has three inputs and seven outputs. The Reset input is an asynchronous input used to reset the controller so that a new computation can begin. The Clock input is used to synchronize the controller's action. All activities are assumed to be synchronized with the trailing edge of the clock pulse.

Step 6: Obtain the state diagram of the controller.

The controller must initiate a set of operations in a specified sequence. Therefore, it is modeled as a sequential circuit. The state diagram of the unsigned multiplier controller is shown in Figure 7.37.

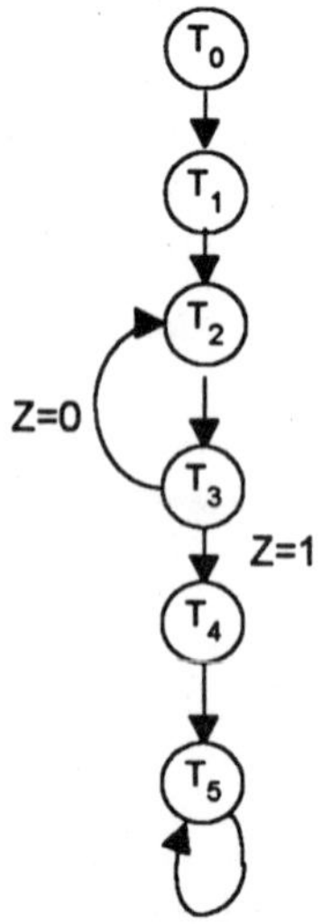

Control State	Operation Performed	Control Signal to be activated
T_0	$R \leftarrow 0$; $M \leftarrow$ inbus	C_0, C_1
T_1	$Q \leftarrow$ inbus	C_2
T_2	$R \leftarrow R + M$, $Q \leftarrow Q - 1$	C_3, C_4, C_6
T_3	None	None
T_4	outbus $\leftarrow R$	C_5
T_5	None	None

(a) State Diagram (b) Controller action

FIGURE 7.37 Controller description

Initially, the controller is in state T_0. At this point, the control signals C_0 and C_1 are HIGH. Operations $R \leftarrow 0$ and $M \leftarrow$ inbus are carried out with the trailing edge of the next clock pulse. The controller moves to state T_1 with this clock pulse. When the controller is in T_2, $R \leftarrow R + M$ and $Q \leftarrow Q - 1$ are performed.

All these operations take place at the trailing edge of the next clock pulse. The controller moves to state T_5 only when the unsigned multiplication is completed. The controller then stays in this state forever. A hardware reset input causes the controller to move to state T_0, and a new computation will start.

In this state diagram, selection of states is made according to the following guidelines:

- If the operations are independent of each other and can be completed within one clock cycle, they are grouped within one control state. For example, in Figure 7.37, operations $R \leftarrow 0$ and $M \leftarrow$ inbus are independent of each other. With this hardware, they can be executed in one clock cycle. That is, they are microoperations. However, if they cannot be completed within T_0 clock cycle, either clock duration must be increased or the operations should be divided into a sequence of microoperations.
- Conditional testing normally implies introduction of new states. For example, in the figure, conditional testing of Z introduces the new state T_3.
- One should not attempt to minimize the number of states. When in doubt, new states must be introduced. The correctness of the control logic is more important than the cost of the circuit.

Step 7: Specify the characteristics of the hardware for generating the required timing signals.

There are six states in the controller state diagram. Six nonoverlapping timing signals (T_0 through T_5) must be generated so that only one will be high for a clock pulse. For example, Figure 7.38 shows the four timing signals T_0, T_1, T_2, and T_3. A mod-8 counter and a 3-to-8 decoder can be used to accomplish this task. Figure 7.39 shows the mod-8 counter.

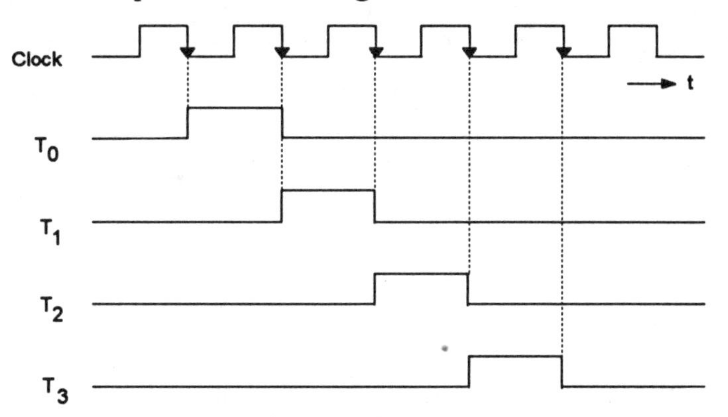

FIGURE 7.38 Timing signals generated by the controller

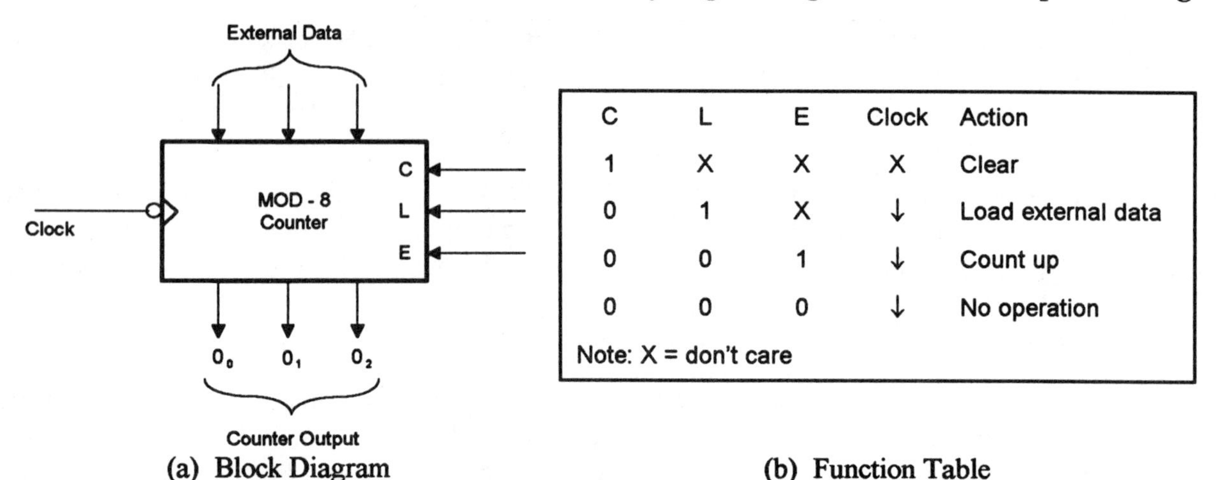

C	L	E	Clock	Action
1	X	X	X	Clear
0	1	X	↓	Load external data
0	0	1	↓	Count up
0	0	0	↓	No operation

Note: X = don't care

(b) Function Table

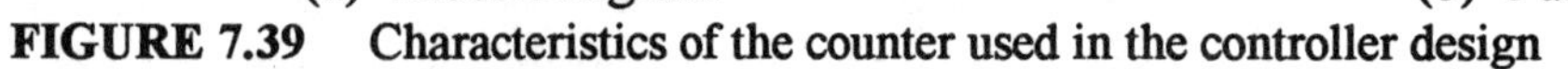
FIGURE 7.39 Characteristics of the counter used in the controller design

Step 8: Draw the logic circuit of the controller.

Figure 7.40 shows the logic circuit of the controller. The key element of the implementation in Figure 7.40 is the sequence controller (SC) hardware, which sequences the controller according to the state diagram of Figure 7.37. Figure 7.41(a) shows the truth table for the SC controller.

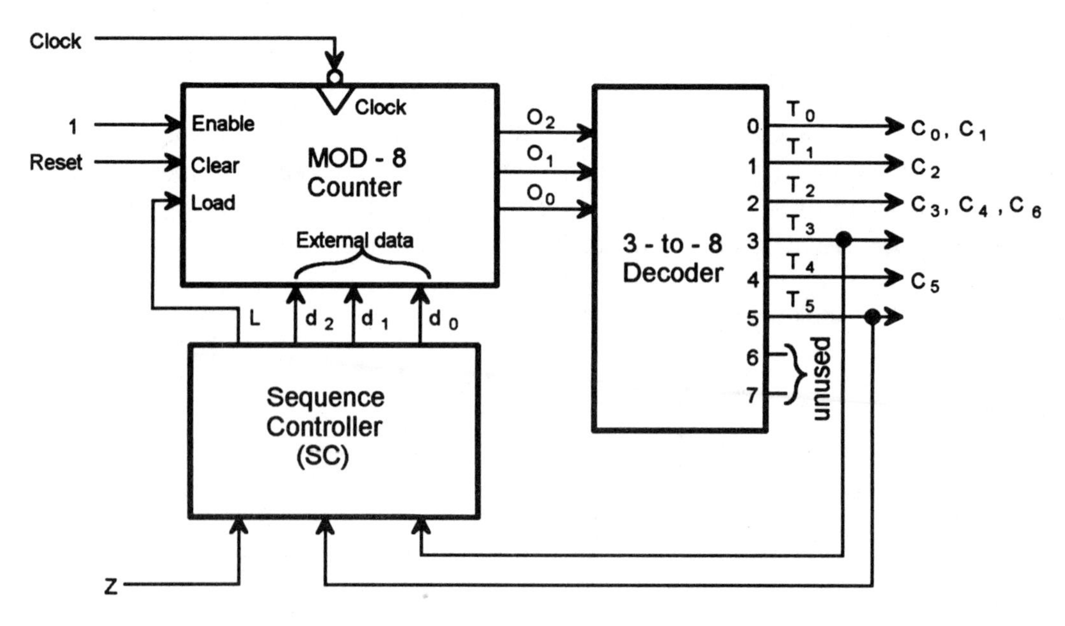

FIGURE 7.40 Logic diagram of the unsigned multiplier controller

Inputs			Outputs			
Z	T_3	T_5	L	d_2	d_1	d_0
0	1	x	1	0	1	0
x	x	1	1	1	0	1

Note: x = don't care

(a) Truth Table

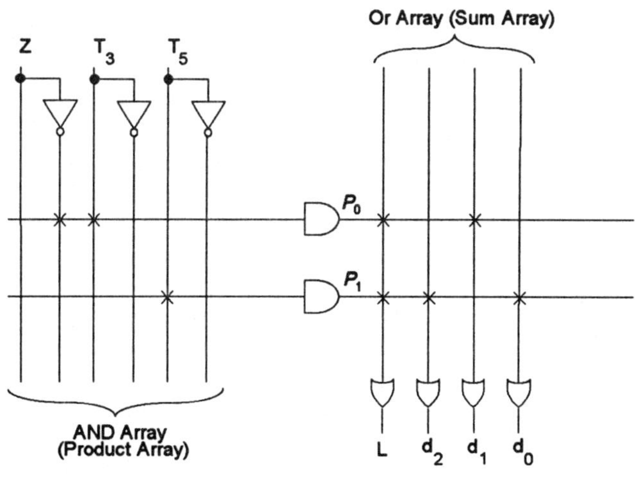

(b) PLA Implementation

FIGURE 7.41 Sequence controller design

Consider the logic involved in deriving the entries of the SC truth table. The mod-8 counter is loaded (or initialized) with the specified external data if the counter control inputs C and L are 0 and 1 respectively from Figure 7.39. In this counter, the counter load control input L overrides the counter enable control input E.

From the controller's state diagram of Figure 7.37, the controller counts up automatically in response to the next clock pulse when the counter load control input $L = 0$ because the enable input E is tied to HIGH. Such normal sequencing activity is desirable for the following situations:

- Present control state is T_0, T_1, T_2, T_4.
- Present control state is T_3 and $Z = 1$; the next state is T_4.

The SC must load the counter with appropriate count when the counter is required to load the count out of its normal sequence.

For example, from the controller's state diagram of Figure 7.37, if the present control state is T_3 (counter output $O_2O_1O_0 = 011$) and if $Z = 0$, the next state is T_2. When these input conditions occur, the counter must be loaded with external value 010 at the trailing edge of the next clock pulse ($T_2 = 1$ only when $O_2O_1O_0$= 010. Therefore, the SC generates $L = 1$ and $d_2d_1d_0 = 010$.

Similarly, from the controller's state diagram of Figure 7.37, if the present state is T_5, the next control state is also T_5. The SC must generate the outputs $L = 1$ and $d_2d_1d_0 = 101$. The SC truth table of Figure 7.41 shows these out-of-sequence counts. For each row of the SC truth table of Figure 7.41(a), a product term is generated in the PLA:

$$P_0 = \overline{Z}T_3 \text{ and } P_1 = T_5.$$

The PLA (Figure 7.41b) generates four outputs: L, d_2, d_1, and d_0. Each output is directly generated by the SC truth table and the product terms. The PLA outputs are as follows:

$$L = P_0 + P_1$$

$$d_2 = P_1$$

$$d_1 = P_0$$

$$d_0 = P_1$$

The controller design is completed by relating the control states (T_0 through T_5) to the control signals (C_0 though C_6) as follows:

$$C_0 = C_1 = T_0$$

$$C_2 = T_1$$

$$C_3 = C_4 = C_6 = T_2$$

$$C_5 = T_4$$

From these equations, when the control is in state T_0 or T_2, multiple micro-operations are performed. Otherwise,when the control is in state T_1 or T_4, a single micro-operation is performed.

The unsigned multiplication algorithm just implemented using hardwired control can be considered as an unsigned multiplication instruction with a microprocessor. To execute this instruction, the microcomputer will read (fetch) this multiplication instruction from external memory into the instruction register located inside the microprocessor. The contents of this instruction register will be input to the control unit for execution. The control unit will generate the control signals C_0 through C_6 as shown in Figure 7.40. These control signals will then be applied to the appropriate components of the processing section in Figure 7.35

at the proper instants of time shown in Figure 7.37. Note that the control signals are physically connected to the hardware elements of Figure 7.35. Thus, the execution of the unsigned multiplication instruction will be completed by the microprocessor.

7.2.3.3 Microprogrammed Control Unit Design

As mentioned earlier, a microprogrammed control unit contains programs written using microinstructions. These programs are stored in a control memory normally in a ROM inside the CPU. To execute instructions, the microprocessor reads (fetches) each instruction into the instruction register from external memory. The control unit translates the instruction for the microprocessor. Each control word contains signals to activate one or more microoperations. A program consisting of a set of microinstructions is executed in a sequence of micro-operations to complete the instruction execution. Generally, all microinstructions have two important fields:

- Control word
- Next address

The control field indicates which control lines are to be activated. The next address field specifies the address of the next microinstruction to be executed. The concept of microprogramming was first proposed by W. V. Wilkes in 1951 utilizing a decoder and an 8×8 ROM with a diode matrix. This concept is extended further to include a control memory inside the CPU. The cost of designing a CPU primarily depends on the size of the control memory. The length of a microinstruction, on the other hand, affects the size of the control memory. Therefore, a major design effort is to minimize cost of implementing a microprogrammed CPU by reducing the length of the microinstruction.

The length of a microinstruction is directly related to the following factors:

- The number of micro-operations that can be activated simultaneously. This is called the "degree of parallelism."
- The method by which the address of the next microinstruction is determined.

All microinstructions executed in parallel can be included in a single microinstruction with a common op-code. The result is a short microprogram. However, the length of the microinstruction increases as parallelism grows.

The control bits in a microinstruction can be organized in several ways. One obvious way is to assign a single bit for each control line. This will provide full parallelism. No decoding of the control field is necessary. For example, consider Figure 7.42 with two registers, X and Y with one outbus. In the figure, the contents of each register are transferred to the outbus when the appropriate control line is activated:

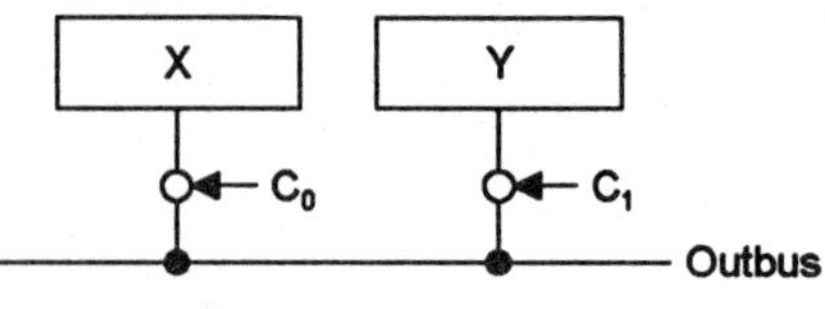

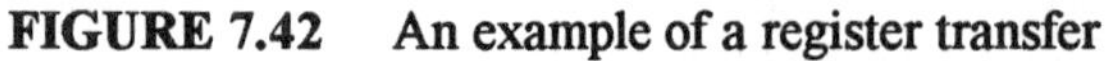
FIGURE 7.42 An example of a register transfer

$$C_0\text{: outbus} \leftarrow X$$
$$C_1\text{: outbus} \leftarrow Y$$

Here, each operation can be performed one at a time because there is only one outbus. A single bit can be assigned to perform each transfer as follows:

Control Bits C_0	Control Bits C_1	Operation Performed
1	0	Outbus ← X
0	1	Outbus ← Y
0	0	No operation

This method is called "unencoded format."

The three operations can be implemented using two bits and a 2-to-4 decoder as shown in Figure 7.43. This is called "encoded format." The relationship between the encoded and actual control information is as follows:

Encoded Bits d_1	Encoded Bits d_0	Operation Performed
0	0	No operation
0	1	Outbus ← x
1	0	Outbus ← y

d_0 d_1

2 to 4 Decoder

3 2 1 0

unused C_1 C_0 No operation

FIGURE 7.43 Encoded format

Note that a 5-bit control field is required for five operations. However, three encoded bits are required for five operations using a 3 to 8 decoder. Hence, the encoded format typically provides a short control field and thus results in short microinstructions. However, the need for a decoder will increase the cost. Therefore, there is a trade-off between the degree of parallelism and the cost.

Microinstructions can be classified into two groups: horizontal and vertical. The horizontal microinstruction mechanism provides long microinstructions, a high degree of parallelism, and little or no encoding. The vertical instruction method, on the other hand, offers short microinstructions, limited parallelism, and considerable decoding.

Microprogramming is the technique of writing microprograms in a microprogrammed control unit. Writing microprograms is similar to writing assembly language programs. Microprograms are basically written in a symbolic language called microassembly language. These programs are translated by a microassembler to generate microcodes, which are then stored in the control memory.

In the early days, the control memory was implemented using ROMs. However, these days control memories are realized in writeable memories. This provides the flexibility of interpreting different instruction set by rewriting the original microprogram, which allows implementation of different control units with the same hardware. Using this approach, one CPU can interpret the instruction set of another CPU. The design of a microprogrammed control unit is considered next. The 4-bit × 4-bit unsigned multiplication using hardwired control (presented earlier) is implemented by microprogramming. The register transfer description shown in Figure 7.33 is rewritten in symbolic microprogram language as shown in Figure 7.44. Note that the unsigned 4-bit × 4-bit multiplication uses repeated addition. The result (product) is assumed to be 4 bits wide.

Control Memory Address		*Control Word*
0	START	R ← 0, M ← inbus;
1		Q ← inbus;
2	LOOP	R ←R + M, Q ← Q - 1;
3		If Z = 0 then goto Loop;
4		outbus ← R;
5	HALT	Go to HALT

FIGURE 7.44 Symbolic microprogram for 4-bit × 4-bit unsigned multiplication using repeated addition

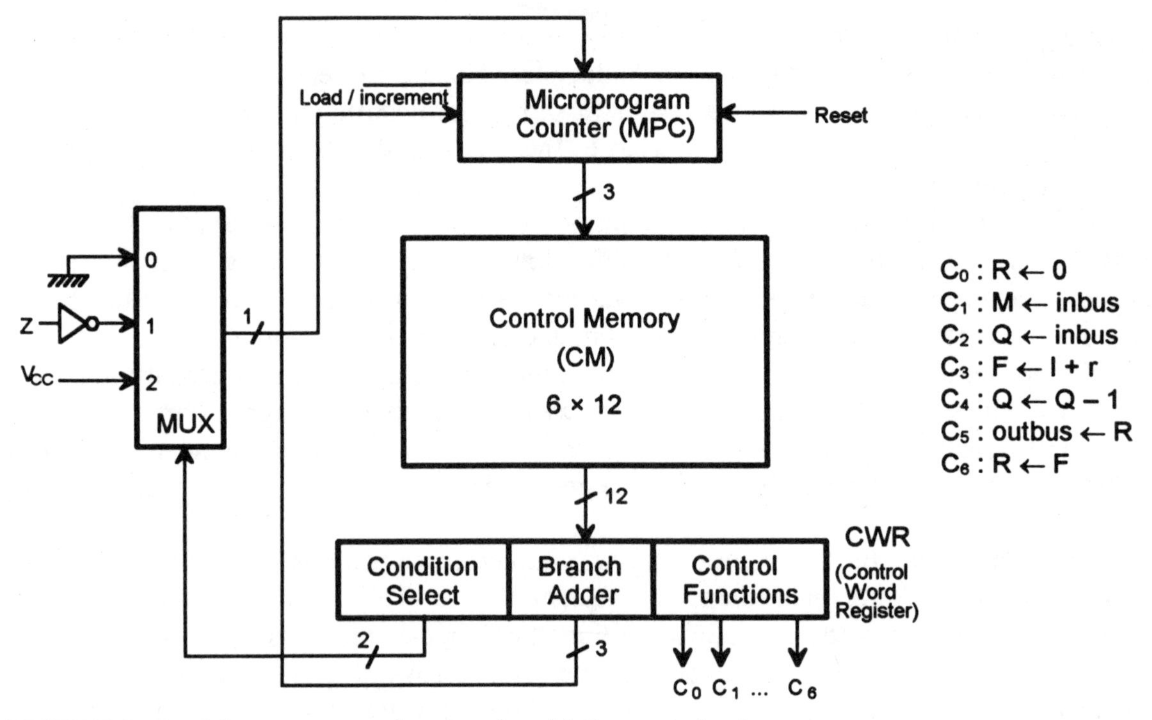

FIGURE 7.45 Microprogrammed unsigned multiplier control unit

To implement the microprogram, the hardware organization of the control unit shown in Figure 7.45 can be used. The various components of the hardware of Figure 7.45 are described in the following:

1. **Microprogram Counter (MPC).** The MPC holds the address of the next microinstruction to be executed. It is initially loaded from an external source to point to the starting address of the microprogram. The MPC is similar to the program counter (PC). The MPC is incremented after each microinstruction fetch. If a branch instruction is encountered, the MPC is loaded with the contents of the branch address field of the microinstruction.

2. **Control Word Register (CWR).** Each control word in the control memory in this example is assumed to contain three fields: condition select, branch address, and control function. Each microinstruction fetched from the Control Memory is loaded into the CWR. The organization of the CWR is same for each control word and contains the three fields just mentioned. In the case of a conditional branch microinstruction, if the condition specified by the condition select field is true, the MPC is loaded with the branch address field of the CWR; otherwise, the MPC is incremented

to point to the next microinstruction. The control function field contains the control signals.

3. **MUX (Multiplexer).** The MUX is a condition select multiplexer. It selects one of the external conditions based on the contents of the condition select field of the microinstruction fetched into the CWR.

In Figure 7.45, a 2-bit condition select field is required as follows:

Condition Select Field		Interpretation
0	0	No branching (no condition)
0	1	Branch if $Z = 0$
1	0	Unconditional branching

From Figure 7.44, six control memory address (addresses 0 through 5) are required for the control memory to store the microprogram. Therefore, a 3-bit address is necessary for each microinstruction. Hence, three bits for the branch address field are required. From Figure 7.45, seven control signals (C_0 through C_6) are required. Therefore, the size of the control function field is 7 bits wide. Thus, the size of each control word can be determined as follows:

size of *a* control word	=	size of the condition select field	+	size of the branch address field	+	number of control signals
	=	2	+	3	+	7
	=	12 bits				

Therefore, the size of the control memory is 6 bits × 12 bits because the microprogram requires six addresses (0 through 5) and each control word is 12 bits wide. The size of the CWR is 12 bits. The complete binary listing of the microprogram is shown in Figure 7.46.

ROM Address		Control Word									Comments
In decimal	In binary	Condition Select	Branch Address	Control Function							
				C_0	C_1	C_2	C_3	C_4	C_5	C_6	
0	0 0 0	0 0	0 0 0	1	1	0	0	0	0	0	R← 0, M ← inbus
1	0 0 1	0 0	0 0 0	0	0	1	0	0	0	0	Q ← inbus
2	0 1 0	0 0	0 0 0	0	0	0	1	1	0	1	R ← R + M, Q ← Q −1, R ← F
3	0 1 1	0 1	0 1 0	0	0	0	0	0	0	0	If Z = 0 then go to address 2 (loop)
4	1 0 0	0 0	0 0 0	0	0	0	0	0	1	0	outbus ← R
5	1 0 1	1 0	1 0 1	0	0	0	0	0	0	0	Go to address 5 (HALT)

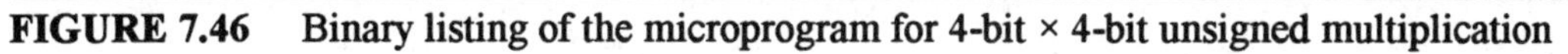

FIGURE 7.46 Binary listing of the microprogram for 4-bit × 4-bit unsigned multiplication

Let us now explain the binary program. Consider the first line of the program. The instruction contains no branching. Therefore, the condition select field is 00. The contents of the branch in this case filled with 000. In the control function field, two micro-operations, C_0 and C_1, are activated. Therefore, both C_0 and C_1 are set to 1; C_2 through C_6 are set to 0. This results in the following binary microinstruction shown in the first line (address 0) of Figure 7.46:

Condition Select	Branch Address	Control Function
00	000	1100000

Next, consider the conditional branch instruction of Figure 7.46. This microinstruction implements the conditional instruction "If $Z = 0$ then go to address 2." In this case, the microinstruction does not have to activate any control signal of the control function field. Therefore, C_0 through C_6 are zero. The condition select field is 01 because the condition is based on $Z = 0$. Also, if the condition is true ($Z = 0$), the program branches to address *2*. Therefore, the branch address field contains 010_2. Thus, the following binary microinstruction is obtained:

Condition Select	Branch Address	Control Function
01	010	000000

The other lines in the binary representation of the microprogram can be explained similarly. To execute an unsigned multiplication instruction implemented using the repeated addition just described, a microprogrammed microprocessor will fetch the instruction from external memory into the instruction register. To execute this instruction, the microprocessor uses the control unit of Figure 7.45 to generate the control word based on the microprogram of Figure 7.46 stored in the control memory. The control signals C_0 through C_6 of the control function field of the CWR will be connected to appropriate components of Figure 7.35. The instruction will thus be executed by the microprocessor.

By inspecting the binary microprogram of Figure 7.46, branch instructions, the control function field is filled with zeros. In a typical microprogram, there may be several conditional and unconditional branch instructions. Therefore, a lot of valuable memory space inside the control unit will be wasted if the control field is filled with zeros. In practice, the format of the control word is organized in a different manner to minimize its size. This reduces the implementation cost of the control unit. *This topic is covered in Chapter 11.*

Nanomemory is another approach for reducing the size of the control memory. This technique contains a two-level memory: control memory and nanomemory. At the outset, are may feel that the two-level memory will increase the overall cost. In fact, it reduces the cost of the system by minimizing the memory size.

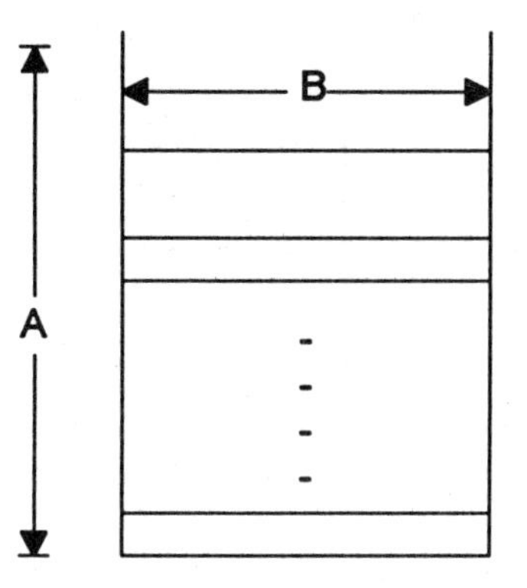

FIGURE 7.47 A microprogram of size A × B

The concept of nanomemory is derived from a combination of horizontal and vertical instructions. However, this method provides trade-offs between them.

Motorola uses nanomemory to design the control units of their popular 16-bit and 32-bit microprocessors, including the 68000, 68020, 68030, and 68040. The nanomemory method provides significant savings in memory when a group of micro-operations occur several times in a microprogram. Consider the microprogram of Figure 7.47, which contains A microinstructions B bits wide.

The size of the control memory to store this microprogram is AB bits. Assume that the microprogram has n ($n < A$) unique microinstructions. These n microinstructions can be held in a separate memory called the "nanomemory" of size nB bits. Each of these n instructions occurs once in the nanomemory. Each microinstruction in the original microprogram is replaced with the address that specifies the location of the nanomemory in which the original B-bit-wide microinstructions are held. Because the nanomemory has n addresses, only the upper integer of $\log_2 n$ bits is required to specify a nanomemory address. This is illustrated in Figure 7.48.

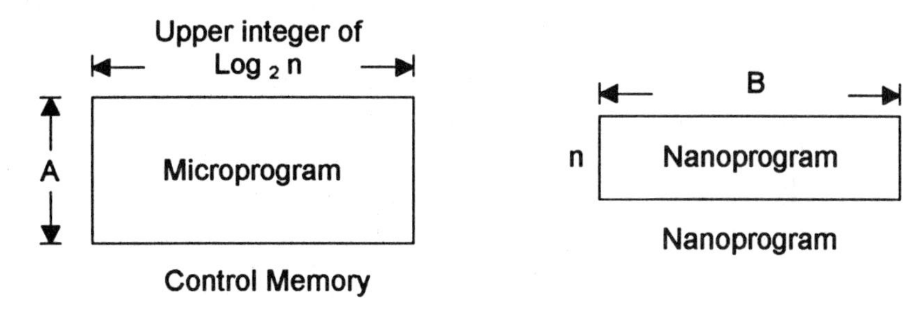

FIGURE 7.48 Nanomemory

Address	Content
000	0100
001	0000
010	0100
011	0100
100	0000
101	1010
110	1010

FIGURE 7.49 7 × 4-bit single control memory

The operation of microprocessor employing a nanomemory can be explained as follows: The microprocessor's control unit reads an address from the microprogram. The content of this address in the nanomemory is the desired control word. The bits in the control word are used by the control unit to accomplish the desired operation. Note that a control unit employing nanomemory (two-level memory) is slower than the one using a conventional control memory (single memory). This is because the nanomemory requires two memory reads (one for the control memory and the other for the nanomemory). For a single conventional control memory, only one memory fetch is necessary. This reduction in control unit speed is offset by the cost of the memory when the same microinstructions occur many times in the microprogram.

Consider the 7 × 4-bit microprogram stored in the single control memory of Figure 7.49. This simplified example is chosen to illustrate the nanomemory concept even though this is not a practical example. In this program, 3 out of 7 microinstructions are unique. Therefore, the size of the microcontrol store is 7 × 2 bits and the size of the nanomemory is 3 × 4 bits. This is shown in Figure 7.50.

Address	Content
000	00
001	01
010	00
011	00
100	01
101	10
110	10

7 × 2-bit microcontrol store

Address	Content
00	0100
01	0000
10	1010

3 × 4 nanocontrol store

FIGURE 7.50 Two-level store (nanomemory)

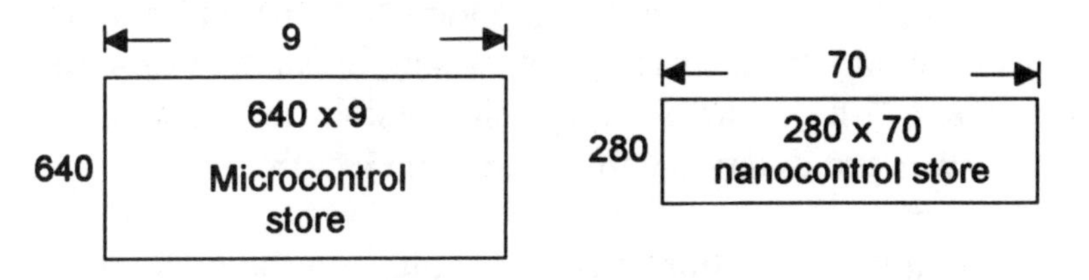

FIGURE 7.51 68000 nanomemory

Memory requirements for the single control memory = 7 × 4 = 28 bits. Memory requirements for nanomemory = (7 × 2 + 3 × 4) bits = 26 bits. Therefore, saving using nanomemory = 28 - 26 = 2 bits. For a simple example like this, 2 bits are saved. The 68000 control unit nanomemory includes a 640 × 9-bit microcontrol store and a 280 × 70-bit nanocontrol store as shown in Figure 7.51. In Figure 7.51, out of 640 microinstructions, 280 are unique. If the 68000 were implemented using a single control memory, the requirements would have been 640 × 70 bits. Therefore,

$$\begin{aligned}\text{Memory savings} &= (640 \times 70) - (640 \times 9 + 280 \times 70) \text{ bits} \\ &= 44{,}800 - 25{,}360 \\ &= 19{,}440 \text{ bits}\end{aligned}$$

This is a tremendous memory savings for the 68000 control unit.

QUESTIONS AND PROBLEMS

7.1 It is desired to implement the following instructions using block code: ADD, SUB, XOR, MOVE, HALT. Draw a block diagram.

7.2 The instruction length and the size of an address field are 9 bits and 3 bits respectively. Is it possible to have
6 two-address instructions
15 one-address instructions
8 zero-address instructions
using expanding op-code technique? Justify your answer.

7.3 Using the instruction format of Problem 7.2, is it possible to have
7 two-address instructions
7 one-address instructions
8 zero-address instructions
using expanding opcode technique?
Justify your answer.

7.4 Assume that it is desired to have 2 two-address, 7 one-address, and 25 zero-address instructions in a computer instruction set. Using expanding op-code technique with a 2-bit op-code and 3-bit address field, is it possible to accomplish the above? If so, justify your answer and determine the instruction length.

7.5 Assume that using an instruction length of 9 bits and the address field size of 3 bits, 5 two-address and 10 one-address instructions have already been designed, using expanding op-code technique. Is it possible to have at least 48 zero-address instructions that can be added to the instruction set?

7.6 Design a combinational logic shifter with 4-bit input and 4-bit output as follows:

	Shift Count		4 - bit output
$\overline{OE}$	S_1	S_0	
1	X	X	High Impedance output lines
0	0	0	No Shift
0	0	1	Right Shift once
0	1	0	Right Shift twice
0	1	1	Right Shift three times

where X means don't care. Using multiplexers and tristate buffers, draw a logic diagram.

7.7 Draw a logic diagram for a 4 × 4 barrel shifter.

7.8 Using a minimum number of full adders and multiplexers, design an incrementer/decrementer circuit as follows: If $S = 0$, output $y = x + 1$; otherwise, $y = x - 1$. Assume x and y are 4-bit signed numbers and the result is 4 bits wide.

7.9 Design a combinational circuit to compute the absolute value of an 8-bit twos complement number. Use 8-bit parallel adder and exclusive-OR gates. Draw a logic circuit.

7.10 Using a 4-bit CLA as the building block, design an 8-bit adder.

7.11 Design an arithmetic logic unit to perform the following functions:

S_1	S_0	F
0	0	A plus B
0	1	A minus B
1	0	A AND B
1	1	A OR B

Use multiplexers, parallel adders, and gates as needed. Assume that A and B are 4-bit numbers. Draw a logic circuit.

7.12 Design a combinational circuit that will perform the following operations:

S_1	S_0	Y
0	0	0
0	1	A
1	0	B
1	1	15_{10}

Assume that A is a 4-bit number and $B = \overline{a_3}\,\overline{a_2}\,\overline{a_1}\,\overline{a_0}$. Draw a logic diagram.

7.13 Design a 4-bit ALU to perform the following operations:

S	F
0	Logical Left Shift A once
1	0

Assume that A is a 4-bit number. Draw a logic diagram using a parallel adder, multiplexers, and inverters as necessary.

7.14 Design a 4-bit arithmetic unit as follows:

S	F
0	A plus B
1	A plus 1

Assume that A and B are 4-bit numbers.

7.15 Design an ALU to perform the following operations:

S_1	S_0	F
0	0	x plus y
0	1	x
1	0	B
1	1	$x \oplus y$

Assume that x and y are 4-bit numbers, and $B = \overline{y_3}\,\overline{y_2}\,\overline{y_1}\,\overline{y_0}$. Draw a logic diagram.

7.16 Assume two 2's complement signed numbers, $M = 11111111_2$ and $Q = 11111100_2$. Perform the signed multiplication using the algorithm described in Section 7.2.2.

7.17 What is the purpose of the control unit in a microprocessor?

7.18 Draw a logic diagram to implement the following register transfers:

(a) If the content of the 8-bit register R is odd, then

$$x \leftarrow x \oplus y$$

$$\text{else } x \leftarrow x \text{ AND } y$$

Assume x and y are 4 bits wide.

(b) If the number in the 8-bit register R is negative, then $x \leftarrow x - 1$ else $x \leftarrow x + 1$. Assume x and y are 4 bits wide.

7.19 Discuss briefly the merits and demerits of single-bus, two-bus, and three-bus architectures inside a control unit.

7.20 What is the basic difference between hardwired control, microprogramming, and nanoprogramming? Name the technique used for designing the control units of the Intel 8086, Motorola 68000, and PowerPC.

7.21 Using the following components: (a) 4-bit general-purpose register, (b) 4-bit adder/subtractor, and (c) tristate buffer, and assuming the inbus and outbus are 4 bits wide, design a control unit using hardwired control to perform the following operations. You may use counters, decoders, and PLAs as required.

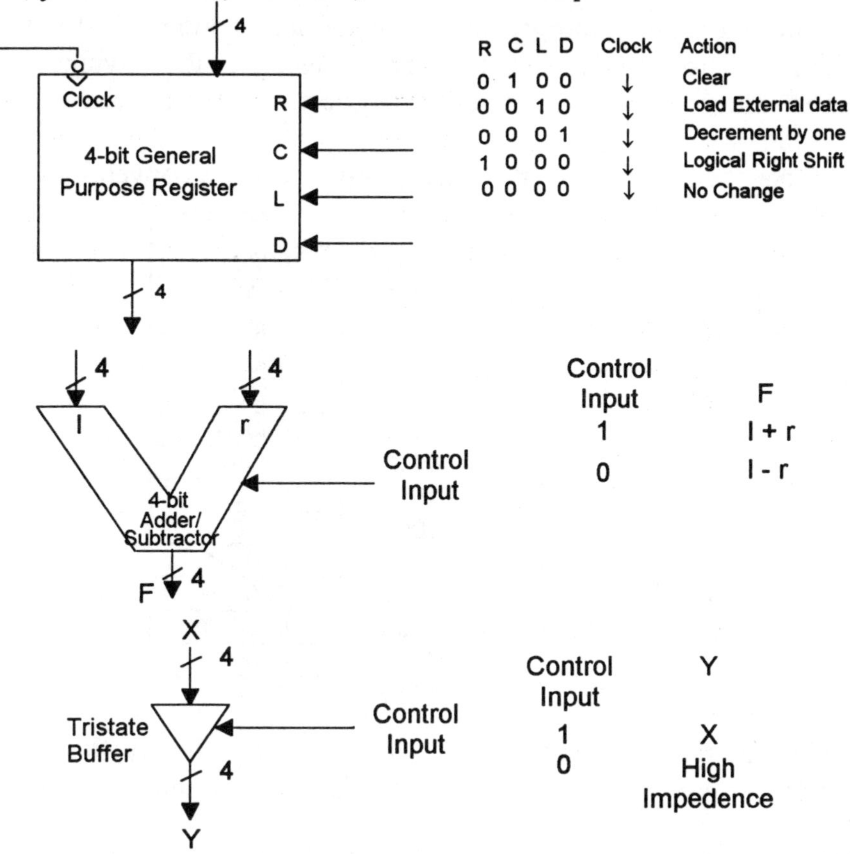

(a) Outbus ← 4 × A. Assume A is a 4-bit unsigned number and the result is 4 bits wide.

(b) If the 4-bit number in register B is odd, outbus ← 0; otherwise outbus ← A + (B / 2). Assume A and B are unsigned 4 bit numbers. Also, assume data is already loaded into B.

(c) If the content of a 4-bit register Q = 0, perform R ← M and then transfer the 4-bit result to outbus. On the other hand, if the content of the 4-bit register Q ≠ 0, perform R ← 0 and then transfer the 4-bit result to the outbus. Assume M and R are 4 bits wide.

7.22 Repeat Problem 7.21 using microprogramming.

7.23 Discuss the basic differences between microprogramming and nanoprogramming.

7.24 A conventional microprogrammed control unit includes 1024 words by 85 bits. Each of 512 microinstructions are unique. Calculate the savings if any by having a nanomemory. Calculate the sizes of microcontrol memory and nanomemory.

7.25 Consider the following 14 × 6 microprogram using a conventional control memory:

Address	Content
0000	000001
0001	000010
0010	000001
0011	000011
0100	000001
0101	000010
0110	000001
0111	000011
1000	000010
1001	000001
1010	000011
1011	000010
1100	000011
1101	000010
1110	000001

Implement this microprogram in a nanomemory. Justify the use of either a single-control memory or a two-level memory for the program.

7.26 Discuss the basic differences between CISC and RISC.

8

MEMORY ORGANIZATION AND INPUT/OUTPUT (I/O) UNIT

This chapter describes the basics of memory organization and input/output techniques. Topics include memory array design, memory management concepts, cache memory organization, and types of input/output methods utilized by typical microprocessors.

8.1 Memory Organization

8.1.1 Introduction

A memory unit is an integral part of any microcomputer system, and its primary purpose is to hold programs and data. The major design goal of a memory unit is to allow it to operate at a speed close to that of the processor. However, the cost of a memory unit is so prohibitive that it is practically not feasible to design a large memory unit with one technology that guarantees a high speed. Therefore, in order to seek a trade-off between the cost and operating speed, a memory system is usually designed with different technologies such as solid state, magnetic, and optical.

In a broad sense, a microcomputer memory system can be logically divided into three groups:

- Processor memory
- Primary or main memory

- Secondary memory

Processor memory refers to a set of microprocessor registers. These registers are used to hold temporary results when a computation is in progress. Also, there is no speed disparity between these registers and the microprocessor because they are fabricated using the same technology. However, the cost involved in this approach limits a microcomputer architect to include only a few registers in the microprocessor. The design of typical registers is described in Chapters 6 and 7.

Main memory *is the storage area in which all programs are executed. The microprocessor can directly access only those items that are stored in main memory.* Therefore, all programs and data must be within the main memory prior to execution. MOS technology is normally used these days in main memory design. Usually the size of the main memory is much larger than processor memory and its operating speed is slower than the processor registers. Main memory normally includes ROMs and RAMs. These are described in Chapter 6.

Electromechanical memory devices such as disks are extensively used as microcomputer's secondary memory and allow storage of large programs and data at a low cost. These secondary memory devices access stored data serially. Hence, they are significantly slower than the main memory. Popular secondary memories include hard disk and floppy disk systems. Data are stored on the disks in files. Note that the floppy disk is removable whereas the hard disk is not. Secondary memory stores programs and data in excess of the main memory. Secondary memory is also referred to as "auxiliary" or "virtual" memory. The microcomputer cannot directly execute programs stored in the secondary memory, so in order to execute these programs, the microcomputer must transfer them to its main memory by a program called the "operating system."

Data in disk memories are stored in tracks. A track is a concentric ring of data stored on the surface of a disk. Each track is further subdivided into several sectors. Each sector typically stores 512 or 1024 bytes of data. All disk memories use magnetic media except the optical memory, which stores data on a plastic disk. CD-ROM is an example of a popular optical memory used with microcomputer systems. The CD-ROM is used to store large programs such as a C++ compiler. Other state-of-the-art optical memories include CD-RAM, DVD-ROM and DVD-RAM. These optical memories were discussed in Chapter 1.

One of the most commonly used disk memory with microcomputer systems is the floppy disk. The floppy disk is a flat, round piece of plastic coated with magnetically sensitive oxide material. The floppy disk is provided with a protective jacket to prevent fingerprint or foreign matter from contaminating the disk's surface. The 3½-inch floppy disk is very popular these days because of its smaller size and because it does not bend easily. All floppy disks are provided with an off-center index hole that allows the electronic system reading the disk to find the start of a track and the first sector.

The storage capacity of a typical hard disk ranges from 5 MB to several gigabytes (GB). The 3½-inch floppy disk, on the other hand, can typically store 1.44 MB. The rotational speeds of the hard disk and the 3½-inch floppy disk are 3600 rpm and 360 rpm respectively. Zip disk is the most recent enhancement in removable disk technology providing storage capacity of 100 MB to 200 MB in a single disk with access speed similar to the hard disk. Zip disk does not use a laser. Rather, it uses a magnetic-coated Myler inside, along with smaller read/write heads, and a rotational speed of 3000 rpm. The smaller heads provide the Zip drive to write data using 2,118 tracks per inch, compared to 135 tracks per inch on a floppy disk.

8.1.2 Main Memory Array Design

From the previous discussions, we notice that the main memory of a microcomputer is realized using solid-state technology. In a typical microcomputer application, a designer has to realize the required capacity by interconnecting several small memory chips. This concept is known as the "memory array design." In this section, we address this topic. We also show how to interface a memory system with a typical microprocessor.

Now let us discuss how to design ROM/RAM arrays. In particular, our discussion is focused on the design of memory arrays for a hypothetical microcomputer. The pertinent signals of a typical microprocessor necessary for main memory interfacing are shown in Figure 8.1. In Figure 8.1, there are 16 address lines, A_{15} through A_0, with A_0 being the least significant bit. This means that this microprocessor can directly address a maximum of $2^{16} = 65,536$ or 64K byte memory locations. The control line $IO/\overline{M}$ goes to LOW if the microprocessor executes a memory instruction, and it is held HIGH if the processor executes an I/O instruction. Similarly, the control line $R/\overline{W}$ goes to HIGH to indicate that the operation is READ and it goes to LOW for WRITE operation. Note that all 16 address lines and the two control lines described so far are unidirectional in nature; that is, in these lines information always travels from the processor to external units. Also, in Figure 8.1 eight bidirectional data lines D_7 through D_0 (with D_0 being the least significant bit) are shown. These lines are used to allow data transfer from the processor to external units and vice versa.

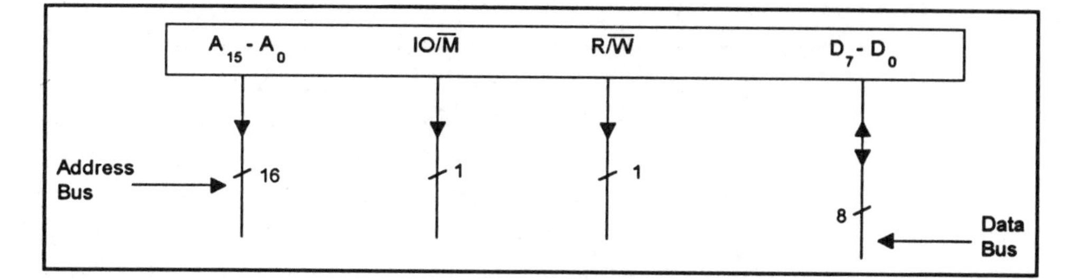

FIGURE 8.1 Pertinent signals of a typical microprocessor required for main memory interfacing

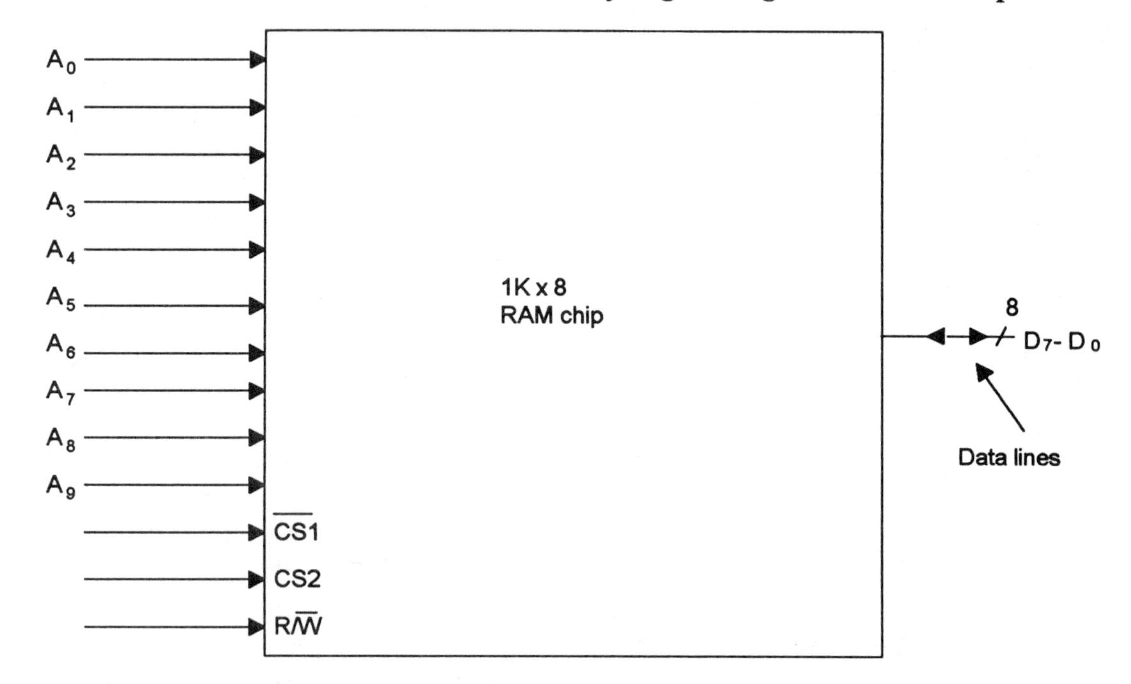

FIGURE 8.2 A typical 1K × 8 RAM chip

In a typical application, the total amount of main memory connected to a microprocessor consists of a combination of both ROMs and RAMs. However, in the following we will illustrate for simplicity how to design memory array using only the RAM chips.

The pin diagram of a typical 1K × 8 RAM chip is shown in Figure 8.2. In this RAM chip there are 10 address lines, A_9 through A_0, so we can access 1024 ($2^{10} = 1024$) different memory words. Also, in this chip there are 8 bidirectional data lines D_7 through D_0 so that data can travel back and forth between the microprocessor and the memory unit. The three control lines $\overline{CS1}$, CS2, and $R/\overline{W}$ are used to control the RAM unit according to the truth table shown in Figure 8.3. From this truth table it can be concluded that the RAM unit is enabled only when $\overline{CS1} = 0$ and CS2 = 1. Under this condition, $R/\overline{W} = 0$ and $R/\overline{W} = 1$ imply write and read operations respectively.

$\overline{CS1}$	CS2	$R/\overline{W}$	Function
0	1	0	Write Operation
0	1	1	Read Operation
1	X	X	The chip is not selected
X	0	X	The chip is not selectd

X means Don't Care

FIGURE 8.3 Truth table for controlling RAM

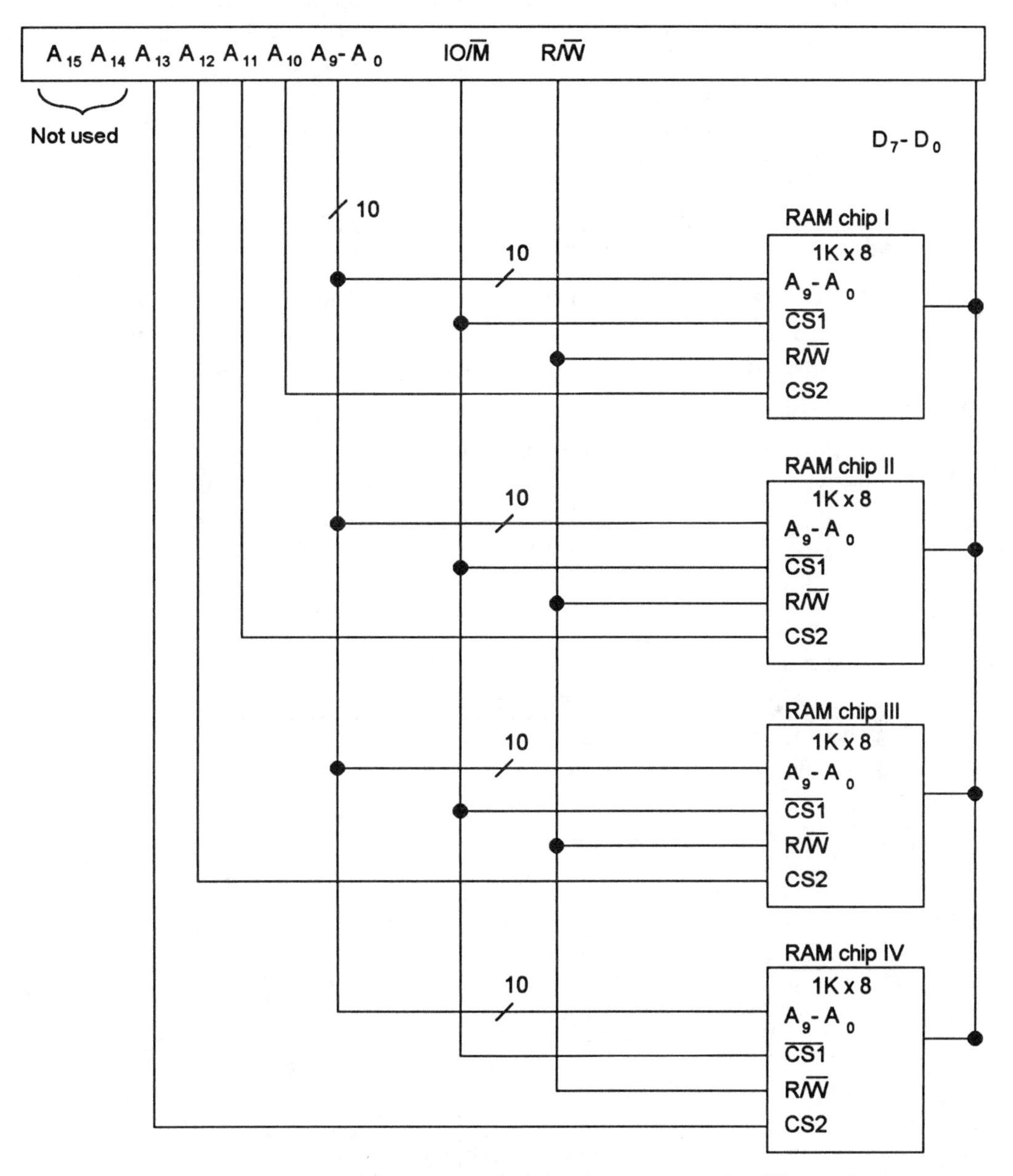

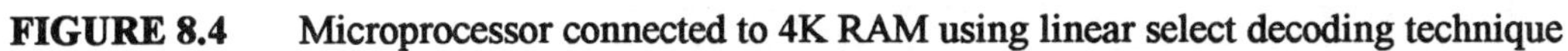
FIGURE 8.4 Microprocessor connected to 4K RAM using linear select decoding technique

To connect a microprocessor to ROM/RAM chips, three address-decoding techniques are usually used: linear decoding, full decoding, and memory decoding using PAL. Let us first discuss how to interconnect a microprocessor with a 4K RAM chip array composed of the four 1K RAM chips of Figure 8.2 using the linear decoding technique.

Address Range in Hexadecimal	RAM Chip Number
0400-07FF	I
0800-0BFF	II
1000-13FF	III
2000-23FF	IV

FIGURE 8.5 Address map of the memory organization of Figure 8.4

Figure 8.4 uses the linear decoding to accomplish this. In this approach, the address lines A_9 through A_0 of the microprocessor are connected to all RAM units. Similarly, the control lines $IO/\overline{M}$ and $R/\overline{W}$ of the microprocessor are connected to the control lines $\overline{CS1}$ and $R/\overline{W}$ respectively of each RAM unit. The high-order address bits A_{10} through A_{13} directly act as chip selects. In particular, the address lines A_{10} and A_{11} select the RAM chips I and II respectively. Similarly, the address lines A_{12} and A_{13} select the RAM chips III and IV respectively. A_{15} and A_{14} are don't cares and are assumed to be 0. Figure 8.5 describes how the addresses are distributed among the four 1K RAM chips. This method is known as "linear select decoding," and its primary advantage is that it does not require any decoding hardware. However, if both A_{10} and A_{11} are high at the same time, both RAM chips I and II are selected, and this causes a bus conflict. Because of this potential problem, the software must be written in such a way that it never reads any address in which more than one of the bits A_{13} through A_{10} are high. Another disadvantage of this method is that it wastes a large amount of address space. For example, whenever the address value is 8400 or 0400, the RAM chip I is selected. In other words, the address 0400 is the mirror reflection of the address 8400 (this situation is also called "memory foldback"). This technique is, therefore, limited to a small system. In particular, we can extend the system of Figure 8.4 up to a total capacity of 6K using A_{14} and A_{15} as chip selects for two more 1K RAM chips.

To resolve the problems with linear decoding, we use the full decoded memory addressing. In this technique, we use a decoder. The same 4K memory system designed using this technique is shown in Figure 8.6. Note that the decoder in the figure is very similar to a practical decoder such as the 74LS138 with three chip enables. In Figure 8.6 the decoder output selects one of the four 1K RAM chips depending on the values of A_{12}, A_{11}, and A_{10}:

A_{12}	A_{11}	A_{10}	Selected RAM Chip
0	0	0	RAM chip I
0	0	1	RAM chip II
0	1	0	RAM chip III
0	1	1	RAM chip IV

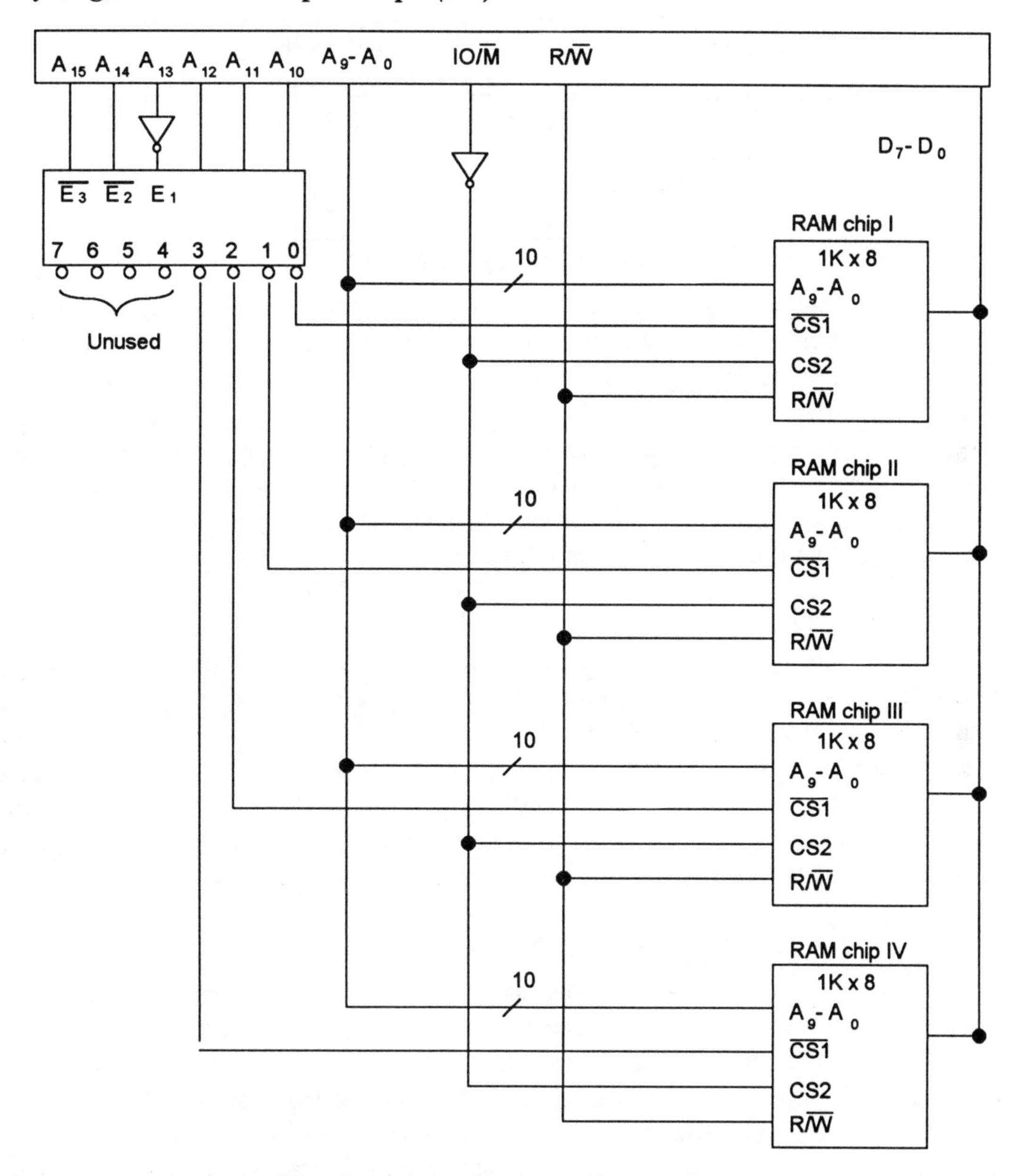

FIGURE 8.6 Interconnecting a microprocessor with a 4K RAM using full decoded memory addressing

Note that the decoder output will be enabled only when $\overline{E3} = \overline{E2} = 0$ and $E1 = 1$. Therefore, in the organization of Figure 8.6, when any one of the high-order bits A_{15}, A_{14}, or A_{13} is 1, the decoder will be disabled, and thus none of the RAM chips will be selected. In this arrangement, the memory addresses are assigned as shown in Figure 8.7.

Address Range in Hexadecimal	RAM Chip Number
0000-03FF	I
0400-07FF	II
0800-0BFF	III
0C00-0FFF	IV

FIGURE 8.7 Address map of the memory organization of Figure 8.6

From the address map of Figure 8.7, it is easy to see that this approach does not waste the address space. In other words, this method does not result in memory foldback. Finally, although the 3-to-8 decoder of Figure 8.6 can select eight 1K RAM chips, its full capability is not used in this example.

As mentioned before, a programmable array logic (PAL) is similar to a ROM in concept except that it does not provide full decoding of the input lines. Instead, a PAL provides a partial sum of products that can be obtained via programming and saves a lot of space on the board. The PAL chip contains a fused programmable AND array and a fixed OR array. Note that in a PLA (programmable logic array) both AND and OR arrays are programmable. The AND and OR gates are fabricated inside the PAL without interconnections. The specific functions desired are implemented during programming via software. Programming of the PAL provides connections of the inputs of the AND gates and the outputs of the AND gates to the inputs of the OR gates. Therefore, the PAL implements the sum of the products of the inputs. PALs are used extensively these days with 32- and 64-bit microprocessors such as the Intel 80386/80486/Pentium and Motorola 68030/68040/PowerPC for performing the memory decode function. PALs connect these microprocessors to memory, I/O devices, and other chips without the use of any additional logic gates or circuits.

8.1.3 Memory Management Concepts

Due to the massive amount of information that must be saved in most systems, the mass storage device is often a disk. If each access is to a disk (even a hard disk), then system throughput will be reduced to unacceptable levels.

An obvious solution is to use a large and fast locally accessed semiconductor memory. Unfortunately the storage cost per bit for this solution is very high. A combination of both off-board disk (secondary memory) and on-board semiconductor main memory must be designed into a system. This requires a mechanism to manage the two-way flow of information between the primary (semiconductor) and secondary (disk) media. This mechanism must be able to transfer blocks of data efficiently, keep track of block usage, and replace them in a nonarbitrary way. The main memory system must, therefore, be able to dynamically allocate memory space.

An operating system must have resource protection from corruption or abuse by users. Users must be able to protect areas of code from each other while maintaining the ability to communicate and share other areas of code. All these requirements indicate the need for a device, located between the microprocessor and memory, to control accesses, perform address mappings, and act as an interface between the logical (Programmer's memory) and and the physical (Microprocessor's directly addressable memory) address spaces. Because this device must manage the memory use configuration, it is appropriately called the "memory management unit (MMU)." Typical 32-bit processors such as the Motorola 68030/68040 and the Intel 80486/Pentium include on-chip MMUs. The MMU reduces the burden of the memory management function on the operating system.

The basic functions provided by the MMU are address translation and protection. The MMU translates logical program addresses to physical memory address. Note that in assembly language programming, addresses are referred to by symbolic names. These addresses in a program are called logical addresses because they indicate the logical positions of instructions and data. The MMU translates these logical addresses to physical addresses provided by the memory chips. The MMU can perform address translation in one of two ways:

1. By using the substitution technique as shown in Figure 8.8(a)
2. By adding an offset to each logical address to obtain the corresponding physical address as shown in Figure 8.8(b)

Address translation using the substitution technique is faster than the offset method. However, the offset method has the advantage of mapping a logical address to any physical address as determined by the offset value.

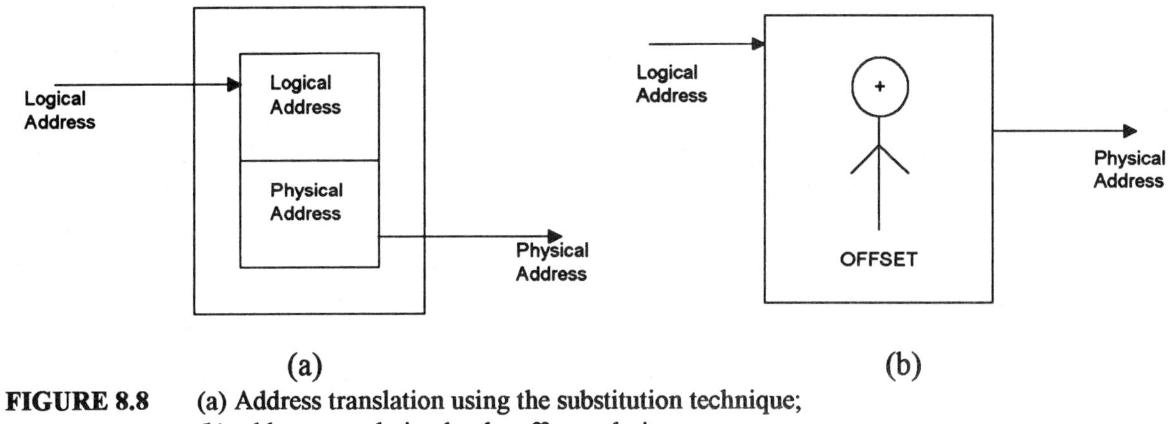

FIGURE 8.8 (a) Address translation using the substitution technique; (b) address translation by the offset technique

Memory is usually divided into small manageable units. The terms "page" and "segment" are frequently used to describe these units. *Paging divides the memory into equal-sized pages; segmentation divides the memory into variable-sized segments.* It is relatively easier to implement the address translation table if the logical and main memory spaces are divided into pages. The term "page" is associated with logical address space, whereas the term "block" usually refers to a page in main memory space.

There are three ways to map logical addresses to physical addresses: paging, segmentation, and combined paging/segmentation. In a paged system, a user has access to a larger address space than physical memory provides. The virtual memory system is managed by both hardware and software. The hardware included in the memory management unit handles address translation. The memory management software in the operating system performs all functions including page replacement policies to provide efficient memory utilization. The memory management software performs functions such as removal of the desired page from main memory to accommodate a new page, transferring a new page from secondary to main memory at the right instant of time, and placing the page at the right location in memory.

If the main memory is full during transfer from secondary to main memory, it is necessary to remove a page from main memory to accommodate the new page. Two popular page replacement policies are first-in–first-out (FIFO) and least recently used (LRU). The FIFO policy removes the page from main memory that has been resident in memory for the longest amount of time. The FIFO replacement policy is easy to implement, but one of its main disadvantages is that it is likely to replace heavily used pages. Note that heavily used pages are resident in main memory for the longest amount of time. Sometimes this replacement policy might be a poor choice. For example, in a time-shared system, several users normally share a copy of the text editor in order to type and correct programs. The FIFO policy on such a system might replace a heavily used editor page to make room for a new page. This editor page might be recalled to main memory immediately. The FIFO, in this case, would be a poor choice. The LRU policy, on the other hand, replaces the page that has not been used for the longest amount of time.

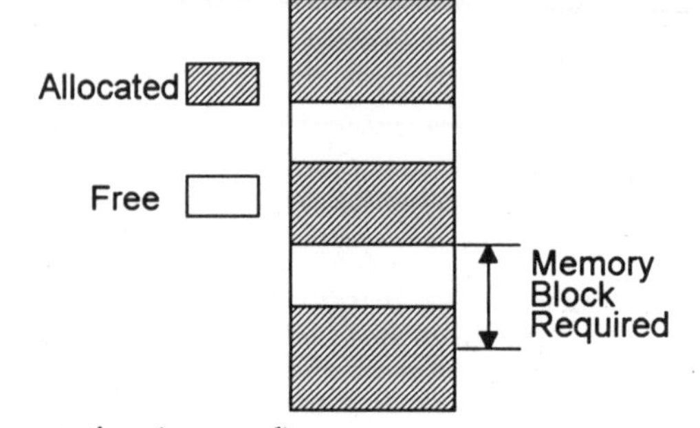

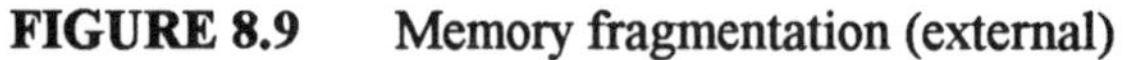
FIGURE 8.9 Memory fragmentation (external)

In the segmentation method, the MMU utilizes the segment selector to obtain a descriptor from a table in memory containing several descriptors. A descriptor contains the physical base address for a segment, the segment's privilege level, and some control bits. When the MMU obtains a logical address from the microprocessor, it first determines whether the segment is already in the physical memory. If it is, the MMU adds an offset component to the segment base component of the address obtained from the segment descriptor table to provide the physical address. The MMU then generates the physical address on the address bus for selecting the memory. On the other hand, if the MMU does not find the logical address in physical memory, it interrupts the microprocessor. The microprocessor executes a service routine to bring the desired program from a secondary memory such as disk to the physical memory. The MMU determines the physical address using the segment offset and descriptor as described earlier and then generates the physical address on the address bus for memory. A segment will usually consist of an integral number of pages, each, say, 256 bytes long. With different-sized segments being swapped in and out, areas of valuable primary memory can become unusable. Memory is unusable for segmentation when it is sandwiched between already allocated segments and if it is not large enough to hold the latest segment that needs to be loaded. This is called "external fragmentation" and is handled by MMUs using special techniques. An example of external fragmentation is given in Figure 8.9. The advantages of segmented memory management are that few descriptors are required for large programs or data spaces and that internal fragmentation (to be discussed later) is minimized. The disadvantages include external fragmentation, the need for involved algorithms for placing data, possible restrictions on the starting address, and the need for longer data swap times to support virtual memory.

Address translation using descriptor tables offers a protection feature. A segment or a page can be protected from access by a program section of a lower privilege level. For example, the selector component of each logical address includes one or two bits indicating the privilege level of the program requesting access to a segment. Each segment descriptor also includes one or two bits providing the privilege level of that segment. When an executing program tries to access a segment, the MMU can compare the selector privilege level with the descriptor privilege level. If the segment selector has the same or higher privilege level, then the MMU permits the access. If the privilege level of the selector is lower than that of the descriptor, the MMU can interrupt the microprocessor, informing it of a privilege-level violation. Therefore, the indirect technique of generating a physical address provides a mechanism of protecting critical program sections in the operating system. Because paging divides the memory into equal-sized pages, it avoids the major problem of segmentation—external fragmentation. Because the pages are of the same size, when a new page is requested and an old one swapped out, the new one will always fit into the vacated space. However, a problem common to both techniques remains—internal fragmentation.

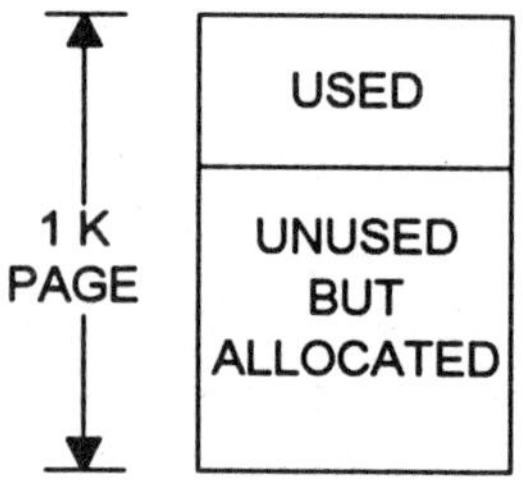

FIGURE 8.10 Memory fragmentation (internal)

Internal fragmentation is a condition where memory is unused but allocated due to memory block size implementation restrictions. This occurs when a module needs, say, 300 bytes and page is 1K bytes, as shown in Figure 8.10

In the paged-segmentation method, each segment contains a number of pages. The logical address is divided into three components: segment, page, and word. The segment component defines a segment number, the page component defines the page within the segment, and the word component provides the particular word within the page. A page component of n bits can provide up to 2^n pages. A segment can be assigned with one or more pages up to maximum of 2^n pages; therefore, a segment size depends on the number of pages assigned to it.

A protection mechanism can be assigned to either a physical address or a logical address. Physical memory protection can be accomplished by using one or more protection bits with each block to define the access type permitted on the block. This means that each time a page is transferred from one block to another, the block protection bits must be updated. A more efficient approach is to provide a protection feature in logical address space by including protection bits in descriptors of the segment table in the MMU. *The concepts associated with virtual memory and memory management are covered in Chapter 11.*

8.1.4 Cache Memory Organization

The performance of a microcomputer system can be significantly improved by introducing a small, expensive, but fast memory between the microprocessor and main memory. This memory is called "cache memory" and this idea was first introduced in the IBM 360/85 computer. Later on, this concept was also implemented in minicomputers such as the PDP-11/70. With the advent of VLSI technology, the cache memory technique is gaining acceptance in the microprocessor world. Studies have shown that typical programs spend most of their execution times in loops. This means that the addresses generated by a microprocessor have a tendency to cluster around a small region in the main memory, a phenomenon known as "locality of reference." Typical 32-bit microprocessors can execute the same instructions in a loop from the on-chip cache rather than reading them repeatedly from the external main memory. Thus, the performance is greatly improved. For example, an on-chip cache memory is implemented in Intel's 32-bit microprocessor, the 80486/Pentium, and Motorola's 32-bit microprocessor, the MC 68030/68040. The 80386 does not have an on-chip cache, but external cache memory can be interfaced to it.

The block diagram representation of a microprocessor system that employs a cache memory is shown in Figure 8.11. Usually, a cache memory is very small in size and its access time is less than that of the main memory by a factor of 5. Typically, the access times of the cache and main memories are 100 and 500 ns, respectively. If a reference is found in the cache, we call it a "cache hit," and the data pertaining to the microprocessor reference is transferred to the microprocessor from the cache. However, if the reference is not found in the cache, we call it a "cache miss." When there is a cache miss, the main memory is accessed by the microprocessor and the data are then transferred to the microprocessor from the main memory. At the same time, a block of data containing the desired data needed by the micro-processor is transferred from the main memory to cache. The block normally contains 4 to 16 words, and this block is placed in the cache using the standard replacement policies such as FIFO or LRU. This block transfer is done with a hope that all future references made by the microprocessor will be confined to the fast cache.

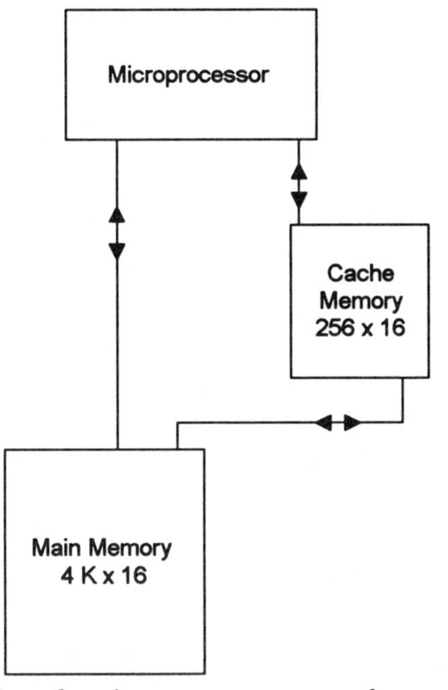

FIGURE 8.11 Memory organization of a microprocessor system that employs a cache memory

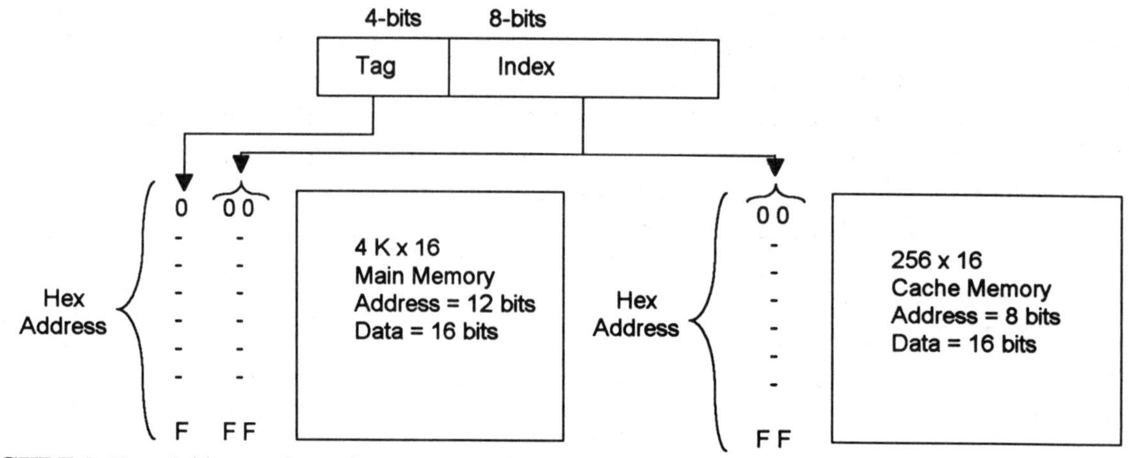

FIGURE 8.12 Addresses for main memory and cache memory

The relationship between the cache and main memory blocks is established using mapping techniques. Three widely used mapping techniques are

1. Direct mapping
2. Fully associative mapping
3. Set-associative mapping

In order to explain these three mapping techniques, the memory organization of Figure 8.12 will be used. The main memory is capable of storing 4K words of 16 bits each. The cache memory on the other hand, can store 256 words of 16 bits each. An identical copy of every word stored in cache exists in main memory. The microprocessor first sends a 12-bit (2^{12} = 4K) address to cache. If there is a hit, the microprocessor accepts the 16-bit word from the cache. In case of a miss, the microprocessor reads the 16-bit data from the main memory and this 16-bit word is then written to the cache.

Direct mapping uses a RAM for the cache. The microprocessor's 12-bit address is divided into two fields, an index field and a tag field. Because the cache address is 8 bits wide ($2^8 = 256$), the low-order 8 bits of the microprocessor's address form the index field, and the remaining 4 bits constitute the tag field. This is illustrated in Figure 8.12.

In general, if the main memory address field is m bits wide and the cache memory address is n bits wide, the index field will then require n bits and the tag field will be $(m - n)$ bits wide. The n-bit address will access the cache. Each word in the cache will include the data word and its associated tag. When the microprocessor generates an address for main memory, the index field is used as the address to access the cache. The tag field of the main memory is compared with the tag field in the word read from cache. A hit occurs if the tags match. This means that the desired data word is in cache. A miss occurs if there is no match, and the required word is read from main memory. It is written in the cache along with the tag. One of the main drawbacks of direct mapping is that numerous misses may occur if two or more words with addresses having the same index but with different tags are accessed several times. This situation should be avoided or can be minimized by having such words far apart in the address lines. Let us now illustrate the concept of direct mapping by means of a numerical example of Figure 8.13. All numbers are in hexadecimal.

The content of index address 00 of cache is tag = 0 and data = 013F. Suppose that the microprocessor wants to access the memory address 100. The index address 00 is used to access the cache. The memory address tag 1 is compared with the cache tag of 0. This does not produce a match. Therefore, the main memory is accessed and the data 2714 is transferred into the microprocessor. The cache word at index address 00 is then replaced with a tag of 1 and data of 2714.

Memory Address	
000	013F
001	1234
002	A370
	-
	-
100	2714
101	23B4
	-
	-
200	7A3F
201	2721
	-
2FF	1523

Main Memory

Index	Tag	Data
00	0	013F
01	0	1234
02	0	A370
	-	-
	-	-
FF	2	1523

Cache Memory

FIGURE 8.13 Direct mapping numerical example

12

12 Bits | Adddress register

Address	Data
010	2F17
247	3245
445	0FA1
731	2A00
-	-

FIGURE 8.14 Associative mapping, numerical example

The fastest cache memory utilizes an associative memory. This method is known as "fully associative mapping." Each element in associative memory contains a main memory address and its content (data). When the microprocessor generates a main memory address, it is compared associatively (simultaneously) with all addresses in the associative memory. If there is a match, the corresponding data word is read from the associative cache memory and sent to the microprocessor. If a miss occurs, the main memory is accessed and the address along with its corresponding data are written to the associative cache memory. If the cache is full, certain policies such as FIFO are used as replacement algorithms for the cache. The associative cache is expensive but provides fast operation. The concept of an associative cache is illustrated by means of a numerical example in Figure 8.14. Assume all numbers are in hexadecimal.

The associative memory stores both the memory address and its contents (data). The figure shows four words stored in the associative cache. Each word in the cache is the 12-bit address along with its 16-bit contents (data). When the microprocessor wants to access memory, the 12-bit address is placed in an address register and the associative cache memory is searched for a matching address. Suppose that the content of the microprocessor address register is 445. Because there is a match, the microprocessor reads the corresponding data 0FA1 into an internal data register.

Set-associative mapping is a combination of direct and associative mapping. Each cache word stores two or more main memory words using the same index address. Each main memory word consists of a tag and its data word. An index with two or more tags and data words forms a set. When the microprocessor generates a memory request, the index of the main memory address is used as the cache address. The tag field of the main memory address is then compared associatively (simultaneously) with all tags stored under the index. If a match occurs, the desired data word is read. If a match does not occur, the data word, along with its tag, is read from main memory and also written into the cache.

Index	Tag	Data	Tag	Data
00	0	013F	2	7A3F
01	1	23B4	2	2721

FIGURE 8.15 Set-associative mapping, numerical example with set size of 2

The hit ratio improves as the set size increases because more words with the same index but different tags can be stored in the cache. The concept of set-associative mapping can be illustrated by a numerical example as shown in Figure 8.15. Assume that all numbers are in hexadecimal.

Each cache word can store two or more memory words under the same index address. Each data item is stored with its tag. The size of a set is defined by the number of tag and data items in a cache word. A set size of two is used in this example. Each index address contains two data words and their associated tags. Each tag includes 4 bits, and each data word contains 16 bits. Therefore, the word length = $2 \times (4 + 16) = 40$ bits. An index address of 8 bits can represent 256 words. Hence, the size of the cache memory is 256×40. It can store 512 main memory words because each cache word includes two data words.

The hex numbers shown in Figure 8.15 are obtained from the main memory contents shown in Figure 8.13. The words stored at addresses 000 and 200 of main memory of figure 8.13 are stored in cache memory (shown in Figure 8.15) at index address 00. Similarly, the words at addresses 101 and 201 are stored at index address 01. When the microprocessor wants to access a memory word, the index value of the address is used to access the cache. The tag field of the microprocessor address is then compared with both tags in the cache associatively (simultaneously) for a cache hit. If there is a match, appropriate data is read into the microprocessor. The hit ratio will improve as the set size increases because more words with the same index but different tags can be stored in the cache. However, this may increase the cost of comparison logic.

There are two ways of writing into cache: the write-back and write-through methods. In the write-back method, whenever the microprocessor writes something into a cache word, a "dirty" bit is assigned to the cache word. When a dirty word is to be replaced with a new word, the dirty word is first copied into the main memory before it is overwritten by the incoming new word. The advantage of this method is that it avoids unnecessary writing into main memory.

In the write-through method, whenever the microprocessor alters a cache address, the same alteration is made in the main memory copy of the altered cache address. This policy can be easily implemented and also ensures that the contents of the main memory are always valid.

This feature is desirable in a multiprocesssor system, in which the main memory is shared by several processors. However, this approach may lead to several unnecessary writes to main memory.

One of the important aspects of cache memory organization is to devise a method that ensures proper utilization of the cache. Usually, the tag directory contains an extra bit for each entry, called a "valid" bit. When the power is turned on, the valid bit corresponding to each cache block entry of the tag directory is reset to zero. This is done in order to indicate that the cache block holds invalid data. When a block of data is first transferred from the main memory to a cache block, the valid bit corresponding to this cache block is set to 1. In this arrangement, whenever the valid bit is zero, it implies that a new incoming block can overwrite the existing cache block. Thus, there is no need to copy the contents of the cache block being replaced into the main memory. *Cache memories are covered in more detail in Chapter 11.*

8.2 Input/Output

One communicates with a microcomputer system via the I/O devices interfaced to it. The user can enter programs and data using the keyboard on a terminal and execute the programs to obtain results. Therefore, the I/O devices connected to a microcomputer system provide an efficient means of communication between the microcomputer and the outside world. These I/O devices are commonly called "peripherals" and include keyboards, CRT displays, printers, and disks.

The characteristics of the I/O devices are normally different from those of the microcomputer. For example, the speed of operation of the peripherals is usually slower than that of the microcomputer, and the word length of the microcomputer may be different from the data format of the peripheral devices. To make the characteristics of the I/O devices compatible with those of the microcomputer, interface hardware circuitry between the microcomputer and I/O devices is necessary. Interfaces provide all input and output transfers between the microcomputer and peripherals by using an I/O bus. An I/O bus carries three types of signals: device address, data, and command.

The microprocessor uses the I/O bus when it executes an I/O instruction. A typical I/O instruction has three fields. When the computer executes an I/O instruction, the control unit decodes the op-code field and identifies it as an I/O instruction. The CPU then places the device address and command from respective fields of the I/O instruction on the I/O bus. The interfaces for various devices connected to the I/O bus decode this address, and an appropriate interface is selected. The identified interface decodes the command lines and determines the function to be performed. Typical functions include receiving data from an input device into the microprocessor or sending data to an output device from the microprocessor. In a typical microcomputer system, the user gets involved with two types of I/O devices: physical I/O and virtual I/O. When the computer has no operating system, the user must work directly with physical I/O devices and perform detailed I/O design.

There are three ways of transferring data between the microcomputer and physical I/O device:

1. Programmed I/O
2. Interrupt I/O
3. Direct memory access (DMA)

The microcomputer executes a program to communicate with an external device via a register called the "I/O port" for programmed I/O. An external device requests the microcomputer to transfer data by activating a signal on the computer's interrupt line during interrupt I/O. In response, the microcomputer executes a program called the interrupt-service routine to carry out the function desired by the external device. Data transfer between the microcomputer's memory and an external device occurs without microprocessor involvement with direct memory access.

In a microcomputer with an operating system, the user works with virtual I/O devices. The user does not have to be familiar with the characteristics of the physical I/O devices. Instead, the user performs data transfers between the microcomputer and the physical I/O devices indirectly by calling the I/O routines provided by the operating system using virtual I/O instructions.

Basically, an operating system serves as an interface between the user programs and actual hardware. The operating system facilitates the creation of many logical or virtual I/O devices, and allows a user program to communicate directly with these logical devices. For example, a user program may write its output to a virtual printer. In reality, a virtual printer may refer to a block of disk space. When the user program terminates, the operating system may assign one of the available physical printers to this virtual printer and monitor the entire printing operation. This concept is known as "spooling" and improves the system throughput by isolating the fast processor from direct contact with a slow printing device. A user program is totally unaware of the logical-to-physical device-mapping process. There is no need to modify a user program if a logical device is assigned to some other available physical device. This approach offers greater flexibility over the conventional hardware-oriented techniques associated with physical I/O.

8.2.1 Programmed I/O

A microcomputer communicates with an external device via one or more registers called "I/O ports" using programmed I/O. I/O ports are usually of two types. For one type, each bit in the port can be individually configured as either input or output. For the other type, all bits in the port can be set up as all parallel input or output bits. Each port can be configured as an input or output port by another register called the "command" or "data-direction register." The port contains the actual input or output data. The data-direction register is an output register and can be used to configure the bits in the port as inputs or outputs.

Each bit in the port can be set up as an input or output, normally by writing a 0 or a 1 in the corresponding bit of the data-direction register. As an example, if an 8-bit data-direction register contains 34H, then the corresponding port is defined as follows:

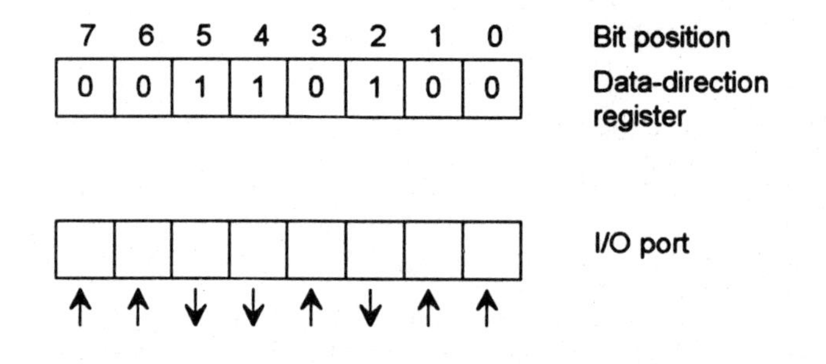

In this example, because 34H (0011 0100) is sent as an output into the data-direction register, bits 0, 1, 3, 6, and 7 of the port are set up as inputs, and bits 2, 4, and 5 of the port are defined as outputs. The microcomputer can then send output to external devices, such as LEDs, connected to bits 2, 4, and 5 through a proper interface. Similarly, the microcomputer can input the status of external devices, such as switches, through bits 0, 1, 3, 6, and 7. To input data from the input switches, the microcomputer assumed here inputs the complete byte, including the bits to which LEDs are connected. While receiving input data from an I/O port, however, the microcomputer places a value, probably 0, at the bits configured as outputs and the program must interpret them as "don't cares." At the same time, the microcomputer's outputs to bits configured as inputs are disabled.

For parallel I/O, there is only one data-direction register, usually known as the "command register" for all ports. A particular bit in the command register configures all bits in the port as either inputs or outputs. Consider two I/O ports in an I/O chip along with one command register. Assume that a 0 or a 1 in a particular bit position defines all bits of ports A or B as inputs or outputs. An example is depicted in the following:

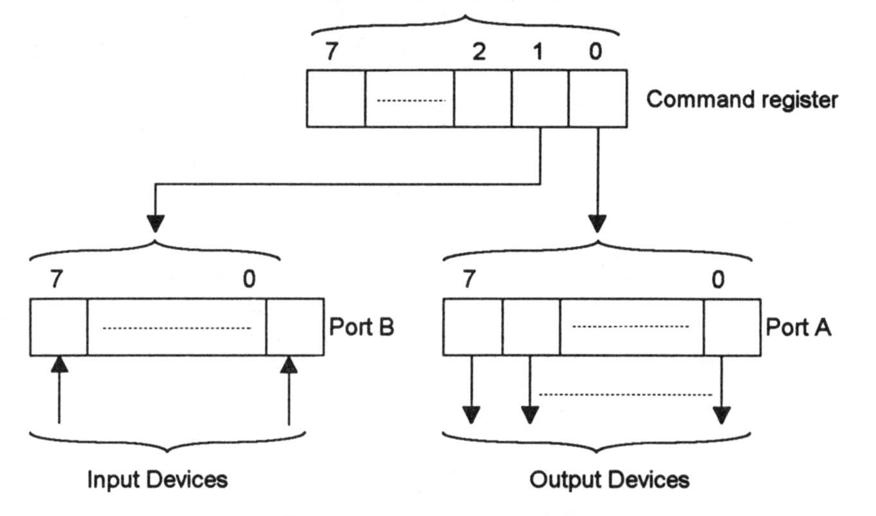

Some I/O ports are called "handshake ports." Data transfer occurs via these ports through exchanging of control signals between the I/O controller and an external device.

I/O ports are addressed using either standard I/O or memory-mapped I/O techniques. The standard I/O, also called "isolated I/O," uses an output pin such as $IO/\overline{M}$ on the microprocessor chip. The processor outputs a HIGH on this pin to indicate to memory and the I/O chips that an I/O operation is taking place. A LOW output from the processor to this pin indicates a memory operation. Execution of IN or OUT instruction makes the $IO/\overline{M}$ HIGH, whereas memory-oriented instructions, such as MOVE, drive the $IO/\overline{M}$ to LOW. In standard I/O, the processor uses the $IO/\overline{M}$ pin to distinguish between I/O and memory. For typical processors, an 8-bit address is commonly used for each I/O port. With an 8-bit I/O port address, these processors are capable of addressing 256 ports. In addition, these processors can also use 16-bit I/O ports. However, in a typical application, four or five I/O ports are usually required. Some of the address bits of the microprocessor are normally decoded to obtain the I/O port addresses. With memory-mapped I/O, the processor does not differentiate between I/O and memory, and therefore does not use the $IO/\overline{M}$ control pin. The processor uses a portion of the memory addresses to represent I/O ports. The I/O ports are mapped into the processor's main memory and, hence, are called "memory-mapped I/O." Each method has its advantages and disadvantages. For example, when standard I/O is used, typical processors normally use 2-byte IN or OUT instruction as follows:

IN port number	2-byte instruction for inputting data from the specified I/O port into the processor's register
OUT port number	2-byte instruction for outputting data from the register into the specified I/O port

With memory-mapped I/O, the processor normally uses instructions, namely, MOVE, as follows:

MOVE M, reg	where M= Port address mapped into memory	instruction for inputting a byte into a register
MOVE reg, M	where M= Port address mapped into memory	instruction for outputting data from a register into the specified port

The processor can send data to an external device at any time during unconditional I/O. The external device must always be ready for data transfer. A typical example is when the processor outputs a 7-bit code through an I/O port to drive a seven-segment display connected to this port.

In conditional I/O, the processor outputs data to an external device via handshaking. Data transfer occurs by the exchanging of control signals between the processor and an external device. The processor inputs the status of the external device to determine whether the device is ready for data transfer. Data transfer takes place when the device is ready. The flow chart in Figure 8.16 illustrates the concept of conditional programmed I/O.

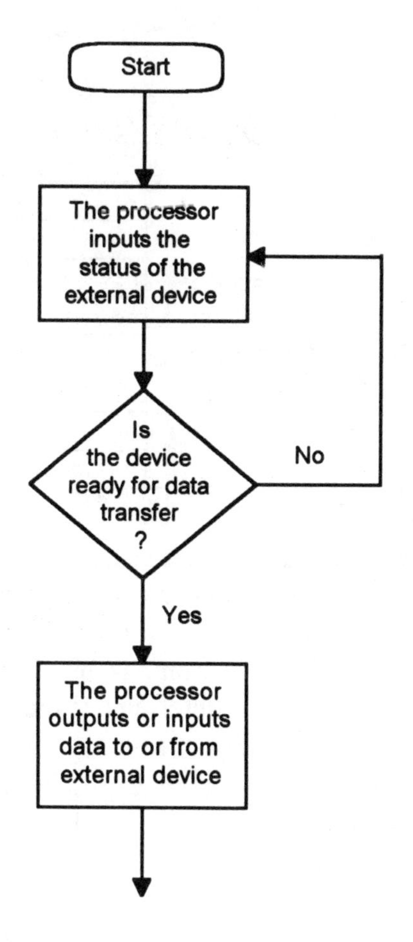

FIGURE 8.16 Flowchart for conditional programmed I/O

The concept of conditional I/O will now be demonstrated by means of data transfer between a processor and an analog-to-digital (A/D) converter. Consider, for example, the A/D converter shown in Figure 8.17. This A/D converter transforms an analog voltage V_x into an 8-bit binary output at pins D_7-D_0. A pulse at the START conversion pin initiates the conversion. This drives the BUSY signal LOW. The signal stays LOW during the conversion process. The BUSY signal goes HIGH as soon as the conversion ends. Because the A/D converter's output is tristated, a LOW on the $\overline{\text{OUTPUT ENABLE}}$ transfers the converter's outputs. A HIGH on the $\overline{\text{OUTPUT ENABLE}}$ drives the converter's outputs to a high impedance state.

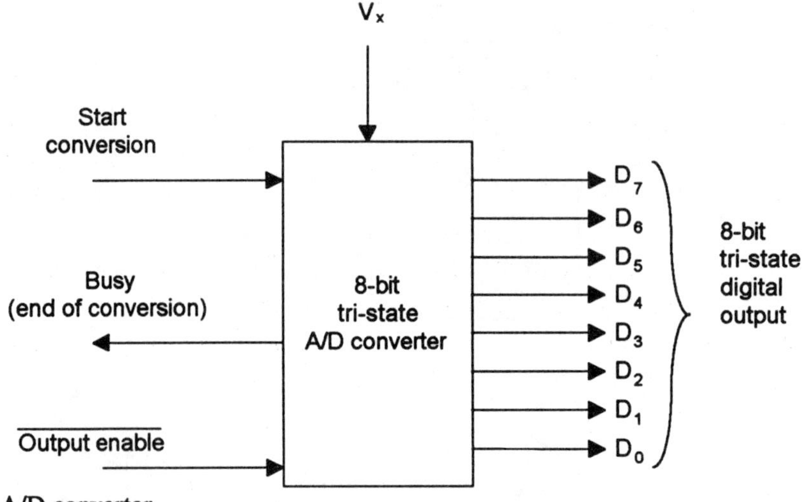

FIGURE 8.17 A/D converter

The concept of conditional I/O can be demonstrated by interfacing the A/D converter to a typical processor. Figure 8.18 shows such an interfacing example. The user writes a program to carry out the conversion process. When this program is executed, the processor sends a pulse to the START pin of the converter via bit 2 of port A. The processor then checks the BUSY signal by bit 1 of port A to determine if the conversion is completed. If the BUSY signal is HIGH (indicating the end of conversion), the processor sends a LOW to the $\overline{\text{OUTPUT ENABLE}}$ pin of the A/D converter. The processor then inputs the converter's D_0-D_7 outputs via port B. If the conversion is not completed, the processor waits in a loop checking for the BUSY signal to go HIGH.

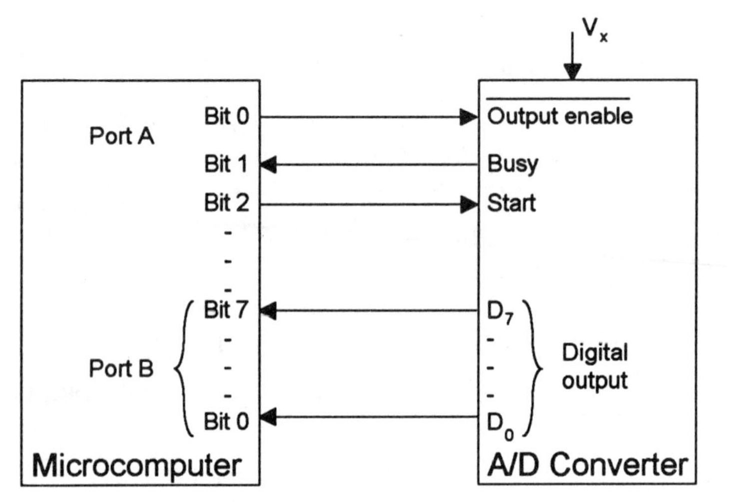

FIGURE 8.18 Interfacing an A/D converter to a microcomputer

8.2.2 Interrupt I/O

A disadvantage of conditional programmed I/O is that the microcomputer needs to check the status bit (BUSY signal for the A/D converter) by waiting in a loop. This type of I/O transfer is dependent on the speed of the external device. For a slow device, this waiting may slow down the microcomputer's capability of processing other data. The interrupt I/O technique is efficient in this type of situation.

Interrupt I/O is a device-initiated I/O transfer. The external device is connected to a pin called the "interrupt (INT) pin" on the processor chip. When the device needs an I/O transfer with the microcomputer, it activates the interrupt pin of the processor chip. The microcomputer usually completes the current instruction and saves at least the contents of the current program counter in the stack.

The microcomputer then automatically loads an address into the program counter to branch to a subroutine-like program called the "interrupt-service routine." This program is written by the user. The external device wants the microcomputer to execute this program to transfer data. The last instruction of the service routine is a RETURN, which is typically the same instruction used at the end of a subroutine. This instruction normally loads the address (saved in the stack before going to the service routine) in the program counter. Then, the microcomputer continues executing the main program. An example of interrupt I/O is shown in Figure 8.19.

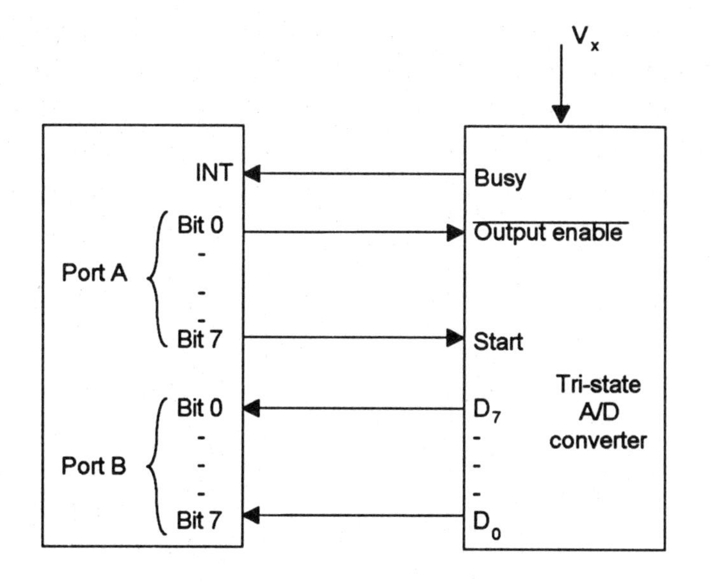

FIGURE 8.19 Microcomputer A/D converter interface via interrupt I/O

Assume the microcomputer is MC68000 based and executing the following program:

```
BEGIN EQU    $3000
      ORG           $2000
      MOVE.B        #$81, DDRA         ;      configure bits 0 and 7
                                       ;      of port A as outputs
      MOVE.B        #$00, DDRB         ;      configure Port B as input
      MOVE.B        #$81, PORTA        ;      send start pulse to A/D
                                       ;      and HIGH to OUTPUT ENABLE
      MOVE.B        #$01, PORTA
      CLR.W         D0                 ;      clear 16-bit register D0 to 0
BEGIN MOVE.W        D1, D2
         :
```

The extensions .B and .W represent byte and word operations. Note that the symbols $ and # indicate hexadecimal number and immediate mode respectively.

The preceding program is arbitrarily written. The program logic can be explained using the MC68000 instruction set. Ports DDRA and DDRB are assumed to be the data-direction registers for ports A and B, respectively. The first four `MOVE` instructions configure bits 0 and 7 of port A as outputs and port B as the input port, and then send a trailing START pulse (HIGH and then LOW) to the A/D converter along with a HIGH to the $\overline{\text{OUTPUT ENABLE}}$. This HIGH $\overline{\text{OUTPUT ENABLE}}$ is required to disable the A/D's output. The microcomputer continues with execution of the `CLR.W D0` instruction. Suppose that the busy signal becomes HIGH, indicating the end of conversion during execution of the `CLR.W D0` instruction. This drives the INT signal to HIGH, interrupting the microcomputer. The microcomputer completes execution of the current instruction, `CLR.W D0`. It then saves the current contents of the program counter (address $3000) and status register automatically and executes a subroutine-like program called the service routine. This program is usually written by the user. The microcomputer manufacturer may define the starting address of the service routine, or it may be provided by the user via external hardware. Assume this address is $4000, where the user writes a service routine to input the A/D converter's output as follows:

```
MOVE.B   #$00, PORTA        ;      Activate OUTPUT ENABLE.
MOVE.B   PORTB, D1          ;      Input A/D
RTE                         ;      Return and restore PC and SR.
```

In this service routine, the microcomputer inputs the A/D converter's output. The return instruction RTE, at the end of the service routine, pops $3000 and the previous status register contents from the stack and loads the program counter and status register with them. The microcomputer executes the `MOVE.W D1,D2` instruction at BEGIN (address $3000) and continues with the main program. The basic characteristics of interrupt I/O have been discussed so far. The main features of interrupt I/O provided with a typical microcomputer are discussed next.

8.2.2.1 Interrupt Types

There are typically three types of interrupts: external interrupts, traps or internal interrupts, and software interrupts. External interrupts are initiated through the microcomputer's interrupt pins by external devices such as A/D converters. A simple example of an external interrupt was given in the previous section.

External interrupts can further be divided into two types: maskable and nonmaskable. A maskable interrupt is enabled or disabled by executing instructions such as CLI or STI. If the computer's interrupt is disabled, the microcomputer ignores the maskable interrupt. Some processors, such as the Intel 8086, have an interrupt-flag bit in the processor status register. When the interrupt is disabled, the interrupt-flag bit is 1, so no maskable interrupts are recognized by the processor. The interrupt-flag bit is reset to zero when the interrupt is enabled.

The nonmaskable interrupt has higher priority than the maskable interrupt. If both maskable and nonmaskable interrupts are activated at the same time, the processor will service the nonmaskable interrupt first. The nonmaskable interrupt is typically used as a power failure interrupt. Processors normally use +5 V DC, which is transformed from 110 V AC. If the power falls below 90 V AC, the DC voltage of +5 V cannot be maintained. However, it will take a few milliseconds before the AC power can drop this low (below 90 V AC). In these few milliseconds, the power-failure-sensing circuitry can interrupt the processor. The interrupt-service routine can be written to store critical data in nonvolatile memory such as battery-backed CMOS RAM, and the interrupted program can continue without any loss of data when the power returns.

Some processors such as the 8086 are provided with a maskable handshake interrupt. This interrupt is usually implemented by using two pins — INTR and $\overline{\text{INTA}}$. When the INTR pin is activated by an external device, the processor completes the current instruction, saves at least the current program counter onto the stack, and generates an interrupt acknowledge ($\overline{\text{INTA}}$). In response to the $\overline{\text{INTA}}$, the external device provides an 8-bit number, using external hardware on the data bus of the microcomputer. This number is then read and used by the microcomputer to branch to the desired service routine.

Internal interrupts, or traps, are activated internally by exceptional conditions such as overflow, division by zero, or execution of an illegal op-code. Traps are handled the same way as external interrupts. The user writes a service routine to take corrective measures and provide an indication to inform the user that an exceptional condition has occurred.

Many processors include software interrupts, or system calls. When one of these instructions is executed, the processor is interrupted and serviced similarly to external or internal interrupts. Software interrupt instructions are normally used to call the operating system. These instructions are shorter than subroutine calls, and no calling program is needed to know the operating system's address in memory. Software interrupt instructions allow the user to switch from user to supervisor mode. For some processors, a software interrupt is the only way to call the operating system, because a subroutine call to an address in the operating system is not allowed.

8.2.2.2 Interrupt Address Vector

The technique used to find the starting address of the service routine (commonly known as the interrupt address vector) varies from one processor to another. With some processors, the manufacturers define the fixed starting address for each interrupt. Other manufacturers use an indirect approach by defining fixed locations where the interrupt address vector is stored.

8.2.2.3 Saving the Microprocessor Registers

When a processor is interrupted, it saves at least the program counter on the stack so that the processor can return to the main program after executing the service routine. Typical processors save one or two registers, such as the program counter and status register, before going to the service routine. The user should know the specific registers the processor saves prior to executing the service routine. This will allow the user to use the appropriate return instruction at the end of the service routine to restore the original conditions upon return to the main program.

8.2.2.4 Interrupt Priorities

A processor is typically provided with one or more interrupt pins on the chip. Therefore, a special mechanism is necessary to handle interrupts from several devices that share one of these interrupt lines. There are two ways of servicing multiple interrupts: polled and daisy chain techniques.

i) Polled Interrupts

Polled interrupts are handled by software and are therefore are slower than daisy chaining. The processor responds to an interrupt by executing one general-service routine for all devices. The priorities of devices are determined by the order in which the routine polls each device. The processor checks the status of each device in the general-service routine, starting with the highest-priority device, to service an interrupt. Once the processor determines the source of the interrupt, it branches to the service routine for the device. Figure 8.20 shows a typical configuration of the polled-interrupt system.

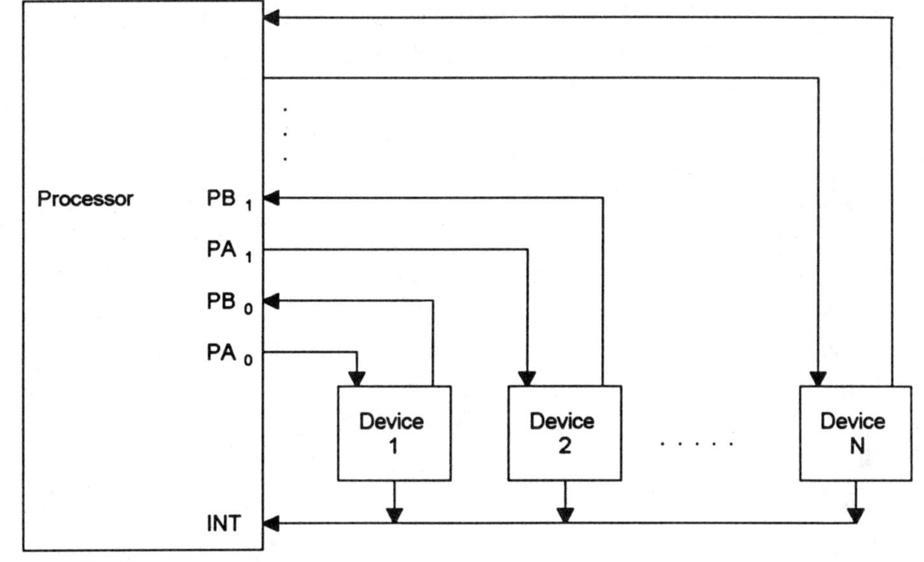

FIGURE 8.20 Polled interrupt

In Figure 8.20, several external devices (device 1, device 2,..., device N) are connected to a single interrupt line of the processor via an OR gate (not shown in the figure). When one or more devices activate the INT line HIGH, the processor pushes the program counter and possibly some other registers onto the stack. It then branches to an address defined by the manufacturer of the processor. The user can write a program at this address to poll each device, starting with the highest-priority device, to find the source of the interrupt. Suppose the devices in Figure 8.20 are A/D converters. Each converter, along with the associated logic for polling, is shown in Figure 8.21.

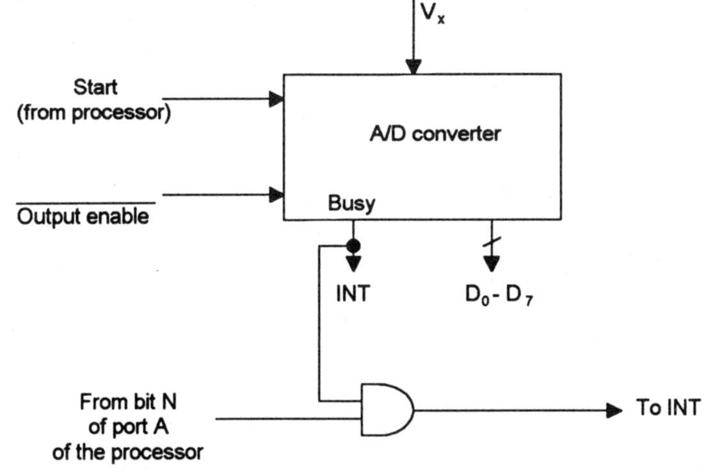

FIGURE 8.21 Device N and associated logic for polled interrupt

Assume that in Figure 8.20 two A/D converters (device 1 and device 2) are provided with the START pulse by the processor at nearly the same time. Suppose the user assigns device 2 the higher priority. The user then sets up this priority mechanism in the general-service routine. For example, when the BUSY signals from device 1 and/or 2 become HIGH, indicating the end of conversion, the processor is interrupted. In response, the processor pushes at least the program counter onto the stack and loads the PC with the interrupt address vector defined by the manufacturer.

The general interrupt-service routine written at this address determines the source of the interrupt as follows: A 1 is sent to PA1 for device 2 because this device has higher priority. If this device has generated an interrupt, the output (PB1) of the AND gate in Figure 8.21 becomes HIGH, indicating to the processor that device 2 generated the interrupt. If the output of the AND gate is 0, the processor sends a HIGH to PA0 and checks the output (PB0) for HIGH. Once the source of the interrupt is determined, the processor can be programmed to jump to the service routine for that device. The service routine enables the A/D converter and inputs the converter's outputs to the processor.

Polled interrupts are slow, and for a large number of devices, the time required to poll each device may exceed the time to service the device. In such a case, a faster mechanism, such as the daisy chain approach, can be used.

ii) Daisy Chain Interrupts

Devices are connected in a daisy chain fashion, as shown in Figure 8.22, to set up priority systems. Suppose one or more devices interrupt the processor. In response, the processor pushes at least the PC and generates an interrupt acknowledge ($\overline{INTA}$) signal to the highest-priority device (device 1 in this case). If this device has generated the interrupt, it will accept the $\overline{INTA}$; otherwise, it will pass the $\overline{INTA}$ onto the next device until the $\overline{INTA}$ is accepted.

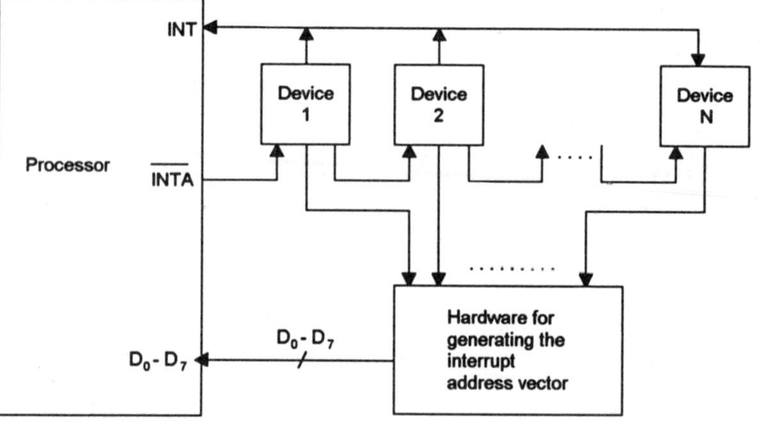

FIGURE 8.22 Daisy chain interrupt

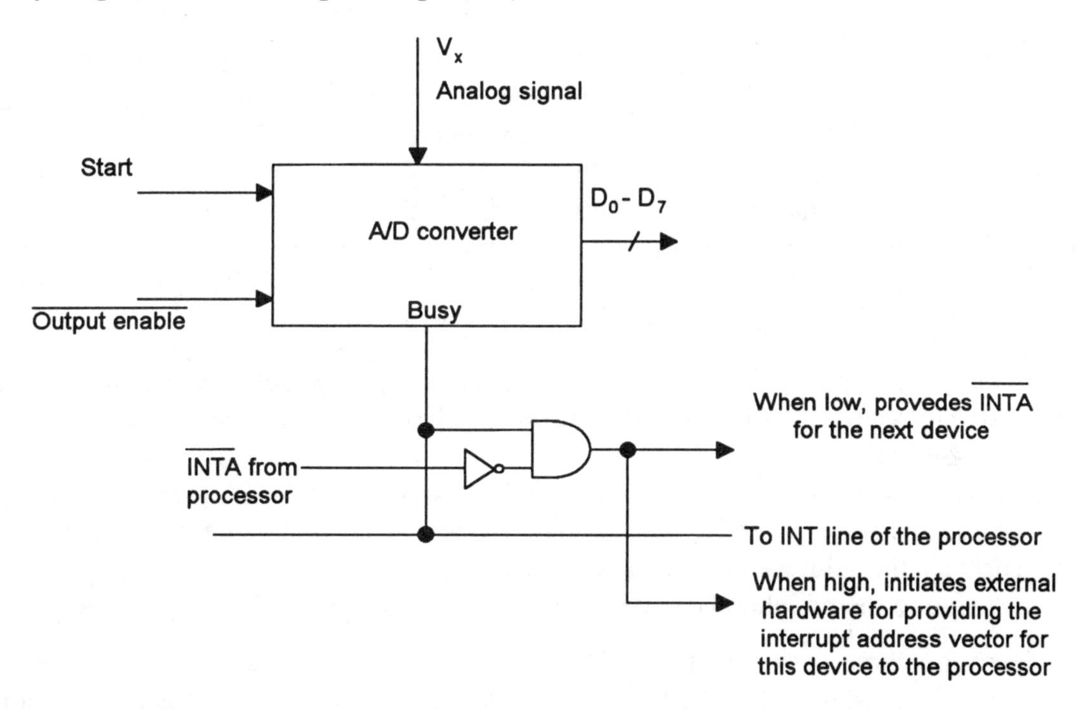

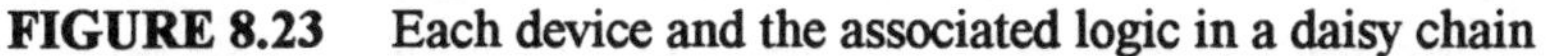

FIGURE 8.23 Each device and the associated logic in a daisy chain

Once accepted, the device provides a means for the processor to find the interrupt-address vector by using external hardware. Assume the devices in Figure 8.22 are A/D converters. Figure 8.23 provides a schematic for each device and the associated logic.

Suppose the processor in Figure 8.22 sends a pulse to start the conversions of the A/D converters of devices 1 and 2 at nearly the same time. When the BUSY signal goes to HIGH, the processor is interrupted through the INT line. The processor pushes the program counter and possibly some other registers. It then generates a LOW at the interrupt-acknowledge ($\overline{\text{INTA}}$) for the highest-priority device (device 1 in Figure 8.22). Device 1 has the highest priority—it is the first device in the daisy chain configuration to receive $\overline{\text{INTA}}$. If A/D converter 1 has generated the BUSY HIGH, the output of the AND gate becomes HIGH. This signal can be used to enable external hardware to provide the interrupt-address vector on the processor's data lines. The processor then branches to the service routine. This program enables the converter and inputs the A/D output to the processor via Port B. If A/D converter #1 does not generate the BUSY HIGH, however, the output of the AND gate in Figure 8.23 becomes LOW (an input to device 2's logic) and the same sequence of operations takes place. In the daisy chain, each device has the same logic with the exception of the last device, which must accept the $\overline{\text{INTA}}$. Note that the outputs of all the devices are connected to the INT line via an OR gate (not shown in Figure 8.22)

8.2.3 Direct Memory Access (DMA)

Direct memory access (DMA) is a technique that transfers data between a microcomputer's memory and an I/O device without involving the microprocessor. DMA is widely used in transferring large blocks of data between a peripheral device and the microcomputer's memory. The DMA technique uses a DMA controller chip for the data-transfer operations. The DMA controller chip implements various components such as a counter containing the length of data to be transferred in hardware in order to speed up data transfer. The main functions of a typical DMA controller are summarized as follows:

- The I/O devices request DMA operation via the DMA request line of the controller chip.
- The controller chip activates the microprocessor HOLD pin, requesting the microprocessor to release the bus.
- The processor sends HLDA (hold acknowledge) back to the DMA controller, indicating that the bus is disabled. The DMA controller places the current value of its internal registers, such as the address register and counter, on the system bus and sends a DMA acknowledge to the peripheral device. The DMA controller completes the DMA transfer.

There are three basic types of DMA: block transfer, cycle stealing, and interleaved DMA. For block-transfer DMA, the DMA controller chip takes over the bus from the microcomputer to transfer data between the microcomputer memory and I/O device. The microprocessor has no access to the bus until the transfer is completed. During this time, the microprocessor can perform internal operations that do not need the bus. This method is popular with microprocessors. Using this technique, blocks of data can be transferred.

Data transfer between the microcomputer memory and an I/O device occurs on a word-by-word basis with cycle stealing. Typically, the microprocessor is generated by ANDing an $\overline{\text{INHIBIT}}$ signal with the system clock. The system clock has the same frequency as the microprocessor clock. The DMA controller controls the $\overline{\text{INHIBIT}}$ line. During normal operation, the $\overline{\text{INHIBIT}}$ line is HIGH, providing the microprocessor clock. When DMA operation is desired, the controller makes the $\overline{\text{INHIBIT}}$ line LOW for one clock cycle. The microprocessor is then stopped completely for one cycle. Data transfer between the memory and I/O takes place during this cycle. This method is called "cycle stealing" because the DMA controller takes away or steals a cycle without microprocessor recognition. Data transfer takes place over a period of time.

With interleaved DMA, the DMA controller chip takes over the system bus when the microprocessor is not using it. For example, the microprocessor does not use the bus while incrementing the program counter or performing an ALU operation. The DMA controller chip identifies these cycles and allows transfer of data between the memory and I/O device. Data transfer takes place over a period of time for this method.

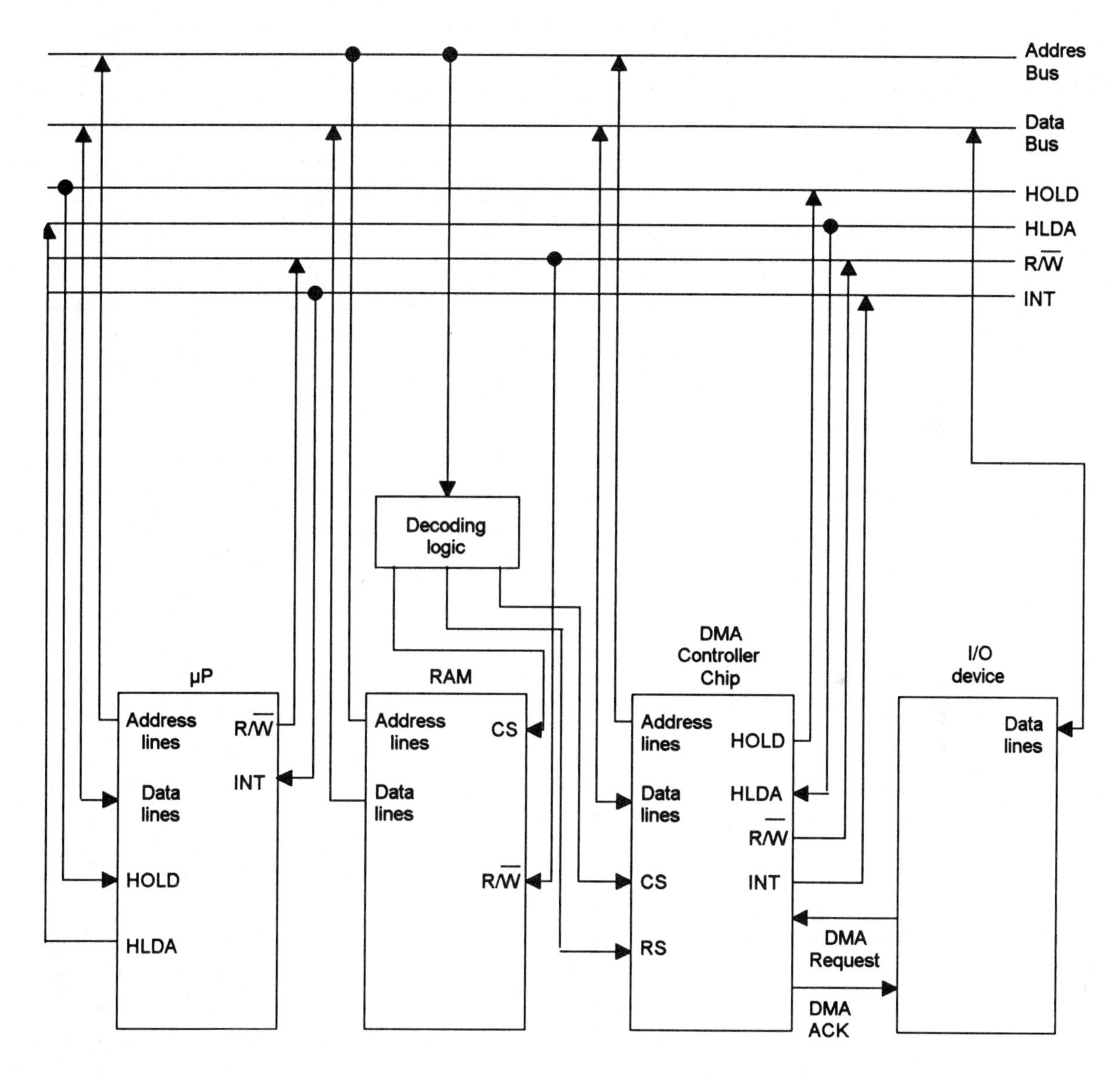

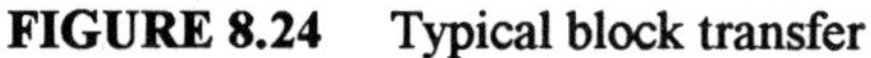
FIGURE 8.24 Typical block transfer

Because block-transfer DMA is common with microprocessors, a detailed description is provided. Figure 8.24 shows a typical diagram of the block-transfer DMA. In the figure, the I/O device requests the DMA transfer via the DMA request line connected to the controller chip. The DMA controller chip then sends a HOLD signal to the microprocessor, and it then waits for the HOLD acknowledge (HLDA) signal from the microprocessor. On receipt of the HLDA, the controller chip sends a DMA ACK signal to the I/O device. The controller takes over the bus and controls data transfer between the RAM and I/O device. On completion of the data transfer, the controller interrupts the microprocessor by the INT line and returns the bus to the microprocessor by disabling the HOLD and DMA ACK signals.

The DMA controller chip usually has at least three registers normally selected by the controller's register select (RS) line: an address register, a terminal count register, and a status register. Both the address and terminal counter registers are initialized by the microprocessor. The address register contains the starting address of the data to be transferred, and the terminal counter register contains the desired block to be transferred. The status register contains information such as completion of DMA transfer. Note that the DMA controller implements logic associated with data transfer in hardware to speed up the DMA operation.

8.3 Summary of I/O

Figure 8.25 summarizes various I/O devices associated with a typical microprocessor.

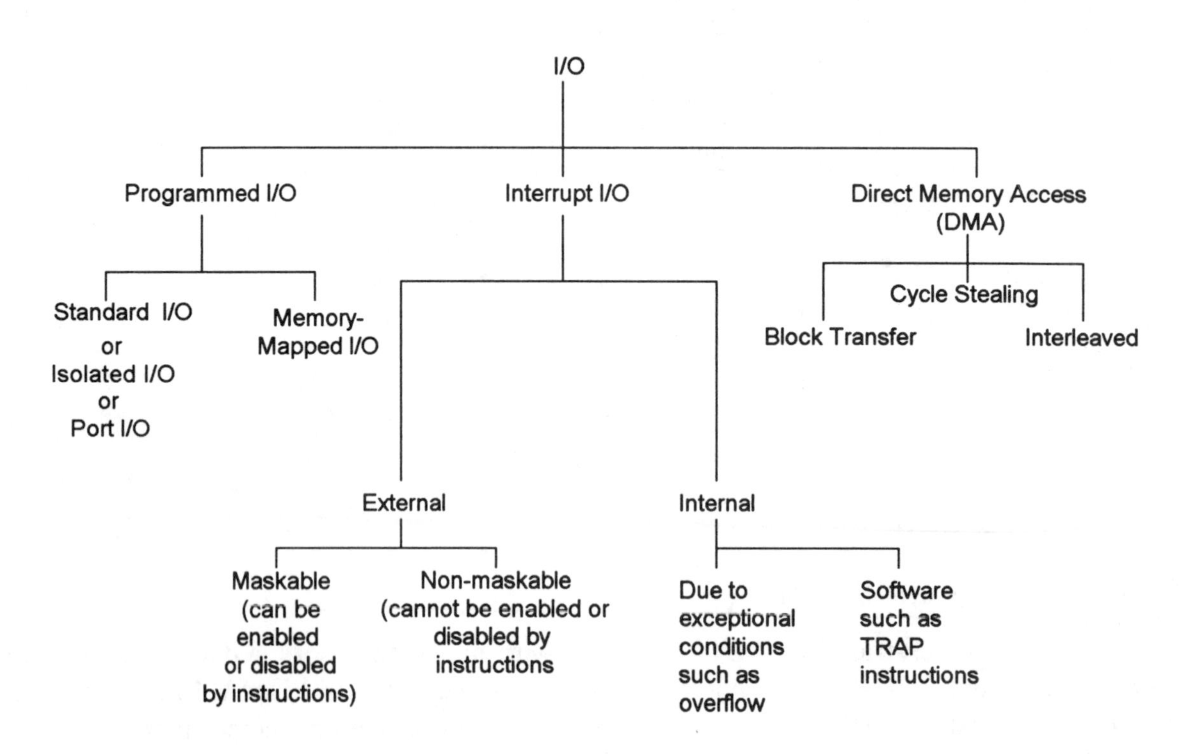

FIGURE 8.25 I/O structure of a typical microcomputer

QUESTIONS AND PROBLEMS

8.1 What is the basic difference between main memory and secondary memory?

8.2 Compare the basic features of hard disk, floppy disk and Zip disk.

8.3 What are the main differences between CD and DVD memories?

8.4 Name the methods used in main memory array design. What are the advantages and disadvantages of each.

8.5 The block diagram of a 512 × 8 RAM chip is shown below. In this arrangement, the memory chip is enabled only when $\overline{CS1}$ = L and CS2 = H.
Design a 1K × 8 RAM system using this chip as the building block. Draw a neat logic diagram of your implementation. Assume that the microprocessor can directly address 64K with a R/$\overline{W}$ and 8 data pins. Using linear decoding and don't-care conditions as 1's, determine the memory map in hex.

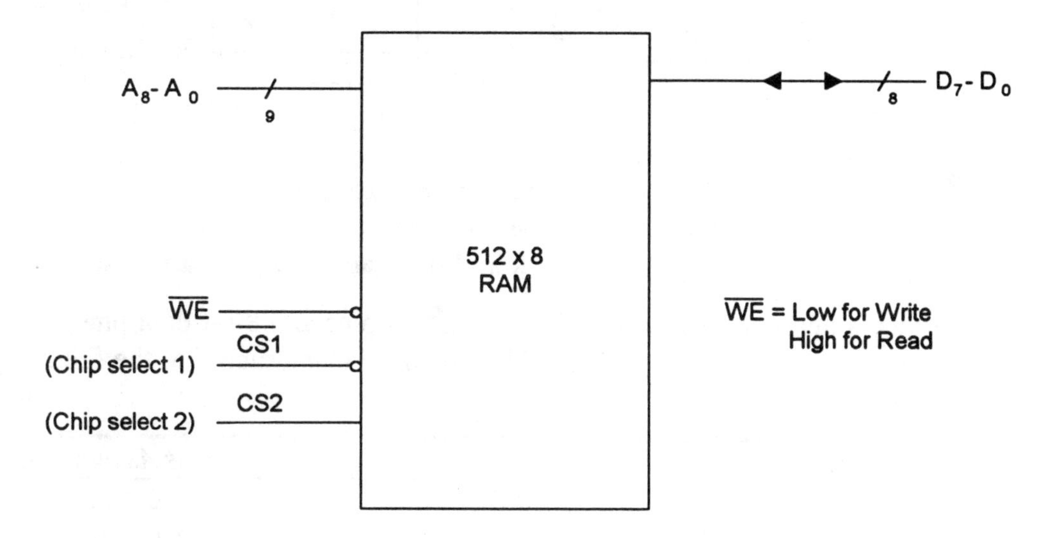

FIGURE P8.5

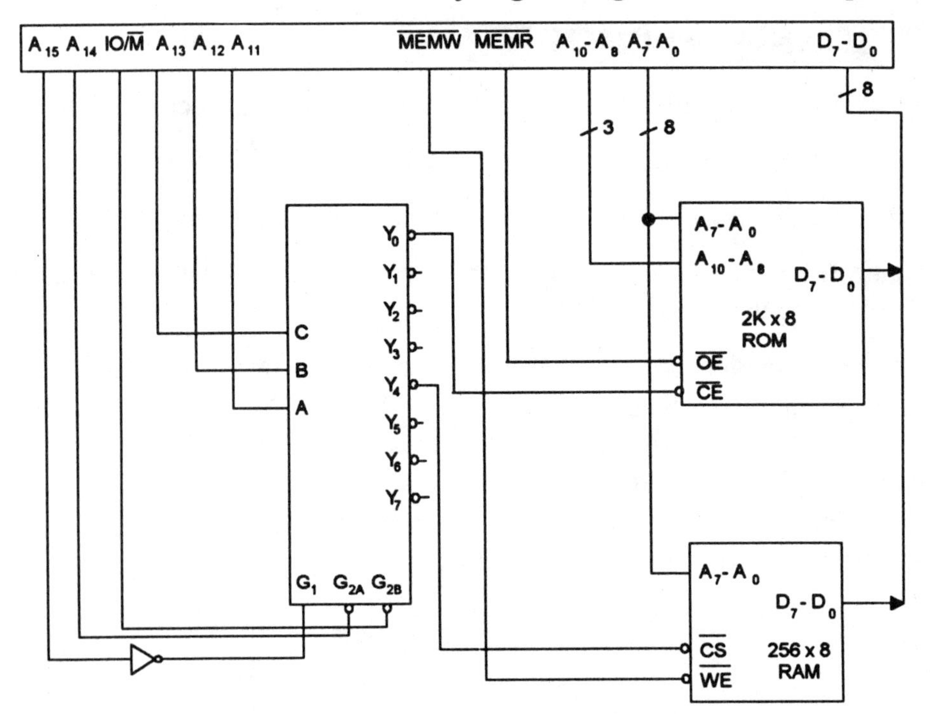

FIGURE P8.6

8.6 Consider the hardware schematic shown in Figure P8.6.

(a) Determine the address map of this system.

(b) Is there any memory foldback in this organization? Clearly justify your answer.

8.7 Interface a microprocessor with 16-bit address pins and 8-bit data pins and a R/$\overline{W}$ pin to a 1K × 8 EPROM chip and two 1K × 8 RAM chips such that the following address map is realized:

Device	*Size*	*Address Assignment (in hex)*
EPROM chip	1K × 8	8000–83FF
RAM chip 0	1K × 8	9000–93FF
RAM chip 1	1K × 8	C000–C3FF

Assume that both EPROM and RAM chips contain two enable pins; $\overline{CE}$ and $\overline{OE}$ for the EPROM, $\overline{CE}$ and $\overline{WE}$ for each RAM. Note that $\overline{WE}$ =1 and $\overline{WE}$ = 0 mean read and write operations for the RAM chip.

8.8 Repeat Problem 8.7, but this time realize the following address map:

Device	*Size*	*Address Assignment in hex*
EPROM chip	1K × 8	7000–73FF
RAM chip 0	1K × 8	D000–D3FF
RAM chip 1	1K × 8	F000–F3FF

8.9 What is meant by “foldback” in linear decoding?

8.10 Comment on the importance of the following features in an operating system implementation:
(a) Address translation
(b) Protection

8.11 Explain briefly the differences between segmentation and paging.

8.12 Draw a block diagram showing the address and data lines for the 2716, 2732, and 2764 EPROM chips.

8.13 How many address and data lines are required for a 1M × 16 memory chip.

8.14 A microprocessor with 24 address pins and 8 data pins is connected to a 1K × 8 memory chip with one-chip enable. How many unused address bits of the microprocessor are available for interfacing other 1K × 8 memory chips. What is the maximum directly addressable memory available with this microprocessor?

8.15 What is the size of a decoder with one chip enable ($\overline{CE}$) to obtain a 64K × 32 memory from the 4K × 8 chips? Where are the inputs and outputs of the decoder connected?

8.16 What is the advantage of having a cache memory? Name a 32-bit microprocessor that does not contain an on-chip cache.

8.17 Discuss the various cache-mapping techniques.

8.18 A microprocessor has a main memory of 8K × 32 and a cache memory of 4K × 32. Using direct mapping, determine the sizes of the tag field, index field, and each word of the cache.

8.19 A microprocessor has a main memory of $4K \times 32$. Using a cache memory address of 8 bits and set-associative mapping with a set size of 2, determine the size of the cache memory.

8.20 A microprocessor can directly address one megabyte of memory with a 16-bit word size. Determine the size of each cache memory word for associative mapping.

8.21 What is the basic difference between:
- (a) Standard I/O and memory-mapped I/O?
- (b) Programmed I/O and virtual I/O?
- (c) Polled I/O and interrupt I/O?
- (d) A subroutine and interrupt I/O?
- (e) Cycle-stealing, block transfer, and interleaved DMA?
- (f) Maskable and nonmaskable interrupts?
- (g) Internal and external interrupts?
- (h) Memory mapping in a microprocessor and memory-mapped I/O?

9

INTEL 16-, 32-,AND 64-BIT MICROPROCESSORS

This chapter covers the Intel 8086 in detail and presents a summary of Intel 32- and 64-bit microprocessors. *Although the Intel 8086 is in the declining stage (almost at the end of its life cycle), Intel 32-bit microprocessors are based on the Intel 8086. Therefore, the 8086 provides an excellent educational tool for understanding Intel 32- and 64-bit microprocessors.* Because the 8086 and its peripheral chips are inexpensive, the implementation costs of 8086-based systems are low. This makes the 8086 appropriate for thorough coverage in a first course on microprocessors. Thus, the 8086 is covered in detail in this chapter. According to Intel, they will support the HCMOS-based 8086 and its peripheral chips directly or through alternate sources in the future. The telephone numbers of Intel, along with the distributors, and alternate sources for Intel and their support chips are

Intel Corporation
Intel Customer Service: 1-800-628-8686
(USA and Canada)
Intel Corporation: 1-408-765-8080
(For local and International users)

The following are some distributors of Intel microprocessors and support chips

Name of Company	Telephone Number for USA and Canada
Arrow Corporation	1-800-777-2776
Avnet Electronics.	1-888-405-2372
Marshall Industries.	1-714-859-5050
Wyle Corp.	1-800-943-7446

These are some alternate sources of Intel microprocessors and support chips.

Name of Company	Telephone Number
NEC	1-408-588-6000
AMD	1-408-732-2400
Harris	1-800-442-7747

Intel's Website address is www.Intel.com. These names of companies along with telephone numbers are provided for the convenience of users of Intel microprocessors. These companies will provide up-to-date information regarding alternative sources for Intel microprocessors and their support chips in case Intel discontinues manufacturing a specific chip. **Note that in the following, symbol [ ] is used to indicate the contents of an 8086 register or a memory location.**

9.1 Introduction

The 8086 was Intel's first 16-bit microprocessor. This means that the 8086 has a 16-bit ALU. The 8086 contains 20 address pins. Therefore, it has a main (directly addressable) memory of one megabyte (2^{20} bytes).

The memory of an 8086-based microcomputer is organized as bytes. Each byte is uniquely addressed with 20-bit addresses of 00000_{16}, 00001_{16}, ... $FFFFF_{16}$. An 8086 word in memory consists of any two consecutive bytes; the low-addressed byte is the low byte of the word and the high-addressed byte contains the high byte as follows:

Low byte of the word	High byte of the word
02_{16}	$A1_{16}$
Address 02000_{16}	Address 02001_{16}

The 16-bit word at the even address 02000_{16} is $A102_{16}$. Next, consider a word stored at an address as follows:

Low byte of the word	High byte of the word
$2E_{16}$	46_{16}
Address 30151_{16}	Address 30152_{16}

The 16-bit word stored at the odd address 30151_{16} is $462E_{16}$.

The 8086 always reads a 16-bit word from memory. This means that a word instruction accessing a word starting at an even address can perform its function with one memory read. A word instruction starting at an odd address, however, must perform two memory accesses to two consecutive memory even addresses, discarding the unwanted bytes of each. For byte read starting at odd address *N*, the byte at the previous even address *N* - 1 is also accessed but discarded. Similarly, for byte read starting at even address N, the byte with odd address $N + 1$ is also accessed but discarded.

For the 8086, register names followed by the letters X, H, or L in an instruction for data transfer between register and memory specify whether the transfer is 16-bit or 8-bit. For example, consider `MOV AX, [START]`. If the 20-bit address `START` is an even number such as 02212_{16}, then this instruction loads the low (AL) and high (AH) bytes of the 8086 16-bit register AX with the contents of memory locations 02212_{16} and 02213_{16}, respectively, in a single access. Now, if `START` is an odd number such as 02213_{16}, then the `MOV AX, [START]` instruction loads AL and AH with the contents of memory locations 02213_{16} and 02214_{16}, respectively, in two accesses. The 8086 also accesses memory locations 02212_{16} and 02215_{16} but ignores their contents.

Next, consider `MOV AL, [START]`. If `START` is an even number such as 30156_{16}, then this instruction accesses both addresses, 30156_{16} and 30157_{16}, but loads AL with the contents of 30156_{16} and ignores the contents of 30157_{16}. However, if `START` is an odd number such as 30157_{16}, then `MOV AL, [START]` loads AL with the contents of 30157_{16}. In this case the 8086 also reads the contents of 30156_{16} but discards it.

The 8086 is packaged in a 40-pin chip. A single +5 V power supply is required. The clock input signal is generated by the 8284 clock generator/driver chip. Instruction execution times vary between 2 and 30 clock cycles.

There are three versions of the 8086: the 8086, 8086-2, and 8086-4. There is no difference between the three versions other than the maximum allowed clock speeds. The 8086 can be operated from a maximum clock frequency of 5 MHz. The maximum clock frequencies of the 8086-2 and 8086-4 are 8 MHz and 4 MHz, respectively.

The 8086 family consists of two types of 16-bit microprocessors, the 8086 and 8088. The main difference is how the processors communicate with the outside world. The 8088 has an 8-bit external data path to memory and I/O; the 8086 has a 16-bit external data path. This means that the 8088 will have to do two READ operations to read a 16-bit word from memory. Similarly, two write operations are required to write a 16-bit word into memory. In most other respects, the processors are identical. Note that the 8088 accesses memory in bytes. No alterations are needed to run software written for one microprocessor on the other. Because of similarities, only the 8086 will be considered here. The 8088 was used in designing IBM's first personal computer.

An 8086 can be configured as a small uniprocessor (minimum mode when the MN/$\overline{\text{MX}}$ pin is tied to HIGH) or as a multiprocessor system (maximum mode when the MN/$\overline{\text{MX}}$ pin is tied to LOW). In a given system, the MN/$\overline{\text{MX}}$ pin is permanently tied to either HIGH or LOW. Some of the 8086 pins have dual functions depending on the selection of the MN/$\overline{\text{MX}}$ pin level.

In the minimum mode (MN/$\overline{MX}$ pin HIGH), these pins transfer control signals directly to memory and I/O devices; in the maximum mode (MN/$\overline{MX}$ pin LOW), these same pins have different functions that facilitate multiprocessor systems. In the maximum mode, the control functions normally present in minimum mode are assumed by a support chip, the 8288 bus controller.

Due to technological advances, Intel introduced the high-performance 80186 and 80188, which are enhanced versions of the 8086 and 8088, respectively. The 8-MHz 80186/80188 provides 2 times greater throughput than the standard 5-MHz 8086/8088. Both have integrated several new peripheral functional units, such as a DMA controller, a 16-bit timer unit, and an interrupt controller unit, into a single chip. Just like the 8086 and 8088, the 80186 has a 16-bit data bus and the 80188 has an 8-bit data bus; otherwise, the architecture and instruction set of the 80186 and 80188 are identical. The 80186/80188 has an on-chip clock generator so that only an external crystal is required to generate the clock. The 80186/80188 can operate at either a 6- or an 8-MHz internal clock frequency. The crystal frequency is divided by 2 internally. In other words, external crystals of 12 or 16 MHz must be connected to generate the 6- or 8-MHz internal clock frequency. The 80186/80188 is fabricated in a 68-pin package. Both processors have on-chip priority interrupt controller circuits to provide five interrupt pins. Like the 8086/8088, the 80186/80188 can directly address one megabyte of memory. The 80186/80188 is provided with 10 new instructions beyond the 8086/8088 instruction set. Examples of these instructions include INS and OUTS for inputting and outputting a string byte or string word.

The 80286, on the other hand, has added memory protection and management capabilities to the basic 8086 architecture. An 8-MHz 80286 provides up to 6 times greater throughput than the 5-MHz 8086. The 80286 is fabricated in a 68-pin package. The 80286 can be operated at a clock frequency of 4, 6, or 8 MHz. An external 82284 clock generator chip is required to generate the clock. The 82284 divides the external clock by 2 to generate the internal clock. The 80286 can be operated in two modes, real address and protected virtual address. Real address mode emulates a very high-performance 8086. In this mode, the 80286 can directly address one megabyte of memory. In virtual address mode, the 80286 can directly address 16 megabytes of memory. Virtual address mode provides (in addition to the real address mode capabilities) virtual memory management as well as task management and protection. The programmer can select one of these modes by loading appropriate data in the 16-bit machine status word (MSW) register by using the load instruction (`LMSW`).

The 80286 was used as the microprocessor of the IBM PC/AT personal computer. An enhanced version of the 80286 is the 32-bit 80386 microprocessor. The 80386 was used as the microprocessor in the IBM 386PC. The 80486 is another 32-bit microprocessor. It is based on the Intel 80386 and includes on-chip floating-point circuitry. IBM's 486 PC contains the 80486 chip. Other 32-bit and 64-bit Intel microprocessors include Pentium, Pentium Pro, Pentium II, and Merced.

Intel plans to replace the 8086 by the 80386. However, the 8086 is expected to be around for some time from second sources. Therefore, a detailed coverage of the 8086 and 80386 is included. A summary of the other 32- and 64-bit microprocessors are then provided.

9.2 8086 Main Memory

The 8086 uses a segmented memory. There are some advantages to working with the segmented memory. First, after initializing the 16-bit segment registers, the 8086 has to deal with only 16-bit effective addresses. That is, the 8086 has to manipulate and store 16-bit address components. Second, because of memory segmentation, the 8086 can be effectively used in time-shared systems. For example, in a time-shared system, several users may share one 8086. Suppose that the 8086 works with one user's program for, say, 5 milliseconds. After spending 5 milliseconds with one of the other users, the 8086 returns to execute the first user's program. Each time the 8086 switches from one user's program to the next, it must execute a new section of code and new sections of data. Segmentation makes it easy to switch from one user program to another.

The 8086's main memory can be divided into 16 segments of 64K bytes each (16×64 KB = 1 MB). A segment may contain codes or data. The 8086 uses 16-bit registers to address segments. For example, in order to address codes, the code segment register must be initialized in some manner (to be discussed later): A 16-bit 8086 register called the "instruction pointer" (IP), which is similar to the program counter of a typical microprocessor, linearly addresses each location in a code segment. Because the size of the IP is 16 bits, the segment size is 64K bytes (2^{16}). Similarly, a 16-bit data segment register must be initialized to hold the segment value of a data segment. The contents of certain 16-bit registers are designed to hold a 16-bit address in a 64-Kbyte data segment. One of these address registers can be used to linearly address each location once the data segment is initialized by an instruction. Finally, in order to access the stack segment, the 8086 16-bit stack segment (SS) register must be initialized; the 64-Kbyte stack is addressed linearly by a 16-bit stack pointer register. Note that the stack memory must be a read/write (RAM) memory. ***Whenever the programmer reads from or writes to the 8086 memory or stack, two components of a memory address must be considered: a segment value and, an address or an offset or a displacement value.*** The 8086 assembly language program works with these two components while accessing memory. These two 16-bit components (the contents of a 16-bit segment register and a 16-bit offset or IP) form a logical address. The programmer writes programs using these logical addresses in assembly language programming.

The 8086 includes on-chip hardware to map or translate these two 16-bit components of a memory address into a 20-bit address called a "physical address" by shifting the contents of a segment register four times to left and then adding the contents of IP or offset. Note that the 8086 contains 20 address pins, so the physical address size is 20 bits wide.

Consider, for example, a logical address with the 16-bit code segment register contents of 2050_{16} and the 16-bit 8086 instruction pointer containing a value of 0004_{16}. Suppose that the programmer writes an 8086 assembly language program using this logical address. The programmer assembles this program and obtains the object or machine code. When the 8086 executes this program and encounters the logical address, it will generate the 20-bit physical address as follows: If 16-bit contents of IP = 0004_{16}, 16-bit contents of code segment = 2050_{16}, 16-bit contents of code segment value after shifting logically 4 times to left = 20500_{16}, then the 20-bit physical address generated by the 8086 on its 20-pin address is 20504_{16}. Note that the 8086 assigns the low address to the low byte of a 16-bit register and the high address to the high byte of the 16-bit register for 16-bit transfers between the 8086 and main memory. This is called *Little-endian byte ordering.*

9.3 8086 Registers

As mentioned in Chapter 8, the 8086 is divided internally into two independent units: the bus interface unit (BIU) and the execution unit (EU). The BIU reads (fetches) instructions, reads operands, and writes results. The EU executes instructions already fetched by the BIU. The 8086 prefetches up to 6 instruction bytes from external memory into a FIFO (first-in–first-out) memory in the BIU and queues them in order to speed up instruction execution.

The BIU contains a dedicated adder to produce the 20-bit address. The bus control logic of the BIU generates all the bus control signals, such as the READ and WRITE signals, for memory and I/O. The BIU also has four 16-bit segment registers: the code segment (CS), data segment (DS), stack segment (SS), and extra segment (ES) registers.

All program instructions must be located in main memory, pointed to by the 16-bit CS register with a 16-bit offset in the segment contained in the 16-bit instruction pointer (IP). Note that immediate data are considered as part of the code segment. The SS register points to the current stack. The 20-bit physical stack address is calculated from the SS and SP (stack pointer) for stack instructions such as PUSH and POP. The programmer can create a programmer's stack with the BP (base pointer) instead of the SP for accessing the stack using the based addressing mode. In this case, the 20-bit physical stack address is calculated from the BP and SS. The *DS register* points to the current data segment; operands for most instructions are fetched from this segment. The 16-bit contents of a register such as the SI (source index) or DI (destination index) or a 16-bit displacement are used as offsets for computing the 20-bit physical address.

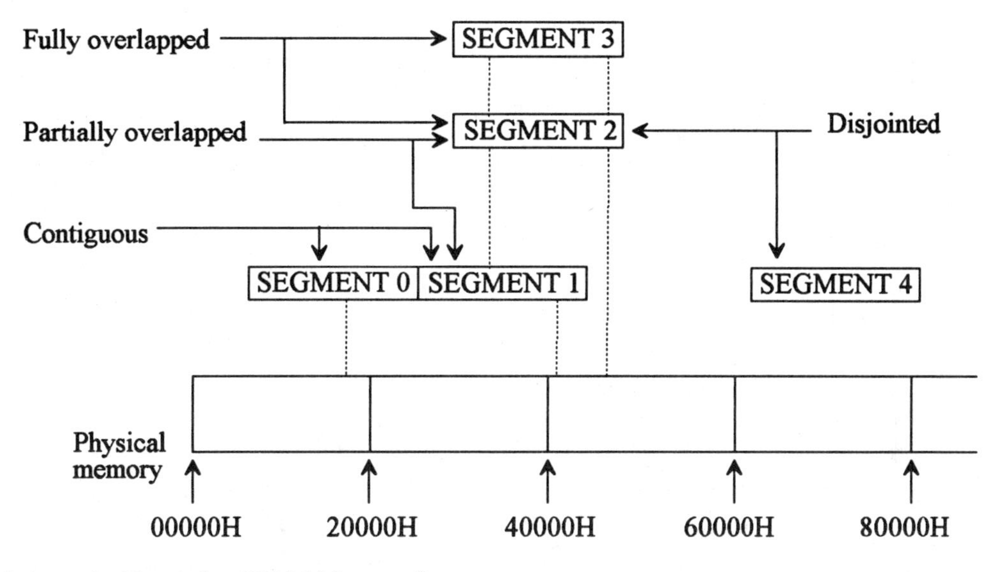

FIGURE 9.1 An Example of 8086 Memory Segments

The *ES register* points to the extra segment in which data (in excess of 64 KB pointed to by the DS) is stored. String instructions always use the ES and DI to determine the 20-bit physical address for the destination.

The segments can be contiguous, partially overlapped, fully overlapped, or disjointed. An example of how five segments (SEGMENT 0 through SEGMENT 4), may be stored in physical memory is shown in Figure 9.1. In this example, SEGMENTs 0 and 1 are contiguous (adjacent), SEGMENTs 1 and 2 are partially overlapped, SEGMENTs 2 and 3 are fully overlapped, and SEGMENTs 2 and 4 are disjointed.

Every segment must start on 16-byte memory boundaries. Typical examples of values of segments should then be selected based on physical addresses starting at 00000_{16}, 00010_{16}, 00020_{16}, 00030_{16}, ..., $FFFF0_{16}$. A physical memory location may be mapped into (contained in) one or more logical segments. Many applications can be written to simply initialize the segment registers and then forget them.

A segment can be pointed to by more than one segment register. For example, the DS and ES may point to the same segment in memory if a string located in that segment is used as a source segment in one string instruction and a destination segment in another string instruction. Note that, for string instructions, a destination segment must be pointed to by the ES. One example of four currently addressable segments is shown in Figure 9.2.

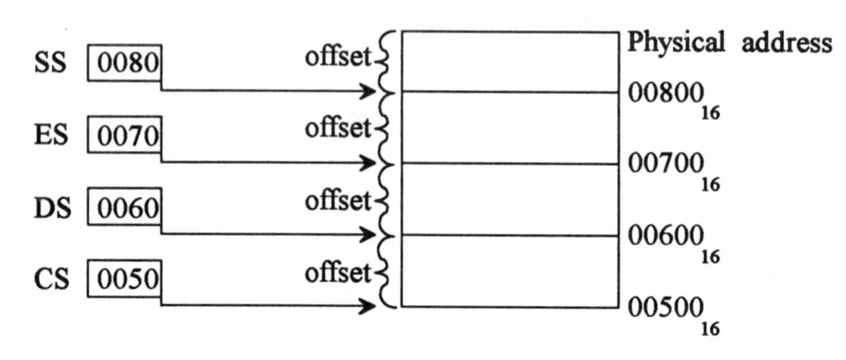

FIGURE 9.2 Four currently addressable 8086 segments

The EU decodes and executes instructions. It has a 16-bit ALU for performing arithmetic and logic operations. The EU has nine 16-bit registers: AX, BX, CX, DX, SP, BP, SI, and DI, and the flag register. The 16-bit general registers AX, BX, CX, and DX can be used as two 8-bit registers (AH, AL; BH, BL; CH, CL; DH, DL). For example, the 16-bit register DX can be considered as two 8-bit registers DH (high byte of DX) and DL (low byte of DX). The general-purpose registers AX, BX, CX, and DX perform the following functions:

- The AX register is 16 bit wide whereas AH and AL are 8 bit wide. The use of AX and AL registers is assumed by some instructions. The I/O (IN or OUT) instructions always use the AX or AL for inputting/outputting 16- or 8-bit data to or from an I/O port. Multiplication and division instructions also use the AX or AL.
- *The BX register is called the "base register." This is the only general-purpose register whose contents can be used for addressing 8086 memory.* All memory references utilizing this register content for addressing use the DS as the default segment register.
- The CX register is known as the *counter* register because some instructions, such as `SHIFT`, `ROTATE`, and `LOOP`, use the contents of CX as a counter, For example, the instruction `LOOP START` will automatically decrement CX by 1 without affecting flags and will check to see if [CX] = 0. If it is zero, the 8086 executes the next instruction; otherwise, the 8086 branches to the label `START`.
- The *DX register*, or *data register,* is used to hold the high 16-bit result (data) (LOW 16-bit data is contained in AX) after 16 × 16 multiplication or the high 16-bit dividend (data) before a 32 ÷ 16 division and the 16-bit remainder after the division (16-bit quotient is contained in AX).
- The two pointer registers, SP (stack pointer) and BP (base pointer), are used to access data in the stack segment. The SP is used as an offset from the current SS during execution of instructions that involve the stack segment in external memory. The SP contents

are automatically updated (incremented or decremented) due to execution of a POP or PUSH instruction. The BP contains an offset address in the current SS. This offset is used by instructions utilizing the based addressing mode.

- The two *index registers,* SI (source index) and DI (destination index), are used in indexed addressing. Note that instructions that process data strings use the SI and DI index registers together with the DS and ES, respectively, in order to distinguish between the source and destination addresses.
- The *flag register* in the EU holds the status flags, typically after an ALU operation. The EU sets or resets these flags to reflect the results of arithmetic and logic operations.

Figure 9.3 depicts the 8086 registers. It shows the nine 16-bit registers in the EU. As described earlier, each one of the AX, BX, CX, and DX registers can be used as two 8-bit registers or as one 16-bit register. The other registers can be accessed as 16-bit registers. Also shown are the four 16-bit segment registers and the 16-bit IP in the BIU. The IP is similar to the program counter. The CS register points to the current code segment from which instructions are fetched. The effective address is derived from the CS and IP. The SS register points to the current stack. The effective address is obtained from the SS and SP. The DS register points to the current data segment. The ES register points to the current extra segment where data is usually stored.

	15 8	7 0
AX	AH	AL
BX	BH	BL
CX	CH	CL
DX	DH	DL

	15 0
Stack pointer	SP
Base pointer	BP
Source index	SI
Destination index	DI
	FLAGS
	EU

15 0	
	Instruction pointer (IP)

15 0	
CS	Code segment
DS	Data segment
SS	Stack segment
ES	Extra segment
BIU	

FIGURE 9.3 8086 Registers

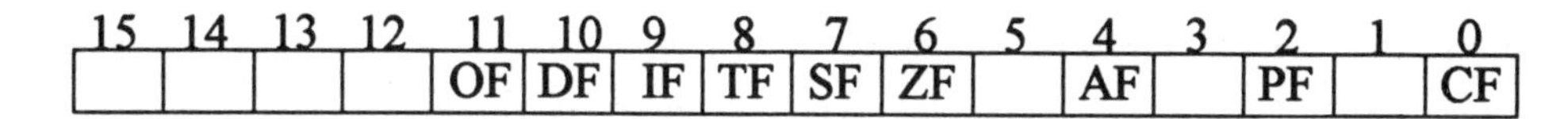

FIGURE 9.4 8086 Flag Register

Figure 9.4 shows the 8086 flag register. The 8086 has six 1-bit status flags. Let us now explain these flags.

- **AF** (auxiliary carry flag) is set if there is a carry due to addition of the low nibble into the high nibble or a borrow due to the subtraction of the low nibble from the high nibble of a number.
  This flag is used by BCD arithmetic instructions; otherwise, AF is zero.
- **CF** (carry flag) is set if there is a carry from addition or a borrow from subtraction.
- **OF** (overflow flag) is set if there is an arithmetic overflow (i.e., if the size of the result exceeds the capacity of the destination location). An interrupt on overflow instruction is available to generate an interrupt in this situation; otherwise, it is zero.
- **SF** (sign flag) is set if the most significant bit of the result is one; otherwise, it is zero.
- **PF** (parity flag) is set if the result has even parity; PF is zero for odd parity of the result.
- **ZF** (zero flag) is set if the result is zero; ZF is zero for a nonzero result.

The 8086 has three control bits in the flag register that can be set or cleared by the programmer:

1. Setting DF (direction flag) causes string instructions to auto-decrement; clearing DF causes string instructions to auto-increment.
2. Setting IF (interrupt flag) causes the 8086 to recognize external maskable interrupts; clearing IF disables these interrupts.
3. Setting TF (trap flag) puts the 8086 in the single-step mode. In this mode, the 8086 generates an internal interrupt after execution of each instruction. The user can write a service routine at the interrupt address vector to display the desired registers and memory locations. The user can thus debug a program.

9.4 8086 Addressing Modes

The 8086 provides various addressing modes to access instruction operands. Operands may be contained in registers, within the instruction op-code, in memory, or in I/O ports. The 8086 has 12 addressing modes, which can be classified into five groups:

1. Register and immediate modes (two modes)
2. Memory addressing modes (six modes)
3. Port addressing mode (two modes)
4. Relative addressing mode (one mode)
5. Implied addressing mode (one mode)

9.4.1 Register and Immediate Modes

Register mode. The addressing modes are illustrated utilizing 8086 instructions with directives of a typical assembler. In register mode, source operands, destination operands, or both may be contained in registers. For example, `MOV AX,BX` moves the 16-bit contents of BX into AX. On the other hand, `MOV AH,BL` moves the 8-bit contents of BL into AH.

Immediate mode. In immediate mode, 8- or 16 bit data can be specified as part of the instruction. For example, `MOV CX,5062H` moves the 16-bit data 5062_{16} into register CX.

9.4.2 Memory Addressing Modes

The EU has direct access to all registers and data for register and immediate modes. However, the EU cannot directly access the memory operands. It must use the BIU to access memory operands. For example, when the EU needs a memory operand, it sends an offset value to the BIU. As mentioned before, this offset is added to the contents of a segment register after shifting it four times to the left, generating a 20-bit physical address. For example, suppose that the contents of a segment register is 2052_{16} and the offset is 0020_{16}. Now, in order to generate the 20-bit physical address, the EU passes this offset to the BIU. The BIU then shifts the segment register four times to the left, obtains 20520_{16} and then adds the 0020_{16} offset to provide the 20-bit physical address 20540_{16}.

Note that the 8086 must use a segment register whenever it accesses the memory. Also, every memory addressing mode has a standard default segment register. However, a segment override instruction can be placed before most of the memory operand instructions whose default segment register is to be overridden. For example, `INC BYTE PTR [START]` will increment the 8-bit contents of a memory location in DS with offset START by 1. However, segment DS can be overridden by ES as follows: `INC ES: BYTE PTR [START]`. Segments cannot be overridden for stack reference instructions (such as `PUSH` and `POP`). The destination segment of a string segment, which must be ES (if a prefix is used with a string instruction, only the source segment DS can be overridden) cannot be overridden. The code segment (CS) register used in program memory addressing cannot be overridden. The EU calculates an offset from the instruction for a memory operand. This offset is called the operand's *effective address,* or EA. It is a 16-bit number that represents the operand's distance in bytes from the start of the segment in which it resides.

The various memory addressing modes will now be described.

1. ***Memory Direct Addressing.*** In this mode, the effective address is taken directly from the displacement field of the instruction. No registers are involved. For example, `MOV BX, [START]`, or `MOV BX, OFFSET START` moves the contents of the 20-bit address computed from DS and START to BX. Some assemblers use square brackets around

START to indicate that the contents of the memory location(s) are at a displacement START from the segment DS. If square brackets are not used, then the programmer may define START as a 16-bit offset by using the assembler directive, `OFFSET`.

2. ***Register Indirect Addressing.*** The effective address of a memory operand may be taken directly from one of the base or index registers (BX, BP, SI, DI). For example, consider `MOV CX,[BX]`. If [DS] = 2000_{16}, [BX] = 0004_{16}, and [20004_{16}] = 0224_{16}, then, after `MOV CX,[BX]`, the contents of CX is 0224_{16}. Note that the segment register used in `MOV CX,[BX]` can be overridden, such as `MOV CX,ES:[BX]`. Now, the `MOV` instruction will use ES instead of DS. If [ES] = 1000_{16} and [10004_{16}] = 0002_{16}, then, after `MOV CX,ES:[BX]`, the register CX will contain 0002_{16}.
3. ***Based Addressing.*** In this mode, the effective address is the sum of a displacement value (signed 8-bit or unsigned 16-bit) and the contents of register BX or BP. For example, `MOV AX,4[BX]` moves the contents of the 20-bit address computed from a segment register and BX + 4 into AX. The segment register is DS or SS. The displacement (4 in this case) can be unsigned 16-bit or sign-extended 8-bit. This means that if the displacement is 8-bit, then the 8086 sign extends this to 16-bit. Segment register SS is used when the stack is accessed; otherwise, this mode uses segment register DS. When memory is accessed, the 20-bit physical address is computed from BX and DS. On the other hand, when the stack is accessed, the 20-bit physical address is computed from BP and SS. Note that BP may be considered as the user stack pointer while SP is the system stack pointer. This is because SP is used by some 8086 instructions (such as CALL subroutine) automatically.

   The based addressing mode with BP is a very convenient way to access stack data. BP can be used as a stack pointer in SS to access local variables. Consider the following instruction sequence (arbitrarily chosen to illustrate the use of BP for stack):

```
PUSH BP              ;    Save BP
MOV  BP,SP           ;    Establish BP
PUSH CX              ;    Save CX
SUB  SP, 6           ;    Allocate 3 words of
                     ;    stack for local variables
MOV  -4[BP], BX      ;    Push BX onto stack using BP
MOV  -6[BP], AX      ;    Push AX onto stack using BP
MOV  -8[BP], DX      ;    Push DX onto stack using BP
ADD  SP, 6           ;    Deallocate stack
POP  CX              ;    Restore CX
POP  BP              ;    Restore BP
```

This instruction sequence can be depicted as follows:

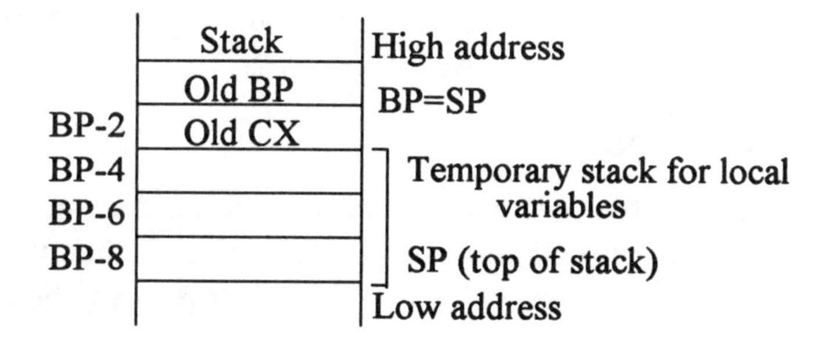

4. ***Indexed Addressing.*** In this mode, the effective address is calculated from the sum of a displacement value and the contents of register SI or DI. For example, `MOV AX, VALUE [SI]` moves the contents of the 20-bit address computed from VALUE, SI and the segment register into AX. The segment register is DS. The displacement (VALUE in this case) can be unsigned 16-bit or sign-extended 8-bit. The indexed mode can be used to access a table.
5. ***Based Indexed Addressing.*** In this mode, the effective address is computed from the sum of a base register (BX or BP), an index register (SI or DI), and a displacement. For example, `MOV AX, 4[BX][SI]` moves the contents of the 20-bit address computed from the segment register and [BX] + [SI] + 4 into AX. The segment register is DS. The displacement can be unsigned 16-bit or sign-extended 8-bit. This mode can be used to access two-dimensional arrays such as matrices.
6. ***String Addressing.*** This mode uses index registers. SI is assumed to point to the first byte or word of the source string, and DI is assumed to point to the first byte or word of the destination when a string instruction is executed. The SI or DI is automatically incremented or decremented to point to the next byte or word depending on DF. The default segment register for source is DS, and it may be overridden; the segment register used for the destination must be ES, and can not be overridden. An example is `MOVS WORD`. If [DF] = 0, [DS] = 3000_{16}, [SI] = 0020_{16}, [ES] 5000_{16}, [DI] = 0040_{16}, [30020] = 30_{16}, [30021] = 05_{16}, [50040] = 06_{16}, and [50041] = 20_{16}, then, after this `MOVS`, [50040] = 30_{16}, [50041] = 05_{16}, [SI] = 0022_{16}, and [DI] = 0042_{16}.

9.4.3 Port Addressing

Two I/O port addressing modes can be used: direct port and indirect port. In either case, 8- or 16-bit I/O transfers must take place via AL or AX respectively.In *direct port mode,* the port number is an 8-bit immediate operand to access 256 ports. For example. `IN AL, 02` moves the contents of port 02 to AL. In *indirect port mode,* the port number is taken from DX, allowing 64K bytes or 32K words of ports. For example, suppose [DX] = 0020, [port

0020] = 02_{16}, and [port 0021] = 03_{16}, then, after `IN AX, DX`, register AX contains 0302_{16}. On the other hand, after `IN AL, DX`, register AL contains 02_{16}.

9.4.4 Relative Addressing Mode

Instructions using this mode specify the operand as a signed 8-bit displacement relative to PC. An example is `JNC START`. This instruction means that if carry = 0, then PC is loaded with the current PC contents plus the 8-bit signed value of START; otherwise, the next instruction is executed.

9.4.5 Implied Addressing Mode

Instructions using this mode have no operands. An example is `CLC`, which clears the carry flag to zero.

9.5 8086 Instruction Set

The 8086 has approximately 117 different instructions with about 300 op-codes. The 8086 instruction set contains no-operand, single-operand, and two-operand instructions. Except for string instructions that involve array operations, 8086 instructions do not permit memory-to-memory operations. Appendices F and H provide 8086 instruction reference data and the instruction set (alphabetical order), respectively.The 8086 instructions can be classified into eight groups:

1. Data Transfer Instructions
2. Arithmetic Instructions
3. Bit Manipulation Instructions
4. String Instructions
5. Unconditional Transfer Instructions
6. Conditional Branch Instructions
7. Interrupt Instructions
8. Processor Control Instructions

Let us now explain some of the 8086 instructions with numerical examples. Note that in the following examples , symbol [ ] is used to indicate the contents of a register or a memory location.

9.5.1 Data Transfer Instructions

Table 9.1 lists the data transfer instructions. Note that LEA is used to load 16-bit offset to a specified register; LDS and LES are similar to LEA except that they load specified register as well as DS or ES. As an example, LEA BX, 3000H has the same meaning as MOV BX,3000H. On the other hand, if [SI]=2000H, then LEA BX,4[SI] will load 2004H into BX while MOV BX,4[SI] will initialize BX with the contents of memory locations computed from 2004H and DS. The LEA instruction can be useful when memory computation is desirable.

TABLE 9.1 8086 Data Transfer Instructions

General Purpose	
MOV d, s	[d] ← [s] MOV byte or word
PUSH d	PUSH word into stack
POP d	POP word off stack
XCHG mem/reg, mem/reg	[mem/reg] ↔ [mem/reg]; No mem to mem.
XLAT	AL ← [20 bit address computed from AL, BX, and DS]
Input / Output	
IN A, DX or Port	Input byte or word
OUT DX or Port, A	Output byte or word
Address Object	
LEA reg, mem	LOAD Effective Address
LDS reg, mem	LOAD pointer using DS
LES reg, mem	LOAD pointer using ES
Flag Transfer	
LAHF	LOAD AH register from flags
SAHF	STORE AH register in flags
PUSHF	PUSH flags onto stack
POPF	POP flags off stack

d = "mem" or "reg" or "segreg," s = "data" or " mem" or "reg" or "segreg," A = AX or AL

In Table 9.1, there are 14 data transfer instructions. These instructions move single bytes and words between a register, a memory location, or an I/O port. Let us explain some of the instructions in Table 9.1.

- `MOV CX,DX` copies the 16-bit contents of DX into CX. `MOV AX,2025H` moves immediate data 2025H into the 16-bit register AX. `MOV CH, [BX]` moves the 8-bit contents of a memory location addressed by BX in segment register DS into CH. If [BX] = 0050H, [DS] 2000H, and [20050H] = 08H, then, after `MOV CH, [BX]`, the contents of CH will be 08H. `MOV START [BP],CX` moves the 16-bit (CL to first location and then CH) contents of CX into two memory locations addressed by the sum of the displacement START and BP in segment register SS. For example, if [CX] = 5009H, [BP]=0030H, [SS] = 3000H, and START = 06H, then, after `MOV START [BP],CX`, [30036H] = 09H and [30037H] = 50H.

- `LDS SI,[0010H]` loads SI and DS from memory. For example, if [DS] = 2000H, [20010] = 0200H, and [20012] = 0100H, then, after `LDS SI,[0010H]`, SI and DS will contain 0200H and 0100H, respectively.
- In the 8086, the SP is decremented by 2 for PUSH and incremented by 2 for POP. For example, consider PUSH [BX]. If [DS] = 2000_{16}, [BX] = 0200_{16}, [SP] = 3000_{16}, [SS] = 4000_{16}, and [20200] = 0120_{16}, then, after execution of `PUSH [BX]`, memory locations 42FFF and 42FFE will contain 01_{16} and 20_{16}, respectively, and the contents of SP will be $2FFE_{16}$.
- `XCHG` has three variations: `XCHG` reg, reg and `XCHG` mem, reg or `XCHG` reg, mem . For example, `XCHG AX, BX` exchanges the contents of 16-bit register BX with the contents of AX. `XCHG` mem, reg exchanges 8- or 16-bit data in mem with 8-or 16-bit reg.
- `XLAT` can be used to employ an index in a table or for code conversion. This instruction utilizes BX to hold the starting address of the table in memory consisting of 8-bit data elements. The index in the table is assumed to be in the AL register. For example, if [BX] = 0200_{16}, [AL] = 04_{16}, and [DS] = 3000_{16}, then, after `XLAT`, the contents of location 30204_{16} will be loaded into AL. Note that the XLAT instruction is the same as `MOV AL, [AL] [BX]`.
- Consider fixed port addressing, in which the 8-bit port address is directly specified as part of the instruction. `IN AL, 38H` inputs 8-bit data from port 38H into AL. `IN AX, 38H` inputs 16-bit data from ports 38H and 39H into AX. `OUT 38H, AL` outputs the contents of AL to port 38H. `OUT 38H, AX`, on the other hand, outputs the 16-bit contents of AX to ports 38H and 39H.
- For variable port addressing, the port address is 16-bit and is specified in the DX register. Assume [DX] = 3124_{16} in all the following examples.

  `IN AL,DX` inputs 8-bit data from 8-bit port 3124_{16} into AL.
  `IN AX,DX` inputs 16-bit data from ports 3124_{16} and 3125_{16} into AX.
  `OUT DX,AL` outputs 8-bit data from AL into port 3124_{16}.
  `OUT DX,AX` outputs 16-bit data from AX into ports 3124_{16} and 3125_{16}.

  Variable port addressing allows up to 65,536 ports with addresses from 0000H to FFFFH. The port addresses in variable port addressing can be calculated dynamically in a program. For example, assume that an 8086-based microcomputer is connected to three printers via three separate ports. Now, in order to output to each one of the printers, separate programs are required if fixed port addressing is used. However, with variable port addressing, one can write a general subroutine to output to the printers and then supply the address of the port for a particular printer in which data output is desired to register DX in the subroutine.

9.5.2 Arithmetic Instructions

Table 9.2 shows the 8086 arithmetic instructions. These operations can be performed on four types of numbers: unsigned binary, signed binary, unsigned packed decimal, and signed packed decimal numbers. Binary numbers can be 8 or 16 bits wide. Decimal numbers are stored in bytes; two digits per byte for packed decimal and one digit per byte for unpacked decimal with the high 4 bits filled with zeros.

TABLE 9.2 8086 Arithmetic Instructions

	Addition	
`ADD` a, b	Add byte or word	
`ADC` a, b	Add byte or word with carry	
`INC` reg/mem	Increment byte or word by one	
`AAA`	ASCII adjust for addition	
`DAA`	Decimal adjust [AL], to be used after ADD or ADC	
	Subtraction	
`SUB` a, b	Subtract byte or word	
`SBB` a, b	Subtract byte or word with borrow	
`DEC` reg/mem	Decrement byte or word by one	
`NEG` reg/mem	Negate byte or word	
`CMP` a, b	Compare byte or word	
`AAS`	ASCII adjust for subtraction	
`DAS`	Decimal adjust [AL], to be used after SUB or SBB	
	Multiplication	
`MUL` reg/mem	Multiply byte or word unsigned	for byte
`IMUL` reg/mem	Integer multiply byte or word (signed)	$[AX] \leftarrow [AL] \cdot [mem/reg]$ for word $[DX][AX] \leftarrow [AX] \cdot [mem/reg]$
	Division	
`DIV` reg/mem	Divide byte or word unsigned	$16 \div 8$ bit; $[AX] \leftarrow \frac{[AX]}{[mem/reg]}$
`IDIV` reg/mem	Integer divide byte or word (signed)	$[AH] \leftarrow$ remainder $[AL] \leftarrow$ quotient $32 \div 16$ bit; $[DX:AX] \leftarrow \frac{[DX:AX]}{[mem/reg]}$ $[DX] \leftarrow$ remainder $[AX] \leftarrow$ quotient
`AAD`	ASCII adjust for division	
`CBW`	Convert byte to word	
`CWD`	Convert word to double word	

a = "reg" or "mem," b = "reg" or "mem" or "data."

Let us explain some of the instructions in Table 9.2.

- Consider `ADC` mem/reg, mem/reg. This instruction adds data with carry from reg to reg, from reg to mem, or from mem to reg. There is no ADC mem, mem instruction. For example, if [AX] = 0020_{16}, [BX] = 0300_{16}, CF = 1, [DS] = 2020_{16}, and [20500] = 0100_{16}, then, after `ADC AX, [BX]`, the contents of register AX = 0020 + 0100 + 1 = 0121_{16}. All flags are affected.
- `DIV` mem/reg performs unsigned division and divides [AX] or [DX:AX] registers by reg or mem. For example, if [AX] = 0005_{16} and [CL] = 02_{16}, then, after `DIV CL`, [AH] = 01_{16} and [AL] = 02_{16}.
- Consider `MUL BL`. If [AL] = 20_{16} and [BL] = 02_{16}, then, after `MUL BL`, register AX will contain 0040_{16}.
- Consider `CBW`. This instruction extends the sign from the AL register to the AH register. For example, if AL = $F1_{16}$, then, after execution of `CBW`, register AH will contain FF_{16} because the most significant bit of $F1_{16}$ is 1. *Note that the sign extension is very useful when one wants to perform an arithmetic operation on two numbers of different lengths.* For example. the 16-bit number 0020_{16} can be added with the 8-bit number $E1_{16}$ by sign-extending E1 as follows:

$$
\begin{array}{lrl}
 & 0020_{16} = & 0000\ 0000\ 0010\ 0000\ (32_{10}) \\
\text{Sign extension} & E1_{16} = & \boxed{1111\ 1111}\ 1110\ 0001\ (-31_{10}) \\
\hline
\text{Ignore carry} \rightarrow & 1 & \underbrace{0000}_{0}\ \underbrace{0000}_{0}\ \underbrace{0000}_{0}\ \underbrace{0001}_{1}\ (+1_{10})
\end{array}
$$

- *Another example of sign extension is that, to multiply a signed 8-bit number by a signed 16-bit number, one must first sign-extend the signed 8-bit into a signed 16-bit number and then the instruction `IMUL` can be used for 16 × 16 signed multiplication.* For unsigned multiplication of a 16-bit number by an 8-bit number, the 8-bit number must be zero extended to 16 bits before using the `MUL` instruction.
- `CWD` sign-extends the AX register into the DX register. That is, if the most significant bit of AX is 1, then $FFFF_{16}$ is stored into DX.
- The distinction between the byte or word operations on memory is sometimes made by some assembler by using B for byte or W for word as trailing characters with the instructions. Typical examples are `MULB` or `MULW`.
- *Numerical data received by an 8086-based microcomputer from a terminal is usually in ASCII code. The ASCII codes for numbers 0 to 9 are 30H through 39H. Two 8-bit data items can be entered into an 8086-based microcomputer via a keyboard. The ASCII*

codes for these data items (with 3 as the upper nibble for each type) can be added. AAA instruction can then be used to provide the correct unpacked BCD. Suppose that ASCII codes for 2 (32_{16}) and 5 (35_{16}) are entered into an 8086-based microcomputer via a keyboard. These ASCII codes can be added and then the result can be adjusted to provide the correct unpacked BCD using the `AAA` *instruction as follows:*

```
ADD     CL, DL    ;    [CL] = 32₁₆ = ACSII for 2
                  ;    [DL] = 35₁₆ = ASCII for 5
                  ;    Result [CL] = 67₁₆
MOV     AL, CL    ;    Move ASCII result
                  ;    into AL because AAA
                  ;    adjusts only [AL]
AAA               ;    [AL] = 07, unpacked
                  ;    BCD for 7
```

Note that, in order to print the unpacked BCD result 07_{16} on an ASCII printer, [AL] = 07 can be ORed with 30H to provide 37H, the ASCII code for 7.

- `DAA` is used to adjust the result of adding two packed BCD numbers in AL to provide a valid BCD number. If, after the addition, the low 4 bits of the result in AL is greater than 9 (or if AF = 1), then the `DAA` adds 6 to the low 4 bits of AL. On the other hand. if the high 4 bits of the result in AL are greater than 9 (or if CF = 1), then `DAA` adds 60H to AL.
- `DAS` may be used to adjust the result of subtraction in AL of two packed BCD numbers to provide the correct packed BCD. While performing these subtractions, any borrows from low and high nibbles are ignored, For example, consider subtracting BCD 55 in DL from BCD 94 in AL:

```
SUB AL, DL ; [AL] = 3FH      low nibble = 1111
DAS        ;  CF = 0                 -6 = 1010
                                          ----
                                         1 1001
           ; [AL] = 39 BCD              1 <------ Ignore 1
```

- `IMUL` mem/reg provides signed 8 × 8 or signed 16 × 16 multiplication. As an example, if [CL] = FDH = -3_{10} and [AL] = FEH = -2_{10}, then, after `IMUL CL`, register AX contains 0006H.
- Consider 16 × 16 unsigned multiplication, `MUL WORD PTR [BX]`. If [BX] = 0050H, [DS] = 3000H, [30050H] = 0002H, and [AX] = 0006H, then, after `MUL WORD PTR [BX]`, [DX] = 0000H and [AX] = 000CH.
- Consider `DIV BL`. If [AX] = 0009H and [BL] = 02H, then, after `DIV BL`,

[AH] = remainder = 01H
[AL] = quotient = 04H

- Consider `IDIV WORD PTR [BX]`. If [BX] = 0020H, [DS] = 2000H, [20020H] = 0004H, and [DX] [AX] = 00000011H, then, after `IDIV WORD PTR [BX]`,

  [DX] = remainder = 0001H
  [AX] = quotient = 0004H

- `AAD` converts two unpacked BCD digits in AH and AL to an equivalent binary number in AL after converting them to packed BCD. `AAD` must be used before dividing two unpacked BCD digits in AX by an unpacked BCD byte. For example, consider dividing [AX] = unpacked BCD 0508 (58 decimal) by [DH] = 07H. [AX] must first be converted to binary by using `AAD`. The register AX will then contain 003AH = 58 decimal. After `DIV DH`, [AL] = quotient = 08 unpacked BCD, and [AH] = remainder 02 unpacked BCD.
- `AAM` adjusts the product of two unpacked BCD digits in AX. If [AL] = BCD3 = 00000011_2 and [CH] = BCD8 = $0000\ 1000_2$, then, after `MUL CH`, [AX] = 0000000000011000_2 = 0018H, and, after using `AAM`, [AX] = 0000001000000100_2 = unpacked 0204. The following instruction sequence accomplishes this:

```
MUL CH
AAM
```

  Note that the 8086 does not allow multiplication of two ASCII codes. Therefore, before multiplying two ASCII bytes received from a terminal, one must make the upper 4 bits of each one of these bytes zero, multiply them as two unpacked BCD digits, and then use `AAM` for adjustment to convert the unpacked BCD product back to ASCII by ORing the product with 3030H. The result in decimal can then be printed on an ASCII printer.

9.5.3 Bit Manipulation Instructions

The 8086 provides three groups of bit manipulation instructions. These are logicals, shifts, and rotates, as shown in Table 9.3. The operand to be shifted or rotated can be either 8- or 16-bit. Let us explain some of the instructions in Table 9.3

- `TEST CL, 05H` logically ANDs [CL] with 00000101_2 but does not store the result in CL. All flags are affected.

- Consider `SHR` mem/reg, `CNT` or `SHL` mem/reg, `CNT`. These instructions are logical right or left shifts, respectively. The CL register contains the number of shifts if the shift is greater than 1. If `CNT` = 1, the shift is immediate data. In both cases, the last bit shifted out goes to CF (carry flag) and 0 is the last bit shifted in.

TABLE 9.3 8086 Bit Manipulation Instructions

	Logicals
NOT mem/reg	NOT byte or word
AND a, b	AND byte or word
OR a, b	OR byte or word
XOR a, b	Exclusive OR byte or word
TEST a, b	Test byte or word
	Shifts
SHL/SAL mem/reg, CNT	Shift logical/arithmetic left byte or word
SHR/SAR mem/reg, CNT	Shift logical/arithmetic right byte or word
	Rotates
ROL mem/reg, CNT	Rotate left byte or word
ROR mem/reg, CNT	Rotate right byte or word
RCL mem/reg, CNT	Rotate through carry left byte or word
RCR mem/reg, CNT	Rotate through carry right byte or word

a = "reg" or "mem," b = "reg" or "mem" or "data," CNT = number of times to be shifted. If CNT > 1, then CNT is contained in CL. Zero or negative shifts and rotates are illegal. If CNT = 1 then CNT is immediate data. Up to 255 shifts are allowed.

- Figure 9.5 shows SAR mem/reg, CNT or SAL mem/reg, CNT. Note that a true arithmetic left shift does not exist in 8086 because the sign bit is not retained after execution of SAL. Note that SAL and SHL perform the same operation except that SAL sets OF to 1 if the sign bit of the number shifted changes during or after shifting.

- ROL mem/reg, CNT rotates [mem/reg] left by the specified number of bits (Figure 9.6). The number of bits to he rotated is either 1 or contained in CL. For example, if CF = 0, [BX] = 0010_{16}, and [CL] = 03_{16} then, after `ROL BX, CL`, register BX will contain 0080_{16} and CF = 0. On the other hand, `ROL BL, 1` rotates the 8-bit contents of BL 1 bit to the left. ROR mem/reg, CNT is similar to ROL except that the rotation is to the right (Figure 9.6).

- Figure 9.7 shows RCL mem/reg, CNT and RCR mem/reg, CNT.

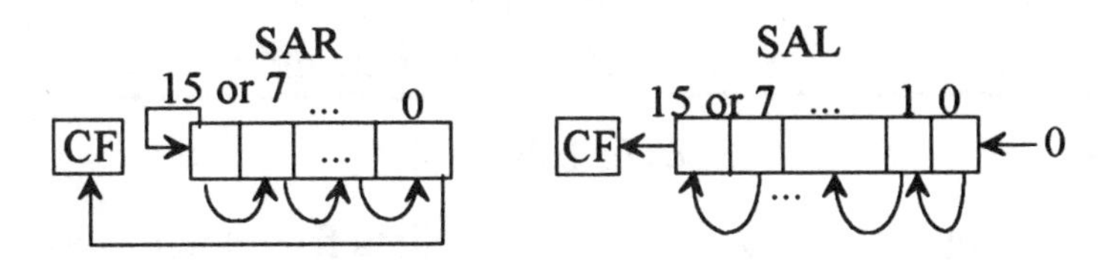

FIGURE 9.5 8086 SAR and SAL instructions

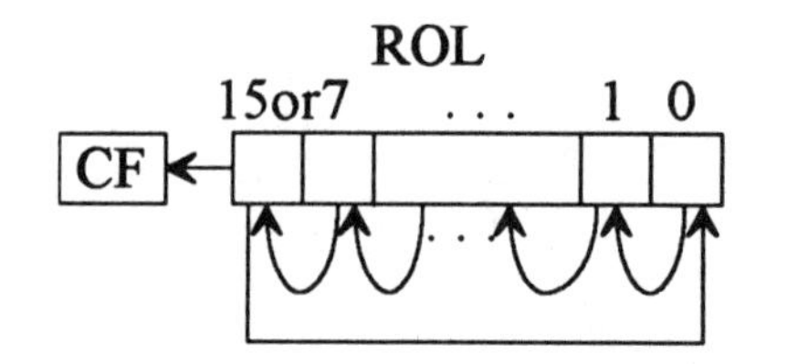

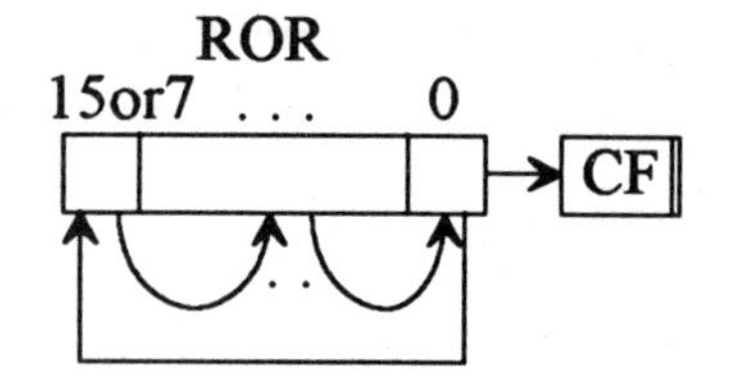

FIGURE 9.6 8086 ROR and ROL instructions

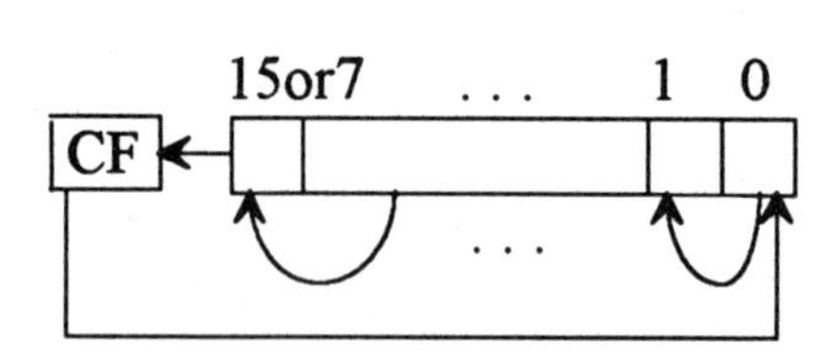

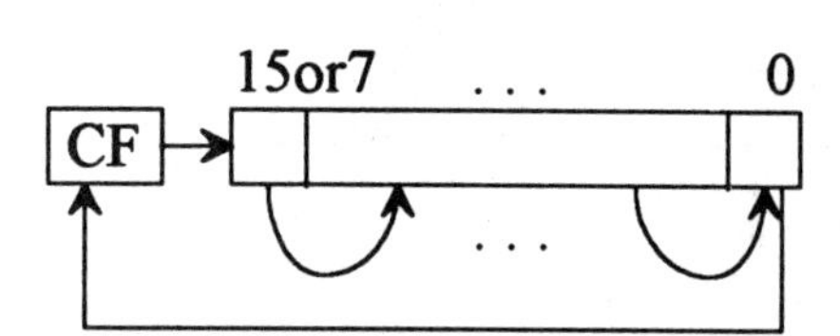

FIGURE 9.7 8086 RCL and RCR instructions

9.5.4 String Instructions

String instructions are available to MOVE, COMPARE, or SCAN for a value as well as to move string elements to and from AL or AX. The instructions, listed in Table 9.4, contain "repeat" prefixes that cause these instructions to be repeated in hardware, allowing long strings to be processed much faster than if done in a software loop.

TABLE 9.4 8086 String Instructions

`REP`	Repeat `MOVS` or `STOS` until CX = 0
`REPE/REPZ.`	Repeat `CMPS` or `SCAS` until ZF = 1 or CX = 0
`REPNE/REPNZ`	Repeat `CMPS` or `SCAS` until ZF = 0 or CX = 0
`MOVS BYTE/WORD`	Move byte or word string
`CMPS BYTE/WORD`	Compare byte or word string
`SCAS BYTE/WORD`	Scan byte or word string
`LODS BYTE/WORD`	Load from memory into AL or AX
`STOS BYTE/WORD`	Store AL or AX into memory

Let us explain some of the instructions in Table 9.4.

- `MOVS WORD` or `BYTE` moves 8- or 16-bit data from the memory location addressed by SI to the memory location addressed by DI. SI and DI are incremented or decremented depending on the DF flag. For example, if [DF] = 0, [DS] = 1000_{16}, [ES] =

3000_{16}, [SI] = 0002_{16}, [DI] = 0004_{16}, and [10002] = 1234_{16}, then, after `MOVS WORD`, [30004] = 1234_{16}, [SI] = 0004_{16}, and [DI] = 0006_{16}.

- `REP` repeats the instruction that follows until the CX register is decremented to 0. A `REPE/REPZ` or `REPNE/REPNZ` prefix can be used with `CMPS` or `SCAS` to cause one of these instructions to continue executing until ZF = 0 (for the `REPNE/REPNZ` prefix) or CX = 0. `REPE` and `REPZ` also provide similar purpose. If `CMPS` is prefixed with `REPE` or `REPZ`, the operation is interpreted as "compare while not end-of-string (CX ≠ 0) or strings are equal (ZF = 1)." If `CMPS` is preceded by `REPNE` or `REPNZ`, the operation is interpreted as "compare while not end-of-string (CX ≠ 0) or strings not equal (ZF = 0)." Thus, repeated `CMPS` instructions can be used to find matching or differing string elements.
- If `SCAS` is prefixed with `REPE` or `REPZ`, the operation is interpreted as "scan while not end-of-string (CX ≠ 0) or string-element = scan-value (ZF = 1)" This form may be used to scan for departure from a given value. If `SCAS` is prefixed with `REPNE` or `REPNZ`, the operation is interpreted as "scan while not end-of-string (CX ≠ 0) or string-element is not equal to scan-value (ZF = 0)." This form may be used to locate a value in a string.
- Consider `SCAS` word or byte. This compares the memory with AL or AX. If [DI] = 0000_{16}, [ES] = 2000_{16}, [DF] = 0, [20000] = 05_{16}, and [AL] = 03_{16}, then, after `SCAS BYTE`, DI will contain 0001_{16} because [DF] = 0 and all flags are affected based on the operation [AL] - [20000].

9.5.5 Unconditional Transfer Instructions

Unconditional transfer instructions transfer control to a location either in the current executing memory segment (intrasegment) or in a different code segment (intersegment). Table 9.5 lists the unconditional transfer instructions. The jump instruction in Table 9.5 has a few variations. Some of these are `JMP` mem, `JMP` disp8, `JMP` disp16, and `JMP` reg. Note that for intersegment jumps, both IP and CS change. For intrasegment jumps, IP changes and CS is fixed. `JMP disp8` adds the second object code byte (signed 8-bit displacement) to [IP + 2], and [CS] is unchanged.

TABLE 9.5 8086 Unconditional Transfers

`CALL` reg/mem/disp 16	Call subroutine
`RET`	Return from subroutine
`JMP` reg/mem/disp8/disp 16	Unconditional jump

`JMP` displ6 adds the next word (16-bit unsigned displacement) to IP + 2, and CS is unchanged. `JMP` mem moves the next two words to IP and CS, respectively. `JMP` reg moves the [reg] into [IP] and [CS] is unchanged. As examples of JMP instruction, consider JMP FAR BEGIN (or some 8086 assembles use JMP FAR PTR BEGIN) unconditionally branches to a label BEGIN in a different code segment.

JMP PROG jumps to a label PROG in the same code segment. CALL instructions can be intersegment and intrasegment. For example, CALL DWORD PTR [BX] pushes CS and IP onto stack and loads IP and CS with the contents of four consecutive bytes pointed to by BX. CALL BX, on the other hand, pushes IP onto the stack, the new value of IP is loaded from BX and CS is unchanged.

9.5.6 Conditional Branch Instructions

All 8086 conditional branch instructions use 8-bit signed displacement. That is, the displacement covers a branch range of -128 to +127, with 0 being positive. The structure of a typical conditional branch instruction is as follows:

If condition is true,
then IP ← IP + disp8,
otherwise IP ← IP + 2 and execute next instruction.

There are two types of conditional branch instructions. In one type, the various relationships that exist between two numbers such as equal, above, below, less than, or greater than can be determined by the appropriate conditional branch instruction after a COMPARE instruction. These instructions can be used for both signed and unsigned numbers. When comparing signed numbers, terms such as "less than" and "greater than" are used. On the other hand, when comparing unsigned numbers, terms such as "below zero" or "above zero" are used.

Table 9.6 lists the 8086 signed and unsigned conditional branch instructions. Note that in Table 9.6 the instructions for checking which two numbers are "equal" or "not equal" are the same for both signed and unsigned numbers. This is because when two numbers are compared for equality, irrespective of whether they are signed or unsigned, they will provide a zero result (ZF = 1) if they are equal and a nonzero result (ZF = 0) if they are not equal. Therefore, the same instructions apply for both signed and unsigned numbers for "equal to" or "not equal to" conditions. The second type of conditional branch instructions is concerned with the setting of flags rather than the relationship between two numbers. Table 9.7 lists these instructions.

TABLE 9.6 8086 Signed and Unsigned Conditional Branch Instructions

Signed		*Unsigned*	
Name	*Alternate Name*	*Name*	*Alternate Name*
`JE` disp 8 (JUMP if equal)	`JZ` disp8 (JUMP if result zero)	`JE` disp8 (JUMP if equal)	`JZ` disp8 (JUMP if zero)
`JNE` disp8 (JUMP if not equal)	`JNZ` disp 8 (JUMP if not zero)	`JNE` disp8 (JUMP if not equal)	`JNZ` disp8 (JUMP if not zero)
`JG` disp8 (JUMP if greater)	`JNLE` disp8 (JUMP if not less or equal)	`JA` disp8 (JUMP if above)	`JNBE` disp8 (JUMP if not below or equal)
`JGE` disp8 (JUMP if greater or equal)	`JNL` disp8 (JUMP if not less)	`JAE` disp8 (JUMP if above or equal)	`JNB` disp8 (JUMP if not below)
`JL` disp8 (JUMP if less than)	`JNGE` disp8 (JUMP if not greater or equal)	`JB` disp8 (JUMP if below)	`JNAE` disp8 (JUMP if not above or equal)
`JLE` disp8 (JUMP if less or equal)	`JNG` disp8 (JUMP if not greater)	`JBE` disp8 (JUMP if below or equal)	`JNA` disp8 (JUMP if not above)

TABLE 9.7 8086 Conditional Branch Instructions Affecting Individual Flags

`JC` disp8	JUMP if carry, i.e., CF = 1
`JNC` disp8	JUMP if no carry, i.e., CF = 0
`JP` disp8	JUMP if parity, i.e., PF = 1
`JNP` disp8	JUMP if no parity. i.e., PF = 0
`JO` disp8	JUMP if overflow, i.e., OF = 1
`JNO` disp8	JUMP if no overflow, i.e., OF = 0
`JS` disp8	JUMP if sign, i.e., SF = 1
`JNS` disp8	JUMP if no sign, i.e.. SF = 0
`JZ` disp8	JUMP if result zero, i.e.. ZF = 1
`JNZ` disp8	JUMP if result not zero, i.e., ZF = 0

Now, in order to check whether the result of an arithmetic or logic operation is zero, nonzero, positive or negative, did or did not produce a carry, did or did not produce parity, or did or did not cause overflow, the following instructions should be used: `JZ`, `JNZ`, `JS`, `JNS`, `JC`, `JNC`, `JP`, `JNP`, `JO`, `JNO`. However, in order to compare two signed or unsigned numbers (a in address A or b in address B) for various conditions, we use `CMP A, B`, which will form $a - b$. and then one of the instructions in Table 9.8.

TABLE 9.8 8086 Instructions To Be Used after `CMP A, B`

Signed "a" and "b"		*Unsigned "a" and "b"*	
JGE disp8	if $a \geq b$	JAE disp8	if $a \geq b$
JL disp8	if $a < b$	JB disp8	if $a < b$
JG disp8	if $a > b$	JA disp8	if $a > b$
JLE disp8	if $a \leq b$	JBE disp8	if $a \leq b$

Now let us illustrate the concept of using the preceding signed or unsigned instructions by an example. Consider clearing a section of memory word starting at B up to and including A, where [A] = 3000_{16} and [B] = 2000_{16} in DS = 1000_{16}, using the following instruction sequence:

```
        MOV     AX, 1000H
        MOV     DS, AX                  ;Initialize DS
        MOV     BX, 2000H
        MOV     CX, 3000H
AGAIN:  MOV     WORD PTR[BX], 0000H
        INC     BX
        INC     BX
        CMP     CX, BX
        JGE     AGAIN
```

JGE treats CMP operands as twos complement numbers. The loop will terminate when BX = 3002H. Now, suppose that the contents of A and B are as follows:

$$[A] = 8500_{16}$$
$$[B] = 0500_{16}$$

In this case, after `CMP CX,BX` is first executed,

$$\begin{aligned}[CX] - [BX] &= 8500 - 0500 \\ &= 8000_{16} \\ &= 1000\ 0000\ 0000\ 0000\end{aligned}$$

↑
SF = 1, i.e., a negative number

Because 8000_{16} is a negative number, the loop terminates.

The correct approach is to use a branch instruction that treats operands as unsigned numbers (positive numbers) and uses the following instruction sequence:

```
        MOV     AX,1000H
        MOV     DS,AX           ;       initialize DS
        MOV     BX,2000H
        MOV     CX,3000H
AGAIN:  MOV     WORD PTR[BX],0000H
        INC     BX
        INC     BX
        CMP     CX,BX
        JAE     AGAIN
```

JAE will work regardless of the values of A and B.

Also, note that addresses are always positive numbers (unsigned). Hence, unsigned conditional jump instruction must be used to obtain correct answer. The above examples are shown for illustrative purposes.

TABLE 9.9 8086 Iteration Control Instructions

`LOOP` disp8	Decrement CX by 1 without affecting flags and LOOP if CX $\neq$ 0; otherwise, go to the next instruction.
`LOOPE/LOOPZ` disp8	Decrement CX by 1 without affecting flags and LOOP if CX not equal / not zero or ZF = 1; otherwise, go to the next instruction.
`LOOPNE/LOOPNZ` disp8	Decrement CX by 1 without affecting flags and LOOP if CX not equal / not zero or ZF = 0; otherwise, go to the next instruction.
`JCXZ` disp8	JMP if register CX =0.

TABLE 9.10 8086 Interrupt Instructions

`INT` interrupt number (can be 0-255_{10})	Software interrupt instructions (`INT` $32_{10} - 255_{10}$ available to the user.)
`INTO`	Interrupt on overflow
`IRET`	Interrupt return

9.5.7 Iteration Control Instructions

Table 9.9 lists *iteration control instructions.* In Table 9.9, `LOOP` disp8 decrements the CX register by 1 without affecting the flags and then acts in the same way as the `JMP` dsp8 instruction except that if CX $\neq$ 0, then the `JMP` is performed: otherwise, the next instruction is executed.

9.5.8 Interrupt Instructions

Table 9.10 shows the *interrupt instructions.*

9.5.9 Processor Control Instructions

Table 9.11 shows the *processor control functions.* Let us explain some of the instructions in Table 9.11.

- `ESC` mem places the contents of the specified memory location on the data bus at the time when the 8086 ready pin is asserted by the addressed memory device. This instruction allows the 8086 to invoke other processors such as the 8087 floating-point processor.

- `LOCK` outputs a LOW on the LOCK pin of the 8086 for the duration of the next instruction. This signal is used in multiprocessing.
- `WAIT` causes the 8086 to enter an idle state if the signal on the $\overline{TEST}$ pin is not asserted. By placing the `WAIT` before the `ESC`, the 8086 can do other things while a subordinate processor such as a floating-point coprocessor is executing an instruction.

TABLE 9.11 8086 Processor Control Instructions

STC	Set carry CF ← 1
CLC	Clear carry CF ← 0
CMC	Complement carry, CF ← $\overline{CF}$
STD	Set direction flag
CLD	Clear direction flag
STI	Set interrupt enable flag
CLI	Clear interrupt enable flag
NOP	No operation
HLT	Halt
WAIT	Wait for $\overline{TEST}$ pin active
ESC mem	Escape to external processor
LOCK	Lock bus during next instruction

9.6 8086 Assembler-Dependent Instructions

Some 8086 instructions do not define whether an 8-bit or a 16-bit operation is to be executed. Instructions with one of the 8086 registers as an operand typically define the operation as 8-bit or 16-bit based on the register size. An example is `MOV CL, [BX]`, which moves an 8-bit number with the offset defined by [BX] in DS into register CL; `MOV CX, [BX]`, on the other hand, moves a 16-bit number from offsets [BX] and [BX + 1] in DS into CX

Instructions with a single-memory operand may define an 8-bit or a 16-bit operation by adding B for byte or W for word with the mnemonic. Typical examples are `MULB [BX]` and `CMPW [ADDR]`. The string instructions may define this in two ways. Typical examples are `MOVSB` or `MOVS BYTE` for 8-bit and `MOVSW` or `MOVS WORD` for 16-bit. Memory offsets can also be specified by including `BYTE PTR` for 8-bit and `WORD PTR` for 16-bit with the instruction. Typical examples are `INC BYTE PTR [BX]` and `INC WORD PTR [BX]`.

9.7 Typical 8086 Assembler Pseudo-Instructions or Directives

One of the requirements of typical 8086 assemblers such as MASM (discussed later) is that a variable's type must be declared as a byte (8-bit), word (16-bit), or double word (4 bytes or 2 words) before using the variable in a program. Some examples are as follows:

`BEGIN DB 0`	BEGIN is declared as a byte offset with contents zero.
`START DW 25F1H`	START is declared as a word offset with contents 25F1H.
`PROG DD 0`	PROG is declared as a double word (4 bytes) offset with zero contents.

Note that the directive `DD` is not used by all assemblers. In that case, one should use the directive DW twice to declare a 32-bit offset.

The `EQU` directive can be used to assign a name to constants. For example, the statement `NUMB EQU 21H` directs the assembler to assign the value 21H every time it finds NUMB in the program. This means that the assembler reads the statement `MOV BH, NUMB` as `MOV BH, 21H`. As mentioned before, `DB`, `DW`, and `DD` are the directives used to assign names and specific data types for variables in a program. For example, after execution of the statement `ADDR DW 2050H` the assembler assigns 50H to the offset name ADDR and 20H to the offset name ADDR + 1. This means that the program can use the instruction `MOV BX, [ADDR]` to load the 16-bit contents of memory starting at the offset ADDR in DS into BX. The `DW` sets aside storage for a word in memory and gives the starting address of this word the name ADDR.

As an example, consider 16 × 16 multiplication. The size of the product should be 32 bits and must be initialized to zero. The following will accomplish this:

```
Multiplicand    DW 2A05H
Multiplier      DW 052AH
Product         DD 0
```

Some versions of MASM assembler use directive `AT` to assign a value to an 8086 segment.

The 8086 addressing mode examples for the typical assemblers are given next:

`MOV AH, BL`	Both source and destination are in register mode.
`MOV CH, 8`	Source is in immediate mode and destination is in register mode.
`MOV AX, [START]`	Source is in memory direct mode and destination is in register mode.
`MOV CH, [BX]`	Source is in register indirect mode and destination is in register mode.
`MOV [SI], AL`	Source is in register mode and destination is in register indirect mode.
`MOV [Dl], BH`	Source is in register mode and destination is in register indirect mode.
`MOV BH, VALUE [Dl]`	Source is in register indirect with displacement mode and destination is in register mode. VALUE is typically defined by the `EQU` directive prior to this instruction.
`MOV AX, 4[DI]`	Source is in indexed with displacement mode and destination is in register mode.
`MOV SI, 2[BP] [DI]`	Source is in based indexed with displacement mode and destination is in register mode.
`OUT 30H, AL`	Source is in register mode and destination is in direct port mode.
`IN AX, DX`	Source is in indirect port mode and destination is in register mode.

In the following paragraphs, more assembler directives such as `SEGMENT`, `ENDS`, `ASSUME`, and `DUP` will be discussed.

9.7.1 SEGMENT and ENDS Directives

A section of a an 8086 program or a data array can be defined by the *SEGMENT and ENDS directives* as follows:

```
START   SEGMENT
X1      DB    0F1H
X2      DB    50H
X3      DB    25H
START   ENDS
```

The segment name is START (arbitrarily chosen). The assembler will assign a numeric value to START corresponding to the base value of the data segment. The programmer must use the 8086 instructions to load START into DS as follows:

```
MOV BX, START
MOV DS, BX
```

Note that the segment registers must be loaded via a 16-bit general purpose register such as BX. A data array or an instruction sequence between the SEGMENT and ENDS directives is called a *logical segment.* These two directives are used to set up a logical segment with a specific name. A typical assembler allows one to use up to 31 characters for the name without any spaces. An underscore is sometimes used to separate words in a name, for example, PROGRAM_BEGIN.

9.7.2 ASSUME Directive

As mentioned before, at any time the 8086 can directly address four physical segments, which include a code segment, a data segment, a stack segment, and an extra segment. The 8086 may contain a number of logical segments containing codes, data, and stack. The ASSUME *directive* assigns a logical segment to a physical segment at any given time. That is, the ASSUME directive tells the assembler what addresses will be in the segment registers at execution time.
For example, the statement ASSUME CS: PROGRAM_1,DS: DATA_1,SS: STACK_1 directs the assembler to use the logical code segment PROGRAM _1 as CS, containing the instructions, the logical data segment DATA_1 as DS, containing data, and the logical stack segment STACK _1 as SS, containing the stack.

9.7.3 DUP Directive

The DUP *directive* can be used to initialize several locations to zero. For example, the statement START DW 4 DUP (0) reserves four words starting at the offset START in DS and initializes them to zero. The DUP directive can also be used to reserve several locations that need not be initialized. A question mark must be used with DUP in this case. For example, the statement BEGIN DB 100 DUP (?) reserves 100 bytes of uninitialized data space to an offset BEGIN in DS. Note that BEGIN should be typed in the label field, DB in the OP code field, and 100 DUP (?) in the operand field.

A typical example illustrating the use of these directives is given next:

```
DATA_1      SEGMENT
ADDR_1      DW 3005H
ADDR_2      DW 2003H
```

```
DATA_1        ENDS
STACK_1       SEGMENT
              DW 60 DUP (0)                      ;     Assign 60 words
                                                 ;     of stack with zeros.
STACK_TOP     LABEL WORD                         ;     Initialize stack
                                                 ;     top to next location
STACK_1       ENDS                               ;     after top of stack.
CODE_1        SEGMENT
              ASSUME CS: CODE_1, DS: DATA_1,  SS: STACK_1
              MOV AX, STACK_1
              MOV SS, AX
              LEA SP, STACK_TOP
              MOV AX, DATA_1
              MOV DS, AX
              LEA SI, ADDR_1
              LEA DI, ADDR_2
                    -      ←      Main program
                    -      ←      body
CODE_1        ENDS  -
```

Note that LABEL is a directive used to initialize STACK_TOP to the next location after the top of the stack. The statement STACK_TOP LABEL WORD gives the name STACK_TOP to the next address after the *60* words are set aside for the stack. The WORD in this statement indicates that PUSH into and POP from the stack are done as words.

Also note that in the above, ASSUME directive tells the assembler to use the logical segment names CODE_1, DATA_1, and STACK_1 as the code segment, data segment, and stack segment, respectively. The extra segment can be assigned a name in a similar manner. When the instructions are executed, the displacements in the instructions along with the segment register contents are used by the assembler to generate the 20-bit physical addresses. The segment register, other than the code segment, must be initialized before it is used to access data. The code segment is typically initialized upon hardware reset or by using ORG.

When the assembler translates an assembly language program, it computes the displacement, or offset, of each instruction code byte from the start of a logical segment that contains it. For example, in the preceding program, the CS: CODE_1 in the ASSUME statement directs the assembler to compute the offsets or displacements by the following instructions from the start of the logical segment CODE_1. This means that when the program is run, the CS will contain the 16-bit value where the logical segment CODE_1 is located in memory. The assembler keeps track of the instruction byte displacements, which are loaded into IP. The 20-bit physical address generated from CS and IP are used to fetch each instruction. Some versions of MASM use directive AT to assign a segment value. Another example to store data bytes in a data segment and to allocate stack is as follows:

```
DSEG        SEGMENT
ARRAY       DB 02H,   0F1H.  0A2H    ;    Store 3 bytes of
DSEG        ENDS                     ;    data in.an ad-
                                     ;    dress defined by
                                     ;    DSEG as DS and
                                     ;    ARRAY as off-
                                     ;    set
SSEG        SEGMENT
            DW 10 DUP (0)            ;    Allocate
                                     ;    10-word stack
STACK _TOP  LABEL WORD               ;    Label initializes
SSEG        ENDS                     ;    top of stack
            -
            -
            -
            MOV AX,  DSEG            ;    Initialize
            MOV DS, AX               ;    DS
            MOV AX, SSEG             ;    Initialize
            MOV SS, AX               ;    SS
            MOV SP,  STACK_TOP       ;    Initialize SP
            -
            -
            -
```

Note that typical 8086 assemblers such as Microsoft and Hewlett-Packard HP64000 use the `ORG` directive to load CS and IP. For example, CS and IP can be initialized with 2000H and 0300H as follows:

For Microsoft 8086 Assembler `ORG 20000300H`
For HP64000 8086 Assembler `ORG 2000H:0300H`

9.8 8086 Assembly Language Program Development Using Microsoft MASM and DEBUG

9.8.1 MASM Assembler

The MASM assembler program developed by Microsoft can be used to translate an 8086 assembly language program into machine language. The 8086 assembly programs are typically developed using a text editor. Each program is then assembled by typing MASM followed by the assembly file which is saved with an extension .ASM. When the file is assembled, MASM creates an .OBJ file, which is used to link the program. By typing `LINK` followed by the file with the extension .OBJ, an executable file is created. This executable file can then be used for DEBUG.

9.8.2 DEBUG

DEBUG is a program included in MS-DOS that enables a programmer to enter and run assembly programs. More specifically, DEBUG allows a programmer to closely observe the immediate results in memory and registers after execution of 8086 instructions. DEBUG allows the programmer the ability to stop programs at specific points during the program execution.

To use the DEBUG program, one needs to type the word "`debug`" at the DOS prompt. Entry into the DEBUG program is denoted by a dash (–) prompt:

`C:\>debug` <return>

At the – prompt, several basic commands can be used throughout programming. First, the command "`a`" is used to begin "assembling" the code. Typing "`a`" at the dash prompt and then pressing <return> results in a new prompt showing the address in memory at which the code is written. An example of this would be: "`103D:0100`", where 103D is the code segment (CS) and 0100 is the offset. The "`a`" command can be used with an offset address as a parameter; however, if no offset is given, one will be assumed.

The following is a simple list of other commands that are typically used during our DEBUG programming:

`u` to "Unassemble"
`f` to "Fill a block of memory"
`d` to "Dump contents of memory"
`e` to "Enter data into memory"
`r` to "view Register contents"
`t` to "Trace"
`g` to "Go"
`q` to "Quit DEBUG"

When writing assembly code in DEBUG, the first parameter following an instruction is the destination, the second is the source. For example, consider `MOV AL, 57`. In the example, the hexadecimal number 57 (source) will be copied to AL (destination), where AL is the low byte of the 16-bit AX register. **Note that in DEBUG, the letter H does not need to be used with a number to define it as a hex number.**

Another convention used in DEBUG is the "Little Endian" convention. This means that when bytes are moved into a register from memory, the bytes are moved from low byte to low byte and high byte to high byte. The same holds true for when data is moved from a register to memory. The following example illustrates the Little Endian convention:

Code:

```
MOV BX,1234
MOV [0200],BX
INT 3
```

Result:

```
CS:0200 34 12
```

What this example shows is that the contents of BL (BL = 34) are moved into the lower byte (offset address 0200) and the contents of BH (BH = 12) are moved into the higher byte (offset address 0201). The result is that while 16-bit register BX contains 1234, offset addresses 0200 and 0201 will contain 3412.

Programs entered using DEBUG can easily be saved in a disk by using a .COM file. The first step is to determine the number of bytes required by the program. The easiest way is to use the offset address for the last line of code (usually the last instruction will be INT 3), and then add 1 to it. This number is stored in register CX using the "r" command so that DEBUG knows how many bytes need to be written. The following is an example of this procedure for a program that requires 12 bytes.

```
-r CX
CX 0000
:12
-n filename.com
-w
Writing 0012 bytes
```

The "n" command is used to give a filename, where filename.com can be any specified filename. The "w" command begins the process of writing to the specified filename. In order to load a program previously created, the "n" command can be used with the filename as the parameter, then pressing <return>; finally, instead of using the "w" (write command), the command "l" (small L) to "Load" the program can be used. By default, the code is written to address CS:0100.

DEBUG is used to determine whether the assembly program is working correctly or not. As mentioned before, once in DEBUG, several commands can be used to test each program at the command prompt (–). Let us explain these commands in detail.

The "u" command is used first to unassemble the program. Once unassembled, both the machine code and the assembly code are displayed with their associated address values. The address values are displayed with the segment followed by the offset.

The "d" command followed by the address is used to dump the data associated with that address. This command is useful in determining the data values in that particular address before the program. Upon executing the program, the dump command, "d", can be used to see whether or not the data at that address location has been modified.

The "g" command is used to execute the program in DEBUG. In order for DEBUG to know when to stop, an "int 3" is used in the assembly program. This command automatically informs DEBUG to stop when it reaches that particular point. Once DEBUG finishes executing the program, all the registers and flags will be displayed. This information can be used to debug the program.

As an alternative, the "t", or "trace," command can be used to execute the program one line at a time. Each time a code is executed, the associated registers and flags are displayed. Also, data stored at a particular address is displayed.

9.8.3 Steps for Developing 8086 Assembly Language Programs Using MASM and DEBUG

The MASM command is used to assemble an 8086 assembly program. One should do the following:

1. Type the 8086 assembly program in any text editor program and save it.
2. After typing the 8086 assembly program, type:
   `MASM /c /la <filename.asm>`
3. Type `LINK <filename.obj>` to save the program as an executable file (filename.exe).

MASM can be used to assemble an 8086 assembly program, and DEBUG can be used to verify the results. As mentioned before, the DEBUG can be used to load and run an 8086 assembly language program using the following steps:

1. Type in "debug <filename.exe> " or just "debug" at the DOS prompt.
2. Press <return>. Entry into the DEBUG program is denoted by the (-) prompt. At this dash prompt, several debug commands can be used.
3. Type in the command "a" after the dash prompt.
4. Type in the 8086 assembly program. (if the assembly language program is not available.)

Using DEBUG, the program can be unassembled by using the "-u <starting address> <ending address>" command to display the unassembled code. To run the program, the command "-g <starting address> <ending address>" is typed. This command will run the program from the starting offset value to an interrupt "INT 3" instruction. At this point, the program will stop and display or print the contents of 8086 registers. If the results are stored in memory, the results can be dumped by typing "-d <starting address> <ending address>." If the program has any bugs in it, one can troubleshoot the program in typically two ways.

First, one can insert an "`INT 3`" instruction, which is the breakpoint. The `INT 3` tells the DEBUG program to stop. At this point, one can check the 8086 registers or memory locations to check the results.

The second way is to use the trace command "`-t` <starting address> <number of instructions>", which will allow one to step through the program and check the contents of all registers and memory locations after execution of each 8086 instruction. To exit DEBUG, type in "q" for "quit" and then press <return>. Finally, note that at the end of some programs the following instructions are used:

```
MOV AH, 4CH
INT 21H
```

This instruction sequence returns control to the DOS operating system.

In order to illustrate the use of MASM, consider the following 8086 assembly program:

```
PROG  SEGMENT
      ASSUME   CS:PROG
      MOV      CL,3
      ADD      CL,2
      HLT
PROG  ENDS
      END
```

Assume that the MASM assembler is located in the DOS directory on the C drive. At the **C:\>** prompt, type `edit` and press Enter. Type in the above program, then save it with a file name (**test.asm**, for example). Exit the editor, and type `masm /c /la test.asm` at the DOS prompt and press Enter. To link the program, type `link test.obj` to generate an executable file, test.exe. To display the assembled code, type at the DOS prompt **edit test.lst**, then press Enter.

```
Microsoft (R) Macro Assembler Version 5.10                 1/7/99 20:59:08

      1 0000            PROG     SEGMENT
      2                          ASSUME  CS:PROG
      3 0000  B1 03              MOV     CL,3
      4 0002  80 C1 02           ADD     CL,2
      5 0005  F4                 HLT
      6 0006            PROG     ENDS
      7                          END
```

The program can be run using DEBUG.

Example 9.1

Determine the effect of each of the following 8086 instructions:

1. `DIV CH`
2. `CBW`
3. `MOVSW`

Assume the following data prior to execution of each of these instructions independently (assume that all numbers are in hexadecimal):

> [DS] = 2000H, [ES] = 4000H, [CX] = 0300H, [AX] = 0091H, [20300H] = 05H, [20301H] = 02H, [40200H] = 06H, [40201H] = 07H, [SI] = 0300H, [DI] = 0200H, DF = 0

Solution

1. Before unsigned division, CH contains 03_{10} and AX contains 145_{10}. Therefore, after `DIV CH`, [AH] = remainder = 01H and [AL] = quotient = 48_{10} = 30H.

2. `CBW` sign-extends the AL register into the AH register. Because the content of AL is 91H, the sign bit is 1. Therefore, after `CBW`, [AX] = FF91H

3. Before `MOVSW`,

Source String	Destination String
[SI] = 0300H, [DS] = 2000H	[DI] = 0200H, [ES] = 4000H
Physical address = 20300H	Physical address = 40200H

After `MOVSW`, [40200H] = 05H, [40201H] = 02H. Because DF = 0, [SI] = 0302H, [DI] = 0202H

Example 9.2

Write an 8086 instruction sequence to move 8-bit data from memory offset 200_{16}–$27F_{16}$ to 300_{16}–$37F_{16}$. Verify the results using DEBUG.

Solution

```
-a
36A1:0100 MOV SI,0200
36A1:0103 MOV DI,0300
36A1:0106 MOV CX,80
36A1:0108 MOV AL, [SI]
36A1:010A MOV [DI],AL
36A1:010C INC SI
36A1:010D INC DI
36A1:010E LOOP 0108
36A1:0110 INT 3
36A1:0111
-u 100 110
36A1:0100 BE0002        MOV     SI,0200
36A1:0103 BF0003        MOV     DI,0300
36A1:0106 B980          MOV     CX,80
36A1:0108 8A04          MOV     AL,[SI]
36A1:010A 8805          MOV     [DI],AL
```

```
36A1:010C 46            INC     SI
36A1:010D 47            INC     DI
36A1:010E E2F8          LOOP    0108
36A1:0110 CC            INT     3
```

Output: ***(Program copies data from memory offset 200–27F to 300–37F)***

```
-g
AX=0050  BX=0000  CX=0000  DX=0000  SP=FFEE  BP=0000  SI=0280  DI=0380
DS=36A1  ES=36A1  SS=36A1  CS=36A1  IP=0110   NV UP EI PL NZ AC PO NC
36A1:0110 CC            INT     3
-d 200
36A1:0200  10 50 E8 06 17 83 C4 02-E8 C1 13 B8 F6 0F 50 E8
36A1:0210  F9 16 83 C4 02 FF 0E F8-0F 83 3E F8 0F 00 7C 0E
36A1:0220  8B 1E F6 0F FF 06 F6 0F-8A 07 2A E4 EB 11 B8 F6
36A1:0230  0F 50 E8 0D 15 83 C4 02-EB 05 C6 06 8A 58 00 B8
36A1:0240  08 00 50 B8 73 27 50 E8-83 DD 83 C4 04 80 3E 8A
36A1:0250  58 00 75 03 E9 B2 00 B8-00 10 50 B8 28 0E 50 E8
36A1:0260  72 19 83 C4 04 B8 00 10-50 B8 49 0E 50 E8 64 19
36A1:0270  83 C4 04 A0 EC 5D 98 50-B8 4C 0E 50 B8 00 10 50
-d 300
36A1:0300  10 50 E8 06 17 83 C4 02-E8 C1 13 B8 F6 0F 50 E8
36A1:0310  F9 16 83 C4 02 FF 0E F8-0F 83 3E F8 0F 00 7C 0E
36A1:0320  8B 1E F6 0F FF 06 F6 0F-8A 07 2A E4 EB 11 B8 F6
36A1:0330  0F 50 E8 0D 15 83 C4 02-EB 05 C6 06 8A 58 00 B8
36A1:0340  08 00 50 B8 73 27 50 E8-83 DD 83 C4 04 80 3E 8A
36A1:0350  58 00 75 03 E9 B2 00 B8-00 10 50 B8 28 0E 50 E8
36A1:0360  72 19 83 C4 04 B8 00 10-50 B8 49 0E 50 E8 64 19
36A1:0370  83 C4 04 A0 EC 5D 98 50-B8 4C 0E 50 B8 00 10 50
-
```

Example 9.3

Write an 8086 assembly language program to add two 16-bit numbers in CX and DX and store the result in location 0500H addressed by DI.

Solution

```
1  0000                   DATA   SEGMENT
2  0004                   DATA   ENDS
3  0000                   CODE   SEGMENT
4                         ASSUME CS:CODE,DS:DATA
5  0000  B8 ---- R               MOV   AX,DATA     ;    Initialize DS
6  0003  8B D0
7  0005  8E D8                   MOV   DS,AX
8  0007  BF 0500                 MOV   DI,0500H
9  000A  03 CA                   ADD   CX,DX       ;    Add
10 000C  89 0D                   MOV   [DI],CX     ;    Store
11 000E  F4                      HLT
12 000F                   CODE   ENDS
13                               END
```

```
Microsoft (R) Macro Assembler Version 5.10                    7/2/98 20:58:07
                                                               Symbols-1
Segments and Groups:
Name                                  Length      Align       Combine Class

CODE . . . . . . . . . . . . . .      000F        PARA        NONE
DATA . . . . . . . . . . . . . .      0000        PARA        NONE

Symbols:

Name                                  Type        Value       Attr

@CPU . . . . . . . . . . . . . .      TEXT        0101h
@FILENAME  . . . . . . . . . . .      TEXT        EX9_3
@VERSION . . . . . . . . . . . .      TEXT        510

     13 Source  Lines
     13 Total   Lines
      7 Symbols

  47176 + 271092 Bytes symbol space free

      0 Warning Errors
      0 Severe  Errors
```

Example 9.4

Write an 8086 assembly language program to add two 64-bit numbers. Assume SI and DI contain the starting address of the numbers. Store the result in memory pointed to by [DI].

Solution

```
1  0000                   DATA_ARRAY   SEGMENT
2  0000  0A71             DATA1        DW     0A71H       ;     DATA1 low
3  0002  F218                          DW     0F218H
4  0004  2F17                          DW     2F17H       ;     DATA1 high
5  0006  6200                          DW     6200H
6  0008  7A24             DATA2        DW     7A24H       ;     DATA2 low
7  000A  1601                          DW     1601H
8  000C  152A                          DW     152AH       ;     DATA2 high
9  000E  671F                          DW     671FH
10 0010                   DATA_ARRAY   ENDS
11 0000                   PROG_CODE    SEGMENT
12                                     ASSUME CS:PROG_CODE, DS:DATA_ARRAY
13 0000  B8 ---- R                     MOV    AX,DATA_ARRAY
14 0003  8E D8                         MOV    DS,AX       ;     Initialize DS
15 0005  BA 0004                       MOV    DX,4        ;     Load 4 into DX
```

```
16 0008  BB 0002                        MOV   BX,2         ;    Initialize BX
17 000B  8B 36 0000 R                   MOV   SI,DATA1     ;    Initialize SI
18 000F  8B 3E 0008 R                   MOV   DI,DATA2     ;    Initialize DI
19 0013  F8                             CLC                ;    Clear Carry
20 0014  8B 04             START:       MOV   AX,[SI]      ;    Load DATA1
21 0016  11 05                          ADC   [DI],AX      ;    Add with carry
22 0018  03 F3                          ADD   SI,BX        ;    Update
23 001A  03 FB                          ADD   DI,BX        ;    pointers
24 001C  4A                             DEC   DX           ;    decrement
25 001D  75 F7                          JNZ   START        ;    branch
26 001F  F4                             HLT
27 0020                    PROG_CODE    ENDS
28                                      END
```

```
Microsoft (R) Macro Assembler Version 5.10              7/2/98 21:22:30
                                                         Symbols-1
Segments and Groups:
Name                                 Length     Align      Combine Class
DATA_ARRAY . . . . . . . . . . .     0010       PARA       NONE
PROG_CODE  . . . . . . . . . . .     0020       PARA       NONE
Symbols:
Name                                 Type       Value      Attr
DATA1  . . . . . . . . . . . . .     L WORD 0000  DATA_ARRAY
DATA2  . . . . . . . . . . . . .     L WORD 0008  DATA_ARRAY
START  . . . . . . . . . . . . .     L NEAR 0014  PROG_CODE
@CPU . . . . . . . . . . . . . .     TEXT  0101h
@FILENAME  . . . . . . . . . . .     TEXT  EX9_4
@VERSION . . . . . . . . . . . .     TEXT  510
     28 Source  Lines
     28 Total   Lines
     10 Symbols
  47138 + 271130 Bytes symbol space free
      0 Warning Errors
      0 Severe  Errors
```

Example 9.5

Write an 8086 assembly language program to multiply two 16-bit unsigned numbers to provide a 32-bit result. Assume that the two numbers are stored in CX and DX.

Solution

```
1 0000             CODE_SEG    SEGMENT
2                              ASSUME CS:CODE_SEG
3 0000   8B C2                 MOV    AX,DX        ;    Move first data
4 0002   F7 E1                 MUL    CX           ;    [DX][AX]<--[AX]*[CX]
5 0004   F4                    HLT
6 0005             CODE_SEG    ENDS
7                              END
```

```
Microsoft (R) Macro Assembler Version 5.10              7/2/98 23:25:00
                                                         Symbols-1
```

```
Segments and Groups:

Name                              Length      Align       Combine Class

CODE_SEG . . . . . . . . . . . .  0005        PARA        NONE

Symbols:

Name                              Type        Value  Attr

@CPU . . . . . . . . . . . . . .  TEXT        0101h
@FILENAME  . . . . . . . . . . .  TEXT        EX9_5
@VERSION . . . . . . . . . . . .  TEXT        510


      7 Source  Lines
      7 Total   Lines
      6 Symbols

  47216 + 271052 Bytes symbol space free

      0 Warning Errors

      0 Severe  Errors
```

Example 9.6

Write an 8086 assembly language program to clear 50_{10} consecutive bytes starting at offset 1000H. Assume DS is already initialized.

Solution

```
1  = 1000              ADDR      EQU    1000H
2  0000                CODE_SEG  SEGMENT
3                                ASSUME    CS:CODE_SEG,DS:DATA_SEG
4  0000  BB 1000                 MOV    BX,ADDR             ;  Initialize BX
5  0003  B9 0032                 MOV    CX,50               ;  Initialize loop count
6  0006  C6 07 00      START:    MOV    BYTE PTR[BX],00H    ;  Clear memory byte
7  0009  43                      INC    BX                  ;  Update pointer
8  000A  E2 FA                   LOOP   START               ;  Decrement CX and loop
9  000C  F4                      HLT
10 000D                CODE_SEG  ENDS
11 0000                DATA_SEG  SEGMENT
12 0000                DATA_SEG  ENDS
13                               END
```

```
Microsoft (R) Macro Assembler Version 5.10                    7/3/98 10:54:05
                                                               Symbols-1
Segments and Groups:
```

```
Name                                   Length    Align     Combine Class

CODE_SEG . . . . . . . . . . . . .     000D      PARA      NONE
DATA_SEG . . . . . . . . . . . . .     0000      PARA      NONE

Symbols:

Name                                   Type      Value     Attr

ADDR . . . . . . . . . . . . . . .     NUMBER    1000

START  . . . . . . . . . . . . . .     L NEAR    0006      CODE_SEG
@CPU . . . . . . . . . . . . . . .     TEXT      0101h
@FILENAME  . . . . . . . . . . . .     TEXT      ex9_6
@VERSION . . . . . . . . . . . . .     TEXT      510

     13 Source  Lines
     13 Total   Lines
      9 Symbols

  47106 + 271482 Bytes symbol space free

      0 Warning Errors
      0 Severe  Errors
```

Example 9.7

Write an 8086 assembly program at CS=1000H to compute $\sum_{i=1}^{100} X_i Y_i$ where X_i and Y_i are signed 8-bit numbers stored at stored at offsets 4000H and 5000H respectively. Initialize DS at 2000H. Store 16-bit result in DX. Assume no overflow.

Solution

```
1  0000               CODE   SEGMENT AT 1000H
2                            ASSUME CS:CODE,DS:DATA
3  0000  B8 2000              MOV    AX,2000H         ;  Initialize
4  0003  8E D8                MOV    DS,AX            ;  Data Segment
5  0005  B9 0064              MOV    CX,100           ;  Initialize loop count
6  0008  BB 4000              MOV    BX,4000H         ;  Initialize pointer of Xi
7  000B  BE 5000              MOV    SI,5000H         ;  Initialize pointer of Yi
8  000E  BA 0000              MOV    DX,0000H         ;  Initialize sum to zero
9  0011  8A 07        START: MOV    AL,[BX]          ;  Load data into AL
10 0013  F6 2C                IMUL   BYTE PTR [SI]    ;  Signed 8x8 multiplication
11 0015  03 D0                ADD    DX,AX            ;  Sum XiYi
12 0017  43                   INC    BX               ;  Update pointer
13 0018  46                   INC    SI               ;  Update pointer
14 0019  E2 F6                LOOP   START            ;  Decrement CX & loop
15 001B  F4                   HLT
16 001C               CODE   ENDS
17 0000               DATA   SEGMENT AT 2000H
18 0000               DATA   ENDS
19                            END                     ;  End program
```

```
Microsoft (R) Macro Assembler Version 5.10                    7/3/98 11:23:33
                                                               Symbols-1
Segments and Groups:
Name                                Length    Align     Combine Class
CODE . . . . . . . . . . . . . . .  001B      AT        1000
DATA . . . . . . . . . . . . . . .  0000      AT        2000

Symbols:
Name                                Type      Value     Attr
ADDR1  . . . . . . . . . . . . . .  L WORD    0000      DATA
ADDR2  . . . . . . . . . . . . . .  L WORD    0002      DATA
START  . . . . . . . . . . . . . .  L NEAR    0013      CODE
@CPU . . . . . . . . . . . . . . .  TEXT      0101h
@FILENAME  . . . . . . . . . . . .  TEXT      ex9_7
@VERSION . . . . . . . . . . . . .  TEXT      510

    20 Source  Lines
    20 Total   Lines
    10 Symbols

  47104 + 271484 Bytes symbol space free

     0 Warning Errors
     0 Severe  Errors
```

Example 9.8

Write an 8086 assembly language program to add two words; each contains two ASCII digits. The first word is stored in two consecutive locations with the low byte pointed to by SI at offset 0300H, while the second byte is stored in two consecutive locations with the low byte pointed to by DI at offset 0700H. Store the unpacked BCD result in memory location pointed to by DI. Assume that each unpacked BCD result of addition is less than or equal to 09H.

Solution

```
1  0000              CODE   SEGMENT AT 1000H
2                           ASSUME CS:CODE,DS:DATA
3  0000  B8 2000            MOV    AX,2000H  ;  Initialize Data Segment
4  0003  8E D8              MOV    DS,AX     ;  at 2000H (arbitrarily chosen)
5  0005  B9 0002            MOV    CX,2      ;  Initialize loop count
6  0008  BE 0300            MOV    SI,0300H  ;  Initialize SI
7  000B  BF 0300            MOV    DI,0700H  ;  Initialize DI
8  000E  8A 04       START: MOV    AL,[SI]   ;  Load data into AL
9  0010  02 05              ADD    AL,[DI]   ;  Perform addition
10 0012  37                 AAA              ;  ASCII adjust
11 0013  88 05              MOV    [DI],AL   ;  Store result
12 0015  46                 INC    SI        ;  Update pointer
13 0016  47                 INC    DI        ;  Update pointer
14 0017  E2 F5              LOOP   START     ;  Decrement CX & loop
```

```
15 0019  F4              HLT              ;  Halt
16 001A           CODE   ENDS
17 0000           DATA   SEGMENT AT 2000H
18 0000           DATA   ENDS
19                       END              ;  End program
```

```
Microsoft (R) Macro Assembler Version 5.10                  7/3/98 11:52:03
                                                              Symbols-1
Segments and Groups:
Name                                   Length    Align    Combine Class
CODE . . . . . . . . . . . . . . .     001A      AT       1000
DATA . . . . . . . . . . . . . . .     0000      AT       2000

Symbols:
Name                                   Type      Value    Attr
START  . . . . . . . . . . . . . .     L NEAR    000E     CODE
@CPU . . . . . . . . . . . . . . .     TEXT      0101h
@FILENAME  . . . . . . . . . . . .     TEXT      ex9_8
@VERSION . . . . . . . . . . . . .     TEXT      510

     20 Source  Lines
     20 Total   Lines
      8 Symbols

  47144 + 271444 Bytes symbol space free

      0 Warning Errors

      0 Severe  Errors
```

Example 9.9

Write an 8086 assembly language program at CS=5000H to compare a source string of 50_{10} words pointed to by an offset of 1000H in the data segment at 2000H with a destination string pointed to by an offset 3000H in another segment at 4000H. The program should be halted as soon as a match is found or the end of string is reached.

Solution

```
1 0000                  CODE   SEGMENT AT 5000H
2                              ASSUME CS:CODE,DS:DATA,ES:DATA1
3  0000  B8 2000               MOV    AX,2000H  ;  Initialize
4  0003  8E D8                 MOV    DS,AX     ;  Data Segment at 2000H
5  0005  B8 4000               MOV    AX,4000H  ;  Initialize
6  0008  8E C0                 MOV    ES,AX     ;  ES at 4000H
7  000A  BE 1000               MOV    SI,1000H  ;  Initialize SI at 1000H
8  000D  BF 3000               MOV    DI,3000H  ;  Initialize DI at 3000H
9  0010  B9 0032               MOV    CX,50     ;  Initialize CX
10 0013  FC                    CLD              ;  Clear DF so that
11                                              ;  SI and DI will
12                                              ;  autoincrement
13                                              ;  after compare
14 0014  F2/ A7                REPNE  CMPSW     ;  Repeat CMPSW until CX=0 or
```

```
15                                          ;   until compared words are equal
16 0016  F4                 HLT             ;   Halt
17 0017              CODE   ENDS
18 0000              DATA   SEGMENT AT 2000H
19 0000              DATA   ENDS
20 0000              DATA1  SEGMENT AT 4000H
21 0000              DATA1  ENDS
22                          END             ;   End program
```

```
Microsoft (R) Macro Assembler Version 5.10                    7/3/98 12:14:11
                                                               Symbols-1
Segments and Groups:
Name                                    Length      Align       Combine Class
CODE . . . . . . . . . . . . . . .      0017        AT          5000
DATA . . . . . . . . . . . . . . .      0000        AT          2000
DATA1 . . . . . . . . . . . . . . .     0000        AT          4000
Symbols:
Name                                    Type        Value       Attr
@CPU . . . . . . . . . . . . . . .      TEXT        0101h
@FILENAME  . . . . . . . . . . . .      TEXT        ex9_9
@VERSION . . . . . . . . . . . . .      TEXT        510

     22 Source  Lines
     22 Total   Lines
      7 Symbols

  47220 + 271368 Bytes symbol space free

      0 Warning Errors
      0 Severe  Errors
```

Example 9.10

Write a subroutine in 8086 assembly language at CS=7000H which can be called by a main program in a different CS=1000H. The subroutine will multiply a signed 16-bit number in CX by a signed 8-bit number in AL. The main program will perform initializations (DS to 5000H, SS to 6000H, SP to 0020H and BX to 2000H), call this subroutine, store the result in two consecutive memory words, and stop. Assume SI and DI are already initialized and contain pointers to the signed 8-bit and 16-bit data respectively.Store 32-bit result pointed to by BX.

Solution

```
1  0000                 CODE   SEGMENT AT 1000H
2                              ASSUME CS:CODE,DS:DATA,SS:STACK
3  0000  B8 5000               MOV    AX,5000H        ;  Initialize Data Segment at
4  0003  8E D8                 MOV    DS,AX           ;  5000H
5  0005  B8 6000               MOV    AX,6000H        ;  Initialize SS at
6  0008  8E D0                 MOV    SS,AX           ;  6000H
7  000A  BC 0020               MOV    SP,0020H        ;  Initialize SP at 0020H
8  000D  BB 2000               MOV    BX,2000H        ;  Initialize BX at 2000H
9  0010  8A 04                 MOV    AL,[SI]         ;  Move 8-bit data
10 0012  8B 0D                 MOV    CX,[DI]         ;  Move 16-bit data
11 0014  9A 0000 ---- R        CALL   FAR PTR MULTI   ;  Call MULTI subroutine
12 0019  89 17                 MOV    [BX],DX         ;  Store high word of result
```

```
13 001B   89 57 02            MOV    [BX+2],AX        ;  Store low word of result
14 001E   F4                  HLT                     ;  Halt
15 001F               CODE    ENDS
16 0000               SUBR    SEGMENT AT 7000H
17                            ASSUME CS:SUBR
18 0000               MULTI   PROC FAR                ;  Must be called from
19                                                    ;  another code segment
20 0000   51
21 0001   50
22 0002   98                  CBW                     ;  Sign extend AL
23 0003   F7 E9               IMUL   CX               ;  [DX][AX]<--[AX]*[CX]
24 0005   58
26 0007   CB                  RET                     ;  Return
27 0008               MULTI   ENDP                    ;  End of procedure
28 0008               SUBR    ENDS                    ;  End subroutine
29 0000               DATA    SEGMENT AT 5000H
30 0000               DATA    ENDS
31 0000               STACK   SEGMENT AT 6000H
32 0000               STACK   ENDS
33                            END                     ;  End program
```

```
Microsoft (R) Macro Assembler Version 5.10                    7/3/98 12:45:17
                                                               Symbols-1
Segments and Groups:
Name                   Length    Align     Combine Class
CODE . . . . . . . . . . . . . . .   001F      AT     1000
DATA . . . . . . . . . . . . . . .   0000      AT     5000
STACK  . . . . . . . . . . . . . .   0000      AT     6000
SUBR . . . . . . . . . . . . . . .   0008      AT     7000
Symbols:
Name                   Type      Value     Attr
MULTI  . . . . . . . . . . . . . .   F PROC    0000      SUBR   Length = 0008
@CPU . . . . . . . . . . . . . . .   TEXT  0101h
@FILENAME  . . . . . . . . . . . .   TEXT  ex9_10
@VERSION . . . . . . . . . . . . .   TEXT  510
    33 Source  Lines
    33 Total   Lines
    10 Symbols
  47086 + 271502 Bytes symbol space free
      0 Warning Errors
      0 Severe  Errors
```

Example 9.11

Write an 8086 assembly program at CS=4000H that converts a temperature (signed) from Fahrenheit degrees stored in a memory location pointed to by SI to Celsius degrees. Store the 8-bit integer part of the result in a memory location pointed to by DI. Also, assume that the temperature can be represented by one byte and DS is already initialized. The source byte is assumed to reside at offset 2000H in the data segment, and the destination byte at an offset of 3000H in the same data segment. Use the formula: $C = \frac{(F-32)}{9} \times 5$

Solution

```
1  0000              CODE   SEGMENT AT 4000H
2                           ASSUME CS:CODE,DS:DATA
3  0000  BE 2000            MOV    SI,2000H       ;  Initialize source pointer
4  0003  BF 3000            MOV    DI,3000H       ;  Init. destination pointer
5  0008  8A 04              MOV    AL,[SI]        ;  Get degrees F
6  000A  98                 CBW                   ;  Sign extend
7  000B  2D 0020            SUB    AX,32          ;  Subtract 32
8  000E  B9 0005            MOV    CX,5           ;  Get multiplier
9  0011  F7 E9              IMUL   CX             ;  Multiply by 5
10 0013  B9 0009            MOV    CX,9           ;  Get divisor
11 0016  F7 F9              IDIV   CX             ;  Divide by 9 to get Celsius
12 0018  88 05              MOV    [DI],AL        ;  Put result in destination
13 001A  F4                 HLT                   ;  Stop
14 001B              CODE   ENDS                  ;  End segment
15 0000              DATA   SEGMENT
16 0000              DATA   ENDS
17                          END
```

```
Microsoft (R) Macro Assembler Version 5.10                        7/3/98 14:16:45
                                                                   Symbols-1
Segments and Groups:
Name                       Length    Align     Combine Class
CODE . . . . . . . . . . . . . . . .  001B      AT      4000
DATA . . . . . . . . . . . . . . . .  0000      PARA    NONE
Symbols:
Name                       Type      Value     Attr
@CPU . . . . . . . . . . . . . . . .  TEXT      0101h
@FILENAME  . . . . . . . . . . . . .  TEXT      ex9_11
@VERSION . . . . . . . . . . . . . .  TEXT      510
     17 Source  Lines
     17 Total   Lines
      7 Symbols
  47124 + 271464 Bytes symbol space free
      0 Warning Errors
      0 Severe Errors
```

8086 CPU

GND	1	40	V_{CC}
AD_{14}	2	39	AD_{15}
AD_{13}	3	38	A_{16}/S_3
AD_{12}	4	37	A_{17}/S_4
AD_{11}	5	36	A_{18}/S_5
AD_{10}	6	35	A_{19}/S_6
AD_9	7	34	$\overline{BHE}/S_7$
AD_8	8	33	$MN/\overline{MX}$
AD_7	9	32	$\overline{RD}$
AD_6	10	31	HOLD
AD_5	11	30	HLDA
AD_4	12	29	$\overline{WR}$
AD_3	13	28	$M/\overline{IO}$
AD_2	14	27	$DT/\overline{R}$
AD_1	15	26	$\overline{DEN}$
AD_0	16	25	ALE
NMI	17	24	$\overline{INTA}$
INTR	18	23	$\overline{TEST}$
CLK	19	22	READY
GND	20	21	RESET

FIGURE 9.8 8086 Pin Diagram

9.9 System Design Using the 8086

This section covers the basic concepts associated with interfacing the 8086 with its support chips such as memory and I/O. Topics such as timing diagrams and 8086 pins and signals will also be included. Appendix E provides data sheets for Intel 8086 and support chips.

9.9.1 8086 Pins and Signals

The 8086 pins and signals are shown in Figure 9.8. As mentioned before, the 8086 can operate in two modes. These are the minimum (uniprocessor systems with a single 8086) and maximum mode (multiprocessor system with more than one 8086). $MN/\overline{MX}$ is an input pin used to select one of these modes.

When $MN/\overline{MX}$ is HIGH, the 8086 operates in the minimum mode. In this mode, the 8086 is configured (that is, pins are defined) to support small single-processor systems using a few devices that use the system bus. When $MN/\overline{MX}$ is LOW, the 8086 is configured (that is, some of the pins are redefined in maximum mode) to support multiprocessor systems. In this case, the Intel 8288 bus controller is added to the 8086 to provide bus control and compatibility with the multibus architecture. Note that, in a particular application, $MN/\overline{MX}$ must be tied to either HIGH or LOW.

The AD_0–AD_{15} lines are a 16-bit multiplexed address/data bus. During the first clock cycle, AD_0–AD_{15} are the low-order 16-bit address. The 8086 has a total of 20 address lines. The upper four lines, A_{16}/S_3, A_{17}/S_4, A_{18}/S_5, and A_{19}/S_6, are multiplexed with the status signals for the 8086. During the first clock period of a bus cycle (read or write cycle), the entire 20-bit address is available on these lines. During all other cycles for memory and I/O, AD_0–AD_{15} lines contain the 16-bit data, and the multiplexed address / status lines become S_3, S_4, S_5, and S_6. S_3 and S_4 are decoded as follows:

A_{17}/S_4	A_{16}/S_3	Function
0	0	Extra segment
0	1	Stack segment
1	0	Code or no segment
1	1	Data segment

Therefore, after the first clock cycle of an instruction execution, the A_{17}/S_4 and A_{16}/S_3 pins specify which segment register generates the segment portion of the 8086 address. Thus, by decoding these pins and then using the decoder outputs as chip selects for memory chips, up to four megabytes (one megabyte per segment) can be included. This provides a degree of protection by preventing erroneous write operations to one segment from overlapping onto another segment and destroying the information in that segment. A_{18}/S_5 and A_{19}/S_6 are used as A_{18} and A_{19}, respectively, during the first clock cycle of an instruction execution. If an I/O instruction is executed, they stay LOW for the first clock period. During all other cycles, A_{18}/S_5 indicates the status of the 8086 interrupt enable flag and A_{19}/S_6 becomes S_6; a LOW S_6 pin indicates that the 8086 is on the bus. During a hold acknowledge clock period, the 8086 tristates the A_{19}/S_6 pin and this allows another bus master to take control of the system bus. The 8086 tristates AD_0–AD_{15} during interrupt acknowledge or hold acknowledge cycles.

$\overline{BHE}/S_7$ is used as $\overline{BHE}$ (bus high enable) during the first clock cycle of an instruction execution. The 8086 outputs a LOW on this pin during the read, write, and interrupt acknowledge cycles in which data are to be transferred in a high-order byte (AD_{15}–AD_8) of the data bus. $\overline{BHE}$ can be used in conjunction with AD_0 to select memory banks. A thorough discussion is provided later. During all other cycles, $\overline{BHE}/S_7$ is used as S_7 and the 8086 maintains the output level ($\overline{BHE}$) of the first clock cycle on this pin. S_7 is the same as $\overline{BHE}$ and does not have any special meaning.

$\overline{RD}$ is low whenever the 8086 is reading data from memory or an I/O location.

$\overline{\text{TEST}}$ is an input pin and is only used by the WAIT instruction. The 8086 enters a wait state after execution of the WAIT instruction until a low is seen on the $\overline{\text{TEST}}$ pin. This input is synchronized internally during each clock cycle on the leading edge of the clock.

INTR is the maskable interrupt input. This line is not latched, so INTR must be held at a HIGH level until it is recognized to generate an interrupt.

NMI is the nonmaskable interrupt pin input activated by a leading edge.

RESET is the system reset input signal. This signal must be HIGH for at least **four** clock cycles to be recognized, except on power-on, which requires a 50-μsec reset pulse. It causes the 8086 to initialize registers DS, ES, SS, IP, and flags to zeros. It also initializes CS to FFFFH. Upon removal of the RESET signal from the RESET pin, the 8086 will fetch its next instruction from a 20-bit physical address FFFF0H (CS = FFFFH, IP = 0000H). When the 8086 detects a positive edge of a pulse on RESET, it stops all activities until the signal goes LOW. Upon hardware reset, the 8086 initializes the system as follows:

8086 Components	Content
Flags	Clear
IP	0000H
CS	FFFFH
DS	0000H
SS	0000H
ES	0000H
Queue	Empty

As mentioned before, the 8086 can be configured in either minimum or maximum mode using the MN/$\overline{\text{MX}}$ input pin. In minimum mode, the 8086 itself generates all bus control signals. These signals are as follows:

- DT/$\overline{\text{R}}$ (data transmit/receive) is an output signal required in a minimum system that uses an 8286/8287 data bus transceiver. It is used to control direction of data flow through the transceiver.
- $\overline{\text{DEN}}$ (data enable) is provided as an output enable for the 8286/8287 in a minimum system that uses the transceiver. $\overline{\text{DEN}}$ is active LOW during each memory and I/O access and for $\overline{\text{INTA}}$ cycles.
- ALE (address latch enable) is an 8086 output signal that can be used to demultiplex the multiplexed 8086 pins including AD_0–AD_{15} into A_0–A_{15} and D_0–D_{15} at the falling edge of ALE.
- M/$\overline{\text{IO}}$ is an 8086 output signal. It is used to distinguish a memory access (M/$\overline{\text{IO}}$ = HIGH) from an I/O access (M/$\overline{\text{IO}}$ = LOW). When the 8086 executes an I/O instruction such as IN or OUT, it outputs a LOW on this pin. On the other hand, the 8086 outputs HIGH on this pin when it executes a memory reference instruction such as `MOV AX, [SI]`.

- $\overline{\text{WR}}$ is used by the 8086 for a write operation. The 8086 outputs a low on this pin to indicate that the processor is performing a write memory or write I/O operation, depending on the M/$\overline{\text{IO}}$ signal.
- For interrupt acknowledge cycles (for the INTR pin), the 8086 outputs LOW on the $\overline{\text{INTA}}$ pin.
- HOLD (input) and HLDA (output) pins are used for DMA. A HIGH on the HOLD pin indicates that another master is requesting to take over the system bus. The processor receiving the HOLD request will output a HIGH on the HLDA as an acknowledgment. At the same time, the processor tristates the system bus. Upon receipt of LOW on the HOLD pin, the processor places LOW on the HLDA pin and takes over the system bus.
- CLK (input) provides the basic timing for the 8086 and bus controller.
- READY (input) pin is used for slow peripheral devices.

The maximum clock frequencies of the 8086-4, 8086, and 8086-2 are 4 MHz, 5 MHz, and 8 MHz, respectively. Because the design of these processors incorporates dynamic cells, a minimum frequency of 2 MHz is required to retain the state of the machine. The 8086-4, 8086, and 8086-2 will be referred to as 8086 in the following discussion.

The reset, clock, and the ready signals of the 8086 can be generated by the Intel 8284. Figure 9.9 shows the pins and signals of the 8284.

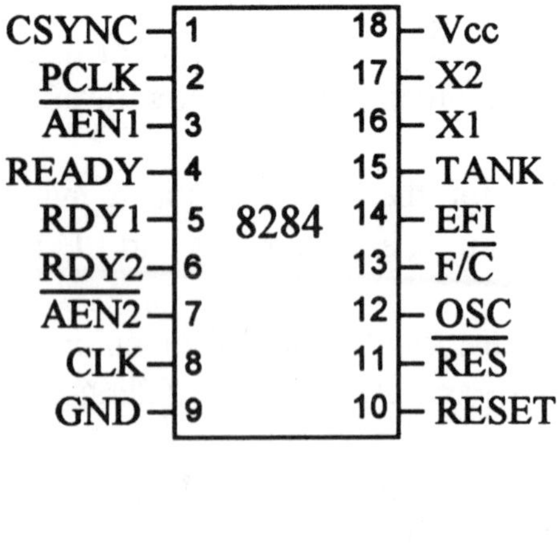

		8284		
CSYNC	1		18	Vcc
PCLK	2		17	X2
$\overline{\text{AEN1}}$	3		16	X1
READY	4		15	TANK
RDY1	5	8284	14	EFI
$\overline{\text{RDY2}}$	6		13	F/$\overline{\text{C}}$
$\overline{\text{AEN2}}$	7		12	OSC
CLK	8		11	$\overline{\text{RES}}$
GND	9		10	RESET

Pin Name	Description
X_1, X_2	Crystal connections
F/$\overline{\text{C}}$	Clock source select
CLK	MOS CLOCK for the 8086
$\overline{\text{RES}}$	Reset input to the 8284 from an RC circuit
RESET	Reset input to the processor
Vcc	+5 V
GND	0V
OSC	Oscillator output
TANK	Used with overtone crystal
EFI	External clock input
CSYNC	Clock synchronization input
RDY1, RDY2	Ready signals from two multibus systems
$\overline{\text{AEN1}}$,$\overline{\text{AEN2}}$	Address enables for ready signals
PCLK	TTL clock for peripherals
READY	Ready output

FIGURE 9.9 8284 pins and signals

The 8284 is an 18-pin chip designed for providing three input signals for the 8086:

1. 8086 CLK input
2. 8086 Reset input
3. 8086 Ready input

The 8284 pins and signals are described in the following.

9.9.1.1 Clock Generation Signals

Because the 8086 has no on-chip clock generator circuitry, the 8284 chip is required to provide the 8086 clock input. The 8284 $F/\overline{C}$ input pin is provided for clock source selection. When the $F/\overline{C}$ pin is connected to LOW, a crystal connected between 8284's X_1 and X_2 pins is used. On the other hand, when $F/\overline{C}$ is connected to HIGH, an external clock source is used; the external clock source is connected to the 8284 EFI (external frequency input) pin. The 8284 divides the clock inputs at the X_1X_2 pins or the EFI pin by 3. This means that if a 15-MHz crystal is connected at the X_1X_2 or EFI pins, the 8284 CLK output pin will be 5 MHz. The 8284 CLK pin will be connected to the 8086 CLK pin. This provides the clock input for the 8086. When selecting a crystal for use with the 8284, the crystal series resistance should be as low as possible. The oscillator delays in the 8284 appear as inductive elements to the crystal and cause the 8284 to run at a frequency below that of the pure series resonance: a capacitor C_L should be placed in series with the crystal and the 8284 X_2 pin. The capacitor cancels the inductive element. The impedance of the capacitor $X_C = 1/(2\pi f C_L)$ where f is the crystal frequency. Intel recommends that the crystal series resistance plus X_c should be kept less than 1 KΩ. Some crystals require resistance to be connected at the 8284 X_1 and X_2 pins as follows:

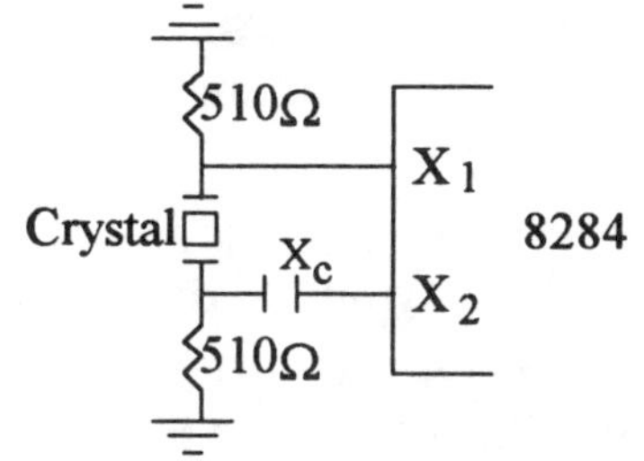

As the crystal frequency increases, C_L should be decreased. For example, a 12-MHz crystal may require C_L = 24 pf whereas a 22-MHz crystal may require C_L = 8pf. C_L values of 12 to 15 pf may be used with a 15-MHz crystal. Two crystal manufacturers recommended by Intel are Crystle Corp., Model CY 15A (15 MHz), and CTS Knight, Inc., Model CY 24A (24 MHz). Note that the 8284 CLK output pin is the MOS clock for the 8086.

There are two more clock outputs on the 8284, the PCLK (peripheral clock) pin and the OSC (oscillator) clock pin. These signals are provided to drive peripheral ICs. The 8284 divides the frequency of the crystal at the X_1X_2 pins or the external clock at the EFI pin by 6 to provide the PCLK. Therefore, the frequency of the PCLK is half the frequency of the 8284 CLK output pin. This means that for a 15-MHz crystal, the PCLK and CLK outputs are 2.5 MHz and 5 MHz respectively. Furthermore, PCLK is provided at the TTL-compatible level rather than at the MOS level. The OSC clock, on the other hand, is derived from the crystal oscillator inside the 8284 and has the same clock frequency as the crystal. Therefore, the OSC output is three times that of the CLK output. The OSC is also TTL compatible. Finally, the CSYNC (clock synchronization) input pin when connected to HIGH provides external synchronization in systems that employ multiple clocks. A typical 8284 interface to the 8086 for providing a 5-MHz clock to the 8086 is shown in the following figure:

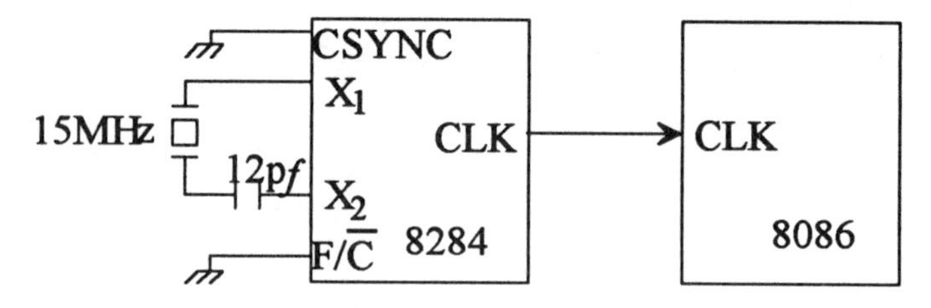

9.9.1.2 Reset Signals

When designing the microprocessor's reset circuit, two types of reset must be considered: power-up reset and manual reset. These reset circuits must be designed using the parameters specified by the manufacturer.

Therefore, a microprocessor must be reset when its Vcc pin is connected to power. This is called "power-up reset." After some time during normal operation the microprocessor can be reset upon activation of a manual switch such as a pushbutton. A reset circuit, therefore, needs to be designed following the timing parameters associated with the microprocessor's reset input pin specified by the manufacturer. The reset circuit, once designed, is connected to the microprocessor's reset pin.

As mentioned before, the 8086 reset input provides a hardware mechanism for initializing the 8086 microprocessor. This is typically done at power-up to provide an orderly start-up of the system. The 8284 $\overline{RES}$ (reset input) pin when driven active LOW generates a HIGH on the 8284 reset output pin. The 8284 reset pin is connected to the 8086 reset (input) pin. As mentioned before, Intel designed the 8086 in such a way that the 8086 requires its reset pin to be HIGH for at least four clock cycles in order to obtain the physical address (FFFF0H) of the first instruction to be executed, except after power-on, which requires a 50-μsec reset pulse.

According to Intel, in order to guarantee a reset from power-up, the 8086 reset input must remain below 1.05 V for 50 μsec after Vcc has reached the minimum supply voltage of 4.5 V. The 8284 $\overline{RES}$ input can be driven by an *RC* circuit as shown in the following figure:

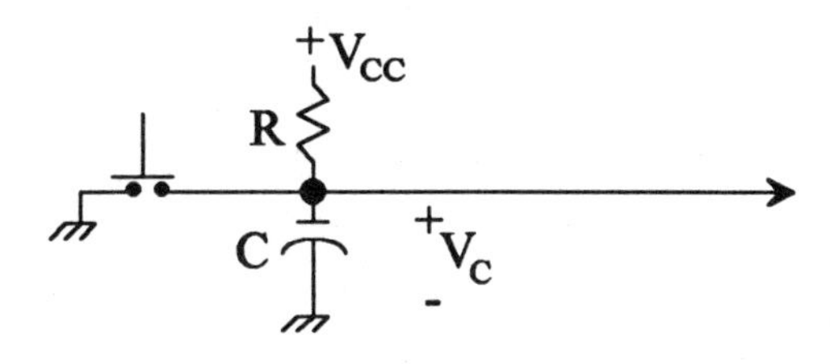

To 8284 $\overline{RES}$ input pin

The voltage across the capacitor initially is zero upon connecting +Vcc to power. If the switch is not depressed, the capacitor charges to +Vcc through the resistor after a definite time determined by the time constant RC.

The charging voltage across the capacitor can be determined from the following equation. Capacitor voltage, $V_c(t) = V_{cc} \times [1 - \exp(-t/RC)]$, where $t = 50$ µsec and $V_c(t) = 1.05$ V, and $V_{cc} = 4.5$ V. Substituting these values in the equation, $RC = 188$ µsec. For example, if C is chosen to be 0.1 µF, then R is 1.88 KΩ.

When the switch is depressed, the 8284 $\overline{RES}$ input pin is short-circuited to ground. This takes the 8284 $\overline{RES}$ pin to LOW and thus discharges the capacitor. As the switch is released, the direct short to ground is broken. However, the 8284 $\overline{RES}$ pin remains effectively short-circuited to ground through the discharged capacitor. The capacitor now starts to recharge with time toward the $+V_{cc}$ voltage level.

The 8284 generates a reset signal from an internal Schmitt trigger input. A Schmitt trigger is a special analog circuit that shifts the switching threshold based on whether the input changes from LOW to HIGH or from HIGH to LOW. To illustrate this, consider a TTL Schmitt trigger inverter. Suppose that the input of this inverter is at 0 V (logic 0). The output will be approximately 3.4 V (logic 1). Now, because of the Schmitt trigger circuit, if the input voltage is increased, the output will not go to low until the value is about 1.7 V. Also, after reaching a low output, the inverter will not produce a HIGH output until the input is decreased to about 0.9 V. Thus, the switching threshold for positive-going input changes is about 1.7 V and for negative-going input changes is about 0.9 V.

The difference between the two thresholds is called "hysteresis." The Schmitt trigger inverter provides 1.7 V - 0.9 V = 0.8 V of hysteresis. Schmitt trigger inputs provide high noise immunity and will normally not respond to the noise encountered in microprocessor systems if its hysteresis is greater than the noise amplitude.

As the voltage across the capacitor increases with time, it remains at logic 0 level as long as the logic 1 threshold of the Schmitt trigger. Thus, the 8284 $\overline{RES}$ input is maintained at logic 0 for at least four clock cycles so that the 8284 RESET output will apply a HIGH at the 8086 reset input for at least four clock cycles. Note that whenever the 8282 $\overline{RES}$ input is at logic 0, the reset output pin of the 8284 is switched to logic 1 according to the timing parameters.

9.9.1.3 Ready Signals

The 8284 Ready (output) pin is connected to the 8086 Ready (input) pin to insert wait states for slow peripheral devices connected to the 8086. There are two main ways to disable this function when not used. One way is to connect the 8086 Ready pin to HIGH, and keep the 8284 Ready output pin floating. The other way is to connect the 8284 RDY1 and RDY2 pins to LOW, and the $\overline{AEN1}$ and $\overline{AEN2}$ to HIGH, which will permanently disable this function. The 8284 Ready (output) pin can then be connected to the 8086 Ready input pin.

The RDY1, $\overline{AEN1}$ and RDY2, $\overline{AEN2}$ input signals provide logic for operation with multiprocessor systems and the 8284 ready output. In multiprocessor systems, these signals are used to control access over the system bus by several 8086's. The 8284 TANK pin is replaced by the $\overline{ASYNC}$ input pin on the newer version of 8284. The $\overline{ASYNC}$ pin can be driven to LOW by a slower device to generate the 8284 READY output pin which can be connected to the 8086 READY pin. This makes it easier for the slower devices to interface to the 8086. Typical 8284 clock (using a 15-MHz crystal), reset, and ready signal (unused) connections to single 8086-appropriate pins are shown in the following figure:

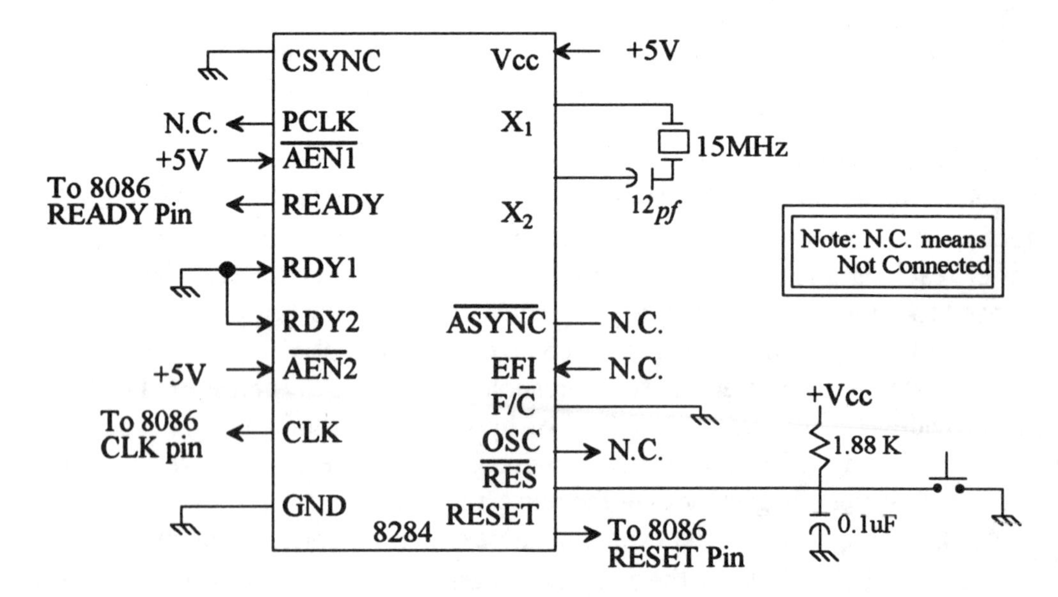

In the maximum mode, some of the 8086 pins in the minimum mode are redefined. For example, pins HOLD, HLDA, $\overline{WR}$, M/$\overline{IO}$, DT/$\overline{R}$, $\overline{DEN}$, ALE, and $\overline{INTA}$ in the minimum mode are redefined as $\overline{RQ}/\overline{GT0}$,$\overline{RQ}/\overline{GT1}$, $\overline{LOCK}$, $\overline{S_2}$, $\overline{S_1}$, $\overline{S_0}$, QS_0, and QS_1, respectively. In maximum mode, the 8288 bus controller decodes the status information from $\overline{S_0}$, $\overline{S_1}$, and $\overline{S_2}$ to generate the bus timing and control signals that are required for a bus cycle. $\overline{S_2}$, $\overline{S_1}$, and $\overline{S_0}$ are 8086 outputs and are decoded as follows:

$\overline{S_2}$	$\overline{S_1}$	$\overline{S_0}$	Function
0	0	0	Interrupt acknowledge
0	0	1	Read I/O port
0	1	0	Write I/O port
0	1	1	Halt
1	0	0	Code access
1	0	1	Read memory
1	1	0	Write memory
1	1	1	Inactive

The $\overline{RQ}/\overline{GT0}$ and $\overline{RQ}/\overline{GT1}$ request/grant pins are used by other local bus masters to force the processor to release the local bus at the end of the processor's current bus cycle. Each pin is bidirectional, with $\overline{RQ}/\overline{GT0}$ having higher priority than $\overline{RQ}/\overline{GT1}$.These pins have internal pull-up resistors so that they may be left unconnected. The request/grant function of the 8086 works as follows:

- A pulse (one clock wide) from another local bus master ($\overline{RQ}/\overline{GT0}$ or $\overline{RQ}/\overline{GT1}$ pin) indicates a local bus request to the 8086.
- At the end of the current 8086 bus cycle, a pulse (one clock wide) from the 8086 to the requesting master indicates that the 8086 has relinquished the system bus and tristates the outputs. Then the new bus master subsequently relinquishes control of the system bus by sending a LOW on $\overline{RQ}/\overline{GT0}$ or $\overline{RQ}/\overline{GT1}$ pin. The 8086 then regains bus control.
- The 8086 outputs LOW on the $\overline{LOCK}$ pin to prevent other bus masters from gaining control of the system bus.

Note that since the 8086 RESET vector is located at the physical address FFFF0H, there may not be enough locations available to write programs. The following 8086 instruction sequence can be used for typical assemblers to jump to a different code segment upon hardware reset to write programs:

```
ORG 0FFFF0000H ; Reset Vector          ORG 10000200H
JMP FAR PTR START                START  —} User
                                        —} Programs
```

The above instruction sequence will allow the 8086 to jump to the offset START (0200H) in code segment 1000H upon hardware reset where the user can write programs.

9.9.2 Basic 8086 System Concepts

This section describes basic concepts associated with the 8086 bus cycles, address and data bus, in minimum mode.

9.9.2.1 8086 Bus Cycle

To communicate with external devices via the system for transferring data or fetching instructions, the 8086 executes a bus cycle. The 8086 basic bus cycle timing diagram is shown in Figure 9.10. The minimum bus cycle contains four microprocessor clock periods or four *T states*. Note that each cycle is called a T state. The bus cycle timing diagram depicted in Figure 9.10 can be described as follows:

1. During the first T state (T_1), the 8086 outputs the 20-bit address computed from a segment register and an offset on the multiplexed address/data/status bus.
2. For the second T state (T_2), the 8086 removes the address from the bus and either tristates or activates the AD_{15}–AD_0 lines in preparation for reading data via the AD_{15}–AD_0 lines during the T_3 cycle. In the case of a write bus cycle, the 8086 outputs data on the AD_{15}–AD_0 lines during the T_3 cycle. Also, during T_2, the upper four multiplexed bus lines switch from address (A_{19}–A_{16}) to bus cycle status (S_6, S_5, S_4, S_3). The 8086 outputs LOW on $\overline{RD}$ (for the read cycle) or $\overline{WR}$ (for the write cycle) during portion of T_2, all of T_3, and portion of T_4.
3. During T_3, the 8086 continues to output status information on the four A_{19}–A_{16}/S_6–S_3 lines and will continue to output write data or input read data to or from the AD_{15}–AD_0 lines.

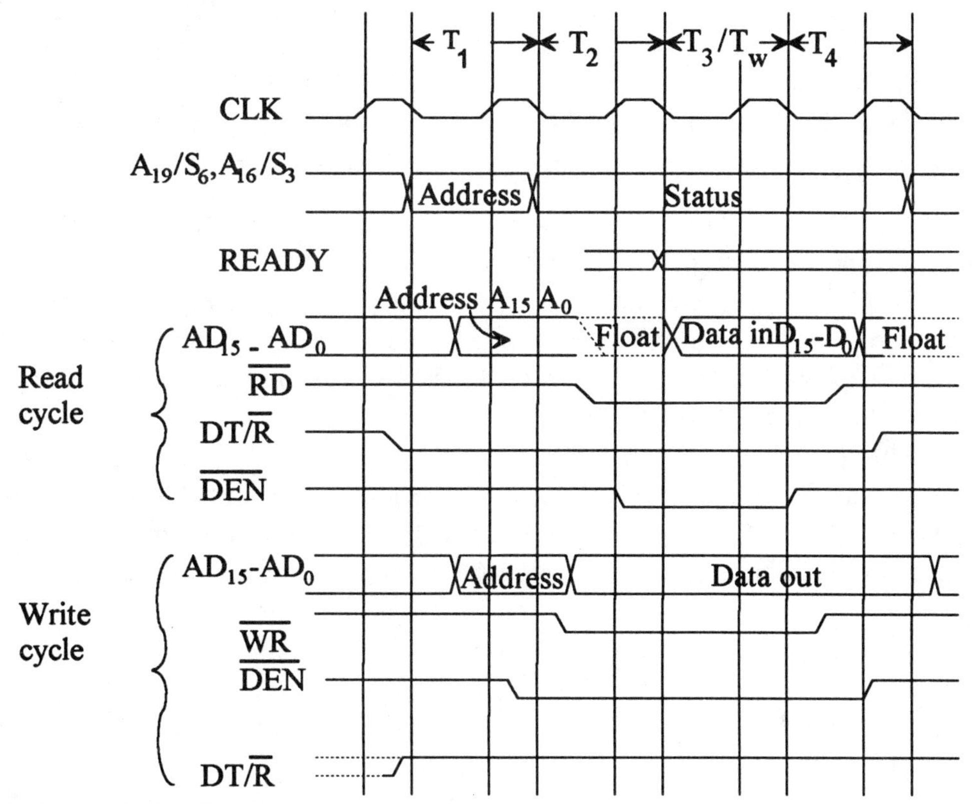

FIGURE 9.10 Basic 8086 bus cycle

If the selected memory or I/O device is not fast enough to transfer data to the 8086, the memory or I/O device activates the 8086's READY input line LOW by the start of T_3. This will force the 8086 to insert additional clock cycles (wait states T_w) after T_2. Bus activity during T_W is the same as that during T_3. When the selected device has had sufficient time to complete the transfer, it must activate the 8086 ready pin HIGH. As soon as the T_w clock period ends, the 8086 executes the last bus cycle (T_4). The 8086 will latch data on the AD_{15}–AD_0 lines during the last wait state or during T_3 if no wait states are requested.

3. During T_4, the 8086 disables the command lines and the selected memory and I/O devices from the bus. Thus, the bus cycle is terminated in T4. The bus cycle appears to devices in the system as an asynchronous event consisting of an address to select the device, a register or memory location within the device, a read strobe, or a write strobe along with data.
4. The $\overline{DEN}$ and $DT/\overline{R}$ pins are used by the 8286/8287 transceiver in a minimum system. During the read cycle, the 8086 outputs $\overline{DEN}$ LOW during part of the T_2 and all of the T_3 cycles. This signal can be used to enable the 8286/8287 transceiver. The 8086 outputs a

LOW on the DT/$\overline{R}$ pin from the start of the T_1 through part of the T_4 cycles. The 8086 uses this signal to receive (read) data from the receiver during T_3–T_4. During a write cycle, the 8086 outputs $\overline{DEN}$ LOW during part of the T_1, all of the T_2, and T_3, and part of the T_4 cycles. The signal can be used to enable the transceiver. The 8086 outputs a HIGH on DT/$\overline{R}$ throughout the 4 bus cycles to transmit (write) data to the transceiver during T_3–T_4.

9.9.2.2 Address and Data Bus Concepts

The majority of memory and I/O chips capable of interfacing to the 8086 require a stable address for the duration of the bus cycle. Therefore, the address on the 8086 multiplexed address/data bus during T_1 should be latched. The latched address is then used to select the desired I/O or memory location.

To demultiplex the bus, the 8086 ALE pin can be used along with three 74LS373 latches. The 74LS373 Output Control ($\overline{OC}$) pin can be connected to ALE with 74LS373 Enable pin (G or C shown as E in Figure 9.11) to HIGH . This will latch the 8086 address and $\overline{BHE}$ pins at the falling edge of ALE. Figure 9.11 shows how this can be accomplished.

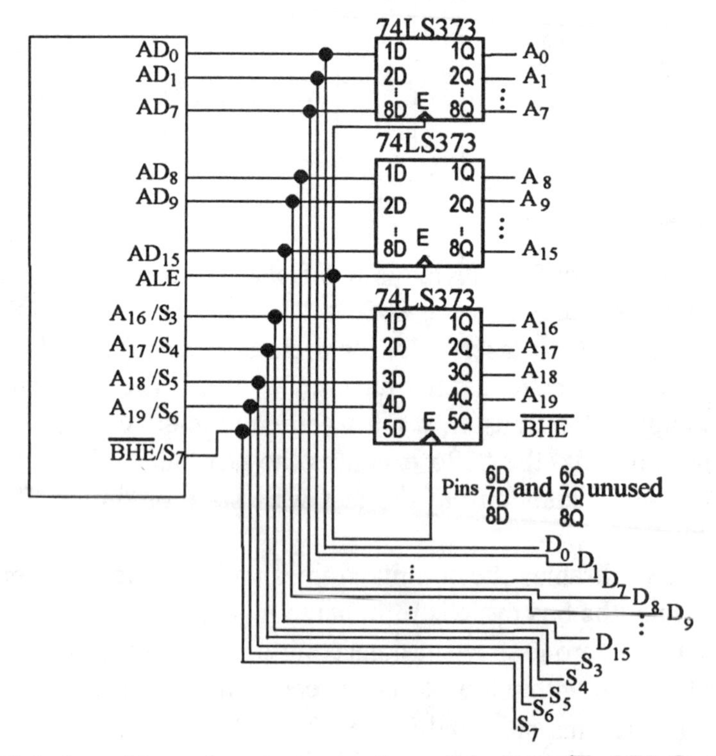

FIGURE 9.11 Demultiplexing address, data, and status lines of the 8086 (E=373 Output Control pin)

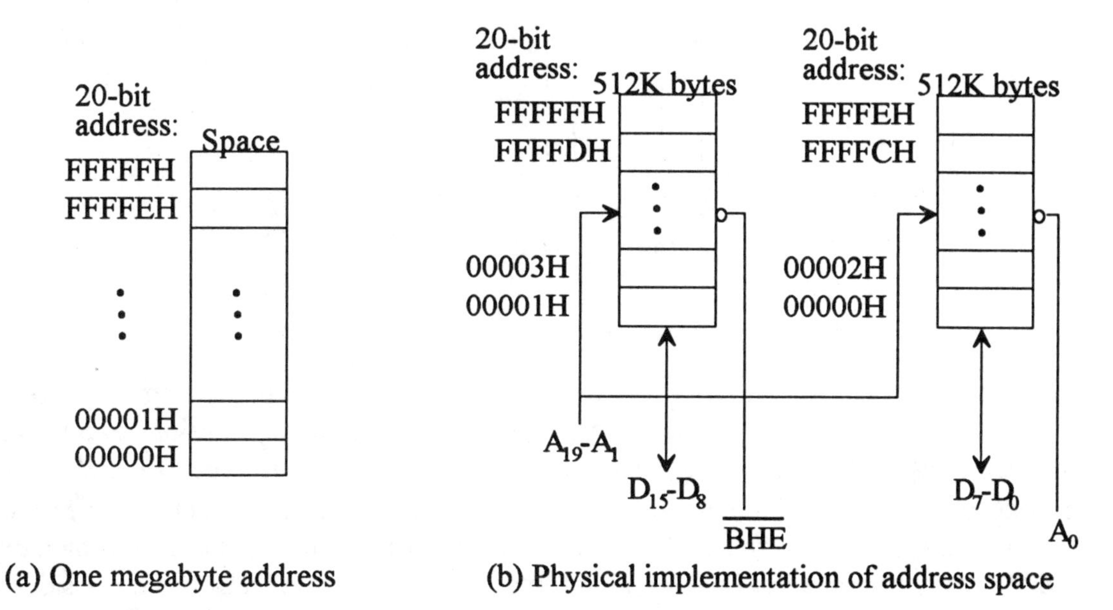

(a) One megabyte address (b) Physical implementation of address space

FIGURE 9.12 8086 Memory

The programmer views the 8086 memory address space as a sequence of one mega bytes in which any byte may contain an 8-bit data element and any two consecutive bytes may contain a 16-bit data element. There is no constraint on byte or word addresses (boundaries). The address space is physically implemented on a 16-bit data bus by dividing the address space into two banks of up to 512K bytes as shown in Figure 9.12. These banks can be selected by $\overline{BHE}$ and A_0 as follows:

$\overline{BHE}$	A_0	Byte transferred
0	0	Both bytes via demultiplexed D_0–D_{15} pins for even address.
0	1	Upper byte to/from odd address via demultiplexed D_8–D_{15} pins.
1	0	Lower byte to/from even address via demultiplexed D_0–D_7 pins.
1	1	None

One bank is connected to D_7–D_0 and contains all even-addressed bytes ($A_0 = 0$). The other bank is connected to D_{15}–D_8 and contains odd-addressed bytes ($A_0 = 1$). A particular byte in each bank is addressed by A_{19}–A_1. The even-addressed bank is enabled by a LOW on A_0, and data bytes are transferred over the D_7–D_0 lines. The 8086 outputs a HIGH on $\overline{BHE}$ (bus high enable) and thus disables the odd-addressed bank. The 8086 outputs a LOW on $\overline{BHE}$ to select the odd-addressed bank and a HIGH on A_0 to disable the even-addressed bank. This directs the data transfer to the appropriate half of the data bus.

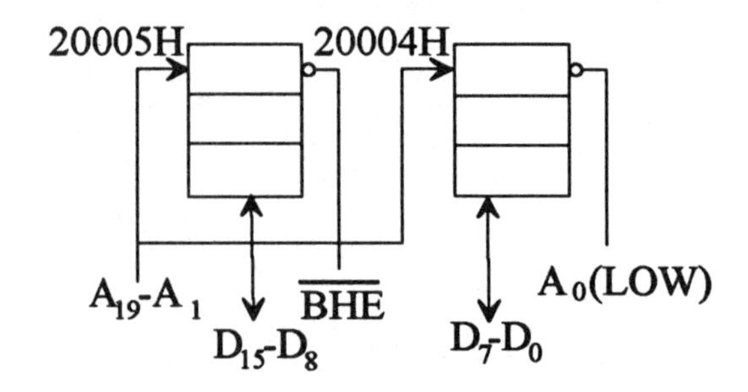

FIGURE 9.13 Even-addressed word transfer

Activation of A_0 and $\overline{BHE}$ is performed by the 8086 depending on odd or even addresses and is transparent to the programmer. As an example, consider execution of the instruction `MOV DH, [BX]`. Suppose the 20-bit address computed by BX and DS is even. The 8086 outputs a LOW on A_0 and a HIGH on $\overline{BHE}$ .This will select the even-addressed bank. The content of the selected memory is placed on the D_7–D_0 lines by a memory chip. The 8086 reads this data via D_7–D_0 and automatically places it in DH. Next, consider writing a 16-bit word by the 8086 with the low byte at an even address as shown in Figure 9.13. For example, suppose that the 8086 executes the instruction `MOV [BX],CX`. Assume [BX] = 0004H and [DS] = 2000H. The 20-bit physical address for the word is 20004H. The 8086 outputs a LOW on both A_0 and $\overline{BHE}$, enabling both banks simultaneously. The 8086 outputs [CL] to the D_7–D_0 lines and [CH] to the D_{15}–D_8 lines, with $\overline{WR}$ = LOW and $M/\overline{IO}$ = HIGH. The enabled memory banks obtain the 16-bit data and write [CL] to location 20004H and [CH] to location 20005H.

Next, consider writing an odd-addressed 16-bit word by the 8086 using `MOV [BX], CX`. For example, suppose the 20-bit physical address computed by the 8086 is 20005H. The 8086 accomplishes this transfer in two bus cycles. In the first bus cycle, the 8086 outputs a HIGH on A_0 and a LOW on $\overline{BHE}$, and thus enables the odd-addressed bank and disables the even-addressed bank. The 8086 also outputs a LOW on the $\overline{WR}$ and a HIGH on the $M/\overline{IO}$ pins. In this bus cycle, the 8086 writes data to odd memory bank via D_{15}–D_8 lines; the 8086 writes the contents of CL to address 20005H. In the second bus cycle, the 8086 outputs a LOW on A_0 and a HIGH on $\overline{BHE}$ and thus enables the even-addressed bank and disables the odd-addressed bank. The 8086 also outputs a LOW on the $\overline{WR}$ and a HIGH on the $M/\overline{IO}$ pins. The 8086 writes data to even memory bank via D_7–D_0 lines; the 8086 writes the contents of CH to address 20006H. This odd-addressed word write is shown in Figure 9.14.

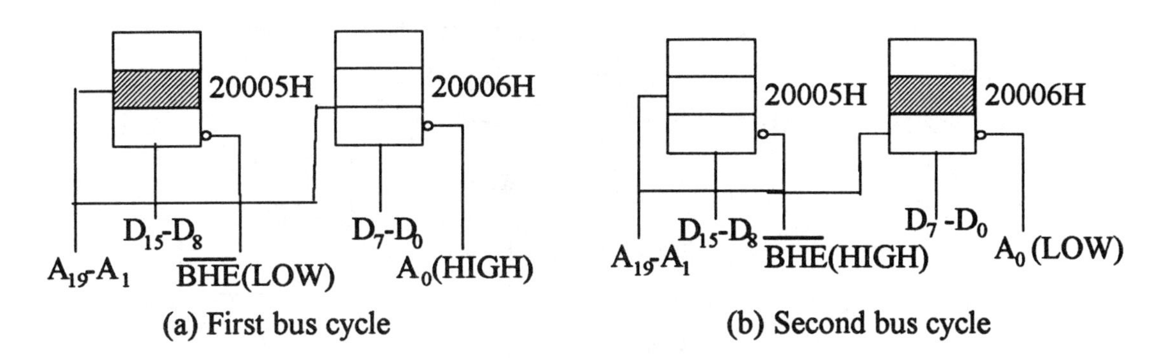

FIGURE 9.14 Odd-addressed word transfer

If memory or I/O devices are directly connected to the multiplexed bus, the designer must guarantee that the devices do not corrupt the address on the bus during T_1. To avoid this, the memory or I/O devices should have an output enable controlled by the 8086 read signal. The 8086 timing guarantees that the read is not valid until after the address is latched by ALE as shown in Figure 9.15.

All Intel peripherals, EPROMs, and RAMs for microprocessors provide output enable for read inputs to allow connection to the multiplexed bus. Several techniques are available for interfacing the devices without output enables to the 8086 multiplexed bus. However, these techniques will not be discussed here.

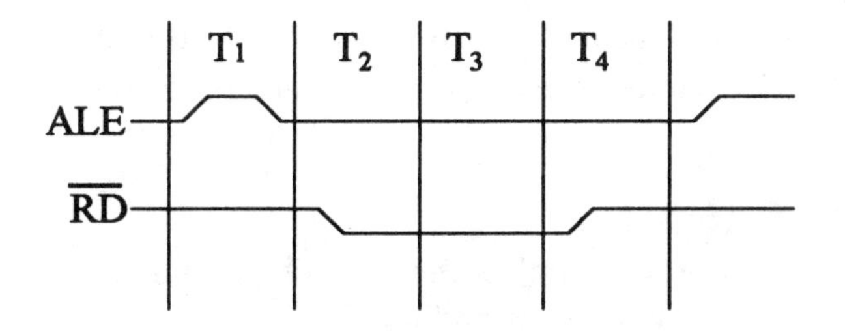

FIGURE 9.15 Relationship of ALE and read

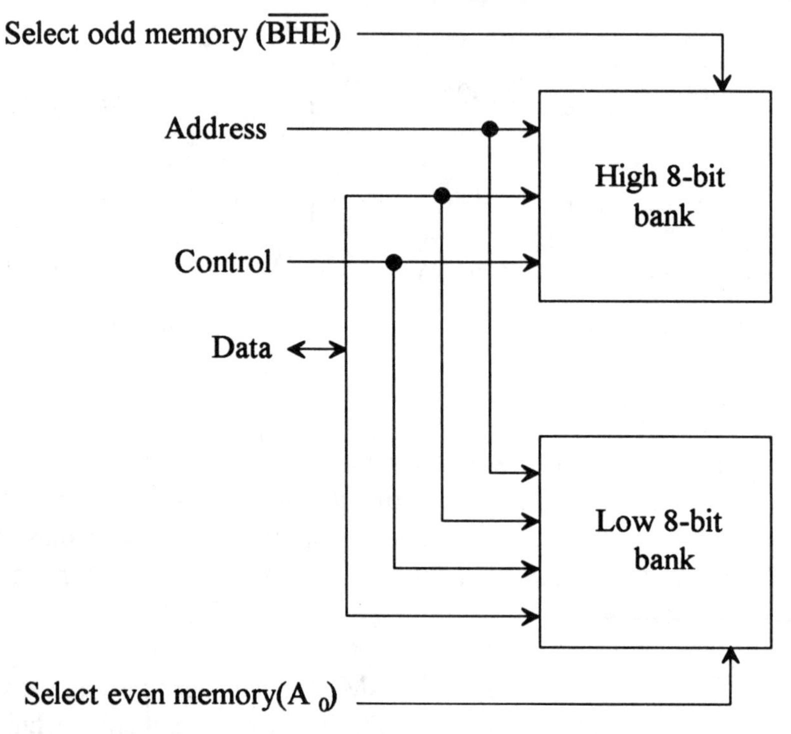

FIGURE 9.16 8086 memory array

9.9.3 Interfacing with Memories

In Figure 9.16, the 16-bit word memory in the 8086 is partitioned into odd and even 8-bit banks on the upper and lower halves of the data bus selected by $\overline{BHE}$ and A_0. This is typically used for RAMs.

9.9.3.1 ROMs and EPROMs

ROMs and EPROMs are the simplest memory chips to interface to the 8086. Because ROMs and EPROMs are read-only devices and the 8086 always reads 16-bit data but discards unwanted bytes (if necessary), A_0 and $\overline{BHE}$ are not required to be part of the chip enable/select decoding (chip enable is similar to chip select decoding except that chip enable also provides whether the chip is in active or standby power mode). The 8086 address lines must be connected to the ROM/EPROM chips starting with A_1 and higher to all the address lines of the ROM/EPROM chips. The 8086 unused address lines can be used as chip enable/select decoding. To interface the ROMs/EPROMs directly to the 8086 multiplexed bus, they must have output enable signals. Figure 9.17 shows the 8086 interfaced to two 2732 chips along with the pin diagram of 2732.

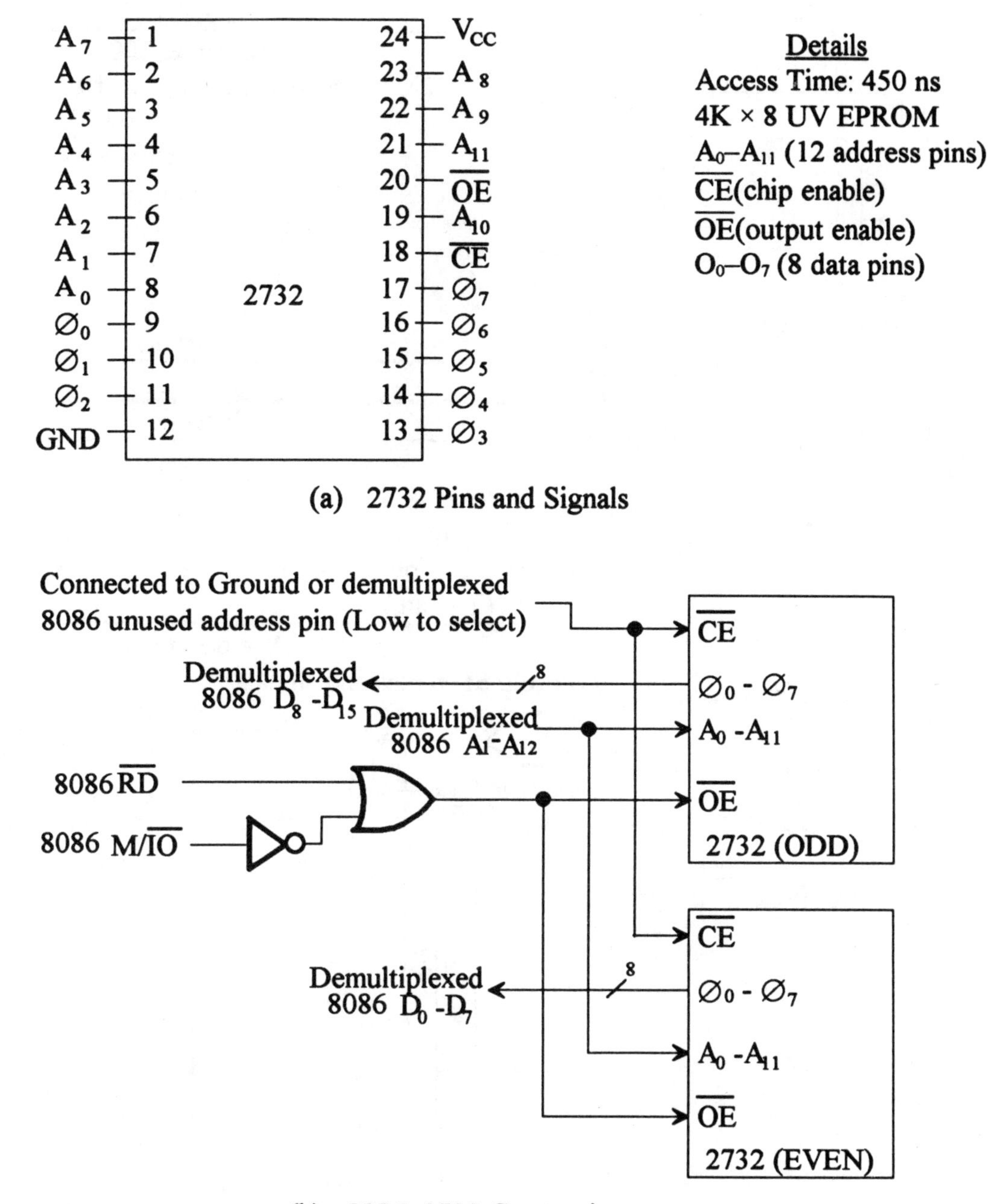

(b) 8086–2732 Connections

FIGURE 9.17 8086–2372 interface along with 2732 pins and signals

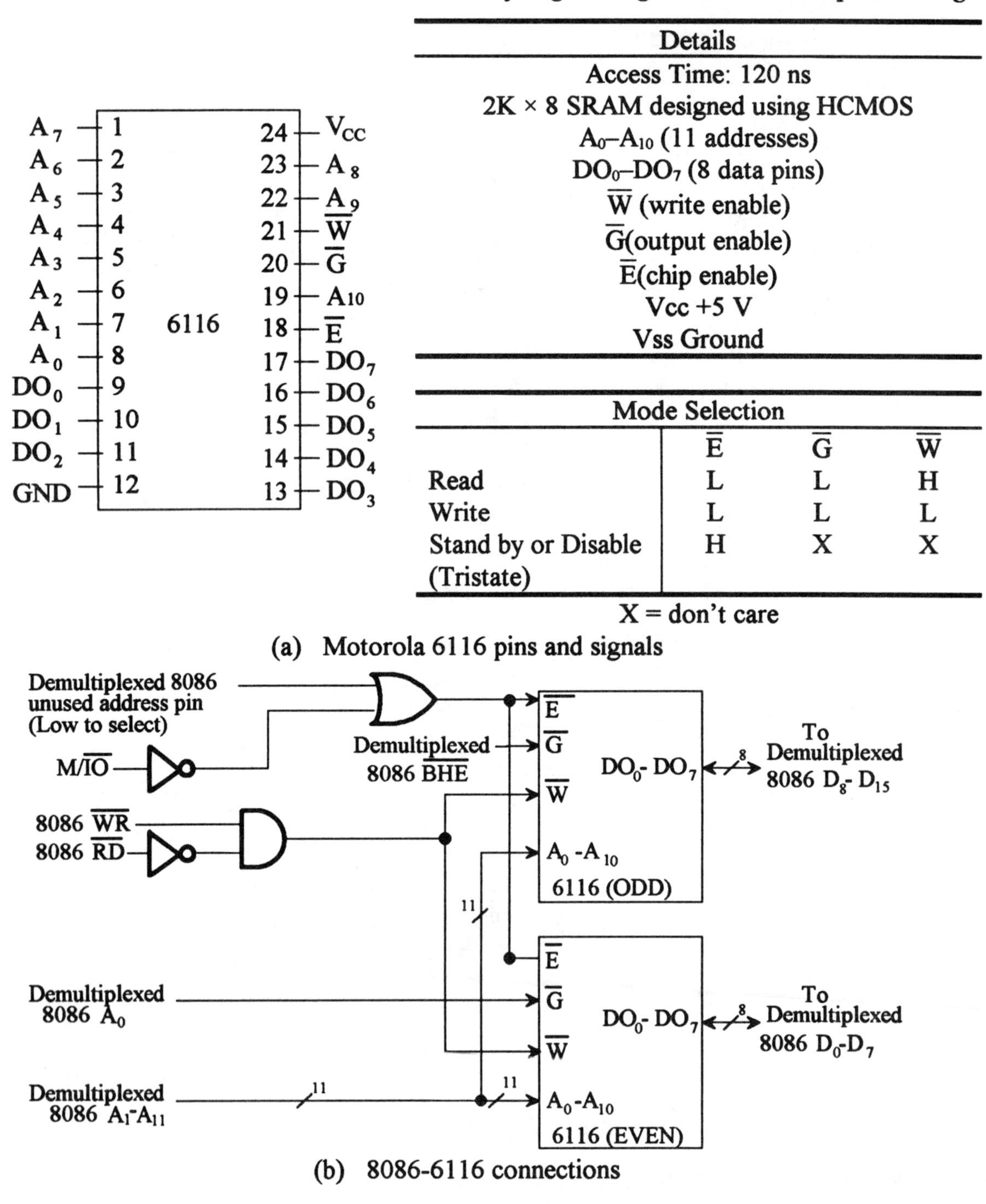

Details
Access Time: 120 ns
2K × 8 SRAM designed using HCMOS
A_0–A_{10} (11 addresses)
DO_0–DO_7 (8 data pins)
$\overline{W}$ (write enable)
$\overline{G}$(output enable)
$\overline{E}$(chip enable)
Vcc +5 V
Vss Ground

Mode Selection	$\overline{E}$	$\overline{G}$	$\overline{W}$
Read	L	L	H
Write	L	L	L
Stand by or Disable (Tristate)	H	X	X

X = don't care

(a) Motorola 6116 pins and signals

(b) 8086-6116 connections

FIGURE 9.18 8086–6116 interface along with 6116 pin diagram

The 8086's interface to 2732 EPROMs in Figure 9.17(b) does not use 8086 $\overline{BHE}$ and A_0 to distinguish between even and odd 2732s. The 8086 $\overline{RD}$ and inverted $M/\overline{IO}$ pins are ORed and connected to the 2732 $\overline{OE}$ pins. The 8086 $\overline{CE}$ can be connected to either ground or an unused 8086 address pin. Note that both 2732's are enabled for all data reads; the odd 2732 places data on the demultiplexed 8086 D_8–D_{15} pins while the even 2732 places data on the demultiplexed 8086 D_0–D_7 pins. The 8086 reads the desired data and discards unwanted data if necessary depending on byte, odd word address or even word address transfers.

9.9.3.2 Static RAMs (SRAMs)

Because static RAMs are read/write memories and data will be written to RAM(s) once selected by the 8086, both A_0 and $\overline{BHE}$ must be included in the chip select logic. For each static RAM, the data lines must be connected to either the upper half (AD_{15}–AD_8) or the lower half (AD_7–AD_0) of the 8086 data lines. Figure 9.18 shows the 8086 interface to two 6116 static RAMs along with the pin diagram of the 6116.

In Figure 9.18, the 8086 demultiplexed $\overline{BHE}$ signal is used to select odd 6116 SRAM chips; the data lines of this odd 6116 are connected to the demultiplexed 8086 D_8–D_{15} pins. The 8086 demultiplexed A_0 signal, on the other hand, is used to select even 6116 SRAM chip; the data lines of this even 6116 are connected to the demultiplexed 8086 D_0–D_7 pins. Note that the 6116 has two chip enables $\overline{E}$ and $\overline{G}$ along with a single read/write pin ($\overline{W}$) .When the 6116 is enabled, $\overline{W} = 1$ for read and $\overline{W} = 0$ for write.

9.9.3.3 Dynamic RAMs (DRAMs)

Dynamic RAMs store information as charges in capacitors. Because capacitors can hold charges for a few milliseconds, refresh circuitry is necessary in dynamic RAMs for retaining these charges. Therefore, dynamic RAMs are complex devices to use to design a system. To relieve the designer of most of these complicated interfacing tasks, Intel provides dynamic RAM controllers to interface with the 8086 to build a dynamic memory system. Dynamic RAMs are used for microcomputers requiring large memories.

9.9.4 8086 I/O Ports

Devices with 8-bit I/O ports can be connected to either the upper or the lower half of the data bus. If the I/O port chip is connected to the lower half of the 8086 data lines (AD_0–AD_7), the port addresses will be even ($A_0 = 0$). On the other hand, the port addresses will be odd ($A_0 = 1$) if the I/O port chip is connected to the upper half of the 8086 data lines (AD_8–AD_{15}). A_0 will always be 1 or 0 for the partitioned I/O chip. Therefore, A_0 cannot be used as an address input to select registers within a particular I/O chip. If two chips are connected to the lower and upper halves of the 8086 address bus that differ only in A_0 (consecutive odd and even addresses), A_0 and $\overline{BHE}$ must be used as conditions of chip select decoding to avoid a write to one I/O chip from erroneously performing a write to the other.

The 8086 uses either standard I/O or memory-mapped I/O. The standard I/O uses the instructions IN and OUT, and is able to provide up to 64K bytes of I/O locations. The standard I/O can transfer either 8-bit data or 16-bit data to or from a peripheral device. The 64-Kbyte I/0 locations can then be configured as 64K 8-bit ports or 32K 16-bit ports. All I/O transfers between the 8086 and peripheral devices take place via AL for 8-bit ports (AH is not involved) and AX for 16-bit ports. The I/O port addressing can be done either directly or indirectly as follows:

- ***Direct***
  `IN AX,PORTA` or `IN AL,PORTA` inputs 16-bit contents of port A into AX or 8-bit contents of port A into AL, respectively.
  `OUT PORTA,AX` or `OUT PORTA,AL` outputs 16-bit contents of AX into port A or 8-hit contents of AL into port A, respectively.
- ***Indirect***
  `IN AX,DX` or `IN AL,DX` inputs 16-bit data into a port addressed by DX into AX or 8-bit data into a port addressed by DX into AL, respectively.
  `OUT DX,AX` or `OUT DX,AL` outputs 16-bit contents of AX into a port addressed by DX or 8-bit contents of AL into a port addressed by DX, respectively.

Memory-mapped I/O is basically accomplished by using the memory instructions such as `MOV AX` or `AL,[BX]` and `MOV [BX],AX` or `AL` for inputting or outputting, 8- or 16-bit data to/from AL or AX addressed by the 20-bit address computed from DS and BX. Note that any 8- or 16-bit general purpose register and memory modes can be used in memory-mapped I/O.

The 8086 programmed I/O capability will be explained in the following paragraphs using the 8255 I/O chip. The 8255 chip is a general-purpose programmable I/O chip. The 8255 has three 8-bit I/O ports: ports A, B, and C. Ports A and B are latched 8-bit ports for both input and output. Port C is also an 8-bit port with latched output, but the inputs are not latched. Port C can be used in two ways: It can be used either as a simple I/0 port or as a control port for data transfer using handshaking via ports A and B.

The 8086 configures the three ports by outputting appropriate data to the 8-bit control register. The ports can be decoded by two 8255 input pins A_0 and A_1, as follows:

A_1	A_0	Port Name
0	0	Port A
0	1	Port B
1	0	Port C
1	1	Control register

The structure of the control register is shown in Figure 9.19.

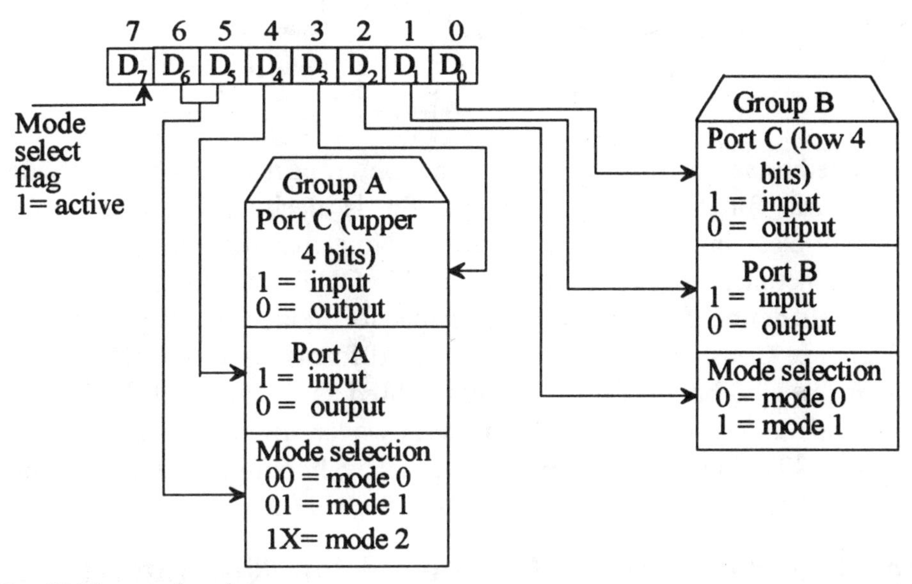

FIGURE 9.19 8255 control register

Bit 7 (D_7) of the control register must be 1 to send the definitions for bits 0–6 (D_0–D_6) as shown in the diagram. In this format, bits D_0–D_6, are divided into two groups: groups A and B. Group A configures all 8 bits of port A and the upper 4 bits of port C; group B defines all 8 bits of port B and the lower 4 bits of port C. All bits in a port can be configured as a parallel input port by writing a 1 at the appropriate bit in the control register by the 8086 OUT instruction, and a 0 in a particular bit position will configure the appropriate port as a parallel output port.

Group A has three modes of operation: modes 0, 1, and 2. Group B has two modes: modes 0 and 1. Mode 0 for both groups provides simple I/O operation for each of the three ports. No handshaking is required. Mode 1 for both groups is the strobed I/O mode used for transferring I/O data to or from a specified port in conjunction with strobes or handshaking signals. Ports A and B use the pins on port C to generate or accept these handshaking signals. Mode 2 of group A is the strobed bidirectional bus I/O and may be used for communicating with a peripheral device on a single 8-bit data bus for both transmitting and receiving data (bidirectional bus I/O). Handshaking signals are required. Interrupt generation and enable/disable functions are also available.

When D_7 = 0, the bit set/reset control word format is used for the control register as follows:

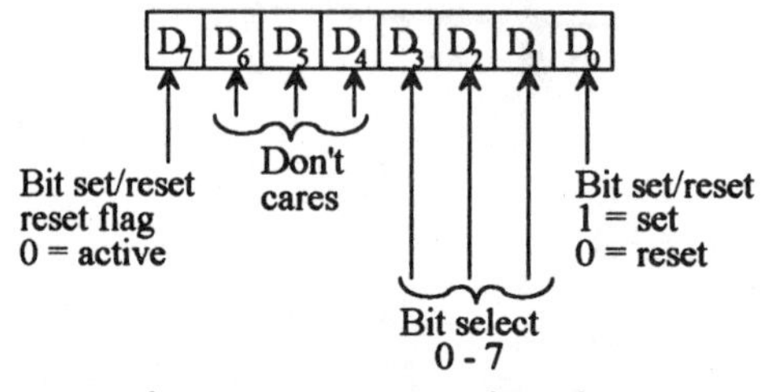

This format is used to set or reset the output on a pin of port C or when enabling of the interrupt output signals for handshake data transfer is desired. For example, the 8 bits (0XXX1100) will clear bit 6 of port C to zero. Note that the control word format can be output to the 8255 control register by using the 8086 OUT instruction.

Now, let us define the control word format for mode 0 more precisely by means of a numerical example. Consider that the control word format is 10000010_2. With this data in the control register, all 8 bits of Port A are configured as outputs and the 8 bits of port C are also configured as outputs. All 8 bits of port B, however, are defined as inputs. On the other hand, outputting 10011011_2 into the control register will configure all three 8-bit ports (ports A, B, and C) as inputs.

9.9.5 Important Points To Be Considered for 8086 Interface to Memory and I/O

From the preceding discussions, the following points can be summarized:

1. For ROMs and EPROMs, $\overline{BHE}$ and A_0 are not required to be part of chip enable/select decoding.
2. For RAMs and I/O port chips, both $\overline{BHE}$ and A_0 must be used in chip select logic.
3. For ROMs/EPROMs and RAMs, both even and odd chips are required. However, for I/O chips, an odd-addressed I/O chip, an even-addressed I/O chip, or both can be used, depending on the number of ports required in an application. 8086 $\overline{BHE}$ and/or A_0 must be used in I/O chip select logic depending on the number of I/O chips used.
4. For interfacing ROMs/EPROMs to the 8086, the same chip select logic must be used for both the even and its corresponding odd memory chip. The same thing applies to RAMs except that $\overline{BHE}$ and A_0 must be used for RAMs.

9.10 <u>8086-Based Microcomputer</u>

In this section, an 8086 will be interfaced in minimum mode to provide 4K × 16 EPROM, 2K × 16 static RAM, and six 8-bit I/O ports. The 2732 EPROM, 6116 static RAM, and 8255 I/O chips are used for this purpose. Memory and I/O maps are determined. Figure 9.20 shows a hardware schematic for accomplishing this.

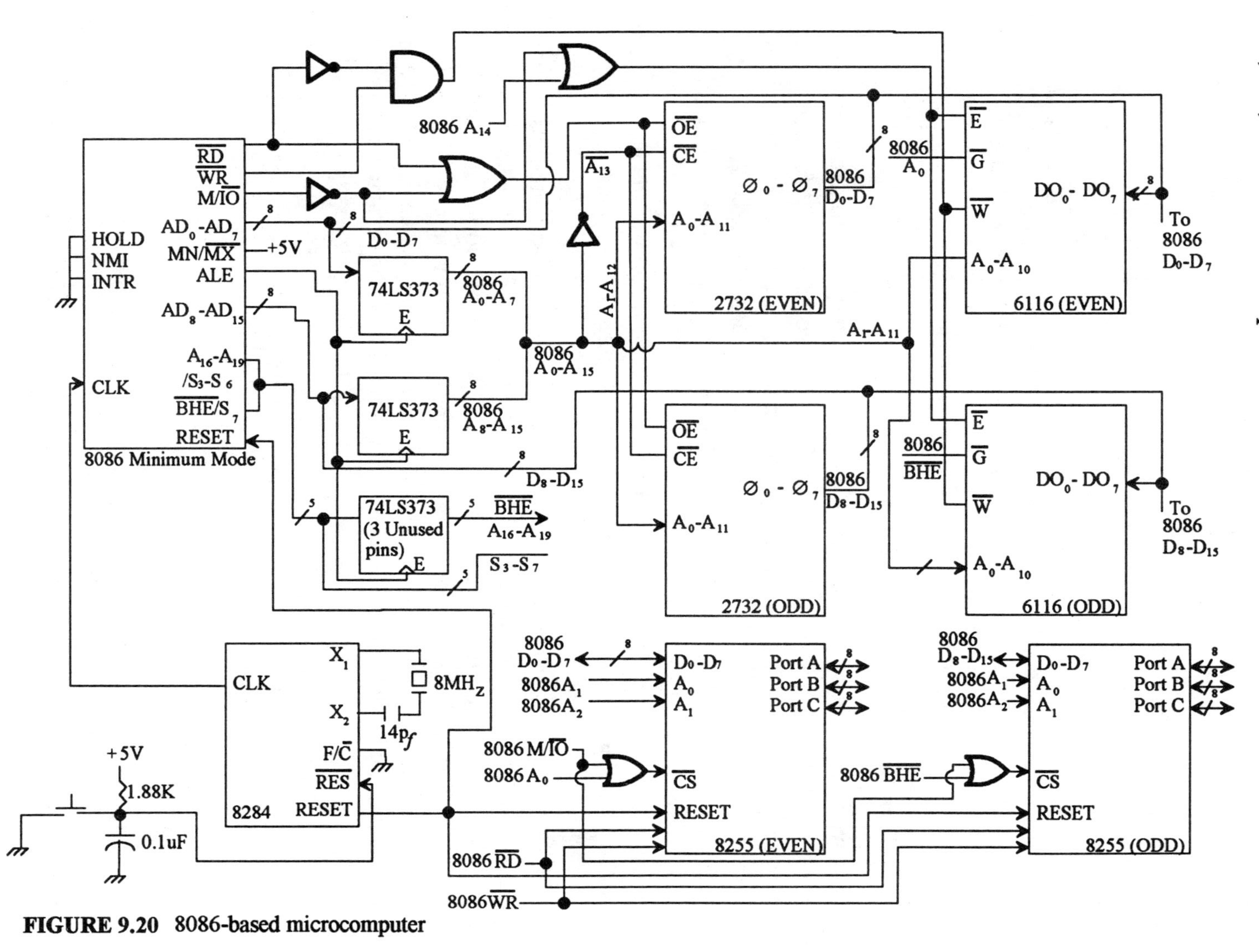

FIGURE 9.20 8086-based microcomputer

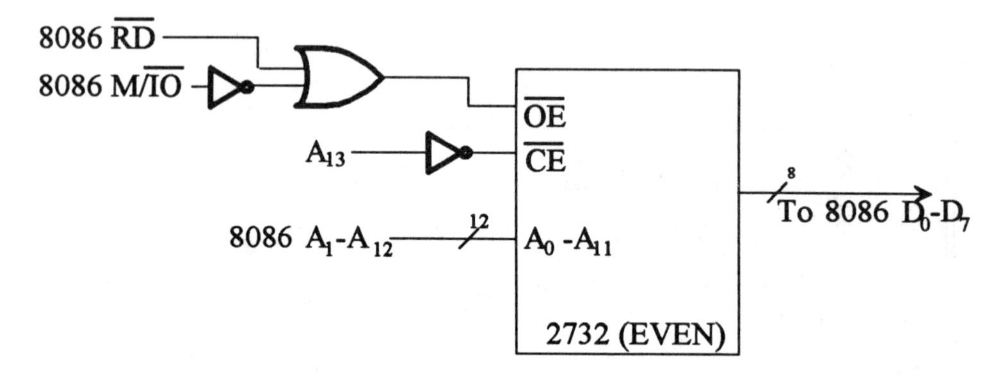

FIGURE 9.21 Even 2732 with pertinent connections

The power and ground pins of all chips must be connected together to the power supply's power and ground pins. The 8086 $MN/\overline{MX}$ is connected to +5 V for minimum mode (single processor) operation. Linear decoding is used to select both EPROMs and SRAMs. 8086 demultiplexed $A_{13} = 1$ is used to select 2732s and 8086 demultiplexed $A_{14} = 0$ is used for 6116s. No unused address pin is used for selecting the 8255s because the 8086 $M/\overline{IO}$ pin distinguishes between memory and I/O.

Let us determine the 8086 memory and I/O maps. To determine the memory map for 2732 EPROMs, consider Figure 9.21 (obtained from Figure 9.20), which shows pertinent connections for the even 2732.

In Figure 9.20, $M/\overline{IO} = 1$ when the 8086 executes a memory-oriented instruction such as `MOV [BX],DL` to access the memory. Also, in the figure, $A_{13} = 1$ is used to select the EPROMs and $A_{14} = 1$ is used to deselect the RAMs. This is done to include the 8086 reset vector $FFFF0_{16}$ in the EPROMs. Therefore, an inverter is used to invert A_{13}. Note that 8086 address pins A_{15}–A_{19} are not used and are, therefore, don't cares. Assume the don't cares to be HIGH. The even memory map for the 2732 in Figure 9.21 can be obtained as follows:

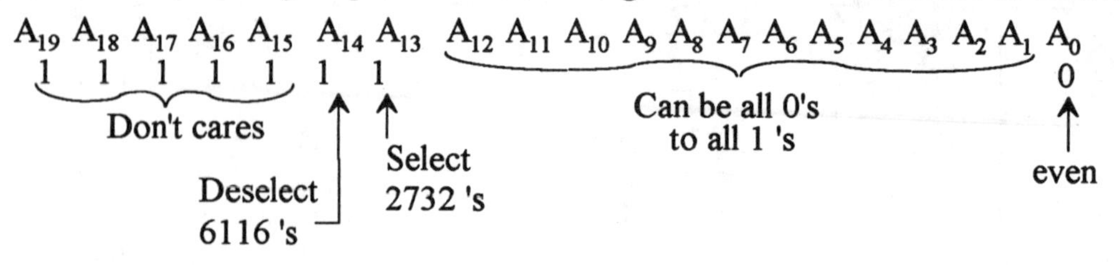

Therefore, the memory map for the even 2732 contains the even addresses FE000H, FE002H, ..., FFFFEH. Similarly, the memory map for the odd 2732 can be determined as: FE001H, FE003H, ..., FFFFFH. Note that the reset vector FFFF0H is included in this map.

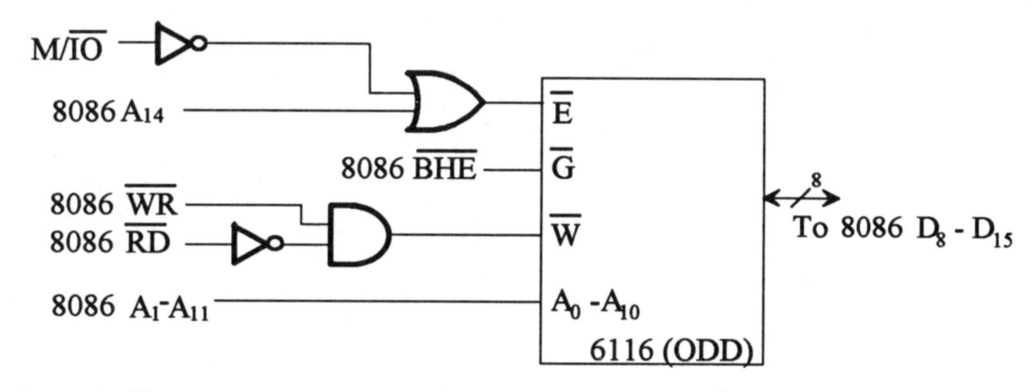

FIGURE 9.22 Odd 6116 with pertinent connections

Let us now determine the memory map for the odd 6116. Consider Figure 9.22 (obtained from Figure 9.20), which shows pertinent connections for the odd 6116.

In Figure 9.20, $A_{13} = 0$ deselects 2732s and $A_{14} = 0$ selects 6116s. Also, the 8086 outputs HIGH on its $M/\overline{IO}$ pin ($M/\overline{IO} = 1$) when it executes a memory-oriented instruction such as `MOV CX, [SI]`. Furthermore, the 8086 outputs a LOW on the $\overline{BHE}$ pin for odd addresses. With don't care addresses, pins A_{15}–A_{19} and A_{12} as ones, the odd memory map for the 6116 in Figure 9.22 can be obtained as follows:

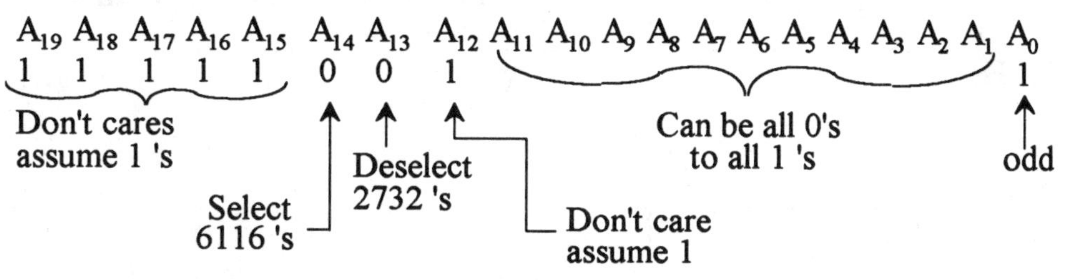

Therefore, the memory for the odd 6116 contains the odd addresses F9001H, F9003H, ..., F9FFFH. Similarly, the memory map for the even 6116 can be obtained as F9000H, F9002H, ..., F9FFEH.

Finally, the I/O map for the 8255s is determined. Consider Figure 9.23 (obtained from Figure 9.20), which shows pertinent connections for the even 8255. The 8086 outputs LOW on its $M/\overline{IO}$ pin ($M/\overline{IO} = 0$) when it executes an IN or OUT instruction. The 8086 outputs LOW ($A_0 = 0$) for an even port address. This will produce a LOW on the $\overline{CS}$ pin of the even 8255. The even 8255 will thus be selected.

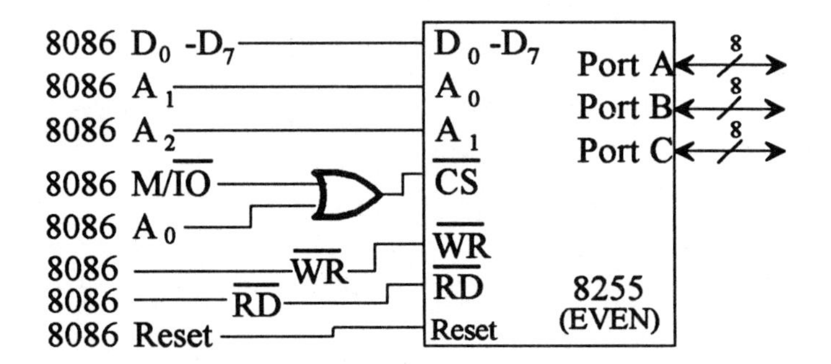

FIGURE 9.23 Even 8255 with pertinent connections

Using 8086 A_1 and A_2 pins for port addresses, the I/O map for the even 8255 chip can be determined as follows:

Port Addresses for even 8255

Port Name	*Address*
Port A	A_7 A_6 A_5 A_4 A_3 A_2 A_1 A_0 = F8H X X X X X 0 0 0 Don't cares assume 1 's (X's); Port A (A_2A_1 = 00); even (A_0 = 0)
Port B	X X X X X 0 1 0 = FAH Don't cares assume 1 's (X's); Port B (A_2A_1 = 01); even (A_0 = 0)
Port C	X X X X X 1 0 0 = FCH Don't cares assume 1 's (X's); Port C (A_2A_1 = 10); even (A_0 = 0)
Control Register	X X X X X 1 1 0 = FEH Don't cares assume 1 's (X's); Control register (A_2A_1 = 11); even (A_0 = 0)

Similarly, the I/O map for the odd 8255 chip is:

Port addresses for the odd 8255		
Port A	=	F9H
Port B	=	FBH
Port C	=	FDH
Control Register	=	FFH

Table 9.12 summarizes the memory and I/O maps.

TABLE 9.12 Memory and I/O Maps for the Microcomputer of Figure 9.20

Memory Map

Chip Number		Physical Address	Logical Address	
			Segment Value	Offset
Even	2732 EPROM	FE000H, FE002H, ... , FFFFEH	FE00H	0000H, 0002H, ... , 1FFEH
Odd	2732 EPROM	FE001H, FE003H, ... , FFFFFH	FE00H	0001H, 0003H, ... , 1FFFH
Even	6116 SRAM	F9000H, F9002H, ... , F9FFEH	F900H	0000H, 0002H, ... , 0FFEH
Odd	6116 SRAM	F9001H, F9003H, ... , F9FFFH	F900H	0001H, 0003H, ... , 0FFFH

I/O Map

Chip Number		Port Address
Even	8255	Port A = F8H, Port B = FAH, Port C = FCH, Control Register = FEH
Odd	8255	Port A = F9H, Port B = FBH, Port C = FDH, Control Register = FFH

Example 9.12

An 8086-8255-2732-6116–based microcomputer is required to drive an LED connected to bit 2 of port B based on two switch inputs connected to bits 6 and 7 of port A. If both switches are either HIGH or LOW, turn the LED ON; otherwise, turn it OFF. Assume a HIGH will turn the LED ON and a LOW will turn it OFF. Write an 8086 assembly language program to accomplish this.

Solution

```
PORTA       EQU     0F8H
PORTB       EQU     0FAH
CNTRL       EQU     0FEH
PROG        SEGMENT
            ASSUME CS: PROG
            MOV     AL, 90H         ;       Configure port A
            OUT     CNTRL, AL       ;       as input and port B
                                    ;       as output
BEGIN:      IN      AL, PORTA       ;       Input port A
```

```
            AND     AL, 0C0H        ;       Retain bits 6 and 7
            JPE     LEDON           ;       If both switches are either
                                    ;       HIGH or LOW, turn the LED ON
            MOV     AL, 00H         ;       Otherwise turn the
            OUT     PORTB, AL       ;       LED OFF
            JMP     BEGIN           ;       Repeat
LEDON:      MOV     AL, 04H         ;       Turn LED
            OUT     PORTB, AL       ;       ON
            JMP     BEGIN
PROG        ENDS
            END
```

Example 9.13

Write an 8086 assembly language program to drive an LED connected to bit 7 of port A based on a switch input at bit 0 of port A. If the switch is HIGH, turn the LED ON; otherwise, turn the LED OFF. Assume an 8086/2732/6116/8255 microcomputer. Also, write a C++ program to accomplish the same task. Compare the 68000 assembly program with the compiled assembly code. Comment on the result.

Solution

The 8086 assembly language program and the C++ program along with the compiled assembly code are shown below. The 8086 assembly program contains 11 instructions whereas the 8086 C++ code generates 16 instructions. This example illustrates that although C++ programming can handle I/O, it generates more codes than assembly language programming. Although programs in C++ are easier to write compared to assembly, the machine code generated by the equivalent assembly language is shorter. Also note that C++ programs are not 100 % portable while the same I/O programs are written using C++ for microprocessors by two different manufactures. This is because of the different hardware configurations (I/O and memory maps) for different manufacturers.

- 8086/8255 Microcomputer Assembly Code for Switch and LED (MASM)

```
= 00F8          PORTA     EQU       0F8H
= 00FE          CTLREG    EQU       0FEH
0000            LAB       SEGMENT
                          ASSUME    CS:LAB
0000  B1 07               MOV       CL,7
0002  B0 90     REPEAT:   MOV       AL,90H
0004  E6 FE               OUT       CTLREG,AL    ;  set PORTA as input
0006  E4 F8               IN        AL,PORTA     ;  read switch
0008  8A D8               MOV       BL,AL        ;  save switch status
000A  B0 80               MOV       AL,80H
000C  E6 FE               OUT       CTLREG,AL    ;  set PORTA as output
000E  8A C3               MOV       AL,BL        ;  get switch status
0010  D2 D0               RCL       AL,CL        ;  rotate switch status
0012  E6 F8               OUT       PORTA,AL     ;  output to LED
0014  EB EC               JMP       REPEAT       ;  repeat
0016            LAB       ENDS
                          END
```

- 8086/8255 Microcomputer C++ program for Switch and LED (C++ Compiler)

```
#include <dos.h>
#define PORTA 0x0F8
#define CNTLREG 0x0FE
int main (){
      int x;
      while(1){
            outportb(CNTLREG, 0x90);      // set PORTA as input
            x = inportb(PORTA);           // read switch
            outportb(CNTLREG, 0x80);      // set PORTA as output
            outportb(PORTA, x << 7);      // output to LED
      }
}
```

• Assembly code generated from C++ code above using Microsoft DEBUG unassembler:

```
-r
AX=0000  BX=0000  CX=022E  DX=0000  SP=FFEE  BP=0000  SI=0000  DI=0000
DS=159B  ES=159B  SS=159B  CS=159B  IP=0100   NV UP EI PL NZ NZ PO NC
159B:0100 800C00        OR     BYTE PTR [SI],00                 DS:0000=CD
-u 2aa 2c8
159B:02AA BAFE00        MOV    DX,00FE
159B:02AD B090          MOV    AL,90
159B:02AF EE            OUT    DX,AL
159B:02B0 BAF800        MOV    DX,00F8
159B:02B3 EC            IN     AL,DX
159B:02B4 B400          MOV    AH,00
159B:02B6 8BD8          MOV    BX,AX
159B:02B8 BAFE00        MOV    DX,00FE
159B:02BB B080          MOV    AL,80
159B:02BD EE            OUT    DX,AL
159B:02BE B107          MOV    CL,07
159B:02C0 8AC3          MOV    AL,BL
159B:02C2 D2E0          SHL    AL,CL
159B:02C4 BAF800        MOV    DX,00F8
159B:02C7 EE            OUT    DX,AL
159B:02C8 EBE0          JMP    02AA
```

9.11 8086 Interrupts

The 8086 assigns every interrupt a type code so that the 8086 can identify it. Interrupts can be initiated by external devices or internally by software instructions or by exceptional conditions such as attempting to divide by zero.

9.11.1 Predefined Interrupts

The first five interrupt types are reserved for specific functions.

Type 0:	INT0	Divide by zero
Type 1:	INT1	Single step
Type 2:	INT2	Nonmaskable interrupt (NMI pin)
Type 3:	INT3	Breakpoint
Type 4:	INT4	Interrupt on overflow

The interrupt vectors for these five interrupts are predefined by Intel. The user must provide the desired IP and CS values in the interrupt pointer table. The user may also initiate these interrupts through hardware or software. If a predefined interrupt is not used in a system, the user may assign some other function to the associated type.

The 8086 is automatically interrupted whenever a division by zero is attempted. This interrupt is nonmaskable and is implemented by Intel as part of the execution of the divide instruction.

When the TF (trap flag) is set by an instruction, the 8086 goes into single-step mode. The TF can be cleared to zero as follows:

```
PUSHF                   ;     Save flags
MOV BP, SP              ;     Move [SP] to [BP]
AND 0[BP], 0FEFFH       ;     Clear TF
POPF                    ;     Pop flags
```

Note here that 0[BP] rather than [BP] is used because BP cannot normally be used without displacement in the 8086 assembler. Now, to set TF, the AND instruction just shown should be replaced by `OR 0[BP],0100H`. Once TF is set to 1, the 8086 automatically generates a type 1 interrupt after execution of each instruction. The user can write a service routine at the interrupt address vector to display memory locations and/or register to debug a program. Single-step mode is nonmaskable and cannot be enabled by the `STI` (enable interrupt) or disabled by the `CLI` (disable interrupt) instruction.

The nonmaskable interrupt is initiated via the 8086 NMI pin. It is edge triggered (LOW to HIGH) and must be active for two clock cycles to guarantee recognition. It is normally used for catastrophic failures such as a power failure. The 8086 obtains the interrupt vector address by automatically executing the `INT2` (type 2) instruction internally.

The type 3 interrupt is used for breakpoints and is nonmaskable. The user inserts the 1-byte instruction `INT3` into a program by replacing an instruction. Breakpoints are useful for program debugging.

The interrupt on overflow is a type 4 interrupt. This interrupt occurs if the overflow flag (OF) is set and the `INTO` instruction is executed. The overflow flag is affected, for example, after execution of a signed arithmetic (such as `IMUL`, signed multiplication) instruction. The user can execute an `INTO` instruction after the `IMUL`. If there is an overflow, an error service routine written by the user at the type 4 interrupt address vector is executed.

9.11.2 Internal Interrupts

The user can generate an interrupt by executing an interrupt instruction `INT`*nn*. The `INT`*nn* instruction is not maskable by the interrupt enable flag (IF). The `INT`*nn* instruction can be used to test an interrupt service routine for external interrupts. Type codes 32–255 can be used; type codes 5 through 31 are reserved by the Intel for future use. If a predefined interrupt is not used in a system, the associate type code can be utilized with the INT*nn* instruction to generate software (internal) interrupts.

9.11.3 External Maskable Interrupts

The 8086 maskable interrupts are initiated via the INTR pin. These interrupts can be enabled or disabled by STI (IF = 1) or CLI (IF = 0), respectively. If IF = 1 and INTR active (HIGH) without occurrence of any other interrupts, the 8086, after completing the current instruction, generates $\overline{\text{INTA}}$ LOW twice, each time for about two cycles.

$\overline{\text{INTA}}$ is only generated by the 8086 in response to INTR, as shown in Figure 9.24. The interrupt acknowledge sequence includes two $\overline{\text{INTA}}$ cycles separated by two clock cycles. ALE is also generated by the 8086 and will load the address latches with indeterminate information. The first $\overline{\text{INTA}}$ bus cycle indicates that an interrupt acknowledge cycle is in progress and allows the system to be ready to place the interrupt type code on the next $\overline{\text{INTA}}$ bus cycle. The 8086 does not obtain the information from the bus during the first cycle. The external hardware must place the type code on the lower half of the 16-bit data bus (D_0–D_7) during the second cycle.

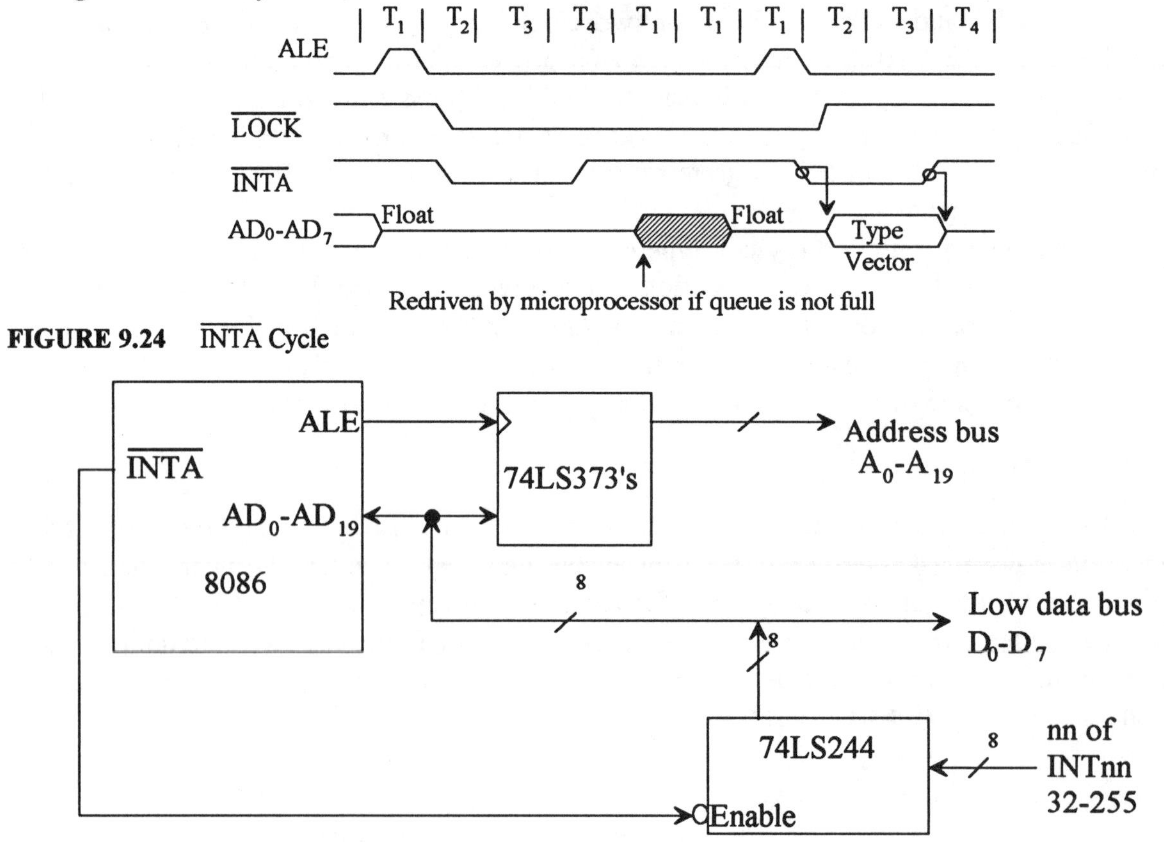

FIGURE 9.24 $\overline{\text{INTA}}$ Cycle

FIGURE 9.25 Servicing the INTR in the minimum mode

In the minimum mode, the $M/\overline{IO}$ is LOW, indicating I/O operation during the $\overline{\text{INTA}}$ bus cycles. The 8086 internal LOCK signal is also LOW from T_2 of the first bus cycle until T_2 of the second bus cycle to keep the BIU from accepting a hold request between the two $\overline{\text{INTA}}$ cycles. Figure 9.25 shows a simplified interconnection between the 8086 and 74LS244 for servicing the INTR. $\overline{\text{INTA}}$ enables the 74LS244 to place type code *nn* on the 8086 data bus. In the maximum mode, the status lines S_0–S_2 will generate the $\overline{\text{INTA}}$ output.

9.11.4 Interrupt Procedures

Once the 8086 has the interrupt type code (via the bus for hardware interrupts, from software interrupt instructions `INT`*nn*, or from the predefined interrupts), the type code is multiplied by 4 to obtain the corresponding interrupt vector in the interrupt vector table. The 4 bytes of the interrupt vector are the least significant byte of the instruction pointer, the most significant byte of the instruction pointer, the least significant byte of the code segment register, and the most significant byte of the code segment register. During the transfer of control, the 8086 pushes the flags and current code segment register and instruction pointer onto the stack. The new CS and IP values are loaded. Flags TF and IF are then cleared to zero. The CS and IP values are read by the 8086 from the interrupt vector table. No segment registers are used when accessing the interrupt pointer table. S_4S_3 has the value 10_2 to indicate no segment register selection.

9.11.5 Interrupt Priorities

As far as the 8086 interrupt priorities are concerned, the single-step interrupt has the highest priority, followed by NMI, followed by the software interrupts. This means that a simultaneous NMI and single-step interrupt will cause the NMI service routine to follow the single step; a simultaneous software interrupt and single step interrupt will cause the software interrupt service routine to follow the single step; and a simultaneous NMI and software interrupt will cause the NMI service routine to be executed prior to the software interrupt service routine. The INTR is maskable and has the lowest priority. A priority interrupt controller such as the 8259A can be used with the 8086 INTR to provide eight levels of interrupts. The 8259A has built-in features for expansion of up to 64 levels with additional 8259s. The 8259A is programmable and can be readily used with the 8086 to obtain multiple interrupts from the single 8086 INTR pin.

9.11.6 Interrupt Pointer Table

The interrupt pointer table provides interrupt address vectors (IP and CS contents) for all the interrupts. There may be up to 256 entries for the 256 type codes. Each entry consists of two addresses, one for storing IP and the other for storing CS. Note that in the 8086 each interrupt address vector is a 20-bit address obtained from IP and CS.

To service an interrupt, the 8086 calculates the two addresses in the pointer table where IP and CS are stored for a particular interrupt type as follows:

For `INT`*nn* ← Type code

The table address for IP = $4 \times nn$ and the table address for CS = $4 \times nn + 2$. For example, consider `INT2`:

Address for IP = 4×2 = 00008H

Address for CS = 00008 + 2 = 0000AH

The values of IP and CS are loaded from location 00008H and 0000AH in the pointer table. Similarly, the IP and CS addresses for other `INT`*nn* are calculated, and their values are obtained from the contents of these addresses in the pointer table (Table 9.13). The 8086 interrupt vectors are defined as follows:

Vectors 0–4	For predefined interrupts
Vectors 5–31	For Intel's future use
Vectors 32–255	For user interrupts

Interrupt service routines should be terminated with an `IRET` (interrupt return) instruction, which pops the top three stack words into the IP, CS, and flags, thus returning control to the right place in the main program.

TABLE 9.13 8086 Interrupt Pointer Table

Interrupt Type Code		*20-Bit Memory Address*
0	IP	00000H
	CS	00002H
1	IP	00004H
	CS	00006H
2		00008H
		0000AH
.		.
.		.
.		.
255	IP	003FCH
	CS	003FEH

9.12 8086 DMA

When configured in minimum mode (MN/$\overline{\text{MX}}$ HIGH) the 8086 provides HOLD and HLDA (hold acknowledge) signals to control the system bus for DMA applications. In this type of DMA, the peripheral device can request the DMA transfer via the DMA request (DRQ) line connected to a DMA controller chip such as the 8257. In response to this request, the 8257 sends a HOLD signal to the 8086. The 8257 then waits for the HLDA signal from the 8086. On receipt of this HLDA, the 8257 sends a $\overline{\text{DMACK}}$ signal to the peripheral device. The 8257 then takes over the bus and controls data transfer between the RAM and peripheral device. On completion of data transfer, the 8257 returns control to the 8086 by disabling the HOLD and $\overline{\text{DMACK}}$ signals.

Example 9.14

In Figure 9.26, an 8086-based microcomputer is required to implement a voltmeter to measure voltage in the range 0 to 5 V and display the result in two decimal digits: one integer part and one fractional part. The microcomputer is required to start the A/D converter at the falling edge of a pulse via bit 0 of Port C. When the conversion is completed, the A/D's "conversion complete" signal will go HIGH. During the conversion, the A/D's "conversion complete" signal stays LOW. Use the 8255 control register = FEH, Port A = F8H, Port B = FAH, and Port C = FCH.

Using programmed I/O, the microcomputer is required to poll the A/D's "conversion complete" signal. When the conversion is completed, the microcomputer will send a LOW of the A/D converter's "output enable" line via bit 1 to port C and then input the 8-bit output from A/D via port B and display the voltage (0 to 5 V) in two decimal digits (one integer and one fractional) via port A on two TIL 311 displays.

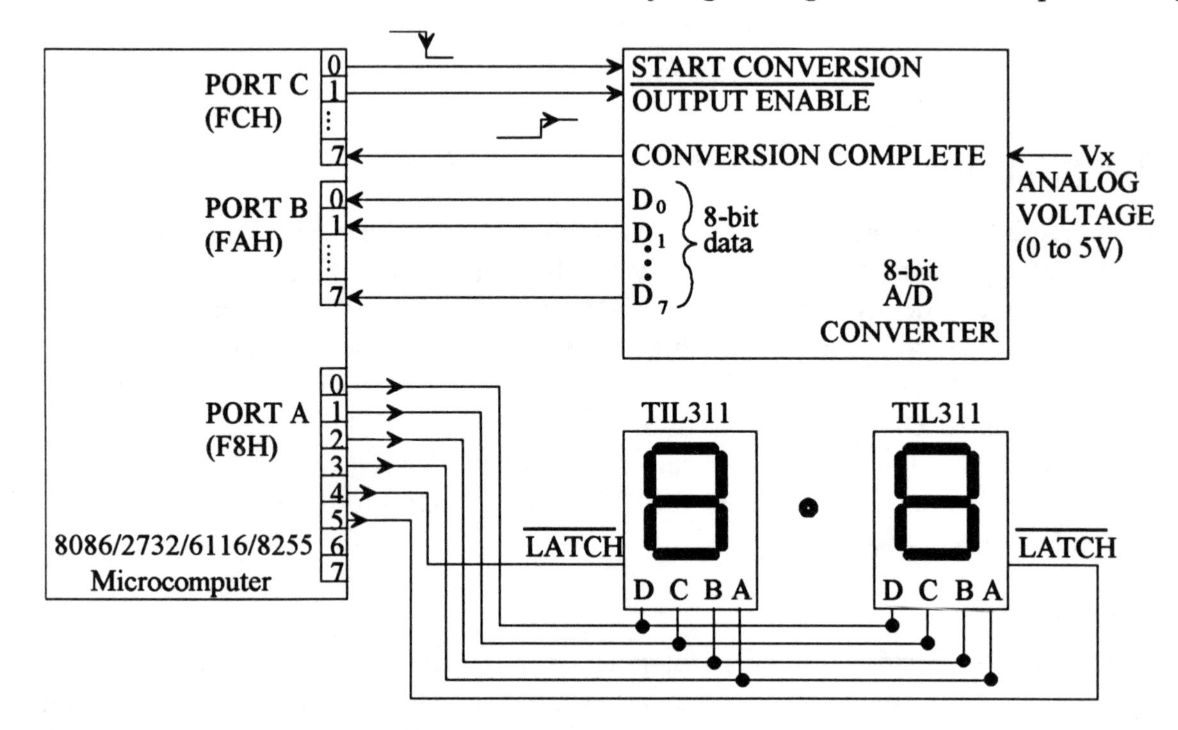

FIGURE 9.26 Figure for Example 9.14

Note that the TIL 311 has an on-chip BCD to seven-segment decoder. The microcomputer will output each decimal digit on the common lines (bits 0–3 of port A) connected to the DCBA inputs of the displays. Each display will be enabled by outputting LOW on each $\overline{\text{LATCH}}$ line in sequence (one after another) so that the input voltage V_x (0 to 5 V) will be displayed with one integer part and fractional part. Write an 8086 assembly language program to accomplish this.

Using interrupt I/O (both NMI and INTR), repeat the task. Write the main program to initialize the 8255 control register and start the A/D. The service routine will input the A/D data, display the result, and return to the main program. Write an 8086 assembly language program for the main program and the service routine. Use the memory map of your choice. Write a service routine for both NMI and INTR at IP = 2000H, CS = 1000H.

Solution

Because the maximum decimal value that can be accommodated in 8 bits is 255_{10} (FF_{16}), the maximum voltage of 5 V will be equivalent to 255_{10}. This means the display in decimal is given by

$$D = 5 \times (\text{Input}/255)$$
$$= \text{Input}/51$$
$$= \underbrace{\text{Quotient}}_{\text{Integer part}} + \text{Remainder}$$

This gives the integer part. The fractional part in decimal is

$$F = (\text{Remainder}/51) \times 10$$
$$\simeq (\text{Remainder})/5$$

For example, suppose that the decimal equivalent of the 8-bit output of A/D is 200.

$$D = 200/51 \Rightarrow \textit{Quotient} = 3, \textit{Remainder} = 47$$
$$\text{Integer part} = 3$$
$$\text{Fractional part}, F = 47/5 = 9$$

Therefore, the display will show 3.9 V.

(a) The 8086 assembly language program using programmed I/O can be written as follows:

```
CDSEG    SEGMENT AT 0FE00H
         ASSUME CS:CDSEG
PORTA    EQU    0F8H
PORTB    EQU    0FAH
PORTC    EQU    0FCH
CNTRL    EQU    0FEH
         MOV    AL,8AH       ;      Configure PORTA, PORTB
         OUT    CNTRL,AL     ;      and PORTC
         MOV    AL,03H       ;      Send 1 to START pin of A/D
         OUT    PORTC,AL     ;      and 1 to (OUTPUT ENABLE-bar)
         MOV    AL,02H       ;      Send 0 to start pin
         OUT    PORTC,AL     ;      of A/D
BEGIN:   IN     AL,PORTC     ;      Check conversion
         ROL    AL,1         ;      Complete bit for HIGH
         JNC    BEGIN
         MOV    AL,00H       ;      Send LOW to (OUTPUT ENABLE-bar)
         OUT    PORTC,AL
         IN     AL,PORTB     ;      Input A/D data
         MOV    AH,0         ;      Convert input data to 16-bit
                             ;      unsigned number in AX
         MOV    DL,51        ;      Convert data to
         DIV    DL           ;      integer part
```

```
        MOV     CL,AL           ;       Save quotient (integer) in CL
        XCHG    AH,AL           ;       Move remainder to AL
        MOV     AH,0            ;       Convert remainder to unsigned
                                ;       16-bit number
        MOV     BL,5            ;       Convert data to
        DIV     BL              ;       fractional part
        MOV     DL,AL           ;       Save quotient (fraction) to DL
        MOV     AL,CL           ;       Move integer part
        OR      AL,20H          ;       Disable fractional display
        AND     AL,2FH          ;       Enable integer display
        OUT     PORTA,AL        ;       Display integer part
        MOV     AL,DL           ;       Move fractional part
        OR      AL,10H          ;       Disable integer display
        AND     AL,1FH          ;       Enable fractional display
        OUT     PORTA,AL        ;       Display fractional part
        HLT
CDSEG   ENDS
        END
```

(b) Using NMI

In Figure 9.26, connect the "conversion complete" to 8086 NMI; all other connections in Figure 9.26 will remain unchanged. Note that all addresses selectable by the user are arbitrarily chosen in the following. The main program in 8086 assembly language is

```
STSEG   SEGMENT AT 0F900H
        DB      32 DUP (?)
STSEG   ENDS
CDSEG   SEGMENT AT 0FE00H
        ASSUME CS:CDSEG,SS:STSEG,DS:DATA
PORTA   EQU     0F8H
PORTB   EQU     0FAH
PORTC   EQU     0FCH
CNTRL   EQU     0FEH
        MOV     AX,0F900H       ;       Initialize
        MOV     SS,AX           ;       stack segment
        MOV     AX,0000H        ;       Initialize
        MOV     DS,AX           ;       data segment
        MOV     SP,0100H        ;       Initialize SP
        MOV     AL,8AH          ;       Configure PORTA, PORTB
        OUT     CNTRL,AL        ;       and PORTC
        MOV     AL,03H          ;       Send 1 to START pin of A/D
        OUT     PORTC,AL        ;       and 1 to (OUTPUT ENABLE)
        MOV     AL,02H          ;       Send 0 to start pin
        OUT     PORTC,AL        ;       of A/D
DELAY:  JMP     DELAY           ;       Wait for interrupt
        HLT
CDSEG   ENDS
        END
```

The NMI Service routine is

```
DATA SEGMENT AT 0000H
     ORG    0008H            ;      Initialize
     DW     2000H            ;      IP = 2000H,
     DW     1000H            ;      CS = 1000H
DATA ENDS                    ;      for Pointer Table
CODE SEGMENT AT 1000H        ;      Start Program at
     ASSUME CS:CODE          ;      CS = 1000H, IP = 2000H
     ORG    2000H
     MOV    AL,00H           ;      Send LOW to (OUTPUT ENABLE)
     OUT    PORTC,AL
     IN     AL,PORTB         ;      Input A/D data
     MOV    AH,0             ;      Convert input to 16-bit unsigned number
     MOV    DL,51            ;      Convert data to
     DIV    DL               ;      integer part
     MOV    CL,AL            ;      Save quotient (integer) in CL
     XCHG   AH,AL            ;      Move remainder to AL
     MOV    AH,0             ;      Convert remainder to unsigned 16-bit
     MOV    BL,5             ;      Convert data to
     DIV    BL               ;      fractional part
     MOV    DL,AL            ;      Save quotient (fraction) to DL
     MOV    AL,CL            ;      Move integer part
     OR     AL,20H           ;      Disable fractional display
     AND    AL,2FH           ;      Enable integer display
     OUT    PORTA,AL         ;      Display integer part
     MOV    AL,DL            ;      Move fractional part
     OR     AL,10H           ;      Disable integer display
     AND    AL,1FH           ;      Enable fractional display
     OUT    PORTA,AL         ;      Display fractional part
     IRET                    ;      Return from interrupt
CODE ENDS
     END
```

(c) Using INTR

All connections in Figure 9.26 will be same except A/D's "conversion complete" to 8086 INTR as shown in Figure 9.27. All other connections in Figure 9.26 will remain unchanged. INT FFH is used. In response to INTR, the 8086 pushes IP and SR onto the stack, and generates LOW on $\overline{\text{INTA}}$. An octal buffer such as 74LS244 can be enabled by this $\overline{\text{INTA}}$ to transfer FF_{16} in this case (can be entered via eight DIP switches connected to + 5 V through a 1 KΩ resistor) to the input of the octal buffer. The output of the octal buffer is connected to the demultiplexed D_0–D_7 lines of the 8086. The 8086 executes INT FFH and goes to the interrupt pointer table to load the contents of physical

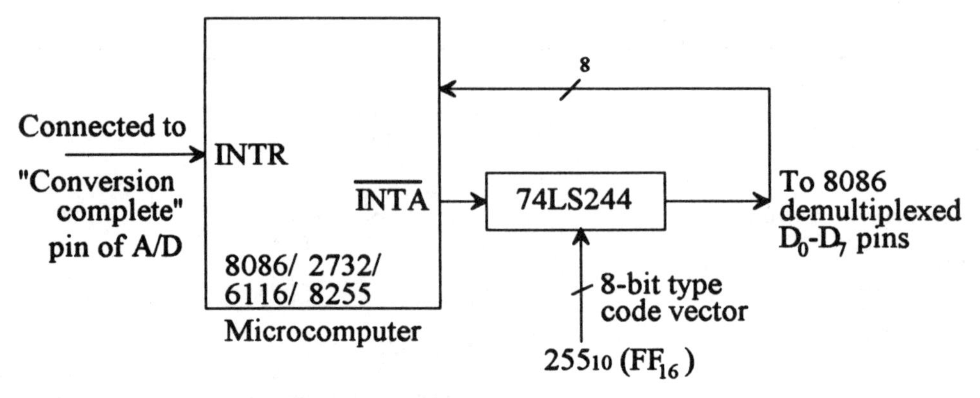

FIGURE 9.27 Hardware interface for 8086 INTR

addresses 003FCH (logical address: CS = 0000H, IP = 03FCH) and 003FEH (logical address: CS = 0000H, IP = 03FEH) to obtain IP and CS for the service routine respectively. Suppose that it is desired to write the service routine at IP = 2000H and CS = 1000H; these IP and CS values must be stored at addresses 003FCH and 003FEH respectively. All user selectable addresses are arbitrarily chosen. The main program in 8086 assembly language is

```
STSEG   SEGMENT AT 0F900H
        ORG     0100H
        DB      32 DUP (?)
STSEG   ENDS
CDSEG   SEGMENT AT 0F300H
        ASSUME CS:CDSEG, SS:STSEG,DS:DATA
PORTA   EQU     0F8H
PORTB   EQU     0FAH
PORTC   EQU     0FCH
CNTRL   EQU     0FEH
        MOV     AX,0F900H       ;       Initialize
        MOV     SS,AX           ;       stack segment
        MOV     AX,0000H        ;       Initialize
        MOV     DS,AX           ;       data segment
        MOV     SP,0100H        ;       Initialize SP
        MOV     AL,8AH          ;       Configure port A, port B,
        OUT     CNTRL,AL        ;       and port C
        STI                     ;       Enable Interrupt
        MOV     AL,03H          ;       Send one to start pin of A/D
        OUT     PORTC,AL        ;       and one to (OUTPUT ENABLE)
        MOV     AL,02H          ;       Send zero to start pin of A/D
        OUT     PORTC,AL
DELAY:  JMP     DELAY           ;       Wait for interrupt
        HLT
CDSEG   ENDS
        END
```

The INTR service routine is

```
DATA SEGMENT AT 0000H
     ORG    03FCH          ;  Initialize
     DW     2000H          ;  IP = 2000H,
     DW     1000H          ;  CS = 1000H
DATA ENDS                  ;  for Pointer Table
CODE SEGMENT AT 1000H
     ASSUME CS:CODE
     ORG    2000H
     MOV    AL,0           ;     Send LOW to
     OUT    PORTC,AL       ;     (OUTPUT ENABLE)
     IN     AL,PORTB       ;     Input A/D data
     MOV    AH,0           ;     Convert input data to
                           ;     16-bit unsigned number in AX
     MOV    DL,51          ;     Convert data
     DIV    DL             ;     to integer part
     MOV    CL,AL          ;     Save quotient (integer) in CL
     XCHG   AH,AL          ;     Move remainder to AL
     MOV    AH,0           ;     Convert remainder to unsigned 16-bit
     MOV    BL,5           ;     Convert data
     DIV    BL             ;     to fractional part
     MOV    DL,AL          ;     Save quotient (fraction) in DL
     MOV    AL,CL          ;     Move integer part
     OR     AL,20H         ;     Disable fractional display
     AND    AL,2FH         ;     Enable integer display
     OUT    PORTA,AL       ;     Display integer part
     MOV    AL,DL          ;     Move fractional part
     OR     AL,10H         ;     Disable integer display
     AND    AL,1FH         ;     Enable fraction display
     OUT    PORTA,AL       ;     Display fractional part
     IRET                  ;     Return from interrupt
CODE ENDS
     END
```

9.13 Interfacing an 8086-Based Microcomputer to a Hexadecimal Keyboard and Seven-Segment Displays

This section describes the characteristics of the 8086-based microcomputer used with a hexadecimal keyboard and a seven-segment display.

9.13.1 Basics of Keyboard and Display Interface to a Microcomputer

A common method of entering programs into a microcomputer is via a keyboard. A popular way of displaying results by the microcomputer is by using seven-segment displays. The main functions to be performed for interfacing a keyboard are:

1. Sense a key actuation.
2. Debounce the key.
3. Decode the key.

Let us now elaborate on keyboard interfacing concepts. A keyboard is arranged in rows and columns. Figure 9.28 shows a 2 × 2 keyboard interfaced to a typical microcomputer. In Figure 9.28, the columns are normally at a HIGH level. A key actuation is sensed by sending a LOW (closing the diode switch) to each row one at a time via PA0 and PA1 of port A. The two columns can then be input via PB2 and PB3 of port B to see whether any of the normally HIGH columns are pulled LOW by a key actuation. If so, the rows can be checked individually to determine the row in which the key is down. The row and column code for the pressed key can thus be found.

The next step is to debounce the key. Key bounce occurs when a key is pressed or released—it bounces for a short time before making the contact. When this bounce occurs, it may appear to the microcomputer that the same key has been actuated several times instead of just once. This problem can be eliminated by reading the keyboard after about 20 ms and then verifying to see if it is still down. If it is, then the key actuation is valid.

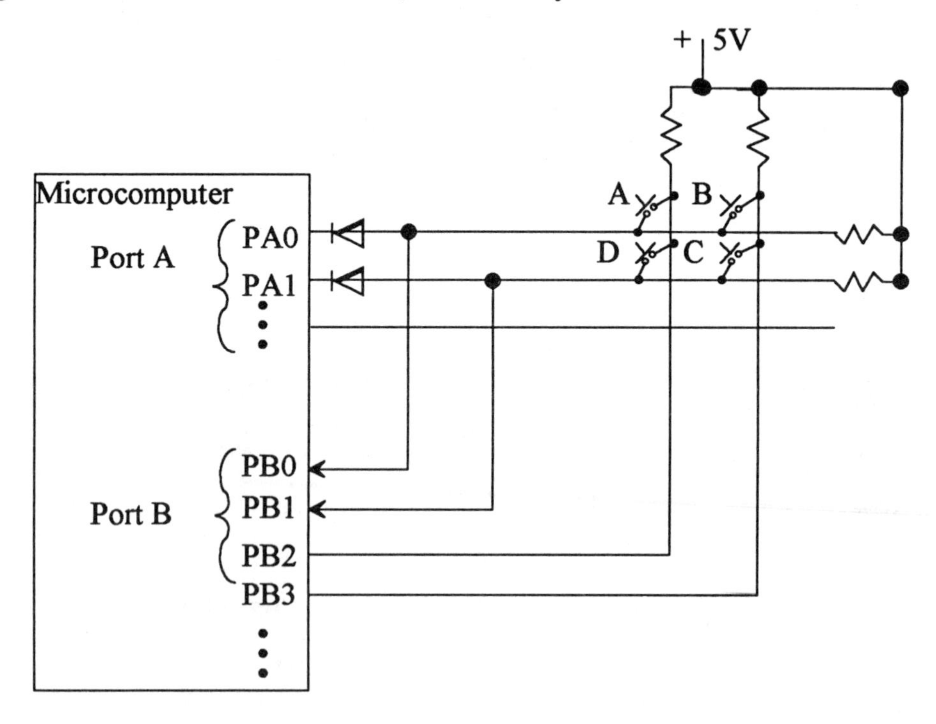

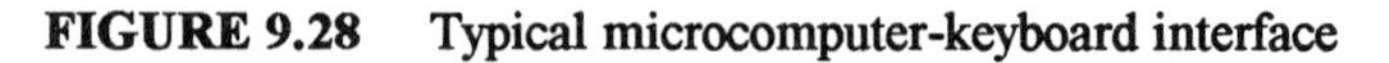
FIGURE 9.28 Typical microcomputer-keyboard interface

The next step is to translate the row and column code into a more popular code such as hexadecimal or ASCII. This can easily be accomplished by a program. Certain characteristics associated with keyboard actuations must be considered while interfacing to a microcomputer. Typically, these are two-key lockout and *N*-key rollover. The two-key lockout ensures that only one key is pressed. An additional key depressed and released does not generate any codes. The system is simple to implement and most often used. However, it might slow down the typing because each key must be fully released before the next one is pressed down. On the other hand, the *N*-key rollover will ignore all keys pressed until only one remains down.

Now let us elaborate on the interfacing characteristics of typical displays. The following functions are typically performed for displays:

1. Output the appropriate display code.
2. Output the code via right entry or left entry into the displays if there are more than one displays.

These functions can easily be realized by a microcomputer program. If there are more than one displays, the displays are typically arranged in rows. A row of four displays is shown in Figure 9.29. In the figure, one has the option of outputting the display code via right entry or left entry. If the code is entered via right entry, the code for the least significant digit of the four-digit display should be output first, then the next digit code, and so on. The program outputs to the displays are so fast that visually all four digits will appear on the display simultaneously. If the displays are entered via left entry, then the most significant digit must be output first and the rest of the sequence is similar to the right entry.

Two techniques are typically used to interface a hexadecimal display to the microcomputer: nonmultiplexed and multiplexed. In nonmultiplexed methods, each hexadecimal display digit is interfaced to the microcomputer via an I/O port. Figure 9.30 illustrates this method. BCD to seven-segment conversion is done in software. The microcomputer can be programmed to output to the two display digits in sequence. However, the microcomputer executes the display instruction sequence so fast that the displays appear to the human eye at the same time.

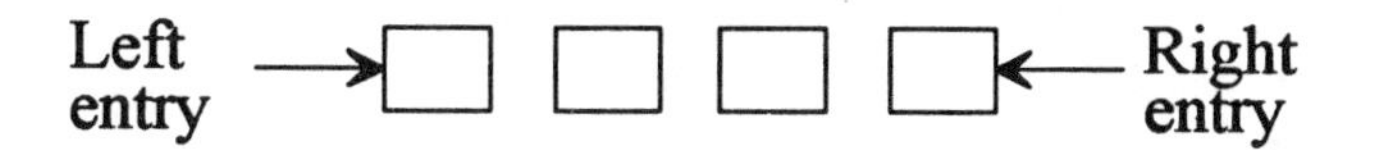

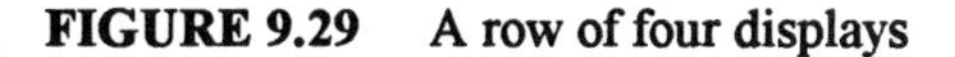

FIGURE 9.29 A row of four displays

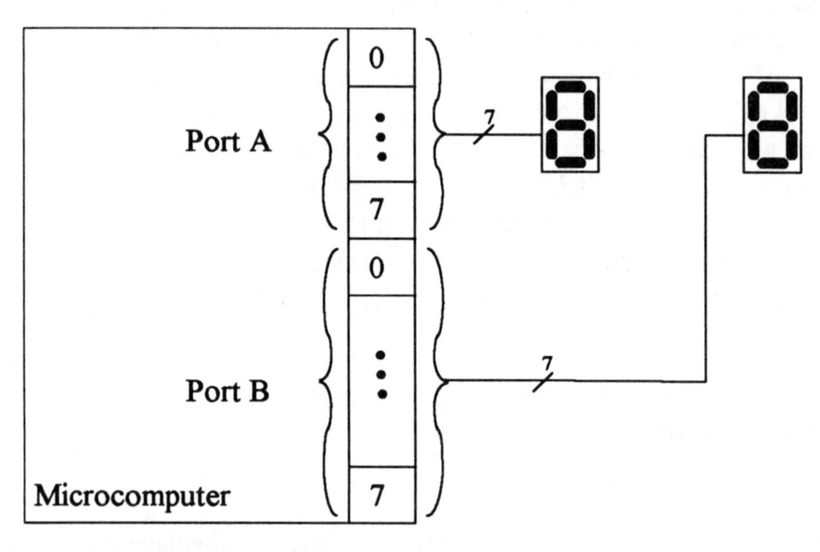

FIGURE 9.30 Nonmultiplexed hexadecimal displays

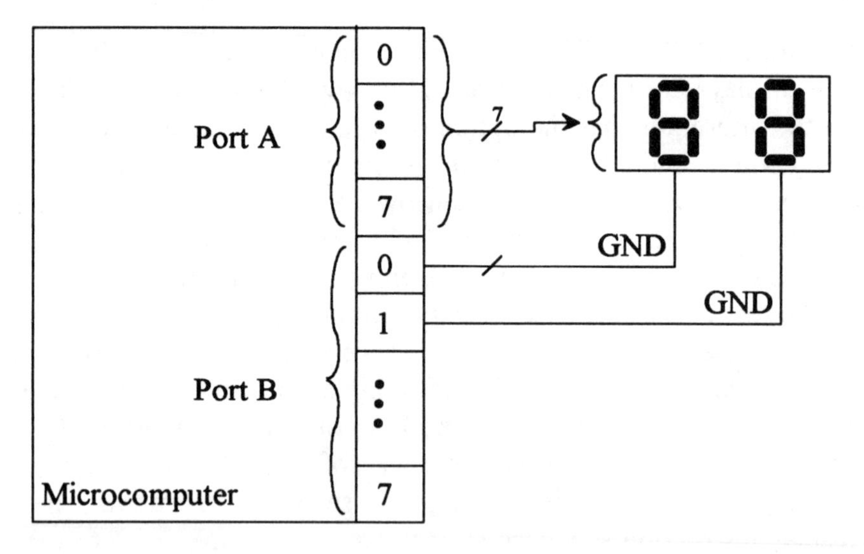

FIGURE 9.31 Multiplexed displays

Figure 9.31 illustrates the multiplexing method of interfacing the two hexadecimal displays to the microcomputer. In the multiplexing scheme, appropriate seven-segment code is sent to the desired displays on seven lines common to all displays. However, the display to be illuminated is grounded. Some displays such as Texas Instrument's TIL 311 have on-chip decoder. In this case, the microcomputer is required to output four bits (decimal) to a display.

The keyboard and display interfacing concepts described here can be realized by either software or hardware. To relieve the microprocessor of these functions, microprocessor manufacturers have developed a number of keyboard/display controller chips. These chips are typically initialized by the microprocessor. The keyboard/display functions are then performed by the chip independent of the microprocessor. The amount of keyboard/display functions performed by the controller chip varies from one manufacturer to another. However, these functions are usually shared between the controller chip and the microprocessor.

9.13.2 Hex Keyboard Interface to an 8086-Based Microcomputer

In this section, an 8086-based microcomputer is designed to display a hexadecimal digit entered via a keypad (16 keys). Figure 9.32 shows the hardware schematic.

1. Port A is configured as an input port to receive the row–column code.
2. Port B is configured as an output port to display the key(s) pressed.
3. Port C is configured as an output port to output zeros to the rows to detect a key actuation.

The system is designed to run at 2 MHz. Debouncing is provided to avoid unwanted oscillation caused by the opening and closing of the key contacts. To ensure stability for the input signal, a delay of 20 ms is used for debouncing the input.

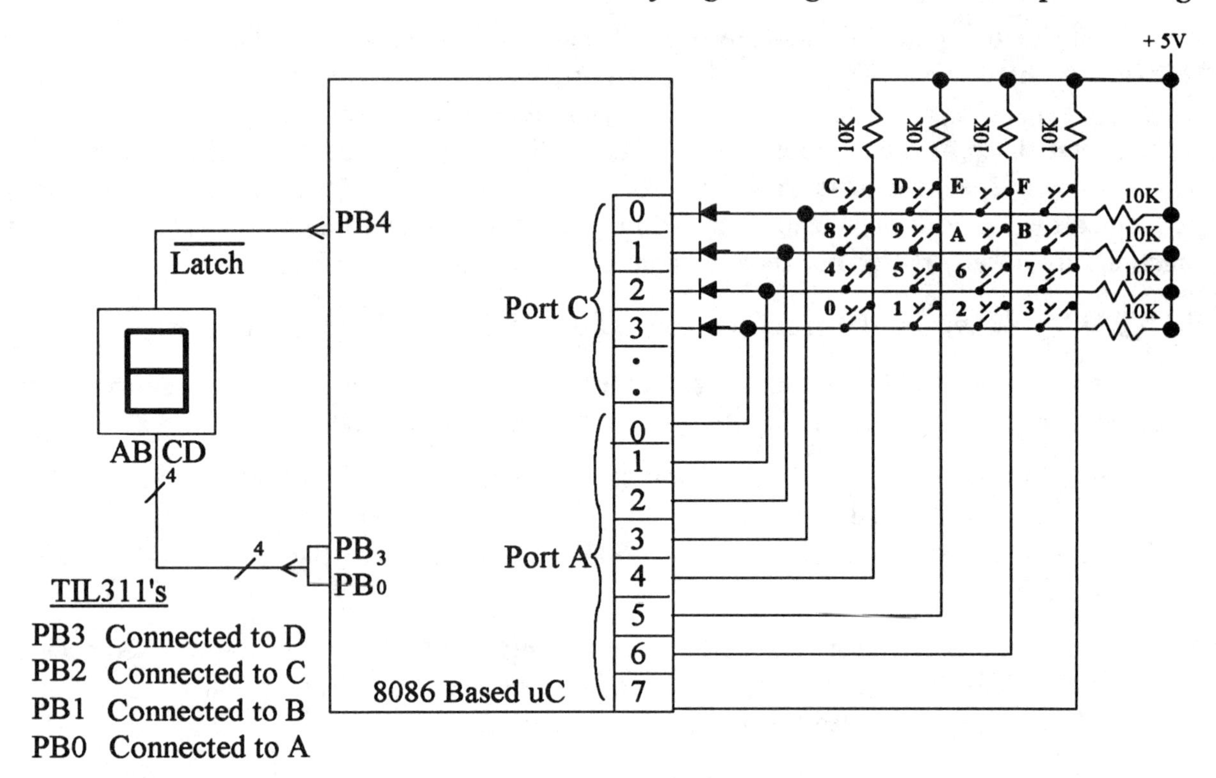

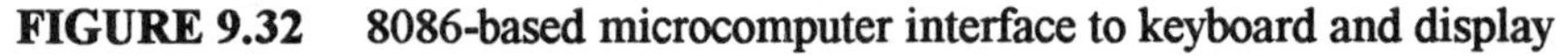
FIGURE 9.32 8086-based microcomputer interface to keyboard and display

The program begins by performing all necessary initializations. Next, it makes sure that all the keys are opened (not pressed). A delay loop of 20 ms is included for debouncing. The initial loop counter values is calculated using the cycles required to execute the following 8086 instructions:

`MOV` reg/imm (4 cycles)
`LOOP` label (19/5 cycles)

Note that the 8086 `LOOP` instructions requires two different execution times. `LOOP` requires 19 cycles when the 8086 branches if the CX is not equal to zero. In the last iteration, the 8086 goes to the next instruction and does not branch when CX = 0, and this requires 5 cycles.

For a 2-MHz clock, each cycle is 500ns. For 20 ms, total cycles = $\frac{20\ msec}{500\ nsec}$ = 40,000. The loop will require 19 cycles for (count - 1) times and the last iteration will take 5 cycles. Thus, total cycles including the `MOV` = 4 + 19 × (count - 1) = 40,000. Hence, count = $38D2_{16}$. Therefore, CX must be loaded with $38D2_{16}$.

The next three lines detect a key closure. If a key closure is detected, it is debounced. It is necessary to determine exactly which key is pressed. To do this, a sequence of row-control codes (0FH, 0EH, 0DH, 0BH, 07H) are output via port C. The row–column code is input via port A to determine if the column code changes corresponding to each different row code. If the column code is not 0FH (changed), the input key is identified. The program then indexes through a look-up table to determine the row–column code saved in DL. If the code is found, the corresponding index value, which equals the input key's value (a single hexadecimal digit) is displayed. The program is written such that it will continuously scan for input key and update the display for each new input. Note that lowercase letters are used to represent the 8086 registers in the program. For example, `al`, `ah`, and `ax` in the program represent the 8086 AL, AH, and AX registers, respectively.

The memory and I/O maps are arbitrarily chosen. A listing of the 8086 assembly language program is given in the following:

```
0000               CDSEG    SEGMENT
                            ASSUME  CS:CDSEG,DS:DTSEG
 = 00F8            PORTA    EQU     0F8h        ;   Hex keyboard input (row/column)
 = 00FA            PORTB    EQU     0FAh        ;   LED displays/controls
 = 00FC            PORTC    EQU     0FCh        ;   Hex keyboard row controls
 = 00FE            CSR      EQU     0FEh        ;   Control status register
 = 00F0            OPEN     EQU     0F0h        ;   Row/column codes if all keys are opened
 0000  BB 0100              mov     bx, 0100h
 0003  8E DB                mov     ds, bx
 0005  B0 90        start:  mov     al, 90h     ;   Config ports A, B, C as i/o/o
 0007  E6 FE                out     CSR, al
 0009  2A C0                sub     al, al      ;   Clear al
 000B  E6 FA                out     PORTB,al    ;   Enable/initialize display
 000D  2A C0     scan_key:  sub     al, al      ;   Clear al
 000F  E6 FC                out     PORTC, al   ;   Set row controls to zero
 0011  E4 F8     key_open:  in      al, PORTA   ;   Read PORTA
 0013  3C F0                cmp     al, OPEN    ;   Are all keys opened?
 0015  75 FA                jnz     key_open    ;   Repeat if closed
 0017  B9 38D2              mov     cx, 38d2h   ;   Delay of 20 ms
 001A  E2 FE       delay1:  loop    delay1      ;   key opened
 001C  E4 F8   key_close:   in      al, PORTA   ;   read PORTA
 001E  3C F0                cmp     al, OPEN    ;   Are all keys closed?
 0020  74 FA                jz      key_close   ;   repeat if opened
 0022  B9 38D2              mov     cx, 38d2h   ;   delay of 20 ms
 0025  E2 FE       delay2:  loop    delay2      ;   Debounce key closed
 0027  B0 FF                mov     al, 0FFh    ;   Set al to all 1's
 0029  F8                   clc                 ;   carry
 002A  D0 D0     next_row:  rcl     al, 1       ;   Set up row mask
 002C  8A C8                mov     cl, al      ;   Save row mask in cl
 002E  E6 FC                out     PORTC, al   ;   Set a row to zero
 0030  E4 F8                in      al, PORTA   ;   Read PORTA
 0032  8A D0                mov     dl, al      ;   Save row/coln codes in dl
 0034  24 F0                and     al, 0F0h    ;   Mask row code
 0036  3C F0                cmp     al, 0F0h    ;   Is coln code affected?
 0038  75 05                jnz     decode      ;   If yes, decode coln code
 003A  8A C1                mov     al, cl      ;   Restore row mask to al
 003C  F9                   stc                 ;   if no, set carry
 003D  EB EB                jmp     next_row    ;   Check next row
 003F  BE FFFF     decode:  mov     si, -1      ;   Initialize index register
```

```
0042  B9 000F              mov     cx, 000Fh    ;  Set up counter
0045  46          search:  inc     si           ;  Increment index
0046  3A 94 0000 R         cmp     dl,[TABLE+si]  ;  Index thru table of codes
004A  E0 F9                loopne  search       ;  Loop if not found
004C  8A C1       done:    mov     al,cl        ;  get character and enable display
004E  E6 FA                out     PORTB,al     ;  display key
0050  EB BB                jmp     scan_key     ;  Return to scan another key input
0052              CDSEG    ends
0000              DTSEG    segment
0000  77          TABLE    DB      77h          ;  Code for F
0001  B7                   DB      0B7h         ;  Code for E
0002  D7                   DB      0D7h         ;  Code for D
0003  E7                   DB      0E7h         ;  Code for C
0004  7B                   DB      7Bh          ;  Code for B
0005  BB                   DB      0BBh         ;  Code for A
0006  DB                   DB      0DBh         ;  Code for 9
0007  EB                   DB      0EBh         ;  Code for 8
0008  7D                   DB      7Dh          ;  Code for 7
0009  BD                   DB      0BDh         ;  Code for 6
000A  DD                   DB      0DDh         ;  Code for 5
000B  ED                   DB      0EDh         ;  Code for 4
000C  7E                   DB      7Eh          ;  Code for 3
000D  BE                   DB      0BEh         ;  Code for 2
000E  DE                   DB      0DEh         ;  Code for 1
000F  EE                   DB      0EEh         ;  Code for 0
0010              DTSEG    ends
                      end
```

In the program, the "Key-open" loop ensures that no keys are closed. On the other hand, the "Key-close" waits in the loop for a key actuation. Note that in this program, the table for the codes for the hexadecimal numbers 0 through F are obtained by inspecting Figure 9.32.

For example, consider key F. When key F is pressed and if a LOW is output by the program to bit 0 of port C, the top row and the rightmost column of the keyboard will be LOW. This will make the content of port A as:

$$\begin{array}{lcccccccc} \text{Bit number:} & 7 & 6 & 5 & 4 & 3 & 2 & 1 & 0 \\ \text{Data:} & \underbrace{0 \quad 1 \quad 1 \quad 1}_{7} & & & & \underbrace{0 \quad 1 \quad 1 \quad 1}_{7} & & & \end{array} = 77_{16}$$

Thus, a code of 77_{16} is obtained at Port A when the key F is pressed. Diodes are connected at the four bits (Bits 0-3) of Port C. This is done to make sure that when a 0 is output by the program to one of these bits (row of the keyboard), the diode switch will close and will generate a LOW on that row.

Now, if a key is pressed on a particular row which is LOW, the column connected to this key will also be LOW. This will enable the programmer to obtain the appropriate key code for each key.

9.14 Intel 32-Bit and 64-Bit Microprocessors

This section provides a summary of Intel 32-bit and 64-bit microprocessors. The Intel line of microprocessors has gone through many changes. The 8080/85 was the first major chip by Intel but did not see major use. In 1978, Intel introduced a more powerful processor called the 8086. The 8086 is covered in detail in earlier sections of this chapter. This chip had many improved features over the 8080/85. As mentioned before, the 8086 is a 16-bit processor and utilizes pipelining. Pipelining allows the processor to execute and fetch instructions at the same time. The Intel line has progressed through the years to the 80286, 80386, 80486, and Pentium. The general trend has been an expansion of the bit width of the processors both internally and externally. The Pentium processor was introduced in 1993, and the name was changed from 80586 to Pentium because of copyright laws. The processor uses more than 3 million transistors and had an initial speed of 60 MHz. The speed has increased over the years to the latest speed of 233 MHz. Table 9.14 compares the basic features of the Intel 80386DX, 80386SX, 80486DX, 80486SX, 80486DX2, and Pentium. These are all 32-bit microprocessors.

Note that the 80386SL (not listed in the table) is also a 32-bit microprocessor with a 16-but data bus like the 80386SX. The 80386SL can run at a speed of up to 25 MHz and has a direct addressing capability of 32 MB. The 80386SL provides virtual memory support along with on-chip memory management and protection. It can be interfaced to the 80387SX to provide floating-point support. The 80386SL includes an on-chip disk controller hardware.

TABLE 9.14 Intel 80386/80486/Pentium Microprocessors

	80386DX	80386SX	80486DX	80486SX	80486DX2	Pentium
• Introduced	October 1985	June 1988	April 1989	April 1991	March 1992	March 1993
• Maximum Clock Speed (MHz)	40	33	50	25	100	233
• MIPS*	6	2.5	20	16.5	54	112
• Transistors	275,000	275,000	1.2 million	1.185 million	1.2 million	3.1 million
• On-chip cache memory	Support chips available	Support chips available	Yes	Yes	Yes	Yes
• Data bus	32-bit	16-bit	32-bit	32-bit	32-bit	64-bit
• Address bus	32-bit	24-bit	32-bit	32-bit	32-bit	32-bit
• Directly addr. memory	4 GB	16MB	4 GB	4 GB	4 GB	4 GB
• Pins	132	100	168	168	168	273
• Virtual memory	Yes	Yes	Yes	Yes	Yes	Yes
• On-chip memory management and protection	Yes	Yes	Yes	Yes	Yes	Yes
• Addressing modes†	11	11	11	11	11	11
• Floating point unit	387DX	387SX	on chip	487SX	on chip	on chip

* MIPS means million of instructions per second that the microprocessor can execute. MIPS is typically used as a measure of performance of a microprocessor. Faster microprocessors have a higher MIPS value.

† Addressing modes include register, immediate, and memory modes.

The Pentium microprocessor uses *superscalar* technology to allow multiple instructions to be executed at the same time. The Pentium uses BICMOS technology, which combines the speed of bipolar transistors and the power efficiency of CMOS technology. The internal registers are only 32 bits even though externally it has a 64-bit data bus. It has a 32-bit address bus, which allows 4 gigabytes of addressable memory space. The math coprocessor is on-chip and is up to ten times faster than the 486 in performing certain instructions. There are two execution units in the Pentium that allow the multiple execution. The multiple execution only works for instructions that are data independent, meaning that an instruction executed immediately after another using the previous result cannot be done. The Pentium uses two execution units called the "U and V pipes." Each has five pipeline stages. The U pipe can execute any of the instructions in the 80x86 set, but the V pipe executes only simple instructions. Another new feature of the Pentium is branch prediction. This feature allows the Pentium to predict and prefetch codes and advance them though the pipeline without waiting for the outcome of the zero flag.

The implementation of virtual memory is an important feature of the Pentium. It allows a total of 64 terabytes of virtual memory. The 386/486 allowed only a 4K page size for virtual memory, but the Pentium allows either 4K or 4M page sizes. The 4K page option makes it backward compatible with the 386/486 processors. The 4M page size option allows mapping of a large program without fragmentation. It reduces the amount of page misses in virtual memory mode.

In this section, the Intel 80386 is covered in detail.

Finally, this section provides an overview of 80486, Pentium, Pentium Pro, Pentium II, Pentium III and Merced/IA-64 microprocessors.

9.14.1 Intel 80386

The Intel 80386 is Intel's first 32-bit microprogrammed microprocessor. Its introduction in 1985 facilitated the introduction of Microsoft's Windows operating systems. The high-speed computer requirement of the graphical interface of Windows operating systems was supplied by the 80386. Also, the on-chip memory management of the 80386 allowed memory to be allocated and managed by the operating system. In the past, memory management was performed by software.

The Intel 80386 is a 32-bit microprocessor and is based on the 8086. A variation of the 80386 (32-bit data bus) is the 80386SX microprocessor, which contains a 16-bit data bus along with all other features of the 80386. The 80386 is software compatible at the object code level with the Intel 8086. The 80386 includes separate 32-bit internal and external data paths along with 8 general-purpose 32-bit registers. The processor can handle 8-, 16-, and 32-bit data types. It has separate 32-bit data and address pins, and generates a 32-bit physical address. The 80386 can directly address up to 4 gigabytes (2^{32}) of physical memory and 64 tetrabytes (2^{46}) of virtual memory. The 80386 can be operated from a 12.5-, 16-, 20-, 25-, 33-, or 40-MHz clock. The chip has 132 pins and is typically housed in a pin grid array (PGA) package. The 80386 is designed using high-speed HCMOS III technology.

The 80386 is highly pipelined and can perform instruction fetching, decoding, execution, and memory management functions in parallel. The on-chip memory management and protection hardware translates logical addresses to physical addresses and provides the protection rules required in a multitasking environment. The 80386 contains a total of 129 instructions. The 80386 protection mechanism, paging, and the instructions to support them are not present in the 8086.

The main differences between the 8086 and the 80386 are the 32-bit addresses and data types and paging and memory management. To provide these features and other applications, several new instructions are added in the 80386 instruction set beyond those of the 8086.

9.14.1.1 Internal 80386 Architecture

The internal architecture of the 80386 includes several functional units that operate in parallel. The parallel operation is known as "pipelined processing." Fetching, decoding, execution, memory management, and bus access for several instructions are performed simultaneously. Typical functional units of the 80386 are these:

- Bus interface unit (BIU)
- Execution unit (EU)
- Segmentation unit
- Paging unit

The 80386 BIU performs similar function as the 8086 BIU. The execution unit processes the instructions from the instruction queue. It contains mainly a control unit and a data unit. The control unit contains microcode and parallel hardware for fast multiplication, division, and effective address calculation. The data unit includes an ALU, 8 general-purpose registers, and a 64-bit barrel shifter for performing multiple bit shifts in one clock cycle. The data unit carries out data operations requested by the control unit. The segmentation unit translates logical addresses into linear addresses at the request of the execution unit. The translated linear address is sent to the paging unit.

Upon enabling of the paging mechanism, the 80386 translates the linear addresses into physical addresses. If paging is not enabled, the physical address is identical to the linear address and no translation is necessary. The 80386 segmentation and paging units support memory management functions. The 80386 does not contain any on-chip cache. However, external cache memory can be interfaced to the 80386 using a cache controller chip.

9.14.1.2 Processing Modes

The 80386 has three processing modes: protected mode, real-address mode, and virtual 8086 mode. Protected mode is the normal 32-bit application of the 80386. All instructions and features of the 80386 are available in this mode. Real-address mode (also known as "real mode") is the mode of operation of the processor upon hardware reset. This mode appears to programmers as a fast 8086 with a few new instructions. This mode is utilized by most applications for initialization purposes only. Virtual 8086 mode (also called "V86 mode") is a mode in which the 80386 can go back and forth repeatedly between V86 mode and protected mode at a fast speed. When entering into V86 mode, the 80386 can execute an 8086 program. The processor can then leave V86 mode and enter protected mode to execute an 80386 program.

As mentioned, the 80386 enters real-address mode upon hardware reset. In this mode, the protection enable (PE) bit in a control register—the control register 0 (CR0)—is cleared to zero. Setting the PE bit in CR0 places the 80386 in protected mode. When the 80386 is in protected mode, setting the VM (virtual mode) bit in the flag register (the EFLAGS register) places the 80386 in V86 mode.

9.14.1.3 Basic 80386 Programming Model

The 80386 basic programming model includes the following aspects:

- Memory organization and segmentation
- Data types
- Registers
- Addressing modes
- Instruction set

I/O is not included as part of the basic programming model because systems designers may select to use I/O instructions for application programs or may select to reserve them for the operating system.

Memory Organization and Segmentation

The 4-gigabyte physical memory of the 80386 is structured as 8-bit bytes. Each byte can be uniquely accessed as a 32-bit address. The programmer can write assembly language programs without knowledge of physical address space. The memory organization model available to applications programmers is determined by the system software designers. The memory organization model available to the programmer for each task can vary between the following possibilities:

An address space includes a single array of up to 4 gigabytes. The 80386 maps the 4-gigabyte space into the physical address space automatically by using an address-translation scheme transparent to the applications programmers.

A segmented address space includes up to 16,383 linear address spaces of up to 4 gigabytes each. In a segmented model, the address space is called the "logical" address space and can be up to 64 terabytes. The processor maps this address space onto the physical address space (up to 4 gigabytes by an address-translation technique).

Data Types

Data types can be byte (8-bit), word (16-bit with the low byte addressed by n and the high byte addressed by $n + 1$), and double word (32-bit with byte 0 addressed by n and byte 3 addressed by $n + 3$). All three data types can start at any byte address. Therefore, the words are not required to be aligned at even-numbered addresses, and double words need not be aligned at addresses evenly divisible by 4. However, for maximum performance, data structures (including stacks) should be designed in such a way that, whenever possible, word operands are aligned at even addresses and double word operands are aligned at addresses evenly divisible by 4. That is, for 32-bit words, addresses should start at 0, 4, 8, ... for the highest speed.

Depending on the instruction referring to the operand, the following additional data types are available: integer (signed 8-, 16-, or 32-bit), ordinal (unsigned 8-, 16-, or 32-bit), near pointer (a 32-bit logical address that is an offset within a segment), far pointer (a 48-bit logical address consisting of a 16-bit selector and a 32-bit offset), string (8-, 16-, or 32-bit from 0 bytes to $2^{32} - 1$ bytes), bit field (a contiguous sequence of bits starting at any bit position of any byte and containing up to 32 bits), bit string (a contiguous sequence of bits starting at any position of any byte and containing up to $2^{32} - 1$ bits), and packed/unpacked BCD. When the 80386 is interfaced to a coprocessor such as the 80287 or 80387, then floating-point numbers are supported.

Registers

Figure 9.33 shows the 80386 registers. As shown in the figure, the 80386 has 16 registers classified as general, segment, status, and instruction pointer. The 8 general registers are the 32-bit registers EAX, EBX, ECX, EDX, EBP, ESP, ESI, and EDI. The low-order word of each of these 8 registers has the 8086 register name AX (AH or AL), BX (BH or BL), CX (CH or CL), DX (DH or DL), BP, SP, SI, and DI. They are useful for making the 80386 compatible with the 8086 processor.

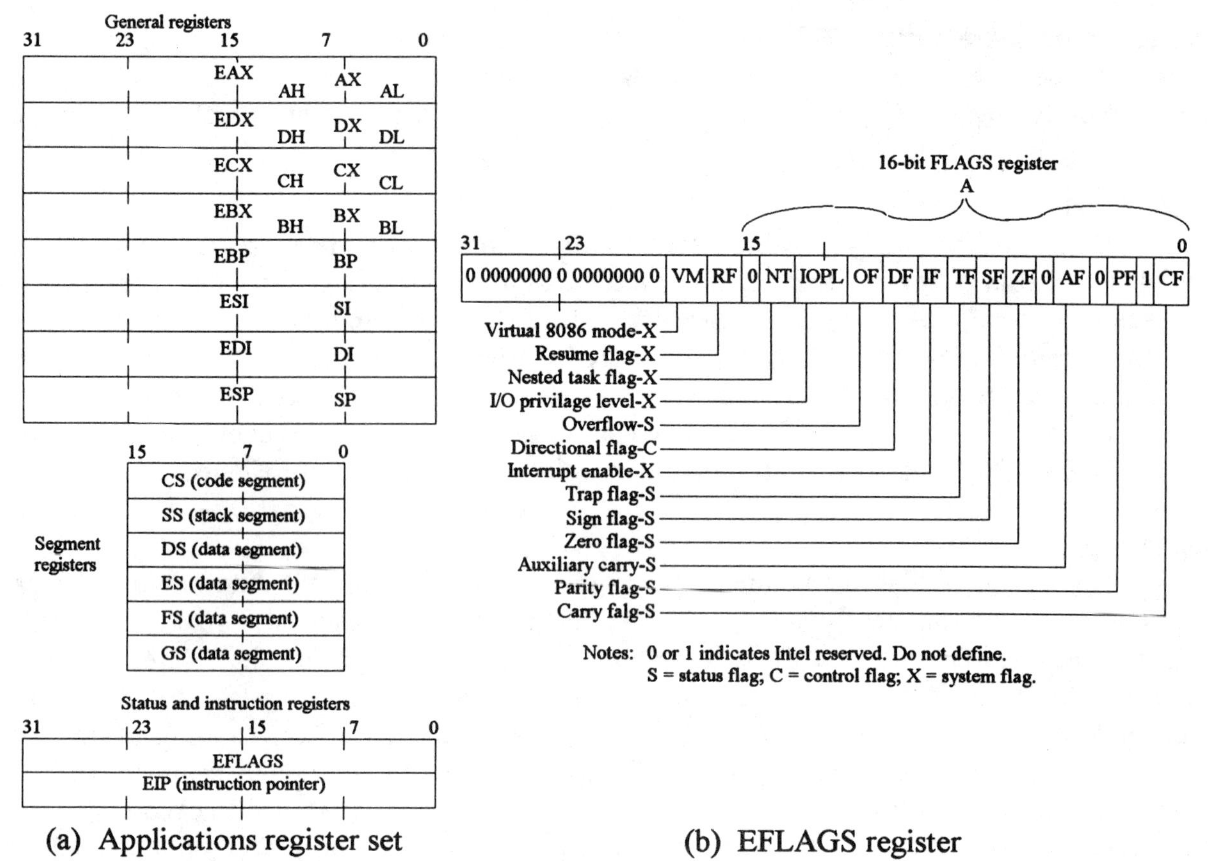

FIGURE 9.33 80386 registers

The six 16-bit segment registers—CS, SS, DS, ES, FS, and GS—allow systems software designers to select either a flat or segmented model of memory organization. The purpose of CS, SS, DS, and ES is same as that of the corresponding 8086 registers. The two additional data segment registers FS and GS are included in the 80386 so that the four data segment registers (DS, ES, FS, and GS) can access four separate data areas and allow programs to access different types of data structures.

The flag register is a 32-bit register, named EFLAGS in Figure 9.33, that shows the meaning of each bit in this register. The low-order 16 bits of EFLAGS is named FLAGS and can be treated as a unit. This is useful when executing 8086 code because this part of EFLAGS is similar to the FLAGS register of the 8086. The 80386 flags are grouped into three types: status flags, control flags, and system flags.

The status flags include CF, PF, AF, ZF, SF, and OF, like the 8086. The control flag DF is used by strings like the 8086. The system flags control I/O, maskable interrupts, debugging, task switching, and enabling of virtual 8086 execution in a protected, multitasking environment. The purpose of IF and TF is identical to the 8086. Let us explain some of the system flags:

- **IOPL** (I/O privilege level). This 2-bit field supports the 80386 protection feature.
- **NT** (nested task). The NT bit controls the IRET operation. If NT = 0, a usual return from interrupt is taken by the 80386 by popping EFLAGS, CS, and EIP from the stack. If NT = 1, the 80386 returns from an interrupt via task switching.
- **RF** (resume flag). is used during debugging.
- **VM** (virtual 8086 mode). When the VM bit is set to 1, the 80386 executes 8086 programs. When the VM bit is 0, the 80386 operates in protected mode.
- The **instruction pointer register** (EIP) contains the offset address relative to the start of the current code segment of the next sequential instruction to be executed. The low-order 16 bits of EIP is named IP and is useful when the 80386 executes 8086 instructions.

80386 Addressing Modes

The 80386 has 11 addressing modes, classified into register/immediate and memory addressing modes. The register/immediate type includes 2 addressing modes, and the memory addressing type contains 9 modes.

Register/Immediate Modes

Instructions using the register or immediate modes operate on either register or immediate operands. In register mode, the operand is contained in one of the 8-, 16-, or 32-bit general registers. An example is `DEC ECX`, which decrements the 32-bit register ECX by 1. In immediate mode, the operand is included as part of the instruction. An example is `MOV EDX,5167812FH`, which moves the 32-bit data $5167812F_{16}$ to the EDX register. Note that the source operand in this case is in immediate mode.

Memory Addressing Modes

The other 9 addressing modes specify the effective memory address of an operand. These modes are used when accessing memory. An 80386 address consists of two parts: a segment base address and an effective address. The effective address is computed by adding any combination of the following four elements:

1. **Displacement.** The 8- or 32-bit immediate data following the instruction is the displacement; 16-bit displacements can be used by inserting an address prefix before the instruction
2. **Base.** The contents of any general-purpose register can be used as a base.

3. **Index.** The contents of any general-purpose register except ESP can be used as an index register. The elements of an array or a string of characters can be accessed via the index register.
4. **Scale.** The index register's contents can be multiplied (scaled) by a factor of 1, 2, 4, or 8. A scaled index mode is efficient for accessing arrays or structures.
   Effective Address, EA = base register + (index register × scale) + displacement

The 9 memory addressing modes are a combination of these four elements. Of the 9 modes, 8 of them are executed with the same number of clock cycles because the effective address calculation is pipelined with the execution of other instructions; the mode containing base, index, and displacement elements requires one additional clock cycle.

1. **Direct mode.**The operand's effective addresses is included as part of the instruction as an 8-, 16-, or 32-bit displacement. An example is `DEC WORD PTR [4000H]`.
2. **Register indirect mode.** A base or index register contains the operand's effective address. An example is `MOV EBX, [ECX]`.
3. **Base mode.** The contents of a base register is added to a displacement to obtain the operand's effective address. An example is `MOV [EDX + 16],EBX`.
4. **Index mode.** The contents of an index register is added to a displacement to obtain the operand's effective address. An example is `ADD START [EDI],EBX`.
5. **Scaled index mode.** The contents of an index register is multiplied by a scaling factor (1, 2, 4, or 8), and the result is added to a displacement to obtain the operand's effective address. An example is `MOV START [EBX * 8],ECX`.
6. **Based index mode.** The contents of a base register is added to the contents of an index register to obtain the operand's effective address. An example is `MOV ECX, [ESI][EAX]`.
7. **Based scaled index mode.** The contents of an index register is multiplied by a scaling factor (1, 2, 4, 8), and the result is added to the contents of a base register to obtain the operand's effective address. An example is `MOV [ECX * 4] [EDX],EAX`.
8. **Based index mode with displacement.** The operand's effective address is obtained by adding the contents of a base register and an index register with a displacement. An example is `MOV [EBX][EBP + 0F24782AH],ECX`.
9. **Based scaled index mode with displacement.** The contents of an index register is multiplied by a scaling factor, and the result is added to the contents of base register and displacement to obtain the operand's effective address. An example is `MOV [ESI * 8] [EBP + 60H],ECX`.

80386 Instruction Set

The 80386 can execute all 16-bit instructions in real and protected modes. This is provided in order to make the 80386 software compatible with the 8086. The 80386 uses either 8- or 32-bit displacements and any register as the base or index register while executing 32-bit code. However, the 80386 uses either 8- or 16-bit displacements with the base and index registers while executing 16-bit code. The base and index registers utilized by the 80386 for 16- and 32-bit addresses are as follows:

	16-Bit Addressing	*32-Bit Addressing*
Base register	BX, BP	Any 32-bit general-purpose register
Index register	SI, DI	Any 32-bit general-purpose register except ESP
Scale factor	None	1, 2, 4, 8
Displacement	0, 8, 16 bits	0, 8, 32 bits

A description of some of the new 80386 instructions is given next.

1. ***Arithmetic Instructions***

There are two new sign extension instructions beyond those of the 8086.

`CWDE`	Sign-extend 16 bit contents of AX to a 32-bit double word in EAX.
`CDQ`	Sign-extend a double word (32 bits) in EAX to a quadword (64 bits) in EDX:EAX

The 80386 includes all of the 8086 arithmetic instructions plus some new ones. Two of the instructions are as follows:

Instruction	*Operation*
`ADC` reg32/mem32, imm32	[reg32 or mem32] ← [reg32 or mem32] + 32-bit immediate data + CF
`ADC` reg32/mem32, imm8	[reg32 or mem32] ← [reg32 or mem32] + 8-bit immediate data sign-extended to 32 bits + CF

Similarly, the other add instructions include the following:

`ADC`	reg32/mem32,	reg32/mem32
`ADD`	reg32/mem32,	imm32
`ADD`	reg32/mem32,	imm8
`ADD`	reg32/mem32,	reg32/mem32

The 80386 `SUB`/`SBB` instructions have the same operands as the `ADD`/`ADC` instructions.

The 80386 multiply instructions include all of the 8086 instructions plus some new ones. Some of them are listed next:

Instruction	*Operation*
IMUL EAX, reg32/mem32	EDX:EAX ← EAX * reg32 or mem32 (signed multiplication). CF and OF flags are cleared to 0 if the EDX value is 0; otherwise, they are set.
IMUL AX, reg16/mem16	DX:AX ← AX * reg16/mem16 (signed multiplication)
IMUL AL, reg8/mem8	(signed multiplication) AX ← AL * reg8/mem8
IMUL reg16, reg16/mem16,imm8	reg16 ← reg16/mem16 * (imm8 sign-extended to 16-bits) (signed multiplication). The result is the low 16 bits of product.
IMUL reg32, reg32/mem32, imm8	reg32 ← reg32/mem32 * (imm8 sign-extended to 32 bits) (signed multiplication). The result is the low 32 bits of product.

The unsigned multiplication `MUL` instruction has the same operands.

The 80386 divide instructions include all of the 8086 instructions plus some new ones. Some of them are listed next:

Instruction	*Operation*
IDIV EAX, reg32/mem32	EDX:EAX ÷ reg32 or mem32 (signed division). EAX = quotient and EDX = remainder.
IDIV AL, reg8/mem8	AX ÷ reg8 or mem8 (signed division) AL = quotient and AH = remainder.
IDIV AX, reg16/mem16	DX:AX ÷ reg16 or mem16 (signed division) AX = quotient and DX = remainder.

The `DIV` instruction performs unsigned division, and the operation is the same as `IDIV`.

2. *Bit Instructions*

The six 80386 bit instructions are as follows:

BSF	Bit scan forward
BSR	Bit scan reverse
BT	Bit test
BTC	Bit test and complement
BTR	Bit test and reset
BTS	Bit test and set

These instructions are discussed separately next.

- **BSF** (bit scan forward) takes the form

```
BSF    d,      s
       reg16,  reg16
       reg16,  mem16
       reg32,  reg32
       reg32,  mem32
```

BSF scans (checks) the 16-bit (word) or 32-bit (double word) number defined by *s* from right to left (bit 0 to bit 15 or bit 31). The bit number of the first 1 found is stored in *d*. If the whole 16-bit or 32-bit number is 0, the ZF flag is set to 1; Otherwise, ZF = 0. For example, consider `BSF EBX, EDX`. If [EDX] = 01241240_{16}, then [EBX] = 00000006_{16} and ZF = 0. The bit number 6 in EDX (contained in the second nibble of EDX) is the first 1 found when [EDX] is scanned from the right.

- **BSR** (bit scan reverse) takes the form

```
BSR    d,      s
       reg16,  reg16
       reg16,  mem16
       reg32,  reg32
       reg32,  mem32
```

BSR scans (checks) the 16-bit or 32-bit number defined by *s* from the most significant bit (bit 15 or bit 31) to the least significant bit (bit 0). The destination operand d is loaded with the bit index (bit number) of the first set bit. If the bits in the number are all 0's, ZF is set to 1 and operand d is undefined; ZF is reset to 0 if a 1 is found.

- **BT** (bit test) takes the form

```
BT     d,      s
       reg16,  reg16
       mem16,  reg16
       reg16,  imm8
       mem16,  imm8
       reg32,  reg32
       mem32,  reg32
       reg32,  imm8
       mem32,  imm8
```

BT assigns the bit value of operand *d* (base) specified by operand *s* (bit offset) to the carry flag. Only CF is affected. If operand *s* is an immediate data, only 8 bits are allowed in the instruction. This operand is taken modulo 32 so that the range of immediate bit offset is from 0 to 31. This permits any bit within a register to be selected. If *d* is a register, the bit value assigned to CF is defined by the value of the bit number defined by *s* taken modulo the register size (16 or 32). If d is a memory bit string, the desired 16 bits or 32 bits can be determined by adding *s* (bit index) divided by the operand size (16 or 32) to the memory address of *d*. The bit within this 16- or 32-bit word is defined by *d* taken modulo the operand size (16 or 32). If d is a memory operand, the 80386 may access 4 bytes in memory starting at effective address plus 4 × [bit offset divided by 32]. As an example, consider BT CX, DX. If [CX] = 081F and [DX] = 0021_{16}, then, because the contents of DX is 33_{10}, the bit number 1 [remainder of 33/16 = 1 of CX (value 1)] is reflected in CF and therefore CF = 1.

- **BTC** (bit test and complement) takes the form

  `BTC` *d*, *s*

  where *d* and *s* have the same definitions as for the BT instruction. The bit of *d* defined by *s* is reflected in CF. After CF is assigned, the same bit of *d* defined by *s* is ones complemented. The 80386 determines the bit number from *s* (whether *s* is immediate data or register) and *d* (whether *d* is register or memory bit string) in the same way as for the `BT` instruction.

- **BTR** (bit test and reset) takes the form

  `BTR` *d*, *s*

  Where *d* and *s* have the same definitions as for the BT instruction. The bit of *d* defined by *s* is reflected in CF. After CF is assigned, the same bit of *d* defined by *s* is reset to 0. Everything else applicable to the BT instruction also applies to BTR.

- **BTS** (bit test and set) takes the form

  `BTS` *d*, *s*

  `BTS` is the same as `BTR` except that the specified bit in *d* is set to 1 after the bit value of *d* defined by *s* is reflected in CF. Everything else applicable to the `BT` instruction also applies to `BTS`.

3. ***Set Byte on Condition Instructions***

These instructions set a byte to 1 or reset a byte to 0 depending on any of the 16 conditions defined by the status flags. The byte may be located in memory or in a 1-byte general register. These instructions are very useful in implementing Boolean expressions in high-level languages. The general structure of these instructions is `SETcc` (set byte on condition *cc*), which sets a byte to 1 if condition *cc* is true or else resets the byte to 0. As an example, consider `SETB BL` (set byte if below; CF = 1). If [BL] = 52_{16} and CF = 1, then, after this instruction is executed, [BL] = 01_{16} and CF remains at 1; all other flags (OF, SF, ZF, AF, PF) are undefined. On the other hand, if CF = 0, then, after execution of this instruction, [BL] = 00_{16}, CF = 0, and ZF = 1; all other flags are undefined. The other `SETcc` instructions can similarly be explained.

4. ***Conditional Jumps and Loops***

`JECXZ` disp8 jumps if [ECX] = 0; disp8 means a relative address. JECXZ tests the contents of the ECX register for zero and not the flags. If [ECX] = 0, then, after execution of the `JECXZ` instruction, the program branches with a signed 8-bit relative offset ($+127_{10}$ to -128_{10} with 0 being positive) defined by disp8. The `JECXZ` instruction is useful at the beginning of a conditional loop that terminates with a conditional loop instruction such as `LOOPNE` *label*. `JECXZ` prevents entering the loop with [ECX] = 0, which would cause the loop to execute up to 2^{32} times instead of zero times.

The loop instructions are listed next:

LOOP disp8	Decrement CX/ECX by 1 and jump if CX/ECX $\neq$ 0
LOOP/LOOPZ disp8	Decrement CX/ECX by 1 and jump if CX/ECX $\neq$ 0 or ZF = 1
LOOPNE/LOOPNZ disp8	Decrement CX/ECX by 1 and jump if CX/ECX $\neq$ 0 or ZF = 0

The 80386 loop instructions are similar to those of the 8086 except that if the counter is more than 16 bits, the ECX register is used as the counter.

5. ***Data Transfer Instructions***

a. ***Move Instructions***

The move instructions are described as follows:

MOVSX	*d*,	*s*	Move and sign-extend
MOVZX	*d*,	*s*	Move and zero-extend
	reg16,	reg8	
	reg16,	mem8	
	reg32,	reg8	
	reg32,	mem8	
	reg32,	reg16	
	reg32,	mem16	

MOVSX reads the contents of the effective address or register as a byte or a word from the source, sign-extends the value to the operand size of the destination (16 or 32 bits), and stores the result in the destination. No flags are affected. MOVZX, on the other hand, reads the contents of the effective address or register as a byte or a word, zero-extends the value to the operand size of the destination (16 or 32 bits), and stores the result in the destination. No flags are affected. For example, consider MOVSX BX, CL. If CL = 81_{16} and [BX] = $21AF_{16}$, then, after execution of this MOVSX, register BX contains $FF81_{16}$ and the contents of CL do not change. Now, consider MOVZX CX, DH. If CX = $F237_{16}$ and [DH] = 85_{16}, then, after execution of this MOVZX, register CX contains 0085_{16} and DH contents do not change.

b. Push and Pop Instructions

There are new push and pop instructions in the 80386 beyond those of the 8086: PUSHAD and POPAD. PUSHAD saves all 32-bit general registers (the order is EAX, ECX, EDX, EBX, original ESP, EBP, ESI, and EDI) onto the 80386 stack. PUSHAD decrements the stack pointer (ESP) by 32_{10} to hold the eight 32-bit values. No flags are affected. POPAD reverses a previous PUSHAD. It pops the eight 32-bit registers (the order is EDI, ESI, EBP, ESP, EBX, EDX, ECX, and EAX). The ESP value is discarded instead of loading onto ESP. No flags are affected. Note that ESP is actually popped but thrown away so that [ESP], after popping all the registers, will be incremented by 32_{10}.

c. Load Pointer Instructions

There are five instructions in the load pointer instruction category: LDS, LES, LFS, LGS, and LSS. The 80386 can have four versions for each one of these instructions as follows:

LDS	reg16,	mem16:mem16
LDS	reg32,	mem16:mem32
LES	reg16,	mem16:mem16
LES	reg32,	mem16:mem32

Note that mem16:mem16 or mem16:mem32 defines a memory operand containing the pointers composed of two numbers. The number to the left of the colon corresponds to the pointer's segment selector; the number to the right corresponds to the offset. These instructions read a full pointer from memory and store it in the selected segment register:specified register. The instruction loads 16 bits into DS (for `LDS`) or into ES (for `LES`). The other register loaded is 32 bits for 32-bit operand size and 16 bits for 16-bit operand size. The 16- and 32-bit registers to be loaded are determined by the reg16 or reg32 register specified.

The three instructions `LFS`, `LGS`, and `LSS` are associated with segment registers FS, GS, and SS can similarly be explained.

6. ***Flag Control Instructions***

There are two new flag control instructions in the 80386 beyond those of the 8086: `PUSHFD` and `POPFD`. `PUSHFD` decrements the stack pointer by 4 and saves the 80386 EFLAGS register to the new top of the stack. No flags are affected. `POPFD` pops the 32 bits (double word) from the top of the stack and stores the value in EFLAGS. All flags except VM and RF are affected.

7. ***Logical Instructions***

There are new logical instructions in the 80386 beyond those of the 8086:

`SHLD`	*d*,	*s*,	count	Shift left double
`SHRD`	*d*,	*s*,	count	Shift right double
	d	*s*	count	
	reg16,	reg16,	imm8	
	mem16,	reg16,	imm8	
	reg16,	reg16,	CL	
	mem16,	reg16,	CL	
	reg32,	reg32,	CL	
	mem32,	reg32,	imm8	
	reg32,	reg32,	CL	
	mem32,	reg32,	CL	

For both `SHLD` and `SHRD`, the shift count is defined by the low 5 bits, so shifts from 0 to 31 can be obtained.

`SHLD` shifts the contents of *d:s* by the specified shift count with the result stored back into *d*; *d* is shifted to the left by the shift count with the low-order bits of *d* filled from the high-order bits of *s*. The bits in *s* are not altered after shifting. The carry flag becomes the value of the bit shifted out of the most significant bit of *d*. If the shift count is zero, this instruction works as an NOP. For the specified shift count, the SF, ZF, and PF flags are set according to the result in *d*. CF is set to the value of the last bit shifted out. OF and AF are undefined.

SHRD shifts the contents of *d*:*s* by the specified shift count to the right with the result stored back into *d*. The bits in *d* are shifted right by the shift count, with the high-order bits filled from the low-order bits of *s*. The bits in *s* are not altered after shifting. If the shift count is zero, this instruction operates as an NOP. For the specified shift count, the SF, ZF, and PF flags are set according to the value of the result. CF is set to the value of the last bit shifted out. OF and AF are undefined.

As an example, consider SHLD BX, DX, 2. If [BX] = $183F_{16}$ and [DX] = $01F1_{16}$, then, after this SHLD, [BX] = $60FC_{16}$, [DX] = $01F1_{16}$, CF = 0, SF = 0, ZF = 0, and PF = 1. Similarly, the SHRD instruction can be illustrated.

8. *String Instructions*

a. *Compare String Instructions*

A new 80386 instruction, CMPS mem32, mem32 (or CMPSD) beyond the compare string instructions available with the 8086 compares 32-bit words ES:EDI (second operand) with DS:ESI and affects the flags. The direction of subtraction of CMPS is [ESI] - [EDI]. The left operand (ESI) is the source, and the right operand (EDI) is the destination. This is a reverse of the normal Intel convention in which the left operand is the destination and the right operand is the source. This is true for byte (CMPSB) or word (CMPSW) compare instructions. The result of subtraction is not stored; only the flags are affected. For the first operand (ESI), DS is used as the segment register unless a segment override byte is present; for the second operand (EDI), ES must be used as the segment register and cannot be overridden. ESI and EDI are incremented by 4 if DF = 0 and are decremented by 4 if DF = 1. CMPSD can be preceded by the REPE or REPNE prefix for block comparison. All flags are affected.

b. *Load and Move String Instructions*

There are new load and move instructions in the 80386 beyond those of 8086. These are LODS mem32 (or LODSD) and MOVS mem32, mem32 (or MOVSD). LODSD loads the (32-bit) double word from a memory location specified by DS:ESI into EAX. After the load, ESI is automatically incremented by 4 if DF = 0 and decremented by 4 if DF = 1. No flags are affected. LODS can be preceded by the REP prefix. LODS is typically used within a loop structure because further processing of the data moved into EAX is normally required. MOVSD copies the (32-bit) double word at the memory location addressed by DS:ESI to the memory location at ES:EDI. DS is used as the segment register for the source and may be overridden. After the move, ESI and EDI are incremented by 4 if DF = 0 and are decremented by 4 if DF = 1. MOVS can be preceded by the REP prefix for block movement of ECX double words. No flags are affected.

c. ***String I/O Instructions***

There are new string I/O instructions in the 80386 beyond those of the 8086: INS mem32, DX (or INSD) and OUTS DX, mem32 (or OUTSD). INSD inputs 32-bit data from a port addressed by the contents of DX into a memory location specified by ES:EDI. ES cannot be overridden. After data transfer, EDI is automatically incremented by 4 if DF = 0 and decremented by 4 if DF = 1. INSD can be preceded by the REP prefix for block input of ECX double words. No flags are affected. OUTSD outputs 32-bit data from a memory location addressed by DS:ESI to a port addressed by the contents of DX. DS can be overridden. After data transfer, ESI is incremented by 4 if DF = 0 and decremented by 4 if DF = 1. OUTSD can be preceded by the REP prefix for block output of ECX double words.

d. ***Store and Scan String Instructions***

There is a new 80386 STOS mem32 (or STOSD) instruction. STOS stores the contents of the EAX register to a double word addressed by ES and EDI. ES cannot be overridden. After the storage, EDI is automatically incremented by 4 if DF = 0 and decremented by 4 if DF = 1. No flags are affected. STOS can be preceded by the REP prefix for a block fill of ECX double words. There is also a new scan instruction, the SCAS mem32 (or SCASD) in the 80386. SCASD performs the 32-bit subtraction [EAX] - [memory addressed by ES and EDI]. The result of subtraction is not stored, and the flags are affected. SCASD can be preceded by the REPE or REPNE prefix for block search of ECX double words. All flags are affected.

e. ***Table Look-Up Translation Instruction***
A modified version of the 8086 XLAT instruction is available in the 80386. XLAT mem8 (XLATB) replaces the AL register from the table index to the table entry. AL should be the unsigned index into a table addressed by DS:BX for a 16-bit address and by DS:EBX for the 32-bit address. DS can be overridden. No flags are affected.

9. ***High-Level Language Instructions***
Three instructions, ENTER, LEAVE, and BOUND, are included in the 80386. The ENTER imm16,imm8 instruction creates a stack frame. The data imm8 defines the nesting depth of the subroutine and can be from 0 to 31. The value 0 specifies the first subroutine only. The data imm8 defines the number of stack frame pointers copied into the new stack frame from the preceding frame. After the instruction is executed, the 80386 uses EBP as the current frame pointer and ESP as the current stack pointer. The data imm16 specifies the number of bytes of local variables for which the stack space is to be allocated. If imm8 is zero, ENTER pushes the frame pointer EBP onto the stack; ENTER then subtracts the first operand imm16 from the ESP and sets EBP to the current ESP.

For example, a procedure with 28 bytes of local variables would have an ENTER 28, 0 instruction at its entry point and a LEAVE instruction before every RET. The 28 local bytes would be addressed as offset from EBP. Note that the LEAVE instruction sets ESP TO EBP and then pops EBP. The 80386 uses BP (low 16 bits of EBP) and SP (low 16 bits of ESP) for 16-bit operands and uses EBP and ESP for 32-bit operands.

The BOUND instruction ensures that a signed array index is within the limits specified by a block of memory containing an upper and lower bound. The 80386 provides two forms of the BOUND instruction:

```
BOUND    reg16, mem32
BOUND    reg32, mem64
```

The first form is for 16-bit operands. The second form is for 32-bit operands and is included in the 80386 instruction set. For example, consider BOUND EDI,ADDR. Suppose [ADDR] = 32-bit lower bound d_l and [ADDR + 4] = 32 bit upper bound d_u. If, after execution of this instruction, [EDI] $<d_l$ or$>d_u$ the 80386 traps to interrupt 5; otherwise, the array is accessed.

The BOUND instruction is usually placed following the computation of an index value to ensure that the limits of the index value are not violated. This permits a check to determine whether or not an address of an array being accessed is within the array boundaries when the register indirect with index mode is used to access an array element. For example, the following instruction sequence will allow accessing an array with base address in ESI, the index value in EDI, and an array lenght 50 bytes; assuming the 32-bit contents of memory location, 20000100_{16} and 20000104_{16} are 0 and 49, respectively:

```
      :
BOUND       EDI, 20000100H
MOV         EAX, [EDI][ESI]
      :
```

Example 9.15

Determine the effect of each of the following 80386 instructions:

```
(a) CDQ
(b) BTC CX, BX
(c) MOVSX ECX, E7H
```

Assume [EAX] = FFFFFFFFH, [ECX] = F1257124H, [EDX] = EEEEEEEEH, and [BX] = 0004H prior to execution of each of these given instructions.

Solution

(a) After `CDQ`,

[EAX] = FFFFFFFFH
[EDX] = FFFFFFFFH

(b) After `BTC CX, BX`, bit 4 of register CX is reflected in CF and then ones complemented in CX, as is shown below.

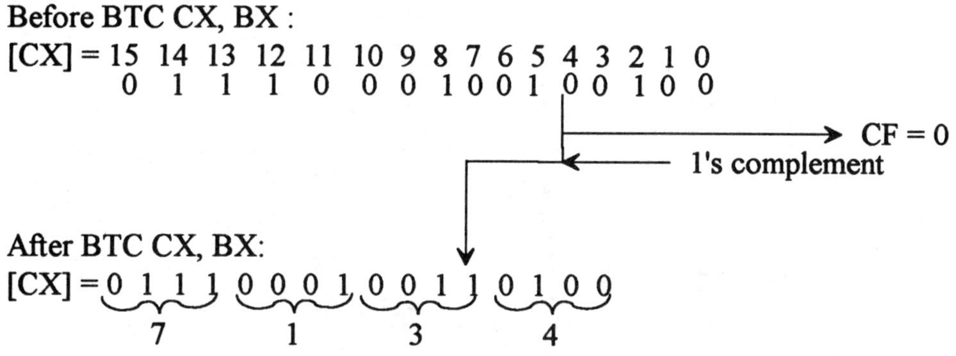

Before BTC CX, BX :
[CX] = 15 14 13 12 11 10 9 8 7 6 5 4 3 2 1 0
0 1 1 1 0 0 0 1 0 0 1 0 0 1 0 0

→ CF = 0
← 1's complement

After BTC CX, BX:
[CX] = 0 1 1 1 0 0 0 1 0 0 1 1 0 1 0 0
7 1 3 4

Hence,

[CX] = 7134H
[BX] = 0004H

(c) `MOVSX ECX,E7H` copies the 8-bit data E7H into the low byte of ECX and then sign-extends to 32 bits. Therefore, after `MOVSX ECX, E7H`,

[ECX] = FFFFFFE7H

Example 9.16

Write an 80386 assembly language program to multiply a signed 8-bit number in AL by a signed 32-bit number in ECX. Assume that the segment registers are already initialized.

Solution

```
CBW                 ;     Sign-extend byte to word
CWDE                ;     Sign-extend word to 32-bit
IMUL EAX, ECX       ;     Perform singed multiplication
HLT                 ;     Stop
```

Example 9.17

Write an 80386 assembly language program to move two columns of ten thousand 32-bit numbers from A (*i*) to B (*i*). In other words, move A (1) to B (1), A (2) to B (2), and so on.

Solution

```
MOV      ECX, 10000          ;     Initialize counter
MOV      BX, SOURCE_SEG      ;     Initialize DS
MOV      DS, BX              ;     register
MOV      BX, DEST_SEG        ;     Initialize ES
MOV      ES, BX              ;     register
MOV      ESI, SOURCE_INDX    ;     Initialize ESI
MOV      EDI, DEST_INDX      ;     Initialize EDI
CLD                          ;     Clear DF to auto-increment
REP MOVSD                    ;     MOV A (i) to
HLT                          ;     B (i) until ECX = 0
```

9.14.1.4 80386 Pins and Signals

The 80386 contains 132 pins in Pin Grid Array (PGA) or other packages. Figure 9.35 shows functional grouping of the 80386 pins. A brief description of the 80386 pins and signals is provided in the following. The # symbol at the end of the signal name or the — symbol above a signal name indicates the active or asserted state when it is low. When the symbol # is absent after the signal name or the symbol — is absent above a signal name, the signal is asserted when high.

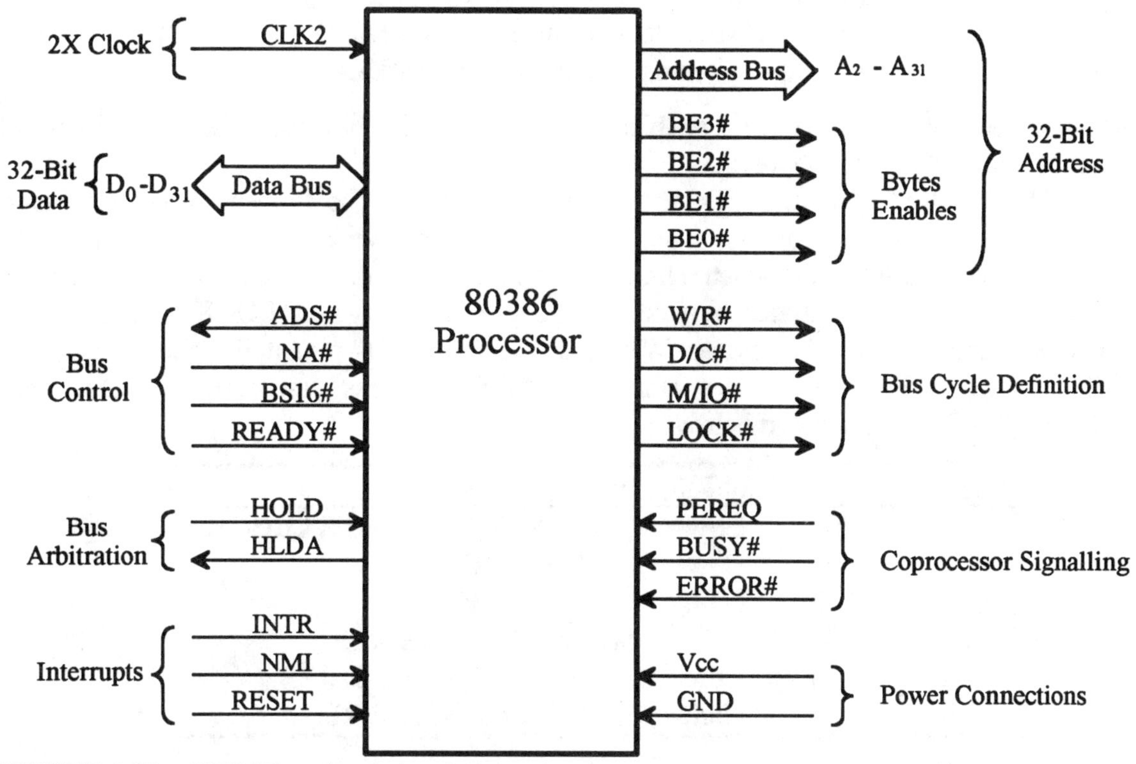

FIGURE 9.35 80386 Functional signal groups

The 80386 has 20 Vcc and 21 GND pins for power distribution. These multiple power and ground pins reduce noise. Preferably, the circuit board should contain Vcc and GND planes.

CLK2 pin provides the basic timing for the 80386. This clock is then divided by 2 by the 80386 internally to provide the clock used for instruction execution. The 80386 is reset by activating the RESET pin for at least 15 CLK2 periods. The RESET signal is level-sensitive. When the RESET pin is asserted, the 80386 will start executing instructions at address FFFF FFF0H. The 82384 clock generator provides system clock and reset signals.

D_0-D_{31} provides the 32-bit data bus. The 80386 can transfer 16- or 32-bit data via the data bus.

The address pins A_2-A_{31} along with the byte enable signals BE0# through BE3# are used to generate physical memory or I/O port addresses. Using the pins, the 80386 can directly address 4 gigabytes by physical memory (00000000H through FFFFFFFFH).

The byte enable outputs, BE0# through BE3# of the 80386, define which bytes of D_0-D_{31} are utilized in the current data transfer. These definations are given below:

BE0# is low when data is transferred via D_0-D_7
BE1# is low when data is transferred via D_8-D_{15}
BE2# is low when data is transferred via D_{16}-D_{23}
BE3# is low when data is transferred via D_{24}-D_{31}

The 80386 asserts one or more byte enables depending on the physical size of the operand being transferred (1, 2, 3, or 4 bytes).

W/R#, D/C#, M/IO#, and LOCK# output pins specify the type of bus cycle being performed by the 80386. W/R# pin, when HIGH, identifies write cycle and, when LOW, indicates read cycle. D/C# pin, when HIGH, identifies data cycle , when LOW, indicates control cycle. M/IO# differentiates between memory and I/O cycles. LOCK# distinguishes between locked and unlocked bus cycles. W/R#, D/C#, and M/IO# pins define the primary bus cycle. This is because these signals are valid when ADS# (address status output) is asserted. Some of these bus cycles are listed below.

M/IO#	D/C#	W/R#	Bus cycle type
Low	Low	Low	INTERRUPT ACKNOWLEDGE
Low	High	Low	I/O DATA READ
Low	High	High	I/O DATA WRITE
High	Low	Low	MEMORY CODE READ
High	High	Low	MEMORY DATA READ
High	High	High	MEMORY DATA WRITE

The 80386 bus control signals include ADS# (address status), READY# (transfer acknowledge), NA# (next address request), and BS16# (bus size 16).

The 80386 outputs LOW on the ADS# pin indicate a valid bus cycle (W/R#, D/C#, M/IO#) and bus enable / address (BE0#-BE3#, A_2-A_{31}) signals.

When READY# input is LOW during a read cycle or an interrupt acknowledge cycle, the 80386 latches the input data on the data pins and ends the cycle. When READY# is low during a write cycle, the 80386 ends the bus cycle.

The NA# input pin is activated low by external hardware to request address pipelining. BS16# input pin permits the 80386 to interface to 32- and 16-bit memory or I/O. For 16-bit memory or I/O, BS16# input pin is asserted low by an external device, the 80386 uses the low-order half (D_0-D_{15}) of the data bus corresponding to BE0# and BE1# for data transfer.

BS16# is asserted high for 32-bit memory or I/O. HOLD (input) and HLDA (output) pins are 80386 bus arbitration signals. These signals are used for DMA transfers. PEREQ, BUSY#, and ERROR# pins are used for interfacing coprocessors such as 80287 or 80387 to the 80386.

There are two interrupt pins or the 80386. These are INTR (maskable) and NMI (nonmaskable) pins. NMI is leading-edge sensitive, whereas INTR is level-sensitive. When INTR is asserted and if the IF bit in the EFLAGS is 1, the 80386 (when ready) responds to the INTR by performing two interrupt acknowledge cycles and at the end of the second cycle latches an 8-bit vector on D_0-D_7 to identify the source of interrupt. Interrupts are serviced in a similar manner as the 8086.

9.14.1.5 80386 Modes

As mentioned before, the 80386 can be operated in real, protected, or virtual 8086 mode. These modes can be selected by some of the bits in the status register. Upon reset or power-up, the 80386 operates in real mode. In real mode, the 80386 can access all the 8086 registers along with the 80386 32-bit register. In real mode, the 80386 can directly address up to one megabyte of memory. The address lines A_2-A_{19}, BE0#-BE3# are used by the 80386 in this mode.

The protected mode provides increased memory space than real mode. Furthermore, this mode supports on-chip memory management and protection features along with multi-tasking operating system. Finally, the virtual 8086 mode permits the execution of 8086 programs, taking full advantage of the 80386 protection mechanism. In particular, in the virtual 8086 mode allows execution of 8086 operating system and application programs concurrently with 80386 operating system and application programs.

9.14.1.6 80386 System Design

In this section, the 80386 is interfaced to typical EPROM chips. As mentioned in the last section the 80386 address and data lines are not multiplexed. There is a total of thirty address pins (A_2-A_{31}) on the chip. A_0 and A_1 are decoded internally to generate four byte enable outputs, BE0#, BE1#, BE2#, and BE3#. In real mode, the 80386 utilizes 20-bit addresses and A_2 through A_{19} address pins are active and the address pins A_{20} through A_{31} are used in real mode at reset, high for code segment (CS)-based accesses, low for others, and always low after CS changes. In the protected mode, on the other hand, all address pins A_2 through A_{31} are active. In both modes, A_0 and A_1 are obtained internally. In all modes, the 80386 outputs on the byte enable pins to activate appropriate portions of the data to transfer byte (8-bit), word (16-bit), and double-word (32-bit) data as follows:

Byte Enable Pins	Data Bus
BE0#	D_0-D_7
BE1#	D_8-D_{15}
BE2#	D_{16}-D_{23}
BE3#	D_{24}-D_{31}

The 80386 supports dynamic bus sizing. This feature connects the 80386 with 32-bit or 16-bit data busses for memory or I/O. The 80386 32-bit data bus can be dynamically switched to a 16-bit bus by activating the BS16# input from high to low by a memory or I/O device. In this case, all data transfers are performed via D_0-D_{15} pins. 32-bit transfers take place as two consecutive 16-bit transfers over data pins D_0 through D_{15}. On the other hand, the 32-bit memory or I/O device can activate the BS16# pin HIGH to transfer data over D_0-D_{31} pins.

The 80386 address pins A_1 and A_0 specify the four addresses of a four byte (32-bit) word. Consider the following :

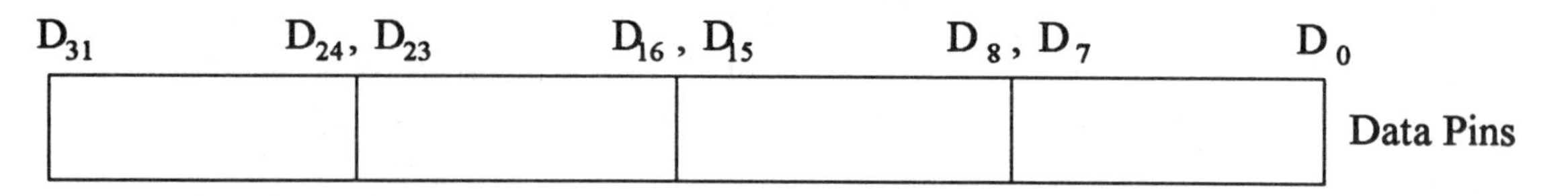

The contents of the memory addresses which include 0, 4, 8, ... with $A_1A_0 = 00_2$ are transferred over D_0-D_7. Similarly, the contents of addresses which include 1,5,9, ..., with $A_1A_0 = 01_2$ are transferred over D_{15}-D_8. On the other hand, the contents of memory addresses 2, 6, 10, ... with $A_1A_0 = 10_2$ are transferred over D_{16}-D_{23} while contents of addresses 3, 7, 11, ... with $A_1A_0 = 11_2$ are transferred over D_{24}-D_{31}. Note that A_1A_0 is encoded from BE3# -BE0#. The following figure depicts this:

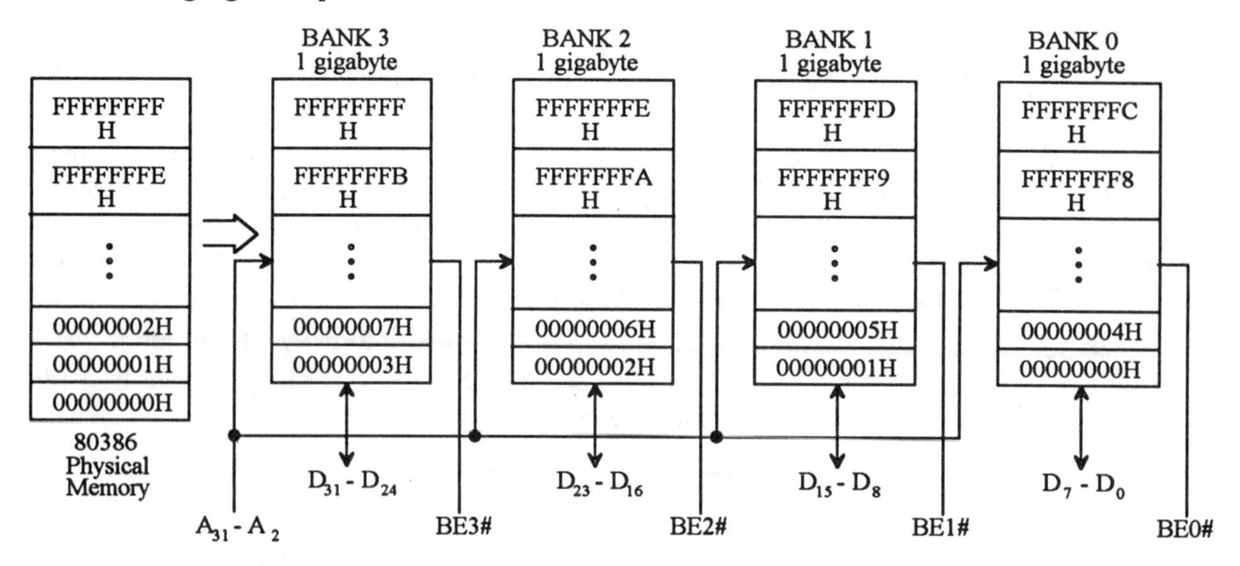

In each bank, a byte can be accessed by enabling one of the byte enables, BE0# -BE3#. For example, in response to execution of a byte-MOVE instruction such as MOV

[00000006H], BL, the 80386 outputs low on BE2# and high on BE0#, BE1# and BE3# and the content of BL is written to address 00000006H. On the other hand, when the 80386 executes a MOVE instruction such as `MOV [00000004H],AX`, the 80386 drives BE0# and BE1# to low. The locations 00000004H and 00000005H are written with the contents of AL and AH via D_0-D_7 and D_8-D_{15} respectively. For 32-bit transfer, the 80386 executing a MOVE instruction from an aligned address such as `MOV [00000004H],EAX`, drives all bus enable pins (BE0# -BE3#) to low and writes four bytes to memory locations 00000004H through 00000007H from EAX. Byte (8-bit), aligned word (16-bit), and aligned double-word (32-bit) are transferred by the 80386 in a single bus cycle.

The 80386 performs misaligned transfers in multiple cycles. For example, the 80386 executing a misaligned word MOVE instruction such as `MOV [00000003H],AX` drives BE3# to low in the first bus cycle and writes into location 00000003H (bank 3) from AL in the first bus cycle. The 80386 then drives BE0# to low in the second bus cycle and writes into location 00000004H (bank 0) from AH. This transfer takes two bus cycles.

A 32-bit misaligned transfer such as `MOV [00000002H],EAX`, on the other hand, takes two bus cycles. In the first bus cycle, the 80386 enables BE2# and BE3#, and writes the contents of low 16-bits of EAX into addresses 00000002H and 00000003H from banks 2 and 3 respectively. In the second cycle, the 80386 enables BE0# and BE1# to low and then writes the contents of upper 16-bits of EAX into addresses 00000004H and 00000005H.

In the following, design concepts associated with the 80386's interface to memory will be discussed. The 80386 device will use 128 Kbyte, 32-bit wide memory. Four 27C256's (32 K x 8 HCMOS EPROMs) are used.

Since the 27C256 chip is 32K x 8 chip, the 80386 address lines A_2-A_{16} are used for addressing the 27C256's. The 80386 M/IO#, D/C#, W/R#, and BE0#-BE3# are also used. Figure 9-36 shows a simplified 80386 - 27C256 interface.

In figure 9.36, A_1 A_0, BE3#-BE0#, D/C#, and ADS# pins of the 80386 are used to generate four byte enable signals, $\overline{E0}$, $\overline{E1}$, $\overline{E2}$, and $\overline{E3}$.

The 80386 outputs low on ADS# (Address status) pin to indicate valid bus cycle (W/R#, D/C#, M/IO#) and address (BE0# -BE3#) signals.

The 80386 A_1 and A_0 bits (obtained internally) indicate which portion of the data bus will be used to transfer data. For example, A_1 A_0 = 11 means that contents of addresses such as 00000003H, 00000007H, ... will be used by the 80386 to transfer data via its D_{31}-D_{24} pins. BE3#-BE0# and D/C# are used to produce the byte enable signals which are connected to the $\overline{CE}$ pin of the appropriate EPROM. The inverted M/IO# is logically ORed with the W/R# pin. The output of this OR gate is connected to the $\overline{OE}$ pin of all four EPROM's.

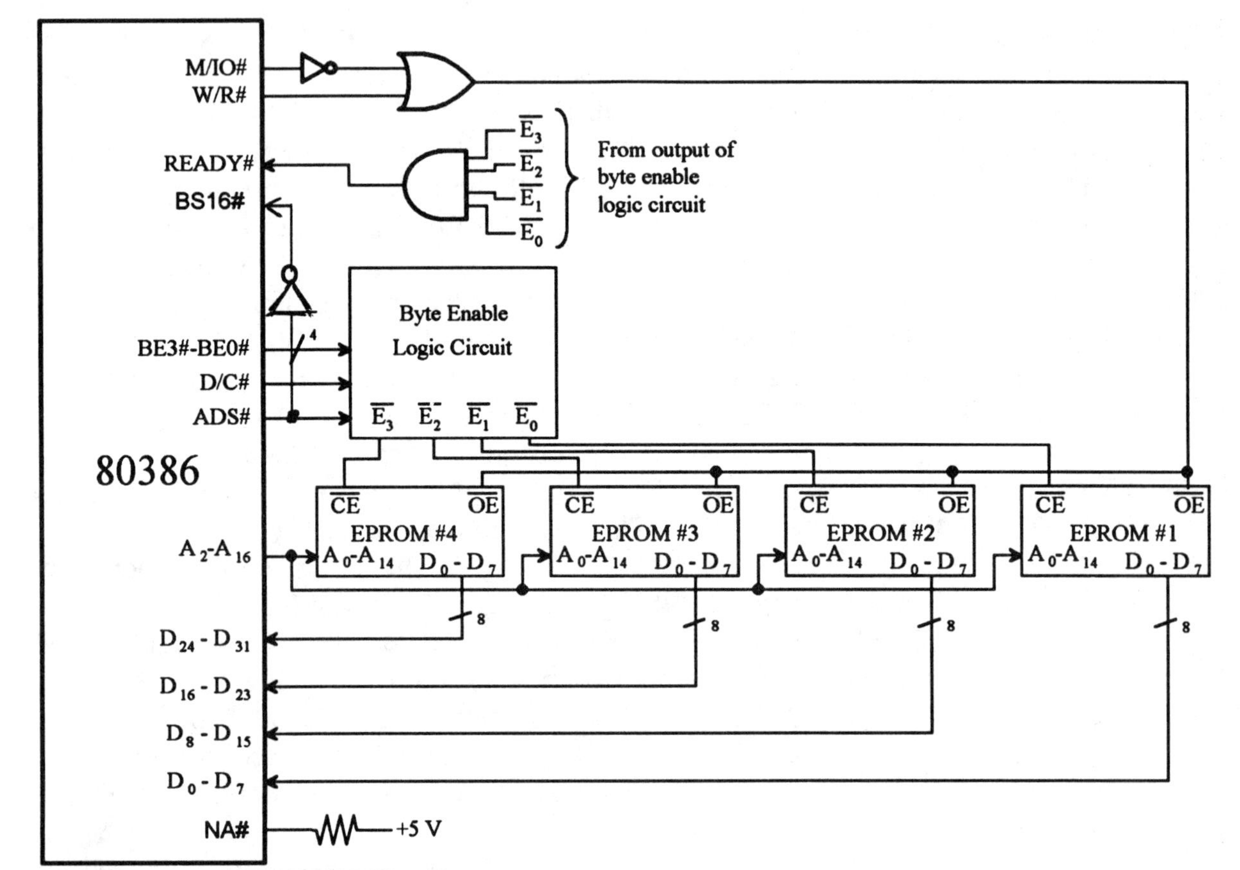

FIGURE 9-36 80386/27C256 Interface.

$\overline{E0}$, $\overline{E1}$, $\overline{E2}$, and $\overline{E3}$ are ANDed and connected to the READY# pin. When the READY# pin is asserted LOW, the 80386 latches or reads data. Until READY# pin is asserted LOW by the external device, the 80386 inserts wait states. One must ensure that the data is ready before READY# is asserted. The BS16# is asserted HIGH by connecting it to inverted ADS# to indicate 32-bit memory. NA# is connected to +5 V to disable pipelining.

The memory map can be determined as follows:

EPROM#1 :

$$\underbrace{A_{31}\,A_{30}\,\cdots\,A_{17}}_{\substack{\text{Don't cares}\\ \text{Assume zeros}}}\;\underbrace{A_{16}\,\cdots\,A_{2}}_{\substack{\text{all zeros}\\ \text{to ones}}}\;\underset{0}{A_1}\;\underset{0}{A_0}$$

= 00000000H, 00000004H, ... , 0001FFFCH

Similarly, the memory maps for other EPROMs are :

EPROM#2: 00000001H, 00000005H, ... , 0001FFFDH
EPROM#3: 00000002H, 00000006H, ... , 0001FFFEH
EPROM#4: 00000003H, 00000007H, ... , 0001FFFFH

9.14.1.7 80386 I/O

The 80386 can use either a standard I/O or a memory-mapped I/O technique.

The address decoding required to generate chip selects for devices using standard I/O is often simpler than that required for memory-mapped devices. But, memory-mapped I/O offers more flexibility in protection than standard I/O does.

The 80386 can operate with 8-, 16-, and 32-bit peripherals. Eight-bit I/O devices can be connected to any of the four 8-bit sections of the data bus. For efficient operation, 32-bit I/O devices should be assigned to addresses that are even multiples of four. For standard I/O, the 80386 includes there types of I/O instructions. These are direct, indirect, and string I/O instructions which include the following:

Direct

For 8-bit :
```
IN AL, PORT
OUT PORT, AL
```
For 16-bit:
```
IN AX, PORT
OUT PORT, AX
```

Indirect

For 8-bit :
```
IN AL, DX
OUT DX, AL
```
For 16-bit:
```
IN AX, DX
OUT DX, AX
```
For 32-bit:
```
IN EAX, DX
OUT DX, EAX
```

String

For 8-bit :

`INSB,` $(ES:DI) \leftarrow ((DX))$
$DI \leftarrow DI \pm 1$

`OUTSB` $((DX)) \leftarrow (ES:SI)$
$SI \leftarrow SI \pm 1$

For 16-bit:

`INSW,` $(ES:DI) \leftarrow ((DX))$
$(DI) \leftarrow DI \pm 2$

`OUTSW,` $(ES:SI) \leftarrow ((DX))$

		(SI) ← SI ± 2
For 32-bit:	INSD,	(ES:EDI) ← ((DX))
		EDI ← EDI ± 4
	OUTSD,	((DX)) ← (ES:ESI)
		ESI ← ESI ± 4

9.14.2 Intel 80486 Microprocessor

The Intel 80486 is an enhanced 80386 microprocessor with on-chip floating-point hardware.

9.14.2.1 Intel 80486/80386 Comparison

Table 9.15 compares the basic features of the 80486 with those of the 80386.

TABLE 9.15 80386 vs. 80486

Characteristic	80386	80486
Introduced in	1985; 386SX in 1988	1989
Main features	Adds paging 32-bit extension, on chip address translation, and greater speed than 8086. 32-bit microprocessor	Adds on-chip cache, floating-point unit, and greater speed than 386. 32-bit microprocessor.
Data bus size accommodated	16-, 32-bit	8-, 16-, 32-bit
On-chip Cache	No; Can be interfaced externally	Yes
Address bus size	32-bit	32-bit
On-chip transistors	275,000	1.2 million
Directly addressable memory	4 Gigabytes	4 Gigabytes
Virtual memory size	64 Terabytes	64 Terabytes
Clock	25 MHz to 50 MHz	25 MHz to 100 MHz
Pins	100 for 80386SX; 168 for other 80386's	168
Address and data buses	non-multiplexed	non-multiplexed
Registers	8 32-bit general purpose registers 32-bit EIP and Flag register 6 16-bit segment registers 6 64-bit segment descriptor registers 4 32-bit system control registers (CR0-CR3)	All registers listed under the 80386 plus the following registers: 8 80-bit 8 2-bit 8 16-bit 3 16-bit 2 48-bit
Address	Defined by A_2-A_{31}; BE0#-BE3#	Same as the 80386

TABLE 9.15 80386 vs. 80486 (Continued)

Characteristic	80386	80486
Address HOLD	Not available	The AHOLD input pin causes the 80486 to float its address bus in the next clock cycle. This allows an external device to drive an address into the 80486 for internal cache line invalidation.
Direct Memory Access (DMA)	Two pins are used: HOLD input pin HLDA output pin	Three pins are used: HOLD input pin HLDA output pin BREQ output
Bus backoff	Not available	The BOFF# input pin indicates that another bus master needs to complete a bus cycle in order for the 80486's current cycle to complete.
On-chip memory management hardware	Yes	Yes
Operating modes: Real, Protected, and Virtual 8086 modes	Yes. Does not support maximum or minimum modes like the 8086	Same as the 80386
Addressing modes	11	11
On-chip floating-point hardware	No	Yes
Instructions	129 including the floating-point instrucions where the 80386 is interfaced to the 80387	All 80386 instructions including the floating-point instructions for the on-chip floating-point hardware plus six new instructions

9.14.2.2 Special Features of the 80486

The Intel 80486 is a 32-bit microprocessor, like the Intel 80386. It executes the complete instruction set of the 80386 and the 80387DX floating-point coprocessor. Unlike the 80386, the 80486 on-chip floating-point hardware eliminates the need for an external floating-point coprocessor chip and the on-chip cache minimizes the need for an external cache and associated control logic.

The 80486 is object code compatible with the 8086, 8088, 80186, 80286, and 80386 processors. It can perform a complete set of arithmetic and logical operations on 8-, 16-, and 32-bit data types using a full-width ALU and eight general-purpose registers. Four gigabytes of physical memory can be addressed directly via its separate 32-bit addresses and data paths. An on-chip memory management unit is added, which maintains the integrity of memory in the multitasking and virtual-memory environments. Both memory segmentation and paging are supported.

The 80486 has an internal 8 Kbyte cache memory. This provides fast access to recently used instructions and data. The internal write-through cache can hold 8 Kbytes of data or instructions. The on-chip floating-point unit performs floating-point operations on the 32-, 64-, and 80- bit arithmetic formats specified in the IEEE standard and is object code compatible with the 8087, 80287, and 80387 coprocessors. The fetching, decoding, execution, and address translation of instructions is overlapped within the 80486 processor using instruction pipelining. This allows a continuous execution rate of one clock cycle per instruction for most instructions.

Like the 80386, the 80486 processor can operate in three modes (set in software): real, protected, and virtual 8086 mode. After reset or power up, the 80486 is initialized in real mode. This mode has the same base architecture as the 8086, but allows access to the 32-bit register set of the 80486 processor. Nearly all of the 80486 processor instructions are available, but the default operand size is 16 bits. The main purpose of real mode is to set up the processor for protected mode.

Protected mode, or protected virtual address mode, is where the complete capabilities of the 80486 become available. Segmentation and paging can both be used in protected mode. All 8086, 80286, and 386 processor software can be run under the 80486 processor's hardware-assisted protection mechanism.

Virtual 8086 mode is a submode for protected mode. It allows 8086 programs to be run but adds the segmentation and paging protection mechanisms of protected mode. It is more flexible to run 8086 in this mode than in real mode because virtual 8086 mode can simultaneously execute the 80486 operating system and both 8086 and 80486 processor applications.

The 80486 is provided with a bus backoff feature. Using this, the 80486 will float its bus signals if another bus master needs control of the bus during a 80486 bus cycle and then restart its cycle when the bus again becomes available. The 80486 includes dynamic bus sizing. Using this feature, external controllers can dynamically alter the effective width of the data bus with 8-, 16-, or 32-bit bus widths.

In terms of programming models, the Intel 80386 has very few differences with the 80486 processor. The 80486 processor defines new bits in the EFLAGS, CR0, and CR3 registers. In the 80386 processor, these bits were reserved, so the new architectural features should be a compatibility issue.

9.14.2.3 80486 New Instructions Beyond Those of the 80386

There are six basic instructions plus floating-point instructions added to the 80486 instruction set beyond those of the 80386 instruction set as follows:

1. Three New Application Instructions
   - `BSWAP`
   - `XADD`
   - `CMPXCHG`
2. Three New System Instructions
   - `INVD`
   - `WBINVD`
   - `INVLPG`

The 80386 can execute all its floating-point instructions when the 80387 is present in the system. The 80486, on the other hand, can directly execute all its floating-point instructions (same as the 80386 floating-point instructions) because it has the on-chip floating-point hardware.

The three new application instructions included with the 80486 are `BSWAP` reg32; `XADD` dest, source; and `CMPXCHG` dest, source. `BSWAP` reg32 reverses the byte order of a 32-bit register, converting a value in little/big endian form to big/little endian form. That is, the `BSWAP` instruction exchanges bits 7–0 with bits 31–24 and bits 15–8 with bits 23–16 of a 32-bit register. Executing this instruction twice in a row leaves the register with the original value. When `BSWAP` is used with a 16-bit operand size, the result left in the destination operand is undefined. Consider an example of a 32-bit operand: If [EAX] = 12345678H, then after `BSWAP EAX`, the contents of EAX are 78563412H. Note that little endian is a byte-oriented method in which the bytes are ordered (left to right) as 3, 2, 1, and 0, with byte 3 being the most significant byte. Big endian on the other hand, is also a byte-oriented method where the bytes are ordered (left to right) as 0, 1, 2, and 3 with byte 0 being the most significant byte. The `BSWAP` instruction speeds up execution of decimal arithmetic by operating on four digits at a time.

`XADD` dest, source has the form

```
XADD   dest,           source
       reg8/mem8,      reg8
       reg16/mem16,    reg16
       reg32/mem32,    reg32
```

The `XADD` dest, source instruction loads the destination into the source and then loads the sum of the destination and the original value of the source into the destination. For example, if [AX] = 0123H, [BX] = 9876H, then after `XADD AX,BX`, the contents of AX and BX are respectively 9999H and 0123H.

`CMPXCHG` dest, source has the form:

```
CMPXCHG  dest,           source
         reg8/mem8,      reg8
         reg16/mem16,    reg16
         reg32/mem32,    reg32
```

The `CMPXCHG` instruction compares the (AL, AX or EAX register) with the destination. If they are equal, the source is loaded into the destination; Otherwise, the destination is loaded into the AL,AX or EAX. For example, if [DX] = 4324H, [AX] = 4532H, and [BX] = 4532H, then after `CMPXCHG BX, DX`, the ZF flag is set to one and [BX] = 4324H.

9.14.3 Intel Pentium Microprocessor

Table 9.16 summarizes the fundamental differences between the basic features of 486 and Pentium families. Microprocessors have served largely separate markets and purposes: business PCs and engineering workstations. The PCs have used Microsoft's DOS and Windows operating systems whereas the workstations have used various features of Unix. The PCs have not been utilized in the workstation market because of their relatively modest performance, especially with regard to complicated graphics display and floating-point calculations. Workstations have been kept out of the PC market partially because of their high prices and hard-to-use system software.

The Pentium has brought the PCs up to workstation-class computational performance with sophisticated graphics. The Intel Pentium is a 32-bit microprocessor with a 64-bit data bus. The Intel Pentium, like its predecessor the Intel 80486, is 100% object code compatible with 8086/80386 systems. BICMOS(Bipolar and CMOS) technology is used for the Pentium.

The Pentium processor has three modes of operation; real-address mode (also called "real mode"), protected mode, and system management mode. The mode determines which instructions and architecture features are accessible. In *real-address mode,* the Pentium processor runs programs written for 8086 or for the real-address mode of an 80386 or 80486.

TABLE 9.16 Basic Differences Between 80486 and Pentium Processor Families

Feature	486 Processor	Pentium Processor
Clock	25 to 100 MHz	60 to 233 MHz
Address and data buses	32-bit data bus 32-bit address bus	64-bit data bus 32-bit address bus
Pipeline model	Single	Dual
Internal cache	8K for both data and instruction	8k for data and 8k for instruction

TABLE 9.16 Basic Differences Between 80486 and Pentium Processor Families (continued)

Feature	486 Processor	Pentium Processor
Number of transistors	1.2 million	3.2 million
Performance at 66 MHZ in MIPS (millions of instructions per second)	54 MIPS	112 MIPS
Number of pins	168	273

The architecture of the Pentium processor in this mode is identical to that of the 8086 microprocessor. In *protected mode,* all instruction and architectural features of the Pentium are available to the programmer. Some of the architectural features of the Pentium processor include memory management, protection, multitasking, and multiprocessing. While in protected mode, the virtual 8086 (v86) mode can be enabled for any task. For the v86 mode, the Pentium can directly execute "real-address-mode" 8086 software in a protected, multitasking environment.

The Pentium processor is also provided with a *system management mode* (SMM) similar to the one used in the 80486SL, which allows to design for low power usage. SMM is entered through activation of an external interrupt pin (system management interrupt, SMI#). In December 1994, Intel detected a flaw in the Pentium chip while performing certain division calculations. The Pentium is not the first chip that Intel has had problems with. The first version of the Intel 80386 had a math flaw that Intel quickly fixed before there were any complaints. Some experts feel that Intel should have acknowledged the math problem in the Pentium when it was first discovered and then have offered to replace the chips. In that case, the problem with the Pentium most likely would have been ignored by the users. However, Intel was heavily criticized by computer magazines when the division flaw in the Pentium chip was first detected.

The flaw in the division algorithm in the Pentium was caused by a problem with a look-up table used in the division. Errors occur in the fourth through the fifteenth significant decimal digits. This means that in a result such as 5.78346, the last three digits could be incorrect. For example, the correct answer for the operation 4,195,835 - (4,195,835 + 3,145,727) + (3,145,727) is zero. The Pentium provided a wrong answer of 256. IBM claimed this problem can occur once every 24 days. It is the author's opinion that the circuitry inside the 32-bit microprocessors is so complex that the math flaw in the Pentium is not unusual. Intel has regained its reputation by fixing the division flaw in the Pentium by shipping replacement chips.

The Pentium microprocessor is based on a superscalar design. This means that the processor includes dual pipelining and executes more than one instruction per clock cycle; note that scalar microprocessors such as the 80486 family have only one pipeline and execute one instruction per clock cycle, and superscalar processors allow more than one instruction to be executed per clock cycle.

The Pentium microprocessor contains the complete 80486 instruction set along with some new ones that are discussed later. Pentium's on-chip memory management unit is completely compatible with that of the 80486.

The Pentium includes faster floating-point on-chip hardware than the 80486. Pentium's on-chip floating-point hardware has been completely redesigned over the 80486. Faster algorithms provide up to ten times speed-up for common operations such as add, multiply, and load. The two instruction pipelines and on-chip floating-point unit are capable of independent operations. Each pipeline issues frequently used instructions in a single clock cycle. The dual pipelines can jointly issue two integer instructions in one clock cycle or one floating-point instruction (under certain circumstances, two floating-point instructions) in one clock cycle.

Branch prediction is implemented in the Pentium by using two prefetch buffers, one to prefetch code in a linear fashion and one to prefetch code according to the contents of the branch target buffer (BTB), so the required code is almost always prefetched before it is needed for execution. Note that the branch addresses are stored in the branch target buffer (BTB).

There are two instruction pipelines, the U pipe and the V pipe, which are not equivalent and interchangeable. The U pipe can execute all integer and floating-point instructions, whereas the V pipe can only execute simple integer instructions and the floating-point exchange register contents (`FXCH`) instructions.

The instruction decode unit decodes the prefetched instructions so that the Pentium can execute them. The control ROM includes the microcode for the Pentium processor and has direct control over both pipelines. A barrel shifter is included in the chip for fast shift operations.

9.14.3.1 Pentium Registers

The Pentium processor includes the same registers as the 80486. Three new system flags are added to the 32-bit EFLAGS register.

9.14.3.2 Pentium Addressing Modes and Instructions

The Pentium includes the same addressing modes as the 80386/80486.

The Pentium microprocessor includes three new application instructions and four new system instructions beyond those of the 80486. One of the new application instruction is the `CMPXCHG8B`. As an example, `CMPXCHG8B` reg64 or mem64 compares the 64-bit value in EDX:EAX with the 64 bit contents of reg64 or mem64. If they are equal, the 64-bit value in

ECX:EBX is stored in reg64 or mem64; otherwise the content of reg64 or mem64 is loaded into EDX:EAX.

Pentium floating-point instructions execute much faster than those of the 80486 instructions. For example, a 66-MHz Pentium microprocessor provides about three times the floating-point performance of a 66-MHz Intel 80486 DX2 microprocessor.

9.14.3.3 Pentium versus 80486: Basic Differences in Registers, Paging, Stack Operations, and Exceptions

Registers of the Pentium Processor versus Those of the 80486

This section discusses the basic differences between the Pentium and 80486 control, debug, and test registers.

One new control register, CR4, is included in the Pentium. CR4 contains bits that enable certain extensions to the 80486 provided in the Pentium processor. These extensions include functions for handling certain hardware error conditions.

The Pentium processor defines the type of breakpoint access by two bits in DR7 to perform breakpoint functions such as break on instruction execution only, break on data writes only, and break on data reads or writes but not instruction fetches. The implementation of test registers on the 80486 used for testing the cache has been redesigned in the Pentium processor.

Paging

The Pentium processor provides an extension to the memory management/paging functions of the 80486 to support larger page sizes.

Stack Operations

The Pentium, 80486, and 80386 microprocessors push a different value of SP on the stack for a PUSH instruction than does the 8086. The 32-bit processors push the value of the SP before it is decremented whereas the 8086 pushes the value of the SP after it is decremented.

Exceptions

The Pentium processor implements new exceptions beyond those of the 80486. For example, a machine check exception is newly defined for reporting parity errors and other hardware errors.

External hardware interrupts on the Pentium may be recognized on different instruction boundaries due to the pipelined execution of the Pentium processor and possibly an extra instruction passing through the V pipe concurrently with an instruction in the U pipe. When the two instructions complete execution, the interrupt is then serviced. Therefore, the EIP pushed onto the stack when servicing the interrupt on the Pentium processor may be different than that for the 80486 (i.e., it is serviced later). The priority of exceptions is the same on both the Pentium and 80486.

9.14.3.4 Pentium Input/Output

The Pentium processor handles I/O in the same way as the 80486. The Pentium can use either standard I/O or memory-mapped I/O. Standard I/O is accomplished by using `IN/OUT` instructions and a hardware protection mechanism. When memory-mapped I/O is used, memory-reference instructions are used for input/output and the protection mechanism is provided via segmentation or paging.

The Pentium can transfer 8, 16, or 32 bits to a device. Like memory-mapped I/O, 16-bit ports using standard I/O should be aligned to even addresses so that all 16 bits can be transferred in a single bus cycle. Like double words in memory-mapped I/O, 32-bit ports in standard I/O should be aligned to addresses that are multiples of four. The Pentium supports I/O transfer to misaligned ports, but there is a performance penalty because an extra bus cycle must be used.

The `INS` and `OUTS` instructions move blocks of data between I/O ports and memory. The `INS` and `OUTS` instructions, when used with repeat prefixes, perform block input or output operations. The string I/O instructions can operate on byte (8-bit) strings, word (16-bit) strings, or double word (32-bit) strings. When the Pentium is running in protected mode, I/O operates as in real address mode with additional protection features.

9.14.3.5 Applications with the Pentium

The performance of the Pentium's floating-point unit (FPU) makes it appropriate for wide areas of numeric applications:

- Pentium's FPU can accept decimal operands and produce extra decimal results of up to 18 digits. This greatly simplifies accounting programming. Financial calculations that use power functions can take advantage of exponential and logarithmic functions.
- Many minicomputer and mainframe large simulation problems can be executed by the Pentium. These applications include complex electronic circuit simulations using SPICE and simulation of mechanical systems using finite element analysis.
- The Pentium's FPU can move and position machine control heads with accuracy in real time. Axis positioning can efficiently be performed by the hardware trigonometric support provided by the FPU. The Pentium can therefore be used for computer numerical control (CNC) machines.
- The pipelined instruction feature of the Pentium processor makes it an ideal candidate for DSP (digital signal processing) and related applications for computing matrix multiplications and convolutions.

- Other possible application areas for the Pentium include robotics, navigation, data acquisition, and process control.

9.14.4 Pentium versus Pentium Pro

The Pentium was first introduced by Intel in March 1993, and the Pentium Pro was introduced in November 1995. The Pentium processor provides pipelined superscalar architecture. The Pentium processor's pipelined implementation uses five stages to extract high throughput and the Pentium Pro utilizes 12-stage, superpipelined implementation, trading less work per pipestage for more stages. The Pentium Pro processor reduced its pipestage time by 33% compared with a Pentium processor, which means the Pentium Pro processor can have a 33% higher clock speed than a Pentium processor and still be equally easy to produce from a semiconductor manufacturing process. A 200-MHz Pentium Pro is always faster than a 200-MHz Pentium for 32-bit applications such as computer-aided design (CAD), 3-D graphics, and multimedia applications.

The Pentium processor's superscalar architecture, with its ability to execute two instructions per clock, was difficult to exceed without a new approach. The new approach used by the Pentium Pro processor removes the constraint of linear instruction sequencing between the traditional "fetch" and "execute" phases, and opens up a wide instruction pool. This approach allows the "execute" phase of the Pentium Pro processor to have much more visibility into the program's instruction stream so that better scheduling may take place. This allows instructions to be started in any order but always be completed in the original program order.

Microprocessor speeds have increased tremendously over the past 10 years, but the speed of the main memory devices has only increased by 60 percent. This increasing memory latency, relative to the microprocessor speed, is a fundamental problem that the Pentium Pro is designed to solve. The Pentium Pro processor "looks ahead" into its instruction pool at subsequent instructions and will do useful work rather than be stalled. The Pentium Pro executes instructions depending on their readiness to execute and not on their original program order. In summary, it is the unique combination of improved branch prediction, choosing the best order, and executing the instructions in the preferred order that enables the Pentium Pro processor to improve program execution over the Pentium processor. This unique combination is called "dynamic execution."

The Pentium Pro does a great job running some operating systems such as Windows NT or Unix. The first release of Windows 95 contains a significant amount of 16-bit code in the graphics subsystem. This causes operations on the Pentium Pro to be serialized instead of taking advantage of the dynamic execution architecture. Nevertheless, the Pentium Pro is up to 30% faster than the fastest Pentium in 32-bit applications. Table 9.17 compares the basic features the Pentium with those of the Pentium Pro.

TABLE 9.17 Pentium vs. Pentium Pro

Pentium	Pentium Pro
First introduced March 1993	Introduced November 1995
2 instructions per clock cycle	3 instructions per clock cycle
Primary cache of 16K	Primary cache of 16K
Current clock speeds of 100, 120, 133, 150, 166, 200, and 233 MHz	Current clock speeds 166, 180, 200 MHz
More silicon is needed to produce the chip	Tighter design reduces silicon needed and makes chip faster (shorter distances between transistors)
Designed for operating systems written in 16-bit code	Designed for operating systems written in 32-bit code.

9.14.5 Pentium II / Celeron / Pentium II Xeon™ / Pentium III / Pentium III Xeon™

The 32-bit Pentium II processor is Intel's latest addition to the Pentium line of microprocessors, which originated form the widely cloned 80x86 line. It basically takes attributes of the Pentium Pro processor plus the capabilities of MMX technology to yield processor speeds of 333, 300, 266, and 233 MHz. The Pentium II processor uses 0.25 micron technology (this refers to the width of the circuit lines on the silicon) to allow increased core frequencies and reduce power consumption. The Pentium II processor took advantage of four new technologies to achieve its performance ratings:

- Dual Independent Bus Architecture (DIB)
- Dynamic Execution
- Intel MMX Technology
- Single-Edge-Contact Cartridge

DIB was first implemented in the Pentium Pro processor to address bandwidth limitations. The DIB architecture consists of two independent buses, an L2 cache bus and a system bus, to offer three times the bandwidth performance of single bus architecture processors. The Pentium II processor can access data from both buses simultaneously to accelerate the flow of information within the system.

Dynamic execution was also first implemented in the Pentium Pro processor. It consists of three processing techniques to improve the efficiency of executing instructions.

These techniques include multiple branch prediction, data flow analysis, and speculative execution. Multiple branch prediction uses an algorithm to determine the next instruction to be executed following a jump in the instruction flow. With data flow analysis, the processor determines the optimum sequence for processing a program after looking at software instructions to see if they are dependent on other instructions. Speculative execution increases the rate of execution by executing instructions ahead of the program counter that are likely to be needed.

MMX (**m**atrix **m**ath e**x**tensions) technology is Intel's greatest enhancement to its microprocessor architecture. MMX technology is intended for efficient multimedia and communications operations. To achieve this, 57 new instructions have been added to manipulate and process video, audio, and graphical data more efficiently. These instructions support single-instruction multiple-data (SIMD) techniques, which enable one instruction to perform the same function on multiple pieces of data. Programs written using the new instructions significantly enhance the capabilities of Pentium II.

The final feature in Intel's Pentium II processor is single-edge-contact (SEC) packaging. In this packaging arrangement, the core and L2 cache are fully enclosed in a plastic and metal cartridge. The components are surface mounted directly to a substrate inside the cartridge to enable high-frequency operation.

Intel Celeron processor utilizes Pentium II as core .The Celeron processor family includes: 333 MHz, 300A MHz, 300 MHz, and 266 MHz processors.The Celeron 266 MHz and 300 MHz processors do not contain any level 2 cache. But the Celeron 300A MHz and 333 MHz processors incorporate an integrated L2 cache. All Celeron processors are based on Intel's 0.25 micron CMOS technology. The Celeron processor is designed for inexpensive or "Basic PC" desktop systems and can run Windows 98. The Celeron processor offers good floating-point (3D geometry calculations) and multimedia (both video and audio) performance.

The Pentium II Xeon processor contains large, fast caches to transfer data at super high speed through the processor core. The processor can run at either 400 MHz or 450 MHz. The Pentium II Xeon is designed for any mid-range or higher Intel-based server or workstation.The 450 MHz Pentium II Xeon can be used in dual-processor (two-way) workstations and servers. The 450 MHz Pentium II Xeon processor with four-way servers is expected to be available in the future.

The Pentium III operates at 450 MHz and 500 MHz. It is designed for desktop PCs. The Pentium III enhances the multimedia capabilities of the PC, including full screen video and graphics. Pentium III Xeon processors run at 500 MHz and 550 MHz. They are designed for mid-range and higher Internet-based servers and workstations. It is compatible with Pentium II Xeon processor-based platforms. Pentium III Xeon is also designed for demanding workstation applications such as 3-D visualization, digital content creation, and dynamic Internet content development. Pentium III-based systems can run applications on Microsoft Windows NT or UNIX-based environments. The Pentium III Xeon is available in a number of L2 cache versions such as 512-Kbytes, 1-Mbyte, or 2-Mbytes (500 MHz); 512 Kbytes (550 MHz) to satisfy a variety of Internet application requirements.

9.14.6 Merced/IA-64

Intel and Hewlett-Packard are collaborating on a new generation of 64-bit microprocessors due by the year 2000 code-named "Merced" and also known as "Intel Architecture–64" (IA-64). The microprocessor will not be simply an extension of Intel's 32-bit 80x86 or Pentium series processors, nor will it be an evolution of HP's 64-bit RISC architecture. IA-64 is a new design that will implement innovative forward-looking features to help improve parallel instruction processing: that is, long instruction words, instruction prediction, branch elimination, and speculative loading. These techniques aren't necessarily new concepts, but they will be implemented in ways that will be much more efficient. We will look at how the current line of 80x86 or Pentium family processors approach these techniques and compare them with the IA-64 implementation.

An 80x86 instruction varies in length from 8 to 108 bits, and the microprocessor spends time and work decoding each instruction while scanning for the instruction boundaries during execution. In addition, Pentium processors frantically try to reorder instructions and group them so that two instructions can be fed into two processing pipelines simultaneously. Although improving performance, this approach is still rather ineffective and has a high cost of logic circuitry in the chip.

The IA-64 packs three instructions into a single 128-bit bundle—something Intel calls "explicitly parallel instruction computing" (EPIC). During compilation of a program, the compiler explicitly tells the microprocessor inside the 128-bit packet which of the instructions can be executed in parallel. Hence, the microprocessor will no longer need to scramble at run-time to discover and reorder instructions for parallel execution because all of this will already have been done at compilation. While trying to keep the instruction pipeline full, 80x86 or Pentium family processors try to predict which way branches will take place and speculatively execute instructions along the predicted path. In case of wrong guesses, the microprocessor must discard the speculative results, flush the pipelines, and reload the correct instructions into the pipe. This results in a large loss of microprocessor cycles.

In dealing with branch prediction, the IA-64 will again put the burden on the compiler. Wherever practical, the compiler will insert flags into the instruction packets to mark separate paths from a branch instruction. These flags, known as "predicates," will allow the microprocessor to funnel instructions for a specific branch into a pipe and execute each branch separately and simultaneously. This will effectively let the microprocessor process different paths of a branch at the same time, then discard the results of the path it doesn't need.

One drawback of the 80x86 processor series is the fact that data is not fetched from memory until the microprocessor needs it and calls for it. The IA-64 will implement speculative loading, which allows the memory and I/O devices to be delivering data to the microprocessor before the processor actually needs it, eliminating some of the delays the 80x86 processor incurs while waiting for data to appear on the bus.

During compilation of a program, the compiler scans the source code and when it sees an upcoming load instruction, removes it and inserts a speculative load instruction a few cycles ahead of it. Also, a speculative check instruction will be placed immediately before the original load instruction. Hence, when the microprocessor receives the speculative load instruction, it informs the memory or I/O systems to start fetching data. In the meantime the microprocessor will continue to execute other instructions until it reaches the speculative check operand. At this point, the microprocessor will check to see if the data it requested is ready. In this manner, the IA-64 will be able to continue executing code while minimizing delay time that the memory or I/O devices inherently incur.

Hopefully, this section provides an overview of the latest Intel microprocessors. Unfortunately, they may be old news within a year. However, once the computer engineers and the scientists understand the fundamental concepts associated with computers, it will be easier to handle the challenging and rapid growth in microprocessor technology.

QUESTIONS AND PROBLEMS

9.1 What is the basic difference between the 8086, 8086-2, and 8086-4?

9.2 Assume [DS]=1000H, [SS]=2000H, [CS]=3000H, [BP]=000FH, [BX]=000AH before execution of the following 8086 instructions:
(a) MOV CX,[BX] (b) MOV DX,[BP]
Which instruction will be executed faster by the 8086, and why ?

9.3 What is the purpose of the 8086 MN/$\overline{\text{MX}}$ pin?

9.4 If [DS] = 205FH and OFFSET = 0052H, what is the 8086 physical address? Does the EU or BIU compute this physical address?

9.5 In an 8086 system, SEGMENT 1 contains addresses 00100H–00200H and SEGMENT 2 also contains addresses 00100H–00200H. What are these segments called?

9.6 Determine the addressing modes for the following 8086 instructions:

(a) `CLC`
(b) `CALL WORDPTR [BX]`
(c) `MOV AX, DX`
(d) `ADD [SI], BX`

9.7 Find the overflow, direction, interrupt, trap, sign, zero, parity, and carry flags after execution of the following 8086 instruction sequence:

```
MOV     AH, 0FH
SAHF
```

9.8 What is the content of AL after execution of the following 8086 instruction sequence?

```
MOV     BH, 33H
MOV     AL, 32H
ADD     AL, BH
AAA
```

9.9 What happens after execution of the following 8086 instruction sequence? Comment.

```
MOV     DX, 001FH
XCHG    DL, DH
MOV     AX, DX
IDIV    DL
```

9.10 What are the remainder, quotient, and registers containing them after execution of the following 8086 instruction sequence?

```
MOV     AH, 0
MOV     AL, 0FFH
MOV     CX, 2
IDIV    CL
```

9.11 Write an 8086 instruction sequence to set the trap flag for single stepping.

9.12 Write an 8086 assembly language program to subtract two 64-bit numbers. Assume SI and DI point to the low words of the numbers.

9.13 Write an 8086 assembly program to add a 16-bit number stored in BX (bits 0 to 7 containing the high-order byte of the number and bits 8 to 15 containing the low-order byte) with another 16-bit number stored in CX (bits 0 to 7 containing the low-order 8 bits of the number and bits 8 thorough 15 containing the high-order 8 bits). Store the result in AX.

9.14 Write an 8086 assembly program to multiply the top two 16-bit unsigned words of the stack. Store the 32-bit result onto the stack.

9.15 Write an 8086 assembly language program to add three 16-bit numbers. Store the 16-bit result in AX. Verify your result using DEBUG.

9.16 Write an 8086 assembly language find the area of a circle with radius 2 meters and save the result in AX. Verify your result using DEBUG.

9.17 Write an 8086 assembly language program to convert 25 degrees Celsius to Fahrenheit degrees and store the value in AX. Use the equation

$$F = (C/5) * 9 + 32$$

9.18 Assume AL, CX and DXBX contain a signed byte, a signed word, and a signed 32-bit number respectively. Write an 8086 assembly language program that will compute the signed 32-bit result: AL – CX + DXBX → DXBX.

9.19 Write an 8086 assembly program to divide an 8-bit signed number in CH by an 8-bit signed number in CL. Store the quotient in CH and the remainder in CL.

9.20 Write an 8086 assembly program to add 25 16-bit numbers stored in consecutive memory locations starting at displacement 0100H in DS = 0020H. Store the 16-bit result onto the stack.

9.21 Write an 8086 assembly program to find the minimum value of a string of 10 signed 8-bit numbers using indexed addressing. Assume Offset 5000H contains the first number.

9.22 Write an 8086 assembly program to move 100 words from a source with offset 0010H in ES = 1000H to a destination with offset 0100H in the same extra segment.

9.23 Write an 8086 assembly program to divide a 28-bit unsigned number in the high 28 bits of DX AX by 8_{10}. Do not use any divide instruction. Store the quotient in the low 28 bits of DX AX. Discard remainder.

9.24 Write an 8086 assembly program to compare two strings of 15 ASCII characters. The first string (string 1) is stored starting at offset 5000H in DS = 0020H followed by the string. The first character of the second string (string 2) is stored starting at 6000H in ES = 1000H. The ASCII character in the first location of string 1 will be compared with the first ASCII character of string 2, and so on. As soon as a match is found, store $00EE_{16}$ onto the stack; otherwise, store 0000_{16} onto the stack.

9.25 Write a subroutine in 8086 assembly language that can be called up by a main program in a different code segment. The subroutine will compute the 16-bit sum

$$\sum_{i=1}^{100} X_i^2$$

Assume the X_i's are signed 8-bit numbers and are stored in consecutive locations starting at displacement 0050H in DS = 2020H. Also, write the main program that will call this subroutine to compute

$$\sum_{i=1}^{100} \frac{X_i^2}{100}$$

and store the 16-bit result (8-bit remainder and 8-bit quotient) in two consecutive memory bytes starting at offset 0400H in DS = 2020H.

9.26 Write a subroutine in 8086 assembly language to convert a 2-digit unpacked BCD number to binary. The most significant digit is stored in a memory location starting at offset 4000H in DS = 0020H, and the least significant is stored at offset 4001H in the same DS. Store the binary result in DX. Use the value of the 2-digit BCD number,

$$V = D_1 \times 10 + D_0$$

9.27 Assume an 8086/2732/6116/8255 microcomputer. Suppose that four switches are connected at bits 0 through 3 of port A and an LED is connected at bit 4 of port B. If the number of LOW switches is even, turn the port B LED ON; otherwise, turn the port B LED OFF. Write an 8086 assembly language program to accomplish this.

9.28 Interface two 2732 and one 8255 odd to an 8086 to obtain even and odd 2732 locations and odd addresses for the 8255's port A, port B, port C, and control registers. Show only the connections for the pins shown in Figure P9.28. Assume all unused address lines to be zeros.

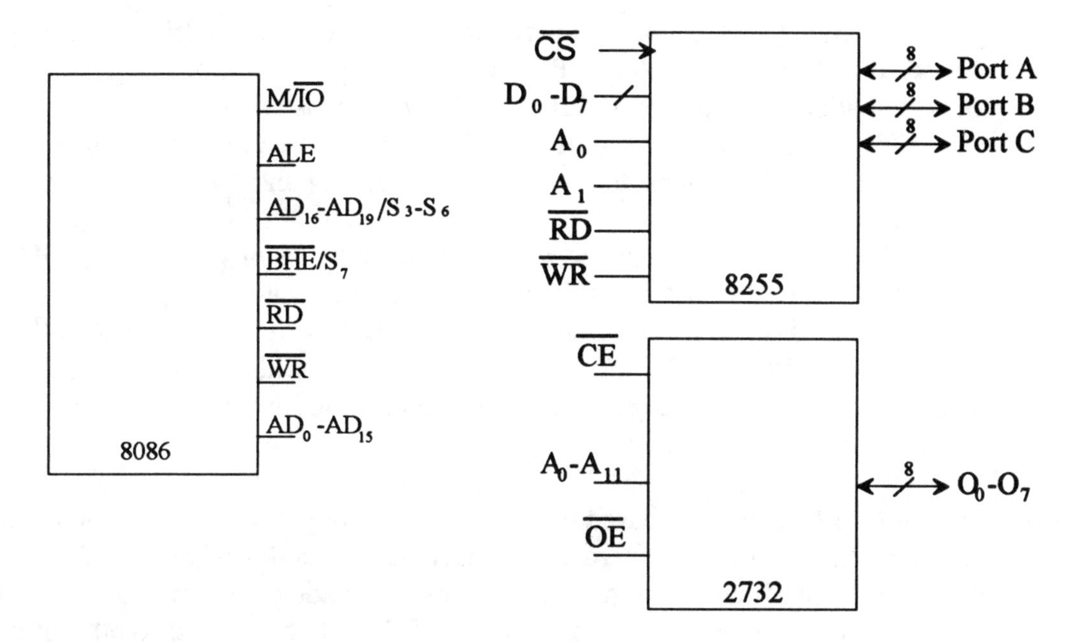

FIGURE P9.28

9.29 In Figure P9.29, if $V_m > 12$ V, turn the LED ON connected at bit 4 of port A. On the other hand, if $V_m < 11$ V, turn the LED OFF. Use ports, registers, and memory locations of your choice. Draw a hardware block diagram showing the microcomputer and the connections of the figure to its ports. Write a service routine in 8086 assembly language. Assume all segment registers are already initialized. The service routine should be written as CS=1000H, IP=2000H. The main program will initialize SP to 2050H, initialize ports, and wait for interrupts.

9.30 Repeat Problem 9.29 using the 8086 NMI interrupt.

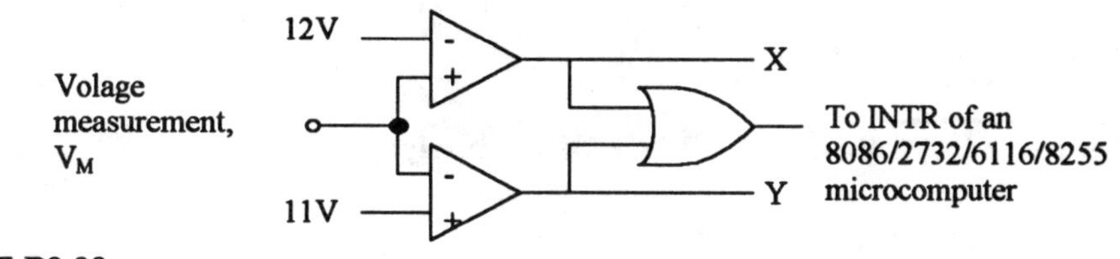

FIGURE P9.29

9.31 An 8086/2732/6116/8255-based microcomputer is required to drive the LEDs connected to bit 0 of ports A and B based on the input conditions set by switches connected at bit 1 of ports A and B. The I/O conditions are as follows:

- If the input to bit 1 of port A is HIGH and the input to bit 1 of port B is low, then the LED connected to port A will be ON and the LED connected to port B will be OFF.
- If the input to bit 1 of port A is LOW and that of port B is HIGH, then the LED of port A will be OFF and that of port B will be ON
- If the bit 1 inputs of both ports A and B are the same (either both HIGH or both LOW), then both LEDs of ports A and B will be ON.

Write an 8086 assembly language program to accomplish this. Do not use any instructions involving the parity flag.

9.32 An 8086/2732/6116/8255-based microcomputer is required to test a NAND gate. Figure P9.32 shows the I/O hardware needed to test the NAND gate. The microcom puter is to be programmed to generate the various logic conditions for the NAND inputs, input the NAND output, and turn the LED ON connected to bit 3 of port A if the NAND gate chip is found to be faulty. Otherwise, turn the LED ON connected to bit 4 of port A. Write an 8086 assembly language program to accomplish this.

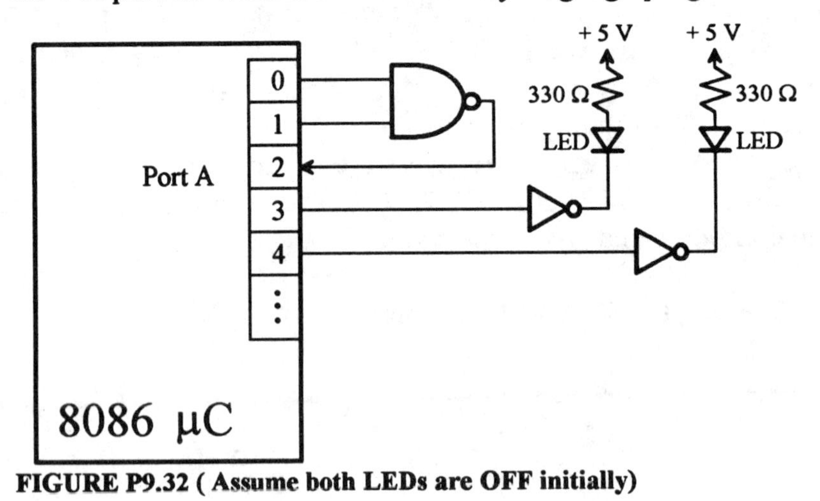

FIGURE P9.32 (Assume both LEDs are OFF initially)

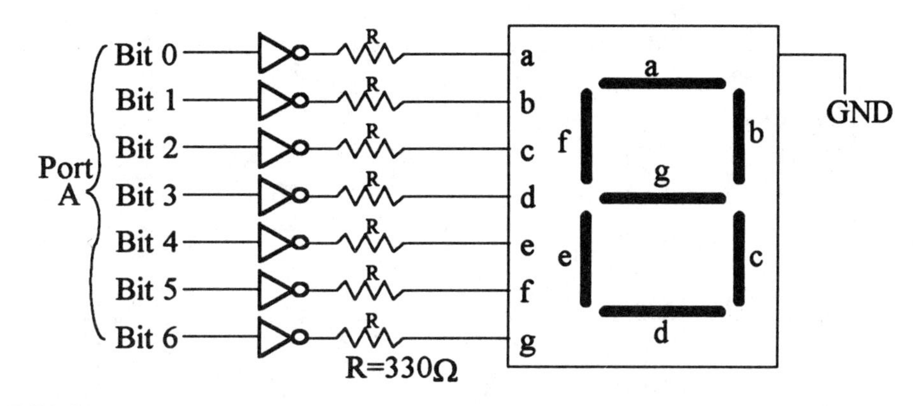

FIGURE P9.33

9.33 An 8086/2732/6116/8255 microcomputer is required to add two 3-bit numbers in AL and BL and output the sum (not to exceed 9) to a common cathode seven-segment display connected to port A as shown in Figure P9.33.Write an 8086 assembly language program to accomplish this by using a look-up table. Do not use XLAT instruction.

9.34 Write an 8086 assembly language program to turn an LED OFF connected to bit 2 of port A of an 8086/2732/6116/8255 microcomputer and then turn it on after delay of 15 s. Assume the LED is ON initially

9.35 What are the factors to be considered for interfacing a hex keyboard to a microcomputer?

9.36 An 8086/2732/6116/8255 microcomputer is required to input a number from 0 to 9 from an ASCII keyboard interfaced to it and output to an EBCDIC printer. Assume that the keyboard is connected to port A and the printer is connected to port B. Write an 8086 assembly language to accomplish this. Use XLAT instruction.

9.37 Will the circuit shown in Figure P9.37 work? If so, determine the I/O map in hex. If not, justify briefly, modify the circuit and determine the I/O map in hex. Use only the pins and signals provided. Assume all don't cares to be zeros. Note that I/O map includes the addresses for port A, port B, port C, and the control register. Using the logical port addresses, write an instruction sequence to configure port A as input and port B as output.

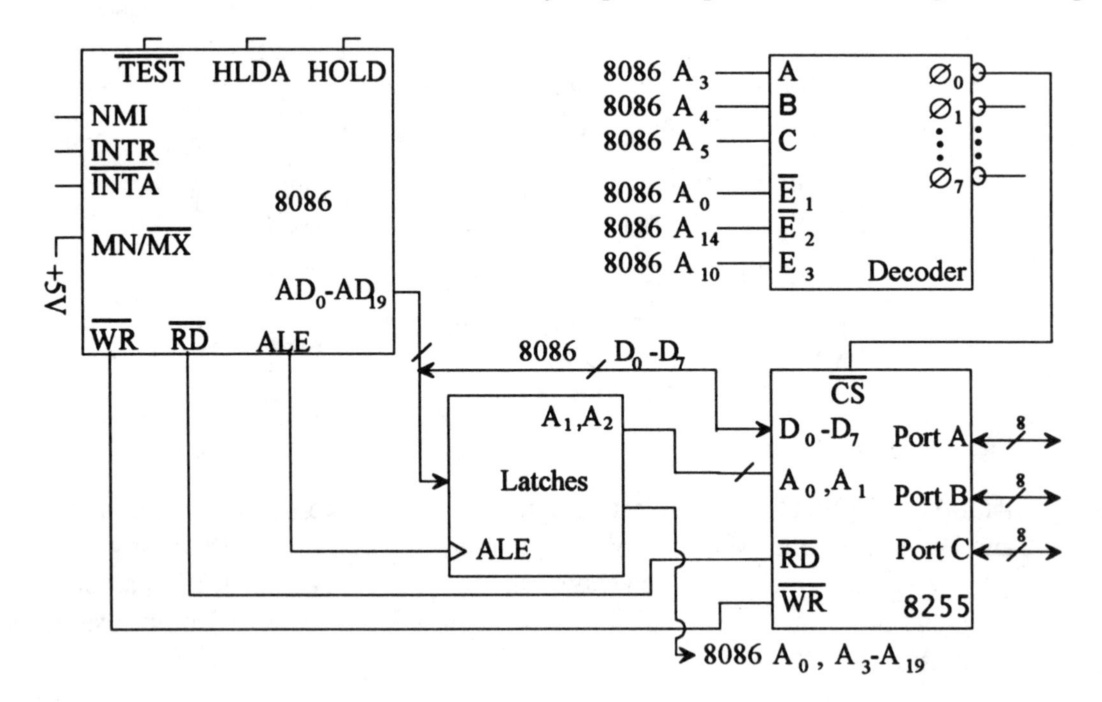

FIGURE P9.37

9.38 (a) What is the basic difference between the 80386 and 80386SX?
(b) What is the basic difference between the 80386 and 80486?

9.39 What is the difference between the 80386 protected, real-address, and virtual 8086 modes?

9.40 Discuss the basic features of the 80486.

9.41 Assume the following 80386 register contents

[EBX] = 00001000H
[ECX] = 04000002H
[EDX] = 20005000H

prior to execution of each of the following 80386 instructions. Determine the effective address after execution of each of the following instructions and identify the addressing modes:

```
MOV [EBX * 4] [ECX], EDX
MOV [EBX * 2] [ECX + 2020H], EDX
```

9.42 Determine the effect of each of the following 80386 instructions:

(a) `MOVZX EAX, CH`

Prior to execution of this `MOVZX` instruction, assume

[EAX] = 80001234H
[ECX] = 00008080H

(b) `MOVSX EDX, BL`

Prior to execution of this `MOVSX` assume

[EDX] = FFFFFFFFH
[EBX] = 05218888H

9.43 Write an 80386 assembly program to add a 64-bit number in [ECX] [EDX] with another 64-bit number in [EAX] [EBX]. Store the result in [EAX] [EBX].

9.44 Write an 80386 assembly program to divide a signed 32-bit number in DX:AX by an 8-bit signed number in BH. Store the 16-bit quotient and 16-bit remainder in AX and DX respectively.

9.45 Write an 80386 assembly program to compute

$$\sum_{i=1}^{N} X_i^2$$

where $N = 1000$ and the X_i's are signed 32-bit numbers.
Assume that ΣX_i^2 can be stored as a 32-bit signed number.

9.46 Discuss 80386 I/O.

9.47 Compare the on-chip hardware features of the 80486 and Pentium microprocessors.

9.48 What are the sizes of the address and data buses of the 80486 and the Pentium?

9.49 Identify the main differences between the 80486 and the Pentium.

9.50 What are the clock speed, pipeline model, number of on-chip transistors, and number of pins on the 80486 and Pentium processors?

9.51 Discuss typical applications of Pentium.

9.52 Identify the main differences between the Intel 80386 and 80486.

9.53 What is meant by the 80486 BUS BACKOFF feature?

9.54 How many pipeline stages are in Pentium and Pentium Pro?

9.55 How many new instructions are added to the 80486 beyond those of the 80386?

9.56 Given the following register contents,

[EBX] = 7F27108AH

[ECX] = 2A157241H

what is the content of ECX after execution of the following 80486 instruction sequence:

```
MOV       EBX,ECX
BSWAP     ECX
BSWAP     ECX
BSWAP     ECX
BSWAP     ECX
```

9.57 If [EBX] = 0123A212H and [EDX] = 46B12310H, then what are the contents of EBX and EDX after execution of the 80486 instruction `XADD EBX,EDX`?

9.58 If [BX] = 271AH, [AX] = 712EH, and [CX] = 1234H, what are the contents of CX after execution of the 80486 instruction `CMPXCHG CX,BX`?

9.59 What are three modes of the Pentium processor? Discuss them briefly.

9.60 What is meant by the statement, "The Pentium processor is based on a superscalar design"?

9.61 What are the purposes of the U pipe and V pipe of the Pentium processor?

9.62 What are the sizes of the data and instruction caches in the Pentium?

9.63 Summarize the basic differences among Pentium, Pentium Pro, and Pentium II, Celeron, Pentium II Xeon, Pentium III, and Pentium III Xeon processors.

9.64 Why are the Pentium Pro's complete capabilities not used by the Windows 95 operating system?

9.65 Summarize the basic features of the Intel/Hewlett-Packard "Merced" microprocessor.

10

MOTOROLA 16-,32-,AND 64- BIT MICROPROCESSORS

This chapter describes the basic features of Motorola's typical 16-, 32-, and 64-bit microprocessors. The MC68000 and MC68020 are covered in detail, and an overview of other Motorola 32-bit and 64-bit microprocessors is provided.

Motorola's 16-bit microprocessor model MC68000 was designed using HMOS technology. Motorola has discontinued manufacturing the MC68000 (end of life cycle) and has replaced it by a lower power MC68HC000, which is designed using HCMOS technology. The MC68HC000 is equivalent to the MC68000 in all aspects except that the MC68HC000 is designed using HCMOS whereas the MC68000 was designed using HMOS technology. This means that unlike the MC68000, the unused inputs of the MC68HC000 should not be kept floating, they should be connected to +5 V, ground, or outputs of other chips as appropriate. Also, note that an HCMOS output can drive 10 LSTTL inputs. However, an LSTTL output is not guaranteed to provide HCMOS input voltage. Hence, the HCT gates may be required when driving HC inputs. The MC 68HC000 has the same registers, addressing modes, instruction set, pins and signals, and I/O capabilities as the MC68000. The term "MC68000" will be used interchangeably with the term "MC68HC000" throughout this chapter.

The MC68HC000, implemented in HCMOS, is applicable to designs for which the following considerations are relevant:

- The MC68HC000 completely satisfies the input/output drive requirements of HCMOS logic devices.
- The MC68HC000 provides an order of magnitude reduction in power dissipation when compared to the HMOS MC68000.
- The minimum operating frequency of the MC68HC000 is 4 MHz.

Although the MC68HC000 is implemented with input protection diodes, care should be exercised to ensure that the maximum input voltage specification (-0.3 V to +6.5 V) is not exceeded.

This chapter also describes the MC68020, Motorola's first 32-bit microprocessor, in detail. An overview of other Motorola 32-bit microprocessors, namely the MC68030, MC68040, and MC68060 is provided. Finally, a summary of the PowerPC and Motorola's future microprocessors is included.

Due to rapid technological advancements, it is likely that Motorola may discontinue manufacturing some microprocessors and/or their support chips in future. Alternate sources may be the only way to obtain these chips. For example, Motorola discontinued production of the 68230 I/O chip. However, this chip is available from a second source, namely ST Microelectronics (SGS-Thompson) Company of Santa Ana, California. The following information is, therefore, provided for the convenience of users of Motorola microprocessors and their support chips:

1. Motorola, Inc. Telephone numbers for the United States and Canada

   Motorola Customer Service: 1-800-521-6274
   Motorola Inc. for Local and International customers: 1-602-244-6900
   Motorola Web site: www.mot.com

2. Distributors of Motorola microprocessors and support chips: telephone numbers for the United States and Canada

   Arrow Co. 1-800-833-3557
   1-800-777-2776
   New York Electronics 1-800-463-9275

3. Second source for Motorola microprocessors and support chips: telephone numbers for local and international users

   ST Microelectronics 1-602-867-6100
   (SGS-Thompson Co.) 1-949-347-0717
   Philips ECG 1-800-526-9354

10.1 Introduction

The MC68000 is Motorola's first 16-bit microprocessor. Its address and data registers are all 32 bits wide, and its ALU is 16 bits wide. The 68000 is designed using HMOS technology. The 68000 requires a single 5-V supply. The processor can be operated from a maximum internal clock frequency of 25 MHz. The 68000 is available in several frequencies, including 4,

6, 8, 10, 12.5, 16.67, and 25 MHz. The 68000 does not have on-chip clock circuitry and therefore, requires an external crystal oscillator or clock generator/driver circuit to generate the clock.

The 68000 has several different versions, which include the 68008, 68010, and 68012. The 68000 and 68010 are packaged either in a 64-pin DIP (dual in-line package) with all pins assigned or in a 68-pin quad pack or PGA (pin grid array) with some unused pins. The 68000 is also packaged in 68-terminal chip carrier. The 68008 is packed in a 48-pin dual in-line package, whereas the 68012 is packed in an 84-pin grid array. The 68008 provides the basic 68000 capabilities with inexpensive packaging. It has an 8-bit data bus, which facilitates the interfacing of this chip to inexpensive 8-bit peripheral chips. The 68010 provides hardware-based virtual memory support and efficient looping instructions. Like the 68000, it has a 16-bit data bus and a 24-bit address bus. The 68012 includes all the 68010 features with a 31-bit address bus. The clock frequencies of the 68008, 68010, and 68012 are the same as those of the 68000. The following table summarizes the basic differences among the 68000 family members:

	68000	68008	68010	68012
Data size (bits)	16	8	16	16
Address bus size (bits)	24	20	24	31
Virtual memory	No	No	Yes	Yes
Control registers	None	None	3	3
Directly addressable memory (bytes)	16 MB	1 MB	16 MB	2 GB

To implement operating systems and protection features, the 68000 can be operated in two modes: supervisor and user. The supervisor mode is also called the “operating system mode.” In this mode, the 68000 can execute all instructions. The 68000 operates in one of these modes based on the S bit of the status register. When the S bit is 1, the 68000 operates in the supervisor mode; when the S bit is 0, the 68000 operates in the user mode.

Table 10.1 lists the basic differences between the 68000 user and supervisor modes. From Table 10.1, it can be seen that the 68000 executing a program in the supervisor mode can enter the user mode by modifying the S bit of the status register to 0 via an instruction. Instructions such as `MOVE` to/from SR, `ORI` to/from SR, and `EORI` to/from SR can be used to accomplish this. On the other hand, the 68000 executing a program in the user mode can enter the supervisor mode only via recognition of a trap, reset, or interrupt. Note that, upon hardware reset, the 68000 operates in the supervisor mode and can execute all instructions. An attempt to execute *privileged instructions* (instructions that can only be executed in the supervisor mode) in the user mode will automatically generate an internal interrupt (trap) by the 68000.

TABLE 10.1 68000 User and Supervisor Modes

	Supervisor Mode	*User Mode*
Enter mode by	Recognition of a trap, reset, or interrupt	Clearing status bit S
System stack pointer	Supervisor stack pointer	User stack pointer
Other stack pointers	User stack pointer and registers A0-A6	registers, A0-A6
Instructions available	All including: STOP RESET MOVE to/from SR ANDI to/from SR ORI to/from SR EORI to/from SR MOVE USP to (An) MOVE to USP RTE	All except those listed under Supervisor mode
Function code pin FC2	1	0

The logical level in the 68000 function code pin (FC2) indicates to the external devices whether the 68000 is currently operating in the user or supervisor mode. The 68000 has three function code pins (FC2, FC1, and FC0), which indicate to the external devices whether the 68000 is accessing supervisor program/data or user program/data or performing an interrupt acknowledge cycle.

The 68000 can operate on five different data types: bits, 4-bit binary-coded decimal (BCD) digits, bytes, 16-bit words, and 32-bit long words. The 68000 instruction set includes 56 basic instruction types. With 14 addressing modes, 56 instructions, and 5 data types, the 68000 contains over 1000 op-codes. The fastest instruction is one that copies the contents of one register into another register. It is executed in 500 ns at an 8-MHz clock rate. The slowest instruction is 32-bit by 16-bit divide, which in executed in 21.25 μs at 8 MHz. The 68000 has no I/O instructions. Thus, the I/O is memory mapped. Hence, MOVE instructions between a register and a memory address are also used as I/O instructions. The MC68000 is a general-purpose register-based microprocessor. Although the 68000 PC is 32 bits wide, only the low-order 24 bits are used. Because this is a byte-addressable machine, it follows that the 68000 microprocessor can directly address 16 MB of memory. **Note that symbol [ ] is used to indicate the contents of a 68000 register or a memory location.**

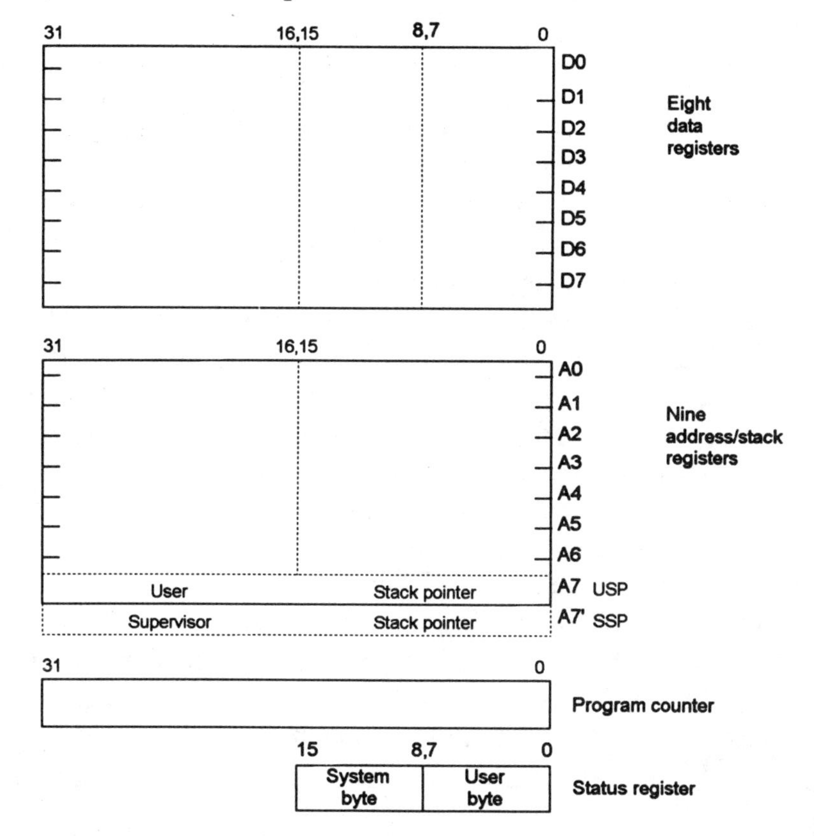

FIGURE 10.1 MC68000 programming model

10.2 68000 Registers

Figure 10.1 shows the 68000 registers. This microprocessor includes eight 32-bit data registers (D0–D7) and nine 32-bit address registers (A0–A7 plus A7'). Data registers normally hold data items such as 8-bit bytes, 16-bit words, and 32-bit long words. An address register usually holds the memory address of an operand; A0-A6 can be used as 16- or 32-bit. Because the 68000 uses 24-bit addresses, it discards the uppermost 8 bits (bits 24–31) while using the address registers to hold memory addresses. The 68000 uses A7 or A7' as the user or supervisor stack pointer (USP or SSP), respectively, depending on the mode of operation.

The 68000 status register is composed of two bytes: a user byte and a system byte (Figure 10.2). The user byte includes typical condition codes such as C, V, N, Z, and X. The meaning of the C, V, N, and Z flags is obvious. Let us explain the meaning of the X bit. Note that the 68000 does not have any `ADDC` or `SUBC` instructions; rather, it has `ADDX` and `SUBX` instructions.

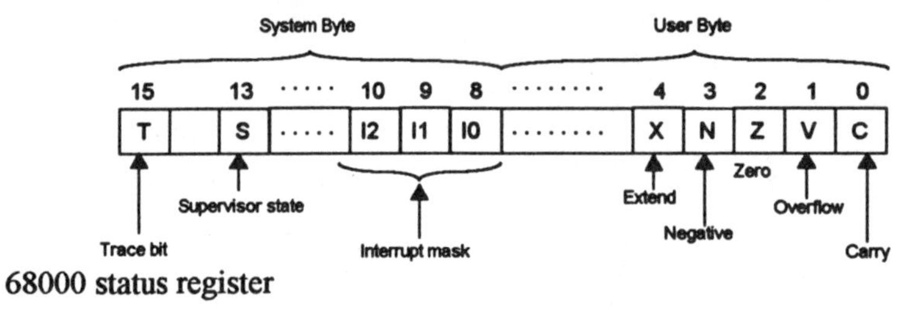

FIGURE 10.2 68000 status register

Because the flags C and X are usually affected in an identical manner, one can use ADDX or SUBX to reflect the carries or borrows in multiprecision arithmetic. The contents of the system byte include a 3-bit interrupt mask (I2, I1, I0), a supervisor flag (S), and a trace flag (T). When the supervisor flag is 1, then the system operates in the supervisor mode; otherwise, the user mode of operation is assumed. When the trace flag is set to 1, the processor generates a trap (internal interrupt) after executing each instruction. A debugging routine can be written at the interrupt address vector to display registers and/or memory after execution of each instruction. Thus, this will provide a single-stepping facility. Note that the trace flag can be set to one in the supervisor mode by executing the instruction ORI# $8000, SR.

The interrupt mask bits (I2, I1, I0) provide the status of the 68000 interrupt pins $\overline{IPL2}$, $\overline{IPL1}$, and $\overline{IPL0}$. I2 I1 I0 = 000 indicates that all interrupts are enabled. I2 I1 I0 = 111 indicates that all maskable interrupts except the nonmaskable interrupt (Level 7) are disabled. The other combinations of I2, I1, and I0 provide the maskable interrupt levels. Note that the signals on the $\overline{IPL2}$, $\overline{IPL1}$, and $\overline{IPL0}$ pins are inverted internally and then compared with I2, I1, and I0, respectively.

10.3 68000 Memory Addressing

The MC68000 supports bytes (8 bits), words (16 bits), and long words (32 bits) as shown in Figure 10.3. Byte addressing includes both odd and even addresses (0, 1, 2, 3, ...), word addressing includes only even addresses in increments of 2 (0, 2, 4, ...), and long word addressing contains even addresses in increments of 4 (0, 4, 8, ...). As an example of 68000 addressing structure, consider MOVE.L D0,$506080. If [D0] = $07F12481, then after this MOVE, [$506080] = $07, [$506081] = $F1, [$506082] = $24, and [$506083] = $81.

In the 68000, all instructions must be located at even addresses for byte, word, and long word instructions; otherwise, the 68000 generates an internal interrupt. The size of each 68000 instruction is even multiples of a byte. This means that once the programmer writes a program starting at an even address, all instructions are located at even addresses after assembling the program. For byte instructions, data can be located at even or odd addresses. On the other hand, data for word and long word instruction must be located at even addresses; otherwise the 68000 generates an internal interrupt.

	15 — 8	7 — 0	
Address = N	Byte 0	Byte 1	$N+1$
$N+2$	Byte 2	Byte 3	$N+3$

(a) 68000 Words Stored in Bytes (4 Bytes)

	15 — 0	
Address = N	Word 0	$N+1$
$N+2$	Word 1	$N+3$
$N+4$	Word 2	$N+5$

(b) 68000 Word Structure (3 Words)

	15 — 0	
Address = N	Long word 0 (H)	$N+1$
$N+2$	Long word 0 (L)	$N+3$
$N+4$	Long word 1 (H)	$N+5$
$N+6$	Long word 1 (L)	$N+7$

(c) 68000 Long Word Structure (2 Long Words)

FIGURE 10.3 68000 addressing structure (N is an even number)

Note that in 68000 for word and long word data, the low-order address stores the high-order byte of a number. This is called *Big-endian byte ordering.*

10.4 68000 Addressing Modes

The 14 addressing modes of the 68000 shown in Table 10.2 can be divided into 6 basic groups: register direct, address register indirect, absolute, program counter relative, immediate, and implied.

As mentioned, the 68000 has three types of instructions: no operand, single operand, and double operand. The single-operand instructions contain the effective address (EA) in the operand field. The EA for these instructions is calculated by the 68000 using the addressing mode used for this operand. In the case of two-operand instructions, one of the operands usually contains the EA and the other operand is usually a register or memory location. The EA in these instructions is calculated by the 68000 based on the addressing mode used for the EA.

Some two-operand instructions have the EA in both operands. This means that the operands in these instructions use two addressing modes. Note that the 68000 address registers do not support byte-sized operands. Therefore, when an address register is used as a source operand, either the low-order word or the entire long word operand is used, depending on the operation size. When an address register is used as the destination operand, the entire register is affected regardless of operation size. If the operation size is a word, an address register in the destination operand is sign-extended to 32 bits after the operation is performed. Data registers, on the other hand, support data operands of byte, word, or long word size.

TABLE 10.2 68000 Addressing Modes

Addressing Mode	*Generation*	*Assembler Syntax*
• Register direct addressing		
Data register direct	EA = Dn	Dn
Address register direct	EA = An	An
• Address register indirect addressing		
Register indirect	EA = (An)	(An)
Postincrement register indirect	EA = (An), An ← An + N	(An)+
Predecrement register indirect	An ← An - N, EA = (An)	-(An)
Register indirect with offset	EA = (An) + d_{16}	d(An)
Indexed register indirect with offset	EA = (An) + (Ri) + d_8	d(An, Ri)
• Absolute data addressing		
Absolute short	EA = (Next word)	xxxx
Absolute long	EA = (Next two words)	xxxxxxxx
• Program counter relative addressing		
Relative with offset	EA = (PC) + d_{16}	d
Relative with index and offset	EA = (PC) + (Ri) + d_8	d(Ri)
• Immediate data addressing		
Immediate	DATA = Next word(s)	#xxxx
Quick immediate	Inherent data	#xx
• Implied addressing		
Implied register	EA = SR, USP, SP, PC	

Notes:

EA	= effective address	USP	= user stack pointer
An	= address register	d_8	= 8-bit signed offset (displacement)
Dn	= data register	d_{16}	= 16-bit signed offset (displacement)
Ri	= address or data register used as index register	N	= 1 for byte, 2 for words, and 4 for long words
SR	= status register	( )	= contents of
PC	= program counter	←	= replaces
SP	= active system stack pointer		

To identify the operand size of an instruction, the following notation is placed after a 68000 mnemonic: .B for byte, .W or none (default) for word, and .L for long word. For example,

`ADD.B D0, D1 ;`	$[D1]_{low\ byte}$	$\leftarrow [D0]_{low\ byte} + [D1]_{low\ byte}$
`ADD.W D0, D1 ;`	$[D1]_{low\ 16\ bit}$	$\leftarrow [D0]_{low\ 16\ bit} + [D1]_{low\ 16\ bit}$
`ADD.L D0, D1 ;`	$[D1]_{32\ bits}$	$\leftarrow [D1]_{32\ bits} + [D0]_{32\ bits}$

10.4.1 Register Direct Addressing

In this mode, the eight data registers (D0–D7) or seven address registers (A0–A6) contain the data operand. For example, consider `ADD $005000,D0`. The destination operand of this instruction is in data register direct mode. Now, if [005000] = 0002_{16} and [D0] = 0003_{16}, then after execution of `ADD $005000,D0`, the contents of D0 = 0002 + 0003 = 0005. Note that in this instruction, the $ symbol is used by Motorola to represent hexadecimal numbers. Also note that instructions are not available for byte operations using address registers.

10.4.2 Address Register Indirect Addressing

There are five different types of address register indirect mode. In this mode, an address register contains the effective address. For example, consider `CLR (A1)`. If [A1] = $00003000, then, after execution of `CLR (A1)`, the 16-bit contents of memory location $003000 will be cleared to zero.

The postincrement address register indirect mode increments an address register by 1 for byte, 2 for word, and 4 for long word after it is used. For example, consider `CLR.L (A0)+`. If [A0] = 00005000_{16}, then after execution of `CLR.L (A0)+`, the contents of memory locations 005000_{16} and 005002_{16} are cleared to zero and [A0] = 00005000 + 4 = 00005004. The postincrement mode is typically used with memory arrays stored from LOW to HIGH memory locations. For example, to clear 1000_{16} words starting at memory location 003000_{16} and above, the following instruction sequence can be used:

```
         MOVE.W    #$1000,D0      ;  Load length of data into D0
         MOVEA.L   #$003000,A0    ;  Load starting address into A0
REPEAT   CLR.W     (A0)+          ;  Clear a location pointed to
                                  ;  by A0 and increment A0 by 2
         SUBQ      #1,D0          ;  Decrement D0 by 1
         BNE       REPEAT         ;  Branch to REPEAT if Z = 0;
         ...                      ;  otherwise, go to next instruction
```

Note that the symbol # in the above is used by the Motorola assember to indicate immediate mode. This will be discussed later in this section. Also, note that `CLR.W (A0)+` automatically points to the next location by incrementing A0 by 2 after clearing a memory location.

The predecrement address register indirect mode, on the other hand, decrements an address register by 1 for byte, 2 for word, and 4 for long word before using a register. For example, consider `CLR.W -(A0)`. If [A0] = $00002004, then the content of A0 is first decremented by 2—that is, [A0] = 00002002_{16}. The content of memory location 002002 is then cleared to zero. The predecrement mode is used with arrays stored from HIGH to LOW memory locations. For example, to clear 1000_{16} words starting at memory location 4000_{16} and below, the following instruction sequence can be used:

```
        MOVE.W    #$1000,D0        ;  Load length of data into D0
        MOVEA.L   #$004002,A0      ;  Load starting address plus 2 into A0
REPEAT  CLR.W     -(A0)            ;  Decrement A0 by 2 and clear memory
                                   ;  location addressed by A0
        SUBQ      #1,D0            ;  Decrement D0 by 1
        BNE       REPEAT           ;  If Z = 0, branch to REPEAT
        ...                        ;  otherwise, go to next instruction
```

In this instruction sequence, `CLR.W -(A0)` first decrements A0 by 2 and then clears the location. Because the starting address is 004000_{16}, A0 must initially be loaded with 00004002_{16}. It should be pointed out that the predecrement and postincrement modes can be combined in a single instruction. A typical example is `MOVE.W (A5)+,-(A3)`.

The two other address register modes provide accessing of the tables by allowing offsets and indexes to be included to an indirect address pointer. The *address register indirect with offset mode* determines the effective address by adding a 16-bit signed integer to the contents of an address register. For example, consider `MOVE.W $10(A5),D3` in which the source operand is in address register indirect with offset mode. If [A5] = 00002000_{16} and $[002010]_{16}$ = 0014_{16}, then, after execution of `MOVE.W $10(A5),D3`, register D3.W will contain 0014_{16}.

The *indexed register indirect with offset mode* determines the effective address by adding an 8-bit signed integer and the contents of a register (data or address register) to the contents of an address (base) register. This mode is usually used when the offset from the base address register needs to be varied during program execution. The size of the index register can be a sign-extended 16-bit integer or a 32-bit value. As an example, consider `MOVE.W $10(A4,D3.W),D4` in which the source is in the indexed register indirect with offset mode. Note that in this instruction A4 is the base register and D3.W is the 16-bit index register (sign-extended to 32 bits). This register can be specified as 32 bits by using D3.L in the instruction, and 10_{16} is the 8-bit offset that is sign-extended to 32 bits. If [A4] = 00003000_{16}, [D3.W] = 0200_{16}, and [003210_{16}] = 0024_{16}, then this `MOVE` instruction will load 0024_{16} into the low 16 bits of register D4.

The address register indirect with offset mode can be used to access a single table. The offset (maximum 16 bits) can be the starting address of the table (fixed number), and the address register can hold the index number in the table to be accessed. Note that the starting address plus the index number provides the address of the element to be accessed in the table. For example, consider `MOVE.W $3400(A5),D1` . If A5 contains 04, then this `MOVE` instruction transfers the contents of 3404 (i.e., the fifth element, 0 being the first element) into the low 16 bits of D1. The indexed register indirect with offset mode, on the other hand, can be used to access multiple tables. Here, the offset (maximum 8 bits) can be the element number to be accessed. The address register pointer can be used to hold the starting address of the table containing the lowest starting address, and the index register can be used to hold the difference between the starting address of the table being accessed and the table with the lowest starting address. For example, consider three tables, with table 1 starting at 002000_{16}, table 2 at 003000_{16}, and table 3 at 004000_{16}. To transfer the seventh element (0 being the first element) in table 2 to the low 16 bits of register D0, the instruction `MOVE.W $06(A2, D1.W),D0` can be used, where [A2] = the starting address of the table with the lowest address (= 002000_{16} in this case) and $[D1]_{low\ 16\ bits}$ = the difference between the starting address of the table being accessed and the starting address of the table with the lowest address = 003000_{16} - 002000_{16} = 1000_{16}. Therefore, this `MOVE` instruction will transfer the contents of address 003006_{16} (the seventh element in table 2) to register D0. The indexed register indirect with offset mode can also be used to access two-dimensional arrays such as matrices.

10.4.3 Absolute Addressing

In this mode, the effective address is part of the instruction. The 68000 has two modes: absolute short addressing, in which a 16-bit address is used (the address is sign-extended to 24 bits before use), and absolute long addressing, in which a 24-bit address is used. For example, consider `ADD $2000,D2` as an example of the absolute short mode. If [2000] = 0012_{16} and [D2] = 0010_{16}, then, after executing `ADD $2000,D2` , register D2 will contain 0022_{16}. The absolute long addressing mode is used when the address size is more than 16 bits. For example, `MOVE.W $240000,D5` loads the 16-bit contents of memory location 240000_{16} into the low 16 bits of D5. The absolute short mode includes an address ADDR in the range of 0 ≤ ADDR ≤ $7FFF or $FF8000 ≤ ADDR ≤ $FFFFFF. Note that a single instruction may use both short and long absolute modes, depending on whether the source or destination address is less than, equal to, or greater than the 16-bit address. A typical example is `MOVE.W $500002,$1000`. Also, note that absolute long mode must be used for `MOVE` to or from address $8000. For example, `MOVE` $8000,D1 will move the contents of location $FF8000 to D1 while `MOVE.L` $00008000,D1 will transfer the contents of address $008000 to D1.

10.4.4 Program Counter Relative Addressing

The 68000 has two program counter relative addressing modes: relative with offset and relative with index and offset. In relative with offset mode, the effective address is obtained by adding the contents of the current PC with a sign-extended 16-bit displacement. This mode can be used when the displacement needs to be fixed during program execution. Typical branch instructions such as `BEQ`, `BRA`, and `BLE` use the relative with offset mode. This mode can also be used by some other instructions. For example, consider `ADD $30(PC),D5`, in which the source operand is in relative with offset mode. Now suppose that the current PC contents is $002000, the content of 002030_{16} is 0005, and the low 16 bits of D5 contain 0010_{16}. Then, after execution of this ADD instruction, D5 will contain 0015_{16}.

In relative with index and offset mode, the effective address is obtained by adding the contents of the current PC, a signed 8-bit displacement (sign-extended to 32 bits), and the contents of an index register (address or data register). The size of the index register can be 16 or 32 bits wide. For example, consider `ADD.W $4(PC,D0.W),D2`. If [D2] = 00000012_{16}, [PC] = 002000_{16}, $[D0]_{low\ 16\ bits} = 0010_{16}$, and [002014] = 0002_{16}, then, after this ADD, $[D2]_{low\ 16\ bits} = 0014_{16}$. This mode is used when the displacement needs to be changed during program execution by modifying the content of the Index register.

10.4.5 Immediate Data Addressing

Two immediate modes are available with the 68000: immediate and quick immediate modes. In immediate mode, the operand data is constant data, which is part of the instruction. For example, consider `ADDI #$0005,D0`. If [D0] = 0002_{16}, then, after this `ADDI` instruction, [D0] = $0002_{16} + 0005_{16} = 0007_{16}$. Note that the # symbol is used by Motorola to indicate the immediate mode. Quick immediate mode allows one to increment or decrement a register by a number from 0 to 7. For example, `ADDQ #1,D0` increments the contents of D0 by 1. Note that immediate data, 1 is inherent in the instruction. That is, data 0 to 7 is contained in the three bits of the instruction. ADDQ #0,Dn is similar to the NOP instruction.

10.4.6 Implied Addressing

The instructions using implied addressing mode do not require any operand, and registers such as PC, SP, or SR are referenced in these instructions. For example, `RTS` returns to the main program from a subroutine by placing the return address into PC using the PC implicitly.

It should be pointed out that in the 68000 the first operand of a two-operand instruction is the source and the second operand is the destination. Recall that in the case of the 8086, the first operand is the destination and the second operand is the source.

10.5 Functional Categories Of 68000 Addressing Modes

All of the 68000 addressing modes in Table 10.2 can be further divided into four functional categories as shown in Table 10.3.

TABLE 10.3 68000 Addressing Modes – Functional Categories

Addressing Modes	*Data*	*Memory*	*Control*	*Alterable*
	Addressing Category			
Data register direct	X	-	-	X
Address register direct	-	-	-	X
Address register indirect	X	X	X	X
Address register indirect with postincrement	X	X	-	X
Address regisiter indirect with predecrement	X	X	-	X
Address register indirect with displacement	X	X	X	X
Address register indirect with index	X	X	X	X
Absolute short	X	X	X	X
Absolute long	X	X	X	X
Program counter with displacement	X	X	X	-
Program counter with index	X	X	X	-
Immediate	X	X	-	-

- ***Data Addressing Mode.*** An addressing mode is said to be a data addressing mode if it references data objects. For example, all 68000 addressing modes except the address register direct mode fall into this category.

- ***Memory Addressing Mode.*** An addressing mode capable of accessing a data item stored in memory is classified as a memory addressing mode. For example, the data and address register direct addressing modes cannot satisfy this definition.

- ***Control Addressing Mode.*** This refers to an addressing mode that has ability to access a data item stored in memory without the need to specify its size. For example, all 68000 addressing modes except the following are classified as control addressing modes: data register direct, address register direct, address register indirect with postincrement, address register indirect with predecrement, and immediate.

- ***Alterable Addressing Mode.*** If the effective address of an addressing mode is written into, then that mode is an alterable addressing mode. For example, the immediate and the program counter relative addressing modes will not satisfy this definition.

10.6 68000 Instruction Set

The 68000 instruction set contains 56 basic instructions. Table 10.4 lists some of the instructions affecting the condition codes. Appendices D and G provide the 68000 instruction execution times and the instruction set (alphabetical order), respectively.

TABLE 10.4 Some of the 68000 Instructions affecting Conditional codes.

Instruction	X	N	Z	V	C
ABCD	✓	U	✓	U	–
ADD, ADDI, ADDQ, ADDX	✓	✓	✓	✓	✓
AND, ANDI	–	✓	✓	0	0
ASL, ASR	✓	✓	✓	✓	✓
BCHG, BCLR, BSET, BTST	–	–	✓	–	–
CHK	–	✓	U	U	U
CLR	–	0	1	0	0
CMP, CMPA, CMPI, CMPM	–	✓	✓	✓	✓
DIVS, DIVU	–	✓	✓	✓	0
EOR, EORI	–	✓	✓	0	0
EXT	–	✓	✓	0	0
LSL, LSR	✓	✓	✓	0	✓
MOVE (ea),(ea)	--	✓	✓	0	0
MOVE TO CCR	✓	✓	✓	✓	✓
MOVE TO SR	✓	✓	✓	✓	✓
MOVEQ	–	✓	✓	0	0
MULS, MULU	–	✓	✓	0	0
NBCD	✓	U	✓	U	✓
NEG, NEGX	✓	✓	✓	✓	✓
NOT	–	✓	✓	0	0
OR, ORI	–	✓	✓	0	0
ROL, ROR	–	✓	✓	0	✓
ROXL, ROXR	✓	✓	✓	0	✓
RTE, RTR	✓	✓	✓	✓	✓
SBCD	✓	U	✓	U	✓
STOP	✓	✓	✓	✓	✓
SUB, SUBI, SUBQ, SUBX	✓	✓	✓	✓	✓
SWAP	–	✓	✓	0	0
TAS	–	✓	✓	0	0
TST	–	✓	✓	0	0

✓ Affected, – Not Affected, U Undefined

Note: ADDA, B_{cc}, and RTS do not affect flags.

TABLE 10.5 68000 Data Movement Instructions

Instruction	*Size*	*Comment*
EXG Rx, Ry	L	Exchange the contents of two registers. Rx or Ry can be any address or data register. No flags are affected.
LEA (EA), An	L	The effective address (EA) is calculated using the particular addressing mode used and then loaded into the address register. (EA) specifies the actual data to be loaded into An.
LINK An, #-displacement	Unsized	The current contents of the specified address register are pushed onto the stack. After the push, the address register is loaded from the updated SP. Finally, the 16-bit sign-extended displacement is added to the SP. A negative displacement is specified to allocate stack.
MOVE (EA), (EA)	B, W,L	(EA)s are calculated by the 68000 using the specific addressing mode used. (EA)s can be register or memory location. Therefore, data transfer can take place between registers, between a register and a memory location, and between different memory locations. Flags are affected. For byte-size operation, address register direct is not allowed. An is not allowed in the destination (EA). The source (EA) can be An for word or long word transfers.
MOVEM reg list, (EA) or (EA), reg list	W, L	Specified registers are transferred to or from consecutive memory locations starting at the location specified by the effective address.
MOVEP Dn, d (Ay) or d (Ay), Dn	W, L	Two (W) or four (L) bytes of data are transferred between a data register and alternate bytes of memory, starting at the location specified and incrementing by 2. The high-order byte of data is transferred first, and the low-order byte is transferred last. This instruction has the address register indirect with displacement only mode.
MOVEQ # data, Dn	L	This instruction moves the 8-bit inherent data into the specified data register. The data is then sign-extended to 32 bits.
PEA (EA)	L	Computes an effective address and then pushes the 32-bit address onto the stack.
SWAP Dn	W	Exchanges 16-bit halves of a data register.
UNLK An	Unsized	An → SP; (SP) + → An

- (EA) in LEA (EA), An can use all addressing modes except Dn, An, (An) +, – (An), and immediate.
- Destination (EA) in MOVE (EA), (EA) can use all modes except An, relative, and immediate.
- Source (EA) in MOVE (EA), (EA) can use all modes.
- Destination (EA) in MOVEM reg list, (EA) can use all modes except, An, (An)+, relative, and immediate.
- Source (EA) in MOVEM (EA), reg list can use all modes except Dn, An, — (An), and immediate.
- (EA) in PEA (EA) can use all modes except, An, (An)+, —(An), and immediate.

The 68000 instructions can be classified into eight groups as follows:

1. Data movement instructions
2. Arithmetic instructions
3. Logical instructions
4. Shift and rotate instructions
5. Bit manipulation instructions

6. Binary-coded decimal instructions
7. Program control instructions
8. System control instructions

10.6.1 Data Movement Instructions

These instructions allow data transfers from register to register, register to memory, memory to register, and memory to memory. In addition, there are also special data movement instructions such as MOVEM (move multiple registers). Typically, byte, word, or long word data can be transferred. A list of the 68000 data movement instructions is given in Table 10.5. Let us now explain the data movement instructions.

10.6.1.1 MOVE Instructions

The format for the basic MOVE instruction is MOVE.S (EA), (EA), where S = L, W, or B. (EA) can be a register or memory location, depending on the addressing mode used. Consider MOVE.B D3,D1, which uses the data register direct mode for both the source and destination. If [D3] = 05_{16} and [D1] = 01_{16}, then, after execution of this MOVE instruction, [D1] = 05_{16} and [D3] = 05_{16}.

There are several variations of the MOVE instruction. For example MOVE.W CCR, (EA) moves the contents of the low-order byte of SR (i.e., CCR) to the low-order byte of the destination operand; the upper byte of SR is considered to be zero. The source operand is a word. Similarly, MOVE.W (EA), CCR moves an 8-bit immediate number, or low-order 8-bit data, from a memory location or register into the condition code register; the upper byte is ignored. The source operand is a word. Data can also be transferred between (EA) and SR or USP using the following privileged instructions:

MOVE.W (EA), SR
MOVE.W SR, (EA)
MOVE.L USP, An
MOVE.L An, USP

MOVEA.W or.L (EA), An can be used to load an address into an address register. Word-size source operands are sign-extended to 32 bits. Note that (EA) is obtained by using an addressing mode. As an example, MOVEA.W #$2000, A5 moves the 16-bit word 2000_{16} into the low 16 bits of A5 and then sign-extends 2000_{16} to the 32-bit number 00002000_{16}. Note that sign extension means extending bit 15 of 2000_{16} from bit 16 through bit 31. As mentioned before, sign extension is required when an arithmetic operation between two signed binary numbers of different sizes is performed. The (EA) in MOVEA can use all addressing modes.

The MOVEM instruction can be used to push or pop multiple registers to or from the stack. For example, MOVEM.L D0-D7/A0-A6,-(SP) saves the contents of all eight data registers and seven address registers in the stack. This instruction stores address registers in the order A6–A0 first, followed by data registers in the order D7–D0, regardless of the order in the register list. `MOVEM.L (SP)+,D0-D7/A0-A6` restores the contents of the registers in the order D0–D7, A0–A6, regardless of the order in the register list. The `MOVEM` instruction can also be used to save a set of registers in memory. In addition to the preceding predecrement and postincrement modes for the effective address, the `MOVEM` instruction allows all the control modes. If the effective address is in one of the control modes, such as absolute short, then the registers are transferred starting at the specified address and up through higher addresses. The order of transfer is from D0 to D7 and then from A0 to A7. For example, `MOVEM.W A5/D1/D3/A1-A3,$2000` transfers the low 16-bit contents of D1, D3, A1, A2, A3, and A5 to locations $2000, $2002, $2004, $2006, $2008, and $200A, respectively.

`The MOVEQ.L $d8,` D*n* instruction moves the immediate 8-bit data into the low byte of D*n*. The 8-bit data is then sign-extended to 32 bits. This is a one-word instruction. For example, `MOVEQ.L #$8F,D5` moves $FFFFFF8F into D5.

To transfer data between the 68000 data registers and 6800 (8-bit) peripherals, the `MOVEP` instruction can be used. This instruction transfers 2 or 4 bytes of data between a data register and alternate byte locations in memory, starting at the location specified and incrementing by 2. Register indirect with displacement is the only addressing mode used with this instruction. If the address is even, all transfers are made on the high-order half of the data bus; if the address is odd, all transfers are made on the low-order half of the data bus. The high-order byte to/from the register is transferred first, and the low-order byte is transferred last. For example, consider `MOVEP.L $0020(A2),D1`. If [A2] = $00002000, [002020] = 02, [002022] = 05, [002024] = 01, and [002026] = 04, then, after execution of this `MOVEP` instruction, D1 will contain 02050104_{16}.

10.6.1.2 EXG and SWAP Instructions

The `EXG` R*x*, R*y* instruction exchanges the 32-bit contents of R*x* with that of R*y*. The exchange is between two data registers, two address registers, or an address register and a data register. The `EXG` instruction exchanges only 32-bit-long words. The data size (L) does not have to be specified after the `EXG` instruction because this instruction has only one data size (L) and it is assumed that the default is this single data size. No flags are affected. The `SWAP` D*n* instruction, on the other hand, exchanges the low 16 bits of D*n* with the high 16 bits of D*n*. All condition codes are affected.

10.6.1.3 LEA and PEA Instructions

The `LEA.L` (EA), A*n* instruction moves an effective address (EA) into the specified address register. The (EA) can be calculated based on the addressing mode of the source. For example, `LEA $00256022,A5` moves $00256022 into A5. This instruction is equivalent to

MOVEA.L #$00256022,A5. Note that $00256022 is contained in PC. It should be pointed out that the LEA instruction is very useful when address calculation is desired during program execution. The (EA) in LEA specifies the actual data to be loaded into An, whereas the (EA) in MOVEA specifies the address of actual data. For example, consider LEA $04(A5, D2.W),A3. If [A5] = 00002000_{16} and [D2] = 0028_{16}, then the LEA instruction moves $0000202C_{16}$ into A3. On the other hand, MOVEA $04(A5, D2.W), A3 moves the contents of $0000202C_{16}$ into A3. Therefore, it is obvious that if address calculation is required, the instruction LEA is very useful.

10.6.1.4 LINK and UNLK Instructions

Before calling a subroutine, the main program quite often transfers the values of certain parameters to the subroutine. It is convenient to save these variables onto the stack before calling the subroutine. These variables can then be read from the stack and used by the subroutine for computations. The 68000 LINK and UNLK instructions are used for this purpose. In addition, the 68000 LINK instruction allows one to reserve temporary storage for the local variables of a subroutine. This storage can be accessed as needed by the subroutine and can be released using UNLK before returning to the main program. The LINK instruction is usually used at the beginning of a subroutine to allocate stack space for storing local variables and parameters for nested subroutine calls. The UNLK instruction is usually used at the end of a subroutine before the RETURN instruction to release the local area and restore the stack pointer contents so that it points to the return address.

The LINK An, #- displacement instruction causes the current contents of the specified An to be pushed onto the system stack. The updated SP contents are then loaded into An. Finally, a sign-extended twos complement displacement value is added to the SP. No flags are affected. For example, consider LINK A5,#-$100. If [A5] = 00002100_{16} and [USP] = 00004104_{16}, then after execution of the LINK instruction, the situation shown in Figure 10.4 occurs. This means that after the LINK instruction, [A5] = $00002100 is pushed onto the stack and the [updated USP] = $004100 is loaded into A5. USP is then loaded with $004000 and therefore 100_{16} locations are allocated to the subroutine at the beginning of which this particular LINK instruction can be used. Note that A5 cannot be used in the subroutine.

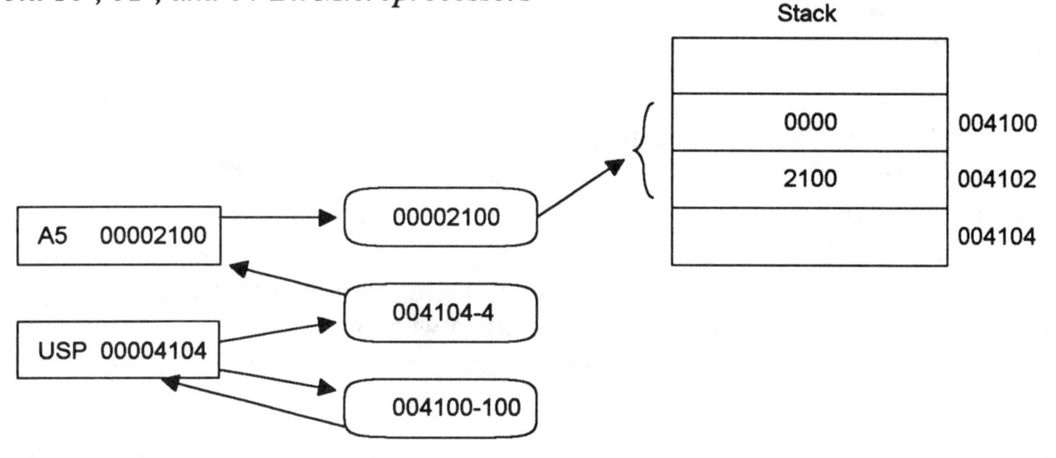

FIGURE 10.4 Execution of the `LINK` instruction

The `UNLK` instruction at the end of this subroutine before the `RETURN` instruction releases the 100 locations and restores the contents of A5 and USP to those prior to using the `LINK` instruction. For example, `UNLK A5` will load [A5] = 00004100 into USP and the two stack words $00002100 into A5. USP is then incremented by 4 to contain $00004104. Therefore, the contents of A5 and USP prior to using the `LINK` instruction are restored.

In this example, after execution of the `LINK`, addresses $0003FF and below can be used as the system stack. One hundred (Hex) locations starting at $004000 and above can be reserved for storing the local variables of the subroutine. These variables can then be accessed with an address register such as A5 as a base pointer using the address register indirect with displacement mode. `MOVE.W d(A5),D1` for read and `MOVE.W D1,d(A5)` for write are typical examples. The use of `LINK` and `UNLK` can be illustrated by the following subroutine structure:

```
SUBR LINK A2, #-50  ;     Allocate 50 bytes
       •
       •
       •
     UNLK A2        ;     Restore original values
     RTS            ;     Return to subroutine
```

The `LINK` instruction is used in this case to allocate 50 bytes for local variables. At the end of the subroutine, `UNLK A2` is used before `RTS` to restore the original values of the registers and the stack. `RTS` returns program execution in the main program.

10.6.2 Arithmetic Instructions

These instructions allow:

- 8-, 16-, or 32-bit additions and subtractions.
- 16-bit by 16-bit multiplication (both signed and unsigned) and 32-bit by 16-bit division (both signed and unsigned)
- Compare, clear, and negate instructions.
- Extended arithmetic instruction for performing multiprecision arithmetic.
- Test (TST) instruction for comparing the operand with zero.
- Test and set (TAS) instruction, which can be used for synchronization in a multiprocessor system.

The 68000 arithmetic instructions are summarized in Table 10.6. Let us now explain the arithmetic instructions.

TABLE 10.6 68000 Arithmetic Instructions

Instruction	*Size*	*Operation*
		Addition and Subtraction Instructions
ADD (EA), (EA)	B, W, L	(EA) + (EA) → (EA)
ADDI #Data, (EA)	B, W, L	(EA) + data → (EA)
ADDQ #d_8, (EA)	B, W, L	(EA) + d_8 → (EA) d_8 can be an integer from 0 to 7
ADDA (EA), An	W, L	An + (EA) → An
SUB (EA), (EA)	B, W, L	(EA) – (EA) → (EA)
SUBI # data, (EA)	B, W, L	(EA) – data → EA
SUBQ #d_8, (EA)	B, W, L	(EA) – d_8 → EA d_8 can be an integer from 0 to 7
SUBA (EA), An	W, L	An – (EA) → An
		Multiplication and Division Instructions
MULS (EA), Dn	W	$(Dn)_{16}$ * $(EA)_{16}$ → $(Dn)_{32}$ (signed multiplication)
MULU (EA), Dn	W	$(Dn)_{16}$ * $(EA)_{16}$ → $(Dn)_{32}$ (unsigned multiplication)
DIVS (EA), Dn	W	$(Dn)_{32}$ / $(EA)_{16}$ → $(Dn)_{32}$ (signed division, high word of Dn contains remainder and low word of Dn contains the quotient)
DIVU (EA), Dn	W	$(Dn)_{32}$ / $(EA)_{16}$ → $(Dn)_{32}$ (unsigned division, remainder is in high word of Dn and quotient is in low word of Dn)

TABLE 10.6 68000 Arithmetic Instructions (continued)

		Compare, Clear, and Negate Instructions
CMP (EA), Dn	B, W, L	Dn - (EA) → No result. Affects flags.
CMPA (EA), An	W, L	An - (EA) → No result. Affects flags.
CMPI # data, (EA)	B, W, L	(EA) – data → No result. Affects flags.
CMPM (Ay) +, (Ax) +	B, W, L	(Ax)+ – (Ay)+ → No result. Affects flags.
CLR (EA)	B,W,L	0 → (EA)
NEG (EA)	B,W,L	0 – (EA) →(EA)
		Extended Arithmetic Instructions
ADDX Dy,Dx	B, W, L	Dx + Dy + X → Dx
ADDX – (Ay), – (Ax)	B, W, L	– (AX) + – (Ay) + X → (Ax)
EXT Dn	W, L	If size is W, then sign extend low byte of Dn to 16 bits. If size is L. then sign extend low 16 bits of Dn to 32 bits.
NEGX (EA)	B, W, L	0 – (EA) – X → (EA)
SUBX Dy,Dx	B, W, L	Dx – Dy – X → Dx
SUBX – (Ay), – (Ax)	B, W, L	– (Ax) – – (Ay) – X → (Ax)
		Test Instruction
TST (EA)	B, W, L	(EA) - 0 ⇒ Flags are affected.
		Test and Set Instruction
TAS (EA)	B	If (EA) = 0, then set Z = 1; else Z = 0, N = 1 and then always set bit 7 of (EA) to 1.

NOTE: If source (EA) in the ADDA or SUBA instruction is an address register, the operand length is WORD or LONG WORD.
(EA) in any instruction is calculated using the addressing mode used.
All instructions except ADDA and SUBA affect condition codes.

- Source (EA) in the above ADD, ADDA, SUB, and SUBA can use all modes. Destination (EA) in the above ADD and SUB instructions can use all modes except An. relative, and immediate.
- Destination (EA) in ADDI and SUBI can use all modes except An. relative, and immediate.
- Destination (EA) in ADDQ and SUBQ can use all modes except relative and immediate.
- (EA) in all multiplication and division instructions can use all modes except An.
- Source (EA) in CMP and CMPA instructions can use all modes.
- Destination (EA) in CMPI can use all modes except An, relative, and immediate.
- (EA) in CLR and NEG can use all modes except An, relative, and immediate.
- (EA) in NEGX can use all modes except An, relative and immediate.
- (EA) in TST can use all modes except An, relative, and immediate.
- (EA) in TAS can use all modes except An, relative, and immediate.

10.6.2.1 Addition and Subtraction Instructions

- Consider `ADD.W $122000,D0`. If $[122000_{16}] = 0012_{16}$ and $[D0] = 0002_{16}$, then, after execution of this `ADD`, the low 16 bits of D0 will contain 0014_{16}.
- The `ADDI` instruction can be used to add immediate data to a register or memory location. The immediate data follows the instruction word. For example, consider `ADDI.W #$0012,$100200`. If $[100200_{16}] = 0002_{16}$, then, after execution of this `ADDI`, memory location 100200_{16} will contain 0014_{16}.
- `ADDQ` adds a number from 0 to 7 to the register or memory location in the destination operand. This instruction occupies 16 bits, and the immediate data 0 to 7 is specified by 3 bits in the instruction word. For example, consider `ADDQ.B #2,D1`. If $[D1]_{\text{low byte}} = 20_{16}$, then, after execution of this `ADDQ`, the low byte of register D1 will contain 22_{16}.
- All subtraction instructions subtract the source from the destination. For example, consider `SUB.W D2,$122200`. If $[D2]_{\text{low word}} = 0003_{16}$ and $[122200_{16}] = 0007_{16}$, then, after execution of this SUB, memory location 122200_{16} will contain 0004_{16}.
- Consider `SUBI.W #3,D0`. If $[D0]_{\text{low word}} = 0014_{16}$, then, after execution of this `SUBI`, D0 will contain 0011_{16}. Note that the same result can be obtained by using a `SUBQ.W #3,D0`. However in this case, the data item 3 is inherent in the instruction word.

10.6.2.2 Multiplication and Division Instructions

The 68000 instruction set includes both signed and unsigned multiplication of integer numbers.

- `MULS` (EA), D*n* multiplies two 16-bit signed numbers and provides a 32-bit result. For example, consider `MULS #-2,D5`. If $[D5] = 0003_{16}$, then, after this `MULS`, D5 will contain the 32-bit result $FFFFFFFA_{16}$, which is -6 in decimal.
- `MULU` (EA), D*n* performs unsigned multiplication. Consider `MULU (A0),D1`. If $[A0] = 00102000_{16}$, $[102000] = 0300_{16}$, and $[D1] = 0200_{16}$, then, after this `MULU`, D1 will contain the 32-bit result 00060000_{16}.
- Consider `DIVS #2, D1`. If $[D1] = -5_{10} = FFFFFFFB_{16}$, then, after this `DIVS`, register D1 will contain

D1	FFFF	FFFE
	16-bit remainder $= -1_{10}$	16-bit quotient $= -2_{10}$

Note that in the 68000, after `DIVS`, the sign of remainder is always the same as the dividend unless the remainder is equal to zero. Therefore, in this example, because the dividend is negative (-5_{10}), the remainder is negative (-1_{10}). Also, division by zero causes an internal interrupt automatically. A service routine can be written by the user to indicate an error. N = 1 if the quotient is negative, and V = 1 if there is an overflow.

- `DIVU` is the same as the `DIVS` instruction except that the division is unsigned. For example, consider `DIVU #4,D5.` If [D5] = 14_{10} = $0000000E_{16}$, then after this `DIVU`, register D5 will contain

D5	0002	0003
	16-bit remainder	16-bit quotient

As with the `DIVS` instruction, division by zero using `DIVU` causes a trap (internal interrupt).

10.6.2.3 Compare, Clear, and Negate Instructions

- The compare instructions affect condition codes but do not provide the substraction result. Consider `CMPM.W (A0)+, (A1)+`. If [A0] = 00100000_{16}, [A1] = 00200000_{16}, [100000] = 0005_{16}, and [200000] = 0006_{16}, then, after this `CMP` instruction, N = 0, C = 0, X = 0, V = 0, Z = 0, [A0] = 00100002_{16}, and [A1] = 00200002_{16}.
- `CLR.L D5` clears all 32 bits of D5 to zero.
- Consider `NEG.W (A0)`. If [A0] = 00200000_{16} and [200000] = 5_{10}, then after this `NEG` instruction, the low 16 bits of location 200000_{16} will contain $FFFB_{16}$.

10.6.2.4 Extended Arithmetic Instructions

- The `ADDX` and `SUBX` instruction can be used in performing multiprecision arithmetic because there are no `ADDC` (add with carry) or `SUBC` (subtract with borrow) instructions. For example, in order to perform a 64-bit addition, the following two instructions can be used:

```
ADD.L D0,D5     Add low 32 bits of data and store in D5.
ADDX.L D1,D6    Add high 32 bits of data along with any
                carry from the low 32-bit addition and
                store result in D6.
```

Note that in this example, D1D0 contain one 64-bit number and D6D5 contain the other 64-bit number. The 64-bit result is stored in D6D5.

- Consider `EXT.W D2`. If $[D2]_{\text{low byte}}$ = $F3_{16}$, then, after the `EXT`, [D2] = $FFF3_{16}$.

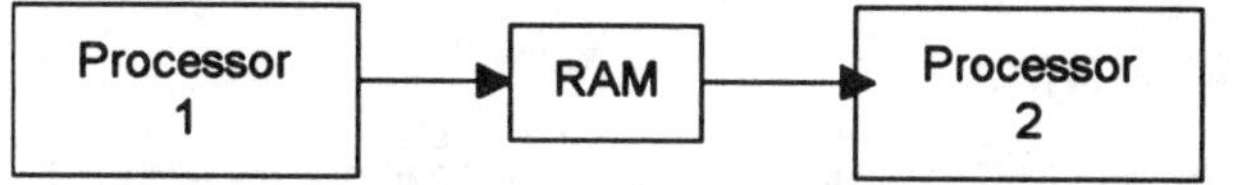

FIGURE 10.5 Two 68000s interfaced via shared RAM

10.6.2.5 Test Instructions

Consider `TST.W (A0)`. If [A0] = 00300000_{16} and [300000] = $FFFF_{16}$, then, after the `TST.W (A0)`, the operation $FFFF_{16} - 0000_{16}$ is performed internally by the 68000, Z is cleared to 0, and N is set to 1. The V and C flags are always cleared to 0.

10.6.2.6 Test and Set Instructions

`TAS (EA)` is usually used to synchronize two processors in multiprocessor data transfers. For example, consider two 68000-based microcomputers with shared RAM as shown in Figure 10.5.

Suppose that it is desired to transfer the low byte of D0 from processor 1 to the low byte of D2 in processor 2. A memory location, namely, TRDATA, can be used to accomplish this. First, processor 1 can execute the TAS instruction to test the byte in the shared RAM with address TEST for zero value. If it is, processor 1 can be programmed to move the low byte of D0 into location TRDATA in the shared RAM. Processor 2 can then execute an instruction sequence to move the contents of TRDATA from the shared RAM into the low byte of D2. The following instruction sequence will accomplish this:

Processor 1 Routine			*Processor 2 Routine*		
`Proc_1`	`TAS`	`TEST`	`Proc_2`	`TAS`	`TEST`
	`BNE`	`Proc_1`		`BNE`	`Proc_2`
	`MOVE.B`	`D0,TRDATA`		`MOVE.B`	`TRDATA,D2`
	`CLR.B`	`TEST`		`CLR.B`	`TEST`
	—			—	
	—			—	
	—			—	

Note that in these instruction sequences, `TAS TEST` checks the byte addressed by TEST for zero. If [TEST] = 0, then Z is set to 1; otherwise, Z = 0 and N = 1. After this, bit 7 of [TEST] is set to 1. Note that a zero value of [TEST] indicates that the shared RAM is free for use, and the Z bit indicates this after the TAS is executed. In each of the instruction sequences, after a data transfer using the `MOVE` instruction, [TEST] is cleared to zero so that the shared RAM is free for use by the other processor. To avoid testing the TEST byte simultaneously by two processors, the `TAS` is executed in a read-modify-write cycle. This means that once the operand is addressed by the 68000 executing the `TAS`, the system bus is not available to the other 68000 until the `TAS` is completed.

10.6.3 Logical Instructions

These instructions include logical `OR`, `EOR`, `AND`, and `NOT` as shown in Table 10.7.

- Consider `AND.W D1,D5`. If [D1] = 0001_{16} and [D5] = $FFFF_{16}$, then, after execution of this AND, the low 16 bits of both D1 and D5 will contain 0001_{16}.
- Consider `ANDI.B #$00,CCR`. If [CCR] = 01_{16}, then, after this ANDI, register CCR will contain 00_{16}.

TABLE 10-7 68000 Logical Instructions

Instruction	*Size*	*Operation*
AND (EA), (EA)	B, W, L	(EA) AND (EA) → (EA); (EA) cannot be address register
ANDI # data, (EA)	B, W, L	(EA) AND # data → (EA); (EA) cannot be address register
ANDI # $data_8$, CCR	B	CCR AND # data → CCR
ANDI # $data_{16}$, SR	W	SR AND# data → SR
EOR Dn, (EA)	B, W, L	Dn ⊕ (EA)→ (EA); (EA) cannot be address register
EORI # data, (EA)	B, W, L	(EA) ⊕ # data → (EA); (EA) cannot be address register
NOT (EA)	B, W, L	One's complement of (EA) → (EA);
OR (EA), (EA)	B, W, L	(EA) OR (EA) → (EA); (EA) cannot be address register
ORI # data, (EA)	B, W, L	(EA) OR # data → (EA); (EA) cannot be address register
ORI # $data_8$, CCR	B	CCR OR # $data_8$ → CCR
ORI # $data_{16}$, SR	W	SR OR # data → SR

Source (EA) in AND and OR can use all modes except An.
Destination (EA) in AND or OR or EOR can use all modes except An, relative, and immediate.
Destination (EA) in ANDI, ORI, and EORI can use all modes except An, relative, and immediate.
(EA) in NOT can use all modes except An, relative, and immediate.

- Consider `EOR.W D1,D2`. If [D1] = $FFFF_{16}$ and [D2] = $FFFF_{16}$, then, after execution of this EOR, register D2 will contain 0000_{16}, and D1 will remain unchanged at $FFFF_{16}$.
- Consider `NOT.B D5`. If [D5] = 02_{16}, then, after execution of this NOT, the low byte of D5 will contain FD_{16}.
- Consider `ORI #$E002,SR`. If [SR] = $111D_{16}$, then after execution of this ORI, register SR will contain F11F. Note that this is a privileged instruction because the high byte of SR containing the control bits is changed and therefore, can be executed only in the supervisor mode.

10.6.4 Shift and Rotate Instructions

The 68000 shift and rotate instruction are listed in Table 10.8.

- All the instructions in Table 10.8 affect N and Z flags according to the result. V is reset to zero except for `ASL`.
- Note that in the 68000 there is no true arithmetic shift left instruction. In true arithmetic shifts, the sign bit of the number being shifted is retained. In the 68000, the instruction `ASL` does not retain the sign bit, whereas the instruction `ASR` retains the sign bit after performing the arithmetic shift operation.

TABLE 10.8 68000 Shift and Rotate Instructions

Instruction	*Size*	*Operation*
`ASL` Dx, Dy	B, W, L	c ← / x ← [D_y ←] ← 0 Shift [Dy] by the number of times to left specified in Dx; the low 6 bits of Dx specify the number of shifts from 0 to 63.
`ASL` # data, Dn	B, W, L	Same as `ASL` Dx, Dy, except that the number of shifts is specified by immediate data from 0 to 7.
`ASL` (EA)	B, W, L	(EA) is shifted one bit to left; the most significant bit of (EA) goes to x and c, and zero moves into the least significant bit.
`ASR` Dx, Dy	B, W, L	[D_y →] → c / → x Arithmetically shift [Dy] to the right by retaining the sign bit; the low 6 bits of Dx specify the number of shifts from 0 to 63.
`ASR` # data, Dn	B, W, L	Same as above except the number of shifts is from 0 to 7.
`ASR` (EA)	B, W, L	Same as above except (EA) is shifted once to the right.
`LSL` Dx, Dy	B, W, L	c ← / x ← [D_y ←] ← 0 Low 6 bits of Dx specify the number of shifts from 0 to 63.
`LSL` # data, Dn	B, W, L	Same as above except that the number of shifts is specified by immediate data from 0 to 7.
`LSL` (EA)	B, W, L	(EA) is shifted one bit to left.
`LSR` Dx, Dy	B, W, L	0 → [D_y →] → c / → Same as `LSL` Dx, Dy, except shift is to the right.
`LSR` # data, Dn	B, W, L	Same as above except shift is to the right by immediate data from 0 to 7.

TABLE 10.8 68000 Shift and Rotate Instructions (continued)

LSR (EA)	B, W, L	Same as LSL (EA) except shift is to the right.
ROL Dx, Dy	B, W, L	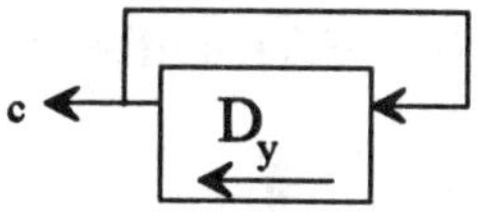 Low 6 bits of Dx specify the number of times [Dy] to be rotated.
ROL # data, Dn	B, W, L	Same as above except that the immediate data specifies that [Dn] to be rotated from 0 to 7.
ROL (EA)	B, W, L	(EA) is rotated one bit to left.
ROR Dx, Dy	B, W, L	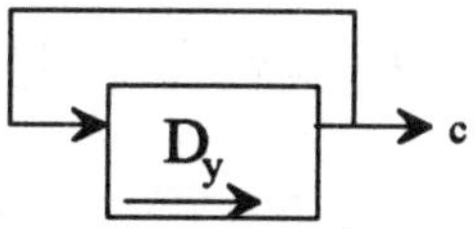
ROR # data, Dn	B, W, L	Same as ROL # data, Dn except the rotate is to the right by immediate data from 0 to 7.
ROR (EA)	B, W, L	(EA) is rotated one bit to right.
ROXL Dx, Dy	B, W, L	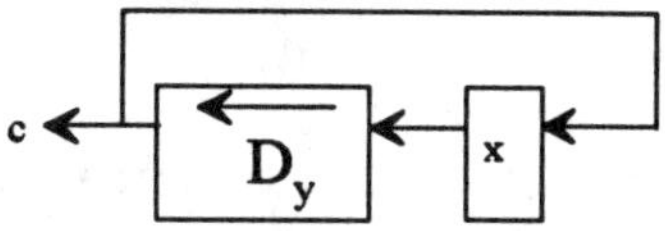 Low 6 bits of Dx contain the number of rotates from 0 to 63.
ROXL # data, Dn	B, W, L	Same as above except that the immediate data specifies number of rotates from 0 to 7.
ROXL (EA)	B, W, L	(EA) is rotated one bit to left.
ROXR Dx, Dy	B, W, L	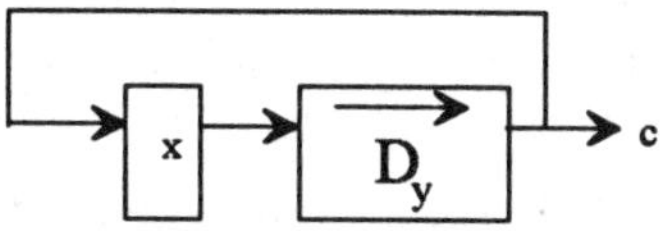 Same as ROXL Dx, Dy except the rotate is to the right.
ROXR # data, Dn	B, W, L	Same as ROXL # data, Dn, except rotate is to the right by immediate data from 0 to 7.
ROXR (EA)	B, W, L	Same as ROXL Dx, Dy except rotate is to the right.

· (EA) in ASL, ASR, LSL, LSR, ROL, ROR, ROXL, and ROXR can use all modes except Dn, An, relative, and immediate.

- Consider `ASL.W D1,D5`. If $[D1]_{\text{low 16 bits}} = 0002_{16}$ and $[D5]_{\text{low 16 bits}} = 9FF0_{16}$, then, after this ASL instruction, $[D5]_{\text{low 16 bits}} = 7FC0_{16}$, C = 0, and X = 0. Note that the sign of the contents of D5 is changed from 1 to 0 and, therefore, the overflow is set. The sign bit of D5 is changed after shifting [D5] twice. For ASL, the overflow flag is set to one if the sign bit changes during or after shifting. The contents of D5 are not updated after each shift.

- ASR retains the sign bit. For example, consider ASR.W #2, D1. If [D1] = $FFE2_{16}$, then, after this ASR, the low 16 bits of [D1] = $FFF8_{16}$, C = 1, and X = 1. Note that the sign bit is retained.
- ASL (EA) or ASR (EA) shifts (EA) 1 bit to left or right, respectively. For example, consider ASL.W (A0). If [A0] = 00002000_{16} and [002000] = 9001_{16}, then, after execution of this ASL, [002000] = 2002_{16}, X = 1, and C = 1. On the other hand, after ASR.W (A0), memory location 002000_{16} will contain $C800_{16}$, C = 1, and X = 1.
- The LSL and ASL instructions are the same in the 68000 except that with the ASL, V is set to 1 if the sign of the result is changed from the sign of the original value during or after shifting.
- Consider LSR.W #3,D1. If [D1] = 8000_{16}, then after this LSR, [D1] = 1000_{16}, X = 0, and C = 0.
- Consider ROL.B #2,D2. If [D2] = $B1_{16}$ and C = 1, then, after this ROL, the low byte of [D2] = $C6_{16}$ and C = 0. On the other hand, with [D2] = $B1_{16}$ and C = 1, consider ROR.B #2,D2. After this ROR, register D2 will contain $6C_{16}$ and C = 0.
- Consider ROXL.W D2,D1. If [D2] = 0003_{16}, [D1] = $F201_{16}$, C = 0, and X = 1, then the low 16 bits after execution of this ROXL are [D1] = $900F_{16}$, C = 1, and X = 1.

10.6.5 Bit Manipulation Instructions

The 68000 has four bit manipulation instructions, and these are listed in Table 10.9.

- In all of the instructions in Table 10.9, the ones complement of the specified bit is reflected in the Z flag. The specified bit is ones complemented, cleared to 0, set to 1, or unchanged by BCHG, BCLR, BSET, or BTST, respectively. In all the instructions in Table 10.9, if (EA) is D*n*, then the length of D*n* is 32 bits; otherwise, the length of the destination is one byte memory.
- Consider BCHG.B #2,$003000. If [003000] = 05_{16}, then, after execution of this BCHG, Z = 0 and [003000] = 01_{16}.
- Consider BCLR.L #3,D1. If [D1] = $F210E128_{16}$, then after execution of this BCLR, register D1 will contain $F210E120_{16}$ and Z = 0.
- Consider BSET.B #0,(A1). If [A1] = 00003000_{16} and [003000] = 00_{16}, then, after execution of this BSET, memory location 003000 will contain 01_{16} and Z = 1.
- Consider BTST.B #2,$002000. If [002000] = 02_{16}, then, after execution of this BTST, Z = 1, and [002000] = 02_{16}; no other flags are affected.

TABLE 10.9 Bit Manipulation Instructions

Instruction	*Size*	*Operation*
BCHG Dn, (EA) BCHG # data, (EA)	B,L	A bit in (EA) specified by Dn or immediate data is tested: the 1's complement of the bit is reflected in both the Z flag and the specified bit position.
BCLR Dn, (EA) BCLR # data, (EA)	B,L	A bit in (EA) specified by Dn or immediate data is tested and the 1's complement of the bit is reflected in the *Z* flag: the specified bit is cleared to zero.
BSET Dn, (EA) BSET # data, (EA)	B,L	A bit in (EA) specified by Dn or immediate data is tested and the 1's complement of the bit is reflected in the Z flag: the specified bit is then set to one.
BTST Dn, (EA) BTST # data, (EA)	B,L	A bit in (EA) specified by Dn or immediate data is tested. The 1's complement of the specified bit is reflected in the Z flag.

•(EA) in the above instructions can use all modes except An, relative,and immediate.
•If (EA) is memory location then data size is byte: if (EA) is Dn then data size is long word.

TABLE 10.10 68000 Binary Coded Decimal Instructions

Instruction	*Operand Size*	*Operation*
ABCD Dy, Dx	B	$(Dx)_{10} + (Dy)_{10} + X \rightarrow Dx$
ABCD - (Ay), -(Ax)	B	$-(Ax)_{10} + -(Ay)_{10} + X \rightarrow (Ax)$
SBCD Dy, Dx	B	$(Dx)_{10} - (Dy)_{10} - X \rightarrow Dx$
SBCD - (Ay), - (Ax)	B	$-(Ax)_{10} - -(Ay)_{10} - X \rightarrow (Ax)$
NBCD (EA)	B	$0 - (EA)_{10} - X \rightarrow (EA)_{10}$

•(EA) in NBCD can use all modes except An, relative, and immediate.

10.6.6 Binary-Coded-Decimal Instructions

The 68000 instruction set contains three BCD instructions, namely, ABCD for adding, SBCD for subtracting, and NBCD for negating. They operate on two packed BCD bytes and provide the result containing one packed BCD byte. These instructions always include the extend (X) bit in the operation. The BCD instructions are listed in Table 10.10.

- Consider `ABCD.B D1,D2`. If [D1] = 25_{10}, [D2] = 15_{10}, and X = 0, then, after execution of this ABCD instruction, [D2] = 40_{10}, X = 0, and Z = 0.
- Consider `SBCD.B -(A2),-(A3)`. If [A2] = 00002004_{16}, [A3] = 00003003_{16}, [002003_{16}] = 05_{10}, [003002_{16}] = 06_{10}, and X = 1, then after execution of this SBCD instruction, [003002_{16}] = 00_{10}, X = 0, and Z = 1.
- Consider `NBCD.B (A1)`. If [A1] = [00003000_{16}], [003000_{16}] =05_{10}, and X = 1, then, after execution of this NBCD instruction, [003000_{16}] = -6_{10}.

  Note that packed BCD subtraction used in the instructions SBCD and NBCD can be obtained by using the concepts discussed in Chapter 2 (Section 2.5.2.2).

10.6.7 Program Control Instructions

These instructions include branches, jumps, and subroutine calls as listed in Table 10.11.

Consider `Bcc` *d*. There are 14 branch conditions. This means that the *cc* in B*cc* can be replaced by 14 conditions providing 14 instructions: `BCC`, `BCS`, `BEQ`, `BGE`, `BGT`, `BHI`, `BLE`, `BLS`, `BLT`, `BMI`, `BNE`, `BPL`, `BVC`, and `BVS`. It should be mentioned that some of these instructions are applicable to both signed and unsigned numbers, some can be used with only signed numbers, and some instructions are applicable to only unsigned numbers.

After signed arithmetic operation such as `ADD` or `SUB`, instructions such as `BEQ`, `BNE`, `BVS`, `BVC`, `BMI`, and `BPL` can be used. On the other hand, after unsigned arithmetic operations, instructions such as `BCC`, `BCS`, `BEQ`, and `BNE` can be used. It should be pointed out that if V = 0, `BPL` and `BGE` have the same meaning, Likewise, if V = 0, `BMI` and `BLT` perform the same function.

TABLE 10.11 68000 Program Control Instructions

Instruction	*Size*	*Operation*
`Bcc` d	B,W	If condition code cc is true, then PC + d → PC. The PC value is current instruction location plus 2. d can be 8- or 16-bit signed displacement. If 8-bit displacement is used, then the instruction size is 16 bits with the 8-bit displacement as the low byte of the instruction word. If 16-bit displacement is used, then the instruction size is two words with 8-bit displacement field (low byte) in the instruction word as zero and the second word following the instruction word as the 16-bit displacement. There are 14 conditions such as BCC (Branch if Carry Clear), BEQ (Branch if result equal to zero, i.e., Z = 1), and BNE (Branch if not equal, i.e., Z = 0). Note that the PC contents will always be even since the instruction length is either one word or two words depending on the displacement widths.
`BRA` d	B,W	Branch always to PC + d where PC value is current instruction location plus 2. As with Bcc, d can be signed 8 or 16 bits. This is an unconditional branching instruction with relative mode. Note that the PC contents are even since the instruction is either one word or two words.
`BSR` d	B,W	PC → – (SP) PC + d → PC The address of the next instruction following PC is pushed onto the stack. PC is then loaded with PC + d. As before, d can be signed 8 or 16 bits. This is a subroutine call instruction using relative mode.
`DB`*cc* Dn, d	W	If *cc* is false, then Dn – 1 → Dn, and if Dn = – 1, then PC + 2 → PC If Dn ≠ – 1 , then PC + d → PC; else PC + 2 → PC.
`JMP` (EA)	unsized	(EA) → PC This is an unconditional jump instruction which uses control addressing mode

TABLE 10.11 68000 Program Control Instructions (continued)

Instruction	*Size*	*Operation*
JSR (EA)	unsized	PC → − (SP) (EA) → PC This is a subroutine call instruction which uses control addressing mode
RTR	unsized	(SP) + → CCR (SP) + → PC Return and restore condition codes
RTS	unsized	Return from subroutine (SP) + → PC
Scc (EA)	B	If *cc* is true, then the byte specified by (EA) is set to all ones; otherwise the byte is cleared to zero.

•(EA) in JMP and JSR can use all modes except Dn, An, (An) +, − (An), and immediate.
•(EA) in Scc can use all modes except An, relative, and immediate.

The conditional branch instruction can be used after typical arithmetic instructions such as subtraction to branch to a location if *cc* is true. For example, consider `SUB.W D1, D2`. Now if [D1] and [D2] are unsigned numbers, then

BCC *d* can be used if [D2] > [D1]
BCS *d* can be used if [D2] ≤ [D1]
BEQ *d* can be used if [D2] = [D1]
BNE *d* can be used if [D2] ≠ [D1]
BHI *d* can be used if [D2] < [D1]
BLS *d* can be used if [D2] ≤ [D1]

On the other hand, if [D1] and [D2] are signed numbers, the after `SUB.W D1, D2`, the following branch instruction can be used:

BEQ *d* can be used if [D2] = [D1]
BNE *d* can be used if [D2] ≠ [D1]
BLT *d* can be used if [D2] < [D1]
BLE *d* can be used if [D2] ≤ [D1]
BGT *d* can be used if [D2] > [D1]
BGE *d* can be used if [D2] ≥ [D1]

Now as a specific example, consider `BEQ BEGIN`. If [PC] = 000200_{16}, and BEGIN=$20 then, after execution of this `BEQ`, program execution starts at 000220_{16} if Z = 1; if Z = 0, program execution continues at 0002000_{16}.

- The instructions BRA and JMP are unconditional jump instructions. BRA uses the relative addressing mode, whereas JMP uses only control addressing mode. For example, consider BRA START. If [PC] = 000200_{16}, and START=$40 then, after execution of this BRA, program execution starts at 000240_{16}.

  Now, consider JMP (A1) . If [A1] = 00000220_{16}, then, after execution of this JMP, program execution starts at 000220_{16}.
- The instructions BSR and JSR are subroutine call instructions. BSR uses the relative mode, whereas JSR uses the control addressing mode. Consider the following program segment: Assume that the main program uses all registers; the subroutine stores the result in memory.

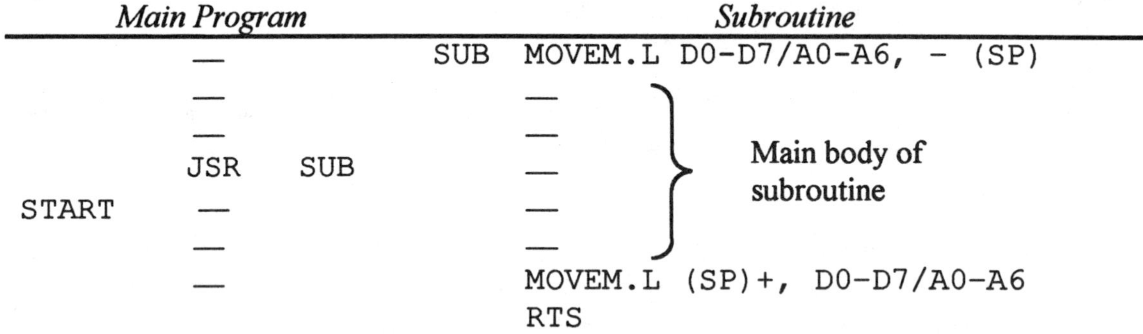
Main Program
Subroutine
JSR SUB
START
SUB MOVEM.L D0-D7/A0-A6, - (SP)
Main body of subroutine
MOVEM.L (SP)+, D0-D7/A0-A6
RTS

  Here, the JSR SUB instruction calls the subroutine SUB. In response to JSR, the 68000 pushes the current PC contents called START onto the stack and loads the starting address SUB of the subroutine into PC. The first MOVEM in the SUB pushes all registers onto the stack and, after the subroutine is executed, the second MOVEM instruction pops all the registers back. Finally, RTS pops the address START from the stack into PC, and program control in returned to the main program. Note that BSR SUB could have been used instead of JSR SUB in the main program. In that case, the 68000 assembler would have considered the SUB with BSR as a displacement rather than as an address with the JSR instruction.
- DB*cc* D*n*, *d* tests the condition codes and the value in a data register. DB*cc* first checks if *cc* (NE, EQ, GT, etc.) is satisfied. If *cc* is satisfied, the next instruction is executed. If *cc* is not satisfied, the specified data register is decremented by 1; if [D*n*] = -1, then the next instruction is executed; on the other hand, if D*n* ≠ -1, then branch to PC + d is performed. For example, consider DBNE D5,BACK with [D5] = 00003002_{16}, BACK= -4 and [PC] = 002006_{16}. If Z = 1, then [D5] = 00003001_{16}. Because [D5] ≠ -1, program execution starts at 002002_{16}.

It should be pointed out that there is a false condition in the DB*cc* instruction and that this instruction is the DBF (some assemblers use DBRA for this). In this case, the condition is always false. This means that, after execution of this instruction, D*n* is decremented by 1 and if [D*n*] = –1, then the next instruction is executed. If [D*n*] ≠ –1, then branch to PC + *d*.

- Consider SPL (A5). If [A5] = 00200020_{16} and N = 0, then, after execution of this SPL, memory location 200020_{16} will contain 11111111_2.

10.6.8 System Control Instructions

The 68000 system control instructions contain certain privileged instructions including RESET, RTE, STOP and instructions that use or modify SR. Note that the privileged instructions can be executed only in the supervisor mode. The system control instructions are listed in Table 10.12.

TABLE 10-12 68000 System Control Instructions

Instruction	*Size*	*Operation*
RESET	Unsized	If supervisor state, then assert reset line; else TRAP
RTE	Unsized	If supervisor state, then restore SR and PC; else TRAP
STOP # data	Unsized	If supervisor state, then load immediate data to SR and then STOP; else TRAP
ORI to SR MOVE USP ANDI to SR EORI to SR MOVE (EA) to SR	}	These instructions were discussed earlier
	Trap and Check Instructions	
TRAP # vector	Unsized	PC → - (SP) SR → - (SP) Vector address → PC
TRAPV	Unsized	TRAP if V = 1
CHK (EA), Dn	W	If Dn < 0 or Dn > (EA), then TRAP; else, go to the next instruction.
	Status Register	
ANDI to CCR EORI to CCR MOVE (EA) to/from CCR ORI to CCR MOVE SR to (EA)	}	Already explained earlier

•(EA) in CHK can use all modes except An.

- The `RESET` instruction when executed in the supervisor mode outputs a low signal on the reset pin of the 68000 in order to initialize the external peripheral chips. The 68000 reset pin is bidirectional. The 68000 can be reset by asserting the reset pin using hardware, whereas the peripheral chips can be reset using the software RESET instruction.
- `MOVE.L A7,An` or `MOVE.L An,A7` can be used to save, restore, or change the contents of the `A7` in supervisor mode. `A7` must be loaded in supervisor mode because `MOVE A7` is a privileged instruction.
- Consider `TRAP` *#n*. There are 16 `TRAP` instructions with *n* ranging from 0 to 15. The hexadecimal vector address is calculated using the equation: Hexadecimal vector address $= 80 + 4 \times n$. The `TRAP` instruction first pushes the contents of the PC and then the SR onto the stack. The hexadecimal vector address is then loaded into PC. `TRAP` is basically a software interrupt. The `TRAP` instruction can be used for service calls to the operating system. For application programs running in the user mode, `TRAP` can be used to transfer control to a supervisor utility program. `RTE` at the end of the `TRAP` routine can be used to return to the application program by placing the saved SR from the stack, thus causing the 68000 to return to the user mode.

  There are other traps that occur due to certain arithmetic errors. For example, division by zero automatically traps to location 14_{16}. On the other hand, an overflow condition (i.e., if V = 1) will trap to address $1C_{16}$ if the instruction `TRAPV` is executed.
- The `CHK` (EA), D*n* instruction compares [D*n*] with (EA). If $[Dn]_{\text{low 16 bits}} < 0$ or if $[Dn]_{\text{low 16 bits}} >$ (EA), then a trap to location 0018_{16} is generated. Also, N is set to 1 if $[Dn]_{\text{low 16 bits}} < 0$, and N is reset to 0 if $[Dn]_{\text{low 16 bits}} >$ (EA). (EA) is treated as a 16-bit twos complement integer. Note that program execution continues if $[Dn]_{\text{low 16 bits}}$ lies between 0 and (EA).

  Consider `CHK (A5),D2`. If $[D2]_{\text{low 16 bits}} = 0200_{16}$, $[A5] = 00003000_{16}$, and $[003000_{16}] = 0100_{16}$, then, after execution of this `CHK`, the 68000 will trap because $[D2] = 0200_{16}$ is greater than $[003000] = 0100_{16}$.

  The purpose of the `CHK` instruction is to provide boundary checking by testing if the content of a data register is in the range from zero to an upper limit. The upper limit used in the instruction can be set equal to the length of the array. Then, every time the array is accessed, the `CHK` instruction can be executed to make sure that the array bounds have not been violated.

  The `CHK` instruction is usually placed after the computation of an index value to ensure that the index value is not violated. This permits a check of whether or not the address of an array being accessed is within array boundaries when address register indirect with index mode is used to access an array element. For example, the following instruction sequence permits accessing of an array with base address in A2 and array length of 50_{10} bytes:

```
—
—
—
CHK #49,D2
MOVE.B 0(A2,D2*W),D3
—
—
```

Here, if the low 16 bits of D2 are greater than 49, the 68000 will trap to location 0018_{16}. It is assumed that D2 is computed prior to execution of the CHK instruction.

10.6.9 68000 Stack

The 68000 supports stacks with the address register indirect postincrement and predecrement addressing modes. In addition to two system stack pointers (A7 and A7'), all seven address registers (A0–A6) can be used as user stack pointers by using appropriate addressing modes. Subroutine calls, traps, and interrupts automatically use the system stack pointers: USP (A7) when S = 0 and SSP (A7') when S = 1. Subroutine calls push the PC onto the system stack; `RTS` pops the PC from the stack. Traps and interrupts push both PC and SR onto the system stack; `RTE` pops PC and SR from the stack.

The 68000 accesses the system stack from the top for operations such as subroutine calls or interrupts. This means that stack operations such as subroutine calls or interrupts access the system stack automatically from HIGH to LOW memory. Therefore, the system SP is decremented by 2 for word or 4 for long word after a push and incremented by 2 for word or 4 for long word after a pop. As an example, suppose that a 68000-`CALL` instruction (JSR or BSR) is executed when PC = $0031F200; then, after execution of the subroutine call, the stack will push the PC as follows:

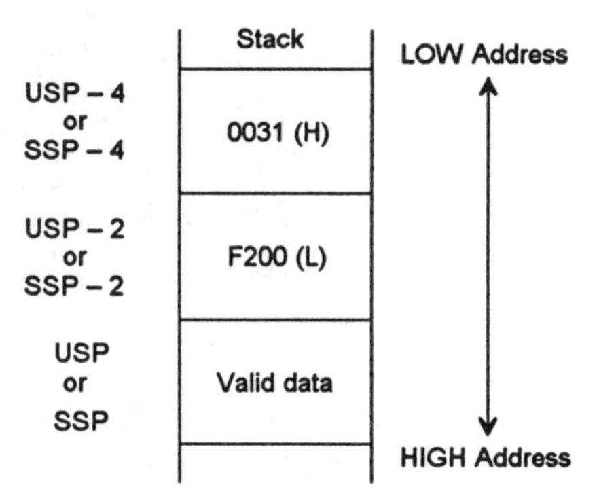

Note that the 68000 SP always points to valid data.

In 68000, stacks can be created by using address register indirect with postincrement or predecrement modes. Typical 68000 memory instructions such as `MOVE` to/from (PUSH/POP in Intel 8086) can be used to access the stack. Also, by using one of the seven address registers (A0–A6) and system stack pointers (A7,A7'), stacks can be filled from either HIGH to LOW memory or vice versa:

1. Filling a stack from HIGH to LOW memory (Top of the stack) is implemented with predecrement mode for push and postincrement mode for pop.
2. Filling a stack from LOW to HIGH (Bottom of the stack) memory is implemented with postincrement for push and predecrement for pop.

For example, consider the following stack growing from HIGH to LOW memory addresses in which A7 is used as the stack pointer:

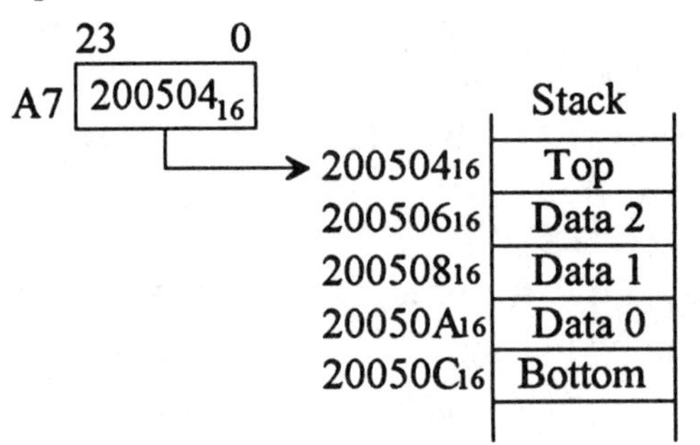

To push the 16-bit contents 0504_{16} of memory location 305016_{16}, the instruction `MOVE.W $305016,-(A7)` can be used as follows:

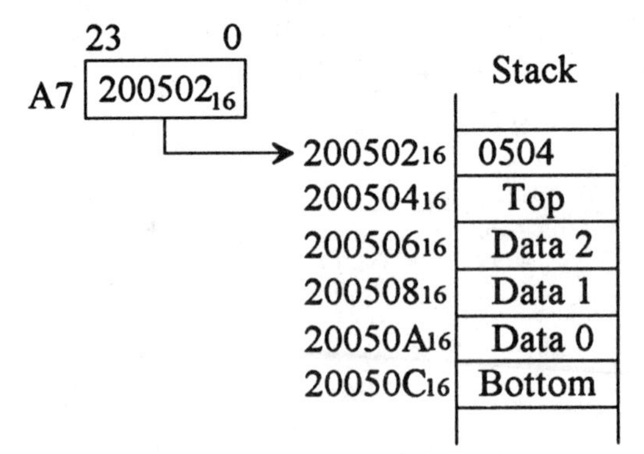

The 16-bit data item 0504_{16} can be popped from the stack into the low 16 bits of D0 by using `MOVE.W (A7)+,D0`. Register A7 will contain 200504_{16} after the pop. Note that, in this case, the stack pointer A7 points to valid data. Next, consider the stack growing from LOW to HIGH memory addresses in which the user utilizes A6 as the stack pointer:

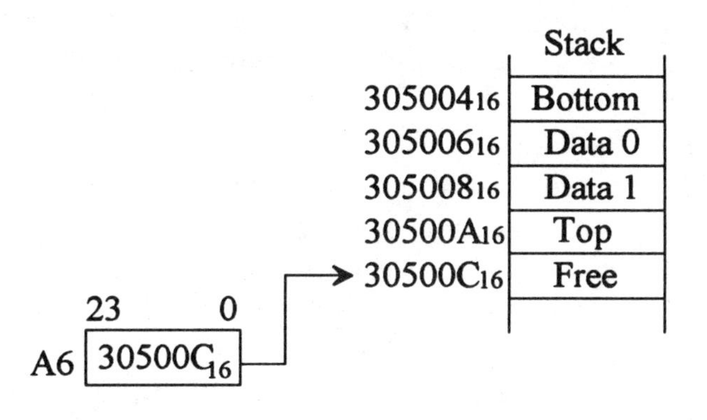

To push the 16-bit contents 2070_{16} of the low 16 bits of D5, the instruction `MOVE.W D5,(A6)+` can be used as follows. The 16-bit data item 2070_{16} can be popped from the stack into the 16-bit contents of memory location 417024_{16} by using `MOVE.W -(A6), $417024`. Note that, in this case, the stack pointer A6 points to the free location above the valid data.

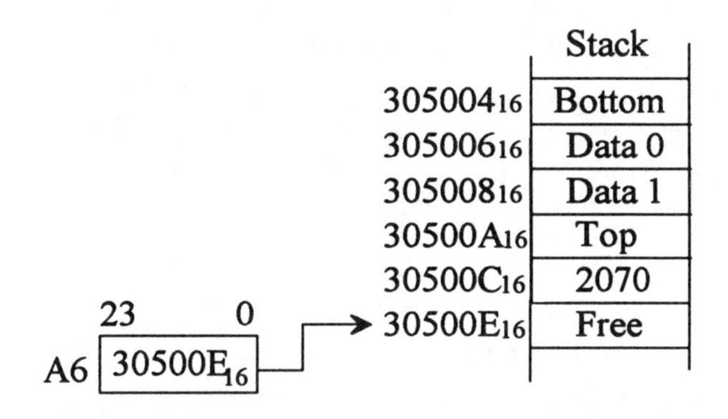

10.7 68000 Assembly Language Program Development Using Motorola ASM68K/68KSIM Assembler/simulator

Motorola ASM68K Assembler can be used to assemble programs written in 68000 assembly language as depicted below:

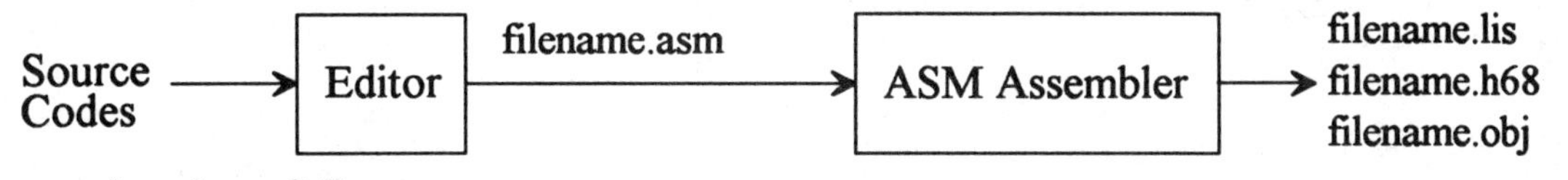

The procedure is as follows:

1. Source codes are entered and saved as *filename.asm* using a text editor.
   At the command prompt, type **edit filename.asm** and press Enter.
   Enter the program codes.
   Select **File** and then **Save As**.
   Specify the name and directory where the program file will be saved. Your file must be saved in the same directory as the ASM assembler.
2. The program file is assembled by the ASM assembler as follows:
   At the command prompt, use the command
   asm -l filename.asm
   The assembler will create *filename.lis*, *filename.h68*, and *filename.obj*

Using this procedure, a 68000 assembly program can be written and assembled using the text editor and the ASM assembler. The program can then be loaded and run using the 68000 simulator program called 68KSIM. The following shows the steps:

1. At the command prompt, type **SIM**. A screen will appear and show the contents of all 8 data registers.
2. At the > prompt, type **ld filename.h68**.
3. To turn on Trace and Single step, at the > prompt type **trace** and then **sstep**.
4. In order to load registers, type **ch** ***register_name value_in_hex***.
5. To run the program, at the > prompt type **go** ***starting_address*** and press Enter.
   As an example, consider the following program:

```
        ORG    $4000
        ADD    D1,D2
FINISH  JMP    FINISH
```

Assume that the operating system is MS-DOS, and the ASM assembler is located in a directory called "68k" on the C: drive. At the **C:\>** prompt, type **cd 68k** and press Enter. At the **C:\68k>** prompt type **edit** and press Enter. Type in the above program, then save it with a file name (**test.asm**, for example). Exit the editor, and type **asm -l test.asm** at the DOS prompt and press Enter. To display the program, enter at the DOS prompt **type test.lis**, then press Enter. In order to execute the program using SIM, follow steps 1 through 3, and change the file name in step 2 to **test.h68**. To enter data (for example, $0002 and $0003) into registers D1 and D2, do the following:

> ch d1 0002
> ch d2 0003

Execute the program, type **go 4000** at the > prompt. See the result (0005) in register D2 at the top of the display. To end execution of the program, type **exit** at the prompt and press Enter.

Example 10.1

Determine the effect of each of the following 68000 instructions:

- `CLR D0`
- `MOVE.L D1, D0`
- `CLR.L (A0) +`
- `MOVE -(A0), D0`
- `MOVE 20(A0), D0`
- `MOVEQ.L #$D7, D0`
- `MOVE 21(A0, A1.L), D0`

Assume the following initial configuration before each instruction is executed; also assume all numbers in hex:

[D0] = 22224444, [D1] = 55556666
[A0] = 00002224, [A1] = 00003333
[002220] = 8888, [002222] = 7777
[002224] = 6666, [002226] = 5555
[002238] = AAAA, [00556C] = FFFF

Instruction	*Effective Address*	*Net Effect (Hex)*
CLR D0	Destination EA = D0	D0 ← 22220000
MOVE.L D1,D0	Destination EA = D0	D0 ← 55556666
CLR.L (A0)+	Destination EA = [A0]	[002224] ← 0000
		[002226] ← 0000
		A0 ← 00002228
MOVE -(A0),D0	Source EA = [A0] – 2	A0 ← 00002222
	Destination EA = D0	D0 ← 22227777
MOVE 20(A0),D0	Source EA = [A0] + 20_{10} (or 14_{16}) = 002238	D0 ← 2222AAAA
	Destination EA = D0	
MOVEQ.L #$D7,D0	Source data = $D7_{16}$	D0 ← FFFFFFD7
	Destination EA = D0	
MOVE 21(A0, A1.L),D0	Source EA = [A0] + [A1] + 21_{10} = 00556C	D0 ← 2222FFFF
	Destination EA = D0	

Example 10.2

Write a 68000 assembly language program that implements each of the following C language program segments:

(a)
```
if (x >= y)
     x = x + 10;
else y = y - 12;
```

(b)
```
sum = 0;
for (i = 0; i <= 9; i= i + 1)
sum = sum + a(i);
```

Assume the following information about the variables involved in this problem:

	Variable	Comments
(a)	x	Address of a 16-bit signed integer
	y	Address of a 16-bit signed integer
(b)	sum	Address of the 16-bit result of addition

Solution

```
(a)     x           EQU     100
        y           EQU     200
                    LEA.L   x,A0         ;  Initialize A0
                    LEA.L   y,A1         ;  Initialize A1
                    MOVE.W  (A0),D0      ;  Move [x] into D0
                    CMP.W   (A1),D0      ;  Compare [x] with [y]
                    BGE.B   THPRT
                    SUBI.W  #12,(A1)     ;  Execute else part
                    BRA.B   STAY
        THPRT       ADDI.W  #10,(A0)     ;  Execute then part
        STAY        JMP     STAY         ;  Halt
(b) Assume register A0 holds the address of the first element of the array.

        SUM         EQU     300          ;  Initialize SUM to 300 for result
                    LEA.L   200,A0       ;  Point A0 to a[0]
                    CLR.W   D0           ;  Clear the sum to zero
                    MOVE.W  #9,D1        ;  Initialize D1 with loop limit
        LOOP        ADD.W   (A0)+,D0     ;  Perform the iterative summation
                    DBF.W   D1,LOOP
                    MOVE.W  D0,SUM       ;  Store 16-bit result in address SUM
        FINISH      JMP     FINISH       ;  Halt
```

Note that, in the above condition F in `DBF` is always false. Hence, the program exits from the LOOP when D1= -1. Therefore, the addition process is performed 10 times.

Example 10.3

Write a 68000 assembly program at address \$2000 to clear 100_{10} consecutive bytes to zero starting at location \$3000.

Solution

```
            ORG         $2000
            MOVEA.L     #$3000,A0     ; Load A0 with $3000
            MOVE.W      #99,D0        ; Move 99₁₀ into D0
LOOP        CLR.B       (A0)+         ; Clear [3000₁₆] and point to next address
            DBF.W       D0,LOOP       ; Decrement and branch
FINISH      JMP         FINISH        ; Halt
```

Note that the 68000 has no HALT instruction.. Therefore, the unconditional jump to the same location such as `FINISH JMP FINISH` is normally used at the end of the program. Because `DBF` is a word instruction and considers D0's low 16-bit word as the loop count, one should be careful about initializing D0 using `MOVEQ.L` #d8,Dn since this instruction sign extends low byte of Dn to 32 bits.

Example 10.4

Write a 68000 assembly language program at address \$1000 to compute $\sum_{i=1}^{N} X_iY_i$, where X_i and Y_i are signed 16-bit numbers and $N = 100$. Store the 32-bit result in D1. Assume that the starting addresses of X_i and Y_i are \$100 and \$200 respectively.

Solution

```
P         EQU       $100
Q         EQU       $200
          ORG       $1000
          MOVE.W    #99,D0        ;  Move 99₁₀ into D0
          LEA.L     P,A0          ;  Load address P into A0
          LEA.L     Q,A1          ;  Load address Q into A1
          CLR.L     D1            ;  Initialize D1 to zero
LOOP      MOVE.W    (A0)+,D2      ;  Move [X] to D2
          MULS.W    (A1)+,D2      ;  D2 ← [X] * [Y]
          ADD.L     D2,D1         ;  D1 ← SUM XiYi
          DBF.W     D0,LOOP       ;  Decrement and branch
FINISH    JMP       FINISH        ;  Halt
```

Example 10.5

Write a 68000 subroutine to compute $Y = \sum_{i=1}^{N} X^2_i/N$. Assume the X_i's are 16-bit signed integers and $N = 100$. The numbers are stored in consecutive locations. Assume A0 points to the X_i's and A7 is already initialized in the main program. Store 32-bit result in D1 (16-bit remainder in high word of D1 and 16-bit quotient in the low word of D1).

Solution

```
SQR       MOVEM.L   D2/D3/A0,-(A7)  ;  Save registers
          CLR.L     D1              ;  Clear sum
          MOVE.W    #99,D2          ;  Initialize loop count
BACK      MOVE.W    (A0)+,D3        ;  Move Xi's into D3
          MULS.W    D3,D3           ;  Compute Xi**2 using MULS.
          ADD.L     D3,D1           ;  Since Xi**2 is always positive,
          DBF.W     D2,BACK         ;   compute
          DIVU.W    #100,D1         ;   ∑Xi²/N using DIVU
          MOVEM.L   (A7)+,D2/D3/A0  ;  Restore registers
          RTS
```

Example 10.6

Write a 68000 assembly language program at address 0 to move a block of data of length 100_{10} from the source block starting at location 002000_{16} to the destination block starting at location 003000_{16} from low to high addresses.

Solution

```
         MOVEA.W  #$2000,A4     ;  Load A4 with source address
         MOVEA.W  #$3000,A5     ;  Load A5 with destination address
         MOVE.W   #99,D0        ;  Load D0 with count -1 = 99
START    MOVE.W   (A4)+,(A5)+   ;  Move source data to destination
         DBF.W    D0,START      ;  Branch if D0≠ -1
STAY     JMP      STAY          ;  Halt
```

Example 10.7

Write a 68000 assembly language program at address 0 to add two words, each containing two ASCII digits. The first word is stored in two consecutive locations (from LOW to HIGH) with the low byte pointed to by A0 at address 000300_{16}, and the second word is stored in two consecutive locations (from LOW to HIGH) with low byte pointed to by A1 at 000700_{16}. Store the packed BCD result in D5.

Solution

```
         MOVEQ.L  #1,D2
         MOVEA.W  #$0300,A0    ;  Initialize A0
         MOVEA.W  #$0700,A1    ;  Initialize A1
START    ANDI.B   #$0F,(A0)+   ;  Convert 1st number to unpacked BCD
         ANDI.B   #$0F,(A1)+   ;  Convert 2nd number to unpacked BCD
         DBF.W    D2,START
         MOVE.B   -(A0),D6     ;  Obtain high unpack. byte of 1st num. in D6
         MOVE.B   -(A0),D7     ;  Obtain low unpack. byte of 1st num. in D7
         LSL.B    #4,D6        ;  Shift 1st num. high byte  4 times to left
         ADD.B    D7,D6        ;  D6 contains packed BCD byte of the 1st num
         MOVE.B   -(A1),D5     ;  Obtain high unpack. byte of 2nd num. in D5
         MOVE.B   -(A1),D4     ;  Obtain low unpack. byte of 2nd num. in D4
         LSL.B    #4,D5        ;  Shift 2nd num. high byte 4 times to left
         ADD.B    D4,D5        ;  D5 contains packed BCD byte of the 2nd num
         ADDI.B   #0,D0        ;  Clear x-bit
         ABCD.B   D6,D5        ;  D5.B contains the packed BCD result
FINISH   JMP      FINISH
```

Note: Typical assemblers assemble a program starting at address 0 if assembler directive ORG is not used at the beginning of the program.

Example 10.8

Write a 68000 assembly language program that will perform : $5 \times X + 6 \times Y + (Y/8) \rightarrow$ D1.L where X is an unsigned 8-bit number stored in the lowest byte of D0 and Y is a 16-bit signed number stored in the upper 16 bits of D1. Neglect the remainder of $Y/8$.

Solution

```
         ANDI.W   #$00FF,D0     ; Convert X to unsigned 16-bit in D0
         MULU.W   #5,D0         ; Compute unsigned 5 * X in D0.L
         SWAP.W   D1            ; Move Y to low 16 bits in D1
         MOVE.W   D1,D2         ; Save Y to low 16 bits in D2
         MULS.W   #6,D1         ; Compute signed 6 * Y in D1.L
         ADD.L    D0,D1         ; Add 5 * X with 6 * Y
         EXT.L    D2            ; Sign extend low 16 bits in D2 to 32 bits
         ASR.L    #3,D2         ; Perform Y/8
                                ; Discard remainder of Y/8
         ADD.L    D2,D1         ; Perform 5 * X + 6 * Y + Y/8
FINISH   JMP      FINISH
```

Example 10.9

Write a 68000 assembly language program to convert temperature from Fahrenheit to Celsius using the following equation: $C = [(F - 32)/9] \times 5$; assume that the low byte of D0 contains the temperature in Fahrenheit. The temperature can be positive or negative. Store result in D0.

Solution

```
         EXT.W    D0       ;  Sign extend (F) low byte of D0 to 16 bits
         SUBI.W   #32,D0   ;  Perform F - 32
         MULS.W   #5,D0    ;  Perform 5 * (F - 32)/9 and store
         DIVS.W   #9,D0    ;  remainder in high word of D0
FINISH   JMP      FINISH   ;  and quotient in low word of D0
```

Example 10.10

Write a 68000 assembly language program to add twenty 32-bit numbers stored in consecutive locations starting at address \$502040. Store the 32-bit result onto the user stack. Assume that no carry is generated due to addition of two cosecutive 32-bit numbers and A7 is already initialized.

Solution

```
         MOVEQ.L  #19,D0         ;  Initialize D0 with count
         MOVEA.L  #$00502040,A0  ;  Initialize A0 as pointer
         CLR.L    D1             ;  Clear sum to 0
START    ADD.L    (A0)+,D1       ;  Add 32-bit data and store in D1
```

```
         DBF.W      D0,START
         MOVE.L     D1,-(A7)       ;  Store result
FINISH   JMP        FINISH
```

Example 10.11

Write a subroutine in 68000 assembly language to implement the C language assignment statement: p = p + q; where addresses p and q hold two 16-digit (64-bit) packed BCD numbers (N1 and N2). The main program will initialize addresses p and q to $002000 and $003000 respectively. Address $002007 will hold the lowest byte of N1 with the highest byte at address $002000 while Address $003007 will contain the lowest byte of N2 with the highest byte at address $003000. Also, write the main program at address $004000 which will perform all initializations including addresses p,q, loop count, and then call the subroutine at $008000 and stop. Assume supervisor mode. Note that the 68000 supervisor stack pointer is initialized upon hardware reset.

Solution

MAIN PROGRAM

```
      ORG       $004000
      MOVEA.W   #$2000,A0
      MOVEA.W   #$3000,A1
      MOVE.W    #7,D1
      JSR       BCDADD
STAY  BRA.B     STAY
```

SUBROUTINE

```
         ORG     $008000
BCDADD   LEA.L   1(A0,D1.W),A0   ;Update A0
         LEA.L   1(A1,D1.W),A1   ;and A1
         ADDI.B  #0,D2           ;X-bit=0
ALOOP    ABCD.B  -(A1),-(A0)     ;Add
         DBF.W   D1,ALOOP
         RTS
```

Example 10.12

Write a 68000 assembly program to multiply an 8-bit signed number in the low byte of D1 by a 16-bit signed number in the high word of D5. Store the result in D3.

Solution

```
         EXT.W      D1        ;  Sign extends low byte of D1
         SWAP.W     D5        ;  Swap low word with high word of D5
         MULS.W     D1,D5     ;  Multiply D1 with D5, store result in D5
         MOVE.L     D5,D3     ;  copy result to D3
FINISH   JMP        FINISH    ;  Halt
```

Example 10.13

Repeat Example 10.10 without using the DBF instruction. Assume that a carry may be generated due to addition of two consecutive 32-bit numbers. Initialize A6 to $00200504 and use low 24 bits of A6 as the stack pointer to push the 32-bit result.

Solution

```
START_ADR    EQU        $00502040
COUNT        EQU        20
             MOVEA.L    #START_ADR,A0    ;  Load starting address in A0
             MOVE.B     #COUNT,D0        ;  Use D0 as a counter
             MOVEA.L    #$00200504,A6    ;  Use A6 as the stack pointer
             CLR.L      D1               ;  Clear D1 register
             ADDI.B     #0,D1            ;  Clear X bit
AGAIN        MOVE.L     (A0)+,D3         ;  Store a 32-bit number in D3
             ADDX.L     D3,D1            ;  Add numbers with carry
             SUBQ.B     #1,D0            ;  Decrement counter
             BNE.B      AGAIN            ;  Repeat until counter expires
             MOVE.L     D1,-(A6)         ;  Store result onto stack
FINISH       JMP        FINISH
```

Note that ADDX adds two data registers or two memory locations using predecrement modes.

Example 10.14

Write a 68000 assembly language program to subtract two 32-bit packed BCD numbers. The BCD number 1 is stored at the locations starting from $003000 through $003003, with the least significant byte at $003003 and the most significant byte at $003000. Similarly, the BCD number 2 is stored at the locations starting from $004000 through $004003, with the least significant byte at $004003 and the most significant byte at $004000. The BCD number 2 is to be subtracted from BCD number 1. Store the packed BCD result at addresses $005000 (Lowest byte of the result) through $005003 (Highest byte of the result). Do not use DBF.

Solution

```
             MOVE.W     #4,D7            ;  The number of bytes to be subtracted
             MOVEA.W    #$3000,A0        ;  Starting address for first number
             MOVEA.W    #$4000,A1        ;  Starting address for second number
             ADDA.W     D7,A0            ;  Move address pointers to the end
             ADDA.W     D7,A1            ;  of each 32-bit packed BCD number
             MOVEA.W    #$5000,A3        ;  Load pointer for destination addr.
             ADDI.B     #0,D7            ;  Clear X bit
   LOOP      MOVE.B     -(A0),D0         ;  Get a byte from first number
             MOVE.B     -(A1),D1         ;  Get a byte from second number
             SBCD.B     D1,D0            ;  BCD subtraction, result in D0
             MOVE.B     D0,(A3)+         ;  Store result in destination addr.
             SUBQ.B     #1,D7            ;  Continue until counter has expired
             BNE.B      LOOP
   FINISH    JMP        FINISH
```

Example 10.15

Write a 68000 assembly program at address $100000 which is equivalent to the following C language segment:

```
sum = 0;
for (i = 0; i <= 9; i = i + 1)
sum = sum + x[i] * y[i];
```

Assume that the arrays, x[i] and y[i] contain signed 16-bit numbers already stored in memory starting at addresses $300000 and $400000 respectively. Store the 32-bit result at address $500000.

Solution

```
            ORG       $100000
x           EQU       $300000
y           EQU       $400000
sum         EQU       $500000
            MOVE.W    #9,D0        ;  Use D0 as a loop counter
            LEA.L     x,A0         ;  Initialize A0 with x
            LEA.L     y,A1         ;  Initialize A1 with y
            LEA.L     sum,A2       ;  Initialize A2 with sum
            CLR.L     (A2)         ;  Clear sum to zero
LOOP        MOVE.W    (A0)+,D2     ;  Move x[i] into D2
            MULS.W    (A1)+,D2     ;  Compute x[i]*y[i]
            ADD.L     D2,(A2)      ;  Update sum
            DBF.W     D0,LOOP      ;  Repeat until D0=-1
FINISH      JMP       FINISH
```

10.8 68000 Pins And Signals

The 68000 is usually packaged in one of the following:

a) 64-pin dual in-line package (DIP)

b) 68-terminal chip carrier

c) 68-pin quad pack

d) 68-pin grid array (PGA)

Figure 10.6 shows the 68000 pin diagram for the DIP. Appendix C provides data sheets for the 68000 and support chips.

The 68000 is provided with two V_{cc} (+5 V) and two ground pins. Power is thus distributed in order to reduce noise problems at high frequencies. Also, to build a prototype to demonstrate that the paper design for the 68000-based microcomputer is correct, one must use either wire-wrap or solder for the actual construction. Prototype board must not be used because, at high frequencies (above 4 MHz), there will be noise problems due to stray capacitances. The 68000 consumes about 1.5 W of power.

Signal	Pin	Pin	Signal
D_4	1	64	D_5
D_3	2	63	D_6
D_2	3	62	D_7
D_1	4	61	D_8
D_0	5	60	D_9
$\overline{AS}$	6	59	D_{10}
$\overline{UDS}$	7	58	D_{11}
$\overline{LDS}$	8	57	D_{12}
$R/\overline{W}$	9	56	D_{13}
$\overline{DTACK}$	10	55	D_{14}
$\overline{BG}$	11	54	D_{15}
$\overline{BGACK}$	12	53	GND
$\overline{BR}$	13	52	A_{23}
Vcc	14	51	A_{22}
CLK	15	50	A_{21}
GND	16	49	Vcc
$\overline{HALT}$	17	48	A_{20}
$\overline{RESET}$	18	47	A_{19}
$\overline{VMA}$	19	46	A_{18}
E	20	45	A_{17}
$\overline{VPA}$	21	44	A_{16}
$\overline{BERR}$	22	43	A_{15}
$\overline{IPL2}$	23	42	A_{14}
$\overline{IPL1}$	24	41	A_{13}
$\overline{IPL0}$	25	40	A_{12}
FC2	26	39	A_{11}
FC1	27	38	A_{10}
FC0	28	37	A_9
A_1	29	36	A_8
A_2	30	35	A_7
A_3	31	34	A_6
A_4	32	33	A_5

FIGURE 10.6 68000 pins and signals

D_0–D_{15} are the 16 data bus pins. All transfers to and from memory and I/O devices are conducted over the 8-bit (LOW or HIGH) or 16-bit data bus depending on the size of the device. A_1–A_{23} are the 23 address lines. A_0 is obtained by encoding the $\overline{UDS}$ (upper data strobe) and $\overline{LDS}$ (lower data strobe) lines.

The 68000 operates on a single-phase TTL-level clock at 4, 6, 8, 10, 12.5, 16.67, or 25 MHz. The clock signal must be generated externally and applied to the 68000 clock input line. An external crystal oscillator chip is required to generate the clock. Figure 10.7 shows the 68000 CLK waveform and clock timing specifications. The clock is at TTL-compatible voltage. The clock timing specifications provide data for three different clock frequencies: 8 MHz, 10 MHz, and 12.5 MHz The 68000 CLK input can be provided by an external crystal oscillator or by designing an external circuit.

Characteristic	Symbol	8 MHz		10 MHz		12.5 MHz		Unit
		Min	Max	Min	Max	Min	Max	
Frequency of operation	f	4.0	8.0	4.0	10.0	4.0	12.5	MHz
Cycle time	t_{CVC}	125	250	100	250	80	250	ns
Clock pulse width	t_{CL}	55	125	45	125	35	125	ns
	t_{CH}	55	125	45	125	35	125	
Rise and fall times	t_{Cr}	—	10	—	10	—	5	ns
	t_{Cf}	—	10	—	10	—	5	

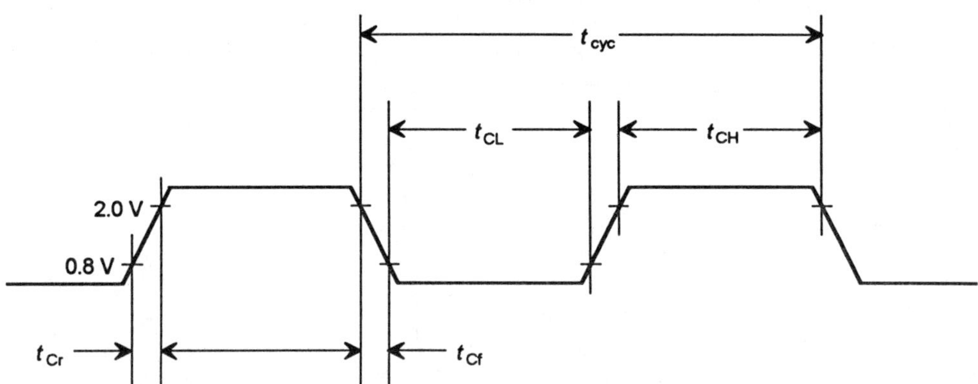

FIGURE 10.7 68000 clock input timing diagram and AC electrical specifications

The 68000 signals can be divided into five functional categories:

1. Synchronous and asynchronous control lines
2. System control lines
3. Interrupt control lines
4. DMA control lines
5. Status lines

10.8.1 Synchronous and Asynchronous Control Lines

The 68000 bus control is asynchronous. This means that once a bus cycle is initiated, the external device must send a signal back to complete it. The 68000 also contains three synchronous control lines that facilitate interfacing to synchronous peripheral devices such as Motorola's inexpensive MC6800 family.

Synchronous operation means that bus control is synchronized or clocked using a common system clock signal. In 6800 family peripherals, this common clock is the E clock signal depending on the particular chip used. With synchronous control, all READ and WRITE operations must be synchronized with the common clock. However, this may create problems when interfacing with slow peripheral devices. This problem does not arise with asynchronous bus control.

Asynchronous operation is not dependent on a common clock signal. The 68000 utilizes the asynchronous control lines to transfer data between the 68000 and peripheral devices via handshaking. Using asynchronous operation, the 68000 can be interfaced to any peripheral chip regardless of the speed.

The 68000 has three control lines to transfer data over its bus in a synchronous manner: E (enable), $\overline{VPA}$ (valid peripheral address), and $\overline{VMA}$ (valid memory address). The E clock corresponds to the clock of the 6800. The E clock is output at a frequency that is one tenth of the 68000 input clock. $\overline{VPA}$ is an input and tells the 68000 that a 6800 device is being addressed and therefore data transfer must be synchronized with the E clock. $\overline{VMA}$ is the processor's response to $\overline{VPA}$. $\overline{VMA}$ is asserted when the memory address is valid. This also tells the external device that the next data transfer over the data bus will be synchronized with the E clock.

$\overline{VPA}$ can be generated by decoding the address pins and address strobe ($\overline{AS}$). Note that the 68000 asserts $\overline{AS}$ LOW when the address on the address bus is valid. $\overline{VMA}$ is typically used as the chip select of the 6800 peripheral. This ensures that the 6800 peripherals are selected and deselected at the correct time. The 6800 peripheral interfacing sequence is as follows:

1. The 68000 initiates a cycle by starting a normal read or write cycle.
2. The 6800 peripheral defines the 68000 cycle by asserting the 68000 $\overline{VPA}$ input. If $\overline{VPA}$ is asserted as soon as possible after assertion of $\overline{AS}$, then $\overline{VPA}$ will be recognized as being asserted after three cycles. If $\overline{VPA}$ is not asserted after three cycles, the 68000 inserts wait states until $\overline{VPA}$ is recognized by the 68000 as asserted. $\overline{DTACK}$ should not be asserted while $\overline{VPA}$ is asserted. The 6800 peripheral must remove $\overline{VPA}$ within 1 clock period after $\overline{AS}$ is negated.
3. The 68000 monitors enable (E) until it is LOW. The 68000 then synchronizes all READ and WRITE operations with the E clock. The $\overline{VMA}$ output pin is asserted LOW by the 68000.
4. The 6800 peripheral waits until E is active (HIGH) and then transfers the data.
5. The 68000 waits until E goes to LOW (on a read cycle, the data is latched as E goes to LOW internally). The 68000 then negates $\overline{VMA}$, $\overline{AS}$, $\overline{UDS}$, and $\overline{LDS}$. The 68000 thus terminates the cycle and starts the next cycle.

The 68000 utilizes five lines to control address and data transfers asynchronously: $\overline{AS}$ (address strobe), R/$\overline{W}$ (read/write), $\overline{DTACK}$ (data acknowledge), $\overline{UDS}$ (upper data strobe), and $\overline{LDS}$ (lower data strobe).

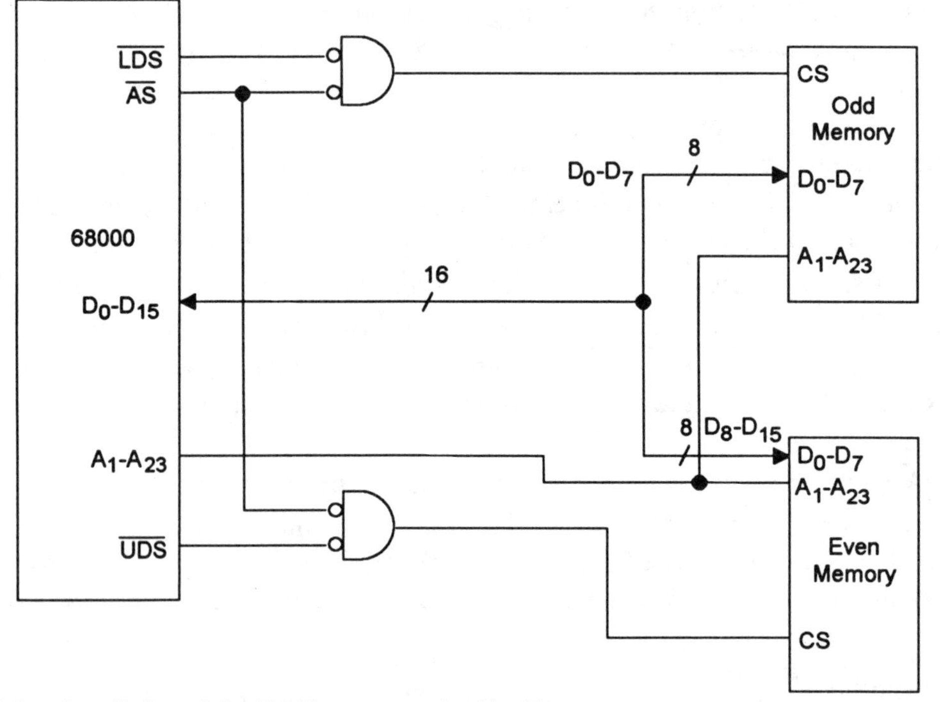

FIGURE 10.8 Interfacing of the 68000 to even and odd addresses

The 68000 outputs $\overline{AS}$ to notify the peripheral device when data is to be transferred. $\overline{AS}$ is active LOW when the 68000 provides a valid address on the address bus. The $R/\overline{W}$ output line indicates whether the 68000 is reading data from or writing data into a peripheral device. $R/\overline{W}$ is HIGH for read and LOW for write. $\overline{DTACK}$ is used to tell the 68000 that a transfer is to be performed. When the 68000 wants to transfer data asynchronously, it first activates the $\overline{AS}$ line and at the same time generates the required address on the address lines to select the peripheral device.

Because the $\overline{AS}$ line tells the peripheral chip when to transfer data, the $\overline{AS}$ line should be part of the address decoding scheme. After enabling $\overline{AS}$, the 68000 enters the wait state until it receives $\overline{DTACK}$ from the selected peripheral device. On receipt of $\overline{DTACK}$, the 68000 knows that the peripheral device is ready for data transfer. The 68000 then utilizes the $R/\overline{W}$ and data lines to transfer data. $\overline{UDS}$ and $\overline{LDS}$ are defined as follows:

$\overline{UDS}$	$\overline{LDS}$	Data Transfer Occurs Via:	Address
1	0	D_0–D_7 pins for byte	Odd
0	1	D_8–D_{15} pins for byte	Even
0	0	D_0–D_{15} pins for word or long word	Even

A_0 is encoded from $\overline{UDS}$ and $\overline{LDS}$. When $\overline{UDS}$ is asserted, the contents of even addresses are transferred on the high-order eight lines of the data bus, D_8–D_{15}. The 68000 internally shifts this data to the low byte of the specified register. When $\overline{LDS}$ is asserted, the contents of odd addresses are transferred on the low-order eight lines of the data bus, D_0–D_7. During word and long word transfers, both $\overline{UDS}$ and $\overline{LDS}$ are asserted and information is transferred on all 16 data lines, D_0–D_{15} pins. Note that during byte memory transfers, A_0 corresponds to $\overline{UDS}$ for even addresses ($A_0 = 0$) and to $\overline{LDS}$ for odd addresses ($A_0 = 1$). The circuit in Figure 10.8 shows how even and odd addresses are interfaced to the 68000.

10.8.2 System Control Lines

The 68000 has three control lines, $\overline{BERR}$ (bus error), $\overline{HALT}$, and $\overline{RESET}$, which are used to control system-related functions. $\overline{BERR}$ is an input to the 68000 and is used to inform the processor that there is a problem with the instruction cycle currently being executed. With asynchronous operation, this problem may arise if the 68000 does not receive $\overline{DTACK}$ from a peripheral device. An external timer can be used to activate the $\overline{BERR}$ pin if the external device does not send $\overline{DTACK}$ within a certain period of time. On receipt of $\overline{BERR}$, the 68000 does one of the following:

- Reruns the instruction cycle that caused the error.
- Executes an error service routine.

The troubled instruction cycle is rerun by the 68000 if it receives a $\overline{HALT}$ signal along with the $\overline{BERR}$ signal. On receipt of LOW on both the $\overline{HALT}$ and $\overline{BERR}$ pins, the 68000 completes the current instruction cycle and then goes into the high-impedance state. On removal of both $\overline{HALT}$ and $\overline{BERR}$ (that is, when both $\overline{HALT}$ and $\overline{BERR}$ are HIGH), the 68000 reruns the troubled instruction cycle. The cycle can be rerun repeatedly if both $\overline{BERR}$ and $\overline{HALT}$ are enabled/disabled continually.

On the other hand, an error service routine is executed only if the $\overline{BERR}$ signal is received without $\overline{HALT}$. In this case, the 68000 will branch to a bus error vector address where the user can write a service routine. If two simultaneous bus errors are received via the $\overline{BERR}$ pin without $\overline{HALT}$, the 68000 automatically goes into the halt state until it is reset.

The $\overline{HALT}$ line can also be used by itself to perform single stepping or to provide DMA. When the $\overline{HALT}$ input is activated, the 68000 completes the current instruction and goes into a high-impedance state until $\overline{HALT}$ is returned to HIGH. By enabling/disabling the $\overline{HALT}$ line continually, the single-stepping debugging can be accomplished. However, because most 68000 instructions consist of more than one clock cycle, single stepping using

$\overline{\text{HALT}}$ is not normally used. Rather, the trace bit in the status register is used to single-step the complete instruction.

One can also use $\overline{\text{HALT}}$ to perform microprocessor-halt DMA. Because the 68000 has separate DMA control lines, DMA using the $\overline{\text{HALT}}$ line will not normally be used. The $\overline{\text{HALT}}$ pin can also be used as an output signal. The 68000 will assert the $\overline{\text{HALT}}$ pin LOW when it goes into a halt state as a result of a catastrophic failure. The double bus error (activation of $\overline{\text{BERR}}$ twice) is an example of this type of error. When this occurs, the 68000 goes into a high-impedance state until it is reset. The $\overline{\text{HALT}}$ line informs the peripheral devices of the catastrophic failure.

The $\overline{\text{RESET}}$ line of the 68000 is also bidirectional. To reset the 68000, both the $\overline{\text{RESET}}$ and $\overline{\text{HALT}}$ pins must be LOW for 10 clock cycles at the same time except when Vcc is initially applied to the 68000. In this case, an external reset must be applied for at least 100 ms. The 68000 executes a reset service routine automatically for loading the PC with the starting address of the program.

The 68000 $\overline{\text{RESET}}$ pin can also be used as an output line. A LOW can be sent to this output line by executing the RESET instruction in the supervisor mode in order to reset external devices connected to the 68000 $\overline{\text{RESET}}$ pin. Upon execution of the RESET instruction, the 68000 drives the $\overline{\text{RESET}}$ pin LOW for 124 clock periods and does not affect any data, address, or status registers. Therefore, the RESET instruction can be placed anywhere in the program whenever the external devices need to be reset.

Upon hardware reset, the 68000 sets S-bit in SR to 1, and then loads the supervisor stack pointer from location $000000 (high 16 bits) and $000002 (low 16 bits) and loads the PC from $000004 (high 16 bits) and $000006 (low 16 bits); but low 24 bits are used. In addition, the 68000 clears the trace bit in SR to 0 and sets bits I2 I1 I0 in SR to 111. All other registers are unaffected.

10.8.3 Interrupt Control Lines

$\overline{\text{IPL0}}$, $\overline{\text{IPL1}}$, and $\overline{\text{IPL2}}$ are the three interrupt control lines These lines provide for seven interrupt priority levels ($\overline{\text{IPL2}}$, $\overline{\text{IPL1}}$, $\overline{\text{IPL0}}$ = 111 means no interrupt, and $\overline{\text{IPL2}}$, $\overline{\text{IPL1}}$, $\overline{\text{IPL0}}$ = 000 means nonmaskable interrupt with the highest priority). The 68000 interrupts will be discussed later in this chapter.

10.8.4 DMA Control Lines

The $\overline{\text{BR}}$ (bus request), $\overline{\text{BG}}$ (bus grant), and $\overline{\text{BGACK}}$ (bus grant acknowledge) lines are used for DMA purposes. The 68000 DMA will be discussed later in this chapter.

TABLE 10.13 Function Code Lines

FC2	*FC1*	*FC0*	*Operation*
0	0	0	Unassigned
0	0	1	User data
0	1	0	User program
0	1	1	Unassigned
1	0	0	Unassigned
1	0	1	Supervisor data
1	1	0	Supervisor program
1	1	1	Interrupt acknowledge

10.8.5 Status Lines

The 68000 has the three output lines called function code pins (output lines) FC2, FC1, and FC0. These lines tell external devices whether user data/program or supervisor data/program is being addressed. These lines can be decoded to provide user or supervisor programs/data and interrupt acknowledge as shown in Table 10.13.

The FC2, FC1, and FC0 pins can be used to partition memory into four functional areas: user data memory, user program memory, supervisor data memory, and supervisor program memory. Each memory partition can directly access up to 16 megabytes, and thus the 68000 can be made to directly address up to 64 megabytes of memory. This is shown in Figure 10.9.

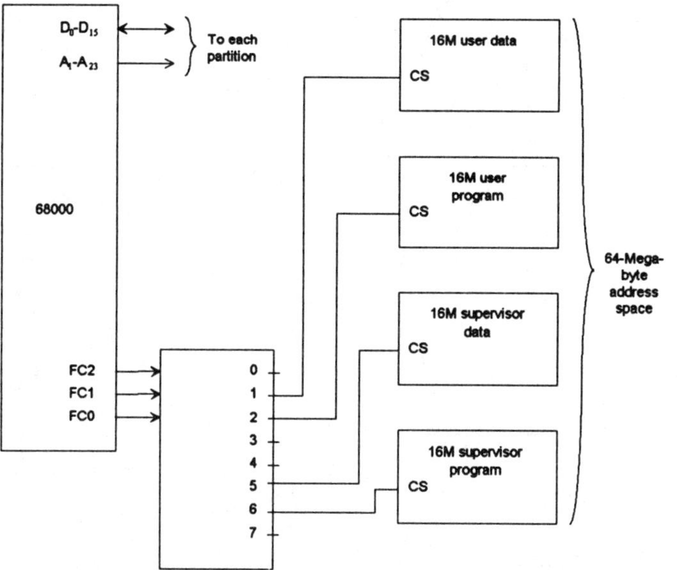

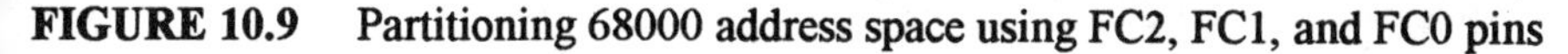

FIGURE 10.9 Partitioning 68000 address space using FC2, FC1, and FC0 pins

10.9 68000 Clock and Reset Signals

This section covers generation of 68000 clock and reset signals in detail because the clock signal and the reset pins are two important signals of any microprocessor.

10.9.1 68000 Clock Signals

As mentioned before, the 68000 does not include an on-chip clock generation circuitry. This means that external crystal oscillator chip is required to generate the clock. The 68000 CLK input can be provided by a crystal oscillator or by designing an external circuit. Figure 10.10 shows a simple oscillator to generate the 68000 CLK input.

This circuit uses two inverters connected in series. Inverter 1 is biased in its transition region by the resistor *R*. Inverter 1 inputs the crystal output (sinusoidal) to provide a logic pulse train at the output of inverter 1. Inverter 2 sharpens the wave and drives the crystal. For this circuit to work, HCMOS logic for the inverters must be used. Therefore, the 74HC04 inverter chip is used. The 74HC04 has high noise immunity and the ability to drive 10 LS-TTL loads. A coupling capacitor should be connected across the supply terminals to reduce the ringing effect during high-frequency switching of the HCMOS devices. Note that the ringing occurs when a circuit oscillates for a short time due to the presence of stray inductance and capacitance. In addition, the output of this oscillator is fed to the CLK input of a D flip-flop (74LS74) to further reduce the ringing. A clock signal of 50% duty cycle at a frequency of ½ the crystal frequency is generated. This means that this circuit with a 16-MHz crystal will generate an 8-MHz clock for the 68000.

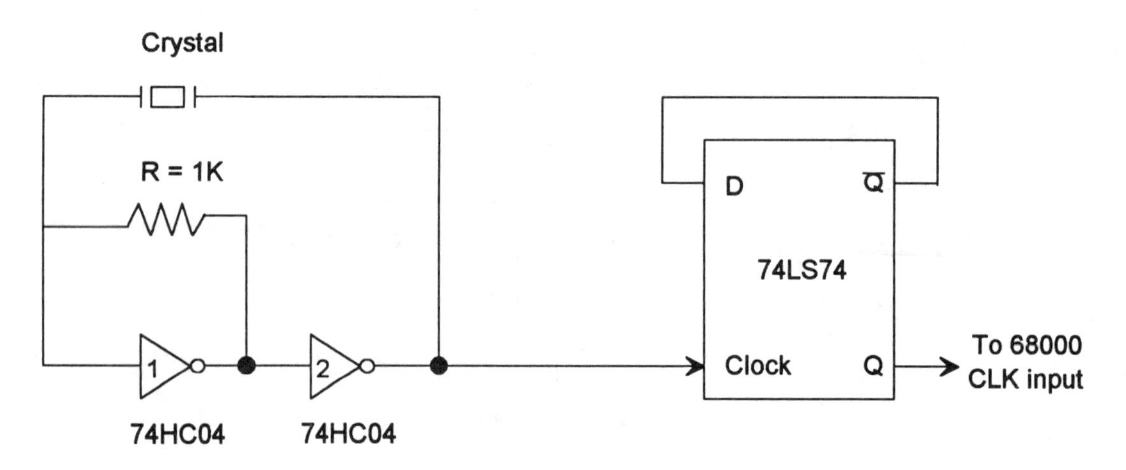

FIGURE 10.10 External clock circuitry

10.9.2 68000 Reset Circuit

When designing the microprocessor's reset circuit, two types of reset must be considered: power-up and manual. These reset circuits must be designed using the parameters specified by the manufacturer. Therefore, a microprocessor must be reset when its Vcc pin is connected to power. This is called "power-up reset." After some time during normal operation, the microprocessor can be reset by the designer upon activation of a manual switch such as a pushbutton. A reset circuit, therefore, needs to be designed following the timing parameters associated typically with the microprocessor's reset input pin specified by the manufacturer. The reset circuit, once designed, is typically connected to the microprocessor's reset pin.

Upon hardware reset, the 68000 sets S-bit in SR to 1 and performs the following:

1. The 68000 loads the supervisor stack pointer from addresses $000000 (high 16 bits) and $000002 (low 16 bits) and loads the PC from $000004 (high 16 bits) and $000006 (low 16 bits). Typical 68000 assembler directives such as DC.L can be used for this purpose. For example, to load $200128 into supervisor SP and $3F1420 into PC, the following instruction sequence can be used:

```
ORG     $00000000
DC.L    $00200128
DC.L    $003F1420
```

2. The 68000 clears the trace bit in SR to 0 and sets the interrupt mask bits I2 I1 I0 in SR to 111. All other registers are unaffected.

To cause a power-up reset, Motorola specifies that both the $\overline{\text{RESET}}$ and $\overline{\text{HALT}}$ pins of the 68000 must be held LOW for at least 100 ms. This means that an external circuit needs to be designed that will generate a negative pulse with a width of at least 100 ms for both $\overline{\text{RESET}}$ and $\overline{\text{HALT}}$. The manual $\overline{\text{RESET}}$ requires both the $\overline{\text{RESET}}$ and $\overline{\text{HALT}}$ pins to be LOW for at least 10 cycles(1.25 microseconds for 8MHz). In general, it is safer to assert $\overline{\text{RESET}}$ and $\overline{\text{HALT}}$ for much longer than the minimum requirements. Figure 10.11 shows a typical 68000 reset circuit that asserts $\overline{\text{RESET}}$ and $\overline{\text{HALT}}$ LOW for approximately 200 ms. The 555 timer is used in the circuit.

The reset circuit in the figure utilizes the 555 timer chip and provides for both power-up and manual resets by asserting the 68000 $\overline{\text{RESET}}$ and $\overline{\text{HALT}}$ pins for at least 200 ms. The computer designer does not have to know about the details of the 555 chip. Instead, the designer should know how to use the 555 chip to generate the 68000 RESET signal.

The 555 is a linear 8-pin chip. The TRIGGER pin is the input signal. When the voltage at the TRIGGER input pin is less than or equal to 1/3 V_{cc}, the OUTPUT pin is HIGH. The DISCHARGE and THRESHOLD pins are tied together to R_A and C. Note that the values of R_A and C determine the output pulse width. The CONTROL input pin controls the

THRESHOLD input voltage. According to manufacturer's data sheets, the control input should be connected to a 0.01-μF capacitor whose other lead should be grounded. Also, from the manufacturer's data sheets, the output pulse width, $t_{pw} = 1.1\ R_A C$ seconds. The values of R_A and C can be chosen for stretching out pulse width. An *RC* circuit is connected at the 555 TRIGGER pin. A slow pulse obtained by charging and discharging the capacitor C_1 is applied at the 555 TRIGGER input pin. The 555 will generate a clean and fast pulse at the output. Capacitor C_1 is at zero voltage upon power-up. This is obviously lower than 1/3 V_{cc} with V_{cc} = 5 V. Thus, the 555 will generate a HIGH at the OUTPUT pin. The OUTPUT pin is connected through a 7404 inverter to provide a LOW at the 68000 $\overline{\text{RESET}}$ and $\overline{\text{HALT}}$ pins. The 7404 output is buffered via two 7407's (noninverting buffers) to ensure adequate currents for the 68000 $\overline{\text{RESET}}$ and $\overline{\text{HALT}}$ pins. Note that the 7407 provides an open collector output. Therefore, a 1-Kohm pull-up is used for each 7407. Now, let us explain how the timing requirements for the 68000 RESET are satisfied.

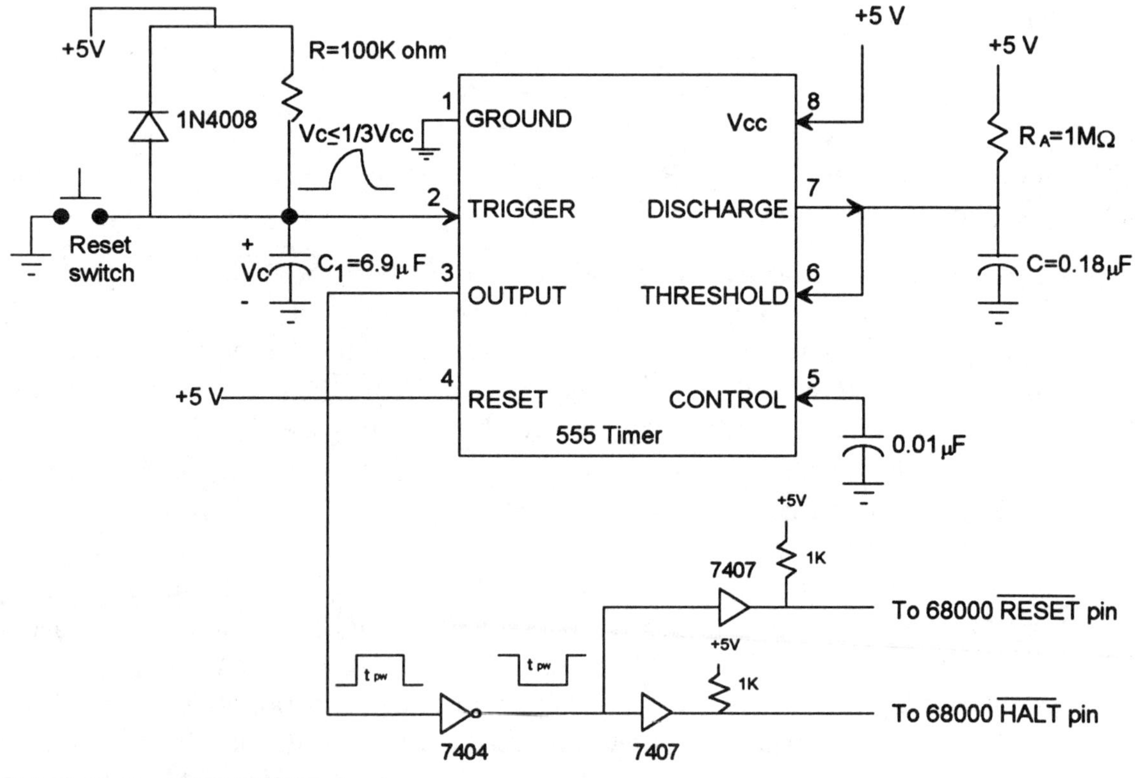

FIGURE 10.11 68000 RESET circuit

As mentioned before, capacitor C_1 is initially at zero voltage upon power-up. C_1 then charges to V_{cc} after a definite time determined by the time constant, RC_1. The charging voltage across the capacitor is

$$Vc(t) = Vcc[1 - e^{-\frac{t}{RC_1}}]$$

$V_c(t)$ must be less than or equal to $V_{cc}/3$ volts (1.7 V). To be on the safe side, let us assume that $V_c = V_{cc}/4 = 5/4 = 1.25$ V.

$$\frac{Vc(t)}{Vcc(t)} = 1 - e^{-\frac{t}{RC_1}}$$

$$\text{Hence, } \frac{1}{4} = 1 - e^{-\frac{t}{RC_1}}$$

$$e^{-\frac{t}{RC_1}} = 0.75$$

$$-\frac{t}{RC_1} = \ln(0.75)$$

$$-\frac{t}{RC_1} = -0.29$$

$$\text{Therefore, } RC_1 = \frac{t}{0.29}$$

As mentioned earlier, it is desired to provide 200 ms (arbitrarily chosen; satisfying the minimum requirements specified by Motorola) reset time for both power-up and manual reset.

$$RC_1 = \frac{200\text{ ms}}{0.29} = 689.65\text{ ms}$$

$$\text{Hence, } RC_1 \cong 0.69\text{ s}$$

If R is arbitrarily chosen as 100 KΩ, then $C_1 = 6.9$ μF.

The 555 output pulse width can be determined using the equation, $t_{pw} = 1.1\ R_A\ C$. Since $t_{pw} = 200$ msec, hence $R_A\ C = 0.18$ seconds.
If $R_A = 1$ MΩ (arbitrarily chosen) then $C = 0.18 / 10^6 = 0.18$ μF.

The reverse-biased diode (1N4008) connected at the 555 TRIGGER input circuit is used to hold the capacitor (C_1 charged to 1.25 V) voltage at 1.25 V in case V_{cc} (obtained using a power supply from AC voltage) drops below 5V to a level such that the capacitor C_1 may discharge through the 100-KΩ resistor. In such a situation, the diode will be forward biased essentially shorting out the 100-Kohm resistor, thus maintaining the capacitor voltage at 1.25 V.

In Figure 10.11, upon power-up, the capacitor C_1 charges to approximately 1.25 V. After some time, if the reset switch is depressed, the capacitor is short-circuited to ground.

The capacitor, therefore, discharges to zero. This logic 0 at the 555 TRIGGER input pin will provide 200 ms LOW at the 68000 $\overline{\text{RESET}}$ and $\overline{\text{HALT}}$ input pins. This will satisfy the minimum requirement of 10 clock cycles(1.25 microseconds for 8MHz clock) at the 68000 $\overline{\text{RESET}}$ and $\overline{\text{HALT}}$ pins for manual reset. The values of R and C_1 at the 555 trigger input should be recalculated for other 68000 clock frequencies for manual reset. Note that the 68000 power-up reset time is fixed with a timing requirement of at least 100 ms whereas the manual reset time depends on the 68000 clock frequency and must be at least 10 clock cycles.

Another way of generating the power-up and manual resets is by using a Schmitt-trigger inverter such as the 7414 chip. Figure 10.12 shows a typical circuit. The purpose of the Schmitt trigger in a microprocessor reset circuit has already been explained in Chapter 9 for 8086 reset using the 8284 chip. The operation of the 68000 power-up and manual resets using the RC circuit in Figure 10.12 has already been described in this section. The purpose of the two 7414 Schmitt-trigger inverters is primarily to shape up a slow pulse generated by the RC circuit to obtain a fast and clean negative pulse. Two 7407 open-collector noninverting buffers are used to amplify currents for the 68000 $\overline{\text{RESET}}$ and $\overline{\text{HALT}}$ pins. Let us now determine the values of R and C.

When the input of the 7414 Schmitt-trigger inverter is low (0 V for example), the output will be HIGH, typically at about 3.7 V. For input voltage from 0 to about 1.7 V, the output of the 7414 will be HIGH. Let us arbitrarily choose V_c = 1.5V to provide a low at the input of the first 7414 in the figure. As before,

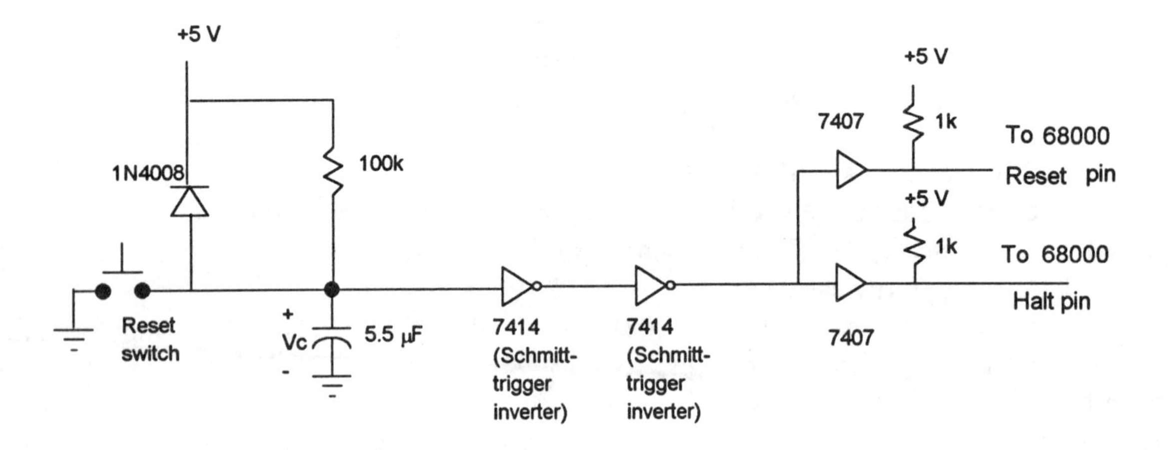

FIGURE 10.12 68000 Reset circuit using a Schmitt trigger

$$Vc = Vcc[1 - e^{-\frac{t}{RC}}]$$

$$\text{Hence, } 1 - e^{-\frac{t}{RC}} = \frac{1.5}{5}$$

$$e^{-\frac{t}{RC}} = 0.7$$

Let us design the reset circuit to provide 200 ms reset time. Therefore, $t = 200$ ms.

$$-\frac{0.2}{RC} = \ln(0.7)$$

$$-\frac{0.2}{RC} = -0.36$$

$$\text{Therefore, } RC = 0.55 \text{ seconds}$$

If R is arbitrarily chosen as 100 KΩ, then $C = 5.5$ μF.

10.10 68000 Read and Write Cycle Timing Diagrams

The 68000 family of processors (68000, 68008, 68010, and 68012) uses a handshaking mechanism to transfer data between the processors and peripheral devices. This means that all these processors can transfer data asynchronously to and from peripherals of varying speeds.

During the read cycle, the 68000 obtains data from a memory location or an I/O port. If the instruction specifies a word (such as `MOVE.W $020504,D1`) or a long word (such as `MOVE.L $030808, D0`), the 68000 reads both upper and lower bytes at the same time by asserting the $\overline{UDS}$ and $\overline{LDS}$ pins. When the instruction is for a byte operation, the 68000 utilizes an internal bit to find which byte to read and then outputs the data strobe required for that byte.

For byte operations, when the address is even ($A_0 = 0$), the 68000 asserts $\overline{UDS}$ and reads data via the D_8–D_{15} pins into the low byte of the specified data register. On the other hand, when the address is odd ($A_0 = 1$), the 68000 outputs a LOW on $\overline{LDS}$ and reads data via the D_0–D_7 pins to the low byte of the specified data register. For example, consider `MOVE.B $507144, D5`. The 68000 outputs a LOW on $\overline{UDS}$ (because $A_0 = 0$) and a HIGH on $\overline{LDS}$. The memory chip's eight data lines must be connected to the 68000 D_8–D_{15} pins. The 68000 reads the data byte via the D_8–D_{15} pins into the low byte of D5. Note that, for reading a byte from an odd address, the data lines of the memory chip must be connected to the 68000 D_0–D_7 pins. In this case, the 68000 outputs a LOW on $\overline{LDS}$ (because $A_0 = 1$) and a HIGH on $\overline{UDS}$, and then reads the data byte into the low byte of the data register.

Figure 10.13 shows the read/write timing diagrams. During S0, address and data signals are in the high-impedance state. At the start of S1, the 68000 outputs the address on its address pins (A_1–A_{23}). During S0, the 68000 outputs FC2–FC0 signals. $\overline{\text{AS}}$ is asserted at the start of S2 to indicate a valid address on the bus. $\overline{\text{AS}}$ can be used at this point to latch the signals on the address pins. The 68000 asserts the $\overline{\text{UDS}}$, $\overline{\text{LDS}}$, and R/$\overline{\text{W}}$ = 1 to indicate a READ operation. The 68000 now waits for the peripheral device to assert $\overline{\text{DTACK}}$. Upon placing data on the data bus, the peripheral device asserts $\overline{\text{DTACK}}$. The 68000 samples the $\overline{\text{DTACK}}$ signal at the end of S4. If $\overline{\text{DTACK}}$ is not asserted by the peripheral device, the processor automatically inserts a wait state(s) (W).

However, upon assertion of $\overline{\text{DTACK}}$, the 68000 negates the $\overline{\text{AS}}$, $\overline{\text{UDS}}$, and $\overline{\text{LDS}}$ signals, and latches the data from the data bus into an internal register at the end of the next cycle. Once the selected peripheral device senses that the 68000 has obtained data from the data bus (by recognizing the negation of $\overline{\text{AS}}$, $\overline{\text{UDS}}$, or $\overline{\text{LDS}}$), the peripheral device must negate $\overline{\text{DTACK}}$ immediately so that it does not interfere with the start of the next cycle.

If $\overline{\text{DTACK}}$ is not asserted by the peripheral at the end of S4 (Figure 10.13, SLOW READ), the 68000 inserts wait states. The 68000 outputs valid addresses on the address pins and keeps asserting $\overline{\text{AS}}$, $\overline{\text{UDS}}$, and $\overline{\text{LDS}}$ until the peripheral asserts $\overline{\text{DTACK}}$. The 68000 always inserts an even number of wait states if $\overline{\text{DTACK}}$ is not asserted by the peripheral because all 68000 operations are performed using the clock with two states per clock cycle. Note in Figure 10.13 that the 68000 inserts 4 wait states or 2 cycles.

As an example of word read, consider that the 68000 is ready to execute the `MOVE.W $602122,D0` instruction. The 68000 performs as follows:

1. At the end of S0 the 68000 places the upper 23 bits of the address 602122_{16} on A_1–A_{23}.
2. At the end of S1, the 68000 asserts $\overline{\text{AS}}$, $\overline{\text{UDS}}$, and $\overline{\text{LDS}}$.
3. The 68000 continues to output a HIGH on the R/$\overline{\text{W}}$ pin from the beginning of the read cycle to indicate a READ operation.
4. At the end of S0, the 68000 places appropriate outputs on the FC2–FC0 pins to indicate either supervisor or user read.
5. If the peripheral asserts $\overline{\text{DTACK}}$ at the end of S4, the 68000 reads the contents of 602122_{16} and 602123_{16} via the D_8–D_{15} and D_0–D_7 pins, respectively, into the high and low bytes of D0.W at the end of S6. If the peripheral does not assert $\overline{\text{DTACK}}$ at the end of S4, the 68000 continues to insert wait states.

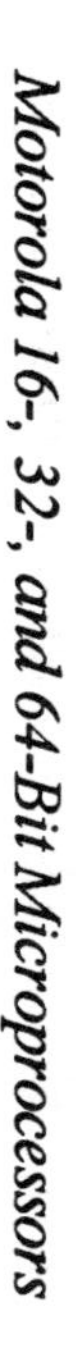

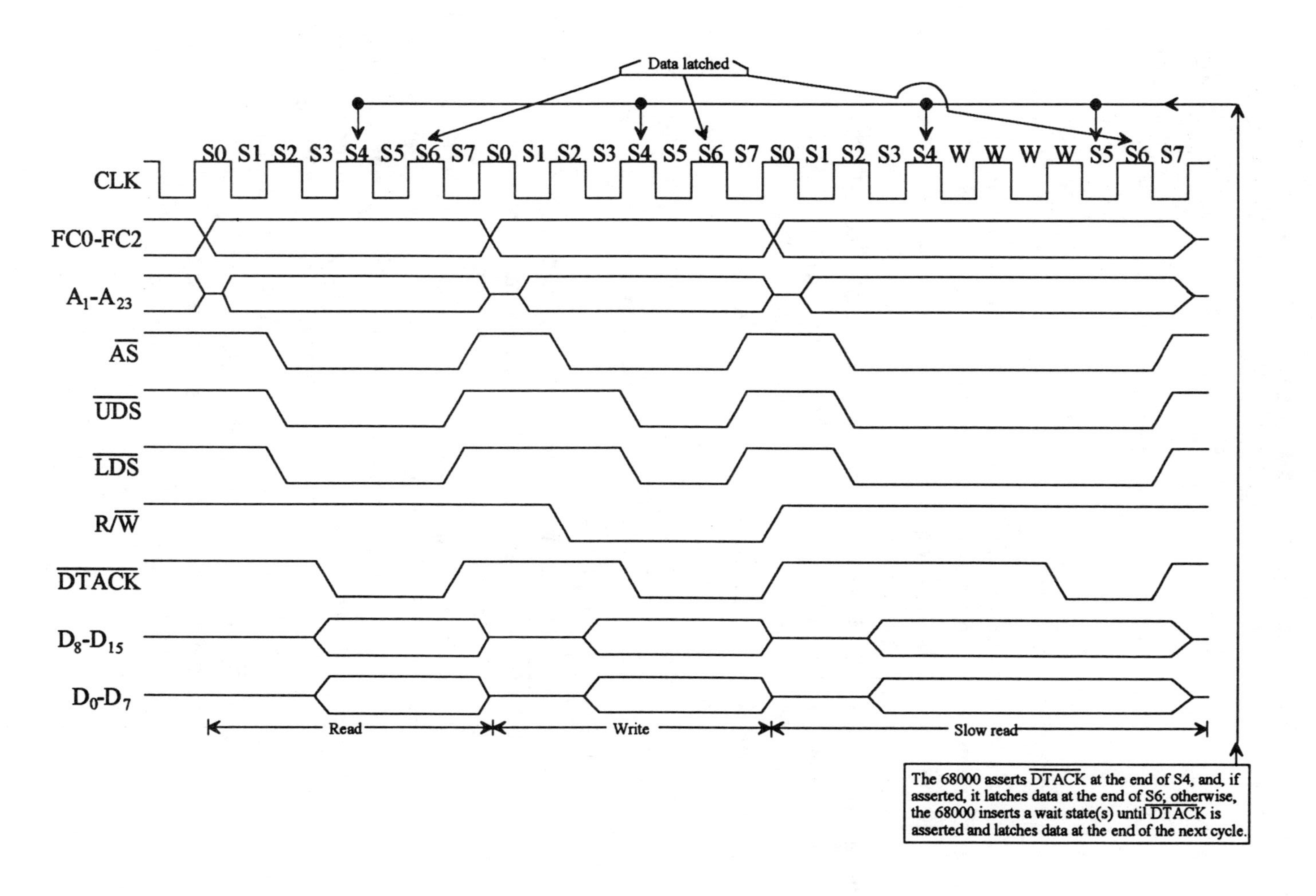

FIGURE 10-13 68000 Read and Write cycle Timing Diagrams

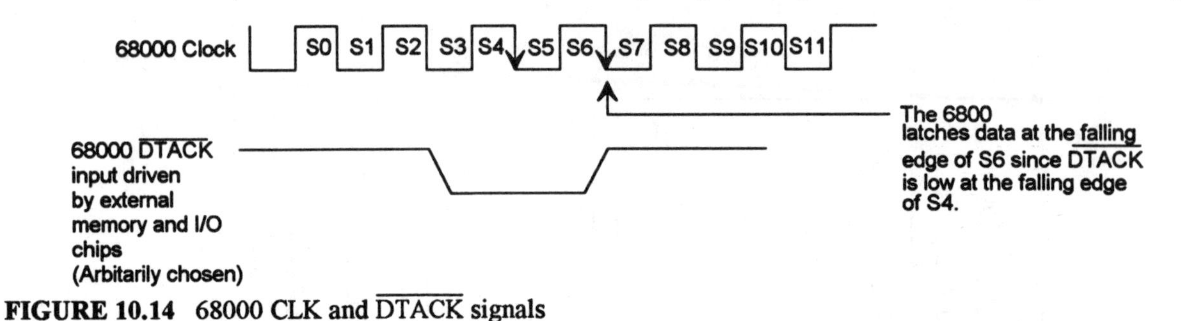

FIGURE 10.14 68000 CLK and $\overline{\text{DTACK}}$ signals

Figure 10.14 shows a simplified timing diagram illustrating the use of $\overline{\text{DTACK}}$ for interfacing external memory and I/O chips to the 68000. As mentioned before, the 68000 checks the $\overline{\text{DTACK}}$ input pin at the following edge of S4 (three cycles), the external memory, or I/O in this case, drives 68000 $\overline{\text{DTACK}}$ input to LOW, and the 68000 waits for one cycle and latches data at the end of S6. However, if the 68000 does not find $\overline{\text{DTACK}}$ LOW at the falling edge of S4, it waits for one clock cycle and then again checks $\overline{\text{DTACK}}$ for LOW. If $\overline{\text{DTACK}}$ is LOW, the 68000 latches data after one cycle (falling edge of S8). If the 68000 does not find $\overline{\text{DTACK}}$ LOW at the falling edge of S6, it checks for $\overline{\text{DTACK}}$ LOW at the falling edge of S8 and the process continues. Note that the minimum time to latch data is four cycles. This means that in the preceding example, if the 68000 clock frequency is 8 MHz, data will be latched after 500 ns because the $\overline{\text{DTACK}}$ is asserted LOW at the end of S4 (375 ns).

10.11 68000 Memory Interface

One of the advantages of the 68000 is that it can easily be interfaced to memory chips because it goes into a wait state if $\overline{\text{DTACK}}$ is not asserted (LOW) by the memory devices at the end of S4. A simplified schematic showing an interface of a 68000 to two 2732's and two 6116's is given in Figure 10.15. As mentioned in Chapter 9, the 2732 is a 4K × 8 EPROM and the 6116 is a 2K × 8 static RAM. The pin diagrams of the 2732 and 6116 are provided in Chapter 9. For a 4-MHz clock, each cycle is 250 ns. Because the 68000 samples data at the falling edge of S4 (750 ns) and latches data at the falling edge of S6 (1000 ns), $\overline{\text{AS}}$ can be used to assert $\overline{\text{DTACK}}$. From the 68000 timing diagram of Figure 10.13, $\overline{\text{AS}}$ goes to LOW after approximately two cycles (500 ns). The time delay between $\overline{\text{AS}}$ going LOW and the falling edge of S6 is 500 ns. Note that $\overline{\text{LDS}}$ and $\overline{\text{UDS}}$ must be used as chip selects as in Figure 10.15. They must not be connected to A0 of the memory chips. Because in that case half of the memory in each memory chip would be wasted. Note that $\overline{\text{LDS}}$ and $\overline{\text{UDS}}$ also go to LOW after about two cycles (500 ns).

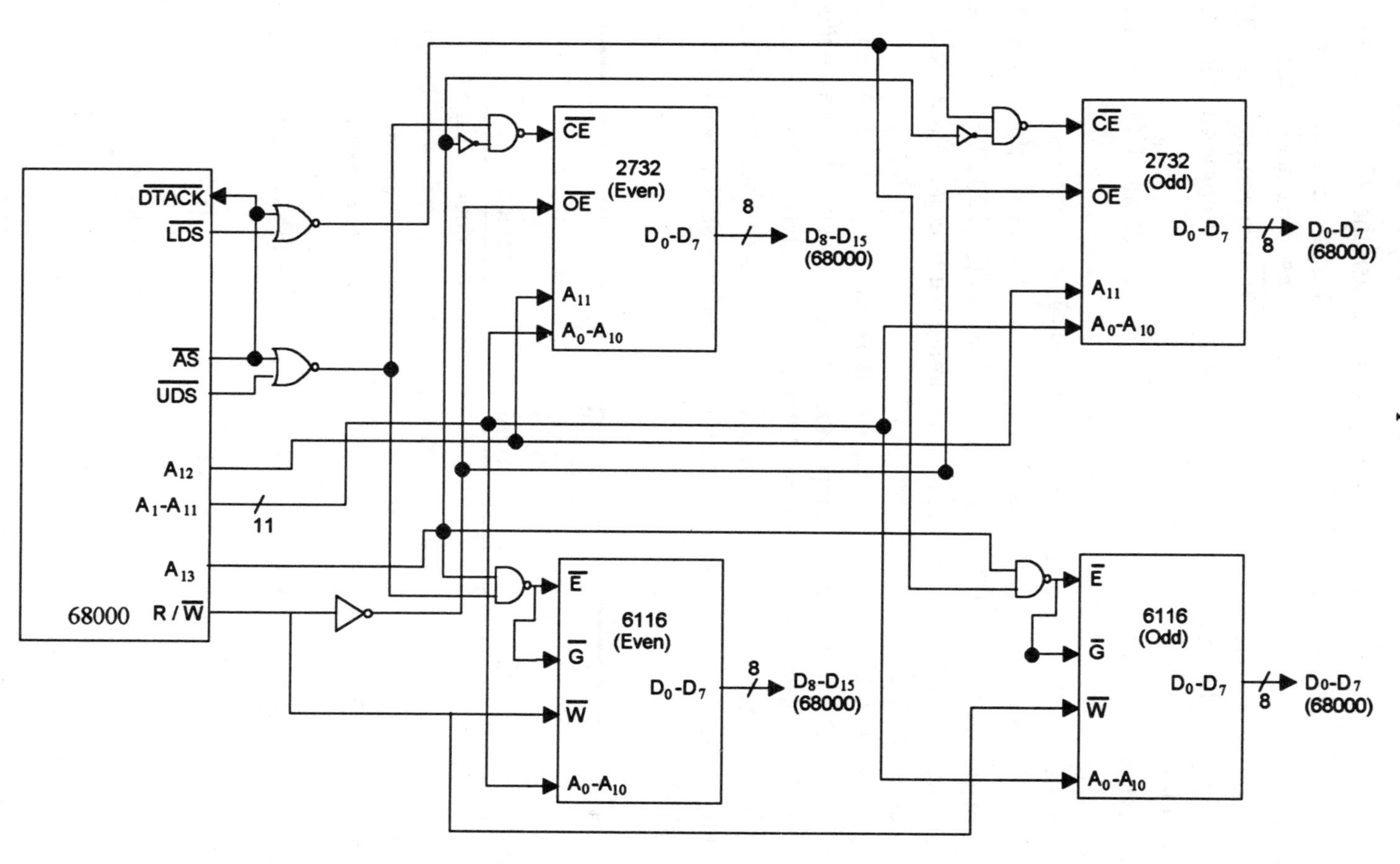

FIGURE 10.15 68000 interface to 2732 / 6116

In Figure 10.15, a delay circuit for $\overline{DTACK}$ is not required because the 2732 and 6116 both place data on the bus lines before the 68000 samples $\overline{DTACK}$. This is due to the 68000 clock frequency of 4 MHz in this case. Thus, each clock cycle is 250 ns. The access times of the 2732 and 6116 are 450 ns and 120 ns respectively. Because $\overline{DTACK}$ is sampled after 3 clock cycles (3×250 ns $= 750$ ns), both the 2732 and 6116 will have adequate time to place data on the bus for the 68000 to latch.

For example, consider the even 2732 EPROM. $\overline{UDS}$ and $\overline{AS}$ are NORed and then NANDed with inverted A_{13} to select this chip. With the 450-ns access time of the 2732, data will be placed on the 68000 D_8–D_{15} pins after approximately 970 nanoseconds (500 ns for $\overline{AS}$ or $\overline{UDS}$ + 10 ns for the NOR gate + 10 ns for the NAND gate + 450 ns for the 2732). Therefore, no delay circuit for the 68000 $\overline{DTACK}$ is required because the 68000 latches data from the D_8–D_{15} pins after 4 cycles (1000 ns in this case). The timing parameters of the 68000-2732 with various 68000 frequencies are shown in Table 10.14.

Next, consider odd 6116 RAM with a 4-MHz 68000. In this case, $\overline{LDS}$ and $\overline{AS}$ are NORed and NANDed with A13 to select this chip. With the 120-ns access time of the 6116 RAM, data will be placed on the 68000 D_0–D_7 pins after approximately 640 ns. Because the 68000 latches data after four cycles (1000 ns in this case), no delay circuit for $\overline{DTACK}$ is required. The requirements for $\overline{DTACK}$ for 68000/6116 for various 68000 clock frequencies can similarly be determined.

TABLE 10.14 68000-2732 Timing Example

Case	68000 Frequency	Clock Cycle	Time before first $\overline{DTACK}$ is sampled	Comment
1	6 MHz	166.7 ns	3(166.7) = 500.1ns	Not enough time for 2732 to place data on bus; needs delay circuit for $\overline{DTACK}$
2	8 MHz	125 ns	3(125) = 375 ns	Same as case 1
3	10 MHz	100 ns	3(100) = 300 ns	Same as case 1
4	12.5 MHz	80 ns	3(80) = 240 ns	Same as case 1
5	16.67 MHz	60 ns	3(60) = 180 ns	Same as case 1
6	25 MHz	40 ns	3(40) = 120 ns	Same as case 1

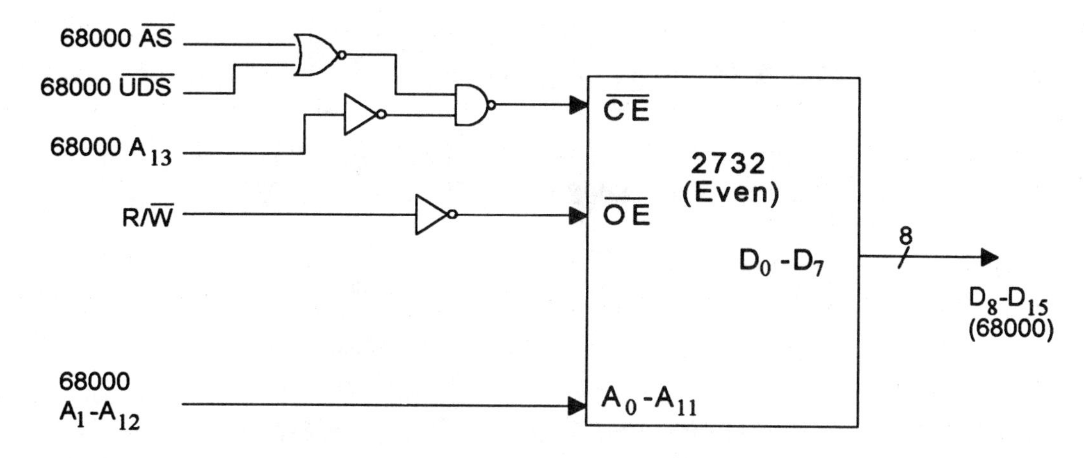

FIGURE 10.16 68000 interface to even 2732

In case a delay circuit for $\overline{DTACK}$ is required, a ring counter with D flip-flops can be used. Let us now determine the memory maps. Figure 10.16 shows the 68000 interface to even 2732 obtained from Figure 10.15. When A_{13} = 0, $\overline{UDS}$ = 0, $\overline{AS}$ = 0, and $R/\overline{W}$ =1, the 2732 will be selected by the 68000 to read data from 68000 D_8–D_{15} pins. The 68000 address pins A_{23}–A_{14} are don't cares (assume 0). The memory map for the even 2732 can be determined as follows:

A_{23} A_{22} · · · A_{14}	A_{13}	A_{12} A_{11} · · · A_1	A_0
0 0 · · · 0	0		0
Don't cares assume 0's	To select 2732	Can be 0's to 1's	even 2732

Address range: \$000000, \$000002, … , \$001FFE

Similarly, the memory for the odd 2732, even 6116, and odd 6116 can be determined as follows:

- **2732 odd**

A_{23} A_{22} · · · A_{14}	A_{13}	A_{12} A_{11} · · · A_1	A_0
0 0 · · · 0	0	Can be 0's to 1's	1

Address range: \$000001, \$000003, … , \$001FFF

- **6116 even**

$A_{23}\ A_{22} \cdots A_{14}\ A_{13}\ A_{12}\ A_{11}\ A_{10} \cdots A_1\ A_0$

0 0 · · · 1 0 Can be 0's to 1's 0

(The brace over A_{23} to A_{14} covers 0 0 · · ·, and the brace over A_{11} to A_1 covers "Can be 0's to 1's".)

Address range: \$002000, \$002002, … , \$002FFE

- **6116 odd**

$A_{23}\ A_{22} \cdots A_{14}\ A_{13}\ A_{12}\ A_{11}\ A_{10} \cdots A_1\ A_0$

0 0 · · · 1 0 Can be 0's to 1's 1

(The brace over A_{23} to A_{14} covers 0 0 · · ·, and the brace over A_{11} to A_1 covers "Can be 0's to 1's".)

Address range: \$002001, \$002003, … , \$002FFF

Note: In the above, for 6116's, A_{12} and A_{14} - A_{23} are don't cares (assume 0's).

10.12 68000 I/O

This section covers the I/O techniques associated with Motorola 68000.

10.12.1 68000 Programmed I/O

As mentioned before, the 68000 uses memory-mapped I/O. Data transfer using I/O ports (programmed I/O) can be achieved in the 68000 in one of the following ways:

- By interfacing the 68000 with an inexpensive slow 6800 I/O chip such as the MC6821.
- By interfacing the 68000 with its own family of I/O chips such as the MC68230.

10.12.1.1 68000/6821 Interface

The Motorola 6821 is a 40-pin peripheral interface adapter (PIA) chip. It is provided with an 8-bit bidirectional data bus (D_0–D_7), two register select lines (RS0, RS1), read/write ($R/\overline{W}$) and reset ($\overline{RESET}$) lines, an enable line (E), two 8-bit I/O ports (PA0–PA7), and (PB0–PB7), and other pins. Figure 10.17 shows the pin diagram of the 6821. There are six 6821 registers. These include two 8-bit ports (ports A and B), two data direction registers, and two control registers. Selection of these registers is controlled by the RS0 and RS1 inputs together with bit 2 of the control register. Table 10.15 shows how the registers are selected. In Table 10.15, bit 2 in each control register (CRA-2 and CRB-2) determines selection of either an I/O port or the corresponding data direction register when the proper register select signals are applied to RS0 and RS1. A 1 in bit 2 in CRA or CRB allows access of I/O ports; a 0 in bit 2 of CRA or CRB selects the data direction registers.

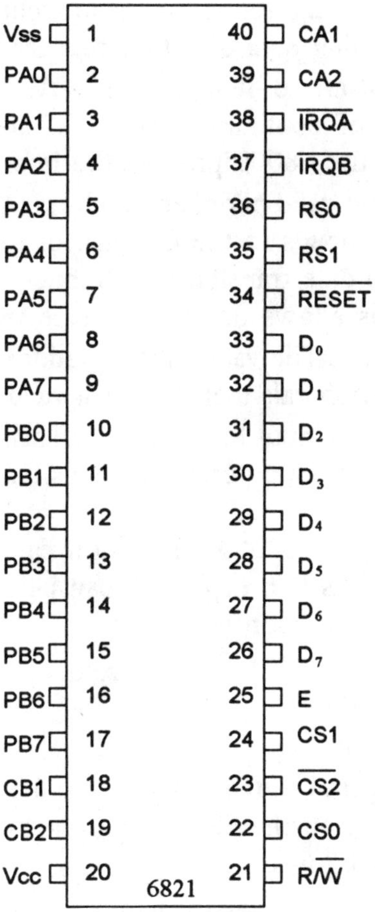

Signal	Pin	Pin	Signal
Vss	1	40	CA1
PA0	2	39	CA2
PA1	3	38	$\overline{IRQA}$
PA2	4	37	$\overline{IRQB}$
PA3	5	36	RS0
PA4	6	35	RS1
PA5	7	34	$\overline{RESET}$
PA6	8	33	D_0
PA7	9	32	D_1
PB0	10	31	D_2
PB1	11	30	D_3
PB2	12	29	D_4
PB3	13	28	D_5
PB4	14	27	D_6
PB5	15	26	D_7
PB6	16	25	E
PB7	17	24	CS1
CB1	18	23	$\overline{CS2}$
CB2	19	22	CS0
Vcc	20	21	$R/\overline{W}$

6821

FIGURE 10.17 6821 pin diagram

TABLE 10.15 6821 Register Definition

		Control Register Bits 2		
RS1	*RS0*	*CRA-2*	*CRB-2*	*Register Selected*
0	0	1	X	I/O port A
0	0	0	X	Data direction register A
0	1	X	X	Control register A
1	0	X	1	I/O port B
1	0	X	0	Data direction register B
1	1	X	X	Control register B

X = Don't care

Each I/O port bit can be configured to act as an input or output. This is accomplished by sending a 1 in the corresponding data direction register bit for those bits that are to be output and a 0 for those bits that are to be inputs. A LOW on the $\overline{\text{RESET}}$ pin clears all PIA registers to 0. This has the effect of configuring PA0–PA7 and PB0–PB7 as inputs.

Three built-in signals in the 68000 provide the interface with the 6821: enable (E), valid memory address ($\overline{\text{VMA}}$), and valid peripheral address ($\overline{\text{VPA}}$). The enable signal (E) is an output from the 68000. It corresponds to the E signal of the 6821. This signal is the clock used by the 6821 to synchronize data transfer. The frequency of the E signal is one tenth of the 68000 clock frequency. This allows one to interface the 68000 (which operates much faster than the 6821) with the 6821. The valid memory address ($\overline{\text{VMA}}$) signal is output by the 68000 to indicate to the 6800 peripherals that there is a valid address on the address bus. The valid peripheral address ($\overline{\text{VPA}}$) is an input to the 68000. This signal is used to indicate that the device addressed by the 68000 is a 6800 peripheral. This tells the 68000 to synchronize data transfer with the enable signal (E).

Let us now discuss how the 68000 instruction can be used to configure the 6821 ports. As an example, bit 7 and bits 0–6 of port A can be configured, respectively, as input and outputs using the following instruction sequence:

```
BCLR.B #$2,CRA        ;     Address DDRA
MOVE.B #$7F,DDRA      ;     Configure port A
BSET.B #$2,CRA        ;     Address port A
```

Once the ports are configured to the designer's specification, the 6821 can be used to transfer data from an input device to the 68000 or from the 68000 to an output device by using the `MOVE.B` instruction as follows:

```
MOVE.B (EA), Dn     ;     Transfer 8-bit data from an input port
                    ;     to the specified data register Dn
MOVE.B Dn, (EA)     ;     Transfer 8-bit data from the specified
                    ;     data register Dn to an output port
```

Figure 10.18 shows a block diagram of how two 6821's are interfaced to the 68000 in order to obtain four 8-bit I/O ports. Note that the least significant bit, A_0, of the 68000 address pin is internally encoded to generate two signals, the upper data strobe ($\overline{\text{UDS}}$) and lower data strobe ($\overline{\text{LDS}}$). For byte transfers, $\overline{\text{UDS}}$ is asserted if an even-numbered byte is being transferred and $\overline{\text{LDS}}$ is asserted for an odd-numbered byte. In Figure 10.18, I/O port addresses can be obtained as follows: When A_{22} = 1 and $\overline{\text{AS}}$= 0, the OR gate output will be LOW. This OR gate output is used to assert $\overline{\text{VPA}}$. The inverted OR gate output, in turn, makes CS1 HIGH on both 6821's. Note that A_{22} is arbitrarily chosen. A_{22} is chosen to be

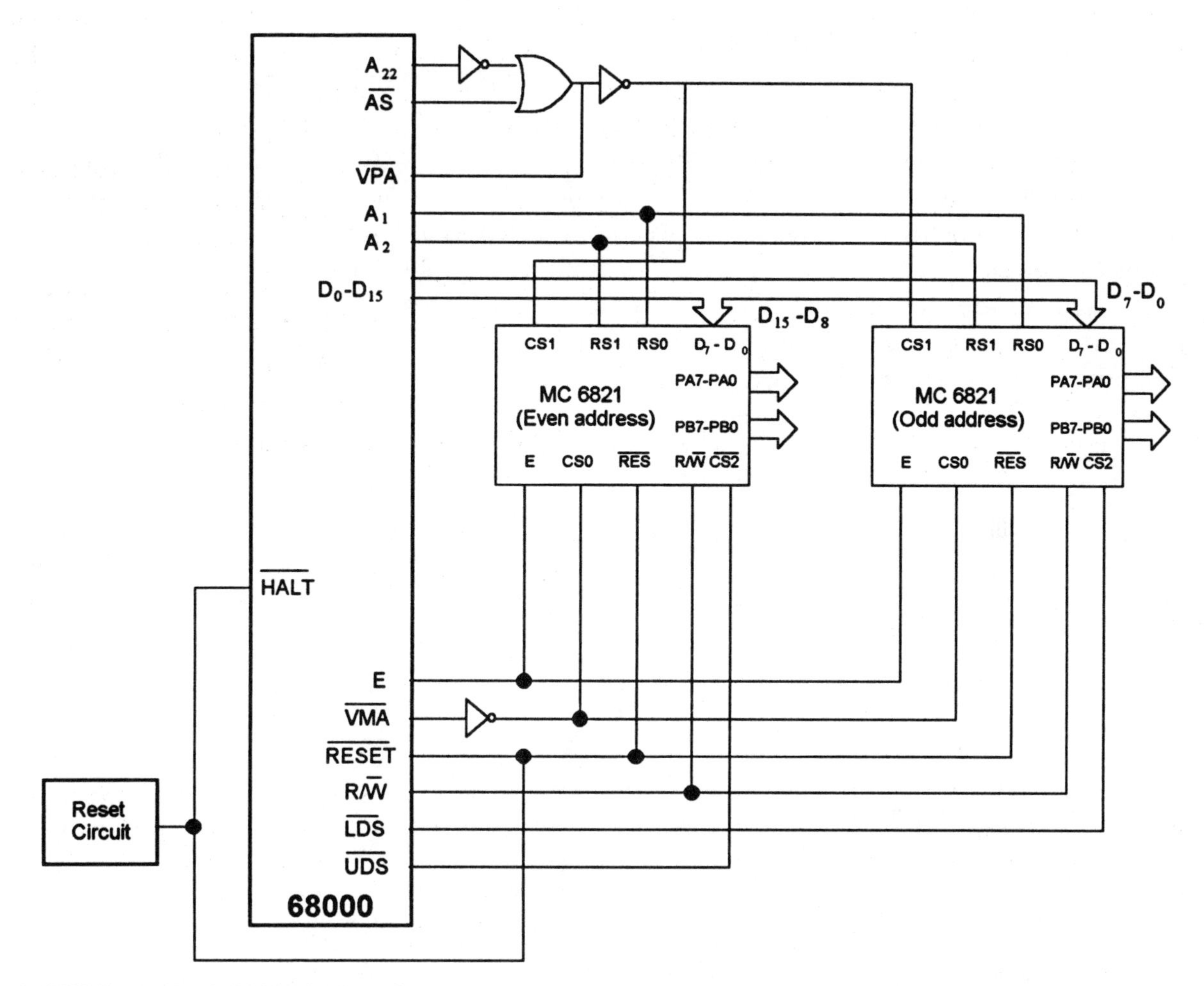

FIGURE 10.18 68000/6821 Interface

HIGH to enable CS1 so that the addresses for the ports and the reset vector are not the same. Assuming that the don't care address lines A_{23} and A_{21}–A_3 are 0's, the addresses for the I/O ports, control registers, and data direction registers for the even 6821 ($A_0 = 0$) can be obtained as shown; similarly, the addresses for the ports, control registers, and data direction registers for the odd 6821 ($A_0 = 1$) can be determined as follows:

	Port A or DDRA	CRA	Port B or DDRB	CRB
6821(even)	$400000	$400002	$400004	$400006
6821(odd)	$400001	$400003	$400005	$400007

10.12.1.2 68000/68230 Interface

The 68230 is a 48-pin I/O chip designed for the 68000 family of microprocessors. The 68230 offers various functions such as programmed I/O, an on-chip timer, and a DMA request pin for connection to a DMA controller. Figure 10.19 shows the 68230 pin diagram.

The 68230 can be configured in two modes of operation: unidirectional and bidirectional. In the unidirectional mode, data direction registers configure the corresponding ports as inputs or outputs. This is programmed I/O mode of operation. Both 8-bit and 16-bit ports can be used. In the bidirectional mode, the 68230 provides data transfer between the 68000 and external devices via exchange of control signals (known as handshaking). This section will only cover the programmed I/O feature of the 68230.

Signal	Pin	Pin	Signal
D_5	1	48	D_4
D_6	2	47	D_3
D_7	3	46	D_2
PA0	4	45	D_1
PA1	5	44	D_0
PA2	6	43	$R/\overline{W}$
PA3	7	42	$\overline{DTACK}$
PA4	8	41	$\overline{CS}$
PA5	9	40	CLK
PA6	10	39	$\overline{RESET}$
PA7	11	38	Vss
Vcc	12	37	PC7/$\overline{TIACK}$
H1	13	36	PC6/$\overline{PIACK}$
H2	14	35	PC5/$\overline{PIRQ}$
H3	15	34	PC4/$\overline{DMAREQ}$
H4	16	33	PC3/TOUT
PB0	17	32	PC2/TIN
PB1	18	31	PC1
PB2	19	30	PC0
PB3	20	29	RS1
PB4	21	28	RS2
PB5	22	27	RS3
PB6	23	26	RS4
PB7	24	25	RS5

FIGURE 10.19 68230 pin diagram

This 68230 ports can be configured in either unidirectional or bidirectional mode by using bits 7 and 6 of the port general control register, PGCR (R0) as follows:

PGCR Bits			
7	6	Mode	
0	0	0	(unidirectional 8-bit)
0	1	1	(unidirectional 16-bit)
1	0	2	(bidirectional 8-bit)
1	1	3	(bidirectional 16-bit)

The other bits of the PGCR are defined for handshaking.

Modes 0 and 2 configure ports A and B as unidirectional or bidirectional 8-bit ports. Modes 1 and 3, on the other hand, combine ports A and B together to form a 16-bit unidirectional or bidirectional port. Ports configured as unidirectional 8-bit must be programmed further as submodes of operation using bits 7 and 6 of PACR (R6) and PBCR (R7) as follows:

Submode	*Bit 7 of PACR or PBCR*	*Bit 6 of PACR or PBCR*	*Comment*
00	0	0	Pin-definable double-buffered input or single-buffered output
01	0	1	Pin-definable double-buffered output or nonlatched input
1X	1	X	Bit I/O (pin-definable single-buffered output or nonlatched input)

Note that X means don't care. Nonlatched inputs are latched internally, but the values are not latched externally by the 68230 at the port. Bit I/O is used for programmed I/O.

The submodes define the ports as parallel input ports, parallel output ports, or bit-configurable I/O ports. In addition to these, the submodes further define the ports as latched input ports, interrupt-driven ports, DMA ports, and ports with various I/O handshake operations. Table 10.16 lists some of the 68230 registers. The registers required for programmed I/O are considered in the following discussion. Note that the 68230 register select pins (RS5–RS1) are used to select the 68230 registers. Figure 10.20 illustrates how to obtain specific addresses for the 68230 I/O ports.

TABLE 10.16 Some of the 68230 Registers

Register Select Bits					
RS5	*RS4*	*RS3*	*RS2*	*RS1*	*Register Selected*
0	0	0	0	0	PGCR, Port General Control Register (R0)
0	0	0	1	0	PADDR, Port A Data Direrction Register (R2)
0	0	0	1	1	PBDDR, Port B Data Direction Register (R3)
0	0	1	0	0	PCDDR, Port C Data Direction Register (R4)
0	0	1	1	0	PACR, Port A Control Register (R6)
0	0	1	1	1	PBCR, Port B Control Register (R7)
0	1	0	0	0	PADR, Port A Data Register (R8)
0	1	0	0	1	PBDR, Port B Data Register (R9)
0	1	1	0	0	PCDR, Port C Data Register (R12)

The hardware schematic for the 68000/68230 interface shown in Figure 10.20 is connected in such a way that each 68230 I/O port has a unique address. A_{23} is chosen to be HIGH to select the 68230 chips so that the port addresses are different from the 68000 reset vector addresses 000000_{16}–000006_{16}. The configuration in the figure will provide even port addresses because $\overline{UDS}$ is used for enabling the 68230 $\overline{CS}$. The 68230 $\overline{DTACK}$ is an open-drain output. Hence, a pull-up resistor is required.

From the figure, addresses for registers PGCR (R0), PADDR (R2), PBDDR (R3), PCDDR (R4), PACR (R6), PBCR (R7), PADR (R8), PBDR (R9) and PCDR (R12) can be obtained. Consider PGCR as follows:

$$\begin{array}{ccccccccccccc} A_{23} & A_{22} & A_{21} & A_{20} & \cdots & A_6 & A_5 & A_4 & A_3 & A_2 & A_1 & A_0 & \\ 1 & 0 & 0 & 0 & \cdots & 0 & \underbrace{0 \quad 0 \quad 0 \quad 0 \quad 0}_{RS5\,-\,RS1} & & & & & \underset{\overline{UDS}}{\overset{\uparrow}{0}} & = \$800000 \end{array}$$

Therefore, Address for PGCR = \$800000

Similarly, Address for PADDR = \$800004, Address for PBDDR = \$800006

Address for PACR = \$80000C, Address for PBCR = \$80000E

Address for PADR = \$800010, Address for PBDR = \$800012

Address for PCDR = \$800018, Address for PCDDR = \$800008

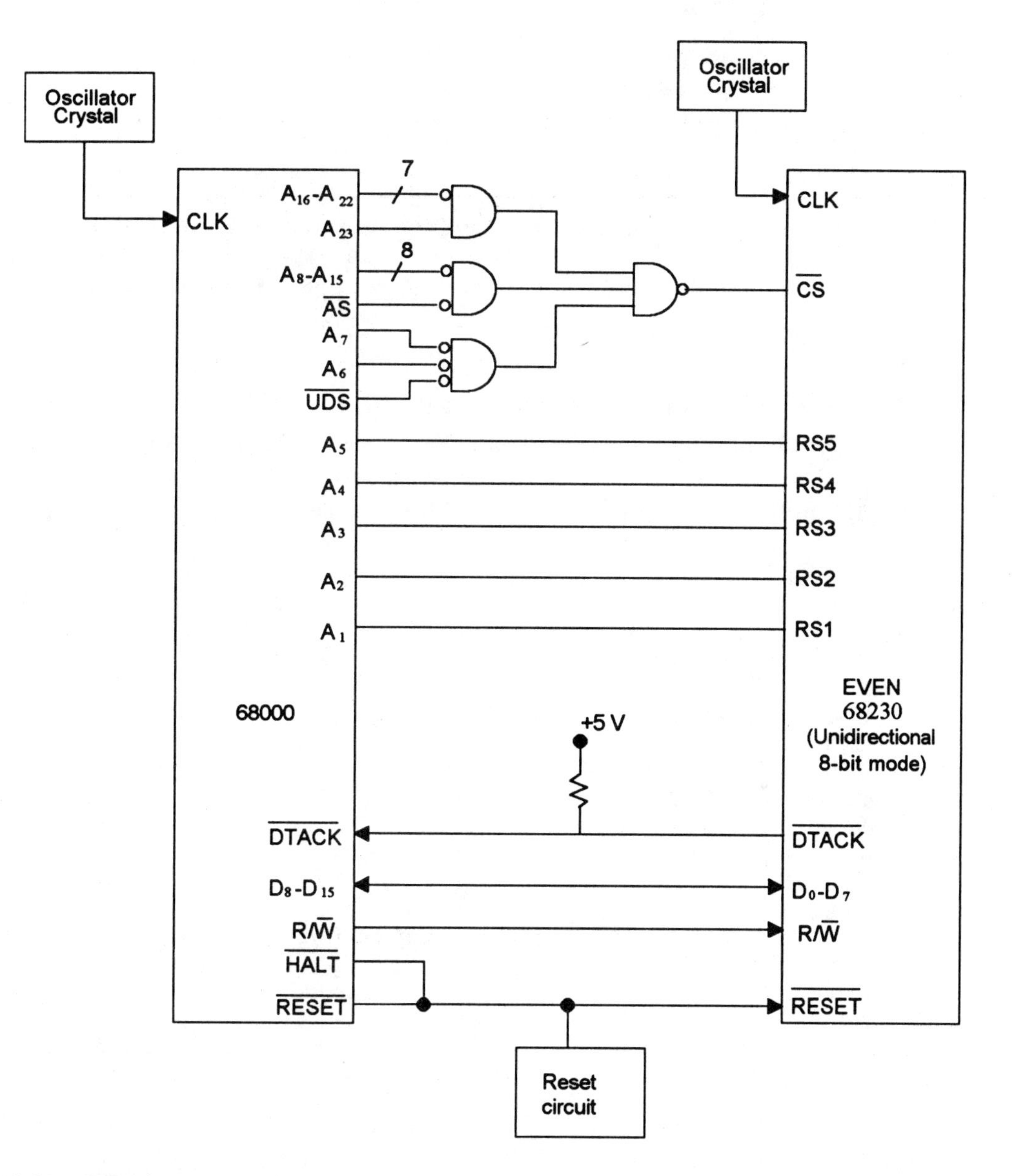

FIGURE 10.20 68000/68230 interface

As an example, the following instruction sequence will select mode 0, submode 1X and configure bits 0–5 of PORT A as outputs, bits 6 and 7 as inputs, and port B as an input port:

```
PGCR    EQU      $800000
PADDR   EQU      $800004
PBDDR   EQU      $800006
PACR    EQU      $80000C
PBCR    EQU      $80000E
        ANDI.B   #$3F,PGCR    ;  Select mode 0
        BSET.B   #7,PACR      ;  Port A bit I/O submode
        BSET.B   #7,PBCR      ;  Port B bit I/O submode
        MOVE.B   #$3F,PADDR   ;  Configure port A bits 0-5 as
                              ;  outputs and bits 6 and 7 as inputs
        MOVE.B   #$00,PBDDR   ;  Configure port B as an input port
```

Example 10.16

A 68000/68230-based microcomputer is required to drive an LED connected at bit 7 of port A based on two switch inputs connected at bits 6 and 7 of port B. If both switches are equal (either HIGH or LOW), turn the LED ON; otherwise turn it OFF. Assume that a HIGH will turn the LED ON and a LOW will turn it OFF. Write a 68000 assembly program to accomplish this.

Solution

```
PGCR    EQU      $800000
PACR    EQU      $80000C
PBCR    EQU      $80000E
PADDR   EQU      $800004
PBDDR   EQU      $800006
PADR    EQU      $800010
PBDR    EQU      $800012
        ANDI.B   #$3F,PGCR    ;  Select mode 0
        BSET.B   #7,PACR      ;  Port A bit I/o submode
        BSET.B   #7,PBCR      ;  Port B bit I/o submode
        MOVE.B   #$80,PADDR   ;  Configure port A bit 7 as output
        MOVE.B   #0,PBDDR     ;  Configure port B bits 6 and 7 as inputs
        MOVE.B   PBDR,D0      ;  Input port B
        ANDI.B   #$0C0,D0     ;  Retain bits 6 and 7
        BEQ.B    LEDON        ;  If both switches LOW, turn LED ON
        CMPI.B   #$0C0,D0     ;  If both switches HIGH, turn LED ON
        BEQ.B    LEDON
        MOVE.B   #$00,PADR    ;  Turn LED OFF
        JMP      FINISH
LEDON   MOVE.B   #$80,PADR    ;  Turn LED ON
FINISH  JMP      FINISH
```

Example 10.17

Write a 68000 assembly language program to drive an LED connected to bit 7 of Port A based on a switch input at bit 0 of Port A. If the switch is HIGH, turn the LED ON; otherwise turn the LED OFF. Assume a 68000/2732/6116/6821 microcomputer. Also, write a C++ program to accomplish the same task. Use port addresses of your choice.

Solution

The 68000 assembly language program and the C++ program follow.

- ***68000/6821 Microcomputer Assembly Code for Switch and LED***

```
PORTA    EQU      $001001
DDRA     EQU      $001001
CRA      EQU      $001003
SETUP    BCLR.B   #2,CRA         ;  address DDRA
         MOVE.B   #$80,DDRA      ;  Configure PORT A
         BSET.B   #2,CRA         ;  Address PORT A
START    MOVE.B   PORTA,D0       ;  Read switch
         ROR.B    #1,D0          ;  Rotate switch status
         MOVE.B   D0,PORTA       ;  Output to LED
         JMP      START          ;  Repeat
```

- ***68000/6821 Microcomputer C++ program for Switch and LED***

```
main()
{
   char *porta, *ddra, *cra;
   porta=0x1001;
   ddra=0x1001;
   cra=0x1003;
   *cra=0;                       /* Address DDRA */
   *ddra=0x80;                   /* Configure Port A */
   *cra=4;                       /* Address Port A */
   while (1)
      *porta=*porta <<7;         /* Read switch and send to LED */
}
```

The C++ compiler will generate more machine codes for the above program compared to the equivalent assembly program. Note that the C++ program is not 100% portable while using I/O. However, it is easier to write programs using C++ than assembly language.

10.12.2 68000 Interrupt System

The 68000 interrupt I/O can be divided into two types: external interrupts and internal interrupts.

10.12.2.1 External Interrupts

The 68000 provides seven levels of external interrupts, 0 through 7. The external hardware provides an interrupt level using the pins $\overline{IPL0}$, $\overline{IPL1}$, and $\overline{IPL2}$. Like other microprocessors, the 68000 checks for and accepts interrupts only between instructions. It compares the value of inverted $\overline{IPL0}$–$\overline{IPL2}$ with the current interrupt mask contained in the bits 10, 9, and 8 of the status register.

If the value of the inverted $\overline{IPL0}$–$\overline{IPL2}$ is greater than the value of the current interrupt mask, then the 68000 acknowledges the interrupt and initiates interrupt processing. Otherwise, the 68000 continues with the current interrupt. Interrupt request level 0 ($\overline{IPL0}$–$\overline{IPL2}$ all HIGH) indicates that no interrupt service is requested. An inverted $\overline{IPL2}$, $\overline{IPL1}$, $\overline{IPL0}$ of 7 is always acknowledged. Therefore, interrupt level 7 is "nonmaskable." Note that the interrupt level is indicated by the interrupt mask bits (inverted $\overline{IPL2}$, $\overline{IPL1}$, $\overline{IPL0}$).

To ensure that an interrupt will be recognized, the following interrupting rules should be considered:

1. The incoming interrupt request level must have a higher priority level than the mask level set in the interrupt mask bits (except for level 7, which is always recognized).
2. The $\overline{IPL2}$–$\overline{IPL0}$ pins must be held at the interrupt request level until the 68000 acknowledges the interrupt by initiating an interrupt acknowledge ($\overline{IACK}$) bus cycle

Interrupt level 7 is edge-triggered. On the other hand, interrupt levels 1–6 are level sensitive. However, as soon as one of them is acknowledged, the processor updates its interrupt mask at the same level.

The 68000 does not have any EI (enable interrupt) or DI (disable interrupt) instructions. Instead, the level indicated by I2 I1 I0 in the SR disables all interrupts below or equal to this value and enables all interrupts above. For example, if I2 I1 I0 = 100, then interrupt levels 1–4 are disabled and 5–7 are enabled. Note that I2 I1 I0 = 000 enables all interrupts and I2 I1 I0 = 111 disables all interrupts except level 7 (nonmaskable).

Once the 68000 has decided to acknowledge an interrupt, it performs several steps:

1. Makes an internal copy of the current status register.
2. Updates the priority mask and address lines A_3–A_1 with the level of the interrupt recognized (inverted $\overline{IPL}$ pins) and then asserts $\overline{AS}$ to inform the external devices that A_1–A_3 has the interrupt level.
3. Enters the supervisor state by setting the S bit in SR to 1.
4. Clears the T bit in SR to inhibit tracing.
5. Pushes the program counter (PC) onto the system stack.
6. Pushes the internal copy of the old SR onto the system stack.

7. Runs an $\overline{\text{IACK}}$ bus cycle for vector number acquisition (to provide the address of the service routine).
8. Multiplies the 8-bit interrupt vector by 4. This points to the location that contains the starting address of the interrupt service routine.
9. Jumps to the interrupt service routine.
10. The last instruction of the service routine should be RTE, which restores the original status word and program counter by popping them from the stack.

External logic can respond to the interrupt acknowledge in one of three ways: by requesting automatic vectoring (autovector), by placing a vector number on the data bus (nonautovector), or by indicating that no device is responding (spurious interrupt).

Autovector (address vectors predefined by Motorola)

If the hardware asserts $\overline{\text{VPA}}$ to terminate the $\overline{\text{IACK}}$ bus cycle, the 68000 directs itself automatically to the proper interrupt vector corresponding to the current interrupt level. No external hardware is inquired for providing the interrupt address vector.

	I2	*I1*	*I0*
Level 1 ← Interrupt vector \$19 for	0	0	1
Level 2 ← Interrupt vector \$1A for	0	1	0
Level 3 ← Interrupt vector \$1B for	0	1	1
Level 4 ← Interrupt vector \$1C for	1	0	0
Level 5 ← Interrupt vector \$1D for	1	0	1
Level 6 ← Interrupt vector \$1E for	1	1	0
Level 7 ← Interrupt vector \$1F for	1	1	1

Nonautovector (user-definable address vectors via external hardware)

The interrupting device uses external hardware to place a vector number on data lines D_0–D_7 and then performs a $\overline{\text{DTACK}}$ handshake to terminate the $\overline{\text{IACK}}$ bus cycle. The vector numbers allowed are \$40 to \$FF, but Motorola has not implemented a protection on the first 64 entries so that user-interrupt may overlap at the discretion of the system designer.

Spurious Interrupt

Another way to terminate an interrupt acknowledge bus cycle is with the $\overline{\text{BERR}}$ (bus error) signal. Even though the interrupt control pins are synchronized to enhance noise immunity, it is possible that external system interrupt circuitry may initiate an $\overline{\text{IACK}}$ bus cycle as a result of noise. Because no device is requesting interrupt service, neither $\overline{\text{DTACK}}$ nor $\overline{\text{VPA}}$ will be asserted to signal the end of the nonexisting $\overline{\text{IACK}}$ bus cycle. When there is no response to an $\overline{\text{IACK}}$ bus cycle after a specified period of time (monitored by the user using an external timer), $\overline{BERR}$ can be asserted by an external timer. This indicates to the processor that it has recognized a spurious interrupt. The 68000 provides 18H as the vector to fetch for the starting address of this exception-handling routine.

It should be pointed out that the spurious interrupt and bus error interrupt due to a troubled instruction cycle (when no $\overline{DTACK}$ is received by the 68000) have two different interrupt vectors. Spurious interrupt occurs when the $\overline{BERR}$ pin is asserted during interrupt processing.

10.12.2.2 Internal Interrupts

The internal interrupt is a software interrupt. This interrupt is generated when the 68000 executes a software interrupt instruction (TRAP) or by some undesirable events such as division by zero or execution of an illegal instruction.

68000 Interrupt Map

The 68000 uses an 8-bit vector n to obtain the interrupt address vector. The 68000 reads the long-word located at memory 4* n. This long word is the starting address of the service routine. Figure 10.21 shows an interrupt map of the 68000. Vector addresses $00 through $2E (not shown in the figure) include vector addresses for reset, bus error, trace, divide by 0, and so on, and addresses $30 through $5C are unassigned. The RESET vector requires four words (addresses 0, 2, 4, and 6); the other vectors require only two words. After hardware reset, the 68000 loads the supervisor SP high and low words, respectively, from addresses 000000_{16} and 000002_{16}, and the PC high and low words, respectively, from 000004_{16} and 000006_{16}. The typical assembler directive DC (define constant) can be used to load the PC and Supervisor SP. For example, the following will load A7′ with $16F128 and PC with $781624:

```
ORG    $000000
DC.L   $0016F128
DC.L   $00781624
```

Vector Address		Vector Number
\$60, \$62	Spurious interrupt	\$18
\$64, \$66	Autovector 1	\$19
\$68, \$6A	Autovector 2	\$1A
\$6C, \$6E	Autovector 3	\$1B
\$70, \$72	Autovector 4	\$1C
\$74, \$76	Autovector 5	\$1D
\$78, \$7A	Autovector 6	\$1E
\$7C, \$7E	Autovector 7	\$1F
\$80 to \$BC	TRAP instructions	\$20 to \$2F
\$C0 to \$FC	Unassigned	\$30 to \$3F
\$100 to \$3FC	User interrupts (nonautovector)	\$40 to \$FF

FIGURE 10.21 68000 interrupt map

68000 Interrupt Address Vector

Suppose that the user decides to write a service routine starting at location \$123456 using autovector 1. Because the autovector 1 address is \$000064 and \$000066, the numbers \$0012 and \$3456 must be stored in locations \$000064 and \$000066, respectively. Note that from Figure 10.21, n = \$19 for autovector 1. Hence, the starting address of the service routine is obtained from the contents of the address 4 x \$19 = \$000064.

An Example of Autovector and Nonautovector Interrupts

As an example to illustrate the concept of autovector and nonautovector interrupts, consider Figure 10.22. In this figure, I/O device 1 uses nonautovector and I/O device 2 uses autovector interrupts. The system is capable of handling interrupts from eight devices because an 8-to-3 priority encoder such as the 74LS148 is used.

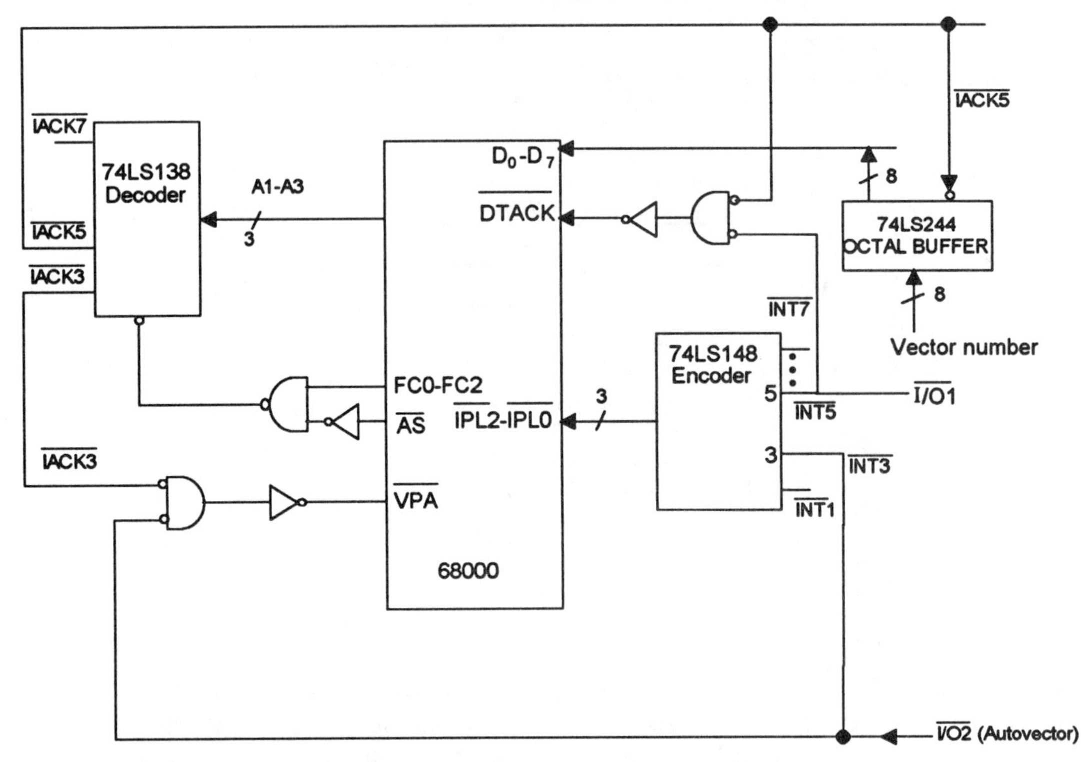

FIGURE 10.22 Autovector and nonautovector interrupts

Suppose that I/0 device 2 drives $\overline{\text{I/O2}}$ LOW in order to activate line 3 of this encoder. This, in turn, interrupts the processor. When the 68000 decides to acknowledge the interrupt, it drives FC0–FC2 HIGH. The interrupt level is reflected on A_1–A_3 when $\overline{\text{AS}}$ is activated by the 68000. The $\overline{\text{IACK3}}$ and $\overline{\text{I/O2}}$ signals are used to generate $\overline{\text{VPA}}$. Once $\overline{\text{VPA}}$ is asserted, the 68000 obtains the interrupt vector address using autovectoring.

In the case of $\overline{\text{I/O1}}$, line 5 of the priority encoder is activated to initiate the interrupt. By using appropriate logic, $\overline{\text{DTACK}}$ is asserted using $\overline{\text{IACK5}}$ and $\overline{\text{I/O1}}$. The vector number is placed on D_0–D_7 by enabling an octal buffer such as the 74LS244 using $\overline{\text{IACK5}}$. The 68000 inputs this vector number and multiplies it by 4 to obtain the interrupt address vector.

Interfacing a Typical A/D Converter to the 68000 Using Autovector and Nonautovector Interrupts

Figure 10.23 shows the interfacing of a typical A/D converter to the 68000-based microcomputer using the autovector interrupt. In the figure, the A/D converter can be started by sending a START pulse. The signal can be connected to line 4 (for example) of the encoder.

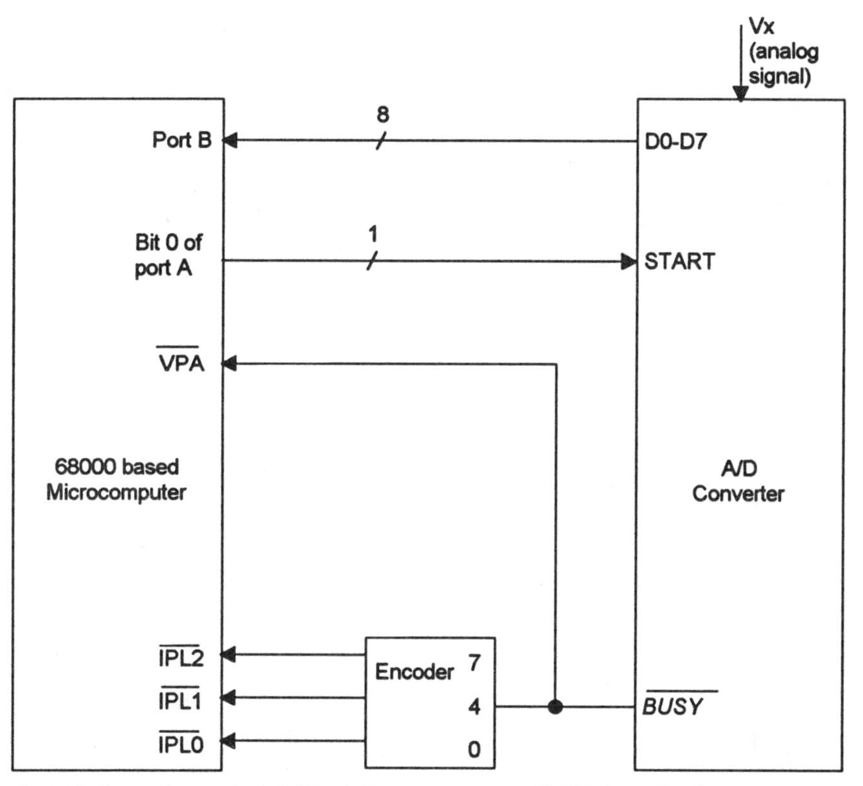

FIGURE 10.23 Interfacing of a typical 8-bit A/D converter to 68000-based microcomputer using autovector interrupt

Note that line 4 is 100_2 for $\overline{IPL2}$, $\overline{IPL1}$, $\overline{IPL0}$, which is a level 3 (inverted 100_2) interrupt. $\overline{BUSY}$ can be used to assert $\overline{VPA}$ so that, after acknowledgment of the interrupt, the 68000 will service the interrupt as a level 3 autovector interrupt. Note that the encoder in Figure 10.23 is used for illustrative purposes. This encoder is not required for a single device such as the A/D converter in the example.

Figure 10.24 shows the interfacing of a typical A/D converter to the 68000-based microcomputer using the nonautovector interrupt. In the figure, the 68000 starts the A/D converter as before. Also, the $\overline{BUSY}$ signal is used to interrupt the microcomputer using line 5 ($\overline{IPL2}$, $\overline{IPL1}$, $\overline{IPL0}$= 101, which is a level 2 interrupt) of the encoder. $\overline{BUSY}$ can be used to assert $\overline{DTACK}$ so that, after acknowledgment of the interrupt, FC2, FC1, FC0 will become 111_2, which can be NANDed to enable an octal buffer such as the 74LS244 in order to transfer an 8-bit vector from the input of the buffer to the D_0–D_7 lines of the 68000. The 68000 can then multiply this vector by 4 to determine the interrupt address vector. As before, the encoder in Figure 10.24 is not required for the single A/D converter.

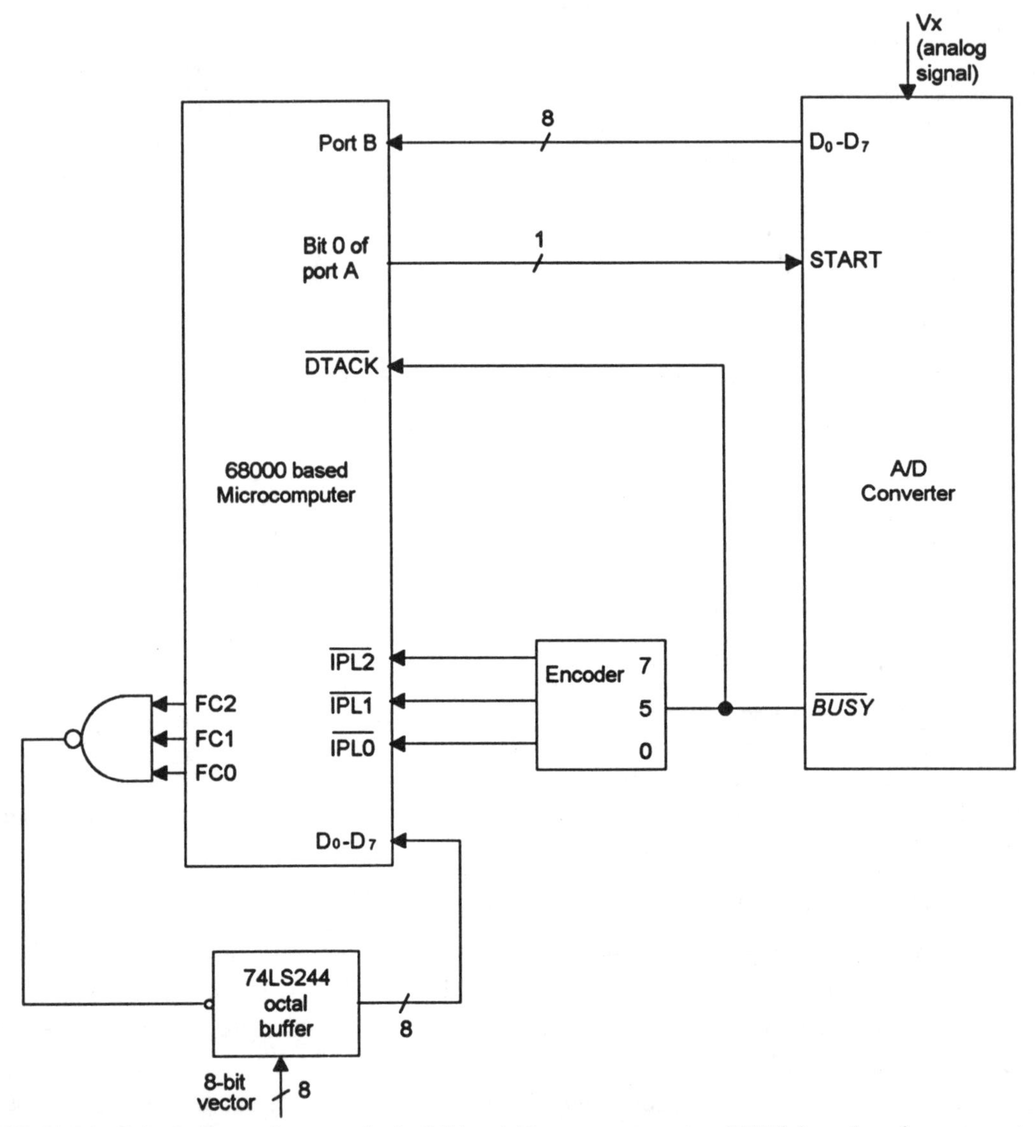

FIGURE 10.24 Interfacing of a typical 8-bit A/D converter to 68000-based microcomputer using nonautovector interrupt

10.12.3 68000 DMA

Three DMA control lines are provided with the 68000. These are $\overline{BR}$ (bus request), $\overline{\text{BG}}$ (bus grant), and $\overline{\text{BGACK}}$ (bus grant acknowledge). The $\overline{\text{BR}}$ line is an input to the 68000. The external device activates this line to tell the 68000 to release the system bus. At least one clock period after receiving $\overline{\text{BR}}$, the 68000 will enable its $\overline{\text{BG}}$ output line to acknowledge the

DMA request. However, the 68000 will not relinquish the bus until it has completed the current instruction cycle. The external device must check the $\overline{AS}$ (address strobe) line to determine the completion of the instruction cycle by the 68000. When $\overline{AS}$ becomes HIGH, the 68000 will tristate its address and data lines and will give up the bus to the external device. After taking over the bus, the external device must enable the $\overline{BGACK}$ line. The $\overline{BGACK}$ line tells the 68000 and other devices connected to the bus that the bus is being used. The 68000 stays in a tristate condition until $\overline{BGACK}$ becomes HIGH.

10.13 68000 Exception Handling

A 16-bit microcomputer is usually capable of handling unusual or exceptional conditions. These conditions include situations such as execution of illegal instruction or division by zero. In this section, the exception-handling capabilities of the 68000 are described.

The 68000 exceptions can be divided into three groups, namely, groups 0, 1, and 2. Group 0 has the highest priority, and group 2 has the lowest priority. Within each group, there are additional priority levels. A list of 68000 exceptions along with individual priorities is as follows:

Group 0	Reset (highest level in this group), address error (next level), and bus error (lowest level)
Group 1	Trace (highest level), interrupt (next level), illegal op-code (next level), and privilege violation (lowest level)
Group 2	TRAP, TRAPV, CHK, and ZERO DIVIDE (no individual priorities assigned in group 2)

Exceptions from group 0 always override an active exception from group 1 or group 2.

Group 0 exception processing begins at the completion of the current bus cycle (2 clock cycles). Note that the number of cycles required for a READ or WRITE operation is called a "bus cycle." This means that during an instruction fetch if there is a group 0 interrupt, the 68000 will complete the instruction fetch and then service the interrupt. Group 1 exception processing begins at the completion of the current instruction. Group 2 exceptions are initiated through execution of an instruction. Therefore, there are no individual priority levels within group 2. Exception processing occurs when a group 2 interrupt is encountered, provided there are no group 0 or group 1 interrupts.

When an exception occurs, the 68000 saves the contents of the program counter and status register onto the stack and then executes a new program whose address is provided by the exception vectors. Once this program is executed, the 68000 returns to the main program using the stored values of program counter and status register.

Exceptions can be of two types: internal or external. The internal exceptions are generated by situations such as division by zero, execution of illegal or unimplemented instructions, and address error. As mentioned before, internal interrupts are called "traps." The external exceptions are generated by bus error, reset, or interrupt instructions. The basic concepts associated with interrupts, relating them to the 68000, have already been described. In this section, we will discuss the other exceptions.

In response to an exceptional condition, the processor executes a user-written program. In some microcomputers, one common program is provided for all exceptions. The beginning section of the program determines the cause of the exception and then branches to the appropriate routine. The 68000 utilizes a more general approach. Each exception can be handled by a separate program.

As mentioned before, the 68000 has two modes of operation: user state and supervisor state. The operating system runs in supervisor mode, and all other programs are executed in user mode. The supervisor state is therefore more privileged. Several privileged instructions such as `MOVE` to SR can be executed only in supervisor mode. Any attempt to execute them in user mode causes a trap.

We will now discuss how the 68000 handles exceptions caused by external resets, trap instructions, bus and address errors, tracing , execution of privileged instructions in user mode, and execution of illegal/unimplemented instructions:

- The reset exception is generated externally. In response to this exception, the 68000 automatically loads the initial starting address into the processor.
- The 68000 has a `TRAP` instruction, which always causes an exception. The operand for this instruction varies from 0 to 15. This means that there are 16 `TRAP` instructions. Each `TRAP` instruction has an exception vector. `TRAP` instructions are normally used to call subroutines in an operating system. Note that this automatically places the 68000 in supervisor state. TRAPs can also be used for inserting breakpoints in a program. Two other 68000 instructions cause traps if a particular condition is true: `TRAPV` and `CHK`. `TRAPV` generates an exception if the overflow flag is set. The `TRAPV` instruction can be inserted after every arithmetic operation in a program in order to cause a trap whenever there is the possibility of an overflow. A routine can be written at the vector address for the `TRAPV` to indicate to the user that an overflow has occurred. The `CHK` instruction is designed to ensure that access to an array in memory is within the range specified by the user. If there is a violation of this range, the 68000 generates an exception.
- A bus error occurs when the 68000 tries to access an address that does not belong to the devices connected to the bus. This error can be detected by asserting the $\overline{\text{BERR}}$ pin on the 68000 chip by an external timer when no $\overline{DTACK}$ is received from the

device after a certain period of time. In response to this, the 68000 executes a user-written routine located at an address obtained from the exception vectors. An address error, on the other hand, occurs when the 68000 tries to read or write a word (16 bits) or long word (32 bits) in an odd address. This address error has a different exception vector from the bus error.

- The trace exception in the 68000 can be generated by setting the trace bit in the status register. In response to the trace exception, the 68000 causes an internal exception after execution of every instruction. The user can write a routine at the exception vectors for the trace instruction to display register and memory contents. The trace exception provides the 68000 with the single-stepping debugging feature.
- As mentioned before, the 68000 has privileged instructions, which must be executed in supervisor mode. An attempt to execute these instructions causes privilege violation.
- Finally, the 68000 causes an exception when it tries to execute an illegal or unimplemented instruction.

10.14 68000/2716-1/6116/6821-Based Microcomputer

Figure 10.25 shows the schematic of a 68000-based microcomputer with a 4K EPROM, a 4K static RAM, and four 8-bit I/O ports. Let us explain the various sections of the hardware schematic. Two 2716-1 and two 6116 chips are required to obtain the 4K EPROM and 4K RAM. The $\overline{\text{LDS}}$ and $\overline{\text{UDS}}$ pins are ORed with the memory select signal to enable the chip selects for the EPROMs and the RAMs. Address decoding is accomplished by using a 3 × 8 decoder. The decoder enables the memory or the I/O chips depending on the status of address lines A_{12}–A_{14} and the $\overline{\text{AS}}$ line of the 68000. $\overline{\text{AS}}$ is used to enable the decoder. $\overline{I_0}$ selects the EPROMs, $\overline{I_1}$ selects the RAMs, and $\overline{I_2}$ selects the I/O ports.

When addressing memory chips, the $\overline{\text{DTACK}}$ input of the 68000 must be asserted for data acknowledge. The 68000 clock in the hardware schematic is 8 MHz. Therefore, each clock cycle is 125 ns. In Figure 10.25, $\overline{\text{AS}}$ is used to enable the 3 × 8 decoder. The outputs of the decoder are gated to assert 68000 $\overline{\text{DTACK}}$. This means that $\overline{\text{AS}}$ is indirectly used to assert $\overline{\text{DTACK}}$. From the 68000 read timing diagram, $\overline{\text{AS}}$ goes to LOW after approximately 2 cycles (250 ns for the 8-MHz clock) from the beginning of the bus cycle. With no wait states, the 68000 samples $\overline{\text{DTACK}}$ at the falling edge of S4 (375 ns) and, if $\overline{\text{DTACK}}$ is recognized, the 68000 latches data at the falling edge of S6 (500 ns). If $\overline{\text{DTACK}}$ is not recognized at the falling edge of S4, the 68000 inserts a 1-cycle (125 ns in this case) wait state, samples $\overline{\text{DTACK}}$ at the end of S6, and, if $\overline{\text{DTACK}}$ is recognized, latches data at the end of S8 (625 ns), and the process continues. Because the access time of the 2716-1 is 350 ns, $\overline{\text{DTACK}}$ recognition by the 68000 at the falling edge of S6 (500 ns) and latching of data at the falling edge of S8 (625 ns) will satisfy the timing requirement. This means that the decoder output $\overline{I_0}$ for EPROM select must go LOW at the end of S6. Therefore, $\overline{DTACK}$ must be delayed by 250 ns (i.e., 2 cycles).

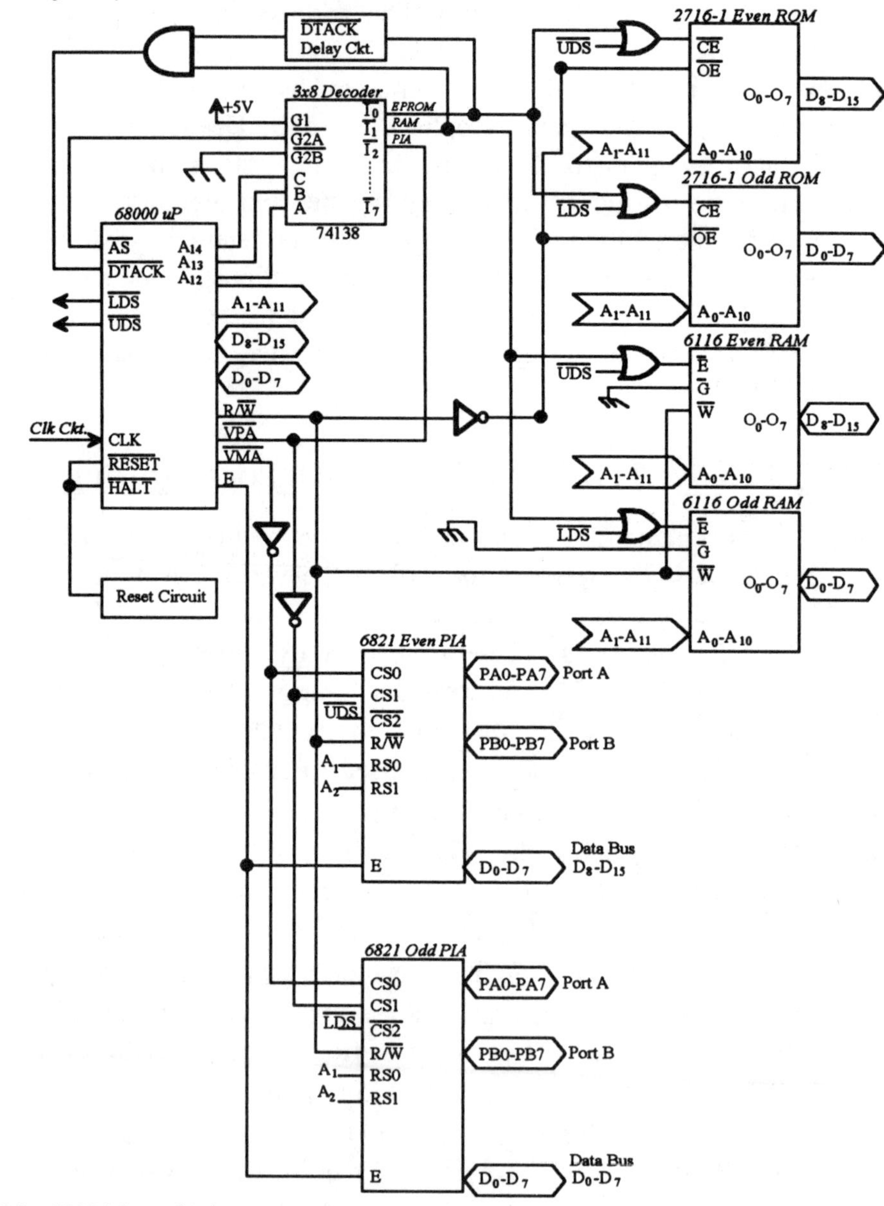

FIGURE 10.25 68000-based microcomputer

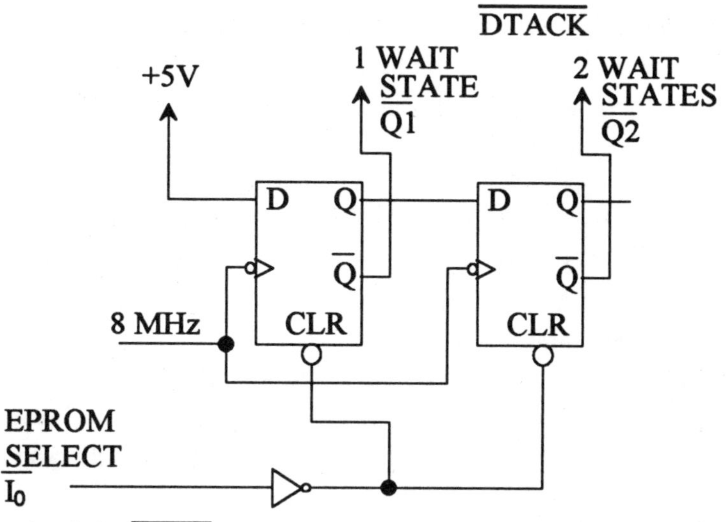

FIGURE 10.26 Delay circuit for DTACK

A delay circuit, as shown in Figure 10.26, is designed using two D flip-flops. EPPOM select activates the delay circuit. The input is then shifted right 2 bits to obtain a 2-cycle wait state to allow sufficient time for data transfer. DTACK assertion and recognition are delayed by 2 cycles during data transfer with EPROMs. Figure 10.27 shows the timing diagram for the DTACK delay circuit. Note that DTACK goes to Low after about 2 cycles if asserted by AS providing erronous result. Therefore, DTACK must be delayed.

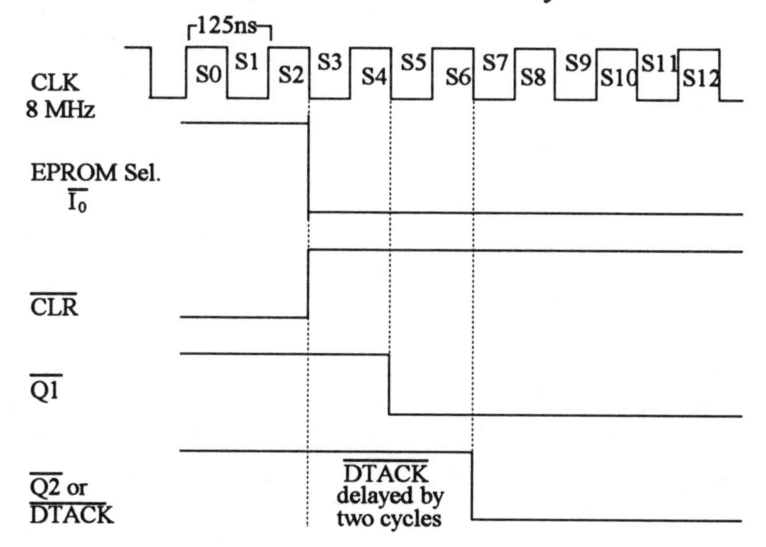

FIGURE 10.27 Timing diagram for the DTACK delay circuit.

When the EPROM is not selected by the decoder, the clear pin is asserted (output of inverter), so Q is forced LOW and $\overline{Q}$ is HIGH. Therefore, $\overline{DTACK}$ is not asserted. When the processor selects the EPROMs, then the output of the inverter is HIGH, so the clear pin is not asserted. The D flip-flop will accept a high at the input, and Q2 will be HIGH and $\overline{Q2}$ will be LOW. Now that $\overline{Q2}$ is LOW, it can assert $\overline{DTACK}$. $\overline{Q1}$ will provide one wait cycle and $\overline{Q2}$ will provide two wait cycles. Because the 2761-1 EPROM has a 350-ns access time and the microprocessor is operating at 8 MHz (125-ns clock cycle), two wait cycles are inserted before asserting $\overline{DTACK}$ ($2 \times 125 = 250$ ns). Therefore, $\overline{Q2}$ can be connected to the $\overline{DTACK}$ pin through an AND gate. No wait state is required for RAMs because the access time for the RAMs is only 120 nanoseconds.

Four 8-bit I/O ports are obtained by using two 6821 chips. When the I/O ports are selected, the $\overline{VPA}$ pin is asserted instead of $\overline{DTACK}$. This will acknowledge to the 68000 that it is addressing a 6800-type peripheral. In response, the 68000 will synchronize all data transfer with the E clock.

The memory and I/O maps for the schematic are as follows:

- *Memory Maps (all numbers in hex)* . A_{23} - A_{15} are don't cares and assumed to be 0's.

A_{23}–A_{15}	A_{14}	A_{13}	A_{12}	A_{11}–A_1	A_0 ($\overline{LDS}$ or $\overline{UDS}$)	
0–0	0	0	0	0–0	0	EPROM(even) = 2K
				:		
0–0	0	0	0	1–1	0	\$000000, \$000002, \$000004, ... , \$000FFE
				:		
0–0	0	0	0	0–0	1	EPROM(odd) = 2K
				:		
0–0	0	0	0	1–1	1	\$000001, \$000003, \$000005, ... , \$000FFF
				:		
0–0	0	0	1	0–0	0	RAM(even) = 2K
				:		
0–0	0	0	1	1–1	0	\$001000, \$001002, ... , \$001FFE
				:		
0–0	0	0	1	0–0	1	RAM(odd) = 2K
				:		
0–0	0	0	1	1–1	1	\$001001, \$001003, ... , \$001FFF

Note that, upon hardware reset, the 68000 loads the supervisor SP high and low words, respectively, from addresses \$000000 and \$000002 and the PC high and low words, respectively, from locations \$000004 and \$000006. The memory map contains these reset vector addresses in the even and odd 2716-1 chips.

- *Memory Mapped I/O (all numbers in hex).* A_{23}-A_{15} and A_{11}-A_3 are don't cares and assumed to be 0's.

					RS1	RS0	$\overline{UDS}$ or $\overline{LDS}$	
A_{23}–A_{15}	A_{14}	A_{13}	A_{12}	A_{11}–A_3	A_2	A_1	A_0	Register Selected (Address) — Even
0–0	0	1	0	0–0	0	0	0	Port A or DDRA = $002000
0–0	0	1	0	0–0	0	1	0	CRA = $002002
0–0	0	1	0	0–0	1	0	0	Port B or DDRB = $002004
0–0	0	1	0	0–0	1	1	0	CRB = $002006
								Register Selected (Address) — Odd
0–0	0	1	0	0–0	0	0	1	Port A or DDRA = $002001
0–0	0	1	0	0–0	0	1	1	CRA = $002003
0–0	0	1	0	0–0	1	0	1	Port B or DDRB = $002005
0–0	0	1	0	0–0	1	1	1	CRB = $002007

10.15 Multiprocessing with the 68000 Using the TAS Instruction and the $\overline{AS}$ Signal

Earlier, the 68000 TAS instruction was discussed. The TAS instruction supports the software aspects of interfacing two or more 68000's via shared RAM. When TAS is executed, the 68000 $\overline{AS}$ pin stays low. During both the read and write portions of the cycle, $\overline{AS}$ remains LOW and the cycle starts as the normal read cycle. However, in the normal read, $\overline{AS}$ going inactive indicates the end of the read. During execution of TAS, $\overline{AS}$ stays LOW throughout the cycle, so $\overline{AS}$ can be used in the design as a bus-locking circuit. Due to the bus locking, only one processor at a time can perform a TAS operation in a multiprocessor system.

The TAS instruction supports multiprocessor operations (globally shared resources) by checking a resource for availability and reserving or locking it for use by a single processor.

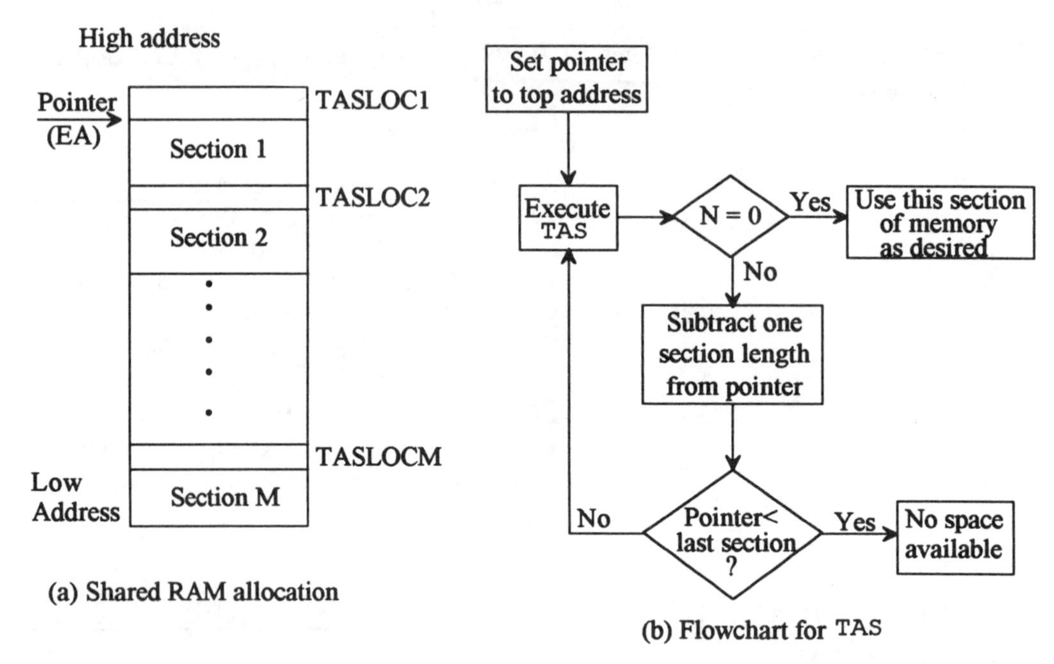

FIGURE 10.28 Memory allocation using TAS

The TAS instruction can, therefore, be used to allocate free memory spaces . The TAS instruction execution flowchart for allocating memory is shown in Figure 10.28. The shared RAM of the Figure 10.28 is divided into *M* sections. The first byte of each section will be pointed to by (EA) of the TAS (EA) instruction. In the flowchart of Figure 10.28, (EA) first points to the first byte of section 1. The instruction TAS (EA) is then executed. The TAS instruction checks the most significant bit (N bit) in (EA). N = 0 indicates that section 1 is free; N = 1 means section 1 is busy. If N = 0, then section 1 will be allocated for use. If N = 1 (section 1 is busy), then a program will be written to subtract one section length from (EA) to check the next section for availability. Also, (EA) must be checked with the value TASLOCM. If (EA) < TASLOCM, then no space is available for allocation. However, if (EA) > TASLOCM, then TAS is executed and the availability of that section is determined.

In a multiprocessor environment, the TAS instruction provides software support for interfacing two or more 68000's via shared RAM. The $\overline{AS}$ signal can be used to provide the bus-locking mechanism.

Example 10.18

Assume that the 68000/2732/6116/6821 microcomputer shown in Figure 10.29 is required to perform the following:

(a) If Vx > Vy , turn the LED ON if the switch is open; otherwise turn the LED OFF. Write a 68000 assembly language program starting at address $000300 to accomplish the above by inputting the comparator output via bit 0 of Port B. Use Port A address = $002000, Port B address = $002004, CRA = $002002, CRB = $002006.

(b) Repeat part (a) using autovector level 7 and nonautovector (Vector $40). Use Port A (address $002000) for LED and switch as above. Assume supervisor mode. Write the main program and service routine in 68000 assembly language starting at addresses $000300 and $000A00 respectively. Also, initialize the supervisor stack pointer at $001200.

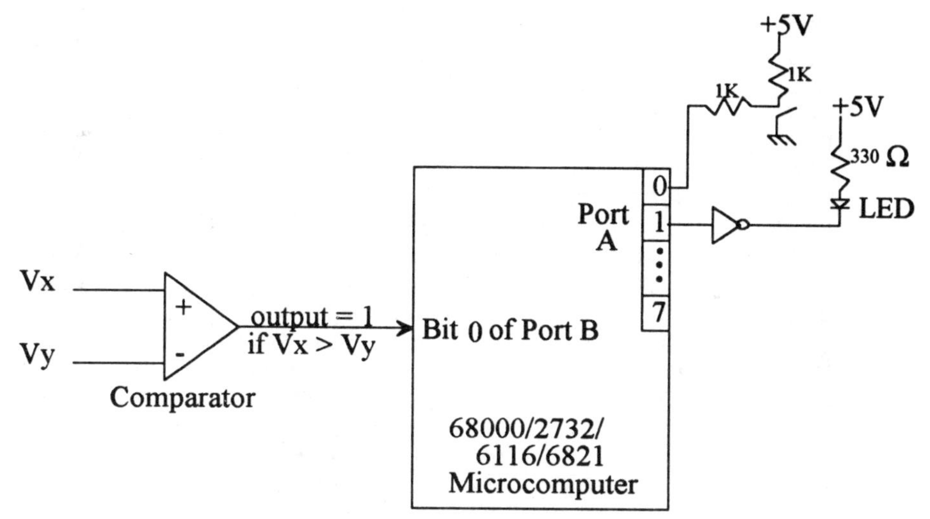

FIGURE 10.29 Figure for Example 10.18

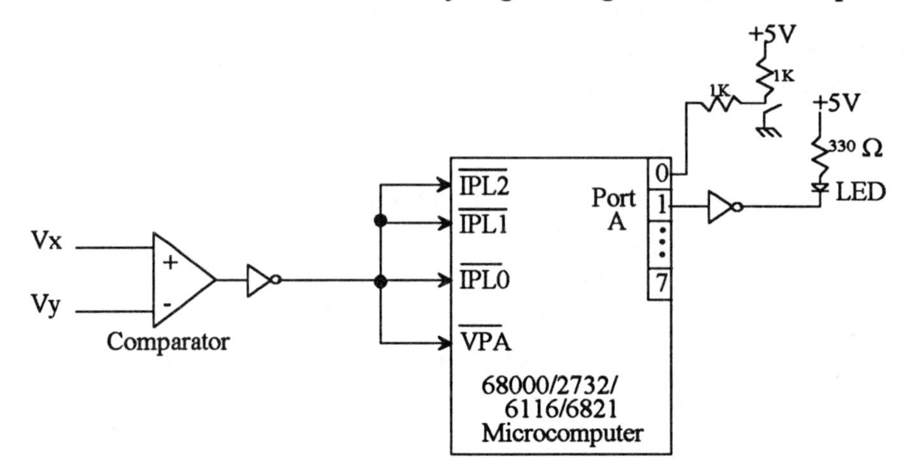

FIGURE 10.30 Example 10.18 using autovectors

Solution

(a)
```
CRA     EQU     $002002
CRB     EQU     $002006
PORTA   EQU     $002000
DDRA    EQU     PORTA
PORTB   EQU     $002004
DDRB    EQU     PORTB
        ORG     $000300
        BCLR.B  #2,CRA     ; Address DDRA
        MOVE.B  #2,DDRA    ; Configure PORTA
        BSET.B  #2,CRA     ; Address PORTA
        BCLR.B  #2,CRB     ; Address DDRB
        MOVE.B  #0,DDRB    ; Configure PORTB
        BSET.B  #2,CRB     ; Address PORTB
COMP    MOVE.B  PORTB,D0   ; Input PORTB
        LSR.B   #1,D0      ; Check
        BCC     COMP       ; Comparator
        MOVE.B  PORTA,D1   ; Input switch
        LSL.B   #1,D1      ; Align LED data
        MOVE.B  D1,PORTA   ; Output to LED
LED     JMP     LED
```

(b) ***Using Autovector Level 7 (nonmaskable interrupt)***

Figure 10.30 shows the pertinent connections for Autovector Level 7 interrupt.

Main Program

```
CRA     EQU     $002002
PORTA   EQU     $002000
DDRA    EQU     PORTA
        ORG     $000300
        BCLR.B  #2,CRA     ; Address DDRA
        MOVE.B  #2,DDRA    ; Configure PORTA
```

```
            BSET.B    #2,CRA       ;  Address PORTA
   WAIT     JMP       WAIT         ;  Wait for interrupts
   FINISH   JMP       FINISH       ;  Halt
```

Service Routine

```
            ORG       $00000A00
            MOVE.B    PORTA, D1    ;  Input switch
            LSL.B     #1, D1       ;  Align LED data
            MOVE.B    D1, PORTA    ;  Output to LED
            RTE
```

Reset Vector

```
            ORG       0
            DC.L      $00001200
            DC.L      $00000300
```

Service Routine Vector

```
            ORG       $0000007C
            DC.L      $00000A00
```

Using Nonautovectoring (vector $40)

Figure 10.31 shows the pertinent connections for nonautovectoring interrupt.

Main Program

```
   CRA      EQU       $002002
   PORTA    EQU       $002000
   DDRA     EQU       PORTA
            ORG       $000300
            BCLR.B    #2,CRA       ;  Address DDRA
            MOVE.B    #2,DDRA      ;  Configure PORTA
```

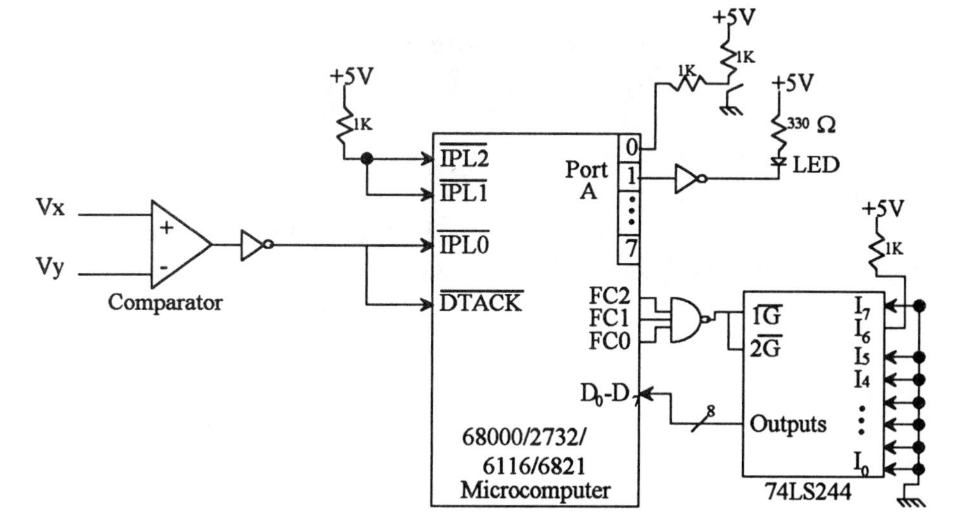

FIGURE 10.31 Example 10.18 using nonautovectors

```
            BSET.B    #2,CRA        ;  Address PORTA
            ANDI.W    #$0F8FF,SR    ;  Enable interrupts
   WAIT     JMP       WAIT          ;  Wait for interrupt
   FINISH   JMP       FINISH        ;  Halt
```

Service Routine

```
            ORG       $000A00
            MOVE.B    PORTA,D1      ;  Input switch
            LSL.B     #$01,D1       ;  Align LED data
            MOVE.B    D1,PORTA      ;  Output to LED
            RTE
```

Reset Vector

```
            ORG       0
            DC.L      $00001200
            DC.L      $00000300
```

Service Routine Vector

```
            ORG       $00000100
            DC.L      $00000A00
```

10.16 Overview of Motorola 32- and 64-bit Microprocessors

This section provides an overview of the state-of-the-art in Motorola's microprocessors. Motorola's 32-bit microprocessors based on 68HC000 architecture include the MC68020, MC68030, MC68040, and MC68060. Table 10.17 compares the basic features of some of these microprocessors with the 68HC000.

The PowerPC family of microprocessors were jointly developed by Motorola, IBM, and Apple. The PowerPC family contains both 32- and 64-bit microprocessors. One of the noteworthy feature of the PowerPC is that it is the first top-of-the-line microprocessor to include an on-chip real-time clock (RTC). The RTC is common in single-chip microcomputers rather than microprocessors. The PowerPC is the first microprocessor to implement this on-chip feature, which makes it easier to satisfy the requirements of time-keeping for task switching and calendar date of modern multitasking operating systems. The PowerPC microprocessor supports both the Power Mac and standard PCs. The PowerPC family is designed using RISC architecture

TABLE 10.17 Motorola MC68HC000 vs. MC68020/68030/68040

	MC68HC000	MC68020	MC68030	MC68040
Comparable Clock Speed	33MHz (4MHz min)*	33 MHz (8 MHz min.)*	33 MHz (8 MHz min.)*	33 MHz (8 MHz min.)*
Pins	64, 68	114	118	118
Address Bus	24-bit	32-bit	32-bit	32-bit
Addressing Modes	14	18	18	18
Maximum Memory	16 Megabytes	4 Gigabytes	4 Gigabytes	4 Gigabytes
Memory Management	NO	By interfacing the 68851 MMU chip	On-chip MMU	On-chip MMU
Cache (on chip)	NO	Instruction cache	Instruction and data cache	Instruction and data cache
Floating Point	NO	By interfacing 68881/68882 floating-point coprocessor chip	By interfacing 68881/68882 floating-point coprocessor chip	On-chip floating point hardware
Total Instructions	56	101	103	103 plus floating-point instructions
ALU size	One 16-bit ALU	Three 32-bit ALU's	Three 32-bit ALU's	Three 32-bit ALU's

*Higher clock speeds available

10.16.1 Motorola MC68020

The MC68020 is Motorola's first 32-bit microprocessor. The design of the 68020 is based on the 68HC000. The 68020 can perform a normal read or write cycle in 3 clock cycles without wait states as compared to the 68HC000, which completes a read or write operation in 4 clock cycles without wait states. As far as the addressing modes are concerned, the 68020 includes new modes beyond those of the 68HC000. Some of these modes are scaled indexing, larger displacements, and memory indirection. Furthermore, several new instructions are added to the 68020 instruction set, including the following:

- Bit field instructions are provided for manipulating a string of consecutive bits with a variable length from 1 to 32 bits.
- Two new instructions are used to perform conversions between packed BCD and ASCII or EBCDIC digits. Note that a packed BCD is a byte containing two BCD digits.
- Enhanced 68000 array-range checking (`CHK2`) and compare (`CMP2`) instructions are included. `CHK2` includes lower and upper bound checking; `CMP2` compares a number with lower and upper values and affects flags accordingly.

- Two advanced instructions, namely, CALLM and RTM, are included to support modular programming.
- Two compare and swap instructions (CAS and CAS2) are provided to support multi-processor systems.

A comparison of the differences between the 68020 and 68HC000 will be provided later in this section.

The 68030 and 68040 are two enhanced versions of the 68020. The 68030 retains most of the 68020 features. It is a virtual memory microprocessor containing an on-chip MMU (memory management unit). The 68040 expands the 68030 on-chip memory management logic to two units: one for instruction fetch and one for data access. This speeds up the 68040's execution time by performing logical-to-physical-address translation in parallel. The on-chip floating-point capability of the 68040 provides it with both integer and floating-point arithmetic operations at a high speed. All 68HC000 programs written in assembly language in user mode will run on the 68020/68030 or 68040. The 68030 and 68040 support all 68020 instructions except CALLM and RTM. Let us now focus on the 68020 microprocessor in more detail.

10.16.1.1 MC68020 Functional Characteristics

The MC68020 is designed to execute all user object code written for the 68HC000. Like the 68HC000, it is manufactured using HCMOS technology. The 68020 consumes a maximum of 1.75 W. It contains 200,000 transistors on a 3/8" piece of silicon. The chip is packaged in a square (1.345" × 1.345") pin grid array (PGA) and other packages. It contains 169 pins (114 pins used) arranged in a 13 × 13 matrix.

The processor speed of the 68020 can be 12.5, 16.67, 20, 25, or 33 MHz. The chip must be operated from a minimum frequency of 8 MHz. Like the 68HC000, it does not have any on-chip clock generation circuitry. The 68020 contains 18 addressing modes and 101 instructions. All addressing modes and instructions of the 68HC000 are included in the 68020. The 68020 supports coprocessors such as the MC68881/MC68882 floating-point and MC68851 MMU coprocessors.

These and other functional characteristics of the 68020 are compared with the 68HC000 in Table 10.18. Some of the 68020 characteristics in Table 10.18 will now be explained.

- Three independent ALUs are provided for data manipulation and address calculations
- A 32-bit barrel shift register (occupies 7% of silicon) is included in the 68020 for very fast shift operations regardless of the shift count.
- The 68020 has three SPs. In the supervisor mode (when S = 1), two SPs can be accessed. These are MSP (when M = 1) and ISP (when M = 0). ISP can be used to simplify and speed up task switching for operating systems.

TABLE 10.18 Functional Characteristics, MC68HC000 vs. MC68HC020

Characteristic	*68HC000*	*68020*
Technology	HCMOS	HCMOS
Number of pins	64, 68	169 (13 × 13 matrix; pins come out at bottom of chip; 114 pins currently used.)
Control unit	Nanomemory (two-level memory)	Nanomemory (two-level memory)
Clock	6 MHz, 10 MHz, 12.5 MHz, 16.67 MHz, 20 MHz, 25 MHz, 33 MHz (4 MHz minimum requirement).	12.5 MHz, 16.67 MHz, 20 MHz, 25 MHz, 33 MHz (8 MHz minimum requirement).
ALU	One 16-bit ALU	Three 32-bit ALUs
Address bus size	24 bits with A_0 encoded from $\overline{UDS}$ and $\overline{LDS}$.	32 bits with no encoding of A_0 is required.
Data bus size	The 68HC000 can only be configured as 16-bit memory (two 8-bit chips) via D_0-D_7 for odd addresses and D_8-D_{15} for even addresses during byte transfers; for word and long word, uses D_0-D_{15}. The I/O can be configured as byte (one 8-bit word) or 16-bit (two 8-bit words).	The 68020 can be configured as 8-bit memory (a single 8-bit chip) via D_{31}-D_{24} pins or 16-bit memory (two 8-bit chips) via D_{31} - D_{16} pins or 32-bit memory (four 8-bit chips) via D_{31}-D_0 pins. I/O can be configured as 8-bit or 16-bit or 32-bit.
Instructions and data access	All word and long word accesses must be at even addresses for both instructions and data.	Instructions must be accessed at even addresses; data accesses can be at any address.
Instruction cache	None	128K 16-bit word cache. At start of an instruction fetch, the 68020 always outputs LOW on $\overline{ECS}$ (early cycle start) pin and acceses the cache. If instruction is found in the cache, the 68020 inhibits outputting LOW on $\overline{AS}$ pin; otherwise, the 68020 sends LOW on $\overline{AS}$ pin and reads instruction from main memory.
Directly addressable memory	16 megabytes	4 gigabytes (4,294,964,296 bytes)
Registers	8 32-bit data registers 7 32-bit address registers 2 32-bit SPs 1 32-bit PC (24 bits used) 1 16-bit SR	8 32-bit data registers 7 32-bit address registers 3 32-bit SPs 1 32-bit PC (all bits used) 1 16-bit SR 1 32-bit VBR (vector base register) 2 3-bit function code registers (SFC and DFC) 1 32-bit CAAR (cache address register) 1 CACR (cache control register)

TABLE 10.18 Functional Characteristics, MC68HC000 vs. MC68HC020 (continued)

Addressing modes	14	18
Instruction set	56 instructions	101 instructions
Barrel shifter	No	Yes. For fast-shift operations.
Stack pointers	USP, SSP	USP, MSP (master SP), ISP (interrupt SP)
Status register	T, S, I0,I1, I2, X, N, Z, V, C	T0, T1, S, M, I0,I1, I2, X, N, Z, V, C
Coprocessor interface	Emulated in software; that is, by writing subroutines, coprocessor functions such as floating-point arithmetic can be obtained.	Can be directly interfaced to coprocessor chips, and coprocessor functions such as floating-point arithmetic can be obtained via 68020 instructions.
FC0, FC1, FC2 pins	FC0, FC1, FC2 = 111 means interrupt acknowledge.	FC0, FC1, FC2 = 111 means CPU space cycle; then by decoding A16-A19, one can obtain breakpoints, coprocessor functions, and interrupt acknowledge.

- The vector base register (VBR) is used in interrupt vector computation. For example, in the 68HC000, the interrupt vector address is obtained by using VBR + 4 × 8-bit vector.
- The SFC (source function code) and DFC (destination function code) registers are 3 bits wide. These registers allow the supervisor to move data between address spaces. In supervisor mode, 3-bit addresses can be written into SFC or DFC using such instructions such as `MOVEC A2,SFC`. The upper 29 bits of SFC are assumed to be zero. The `MOVES.W(A0),D0` can then be used to move a word from a location within the address space specified by SFC and [A0] to D0. The 68020 outputs [SFC] to the FC2, FC1, and FC0 pins. By decoding these pins via an external decoder, the desired source memory location addressed by [A0] can be accessed.
- The new addressing modes in the 68020 include scaled indexing, 32-bit displacements, and memory indirection. To illustrate the concept of scaling, consider moving the contents of memory location 50_{10} to A1. Using the 68000, the following instruction sequence will accomplish this

```
MOVEA.W #10, A0
MOVE.W #10, D0
ASL #2, D0
MOVEA.L 0 (A0, D0.W), A1
```

The scaled indexing mode can be used with the 68020 to perform the same as follows:

```
MOVEA.W #10, A0
MOVE.W #10, D0
MOVEA.L (0, A0, D0.W * 4), A1
```

Note that [D0] here is scaled by 4. Scaling by 1, 2, 4, or 8 can be obtained.

- The new 68020 instructions include bit field instructions to better support compilers and certain hardware applications such as graphics, 32-bit multiply and divide instructions, pack and unpack instructions for BCD, and coprocessor instructions. Bit field instructions can be used to input A/D converters and eliminate wasting main memory space when the A/D converter is not 32 bits wide. For example, if the A/D is 12 bits wide, then the instruction `BFEEXTU $22320000 {2:13},D0` will input bits 2-13 of memory location $22320000 into D0. Note that $22320000 is the memory-mapped port, where the 12-bit A/D is connected at bits 2–13. The next A/D can be connected at bits 14–25, and so on.
- FC2, FC1, FC0 = 111 means CPU space cycle. The 68020 makes CPU space access for breakpoints, coprocessor operations, or interrupt acknowledge cycles. The CPU space classification is generated by the 68020 based upon execution of breakpoint instructions or coprocessor instructions, or during an interrupt acknowledge cycle. The 68020 then decodes A_{16}–A_{19} to determine the type of CPU space. For example, FC2, FC1, FC0 = 111 and A_{19}, A_{18}, A_{17}, A_{16} = 0010 mean coprocessor instruction.
- For performing floating-point operation, the 68HC000 user must write subroutines using the 68HC000 instruction set. The floating-point capability in the 68020 can be obtained by connecting a floating-point coprocessor chip such as the Motorola 68881. The 68020 has two coprocessor chips: the 68881 (floating point) and the 68851 (memory management). The 68020 can have up to eight coprocessor chips. When a coprocessor is connected to the 68020, the coprocessor instructions are added to the 68020 instruction set automatically, and this is transparent to the user. For example, when the 68881 floating-point coprocessor is added to the 68020, instructions such as `FADD` (floating-point add) are available to the user. The programmer can then execute the instruction `FADD FD0,FD1`. Note that registers FD0 and FD1 are in the 68881. When the 68020 encounters the `FADD` instruction, it writes a command in the command register in the 68881, indicating that the 68881 has to perform this operation. The 68881 then responds to this by writing in the 68881 response register. Note that all coprocessor registers are memory mapped. Hence, the 68020 can read the response register and obtain the result of the floating-point add from the appropriate locations.
- The 68HC000 $\overline{DTACK}$ pin is replaced by two pins on the 68020: $\overline{DSACK1}$ and $\overline{DSACK0}$. These pins are defined as follows:

$\overline{DSACK1}$	$\overline{DSACK0}$	Device Size
0	0	32-bit device
0	1	16-bit device
1	0	8-bit device
1	1	Data not ready; insert wait states

The 68020 can be configured as a byte, 16-bit, or 32-bit memory system. As a byte memory system, the data pins of a single 8-bit memory containing all addresses in increments of one can be connected to the 68020 D_{31}–D_{24} pins. All data transfers occur via pins D_{31}–D_{24}. The byte memory chip informs the 68020 of its size by activating $\overline{DSACK1}$ = 1 and $\overline{DSACK0}$ = 0 so that the 68020 transfers data via its D_{31}–D_{24} pins. For byte instructions, one byte is transferred via these pins; for word (16-bit) instructions, two consecutive bytes are transferred via these pins; for long word (32-bit) instructions, four consecutive bytes are transferred via these pins.

When the 68020 is configured as a word (16-bit) memory system, two byte memory chips are interfaced to the 68020 via its D_{31}– D_{16} pins. The data pins of the byte memory chips containing even and odd addresses are connected to the 68020 pins D_{31}–D_{24} and D_{23}-D_{16}, respectively. The memory chips inform the 68020 of the 16-bit memory configuration by activating $\overline{DSACK1}$ = 0 and $\overline{DSACK0}$ = 1. The 68020 then uses D_{31}–D_{16} to transfer data for byte, word, or long word instructions. For byte instructions, one byte is transferred via pins D_{31}–D_{24} or D_{23}–D_{16} depending on whether the address is even or odd. For word instructions, the contents of both even and odd addresses are transferred via pins D_{31}–D_{16} with even-address byte via D_{31}-D_{24} pins and odd-addressed byte via D_{23}-D_{16} pins; for long word instructions, four consecutive bytes are transferred via pins D_{31}–D_{16} with the contents of even addresses via pins D_{31}–D_{16} using additional cycles. Data transfer can be aligned or misaligned. For 16-bit memory systems, a word or long word instruction with data transfer starting at an even address is called an "aligned transfer." For example, the instruction `MOVE.W D1,$30000000` will store one data byte at the even address $30000000 via pins D_{31}–D_{24} and one data byte at the odd address $30000001 via pins D_{23}–D_{16} in one cycle. On the other hand, `MOVE.W D0,$30000001` is a misaligned transfer. The 68020 transfers one byte to $30000001 via pins D_{23}–D_{16} in the first cycle and another byte to $30000002 via pins D_{31}–D_{24} in the second cycle. Thus, the misaligned transfer for word instruction takes two cycles in a 16-bit memory configuration. For 32-bit transfers, `MOVE.L D1,$30000000` is an aligned transfer. During the first cycle, the 68020 transfers 8-bit contents of the highest byte of D0 to $30000000 via pins D_{31}–D_{24}, and the next 8-bit contents of D0 to $30000001 via pins D_{23}–D_{16}. During the second cycle, the 68020 transfers next byte of D0 to $30000002 via pins D_{31}–D_{24} and the lowest byte of register D0 to $30000003 via pins D_{23}–D_{16}. Thus, for aligned transfer with 16-bit memory configuration, the 68020 transfers data in two cycles for 32-bit transfers. Next, consider the instruction, `MOVE.L D0,$30000001`. This is a misaligned transfer. The 68020 transfers the most significant byte of D0 to $30000001 via pins D_{23}–D_{16} in the first cycle, the next byte of register D0 to $30000002 via pins D_{31}–D_{24}, and the next byte of D0 to $30000003 via pins D_{23}-D_{16} in the second cycle and finally, the lowest byte of D0 to address

\$30000004 via pins D_{31}–D_{24} in the third cycle. Thus, for misaligned transfers in a 16-bit memory configuration, the 68020 requires 3 cycles to transfer data for long word instructions.

When the 68020 is configured as a 32-bit memory system, four byte memory chips are connected to D_{31}–D_0. The memory chip with data pins connected to D_{31}–D_{24} contains addresses 0, 4, 8, …; the memory chip with data pins connected to D_{23}–D_{16} contains addresses 1, 5, 9, …; the memory chip with data pins connected to D_{15}–D_8 includes addresses 2, 6, 10, …; and the memory chip with data pins connected to D_7–D_0 contains addresses 3, 7, 11, …. The memory chips inform the 68020 of the 32-bit memory configuration by activating $\overline{DSACK1}$ = 0 and $\overline{DSACK0}$ = 0. The 68020 then uses pins D_{31}–D_0 to transfer data for byte, word, or long word instructions. For byte instructions, data is transferred via the appropriate 8 data pins of the 68020 depending on the address in one cycle. For word instructions starting at addresses 0, 4, 8, …, addresses 1, 5, 9, …, and addresses 2, 6, 10, …, data are aligned, and will be transferred in one cycle. For example, consider `MOVE.W D1,$20000005`. The 68020 transfers the contents of D1 (bits 15-8) to address \$20000005 via pins D_{23}–D_{16} and contents of register D1 (bits 7-0) to address \$20000006 via pins D_{15}–D_8 in one cycle. On the other hand, `MOVE.W D1,$20000007` is a misaligned transfer. In this case, the 68020 transfers the contents of register D1 (bits 15-8) to address \$20000007 via pins D_7–D_0 in the first cycle and the contents of D1 (bits 7-0) to address \$20000008 via pins D_{31}–D_{24} in the second cycle.

For long word instructions, data transfers with addresses starting at 0, 4, 8, … are aligned transfers. They will be performed in one cycle. Data with addresses in all other three chips are misaligned and will require additional cycles. For I/O configuration, one to four chips can be connected to the appropriate D_{31}–D_0 pins as required by an application. The addresses in the I/O chips will be memory mapped and connected to the appropriate portions of pins D_{31}–D_0 in the same way as the memory chips.

10.16.1.2 MC68020 Programmer's Model

The MC68020 programmer's model is based on sequential, nonconcurrent instruction execution. This implies that each instruction is completely executed before the next instruction is executed. Although instructions might operate concurrently in actual hardware, they do not operate concurrently in the programmer's model.

Figure 10.32 shows the MC68020 user and supervisor programming models. The user model has fifteen 32-bit general-purpose registers (D0–D7 and A0–A6), a 32-bit program counter (PC), and a condition code register (CCR) contained within the supervisor status register (SR). The supervisor model has two 32-bit supervisor stack pointers (ISP and MSP), a 16-bit status register (SR), a 32-bit vector base register (VBR), two 3-bit alternate function code registers (SFC and DFC), and two 32-bit cache-handling (address and control) registers (CAAR and CACR). The user stack pointer (USP) A7, interrupt stack pointer (ISP) A7', and master stack pointer (MSP) A7" are system stack pointers.

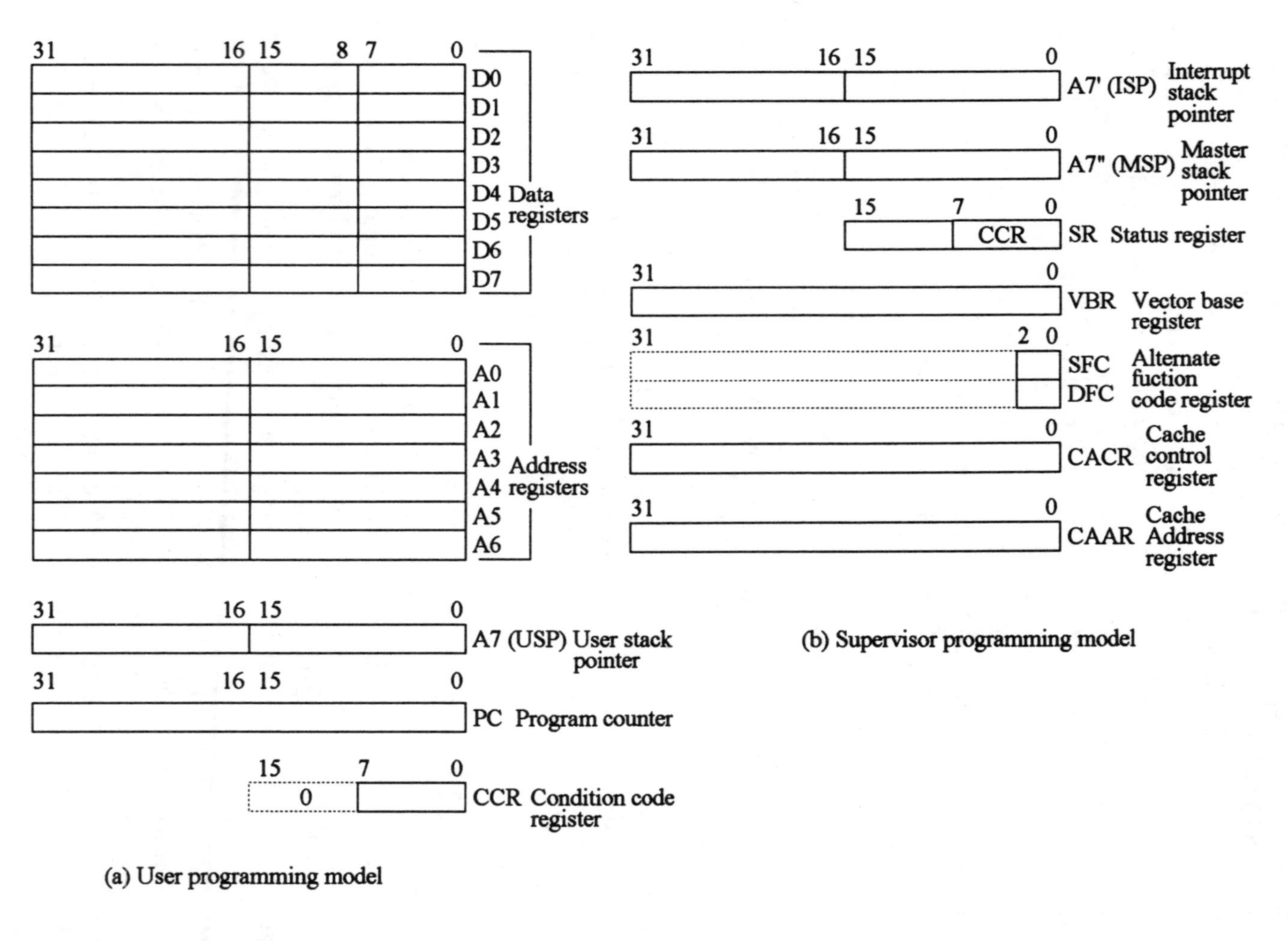

FIGURE 10.32 MC68020 programming model

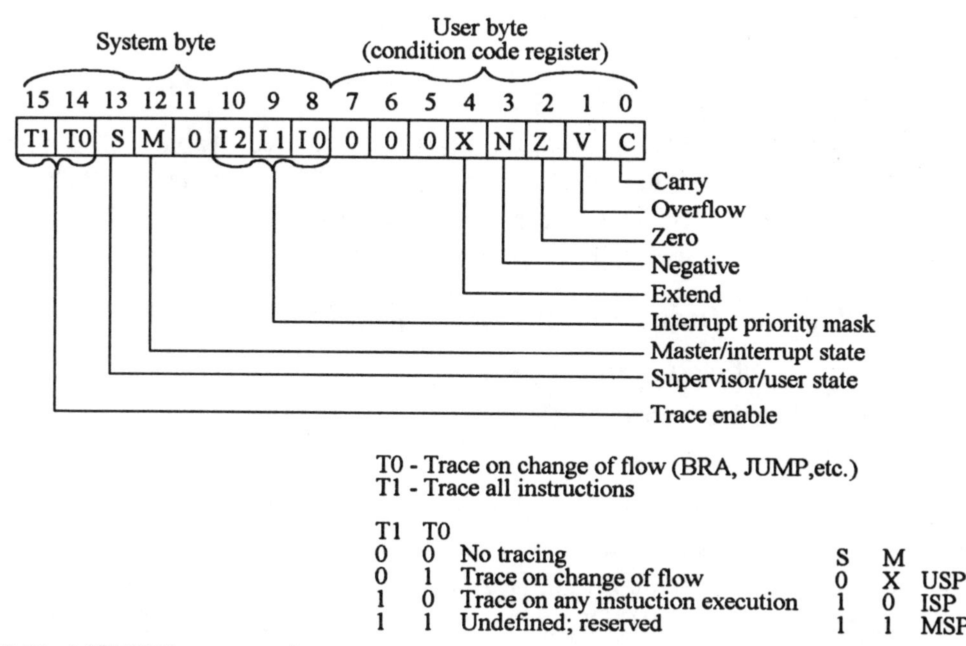

FIGURE 10.33 MC68020 status register

The status register, as shown in Figure 10.33, consists of a user byte (condition code register, CCR) and a system byte. The system byte contains control bits to indicate that the processor is in the trace mode (T1, T0), supervisor/user state (S), and master/interrupt state (M). The user byte consists of the following condition codes: carry (C), overflow (V), zero (Z), negative (N), and extend (X).

The bits in the 68020 user byte are set or reset in the same way as those of the 68HC000 user byte. Bits I2, I1, I0, and S have the same meaning as those of the 68HC000. In the 68020, two trace bits (T1, T0) are included as opposed to one trace bit (T) in the 68HC000. These two bits allow the 68020 to trace on both normal instruction execution and jumps. The 68020 M bit is not included in the 68HC000 status register.

The vector base register (VBR) is used to allocate the exception processing vector table in memory. VBR supports multiple vector tables so that each process can properly manage independent exceptions. The 68020 distinguishes address spaces as supervisor/user and program/data. To support full access privileges in the supervisor mode, the alternate function code registers (SFC and DFC) allow the supervisor to access any address space by preloading the SFC/DFC registers appropriately. The cache registers (CACR and CAAR) allow software manipulation of the instruction code. The CACR provides control and status accesses to the instruction cache; the CAAR holds the address for those cache control functions that require an address.

10.16.1.3 MC68020 Addressing Modes

Table 10.19 lists the MC68020's 18 addressing modes. Table 10.20 compares the addressing modes of the 68HC000 with those of the MC68020. Because 68HC000 addressing modes were covered earlier in this chapter in detail with examples, the 68020 modes not available in the 68HC000 will be covered in the following discussion.

ARI (Address Register Indirect) with Index (Scaled) and 8-Bit Displacement

- Assembler syntax: (d8, A*n*, X*n*.size * scale)
- EA = (A*n*) + (X*n*.size * scale) + d8
- Xn can be W or L.

If the index register (A*n* or D*n*) is 16 bits, then it is sign-extended to 32 bits and multiplied by 1, 2, 4 or 8 to be used in EA calculations. An example is `MOVE.W (0, A2, D2.W * 2),D1`. Suppose that [A2] = $50000000, [D2.W] = $1000, and [$50002000] = $1571; then, after the execution of this `MOVE`, $[D1]_{\text{low 16 bits}}$ = $1571 because EA = $5000000 + $1000 * 2 + 0 = $50002000.

ARI (Address Register Indirect) with Index and Base Displacement

- Assembler syntax: (bd, A*n*, X*n*.size * scale)
- EA = (A*n*) + (X*n*.size * scale) + bd
- Base displacement, bd, has value 0 when present or can be 16 or 32 bits.

The following figure shows the use of ARI with index, X*n*, and base displacement, bd, for accessing tables or arrays:

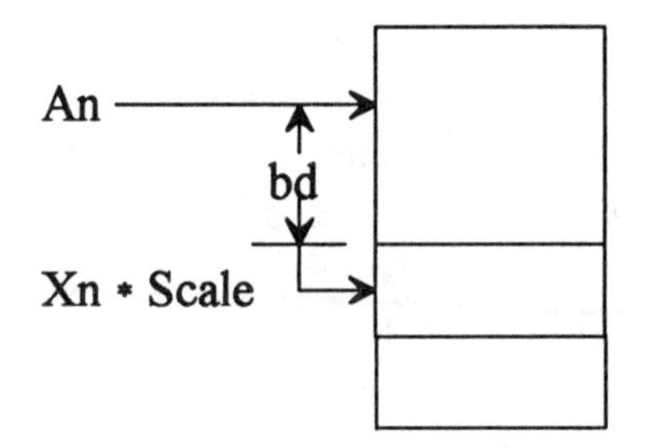

An example is MOVE.W ($5000, A2, D1.W * 4), D5. If [A2] = $30000000, [D1.W] = $0200, and [$30005800] = $0174, then, after execution of this MOVE, $[D5]_{\text{low 16 bits}}$ = $0174 because EA = $5000 + $30000000 + $0200 * 4 = $30005800.

TABLE 10.19 MC68020 Addressing Modes

Mode	*Syntax*
• Register direct	
Data register direct	Dn
Address register direct	An
• Register indirect	
Address register indirect (ARI)	(An)
Address register indirect with postincrement	(An)+
Address register indirect with predecrement	–(An)
Address register indirect with displacement	(d16, An)
• Register indirect with index	
Address register indirect with index (8-bit displacement)	(d8, An, Xn)
Address register indirect with index (base displacement)	(bd, An, Xn)
• Memory indirect	
Memory indirect, postindexed	([bd, An], Xn, od)
Memory indirect, preindexed	([bd, An, Xn], od)
• Program counter indirect with displacement	(d16,PC)
• Program counter indirect with index	
PC indirect with index (8-bit displacement)	(d8, PC, Xn)
PC indirect with index (base displacement)	(bd, PC, Xn)
• Program counter memory indirect	
PC memory indirect, postindexed	([bd, PC], Xn, od)
PC memory indirect, preindexed	([bd, PC, Xn], od)
• Absolute	
Absolute short	(xxx).W
Absolute long	(xxx).L
• Immediate	#data

Notes:

Dn = data register, D0 -D7
An = address register, A0-A6
d8, d16 = 2's complement or sign-extended displacement; added as part of effective address calculation; size is 8 (d8) or 16 (d16) bits; when omitted, assemblers use a value of 0
Xn = address or data register used as an index register; form is Xn.size * scale, where size is .W or .L (indicates index register size) and scale is 1, 2, 4, or 8 (index register is multiplied by scale); use of size and/or scale is optional
bd = 2's complement base displacement; when present, size can be 16 or 32 bits
od = outer displacement, added as part of effective address calculation after any memory indirection; use is optional with a size of 16 or 32 bits
PC = program counter
<data> = immediate value of 8, 16, or 32 bits
( ) = effective address
[ ] = use as indirect address to long word address
ARI = Address Register Indirect

TABLE 10.20 Addressing Modes, MC68HC000 vs. MC68020

Addressing Modes Available	*Syntax*	*68HC000*	*68020*
Data register direct	Dn	Yes	Yes
Address register direct	An	Yes	Yes
Address register indirect (ARI)	(An)	Yes	Yes
ARI with postincrement	(An)+	Yes	Yes
ARI with predecrement	–(An)	Yes	Yes
ARI with displacement (16-bit disp)	(d, An)	Yes	Yes
ARI with index (8-bit disp)	(d, An, Xn)	Yes*	Yes*
ARI with index (base disp; 0, 16, 32)	(bd, An, Xn)	No	Yes
Memory indirect (postindexed)	([bd, An], Xn, od)	No	Yes
Memory indirect (preindexed)	([bd, An, Xn], od)	No	Yes
PC indirect with disp. (16-bit)	(d, PC)	Yes	Yes
PC indirect with index (8-bit disp)	(d, PC, Xn)	Yes*	Yes*
PC indirect with index (base disp)	(bd, PC, Xn)	No	Yes
PC memory indirect (postindexed)	([bd, PC], Xn, od)	No	Yes
PC memory indirect (preindexed)	([bd, PC, Xn], od)	No	Yes
Absolute short	(xxxx).W	Yes	Yes
Absolute long	(xxxxxxxx).L	Yes	Yes
Immediate	#<data>	Yes	Yes

*68HC000 has no scaling capability; 68020 can scale Xn by 1,2,4,or 8.

Memory Indirect

Memory indirect mode is distinguished from address register indirect mode by the use of square brackets in the assembler notation. The concept of memory indirect mode is depicted in the following figure:

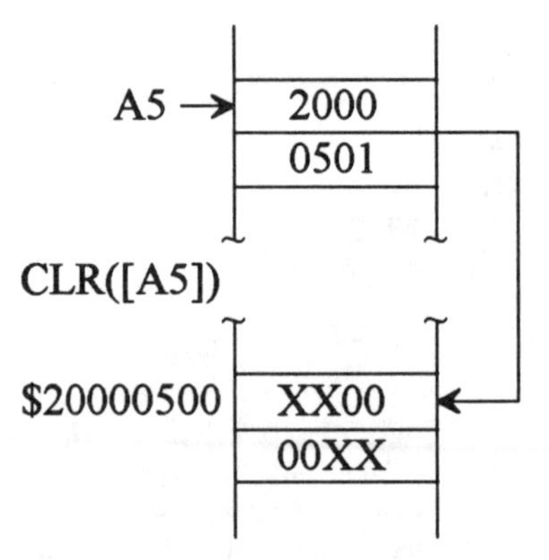

Here, register A5 points to the effective address $20000501. Because `CLR ([A5])` is a 16-bit clear instruction, 2 bytes in location $20000501 and $20000502 are cleared to 0.

Memory indirect mode can be indexed with scaling and displacements. There are two types of memory indirect mode with scaled indexing and displacements: postindexed memory indirect mode and preindexed memory indirect mode. For postindexed memory indirect mode, an indirect memory address is first calculated using the base register (A*n*) and base displacement (bd). This address is used for an indirect memory access of a long word followed by adding a scaled indexed operand and an optional outer displacement (od) to generate the effective address. Note that bd and od can be zero, 16 bits, or 32 bits. In postindexed memory indirect mode, indexing occurs after memory indirection.

- Assembler syntax: ([bd, A*n*], X*n*.size * scale, od)
- EA = ([bd + A*n*]) + (X*n*.size * scale + od)

An example is `MOVE.W ([$0004, A1], D1.W * 2, 2), D2`. If [A1] = $20000000, [$2000004] = $00003000, [D1.W] = $0002, and [$00003006] = $1A40, then, after execution of this `MOVE`, intermediate pointer = (4 + $20000000) = $20000004, [$2000004], which is $00003000 used as a pointer. Therefore, EA = $00003000 + $00000004 + 2 = $00003006. Hence, $[D2]_{\text{low 16 bits}}$ = $1A40.

For memory indirect preindexed mode, the scaled index operand is added to the base register (A*n*) and base displacement (bd). This result is then used as an indirect address into the data space. The 32-bit value at this address is read and an optional outer displacement (od) is added to generate the effective address. The indexing, therefore, occurs before indirection.

- Assembler syntax: ([bd, A*n*, X*n*.size * scale], od)
- EA = (bd, A*n* + X*n*.size * scale) + od

As an example of the preindexed mode, consider several I/O devices in a system. The addresses of these devices can be held in a table pointed to by An, bd, and Xn. The actual programs for these devices can be stored in memory pointed to by the respective device addresses plus od.

The memory indirect preindexed mode will now be illustrated by a numerical example. Consider

```
MOVE.W ([$0002, A1,D0.W*2], 2), D1
```

If [A1] = $20000000, [D0.W] = $0004, [$2000000A] = $00121502, [$00121504] = $F124, then after execution of this `MOVE`, intermediate pointer = $20000000 + $0002 + $0004*2 = $2000000A. Therefore, [$2000000A], which is $00121502, is used as a memory pointer. Hence, [D1] low 16 bits = $F124.

TABLE 10.21 68020 New Instructions

Instruction	*Description*
BFCHG	Bit field change
BFCLR	Bit field clear
BFEXTS	Bit field signed extract
BFEXTU	Bit field unsigned extract
BFFFO	Bit field find first one set
BFINS	Bit field insert
BFSET	Bit field set
BFTST	Bit field test
CALLM	Call module
CAS	Compare and swap
CAS2	Compare and swap (two operands)
CHK2	Check register against upper and lower bounds
CMP2	Compare register against upper and lower bounds
cpBcc	Coprocessor branch on coprocessor condition
cpDBcc	Coprocessor test condition, decrement, and branch
cpGEN	Coprocessor general function
cpRESTORE	Coprocessor restore internal state
cpSAVE	Coprocessor save internal state
cpSETcc	Coprocessor set according to coprocessor condition
cpTRAPcc	Coprocessor trap on coprocessor condition
PACK	Pack BCD
RTM	Return from module
UNPK	Unpack BCD

10.16.1.4 MC68020 Instruction Set

The MC68020 instruction set includes all 68HC000 instructions plus some new ones. Some of the 68HC000 instructions are enhanced. Over 20 new instructions are added to provide new functionality. A list of these instructions is given in Table 10.21.

Succeeding sections will discuss the 68020 instructions listed next:

- 68020 new priveleged move instructions
- RTD instruction
- CHK/CHK2 and CMP/CMP2 instructions
- TRAP*cc* instructions
- Bit field instructions
- PACK and UNPK instructions
- Multiplication and division instructions
- 68HC000 enhanced instructions

68020 New Priveleged Move Instructions

The 68020 new priveleged move instructions can be executed by the 68020 in the supervisor mode. They are listed below:

Instruction	*Operand Size*	*Operation*	*Notation*
`MOVE`	16	SR → destination	`MOVE SR,` (EA)
`MOVEC`	32	Rc → Rn	`MOVEC.L` Rc, Rn
		Rn → Rc	`MOVEC.L` Rn, Rc
`MOVES`	8, 16, 32	Rn → destination using DFC	`MOVES.S` Rn, (EA)
		Source using SFC → Rn	`MOVES.S` (EA),Rn

Note that Rc includes VBR, SFC, DFC, MSP, ISP, USP, CACR, and CAAR. Rn can be either an address or a data register.

The operand size (.L) indicates that the `MOVEC` operations are always long word. Notice that only register to register operations are allowed. A control register (Rc) can be copied to an address or a data register (Rn) or vice versa. When the 3 bit SFC or DFC register is copied into Rn, all 32 bits of the register are overwritten and the upper 29 bits are "0."

The `MOVES` (move to alternate space) instruction allows the operating system to access any addressed space defined by the function codes. It is typically used when an operating system running in the supervisor mode must pass a pointer or value to a previously defined user program or data space. The operand size (.S) indicates that the `MOVES` instruction can be byte (.B), word (.W), or long word (.L). The `MOVES` instruction allows register to memory or memory to register operations. When a memory to register move occurs, this instruction causes the contents of the source function code register to be placed on the external function hardware pins. For a register to memory move, the processor places the destination function code register on the function code pins. The `MOVES` instruction can be used to move information from one space to another.

Example 10.19

(a) Find the contents of address $70000023 and the function code pins FC2, FC1, and FC0 after execution of `MOVES.B D5, (A5)`. Assume the following data prior to execution of this `MOVES` instruction: [SFC] = 101_2, [DFC] = 100_2 , [A5] = $70000023, [D5] = $718F2A05, [$70000020] = $01, [$70000021] = $F1, [$70000022] = $A2, [$70000023] = $2A

Solution

After execution of this `MOVES` instruction,

FC2 FC1 FC0 = 100_2 , [$70000023] = $05

(b) The following 68000 instruction sequence:

```
MOVEA.L 8(A7),A0
MOVE.W  (A0),D3
```

is used by a subroutine to access a parameter whose address has been passed into A0 and then moves the parameter to D3. Find the equivalent 68020 instruction.

Solution `MOVE.W ([8,A7]),D3`

Return and Delocate Instruction

The return and delocate (RTD) instruction is useful when a subroutine has the responsibility to remove parameters off the stack that were pushed onto the stack by the calling routine. Note that the calling routine's JSR (jump to subroutine) or BSR (branch to subroutine) instructions do not automatically push parameters onto the stack prior to the call as do the CALLM instructions. Rather, the pushed parameters must be placed there using the MOVE instruction. The format of the RTD instruction is shown next:

Instruction	*Operand Size*	*Operation*	*Notation*
RTD	Unsized	(SP) → PC, SP + 4 + d → SP	RTD # <disp>

As an example, consider RTD #8, which, at the end of a subroutine, deallocates 8 bytes of unwanted parameters off the stack by adding 8 to the stack pointer and returns to the main program. The size of the displacement is 16-bit.

CHK/CHK2 and CMP/CMP2 Instructions

The 68020 check instruction (CHK) compares a 32-bit twos complement integer value residing in a data register (D*n*) against a lower bound (LB) value of zero and against an upper bound (UB) value of the programmer's choice. The upper bound value is located at the effective address (EA) specified in the instruction format. The CHK instruction has the following format: `CHK.S` (EA), D*n* where the operand size (.S) designates word (.W) or long word (.L).

If the data register value is less than zero (D*n* < 0) or if the data register is greater than the upper bound (D*n* > UB), then the processor traps through exception vector 6 (offset $18) in the exception vector table. Of course, the operating system or the programmer must define a check service handler routine at this vector address. The condition codes after execution of the CHK are affected as follows: If D*n* < 0 then N = 1; if D*n* > UB (upper bound) then N =0.

If 0 ≤ D*n* ≤ UB then N is undefined. X is unaffected and all other flags are undefined and program execution continues with the next instruction.

The CHK instruction can be used for maintaining array subscripts because all subscripts can be checked against an upper bound (i.e., UB = array size - 1). If the compared subscript is within the array bounds (i.e., 0 ≤ subscript value ≤ UB value), then the subscript is valid, and the program continues normal instruction execution. If the subscript value is out of array limits (i.e., 0 > subscript value or subscript value > UB value), then the processor traps through the CHK exception.

Example 10.20

Determine the effects of execution of CHK.L (A5), D3, where A5 represents a memory pointer to the array's upper bound value. Register D3 contains the subscript value to be checked against the array bounds. Assume the following data prior to execution of this CHK instruction:

[D3] = $01507126
[A5] = $00710004
[$00710004] = $01500000

Solution

The long word array subscript value $01507126 contained in data register D3 is compared against the long word UB value $01500000 pointed to by address register A5. Because the value $01507126 contained in D3 exceeds the UB value $01500000 pointed to by A5, the N bit is cleared. (X is unaffected and the remaining CCR bits are not affected.) This out-of-bounds condition causes the program to trap to a check exception service routine.

Before CHK.L(A5), D3	Operation	After
D3: 01507126	0 < D3.L > $01500000 ∴ N = 0, TRAP	Enter check exception service routine
Memory (bits 31–0) A5 = $00710004: 01500000		CCR X N Z V C X 0 U U U

The operation of the CHK instruction can be summarized as follows:

Instruction	*Operand Size*	*Operation*	*Notation*
CHK	16, 32	If D*n* < 0 or D*n* > source, then TRAP	CHK (EA), D*n*

The 68020 CMP.S (EA), D*n* instruction subtracts (EA) from D*n* and affects the condition codes without any result. The operand size designator (.S) is either byte (.B) or word (.W) or long word (.L).

Both the CHK2 and the CMP2 instructions have similar formats:

CHK2.S (EA), R*n*

and

CMP2.S (EA), R*n*

They compare a value contained in a data or address register (designated by R*n*) against two (2) bounds chosen by the programmer. The size of the data to be compared (.S) may be specified as byte (.B), word (.W), or long word (.L). As shown in the following figure, the lower bound (LB) value must be located in memory at the effective address (EA) specified in the instruction, and the upper bound (UB) value must follow immediately at the next higher memory address. That is, UB addr = LB addr + size, where size = B (+1), W (+2), or L (+4).

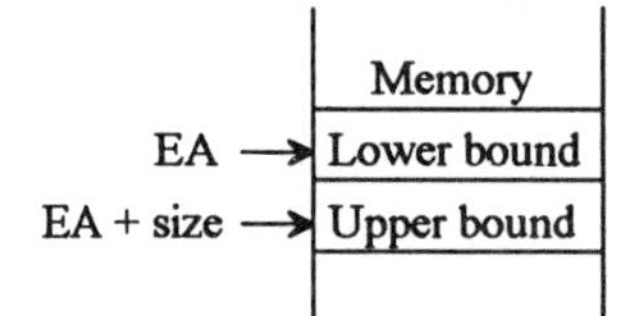

If the compared register is a data register (i.e., R*n* = D*n*) and the operand size (.S) is a byte or word, then only the appropriate low-order part of the data register is checked. If the compared register is an address register (i.e., R*n* = A*n*) and the operand size (.S) is a byte or word, then the bound operands are sign-extended to 32 bits and the extended operands are compared against the full 32 bits of the address register. After execution of CHK2 and CMP2, the condition codes are affected as follows:

carry	=	1	if the contents of Dn are out of bounds
	=	0	otherwise.
Z	=	1	if the contents of Dn are equal to either bound
	=	0	otherwise.

In the case where an upper bound equals the lower bound, the valid range for comparison becomes a single value. The only difference between the CHK2 and CMP2 instructions is that, for comparisons determined to be out of bounds, CHK2 causes exception processing utilizing the same exception vector as the CHK instructions, whereas the CMP2 instruction execution affects only the condition codes.

In both instructions, the compare is performed for either signed or unsigned bounds. The 68020 automatically evaluates the relationship between the two bounds to determine which kind of comparison to employ. If the programmer wishes to have the bounds evaluated as signed values, the arithmetically smaller value should be the lower bound. If the bounds are to be evaluated as unsigned values, the programmer should make the logically smaller value the lower bound.

The following CMP2 and CHK2 instruction examples are identical in that they both utilize the same registers, comparison data, and bound values. The difference is how the upper and lower bounds are arranged.

Example 10.21

Determine the effects of execution of `CMP2.W (A2),D1`. Assume the following data prior to execution of this `CMP2` instruction:

[D1] = $50000200, [A2] = $00007000
[$00007000] = $B000, [$00007002] = $5000

Solution

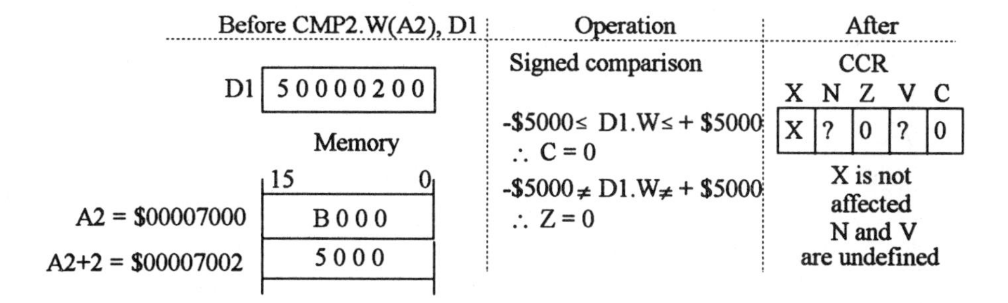

In this example, the word value $B000 contained in memory (as pointed to by address register A2) is the lower bound and the word value $5000 immediately following $B000 is the upper bound. Because the lower bound is the arithmetically smaller value, the programmer is indicating to the 68020 to interpret the bounds as signed numbers. The twos complement value $B000 is equivalent to an actual value of –$5000. Therefore, the instruction evaluates the word contained in data register D1 ($0200) to determine whether it is greater than or equal to the upper bound, +$5000, or less than or equal to the lower bound, -$5000. Because the compared value $0200 is within bounds, the carry bit (C) is cleared to 0. Also, because $0200 is not equal to either bound, the zero bit (Z) is cleared. The following figure shows the range of valid values that D1 could contain:

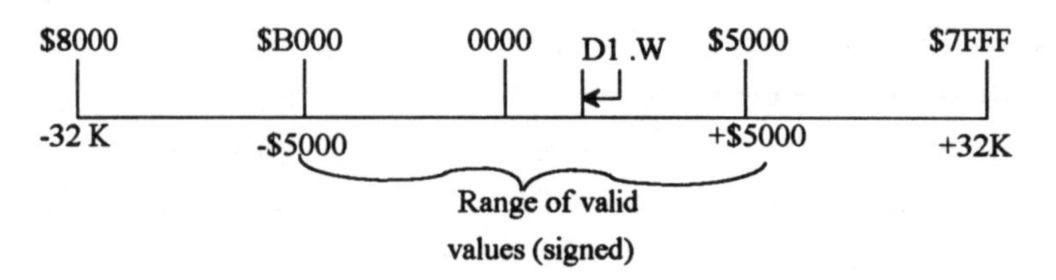

A typical application for the `CMP2` instruction would be to read in a number of user entries and verify that each entry is valid by comparing it against the valid range bounds. In the preceding `CMP2` example, the user-entered value would be in register D1 and register A2 would point to a range for that value. The `CMP2` instruction would verify whether the entry is in range by clearing the CCR carry bit if it is in bounds and setting the carry bit if it is out of bounds.

Example 10.22

Determine the effects of execution of `CHK2.W (A2), D1`. Assume the following data prior to execution of this `CHK2` instruction:

[D1] = $50000200, [A2] = $00007000
[$00007000] = $5000, [$00007002] = $B000

Solution

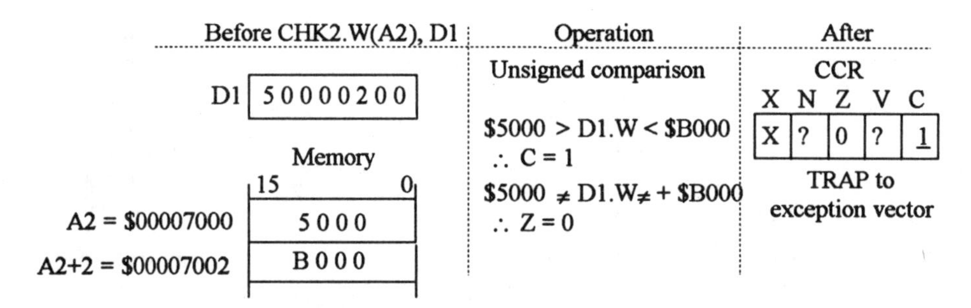

This time, the value $5000 located in memory is the lower bound and the value $B000 is the upper bound.

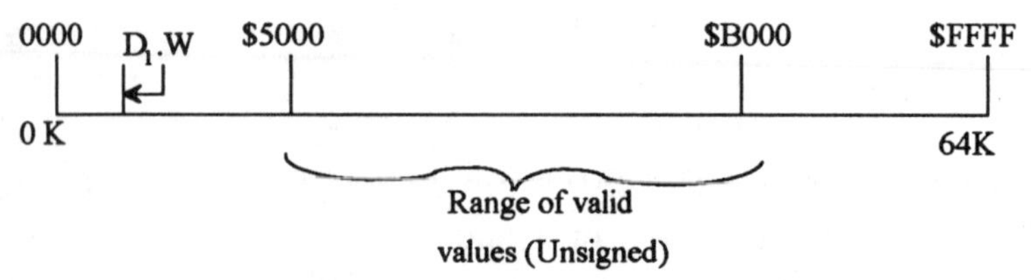

Now, because the lower bound contains the logically smaller value, the programmer is indicating to the 68020 to interpret the bounds as unsigned numbers, representing only a magnitude.

Therefore, the instruction evaluates the word contained in register D1 ($0200) to determine whether it is greater than or equal so the lower bound, $5000, or less than or equal to the upper bound, $B000. Because the compared value $0200 is less than $5000, the carry bit is set to indicate an out of bounds condition and the program traps to the `CHK/CHK2` exception vector service routine. Also, because $0200 is not equal to either bound, the zero bit (Z) is cleared. The figure above shows the range of valid values that D1 could contain.

A typical application for the `CHK2` instruction would be to cause a trap exception to occur if a certain subscript value is not within the bounds of some defined array. Using the `CHK2` example format just given, if we define an array of 100 elements with subscripts ranging from 0- 99_{10}, and if the two words located at (A2) and (A2 + 2) contain 50 and 99, respectively, and register D1 contains 100_{10}, then execution of the `CHK2` instruction would cause a trap through the `CHK/CHK2` exception vector. The operation of the `CMP2` and `CHK2` instructions are summarized as follows:

Instruction	*Operand Size*	*Operation*	*Notation*
`CMP2`	8,16, 32	Compare R*n* < source – lower bound or R*n* > source – upper bound and set CCR	`CMP2` (EA), R*n*
`CHK2`	8, 16, 32	If R*n* < source – lower bound or R*n* > source – upper bound, then TRAP	`CHK2` (EA), R*n*

Trap-on-Condition Instructions

The new trap condition (`TRAP`*cc*) instruction allows a conditional trap exception on any of the condition codes shown in Table 10.22. These are the same conditions that are allowed for the set-on-condition (S*cc*) and the branch-on-condition (B*cc*) instructions. The `TRAP`*cc* instruction evaluates the selected test condition based on the state of the condition code flags, and if the test is true, the 68020 initiates exception processing by trapping through the same exception vector as the `TRAPV` instruction (vector 7, offset $1C, VBR = VBR + offset). The trap-on-condition instruction format is

`TRAP`*cc* or `TRAP`*cc*.S #<data>

where the operand size (.S) designates word (.W) or long word (.L).

If either a word or long word operand is specified, a 1- or 2-word immediate operand is placed following the instruction word. The immediate operand(s) consists of argument parameters that are passed to the trap handler to further define requests or services it should perform. If *cc* is false, the 68020 does not interpret the immediate operand(s) but instead adjusts the program counter to the beginning of the following instruction. The exception handler can access this immediate data as an offset to the stacked PC. The stacked PC is the next instruction to be executed.

A summary of the TRAP*cc* instruction operation is shown next:

Instruction	*Operand Size*	*Operation*	*Notation*
TRAP*cc*	None	If *cc*, then TRAP	TRAP*cc*
	16	Same	TRAPcc.W #<data>
	32	Same	TRAPcc.L #<data>

TABLE 10.22 Conditions for TRAP*cc*

Code	*Description*	*Result*
CC	Carry clear	$\overline{C}$
CS	Carry set	C
EQ	Equal	Z
F	Never true	0
GE	Greater or equal	$N \cdot V + \overline{N} \cdot \overline{V}$
GT	Greater than	$N \cdot V \cdot Z + \overline{N} \cdot \overline{V} \cdot \overline{Z}$
HI	High	$\overline{C} \cdot \overline{Z}$
LE	Less or equal	$Z + N \cdot \overline{V} + \overline{N} \cdot V$
LS	Low or same	$C + Z$
LT	Less than	$N \cdot \overline{V} + \overline{N} \cdot V$
MI	Minus	N
NE	Not equal	$\overline{Z}$
PL	Plus	N
T	Always true	1
VC	Overflow clear	$\overline{V}$
VS	Overflow set	V

Bit Field Instructions

The bit field instructions, which allow operations to clear, set, ones complement, input, insert, and test one or more bits in a string of bits (bit field), are listed on the next page. Note that the condition codes are affected according to the value in the field before execution of the instruction. All bit field instructions affect the N and Z bits as shown for BFTST. That is, for all instructions, Z = 1 if all bits in a field prior to execution of the instruction are zero; Z = 0 otherwise. N = 1 if the most significant bit of the field prior to execution of the instruction is one; N = 0 otherwise. C and V are always cleared. X is always unaffected. Next, consider BFFFO. The offset of the first bit set 1 in a bit field is placed in D*n*; if no set bit is found, D*n* contains the offset plus the field width.

Immediate offset is from 0 to 31, whereas offset in D*n* can be specified from -2^{31} to $2^{31} - 1$. All instructions are unsized. They are useful for memory conservation, graphics, and communications.

Instruction	*Operand Size*	*Operation*	*Notation*
`BFTST`	1-32	Field MSB → N, Z = 1 if all bits in field are zero; Z = 0 otherwise	`BFTST` (EA) {offset:width}
`BFCLR`	1-32	0's → Field	`BFCLR` (EA) {offset:width}
`BFSET`	1-32	1's → Field	`BFSET` (EA) {offset:width}
`BFCHG`	1-32	$\overline{\text{Field}}$ → Field	`BFCHG` (EA) {offset:width}
`BFEXTS`	1-32	Field → D*n*; sign-extended	`BFEXTS` (EA) {offset:width}, D*n*
`BFEXTU`	1-32	Field → D*n*; Zero-extended	`BFEXTU` (EA) {offset:width}, D*n*
`BFINS`	1-32	D*n* → field	`BFINS Dn,` (EA) {offset:width}
`BFFFO`	1-32	Scan for first bit-set in field	`BFFFO` (EA) {offset:width}, D*n*

As an example, consider `BFCLR $5002{4:12}`. Assume the following memory contents:

	7	6	5	4	3	2	1	0 ← Bit number
\$5001	1	0	1	0	0	0	0	1
\$5002 (Base address) →	1	0	0	1	1	1	0	0
\$5003	0	1	1	1	0	0	0	1
\$5004	0	0	0	1	0	0	1	0

Bit 7 of the base address \$5002 has the offset 0. Therefore, bit 3 of \$5002 has the offset value of 4. Bit 0 of location \$5001 has offset value -1, bit 1 of \$5001 has offset value -2, and so on. The example `BFCLR` instruction just given clears 12 bits starting with bit 3 of \$5002. Therefore, bits 0–3 of location \$5002 and bits 0–7 of location \$5003 are cleared to 0. Therefore, the memory contents change as follows:

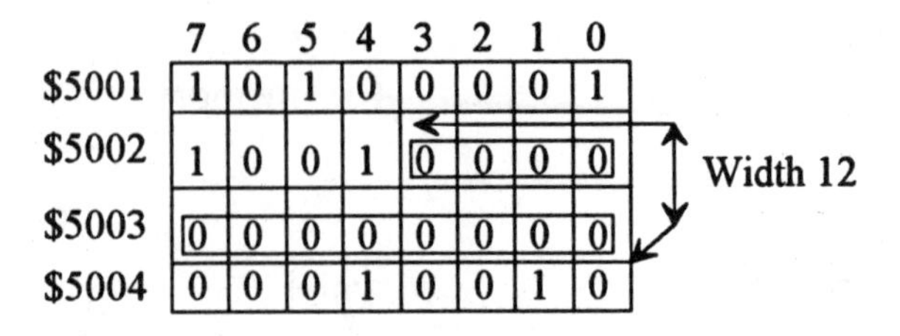

The use of bit field instructions may result in memory savings. For example, assume that an input device such as a 12-bit A/D converter is interfaced via a 16-bit port of a MC68020 based microcomputer. Now, suppose that 1 million pieces of data are to be collected from this port. Each 12 bits can be transferred to a 16-bit memory location or bit field instructions can be used.

- Using a 16-bit location for each 12 bits:

$$\text{Memory requirements} = 2 \times 1 \text{ million}$$
$$= 2 \text{ million bytes}$$

- Using bit fields:

$$12 \text{ bits} = 1.5 \text{ bytes}$$
$$\text{Memory requirements} = 1.5 \times 1 \text{ million}$$
$$= 1.5 \text{ million bytes}$$
$$\text{Savings} = 2 \text{ million bytes} - 1.5 \text{ million bytes}$$
$$= 500,000 \text{ bytes}$$

Example 10.23

Determine the effect of each of the following bit field instructions:

```
BFCHG $5004{D5:D6}
BFEXIU $5004{2:4},D5
BFINS D4,(A0){D5:D6}
BFFFO $5004{D6:4},D5
```

Assume the following data prior to execution of each of the given instructions. Register contents are given in hex, CCR and memory contents in binary, and offset to the left of memory in decimal.

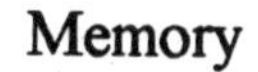

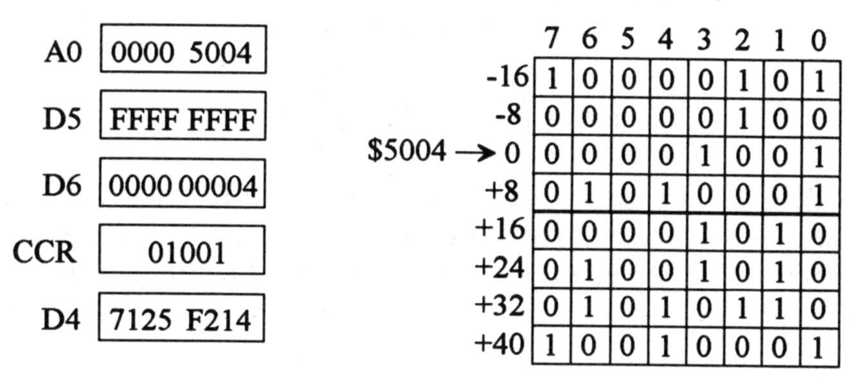

Register	Contents
A0	0000 5004
D5	FFFF FFFF
D6	0000 00004
CCR	01001
D4	7125 F214

	7	6	5	4	3	2	1	0
-16	1	0	0	0	0	1	0	1
-8	0	0	0	0	0	1	0	0
$5004 → 0	0	0	0	0	1	0	0	1
+8	0	1	0	1	0	0	0	1
+16	0	0	0	0	1	0	1	0
+24	0	1	0	0	1	0	1	0
+32	0	1	0	1	0	1	1	0
+40	1	0	0	1	0	0	0	1

Solution

- `BFCHG $5004 {D5:D6}`
  Offset = - 1, Width = 4

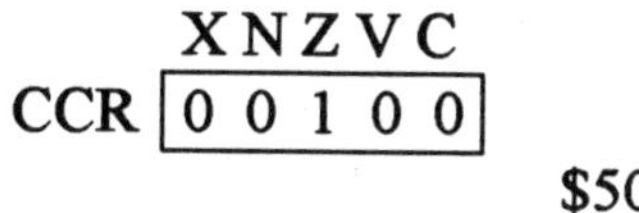

Memory

				1
$5004	1	1	1	

- `BFEXTU $5004 {2:4},D5`
  Offset = 2, Width = 4

	X	N	Z	V	C
CCR	0	0	0	0	0

D5 0 0 0 0 0 0 0 2

- `BFINS D4,(A0) {D5:D6}`
  Offset = - 1, Width = 4

Memory

				0
5004	1	0	0	

	X	N	Z	V	C
CCR	0	0	1	0	0

- `BFFFO $5004 {D6:4},D5`
  Offset = 4, Width = 4

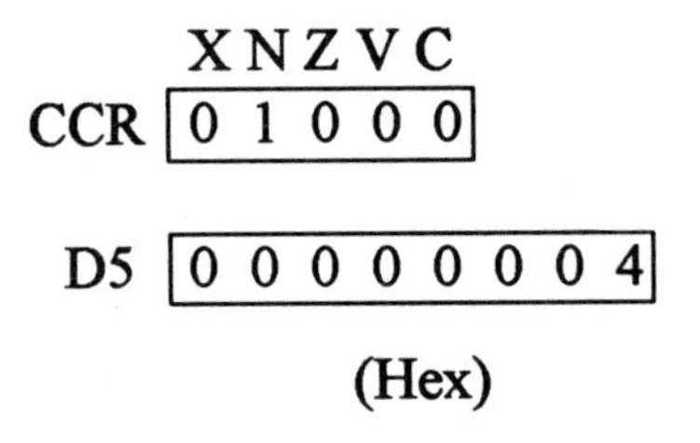

Pack and Unpack Instructions

The details of the PACK and UNPK instructions are listed next:

Instruction	*Operand Size*	*Operation*	*Notation*
PACK	16 → 8	Unpacked source + #data → packed destination	PACK –(A*n*), –(A*n*), #<data>
			PACK D*n*, D*n*,#<data>
UNPK	8 → 16	Packed source → unpacked source	UNPK –(A*n*), –(A*n*), #<data>
		unpacked source + #data → unpacked destination	UNPK D*n*, D*n*,#<data>

Both instructions have three operands and are unsized. They do not affect the condition codes. The PACK instruction converts two unpacked BCD digits to two packed BCD digits:

Unpacked BCD:

15–12	11–8	7–4	3–0
0 0 0 0	BCD0	0 0 0 0	BCD1

Packed BCD:

7–4	3–0
BCD0	BCD1

The UNPK instruction reverses the process and converts two packed BCD digits to two unpacked BCD digits. Immediate data can be added to convert numbers from one code to another. That is, these instructions can be used to translate codes such as ASCII or EBCDIC to a BCD and vice versa.

The PACK and UNPK instructions are useful when I/O devices such as an ASCII keyboard and an ASCII printer are interfaced to an MC68020-based microcomputer. Data can be entered into the microcomputer via the keyboard in ASCII codes. The PACK instruction can be used with appropriate adjustments to convert these ASCII codes into packed BCD. Arithmetic operations can be performed inside the microcomputer, and the result will be in packed BCD. The UNPK instruction can similarly be used with appropriate adjustments to convert packed BCD to ASCII codes for outputting to the ASCII printer.

Example 10.24

Determine the effect of execution of each of the following `PACK` and `UNPK` instructions:

- `PACK D0,D5,#$0000`
- `PACK-(A1),-(A4),#$0000`
- `UNPK D4,D6,#$3030`
- `UNPK-(A3),-(A2),#$3030`

Assume the following data prior to execution of each of the above instructions:

Register	31 … 0
D0	X X X X 32 37
D5	X X X X X 26
D4	X X X X X 35
D6	X X X X X 27
A2	3 0 0 5 0 0 A3
A3	5 0 7 1 2 4 B9
A1	5 0 7 1 2 4 B3
A4	3 0 0 5 0 0 A1

Memory

Address	7 … 0
$507124B1	32
$507124B2	37
$507124B3	00
$507124B4	27
$507124B5	02
$507124B6	07
$507124B7	27
$507124B8	27

Solution

- `PACK D0,D5,#$0000`

[D0] = 32 37
low word
+ 00 00
32 37
[D5] = 27

Note that ASCII code for 2 is $32 and for 7 is $37. Hence, this pack instruction converts ASCII code to packed BCD.

- `PACK -(A1),-(A4),$0000`

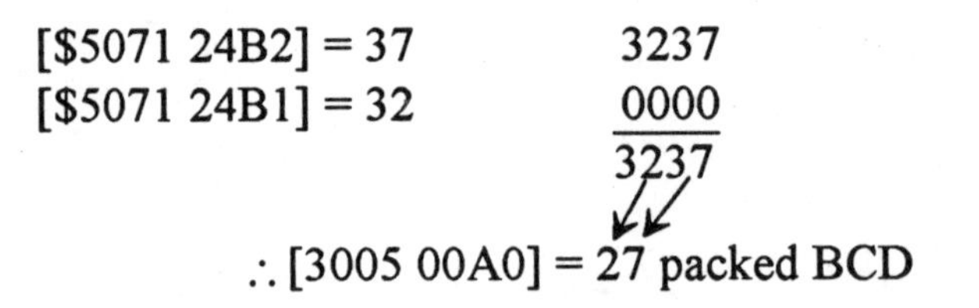

Hence, this pack instruction with the specified data converts two ASCII digits to their equivalent packed BCD form.

- `UNPK D4,D6,#$3030`

[D4] = XXXXXX 35
03 05
+ 30 30
33 35
∴ [D6] = XXXX 33 35
[D4] = XXXXXX 35

Therefore, this UNPK instruction with the assumed data converts from packed BCD in D4 to ASCII code in D6; the contents of D4 are not changed.

- `UNPK -(A3),-(A2),#$3030`

[$5071 24B8] = 27

```
  02 07
  30 30
  -----
  32 37
```

∴ [$300500A2] = 37
[$300500A1] = 32

This UNPK instruction with the assumed data converts two packed BCD digits to their equivalent ASCII digits.

Multiplication and Division Instructions

The 68020 includes the following signed and unsigned multiplication instructions:

Instruction	*Operand Size*	*Operation*
`MULS.W` (EA), D*n* or `MULU`	$16 \times 16 \rightarrow 32$	(EA)16 * (D*n*)16 → (D*n*)32
`MULS.L` (EA), D*n* or `MULU`	$32 \times 32 \rightarrow 32$	(EA) * D*n* → D*n*. D*n* holds 32 bits of the result after multiplication. Upper 32 bits of the result are discarded.
`MULS.L` (EA),D*h*:D*n* or `MULU`	$32 \times 32 \rightarrow 64$	(EA) * D*n* → D*h*:D*n*. (EA) holds 32-bit multiplier before multiplication. D*h* holds high 32 bits of product after multiplication. Dn holds 32-bit multiplicand before multiplication and low 32 bits of product after multiplication.

(EA) can use all modes except A*n*. The condition codes N. Z. and V are affected; C is always cleared to 0, and X is unaffected for both `MULS` and `MULU`. For signed multiplication, overflow (V = 1) can only occur for 32×32 multiplication, producing a 32-bit result if the high-order 32 bits of the 64-bit product are not the sign extension of the low-order 32 bits. In the case of unsigned multiplication, overflow (V = 1) can occur for 32×32 multiplication, producing a 32-bit result if the high-order 32 bits of the 64-bit product are not zero.

Both MULS and MULU have a word form and a long word form. For the word form (16×16), the multiplier and multiplicand are both 16 bits and the result is 32 bits. The result is saved in the destination data register. For the long word form (32×32), the multiplier and multiplicand are both 32 bits and the result is either 32 bits or 64 bits. When the result is 32 bits for a 32-bit × 32-bit operation, the low-order 32 bits of the 64-bit product are provided.

The signed and unsigned division instructions of the 68020 include the following, in which the source is the divisor, the destination is the dividend.

Instruction	*Operation*
DIVS.W (EA), D*n*	$32/16 \rightarrow 16r{:}16q$
or	
DIVU	
DIVS.L (EA), D*q*	$32/32 \rightarrow 32q$
or	No remainder is provided.
DIVU	
DIVS.L (EA),D*r*:D*q*	$64/32 \rightarrow 32r{:}32q$
or	
DIVU	
DIVSL.L (EA),D*r*:D*q*	D*r*/(EA) $\rightarrow 32r{:}32q$
or	D*r* contains 32-bit dividend
DIVUL	

(EA) can use all modes except A*n*. The condition codes for either signed or unsigned division are affected as follows: N = 1 if the quotient is negative; N = 0 otherwise. N is undefined for overflow or divide by zero. Z = 1 if the quotient is zero; Z = 0 otherwise. Z is undefined for overflow or divide by zero. V = 1 for division overflow; V = 0 otherwise. X is unaffected. Division by zero causes a trap. If overflow is detected before completion of the instruction, V is set to 1, but the operands are unaffected.

Both signed and unsigned division instructions have a word form and three long word forms. For the word form, the destination operand is 32 bits and the source operand is 16 bits. The 32-bit result in D*n* contains the 16-bit quotient in the low word and the 16-bit remainder in the high word. The sign of the remainder is the same as the sign of the dividend.

For the instruction

DIVS.L (EA), D*q*
or
DIVU

both destination and source operands are 32 bits. The result in D*q* contains the 32-bit quotient and the remainder is discarded.

For the instruction

`DIVS.L` (EA), D*r*:D*q*
or
`DIVU`

the destination is 64 bits contained in any two data registers and the source is 32 bits. The 32-bit register D*r* (D0–D7) contains the 32-bit remainder and the 32-bit register D*q* (D0–D7) contains the 32-bit quotient.

For the instruction

`DIVSL.L` (EA), D*r*:D*q*
or
`DIVUL`

the 32-bit register D*r* (D0–D7) contains the 32-bit dividend and the source is also 32 bits. After division, D*r* contains the 32-bit remainder and D*q* contains the 32-bit quotient.

Example 10.25

Determine the effect of execution of each of the following multiplication and division instructions:

- `MULU.L #$2,D5` if [D5] = $FFFFFFFF
- `MULS.L #$2,D5` if [D5] = $FFFFFFFF
- `MULU.L #$2,D5:D2` if [D5] = $2ABC1800 and [D2] = $FFFFFFFF
- `DIVS.L #$2,D5` if [D5] = $FFFFFFFC
- `DIVS.L #$2,D2:D0` if [D2] = $FFFFFFFF and [D0] = $FFFFFFFC
- `DIVSL.L #$2,D6:D1` if [D1] = $00041234 and [D6] = $FFFFFFFD

Solution

- `MULU.L #$2,D5` if [D5] = $FFFFFFFF

```
                $FFFFFFFF
              * $00000002
    00000001    FFFFFFFE
    V = 1       Low 32-bit
    since       result in D5
    this is
    nonzero
```

Therefore, [D5] = $FFFFFFFE, N = 0 since the most significant bit of the result is 0, Z = 0 because the result is nonzero, V = 1 because the high 32 bits of the 64-bit product are not zero, C = 0 (always), and X is not affected.

- `MULS.L #$2,D5` if [D5] = $FFFFFFFF

```
                  $FFFFFFFF  (-1)
                * $00000002  (+2)
      $FFFFFFFF   $FFFFFFFE  (-2)
                  \_________/
                  Result in D5
```

Therefore, [D5] = $FFFFFFFE, X is unaffected, C = 0, N = 1, V = 0, and Z = 0.

- `MULU.L #$2,D5:D2` if [D5] = $2ABC1800 and D2 = $FFFFFFFF

```
                  $FFFFFFFF
                * $00000002
                 ----------
      00000001    FFFFFFFE
      \______/    \______/
         D5          D2
```

Here N = 0, Z = 0, V = 0, C = 0, and X is not affected.

- `DIVS.L #$2,D5` if [D5] = $FFFFFFFC

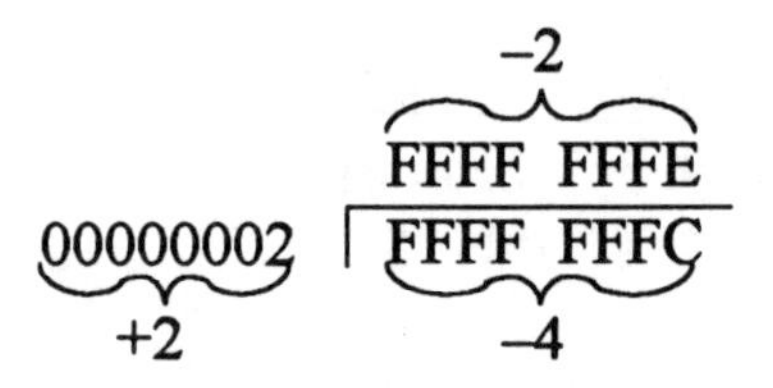

[D5] = $FFFFFFFE, X is unaffected, N = 1, Z = 0, V = 0, and C = 0 (always).

- `DIVS.L #$2,D2:D0` if [D2] = $FFFFFFFF and [D0] = $FFFFFFFC

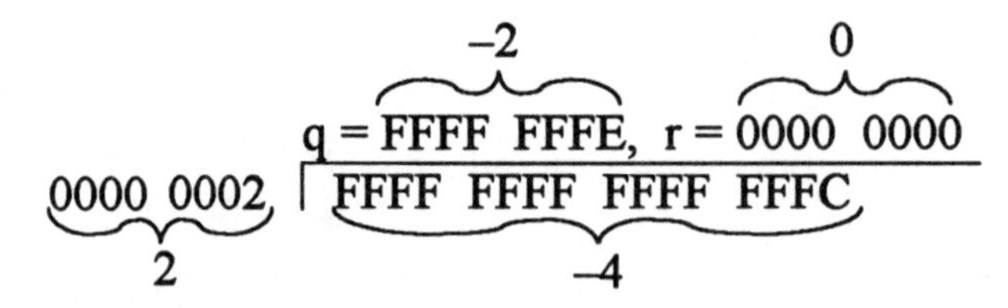

[D2] = $00000000 = remainder, [D0] = $FFFFFFFE = quotient, X is unaffected, Z = 0, N = 1, V = 0, and C = 0 (always).

- `DIVSL.L #$2,D6:D1` if [D1] = $00041234 and [D6] = $FFFFFFFD

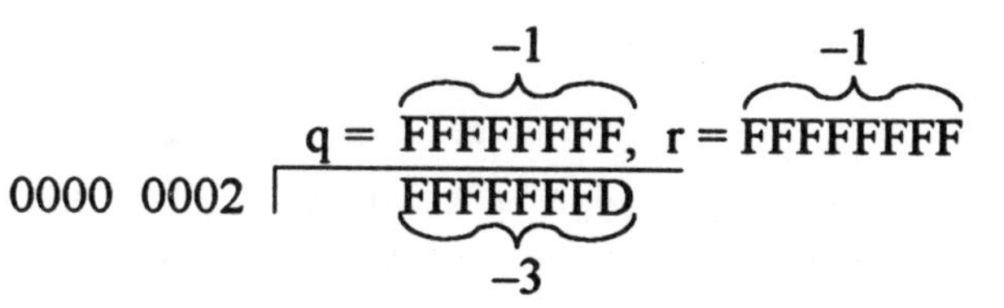

[D6] = \$FFFFFFFF = remainder, [D1] = \$FFFFFFFF = quotient, X is unaffected, N = 1, Z = 0, V = 0, and C = 0 (always).

MC68HC000 Enhanced Instructions

The MC68020 includes the enhanced version of the instructions as listed next:

Instruction	*Operand Size*	*Operation*
`BRA` *label*	8, 16, 32	PC + d → PC
`Bcc` *label*	8, 16, 32	If *cc* is true, then PC + d → PC; else next instruction
`BSR` *label*	8, 16, 32	PC → -(SP); PC + d → PC
`CMPI.S` #data, (EA)	8, 16, 32	Destination – #data → CCR is affected
`TST.S` (EA)	8, 16, 32	Destination – 0 → CCR is affected
`LINK.S` An, -d	16, 32	An → -(SP); SP → An; SP + d → SP
`EXTB.L` Dn	32	Sign-extend byte to long word

Note that S can be B, W, or L. In addition to 8- and 16-bit signed displacements for BRA, B*cc*, and BSR like the 68HC000, the 68020 also allows signed 32-bit displacements. LINK is unsized in the 68HC000. (EA) in CMPI and TST supports all 68HC000 modes plus PC relative. An example is CMPI.W #\$2000, (START, PC). In addition to EXT.W D*n* and EXT.L D*n* like the 68HC000, the 68020 also provides an EXTB.L instruction.

Example 10.26

Write a program in 68020 assembly language to multiply a 32-bit signed number in D2 by a 32-bit signed number in D3 by storing the multiplication result in the following manner:

(a) Store the 32-bit result in D2.

(b) Store the high 32 bits of the result in D3 and the low 32 bits of the result in D2.

Solution

(a)
```
        MULS.L  D3,D2
FINISH  JMP     FINISH
```

(b)
```
        MULS.L  D3,D3:D2
FINISH  JMP     FINISH
```

Example 10.27

Write a program in 68020 assembly language to convert 10 packed BCD bytes (20 BCD digits) stored in memory starting at address $2000 and above, to their ASCII equivalents and, store the result in memory locations starting at $FFFF8000.

Solution

```
       MOVEA.L  #$2000,A0       ;  Load starting addr. of BCD array into A0
       MOVEA.L  #$8000,A1       ;  Load starting addr. of ASCII array into A1
       MOVEQ.L  #9,D0           ;  Load data length into D0
START  MOVE.B   (A0)+,D1        ;  Load a packed BCD byte
       UNPK     D1,D2,#$3030    ;  Convert to ASCII
       MOVE.W   D2,(A1)+        ;  Store ASCII data to addr. pointed to by A1
       DBF      D0,START        ;  Decrement and branch if false
FINISH JMP      FINISH          ;  otherwise stop
```

10.16.1.5 M68020 Pins and Signals

The 68020 is arranged in a 13 × 13 matrix array (114 pins defined) and fabricated in a pin grid array (PGA) or other packages such as RC suffix package. Both the 32-bit address (A_0–A_{31}) and data (D_0–D_{31}) pins of the 68020 are nonmultiplexed. The 68020 transfers data with an 8-bit device via D_{31}–D_{24}, with a 16-bit device via D_{16}–D_{31}, and with a 32-bit device via D_{31}–D_0. Figure 10.34 shows the MC68020 functional signal group. Table 10.23 lists these signals along with a description of each. There are 10 Vcc (+5 V) and 13 ground pins to distribute power in order to reduce noise.

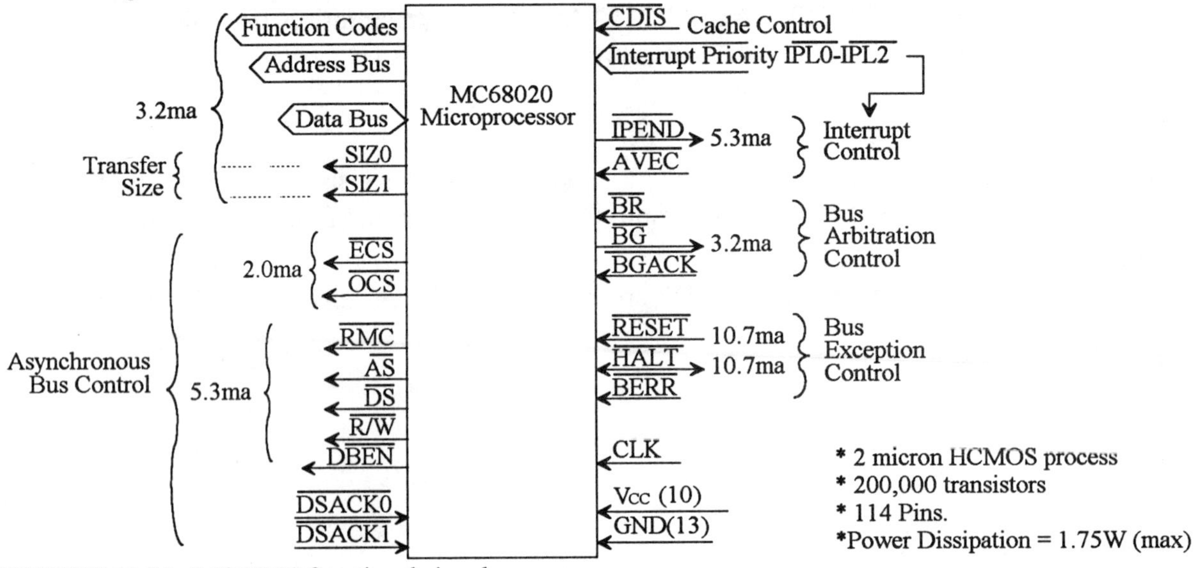

FIGURE 10.34 MC68020 functional signal groups

TABLE 10.23 Hardware Signal Index

Signal Name	*Mnemonic*	*Function*
Address bus	A_0-A_{31}	32-bit address bus used to address any of 4,294,967,296 bytes
Data bus	D_0-D_{31}	32-bit data bus used to transfer 8,16,24, or 32 bits of data per bus cycle
Function codes	FC0-FC2	3-bit function code used to identify the address space of each bus cycle
Size	SIZ0/SIZ1	Indicates the number of bytes remaining to be transferred for this cycle; these signals, together with A0 and A1, define the active sections of the data bus.
Read-modify-write cycle	$\overline{RMC}$	Provides an indicator that the current bus cycle is part of an indivisible read-modify-write operation
External cycle start	$\overline{ECS}$	Provides an indication that a bus cycle is beginning
Operand cycle start	$\overline{OCS}$	Identical operation to that of $\overline{ECS}$ except that $\overline{OCS}$ is asserted only during the first bus cycle of an operand transfer
Address strobe	$\overline{AS}$	Indicates that a valid address is on the bus
Data strobe	$\overline{DS}$	Indicates that valid data is to be placed on the data bus by an external device or has been placed on the data bus by the MC68020
Read/write	$R/\overline{W}$	Defines the bus transfer as an 68020 read or write
Data buffer enable	$\overline{DBEN}$	Provides an enable signal for external data buffers
Data transfer and size acknowledge	$\overline{DSACK0}$/ $\overline{DSACK1}$	Bus response signals that indicate the requested data transfer operation are completed; in addition, these two lines indicate the use of the external bus port on a cycle-by-cycle basis
Cache disable	$\overline{CDIS}$	Dynamically disables the on-chip cache
Interrupt priority level	$\overline{IPL0}$-$\overline{IPL2}$	Provides an encoded interrupt level to the processor
Autovector	$\overline{AVEC}$	Requests an autovector during an interrupt acknowledge cycle
Interrupt pending	$\overline{IPEND}$	Indicates that an interrupt is pending
Bus request	$\overline{BR}$	Indicates that an external device requires bus mastership
Bus grant	$\overline{BG}$	Indicates that an external device may assume bus mastership
Bus grant acknowledge	$\overline{BGACK}$	Indicates that an external device has assumed bus control
Reset	$\overline{RESET}$	System reset
Halt	$\overline{HALT}$	Indicates that the processor should suspend bus activity
Bus error	$\overline{BERR}$	Indicates that an illegal bus operation is being attempted
Clock	CLK	Clock input to the processor
Power supply	VCC	+5 volt ± 5% power supply
Ground	GND	Ground connection

Like the MC68HC000, the three function code signals FC2, FC1, and FC0 identify the processor state (supervisor or user) and the address space of the bus cycle currently being executed except that the 68020 defines the CPU space cycle as follows:

FC2	*FC1*	*FC0*	*Cycle type*
0	0	0	Undefined, reserved
0	0	1	User data space
0	1	0	User program space
0	1	1	Undefined, reserved
1	0	0	Undefined, reserved
1	0	1	Supervisor data space
1	1	0	Supervisor program space
1	1	1	CPU space

Note that in the 68HC000, FC2, FC1, FC0 = 111 indicates the interrupt acknowledge cycle. In the MC68020, it indicates the CPU space cycle. In this cycle, by decoding the address lines A_{19}–A_{16}, the MC68020 can perform various types of functions such as coprocessor communication, breakpoint acknowledge, interrupt acknowledge, and module operations as follows:

A_{19}	A_{18}	A_{17}	A_{16}	*Function performed*
0	0	0	0	Breakpoint acknowledge
1	0	0	1	Module operations
0	0	1	0	Coprocessor communication
1	1	1	1	Interrupt acknowledge

Note that A_{19}, A_{18}, A_{17}, A_{16} = 0011_2 to 1110_2 is reserved by Motorola. In the coprocessor communication CPU space cycle, the MC68020 determines the coprocessor type by decoding A_{15}–A_{13} as follows:

A_{15}	A_{14}	A_{13}	*Coprocessor Type*
0	0	0	MC68851 paged memory management unit
0	0	1	MC68881 floating-point coprocessor

The 68020 offers a feature called "dynamic bus sizing," which enables designers to use 8-bit and 16- and 32-bit memory and I/O devices without sacrificing system performance. The SIZ0, SIZ1, $\overline{\text{DSACK0}}$ and $\overline{\text{DSACK1}}$ pins are used to implement this. These pins are defined as follows:

SIZ1	*SIZ0*	*Number of Bytes Remaining to be Transferred*
0	1	Byte
1	0	Word
1	1	3 bytes
0	0	Long words

$\overline{DSACK1}$	$\overline{DSACK0}$	*Device Size*
0	0	32-bit device
0	1	16-bit device
1	0	8-bit device
1	1	Data not ready; insert wait states

During each bus cycle, the external device indicates its width via $\overline{DSACK0}$ and $\overline{DSACK1}$. The $\overline{DSACK0}$ and $\overline{DSACK1}$ pins are used to indicate completion of bus cycle. At the start of a bus cycle, the 68020 always transfers data to lines D_0–D_{31}, taking into consideration that the memory or I/O device may be 8, 16, or 32 bits wide. After the first bus cycle, the 68020 knows the device size by checking the $\overline{DSACK0}$ and $\overline{DSACK1}$ pins and generates additional bus cycles if needed to complete the transfer.

Unlike the 68HC000, the 68020 permits word and long word operands to start at an odd address. However, if the starting address is odd, additional bus cycles are required to complete the transfer. For example, for a 16-bit device, the 68020 requires 2 bus cycles for a write to an even address such as `MOVE.L D1,$40002050` to complete the operation. On the other hand, the 68020 requires 3 bus cycles for `MOVE.L D1,$40002051` for a 16-bit device to complete the transfer. Note that, as in the 68HC000, instructions in the 68020 must start at even addresses.

Next, consider an example of dynamic bus sizing. The four bytes of a 32-bit data can be defined as follows:

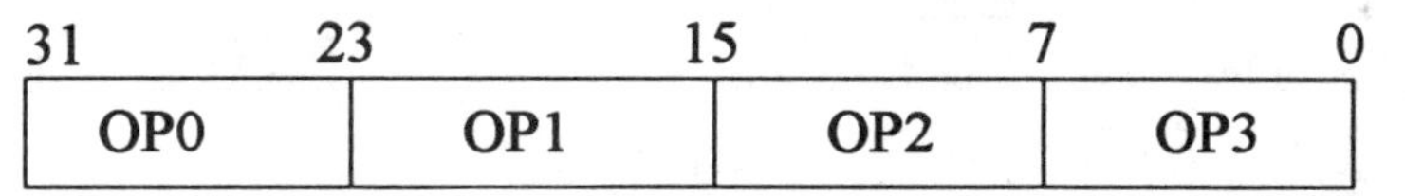

If this data is held in a data register D*n* and is to be written to a memory or 1/0 location, then the address lines A_1 and A_0 define the byte position of data. For a 32-bit device, A_1A_0 = 00 (addresses 0, 4, 8, ...), A_1A_0 = 01 (addresses 1, 5, 9, ...), A_1A_0 = 10 (addresses 2, 6, 10, ...), and A_1A_0 = 11 (addresses 3, 7, 11, ...) will store OP0, OP1, OP2, and OP3, respectively. This data is written via the 68020 D_{31}–D_0 pins. However, if the device is 16-bit, data is always transferred as follows:

All even-addressed bytes via pins D_{31}–D_{24}.
All odd-addressed bytes via pins D_{23}–D_{16}.

Finally, for an 8-bit device, both even- and odd-addressed bytes are transferred via pins D_{31}–D_{24}.

The 68020 always starts transferring data with the most significant byte first. As an example, consider `MOVE.L D1,$20107420`. In the first bus cycle, the 68020 does not know the size of the device and, hence, outputs all combinations of data on pins D_{31}–D_0, taking into consideration that the device may be 8, 16, or 32 bits wide. Assume that the content of D1 is \$02A10512 (OP0 = \$02, OP1 = \$A1, OP2 = \$05, and OP3 = \$12). In the first bus cycle, the 68020 sends SIZ1 SIZ0 = 00, indicating a 32-bit transfer, and then outputs data on its D_{31}–D_0 pins as follows:

$D_{31}:D_{24}$	$D_{23}:D_{16}$	$D_{15}:D_8$	$D_7:D_0$
\$02	\$A1	\$05	\$12

If the device is 8-bit, it will take data \$02 from pins D_{31}–D_{24} in the first cycle and will then assert $\overline{DSACK1}$ and $\overline{DSACK0}$ as 10, indicating an 8-bit device. The 68020 then transfers the remaining 24 bits (\$A1 first, \$05 next, and \$12 last) via pins D_{31}–D_{24} in three consecutive cycles, with a total of four cycles being necessary to complete the transfer.

However, if the device is 16-bit, in the first cycle the device will take the 16-bit data \$02A1 via pins D_{31}–D_{16} and will then assert $\overline{DSACK1}$ and $\overline{DSACK0}$ as 01, indicating a 16-bit device. The 68020 then transfers the remaining 16 bits (\$0512) via pins D_{31}–D_{16} in the next cycle, requiring a total of two cycles for the transfer.

Finally, if the device is 32-bit, the device receives all 32-bit data \$02A10512 via pins D_{31}–D_0 and asserts $\overline{DSACK1}$ $\overline{DSACK0}$ = 00 to indicate completion of the transfer. Aligned data transfers for various devices are as follows :

For 8-bit device:

31. . . . . . . 0 ← Bit number

Register D1 | 02 | A1 | 05 | 12 |

68020 pins	D_{31} D_{24}	SIZ1	SIZ0	A_1	A_0	$\overline{DSACK1}$	$\overline{DSACK0}$
First cycle	02	0	0	0	0	1	0
Second cycle	A1	1	1	0	1	1	0
Third cycle	05	1	0	1	0	1	0
Fourth cycle	12	0	1	1	1	1	0

For 16-bit device:

68020 pins	D_{31} D_{24}	D_{23} D_{16}	SIZ1	SIZ0	A_1	A_0	$\overline{DSACK1}$	$\overline{DSACK0}$
First cycle	02	A1	0	0	0	0	0	1
Second cycle	05	12	1	0	1	0	0	1

For 32-bit device:

68020 pins	D_{31} D_0	SIZ1	SIZ0	A_1	A_0	$\overline{DSACK1}$	$\overline{DSACK0}$
First cycle	02 A1 05 12	0	0	0	0	0	0

Next, consider a misaligned transfer such as MOVE.W D1, \$02010741 with [D1] = \$20F107A4. The 68020 outputs \$0707A4XX on its D_{31}-D_0 pins in its first cycle where XX are don't cares. Data transfers to various devices are summarized below:

For 8-bit device:

	31–24	23–16	15–8	7–0
Register D1	20	F1	07	A4

(← Bit number: 31, 23, 15, 7, 0)

68020 pins	D_{31} D_{24}	SIZ1	SIZ0	A_1	A_0	$\overline{DSACK1}$	$\overline{DSACK0}$
First cycle	07	1	0	0	1	1	0
Second cycle	A4	0	1	1	0	1	0

For 16-bit device:

68020 pins	D_{31} D_{24}	D_{23} D_{16}	SIZ1	SIZ0	A_1	A_0	$\overline{DSACK1}$	$\overline{DSACK}$
First cycle		07	1	0	0	1	0	1
Second cycle	A4		0	1	1	0	0	1

For 32-bit device:

68020 pins	D_{31} D_{24}	D_{23} D_{16}	D_{15} D_8	D_7 D_0	SIZ1	SIZ0	A_1	A_0	$\overline{DSACK1}$	$\overline{DSACK}$
First cycle		07	A4		1	0	0	1	0	0

Let us explain some of the other 68020 pins.
The $\overline{ECS}$ (external cycle start) pin is an MC68020 output pin. The MC68020 asserts this pin during the first one half clock of every bus cycle to provide the earliest indication of the start of a bus cycle. The use of $\overline{ECS}$ must be validated later with $\overline{AS}$, because the MC68020 may start an instruction fetch cycle and then abort it if the instruction is found in the cache. In the case of a cache hit, the MC68020 does not assert $\overline{AS}$, but provides A_{31}–A_0, SIZ1, SIZ0, and FC2–FC0 outputs.

The MC68020 $\overline{AVEC}$ input is activated by an external device to service an autovector interrupt. The $\overline{AVEC}$ has the same function as $\overline{VPA}$ on the 68HC000. The functions of the other signals, such as $\overline{AS}$, R/$\overline{W}$, $\overline{IPL2}$ - $\overline{IPL0}$, $\overline{BR}$, $\overline{BG}$, and $\overline{BGACK}$, are similar to those of the MC68HC000.

The MC68020 system control pins are functionally similar to those of the MC68HC000. However, there are some minor differences. For example, for hardware reset, $\overline{RESET}$ and $\overline{HALT}$ pins need not be asserted simultaneously. Therefore, unlike the 68HC000, the $\overline{RESET}$ and $\overline{HALT}$ pins are not required to be tied together in the MC68020 system.

The $\overline{\text{RESET}}$ and $\overline{\text{HALT}}$ pins are bidirectional and open drain (external pull-up resistances are required), and their functions are independent. The $\overline{\text{RESET}}$ signal is a bidirectional signal. The $\overline{\text{RESET}}$ pin, when asserted by an external circuit for a minimum of 520 clock periods, the $\overline{\text{RESET}}$ pin resets the entire system including the MC68020. Upon hardware reset, the MC68020 completes any active bus cycle in an orderly manner and then performs the following:

- Reads the 32-bit content of address $00000000 and loads it into the ISP (the contents of $00000000 are loaded to the most significant byte of the ISP and so on).
- Reads the 32-bit contents of address $00000004 into the PC (contents of $00000004 to most significant byte of the PC and so on).
- Sets the I2 I1 I0 bits of the SR to 1 1 1, sets the S bit in the SR to 1, and clears the T1, T0, and M bits in the SR.
- Clears the VBR to $00000000.
- Clears the cache enable bit in the CACR.
- All other registers are unaffected by hardware reset.

When the RESET instruction is executed, the MC68020 asserts the $\overline{\text{RESET}}$ pin for 512 clock cycles and the processor resets all the external devices connected to the $\overline{\text{RESET}}$ pin. Software reset does not affect any internal register.

As mentioned earlier while describing dynamic bus sizing, the 68020 always drives all data lines during a write operation. Furthermore, for all inputs there is a sample window of at least 20 ns during which the 68020 latches the input level. To guarantee the recognition of a certain level on a particular falling edge of the clock, the input level must be held stable throughout this sample window, 20 ns; otherwise, the level recognized by the MC68020 is unknown or legal.

During data transfer operations, the 68020 can use either synchronous or asynchronous operation. In synchronous operation, the 68020 clock is used to generate $\overline{\text{DSACK1}}$, $\overline{\text{DSACK0}}$, and other asynchronous inputs. Also, in synchronous operation, if the $\overline{\text{DSACK1}}$ and $\overline{\text{DSACK0}}$ are asserted for the required window of at least 20 ns (at least 5 ns before and at least15 ns after the falling edge of S2) on the falling edge S2, the 68020 latches valid data on the falling edge of S4 on a read cycle. The 68020 does not generate any wait states if $\overline{\text{DSACK1}}$ and $\overline{\text{DSACK0}}$ are asserted at the falling edge of S2; otherwise the 68020 inserts wait cycles like the 68HC000 and latches data at the falling edge of the following cycle as soon as $\overline{\text{DSACK1}}$ and $\overline{\text{DSACK0}}$ are asserted. A minimum of three clock cycles are required for a read operation.

In asynchronous operation, clock frequency independence at a system level is achieved and the 68020 is used in an asynchronous manner. This typically requires using the bus signals such as $\overline{AS}$, $\overline{DS}$, $\overline{DSACK1}$, and $\overline{DSACK0}$ to control data transfer. Using asynchronous operation, $\overline{AS}$ starts the bus cycle and $\overline{DS}$ is used as a condition of valid data on a write cycle. Decoding of SIZ1, SIZ0, A_1, and A_0 provides enable signals, which indicate the portion of the data bus is used in data transfer. The memory or I/O chip then responds by placing the requested data on the correct portion of the data bus for a read cycle or latching the data on a write cycle and asserting $\overline{DSACK1}$ and $\overline{DSACK0}$, corresponding to the memory or I/O port size (8-bit, 16-bit, or 32-bit), to terminate the bus cycle. If no memory or I/O device responds or the address is invalid, the external control logic asserts the $\overline{BERR}$ or $\overline{BERR}$ and $\overline{HALT}$ signal(s) to abort or retry the bus cycle or retries the bus cycle.

In asynchronous operation, the $\overline{DSACK1}$ and $\overline{DSACK0}$ signals are allowed to be asserted before the data from memory or an I/O device is valid on a read cycle. The 68020 latches data according to Parameter #31 provided in Motorola manuals. (Parameter #31 is a maximum of 60 ns for the 12.5-MHz 68020, a maximum of 50 ns for the 16.67-MHz 68020, and a maximum of 43 ns for the 20-Mhz 68020, and maximum time is specified from the assertion of $\overline{AS}$ to the assertion of $\overline{DSACK1}$ and $\overline{DSACK0}$. This is because the 68020 will insert wait cycles in one-clock-cycle increments until $\overline{DSACK1}$ and $\overline{DSACK0}$ are recognized as asserted.)

10.16.1.6 MC68020 System Design

The following 8-MHz 68020 system design will use a 128 KB 32-bit wide supervisor data memory. Four 27C256's (32K × 8 HCMOS EPROM with 120-ns access time) are used for this purpose. Because the memory is 32 KB, the 68020 address lines A_2-A_{16} are used for addressing the 27C256's. The 68020 SIZ1, SIZ0, A_1, A_0, $\overline{DSACK1}$, and $\overline{DSACK0}$ pins are utilized for selecting the memory chips.

Table 10.24 shows the table for designing the enable logic for the four 27C256 chips. The 68020 A_{17} pin is used to distinguish between memory and I/O. $A_{17} = 0$ is used to select the memory chips; $A_{17} = 1$ is used to select I/O chips (not shown in the design). Table 10.25 shows the K-maps for the enable logic. A logic diagram can be drawn for generating the memory byte enable signals $\overline{DBBE1}$, $\overline{DBBE2}$, $\overline{DBBE3}$, and $\overline{DBBE4}$.

The 68020 system with 32-bit memory consists of four 27C256's, each connected to its associated portion of the system data bus (D_{31}–D_{24}, D_{23}–D_{16}, D_{15}–D_8, and D_7–D_0). To manipulate this memory configuration, 32-bit data bus control byte enable logic is incorporated to generate byte enable signals ($\overline{DBBE1}$, $\overline{DBBE2}$, $\overline{DBBE3}$, and $\overline{DBBE4}$). These byte enables are generated by using 68020's SIZ1, SIZ0, A_1, A_0, A_{17}, and $\overline{DS}$ pins as shown in the individual logic diagrams of the byte enable logic. A PAL can be programmed to implement this logic. A schematic of the 68020–27C256 interface is shown in Figure 10.35.

TABLE 10.24 Table for memory enables for 32-bit memory

SIZ1	*SIZ0*	A_1	A_0	*DBBE11*	*DBBE22*	*DBBE33*	*DBBE44*
0	1	0	0	1	0	0	0
		0	1	0	1	0	0
		1	0	0	0	1	0
		1	1	0	0	0	1
1	0	0	0	1	1	0	0
		0	1	0	1	1	0
		1	0	0	0	1	1
		1	1	0	0	0	1
1	1	0	0	1	1	1	0
		0	1	0	1	1	1
		1	0	0	0	1	1
		1	1	0	0	0	1
0	0	0	0	1	1	1	1
		0	1	0	1	1	1
		1	0	0	0	1	1
		1	1	0	0	0	1

TABLE 10.25 K-maps for Enable Signals for Memory

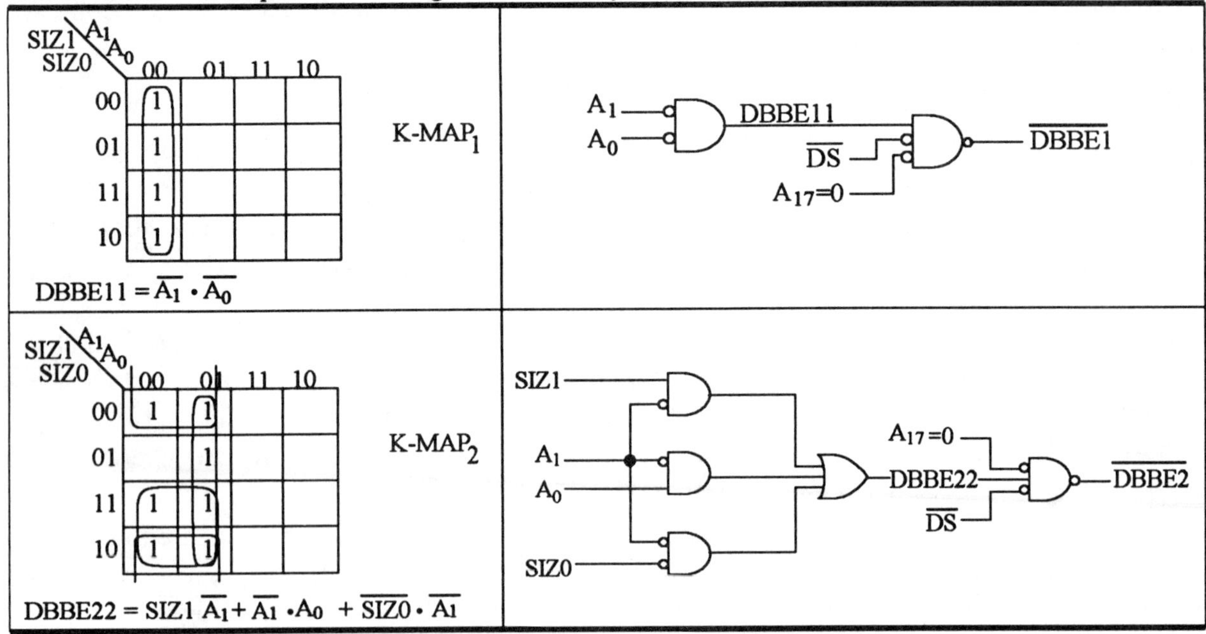

TABLE 10.25 K-maps for Enable Signals for Memory (continued)

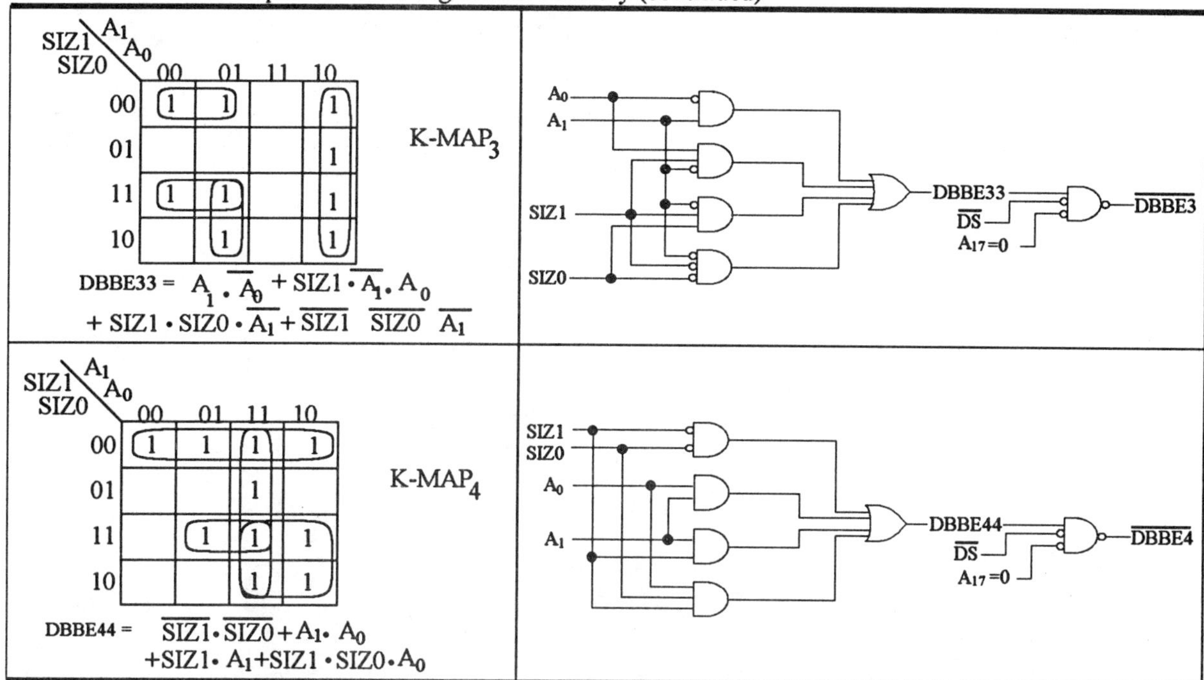

Because the 68020 clock is used to generate $\overline{\text{DSACK1}}$ and $\overline{\text{DSACK0}}$, the 68020 operates in synchronous mode.

A 74HC138 decoder is used for selecting memory banks to enable the appropriate memory chips. The 74HC138 is enabled by $\overline{\text{AS}} = 0$. The output line 5 (FC2FC1FC0 = 101 for supervisor data) is used to select the memory chips. Assuming don't cares to be zeros and also note that $A_{17} = 0$ for memory, the supervisor data memory map is obtained as follows:

EPROM #1 $00000000, $00000004, ..., $0001FFFC
EPROM #2 $00000001, $00000005, ..., $0001FFFD
EPROM #3 $00000002, $00000006, ..., $0001FFFE
EPROM #4 $00000003, $00000007, ..., $0001FFFF

$\overline{\text{DSACK1}}$ and $\overline{\text{DSACK0}}$ are generated by ANDing the $\overline{\text{DBBE1}}$, $\overline{\text{DBBE2}}$, $\overline{\text{DBBE3}}$, and $\overline{\text{DBBE4}}$ outputs of the byte enable logic circuit. When one or more EPROM chips are selected, the appropriate enables ($\overline{\text{DBBE1}}$–$\overline{\text{DBBE4}}$) will be low, thus asserting $\overline{\text{DSACK1}} = 0$ and $\overline{\text{DSACK0}} = 0$. This will tell the 68020 that the memory is 32 bits wide. Data from the selected memory chip(s) will be placed on the appropriate data pins of the 68020. For example, in response to execution of the instruction `MOVE.W $00000001,D0` in the supervisor mode, the 68020 will generate appropriate signals to generate $\overline{\text{DBBE1}}= 1$, $\overline{\text{DBBE2}}= 0$, $\overline{\text{DBBE3}}= 0$, $\overline{\text{DBBE4}}= 1$, $R/\overline{W} = 1$, and output 5 of the decoder = 0.

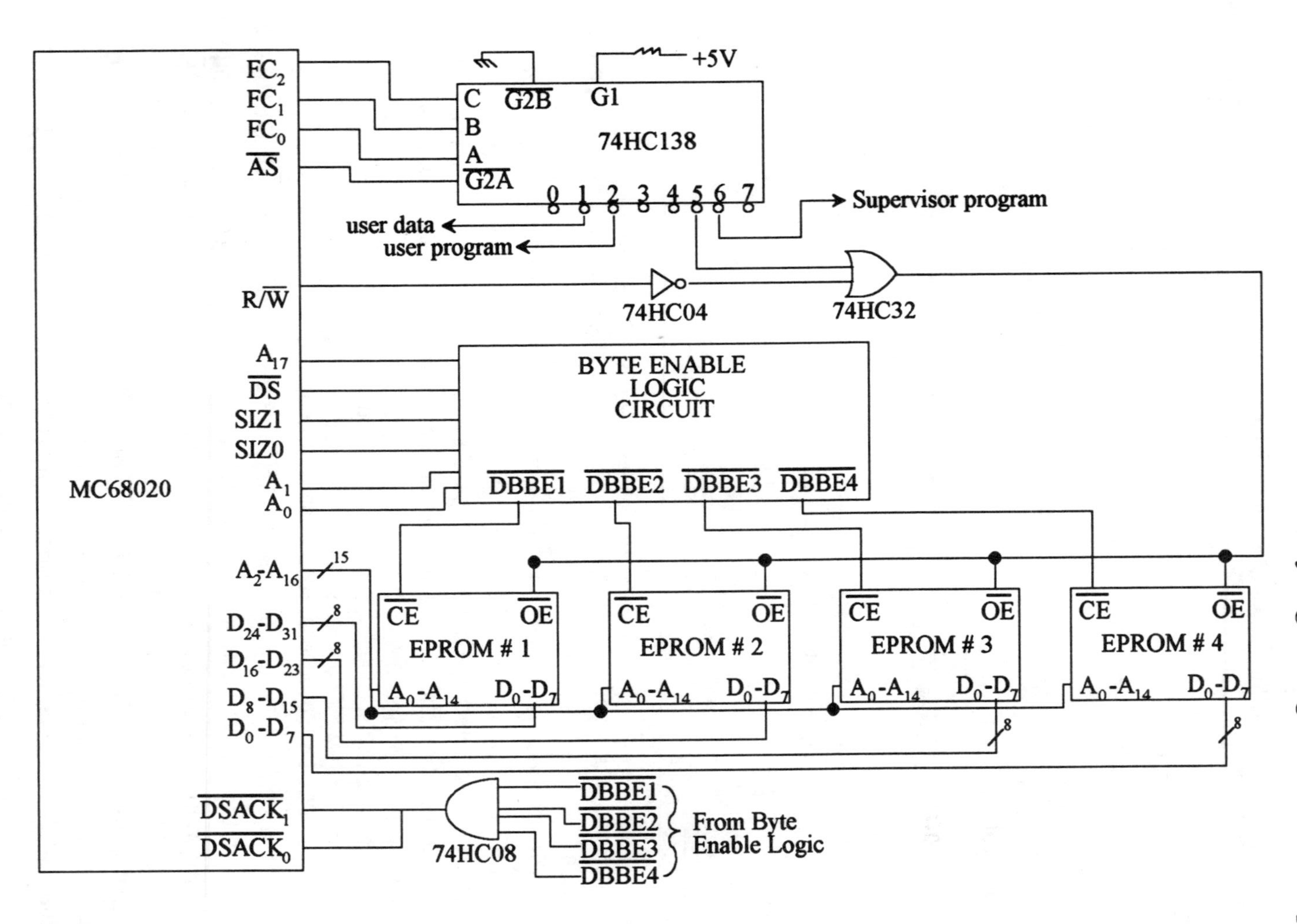

FIGURE 10.35 68020 / 27C256 System

This will select EPROM #2 and EPROM #3 chips. Thus, the contents of address $00000001 are transferred to D0 (bits 8–15) and the contents of address $00000002 are moved to D0 (bits 0–7). The supervisor program, user program, and user data memories can be connected in a similar way (not shown in the figure). For each memory space, four memory chips are required.

Let us discuss the timing requirements of the 68020/27C256 system. Because the 68020 clock is used to generate $\overline{\text{DSACK1}}$ and $\overline{\text{DSACK0}}$, the 68020 operates in synchronous mode. This means that the 68020 checks $\overline{\text{DSACK1}}$ and $\overline{\text{DSACK0}}$ for LOW at the falling edge of S2 (two cycles). From the 68020 timing diagram (Motorola manual), $\overline{\text{AS}}$, $\overline{\text{DS}}$, and all other output signals used in memory decoding go to LOW at the end of approximately one clock cycle. For an 8-MHz 68020 clock, each cycle is 125 ns. From byte enable logic diagrams, a maximum of four gate delays (40 ns) are required. Therefore, the selected EPROM(s) will be enabled after 165 ns (125 ns + 40 ns). With 120-ns access time, the EPROM(s) will place data on the output lines after approximately 285 ns (165 ns + 120 ns). With an 8-MHz 68020 clock, $\overline{\text{DSACK1}}$ and $\overline{\text{DSACK0}}$ will be checked for LOW (32-bit memory) after two cycles (250 ns) and if LOW, the 68020 will latch data after three cycles (375 ns). Hence, no delay circuit is required for $\overline{\text{DSACK1}}$ and $\overline{\text{DSACK0}}$. In case a delay circuit is required, a ring counter can be used. Note that the 20-ns window requirement for $\overline{\text{DSACK1}}$ and $\overline{\text{DSACK0}}$ inputs (5 ns before and 15 ns after the falling edge of S2) is satisfied.

MC68020 I/O

The 68020 I/O handling features are very similar to those of the 68000. This means that the 68020 uses memory-mapped I/O, and the 68230 I/O chip can be used for programmed I/O. The external interrupts are handled via the 68020 $\overline{\text{IPL2}}$, $\overline{\text{IPL1}}$, and $\overline{\text{IPL0}}$ pins using autovectoring and nonautovectoring pins. However, the 68020 uses a new pin called $\overline{\text{AVEC}}$ rather than $\overline{\text{VPA}}$ (68HC000) for autovectoring. Nonautovectoring is handled using $\overline{\text{DSACK0}}$ = 0 and $\overline{\text{DSACK1}}$ = 0 rather than $\overline{\text{DTACK}}$= 0 (as with the 68HC000). Note that the 68020 does not have the $\overline{\text{VPA}}$ pin. Like the 68HC000, the 68020 uses the $\overline{\text{BR}}$, $\overline{\text{BG}}$, and $\overline{\text{BGACK}}$ pins for DMA transfer. The 68020 exceptions are similar to those of the 68000 with some variations such as coprocessor exceptions.

10.16.2 Motorola MC68030

The MC68030 is a virtual memory microprocessor based on the MC68020 with additional features. The MC68030 is designed by using HCMOS technology and can be operated at clock rates of 16.67 and 33 MHz. The MC68030 contains all features of the MC68020, plus some additional ones. The basic differences between the MC68020 and MC68030 are as follows:

Characteristics	*MC 68020*	*MC68030*
On-chip cache	256-byte instruction cache	256-byte instruction cache and 256 byte data cache
On-chip memory management unit (MMU)	None	Paged data memory management (demand page of the MC68851)
Instruction set	101	103 (four new instructions are for on-chip MMU); CALLM and RTM instructions are not supported.

Like the MC68020, the MC68030 also supports 7 data types and 18 addressing modes. The MC68030 I/O is identical to the MC68020.

10.16.3 Motorola MC68040 / MC68060

This section presents an overview of the Motorola MC68040 and MC 68060 32-bit microprocessors. The MC68040 is Motorola's enhanced 68030, 32-bit microprocessor, implemented in HCMOS technology. Providing balance between speed, power, and physical device size, the MC68040 integrates on-chip MC68030-compatible integer unit, an MC68881/ MC68882-compatible floating-point unit (FPU), dual independent demand-paged memory management units (MMUs) for instruction and data stream accesses, and an independent 4 KB instruction and data cache. A high degree of instruction execution parallelism is achieved through the use of multiple independent execution pipelines, multiple internal buses, and separate physical caches for both instruction and data accesses. The MC68040 also includes 32-bit nonmultiplexed external address and data buses .

The MC68060 is a superscalar (two instructions per cycle) 32-bit microprocessor. For some reason, Motorola does not offer MC68050 microprocessor. The 68060 is fully compatible with the 68040 in the user mode. The 68060 can operate at 50- and 66-MHz clocks with performance much faster than the 68040. An striking feature of the 68060 is the power consumption control. The 68060 is designed using static HCMOS to reduce power during normal operation.

10.16.4 PowerPC Microprocessor

This section provides an overview of the hardware, software, and interfacing features associated with the RISC microprocessor called the PowerPC. Finally, the basic features of both 32-bit and 64-bit PowerPC microprocessors are discussed.

10.16.4.1 Basics of RISC

RISC is an acronym for Reduced Instruction Set Computer. This type of microprocessor emphasizes simplicity and efficiency. RISC designs start with a necessary and sufficient instruction set. The purpose of using RISC architecture is to maximize speed by reducing clock cycles per instruction. Almost all computations can be obtained from a few simple operations. *The goal of RISC architecture is to maximize the effective speed of a design by performing infrequent operations in software and frequent functions in hardware, thus obtaining a net performance gain.* The following summarizes the typical features of a RISC microprocessor:

1. *The RISC microprocessor is designed using hardwired control with little or no microcode. Note that variable-length instruction formats generally require microcode design. All RISC instructions have fixed formats, so microcode design is not necessary.*
2. *A RISC microprocessor executes most instructions in a single cycle.*
3. The instruction set of a RISC microprocessor typically includes only register, load, and store instructions. All instructions involving arithmetic operations use registers, and load and store operations are utilized to access memory.
4. The instructions have a simple fixed format with few addressing modes.
5. A RISC microprocessor has several general-purpose registers and large cache memories.
6. A RISC microprocessor processes several instructions simultaneously and thus includes pipelining.
7. Software can take advantage of more concurrency. For example, Jumps occur after execution of the instruction that follows. This allows fetching of the next instruction during execution of the current instruction.

RISC microprocessors are suitable for embedded applications. An embedded application is one in which the processor monitors and analyzes signals from one segment of the system and produces output required by another segment of the system; thus, it behaves as a controller that bridges various parts of the entire system. It performs all its functions without any user input.

RISC microprocessors are well suited for applications such as image processing, robotics, graphics, and instrumentation. The key features of the RISC microprocessors that make them ideal for these applications are their relatively low level of integration in the chip and instruction pipeline architecture. These characteristics result in low power consumption, fast instruction execution, and fast recognition of interrupts. Typical 32- and 64-bit RISC microprocessors include PowerPC microprocessors.

10.16.4.2 IBM/Motorola/Apple PowerPC 601

This section provides an overview of the basic features of PowerPC microprocessors. The PowerPC 601 is jointly developed by Apple, IBM, and Motorola. It is available from IBM as PP 601 and from Motorola as MPC 601. The PowerPC 601 is the first implementation of the PowerPC family of Reduced Instruction Set Computer (RISC) microprocessors. There are two types of PowerPC implementations: 32-bit and 64-bit. The PowerPC 601 implements the 32-bit portion of the IBM PowerPC architectures and Motorola 88100 bus control logic. It includes 32-bit effective (logical) addresses, integer data types of 8, 16, and 32 bits, and floating-point data types of 32 and 64 bits. For 64-bit PowerPC implementations, the PowerPC architecture provides 64-bit integer data types, 64-bit addressing, and other features necessary to complete the 64-bit architecture.

The 601 is a pipelined superscalar processor and is capable of executing three instructions per clock cycle. A pipelined processor is one in which the processing of an instruction is broken down into discrete stages, such as decode, execute, and write-back (the result of the operation is written back in the register file).

Because the tasks required to process an instruction are broken into a series of tasks, an instruction does not require the entire resources of an execution unit. For example, after an instruction completes the decode stage, it can pass on to the next stage, and the subsequent instruction can advance into the decode stage. This improves the throughput of the instruction flow. For example, it may take three cycles for an integer instruction to complete, but if there are no stalls in the integer pipeline, a series of integer instructions can have a throughput of one instruction per cycle. Each unit is kept busy in each cycle.
A superscalar processor is one in which multiple pipelines are provided to allow instructions to execute in parallel. The PowerPC 601 includes three execution units: a 32-bit integer unit (IU), a branch processing unit (BPU), and a pipelined floating-point unit (FPU).

The PowerPC 601 contains an on-chip, 32 KB unified cache (combined instruction and data cache) and an on-chip memory management unit (MMU). It has a 64-bit data bus and a 32-bit address bus. The 601 supports single-beat and four-beat burst data transfer for memory accesses. Note that a single-beat transaction indicates data transfer of up to 64 bits. The PowerPC 601 uses memory-mapped I/O. Input/output devices can also be interfaced to the PowerPC 601 by using the I/O controller. The 601 is designed by using an advanced, CMOS process technology and maintains full compatibility with TTL devices.

The PowerPC 601 contains an on-chip real-time clock (RTC). The RTC was normally an I/O device completely outside the CPU in earlier microcomputers. Although the RTC appearing inside the microcomputer chip is common on single-chip microcomputers, this is the first time the RTC is implemented inside a top-of-the-line microprocessor such as the PowerPC. This implication is that modern multitasking operating systems require time keeping for task switching as well as keeping the calendar date. The 601 real-time clock (RTC) on-chip hardware provides a measure of real time in terms of time of day and date, with a calendar range of 136.19 years.

To specify the ordering of four bytes (ABCD) within 32 bits, the 601 can use either the ABCD (big-endian) or DCBA (little-endian) ordering. The 601 big- or little-endian modes can be selected by setting the LM bit (bit 28) in the HID0 register. Note that big-endian ordering (ABCD) assigns the lowest address to the highest-order eight bits of the multibyte data. On the other hand, little-endian byte ordering (DCBA) assigns the lowest address to the lowest order (rightmost) 8 bits of the multibyte data.

Note that Motorola 68XXX microprocessors support big-endian byte ordering whereas Intel 80XXX microprocessors support little-endian byte ordering.

PowerPC 601 Registers

PowerPC 601 registers can be accessed depending on the program's access privilege level (supervisor or user mode). The privilege level is determined by the privilege level (PR) bit in the machine status register (MSR). The supervisor mode of operation is typically used by the operating system, and user mode is used by the application software. The PowerPC 601 programming model contains user- and supervisor-level registers. Some of these are

- The user-level register can be accessed by all software with either user or supervisor privileges.
- The 32-bit GPRs (general-purpose registers, GPR0–GPR31) can be used as the data source or destination for all integer instructions. They can also provide data for generating addresses.
- The 32-bit FPRs (floating-point registers, FPR0–FPR31) can be used as data sources and destinations for all floating-point instructions.
- The floating-point status and control register (FPCSR) is a user control register in the floating-point unit (FPU). It contains floating-point status and control bits such as floating-point exception signal bits, exception summary bits, and exception enable bits.
- The condition register (CR) is a 32-bit register, divided into eight 4-bit fields, CR0–CR7. These fields reflect the results of certain arithmetic operations and provide mechanisms for testing and branching.

  The remaining user-level registers are 32-bit special purpose registers—SPR0, SPR1, SPR4, SPR5, SPR8, and SPR9.

- SPR0 is known as the MQ register and is used as a register extension to hold the product for the multiplication instructions and the dividend for the divide instructions. The MQ register is also used as an operand of long shift and rotate instructions.

- SPR1 is called the integer exception register (XER). The XER is a 32-bit register that indicates carries and overflow bits for integer operations. It also contains two fields for load string and compare byte indexed instructions.
- SPR4 and SPR5 respectively represent two 32-bit read only registers and hold the upper (RTCU) and lower (RTCL) portions of the real-time clock (RTC). The RTCU register maintains the number of seconds from a time specified by software. The RTCL register maintains the fraction of the current second in nanoseconds.
- SPR8 is the 32-bit link register (LR). The link register can be used to provide the branch target address and to hold the return address after branch and link instructions.
- SPR9 represents the 32-bit count register (CTR). The CTR can be used to hold a loop count that can be decremented during execution of certain branch instructions. The CTR can also be used to hold the target address for the branch conditional to count register instruction.

PowerPC 601 Addressing Modes

The effective address (EA) is the 32-bit address computed by the processor when executing a memory access or branch instruction or when fetching the next sequential instruction. *Since the PowerPC is based on the RISC architecture, arithmetic and logical instructions do not read or modify memory.*

Load and store operations have two types of effective address generation:

i) ***Register Indirect with Immediate Index Mode***

Instructions using this mode contain a signed 16-bit index (d operand in the 32-bit instruction) which is sign extended to 32-bits, and added to the contents of a general-purpose register specified by five bits in the 32-bit instruction (rA operand) to generate the effective address. A zero in the rA operand causes a zero to be added to the immediate index (d operand). The option to specify rA or 0 is shown in the instruction descriptions of the 601 user's manual as the notation (rA|0).

An example is `lbz` rD,d (rA) where rA specifies a general-purpose register (GPR) containing an address, d is the the 16-bit immediate index and rD specifies a general-purpose register as destination. Consider `lbz r1,20(r3)`. The effective address (EA) is the sum r3+20. The byte in memory addressed by the EA is loaded into bits 31 through 24 of register r1. The remaining bits in r1 are cleared to zero. Note that the registers r1 and r3 represent GPR1 and GPR3 respectively.

ii) ***Register Indirect with Index Mode***

Instructions using this addressing mode add the contents of two general-purpose registers (one GPR holds an address and another holds the index). An example is `lbzx` rD, rA, rB where rD specifies a GPR as destination, rA specifies a GPR as the index, and rB specifies a GPR holding an address. Consider `lbzx r1,r4,r6`. The effective address (EA) is the sum (r4|0)+(r6). The byte in memory adressed by the EA is loaded into register r1 (24-31). The remaining bits in register rD are cleared to zero.

PowerPC 601 conditional and unconditional branch instructions compute the effective address (EA) or the next instruction address using various addressing modes A few of them are described below:

- ***Branch Relative*** Branch instructions (32-bit wide) using the relative mode generate the address of the next instruction by adding an offset and the current program counter contents. An example of this mode is an instruction `be start` unconditionally jumps to the address PC + start.

- ***Branch Absolute*** Branch instructions using this mode include the address of the next instruction to be executed. For example, the instruction `ba begin` unconditionally branches to the absolute address "begin" specified in the instruction.

- ***Branch to Link Register*** Branch instructions using this mode branch to the address computed as the sum of the immediate offset and the address of the current instruction. The instruction address following the instruction is placed into the link register. For example, the instruction `bl,start` unconditionally jumps to the address computed from current PC contents plus start. The return address is placed in the link register.

- ***Branch to Count Register*** Instructions using this mode branch to the address contained in the current register. Consider `bcttr BO, BI` means branch conditional to count register. This instruction branches conditionally to the address specified in the count register.

  The BI operand specifies the bit in the condition register to be used as the condition of the branch. The BO operand specifies how the branch is affected by or affects condition or count registers. Numerical values specifying BI and BO can be obtained from the 601 manual.

Note that some instructions combine the link register and count register modes. An example is `bcctr BO, BI`. This instruction first performs the same operation as the bcttr and then places the instruction address following the instruction into the link register. This instruction is a form of "conditional call" because the return address is saved in the link register.

Typical PowerPC 601 Instructions

The 601 instructions are divided into the following categories:

1. Integer Instructions
2. Floating-point Instructions

3. Load/store Instructions
4. Flow Control Instructions
5. Processor Control Instructions

Integer instructions operate on byte (8-bit), half-word (16-bit), and word (32-bit) operands. Floating-point instructions operate on single-precision and double-precision floating-point operands.

Integer Instructions

The integer instructions include integer arithmetic, integer compare, integer rotate and shift, and integer logical instructions. The integer arithmetic instructions always set the integer exception register bit, CA, to reflect the carry out of bit 7. Integer instructions with the overflow enable (OE) bit set will cause the XER bits SO (summary overflow — overflow bit set due to exception) and OV (overflow bit set due to instruction execution) to be set to reflect overflow of the 32-bit result. Some examples of integer instructions are provided in the following. Note that rS, rD, rA, and rB in the following examples are 32-bit general purpose registers (GPRs) of the 601 and SIMM is 16-bit signed immediate number.

- `add rD,rA,SIMM` performs the following immediate operation: rD ← (rA|0) + SIMM; rA|0) can be either (rA) or 0. An example is `add rD, rA, SIMM` or `add rD, 0, SIMM`.
- `add rD,rA,rB` performs rD ← rA + rB.
- `add.rD,rA,rB` adds with CR update as follows: rD ← rA + rB. The dot suffix enables the update of the condition register.
- `subf rD,rA,rB` performs rD ← rB - rA.
- `sub rD,rA,rB` performs the same operation as subf but updates the condition code register.
- `addme rD,rA` performs the (add to minus one extended) operation: rD ← (rA) + FFFF FFFFH + CA bit in XER.
- `subfme rD,rA` performs the (subtract from minus one extended) operation: rD ← $(\overline{rA})$ + FFFF FFFFH + CA bit in XER, where $(\overline{rA})$ represents the ones complement of the contents of rA.
- `mulhwu rD,rA,rB` performs an unsigned multiplication of two 32-bit numbers in rA and rB. The high-order 32 bits of the 64-bit product are placed in rD.
- `mulhw rD,rA,rB` performs the same operation as the mulhwu except that the multiplication is for signed numbers.
- `mullw rD,rA,rB` places the low order 32-bits of the 64-bit product (rA)*(rB) into rD. The low-order 32-bit products are independent whether the operands are treated as signed or unsigned integers.

- `mulli rD,rA,SIMM` places the low-order 32 bits of the 48-bit product (rA)*$SIMM_{16}$ into rD. The low-order bits of the 32-bit product are independent whether the operands are treated as signed or unsigned integers.
- `divw rD,rA,rB` divides the 32-bit signed dividend in rA by the 32-bit signed divisor in rB. The 32-bit quotient is placed in rD and the remainder is discarded.
- `divwu rD,rA,rB` is the same as the divw instruction except that the division is for unsigned numbers.
- `cmpi crfD,L,rA,SIMM` compares 32 bits in rA with immediate SIMM treating operands as signed integer. The result of comparison is placed in crfd field (0 for CR0, 1 for CR1, and so on) of the condition register. L=0 indicates 32-bit operands while L=1 represents the 64-bit operands. For example, `cmpi 0,0, rA, 200` compares 32 bits in register rA with immediate value 200 and CR0 is affected according to the comparison.
- `xor rA,rS,rB` performs exclusive-or operation between the contents of rS and rB. The result is placed into register rA.
- `extsb rA,rS` places bits 24-31 of rS into bits 24-31 of rA. Bit 24 of rS is then sign extended through bits 0-23 of rA.
- `slw rA,rS,rB` shifts the contents of rS left the shift count specified by rB [27-31]. Bits shifted out of position 0 are lost. Zeros are placed in the vacated positions on the right. The 32-bit result is placed into rA.
- `srw rA,rS,rB` is similar to `slw rA,rS,rB` except that the operation is for right shift.

Floating-Point Instructions

Some of the 601 floating-point instructions are provided below:

- `fadd frD,frA,frB` adds the contents of the floating-point register, frA to the contents of the floating-point register frB. If the most significant bit of the resultant significand is not a one, then the result is normalized. The result is rounded to the specified position under control of the FPSCR register. The result is rounded to the specified precision under control of the FPSCR register. The result is then placed in frD.

  Note that this fadd instruction requires one cycle in execute stage, assuming normal operations; however, there is an execute stage delay of three cycles if the next instruction is dependent.

  The 601 floating point addition is based on "exponent comparison and add by one" for each bit shifted, until the two exponents are equal. The two significands are then added algebraically to form an intermediate sum. If a carry occurs, the sum's significand is shifted right one bit position and the exponent is increased by one.

- `fsub frD,frA,frB` performs frA – frB, normalization, and rounding of the result are performed in the same way as the `fadd` instruction.
- `fmul frD,frA,frC` performs frD ← frA * frC.
  Normalization and rounding of the result are performed in the same way as the fadd. Floating-point multiplication is based on exponent addition and multiplication of the significands.
- `fdiv frD,frA,frB` performs the floating-point division frD ← frA/frB. No remainder is provided. Normalization and rounding of the result are performed in the same way as the `fadd` instruction.
- `fmsub frD,frA,FrC,frB` performs frD ← frA * frC – frB. Normalization and rounding of the result are performed in the same way as the `fadd` instruction.

Load/Store Instructions

Some examples of the 601 load and store instructions are

- `lhzx rD,rA,rB` loads the half word (16 bits) in memory addressed by the sum (rA|0) + (rB) into bits 16 through 31 of rD. The remaining bits of rD are cleared to zero.
- `sthux rS,rA,rB` stores the 16-bit half word from bits 16–31 of register rS in memory addressed by the sum (rA|0) + (rB). The value (rA|0) + rB is placed into register rA.
- `lmw rD,d(rA)` loads n (where $n = 32 - D$ and $D = 0$ through 31) consecutive words starting at memory location addressed by the sum (r|0) + d into the general-purpose register specified by rD through r31.
- `stmu rS,d(rA)` is similar to `lmw` except that `stmw` stores n consecutive words.

Flow Control Instructions

Flow control instructions include conditional and unconditional branch instructions. An example of one of these instructions is

- `bc` (branch conditional) `BO, BI,` target branch with offset target if the condition bit in CR specified by bit number BI is true (The condition "true" is specified by a value in BO).
  For example, `bc 12,0,target` means that branch with offset target if the condition specified by bit 0 in CR (BI = 0 indicates the result is negative) is true (specified by the value BO = 12 according to Motorola PowerPC 601 manual).

Processor Control Instructions

Processor control instructions are used to read from and write to the machine state register (MSR), condition register (CR), and special status register (SPRs). Some examples of these instructions are

- `mfcr rD` places the contents of the condition register into rD.
- `mtmsr rS` places the contents of rS into the MSR. This is a supervisor-level instruction.

- `mfimsr rD` places the contents of MSR into rD. This is a supervisor-level instruction.

PowerPC 601 Exception Model

All 601 exceptions can be described as either precise or imprecise and either synchronous or asynchronous. Asynchronous exceptions are caused by events external to the processor's execution. Synchronous exceptions, on the other hand, are handled precisely by the 601 and are caused by instructions; precise exception means that the machine state at the time the exception occurs is known and can be completely restored. That is, the instructions that invoke trap and system call exceptions complete execution before the exception is taken. When exception processing completes, execution resumes at the address of the next instruction.

An example of a maskable asynchronous, precise exception is the external interrupt. When an asynchronous, precise exception such as the external interrupt occurs, the 601 postpones its handling until all instructions and any exceptions associated those instructions complete execution. System reset and machine check exceptions are two nonmaskable exceptions that are asynchronous and imprecise. These exceptions may not be recoverable or may provide a limited degree of recoverability for diagnostic purpose.

Asynchronous, imprecise exceptions have the highest priority with the synchronous, precise exceptions having the next priority and the asynchronous, precise exceptions the lowest priority.

The 601 exception mechanism allows the processor to change automatically to supervisor state as a result of exceptions. When exceptions occur, information about the state of the processor is saved to certain registers rather than in memory as is usually done with other processors in order to achieve high speeds. The processor then begins execution at an address (exception vector) predetermined for each exception. The exception handler at the specified vector is then processed with processor in supervisor mode.

601 System Interface

The pins and signals of the PowerPC 601 include a 32-bit address bus and 52 control and information signals. Memory access allows transfer sizes of 8, 16, 24, 32, 40, 48, 56, or 64 bits in one bus clock cycle. Data transfer occurs in either single-beat transactions or four-beat burst transactions. Both memory and I/O accesses can use the same bus transfer protocols. The 601 also has the ability to define memory areas as I/O controller interface areas. The 601 uses the $\overline{TS}$ pin for memory-mapped accesses and the XATS pin for I/O controller interface accesses.

Summary of PowerPC 601 Features

The PowerPC 601 is a RISC-based superscalar microprocessor. That is, it can execute two or more instructions per cycle. The PowerPC 601 is based on load/store architectures. This means that all instructions that access memory are either loads or stores, and all operate instructions are from register to register. Both load and store instructions have 32-bit fixed-length instructions along with 32-bit integer and 32-bit floating-point registers.

The PowerPC 601 includes two primary addressing modes: register plus displacement and register plus register. In addition, the 601 load and store instructions perform the load or store operation and also modify the index register by placing the effective address just computed. In the PowerPC 601, Branch target addresses are normally determined by using program counter relative mode. That is, the branch target address is determined by adding a displacement to the program counter. However, as mentioned before, conditional branches in the 601 may test fields in the condition code register and the contents of a special register called the count register (CTR). A single 601 branch instruction can implement a loop-closing branch by decrementing the CTR, testing its value, and branching if it is nonzero.

The PowerPC 601 saves the return address for certain control transfer instructions such as subroutine call in a general-purpose register. The 601 does this in any branch by setting the link (LK) bit to one. The return address is saved in the link register. The PowerPC 601 utilizes sophisticated pipelines. The 601 uses relatively short independent pipelines with more buffering. The 601 does a lot of computation in each pipe stage. The 601 has a unified (combined) 32 KB cache. That is, instructions and data reside in the same cache in the 601. Finally, the 601 offers high performance by utilizing sophisticated design tricks. For example, the 601 includes powerful instructions such as floating-point multiply-add and update load/store that perform more tasks with fewer instructions.

10.16.4.3 PowerPC 64-Bit Microprocessors

PowerPC 64-bit microprocessors include the PowerPC 620, 603e, 750/740, and 604e. These microprocessors are 64-bit superscalar processors. This means that they can execute more than one instruction in a cycle. Table 10.26 compares the basic features of the 32-bit PowerPC 601 with the 64-bit PowerPC 620.

TABLE 10.26 PowerPC 601 vs. 620

Features	*PowerPC 601*	*PowerPC 620*
Technology	HCMOS	HCMOS
Transistor count	2.8 million	7 million
Clock speed	50 MHz, 66 MHz	133 MHz
Size of the microprocessor	32-bit	64-bit
Address bus	32-bit	40-bit
Data bus	64-bit	128-bit

There are a few versions of the 64-bit PowerPC available: PowerPC 603e, PowerPC 750/740, and PowerPC 604e. The PowerPC 603e microprocessor is available at speeds of 250, 275, and 300 MHz. The 603e has high performance and low power consumption, which makes it suited for applications found in the embedded system market. The PowerPC 603e is used in the Power Macintosh C500 series, which offers features such as accelerated multimedia, advanced video capture, and publishing. The PowerPC 750/740 is available at speeds up to 266 MHz and uses only 5 watts of power. The unique features offered by this microprocessor are built-in power-saving modes, an on-chip thermal sensor to regulate processor temperature, and a choice of packaging configurations. The PowerPC 604e microprocessor, another member of the PowerPC family, provides speeds of 350 MHz and using 8.0 watts of power. Like Intel, Motorola used the 0.25 μ process technology to achieve this speed. The PowerPC 604e is intended for high-end Macintosh and Mac-compatible systems.

Apple Computer's original G3 (Marketing name used by Apple) utilized PowerPC 750 for Apple's iMac and Power Macintosh personal computers. Apple's G3 (later version) used Motorola's copper-based PowerPC microprocessor, providing speed of up to 400 MHz.

10.17 Motorola's State-of-the-art Microprocessors

As part of their plans to carry the PowerPC architecture into the next century, Motorola /IBM/Apple already announced AltiVec extensions for the PowerPC family. The result is the MPC7400 PowerPC microprocessor. This microprocessor is avilable in 400 MHz, 450 MHz and 500 MHz clock speeds. Motorola's AltiVec technology is the foundation for the Velocity Engine of Apple Cmputer's next generation desktop computers. For example, Apple rececently announced Power Mac G4 which uses the MPC7400.

AltiVec extensions are somewhat comparable to the MMX extensions in Intel's Pentium family. AltiVec has independent processing units while Intel tied MMX to the floating-point unit. Both utilize SIMD (Chapter 11). A comparison of some of the features of AltiVec vs. MMX is provided below:

Features	AltiVec	MMX
Size	128 bits at a time	64 bits at a time
Instructions	162 instructions	57 instructions
Registers	32 registers	8 registers
Unit	Independent	tied to Floating-point Unit

In AltiVec, each processing unit can work independent of the others. This provides more parallelism by separate units. Since Intel tied MMX to floating-point unit, Pentiums can perform either floating-point math or switch over to MMX, but not both simultaneously. The switch requires a mode change that can cost hundreds of cycles, both going into and coming out of MMX mode. It may be very tricky with Pentiums to write good and efficient codes when mixing of modes are required in some computing algorithms.

AltiVec can vetorize (Chapter 11) the floating-point operations. This means that one can use AltiVec to work on some data in the Floating-point Unit, then load the data in the AltiVec side (Vector Unit) without any significant mode switch. This may save hundreds of cycles . Also, this allows programmers to do more with the Vector Unit since they can go back and forth to mix and match.

The biggest drawback with MMX or AltiVec is getting programmers to use them. Programmers are required to use assembly language for MMX. Therefore, a few programmers used MMX for dedicated applications. For example, Intel hand tuned some photoshop filters for Adobe. Programmers can use C language with AltiVec. Therefore, it is highly likely that more programmers will use AltiVec than MMX.

By the year 2001, Motorola and IBM plan to introduce the PowerPC series 2K. It is expected that the chip will contain 100 million transistors and have clock speeds greater than 1 GHz. The result should be a microprocessor that will run much faster than today's fastest microprocessor.

QUESTIONS AND PROBLEMS

10.1 What are the basic differences between the 68000, 68008, 68010, and 68012?

10.2 What does a HIGH on the 68000 FC2 pin indicate?

10.3 (a) If a 68000-based system operates in the user mode and an interrupt occurs, what will the 68000 mode be?

(b) If a 68000-based system operates in the supervisor mode, how can the mode be changed to user mode?

10.4 (a) What is the purpose of 68000 trace and X flags?

(b) How can you set or reset them?

10.5 Indicate whether the following 68000 instructions are valid or not valid. Justify your answers.

(a) `MOVE.B D0,(A1)`

(b) `MOVE.B D0,A1`

10.6 How many addressing modes and instructions does the 68000 have?

10.7 What happens after execution of the following 68000 instruction?

```
MOVE.L D0,$03000013
```

10.8 What is meant by 68000 privileged instructions?

10.9 Identify the following 68000 instructions as privileged or nonprivileged:

(a) `MOVE (A2),SR`

(b) `MOVE CCR,(A5)`

(c) `MOVE.L A7,A2`

10.10 (a) Find the contents of locations $305020 and $305021 after execution of the `MOVE D5,$305020`. Assume [D5] = $6A2FA150 prior to execution of this 68000 MOVE instruction.

(b) If [A0] = $203040FF and [D0] = $40F12560, what happens after execution of the 68000 instruction: `MOVE (A0),D0`?

10.11 Identify the addressing modes for each of the following 68000 instructions:

(a) `CLR D0`

(b) `MOVE.L (A1)+,-(A5)`

(c) `MOVE $2000(A2),D1`

10.12 Determine the contents of registers / memory locations affected by each of the following 68000 instructions:

(a) `MOVE (A0)+,D1`

Assume the following data prior to execution of this MOVE:

[A0] = $50105020 [$105021] = $51

[D1] = $70801F25 [$105022] = $52

[$105020] = $50 [$105023] = $7F

(b) `MOVEA D5,A2`

Assume the following data prior to execution of this MOVEA:

[D5] = $A725B600

[A2] = $5030801F

10.13 Find the contents of register D0 after execution of the following 68000 instruction sequence:

```
EXT D0
EXT.L D0
```

Assume [D0] = $F215A700 prior to execution of the instruction sequence.

10.14 Find the contents of D1 after execution of `DIVS #6,D1`. Assume [D1] = $FFFFFFF7 prior to execution of the 68000 instruction. Identify the quotient and remainder. Comment on the sign of the remainder.

10.15 Write a 68000 assembly program to multiply a 16-bit signed number in the low word of D0 by an 8-bit signed number in the highest byte (bits 31–24) of D0.

10.16 Write a 68000 assembly program to divide a 16-bit signed number in the high word of D1 by an 8-bit signed number in the lowest byte of D1.

10.17 Write a 68000 assembly program to add the top two 16 bits of the stack. Store the 16-bit result onto the stack. Assume supervisor mode.

10.18 Write a 68000 assembly program to add a 16-bit number in the low word (bits 0–15) of D1 with another 16-bit number in the high word (bits 16–31) of D1. Store the result in the high word of D1.

10.19 Write a 68000 assembly program to add two 48-bit data items in memory as shown in Figure P10.19. Store the result pointed to by A1. The operation is given by

```
  $00 02 03 A1 07 20
  $07 03 02 02 03 1A
  ------------------
  $07 05 05 A3 0A 3A
```

Assume that the data pointers and the data are already initialized.

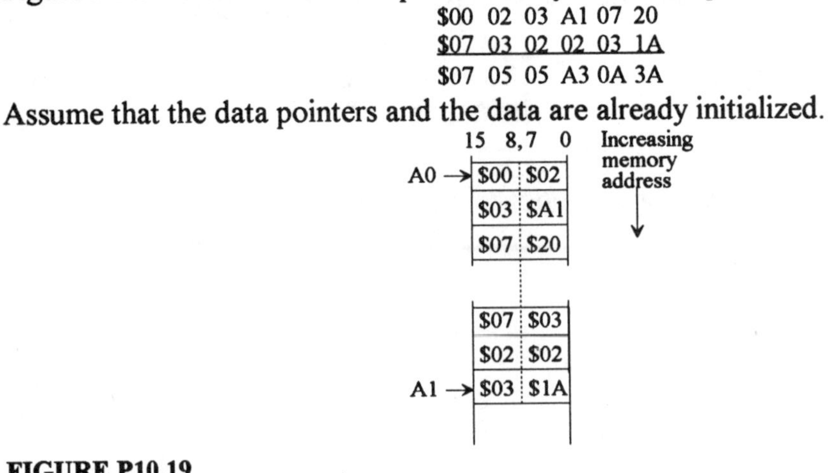

FIGURE P10.19

10.20 Write a 68000 assembly program to divide a 9-bit unsigned number in the high 9 bits (bits 31–23) of D0 by 8_{10}. Do not use any division instruction. Store the result in D0. Neglect the remainder.

10.21 Write a 68000 assembly program to compare two strings of 15 ASCII characters. The first string is stored starting at $502030. The second string is stored at location $302510. The ASCII character in location $502030 of string 1 will be compared with the ASCII character in location $302510 of string 2, [$502031] will be compared with [$302511], and so on. Each time there is a match, store $EEEE onto the stack; otherwise, store $0000 onto the stack. Assume user mode.

10.22 Write a subroutine in 68000 assembly language to subtract two 32-bit packed BCD numbers. BCD number 1 is stored at a location starting from $500000 through $500003, with the least significant digit at $500003 and the most significant digit at $500000. BCD number 2 is stored at a location starting from $700000 through $700003, with the least significant digit at $700003 and the most significant digit at $700000. BCD number 2 is to be subtracted from BCD number 1. Store the result as packed BCD digits in D5.

10.23 . Write a subroutine in 68000 assembly language to compute

$$Z = \sum_{i=1}^{100} Xi$$

Assume the X_i's are signed 8-bit and stored in consecutive locations starting at $504020. Assume A0 points to the X_i's. Also, write the main program in 68000 assembly language to perform all initializations, call the subroutine, and then compute $Z/100$.

10.24 Write a subroutine in 68000 assembly language to convert a 3-digit unpacked BCD number to binary using unsigned multiplications by 10, and additions. The most significant digit is stored in a memory location starting at $3000, the next digit is stored at $3001, and so on. Store the binary result in D3. Use the value of the 3-digit BCD number,

$$N = N2 \times 10^2 + N1 \times 10^1 + N0$$

$$= ((10 \times N2) + N1) \times 10 + N0$$

10.25 Write a 68000 assembly program to compute the following:

$$I = 6 \times J + K/M$$

where the locations \$6000, \$6002, & \$6004 contain the 16-bit signed integers J, K, and M. Store the result into a long word starting at \$6006. Discard the remainder of K/M.

10.26 Determine the status of $\overline{AS}$, FC2–FC0, $\overline{LDS}$, $\overline{UDS}$, and address lines immediately after execution of the following instruction sequence (before the 68000 tristates these lines to fetch the next instruction):

```
MOVE #$2050,SR
MOVE.B D0,$405060
```

Assume the 68000 is in the supervisor mode prior to execution of the instructions.

10.27 Suppose that three switches are connected to bits 0–2 of port A and an LED to bit 6 of port B. If the number of HIGH switches is even, turn the LED ON; otherwise, turn the LED OFF. Write a 68000 assembly language program to accomplish this.

(a) Assume a 68000/6821 system.

(b) Assume a 68000/68230 system.

10.28 Assume the pins and signal shown in Figure P10.28 for the 68000, 68230 (ODD), 2764 (ODD and EVEN). Connect the chips and draw a neat schematic. Determine the memory and I/O maps. Assume a 16.67-MHz internal clock on the 68000.

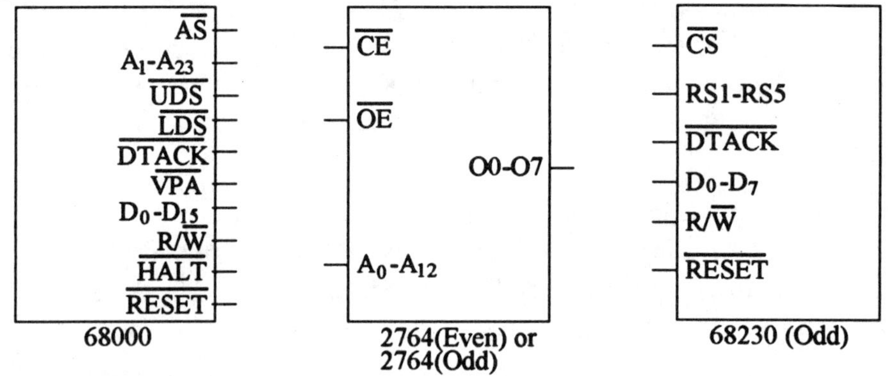

FIGURE P10.28

10.29 Find $\overline{LDS}$ and $\overline{UDS}$ after execution of the following 68000 instruction sequence:

```
MOVEA.L #$0005A123,A2
MOVE.B (A2),D0
```

10.30 Write a 68000 service routine at address \$1000 for a hardware reset that will initialize all data registers to zero, address registers to \$FFFFFFFF, supervisor SP to \$502078, and user SP to \$1F0524, and then jump to \$7020F0.

10.31 Assume the 68000 stack and register values shown in Figure P10.31 before occurrence of an interrupt. If an external device requests an interrupt by asserting the $\overline{IPL2}$, $\overline{IPL1}$, and $\overline{IPL0}$ pins with the value 000_2, determine the contents of A7′ and SR during interrupt and after execution of RTE at the end of the service routine of the interrupt. Draw the memory layouts and show where A7′ points to and the stack contents during and after interrupt. Assume that the stack is not used by the service routine.

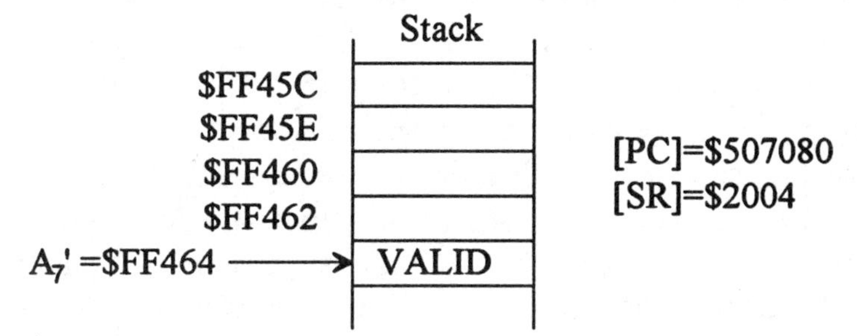

FIGURE P10.31

10.32 Consider the following data prior to a 68000 hardware reset:

[D0] = \$7F2A1620
[A1] = \$6AB11057
[SR] = \$001F

What are the contents of D0, A1, and SR after hardware reset?

10.33 In Figure P.10.33, if VM > 12 V, turn an LED ON connected at bit 3 of port A. If VM < 11 V, turn the LED OFF. Using ports, registers, and memory locations as needed and level 1 autovectored interrupt:

(a) Draw a neat block diagram showing the 68000/6821 microcomputer and the connections to the diagram in Figure P10.33 to ports.

(b) Write a service routine in 68000 assembly language.

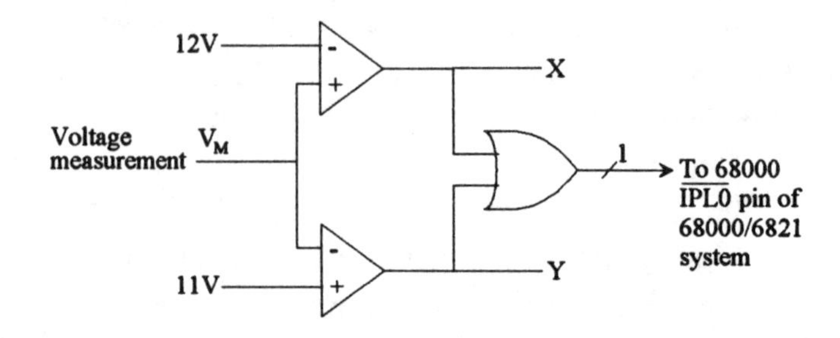

FIGURE P10.33

10.34 Write a subroutine in 68000 assembly language using the `TAS` instruction to find, reserve, and lock a memory segment for the main program. The memory is divided into three segments (0, 1, 2) of 16 bytes each. The first byte of each segment includes a flag byte to be used by the `TAS` instruction. In the subroutine, a maximum of three 16-byte memory segments must be checked for a free segment (flag byte = 0). The `TAS` instruction should be used to find a free segment. The starting address of the free segment (once found) must be stored in A0 and the low byte D0 must be cleared to zero to indicate a free segment and the program control should return to the main program.. If no free block is found, \$FF must be stored in the low byte of D0 and the control should return to the main program.

10.35 Will the circuit in Figure P10.35 work? If so, determine the I/O port addresses for PGCR, PADR, PADDR, PBDR, PBDDR, PCDR and PCDDR. If not, comment briefly, modify the circuit, and then determine the port addresses. Use only the pins and the signals shown. Assume all don't cares to be zeros.

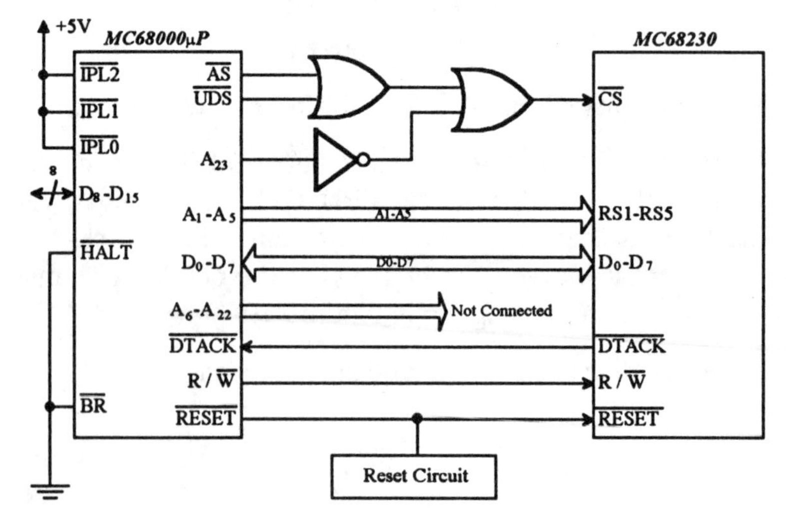

FIGURE P10.35

10.36 Summarize the basic differences between the 68000, 68020, 68030, 68040 and 68060.

10.37 What is the unique feature of the Power PC 601?

10.38 Name three new 68020 instructions that are not provided with the 68000.

10.39 Find the contents of the affected registers and memory locations after execution of the 68020 instruction `MOVE ($1000,A5,D3.W*4),D1`. Assume the following data prior to execution of this `MOVE`:

[A5] = $0000F210 , [$ 00014218] = $4567

[D3] = $00001002 , [$ 0001421A] = $2345

[D1] = $F125012A

10.40 Assume the following 68020 memory configuration:

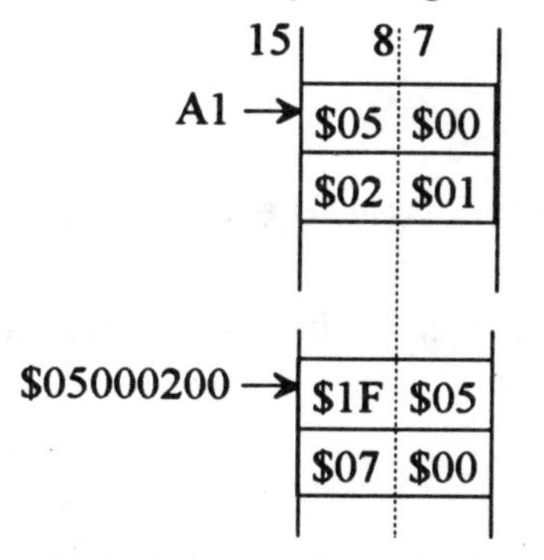

Find the contents of the affected memory locations after execution of `MOVE.W #$1234,([A1])`.

10.41 Find the 68020 compare instruction with the appropriate addressing mode to replace the following 68000 instruction sequence:

```
ASL.L #1,D5
CMP.L 0 (A0,D5.L),D0
```

10.42 Find the contents of D1, D2, A4, and CCR and the memory locations after execution of each of the following 68020 instructions:

```
(a) BFSET $5000 {D1:10}
(b) BFINS D2, (A4) {D1:D4}
```

Assume the data given in Figure P10.42 prior to execution of each of these instructions.

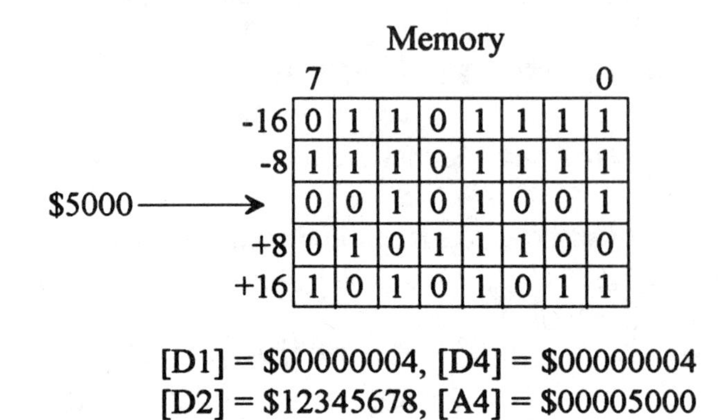

FIGURE P10.42

10.43 Identify the following 68020 instructions as valid or invalid. Justify your answers.
(a) `DIVS A0,D1`
(b) `CHK.B D0,(A0)`
(c) `MOVE.L D0,(A0).`
It is given that [A0] = $1025671A prior to execution of the `MOVE`.

10.44 Determine the values of the Z and C flags after execution of each of the following 68020 instructions:
(a) `CHK2.W (A5),D3`
(b) `CMP2.L $2001,A5`
Assume the following data prior to execution of each of these instructions:

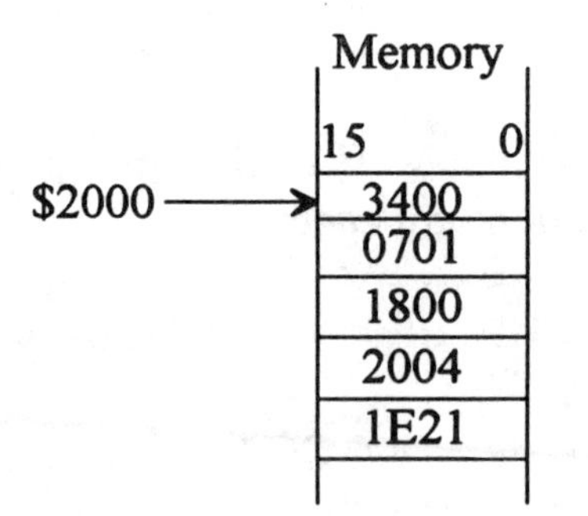

[D3] = $02001740, [A5] = $0002004

10.45 Write a 68020 assembly program to add two 64-bit numbers in D1D0 with another 64-bit number in D2D3. Store the result in D1D0.

10.46 Write a 68020 assembly program to multiply a 32-bit signed number in D5 by another 16-bit signed number in D1. Store the 64-bit result in D5D1.

10.47 Write a subroutine in 68020 assembly language to compute

$$Y = \sum_{i=1}^{50} \frac{Xi^2}{50}$$

Assume the X_i's are signed 32-bit numbers and the array starts at $50000021. Neglect overflow.

10.48 What is meant by 68020 dynamic bus sizing?

10.49 Consider the 68020 instruction `MOVE.B D1,$00000016`. Find the 68020 data pins over which data will be transferred if $\overline{DSACK1}\ \overline{DSACK0}$ = 00. What are the 68020 data pins if $\overline{DSACK1}\ \overline{DSACK0}$ = 10?

10.50 If a 32-bit data is transferred using 68020 `MOVE.L D0,$50607011` instruction to a 32-bit memory with [D0] = $81F27561, how many bus cycles are needed to perform the transfer? What are A_1A_0 equal to during each cycle? What is the SIZ1 SIZ0 code during each cycle? What bytes of data are transferred during each bus cycle?

10.51 Discuss 68020 I/O.

10.52 What do you mean by the unified cache of the 601? What is its size?

10.53 List the user-level and general-purpose registers of the 601.

10.54 Name one supervisor-level register in the 601. What is its purpose?

10.55 How does the 601 MSR indicate the following:
(a) The 601 executes both the user- and supervisor- level instructions.
(b) The 601 executes only the user-level instructions.

10.56 Explain the operation performed by each of the following 601 instructions:

```
(a) add.r1,r2,r3
(b) divwu r2,r3,r4
(c) extsb r1,r2
```

10.57 Discuss briefly the exceptions included in the PowerPC 601.

10.58 Compare the basic features of the 601 with the 620. Discuss PowerPC 64-bit $\mu p's$.

10.59 Write a subroutine in 68000 assembly to recursively compute n!, where n is positive.

10.60 Summarize the basic features of Motorola's state-of-the-art microprocessors.

10.61 Write a program in 68020 assembly language to find the first one in a bit field which is greater than or equal to 16 bits and less than or equal to 512 bits. Assume that the number of bits to be checked is divisible by 16. If no ones are found, store zero in D3; otherwise store the offset of the first set bit in D3, and then stop. Assume A2 contains the starting address of the array, and D2 contains the number of bits in the array.

10.62 Write a program in 68020 assembly language to multiply a signed byte by a 32-bit signed number to obtain a 64-bit result. Assume that the numbers are respectively pointed to by the addresses that are passed on to the user stack by a subroutine pointed to by (A7+6) and (A7+8). Store the 64-bit result in D2:D1.

10.63 Repeat problem 9.31 using a 68000/2732/6116/68230 microcomputer.

10.64 Repeat problem 9.32 using a 68000/2732/6116/6821 microcomputer.

10.65 Repeat problem 9.33 using a 68000/2732/6116/68230 microcomputer. Note: 68000 does not have any instruction similar to 8086 XLAT instruction.

10.66 Repeat problem 9.36 using a 68000/2732/6116/68230 microcomputer. Note: 68000 does not have any instruction similar to 8086 XLAT instruction.

10.67 Write a subroutine in 68000 assembly language program to compute the trace of a 4X4 matrix containing 8-bit unsigned integers. Assume that each element is stored in memory as a 16-bit number with upper byte as zero in the row-major order form; that is, elements are stored in memory as row by row and within a row, elements are stored as column by column. Note that the trace of a matrix is the sum of the elements of the leading diagonal.

11

SELECTED TOPICS IN COMBINATIONAL/SEQUENTIAL LOGIC AND COMPUTER ORGANIZATION

This chapter provides a more detailed coverage of some of the special topics on combinational/ sequential logic design and computer organization. These include state machine design using ASM (Algorithmic State Machine) chart, an overview of VHDL, array and ROM-based multipliers, microprogramming, RISC, cache memory, virtual memory, and pipeline processing.

11.1 State Machine Design using ASM chart

An overview of the ASM chart is provided in Chapter 5. As mentioned before, an ASM chart is used to define digital hardware algorithms. This can be used to design and implement state machines. This section describes a procedure for designing state machines using the ASM chart. This is a three step process as follows:

1. Draw the ASM chart from problem definition.
2. Derive the state transition table representing the sequence of operations to be performed.
3. Derive the logic equations and draw the hardware schematic. The hardware can be designed using either classical sequential design or PLAs. Examples 11.1 and 11.2 illustrate this.

The ASM chart described in section 7.2.3 (Page 317) can be used to design the control unit. This can be accomplished using an input (I) for the system to return to the initial state T_0 *when I equals to one (for example); the system stays in the last state* T_n *if* $I=0$*. The flowcharts and the state diagrams can be converted to ASM charts following the concepts described in this section. Problem 11.8 (page 771) is included at the end of this chapter for this purpose.*

Example 11.1

Design a digital system using an ASM chart that will operate as follows:
The system will contain a 2-bit binary counter. The binary counter will count in the sequence 00, 01, 10, and 11. The most significant bit of the binary count XY is X while Y is the least significant bit. The system starts with an initial count of 3. A start signal I (represented by a switch) initiates a sequence of operations. If I = 0, the system stays in the initial state T_0 with count of 3. On the other hand, I = 1 starts the sequence.

When I = 1, counter Z (represented by XY) is first cleared to zero. The system then moves to state T_1. In this state, counter Z is incremented by 1 at the leading edge of each clock pulse. When the counter reaches 3, the system goes back to the initial state T_0, and the process continues depending on the status of the start switch I. The counter output will be displayed on a seven-segment display. An LED will be connected at the output of flip-flop W. The system will turn the LED ON for the count sequence 1, 2 by clearing flip-flop W to 0. The flip-flop W will be preset to 1 in the initial state to turn the LED OFF. This can be accomplished by using input I as the PRESET input of flip-flop W. Use D flip-flops for the system.

Solution

Step 1: **Draw the ASM chart.** Figure 11.1 shows the ASM chart.

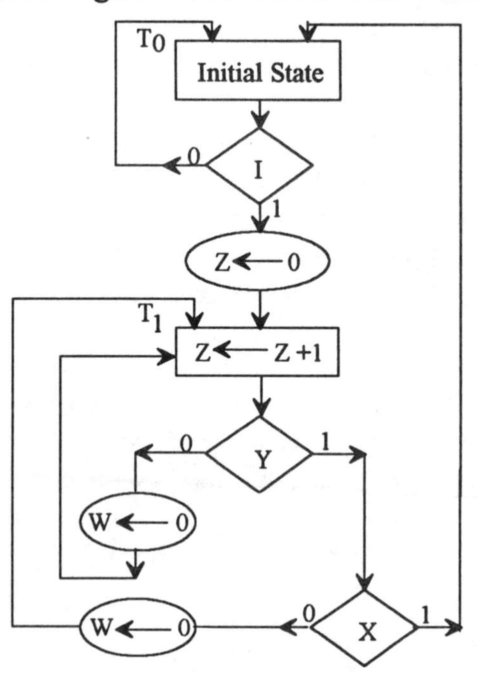

FIGURE 11.1 ASM Chart for Example 11.1

The symbol T_n is used without its binary value for the state boxes in all ASM charts in this section. In the ASM chart of Figure 11.1, when the system is in initial state T_0, it waits for the start signal (I) to become HIGH. When I=1, Counter Z is cleared to zero and the system goes to state T_1. The counter is incremented at the leading edge of each clock pulse. In state T_1, one of the following possible operations occurs after the next clock pulse transition:
Either, if counter Z is 1 or 2, flip-flop W is cleared to zero and control stays in state T_1;

or

If the Counter Z counts to 3, the system goes back to initial state T_0.

The ASM chart consists of two states and two blocks. The block associated with T_0 includes one state box, one decision box, and one conditional box. The block in T_1 consists of one state box, two decision boxes and two conditional boxes.

***Step 2:* Derive the state transition table representing the sequence of operations.**
One common clock pulse specifies the operations to be performed in every block of an ASM chart. Table 11.1 shows the State Transition Table.

TABLE 11.1 State Transition Table

COUNTER X	COUNTER Y	FLIP-FLOP W (Q)	CONDITIONS	STATE
0	0	1	X = 0, Y = 0	T_0
0	1	0	X = 0, Y = 1	T_1
1	0	0	X = 1, Y = 0	T_1
1	1	1	X = 1, Y = 1	T_0

The binary values of the counter along with the corresponding outputs of flip-flop W is shown in the transition table. In state T_0, if I = 1, Counter Z is cleared to zero (XY = 00) and the system moves from state T_0 to T_1. In state T_1, Counter Z is first incremented to XY = 01 at the leading edge of the clock pulse; Counter Z then counts to XY = 10 at the leading edge of the following clock pulse. Finally, when XY = 11, the system moves to state T_0. The system stays in the initial state T_0 as long as I = 0; otherwise the process continues.
The operations that are performed in the digital hardware as specified by a block in the ASM chart occur during the same clock period and not in a sequence of operations following each other in time, as is usually interpreted in a conventional flowchart. For example, consider state T_1. The value of Y to be considered in the decision box is taken from the value of the counter in the present state T_1. This is because the decision boxes for Flip-flop W belong to the same block as state T_1. The digital hardware generates the signals for all operations specified in the present block before arrival of the next clock pulse.

Step 3: **Derive the logic equations and draw the hardware.**

The system can be divided into two sections. These are data processor and controller. The requirements for the design of the data processor are defined inside the state and conditional boxes. The logic for the controller, on the other hand, is determined from the decision boxes and the necessary state transitions.

The design of the data processor is typically implemented by using digital components such as registers, counters, multiplexers, and adders. The system can be designed using the theory of sequential logic discussed in Chapter 5. Figure 11.2 shows the hardware block diagram. The Controller is shown with the required inputs and outputs. The data processor includes a 2-bit counter, one flip-flop, and one AND gate. The counter is incremented by one at the leading edge of every clock pulse when control is in state T_1. The counter is assumed to be in count 3 initially. It is cleared to zero only when control is in state T_0 and I=1. Therefore, T_0 and I are logically ANDed. The D-input of Flip-flop W is connected to output X of the counter to clear Flip-flop W during state T_1. This is because if present count is 00 (X=0), the counter will be 01 after the next clock. On the other hand, if the present count is 01 (X=0), the count will be 10 after the next clock. Hence, X is connected to the D-input of Flip-flop W to turn the

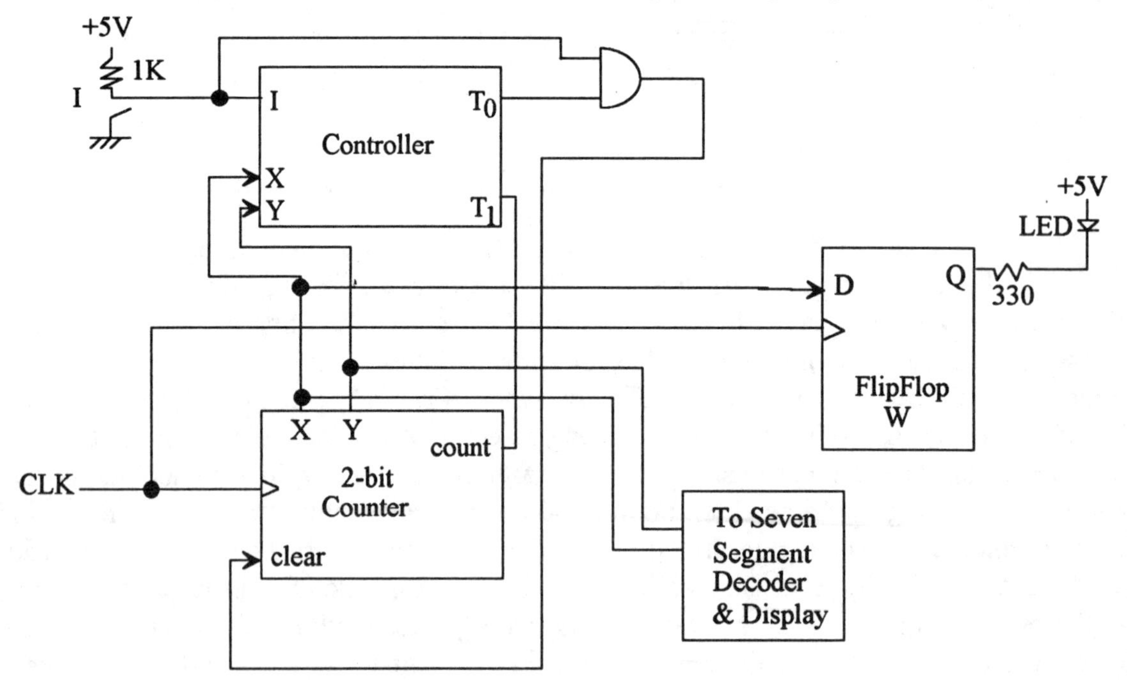

FIGURE 11.2 Hardware Schematic for Example 11.1

LED ON for count sequence 1, 2. A common clock is used for all flip-flops in the system including the flip-flops in the counter and Flip-flop W.
This example illustrates a technique of designing digital systems using the ASM chart. The 2-bit counter can be designed using the concepts described in Chapter 5. In order to design the Controller, a state table for the controller must be derived. Table 11.2 shows the state table for the Controller.

TABLE 11.2 State Table for the Controller

Present State (Controller)	Present States (counter)		Inputs (Controller)			Next States (counter)		Next Output States (controller)	
	X	Y	I	X	Y	X+	Y+	T_1	T_0
T_0	1	1	0	1	1	1	1	0	1
T_0	1	1	1	1	1	0	0	0	1
T_0	0	0	1	0	0	0	1	1	0
T_1	0	1	1	0	1	1	0	1	0
T_1	1	0	1	1	0	1	1	0	1

There is a row in the table for each possible transition between states. Initial state T_0 stays in T_0 or goes from T_0 to T_1 depending on the status of the switch input (I). The same procedure for designing a sequential circuit described in Chapter 5 can be utilized. Since there are two controller outputs (T_1,T_0) and three inputs (I, X, Y), a three-variable K-map is required. The design of the final hardware schematic is left as an exercise to the reader.The system will contain D flip-flops with the same common clock and a combinational circuit.

The design of the system using classical sequential design method may be cumbersome. Hence, other simplified methods using PLAs can be used as illustrated in Example 11.2. In the following example, one PLA with a MOD-4 counter, and a 2 to 4 decoder is used. The counter and the decoder can be replaced by using two D flip-flips. In this case, the present outputs (Q1, Q0) of the flip-flops can be connected as two inputs to the PLA. A common clock pulse can be used for both the D flip-flips. In example 11.2, the PLA will have five inputs with *X, Y, and Z* as the other three inputs. The PLA will have six outputs representing the two next outputs (Q1+, Q0+) of the D flip-flops and four timing signals T_0, T_1, T_2, and T_3. The two outputs Q1+, Q0+ from the PLA outputs can be fed back and connected to the D inputs of the corresponding flip-flops. The *C* input can be connected to the CLEAR inputs of the two flip-flops. The PLA can be programmed with five inputs (Q1,Q0, *X,Y,Z)* to generate six outputs (Q1+,Q0+,T_0, T_1, T_2,. T_3).

Example 11.2

A digital system has three inputs (*X, Y, Z*) and a 2-bit MOD-4 counter (W) to count from 0 to 3. The four counter states are T_0, T_1, T_2, and T_3. The operation of the system is initiated by the counter clear input, *C*. When $C = 0$, the system stays in initial state T_0. On the other hand, when $C = 1$, state transitions to be handled by the system are as follows:

INPUTS	STATE TRANSITIONS
$X = 0$	The system moves from T_0 to T_1
$X = 1$	The system stays in T_0
$Y = 0$	The system moves back from T_1 to T_0
$Y = 1$	The system goes from T_1 to T_2
$Z = 0$	The system stays in T_2
$Z = 1$	The system moves from T_2 to T_3 and then stays in T_3 indefinitely (for counter clear input C=1) until counter W is reset to zero (state T_0) by activating the counter clear input *C* to 0 to start a new sequence.

Design the digital system using an ASM chart. Use counter, decoder, and a PLA. Figure 11.3 shows the block diagram of the MOD-4 counter to be used in the design.

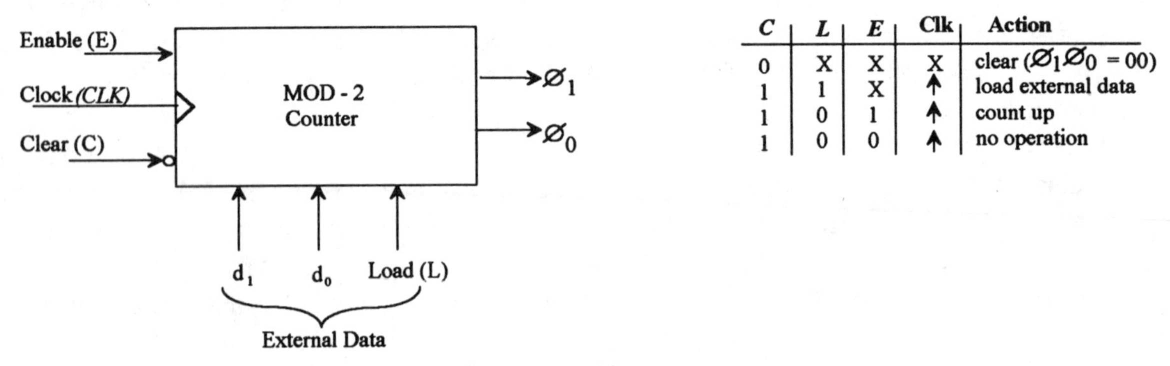

C	*L*	*E*	Clk	Action
0	X	X	X	clear ($Ø_1Ø_0 = 00$)
1	1	X	↑	load external data
1	0	1	↑	count up
1	0	0	↑	no operation

FIGURE 11.3 Block diagram and truth table of the 2-bit counter

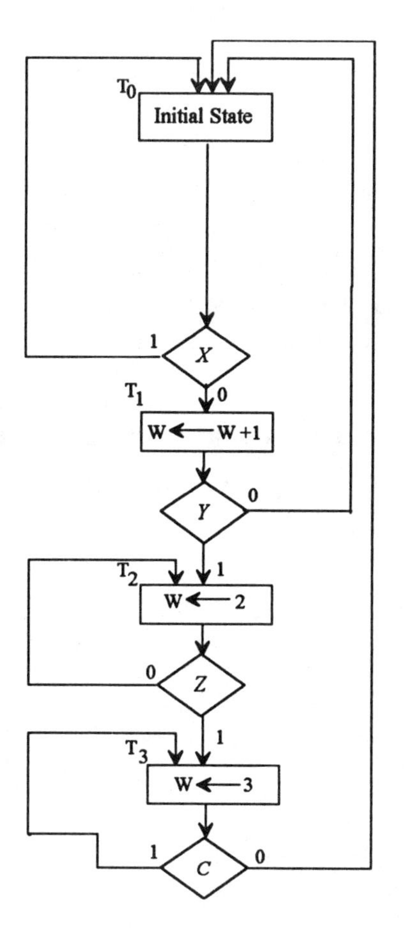

FIGURE 11.4 ASM Chart for Example 11.2

Solution:

Step 1: **Draw an ASM chart.**
The ASM chart is shown in Figure 11.4

Step 2: **Derive the inputs, outputs, and a sequence of operations.**
The system will be designed using a PLA, a MOD-4 counter, and a 2 to 4 decoder. The MOD-4 counter is loaded or initialized with the external data if the counter control inputs C and L are both ones. The counter load control input L overrides the counter enable control input E.
The counter counts up automatically in response to the next clock pulse when the counter load control input L = 0 and the enable input E is tied to HIGH. Such normal activity is

desirable for the situation (obtained from the ASM chart) when the counter goes through the sequence T_0, T_1, T_2, T_3 for the specified inputs.
However, if the following situations occur, the counter needs to be loaded with data out of its normal sequence:
If the counter is in initial state T_0 (Counter W=0 with C= 0) , it stays in T_0 for X = 1. This means that if the counter output is 00 and if X = 1, the counter must be loaded with external data d_1d_0 = 00. Similarly, the other out of normal sequence count includes transitions from T_1 to T_0 (for Y = 0), T_2 to T_2 (for Z = 0) with count 2, and T_3 to T_3 (for C=1). Finally, if C = 0, transition from T_3 to T_0 occurs and the process continues. The appropriate external data must be loaded into the counter for out of normal count sequence by the PLA using the L input of the counter.

Step 3: **Derive the logic equations and draw a hardware schematic.**

Figure 11.5 depicts the logic diagram. Figure 11.6 shows the truth table and hardware schematic for PLA-based implementation.

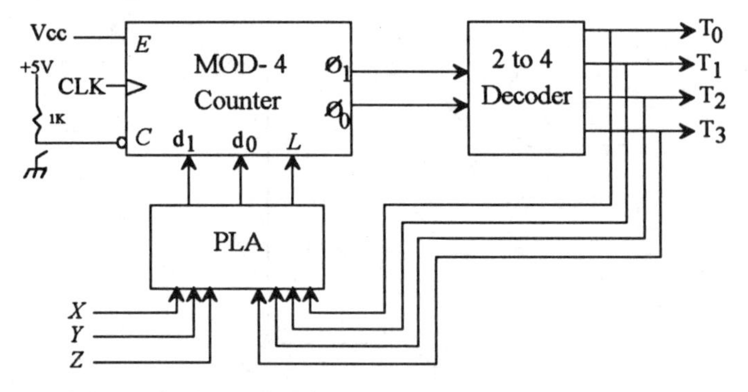

FIGURE 11.5 Hardware Schematic for Example 11.2

Inputs								Outputs		
C	X	Y	Z	T_0	T_1	T_2	T_3	L	d_1	d_0
1	1	X	X	1	X	X	X	1	0	0
1	X	0	X	X	1	X	X	1	0	0
1	X	X	0	X	X	1	X	1	1	0
1	X	X	X	X	X	X	1	1	1	1
0	X	X	X	X	X	X	1	1	0	0

X = don't cares

(a) Truth Table for out of normal Count sequence

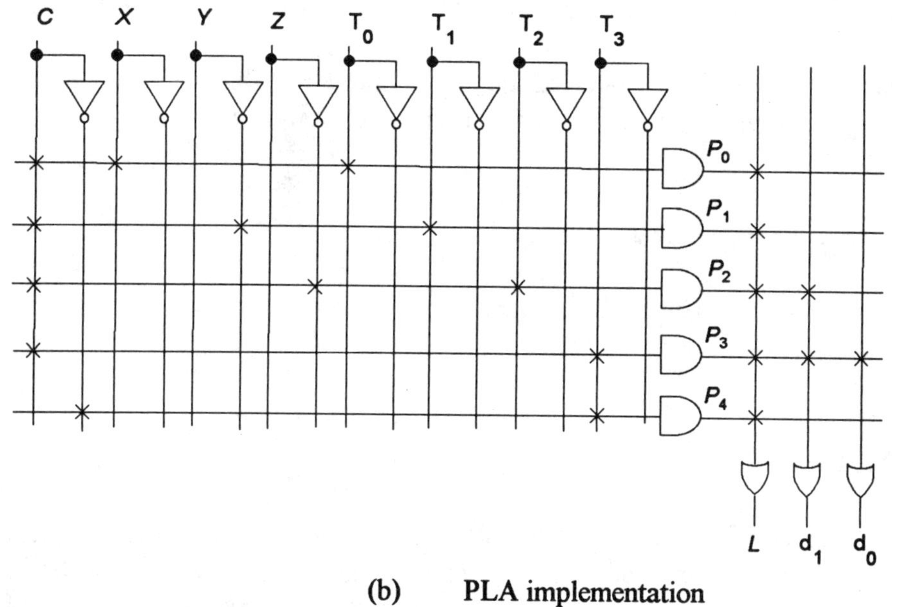

(b) PLA implementation

FIGURE 11.6 PLA-based System

The equations for the product terms are:

$P_0 = X T_0 C \quad P_1 = \overline{Y} T_1 C \quad P_2 = \overline{Z} T_2 C \quad P_3 = T_3 C \quad P_4 = T_3 \overline{C}$

$L = P_0 + P_1 + P_2 + P_3 + P_4 \quad d_1 = P_2 + P_3 \quad d_0 = P_3$

11.2 Overview of VHDL

VHDL is a PLD programming language like ABEL. VHDL is an acronym for VHSIC Hardware Description Language. VHSIC stands for Very High Speed Integrated Circuits. The design of VHDL evolved from the United States Department of Defense (DOD) VHSIC program. VHDL is based on ADA programming language.

The design of VHDL started in 1983 and after going through several versions was formally accepted as an IEEE (Institute of Electrical and Electronics Engineers) standard in 1987. VHDL has been gaining wide acceptance as a hardware design language for PLDs. An overview of VHDL is presented in this section.

Each VHDL description contains two blocks. These are input/output and architectural components. The input/output description specifies the input and output connections (ports)

to the hardware. The architectural component defines the behavior of the hardware entity being designed.
A typical VHDL description includes a port statement contained within an entity statement. All keywords in VHDL are reserved. This means that they cannot be used for any other purpose.

A typical VHDL structure is given below:

```
entity EXAMPLE is  -- Entity Statement
port               -- port   Statement
 (X,Y,Z : in  BIT;
  W     : out BIT);
end    EXAMPLE
```

The **entity** statement begins with the keyword **entity** followed by the name of the entity EXAMPLE followed by the word **is**. The **port** statement is contained within an **entity** statement. The VHDL design entity is comprised of two parts: an interface and a body. The interface is specified by the keyword **entity** and the body is denoted by the keyword **architecture.** Typical logic and arithmetic operators are listed below:

LOGIC OPERATORS

and	AND Operation
or	OR Operation
xor	Exclusive-OR Operation
xnor	Exclusive-NOR Operation
nand	NAND Operation
nor	NOR Operation
not	NOT Operation

ARITHMETIC OPERATORS

+	Positive sign or addition

–	Negative sign or subtraction
*	Multiplication
/	Division
mod	Modulus
rem	Remainder
abs	Absolute value
**	Exponential

TYPICAL PORT MODES

in	Information flows into the design entity
out	Information flows out of the design entity

In the following, typical programs will be provided. The keywords will be indented and other words will be written in uppercase. A comment is indicated by the symbol – – before a statement.

A VHDL program for an Exclusive-NOR operation between two Boolean variables X and Y is provided below:

```
-- Exclusive-NOR Operation
entity  XNOR  is
port
(X,Y : in   BIT;
Z      : out BIT);
end     XNOR;
-- Body
architecture BEHAVIOR of XNOR is
begin
Z<=X xnor Y;
end  BEHAVIOR;
```

The following VHDL program uses the equation, $Z = XY + \bar{X}\,\bar{Y}$ to implement the Exclusive-NOR operation:

```
-- Exclusive NOR using Equation
entity  XNOR  is
port
(X,Y: in   BIT;
Z    :  out  BIT);
end    XNOR;
```

```
--      Body
architecture EQUATION of XNOR is
begin
Z<= (X and Y) or (not X and not Y);
end EQUATION;
```

Note that in the above examples **architecture** declares the name XNOR to associate the architecture with the XNOR design entity interface.

11.3 Array and ROM-Based Multipliers

In a typical 8-bit processor such as the Intel 8085, the processor hardware does not include any provision to multiply two numbers, due to a limitation on the component density. Hence, the processor's instruction set does not include a multiplication instruction. In this situation, however, one can write a program that multiplies two numbers. Although this solution seems viable, the operational speed is unsatisfactory.

For application environments such as real-time digital filtering, in which the processor is expected to perform 32 to 64 eight-bit multiplication operations within 100 μsec (sampling frequency = 10 kHz), speed is an important factor.

New device technologies such as BICMOS and HCMOS, allow manufacturers to pack millions of transistors in a chip. Consequently, state-of-the-art 32-bit microprocessors such as the Motorola 68060 (HCMOS) and Intel Pentium (BICMOS) designed using these technologies, have a larger instruction set than their predecessors, which includes multiplication and division instructions. In this section, multiplier design principles are discussed. Two unsigned integers can be multiplied in the same way as two decimal numbers are multiplied by paper and pencil method. Consider the multiplication of two unsigned integers, where the multiplier Q = 15 and the multiplicand is M = 14, as illustrated:

M ⟶ 1110 — Multiplicand (14_{10})

Q ⟶ 1111 — Multiplier (15_{10})

1110 ⟵

1110 ⟵

1110 ⟵ Partial products

1110 ⟵

P ⟶ 11010010 ⟵ Final product

In the paper and pencil algorithm, shifted versions of multiplicands are added. This procedure can be implemented by using combinational circuit elements such as AND gates and FULL adders.

Generally, a 4-bit unsigned multiplier Q and a 4-bit unsigned multiplicand M can be written as

$$M\colon m_3\, m_2\, m_1\, m_0$$

$$Q\colon q_3\, q_2\, q_1\, q_0$$

The process of generating the partial products and the final product can also be generalized as shown in Figure 11.7.

Each cross-product term ($m_i\, q_j$) in this figure can be generated using an AND gate. This requires 16 AND gates to generate all cross-product terms that are summed by full adder arrays, as shown in Figure 11.8.

				m_3	m_2	m_1	m_0	
				q_3	q_2	q_1	q_0	
				m_3q_0	m_2q_0	m_1q_0	m_0q_0	⟵ *Partial product PR_0*
			m_3q_1	m_2q_1	m_1q_1	m_0q_1		⟵ *Partial product PR_1*
		m_3q_2	m_2q_2	m_1q_2	m_0q_2			⟵ *Partial product PR_2*
	m_3q_3	m_2q_3	m_1q_3	m_0q_3				⟵ *Partial product PR_3*
P_7	P_6	P_5	P_4	P_3	P_2	P_1	P_0	

FIGURE 11.7 Generalized Version of the Multiplication of Two 4-bit Numbers Using the Paper and Pencil Algorithm

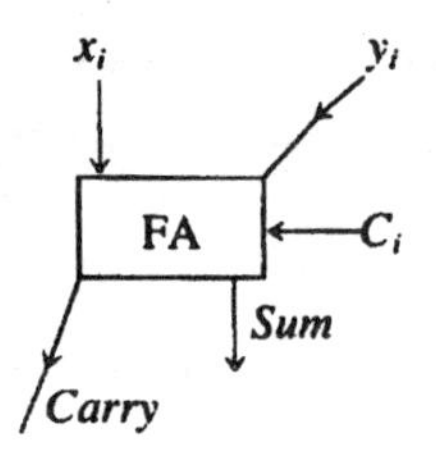

a. Basic Cell

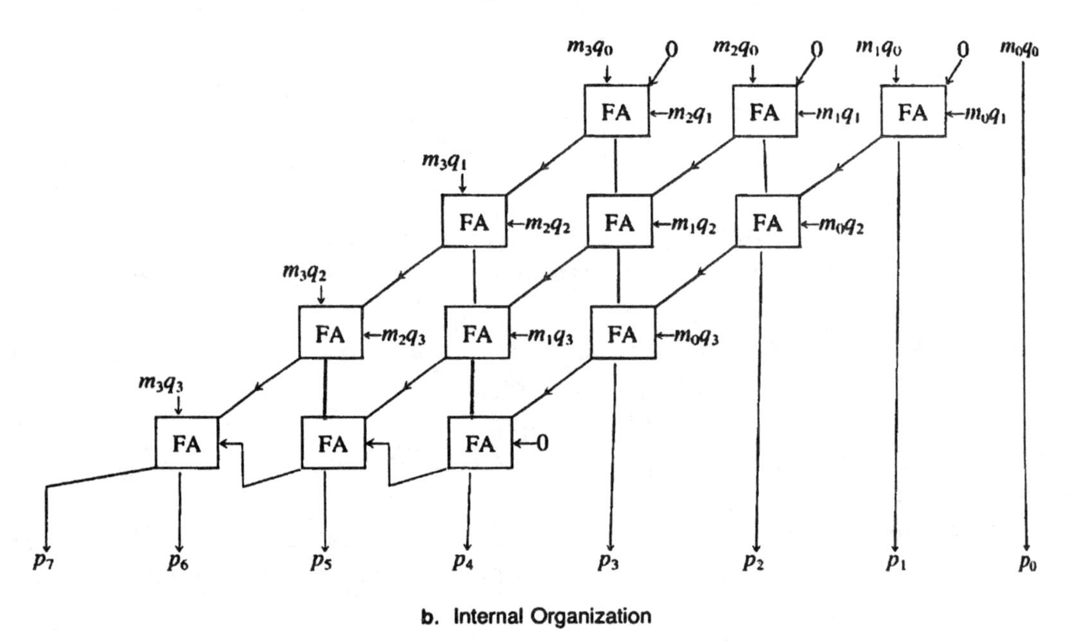

b. Internal Organization

FIGURE 11.8 4 ×4 Array Multiplier

Consider the generation of p_2 in Figure 11.8(b). From Figure 11.7, p_2 is the sum of m_2q_0, m_1q_1 and m_0q_2. The sum of these three elements is obtained by using two full adders. (See column for p_2 in Figure 11.8). The top full-adder in this column generates the sum $m_2q_0 + m_1q_1$. This sum is then added to m_0q_2 by the bottom full-adder along with any carry from the previous full-adder for p_1.

The time required to complete the multiplication can be estimated by considering the longest carry propagation path comprising of the rightmost diagonal (which includes the full-adder for p_1 and the bottom full-adders for p_2 and p_3), and the last row (which includes the full-adder for p_6 and the bottom full-adders for p_4 and p_5). The time taken to multiply two n-bit numbers can be expressed as follows:

$$T(n) = \Delta_{AND\,gate} + (n-1)\,\Delta_{carry\ propagation} + (n-1)\,\Delta_{carry\ propagation}$$

In this equation, all cross-product terms m_iq_i can be generated simultaneously by an array of AND gates. Therefore, only one AND gate delay is included in the equation. Also, the right-most diagonal and the bottom row contain (n - 1) full-adders each for the n × n multiplier. Assuming that $\Delta_{AND\ gate} = \Delta_{carry propogation} = 2\ gate\ delays = 2\Delta$, the preceding expression can be simplified as shown:

$T(n) = 2\Delta + (2n - 2)2\Delta = (4n - 2)\Delta$.

The array multiplier that has been considered so far is known as Braun's multiplier. The hardware is often called a nonadditive multiplier (NM), since it does not include any additive inputs. An additive multiplier (AM) includes an extra input R; it computes products of the form

$P = M * Q + R$

This type of multiplier is useful in computing the sum of products of the form ΣXiYi.

Both an NM and an AM are available as standard 1C blocks. Since these systems require more components, they are available only to handle 4- or 8-bit operands.

Alternatively, the same 4x4 NM discussed earlier can be obtained using a 256 × 8 ROM as shown in Figure 11.9.

It can be seen that a given MQ pair defines a ROM address, where the corresponding 8-bit product is held. The ROM approach can be used for small-scale multipliers because:

- The technological advancements allow the manufacturers to produce low-cost ROMs.
- The design effort is minimum.

In case of large multipliers, ROM implementation is unfeasible, since large-size ROMs are required. For example, in order to implement an 8 × 8 multiplier, a $2^{16} \times 16$ ROM is required. If the required 8 × 8 product is decomposed into a linear combination of four 4x4 products, an 8 × 8 multiplier can be implemented using four 256 × 8 ROMs and a few 4-bit parallel adders. However, PLDs can be used to accomplish this.

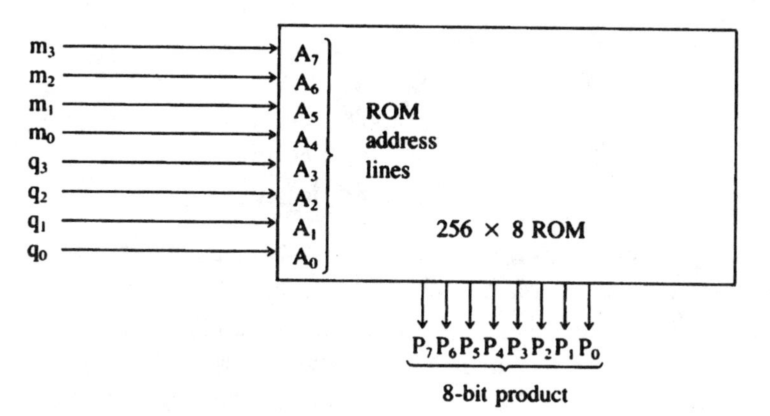

FIGURE 11.9 ROM-based 4x4 Multiplier

11.4 Microprogramming – More Details

The basic concepts associated with microprogramming are described in Chapter 7. More coverage is provided in this section. Design of a microprogrammed CPU is also included.

11.4.1 Reduction of the length of a Microprogram

By examining the microprograms in Figure 7.46 of Chapter 7, it is obvious that the control function field contains all zeros in case of branch instructions. Whenever there are several branch instructions, the microinstructions, can be formatted by using a method called multiple microinstruction format. In this approach, the microinstructions are divided into two groups: operate and branch instructions.

An operate instruction initiates one or more microoperations. For example, after the execution of an operate instruction, the MPC will be incremented by 1. In the case of a branch instruction, no microoperation will usually be initiated, and the MPC may be loaded with a new value.

This means that the branch address field can be removed from the microinstruction format. Therefore, the control function field is used to specify the branch address itself. Typically, each microinstruction will have two fields, as shown next:

CONDITION-SELECT FIELD		CONTROL FUNCTION FIELD						
S_1	S_0	C_6	C_5	C_4	C_3	C_2	C_1	C_0

If S_1 S_0 = 00, the microinstruction is considered as an operate instruction, and the contents of the control function field are treated as the control signals. Assume the Condition Select Field is encoded as follows:

S_1	S_0	
0	0	No branch
0	1	Branch if cond-1 = 1
1	1	Branch if cond-2 = 1
1	0	Unconditional branch

If S_1 S_0 = 01, the instruction is regarded as a branch instruction, and the contents of the control field are assumed to be a 7-bit branch address. In this example, it is assumed that when S_1 S_0 = 01, the MPC will be loaded with the appropriate address specified by C_6 C_5 C_4 C_3 C_2 C_1 C_0 if the condition Z = 0 is satisfied; on the other hand, if S_1 S_0 = 10, an unconditional branch to the address specified by the Control Function / Branch Address Field occurs.

ROM Address	Control Word									Comments	
	Condition select field		Control Function / Branch Address Field							Instruction Type	Operation Performed
	S_1	S_0	C_6	C_5	C_4	C_3	C_2 / br_2	C_1 / br_1	C_0 / br_0		
0 0 0	0	0	0	0	0	0	0	1	1	Operate	R ← 0 M ← Inbus
0 0 1	0	0	0	0	0	0	1	0	0	Operate	Q ← Inbus
0 1 0	0	0	1	0	1	1	0	0	0	Operate	R ← R+M Q ← Q-1 R ← F
0 1 1	0	1	0	0	0	0	0	1	0	Branch	If Z=0 Then go to address 2 (loop)
1 0 0	0	0	0	1	0	0	0	0	0	Operate	Outbus ← R
1 0 1	1	0	0	0	0	0	1	0	1	Branch	Go to Address 5(halt)

FIGURE 11.10 Reduction of the length of microinstruction of Figure 7.46

In order to illustrate this concept, the microprogram for 4-bit by 4-bit unsigned multiplication of Figure 7.46 is rewritten using the multiple instruction format as shown in Figure 11.10.

It can be seen from the figure 11.10 that the total size of the control store is 54 bits ($6 \times 9 = 54$). In contrast, the control store of figure 7.46 contains 72 bits. For large microprograms with many branch instructions, tremendous memory savings can be accomplished using the multiple microinstructon format. Addresses 0, 1, 2, and 4 contain microinstructions with the contents of the conditional select field as 00, and are considered as operate instructions. In this case, the contents of the control function field are directed to the processing hardware.

Address 3 contains a conditional branch instruction since the contents of the condition select field are 01; while address 5 contains an unconditional branch instruction (halt instruction; that is, jump to the same address) since the condition select field is 10. Hence, the 7-bit control function field directly specifies the desired branch addresses 2 and 5, respectively. Figure 11.11 shows the hardware schematic.

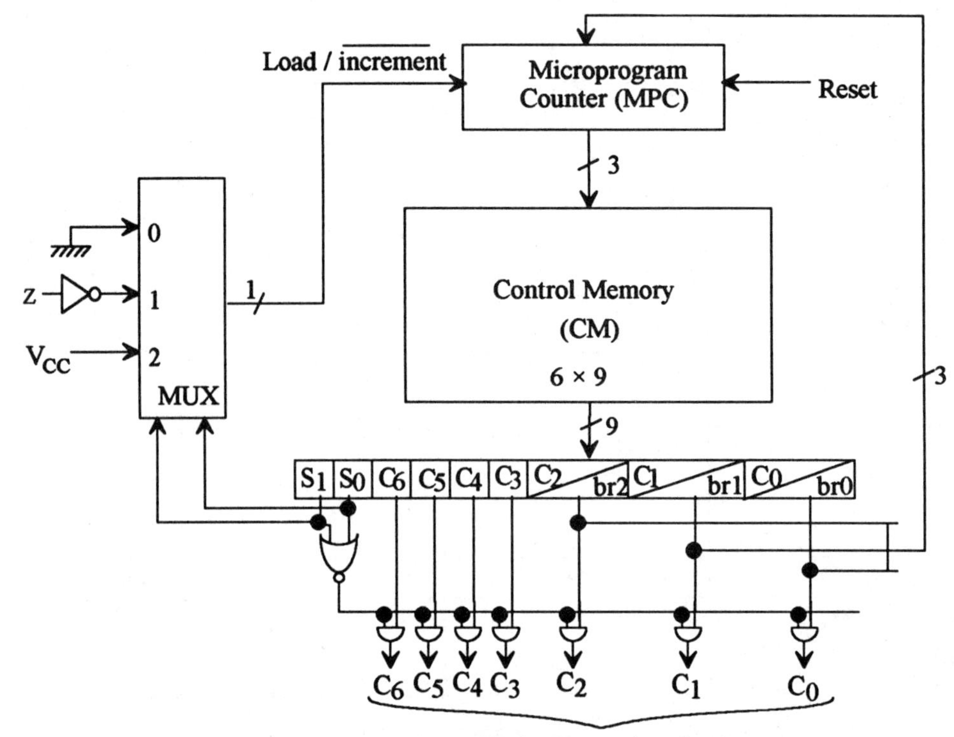

FIGURE 11.11 Microprogrammed Controller for the Microprogram of Figure 11-10.

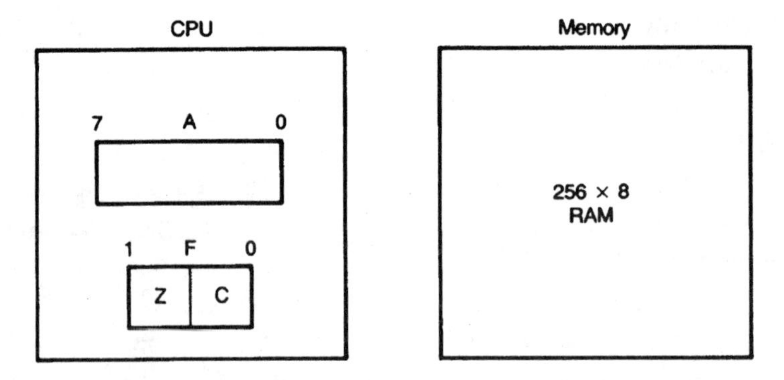

FIGURE 11. 12 Programming Model of a Simple Processor

11.4.2 Design of a Microprogrammed CPU

Next, the design of a microprogrammed processor is illustrated. The programming model of this processor is shown in Figure 11.12.

The CPU contains two registers:

1. An 8-bit register A
2. A 2-bit flag register F

The flag register holds only zero (Z) and carry (C) flags. All programs and data are stored in the 256 × 8 RAM. The detailed hardware schematic of the data-flow part of this processor is shown in Figure 11.13.

From Figure 11.13, it can be seen that the hardware organization includes four more 8-bit registers, PC, IR, MAR, and BUFFER. These registers are transparent to a programmer. The 8-bit register BUFFER is used to hold the data that is retrieved from memory. In this system, only a restricted number of data paths are available. These paths are controlled by the control inputs C_0 through C_9, as defined in Table 11.3.

TABLE 11. 3 Definitions of the Control Inputs C_0-C_9

MICROOPERATION	COMMENT
C_0: PC $\leftarrow$ 0	Clear PC to zero.
C_1: PC $\leftarrow$ PC + 1	Advance the PC.
$C_2C_5\overline{C_6}$: PC $\leftarrow$ M ((MAR))	Read the data from the memory and save it in the PC.
$\overline{C_3}C_4$: MAR $\leftarrow$ PC	Transfer the contents of the PC into MAR.
$C_5\,\overline{C_6}\,C_7$: BUFFER $\leftarrow$ M ((MAR))	Read the data from the memory and save the result in BUFFER.
C_3C_4: MAR $\leftarrow$ BUFFER	Transfer the content of the BUFFER into MAR.
$C_5\,\overline{C_6}\,C_8$: IR $\leftarrow$ M ((MAR))	Read the data from memory and save the result into IR.
C_9: A $\leftarrow$ F	Transfer the ALU output into the A register.
$\overline{C_5}\,\overline{C_6}$: M ((MAR)) $\leftarrow$ A	Save contents of register A into memory.

The eight ALU operations performed by the CPU are defined by $C_{10}C_{11}C_{12}$ as follows:

C_{10}	C_{11}	C_{12}	F
0	0	0	0
0	0	1	R
0	1	0	L+R
0	1	1	L-R
1	0	0	L+1
1	0	1	L-1
1	1	0	L AND R
1	1	1	NOT L

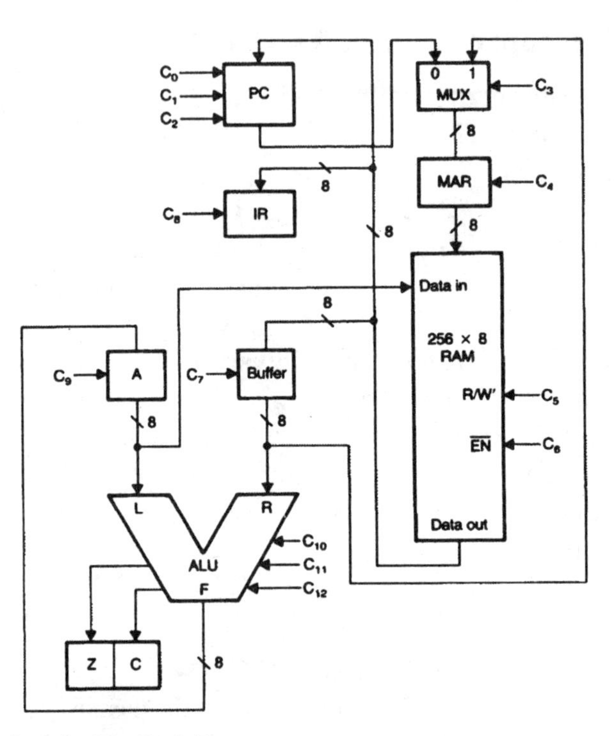

FIGURE 11. 13 Hardware Schematic of the Simple Processor

General Format	Instruction Length in Bytes	Object Code		Instruction Type	Operation	Comment
		In binary	In hex			
LDA ‹addr›	2	0000 1000	08	MRI	A ← M (‹addr›)	Load register A direct
		‹addr8›	‹addrH›			
STA ‹addr›	2	0000 1001	09	MRI	M (‹addr›) ← A	Store register A direct
		‹addr8›	‹addrH›			
ADD ‹addr›	2	0000 1010	0A	MRI	A ← A + M (‹addr›)	Add register A direct
		‹addr8›	‹addrH›			
SUB ‹addr›	2	0000 1011	0B	MRI	A ← A − M (‹addr›)	Subtract register A direct
		‹addr8›	‹addrH›			
JZ ‹addr›	2	0000 1100	0C	MRI	If Z=1 then PC ← ‹addr› else PC ← PC + 1	Jump on zero flag set
		‹addr8›	‹addrH›			
JC ‹addr›	2	0000 1101	0D	MRI	If C = 1 then PC ← ‹addr› else PC ← PC + 1	Jump on carry flag set
		‹addr8›	‹addrH›			
AND ‹addr›	2	0000 1110	0E	MRI	A ← A ^ M (‹addr›)	And register A direct
		‹addr8›	‹addrH›			
CMA	1	0000 0000	00	NMRI	A ← $\overline{A}$	Complement register A
INCA	1	0000 0010	02	NMRI	A ← A + 1	Increment register A
DCRA	1	0000 0100	04	NMRI	A ← A − 1	Decrement register A
HLT	1	0000 0110	06	NMRI	Halt	Halt CPU.

‹addr8›: 8-bit memory address in binary
‹addrH›: 8-bit memory address in hex
MRI: memory reference instruction
NMRI: nonmemory reference instruction.

FIGURE 11.14 Instruction Set to be Implemented

From Figure 11.14, notice that the proposed instruction set contains 11 instructions. The first 7 instructions are classified as memory reference instructions, since they all require a memory address (which is an 8-bit number in this case). The last 4 instructions do not require any memory address; they are called nonmemory reference instructions. Each memory reference instruction is assumed to occupy 2 consecutive bytes in the RAM. The first byte is reserved for the op-code, and the second byte indicates the 8-bit memory address. In contrast, a nonmemory reference instruction takes only one byte of storage. This instruction set supports only two addressing modes: implicit and direct. Both branch instructions are assumed to be absolute mode branch instructions. The op-code encoding for this instruction set is carried out in a logical manner, as explained in Figure 11.15.

The bit I3 of Figure 11.15 decides the instruction type. If I3 = 1, it is a memory reference instruction (MRI), otherwise it is a nonmemory reference instruction (NMRI).

Within the memory reference category, instructions are classified into four groups, as follows:

GROUP NO.	INSTRUCTIONS
0	Load and store
1	Add and subtract
2	Jumps
3	Logical

There are two instructions in the first three groups. Bit I_0 is used to determine the desired instruction of a particular group. If Io of group 0 equals zero, it is the load (LDA) instruction; otherwise it is the store (STA) instruction. Nevertheless, no such classification is required for group 3 and the nonmemory reference instructions.

As mentioned before, the instruction execution involves the following steps:

Step 1: Fetch the instruction.
Step 2: Decode the instruction to find out the required operation.
Step 3: If the required operation is a halt operation, then go to Step 6; otherwise continue.
Step 4: Retrieve the operands and perform the desired operation.
Step 5: Go to Step 1.
Step 6: Execute an infinite LOOP.

The first step is known as the fetch cycle, and the rest are collectively known as the execution cycle. To decode the instruction, the hardware shown in Figure 11.16 is used.

With this hardware and the status flags (Z and C), a microprogram to implement the instruction set can be written. The symbolic version of this microprogram is shown in Figure 11.17.

Mnemonic	Op-code Bit of Their Interpretations							
					TC	GN		SC
	I7	I6	I5	I4	I3	I2	I1	I0
LDA	0	0	0	0	1	0	0	0
STA	0	0	0	0	1	0	0	1
ADD	0	0	0	0	1	0	1	0
SUB	0	0	0	0	1	0	1	1
JZ	0	0	0	0	1	1	0	0
JC	0	0	0	0	1	1	0	1
AND	0	0	0	0	1	1	1	0
CMA	0	0	0	0	0	0	0	0
INCA	0	0	0	0	0	0	1	0
DCRA	0	0	0	0	0	1	0	0
HLT	0	0	0	0	0	1	1	0

Note:

TC: Type classifier (if I3 = 1, then it is a MRI; otherwise it is a NMRI)

GN: Group number within a type

(I2 I1 Group no.
0 0 0
0 1 1
1 0 2
1 1 3)

SC: Subcategory within a group

FIGURE 11.15 Op-code Encoding Logic

The hardware organization of the microprogrammed control unit for this situation shown in Figure 11.18 directly follows the symbolic listing shown in Figure 11.17. No attempt has been made toward arriving at a minimal microprogram. Rather, the concept was presented. The task of translating the symbolic microprogram of Figure 11.17 into a binary microprogram is left as an exercise.

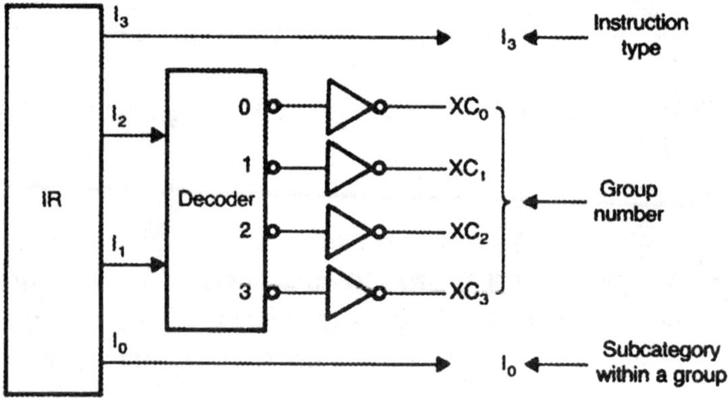

FIGURE 11.16 Instruction-decoding Hardware

Symbolic Microprogram:

ROM Address			
0		PC $\leftarrow$ 0;	These operations constitute the fetch cycle.
1	FETCH	MAR $\leftarrow$ PC;	
2		IR $\leftarrow$ M ((MAR)), PC $\leftarrow$ PC + 1;	
3		IF I_3 = 1 then go to MEMREF;	
4		IF XC_0 = 1 then go to CMA;	Here we decode the instructions.
5		IF XC_1 = 1 then go to INCA;	
6		IF XC_2 = 1 then go to DCRA;	
7		Go to HALT;	
8	CMA	A $\leftarrow$ $\overline{A}$;	Execute CMA instructions.
9		Go to FETCH;	
10	INCA	A $\leftarrow$ A + 1;	Execute INCA instruction.
11		Go to FETCH;	
12	DCRA	A $\leftarrow$ A - 1;	Execute DCRA instruction.
13		Go to FETCH;	
14	MEMREF	IF XC_0 = 1 then go to LDSTO;	Here we branch to the various groups of the memory reference instruction.
15		IF XC_1 = 1 then go to ADSUB;	
16		IF XC_2 = 1 then go to JMPS;	
17	AND	MAR $\leftarrow$ PC;	
18		BUFFER $\leftarrow$ M ((MAR)), PC $\leftarrow$ PC + 1;	Execute AND instruction.
19		MAR $\leftarrow$ BUFFER;	
20		BUFFER $\leftarrow$ M ((MAR));	
21		A $\leftarrow$ A $\wedge$ BUFFER;	
22		Go to FETCH;	
23	LDSTO	MAR $\leftarrow$ PC;	
24		BUFFER $\leftarrow$ M ((MAR)), PC $\leftarrow$ PC + 1;	
25		MAR $\leftarrow$ BUFFER;	
26		IF I_0 = 1 then go to STO;	
27	LOAD	BUFFER $\leftarrow$ M ((MAR));	
28		A $\leftarrow$ BUFFER;	
29		Go to FETCH;	
30	STO	M ((MAR)) $\leftarrow$ A;	
31		Go to FETCH;	

FIGURE 11.17 Symbolic Microprogram that Implements the Instruction Set Shown in Figure 11.14 for the Hardware Shown in Figure 11.13 *(Continued)*

32	ADSUB	MAR ← PC;	
33		BUFFER ← M ((MAR)), PC ← PC + 1;	
34		MAR ← BUFFER ;	
35		BUFFER ← M ((MAR));	
36		IF I_0 = 1 then go to SUB;	
37	ADD	A ← A + BUFFER;	Execute ADD instruction
38		Go to FETCH;	
39	SUB	A ← A – BUFFER;	Execute SUB instruction
40		Go to FETCH;	
41	JMPS	MAR ← PC;	
42			
43		IF I_0 = 1 then go to JOC;	
44	JOZ	IF Z = 1 then go to LOADPC;	Execute JZ instruction
45		Go to FETCH;	
46	JOC	IF C = 1 then go to LOADPC;	Execute JC instruction
47		Go to FETCH;	
48	LOADPC	PC ← M((MAR));	
49		Go to FETCH;	
50	HALT	Go to HALT;	Execute HALT instruction

FIGURE 11.17 *Continued*

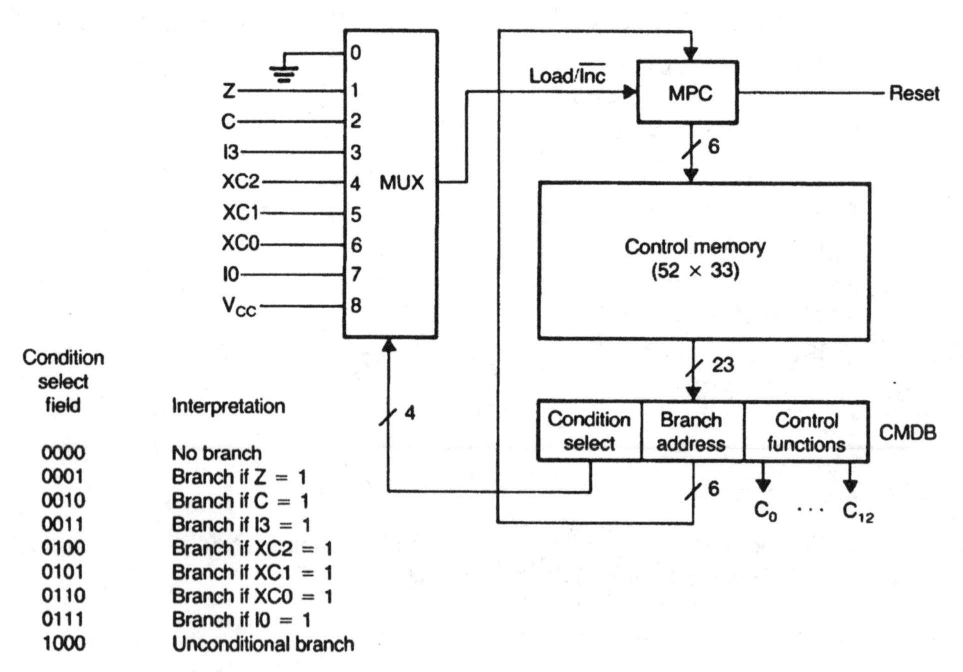

FIGURE 11.18 Microprogrammed Controller for the CPU

Example 11.3

If the following two instructions are to be added to the instruction set of Figure 11.14, write a symbolic microprogram for the CPU of section 11.4.2 that describes the execution of each instruction:

	GENERAL FORMAT	OPERATION	DESCRIPTION
(a)	**CLRA**	$A \leftarrow 0$	**Clear register A**
(b)	**PRSA**	$A \leftarrow 11111111$	**Set register A to all ones**

Solution:

(a) CLRA: $A \leftarrow 0$; Use ALU's zero output ($C_{10}C_{11}C_{12}$ =000)
go to FETCH ;

(b) PRSA: $A \leftarrow 0$; Use ALU's zero output ($C_{10}C_{11}C_{12}$ =000)
$A \leftarrow \overline{A}$;
go to FETCH ;

11.5 Reduced Instruction Set Computer (RISC) - More Details

RISC, which stands for reduced instruction set computer, is a new generation of faster and inexpensive machines. The initial application of RISC principles has been in desktop workstations. As mentioned before, PowerPC is a RISC microprocessor.

The basic idea behind RISC is for machines to cost less yet run faster, by using a small set of simple instructions for their operations. Also, RISC allows a balance between hardware and software based on functions to be achieved to make a program run faster and more efficiently. The philosophy of RISC is based on six principles: reliance on optimizing compilers, few instruction and addressing modes, fixed instruction format, instructions executed in one machine cycle, only call/return instructions accessing memory, and hardwired control.

The trend has always been to build CISCs (complex instruction set computers), which use many detailed instructions. However, because of their complexity, more hardware would have to be used. The more instructions, the more hardware logic is needed to implement and support them. For example, in a RISC machine, an ADD instruction takes its data from registers. On a CISC, each operand can be stored in any of many different forms, so the compiler must check several possibilities. Thus, both RISC and CISC have advantages and disadvantages. However, the principles of understanding optimizing compilers and what actually happens when a program is executed lead to RISC.

11.5.1 Case Study: RISC I (University of California, Berkeley)

The RISC machine presented in this section is the one investigated at the University of California, Berkeley. The RISC I is designed with the following design constraints:

1. Only one instruction is executed per cycle.
2. All instructions have the same size.
3. Only load and store instructions can access memory.
4. High-level languages (HLL) are supported.

Two high level Languages (C and Pascal) were supported by RISC I. A simple architecture implies a fewer transistors, and this leads to the fact that most pieces of a RISC HLL system are in software. Hardware is utilized for time-consuming operations. Using C and Pascal, a comparison study was made to determine the frequency of occurrence of particular variable and statement types. Studies revealed that integer constants appeared most frequently, and a study of the code produced revealed that the procedure calls are the most time-consuming operations.

i) Basic RISC Architecture

The RISC I instruction set contains a few simple operations (arithmetic, logical, and shift). These instructions operate on registers. Instruction, data, addresses and registers are all 32 bits long. RISC instructions fall in four categories: ALU, memory access, branch, and miscellaneous. The execution time is given by the time taken to read a register, perform an ALU operation, and store the result in a register. Register 0 always contains a 0. Load and store instructions move data between registers and memory. These instructions use two CPU cycles. Variations of memory-access instructions exist in order to accommodate sign-extended or zero-extended 8-bit, 16-bit and 32-bit data. Though absolute and register indirect addressing are not directly available, they may be synthesized using register 0. Branch instructions include CALL, RETURN, and conditional and unconditional jumps. The following instruction format is used:

opcode(7)	scc(l)	dest(5)	source1(5)	imm(l)	source2(13)

For register-to-register instructions, dest selects one of the 32 registers as destination of the result of the operation that is itself performed on registers source 1 and source2. If imm equals 0, the low-order 5 bits of source2 specify another register. If imm equals 1, then source2 is regarded as a sign-extended 13-bit constant. Since the frequency of integer constants is high, the immediate field has been made an option in every instruction. Also, Scc determines whether the condition codes are set. Memory-access instructions use source 1 to specify the index register and source2 to specify offset.

ii) Register Windows

The procedure-call statements take the maximum execution time. A RISC program has more call statements, since the complex instructions available in CISC are subroutines in RISC. The RISC register window scheme strives to make the call operation as fast as possible and also to reduce the number of accesses to data memory. The scheme works as follows.

Using procedures involve two groups of time-consuming operations, namely, saving or restoring registers on each call/return and passing parameters and results to and from the procedure. Statistics indicate that local variables are the most frequent operands.

This creates a need to support the allocation of locals in the registers. One available scheme is to provide multiple banks of registers on the chip to avoid saving and restoring of registers. Thus each procedure call results in a new set of registers being allocated for use by that procedure. The return alters a pointer that restores the old set. A similar scheme is adopted by RISC. However, there are some registers that are not saved or restored; these are called global registers. In addition, the sets of registers used by different processes are overlapped in order to allow parameters to be passed. In other machines, parameters are usually passed on the stack with the calling procedure using a register to point to the beginning of the parameters (and also to the end of the locals). Thus all references to parameters are indexed references to memory. In RISC I the set of window registers (r10 to r31) is divided into three parts. Registers r26 to r31 (HIGH) contain parameters passed from the calling procedure. Registers rl6 to r25 (LOCAL) are for local storage. Registers r10 to rl5 (LOW) are for local storage and for parameters to be passed to the called procedure. On each call, a new set of r10 to r31 registers is allocated. The LOW registers of the caller are required to become the HIGH registers of the called procedure. This is accomplished by having the hardware overlap the LOW registers of the calling frame with the HIGH registers of the called frame. Thus without actually moving the information, parameters are transferred.

Multiple register banks require a mechanism to handle the case in which there are no free register banks available. RISC handles this problem with a separate register-overflow stack in memory and a stack pointer to it. Overflow and underflow are handled with a trap to a software routine that adjusts the stack. The final step in allocating variables in registers is handling the problem of pointers. RISC resolves this by giving addresses to the window registers. If a portion of the address space is reserved, we can determine with one comparison whether an address points to a register or to memory. Load and store are the only instructions that access memory and they take an extra cycle already. Hence this feature may be added without reducing the performance of the load and store instructions. This permits the use of straightforward computer technology and still leaves a large fraction of the variables in registers.

iii) Delayed Jump

A normal RISC I instruction cycle is long enough to execute the following sequence of operations:

1. Read a register.
2. Perform an ALU operation.
3. Store the result back into a register.

Performance is increased by prefetching the next instruction during the current instruction. To facilitate this, jumps are redefined such that they do not occur until after the following instruction. This is called delayed jump.

11.6 Cache Memories - More Details

Cache Memory is discussed in Chapter 8. This section provides a more detailed coverage.

The performance of a system that employs a cache can be formally analyzed as follows: If t_c, h, and t_m specify the cache-access time, hit ratio, and the main memory access time, respectively; then the average access time can be determined as shown in the equation below:

$$t_{av} = ht_c + (1 - h)(t_c + t_m)$$

The hit ratio h always lies in the closed interval 0 and 1, and it specifies the relative number of successful references to the cache. In the above equation, when there is a cache hit, the main memory will not be accessed; and in the event of a cache miss, both main memory and cache will be accessed. Suppose the ratio of main memory access time to cache access time is γ, then an expression for the efficiency of a system that employs a cache can be derived as follows:

$$\begin{aligned} \mathit{Efficieny} = E &= \frac{t_c}{t_{av}} \\ &= \frac{t_c}{ht_c + (1-h)(t_c + t_m)} \\ &= \frac{1}{h + (1-h)(1 + \frac{t_m}{t_c})} \\ &= \frac{1}{h + (1-h)(1+\gamma)} \\ &= \frac{1}{1 + \gamma(1-h)} \end{aligned}$$

Note that E is maximum when $h = 1$ (when all references are confined to the cache). A hit ratio of 90% (h = 0.90) is not uncommon with many contemporary systems.

Example 11.4

Calculate t_{av}, γ, and E of a memory system whose parameters are as indicated:

$t_c = 160$ ns

$t_m = 960$ ns

$h = 0.90$

Solution

$$t_{av} = h t_c + (1 - h)(t_c + t_m)$$
$$= 0.9\,(160) + (0.1)\,(960 + 160)$$
$$= 144 + 112$$
$$= 256 \text{ ns}$$

$$\gamma = \frac{t_m}{t_c} = \frac{960}{160} = 6$$

$$E = \frac{1}{1+\gamma(1-h)} = \frac{1}{1+6(0.1)} = 0.625$$

This result indicates that by employing a cache, efficiency is improved by 62.5%. Assume the unit of mapping is a block; then the relationship between the main and cache memory blocks can be established by using a specific mapping technique. Three mapping techniques are already discussed in section 8.1.4.

In fully associative mapping, a main memory block i can be mapped to any cache block j, where $0 \leqslant i \leqslant M-1$ *and* $0 \leqslant j \leqslant N-1$. Note that the main memory has M blocks and the cache is divided into N blocks. To determine which block of main memory is stored into the cache, a tag is required for each block. Hence,

Tag (j) = address of the main memory block stored in the cache block j.

Suppose $M = 2^m$ and $N = 2^n$; then m and n bits are required to specify the addresses of a main and cache memory block, respectively. Also, block size = 2^w, where w bits are required to specify a word in a block.

For Associative mapping : m bits of the main memory are used as a tag; and N tags are needed since there are N cache blocks.

Main memory address = (Tag + w)bits.

For Direct mapping: High order (m-n) bits are used as a tag.

Main memory address = (Tag + n + w)bits

For Set-associative mapping:

Tag field = (m - n + s) bits, where Blocks/set = 2^s

Cache set number = (n - s) bits

Main memory address = (Tag size + cache set number + w) bits.

Example 11.5

The parameters of a computer memory system are specified as follows:

- Main memory size = 8K blocks
- Cache memory size =512 blocks
- Block size = 8 words

Determine the sizes of the tag field along with the main memory address using each of the following methods:

(a) Fully associative mapping
(b) Direct mapping
(c) Set associative mapping with 16 blocks/set

Solution

With the given data, compute the following:

- M = 8K = 8192 = 2^{13}, and thus m = 13.
- N = 512 = 2^9, and thus n = 9.
- Block size = 8 words = 2^3 words, and thus we require 3 bits to specify a word within a block.

Using this information, we can determine the main and cache memory address formats as shown next:

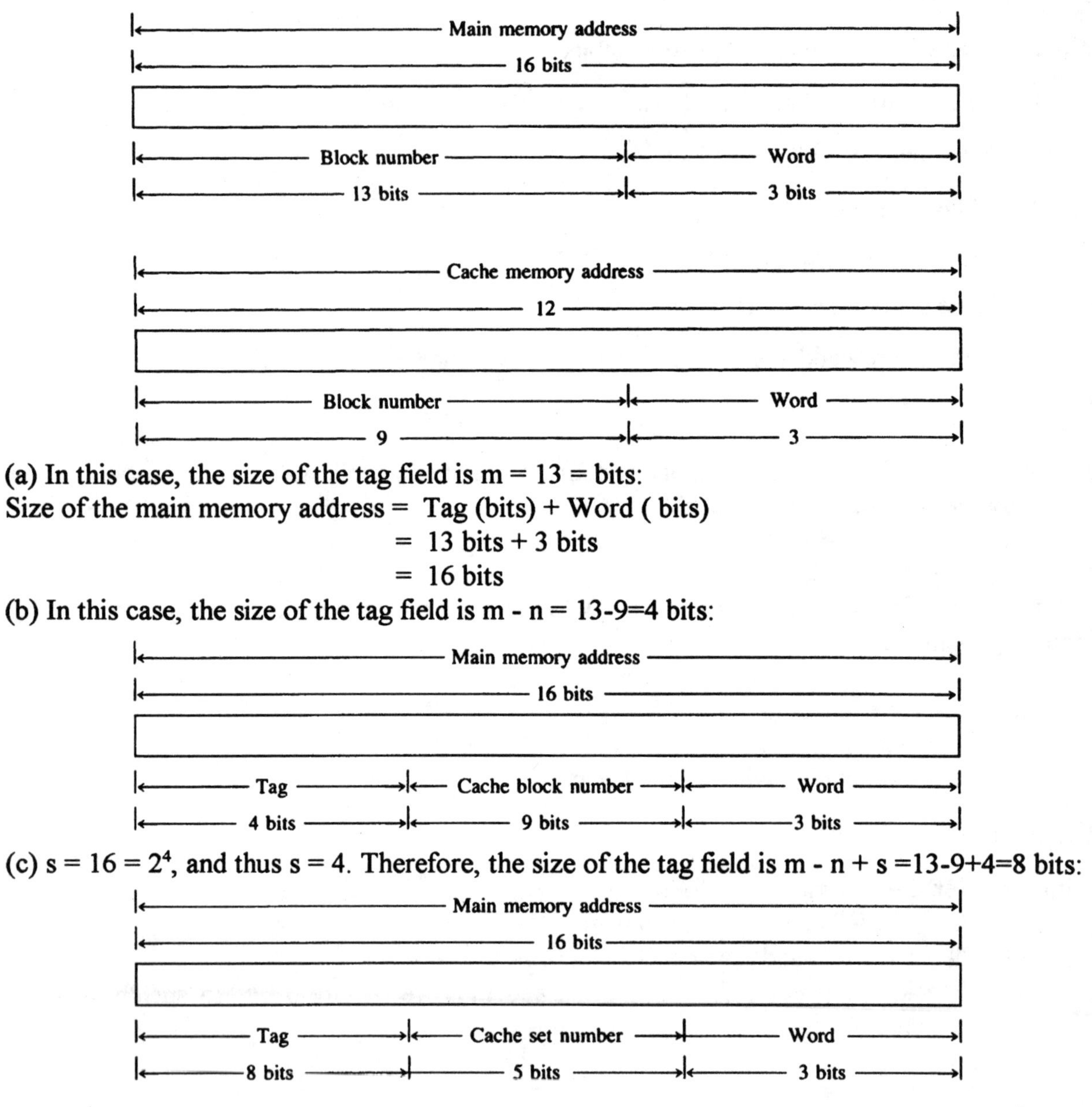

(a) In this case, the size of the tag field is m = 13 = bits:

Size of the main memory address = Tag (bits) + Word (bits)
= 13 bits + 3 bits
= 16 bits

(b) In this case, the size of the tag field is m - n = 13-9=4 bits:

(c) $s = 16 = 2^4$, and thus s = 4. Therefore, the size of the tag field is m - n + s =13-9+4=8 bits:

Example 11.6

The access time of a cache memory is 50 ns and that of the main memory is 500 ns. It is estimated that 80% of the main memory requests are for read and the remaining are for write. The hit ratio for read access only is 0.9 and a write-through policy is used.

(a) Determine the average access time considering only the read cycles.

(b) What is the average time if the write requests are also taken into consideration

Solution

(a)

$$t_{av} = ht_c + (1-h)(t_c + t_m)$$
$$= 0.9 \times 50 + (0.1)(550)$$
$$= 45 + 55 \text{ ns}$$
$$= 100 \text{ ns}$$

(b)

$$t_{read/write} = (\textit{read request probability}) \times t_{av\,read} + (1 - \textit{read request probability}) \times t_{av\,write}$$

$$\textit{read request probability} = 0.8$$

$$\textit{write request probability} = 0.2$$

$$t_{av\,read} = t_{av} = 100 \text{ ns (result of part (a))}$$

$$t_{av\,write} = 500 \text{ ns (because both the main and cache memories are updated at the same time)}$$

$$t_{read/write} = 0.8 \times 100 + 0.2 \times 500$$
$$= 80 + 100 \text{ ns}$$
$$= 180 \text{ ns}$$

The growth in 1C technology has allowed manufacturers to fabricate a small cache on the CPU chip. The on-chip cache of Motorola's 32-bit microprocessor, the MC68020, is discussed next.

The MC68020 on-chip cache is a direct mapped instruction cache. Only instructions are cached; data items are not. This cache is a collection of 64 entries, where each cache entry consists of a 26-bit tag field and 32-bit instruction data. The tag field includes the following components:

- High-order 24 bits of the memory address.
- The most-significant bit FC2 of the function code. In the MC68020 processor, the 3-bit function code combination FC2 FC1 FC0 is used to identify the status of the

processor and the address space (discussed in Chapter 10) of the bus cycle. For example, FC2 = 1 means the processor operates in the supervisory or privileged mode. Otherwise, it operates in the user mode. Similarly, when FC1 FC0 = 01, the bus cycle is made to access data. When FC1 FC0 = 10, the bus cycle is made to access code.
- Valid bit.

A block diagram of the MC68020 on chip cache is shown in Figure 11.19.

If an instruction fetch occurs when the cache is enabled, the cache is first checked to determine if the word requested is in the cache. This is achieved by first using 6 bits of the memory address (A7-A2) to select one of the 64 entries of the cache. Next, address bits A31-A8 and function bit FC2 are compared to the corresponding values of the selected cache entry. If there is a match and the valid bit is set, a cache bit is said to occur. In this case, the address bit Al is used to select the proper instruction word stored in the cache and the cycle ends.

If there is no match or the valid bit is cleared, and a cache miss occurs. In this case, the instruction is fetched from external memory. This new instruction is automatically written into the cache and the valid bit is set. Since the processor always pre fetches instructions from the external memory in the form of long words, both instruction data words of the cache will be updated regardless of which word caused the miss.

The MC68020 on-chip instruction cache obtains a significant increase in performance by reducing the number of fetches required to external memory.

Typically, this cache reduces the instruction execution time in two ways. First, it provides a two-clock-cycle access time for an instruction that hits in the cache (see Figure 11.20); second, if the access hits in the cache, it allows simultaneous instruction and data access to occur. Of these two benefits, simultaneous access is more significant, since it allows 100% reduction in the time required to access the instruction rather than the 33% reduction afforded by going from three to two clocks.

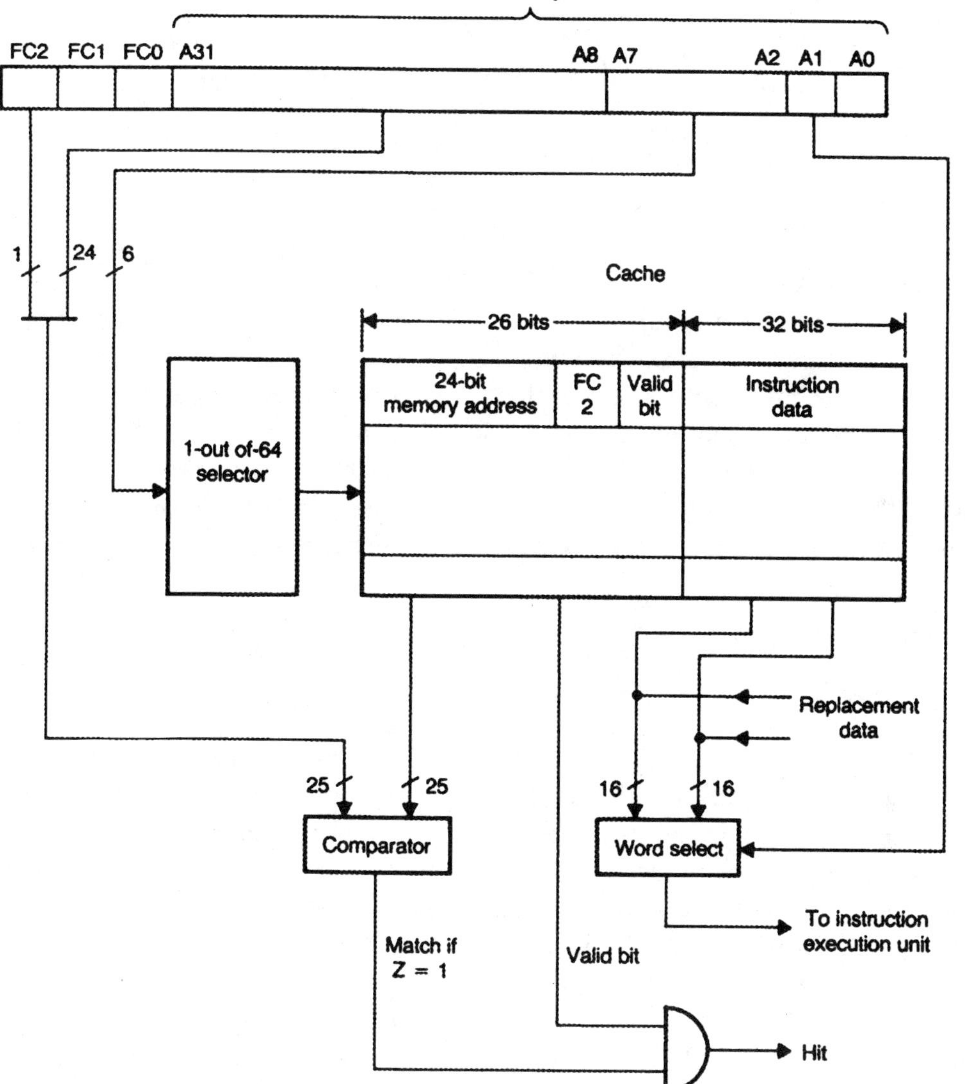

FIGURE 11.19 MC68020 On-chip Cache Organization

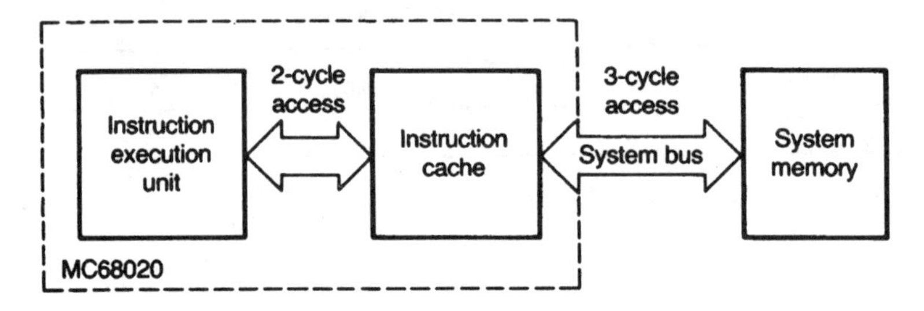

FIGURE 11.20 Mechanics of the MC68020 Instruction Cache.

Finally, microprocessors such as Intel Pentium II support two-levels of cache. These are L1 (Level 1) and L2 (Level 2) cache memories. The L1 cache (Smaller in size) is contained inside the processor chip while the L2 cache (Larger in size) is interfaced external to the microprocessor. The L1 cache normally provides separate instruction and data caches. The processor can directly access the L1 cache while the L2 cache normally supplies instructions and data to the L1 cache. The L2 cache is usually accessed by the microprocessor only if L1 misses occur. This two-level cache memory enhances the performance of the microprocessor.

11.7 Virtual Memory and Memory-Management Concepts - More Details.

This topic is discussed briefly in chapter 8. This section describes basic memory-management concepts in more detail. Topics include virtual memory, paging, segmentation, and address mapping schemes. Virtual memory is the most fundamental concept implemented by a system that performs memory-management functions such as space allocation, program relocation, code sharing and protection.The key idea behind this concept is to allow a user program to address more locations than those available in a physical memory. An address generated by a user program is called a virtual address. The set of virtual addresses constitutes the virtual address space. Similarly, the main memory of a computer contains a fixed number of addressable locations and a set of these locations forms the physical address space. The basic hardware for virtual memory is implemented in modern microprocessors as an on-chip feature. These contemporary processors support both cache and virtual memories. The virtual addresses are typically converted to physical addresses and then applied to cache.

In the early days, when a programmer used to write a large program that could not fit into the main memory, it was necessary to divide the program into small portions so each one could fit into the primary memory. These small portions are called overlays. A programmer has to design overlays so they are independent of each other. Under these circumstances, one can successively bring each overlay into the main memory and execute them in a sequence.

Although this idea appears to be simple, it increases the program-development time considerably.

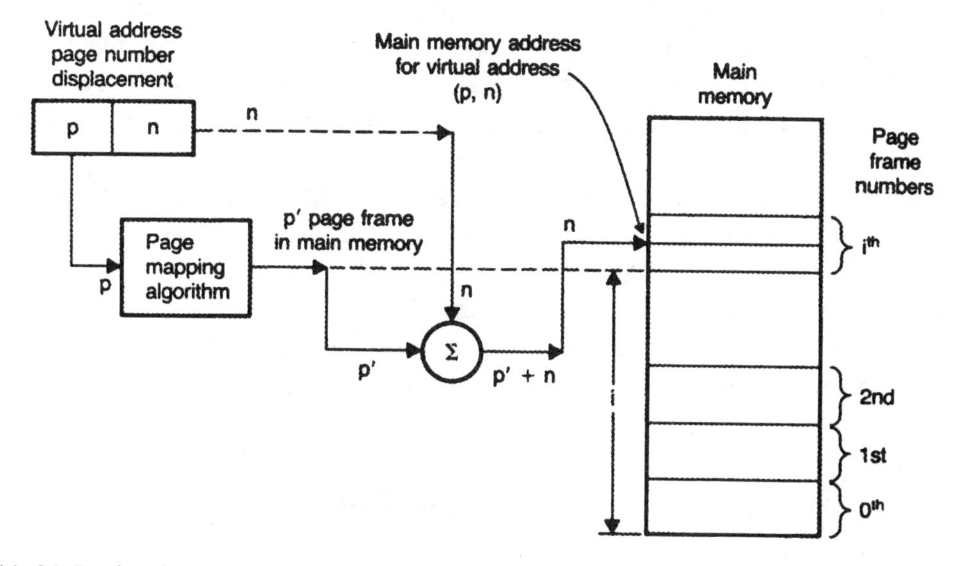

FIGURE 11. 21 Paging Systems—Virtual versus Main Memory Mapping

However, in a system that uses a virtual memory, the size of the virtual address space is usually much larger than the available physical address space. In such a system, a programmer does not have to worry about overlay design, and thus a program can be written assuming a huge address space is available. In a virtual memory system, the programming effort can be greatly simplified. However, in reality, the actual number of physical addresses available is considerably less than the number of virtual addresses provided by the system. There should be some mechanism for dividing a large program into small overlays automatically. A virtual memory system is one that mechanizes the process of overlay generation by performing a series of mapping operations.

A virtual memory system may be configured in one of the following ways:

- Paging systems
- Segmentation systems

In a paging system, the virtual address space is divided into equal-size blocks called pages. Similarly, the physical memory is also divided into equal-size blocks called frames. The size of a page is the same as the size of a frame. The size of a page may be 512, 1024 or 2048 words.

In a paging system, each virtual address may be regarded as an ordered pair (p, n), where p is the page number and n is the word number within the page p. Sometimes the quantity n is referred to as the displacement, or offset. A user program may be regarded as a sequence of pages, and a complete copy of the program is always held in a backup store such as a disk. A page p of the user program can be placed in any available page frame p' of the main memory. A program may access a page if the page is in the main memory. In a paging scheme, pages are brought from secondary memory and are stored in main memory in a dynamic manner. All virtual addresses generated by a user program must be translated into physical memory addresses. This process is known as dynamic address translation and is shown in Figure 11.21.

When a running program accesses a virtual memory location v = (p, n), the mapping algorithm finds that the virtual page p is mapped to the physical frame p'. The physical address is then determined by appending p' to n.

This dynamic address translator can be implemented using a page table. In most systems, this table is maintained in the main memory. It will have one entry for each virtual page of the virtual address space. This is illustrated in the following example.

Example 11.7

Design a mapping scheme with the following specifications:

- Virtual address space = 32K words
- Main memory size = 8K words
- Page size = 2K words
- Secondary memory address = 24 bits

Solution

32K words can be divided into 16 virtual pages with 2K words per page, as follows:

VIRTUAL ADDRESS	PAGE NUMBER
0-2047	0
2048-4095	1
4096-6143	2
6144-8191	3
8192-10239	4
10240-12287	5
12288-14335	6
14336-16383	7
16384-18431	8
18432-20479	9
20480-22527	10
22528-24575	11
24576-26623	12
26624-28671	13

28672-30719	14
30720-32767	15

Since there are 8K words in the main memory, 4 frames with 2K words per frame are available:

PHYSICAL ADDRESS	FRAME NUMBER
0-2047	0
2048-4095	1
4096-6143	2
6144-8191	3

Since there are 32K addresses in the virtual space, 15 bits are required for the virtual address. Because there are 16 virtual pages, the page map table contains 16 entries. The 4 most-significant bits of the virtual address are used as an index to the page map table, and the remaining 11 bits of the virtual address are used as the displacement to locate a word within the page frame. Each entry of the page table is 32 bits long. This can be obtained as follows:

1 bit for determining whether the page table is in main memory or not (residence bit).
2 bits for main memory page frame number.
24 bits for secondary memory address
5 bits for future use. (Unused)
32 bits total

The complete layout of the page table is shown in Figure 11.22. Assume the virtual address generated is 0111 000 0010 1101. From this, compute the following:

Virtual page number = 7_{10}
Displacement = 43_{10}

From the page-map table entry corresponding to the address 0111, the page can be found in the main memory (since the page resident bit is 1).

The required virtual page is mapped to main memory page frame number 2. Therefore, the actual physical word is the 43rd word in the second page frame of the main memory.

So far, a page referenced by a program is assumed always to be found in the main memory. In practice, this is not necessarily true. When a page needed by a program is not assigned to the main memory, a page fault occurrs. A page fault is indicated by an interrupt, and when this interrupt occurs, control is transferred to a service routine of the operating system called the page-fault handler. The sequence of activities performed by the page-fault handler are summarized as follows:

- The secondary memory address of the required page p is located from the page table.
- Page p from the secondary memory is transferred into one of the available main memory frames by performing a block-move operation.
- The page table is updated by entering the frame number where page p is loaded and by setting the residence bit to 1 and the change bit to 0.

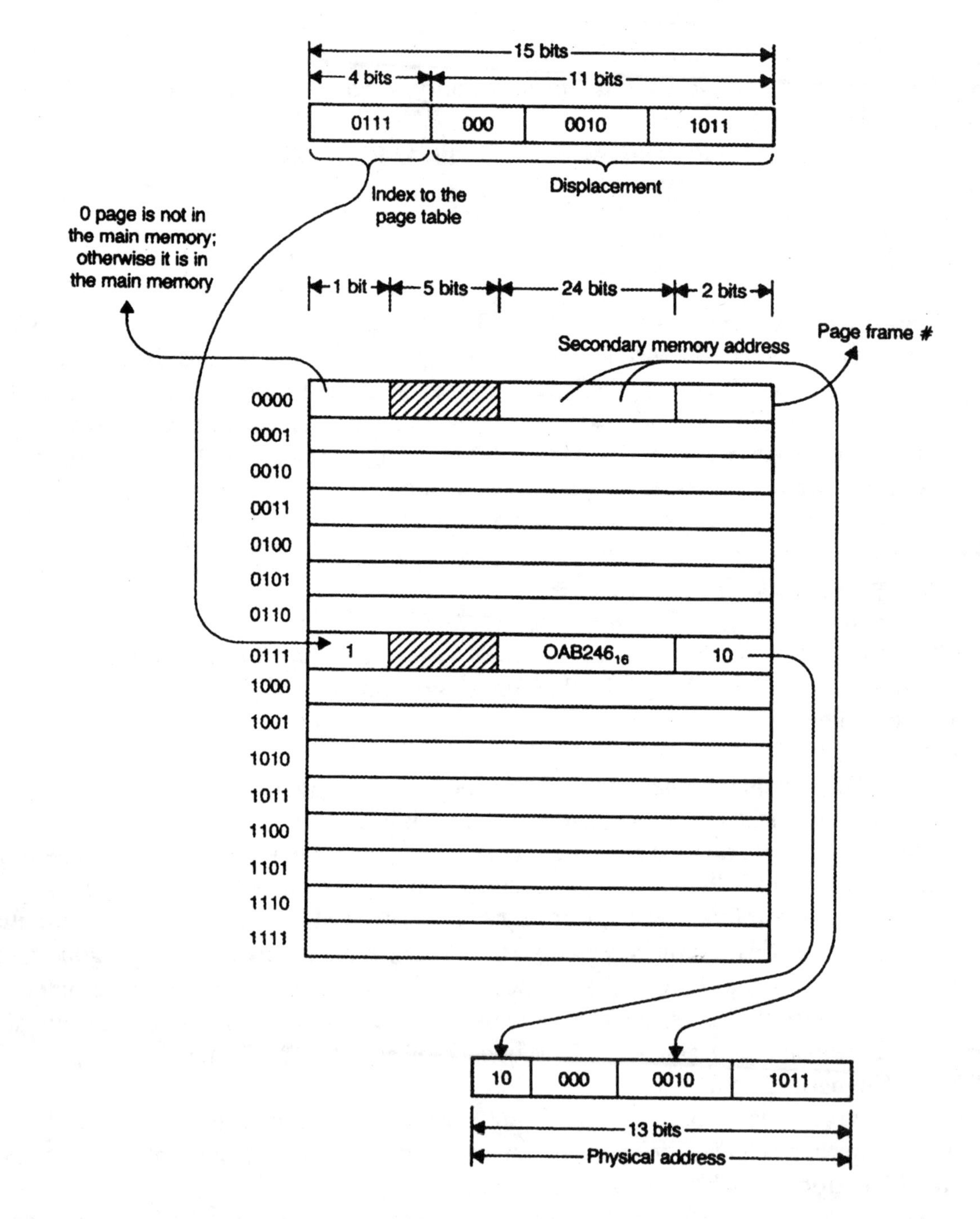

FIGURE 11.22 Mapping Scheme for the Paging System of Example 11.7

When a page-fault handler completes its task, control is transferred to the user program, and the main memory is accessed again for the required data or instruction. All these activities are kept hidden from a user. Pages are transferred to main memory only at specified times. The policy that governs this decision is known as the fetch policy. Similarly, when a page is to be transferred from the secondary memory to main memory, all frames may be full. In such a situation, one of the frames has to be removed from the main memory to provide room for an incoming page. The frame to be removed is selected using a replacement policy. The performance of a virtual memory system is dependent upon the fetch and replacement strategies. These issues are discussed later.

The paging concept covered so far is viewed as a one-dimensional technique because the virtual addresses generated by a program may linearly increase from 0 to some maximum value M. There are many situations where it is desirable to have a multidimensional virtual address space. This is the key idea behind segmentation systems.

Each logical entity such as a stack, an array, or a subroutine has a separate virtual address space in segmentation systems. Each virtual address space is called a segment, and each segment can grow from zero to some maximum value. Since each segment refers to a separate virtual address space, it can grow or shrink independently without affecting other segments.

In a segmentation system, the details about segments are held in a table called a segment table. Each entry in the segment table is called a segment descriptor, and it typically includes the following information:

- Segment base address b (starting address of the segment in the main memory)
- Segment length l (size of a segment)
- Segment presence bit
- Protection bits

From the structure of a segment descriptor, it is possible to create two or more segments whose sizes are different from one another. In a sense, a segmentation system becomes a paging system if all segments are of equal length. Because of this similarity, there is a close relationship between the paging and segmentation systems from the viewpoint of address translation.

A virtual address, V, in a segmentation system is regarded as an ordered pair (s, d), where s is the segment number and d is the displacement within segment s. The address translator for a segmentation system can be implemented using a segment table, and its organization is shown in Figure 11.23.

The details of the address translation process is briefly discussed next.

Let V be the virtual address generated by the user program. First, the segment number field, s, of the virtual address V is used as an index to the segment table. The base address and length of this segment are b_s and l_s, respectively. Then, the displacement d of the virtual address V is comared with the length of the segment l_s to make sure that the required address lies within the segment. If d is less than or equal to l_s, then the comparator output Z will be high. When $d \leq l_s$, the physical address is formed by adding b_s and d. From this physical address, data is retrieved and transferred to the CPU. However, when $d > l_s$, the required address lies out of the segment range, and thus an address out of range trap will be generated. A trap is a nonmaskable interrupt with highest priority.

In a segmentation system, a segment needed by a program may not reside in main memory. This situation is indicated by a bit called a valid bit. A valid bit serves the same purpose as that of a page resident bit, and thus it is regarded as a component of the segment descriptor. When the valid bit is reset to 0, it may be concluded that the required segment is not in main memory.

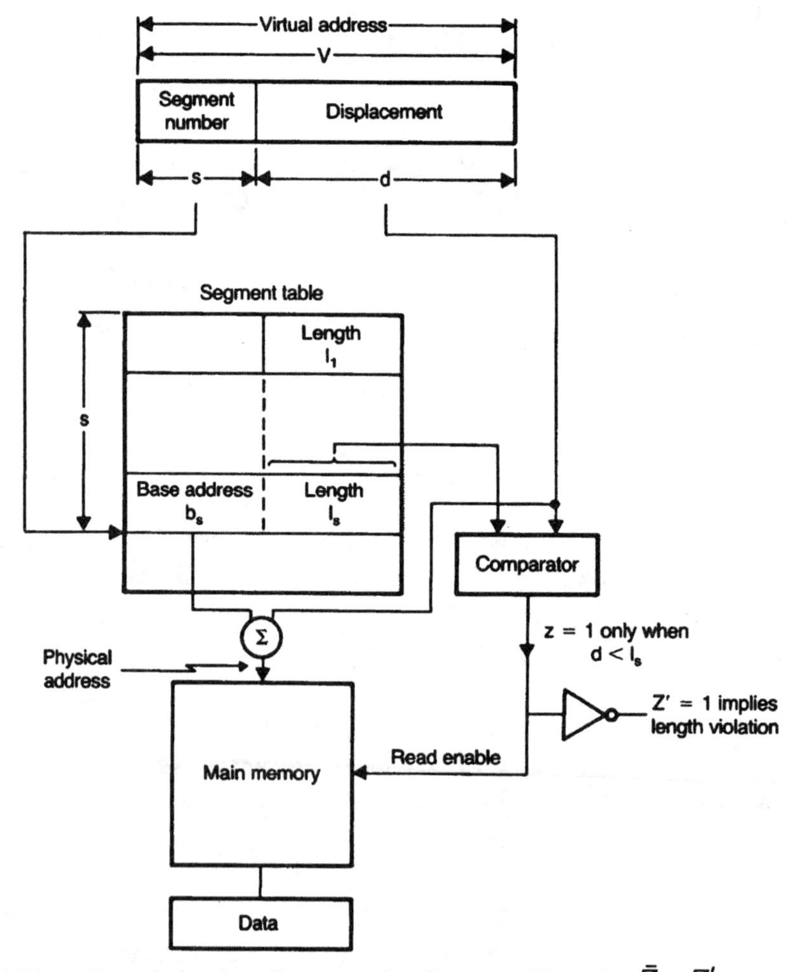

FIGURE 11.23 Address Translation in a Segmentation System. (Note that $\bar{Z} = Z'$)

This means that its secondary memory address must be included in the segment descriptor. Recall that each segment represents a logical entity. This implies that we can protect segments with different protection protocols based on the logical contents of the segment. The following are the common protection protocols used in a segmentation system:

- Read only
- Execute only
- Read and execute only
- Unlimited access
- No access

Thus it follows that these protection protocols have to be encoded into some protection codes and these codes have to be included in a segment descriptor.

In a segmented memory system, when a virtual address is translated into a physical address, one of the following traps may be generated:

- Segment fault trap is generated when the required segment is not in the main memory.
- Address violation trap occurs when $d > l_s$.
- Protection violation trap is generated when there is a protection violation.

When a segment fault occurs, control will be transferred to the operating system. In response, the operating system has to perform the following activities:

- First, it finds the secondary memory address of the required segment from its segment descriptor.
- Next, it transfers the required segment from the secondary to primary memory.
- Finally, it updates the segment descriptor to indicate that the required segment is in the main memory.

After performing the preceding activities, the operating system transfers control to the user program and the data or instruction retrieval or write operation is repeated.

A comparison of the paging and segmentation systems is provided next. The primary idea behind a paging system is to provide a huge virtual space to a programmer, allowing a programmer to be relieved from performing tedious memory-management tasks such as overlay design. The main goal of a segmentation system is to provide several virtual address spaces, so the programmer can efficiently manage different logical entities such as a program, data, or a stack.

The operation of a paging system can be kept hidden at the user level. However, a programmer is aware of the existence of a segmented memory system.

To run a program in a paging system, only its current page is needed in the main memory. Several programs can be held in the main memory and can be multiplexed. The paging concept improves the performance of a multiprogramming system. In contrast, a segmented memory system can be operated only if the entire program segment is held in the main memory.

In a paging system, a programmer cannot efficiently handle typical data structures such as stacks or symbol tables because their sizes vary in a dynamic fashion during program execution. Typically, large pages for a symbol table or small pages for a stack cannot be created. In a segmentation system, a programmer can treat these two structures as two logical entities and define the two segments with different sizes.

The concept of segmentation encourages people to share programs efficiently. For example, assume a copy of a matrix multiplication subroutine is held in the main memory. Two or more users can use this routine if their segment tables contain copies of the segment descriptor corresponding to this routine. In a paging system, this task cannot be accomplished efficiently because the system operation is hidden from the user. This result also implies that in a segmentation system, the user can apply protection features to each segment in any desired manner. However, a paging system does not provide such a versatile protection feature.

Since page size is a fixed parameter in a paging system, a new page can always be loaded in the space used by a page being swapped out. However, in a segmentation system with uneven segment sizes, there is no guarantee that an incoming segment can fit into the free space created by a segment being swapped out.

In a dynamic situation, several programs may request more space, whereas some other programs may be in the process of releasing the spaces used by them. When this happens in a segmented memory system, there is a possibility that uneven-sized free spaces may be sparsely distributed in the physical address space. These free spaces are so irregular in size that they cannot normally be used to satisfy any new request. This is called an external fragmentation, and an operating system has to merge all free spaces to form a single large useful segment by moving all active segments to one end of the memory. This activity is known as memory compaction. This is a time-consuming operation and is a pure overhead. Since pages are of equal size, no external fragmentation can occur in a paging system.

In a segmented memory system, a programmer defines a segment, and all segments are completely filled.

The page size is decided by the operating system, and the last page of a program may not be filled completely when a program is stored in a sequence of pages. The space not filled in the last page cannot be used for any other program. This difficulty is known as internal fragmentation—a potential disadvantage of a paging system.

In summary, the paging concept simplifies the memory-management tasks to be performed by an operating system and therefore, can be handled efficiently by an operating system. The segmentation approach is desirable to programmers when both protection and sharing of logical entities among a group of programmers are required.

To take advantage of both paging and segmentation, some systems use a different approach, in which these concepts are merged. In this technique, a segment is viewed as a collection of pages. The number of pages per segment may vary. However, the number of words per page still remains fixed. In this situation, a virtual address V is an ordered triple (s, p, d), where s is the segment number and p and d are the page number and the displacement within a page, respectively.

The following tables are used to translate a virtual address into a physical address:

- Page table: This table holds pointers to the physical frames.
- Segment table: Each entry in the segment table contains the base address of the page table that holds the details about the pages that belong to the given segment.

The address-translation scheme of such a paged-segmentation system is shown in Figure 11.24:

- First, the segment number s of the virtual address is used as an index to the segment table, which leads to the base address b_p of the page table.
- Then, the page number p of the virtual address is used as an index to the page table, and the base address of the frame number p' (to which the page p is mapped) can be found.
- Finally, the physical memory address is computed by adding the displacement d of the virtual address to the base address p' obtained before.

To illustrate this concept, the following numerical example is provided.

Example 11.8

Assume the following values for the system of Figure 11.24:

- Length of the virtual address field =32 bits
- Length of the segment number field =12 bits
- Length of the page number field = 8 bits
- Length of the displacement field =12 bits

Now, determine the value of the physical address using the following information:

- Value of the virtual address field = $000FA0BA_{16}$
- Contents of the segment table address $(000)_{16} = 0FF_{16}$
- Contents of the page table address $(1F9_{16}) = AC_{16}$

Solution

From the given virtual address, the segment table address is 000_{16} (three high-order hexadecimal digits of the virtual address). It is given that the contents of this segment-able address is $0FF_{16}$. Therefore, by adding the page number p (fourth and fifth hexadecimal digits of the virtual address) with $0FF_{16}$, the base address of the page table can be determined as:

$$0FF_{16} + FA_{16} = 1F9_{16}$$

Since the contents of the page table address $1F9_{16}$ is AC_{16}, the physical address can be obtained by adding the displacement (low-order three hexadecimal digits of the virtual address) with AC_{16} as follows:

$AC000_{16} + 000BA_{16} = AC0BA_{16}$

In this addition, the displacement value 0BA is sign-extended to obtain a 20-bit number that can be directly added to the base value p'. The same final answer can be obtained if p' and d are first concatenated. Thus, the value of the physical address is $AC0BA_{16}$.

The virtual space of some computers use both paging and segmentation, and it is called a linear segmented virtual memory system. In this system, the main memory is accessed three times to retrieve data (one for accessing the page table; one for accessing the segment table; and one for accessing the data itself).

Accessing the main memory is a time-consuming operation. To speed up the retrieval operation, a small associative memory (implemented as an on-chip hardware in modern microprocessors) called the translation lookaside buffer (TLB) is used. The TLB stores the translation information for the 8 or 16 most recent virtual addresses. The organization of a address translation scheme that includes a TLB is shown in Figure 11.25.

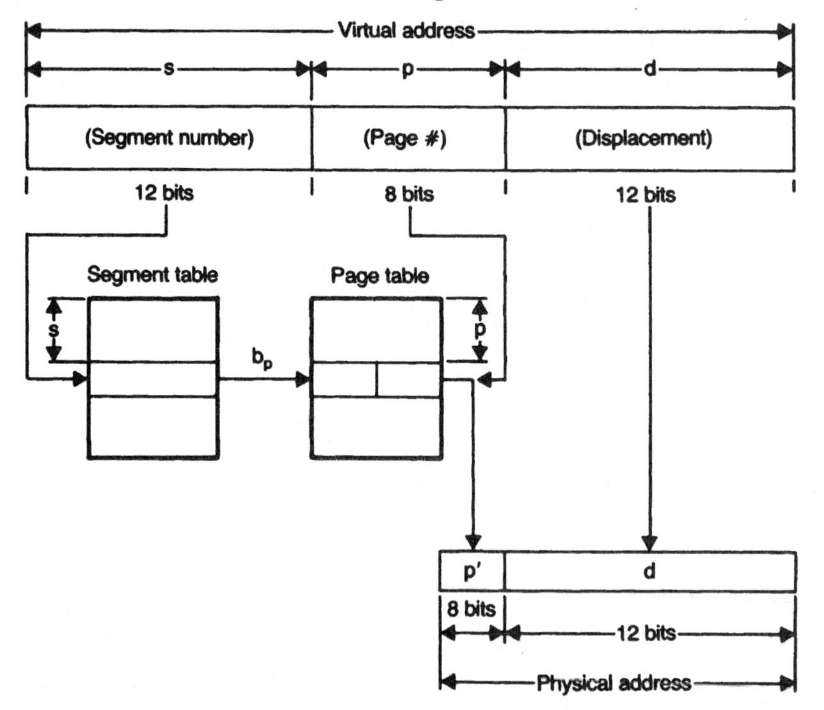

FIGURE 11.24 Address-translation Scheme for a Paged-segmentation System

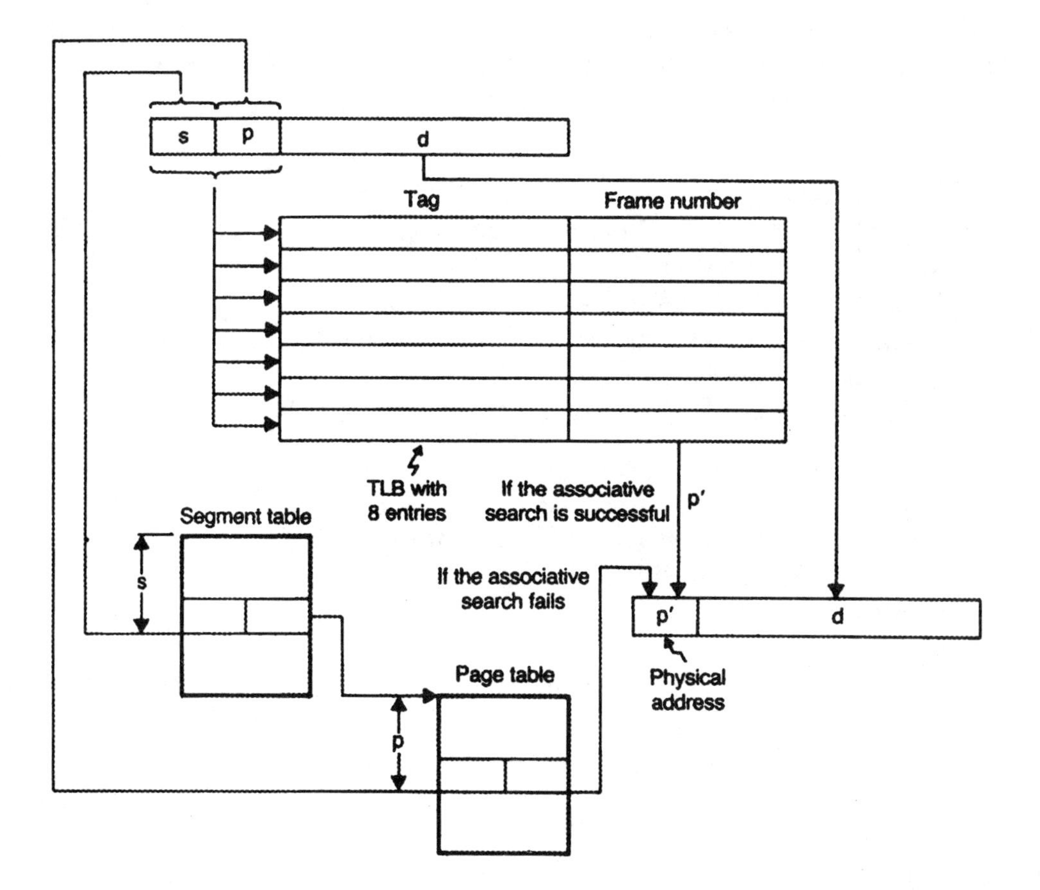

FIGURE 11.25 Address Translation Using a TLB

In this scheme, assume the TLB is capable of holding the translation information about the 8 most recent virtual addresses.

The pair (s, p) of the virtual address is known as a tag, and each entry in the TLB is of the form:

(s,p) or tag	Base address of the frame p'

When a user program generates a virtual address, the (s, p) pair is associatively compared with all tags held in the TLB for a match. If there is a match, the physical address is formed by retrieving the base address of the frame p' from the TLB and concatenating this with the displacement d. However, in the event of a TLB miss, the physical address is generated after accessing the segment and page tables, and this information will also be loaded in the TLB. This ensures that translation information pertaining to a future reference is confined to the TLB. To illustrate the effectiveness of the TLB, the following numerical example is provided.

Example 11.9

The following measurements are obtained from a computer system that uses a linear segmented memory system with a TLB:

- Number of entries in the TLB =16
- Time taken to conduct an associative search in the TLB = 160 ns
- Main memory access time = 1 μs

Determine the average access time assuming a TLB hit ratio of 0.75.

Solution

In the event of a TLB hit, the time needed to retrieve the data is:

t1 = TLB search time + time for one memory access
= 160 ns + 1 μs
= 1.160 μs

However, when a TLB miss occurs, the main memory is accessed three times to retrieve the data. Therefore, the retrieval time t2 in this case is

t2 = TLB search time + 3 (time for one memory access)
= 160 ns + 3 μs
= 3.160 μs

The average access time,

$t_{av} = ht1 + (1 - h)t2$

where h is the TLB hit ratio.

The average access time t_{av} = 0.75 (1.6) + 0.25 (3.160) μsec
= 1.2 + 0.79 μsec
=1.99 μsec

This example shows that the use of a small TLB significantly improves the efficiency of the retrieval operation (by 33%). There are two main reasons for this improvement. First, the TLB is designed using the associated memory. Second, the TLB hit ratio may be attributed to the locality of reference. Simulation studies indicate that it is possible to achieve a hit ratio in the range of 0.8 to 0.9 by having a TLB with 8 to 16 entries.

In a computer based on a linear segmented virtual memory system, the performance parameters such as storage use are significantly influenced by the page size p. For instance, when p is very large, excessive internal fragmentation will occur. If p is small, the size of the page table becomes large. This results in poor use of valuable memory space. The selection of the page size p is often a compromise. Different computer systems use different page sizes. In the following, important memory-management strategies are described. There are three major strategies associated with the management:

- Fetch strategies
- Placement strategies
- Replacement strategies

All these strategies are governed by a set of policies conceived intuitively. Then they are validated using rigorous mathematical methods or by conducting a series of simulation experiments. A policy is implemented using some mechanism such as hardware, software, or firmware.

Fetch strategies deal with when to move the next page to main memory. Recall that when a page needed by a program is not in the main memory, a page fault occurs. In the event of a page fault, the page-fault handler will read the required page from the secondary memory and enter its new physical memory location in the page table, and the instruction execution continues as though nothing has happened.

In a virtual memory system, it is possible to run a program without having any page in the primary memory. In this case, when the first instruction is attempted, there is a page fault. As a consequence, the required page is brought into the main memory, where the instruction execution process is repeated again. Similarly, the next instruction may also cause a page fault. This situation is handled exactly in the same manner as described before. This strategy is referred to as demand paging because a page is brought in only when it is needed. This idea is useful in a multiprogramming environment because several programs can be kept in the main memory and executed concurrently.

However, this concept does not give best results if the page fault occurs repeatedly. For instance, after a page fault, the page-fault handler has to spend a considerable amount of time to bring the required page from the secondary memory. Typically, in a demand paging system, the effective access time t_{av} is the sum of the main memory access time t and μ, where μ is the time taken to service a page fault. Example 11.10 illustrates the concept.

Example 11.10

(a) Assuming that the probability of a page fault occurring is p, derive an expression for t_{av} in terms of t, μ, and p.

(b) Suppose that t = 500 ns and μ = 30 ms, calculate the effective access time t_{av} if it is given that on the average, one out of 200 references results in a page fault.

Solution

(a) If a page fault does not occur, then the desired data can be accessed within a time t. (From the hypothesis the probability for a page fault not to occur is $1 - p$). If the page fault occurs, then μ time units are required to access the data. The effective access time is

$$t_{av} = (1 - p)t + p\mu$$

(b) Since it is given that one out of every 200 references generates a page fault, p = 1/200. Using the result derived in part (a):

$$\begin{aligned} t_{av} &= [(1- 0.005) \times 0.5 + 0.005 \times 30{,}000]\ \mu s \\ &= [0.995 \times 0.5 + 150]\ \mu s = [0.4975 + 150]\ \mu s \\ &= 150.4975\ \mu s \end{aligned}$$

These parameters have a significant impact on the performance of a time-sharing system.

As an alternative approach, anticipatory fetching can be adapted. This conclusion is based on the fact that in a short period of time addresses referenced by a program are clustered around a particular region of the address space. This property is known as locality of reference.

The working set of a program $W(m, t)$ is defined as the set of m most recently needed pages by the program at some instant of time t. The parameter m is called the window of the working set. For example, consider the stream of references shown in Figure 11.26:

From this figure, determine that:

$$W(4, t_1) = (2, 3) \qquad W(4, t_2) = \{1, 2, 3\} \quad W(5, t_2) = \{1,2,3,4\}$$

In general, the cardinality of the set $W(0, t)$ is zero, and the cardinality of the set $W(\infty, t)$ is equal to the total number of distinct pages in the program. Since $m + 1$ most-recent page references include m most-recent page references:

$$\#[W(m + 1, t)] \subseteq \#[W(m, t)]$$

In this equation, the symbol # is used to indicate the cardinality of the set $W(m, t)$. When m is varied from 0 to ∞, $\#W(m, t)$ increases exponentially. The relationship between m and $\#W(m, t)$ is shown in Figure 11.27.

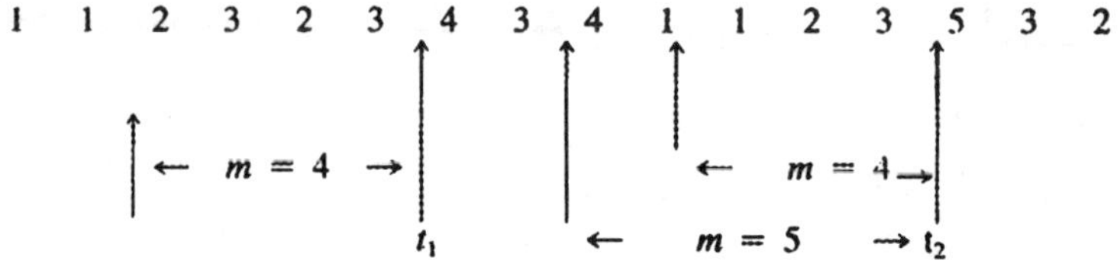

1 1 2 3 2 3 4 3 4 1 1 2 3 5 3 2

← $m = 4$ → t_1 ← $m = 4$ → ← $m = 5$ → t_2

FIGURE 11.26 Stream of Page References

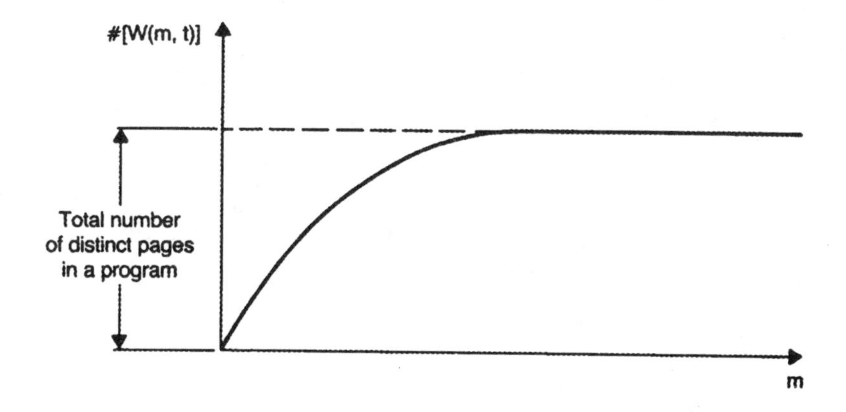

FIGURE 11.27 Relationship between One Cardinality of the Working Set and the Window Size m

In practice, the working set of program varies slowly with respect to time. Therefore, the working set of a program can be predicted ahead of time. For example, in a multiprogramming system, when the execution of a suspended program is resumed, its present working set can be reasonably estimated based on the value of its working set at the time it was suspended. If this estimated working set is loaded, page faults are less likely to occur. This anticipatory fetching further improves the system performance because the working set of a program can be loaded while another program is being executed by the CPU.

However, the accuracy of a working set model depends on the value of m. Larger values of m result in more-accurate predictions. Typical values of m lie in the range of 5000 to 10,000.

To keep track of the working set of a program, the operating system has to perform time-consuming housekeeping operations. This activity is pure overhead, and thus the system performance may be degraded.

Placement strategies are significant with segmentation systems, and they are concerned with where to place an incoming program or data in the main memory. The following are the three widely used placement strategies:

- First-fit technique
- Best-fit technique
- Worst-fit technique

The first-fit technique places the program in the first available free block or hole that is adequate to store it. The best-fit technique stores the program in the smallest free hole of all the available holes able to store it. The worst-fit technique stores the program in the largest free hole.

The first-fit technique is easy to implement and does not have to scan the entire space to place a program. The best-fit technique appears to be efficient because it finds an optimal hole size. However, it has the following drawbacks:

- It is very difficult to implement.
- It may have to scan the entire free space to find the smallest free hole that can hold the incoming program. Therefore, it may be time-consuming.
- It has the tendency continuously to divide the holes into smaller sizes. These smaller holes may eventually become useless.

Worst-fit strategy is sometimes used when the design goal is to avoid creating small holes. In general, the operating system maintains a list known as the available space list (ASL) to indicate the free memory space. Typically, each entry in this list includes the following information:

- Starting address of the free block
- Size of the free block

After each allocation or release, the operating system updates the ASL. In the following example, the mechanics of the various placement strategies presented earlier are explained.

Example 11.11

The available space list of a computer memory system is specified as follows:

STARTING ADDRESS	BLOCK SIZE (IN WORDS)
100	50
200	150
450	600
1,200	400

Determine the available space list after allocating the space for the stream of requests consisting of the following block sizes:

25, 100, 250, 200, 100, 150

a) Use the first-fit method.
b) Use the best-fit method.
c) Use the worst-fit method.

Solution

a) First-fit method. Consider the first request with a block size of 25. Examination of the block sizes of the available space list reveals that this request can be satisfied by allocating from the first available block. The block size (50) is the first of the available space list and is adequate to hold the request (25 blocks). Therefore, the first request with 25 blocks will be allocated from the available space list starting at address 100 with a block size of 50. Request 1 will be allocated starting at an address of 100 ending at an address 100 + 24 = 124 (25 locations including 100). Therefore, the first block of the available space list will start at 125 with a block size of 25. The starting address and block size of each request can be calculated similarly.

b) Best-fit method. Consider request 1. Examination of the available block size reveals that this request can be satisfied by allocating from the first smallest available block capable of holding it. Request 1 will be allocated starting at address 100 and ending at 124. Therefore, the available space list will start at 125 with a block size of 25.

c) Worst-fit method. Consider request 1. Examination of the available block sizes reveals that this request can be satisfied by allocating from the third block (largest) starting at 450. After this allocation the starting address of the available list will be 500 instead of 450 with a block size of 600 - 25 = 575. Various results for all the other requests are shown in Figure 11.28.

	Request 1 (25)		Request 2 (100)		Request 3 (250)		Request 4 (200)		Request 5 (100)		Request 6 (150)	
	Start address	Block size	Start address	Block size	Start address	Block size	Start address	Block size	Start address	Block size	Start address	Block size
First fit	125	25	125	25	125	25	125	25	125	25	125	25
	200	150	300	50	300	50	300	50	300	50	300	50
	450	600	450	600	700	350	900	150	1000	50	1000	50
	1200	400	1200	400	1200	400	1200	400	1200	400	1350	250
Best fit	125	25	125	25	125	25	125	25	125	25	125	25
	200	150	300	50	300	50	300	50	300	50	300	50
	450	600	450	600	450	600	650	400	650	400	800	250
	1200	400	1200	400	1450	150	1450	150	1550	50	1550	50
Worst fit	100	50	100	50	100	50	100	50	100	50	100	50
	200	150	200	150	200	150	200	150	200	150	200	150
	500	575	600	475	850	225	850	225	950	125	850	125
	1200	400	1200	400	1200	400	1400	200	1400	200	1550	50

FIGURE 11.28 Memory Map after Allocating Space for All Requests Given Example Using Different Placement Strategies

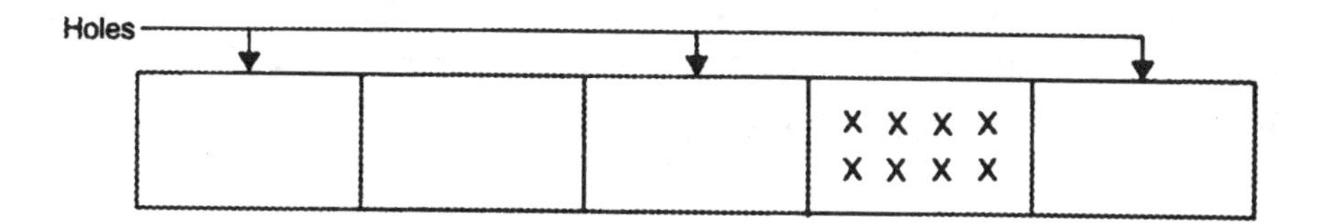

FIGURE 11.29 Memory Status before Compaction

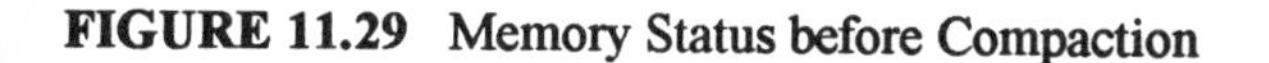

FIGURE 11.30 Memory Status after Compaction

In a multiprogramming system, programs of different sizes may reside in the main memory. As these programs are completed, the allocated memory space becomes free. It may happen that these unused free spaces, or holes, become available between two allocated blocks, or partitions. Some of these holes may not be large enough to satisfy the memory request of a program waiting to run. Thus valuable memory space may be wasted. One way to get around this problem is to combine adjacent free holes to make the hole size larger and usable by other jobs. This technique is known as coalescing of holes.

It is possible that the memory request made by a program may be larger than any free hole but smaller than the combined total of all available holes. If the free holes are combined into one single hole, the request can be satisfied. This technique is known as memory compaction. For example, the status of a computer memory before and after memory compaction is shown in Figures 11.29 and 11.30, respectively.

Placement strategies such as first-fit and best-fit are usually implemented as software procedures. These procedures are included in the operating system's software. The advent of high-level languages such as Pascal and C greatly simplify the programming effort because they support abstract data objects such as pointers. The available space list discussed in this section can easily be implemented using pointers.

The memory compaction task is performed by a special software routine of the operating system called a garbage collector. Normally, an operating system runs the garbage collector routine at regular intervals.

In a paged virtual memory system, when no frames are vacant, it is necessary to replace a current main memory page to provide room for a newly fetched page. The page for replacement is selected using some replacement policy. An operating system implements the chosen replacement policy. In general, a replacement policy is considered efficient if it guarantees a high hit ratio. The hit ratio h is defined as the ratio of the number of page references that did not cause a page fault to the total number of page references.

The simplest of all page replacement policies is the FIFO policy. This algorithm selects the oldest page (or the page that arrived first) in the main memory for replacement. The hit ratio h for this algorithm can be analytically determined using some arbitrary stream of page references as illustrated in the following example.

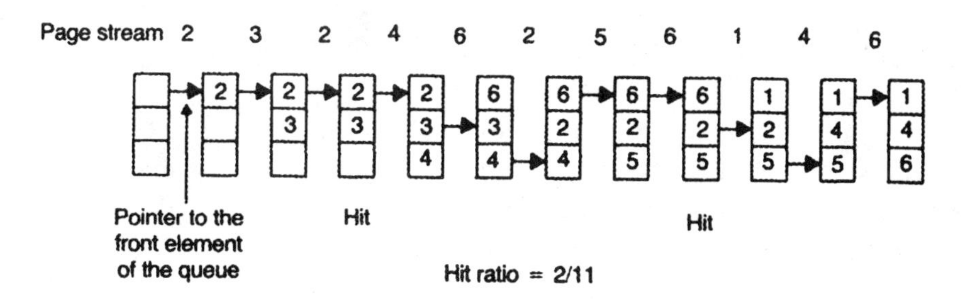

FIGURE 11.31 Hit Ratio Computation for Example 11.12

Example 11.12

Consider the following stream of page requests.

2, 3, 2, 4, 6, 2, 5, 6, 1, 4, 6

Determine the hit ratio h for this stream using the FIFO replacement policy. Assume the main memory can hold 3 page frames and initially all of them are vacant.

Solution

The hit ratio computation for this situation is illustrated in Figure 11.31.

From Figure 11.31, it can be seen that the first two page references cause page faults. However, there is a hit with the third reference because the required page (page 2) is already in the main memory. After the first four references, all main memory frames are completely used. In the fifth reference, page 6 is required. Since this page is not in the main memory, a page fault occurs. Therefore, page 6 is fetched from the secondary memory. Since there are no vacant frames in the main memory, the oldest of the current main memory pages is selected for replacement. Page 6 is loaded in this position. All other data tabulated in this figure are obtained in the same manner. Since 9 out of 11 references generate a page fault, the hit ratio is 2/11.

The primary advantage of the FIFO algorithm is its simplicity. This algorithm can be implemented by using a FIFO queue. FIFO policy gives the best result when page references are made in a strictly sequential order. However, this algorithm fails if a program loop needs a variable introduced at the beginning. Another difficulty with the FIFO algorithm is it may give anomalous results.

Intuitively, one may feel that an increase in the number of page frames will also increase the hit ratio. However, with FIFO, it is possible that when the page frames are increased, there is a drop in the hit ratio. Consider the following stream of requests:

1, 2, 3, 4, 5, 1, 2, 5, 1, 2, 3, 4, 5, 6, 5

Assume the main memory has 4 page frames; then using the FIFO policy there is a hit ratio of 4/15. However, if the entire computation is repeated using 5 page frames, there is a hit ratio of 3/15. This computation is left as an exercise.

Another replacement algorithm of theoretical interest is the optimal replacement policy. When there is a need to replace a page, choose that page which may not be needed again for the longest period of time in the future.

The following numerical example explains this concept.

Example 11.13

Using the optimal replacement policy, calculate the hit ratio for the stream of page references specified in Example 11.12. Assume the main memory has three frames and initially all of them are vacant.

Solution

The hit ratio computation for this problem is shown in Figure 11.32.

From Figure 11.32, it can be seen that the first two page references generate page faults. There is a hit with the sixth page reference, because the required page (page 2) is found in the main memory. Consider the fifth page reference. In this case, page 6 is required. Since this page is not in the main memory, it is fetched from the secondary memory. Now, there are no vacant page frames. This means that one of the current pages in the main memory has to be selected for replacement. Choose page 3 for replacement because this page is not used for the longest period of time. Page 6 is loaded into this position. Following the same procedure, other entries of this figure can be determined. Since 6 out of 11 page references generate a page fault, the hit ratio is 5/11.

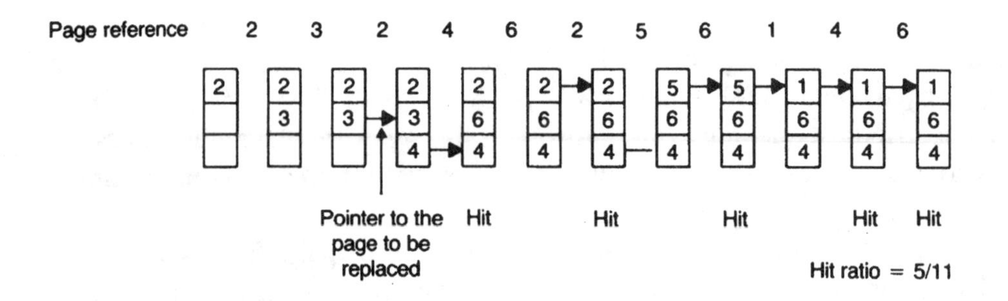

FIGURE 11.32 Hit Ratio Computation for Example 11.13

The decision made by the optimal replacement policy is optimal because it makes a decision based on the future evolution. It has been proven that this technique does not give any anomalous results when the number of page frames is increased. However, it is not possible to implement this technique because it is impossible to predict the page references well ahead of time. Despite this disadvantage, this procedure is used as a standard to determine the efficiency of a new replacement algorithm. Since the optimal replacement policy is practically unfeasible, some method that approximates the behavior of this policy is desirable. One such approximation is the least recently used (LRU) policy.

According to the LRU policy, the page that is selected for replacement is that page that has not been referenced for the longest period of time. To see how this idea works, examine Example 11.14.

Example 11.14

Solve Example 11.13 using the LRU policy.

Solution

The hit ratio computation for this problem is shown in Figure 11.33.

In the figure we again notice that the first two references generate a page fault, whereas the third reference is a hit because the required page is already in the main memory. Now, consider what happens when the fifth reference is made. This reference requires page 6, which is not in the memory.

Also, we need to replace one of the current pages in the main memory because all frames are filled. According to the LRU policy, among pages 2, 3, and 4, page 3 is the page that is least recently referenced. Thus we replace this page with page 6. Following the same reasoning the other entries of Figure 11.33 can be determined. Note that 7 out of 11 references generate a page fault; therefore, the hit ratio is 4/11. From the results of the example, we observe that the performance of the LRU policy is very close to that of the optimal replacement policy. Also, the LRU obtains a better result than the FIFO because it tries to retain the pages that are used recently.

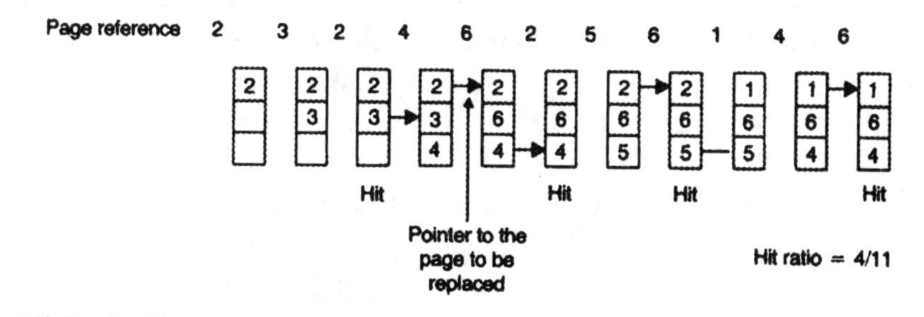

FIGURE 11.33 Hit Ratio Computation for Example 11.14

Now, let us summarize some important features of the LRU algorithm.

- In principle, the LRU algorithm is similar to the optimal replacement policy except that it looks backward on the time axis. Note that the optimal replacement policy works forward on the time axis.
- If the request stream is first reversed and then the LRU policy is applied to it, the result obtained is equivalent to the one that is obtained by the direct application of the optimal replacement policy to the original request stream.
- It has been proven that the LRU algorithm does not exhibit Belady's anamoly. This is because the LRU algorithm is a stack algorithm. A page-replacement algorithm is said to be a stack algorithm if the following condition holds:

  $$P_t(i) \subset P_t(i + 1)$$

  In the preceding relation the quantity Pt(i) refers to the set of pages in the main memory whose total capacity is i frames at some time t. This relation is called the inclusion property. One can easily demonstrate that FIFO replacement policy is not a stack algorithm. This task is left as an exercise.
- The LRU policy can be easily implemented using a stack. Typically, the page numbers of the request stream are stored in this stack. Suppose that p is the page number being referenced. If p is not in the stack, then p is pushed into the stack. However, if p is in the stack, p is removed from the stack and placed on the top of the stack. The top of the stack always holds the most recently referenced page number, and the bottom of the stack always holds the least-recent page number. To see this clearly, consider Figure 11.34, in which a stream of page references and the corresponding stack instants are shown. The principal advantage of this approach is that there is no need to search for the page to be replaced because it is always the bottom most element of the stack. This approach can be implemented using either software or microcodes. However, this method takes more time when a page number is moved from the middle of the stack.
- Alternatively, the LRU policy can be implemented by adding an age register to each entry of the page table and a virtual clock to the CPU. The virtual clock is organized so that it is incremented after each memory reference. When a page is referenced, its age register is loaded with the contents of the virtual clock. The age register of a page holds the time at which that page was most recently referenced. The least-recent page is that page whose age register value is minimum. This approach requires an operating system to perform time-consuming housekeeping operations. Thus the performance of the system may be degraded.
- To implement these methods, the computer system must provide adequate hardware support. Incrementing the virtual clock using software takes more time. Thus the operating speed of the entire system is reduced. The LRU policy can not be implemented in

systems that do not provide enough hardware support. To get around this problem, some replacement policy is employed that will approximate the LRU policy.

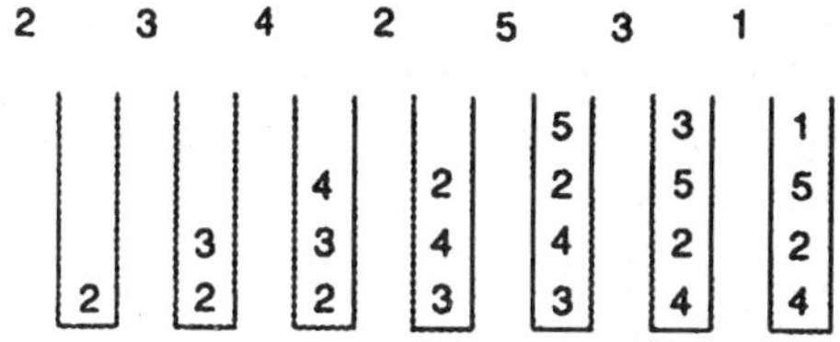

2	3	4	2	5	3	1
				5	3	1
		4	2	2	5	5
	3	3	4	4	2	2
2	2	2	3	3	4	4

FIGURE 11.34 Implementation of the LRU Algorithm Using a Stack

- The LRU policy can be approximated by adding an extra bit called an activity bit to each entry of the page table. Initially all activity bits are cleared to 0. When a page is referenced, its activity bit is set to 1. Thus this bit tells whether or not the page is used. Any page whose activity bit is 0 may be a candidate for replacement. However, the activity bit cannot determine how many times a page has been referenced.
- More information can be obtained by adding a register to each page table entry. To illustrate this concept, assume a 16-bit register has been added to each entry of the page table. Assume that the operating system is allowed to shift the contents of all the registers 1 bit to the right at regular intervals. With one right shift, the most-significant bit position becomes vacant. If it is assumed that the activity bit is used to fill this vacant position, some meaningful conclusions can be derived. For example, if the content of a page register is 0000_{16}, then it can be concluded that this page was not in use during the last 16 time-interval periods. Similarly, a value $FFFF_{16}$ for page register indicates that the page should have been referenced at least once in the last 16 time-interval periods. If the content of a page register is $FF00_{16}$ and the content of another one is $00F0_{16}$, the former was used more recently.
- If the content of a page register is interpreted as an integer number, then the least-recent page has a minimum page register value and can be replaced. If two page registers hold the minimum value, then either of the pages can be evicted, or one of them can be chosen on a FIFO basis.
- The larger the size of the page register, the more time is spent by the operating system in the update operations. When the size of the page register is 0, the history of the system can only be obtained via the activity bits. If the proposed replacement procedure is applied on the activity bits alone, the result is known as the second-chance replacement policy.
- Another bit called a dirty bit may be appended to each entry of the page table. This bit is initially cleared to 0 and set to 1 when a page is modified.
- This bit can be used in two different ways:

- The idea of dirty bit reduces the swapping overhead because when the dirty bit of a page to be replaced is zero, there is no need to copy this page into the secondary memory, and it can be overwritten by an incoming page. A dirty bit can be used in conjunction with any replacement algorithm.
- A priority scheme can be set up for replacement using the values of the dirty and activity bits, as described next.

PRIORITY LEVEL	ACTIVITY BIT	DIRTY BIT	MEANING
0	0	0	Neither used nor modified.
1	0	1	Not recently used but modified.
2	1	0	Used but not modified.
3	1	1	Used as well as dirty.

Using the priority levels just described, the following replacement policy can be formulated: When it is necessary to replace a page, choose that page whose priority level is minimum. In the event of a tie, select the victim on a FIFO basis.

In some systems, the LRU policy is approximated using the least frequently used (LFU) and most frequently used (MFU) algorithms. A thorough discussion of these procedures is beyond the scope of this book.

- One of the major goals in a replacement policy is to minimize the page-fault rate. A program is said to be in a thrashing state if it generates excessive numbers of page faults. Replacement policy may not have a complete control on thrashing. For example, suppose a program generates the following stream of page references:

  1,2,3,4, 1,2,3,4, 1,2,3,4, . . .

  If it runs on a system with three frames it will definitely enter into thrashing state even if the optimal replacement policy is implemented.
- There is a close relationship between the degree of multiprogramming and thrashing. In general, the degree of multiprogramming is increased to improve the CPU use. However, in this case more thrashing occurs. Therefore, to reduce thrashing, the degree of multiprogramming is reduced. Now the CPU utilization drops. CPU utilization and thrashing are conflicting performance issues.

11.8 Pipeline Processing

The purpose of this section is to provide a brief overview of pipelining.

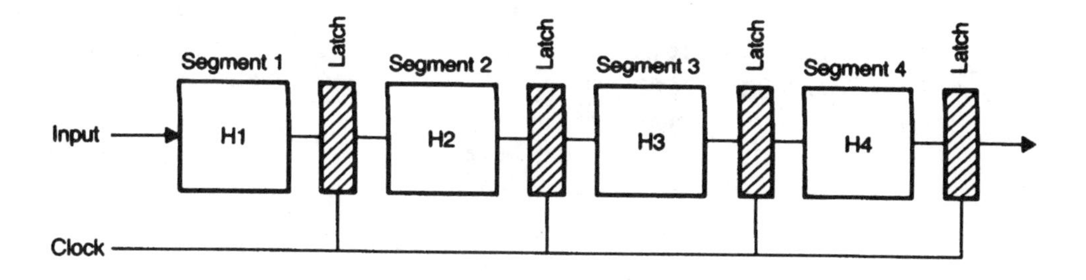

FIGURE 11.35 A Four-segment Pipeline

11.8.1 Basic Concepts

Assume a task T is carried out by performing four activities: Al, A2, A3, and A4, in that order. Hardware Hi is designed to perform the activity Ai. Hi is referred to as a segment, and it essentially contains combinational circuit elements. Consider the arrangement shown in Figure 11.35.

In this configuration, a latch is placed between two segments so the result computed by one segment can serve as input to the following segment during the next clock period.
The execution of four tasks Tl, T2, T3, and T4 using the hardware of Figure 11.35 is described using a space-time chart shown in Figure 11.36.
Initially, task Tl is handled by segment 1. After the first clock, segment 2 is busy with Tl while segment 1 is busy with T2. Continuing in this manner, the task Tl is completed at the end of the fourth clock. However, following this point, one task is shipped out per clock. This is the essence of the pipeline concept. A pipeline gains efficiency for the same reason as an assembly line does: Several activities are performed but not on the same material.
Suppose ti and L denote the propagation delays of segment i and the latch, respectively. Then the pipeline clock period T can be expressed as follows:

$$T = \max (Tl, T2, \ldots Tn) + L$$

The segment with the maximum delay is known as the bottleneck, and it decides the pipeline clock period T. The reciprocal of T is referred to as the pipeline frequency.

Consider the execution of m tasks using an n-segment pipeline. In this case, the first task will be completed after n clocks (because there are n segments) and the remaining m-1 tasks are shipped out at the rate of one task per pipeline clock.

Therefore, n + (m − 1) clock periods are required to complete m tasks using an n-segment pipeline. If all m tasks are executed without any overlap, mn clock periods are needed because each task has to pass through all n segments. Thus speed gained by an n segment pipeline can be shown as follows:

$$\text{speedup } P(n) = \frac{\text{number of clocks required when there is no overlap}}{\text{number of clocks required when tasks are overlapped in time}} = \frac{mn}{n + m - 1}$$

P(n) approaches n when m approaches infinity. This implies that when a large number of tasks are carried out using an *n*-segment pipeline, an *n*-fold increase in speed can be expected.

The previous result shows that the pipeline completes m tasks in the m + n - 1 clock periods. Therefore, its throughput can be defined as follows:

$$\text{throughput of an } n\text{-segment pipeline} = U(n) = \text{number of tasks computed per unit time} = \frac{m}{(n + m - 1)T}$$

For a large value of m, U(n) approaches 1/T, which is the pipeline frequency. Thus the throughput of an ideal pipeline is equal to the reciprocal of its clock period. The efficiency of an *n*-segment pipeline is defined as the ratio of the actual speedup to the maximum speedup realized.

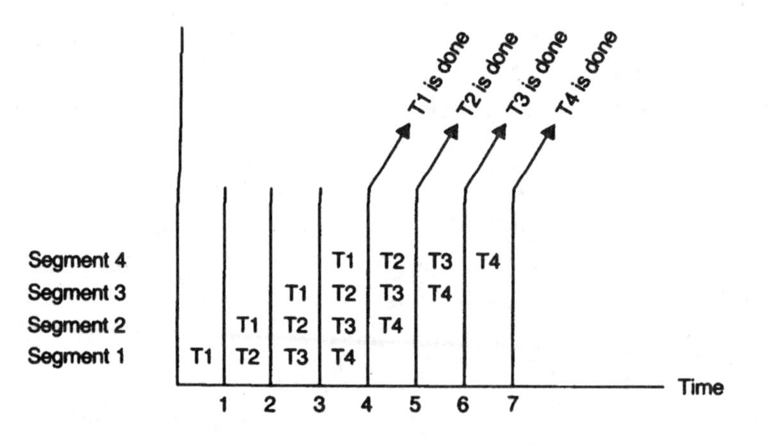

FIGURE 11.36 Overlapped Execution of Four Tasks Using a Pipeline

$$\begin{array}{l}\text{efficiency}\\ \text{of an } n\text{-}\\ \text{segment}\\ \text{pipeline}\end{array} = E(n) = \frac{\text{actual speedup}}{\text{maximum speedup}} = \frac{P(n)}{n}$$

This illustrates that when m is very large, E(n) approaches 1 as expected.

In many modem computers, the pipeline concept is used in carrying out two tasks: arithmetic operations and instruction execution.

11.8.2 Arithmetic Pipelines

The pipeline concept can be used to build high-speed multipliers. Consider the multiplication P = M * Q, where M and Q are 8-bit numbers. The 16-bit product P can be expressed as:

$P = M(q_7 2^7 + q_6 2^6 + q_5 2^5 + q_4 2^4 + q_3 2^3 + q_2 2^2 + q_1 2^1 + q_0 2^0)$. Hence, $P = \sum_{i=0}^{7} Mq_i 2^i$. This result can also be rewritten as:

$$P = \sum_{i=0}^{7} S_i$$

where, $S_i = Mq_i 2^i$ and each S_i represents a 16-bit partial product. Each partial product is the shifted multiplicand. All 8 partial products can be added using several carry-save adders.

This concept can be extended to design an n × n pipelined multiplier. Here n partial products must be summed with 2n bits per partial product. So, as n increases, the hardware cost associated with a fully combinational multiplier increases in an exponential fashion. To reduce the hardware cost, large multipliers are designed.

The pipeline concept is widely used in designing floating-point arithmetic units. Consider the process of adding two floating point numbers A = 0.9234 * 10^4 and B = 0.48 * 10^2. First, notice that the exponents of A and B are unequal. Therefore, the smaller number should be modified so that its exponent is equal to the exponent of the greater number. For this example, modify B to 0.0048 * 10^4. This modification step is known as exponent alignment. Here the decimal point of the significand 0.48 is shifted to the right to obtain the desired result. After the exponent alignment, the significands 0.9234 and 0.0048 are added to obtain the final solution of 0.9282 * 10^4.

For a second example, consider the operation A - B, where A = 0.9234 * 10^4 and B = 0.9230 * 10^4. In this case, no exponent alignment is necessary because the exponent of A equals to the exponent of B. Therefore, the significand of B is subtracted from the significand of A to obtain 0.9234 - 0.9230 = 0.0004. However, $0.0004 * 10^4$ cannot be the final answer because the significand, 0.0004, is not normalized. A floating-point number with base b is said to be normalized if the magnitude of its significand satisfies the following inequality: $\frac{1}{b} \leq |\text{significand}| < 1$.

In this example, since b = 10, a normalized floating-point number must satisfy the condition:

$$0.1 \leq |\text{significand}| < 1$$

(Note that normalized floating-point numbers are always considered because for each real-world number there exists one and only one floating-point representation. This uniqueness property allows processors to make correct decisions while performing compare operations).

The final answer is modified to $0.4 * 10^1$. This modification step is known as postnormalization, and here the significand is shifted to the left to obtain the correct result.

In summary, addition or subtraction of two floating-point numbers calls for four activities:

1. Exponent comparison
2. Exponent alignment
3. Significand addition or subtraction
4. Postnormalization

Based on this result, a four-segment floating-point adder/subtracter pipeline can be built, as shown in Figure 11.37.

It is important to realize that each segment in this pipeline is primarily composed of combinational components such as multiplexers. The shifter used in this system is the barrel shifter discussed earlier. Modern microprocessors such as Motorola MC 68040 include a 3-stage floating-point pipeline consisting of operand conversion, execute, and result normalization.

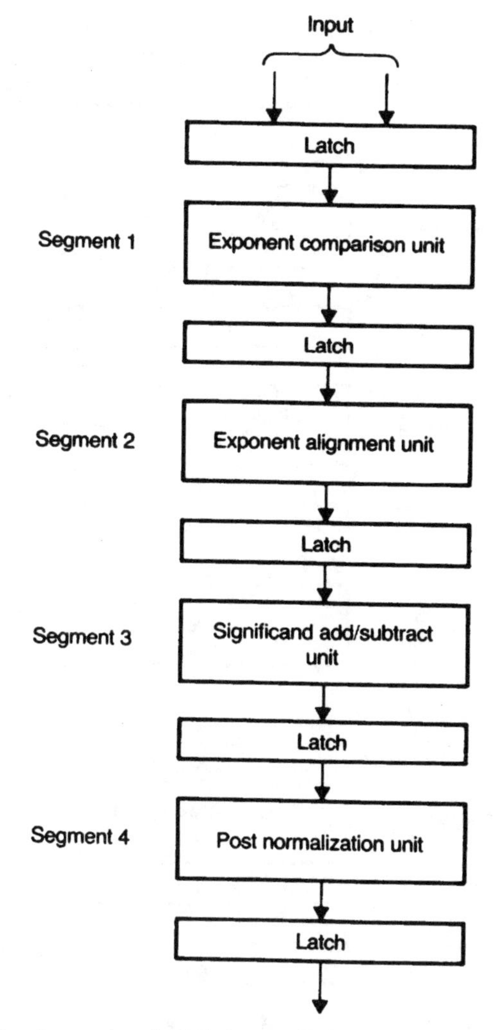

FIGURE 11.37 A Pipelined Floating-point Add/Subtract Unit

11.8.3 Instruction Pipelines

Modern microprocessors such as Motorola MC 68020 contain a 3-stage instruction pipeline. Recall that an instruction cycle typically involves the following activities:

1. Instruction fetch 2. Instruction decode 3. Operand fetch 4. Operation execution 5. Result routing.

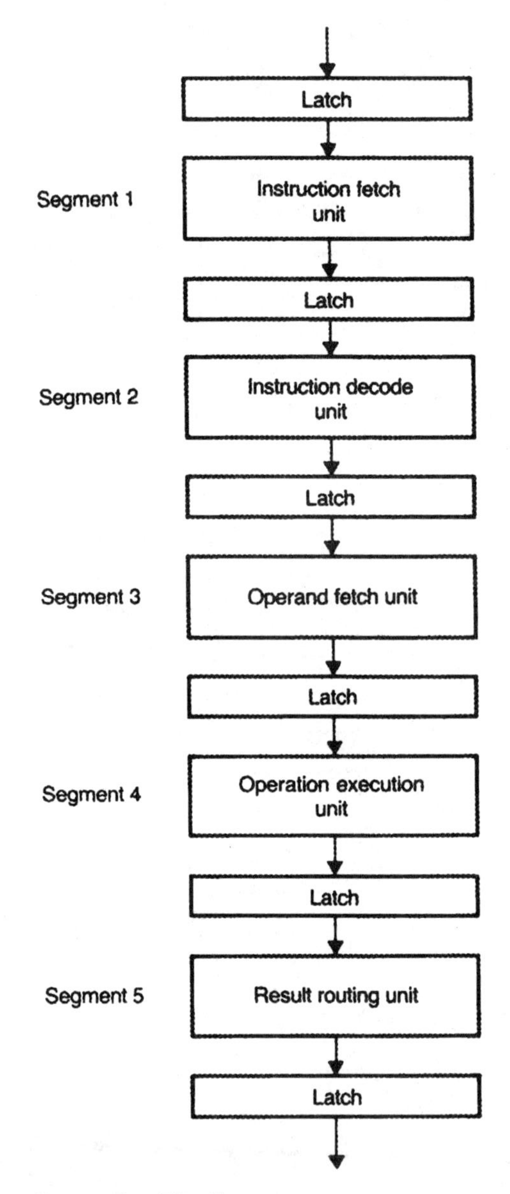

FIGURE 11.38 A Five-segment Instruction Pipeline

This process can be effectively carried out by using the pipeline shown in Figure 11.38.

As mentioned earlier, in such a pipelined scheme the first instruction requires five clocks to complete its execution. However, the remaining instructions are completed at a rate of one per pipeline clock. Such a situation prevails as long as all the segments are busy.

In practice, the presence of branch instructions and conflicts in memory accesses poses a great problem to the efficient operation of an instruction pipeline.
For example, consider the execution of a stream of five instructions: I1, I2, I3, I4, and I5 in which I3 is a conditional branch instruction. This stream is processed by the instruction pipeline (Figure 11.38) as depicted in Figure 11.39.

When a conditional branch instruction is fetched, the next instruction cannot be fetched because the exact target is not known until the conditional branch instruction has been executed. The next fetch can occur once the branch is resolved. Four additional clocks are required due to I3.

Suppose a stream of s instructions is to be executed using an n-segment pipeline. If c is the probability for an instruction to be a conditional branch instruction, there will be sc conditional branch instructions in a stream of s instructions. Since each branch instruction requires $n - 1$ additional clocks, the total number of clocks required to process a stream of s instructions is

$$(n + s - 1) + sc(n - 1)$$

An instruction cycle constitutes n pipeline clocks. Therefore, the total number of instruction cycles required to execute an instruction is

$$I = \frac{(n + s - 1) + sc(n - 1)}{n}$$

The average number of instructions executed per instruction cycle is

$$\frac{s}{I} = \frac{sn}{(n + s - 1) + sc(n - 1)} = \frac{n}{\frac{n}{s} + \frac{(s - 1)}{s} + c(n - 1)}$$

For a large value of s, the preceding result can be simplified as shown on the following page:

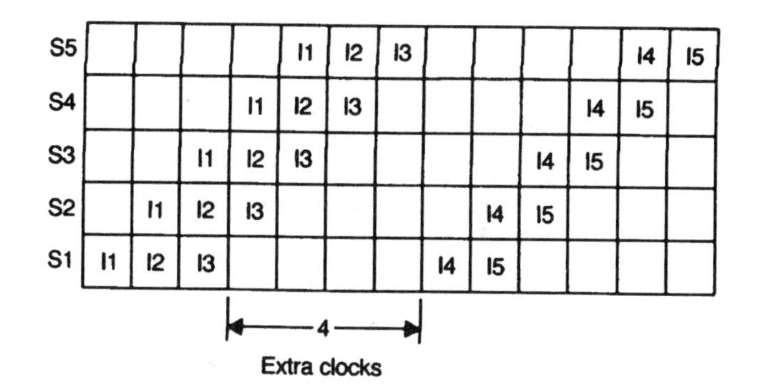

FIGURE 11.39 Pipelined Execution Of A Stream of Five instructions that Includes a Branch Instruction

$$\lim_{s \to \infty} \frac{S}{I} = \frac{n}{1 + c(n-1)}$$

For n = 5, the equation becomes:

$$\frac{5}{1 + 4c}$$

For no conditional branch instructions (c = 0), 5 instructions per instruction cycle are executed. This is the best result produced by a five-segment pipeline. If 25% of the instructions are branch instructions only,

$$\frac{5}{1 + 4 * 0.25} = 2.5 \text{ instructions}$$

per instruction cycle can be executed. This shows how pipeline efficiency is significantly decreased even with a small percentage of branch instructions.

In many contemporary systems, branch instructions are handled using a strategy called Target Prefetch. When a conditional branch instruction is recognized, the immediate successor of the branch instructions and the target of the branch are prefetched. The latter is saved in a buffer until the branch is executed. If the branch condition is successful, one pipeline is still busy because the branch target is in the buffer.

Another approach to handle branch instructions is the use of the delayed branch concept. In this case, the branch does not take place until after the following instruction. To illustrate this, consider the instruction sequence shown in Figure 11.40.

Suppose the compiler inserts a NOP instruction and changes the branch instruction to JMP 2051. The program semantics remain unchanged. This is shown in Figure 11.41.

MEMORY ADDRESS	INSTRUCTION
2000	LDA X
2001	INC Y
2002	JMP 2050
2003	SUB Z
.	
.	
.	
.	
.	
2050	STA W
.	
.	
.	
.	
.	

FIGURE 11.40 A Hypothetical Program

MEMORY ADDRESS	INSTRUCTION
2000	LDA X
2001	INC Y
2002	JMP 2051
2003	NOP
2004	SUB Z
.	
.	
.	
2051	STA W

FIGURE 11.41 Modified Sequence

Instruction fetch	LDA X	INC Y	JMP 2051	NOP	STA W
Instruction execute		LDA X	INC Y	JMP 2051	NOP

FIGURE 11.42 Pipelined Execution of a Hypothetical Instruction Sequence

This modified sequence will be executed by a two-segment pipeline, as shown in Figure 11.42:

- Instruction fetch
- Instruction execute

Because of the delayed branch concept, the pipeline still functions correctly without damage.

The efficiency of this pipeline can be further improved if the compiler produces a new sequence as shown in Figure 11.43.

In this case, the compiler has reversed the instruction sequence. The JMP instruction is placed in the location 2001, and the INC instruction is moved to memory location 2002. This reversed sequence is executed by the same 2-segment pipeline, as shown in Figure 11.44.

MEMORY ADDRESS	INSTRUCTION
2000	LDA X
2001	JMP 2050
2002	INC Y
2003	SUB Z
.	.
.	.
.	.
2050	STA W

FIGURE 11.43 Instruction Sequence with Branch Instruction Reversed

Instruction fetch	LDA X	JMP 2050	INC Y	STA W
Instruction execute		LDA X	JMP 2050	INC Y

FIGURE 11.44 Execution of the Reversed-instruction Sequence

It is important to understand that due to the delayed branch rule, the INC Y instruction is fetched before the execution of JMP 2050 instruction; therefore, there is no change in the order of instruction execution. This implies that the program will still produce the same result. Since the NOP instruction was eliminated, the program is executed more efficiently. The concept of delayed branch is one of the key characteristics of RISC as it makes concurrency visible to a programmer.

As does the presence of branch instructions, memory-access conflicts cause damage to pipeline performance. For example, if the instructions in the operand fetch and result-saving units refer to the same memory address, these operations cannot be overlapped.

To reduce such memory conflicts, a new approach called memory interleaving is often employed. For this case, the memory addresses are distributed among a set of memory modules, as shown in Figure 11.45.

In this arrangement, memory is distributed among many modules. Since consecutive addresses are placed into different modules, the CPU can access several words in one memory access.

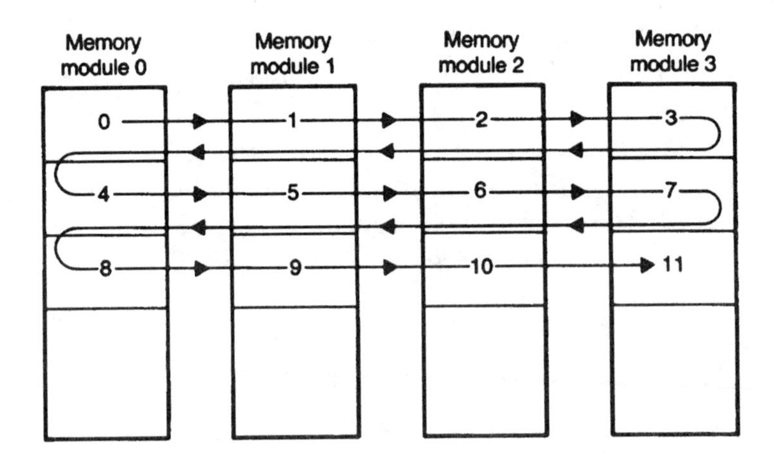

FIGURE 11.45 Memory Interleaving

11.9 General Classifications of Computer Architectures

Over the last two decades, parallel processing has drawn the attention of many research workers, and several high-speed architectures have been proposed. To present these results in a concise manner, different architectures must be classified in well defined groups.

All computers may be categorized into different groups using one of three classification methods:

1. Flynn
2. Feng
3. Handler

The two principal elements of a computer are the processor and the memory. A processor manipulates data stored in the memory as dictated by the instruction. Instructions are stored in the memory unit and always flow from memory to processor. Data movement is bidirectional, meaning data may be read from or written into the memory. Figure 11.46 shows the processor-memory interaction.

The number of instructions read and data items manipulated simultaneously by the processor form the basis for Flynn's classification.

Figure 11.47 shows the four types of computer architectures that are defined using Flynn's method.

The SISD computers are capable of manipulating a single data item by executing one instruction at a time. The SISD classification covers the conventional uniprocessor systems such as the VAX-11, IBM 370, Intel 8085, and Motorola 6809. The processor unit of a SISD machine may have one or many functional units. For example, the VAX-11/780 is a SISD machine with a single functional unit. CDC 6600 and IBM 370/168 computers are typical examples of SISD systems with multiple functional units. In a SISD machine, instructions are executed in a strictly sequential fashion.

The SIMD system allows a single instruction to manipulate several data elements. These machines are also called vector machines or array processors. Examples of this type of computer are the ILLIAC-IV and Burroughs Scientific Processor (BSP). The ILLIAC-IV was an experimental parallel computer proposed by the University of Illinois and built by the Burroughs Corporation. In this system, there are 64 processing elements. Each processing element has its own small local memory unit. The operation of all the processing elements is under the control of a central control unit (CCU). Typically, the CCU reads an instruction from the common memory and broadcasts the same to all processing units so the processing units can all operate on their own data at the same time. This configuration is very useful for carrying out a high volume of computations that are encountered in application areas such as finite-element analysis, logic simulation, and spectral analysis. Modern microprocessors such as Intel Pentium II use the SIMD architecture.

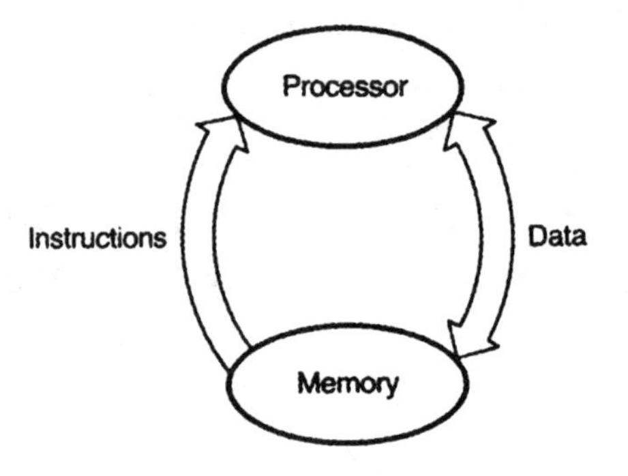

FIGURE 11.46 Processor-Memory Interaction

NAME OF THE ARCHITECTURE	NAME OF THE ARCHITECTURE IN ABBREVIATED FORM
Single-instruction stream-single-data stream	SISD
Single-instruction stream-multiple-data stream	SIMD
Multiple-instruction stream-single-data stream	MISD
Multiple-instruction stream-multiple-data stream	MIMD

FIGURE 11.47 Classification of Computers Using Flynn's Method

By definition, MISD refers to a computer in which several instructions manipulate the same data stream concurrently. The notion of pipelining is very close to the MISD definition.

A set of instructions constitute a program, and a program operates on several data elements. MIMD organization refers to a computer that is capable of processing several programs simultaneously. MIMD systems include all multiprocessing systems. Based on the degree of processor interaction, multiprocessor systems may be further divided into two groups: loosely coupled and tightly coupled. A tightly coupled system has high interaction between processors. Multiprocessor systems with low interprocessor communications are referred to as loosely coupled systems.

In Feng's approach, computers are classified according to the number of bits processed within a unit time. However, Handler's classification scheme categorizes computers on the basis of the amount of parallelism found at the following levels:

- CPU
- ALU
- Bit

A thorough discussion of these schemes is beyond the scope of this book. Since contemporary microprocessors such as Intel Pentium II use SIMD architechture, a basic coverage of SIMD is provided next. The SIMD computers are also called array processors.

A synchronous array processor may be defined as a computer in which a set of identical processing elements act under the control of a master controller (MC). A command given by the MC is simultaneously executed by all processing elements, and a SIMD system is formed. Since all processors execute the same instruction, this organization offers a great attraction for vector processing applications such as matrix manipulation.

A conceptual organization of a typical array processor is shown in Figure 11.48. The Master Controller (MC) controls the operation of the processor array. This array consists of N identical processing elements (P_0 through P_{n-1}). Each processing element P_i is assumed to have its own memory, PM_i, to store its data. The MC of Figure 11.48 contains two major components:

- The master control unit (MCU)
- The master control memory (MCM)

The MCU is the CPU of the master controller and includes an ALU and a set of registers. The purpose of the MCM is to hold the instructions and common data.
Each instruction of a program is executed under the supervision of the MCU in a sequential fashion. The MCU fetches the next instruction, and the execution of this instruction will take place in one of the following ways:

- If the instruction fetched is a scalar or a branch instruction, it is executed by the MC itself.
- If the instruction fetched is a vector instruction, such as vector add or vector multiply, then the MCU broadcasts the same instruction to each P_i, of the processor array, allowing all P_i's to execute this instruction simultaneously.

Assume the required data is already within the processing element's private memory. Before execution of a vector instruction, the system ensures that appropriate data values are routed to each processing element's private memory. Such an operation can be performed in two ways:

- All data values can be transferred to the private memories from an external source via the system data bus.
- The MCU can transfer the data values to the private memories via the control bus.

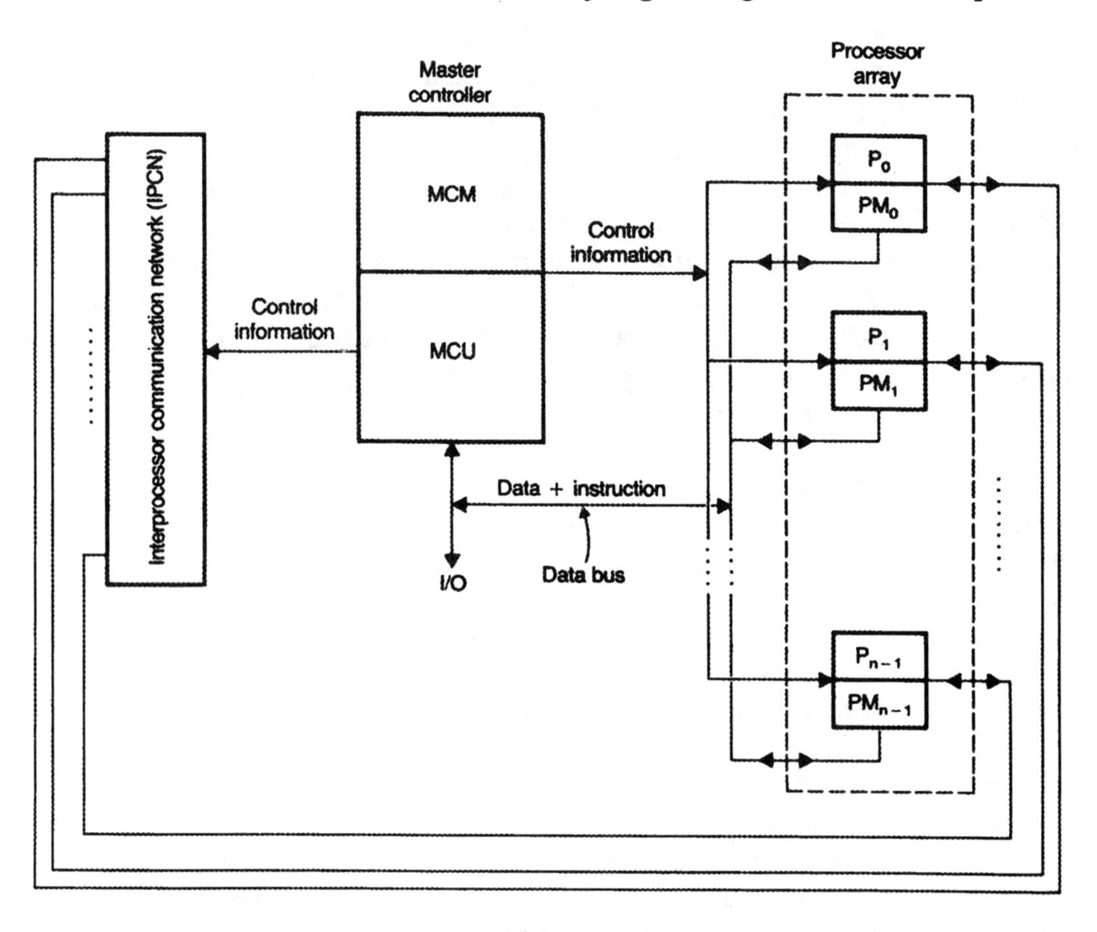

FIGURE 11.48 A Typical Array Processor Organization

In an array processor like the one shown in Figure 11.48, it may be necessary to disable some processing elements during a vector operation. This is accomplished by including a mask register, M, in the MCU. The mask register contains a bit, m_i, for each processing element, p_i. A particular processing element, p_i, will respond to a vector instruction broadcast by the MCU only when its mask bit, m_i, is set to 1; otherwise, the processing element. P_i, will not respond to the vector instruction and is said to be disabled.

In an array processor, it may be necessary to exchange data between processing elements. Such an exchange of data between processing elements takes place through the path provided by the interprocessor communication network (IPCN). Data exchanges refers to exchanges between scratchpad registers of the processing elements and exchanges between private memories of the processing elements.

QUESTIONS AND PROBLEMS

11.1 Draw an ASM chart for each of the following sequence of operations:

(a) The ASM chart will define a conditional operation to perform the operation $R_2 \leftarrow R_2 - R_1$ during State T_0 and will transfer control to State T_1 if the control input c is 1; if c=0, the system will stay in T_0. Assume that R_1 and R_2 are 8-bit registers.

(b) The ASM chart in which the system is initially in State T_0 and then checks a control input c. If c=1, control will move from State T_0 to State T_1; if c=0, the system will increment an 8-bit register R by 1 and control will return to the initial state.

11.2 Draw an ASM chart for the following state diagram of Figure P11.2:

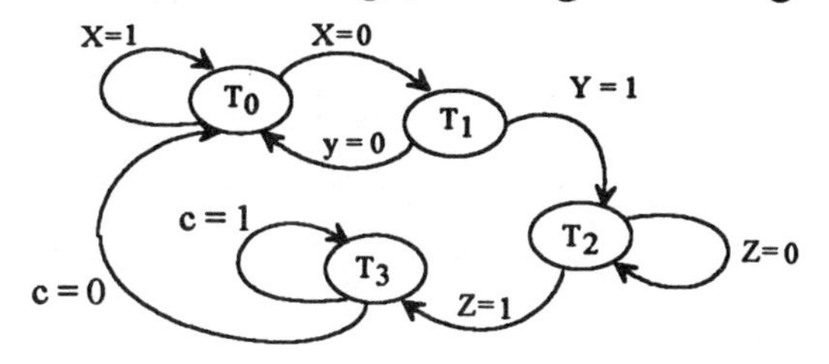

FIGURE P11.2

Assume that the system stays in initial state T_0 when control input c = 0 and input X = 1. The sequence of operations is started from T_0 when X = 0 . When the system reaches state T_3, it stays in T_3 indefinitely as long as c = 1; the system returns to state T_0 when c = 0.

11.3 What does VHDL stand for? Write a VHDL description for inverting a Boolean variable X without using the not instruction.

11.4 Using the concepts described in Chapters 3, 4, and 7, design the following:

(a) Using a 4-bit CLA as a building block, design a 16-bit adder whose worst-case add-time is 10Δ.

(b) Using a 4-bit CLA as the building block, design the fastest 64-bit adder. Estimate the worst-case add-time of your design.

(c) Design a combinational circuit to compute the function $f(x) = (\frac{3}{8}) * x$ where x is a 4-bit 2's complement number.

(d) Design and implement a combinational circuit that will work as follows:

S1	S0	F
0	0	A plus B
0	1	Shift left (A)
1	0	A plus B plus 1
1	1	Shift left (A) + 1

Note that A and B are 4-bit operands

(e) i) Design a combinational circuit that will satisfy the following specification.

S1	S0	Y_i
0	0	0
0	1	X_i
1	0	$\overline{X_i}$
1	1	1

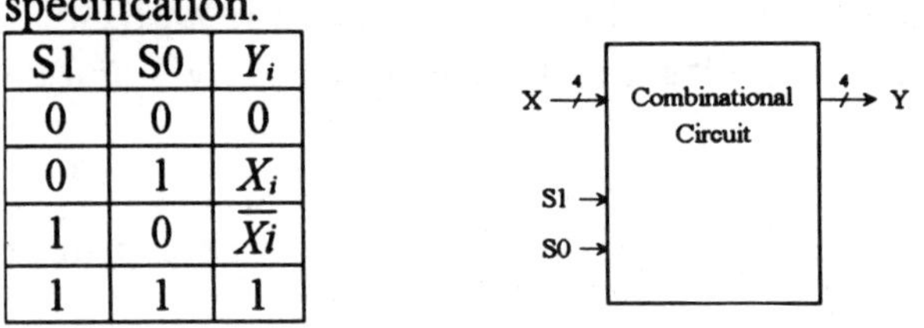

ii) Using the results of part i), design a 4-bit, 8-function arithmetic unit that ii) will function as described next:

S2	S1	S0	F
0	0	0	A
0	0	1	A plus B
0	1	0	A plus $\overline{B}$
0	1	1	A minus 1
1	0	0	A plus 1
1	0	1	A plus B plus 1
1	1	0	A plus $\overline{B}$ plus 1
1	1	1	A

(f) Design a 4-bit, 8-function arithmetic unit that will meet the following specifications:

S2	S1	S0	F
0	0	0	2A
0	0	1	A plus $\overline{B}$
0	1	0	A plus B
0	1	1	A minus 1
1	0	0	2A plus 1
1	0	1	A plus $\overline{B}$ plus 1
1	1	0	A plus B plus 1
1	1	1	A

(g) i) Using a 4-bit parallel adder with inputs (A, B, and C_{in}), outputs (F and C_{out}), and one selection bit (S0), design an arithmetic circuit as follows:

S0	FUNCTION TO BE PERFORMED
0	A plus B
1	B plus 1

ii) Using another selection bit S1, modify the circuit of i) to include the arithmetic and logic functions as follows:

S1	S0	FUNCTION TO BE PERFORMED
0	0	F = A plus B
0	1	F = B
1	0	F = shift left (logical) A
1	1	$F = \overline{A}$

(h) Design a 4-bit logic unit that will function as follows:

S1	S0	F
0	0	A + B
0	1	A • B
1	0	$\overline{A}$
1	1	$A \oplus B$

11.5 Design and implement a 6 × 6 array multiplier.

11.6 Design an unsigned 8 × 4 non-additive multiplier using additive-multiplier-module whose block diagram representation is as follows:

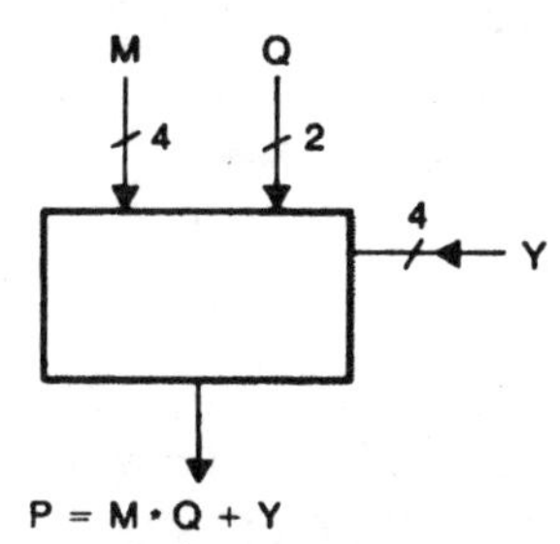

Assume that M, Q, and Y are unsigned integers.

11.7 Using four 256 × 8 ROMS and 4-bit parallel adders, design a 8 × 8 unsigned, nonadditive multiplier. Draw a logic diagram of your implementation.

11.8 (a) Repeat Example 11.2 using the ASM chart shown in the solution of the example with D flip-flops and a PLA.

(b) Repeat Problem 7.21 by drawing an ASM chart for each case.

11.9 Consider the registers and ALU shown in Figure P11.9:

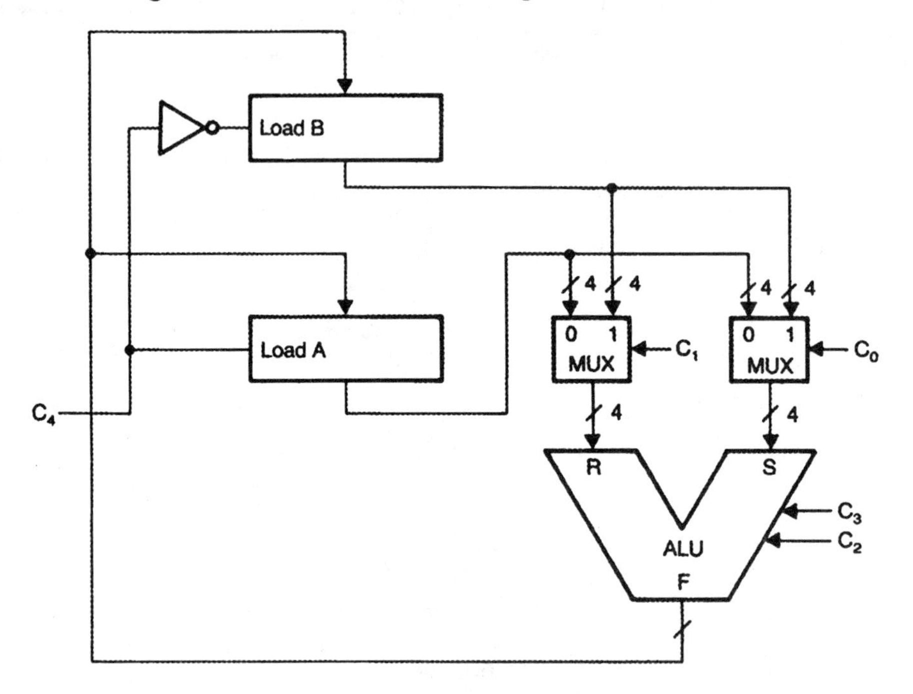

The interpretation of various control points are summarized as follows:

C_3	C_2	F
0	0	R plus S
0	1	R minus S
1	0	R and S
1	1	R EX-OR S

C_1	C_0	R-INPUT	S-INPUT
0	0	A	A
0	1	A	B
1	0	B	A
1	1	B	B

C_4	ACTION
0	B ← F
1	A ← F

FIGURE P11.9

Answer the following questions by writing suitable control word(s). Each control word must be specified according to the following format:

$$C_4\ C_3\ C_2\ C_1\ C_0$$

For example:

C_4 C_3 C_2 C_1 C_0

1 0 0 0 1 ;A ← A plus B

(a) How will the A register be cleared? (Suggest at least two possible ways.) DIRECT CLEAR input is not available.

(b) Suggest a sequence of control words that exchanges the contents of A and B registers (exchange means A ← B and B ← A).

11.10 Consider the following algorithm:

```
Declare registers A [8], B [8], C [8];
START: A ← 0; B ← 00001010;
LOOP:          A ← A + B; B ← B – 1;
               If B < > 0 then go to LOOP
               C ← A;
HALT:          Go to HALT
```

Design a hardwired controller that will implement this algorithm.

11.11 It is desired to build an interface in order establish communication between a 32-bit host computer and a front end 8-bit microcomputer (See Figure P11.11). The operation of this system is described as follows:

Step 1: First the host processor puts a high signal on the line "want" (saying that it needs a 32-bit data) for one clock period.

Step 2: The interface recognizes this by polling the want line.

Step 3: The interface unit puts a high signal on the line "fetch" for one clock period (that is it instructs the microcomputer to fetch an 8-bit data).

Step 4: In response to this, the microcomputer samples the speech signal, converts it into an 8-bit digital data and informs the interface that the data is ready by placing a high signal on the "ready" line for one clock period.

Step 5: The interface recognizes this (by polling the ready line), and it reads the 8-bit data into its internal register.

Step 6: The interface unit repeats the steps 3 through 5 for three more times (so that it acquires 32-bit data from the microcomputer).

Step 7: The interface informs the host computer that the latter can read the 32-bit data by placing a high signal on the line "takeit" for one clock period.

Step 8: The interface unit maintains a valid 32-bit data on the 32-bit output bus until the host processor says that it is done (the host puts a high signal on the line "done" for one clock period). In this case, the interface proceeds to step 1 and looks for a high on the "want" line.

(a) Provide a Register Transfer Language description of the interface.

(b) Design the processing section of the interface.

(c) Draw a block diagram of the interface controller.

(d) Draw a state diagram of the interface controller.

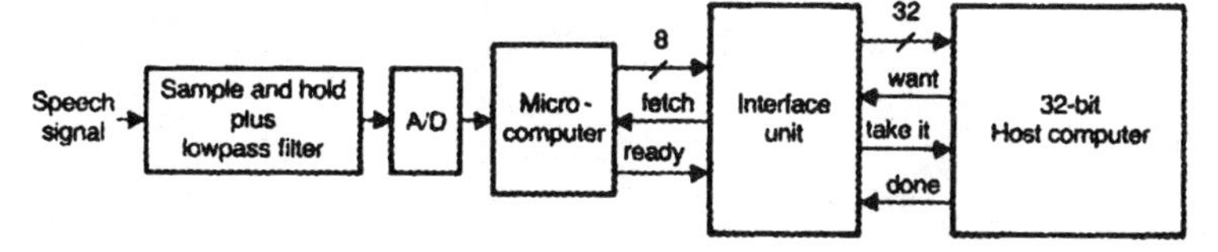

FIGURE P11.11

11.12 Solve Problem 11.10 using the microprogrammed approach.

11.13 Design a microprogrammed system to add numbers stored in the register pair AB and CD. A, B, C, and D are 8-bit registers. The sum is to be saved in the register pair AB. Assume that only an 8-bit adder is available.

11.14 The goal of this problem is to design a microprogrammed 3rd order FIR (Finite impulse response) digital filter. In this system, there are 4 coefficients w_0, w_1, w_2, and w_3. The output y_k (at the kth clock period) is the discrete convolution product of the inputs ($x_k s$) and the filter coefficients. This is formally expressed as follows:

$$y_k = w_0 * x_k + w_1 * x_{k-1} + w_2 * x_{k-2} + w_3 * x_{k-3}$$

In the above summation, x_k represents the input at the kth clock period while x_{k-i} represents input at $(k - i)$th sample period. For all practical purposes, we assume that our system is causal and so $x_i = 0$ for $i < 0$. The processing hardware is shown in Figure P11.14. This unit includes 8 eight-bit registers (to hold data and coefficients), A/D (Analog digital converter), MAC (multiplier accumulator), and a D/A (Digital analog converter). The processing sequence is shown below:

1. Initialize coefficient registers
2. Clear all data registers except x_i
3. Start A/D conversion (first make $sc = 1$ and then retract it to 0)
4. Wait for one control state (To make sure that the conversion is complete)
5. Read the digitized data into the register x_k
6. Iteratively calculate filter output y_k (use MAC for this)
7. Pass y_k to D/A (Pass Accumulator's output to D/A via Rounding ROM)
8. Move the data to reflect the time shift ($x_{k-3} = x_{k-2}$, $x_{k-2} = x_{k-1}$, $x_{k-1} = x_k$)
9. Go to 3

(a) Specify the controller organization.

(b) Produce a well documented listing of the binary microprogram

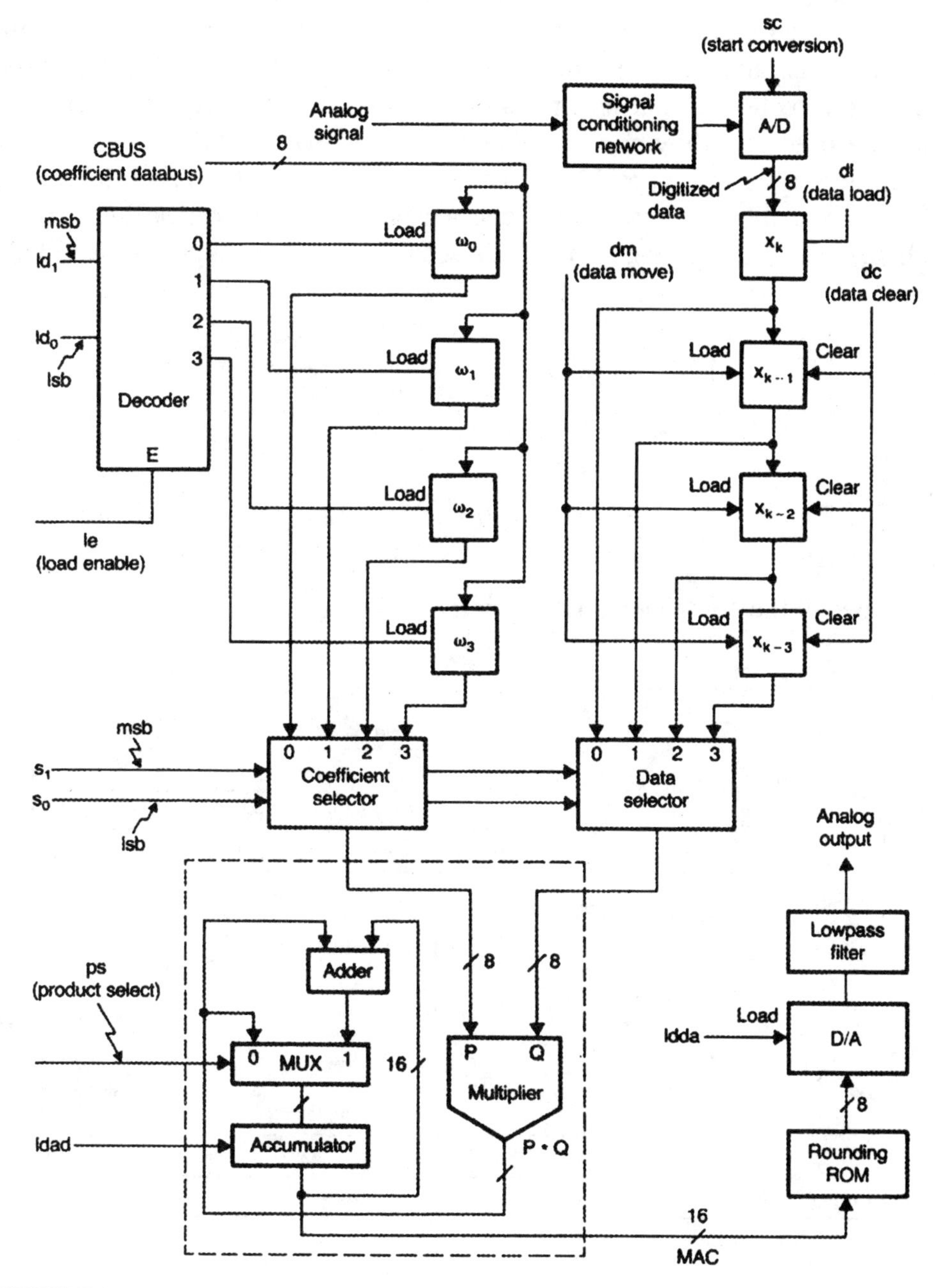

FIGURE P11.14

11.15 Your task is to design a microprogrammed controller for a simple robot with 4 sensors (see Fig. P11.15a). The sensor output will go high only if there is a wall or an obstruction within a certain distance. For example, if F= 1, there is an obstruction or wall in the forward direction. In particular, your controller is supposed to communicate with a motor controller unit shown in Fig. P11.15b. The flow chart that describes the control algorithm is shown in Fig. P11.15c. The outputs such as MFTS, MRT, MLT, MUT, and STP, andd the status signals such as FMC, and TC will be high for one clock period. Assume that a power on reset causes the controller to go the WAIT STATE 0.

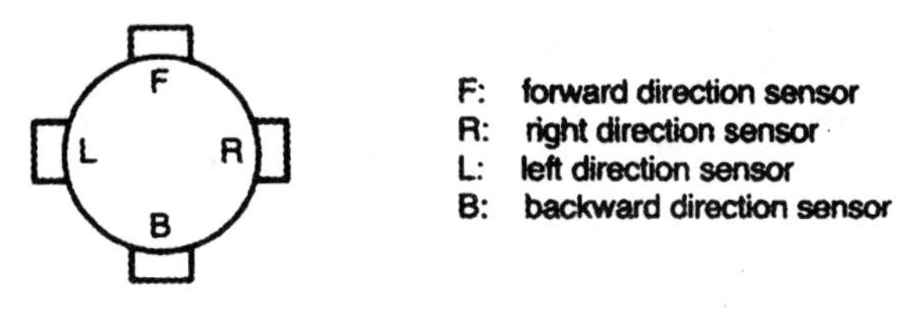

Figure A

FIGURE P11.15a

(a) Specify the controller organization.

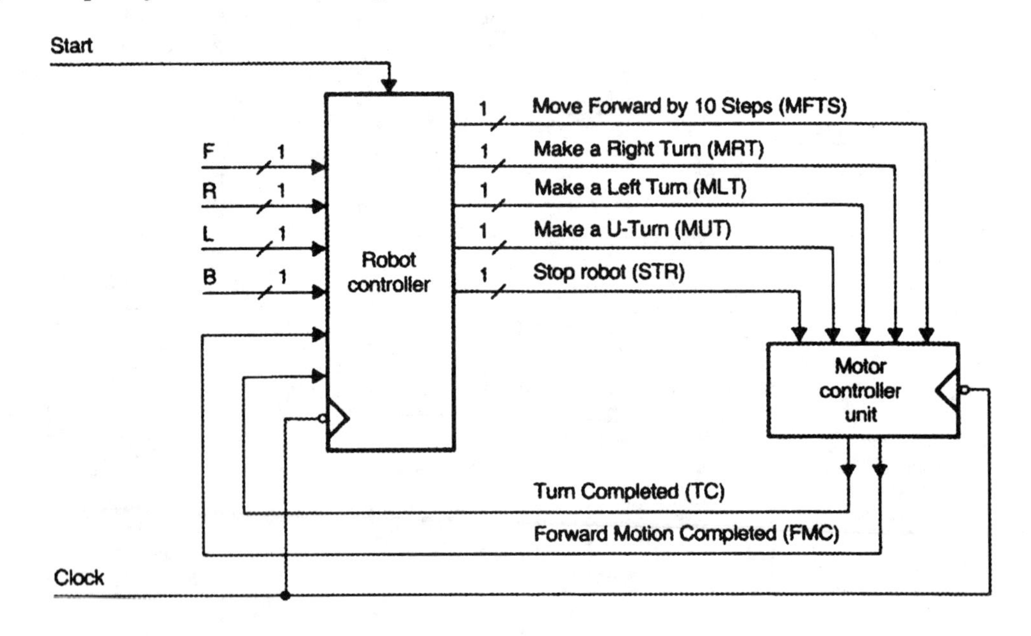

Figure B

FIGURE P11.15b

(b) Provide a well documented listing of the binary microprogram.

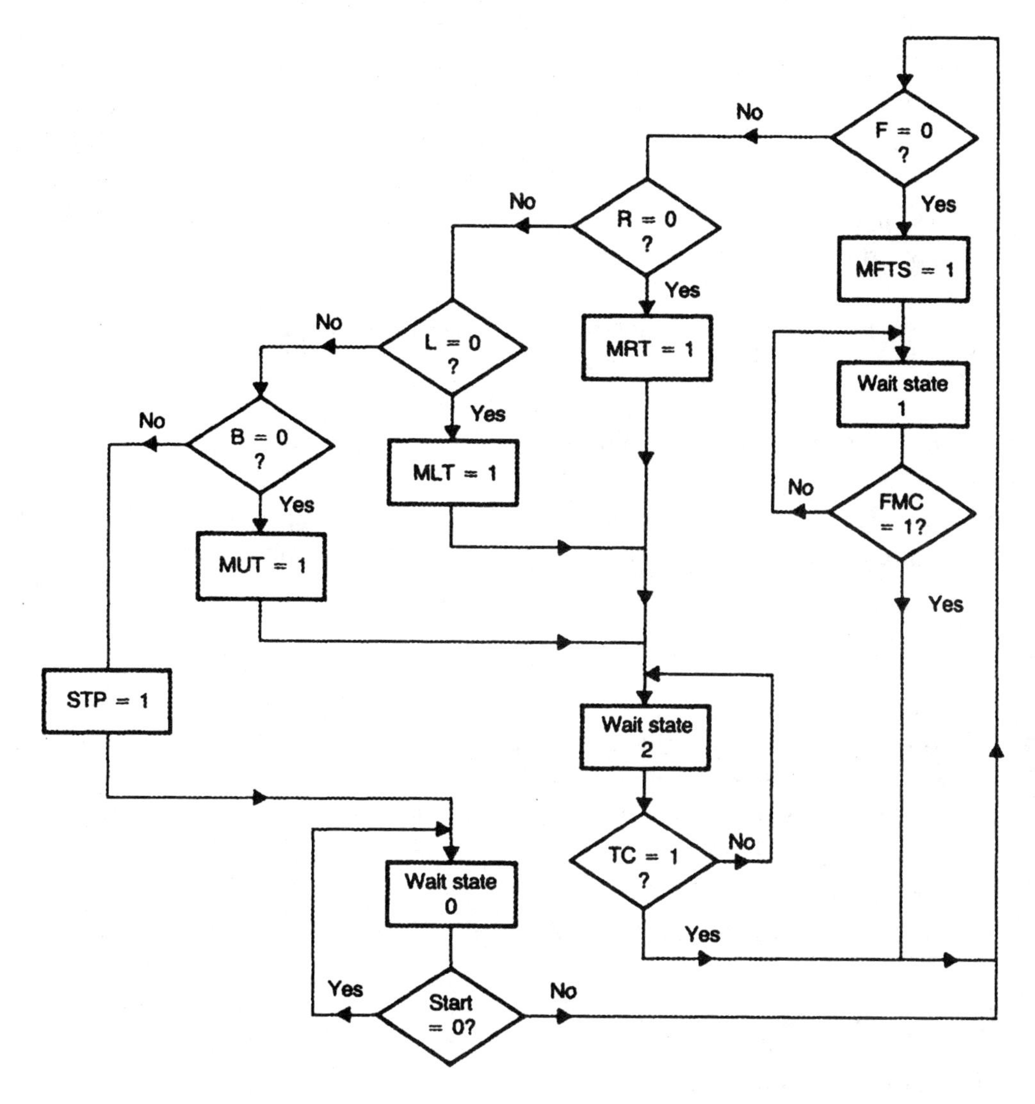

Figure C

FIGURE P11.15c

11.16 It is desired to add the following instructions to the instruction set shown in Figure 11.14.

	GENERAL FORMAT	OPERATION	DESCRIPTION
(a)	MVIA ‹data8›	A ← ‹data8›	This is an immediate mode move instruction. The first byte contains the op-code while the second byte contains the 8-bit data.
(b)	NEGA	A ← − A	This instruction negates the contents of A

Write a symbolic microprogram that describes the execution of each instruction.

11.17 Explain how the effect of an unconditional branch instruction of the following form is simulated:

JP ‹addr›

Use the instruction set shown in Figure 11.14.

11.18 Using the instruction set shown in Figure 11.14, write a program to add the contents of the memory locations 64_{16} through $6D_{16}$ and save the result in the address $6E_{16}$.

11.19 Show that it is possible to specify 675 microoperations using a 10 bit control function field.

11.20 A microprogram occupies 100 words and each word typically emits 70 control signals. The architect claims that by using a $2^i \times 70$ nanomemory (for some $i > 0$), it is possible to save 4260 bits. If this were true, determine the number of distinct control states in the original microprogram (Note that here when we say a control state we refer only to the control function field).

Hint: You may have to employ a trial and error approach to solve this problem.

11.21 A typical computer system has a 32K main memory and a 4K fully associative cache memory. The cache block size is 8 words. The access time for the main memory is 10 times that of the cache memory.

(a) How many hardware comparators are needed?

(b) What is the size of the tag field?

(c) If a direct mapping scheme were used instead, what would be the size of the tag field?

(d) Suppose the access efficiency is defined as the ratio of the average access time with a cache to the average access time without a cache, determine the access efficiency assuming a cache hit ratio h of 0.9.

(e) If the cache access time is 200 nanoseconds, what hit ratio would be required to achieve an average access time equal to 500 nanoseconds?

11.22 A set associative cache has a total of 64 blocks divided into sets of 4 blocks each.

(a) Main memory has 1024 blocks with 16 words per block. How many bits are needed in each of the tag, set, and word fields of the main memory address?

(b) A computer system has 32K words of main memory and a set associative cache. The block size is 16 words and the TAG field of the main memory address is 5-bit wide. If the same cache were direct mapped, the main memory will have a 3-bit TAG field. How many words are there in the cache? How many blocks are there in a cache set?

11.23 Under what condition does the set associative mapping method become one of the following?

(a) Direct mapping

(b) Fully associative mapping

11.24 Discuss the main features of Motorola 68020 on-chip cache.

11.25 Design a direct mapped virtual memory system with the following specifications:

- Size of the virtual address space = 64K
- Size of the physical address space = 8K
- Page size = 512 words
- Total length of a page table entry = 24 bits

11.26 A virtual memory system has the following specifications:

- Size of the virtual address space = 64K
- Size of the physical address space = 4K
- Page size = 512

From the page table the following mapping is recognized:

VIRTUAL PAGE NUMBER	PHYSICAL PAGE FRAME NUMBER
0	0
3	1
7	2
4	3
10	4
12	5
24	6
30	7

(a) Find all virtual addresses that will generate a page fault.
(b) Compute the main memory address for the following virtual addresses:
24, 3784, 10250, 30780

11.27 Assume a computer has a segmented memory with paged segments. (Fig. P11.27) The instruction format of this machine is as shown:

Op-code	BR	IR	Displacement
4 bits	2 bits	2 bits	4 bits

This format has the following fields:

- Op-code field
- 2-bit base register field BR
- 2-bit index register field IR
- 4-bit displacement field

The contents of the specified base and index registers are added with the displacement to produce a virtual address whose format is shown next:

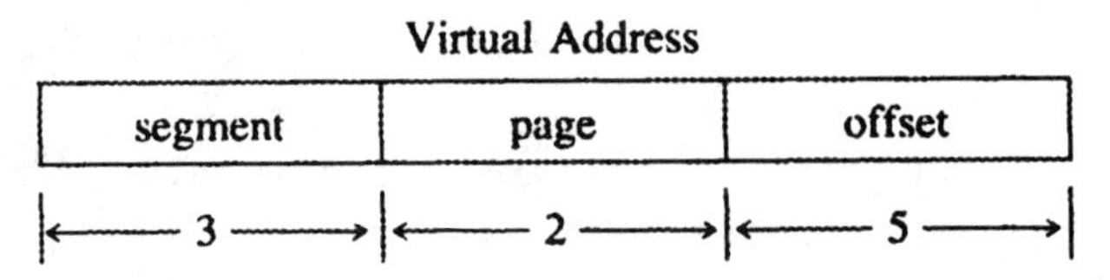

The virtual address is translated into a physical address by means of segment and page tables, which are stored in the main memory. The segment table entry contains the starting address of its page table and the page table entry contains the address of the location which holds the page frame number. The segment table base address register contains the start address of the segment table. The final physical address is the sum of the page table entry and the offset from the virtual address. Consider the following situation:

(a) Compute the physical address needed by the given situation
(b) How many two-operand summations are required to compute one physical address?

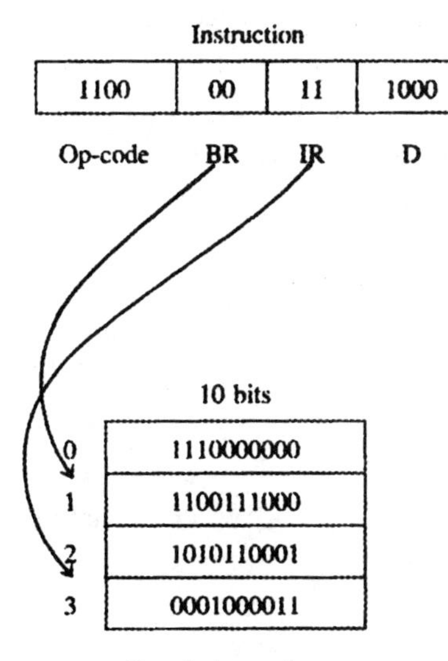

Segment table base address

14_{10}

Main memory

Address	Contents
12	38
13	46
14	57
15	68
16	79
17	106
18	114
19	320
20	40
21	27
22	380
23	440
24	606
25	639
26	900
27	748
28	450
29	121
30	836
31	488
32	594
33	638
34	798
35	868
36	997
37	1048
38	2141
39	2284

FIGURE P11.27

11.28 Assume a main memory has 4 page frames and initially all page frames are empty. Consider the following stream of references;

1, 2, 3, 4, 5, 1, 2, 6, 1, 2, 3, 4, 5, 6, 5

Calculate the hit ratio if the replacement policy used is as follows.

(a) FIFO

(b) LRU

11.29 Repeat Problem 11.28 when the main memory has 5 page frames instead of 4. Comment on your results.

11.30 Consider the stream of references given in Problem 11.28. Plot a graph between the hit ratio and the number of frames *(f)* in the main memory after computing the hit ratio for all values *f* in the range of 1 to 8. Assume LRU policy is used. (Hint: Use the stack algorithm.)

11.31 Consider the floating-point pipeline discussed in section 11.8. Assume:

$T_1 = 40$ ns $\quad T_2 = 100$ ns

$T_3 = 180$ ns $\quad T_4 = 60$ ns

$T_1 = 20$ ns

(a) Determine the pipeline clock rate.

(b) Find the time taken to add 1000 pairs of floating-point numbers using this pipeline.

(c) What is the efficiency of the pipeline when 2000 pairs of floating-point numbers are added?

11.32 Design a pipeline multiplier using carry/save adders (CSA) and carry-look-ahead adders to multiply a stream of input numbers X0, X1, X2, by a fixed number Y. Assume all Xs and Ys are 6-bit numbers. The output should be a stream of 12-bit products YX0, YX1, YX2. Draw a neat schematic diagram of your design.

11.33 Consider the execution of 1000 instructions using a 6-segment pipeline.

(a) What is the average number of instructions executed per instruction cycle when C = 0.2?

(b) What must be the value of C so execution of at least 4 instructions per instruction cycle is always allowed.

11.34 Describe the methods used to handle branches in a pipeline instruction execution unit.

11.35 Modify each of the following programs so the data flow in the 2-segment pipeline (Figure 11.42) is properly regularized:

(a)

MEMORY ADDRESS	INSTRUCTION
2000	LDA X
2001	DCR Y
2002	JMP 2040
2003	SUB Z
:	:
2040	STA W

(b)

MEMORY ADDRESS	INSTRUCTION
2000	LDA X
2001	DCR Y
2002	JNZ 2040
2003	SUB Z
2004	:
:	STAW
2040	

11.36 Explain the significance of interleaved memory organization in pipelined computers .

11.37 Discuss the basic differences between SISD and SIMD.

11.38 The Cray - I computer has one CPU, and 12 functional units. Up to a maximum of 8 functional units can be cascaded to form a chain. Each functional unit is pipelined and the number of pipeline segments vary from 1 to 14. Each functional unit is capable of manipulating 64-bit data. Is it possible to describe this machine using Flynn's approach? Explain.

11.39 Consider a processor array with 4 floating-point processors (FPP). Suppose that each FPP takes 4 time units to produce one result, how long it would take to carry out 100 floating point operations? Is there any performance improvement if the same 100 floating-point operations are carried out using a 4-segment pipelined processor in which each segment takes 1 time unit to produce the result (Ignore latch delay)?

11.40 Explain the significance of masking in array processors.

APPENDIX

ANSWERS TO SELECTED PROBLEMS

Chapter 2

2.1 $1101.101_2 = 13.625_{10}$
2.2(b) $343_{10} = 101010111_2$
2.3(a) $1843_{10} = 3463_8$
2.4(b) $3072_{10} = C00_{16}$
2.6(c) $-48_{10} = 1101\ 0000_2$
2.11(c) $61440_{10} = 1001\ 0100\ 0111\ 0111\ 0011_2$
2.16(b) $0011\ 1110_2$
2.19(b) 0; no overflow
2.19(d) overflow
2.22(a) $0001\ 0000\ 0010_2$ = 102 in BCD

Chapter 3

3.1 $36_{16} \oplus 2A_{16} = 1C_{16}$
3.3 1's Complement of $A7_{16}$
3.4(d)
$$\overline{(A + \overline{A}B)} = \overline{A}(\overline{\overline{A}B}) = \overline{A}(A + \overline{B}) = \overline{A}\,\overline{B}$$

3.4(f)
$$\overline{B}\,\overline{C} + ABC + \overline{A}\,\overline{C} = \overline{C}(\overline{B} + \overline{A}) + ABC = \overline{C}(\overline{AB}) + (AB)C = \overline{C \oplus (AB)}$$

3.5(c) BC

3.7(a) $\overline{F} = \Pi M(0, 1, 5, 7, 10, 14, 15)$
3.9(c) $F = \overline{Z}$
3.10(b) $F = BC + \overline{A}B$
3.11(d) $F = W \oplus Y$
3.11(e) $F = Z$
3.14(a) $f = A + \overline{B}C + B\overline{C}$
3.14(c) $f = \overline{B}$
3.15 $F = \overline{A}\,\overline{C} + \overline{C}\,\overline{D}$
3.17(b) $F = \overline{\overline{(A + C)} + \overline{(B + \overline{C})}}$

Chapter 4

4.1 $F = 0$
4.3(c) $F = A\overline{C} + BC$
4.7 $f = \overline{A \oplus C}$
4.10 $f_3 = \overline{A}\,\overline{B}\,\overline{C}, \quad f_1 = C$

$f_2 = B \oplus C, \quad f_0 = \overline{D}$

4.13 Add the 4-bit unsigned number to itself using full-adders.
4.16 $Z = 1$

$Y = 0$

$X = m5$

$W = m9$

4.20 For 4-Bit signed number, A
$A + 1111_2 = A - 1$, decrement by 1.
$A + 0001_2 = A + 1$, increment by 1.
Manipulate C_{in} to accomplish the above.

Chapter 5

5.5 $A = 1, \ B = 0$
5.7 $A = 1, \ B = 1$

5.9

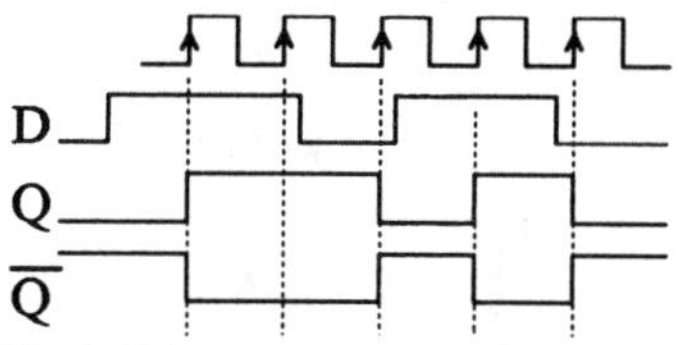

D

Q

$\overline{Q}$

Figure for solution 5.9

5.13 Tie JK inputs to HIGH ; Clock is the T input.

5.15

$$B_+ = A, output \ \ y = \overline{B}$$

$$A += B \oplus x$$

5.17(b)

$$Jx = z, \quad kx = y$$

$$Jy = 1, \quad ky = x + z$$

$$Jz = \overline{x}y, \ kz = x$$

5.19

$$D_A = (A \oplus x) + \overline{B}x$$

$$D_B = \overline{x}(\overline{A \oplus B}) + A\overline{B}x$$

Where x is the input

5.20(c)

$$Tx = \overline{y}$$

$$Ty = 1$$

5.23

$$T_3 = Q_3Q_0 + Q_2Q_1Q_0$$

$$T_2 = Q_1Q_0$$

$$T_1 = \overline{Q_3}Q_0$$

$$T_0 = 1$$

5.24(a) $J_A = B$, $K_A = BC$, $J_B = C$, $K_B = C$, $J_C = 1$, $K_C = A + B$
self correcting

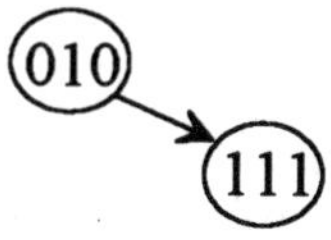

010 → 111

Chapter 6

6.4(a) sign = 0, carry = 0, zero = 0, overflow = 0.
(d) sign = 1, carry = 0, zero = 0, overflow = 1.

6.6(a) 20BE
(b) (20BE) = 05, (20BF) = 02

6.13(a) 16,384

(b) 128 chips

(c) 4 bits

6.18 Use the following identities: $a \oplus a = 0$ and $a \oplus 0 = a$ and $(a \oplus b) \oplus a = b$

Chapter 7

7.2 Yes, it is possible

7.5 Yes, it is possible

7.6 Use four mux's. Manipulate inputs of the mux's to obtain the desired outputs. Use the tristate buffers at the outputs of the mux's.

7.9 $y = |x|$

If $x_7 = 0$, then $y_7....y_2y_1y_0 = x_7....x_2x_1x_0$

else $y_7....y_2y_1y_0 = \bar{x}_7.....\bar{x}_2\bar{x}_1\bar{x}_0 + 1$

use XOR gates for finding 1's complement of x .

7.13 Refer to figure below:

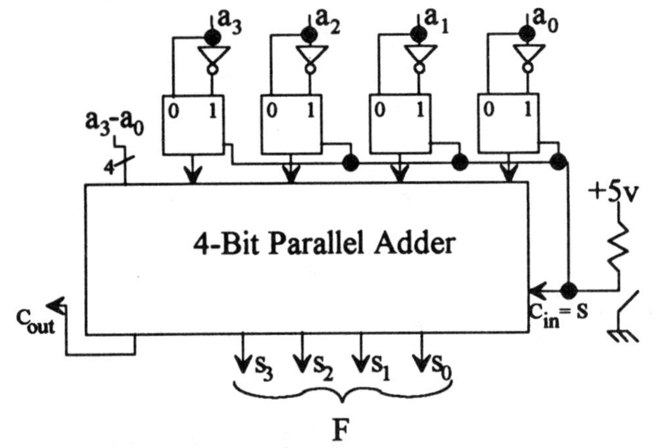

7.16 Product = $0000\ 0000\ 0000\ 0100_2$

7.21(a) 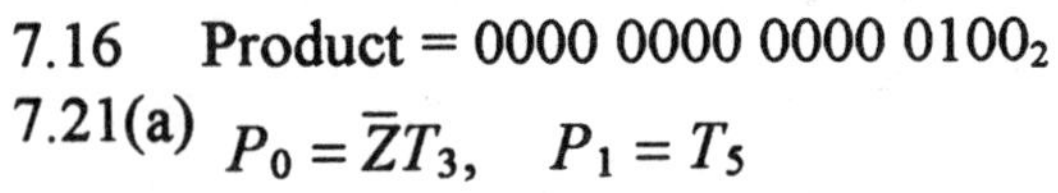
$P_0 = \bar{Z}T_3, \quad P_1 = T_5$

$L = P_0 + P_1, \quad d_2 = P_1, \quad d_1 = P_0, \quad d_0 = P_1$

$C_0 = C_I = T_0, \quad C_2 = T_1, \quad C_3 = C_4 = C_6 = T_2, \quad C_5 = T_4$

7.24 Savings = 34,304 bits

Chapter 8

8.5 Memory Chip #1 EC00H - EDFFH

Memory Chip #2 F200H - F3FFH

8.6(a) ROM Map: 0000H - 07FFH
RAM Map: 2000H - 27FFH

8.13 20

8.14 Maximum Directly Addressable Memory = 16 Megabytes;
14 unused address pins Available.

8.15 6 x 64 decoder

8.18 Cache Tag Field = 1-bit
Cache Index Field = 12-bits
Cache Data Field = 32-bits

8.20 Cache word size = 36 bits.

Chapter 9

9.4 20642H

9.6(a) Implied

9.8 [AL] = 5

9.13

```
XCHG  BL,BH
MOV   AX,BX
ADD   AX,CX
HLT
```

9.19

```
MOV   AL,CH
CBW
IDIV  CL
MOV   CL,AH
MOV   CH,AL
HLT
```

9.26

```
CONV        SEGMENT
            ASSUME   CS:CONV
BCD2BIN     PROC   FAR
            MOV    BX,4000H
            MOV    CL,10
            MOV    DX,0
            MOV    AL,[BX]
            MUL    CL
            ADD    DX,AX
            INC    BX
            ADD    DL,[BX]
            RET
BC2BIN      ENDP
CONV        ENDS
            END
```

9.27

```
            MOV   CL,4
            MOV   AL,90H
            OUT   CNTRL,AL
            MOV   BL,0
BACK        IN    AL,PORTA
            RCR   AL,1
            JC    START
            INC   BL
START       DEC   CL
            JNZ   BACK
            RCR   BL,1
            JNC   LEDON
            MOV   AL,0
            OUT   PORTB,AL
            HLT
LEDON       MOV   AL,10H
            OUT   PORTB,AL
            HLT
```

9.28 Port A = 01H, Port B = 03H, Port C = 05H, CNTRL = 07H
2732 ODD = 00001H,00003H,...,01FFFH
2732 EVEN = 00000H,00002H,...,01FFEH

9.34 For 15 sec. delay: a count of 38D2H provides a delay of 20 msec; this loop needs to be executed 750 times.

9.42(a) [EAX] = 0000 0080H

9.44
```
MOVSX CX, BH
IDIV  AX,CX
HLT
```

9.56 [ECX] = 2A157241H

9.58 [AX] = 1234H

Chapter 10

10.7 TRAP occurs since odd address.

10.9(c) Privileged

10.13 $0000 0000

10.16
```
         SWAP     D1
         MOVE     D1,D0
         EXT.L    D0
         SWAP     D1
         EXT.W    D1
         DIVS     D1,D0
FINISH   JMP      FINISH
```

10.18

```
            MOVE.W   D1,D0
            SWAP     D1
            ADD      D0,D1
            SWAP     D1
FINISH      JMP      FINISH
```

10.26

$$\overline{AS} = 0, \quad FC2FC1FC0 = 101$$

$$\overline{LDS} = 1, \quad \overline{UDS} = 0$$

10.28 Memory map:
even 2764 $000000,$000002,...,$003FFE
odd 2764 $000001,$000003,...,$003FFF
68230 I/O map:
PGCR = $004001, PADDR = $004005
PBDDR = $004007, PACR = $00400D
PBCR = $00400F, PADR = $004011
PBDR = $004013

10.39 [D1.W] = $4567

10.41 `CMP.L (0,A0,D5.L*2),D0`

10.45

```
            ADD.L    D3,D0
            ADDX.L   D2,D1
FINISH      JMP      FINISH
```

10.49 *32-bit device: Byte data will be transferred via 68020 D_{15} - D_8 pins.
*8-bit device: Byte data will be transferred via D_{31} - D_{24} pins.

10.53 GPR0 - GPR31

10.55(b) The PR bit in MSR is 1.

10.56(a) The 32-bit contents of r2 and r3 are added; the result is stored in r1. The dot suffix enables the update of the condition register.

Chapter 11

11.2 See the ASM chart of example 11.2 in the text.

11.4(a) $s_{15} \xleftarrow[3\Delta]{} c_{15} \xleftarrow[2\Delta]{} c_{12} \xleftarrow[2\Delta]{} g_i p_i \xleftarrow[2\Delta]{} G_i P_i \xleftarrow[\Delta]{} \begin{matrix} x_i \\ y_i \\ c_0 \end{matrix}$; worst case add-time: 10Δ

11.9(a)

	C_4	C_3	C_2	C_1	C_0	
Solution 1	1	0	1	0	0	; A ← A minus A
Solution 2	1	1	1	0	0	; A ← A ex-or A

11.17 *Step 1:* Make F=0 (set $c_{10}c_{11}c_{12}$ to 000) and set the zero flag to 1.
Step 2: Execute JZ instruction.

11.21(a) 512 (e) $h = 0.85$

11.22(b) Cache size is 4K words.
4 blocks per set.

11.26(b)

Virtual address	Physical address
24	24
3784	1224
10250	page fault
30780	page fault

11.28(a) 4/15

11.31(a) Pipeline clock rate = 5 MHz

(c) Efficiency = 99.8%

11.33(a) Avg. number of instructions executed per instruction cycle ≅ 4.98

11.35(a)

```
        LDA   X
        JMP   2040
        DCR   Y
        SUB   Z
           :
2040    STA   W
```

The above program assumes that the system supports delayed branch.

APPENDIX

B

GLOSSARY

ABEL: A programming language for PLDs developed by Data I/O Corporation.

Absolute Addressing: This addressing mode specifies the address of data with the instruction.

Accumulator: Used for storing the result after most ALU operations; available with 8-bit microprocessors.

Address: A unique identification number (or locator) for source or destination of data. An address specifies the register or memory location of an operand involved in the instruction.

Addressing Mode: The manner in which a microprocessor determines the effective address of source and destination operands in an instruction.

Address Register: A register used to store the address (memory location) of data.

Address Space: The number of storage location in a microcomputer's memory that can be directly addressed by the microprocessor. The addressing range is determined by the number of address pins provided with the microprocessor chip.

American Standard Code for Information Interchange (ASCII): An 8-bit code commonly used with microprocessors for representing alphanumeric codes.

Analog-to-Digital (A/D) Converter: Transforms an analog voltage into its digital equivalent.

AND gate: The output is 1, if all inputs are 1; otherwise the output is 0.

Architecture: The organizational structure or hardware configuration of a computer system.

Arithmetic and Logic Unit (ALU): A digital circuit which performs arithmetic and logic operations on two n-bit numbers.

Assembler: A program that translates and assembly language program into a machine language program.

Assembly Language: A type of microprocessor programming language that uses a semi-English-language statement.

Asynchronous Operation: The execution of a sequence of steps such that each step is initiated upon completion of the previous step.

Asynchronous Sequential Circuit: Completion of one operation starts the next operation in sequence. Time delay devices (logic gates) are used as memory.

Asynchronous Serial Data Transmission: The transmitting device does not need to be synchronized with the receiving device.

Autodecrement Addressing Mode: The contents of the specified microprocessor register are first decremented by K (1 for byte, 2 for 16-bit, and 4 for 32-bit) and then the resulting value is used as the address of the operand.

Autoincrement Addressing Mode: The contents of a specified microprocessor register are used as the address of the operand first and then the register contents are automatically incremented by K (1 for byte, 2 for 16-bit, and 4 for 32-bit).

Bandwidth: Bandwidth of a bus or memory is a measure of communications throughput and can be represented as the product of the maximum number of transactions per second and number of data bits per transaction.

Barrel Shifter: A specially configured shift register that is normally included in 32-bit microprocessors for cycle rotation. That is , the barrel shifter shifts data in one direction.

Base address: An address that is used to convert all relative addresses in a program to absolute (machine) addresses.

Baud Rate: Rate of data transmission in bits per second.

Binary-Coded Decimal (BCD): The representation of 10 decimal digits, 0 through 9, by their corresponding 4-bit binary number.

Bit: An abbreviation for a binary digit. A unit of information equal to one of two possible states (one or zero, on or off, true or false).

Block Transfer DMA: A peripheral device requests the DMA transfer via the DMA request line, which is connected directly or through a DMA controller chip to the microprocessor. The DMA controller chip completes the DMA transfer and transfers the control of the bus to the microprocessor.

Branch: The branch instruction allows the computer to skip or jump out of program sequence to a designated instruction either unconditionally or conditionally (based on conditions such as carry or sign).

Breakpoint: Allows the user to execute the section of a program until one of the breakpoint conditions is met. It is then halted. The designer may then single step or examine memory and registers. Typically breakpoint conditions are program counter address or data references. Breakpoints are used in debugging assembly language programs.

Browser: Program in the personal computer to see contents on the web via http protocol.

Buffer: A temporary memory storage device deigned to compensate for the different data rates between a transmitting device and a receiving device (for example, between a CPU and a peripheral). Current amplifiers are also referred to as buffers.

Bus: A collection of wires that interconnects computer modules. The typical microcomputer interface includes separate buses for address, data, control, and power functions.

Bus Arbitration: Bus operation protocols (rules) that guarantee conflict-free access to a bus. Arbitration is the process of selecting one respondent from a collection of several candidates that concurrently request service.

Bus Cycle: The period of time in which a microprocessor carries out read or write operations.

Cache Memory: A high speed, directly accessible, relatively small, semiconductor read/write memory block used to store data/instructions that the microcomputer may need in the immediate future. Increases speed by reducing the number of external memory reads required by the processor. Typical 32 and 64-bit microprocessors are normally provided with on-chip cache memory.

Central Processing Unit (CPU): The brain of a computer containing the ALU, register section, and control unit. CPU on a single chip is called microprocessor.

Chip: An integrated package (IC) containing digital circuits.

Clock: Timing signals providing synchronization among the various components in a microcomputer system.

Combinational Circuit: Output is provided upon application of inputs; contains no memory.

Code: A system of symbols for representation of data in a digital computer. Examples include ASCII and EBCDIC.

Compiler: A software program which translates the source code written in a high-level programming language into machine language that is understandable to the processor.

Conditional Branching: Conditional branch instructions are used to change the order of execution of a program based on the conditions set by the status flags.

Condition Code Register: contains information such as carry, sign, zero, and overflow based on ALU operations.

Control Store: Used to contain microcode (usually in ROM) in order to provide for microprogrammed "firmware" control functions. An integral part of a microprogrammed CPU.

Control Unit: Part of the CPU; its purpose is to read and decode instructions from the memory.

Controller/ Sequencer: The hardware circuits which provides signals to carry out selection and retrieval of instructions from storage in sequence, interpret them, and initiate the required operation. The system functions may be implemented by hardware control, firmware control, or software control.

Coprocessor: A companion microprocessor that performs specific functions such as floating-point operations independently from the microprocessor to speed up overall operations.

Cycle Stealing DMA: The DMA controller transfers a byte of data between the microcomputer's memory and a peripheral device such as the disk by stealing a clock cycle of microprocessor.

Data: Basic elements of information represented in binary form (that is, digits consisting of bits) that can be processed or produced by a microcomputer. Data represents any group of operands made up of numbers, letters, or symbols denoting any condition, value, or state. Typical microcomputer operand sizes include: a word, which typically contains 2 bytes or 16-bits; a long word, which contains 4 bytes or 32 bits; a quad word, which contains 8 bytes or 64 bits.

Data Register: A register used to temporarily hold operational data begin sent to and from a peripheral device.

Debugger: A program that executes and debugs the object program generated by the assembler or compiler. The debugger provides a single stepping, breakpoints, and program tracing.

Decoder: A device capable of generating 2^n output lines based on n inputs.

Demultiplexer: Performs reverse operation of a multiplexer.

Digital to Analog (D/A) Converter: Converts binary number to analog signal.

Diode: Two terminal electronic switch.

Direct Memory Access (DMA): A type of input/output technique in which data can be transferred between the microcomputer memory and external devices without the microprocessor's involvement.

Directly Addressable Memory: The memory address space in which the microprocessor can directly execute programs. The maximum directly addressable memory is determined by the number of the microprocessor's address pins.

Dynamic RAM: Stores data in capacitors and , therefore, must be refreshed, uses refresh circuitry.

EAROM (Electrically Alterable Read-Only Memory): Can be programmed without removing the memory from its sockets. This memory is also called read-mostly memory since it has much slower write times than read times.

Editor: A program that produces an error-free source program, written in assembly or high-level languages.

EEROM or **E^2ROM:** Same as EAROM (see EAROM).

Effective Address: The final address used to carry out an instruction. Determined by the addressing mode.

Emulator: A hardware device that allows a microcomputer system to emulate (that is, mimic the procedures or protocols) another microcomputer system.

Encoder: Performs reverse operation of a decoder. Contains a maximum of 2^n inputs and n outputs.

EPROM (Erasable Programming Read-Only Memory): Can be programmed and erased using ultraviolet light. The chip must be removed from the microcomputer system for programming.

Equivalence: Same as Exclusive-NOR.

Exception Processing: The CPU processing state associated with interrupts, trap instructions, tracing, and other exceptional conditions, whether they are initiated internally or externally.

Exclusive-OR: The output is 0, if inputs are same; otherwise; the output is 1.

Exclusive-NOR: The output is 1, if inputs are same; otherwise, the output is 0.

Extended Binary-Coded Decimal Interchange Code (EBCDIC): An 8-bit code commonly used with microprocessors for representing character codes. Normally used by IBM.

Firmware: Microprogram is sometimes referred to as firmware to distinguish it from hardwired control (purely hardware method).

Flag(s): An indicator, often a single bit, to indicate some conditions such as trace, carry, zero, and overflow.

Flip-Flop: One-bit memory.

Flowchart: Representation of a program in a schematic form. It is convenient to flowchart a problem before writing the actual programs.

Full-Adder: Adds three bits.

Gate: Digital circuits which perform logic operations.

Half-Adder: Adds two bits.

Handshaking: Data transfer via exchange of control signals between the microprocessor and an external device.

Hardware: The physical electronic circuits (chips) that make up the microcomputer system.

Hardwired Control: Used for designing the control unit.

HCMOS: Low-power HMOS. Technology of future.

Hexadecimal Number System: Base-16 number system.

High-Level Language: A type of programming language the uses a more understandable human-oriented language such as C.

HMOS: High-performance MOS reduces the channel length of the NMOS transistor and provides increased density and speed in VLSI circuits.

Immediate Address: An address that is used as an operand by the instruction itself.

Implied Address: An address is not specified, but is contained implicitly in the instruction.

In-Circuit Emulation: The most powerful hardware debugging technique; especially valuable when hardware and software are being debugged simultaneously.

Index: A symbol used to identify or place a particular quantity in an array (list) of similar quantities. Also, and ordered list of references to the contents of a larger body of data such as a file or record.

Indexed Addressing: The effective address of the instruction is determined by the sum of the 16-bit address and the contents of the index register. Used to access arrays.

Index Register: A register used to hold a value used in indexing data, such as when a value is used in indexed addressing to increment a base address contained within an instruction.

Indirect Address: A register holding a memory address to be accessed.

Instruction: A program statement (step) that causes the microcomputer to carry out an operation, and specifies the values or locations of all operands.

Instruction Cycle: The sequence of operations that a microprocessor has to carry out while executing an instruction.

Instruction Register (IR): A register storing instructions; typically 32 bits long for a 32-bit microprocessor.

Instruction Set: Lists all the instructions that the microcomputer can execute.

Interleaved DMA: Using this technique, the DMA controller takes over the system bus when the microprocessor is not using it.

Internal Interrupt: Activated internally by exceptionally conditions such as overflow and division by zero.

Internet: Connects users from around the world via a web of data transmission lines.

Interpreter: A program that executes a set of machine language instructions in response to each high-level statement in order to carry out the function.

Interrupt I/O: An external device can force the microcomputer system to stop executing the current program temporarily so that it can execute another program known as the interrupt service routine.

Interrupts: A temporary break in a sequence of a program, initiated externally or internally, causing control to pass to a routine, which performs some action while the program is stopped.

I/O (Input/Output): Describes that portion of a microcomputer system that exchanges data between the microcomputer system and an external device.

I/O Port: A register that contains control logic and data storage used to connect a microcomputer to external peripherals.

Inverting Buffer: Performs NOT operation. Current amplifier.

Karnaugh Map: Simplifies Boolean expression by a mapping mechanism.

Keyboard: Has a number of push button-type switches configured in a matrix form (rows x columns).

Keybounce: When a mechanical switch opens or closes, it bounces (vibrates) for a small period of time (about 10-20 ms) before settling down.

Large-Scale Integration (LSI): An LSI chip contains 100 to 1000 gates.

LED: Light Emitting Diode. Typically, a current of 10 ma to 20 ma flows at 1.7v drop across it.

Local Area Network: A collection of devices and communication channels that connect a group of computers and peripherals devices together within a small area so that they can communicate with each other.

Logic Analyzer: A hardware development aid for microprocessor-based design; gathers data on the fly and displays it.

Logical Address Space: All storage locations with a programmer's addressing range.

Loops: A programming control structure where a sequence of microcomputer instructions are executed repeatedly (looped) until a terminating condition (result) is satisfied.

Machine Code: A binary code (composed of 1's and 0's) that a microcomputer understands.

Machine Language: A type of microprocessor programming language that uses binary or hexadecimal numbers.

Macroinstruction: Commonly known as an instruction; initiates execution of a complete microprogram. Example include assembly language instructions.

Macroprogram: The assembly language program.

Mask: A pattern of bits used to specify (or mask) which bit parts of another bit pattern are to be operated on and which bits are to be ignored or "masked" out. Uses logical AND operations.

Mask ROM: Programmed by a masking operation performed on the chip during the manufacturing process; its contents cannot be changed by user.

Maskable Interrupt: Can be enabled or disabled by executing typically the interrupt instructions.

Memory: Any storage device which can accept, retain, and read back data.

Memory Access Time: Average time taken to read a unit of information from the memory.

Memory Address Register (MAR): Stores the address of the data.

Memory Cycle Time: Average time lapse between two successive read operations.

Memory Management Unit (MMU): Hardware that performs address translation and protection functions.

Memory Map: A representation of the physical locations within a microcomputer's addressable main memory.

Memory-Mapped I/O: I/O ports are mapped as memory locations, with every connected device treated as if it were a memory location with a specific address. Manipulation of I/O data occurs in "interface registers" (as opposed to memory locations); hence there are no input (read) or output (write) instructions used in memory-mapped I/O.

Microcode: A set of instructions called "microinstructions" usually stored in a ROM in the control unit of a microprocessor to translate instructions of a higher-level programming language such as assembly language programming.

Microcomputer: Consists of a microprocessor, a memory unit, and an input/output unit.

Microcontroller: Typically includes a microcomputer, timer, A/D (Analog to Digital) and D/A (Digital to Analog) converters in the same chip.

Microinstruction: Most microprocessors have an internal memory called control memory. This memory is used to store a number of codes called microinstructions. These microinstructions are combined to design the instruction set of the microprocessor.

Microprocessor: The Central Processing Unit (CPU) of a microcomputer.

Microprocessor Development System: A tool for designing and debugging both hardware and software for microcomputer-based system.

Microprocessor-Halt DMA: Data transfer is performed between the microcomputer's memory and a peripheral device either by completely stopping the microprocessor or by a technique called cycle stealing.

Microprogramming: The microprocessor can use microprogramming to design the instruction set. Each instruction in the register initiates execution of a microprogram in the control unit to perform the operation required by the instruction.

Monitor: Consists of a number of subroutines grouped together to provide "intelligence" to a microcomputer system. This intelligence gives the microcomputer system the capabilities for debugging a user program, system design, and displays.

Multiplexer: A hardware device which selects one of n input lines and produces it on the output.

Multiprocessing: The process of executing two or more programs in parallel, handled by multiple processors all under common control. Typically each processor will be assigned specific processing tasks.

Multitasking: Operating system software that permits more than one program to run on a single microprocessor. Even though each program is given a small time slice in which to execute, the user has the impression that all tasks (different programs) are executing at the same time.

Multiuser: Describes a computer operation system that permits a number of users to access the system on a time-sharing basis.

NAND: The output is 0, if all inputs are 1; otherwise, the output is 1.

NOR: The output is 1, if all inputs are 0's; otherwise, the output is 0.

Nanomemory: Two-level ROM used in designing the control unit.

Nested Subroutine: A commonly used programming technique that includes one subroutine entirely embedded within the "scope" of another subroutine.

Nibble: A 4-bit word.

Non-inverting Buffer: Input is same as output. Current amplifier.

Nonmaskable Interrupt: Occurrence of this type of interrupt cannot be ignored by microcomputer and even though interrupt capability of the microprocessor is disabled. Its effect cannot be disabled by instruction.

Non-Multiplexed: A non-multiplexed system indicates a direct single communication channel (that is, electrical wires).

NOT gate: If the input is 1, the output is 0, and vice versa.

Object Code: The binary (machine) code into which a source program is translated by a compiler, assembler, or interpreter.

Octal Number System: Base 8-number system.

Ones Complement: Obtained by changing 1's to ' 0's, and 0's to 1's to of a binary number.

One-Pass Assembler: This assembler goes through the assembly language program once and translates the assembly language program into a machine language program. This assembler has the problem of defining forward references. See Two-Pass Assembler.

Op Code (Operation Code): The instruction represented in binary form.

Operand: A datum or information item involved in an operation from which the result is obtained as a consequence of defined address modes. Various operand types contain information, such as source address, destination address, or immediate data.

Operating System: Consists of a number of program modules to provide resource management. Typical resources include microprocessors, disks, and printers.

OR Gate: The output is 0, if all inputs are 0; otherwise, the output is 1.

Page: Some microprocessors, divide the memory locations into equal blocks. Each of these blocks is called a page and contains several addresses.

Parallel Operation: Any operation carried out simultaneously with a related operation.

Parallel Transmission: Each bit of binary data is transmitted over a separate wire.

Parity: The number of 1's in a word is odd for odd parity and even for even parity.

Peripheral: An I/O device capable of being operated under the control of a CPU through communication channels. Examples include disk drives, keyboards, CRT's, printers, and modems.

Personal Computer: Low-cost, affordable microcomputer normally used by an individual for word processing and Internet applications.

Physical Address Space: Address space is defined by the address pins of the microprocessor.

Pipeline: A technique that allows a microcomputer processing operation to be broken down into several steps (dictated by the number of pipeline levels or stages) so that the individual step outputs can be handled by the computer in parallel. Often used to fetch the processor's next instruction while executing the current instruction, which considerably speeds up the overall operation of the microcomputer. Overlaps instruction fetch with execution.

Programmable Array Logic (PAL): Contains programmable AND gates and fixed OR gates. Similar to a ROM in concept except that it does not provide full decoding of the input lines. PAL's are used with 32-bit microprocessors for performing the memory decode function.

Programmable Logic Array (PLA): Contains programmable AND and Programmable OR gates.

Programmable Logic Device (PLD): Contains AND gates and OR gates.

Pointer: A storage location (usually a register within a microprocessor) that contains the address of (or points to) a required item of data or subroutine.

Polled Interrupt: A software approach for determining the source of interrupt in a multiple interrupt system.

POP Operation: Reading from the top or bottom of stack.

Port: A register through which the microcomputers communicate with peripheral devices.

Primary or Main Memory: That memory storage which is considered main, integral, or internal to the computing system. The microcomputer can directly execute all instructions in the main memory.

Privileged Instructions: An instruction which is reserved for use by a computer's operating system.

Processor Memory: A set of microprocessor registers for holding temporary results when a computation is in progress.

Program: A self-contained sequence of computer software instructions (source code) that, when converted into machine code, directs the computer to perform specific operations for the purpose of accomplishing some processing task.

Program Counter (PC): A register that normally contains the address of the next instruction in the sequence of operations.

Programmed I/O: The microprocessor executes a program to perform all data transfers between the microcomputer system and external devices.

PROM (Programmable Read-Only Memory): Can be programmed by the user by using proper equipment. Once programmed, its contents cannot be altered.

Protocol: A list of data transmission rules or procedures that encompass the timing, control, formatting, and data representations by which two devices are to communicate. Also known as hardware "handshaking", which is used to permit asynchronous communication.

PUSH Operation: Writing to the top or bottom of stack.

Random Access Memory (RAM): A read/write memory. RAMs (static or dynamic) are volatile in nature (in other words, information is lost when power is removed).

Read-Only-Memory (ROM): A memory in which any addressable operand can be read from, but not written to, after initial programming. ROM storage is nonvolatile (information is not lost at the removal of power).

Reduced Instruction Set Computer (RISC): A necessary and sufficient instruction set is included. The RISC architecture maximizes speed by reducing clock cycles per instruction. Performs infrequent operations in software and frequent functions in hardware.

Register: A one-word, high-speed memory device usually constructed from flip-flops (electronic switches) that are directly accessible to the processor. It can also refer to a specific location in memory that contains word(s) used during arithmetic, logic, and transfer operations.

Register Indirect: Uses a register which contains the address of data.

Relative Address: An address used to designate the position of a memory location in a routine or program.

Rollover: Occurs when more than one key is pushed simultaneously.

Routine: A group of instructions for carrying out a specific processing operation. Usually refers to part of a larger program. A routine and subroutine have essentially the same meaning, but a subroutine could be interpreted as a self-contained routine nested within a routine or program.

Scalar Microprocessor: Provided with one pipeline. Can execute one instruction per clock cycle. The 80486 is a scalar microprocessor.

Schmitt Trigger: An analog circuit that provides high noise immunity.

Scaling: Multiplying an index register by 1,2,4 or 8. Used by the addressing modes of typical 32- and 64-bit microprocessors.

Secondary Memory Storage: An auxiliary data storing device that supplements the main (primary) internal memory of a microcomputer. It is used to hold programs and data that would otherwise exceed the capacity of the main memory. Although it has a much slower access time, secondary storage is less expensive, common devices include magnetic disk (floppy and hard disks).

Server: Large computer performing actual work on the Internet.

Serial Transmission: Only one line is used to transmit the complete binary data bit by bit.

Sequential Circuit: Combinational circuit with memory.

Seven-Segment LED: Can display numbers.

Single-Chip Microcomputer: Microcomputer (CPU, memory, and input/output) on a chip.

Single-chip Microprocessor: Microcomputer CPU (microprocessor) on a chip.

Single Step: Allows the user to execute a program one instruction at a time and examine memory and registers.

Software: Programs in a microcomputer.

Source Code: The high-level language code used by a programmer to write computer instructions. This code must be translated to the object (machine) code to be usable to the microcomputer.

Stack: An area of read/write memory reserved to hold information about the status of a microcomputer the instant an interrupt occurs so that the microcomputer can continue processing after the interrupt has been handled. Another common use in handling the accessing sequence of "nested" subroutines. The stack is the last in/first out (LIFO) read/write memory (RAM) that is manipulated by using PUSH or POP instructions.

Stack Pointer: A register used to address the stack.

Standard I/O: Utilizes a control pin on the microprocessor chip called the $M/\overline{IO}$ pin, in order to distinguish between input/output and memory; typically, IN and OUT instructions are used for input/output operations.

Static RAM: Stores data in flip-flops; does not need to be refreshed. Information is lost upon power failure unless backed by battery.

Status Register: A register which contains information concerning the flags in a processor.

Subroutine: A program carrying out a particular function and which can be called by another program known as the main program. A subroutine needs to be placed only once in memory and can be called by the main program as many times as the programmer wants.

Superscalar Microprocessor: Provided with more than one pipeline and executes more than one instruction per clock cycle. The Pentium is a superscalar microprocessor.

Supervisor State: When the microprocessor processing operations are conducted at a higher privilege level, it is usually in the supervisor state. An operating system typically executes in the supervisor state to protect the integrity of "basic" system operations from user influences.

Synchronous Operation: Operations that occur at intervals directly related to a clock period.

Synchronous Sequential Circuit: The present outputs depend on the present inputs and the previous states stored in flip-flops.

Synchronous Serial Data Transmission: Data is transmitted or received based on a clock signal.

TCP/IP: Protocol used on the Internet.

Tracing: A dynamic diagnostic technique in which a record of internal counter events is made to permit analysis (debugging) of the program's execution.

Transistor: Electronic switch; performs NOT; current amplifier.

Tristate Buffer: Has three output states: logic 0, 1, and a high-impedance state. It is typically enabled by a control signal to provide logic 0 or 1 outputs. This type of buffer can also be disabled by the control signal to place it in a high-impedance state.

Two's Complement: The two's complement of a binary number is obtained by replacing each 0 with a 1 and each 1 with a 0 and adding one to the resulting number.

Two-Pass Assembler: This assembler goes through the assembly language program twice. In the first pass, the assembler translates the assembly language program to the machine language. No problem with forward branching. See One-Pass Assembler.

UART (Universal Asynchronous Receiver Transmitter): A chip that provides all the interface functions when a microprocessor transmits or receives data to or from a serial device. Converts serial data to parallel and vice versa. Also called ACIA (Asynchronous Communications Interface Adapter) by Motorola.

User State: Typical microprocessor operations processing conducted at the user level. The user state is usually at lower privilege level than the supervisor state. In the user mode, the microprocessor can execute a subset of its instruction set, and allows protection of basic system resources by providing use of the operating system in the supervisor state. This is very useful in multiuser/multitasking systems.

Vectored Interrupts: A device identification technique in which the highest priority device with a pending interrupt request forces program execution to branch to an interrupt routine to handle exception processing for the device.

Very Large Scale Integration (VLSI): a VLSI chip contains more than 1000 gates.

Virtual Memory: An operating system technique that allows programs or data to exceed the physical size of the main, internal, directly accessed memory. Program or data segments/pages are swapped from external disk storage as needed. The swapping is invisible (transparent) to the programmer. Therefore, the programmer need not be concerned with the actual physical size of internal memory while writing the code.

Web: All the interconnected data sources that can be accessed by the personal computers on the Internet.

Wide Area Network: Data network connecting systems within a large area.

Word: The bit size of a microprocessor refers to the number of bits that can be processed simultaneously by the basic arithmetic circuits of the microprocessor. A number of bits taken as a group in this manner is called a word.

APPENDIX C

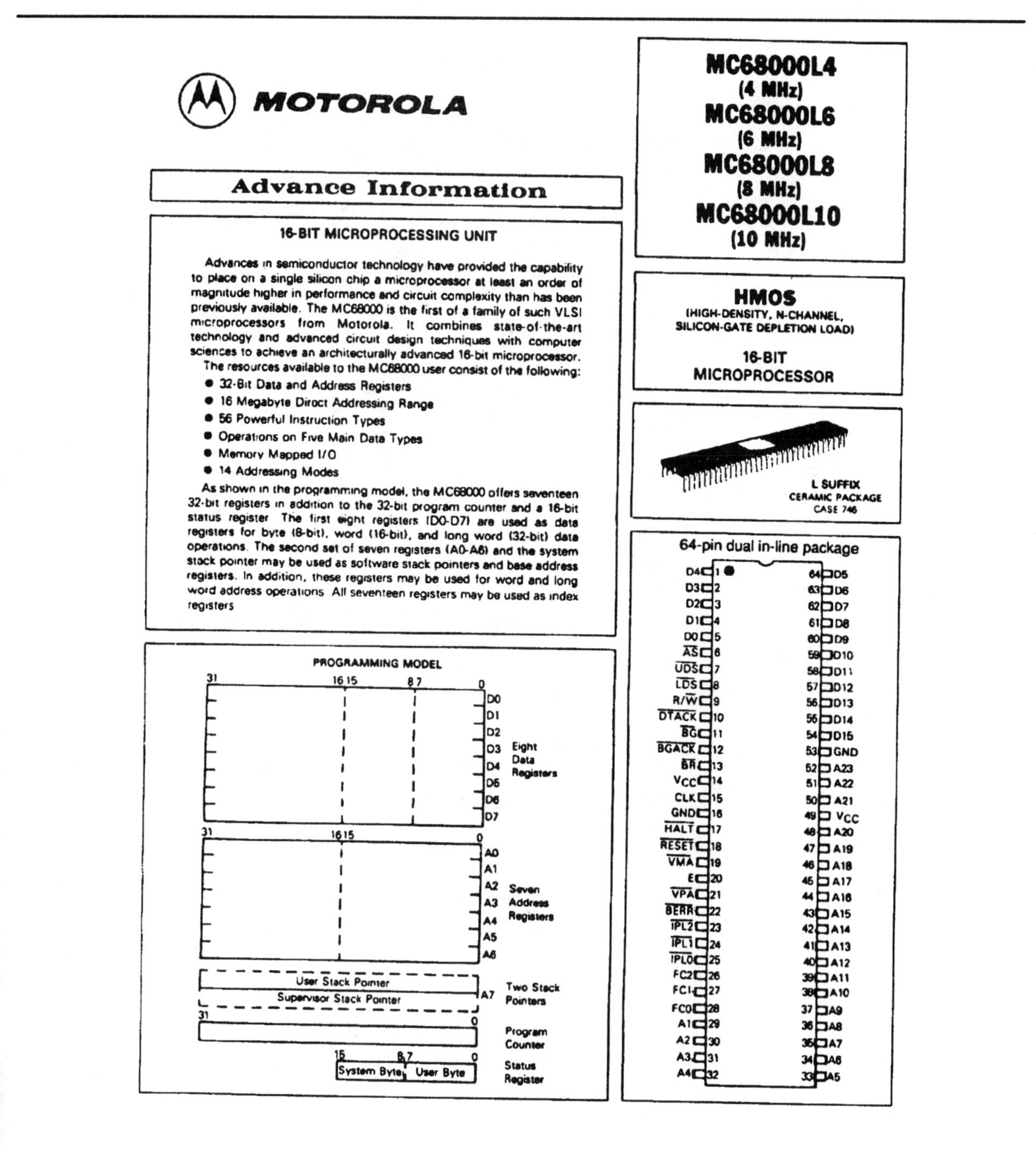

MOTOROLA

Advance Information

MC68000L4 (4 MHz)
MC68000L6 (6 MHz)
MC68000L8 (8 MHz)
MC68000L10 (10 MHz)

16-BIT MICROPROCESSING UNIT

Advances in semiconductor technology have provided the capability to place on a single silicon chip a microprocessor at least an order of magnitude higher in performance and circuit complexity than has been previously available. The MC68000 is the first of a family of such VLSI microprocessors from Motorola. It combines state-of-the-art technology and advanced circuit design techniques with computer sciences to achieve an architecturally advanced 16-bit microprocessor.

The resources available to the MC68000 user consist of the following:

- 32-Bit Data and Address Registers
- 16 Megabyte Direct Addressing Range
- 56 Powerful Instruction Types
- Operations on Five Main Data Types
- Memory Mapped I/O
- 14 Addressing Modes

As shown in the programming model, the MC68000 offers seventeen 32-bit registers in addition to the 32-bit program counter and a 16-bit status register. The first eight registers (D0-D7) are used as data registers for byte (8-bit), word (16-bit), and long word (32-bit) data operations. The second set of seven registers (A0-A6) and the system stack pointer may be used as software stack pointers and base address registers. In addition, these registers may be used for word and long word address operations. All seventeen registers may be used as index registers.

HMOS
(HIGH-DENSITY, N-CHANNEL, SILICON-GATE DEPLETION LOAD)

16-BIT MICROPROCESSOR

L SUFFIX
CERAMIC PACKAGE
CASE 746

PROGRAMMING MODEL

64-pin dual in-line package

Pin	Signal	Pin	Signal
1	D4	64	D5
2	D3	63	D6
3	D2	62	D7
4	D1	61	D8
5	D0	60	D9
6	AS	59	D10
7	UDS	58	D11
8	LDS	57	D12
9	R/W	56	D13
10	DTACK	55	D14
11	BG	54	D15
12	BGACK	53	GND
13	BR	52	A23
14	V_{CC}	51	A22
15	CLK	50	A21
16	GND	49	V_{CC}
17	HALT	48	A20
18	RESET	47	A19
19	VMA	46	A18
20	E	45	A17
21	VPA	44	A16
22	BERR	43	A15
23	IPL2	42	A14
24	IPL1	41	A13
25	IPL0	40	A12
26	FC2	39	A11
27	FC1	38	A10
28	FC0	37	A9
29	A1	36	A8
30	A2	35	A7
31	A3	34	A6
32	A4	33	A5

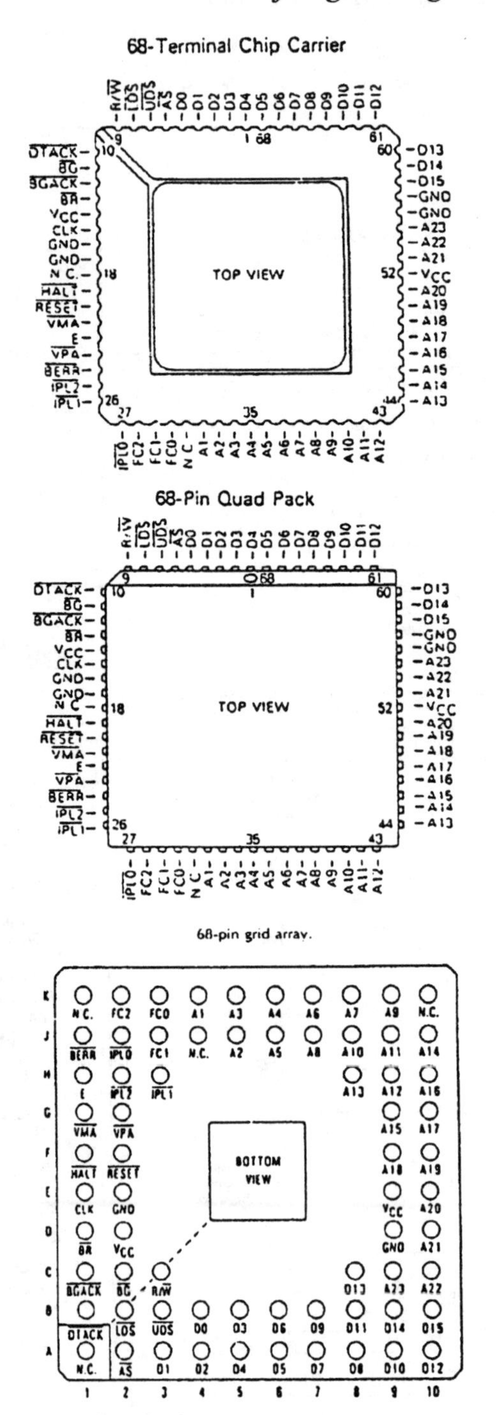
68-Terminal Chip Carrier
TOP VIEW
68-Pin Quad Pack
TOP VIEW
68-pin grid array.
BOTTOM VIEW

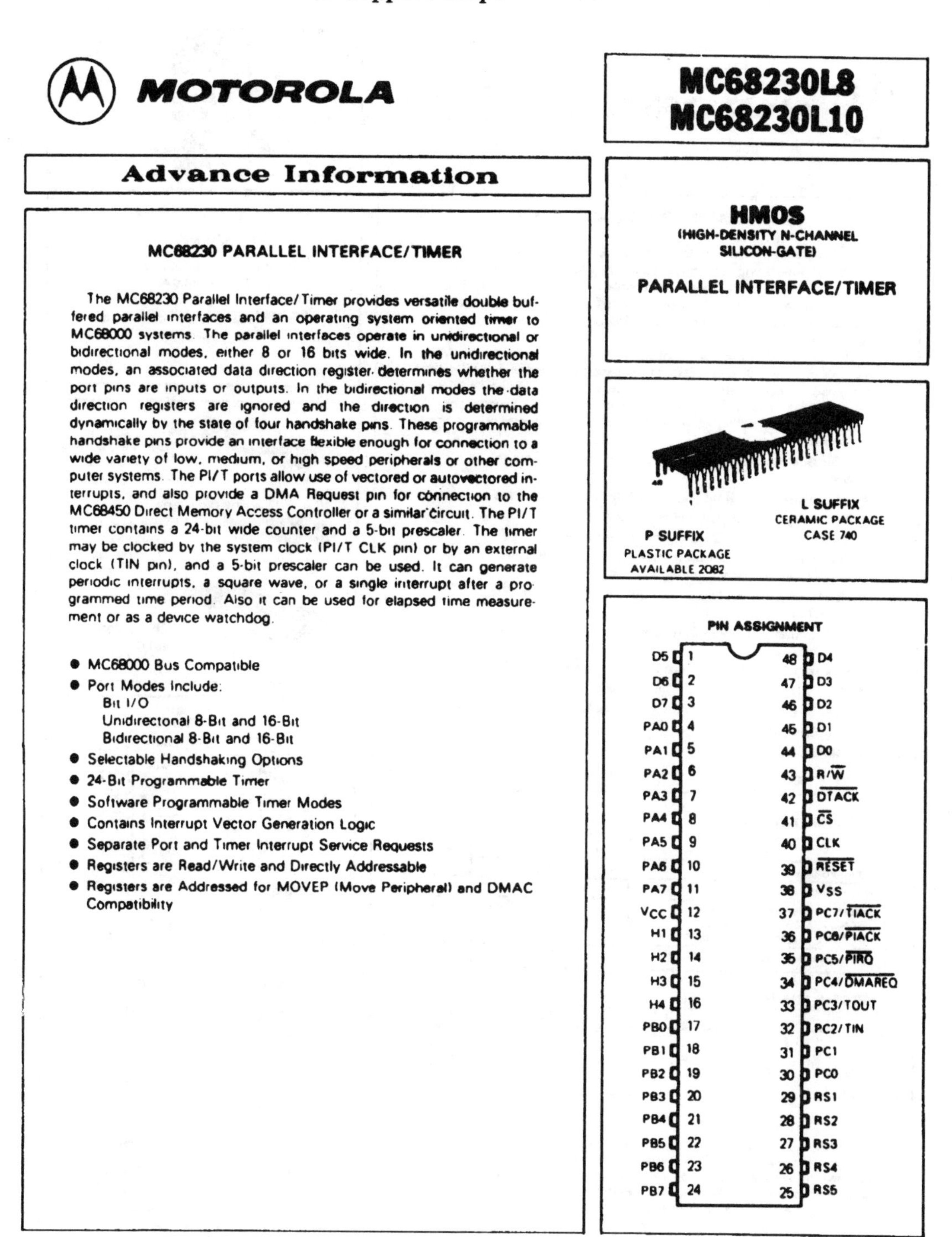

MOTOROLA

MC68230L8
MC68230L10

Advance Information

HMOS
(HIGH-DENSITY N-CHANNEL SILICON-GATE)

PARALLEL INTERFACE/TIMER

MC68230 PARALLEL INTERFACE/TIMER

The MC68230 Parallel Interface/Timer provides versatile double buffered parallel interfaces and an operating system oriented timer to MC68000 systems. The parallel interfaces operate in unidirectional or bidirectional modes, either 8 or 16 bits wide. In the unidirectional modes, an associated data direction register determines whether the port pins are inputs or outputs. In the bidirectional modes the data direction registers are ignored and the direction is determined dynamically by the state of four handshake pins. These programmable handshake pins provide an interface flexible enough for connection to a wide variety of low, medium, or high speed peripherals or other computer systems. The PI/T ports allow use of vectored or autovectored interrupts, and also provide a DMA Request pin for connection to the MC68450 Direct Memory Access Controller or a similar circuit. The PI/T timer contains a 24-bit wide counter and a 5-bit prescaler. The timer may be clocked by the system clock (PI/T CLK pin) or by an external clock (TIN pin), and a 5-bit prescaler can be used. It can generate periodic interrupts, a square wave, or a single interrupt after a programmed time period. Also it can be used for elapsed time measurement or as a device watchdog.

- MC68000 Bus Compatible
- Port Modes Include:
  - Bit I/O
  - Unidirectonal 8-Bit and 16-Bit
  - Bidirectional 8-Bit and 16-Bit
- Selectable Handshaking Options
- 24-Bit Programmable Timer
- Software Programmable Timer Modes
- Contains Interrupt Vector Generation Logic
- Separate Port and Timer Interrupt Service Requests
- Registers are Read/Write and Directly Addressable
- Registers are Addressed for MOVEP (Move Peripheral) and DMAC Compatibility

L SUFFIX
CERAMIC PACKAGE
CASE 740

P SUFFIX
PLASTIC PACKAGE
AVAILABLE 2082

PIN ASSIGNMENT

Signal	Pin	Pin	Signal
D5	1	48	D4
D6	2	47	D3
D7	3	46	D2
PA0	4	45	D1
PA1	5	44	D0
PA2	6	43	R/$\overline{W}$
PA3	7	42	$\overline{DTACK}$
PA4	8	41	$\overline{CS}$
PA5	9	40	CLK
PA6	10	39	$\overline{RESET}$
PA7	11	38	V_{SS}
V_{CC}	12	37	PC7/$\overline{TIACK}$
H1	13	36	PC6/$\overline{PIACK}$
H2	14	35	PC5/$\overline{PIRQ}$
H3	15	34	PC4/$\overline{DMAREQ}$
H4	16	33	PC3/TOUT
PB0	17	32	PC2/TIN
PB1	18	31	PC1
PB2	19	30	PC0
PB3	20	29	RS1
PB4	21	28	RS2
PB5	22	27	RS3
PB6	23	26	RS4
PB7	24	25	RS5

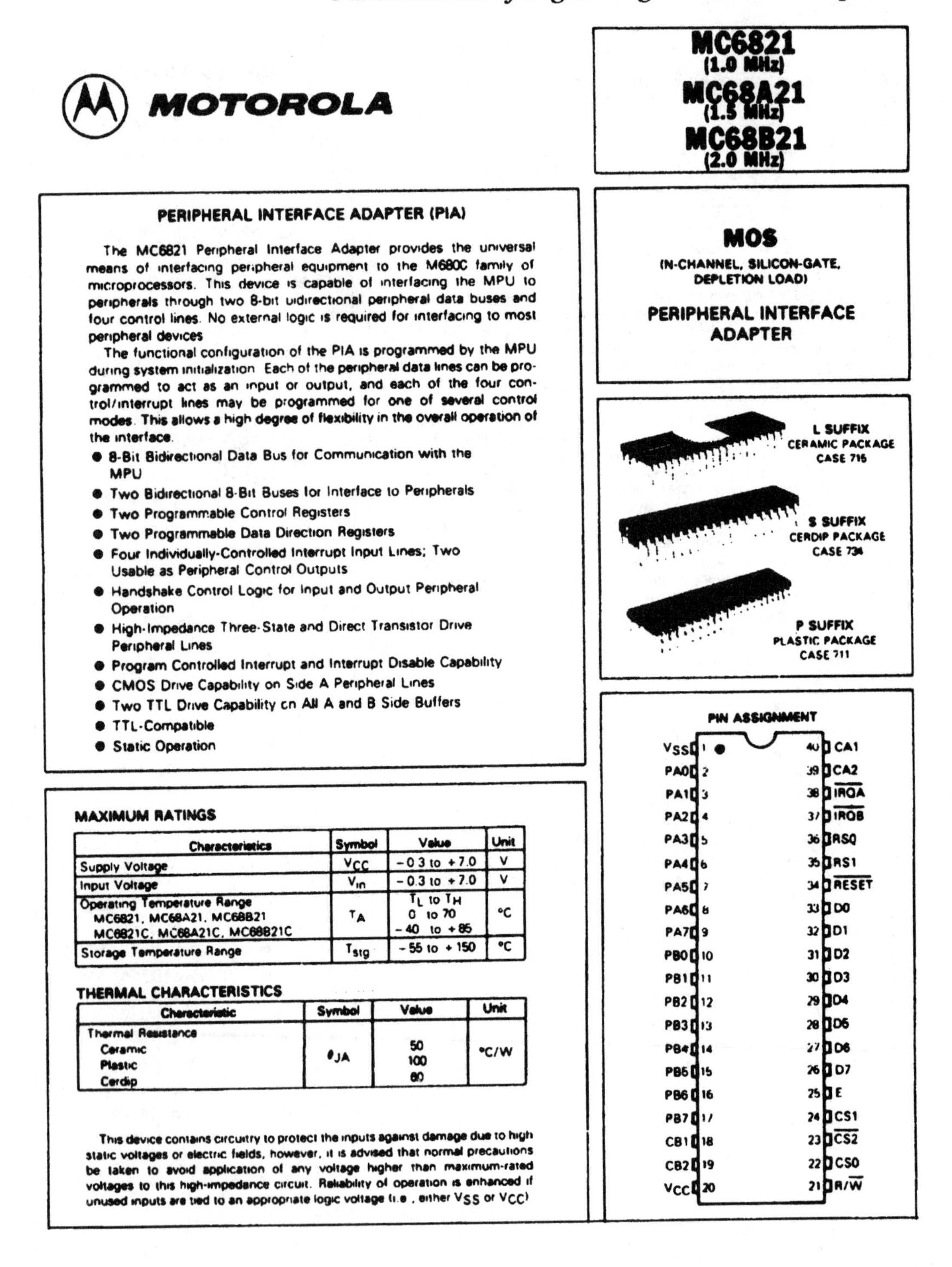

MOTOROLA

MC6821
(1.0 MHz)
MC68A21
(1.5 MHz)
MC68B21
(2.0 MHz)

PERIPHERAL INTERFACE ADAPTER (PIA)

The MC6821 Peripheral Interface Adapter provides the universal means of interfacing peripheral equipment to the M6800 family of microprocessors. This device is capable of interfacing the MPU to peripherals through two 8-bit bidirectional peripheral data buses and four control lines. No external logic is required for interfacing to most peripheral devices.

The functional configuration of the PIA is programmed by the MPU during system initialization. Each of the peripheral data lines can be programmed to act as an input or output, and each of the four control/interrupt lines may be programmed for one of several control modes. This allows a high degree of flexibility in the overall operation of the interface.

- 8-Bit Bidirectional Data Bus for Communication with the MPU
- Two Bidirectional 8-Bit Buses for Interface to Peripherals
- Two Programmable Control Registers
- Two Programmable Data Direction Registers
- Four Individually-Controlled Interrupt Input Lines; Two Usable as Peripheral Control Outputs
- Handshake Control Logic for Input and Output Peripheral Operation
- High-Impedance Three-State and Direct Transistor Drive Peripheral Lines
- Program Controlled Interrupt and Interrupt Disable Capability
- CMOS Drive Capability on Side A Peripheral Lines
- Two TTL Drive Capability on All A and B Side Buffers
- TTL-Compatible
- Static Operation

MOS
(N-CHANNEL, SILICON-GATE, DEPLETION LOAD)

PERIPHERAL INTERFACE ADAPTER

L SUFFIX
CERAMIC PACKAGE
CASE 715

S SUFFIX
CERDIP PACKAGE
CASE 734

P SUFFIX
PLASTIC PACKAGE
CASE 711

MAXIMUM RATINGS

Characteristics	Symbol	Value	Unit
Supply Voltage	V_{CC}	−0.3 to +7.0	V
Input Voltage	V_{in}	−0.3 to +7.0	V
Operating Temperature Range MC6821, MC68A21, MC68B21 MC6821C, MC68A21C, MC68B21C	T_A	T_L to T_H 0 to 70 −40 to +85	°C
Storage Temperature Range	T_{stg}	−55 to +150	°C

THERMAL CHARACTERISTICS

Characteristic	Symbol	Value	Unit
Thermal Resistance Ceramic Plastic Cerdip	θ_{JA}	 50 100 60	°C/W

This device contains circuitry to protect the inputs against damage due to high static voltages or electric fields; however, it is advised that normal precautions be taken to avoid application of any voltage higher than maximum-rated voltages to this high-impedance circuit. Reliability of operation is enhanced if unused inputs are tied to an appropriate logic voltage (i.e., either V_{SS} or V_{CC}).

PIN ASSIGNMENT

1	V_{SS}	40	CA1
2	PA0	39	CA2
3	PA1	38	$\overline{IRQA}$
4	PA2	37	$\overline{IRQB}$
5	PA3	36	RS0
6	PA4	35	RS1
7	PA5	34	$\overline{RESET}$
8	PA6	33	D0
9	PA7	32	D1
10	PB0	31	D2
11	PB1	30	D3
12	PB2	29	D4
13	PB3	28	D5
14	PB4	27	D6
15	PB5	26	D7
16	PB6	25	E
17	PB7	24	CS1
18	CB1	23	$\overline{CS2}$
19	CB2	22	CS0
20	V_{CC}	21	R/$\overline{W}$

Expanded block diagram of the MC6821

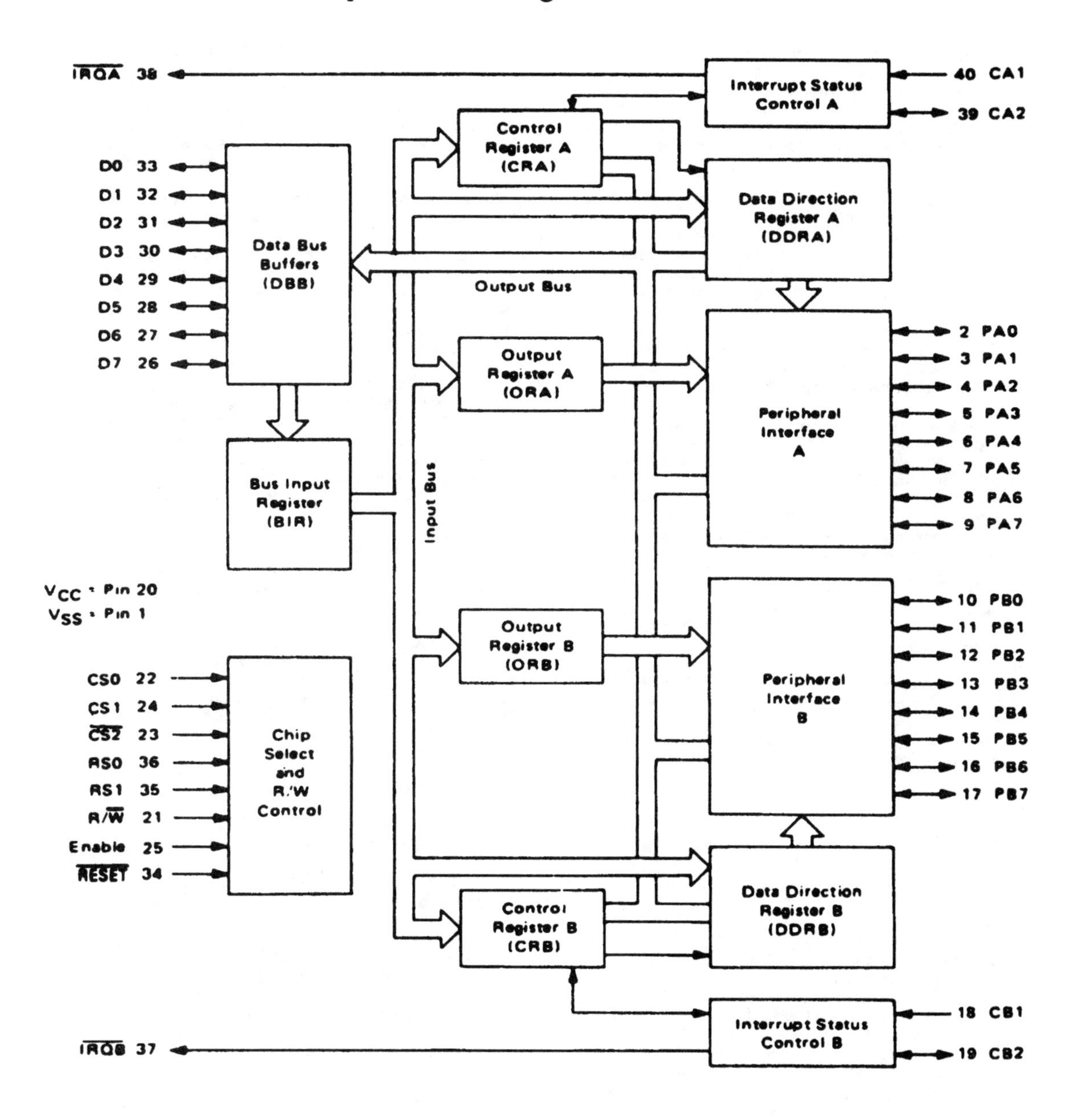

PIA INTERFACE SIGNALS FOR MPU

The PIA interfaces to the M6800 bus with an 8-bit bidirectional data bus, three chip select lines, two register select lines, two interrupt request lines, a read/write line, an enable line and a reset line. To ensure proper operation with the MC6800, MC6802, or MC6808 microprocessors, VMA should be used as an active part of the address decoding.

Bidirectional Data (D0-D7) – The bidirectional data lines (D0-D7) allow the transfer of data between the MPU and the PIA. The data bus output drivers are three-state devices that remain in the high-impedance (off) state except when the MPU performs a PIA read operation. The read/write line is in the read (high) state when the PIA is selected for a read operation.

Enable (E) – The enable pulse, E, is the only timing signal that is supplied to the PIA. Timing of all other signals is referenced to the leading and trailing edges of the E pulse.

Read/Write (R/$\overline{\text{W}}$) – This signal is generated by the MPU to control the direction of data transfers on the data bus. A low state on the PIA read/write line enables the input buffers and data is transferred from the MPU to the PIA on the E signal if the device has been selected. A high on the read/write line sets up the PIA for a transfer of data to the bus. The PIA output buffers are enabled when the proper address and the enable pulse E are present.

$\overline{\text{RESET}}$ – The active low $\overline{\text{RESET}}$ line is used to reset all register bits in the PIA to a logical zero (low). This line can be used as a power-on reset and as a master reset during system operation.

Chip Selects (CS0, CS1, and $\overline{\text{CS2}}$) – These three input signals are used to select the PIA. CS0 and CS1 must be high and $\overline{\text{CS2}}$ must be low for selection of the device. Data transfers are then performed under the control of the enable and read/write signals. The chip select lines must be stable for the duration of the E pulse. The device is deselected when any of the chip selects are in the inactive state.

Register Selects (RS0 and RS1) – The two register select lines are used to select the various registers inside the PIA. These two lines are used in conjunction with internal Control Registers to select a particular register that is to be written or read.

The register and chip select lines should be stable for the duration of the E pulse while in the read or write cycle.

Interrupt Request ($\overline{\text{IRQA}}$ and $\overline{\text{IRQB}}$) – The active low Interrupt Request lines ($\overline{\text{IRQA}}$ and $\overline{\text{IRQB}}$) act to interrupt the MPU either directly or through interrupt priority circuitry. These lines are "open drain" (no load device on the chip). This permits all interrupt request lines to be tied together in a wire-OR configuration.

Each Interrupt Request line has two internal interrupt flag bits that can cause the Interrupt Request line to go low. Each flag bit is associated with a particular peripheral interrupt line. Also, four interrupt enable bits are provided in the PIA which may be used to inhibit a particular interrupt from a peripheral device.

Servicing an interrupt by the MPU may be accomplished by a software routine that, on a prioritized basis, sequentially reads and tests the two control registers in each PIA for interrupt flag bits that are set.

The interrupt flags are cleared (zeroed) as a result of an MPU Read Peripheral Data Operation of the corresponding data register. After being cleared, the interrupt flag bit cannot be enabled to be set until the PIA is deselected during an E pulse. The E pulse is used to condition the interrupt control lines (CA1, CA2, CB1, CB2). When these lines are used as interrupt inputs, at least one E pulse must occur from the inactive edge to the active edge of the interrupt input signal to condition the edge sense network. If the interrupt flag has been enabled and the edge sense circuit has been properly conditioned, the interrupt flag will be set on the next active transition of the interrupt input pin.

PIA PERIPHERAL INTERFACE LINES

The PIA provides two 8-bit bidirectional data buses and four interrupt/control lines for interfacing to peripheral devices.

Section A Peripheral Data (PA0-PA7) – Each of the peripheral data lines can be programmed to act as an input or output. This is accomplished by setting a "1" in the corresponding Data Direction Register bit for those lines which are to be outputs. A "0" in a bit of the Data Direction Register causes the corresponding peripheral data line to act as an input. During an MPU Read Peripheral Data Operation, the data on peripheral lines programmed to act as inputs appears directly on the corresponding MPU Data Bus lines. In the input mode, the internal pullup resistor on these lines represents a maximum of 1.5 standard TTL loads.

The data in Output Register A will appear on the data lines that are programmed to be outputs. A logical "1" written into the register will cause a "high" on the corresponding data line while a "0" results in a "low." Data in Output Register A may be read by an MPU "Read Peripheral Data A" operation when the corresponding lines are programmed as outputs. This data will be read property if the voltage on the peripheral data lines is greater than 2.0 volts for a logic "1" output and less than 0.8 volt for a logic "0" output. Loading the output lines such that the voltage on these lines does not reach full voltage causes the data transferred into the MPU on a Read operation to differ from that contained in the respective bit of Output Register A.

Section B Peripheral Data (PB0-PB7) – The peripheral data lines in the B Section of the PIA can be programmed to act as either inputs or outputs in a similar manner to PA0-PA7. They have three-state capabiity, allowing them to enter a high-impedance state when the peripheral data line is used as an input. In addition, data on the peripheral data lines

PB0-PB7 will be read properly from those lines programmed as outputs even if the voltages are below 2.0 volts for a "high" or above 0.8 V for a "low". As outputs, these lines are compatible with standard TTL and may also be used as a source of up to 1 milliampere at 1.5 volts to directly drive the base of a transistor switch.

Interrupt Input (CA1 and CB1) – Peripheral input lines CA1 and CB1 are input only lines that set the interrupt flags of the control registers. The active transition for these signals is also programmed by the two control registers.

Peripheral Control (CA2) – The peripheral control line CA2 can be programmed to act as an interrupt input or as a peripheral control output. As an output, this line is compatible with standard TTL; as an input the internal pullup resistor on this line represents 1.5 standard TTL loads. The function of this signal line is programmed with Control Register A.

Peripheral Control (CB2) – Peripheral Control line CB2 may also be programmed to act as an interrupt input or peripheral control output. As an input, this line has high input impedance and is compatible with standard TTL. As an output it is compatible with standard TTL and may also be used as a source of up to 1 milliampere at 1.5 volts to directly drive the base of a transistor switch. This line is programmed by Control Register B.

INTERNAL CONTROLS

INITIALIZATION

A $\overline{\text{RESET}}$ has the effect of zeroing all PIA registers. This will set PA0-PA7, PB0-PB7, CA2 and CB2 as inputs, and all interrupts disabled. The PIA must be configured during the restart program which follows the reset.

There are six locations within the PIA accessible to the MPU data bus: two Peripheral Registers, two Data Direction Registers, and two Control Registers. Selection of these locations is controlled by the RS0 and RS1 inputs together with bit 2 in the Control Register, as shown in Table B.1

Details of possible configurations of the Data Direction and Control Register are as follows:

TABLE B.1 INTERNAL ADDRESSING

RS1	RS0	Control Register Bit CRA-2	Control Register Bit CRB-2	Location Selected
0	0	1	X	Peripheral Register A
0	0	0	X	Data Direction Register A
0	1	X	X	Control Register A
1	0	X	1	Peripheral Register B
1	0	X	0	Data Direction Register B
1	1	X	X	Control Register B

X = Don't Care

PORT A-B HARDWARE CHARACTERISTICS

As shown in Figure 1-7, the MC6821 has a pair of I/O ports whose characteristics differ greatly. The A side is designed to drive CMOS logic to normal 30% to 70% levels, and incorporates an internal pullup device that remains connected even in the input mode. Because of this, the A side requires more drive current in the input mode than Port B. In contrast, the B side uses a normal three-state NMOS buffer which cannot pullup to CMOS levels without external resistors. The B side can drive extra loads such as Darlingtons without problem. When the PIA comes out of reset, the A port represents inputs with pullup resistors, whereas the B side (input mode also) will float high or low, depending upon the load connected to it.

Notice the differences between a Port A and Port B read operation when in the output mode. When reading Port A, the actual pin is read, whereas the B side read comes from an output latch, ahead of the actual pin.

CONTROL REGISTERS (CRA and CRB)

The two Control Registers (CRA and CRB) allow the MPU to control the operation of the four peripheral control lines CA1, CA2, CB1, and CB2. In addition they allow the MPU to enable the interrupt lines and monitor the status of the interrupt flags. Bits 0 through 5 of the two registers may be written or read by the MPU when the proper chip select and register select signals are applied. Bits 6 and 7 of the two registers are read only and are modified by external interrupts occurring on control lines CA1, CA2, CB1, or CB2. The format of the control words is shown in Figure B.3

DATA DIRECTION ACCESS CONTROL BIT (CRA-2 and CRB-2)

Bit 2, in each Control Register (CRA and CRB), determines selection of either a Peripheral Output Register or the corresponding Data Direction E Register when the proper register select signals are applied to RS0 and RS1. A "1" in bit 2 allows access of the Peripheral Interface Register, while a "0" causes the Data Direction Register to be addressed.

Interrupt Flags (CRA-6, CRA-7, CRB-6, and CRB-7) – The four interrupt flag bits are set by active transitions of signals on the four Interrupt and Peripheral Control lines when those lines are programmed to be inputs. These bits cannot be set directly from the MPU Data Bus and are reset indirectly by a Read Peripheral Data Operation on the appropriate section.

Control of CA2 and CB2 Peripheral Control Lines (CRA-3, CRA-4, CRA-5, CRB-3, CRB-4, and CRB-5) – Bits 3, 4, and 5 of the two control registers are used to control the CA2 and CB2 Peripheral Control lines. These bits determine if the control lines will be an interrupt input or an output control signal. If bit CRA-5 (CRB-5) is low, CA2 (CB2) is an interrupt input line similar to CA1 (CB1). When CRA-5 (CRB-5) is high, CA2 (CB2) becomes an output signal that may be used to control peripheral data transfers. When in the output mode, CA2 and CB2 have slightly different loading characteristics.

Control of CA1 and CB1 Interrupt Input Lines (CRA-0, CRB-1, CRA-1, and CRB-1) – The two lowest-order bits of the control registers are used to control the interrupt input lines CA1 and CB1. Bits CRA-0 and CRB-0 are used to enable the MPU interrupt signals $\overline{IRQA}$ and $\overline{IRQB}$, respectively. Bits CRA-1 and CRB-1 determine the active transition of the interrupt input signals CA1 and CB1.

FIGURE B.2 PORT A AND PORT B EQUIVALENT CIRCUITS

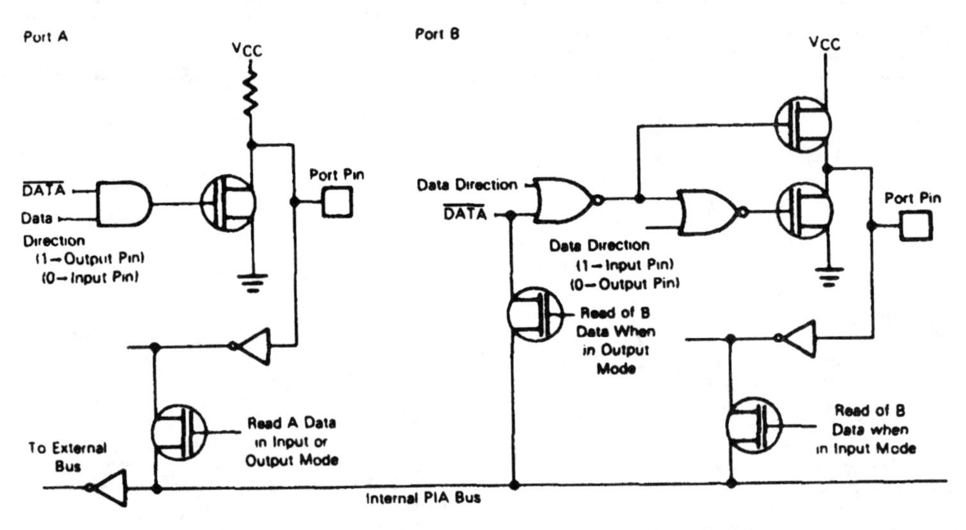

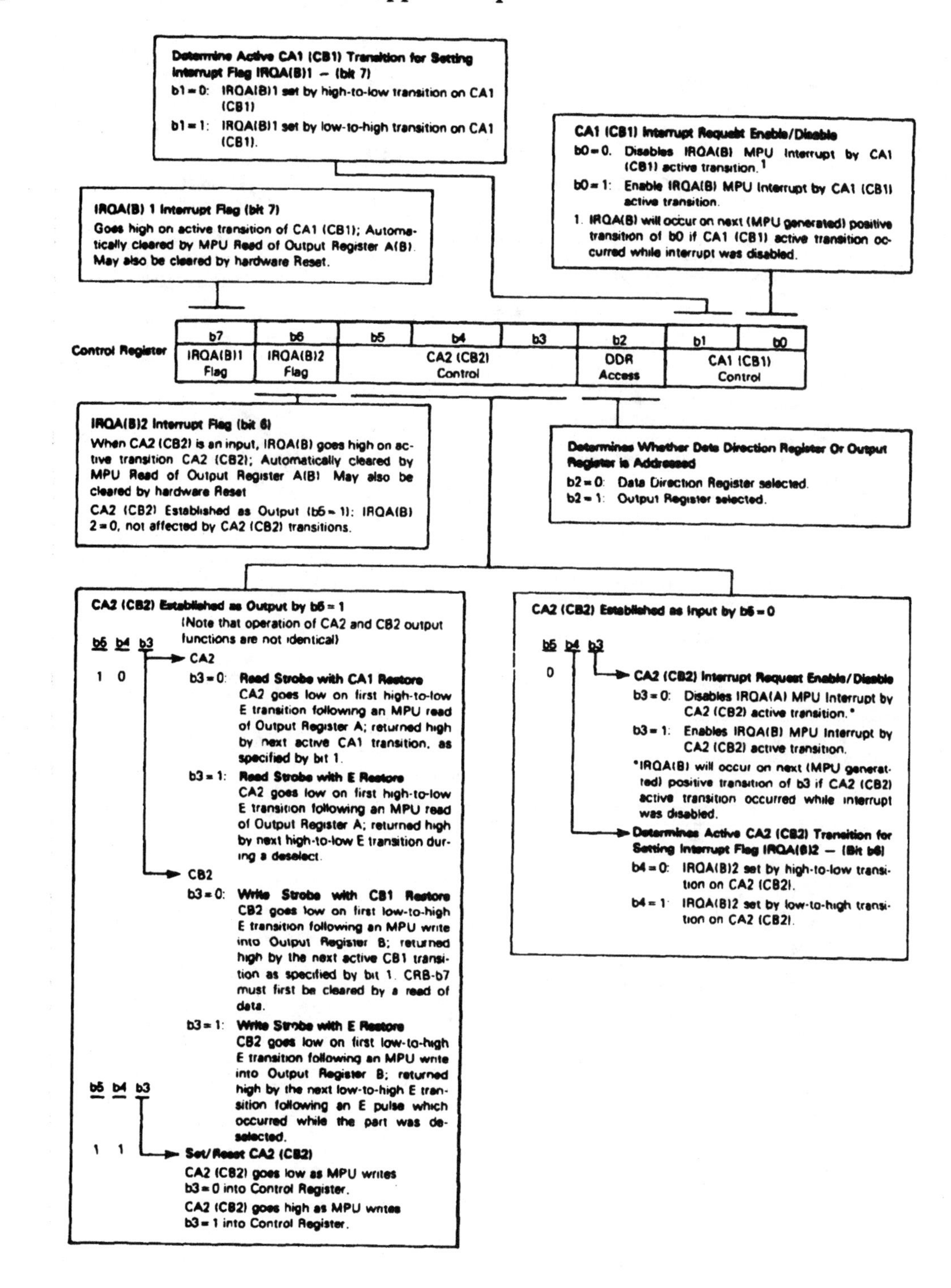
Determine Active CA1 (CB1) Transition for Setting Interrupt Flag IRQA(B)1 — (bit 7)
b1 = 0: IRQA(B)1 set by high-to-low transition on CA1 (CB1)
b1 = 1: IRQA(B)1 set by low-to-high transition on CA1 (CB1).
CA1 (CB1) Interrupt Request Enable/Disable
b0 = 0. Disables IRQA(B) MPU Interrupt by CA1 (CB1) active transition.1
b0 = 1: Enable IRQA(B) MPU Interrupt by CA1 (CB1) active transition.
1. IRQA(B) will occur on next (MPU generated) positive transition of b0 if CA1 (CB1) active transition occurred while interrupt was disabled.
IRQA(B) 1 Interrupt Flag (bit 7)
Goes high on active transition of CA1 (CB1); Automatically cleared by MPU Read of Output Register A(B). May also be cleared by hardware Reset.
Control Register
b7 b6 b5 b4 b3 b2 b1 b0
IRQA(B)1 Flag
IRQA(B)2 Flag
CA2 (CB2) Control
DDR Access
CA1 (CB1) Control
IRQA(B)2 Interrupt Flag (bit 6)
When CA2 (CB2) is an input, IRQA(B) goes high on active transition CA2 (CB2); Automatically cleared by MPU Read of Output Register A(B) May also be cleared by hardware Reset
CA2 (CB2) Established as Output (b5 = 1): IRQA(B) 2 = 0, not affected by CA2 (CB2) transitions.
Determines Whether Data Direction Register Or Output Register is Addressed
b2 = 0: Data Direction Register selected.
b2 = 1: Output Register selected.
CA2 (CB2) Established as Output by b5 = 1
(Note that operation of CA2 and CB2 output functions are not identical)
b5 b4 b3
1 0
CA2
b3 = 0: Read Strobe with CA1 Restore
CA2 goes low on first high-to-low E transition following an MPU read of Output Register A; returned high by next active CA1 transition, as specified by bit 1.
b3 = 1: Read Strobe with E Restore
CA2 goes low on first high-to-low E transition following an MPU read of Output Register A; returned high by next high-to-low E transition during a deselect.
CB2
b3 = 0: Write Strobe with CB1 Restore
CB2 goes low on first low-to-high E transition following an MPU write into Output Register B; returned high by the next active CB1 transition as specified by bit 1. CRB-b7 must first be cleared by a read of data.
b3 = 1: Write Strobe with E Restore
CB2 goes low on first low-to-high E transition following an MPU write into Output Register B; returned high by the next low-to-high E transition following an E pulse which occurred while the part was deselected.
b5 b4 b3
1 1
Set/Reset CA2 (CB2)
CA2 (CB2) goes low as MPU writes b3 = 0 into Control Register.
CA2 (CB2) goes high as MPU writes b3 = 1 into Control Register.
CA2 (CB2) Established as Input by b5 = 0
b5 b4 b3
0
CA2 (CB2) Interrupt Request Enable/Disable
b3 = 0: Disables IRQA(A) MPU Interrupt by CA2 (CB2) active transition.*
b3 = 1: Enables IRQA(B) MPU Interrupt by CA2 (CB2) active transition.
*IRQA(B) will occur on next (MPU generated) positive transition of b3 if CA2 (CB2) active transition occurred while interrupt was disabled.
Determines Active CA2 (CB2) Transition for Setting Interrupt Flag IRQA(B)2 — (Bit b6)
b4 = 0: IRQA(B)2 set by high-to-low transition on CA2 (CB2).
b4 = 1: IRQA(B)2 set by low-to-high transition on CA2 (CB2).

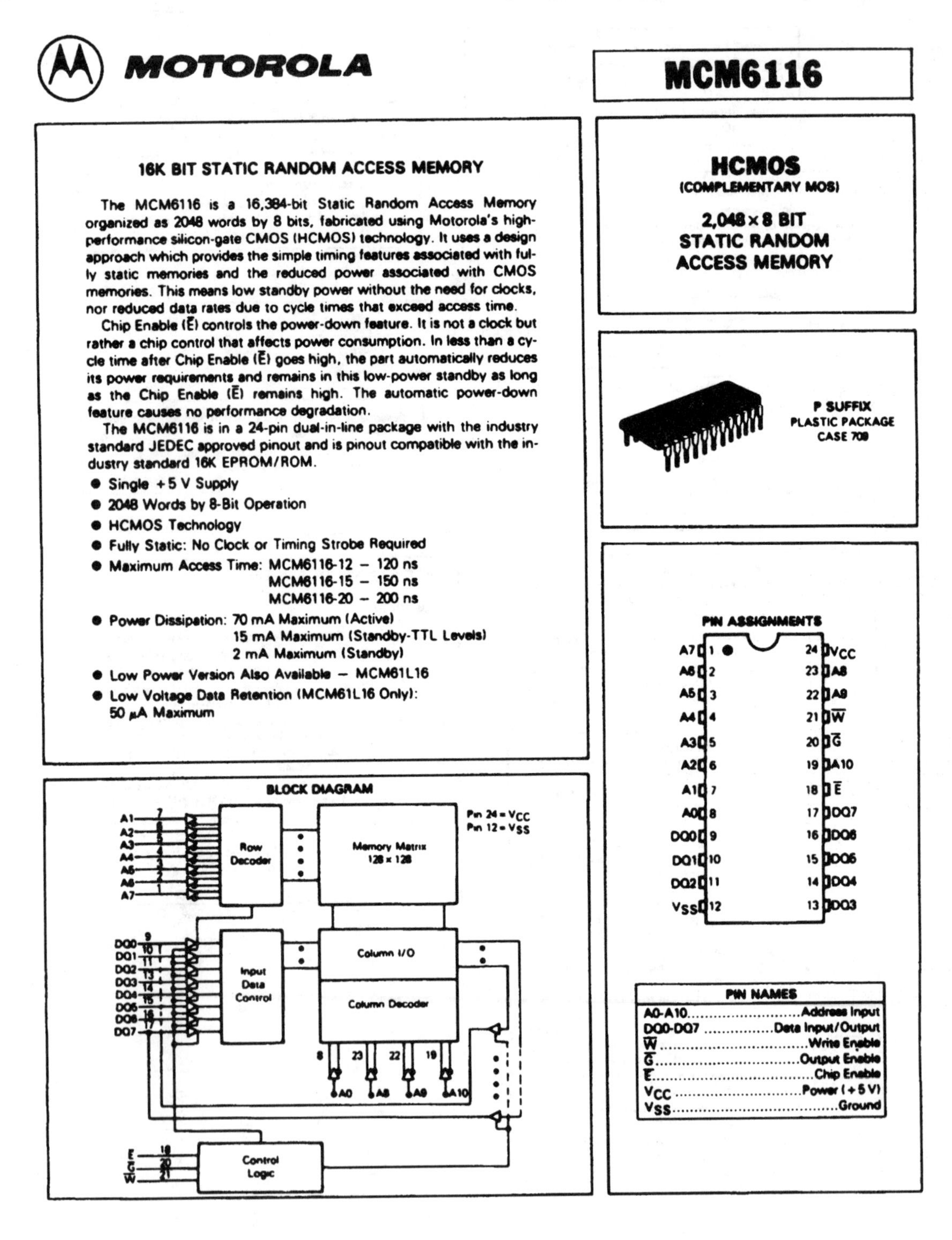

MOTOROLA

MCM6116

HCMOS
(COMPLEMENTARY MOS)

2,048 × 8 BIT STATIC RANDOM ACCESS MEMORY

16K BIT STATIC RANDOM ACCESS MEMORY

The MCM6116 is a 16,384-bit Static Random Access Memory organized as 2048 words by 8 bits, fabricated using Motorola's high-performance silicon-gate CMOS (HCMOS) technology. It uses a design approach which provides the simple timing features associated with fully static memories and the reduced power associated with CMOS memories. This means low standby power without the need for clocks, nor reduced data rates due to cycle times that exceed access time.

Chip Enable ($\overline{E}$) controls the power-down feature. It is not a clock but rather a chip control that affects power consumption. In less than a cycle time after Chip Enable ($\overline{E}$) goes high, the part automatically reduces its power requirements and remains in this low-power standby as long as the Chip Enable ($\overline{E}$) remains high. The automatic power-down feature causes no performance degradation.

The MCM6116 is in a 24-pin dual-in-line package with the industry standard JEDEC approved pinout and is pinout compatible with the industry standard 16K EPROM/ROM.

- Single +5 V Supply
- 2048 Words by 8-Bit Operation
- HCMOS Technology
- Fully Static: No Clock or Timing Strobe Required
- Maximum Access Time: MCM6116-12 – 120 ns
  MCM6116-15 – 150 ns
  MCM6116-20 – 200 ns
- Power Dissipation: 70 mA Maximum (Active)
  15 mA Maximum (Standby-TTL Levels)
  2 mA Maximum (Standby)
- Low Power Version Also Available – MCM61L16
- Low Voltage Data Retention (MCM61L16 Only):
  50 μA Maximum

P SUFFIX
PLASTIC PACKAGE
CASE 709

PIN ASSIGNMENTS

1	A7	24	V_{CC}
2	A6	23	A8
3	A5	22	A9
4	A4	21	$\overline{W}$
5	A3	20	$\overline{G}$
6	A2	19	A10
7	A1	18	$\overline{E}$
8	A0	17	DQ7
9	DQ0	16	DQ6
10	DQ1	15	DQ5
11	DQ2	14	DQ4
12	V_{SS}	13	DQ3

PIN NAMES

A0-A10	Address Input
DQ0-DQ7	Data Input/Output
$\overline{W}$	Write Enable
$\overline{G}$	Output Enable
$\overline{E}$	Chip Enable
V_{CC}	Power (+5 V)
V_{SS}	Ground

BLOCK DIAGRAM

ABSOLUTE MAXIMUM RATINGS (See Note)

Rating	Value	Unit
Temperature Under Bias	−10 to +80	°C
Voltage on Any Pin With Respect to V_{SS}	−1.0 to +7.0	V
DC Output Current	20	mA
Power Dissipation	1.2	Watt
Operating Temperature Range	0 to +70	°C
Storage Temperature Range	−65 to +150	°C

This device contains circuitry to protect the inputs against damage due to high static voltages or electric fields, however, it is advised that normal precautions be taken to avoid application of any voltage higher than maximum rated voltages to this high-impedance circuit.

NOTE: Permanent device damage may occur if ABSOLUTE MAXIMUM RATINGS are exceeded. Functional operation should be restricted to RECOMMENDED OPERATING CONDITIONS. Exposure to higher than recommended voltages for extended periods of time could affect device reliability.

DC OPERATING CONDITIONS AND CHARACTERISTICS

(Full operating voltage and temperature ranges unless otherwise noted.)

RECOMMENDED OPERATING CONDITIONS

Parameter	Symbol	Min	Typ	Max	Unit
Supply Voltage	V_{CC}	4.5	5.0	5.5	V
	V_{SS}	0	0	0	V
Input Voltage	V_{IH}	2.2	3.5	6.0	V
	V_{IL}	−1.0*	−	0.8	V

*The device will withstand undershoots to the −1.0 volt level with a maximum pulse width of 50 ns at the −0.3 volt level. This is periodically sampled rather than 100% tested.

RECOMMENDED OPERATING CHARACTERISTICS

Parameter	Symbol	MCM6116			MCM61L16			Unit
		Min	Typ*	Max	Min	Typ*	Max	
Input Leakage Current (V_{CC} = 5.5 V, V_{in} = GND to V_{CC})	$\|I_{LI}\|$	−	−	1	−	−	1	μA
Output Leakage Current ($\overline{E}$ = V_{IH} or $\overline{G}$ = V_{IH} $V_{I/O}$ = GND to V_{CC})	$\|I_{LO}\|$	−	−	1	−	−	1	μA
Operating Power Supply Current ($\overline{E}$ = V_{IL}, $I_{I/O}$ = 0 mA)	I_{CC}	−	35	70	−	35	55	mA
Average Operating Current Minimum cycle, duty = 100%	I_{CC2}	−	35	70	−	35	55	mA
Standby Power Supply Current ($\overline{E}$ = V_{IH})	I_{SB}	−	5	15	−	5	12	mA
Standby Power Supply Current ($\overline{E}$ ≥ V_{CC} − 0.2 V, V_{in} ≥ V_{CC} − 0.2 V or V_{in} ≤ 0.2 V)	I_{SB1}	−	20	2000	−	4	100	μA
Output Low Voltage (I_{OL} = 2.1 mA)	V_{OL}	−	−	0.4	−	−	0.4	V
Output High Voltage (I_{OH} = −1.0 mA)**	V_{OH}	2.4	−	−	2.4	−	−	V

*V_{CC} = 5 V, T_A = 25°C

**Also, output voltages are compatible with Motorola's new high-speed CMOS logic family if the same power supply voltage is used.

CAPACITANCE (f = 1.0 MHz, T_A = 25°C, periodically sampled rather than 100% tested.)

Characteristic	Symbol	Typ	Max	Unit
Input Capacitance except $\overline{E}$	C_{in}	3	5	pF
Input/Output Capacitance and $\overline{E}$ Input Capacitance	$C_{I/O}$	5	7	pF

MODE SELECTION

Mode	$\overline{E}$	$\overline{G}$	$\overline{W}$	V_{CC} Current	DQ
Standby	H	X	X	I_{SB}, I_{SB1}	High Z
Read	L	L	H	I_{CC}	Q
Write Cycle (1)	L	H	L	I_{CC}	D
Write Cycle (2)	L	L	L	I_{CC}	D

AC OPERATING CONDITIONS AND CHARACTERISTICS

(Full operating voltage and temperature unless otherwise noted.)

Input Pulse Levels .... 0 Volt to 3.5 Volts
Input Rise and Fall Times .... 10 ns
Input and Output Timing Reference Levels .... 1.5 Volts
Output Load .... 1 TTL Gate and C_L = 100 pF

READ CYCLE

Parameter	Symbol	MCM6116-12 MCM61L16-12		MCM6116-15 MCM61L16-15		MCM6116-20 MCM61L16-20		Unit
		Min	Max	Min	Max	Min	Max	
Address Valid to Address Don't Care (Cycle Time when Chip Enable is Held Active)	t_{AVAX}	120	–	150	–	200	–	ns
Chip Enable Low to Chip Enable High	t_{ELEH}	120	–	150	–	200	–	ns
Address Valid to Output Valid (Access)	t_{AVQV}	–	120	–	150	–	200	ns
Chip Enable Low to Output Valid (Access)	t_{ELQV}	–	120	–	150	–	200	ns
Address Valid to Output Invalid	t_{AVQX}	10	–	15	–	15	–	ns
Chip Enable Low to Output Invalid	t_{ELQX}	10	–	15	–	15	–	ns
Chip Enable High to Output High Z	t_{EHQZ}	0	40	0	50	0	60	ns
Output Enable to Output Valid	t_{GLQV}	–	80	–	100	–	120	ns
Output Enable to Output Invalid	t_{GLQX}	10	–	15	–	15	–	ns
Output Enable to Output High Z	t_{GLQZ}	0	40	0	50	0	60	ns
Address Invalid to Output Invalid	t_{AXQX}	10	–	15	–	15	–	ns
Address Valid to Chip Enable Low (Address Setup)	t_{AVEL}	0	–	0	–	0	–	ns
Chip Enable to Power-Up Time	t_{PU}	0	–	0	–	0	–	ns
Chip Disable to Power-Down Time	t_{PD}	–	30	–	30	–	30	ns

WRITE CYCLE

Parameter	Symbol	MCM6116-12 MCM61L16-12		MCM6116-15 MCM61L16-15		MCM6116-20 MCM61L16-20		Unit
		Min	Max	Min	Max	Min	Max	
Chip Enable Low to Write High	t_{ELWH}	70	–	90	–	120	–	ns
Address Valid to Write High	t_{AVWH}	105	–	120	–	140	–	ns
Address Valid to Write Low (Address Setup)	t_{AVWL}	20	–	20	–	20	–	ns
Write Low to Write High (Write Pulse Width)	t_{WLWH}	70	–	90	–	120	–	ns
Write High to Address Don't Care	t_{WHAX}	5	–	10	–	10	–	ns
Data Valid to Write High	t_{DVWH}	35	–	40	–	60	–	ns
Write High to Data Don't Care (Data Hold)	t_{WHDX}	5	–	10	–	10	–	ns
Write Low to Output High Z	t_{WLQZ}	0	50	0	60	0	60	ns
Write High to Output Valid	t_{WHQV}	5	–	10	–	10	–	ns
Output Disable to Output High Z	t_{GHQZ}	0	40	0	50	0	60	ns

TIMING PARAMETER ABBREVIATIONS

t X X X X

- signal name from which interval is defined (first X)
- transition direction for first signal (second X)
- signal name to which interval is defined (third X)
- transition direction for second signal (fourth X)

The transition definitions used in this data sheet are:

H = transition to high
L = transition to low
V = transition to valid
X = transition to invalid or don't care
Z = transition to off (high impedance)

TIMING LIMITS

The table of timing values shows either a minimum or a maximum limit for each parameter. Input requirements are specified from the external system point of view. Thus, address setup time is shown as a minimum since the system must supply at least that much time (even though most devices do not require it). On the other hand, responses from the memory are specified from the device point of view. Thus, the access time is shown as a maximum since the device never provides data later than that time.

APPENDIX D

68000 EXECUTION TIMES

D.1 INTRODUCTION

This Appendix contains listings of the instruction execution times in terms of external clock (CLK) periods. In this data, it is assumed that both memory read and write cycle times are four clock periods. A longer memory cycle will cause the generation of wait states which must be added to the total instruction time.

The number of bus read and write cycles for each instruction is also included with the timing data. This data is enclosed in parenthesis following the number of clock periods and is shown as: (r/w) where r is the number of read cycles and w is the number of write cycles included in the clock period number. Recalling that either a read or write cycle requires four clock periods, a timing number given as 18(3/1) relates to 12 clock periods for the three read cycles, plus 4 clock periods for the one write cycle, plus 2 cycles required for some internal function of the processor.

NOTE

The number of periods includes instruction fetch and all applicable operand fetches and stores.

D.2 OPERAND EFFECTIVE ADDRESS CALCULATION TIMING

Table D-1 lists the number of clock periods required to compute an instruction's effective address. It includes fetching of any extension words, the address computation, and fetching of the memory operand. The number of bus read and write cycles is shown in parenthesis as (r/w). Note there are no write cycles involved in processing the effective address.

Table D-1. Effective Address Calculation Times

Addressing Mode		Byte, Word	Long
	Register		
Dn	Data Register Direct	0(0/0)	0(0/0)
An	Address Register Direct	0(0/0)	0(0/0)
	Memory		
(An)	Address Register Indirect	4(1/0)	8(2/0)
(An)+	Address Register Indirect with Postincrement	4(1/0)	8(2/0)
−(An)	Address Register Indirect with Predecrement	6(1/0)	10(2/0)
d(An)	Address Register Indirect with Displacement	8(2/0)	12(3/0)
d(An, ix)*	Address Register Indirect with Index	10(2/0)	14(3/0)
xxx.W	Absolute Short	8(2/0)	12(3/0)
xxx.L	Absolute Long	12(3/0)	16(4/0)
d(PC)	Program Counter with Displacement	8(2/0)	12(3/0)
d(PC, ix)*	Program Counter with Index	10(2/0)	14(3/0)
#xxx	Immediate	4(1/0)	8(2/0)

*The size of the index register (ix) does not affect execution time.

D.3 MOVE INSTRUCTION EXECUTION TIMES

Tables D-2 and D-3 indicate the number of clock periods for the move instruction. This data includes instruction fetch, operand reads, and operand writes. The number of bus read and write cycles is shown in parenthesis as (r/w).

Table D-2. Move Byte and Word Instruction Execution Times

Source	Destination								
	Dn	An	(An)	(An)+	−(An)	d(An)	d(An, ix)*	xxx.W	xxx.L
Dn	4(1/0)	4(1/0)	8(1/1)	8(1/1)	8(1/1)	12(2/1)	14(2/1)	12(2/1)	16(3/1)
An	4(1/0)	4(1/0)	8(1/1)	8(1/1)	8(1/1)	12(2/1)	14(2/1)	12(2/1)	16(3/1)
(An)	8(2/0)	8(2/0)	12(2/1)	12(2/1)	12(2/1)	16(3/1)	18(3/1)	16(3/1)	20(4/1)
(An)+	8(2/0)	8(2/0)	12(2/1)	12(2/1)	12(2/1)	16(3/1)	18(3/1)	16(3/1)	20(4/1)
−(An)	10(2/0)	10(2/0)	14(2/1)	14(2/1)	14(2/1)	18(3/1)	20(3/1)	18(3/1)	22(4/1)
d(An)	12(3/0)	12(3/0)	16(3/1)	16(3/1)	16(3/1)	20(4/1)	22(4/1)	20(4/1)	24(5/1)
d(An, ix)*	14(3/0)	14(3/0)	18(3/1)	18(3/1)	18(3/1)	22(4/1)	24(4/1)	22(4/1)	26(5/1)
xxx.W	12(3/0)	12(3/0)	16(3/1)	16(3/1)	16(3/1)	20(4/1)	22(4/1)	20(4/1)	24(5/1)
xxx.L	16(4/0)	16(4/0)	20(4/1)	20(4/1)	20(4/1)	24(5/1)	26(5/1)	24(5/1)	28(6/1)
d(PC)	12(3/0)	12(3/0)	16(3/1)	16(3/1)	16(3/1)	20(4/1)	22(4/1)	20(4/1)	24(5/1)
d(PC, ix)*	14(3/0)	14(3/0)	18(3/1)	18(3/1)	18(3/1)	22(4/1)	24(4/1)	22(4/1)	26(5/1)
#xxx	8(2/0)	8(2/0)	12(2/1)	12(2/1)	12(2/1)	16(3/1)	18(3/1)	16(3/1)	20(4/1)

*The size of the index register (ix) does not affect execution time.

Table D-3. Move Long Instruction Execution Times

Source	Destination								
	Dn	An	(An)	(An)+	−(An)	d(An)	d(An, ix)*	xxx.W	xxx.L
Dn	4(1/0)	4(1/0)	12(1/2)	12(1/2)	12(1/2)	16(2/2)	18(2/2)	16(2/2)	20(3/2)
An	4(1/0)	4(1/0)	12(1/2)	12(1/2)	12(1/2)	16(2/2)	18(2/2)	16(2/2)	20(3/2)
(An)	12(3/0)	12(3/0)	20(3/2)	20(3/2)	20(3/2)	24(4/2)	26(4/2)	24(4/2)	28(5/2)
(An)+	12(3/0)	12(3/0)	20(3/2)	20(3/2)	20(3/2)	24(4/2)	26(4/2)	24(4/2)	28(5/2)
−(An)	14(3/0)	14(3/0)	22(3/2)	22(3/2)	22(3/2)	26(4/2)	28(4/2)	26(4/2)	30(5/2)
d(An)	16(4/0)	16(4/0)	24(4/2)	24(4/2)	24(4/2)	28(5/2)	30(5/2)	28(5/2)	32(6/2)
d(An, ix)*	18(4/0)	18(4/0)	26(4/2)	26(4/2)	26(4/2)	30(5/2)	32(5/2)	30(5/2)	34(6/2)
xxx.W	16(4/0)	16(4/0)	24(4/2)	24(4/2)	24(4/2)	28(5/2)	30(5/2)	28(5/2)	32(6/2)
xxx.L	20(5/0)	20(5/0)	28(5/2)	28(5/2)	28(5/2)	32(6/2)	34(6/2)	32(6/2)	36(7/2)
d(PC)	16(4/0)	16(4/0)	24(4/2)	24(4/2)	24(4/2)	28(5/2)	30(5/2)	28(5/2)	32(5/2)
d(PC, ix)*	18(4/0)	18(4/0)	26(4/2)	26(4/2)	26(4/2)	30(5/2)	32(5/2)	30(5/2)	34(6/2)
#xxx	12(3/0)	12(3/0)	20(3/2)	20(3/2)	20(3/2)	24(4/2)	26(4/2)	24(4/2)	28(5/2)

*The size of the index register (ix) does not affect execution time.

D.4 STANDARD INSTRUCTION EXECUTION TIMES

The number of clock periods shown in Table D-4 indicates the time required to perform the operations, store the results, and read the next instruction. The number of bus read and write cycles is shown in parenthesis as (r/w). The number of clock periods and the number of read and write cycles must be added respectively to those of the effective address calculation where indicated.

In Table D-4 the headings have the following meanings: An = address register operand, Dn = data register operand, ea = an operand specified by an effective address, and M = memory effective address operand.

Table D-4. Standard Instruction Execution Times

Instruction	Size	op<ea>, An†	op<ea>, Dn	op Dn, <M>
ADD	Byte, Word	8(1/0)+	4(1/0)+	8(1/1)+
	Long	6(1/0)+ **	6(1/0)+ **	12(1/2)+
AND	Byte, Word	–	4(1/0)+	8(1/1)+
	Long	–	6(1/0)+ **	12(1/2)+
CMP	Byte, Word	6(1/0)+	4(1/0)+	–
	Long	6(1/0)+	6(1/0)+	–
DIVS	–	–	158(1/0)+ *	–
DIVU	–	–	140(1/0)+ *	–
EOR	Byte, Word	–	4(1/0)***	8(1/1)+
	Long	–	8(1/0)***	12(1/2)+
MULS	–	–	70(1/0)+ *	–
MULU	–	–	70(1/0)+ *	–
OR	Byte, Word	–	4(1/0)+	8(1/1)+
	Long	–	8(1/0)+ **	12(1/2)+
SUB	Byte, Word	8(1/0)+	4(1/0)+	8(1/1)+
	Long	6(1/0)+ **	6(1/0)+ **	12(1/2)+

NOTES:

+ add effective address calculation time

† word or long only

* indicates maximum value

** The base time of six clock periods is increased to eight if the effective address mode is register direct or immediate (effective address time should also be added).

*** Only available effective address mode is data register direct.

DIVS, DIVU – The divide algorithm used by the MC68000 provides less than 10% difference between the best and worst case timings.

MULS, MULU – The multiply algorithm requires 38+2n clocks where n is defined as:

MULU: n = the number of ones in the <ea>

MULS: n = concatanate the <ea> with a zero as the LSB; n is the resultant number of 10 or 01 patterns in the 17-bit source; i.e., worst case happens when the source is $5555.

D.5 IMMEDIATE INSTRUCTION EXECUTION TIMES

The number of clock periods shown in Table D-5 includes the time to fetch immediate operands, perform the operations, store the results, and read the next operation. The number of bus read and write cycles is shown in parenthesis as (r/w). The number of clock periods and the number of read and write cycles must be added respectively to those of the effective address calculation where indicated.

In Table D-5, the headings have the following meanings: # = immediate operand, Dn = data register operand, An = address register operand, and M = memory operand. SR = status register.

Table D-5. Immediate Instruction Execution Times

Instruction	Size	op #, Dn	op #, An	op #, M
ADDI	Byte, Word	8(2/0)	–	12(2/1) +
	Long	16(3/0)	–	20(3/2) +
ADDQ	Byte, Word	4(1/0)	8(1/0)*	8(1/1) +
	Long	8(1/0)	8(1/0)	12(1/2) +
ANDI	Byte, Word	8(2/0)	–	12(2/1) +
	Long	16(3/0)	–	20(3/1) +
CMPI	Byte, Word	8(2/0)	–	8(2/0) +
	Long	14(3/0)	–	12(3/0) +
EORI	Byte, Word	8(2/0)	–	12(2/1) +
	Long	16(3/0)	–	20(3/2) +
MOVEQ	Long	4(1/0)	–	–
ORI	Byte, Word	8(2/0)	–	12(2/1) +
	Long	16(3/0)	–	20(3/2) +
SUBI	Byte, Word	8(2/0)	–	12(2/1) +
	Long	16(3/0)	–	20(3/2) +
SUBQ	Byte, Word	4(1/0)	8(1/0)*	8(1/1) +
	Long	8(1/0)	8(1/0)	12(1/2) +

+ add effective address calculation time
*word only

D.6 SINGLE OPERAND INSTRUCTION EXECUTION TIMES

Table D-6 indicates the number of clock periods for the single operand instructions. The number of bus read and write cycles is shown in parenthesis as (r/w). The number of clock periods and the number of read and write cycles must be added respectively to those of the effective address calculation where indicated.

Table D-6. Single Operand Instruction Execution Times

Instruction	Size	Register	Memory
CLR	Byte, Word	4(1/0)	8(1/1)+
	Long	6(1/0)	12(1/2)+
NBCD	Byte	6(1/0)	8(1/1)+
NEG	Byte, Word	4(1/0)	8(1/1)+
	Long	6(1/0)	12(1/2)+
NEGX	Byte, Word	4(1/0)	8(1/1)+
	Long	6(1/0)	12(1/2)+
NOT	Byte, Word	4(1/0)	8(1/1)+
	Long	6(1/0)	12(1/2)+
Scc	Byte, False	4(1/0)	8(1/1)+
	Byte, True	6(1/0)	8(1/1)+
TAS	Byte	4(1/0)	10(1/1)+
TST	Byte, Word	4(1/0)	4(1/0)+
	Long	4(1/0)	4(1/0)+

+ add effective address calculation time

D.7 SHIFT/ROTATE INSTRUCTION EXECUTION TIMES

Table D-7 indicates the number of clock periods for the shift and rotate instructions. The number of bus read and write cycles is shown in parenthesis as (r/w). The number of clock periods and the number of read and write cycles must be added respectively to those of the effective address calculation where indicated.

Table D-7. Shift/Rotate Instruction Execution Times

Instruction	Size	Register	Memory
ASR, ASL	Byte, Word	6 + 2n(1/0)	8(1/1)+
	Long	8 + 2n(1/0)	–
LSR, LSL	Byte, Word	6 + 2n(1/0)	8(1/1)+
	Long	8 + 2n(1/0)	–
ROR, ROL	Byte, Word	6 + 2n(1/0)	8(1/1)+
	Long	8 + 2n(1/0)	–
ROXR, ROXL	Byte, Word	6 + 2n(1/0)	8(1/1)+
	Long	8 + 2n(1/0)	–

+ add effective address calculation time
n is the shift count

D.12 MISCELLANEOUS INSTRUCTION EXECUTION TIMES

Tables D-12 and D-13 indicate the number of clock periods for the following miscellaneous instructions. The number of bus read and write cycles is shown in parenthesis as (r/w). The number of clock periods plus the number of read and write cycles must be added to those of the effective address calculation where indicated.

Table D-12. Miscellaneous Instruction Execution Times

Instruction	Size	Register	Memory
ANDI to CCR	Byte	20(3/0)	–
ANDI to SR	Word	20(3/0)	–
CHK	–	10(1/0) +	–
EORI to CCR	Byte	20(3/0)	–
EORI to SR	Word	20(3/0)	–
ORI to CCR	Byte	20(3/0)	–
ORI to SR	Word	20(3/0)	–
MOVE from SR	–	6(1/0)	8(1/1) +
MOVE to CCR	–	12(2/0)	12(2/0) +
MOVE to SR	–	12(2/0)	12(2/0) +
EXG	–	6(1/0)	–
EXT	Word	4(1/0)	–
	Long	4(1/0)	–
LINK	–	16(2/2)	–
MOVE from USP	–	4(1/0)	–
MOVE to USP	–	4(1/0)	–
NOP	–	4(1/0)	–
RESET	–	132(1/0)	–
RTE	–	20(5/0)	–
RTR	–	20(5/0)	–
RTS	–	16(4/0)	–
STOP	–	4(0/0)	–
SWAP	–	4(1/0)	–
TRAPV	–	4(1/0)	–
UNLK	–	12(3/0)	–

+ add effective address calculation time

Table D-13. Move Peripheral Instruction Execution Times

Instruction	Size	Register → Memory	Memory → Register
MOVEP	Word	16(2/2)	16(4/0)
	Long	24(2/4)	24(6/0)

D.13 EXCEPTION PROCESSING EXECUTION TIMES

Table D-14 indicates the number of clock periods for exception processing. The number of clock periods includes the time for all stacking, the vector fetch, and the fetch of the first two instruction words of the handler routine. The number of bus read and write cycles is shown in parenthesis as (r/w).

Table D-14. Exception Processing Execution Times

Exception	Periods
Address Error	50(4/7)
Bus Error	50(4/7)
CHK Instruction	44(5/4)+
Divide by Zero	42(5/4)
Illegal Instruction	34(4/3)
Interrupt	44(5/3)*
Privilege Violation	34(4/3)
$\overline{\text{RESET}}$**	40(6/0)
Trace	34(4/3)
TRAP Instruction	38(4/4)
TRAPV Instruction	34(4/3)

+ add effective address calculation time

*The interrupt acknowledge cycle is assumed to take four clock periods.

**Indicates the time from when $\overline{\text{RESET}}$ and $\overline{\text{HALT}}$ are first sampled as negated to when instruction execution starts.

APPENDIX E

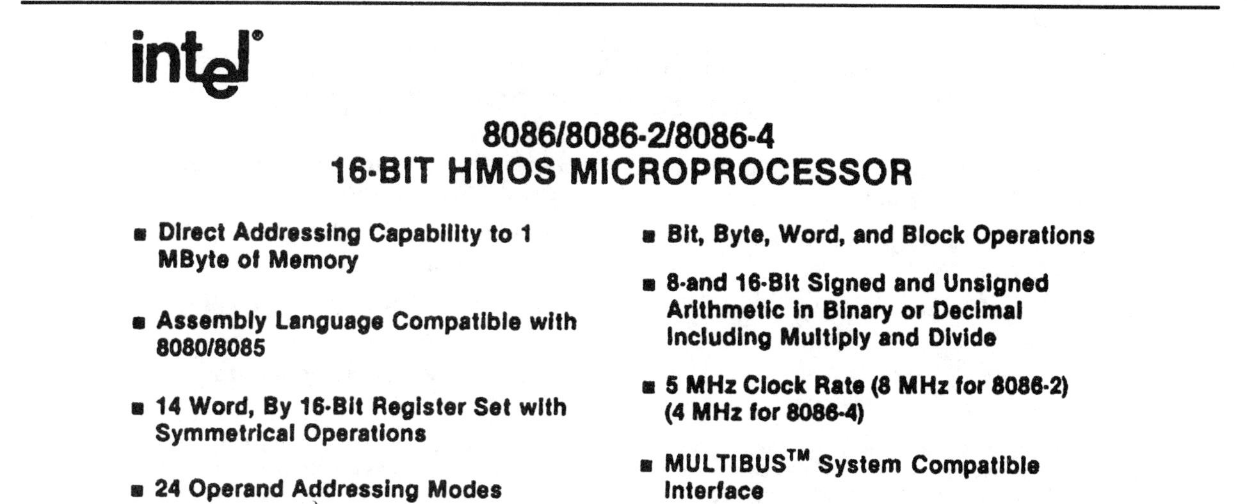

intel

8086/8086-2/8086-4
16-BIT HMOS MICROPROCESSOR

- **Direct Addressing Capability to 1 MByte of Memory**
- **Assembly Language Compatible with 8080/8085**
- **14 Word, By 16-Bit Register Set with Symmetrical Operations**
- **24 Operand Addressing Modes**
- **Bit, Byte, Word, and Block Operations**
- **8-and 16-Bit Signed and Unsigned Arithmetic in Binary or Decimal Including Multiply and Divide**
- **5 MHz Clock Rate (8 MHz for 8086-2) (4 MHz for 8086-4)**
- **MULTIBUS™ System Compatible Interface**

The Intel® 8086 is a new generation, high performance microprocessor implemented in N-channel, depletion load, silicon gate technology (HMOS), and packaged in a 40-pin CerDIP package. The processor has attributes of both 8- and 16-bit microprocessors. It addresses memory as a sequence of 8-bit bytes, but has a 16-bit wide physical path to memory for high performance.

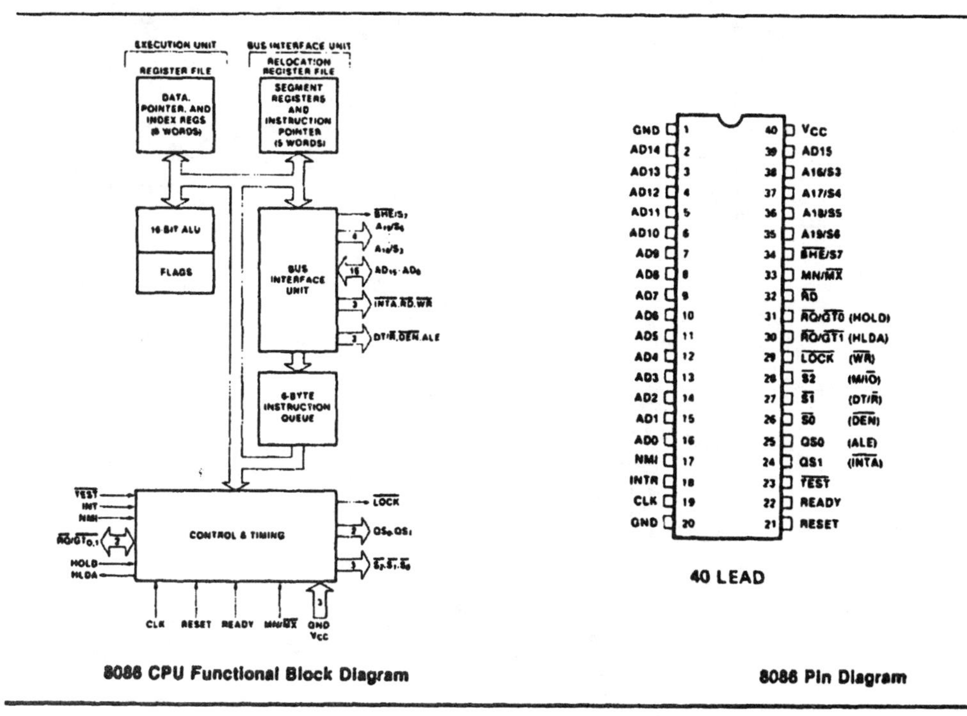

8086 CPU Functional Block Diagram

8086 Pin Diagram

intel®

I8284
CLOCK GENERATOR AND DRIVER FOR 8086, 8088, 8089 PROCESSORS

- **Generates the System Clock for the 8086, 8088 and 8089**
- **Uses a Crystal or a TTL Signal for Frequency Source**
- **Single +5V Power Supply**
- **18-Pin Package**
- **Generates System Reset Output from Schmitt Trigger Input**
- **Provides Local Ready and MULTIBUS™ Ready Synchronization**
- **Capable of Clock Synchronization with other 8284's**
- **Industrial Temperature Range −40° to +85°C**

The I8284 is a bipolar clock generator/driver designed to provide clock signals for the 8086, 8088 & 8089 and peripherals. It also contains READY logic for operation with two MULTIBUS™ systems and provides the processors required READY synchronization and timing. Reset logic with hysteresis and synchronization is also provided.

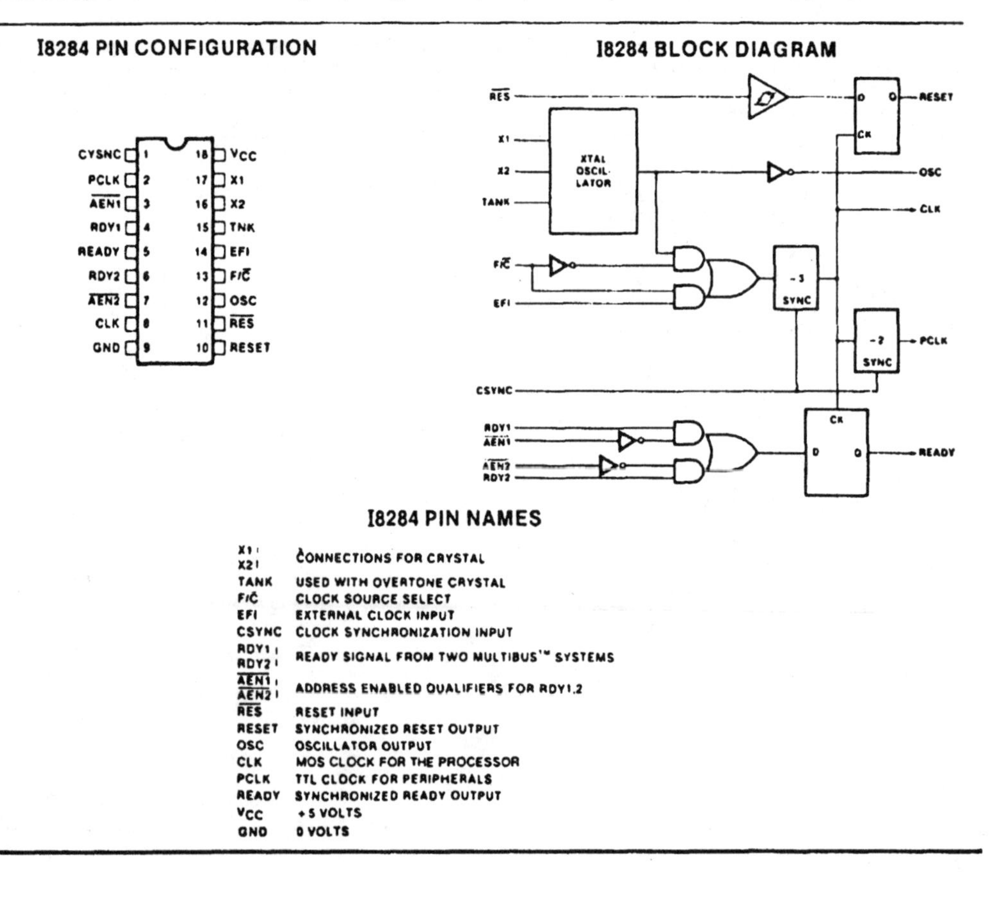

I8284 PIN NAMES

X1, X2	CONNECTIONS FOR CRYSTAL
TANK	USED WITH OVERTONE CRYSTAL
F/C	CLOCK SOURCE SELECT
EFI	EXTERNAL CLOCK INPUT
CSYNC	CLOCK SYNCHRONIZATION INPUT
RDY1, RDY2	READY SIGNAL FROM TWO MULTIBUS™ SYSTEMS
AEN1, AEN2	ADDRESS ENABLED QUALIFIERS FOR RDY1,2
RES	RESET INPUT
RESET	SYNCHRONIZED RESET OUTPUT
OSC	OSCILLATOR OUTPUT
CLK	MOS CLOCK FOR THE PROCESSOR
PCLK	TTL CLOCK FOR PERIPHERALS
READY	SYNCHRONIZED READY OUTPUT
Vcc	+5 VOLTS
GND	0 VOLTS

intel®

8288 BUS CONTROLLER FOR 8086, 8088, 8089 PROCESSORS

- **Bipolar Drive Capability**
- **Provides Advanced Commands**
- **Provides Wide Flexibility in System Configurations**
- **3-State Command Output Drivers**
- **Configurable for Use with an I/O Bus**
- **Facilitates Interface to One or Two Multi-Master Busses**

The Intel® 8288 Bus Controller is a 20-pin bipolar component for use with medium-to-large 8086 processing systems. The bus controller provides command and control timing generation as well as bipolar bus drive capability while optimizing system performance.

A strapping option on the bus controller configures it for use with a multi-master system bus and separate I/O bus.

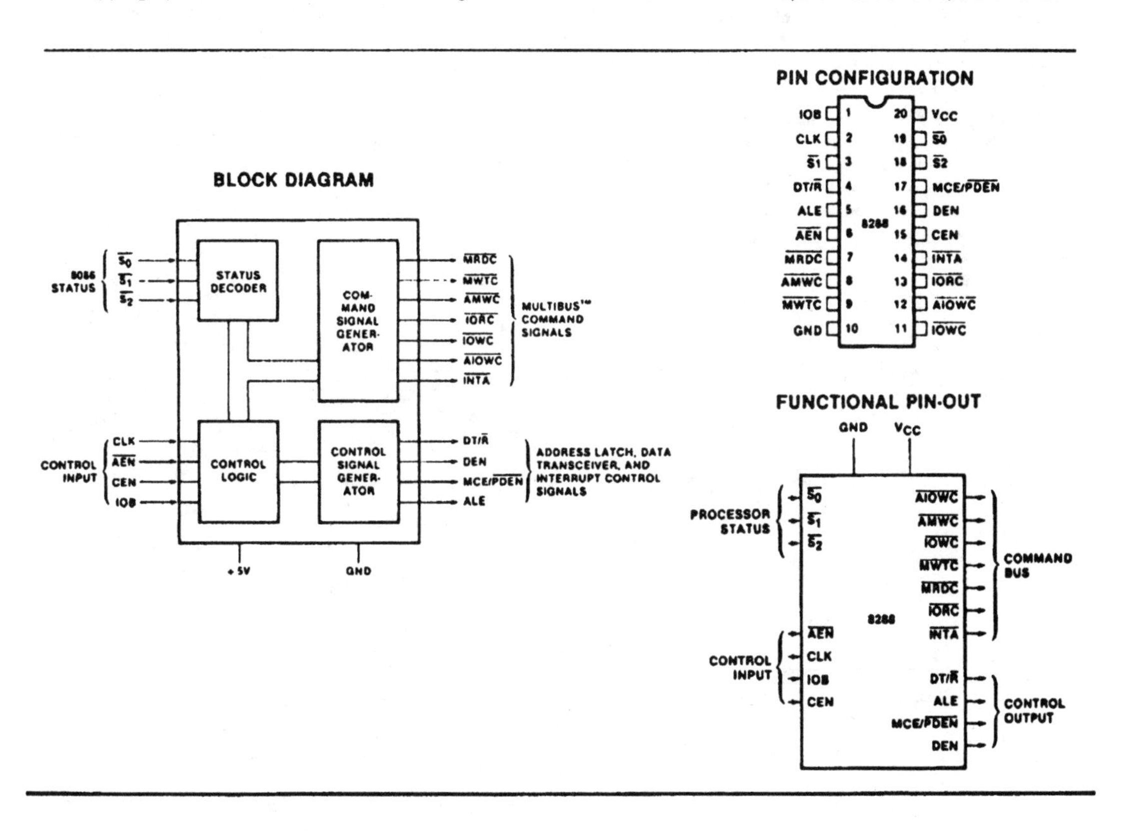

intel®

2716
16K (2K × 8) UV ERASABLE PROM

- **Fast Access Time**
  - **— 350 ns Max. 2716-1**
  - **— 390 ns Max. 2716-2**
  - **— 450 ns Max. 2716**
  - **— 650 ns Max. 2716-6**
- **Single +5V Power Supply**
- **Low Power Dissipation**
  - **— 525 mW Max. Active Power**
  - **— 132 mW Max. Standby Power**
- **Pin Compatible to Intel® 2732 EPROM**
- **Simple Programming Requirements**
  - **— Single Location Programming**
  - **— Programs with One 50 ms Pulse**
- **Inputs and Outputs TTL Compatible during Read and Program**
- **Completely Static**

The Intel® 2716 is a 16,384-bit ultraviolet erasable and electrically programmable read-only memory (EPROM). The 2716 operates from a single 5-volt power supply, has a static standby mode, and features fast single address location programming. It makes designing with EPROMs faster, easier and more economical.

The 2716, with its single 5-volt supply and with an access time up to 350 ns, is ideal for use with the newer high performance +5V microprocessors such as Intel's 8085 and 8086. The 2716 is also the first EPROM with a static standby mode which reduces the power dissipation without increasing access time. The maximum active power dissipation is 525 mW while the maximum standby power dissipation is only 132 mW, a 75% savings.

The 2716 has the simplest and fastest method yet devised for programming EPROMs – single pulse TTL level programming. No need for high voltage pulsing because all programming controls are handled by TTL signals. Program any location at any time—either individually, sequentially or at random, with the 2716's single address location programming. Total programming time for all 16,384 bits is only 100 seconds.

PIN CONFIGURATION

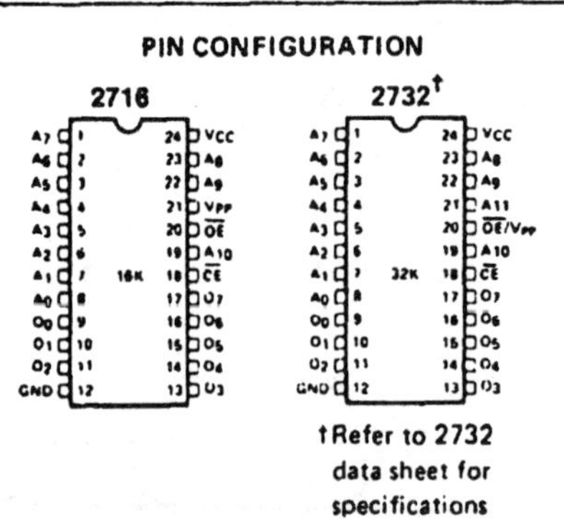

†Refer to 2732 data sheet for specifications

PIN NAMES

A_0–A_{10}	ADDRESSES
$\overline{CE}$/PGM	CHIP ENABLE/PROGRAM
$\overline{OE}$	OUTPUT ENABLE
O_0–O_7	OUTPUTS

MODE SELECTION

MODE \ PINS	$\overline{CE}$/PGM (18)	$\overline{OE}$ (20)	V_{PP} (21)	V_{CC} (24)	OUTPUTS (9-11, 13-17)
Read	V_{IL}	V_{IL}	+5	+5	D_{OUT}
Standby	V_{IH}	Don't Care	+5	+5	High Z
Program	Pulsed V_{IL} to V_{IH}	V_{IH}	+25	+5	D_{IN}
Program Verify	V_{IL}	V_{IL}	+25	+5	D_{OUT}
Program Inhibit	V_{IL}	V_{IH}	+25	+5	High Z

BLOCK DIAGRAM

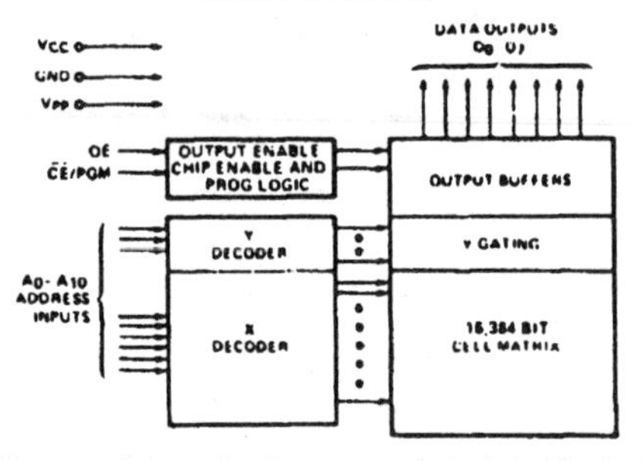

intel®

2732
32K (4K x 8) UV ERASABLE PROM

- **Fast Access Time:**
  - **— 450 ns Max. 2732**
  - **— 550 ns Max. 2732-6**
- **Single +5V ± 5% Power Supply**
- **Output Enable for MCS-85™ and MCS-86™ Compatibility**
- **Low Power Dissipation: 150mA Max. Active Current 30mA Max. Standby Current**
- **Pin Compatible to Intel® 2716 EPROM**
- **Completely Static**
- **Simple Programming Requirements**
  - **— Single Location Programming**
  - **— Programs with One 50ms Pulse**
- **Three-State Output for Direct Bus Interface**

The Intel® 2732 is a 32,768-bit ultraviolet erasable and electrically programmable read-only memory (EPROM). The 2732 operates from a single 5-volt power supply, has a standby mode, and features an output enable control. The total programming time for all bits is three and a half minutes. All these features make designing with the 2732 in microcomputer systems faster, easier, and more economical.

An important 2732 feature is the separate output control, Output Enable ($\overline{OE}$), from the Chip Enable control ($\overline{CE}$). The $\overline{OE}$ control eliminates bus contention in multiple bus microprocessor systems. Intel's Application Note AP-30 describes the microprocessor system implementation of the $\overline{OE}$ and $\overline{CE}$ controls on Intel's 2716 and 2732 EPROMs. AP-30 is available from Intel's Literature Department.

The 2732 has a standby mode which reduces the power dissipation without increasing access time. The maximum active current is 150mA, while the maximum standby current is only 30mA, an 80% savings. The standby mode is achieved by applying a TTL-high signal to the $\overline{CE}$ input.

PIN CONFIGURATION

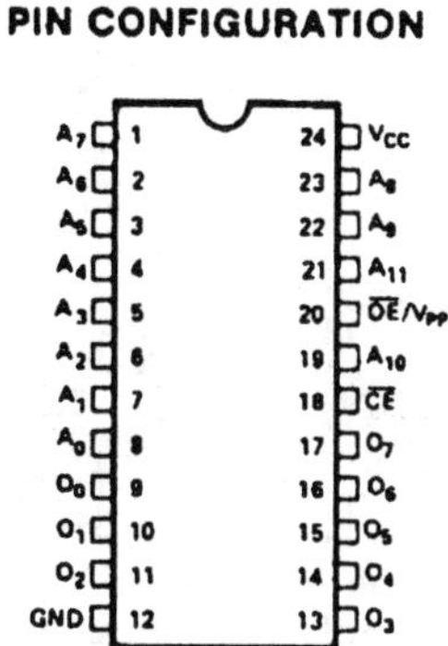

PIN NAMES

A_0-A_{11}	ADDRESSES
$\overline{CE}$	CHIP ENABLE
$\overline{OE}$	OUTPUT ENABLE
O_0-O_7	OUTPUTS

MODE SELECTION

MODE \ PINS	$\overline{CE}$ (18)	$\overline{OE}/V_{PP}$ (20)	V_{CC} (24)	OUTPUTS (9-11,13-17)
Read	V_{IL}	V_{IL}	+5	D_{OUT}
Standby	V_{IH}	Don't Care	+5	High Z
Program	V_{IL}	V_{PP}	+5	D_{IN}
Program Verify	V_{IL}	V_{IL}	+5	D_{OUT}
Program Inhibit	V_{IH}	V_{PP}	+5	High Z

BLOCK DIAGRAM

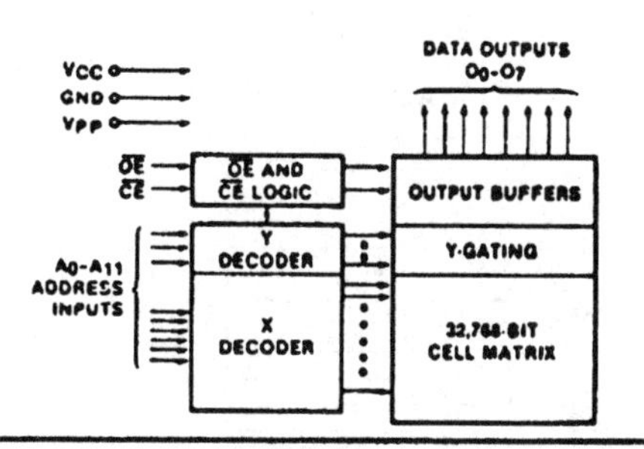

8255A/8255A-5
PROGRAMMABLE PERIPHERAL INTERFACE

- **MCS-85™ Compatible 8255A-5**
- **24 Programmable I/O Pins**
- **Completely TTL Compatible**
- **Fully Compatible with Intel® Microprocessor Families**
- **Improved Timing Characteristics**
- **Direct Bit Set/Reset Capability Easing Control Application Interface**
- **40-Pin Dual In-Line Package**
- **Reduces System Package Count**
- **Improved DC Driving Capability**

The Intel® 8255A is a general purpose programmable I/O device designed for use with Intel® microprocessors. It has 24 I/O pins which may be individually programmed in 2 groups of 12 and used in 3 major modes of operation. In the first mode (MODE 0), each group of 12 I/O pins may be programmed in sets of 4 to be input or output. In MODE 1, the second mode, each group may be programmed to have 8 lines of input or output. Of the remaining 4 pins, 3 are used for handshaking and interrupt control signals. The third mode of operation (MODE 2) is a bidirectional bus mode which uses 8 lines for a bidirectional bus, and 5 lines, borrowing one from the other group, for handshaking.

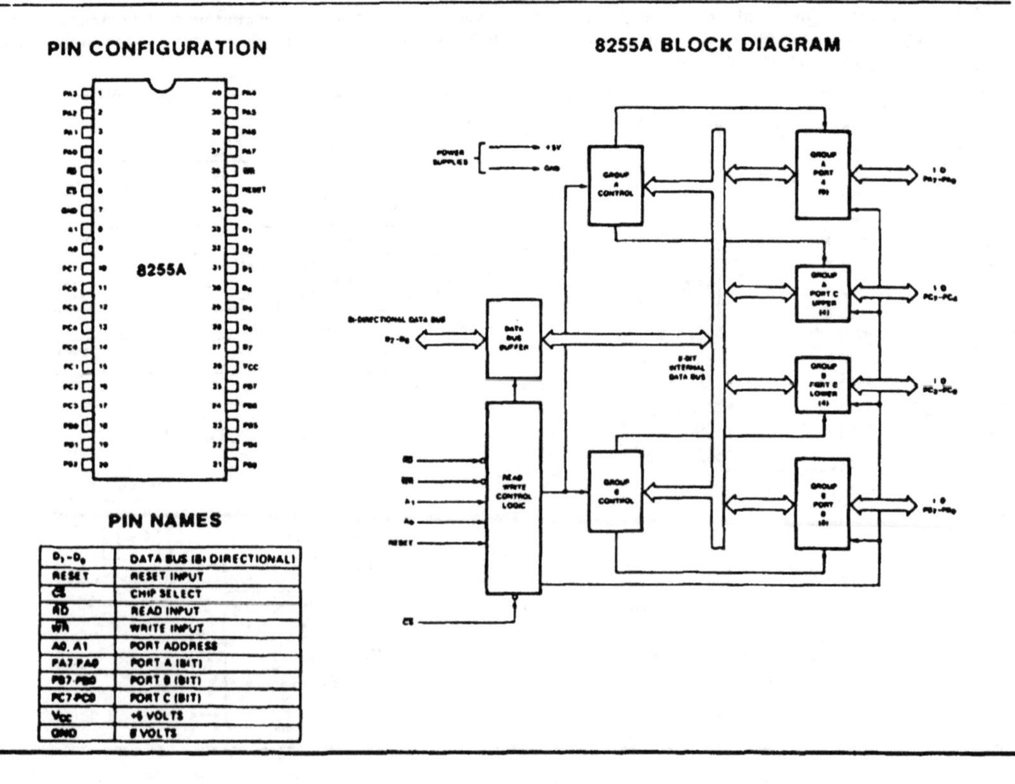

D_7-D_0	DATA BUS (BI DIRECTIONAL)
RESET	RESET INPUT
$\overline{CS}$	CHIP SELECT
$\overline{RD}$	READ INPUT
$\overline{WR}$	WRITE INPUT
A0, A1	PORT ADDRESS
PA7-PA0	PORT A (BIT)
PB7-PB0	PORT B (BIT)
PC7-PC0	PORT C (BIT)
V_{CC}	+5 VOLTS
GND	0 VOLTS

APPENDIX F

8086 INSTRUCTION SET REFERENCE DATA

AAA	AAA (no operands) ASCII adjust for addition	Flags	O	D	I	T	S	Z	A	P	C
			U				U	U	X	U	X

Operands	Clocks	Transfers*	Bytes	Coding Example
(no operands)	4	—	1	AAA

AAD	AAD (no operands) ASCII adjust for division	Flags	O	D	I	T	S	Z	A	P	C
			U				X	X	U	X	U

Operands	Clocks	Transfers*	Bytes	Coding Example
(no operands)	60	—	2	AAD

AAM	AAM (no operands) ASCII adjust for multiply	Flags	O	D	I	T	S	Z	A	P	C
			U				X	X	U	X	U

Operands	Clocks	Transfers*	Bytes	Coding Example
(no operands)	83	—	1	AAM

AAS	AAS (no operands) ASCII adjust for subtraction	Flags	O	D	I	T	S	Z	A	P	C
			U				U	U	X	U	X

Operands	Clocks	Transfers*	Bytes	Coding Example
(no operands)	4	—	1	AAS

*For the 8086, add four clocks for each 16-bit word transfer with an odd address. For the 8088, add four clocks for each 16-bit word transfer.

ADC

ADC destination,source
Add with carry

Flags	O	D	I	T	S	Z	A	P	C
	X				X	X	X	X	X

Operands	Clocks	Transfers*	Bytes	Coding Example
register, register	3	—	2	ADC AX, SI
register, memory	9+EA	1	2-4	ADC DX, BETA [SI]
memory, register	16+EA	2	2-4	ADC ALPHA [BX] [SI], DI
register, immediate	4	—	3-4	ADC BX, 256
memory, immediate	17+EA	2	3-6	ADC GAMMA, 30H
accumulator, immediate	4	—	2-3	ADC AL, 5

ADD

ADD destination,source
Addition

Flags	O	D	I	T	S	Z	A	P	C
	X				X	X	X	X	X

Operands	Clocks	Transfers*	Bytes	Coding Example
register, register	3	—	2	ADD CX, DX
register, memory	9+EA	1	2-4	ADD DI, [BX].ALPHA
memory, register	16+EA	2	2-4	ADD TEMP, CL
register, immediate	4	—	3-4	ADD CL, 2
memory, immediate	17+EA	2	3-6	ADD ALPHA, 2
accumulator, immediate	4	—	2-3	ADD AX, 200

AND

AND destination,source
Logical and

Flags	O	D	I	T	S	Z	A	P	C
	0				X	X	U	X	0

Operands	Clocks	Transfers*	Bytes	Coding Example
register, register	3	—	2	AND AL,BL
register, memory	9+EA	1	2-4	AND CX,FLAG_WORD
memory, register	16+EA	2	2-4	AND ASCII [DI],AL
register, immediate	4	—	3-4	AND CX,0F0H
memory, immediate	17+EA	2	3-6	AND BETA, 01H
accumulator, immediate	4	—	2-3	AND AX, 01010000B

CALL

CALL target
Call a procedure

Flags	O	D	I	T	S	Z	A	P	C

Operands	Clocks	Transfers*	Bytes	Coding Examples
near-proc	19	1	3	CALL NEAR_PROC
far-proc	28	2	5	CALL FAR_PROC
memptr 16	21+EA	2	2-4	CALL PROC_TABLE [SI]
regptr 16	16	1	2	CALL AX
memptr 32	37+EA	4	2-4	CALL [BX].TASK [SI]

CBW

CBW (no operands)
Convert byte to word

Flags	O	D	I	T	S	Z	A	P	C

Operands	Clocks	Transfers*	Bytes	Coding Example
(no operands)	2	—	1	CBW

*For the 8086, add four clocks for each 16-bit word transfer with an odd address. For the 8088, add four clocks for each 16-bit word transfer.

CLC	**CLC** (no operands) Clear carry flag			Flags: O D I T S Z A P C C = 0
Operands	**Clocks**	**Transfers***	**Bytes**	**Coding Example**
(no operands)	2	—	1	CLC

CLD	**CLD** (no operands) Clear direction flag			Flags: O D I T S Z A P C D = 0
Operands	**Clocks**	**Transfers***	**Bytes**	**Coding Example**
(no operands)	2	—	1	CLD

CLI	**CLI** (no operands) Clear interrupt flag			Flags: O D I T S Z A P C I = 0
Operands	**Clocks**	**Transfers***	**Bytes**	**Coding Example**
(no operands)	2	—	1	CLI

CMC	**CMC** (no operands) Complement carry flag			Flags: O D I T S Z A P C C = X
Operands	**Clocks**	**Transfers***	**Bytes**	**Coding Example**
(no operands)	2	—	1	CMC

CMP	**CMP** destination,source Compare destination to source			Flags: O D I T S Z A P C O = X, S Z A P C = X X X X X
Operands	**Clocks**	**Transfers***	**Bytes**	**Coding Example**
register, register	3	—	2	CMP BX, CX
register, memory	9+EA	1	2-4	CMP DH, ALPHA
memory, register	9+EA	1	2-4	CMP [BP+2], SI
register, immediate	4	—	3-4	CMP BL, 02H
memory, immediate	10+EA	1	3-6	CMP [BX].RADAR [DI], 3420H
accumulator, immediate	4	—	2-3	CMP AL, 00010000B

CMPS	**CMPS** dest-string,source-string Compare string			Flags: O D I T S Z A P C O = X, S Z A P C = X X X X X
Operands	**Clocks**	**Transfers***	**Bytes**	**Coding Example**
dest-string, source-string	22	2	1	CMPS BUFF1, BUFF2
(repeat) dest-string, source-string	9+22/rep	2/rep	1	REPE CMPS ID, KEY

*For the 8086, add four clocks for each 16-bit word transfer with an odd address. For the 8088, add four clocks for each 16-bit word transfer.

CWD

CWD (no operands)
Convert word to doubleword

Flags	O	D	I	T	S	Z	A	P	C

Operands	Clocks	Transfers*	Bytes	Coding Example
(no operands)	5	—	1	CWD

DAA

DAA (no operands)
Decimal adjust for addition

Flags	O	D	I	T	S	Z	A	P	C
	X				X	X	X	X	X

Operands	Clocks	Transfers*	Bytes	Coding Example
(no operands)	4	—	1	DAA

DAS

DAS (no operands)
Decimal adjust for subtraction

Flags	O	D	I	T	S	Z	A	P	C
	U				X	X	X	X	X

Operands	Clocks	Transfers*	Bytes	Coding Example
(no operands)	4	—	1	DAS

DEC

DEC destination
Decrement by 1

Flags	O	D	I	T	S	Z	A	P	C
	X				X	X	X	X	

Operands	Clocks	Transfers*	Bytes	Coding Example
reg16	2	—	1	DEC AX
reg8	3	—	2	DEC AL
memory	15+EA	2	2-4	DEC ARRAY [SI]

DIV

DIV source
Division, unsigned

Flags	O	D	I	T	S	Z	A	P	C
	U				U	U	U	U	U

Operands	Clocks	Transfers*	Bytes	Coding Example
reg8	80-90	—	2	DIV CL
reg16	144-162	—	2	DIV BX
mem8	(86-96) +EA	1	2-4	DIV ALPHA
mem16	(150-168) +EA	1	2-4	DIV TABLE [SI]

ESC

ESC external-opcode,source
Escape

Flags	O	D	I	T	S	Z	A	P	C

Operands	Clocks	Transfers*	Bytes	Coding Example
immediate, memory	8+EA	1	2-4	ESC 6,ARRAY [SI]
immediate, register	2	—	2	ESC 20,AL

*For the 8086, add four clocks for each 16-bit word transfer with an odd address. For the 8088, add four clocks for each 16-bit word transfer.

HLT

HLT (no operands)
Halt

Flags	O	D	I	T	S	Z	A	P	C

Operands	Clocks	Transfers*	Bytes	Coding Example
(no operands)	2	—	1	HLT

IDIV

IDIV source
Integer division

Flags	O	D	I	T	S	Z	A	P	C
	U				U	U	U	U	U

Operands	Clocks	Transfers*	Bytes	Coding Example
reg8	101-112	—	2	IDIV BL
reg16	165-184	—	2	IDIV CX
mem8	(107-118) + EA	1	2-4	IDIV DIVISOR_BYTE [SI]
mem16	(171-190) + EA	1	2-4	IDIV [BX].DIVISOR_WORD

IMUL

IMUL source
Integer multiplication

Flags	O	D	I	T	S	Z	A	P	C
	X				U	U	U	U	X

Operands	Clocks	Transfers*	Bytes	Coding Example
reg8	80-98	—	2	IMUL CL
reg16	128-154	—	2	IMUL BX
mem8	(86-104) + EA	1	2-4	IMUL RATE_BYTE
mem16	(134-160) + EA	1	2-4	IMUL RATE_WORD [BP] [DI]

IN

IN accumulator,port
Input byte or word

Flags	O	D	I	T	S	Z	A	P	C

Operands	Clocks	Transfers*	Bytes	Coding Example
accumulator, immed8	10	1	2	IN AL, 0FFEAH
accumulator, DX	8	1	1	IN AX, DX

INC

INC destination
Increment by 1

Flags	O	D	I	T	S	Z	A	P	C
	X				X	X	X	X	

Operands	Clocks	Transfers*	Bytes	Coding Example
reg16	2	—	1	INC CX
reg8	3	—	2	INC BL
memory	15 + EA	2	2-4	INC ALPHA [DI] [BX]

*For the 8086, add four clocks for each 16-bit word transfer with an odd address. For the 8088, add four clocks for each 16-bit word transfer.

INT

INT interrupt-type
Interrupt

Flags	O	D	I	T	S	Z	A	P	C
			0	0					

Operands	Clocks	Transfers*	Bytes	Coding Example
immed8 (type = 3)	52	5	1	INT 3
immed8 (type ≠ 3)	51	5	2	INT 67

INTR†

INTR (external maskable interrupt)
Interrupt if INTR and IF=1

Flags	O	D	I	T	S	Z	A	P	C
			0	0					

Operands	Clocks	Transfers*	Bytes	Coding Example
(no operands)	61	7	N/A	N/A

INTO

INTO (no operands)
Interrupt if overflow

Flags	O	D	I	T	S	Z	A	P	C
			0	0					

Operands	Clocks	Transfers*	Bytes	Coding Example
(no operands)	53 or 4	5	1	INTO

IRET

IRET (no operands)
Interrupt Return

Flags	O	D	I	T	S	Z	A	P	C
	R	R	R	R	R	R	R	R	R

Operands	Clocks	Transfers*	Bytes	Coding Example
(no operands)	24	3	1	IRET

JA/JNBE

JA/JNBE short-label
Jump if above/Jump if not below nor equal

Flags	O	D	I	T	S	Z	A	P	C

Operands	Clocks	Transfers*	Bytes	Coding Example
short-label	16 or 4	—	2	JA ABOVE

JAE/JNB

JAE/JNB short-label
Jump if above or equal/Jump if not below

Flags	O	D	I	T	S	Z	A	P	C

Operands	Clocks	Transfers*	Bytes	Coding Example
short-label	16 or 4	—	2	JAE ABOVE_EQUAL

JB/JNAE

JB/JNAE short-label
Jump if below/Jump if not above nor equal

Flags	O	D	I	T	S	Z	A	P	C

Operands	Clocks	Transfers*	Bytes	Coding Example
short-label	16 or 4	—	2	JB BELOW

*For the 8086, add four clocks for each 16-bit word transfer with an odd address. For the 8088, add four clocks for each 16-bit word transfer.

†INTR is not an instruction; it is included in table 2-21 only for timing information.

JBE/JNA	JBE/JNA short-label Jump if below or equal/Jump if not above			Flags O D I T S Z A P C
Operands	**Clocks**	**Transfers***	**Bytes**	**Coding Example**
short-label	16 or 4	—	2	JNA NOT_ABOVE

JC	JC short-label Jump if carry			Flags O D I T S Z A P C
Operands	**Clocks**	**Transfers***	**Bytes**	**Coding Example**
short-label	16 or 4	—	2	JC CARRY_SET

JCXZ	JCXZ short-label Jump if CX is zero			Flags O D I T S Z A P C
Operands	**Clocks**	**Transfers***	**Bytes**	**Coding Example**
short-label	18 or 6	—	2	JCXZ COUNT_DONE

JE/JZ	JE/JZ short-label Jump if equal/Jump if zero			Flags O D I T S Z A P C
Operands	**Clocks**	**Transfers***	**Bytes**	**Coding Example**
short-label	16 or 4	—	2	JZ ZERO

JG/JNLE	JG/JNLE short-label Jump if greater/Jump if not less nor equal			Flags O D I T S Z A P C
Operands	**Clocks**	**Transfers***	**Bytes**	**Coding Example**
short-label	16 or 4	—	2	JG GREATER

JGE/JNL	JGE/JNL short-label Jump if greater or equal/Jump if not less			Flags O D I T S Z A P C
Operands	**Clocks**	**Transfers***	**Bytes**	**Coding Example**
short-label	16 or 4	—	2	JGE GREATER_EQUAL

JL/JNGE	JL/JNGE short-label Jump if less/Jump if not greater nor equal			Flags O D I T S Z A P C
Operands	**Clocks**	**Transfers***	**Bytes**	**Coding Example**
short-label	16 or 4	—	2	JL LESS

*For the 8086, add four clocks for each 16-bit word transfer with an odd address. For the 8088, add four clocks for each 16-bit word transfer.

JLE/JNG	JLE/JNG short-label Jump if less or equal/Jump if not greater			Flags O D I T S Z A P C
Operands	**Clocks**	**Transfers***	**Bytes**	**Coding Example**
short-label	16 or 4	—	2	JNG NOT_GREATER

JMP	JMP target Jump			Flags O D I T S Z A P C
Operands	**Clocks**	**Transfers***	**Bytes**	**Coding Example**
short-label	15	—	2	JMP SHORT
near-label	15	—	3	JMP WITHIN_SEGMENT
far-label	15	—	5	JMP FAR_LABEL
memptr16	18+EA	1	2-4	JMP [BX].TARGET
regptr16	11	—	2	JMP CX
memptr32	24+EA	2	2-4	JMP OTHER.SEG [SI]

JNC	JNC short-label Jump if not carry			Flags O D I T S Z A P C
Operands	**Clocks**	**Transfers***	**Bytes**	**Coding Example**
short-label	16 or 4	—	2	JNC NOT_CARRY

JNE/JNZ	JNE/JNZ short-label Jump if not equal/Jump if not zero			Flags O D I T S Z A P C
Operands	**Clocks**	**Transfers***	**Bytes**	**Coding Example**
short-label	16 or 4	—	2	JNE NOT_EQUAL

JNO	JNO short-label Jump if not overflow			Flags O D I T S Z A P C
Operands	**Clocks**	**Transfers***	**Bytes**	**Coding Example**
short-label	16 or 4	—	2	JNO NO_OVERFLOW

JNP/JPO	JNP/JPO short-label Jump if not parity/Jump if parity odd			Flags O D I T S Z A P C
Operands	**Clocks**	**Transfers***	**Bytes**	**Coding Example**
short-label	16 or 4	—	2	JPO ODD_PARITY

JNS	JNS short-label Jump if not sign			Flags O D I T S Z A P C
Operands	**Clocks**	**Transfers***	**Bytes**	**Coding Example**
short-label	16 or 4	—	2	JNS POSITIVE

*For the 8086, add four clocks for each 16-bit word transfer with an odd address. For the 8088, add four clocks for each 16-bit word transfer.

JO	JO short-label Jump if overflow			Flags O D I T S Z A P C
Operands	**Clocks**	**Transfers***	**Bytes**	**Coding Example**
short-label	16 or 4	—	2	JO SIGNED_OVRFLW

JP/JPE	JP/JPE short-label Jump if parity/Jump if parity even			Flags O D I T S Z A P C
Operands	**Clocks**	**Transfers***	**Bytes**	**Coding Example**
short-label	16 or 4	—	2	JPE EVEN_PARITY

JS	JS short-label Jump if sign			Flags O D I T S Z A P C
Operands	**Clocks**	**Transfers***	**Bytes**	**Coding Example**
short-label	16 or 4	—	2	JS NEGATIVE

LAHF	LAHF (no operands) Load AH from flags			Flags O D I T S Z A P C
Operands	**Clocks**	**Transfers***	**Bytes**	**Coding Example**
(no operands)	4	—	1	LAHF

LDS	LDS destination,source Load pointer using DS			Flags O D I T S Z A P C
Operands	**Clocks**	**Transfers**	**Bytes**	**Coding Example**
reg16, mem32	16+EA	2	2-4	LDS SI,DATA.SEG [DI]

LEA	LEA destination,source Load effective address			Flags O D I T S Z A P C
Operands	**Clocks**	**Transfers***	**Bytes**	**Coding Example**
reg16, mem16	2+EA	—	2-4	LEA BX, [BP] [DI]

LES	LES destination,source Load pointer using ES			Flags O D I T S Z A P C
Operands	**Clocks**	**Transfers***	**Bytes**	**Coding Example**
reg16, mem32	16+EA	2	2-4	LES DI, [BX].TEXT_BUFF

*For the 8086, add four clocks for each 16-bit word transfer with an odd address. For the 8088, add four clocks for each 16-bit word transfer.

LOCK	LOCK (no operands) Lock bus	Flags O D I T S Z A P C

Operands	Clocks	Transfers*	Bytes	Coding Example
(no operands)	2	—	1	LOCK XCHG FLAG,AL

LODS	LODS source-string Load string	Flags O D I T S Z A P C

Operands	Clocks	Transfers*	Bytes	Coding Example
source-string (repeat) source-string	12 9+13/rep	1 1/rep	1 1	LODS CUSTOMER_NAME REP LODS NAME

LOOP	LOOP short-label Loop	Flags O D I T S Z A P C

Operands	Clocks	Transfers*	Bytes	Coding Example
short-label	17/5	—	2	LOOP AGAIN

LOOPE/LOOPZ	LOOPE/LOOPZ short-label Loop if equal/Loop if zero	Flags O D I T S Z A P C

Operands	Clocks	Transfers*	Bytes	Coding Example
short-label	18 or 6	—	2	LOOPE AGAIN

LOOPNE/LOOPNZ	LOOPNE/LOOPNZ short-label Loop if not equal/Loop if not zero	Flags O D I T S Z A P C

Operands	Clocks	Transfers*	Bytes	Coding Example
short-label	19 or 5	—	2	LOOPNE AGAIN

NMI†	NMI (external nonmaskable interrupt) Interrupt if NMI = 1	Flags O S I T S Z A P C 0 0

Operands	Clocks	Transfers*	Bytes	Coding Example
(no operands)	50*	5	N/A	N/A

*For the 8086, add four clocks for each 16-bit word transfer with an odd address. For the 8088, add four clocks for each 16-bit word transfer.

†NMI is not an instruction; it is included in table 2-21 only for timing information.

MOV

MOV destination,source
Move

Flags	O	D	I	T	S	Z	A	P	C

Operands	Clocks	Transfers*	Bytes	Coding Example
memory, accumulator	10	1	3	MOV ARRAY [SI], AL
accumulator, memory	10	1	3	MOV AX, TEMP_RESULT
register, register	2	—	2	MOV AX,CX
register, memory	8+EA	1	2-4	MOV BP, STACK_TOP
memory, register	9+EA	1	2-4	MOV COUNT [DI], CX
register, immediate	4	—	2-3	MOV CL, 2
memory, immediate	10+EA	1	3-6	MOV MASK [BX] [SI], 2CH
seg-reg, reg16	2	—	2	MOV ES, CX
seg-reg, mem16	8+EA	1	2-4	MOV DS, SEGMENT_BASE
reg16, seg-reg	2	—	2	MOV BP, SS
memory, seg-reg	9+EA	1	2-4	MOV [BX].SEG_SAVE, CS

MOVS

MOVS dest-string,source-string
Move string

Flags	O	D	I	T	S	Z	A	P	C

Operands	Clocks	Transfers*	Bytes	Coding Example
dest-string, source-string	18	2	1	MOVS LINE EDIT_DATA
(repeat) dest-string, source-string	9+17/rep	2/rep	1	REP MOVS SCREEN, BUFFER

MOVSB/MOVSW

MOVSB/MOVSW (no operands)
Move string (byte/word)

Flags	O	D	I	T	S	Z	A	P	C

Operands	Clocks	Transfers*	Bytes	Coding Example
(no operands)	18	2	1	MOVSB
(repeat) (no operands)	9+17/rep	2/rep	1	REP MOVSW

MUL

MUL source
Multiplication, unsigned

Flags	O	D	I	T	S	Z	A	P	C
	X				U	U	U	U	X

Operands	Clocks	Transfers*	Bytes	Coding Example
reg8	70-77	—	2	MUL BL
reg16	118-133	—	2	MUL CX
mem8	(76-83) +EA	1	2-4	MUL MONTH [SI]
mem16	(124-139) +EA	1	2-4	MUL BAUD_RATE

*For the 8086, add four clocks for each 16-bit word transfer with an odd address. For the 8088, add four clocks for each 16-bit word transfer.

NEG

NEG destination
Negate

Flags	O	D	I	T	S	Z	A	P	C
	X				X	X	X	X	1*

Operands	Clocks	Transfers*	Bytes	Coding Example
register	3	—	2	NEG AL
memory	16+EA	2	2-4	NEG MULTIPLIER

*0 if destination = 0

NOP

NOP (no operands)
No Operation

Flags	O	D	I	T	S	Z	A	P	C

Operands	Clocks	Transfers*	Bytes	Coding Example
(no operands)	3	—	1	NOP

NOT

NOT destination
Logical not

Flags	O	D	I	T	S	Z	A	P	C

Operands	Clocks	Transfers*	Bytes	Coding Example
register	3	—	2	NOT AX
memory	16+EA	2	2-4	NOT CHARACTER

OR

OR destination,source
Logical inclusive or

Flags	O	D	I	T	S	Z	A	P	C
	0				X	X	U	X	0

Operands	Clocks	Transfers*	Bytes	Coding Example
register, register	3	—	2	OR AL, BL
register, memory	9+EA	1	2-4	OR DX, PORT_ID [DI]
memory, register	16+EA	2	2-4	OR FLAG_BYTE, CL
accumulator, immediate	4	—	2-3	OR AL, 01101100B
register, immediate	4	—	3-4	OR CX,01H
memory, immediate	17+EA	2	3-6	OR [BX].CMD_WORD,0CFH

OUT

OUT port,accumulator
Output byte or word

Flags	O	D	I	T	S	Z	A	P	C

Operands	Clocks	Transfers*	Bytes	Coding Example
immed8, accumulator	10	1	2	OUT 44, AX
DX, accumulator	8	1	1	OUT DX, AL

POP

POP destination
Pop word off stack

Flags	O	D	I	T	S	Z	A	P	C

Operands	Clocks	Transfers*	Bytes	Coding Example
register	8	1	1	POP DX
seg-reg (CS illegal)	8	1	1	POP DS
memory	17+EA	2	2-4	POP PARAMETER

*For the 8086, add four clocks for each 16-bit word transfer with an odd address. For the 8088, add four clocks for each 16-bit word transfer.

POPF — POPF (no operands) Pop flags off stack

Flags	O	D	I	T	S	Z	A	P	C
	R	R	R	R	R	R	R	R	R

Operands	Clocks	Transfers*	Bytes	Coding Example
(no operands)	8	1	1	POPF

PUSH — PUSH source Push word onto stack

Flags	O	D	I	T	S	Z	A	P	C

Operands	Clocks	Transfers*	Bytes	Coding Example
register	11	1	1	PUSH SI
seg-reg (CS legal)	10	1	1	PUSH ES
memory	16+EA	2	2-4	PUSH RETURN_CODE [SI]

PUSHF — PUSHF (no operands) Push flags onto stack

Flags	O	D	I	T	S	Z	A	P	C

Operands	Clocks	Transfers*	Bytes	Coding Example
(no operands)	10	1	1	PUSHF

RCL — RCL destination,count Rotate left through carry

Flags	O	D	I	T	S	Z	A	P	C
	X								X

Operands	Clocks	Transfers*	Bytes	Coding Example
register, 1	2	—	2	RCL CX, 1
register, CL	8+4/bit	—	2	RCL AL, CL
memory, 1	15+EA	2	2-4	RCL ALPHA, 1
memory, CL	20+EA+ 4/bit	2	2-4	RCL [BP].PARM, CL

RCR — RCR designation,count Rotate right through carry

Flags	O	D	I	T	S	Z	A	P	C
	X								X

Operands	Clocks	Transfers*	Bytes	Coding Example
register, 1	2	—	2	RCR BX, 1
register, CL	8+4/bit	—	2	RCR BL, CL
memory, 1	15+EA	2	2-4	RCR [BX].STATUS, 1
memory, CL	20+EA+ 4/bit	2	2-4	RCR ARRAY [DI], CL

REP — REP (no operands) Repeat string operation

Flags	O	D	I	T	S	Z	A	P	C

Operands	Clocks	Transfers*	Bytes	Coding Example
(no operands)	2	—	1	REP MOVS DEST, SRCE

*For the 8086, add four clocks for each 16-bit word transfer with an odd address. For the 8088, add four clocks for each 16-bit word transfer.

REPE/REPZ

REPE/REPZ (no operands)
Repeat string operation while equal/while zero

Flags

O	D	I	T	S	Z	A	P	C

Operands	Clocks	Transfers*	Bytes	Coding Example
(no operands)	2	—	1	REPE CMPS DATA, KEY

REPNE/REPNZ

REPNE/REPNZ (no operands)
Repeat string operation while not equal/not zero

Flags

O	D	I	T	S	Z	A	P	C

Operands	Clocks	Transfers*	Bytes	Coding Example
(no operands)	2	—	1	REPNE SCAS INPUT_LINE

RET

RET optional-pop-value
Return from procedure

Flags

O	D	I	T	S	Z	A	P	C

Operands	Clocks	Transfers*	Bytes	Coding Example
(intra-segment, no pop)	8	1	1	RET
(intra-segment, pop)	12	1	3	RET 4
(inter-segment, no pop)	18	2	1	RET
(inter-segment, pop)	17	2	3	RET 2

ROL

ROL destination,count
Rotate left

Flags

O	D	I	T	S	Z	A	P	C
X								X

Operands	Clocks	Transfers	Bytes	Coding Examples
register, 1	2	—	2	ROL BX, 1
register, CL	8 + 4/bit	—	2	ROL DI, CL
memory, 1	15 + EA	2	2-4	ROL FLAG_BYTE [DI],1
memory, CL	20 + EA + 4/bit	2	2-4	ROL ALPHA , CL

ROR

ROR destination,count
Rotate right

Flags

O	D	I	T	S	Z	A	P	C
X								X

Operand	Clocks	Transfers*	Bytes	Coding Example
register, 1	2	—	2	ROR AL, 1
register, CL	8 + 4/bit	—	2	ROR BX, CL
memory, 1	15 + EA	2	2-4	ROR PORT_STATUS, 1
memory, CL	20 + EA + 4/bit	2	2-4	ROR CMD_WORD, CL

SAHF

SAHF (no operands)
Store AH into flags

Flags

O	D	I	T	S	Z	A	P	C
				R	R	R	R	R

Operands	Clocks	Transfers*	Bytes	Coding Example
(no operands)	4	—	1	SAHF

*For the 8086, add four clocks for each 16-bit word transfer with an odd address. For the 8088, add four clocks for each 16-bit word transfer.

SAL/SHL	**SAL/SHL** destination,count Shift arithmetic left/Shift logical left			Flags O D I T S Z A P C X X
Operands	**Clocks**	**Transfers***	**Bytes**	**Coding Examples**
register,1	2	—	2	SAL AL,1
register, CL	8+4/bit	—	2	SHL DI, CL
memory,1	15+EA	2	2-4	SHL [BX].OVERDRAW, 1
memory, CL	20+EA+4/bit	2	2-4	SAL STORE_COUNT, CL

SAR	**SAR** destination,source Shift arithmetic right			Flags O D I T S Z A P C X . . . X X U X X
Operands	**Clocks**	**Transfers***	**Bytes**	**Coding Example**
register, 1	2	—	2	SAR DX, 1
register, CL	8+4/bit	—	2	SAR DI, CL
memory, 1	15+EA	2	2-4	SAR N_BLOCKS, 1
memory, CL	20+EA+4/bit	2	2-4	SAR N_BLOCKS, CL

SBB	**SBB** destination,source Subtract with borrow			Flags O D I T S Z A P C X . . . X X X X X
Operands	**Clocks**	**Transfers***	**Bytes**	**Coding Example**
register, register	3	—	2	SBB BX, CX
register, memory	9+EA	1	2-4	SBB DI, [BX].PAYMENT
memory, register	16+EA	2	2-4	SBB BALANCE, AX
accumulator, immediate	4	—	2-3	SBB AX, 2
register, immediate	4	—	3-4	SBB CL, 1
memory, immediate	17+EA	2	3-6	SBB COUNT [SI], 10

SCAS	**SCAS** dest-string Scan string			Flags O D I T S Z A P C X . . . X X X X X
Operands	**Clocks**	**Transfers***	**Bytes**	**Coding Example**
dest-string	15	1	1	SCAS INPUT_LINE
(repeat) dest-string	9+15/rep	1/rep	1	REPNE SCAS BUFFER

SEGMENT†	**SEGMENT** override prefix Override to specified segment			Flags O D I T S Z A P C
Operands	**Clocks**	**Transfers***	**Bytes**	**Coding Example**
(no operands)	2	—	1	MOV SS:PARAMETER, AX

*For the 8086, add four clocks for each 16-bit word transfer with an odd address. For the 8088, add four clocks for each 16-bit word transfer.

†ASM-86 incorporates the segment override prefix into the operand specification and not as a separate instruction. SEGMENT is included in table 2-21 only for timing information.

SHR	SHR destination,count Shift logical right			Flags O D I T S Z A P C X X
Operands	**Clocks**	**Transfers***	**Bytes**	**Coding Example**
register, 1	2	—	2	SHR SI, 1
register, CL	8 + 4/bit	—	2	SHR SI, CL
memory, 1	15 + EA	2	2-4	SHR ID_BYTE [SI] [BX], 1
memory, CL	20 + EA + 4/bit	2	2-4	SHR INPUT_WORD, CL

SINGLE STEP†	SINGLE STEP (Trap flag interrupt) Interrupt if TF = 1			Flags O D I T S Z A P C 0 0
Operands	**Clocks**	**Transfers***	**Bytes**	**Coding Example**
(no operands)	50	5	N/A	N/A

STC	STC (no operands) Set carry flag			Flags O D I T S Z A P C 1
Operands	**Clocks**	**Transfers***	**Bytes**	**Coding Example**
(no operands)	2	—	1	STC

STD	STD (no operands) Set direction flag			Flags O D I T S Z A P C 1
Operands	**Clocks**	**Transfers***	**Bytes**	**Coding Example**
(no operands)	2	—	1	STD

STI	STI (no operands) Set interrupt enable flag			Flags O D I T S Z A P C 1
Operands	**Clocks**	**Transfers***	**Bytes**	**Coding Example**
(no operands)	2	—	1	STI

STOS	STOS dest-string Store byte or word string			Flags O D I T S Z A P C
Operands	**Clocks**	**Transfers***	**Bytes**	**Coding Example**
dest-string	11	1	1	STOS PRINT_LINE
(repeat) dest-string	9 + 10/rep	1/rep	1	REP STOS DISPLAY

*For the 8086, add four clocks for each 16-bit word transfer with an odd address. For the 8088, add four clocks for each 16-bit word transfer.

†SINGLE STEP is not an instruction; it is included in table 2-21 only for timing information.

SUB

SUB destination,source
Subtraction

Flags	O	D	I	T	S	Z	A	P	C
	X				X	X	X	X	X

Operands	Clocks	Transfers*	Bytes	Coding Example
register, register	3	—	2	SUB CX, BX
register, memory	9+EA	1	2-4	SUB DX, MATH_TOTAL [SI]
memory, register	16+EA	2	2-4	SUB [BP+2], CL
accumulator, immediate	4	—	2-3	SUB AL, 10
register, immediate	4	—	3-4	SUB SI, 5280
memory, immediate	17+EA	2	3-6	SUB [BP].BALANCE, 1000

TEST

TEST destination,source
Test or non-destructive logical and

Flags	O	D	I	T	S	Z	A	P	C
	0				X	X	U	X	0

Operands	Clocks	Transfers*	Bytes	Coding Example
register, register	3	—	2	TEST SI, DI
register, memory	9+EA	1	2-4	TEST SI, END_COUNT
accumulator, immediate	4	—	2-3	TEST AL, 00100000B
register, immediate	5	—	3-4	TEST BX, 0CC4H
memory, immediate	11+EA	—	3-6	TEST RETURN_CODE, 01H

WAIT

WAIT (no operands)
Wait while TEST pin not asserted

Flags	O	D	I	T	S	Z	A	P	C

Operands	Clocks	Transfers*	Bytes	Coding Example
(no operands)	3 + 5n	—	1	WAIT

XCHG

XCHG destination,source
Exchange

Flags	O	D	I	T	S	Z	A	P	C

Operands	Clocks	Transfers*	Bytes	Coding Example
accumulator, reg16	3	—	1	XCHG AX, BX
memory, register	17+EA	2	2-4	XCHG SEMAPHORE, AX
register, register	4	—	2	XCHG AL, BL

XLAT

XLAT source-table
Translate

Flags	O	D	I	T	S	Z	A	P	C

Operands	Clocks	Transfers*	Bytes	Coding Example
source-table	11	1	1	XLAT ASCII_TAB

*For the 8086, add four clocks for each 16-bit word transfer with an odd address. For the 8088, add four clocks for each 16-bit word transfer.

XOR	XOR destination,source Logical exclusive or			Flags O D I T S Z A P C 0 X X U X 0
Operands	**Clocks**	**Transfers***	**Bytes**	**Coding Example**
register, register	3	—	2	XOR CX, BX
register, memory	9+EA	1	2-4	XOR CL, MASK_BYTE
memory, register	16+EA	2	2-4	XOR ALPHA [SI], DX
accumulator, immediate	4	—	2-3	XOR AL, 01000010B
register, immediate	4	—	3-4	XOR SI, 00C2H
memory, immediate	17+EA	2	3-6	XOR RETURN_CODE, 0D2H

*For the 8086, add four clocks for each 16-bit word transfer with an odd address. For the 8088, add four clocks for each 16-bit word transfer.

APPENDIX G

68000 INSTRUCTION SET

Instruction	Size	Length (words)	Operation
ABCD – (Ay), – (Ax)	B	1	– [Ay] 10 + – [Ax] 10 + X → [Ax]
ABCD Dy, Dx	B	1	[Dy]10 + [Dx]10 +X → Dx
ADD (EA), (EA)	B,W, L	1	[EA] + [EA] → EA
ADDA (EA), An	W, L	1	[EA] + An → An
ADDI #data, (EA)	B,W, L	2 for B, W 3 for L	data + [EA] → EA
ADDQ #data, (EA)	B, W, L	1	data + [EA] → EA
ADDX – (Ay), – (Ax)	B, W, L	1	– [Ay] + – [Ax] + X → [Ax]
ADDX Dy, Dx	B, W, L	1	Dy + Dx + X → Dx
AND (EA), (EA)	B, W, L	1	[EA] ^ [EA] → EA
ANDI #data, (EA)	B,W, L	2 for B, W 3 for L	data ^ [EA] → EA
ANDI #data8, CCR	B	2	data8 ^ [CCR] → CCR
ANDI #data16, SR	W	2	data16 ^ [SR] → SR if s = 1; else trap
ASL Dx, Dy	B, W, L	1	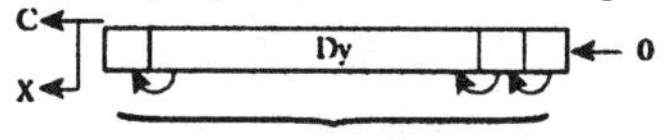 C, X ← Dy ← 0 number of shifts determined by [Dx]
ASL #data, Dy	B, W, L	1	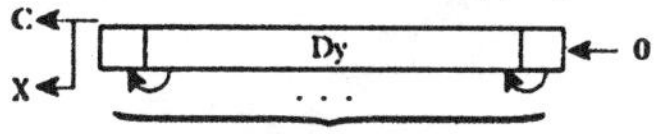 C, X ← Dy ← 0 number of shifts determined by # data
ASL (EA)	B, W, L	1	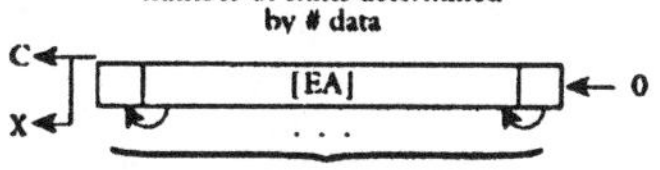 C, X ← [EA] ← 0 shift once
ASR Dx, Dy	B, W, L	1	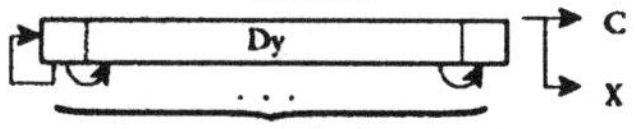 Dy → C, X number of shifts determined by [Dx]
ASR #data, Dy	B, W, L	1	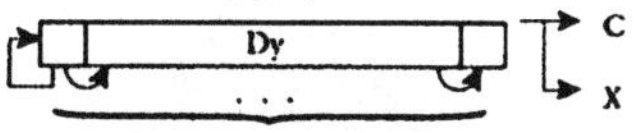 Dy → C, X number of shifts determined by immediate data
ASR (EA)	B,W, L	1	 [EA] → C, X shift once
BCC d	B, W	1 for B 2 for W	Branch to PC + d if carry = 0; else next instruction

Instruction	Size	Length (words)	Operation
BCHG Dn, (EA)	B, L	1	[bit of [EA], specified by Dn]' → Z [bit of [EA] specified by Dn]' → bit of [EA]
BCHG #data. (EA)	B, L	2	Same as BCHG Dn, [EA] except bit number is specified by immediate data
BCLR Dn (EA)	B, L	1	[bit of [EA]]' → Z 0 → bit of [EA] specified by Dn
BCLR #data, (EA)	B, L	2	Same as BCLR Dn, [EA] except the bit is specified by immediate data
BCS d	B, W	1 for B 2 for W	Branch to PC + d if carry = 1; else next instruction
BEQ d	B, W	1 for B 2 for W	Branch to PC + d if Z = 1; else next instruction
BGE d	B, W	1 for B 2 for W	Branch to PC + d if greater than or equal; else next instruction
BGT d	B, W	1 for B 2 for W	Branch to PC + d if greater than; else next instruction
BHI d	B, W	1 for B 2 for W	Branch to PC + d if higher; else next instruction
BLE d	B, W	1 for B 2 for W	Branch to PC + d if less or equal; else next instruction
BLS d	B, W	1 for B 2 for W	Branch to PC + d if low or same; else next instruction
BLT d	B, W	1 for B 2 for W	Branch to PC + d if less than; else next instruction
BMI d	B,W	1 for B 2 for W	Branch to PC +d if N = 1; else next instruction
BNE d	B, W	1 for B 2 for W	Branch to PC +d if Z = 0; else next instruction
BPL d	B, W	1 for B 2 for W	Branch to PC + d if N = 0; else next instruction
BRA d	B, W	1 for B 2 for W	Branch always to PC + d
BSET Dn, (EA)	B, L	1	[bit of [EA]]' → Z 1 → bit of [EA] specified by Dn
BSET #data, (EA)	B, L	2	Same as BSET Dn, [EA] except the bit is specified by immediate data
BSR d	B, W	1 for B 2 for W	PC → – [SP] PC + d → PC
BTST Dn, (EA)	B, L	1	[bit of [EA] specified by Dn]' → Z
BTST #data, (EA)	B, L	2	Same as BTST Dn, [EA] except the bit is specified by data
BVC d	B,W	1 for B 2 for W	Branch to PC + d if V = 0; else next instruction
BVS d	B, W	1 for B 2 for W	Branch to PC + d if V = 1; else next instruction
CHK (EA), Dn	W	1	If Dn < 0 or Dn > [EA], then trap
CLR(EA)	B, W, L	1	0 → EA
CMP (EA), Dn	B, W, L	1	Dn – [EA] → Affect all condition codes except X
CMP (EA), An	W, L	1	An – [EA] → Attect all condition codes except X
CMPI #data, (EA)	B, W, E	2 for B, W 3 for L	[EA] – data → Affect all flags except X-bit
CMPM (Ay) +, (Ax) +	B, W, E	1	[Ax]+ - [Ay]+ ⟶ Affect all flags except X; update Ax and Ay

Instruction	Size	Length (words)	Operation
DBCC Dn, d	W	2	If condition false, i.e., C = 1, then Dn - 1 —> Dn; if Dn ≠ – 1, then PC + d → PC; else PC + 2 → PC
DBCS Dn, d	W	2	Same as DBCC except condition is C = 1
DBEQ Dn, d	W	2	Same as DBCC except condition is Z = 1
DBF Dn, d	W	2	Same as DBCC except condition is always false
DBGE Dn, d	W	2	Same as DBCC except condition is greater or equal
DBGT Gn, d	W	2	Same as DBCC except condition is greater than
DBHIDn, d	\V	2	Same as DBCC except condition is high
DBLE Dn, d	W	2	Same as DBCC except condition is less than or equal
DBLS Dn, d	W	2	Same as DBCC except condition is low or same
DBLT Dn, d	W	2	Same as DBCC except condition is less than
DBM1 Dn, d	W	2	Same as DBCC except condition is N = 1
DBNE Dn, d	W	2	Same as DBCC except condition Z = 0
DBPL Dn, d	W	2	Same as DBCC except condition N = 0
DBT Dn, d	W	2	Same as DBCC except condition is always true
DBVC Dn, d	W	2	Same as DBCC except condition is V = 0
DBVS Dn, d	W	2	Same as DBCC except condition is V = 1
DIVS (EA), Dn	W	1	Signed division [Dn]32/[EA]16 → [Dn] 0-15 = quotient [Dn] 16-31 = remainder
DIVU (EA), Dn	W	1	Same as DIVS except division is unsigned
EOR Dn, (EA)	B,W, L	1	Dn ⊕ [EA] → EA
EORI #data, (EA)	B,W, L	2 for B, W 3 for L	data ⊕ [EA] → EA
EORI #d8, CCR	B	2	d8 ⊕ CCR → CCR
EORI #dl6, SR	W	2	dl6 ⊕ SR → SR if S = 1; else trap
EXG Rx, Ry	L	1	Rx ↔ Ry
EXTDn	W, L	1	Extend sign bit of Dn from 8-bit to 16-bit or from 16-bit to 32-bit depending on whether the operand size is B or W
JMP (EA)	Unsized	1	[EA] → PC Unconditional jump using address in operand
JSR (EA)	Unsized	1	PC → – [SP]; [EA] → PC Jump to subroutine using address in operand
LEA (EA), An	L	1	[EA] → An
LINK An, # -d	Unsized	2	An ← – [SP]; SP → An; SP – d → SP
LSL Dx, Dy	B,W, L	1	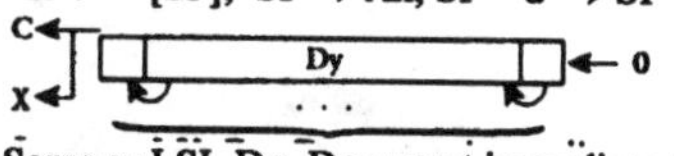 C ← X ← Dy ← 0
LSL #data, Dy	B,W, L	1	Same as LSL Dx, Dy except immediate data specify the number of shifts from 0 to 7
LSL (EA)	B,W, L	1	Same as LSL Dx, Dy except left shift is performed only once
LSR Dx, Dy	B,W, L	1	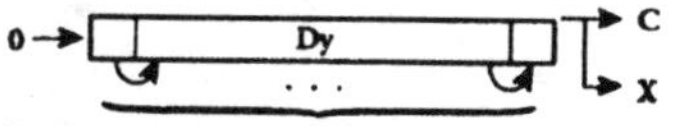 0 → Dy → C, X
LSR #data, Dy	B,W, L	1	Same as LSR except immediate data specifies the number of shifts from 0 to 7
LSR (EA)	B,W,L	1	Same as LSR, Dx, Dy except the right shift is performed only once
MOVE (EA), (EA)	B,W, L	1	[EA] source → [EA] destination
MOVE (EA), CCR	W	1	[EA] → CCR
MOVE CCR, (EA)	W	1	CCR → [EA]
MOVE (EA), SR	W	1	If S = l, then [EA] → SR; else TRAP
MOVE SR, (EA)	W	1	If S = l, then SR → [EA]; else TRAP

Instruction	Size	Length (words)	Operation
MOVE An, USP	L	1	If S = l, then An → USP; else TRAP
MOVE USP, An	W, L	1	[USP] → An
MOVEM register list, (EA)	W,L	2	Register list → [EA]
MOVEM (EA), register list	W, L	2	[EA] → register list
MOVEP Dx, d (Ay)	W, L	2	Dx → d[Ay]
MOVEP d (Ay), Dx	W, L	2	d[Ay] → Dx
MOVEQ #d8, Dn	L	1	d8 sign extended to 32-bit → Dn
MULS(EA)16,(Dn)16	W	1	Signed 16 × 16 multiplication [EA]16 * [Dn]16 → [Dn]32
MULU(EA)16,(Dn)16	W	1	Unsigned 16 × 16 multiplication [EA]16 * [Dn]16 → [Dn]32
NBCD (EA)	B	1	0 – [EA]10 – X → EA
NEC (EA)	B,W, L	1	0 – [EA] → EA
NEGX (EA)	B,W, L	1	0 – [EA] – X → EA
NOP	Unsized	1	No operation
NOT (EA)	B,W, L	1	[EA]' → EA
OR (EA), (EA)	B,W, L	1	[EA]V[EA] → EA
ORI #data, (EA)	B,W, L	2 for B, W 3 for L	data V[EA] → EA
ORI #d8, CCR	B	2	d8VCCR → CCR
ORI #dl6, SR	W	2	If S = 1, then dl6VSR -> SR; else TRAP
PEA (EA)	L	1	[EA] 16 sign extend to 32 bits → – [SP]
RESET	Unsized	1	If S = l, then assert RESET line; else TRAP
ROL Dx, Dy	B, W, L	1	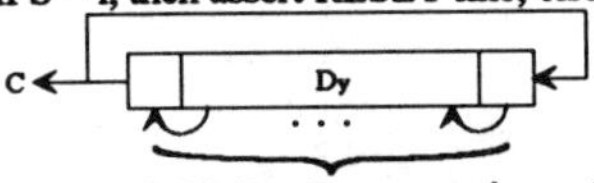
ROL #data, Dy	B, W, L	1	Same as ROL Dx, Dy except immediate data specifies number of times to be rotated from 0 to 7
ROL (EA)	B, W, L	1	Same as ROL Dx, Dy except [EA] is rotated once
ROR Dx, Dy	B, W, L	1	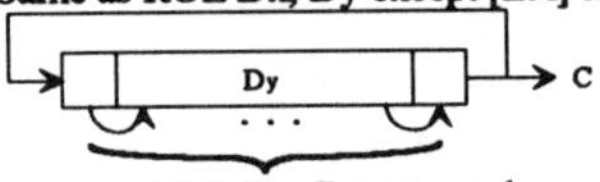
ROR #data, Dy	B, W, L	1	Same as ROR Dx, Dy except the number of rotates is specified by immediate data from 0 to 7
ROR (EA)	B, W, L	1	Same as ROR Dx, Dy except [EA] is rotated once
ROXL Dx, Dy	B, W, L	1	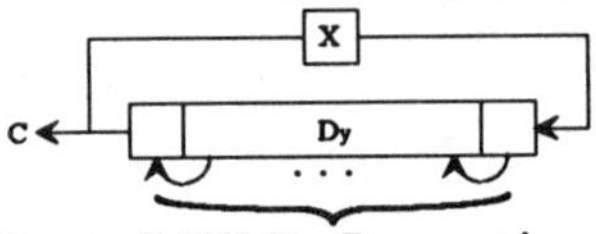
ROXL #data, Dy	B, W, L	1	Same as ROXL Dx, Dy except immediate data specifies number of rotates from 0 to 7
ROXL (EA)	B, W, L	1	Same as ROXL Dx, Dy except [EA] is rotated once
ROXR Dx, Dy	B, W, L	1	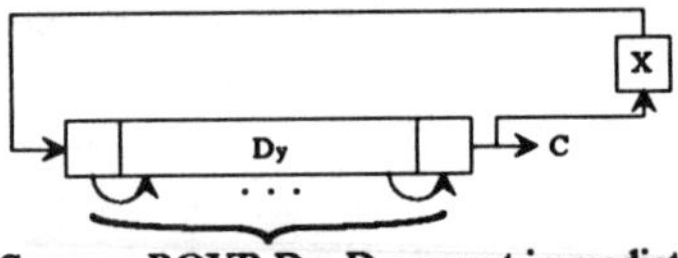
ROXR #data, Dy	B,W, L	1	Same as ROXR Dx, Dy except immediate data specifies number of rotates from 0 to 7
ROXR (EA)	B,W, L	1	Same as ROXR Dx, Dy except [EA] is rotated once
RTE	Unsized	1	If S = l, then [SP] + → SR; [SP] + → PC, else TRAP
RTR	Unsized	1	[SP] + → CC; [SP] + → PC
RTS	Unsized	1	[SP] + → PC
SBCD -(Ay), -(Ax)	B	1	– (Ax)10 – (Ay)10 – X → (Ax)
SBCD Dy, Dx	B	1	[Dx]10 – [Dy]10 – X → Dx

Instruction	Size	Length (words)	Operation
SCC (EA)	B	1	If C = 0, then 1s → [EA] else 0s → [EA]
SCS (EA)	B	1	Same as SCC except the condition is C = 1
SEQ (EA)	B	1	Same as SCC except if Z = 1
SF (EA)	B	1	Same as SCC except condition is always false
SGE (EA)	B	1	Same as SCC except if greater or equal
SGT (EA)	B	1	Same as SCC except if greater than
SHI (EA)	B	1	Same as SCC except if high
SLE (EA)	B	1	Same as SCC except if less or equal
SLS(EA)	B	1	Same as SCC except if low or same
SLT (EA)	B	1	Same as SCC except if less than
SMI (EA)	B	1	Same as SCC except if N = 1
SNE (EA)	B	1	Same as SCC except if Z = 0
SPL(EA)	B	1	Same as SCC except if N = 0
ST (EA)	B	1	Same as SCC except condition always true
STOP #data	Unsized	2	If S= 1, then data → SR and stop; TRAP if executed in user mode
SUB (EA), (EA)	B, W, L	1	[EA] – [EA] → EA
SUBA (EA), An	W,L	1	An – [EA] → An
SUBI #data, (EA)	B, W, L	2 for B, W 3 for L	[EA] – data → EA
SUBQ #data, (EA)	B, W, L	1	[EA] – data → EA
SUBX – (Ay), – (Ax)	B, W, L	1	– [Ax] – [Ay] – X → [Ax]
SUBX Dy, Dx	B, W, L	1	Dx – Dy – X → Dx
SVC (EA)	B	1	Same as SCC except if V = 0
SVS (EA)	B	1	Same as SCC except if V = 1
SWAP Dn	W	1	Dn [31:16] ↔ Dn [15:0]
TAS (EA)	B	1	[EA] tested; N and Z are affected accordingly; 1 → bit 7 of [EA]
TRAP #vector	Unsized	1	PC → – [SSP], SR → – [SSP], (vector) → PC; 16 TRAP
TRAPV	Unsized	1	If V = 1, then TRAP; else next instruction
TST (EA)	B,W, L	1	[EA] – 0 → condition codes affected; no result provided
UNLK An	Unsized	1	An → SP; [SP]+ → An

APPENDIX H

8086 INSTRUCTION SET

Instructions	Interpretation	Comments
AAA	ASCII adjust [AL] after addition	This instruction has implied addressing mode; this instruction is used to adjust the content of AL after addition of two ASCII characters
AAD	ASCII adjust for division	This instruction has implied addressing mode; converts two unpacked BCD digits in AX into equivalent binary numbers in AL; AAD must be used before dividing two unpacked BCD digits by an unpacked BCD byte
AAM	ASCII adjust after multiplication	This instruction has implied addressing mode; after multiplying two unpacked BCD numbers, adjust the product in AX to become an unpacked BCD result; ZF, SF, and PF are affected
AAS	ASCII adjust [AL] after subtraction	This instruction has implied addressing mode used to adjust [AL] after subtraction of two ASCII characters
ADC mem/reg 1, mem/reg 2	[mem/reg 1] ← [mem/reg 1] + [mem/reg 2] + CY	Memory or register can be 8- or 16-bit; all flags are affected; no segment registers are allowed; no memory-to-memory ADC is permitted
ADC mem, data	[mem] ← [mem] + data + CY	Data can be 8- or 16-bit; mem uses DS as the segment register; all flags are affected
ADC reg, data	[reg] ←[reg] + data + CY	Data can be 8- or 16-bit; register cannot be segment register; all flags are affected
ADD mem/reg 1, mem/reg 2	[mem/reg 1] ← [mem/reg 2] + [mem/reg 1]	Add two 8- or 16-bit data; no memory-to-memory ADD is permitted; all flags are affected; mem uses DS as the segment register; reg 1 or reg 2 cannot be segment register
ADD mem, data	[mem] ← [mem] + data	Mem uses DS as the segment register; data can be 8-or 16-bit; all flags are affected
ADD reg, data	[reg] ← [reg] + data	Data can be 8- or 16-bit; no segment registers are allowed; all flags are affected
AND mem/reg 1, mem/reg 2	[mem/reg 1] ← [mem/reg 1] ^ [mem/reg 2]	This instruction logically ANDs 8- or 16-bit data in [mem/reg 1] with 8- or 16-bit data in [mem/reg 2]; all flags are affected; OF and CF are cleared to zero; no segment registers are allowed; no memory-to-memory operation is allowed; mem uses DS as the segment register
AND mem, data	[mem] ← [mem] ^ data	Data can be 8- or 16-bit; mem uses DS as the segment register; all flags are affected with OF and CF always cleared to zero
AND reg, data	[reg] ← [reg] + data	Data can be 8- or 16-bit; reg cannot be segment register; all flags are affected with OF and CF cleared to zero
CALL PROC (NEAR)	Call a subroutine in the same segment with signed 16-bit displacement (to CALL a subroutine in ±32K)	NEAR in the statement BEGIN PROC NEAR indicates that the subroutine 'BEGIN' is in the same segment and BEGIN is 16-bit signed; CALL BEGIN instruction decrements SP by 2 and then pushes IP onto the stack and then adds the signed 16-bit value of BEGIN to IP and CS is unchanged; thus, a subroutine is called in the same segment (intrasegment direct)
CALL reg 16	CALL a subroutine in the same segment addressed by the contents of a 16-bit general register	The 8086 decrements SP by 2 and then pushes IP onto the stack, then specified 16-bit register contents (such as BX, SI, and DI) provide the new value for IP; CS is unchanged (intrasegment indirect)

Instructions	Interpretation	Comments
CALL mem 16	CALL a subroutine addressed by the content of a memory location pointed to by 8086 16-bit register such as BX, SI, and DI	The 8086 decrements SP by 2 and pushes IP onto the stack; the 8086 then loads the contents of a memory location addressed by the content of a 16-bit register such as BX, SI, and DI into IP; [CS] is unchanged (intrasegment indirect)
CALL subroutine in another segment	CALL a subroutine in another segment	FAR in the statement BEGIN PROC FAR indicates that the subroutine 'BEGIN' is in another segment and the value of BEGIN is 32 bit wide The 8086 decrements SP by 2 and pushes CS onto the stack and moves the low 16-bit value of the specified 32-bit number such as 'BEGIN' in CALL BEGIN into CS; SP is again decremented by 2; IP is pushed onto the stack; IP is then loaded with high 16-bit value of BEGIN; thus, this instruction CALLS a subroutine in another code segment (intersegment direct)
CALL DWORDPTR [reg 16]	CALL a subroutine in another segment	This instruction decrements SP by 2, and pushes CS onto the stack; CS is then loaded with the contents of memory locations addressed by [reg 16+2] and [reg 16 + 3] in DS; the SP is again decremented by 2; IP is pushed onto the stack; IP is then loaded with the contents of memory locations addressed by [reg 16] and [reg 16 + 1] in DS; typical 8086 registers used for reg 16 are BX, SI, and DI (intersegment indirect)
CBW	Convert a byte to a word	Extend the sign bit (bit 7) ofAL register into AH
CLC	CF ← 0	Clear carry to zero
CLD	DF ← 0	Clear direction flag to zero
CLI	IF ← 0	Clear interrupt enable flag to zero to disable maskable interrupts
CMC	CF ← $\overline{CF}$	One's complement carry
CMP mem/reg 1, mem/reg 2	[mem/reg 1] – [mem/reg 2], flags are affected	mem/reg can be 8- or 16-bit; no memory-to-memory comparison allowed; result of subtraction is not provided; all flags are affected
CMP mem/reg, data	[mem/reg] – data, flags are affected	Subtracts 8- or 16-bit data from [mem or reg] and affects flags; no result is provided
CMPS BYTE or CMPSB	FOR BYTE [[SI]] – [[DI]], flags are affected [SI] ← [SI] ± 1 [DI] ← [DI] ± 1	8- or 16-bit data addressed by [DI] in ES is subtracted from 8- or 16-bit data addressed by SI in DS and flags are affected without providing any result; if DF = 0, then SI and DI are incremented by one for byte and two for word; if DF = 1, then SI and DI are decremented by one for byte and two for word; the segment register ES in destination cannot be overridden
CMPS WORD or CPSW	FOR WORD [[SI]] – [[DI]], flags are affected [SI] ← [SI] ± 2 [DI] ← [DI] ± 2	
CWD	Convert a word to 32 bits	Extend the sign bit of AX (bit 15) into DX
DAA	Decimal adjust [AL] after addition	This instruction uses implied addressing mode; this instruction converts [AL] into BCD; DAA should be used after BCD addition
DAS	Decimal adjust [AL] after subtraction	This instruction uses implied addressing mode; converts [AL] into BCD; DAS should be used after BCD subtraction
DEC reg 16	[reg 16] ← [reg 16] – 1	This is a one-byte instruction; used to decrement a 16-bit register except segment register; does not affect the carry flag
DEC mem/reg 8	[mem] ← [mem] – 1 or [reg 8] ← [reg 8] – 1	Used to decrement a byte or a word in memory or an 8-bit register content; segment register cannot be decremented by this instruction; does not affect carry flag
DIV mem/reg	16/8 bit divide: $\frac{[AX]}{[mem8 / reg8]}$ [AH] ← Remainder [AL] ← Quotient 32/16 bit divide: $\frac{[DX][AX]}{[mem16 / reg16]}$ [DX] ← Remainder [AX] ← Quotient	Mem/reg is 8-bit for 16-bit by 8-bit divide and 16-bit for 32-bit by 16-bit divide; this is an unsigned division; no flags are affected; division by zero automatically generates an internal interrupt

Instructions	Interpretation	Comments
ESC external OP code, source	ESCAPE to external processes	This instruction is used to pass instructions to a coprocessor such as the 8087 floating point coprocessor which simultaneously monitors the system bus with the 8086; the coprocessor OP codes are 6-bit wide; the coprocessor treats normal 8086 instructions as NOP's; the 8086 fetches all instructions from memory; when the 8086 encounters an ESC instruction, it usually treats it as NOP; the coprocessor decodes this instruction and carries out the operation using the 6-bit OP code independent of the 8086; for ESC OP code, memory, the 8086 accesses data in memory for the coprocessor; for ESC data, register, the coprocessor operates on 8086 registers; the 8086 treats this as an NOP
HLT	HALT	Halt
IDIV mem/reg	Same as DIV mem/reg	Signed division. No flags are affected.
IMUL mem/reg	For 8 x 8 [AX] ← [AL] * [mem 8 / reg 8] For 16 x 16 [DX][AX]← [AX] * [mem 16 / reg 16]	Mem/reg can be 8- or 16-bit; only CF and OF are affected; signed multiplication
IN AL, DX	[AL] ← PORT [DX]	Input AL with the 8-bit content of a port addressed by DX; this is a one-byte instruction
IN AX, DX	[AX] ← PORT [DX]	Input AX with the 16-bit content of a port addressed by DX and DX + 1; this is a one-byte instruction
IN AL, PORT	[AL] ← [PORT]	Input AL with the 8-bit content of a port addressed by the second byte of the instruction
IN AX, PORT	[AX] ← [PORT]	Input AX with the 16-bit content of a port addressed by the 8-bit address in the second byte of the instruction
INC reg 16	[reg 16] ← [reg 16] + 1	This is a one-byte instruction; used to increment a 16-bit register except the segment register; does not affect the carry flag
INC mem/reg 8	[mem] ← [mem] + 1 or [reg 8] ← [reg 8] + 1	This is a two-byte instruction; can be used to increment a byte or word in memory or an 8-bit register content; segment registers cannot be incremented by this instruction; does not affect the carry flag
INT n (n can be zero thru 255)	[SP] ← [SP] – 2 [[SP]] ← Flags IF ← 0 TF ← 0 [SP] ← [SP] – 2 [[SP]] ← [CS] [CS] ← 4n + 2 [SP] ← [SP] – 2 [[SP]] ← [IP] [IP] ← 4n	Software interrupts can be used as supervisor calls; that is, request for service from an operating system; a different interrupt type can be used for each type of service that the operating system could supply for an application or program; software interrupt instructions can also be used for checking interrupt service routines written for hardware-initiated interrupts
INTO	Interrupt on Overflow	Generates an internal interrupt if OF = 1; executes INT 4; can be used after an arithmetic operation to activate a service routine if OF = 1; when INTO is executed and if OF = 1, operations similar to INT n take place
IRET	Interrupt Return	POPS IP, CS and Flags from stack; IRET is used as return instruction at the end of a service routine for both hardware and software interrupts
JA/JNBE disp 8	Jump if above/jump if not below or equal	Jump if above/jump if not below or equal with 8-bit signed displacement; that is, the displacement can be from -128_{10} to $+127_{10}$, zero being positive; JA and JNBE are the mnemonic which represent the same instruction; Jump if both CF and ZF are zero; used for unsigned comparison
JAE/JNB/JNC disp 8	Jump if above or equal/jump if not below/jump if no carry	Same as JA/JNBE except that the 8086 Jumps if CF = 0; used for unsigned comparison
JB/JC/JNAE disp 8	Jump if below/jump if carry/jump if not above or equal	Same as JA/JNBE except that the jump is taken CF = 1, used for unsigned comparison
JBE/JNA disp 8	Jump if below or equal/jump if not above	Same as JA/JNBE except that the jump is taken if CF = 1 or ZF = 0; used for unsigned comparison

Instructions	Interpretation	Comments
JCXZ disp 8	Jump if CX = 0	Jump if CX = 0; this instruction is useful at the beginning of a loop to bypass the loop if CX = 0
JE/JZ disp 8	Jump if equal/jump if zero	Same as JA/JNBE except that the jump is taken if ZF = 1; used for both signed and unsigned comparison
JG/JNLE disp 8	Jump if greater/jump if not less or equal	Same as JA/JNBE except that the jump is taken if ((SF ⊕ OF) or ZF) = 0; used for signed comparison
JGE/JNL disp 8	Jump if greater or equal/ jump if not less	Same as JA/JNBE except that the jump is taken if (SF ⊕ OF) = 0; used for signed comparison
JL/JNGE disp 8	Jump if less/Jump if not greater nor equal	Same as JA/JNBE except that the jump is taken if (SF ⊕ OF) = 1; used for signed comparison
JLE/JNG disp 8	Jump if less or equal/ jump if not greater	Same as JA/JNBE except that the jump is taken if ((SF ⊕ OF) or ZF) = 1; used for signed comparison
JMP Label	Unconditional Jump with a signed 8-bit (SHORT) or signed 16-bit (NEAR) displacement in the same segment	The label START can be signed 8-bit (called SHORT jump) or signed 16-bit (called NEAR jump) displacement; the assembler usually determines the displacement value; if the assembler finds the displacement value to be signed 8-bit (–128 to +127, 0 being positive), then the assembler uses two bytes for the instruction: one byte for the OP code followed by a byte for the displacement; the assembler sign extends the 8-bit displacement and then adds it to IP; [CS] is unchanged; on the other hand, if the assembler finds the displacement to be signed 16-bit (±32 K), then the assembler uses three bytes for the instruction: one byte for the OP code followed by 2 bytes for the displacement; the assembler adds the signed 16-bit displacement to IP; [CS] is unchanged; therefore, this JMP provides a jump in the same segment (intrasegment direct jump)
JMP regl6	[IP] ← [reg 16]; [CS] is unchanged	Jump to an address specified by the contents of a 16-bit register such as BX, SI, and DI in the same code segment; in the example JMP BX, [BX] is loaded into IP and [CS] is unchanged (intrasegment memory indirect jump)
JMP mem 16	[IP] ← [mem]; [CS] is unchanged	Jump to an address specified by the contents of a 16-bit memory location addressed by 16-bit register such as BX, SI, and DI; in the example, JMP [BX] copies the content of a memory location addressed by BX in DS into IP; CS is unchanged (intrasegment memory indirect jump)
JMP Label (to another segment)	Unconditionally jump to another segment	This is a 5-byte instruction: the first byte is the OP code followed by four bytes of 32-bit immediate data; bytes 2 and 3 are loaded into IP; bytes 4 and 5 are loaded into CS to JUMP unconditionally to another segment (intersegment direct)
JMP DWORDPTR [reg 16]	Unconditionally jump to another segment	This instruction loads the contents of memory locations addressed by [reg 16] and [reg 16 + 1] in DS into IP; it then loads the contents of memory locations addressed by [reg 16 + 2] and [reg 16 + 3] in DS into CS; typical 8086 registers used for reg 16 are BX, SI, and DI (intersegment indirect)
JNE/JNZ disp 8	Jump if not equal/jump if not zero	Same as JA/JNBE except that the jump is taken if ZF = 0; used for both signed and unsigned comparison
JNO disp 8	Jump if not overflow	Same as JA/JNBE except that the jump is taken if OF = 0
JNP/JPO disp 8	Jump if no parity/jump if parity odd	Same as JA/JNBE except that the jump is taken if PF = 0
JNS disp 8	Jump if not sign	Same as JA/JNBE except that the jump is taken if SF = 0
JO disp 8	Jump if overflow	Same as JA/JNBE except that the jump is taken if OF = 1
JP/JPE disp 8	Jump if parity/jump if parity even	Same as JA/JNBE except that the jump is taken if PF = 1
JS disp 8	Jump if sign	Same as JA/JNBE except that the jump is taken if SF = 1
LAHF	[AH] ← Flag low-byte	This instruction has implied addressing mode; it loads AH with the low byte of the flag register; no flags are affected
LDS reg, mem	[reg] ← [mem] [DS] ← [mem + 2]	Load a 16-bit register (AX, BX, CX, DX, SP, BP, SI, DI) with the content of specified memory and load DS with the content of the location that follows; no flags are affected; DS is used as the segment register for mem
LEA reg, mem	[reg] ← [offset portion of address]	LEA (load effective address) loads the value of the source operand rather than its content to register (such as SI, DI, BX) which are allowed to contain offset for accessing memory; no flags are affected

Instructions	Interpretation	Comments
LES reg, mem	[reg] ← [mem] [ES] ← [mem+ 2]	DS is used as the segment register for mem; in the example LES DX, [BX], DX is loaded with 16-bit value from a memory location addressed by 20-bit physical address computed from DS and BX; the 16-bit content of the next memory is loaded into ES; no flags are affected
LOCK	LOCK bus during next instruction	Lock is a one-byte prefix that causes the 8086 (configured in maximum mode) to assert its bus LOCK signal while following instruction is executed; this signal is used in multiprocessing; the LOCK pin of the 8086 can be used to LOCK other processors off the system bus during execution of an instruction; in this way, the 8086 can be assured of uninterrupted access to common system resources such as shared RAM
LODS BYTE or LODSB	FOR BYTE [AL] ← [[SI]] [SI] ← [SI] ± 1	Load 8-bit data into AL or 16-bit data into AX from a memory location addressed by SI in segment DS; if DF = 0, then SI is incremented by 1 for byte or incremented by 2 for word after the load; if DF = 1, then SI is decremented by 1 for byte or decremented by 2 for word; LODS affects no flags
LODS WORD or LODSW	FOR WORD [AX] ← [[SI]] [SI] ← [SI] ± 2	
LOOP disp 8	Loop if CX not equal to zero	Decrement CX by one, without affecting flags and loop with signed 8-bit displacement (from –128 to +127, zero being positive) if CX is not equal to zero
LOOPE/LOOPZ disp 8	Loop while equal/loop while zero	Decrement CX by one without affecting flags and loop with signed 8-bit displacement if CX is equal to zero, and if ZF = 1 which results from execution of the previous instruction
LOOPNE/LOOPNZ disp 8	Loop while not equal/loop while not zero	Decrement CX by one without affecting flags and loop with signed 8-bit displacement if CX is not equal to zero and ZF = 0 which results from execution of previous instruction
MOV mem/reg 2, mem/reg 1	[mem/reg 2] ← [mem/reg 1]	mem uses DS as the segment register; no memory-to-memory operation allowed; that is, MOV mem, mem is not permitted; segment register cannot be specified as reg or reg; no flags are affected; not usually used to load or store 'A' from or to memory
MOV mem, data	[mem] ← data	mem uses DS as the segment register; 8- or 16-bit data specifies whether memory location is 8- or 16-bit; no flags are affected
MOV reg, data	[reg] ← data	Segment register cannot be specified as reg; data can be 8- or 16-bit; no flags are affected
MOV segreg, mem/reg	[segreg] ← [mem/reg]	mem uses DS as segment register; used for initializing CS, DS, ES, and SS; no flags are affected
MOV mem/reg, segreg	[mem/reg] ← [segreg]	mem uses DS as segment register; no flags are affected
MOVS BYTE or MOVSB	FOR BYTE [[DI]]← [[SI]] [SI] ← [SI] ± 1	Move 8-bit or 16-bit data from the memory location addressed by SI in segment DS location addressed by DI in ES; segment DS can be overridden by a prefix but destination segment must be ES and cannot be overridden; if DF = 0, then SI is incremented by one for byte or incremented by two for word; if DF = 1, then SI is decremented by one for byte or by two for word
MOVS WORD or MOVSW	FOR WORD [[DI]]← [[SI]] [SI] ← [SI] ± 2	
MUL mem/reg	FOR 8 × 8 [AX] ← [AL] ∗ [mem/reg] FOR 16 × 16 [DX] [AX] ← [AX] ∗ [mem/reg]	mem/reg can be 8- or 16-bit; only CF and OF are affected; unsigned multiplication
NEG mem/reg	[mem/reg] ← $\overline{\text{[mem/reg]}}$ + 1	mem/reg can be 8- or 16-bit; performs two's complement subtraction of the specified operand from zero, that is, two's complement of a number is formed; all flags are affected except CF = 0 if [mem/reg] is zero; otherwise CF = 1
NOP	No Operation	8086 does nothing
NOT reg	[reg] ← $\overline{\text{[reg]}}$	mem and reg can be 8- or 16-bit; segment registers are not allowed; no flags are affected; ones complement reg
NOT mem	[mem] ← $\overline{\text{[mem]}}$	mem uses DS as the segment register; no flags are affected; ones complement mem

Instructions	Interpretation	Comments
OR Mem/reg 1, Mem/reg 2	[mem/reg 1] ← [mem/reg 1] ∨ [mem/reg 2]	No memory-to-memory operation is allowed; [mem] or [reg 1] or [reg 2] can be 8- or 16-bit; all flags are affected with OF and CF cleared to zero; no segment registers are allowed; mem uses DS as segment register
OR mem, data	[mem] ← [mem] ∨ data	mem and data can be 8- or 16-bit; mem uses DS as segment register; all flags are affected with CF and OF cleared to zero
OR reg, data	[reg] ← [reg] ∨ data	reg and data can be 8- or 16-bit; no segment registers are allowed; all flags are affected with CF and OF cleared to zero
OUT DX, AL	PORT [DX] ← [AL]	Output the 8-bit contents of AL into an I/O Port addressed by the 16-bit content of DX; this is a one-byte instruction
OUT DX, AX	PORT [DX] ← [AX]	Output the 16-bit contents of AX into an I/O Port addressed by the 16-bit content of DX; this is a one-byte instruction
OUT PORT, AL	PORT ← [AL]	Output the 8-bit contents of AL into the Port specified in the second byte of the instruction
OUT PORT, AX	PORT ← [AX]	Output the 16-bit contents of AX into the Port specified in the second byte of the instruction
POP mem	[mem] ← [[SP]] [SP] ← [SP] + 2	mem uses DS as the segment register; no flags are affected
POP reg	[reg]← [[SP]] [SP] ← [SP] + 2	Cannot be used to POP segment registers or flag register
POP segreg	[segreg] ← [[SP]] [SP] ← [SP] + 2	POP CS is illegal
POPF	[Flags] ← [[SP]] [SP] ← [SP] + 2	This instruction pops the top two stack bytes in the16-bit flag register
PUSH mem	[SP] ← [SP] – 2 [[SP]] ← [mem]	mem uses DS as segment register; no flags are affected; pushes 16-bit memory contents
PUSH reg	[SP] ← [SP] – 2 [[SP]] ← [reg]	reg must be a 16-bit register; cannot be used to PUSH segment register or Flag register
PUSH segreg	[SP] ← [SP] – 2 [[SP]] ← [segreg]	PUSH CS is illegal
PUSHF	[SP] ← [SP] – 2 [[SP]] ← [Flags]	This instruction pushes the 16-bit Flag register onto the stack
RCL mem/reg, 1	ROTATE through carry left once byte or word in mem/reg	FOR BYTE CF 7 0 FOR WORD 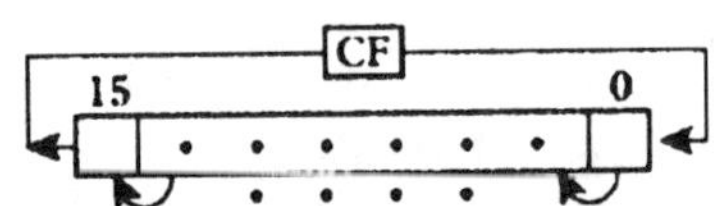
RCL mem/reg, CL	ROTATE through carry left byte or word in mem/reg by [CL]	Operation same as RCL mem/reg, 1 except the number of rotates is specified in CL for rotates up to 255; zero or negative rotates are illegal
RCR mem/reg, 1	ROTATE through carry right once byte or word in mem/reg	FOR BYTE 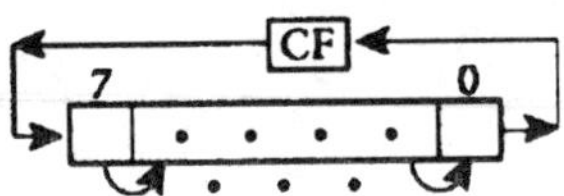 FOR WORD 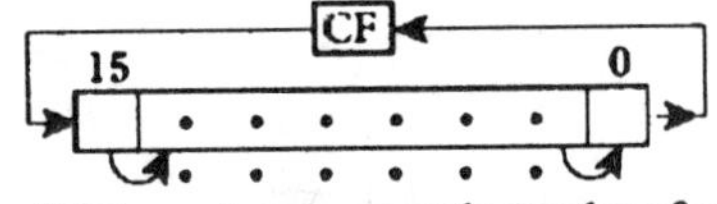
RCR mem/reg, CL	ROTATE through carry right byte or word in mem/reg by [CL]	Operation same as RCR mem/reg, 1 except the number of rotates is specified in CL for rotates up to 255; zero or negative rotates are illegal

Instructions	Interpretation	Comments
RET	.POPS IP for intrasegment CALLS .POPS IP and CS for intersegment CALLS	The assembler generates an intrasegment return if the programmer has defined the subroutine as NEAR; for intrasegment return, the following operations take place: [IP] ← [[SP]], [SP] ← [SP] + 2; on the other hand, the assembler generates an intersegment return if the subroutine has been defined as FAR; in this case, the following operations take place: [IP] ← [[SP]], [SP] ← [SP] + 2, [CS] ← [[SP]], [SP] ← [SP] + 2; an optional 16-bit displacement 'START' can be specified with the intersegment return such as RET START; in this case, the 16-bit displacement is added to the SP value; this feature may be used to discard parameter pushed onto the stack before the execution of the CALL instruction
ROL mem/reg, 1	ROTATE left once byte or word in mem/reg	FOR BYTE 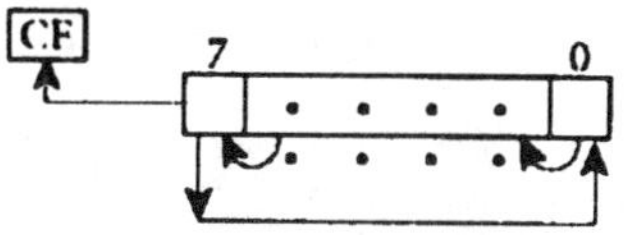 [diagram: CF, bits 7 … 0] FOR WORD [diagram: CF, bits 15 … 0] 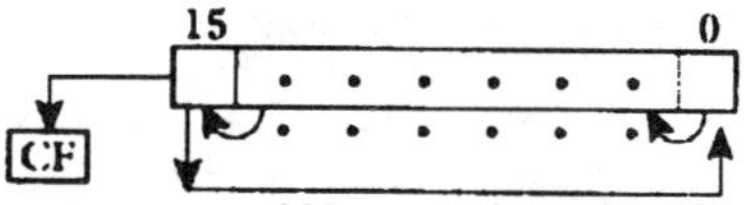
ROL mem/reg, CL	ROTATE left byte or word by the content of CL	[CL] contains rotate count up to 255; zero and negative shifts are illegal; CL is used to rotate count when the rotate is greater than once; mem uses DS as the segment register
ROR mem/reg, 1	ROTATE right once byte or word in mem/reg	FOR BYTE 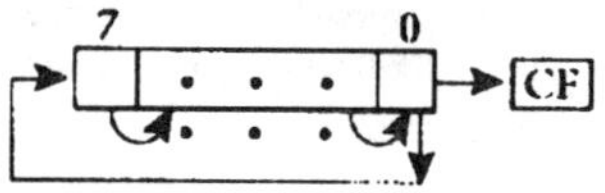 [diagram: bits 7 … 0, CF] FOR WORD [diagram: bits 15 … 0, CF] 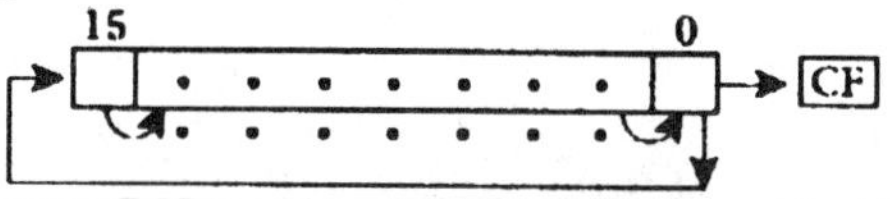
ROR mem/reg, CL	ROTATE right byte or word in mem/reg by [CL]	Operation same as ROR mem/reg, 1; [CL] specifics the number of rotates for up to 255; zero and negative rotates are illegal; mem uses DS as the segment register
SAHF	[Flags, low-byte] ← [AH]	This instruction has the implied addressing mode; the content of the AH register is stored into the low-byte of the flag register; no flags are affected
SAL mem/reg, 1	Shift arithmetic left once byte or word in mem or reg	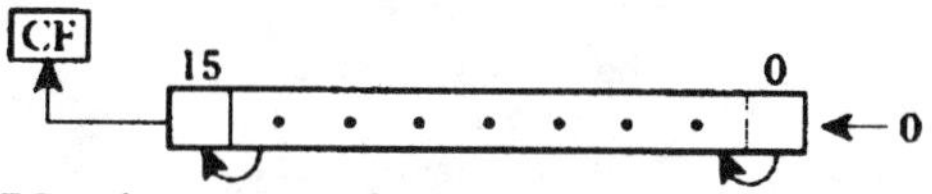 FOR BYTE 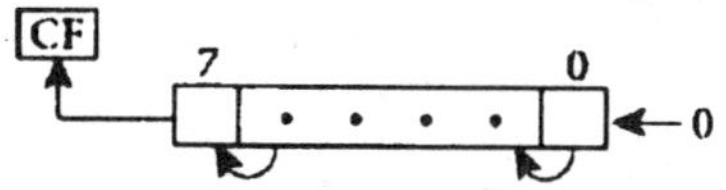 [diagram: CF, bits 7 … 0, ← 0] FOR WORD [diagram: CF, bits 15 … 0, ← 0] Mem uses DS as the segment register; reg cannot be segment registers; OF and CF are affected; if sign bit is changed during or after shifting, the OF is set to one
SAL mem/reg, CL	Shift arithmetic left byte or word by shift count on CL	Operation same as SAL mem/reg, 1; CL contains shift count for up to 255; zero and negative shifts are illegal; [CL] is used as shift count when shift is greater than one; OF and SF are affected; if sign bit of [mem] is changed during or after shifting, the OF is set to one; mem uses DS as segment register

Instructions	Interpretation	Comments
SAR mem/reg, 1	SHIFT arithmetic right once byte or word in mem/reg	FOR BYTE 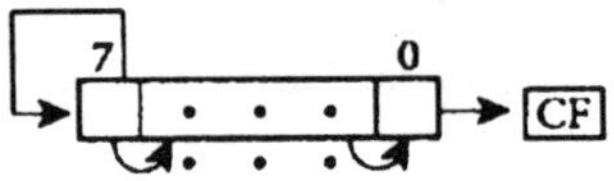 FOR WORD 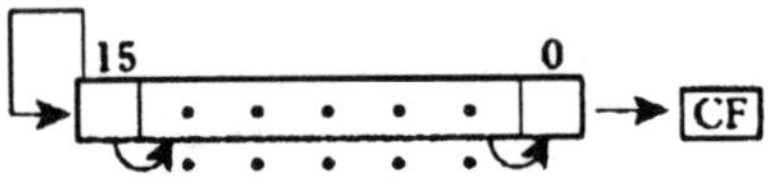
SAR mem/reg, CL	SHIFT arithmetic right byte or word in mem/reg by [CL]	Operation same as SAR mem/reg, 1; however, shift count is specified in CL for shifts up to 255; zero and negative shifts are illegal
SBB mem/reg 1, mem/reg 2	[mem/reg 1] ← [mem/reg 1] - [mem/reg 2] – CY	Same as SUB mem/reg 1, mem/reg 2 except this is a subtraction with borrow
SBB mem, data	[mem] ← [mem] - data - CY	Same as SUB mem, data except this is a subtraction with borrow
SBB reg, data	[reg] ∧ [reg] – data – CY	Same as SUB reg, data except this is a subtraction with borrow
SBB A, data	[A] ← [A] – data – CY	Same as SUB A, data except this is a subtraction with borrow
SCAS BYTE or SCASB SCAS WORD or SCASW	FOR BYTE [AL] – [[DI]], flags are affected,[DI] ← [DI] ± 1 FOR WORD[AX] - [[DI]], flags are affected,[DI] ← [DI] ± 2	8- or 16-bit data addressed by [DI] in ES is subtracted from 8- or 16-bit data in AL or AX and flags are affected without affecting [AL] or [AX] or string data; ES cannot be overridden; if DF = 0, then DI is incremented by one for byte and two for word; if DF = 1, then DI is decremented by one for byte or decremented by two for word
SHL mem/reg, 1	SHIFT logical left once byte or word in mem/reg	Same as SAL mem/reg, 1
SHL mem/reg, CL	SHIFT logical left byte or word in mem/reg by the shift count in CL	Same as SAL mem/reg, CL except overflow is cleared to zero
SHR mem/reg, 1	SHIFT right logical once byte or word in mem/reg	FOR BYTE 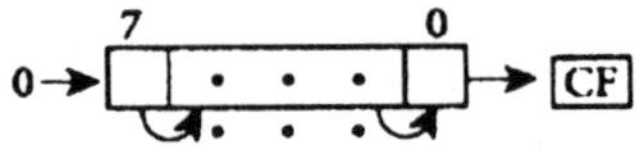 FOR WORD 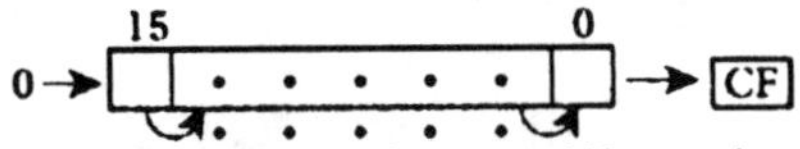
SHR mem/reg, CL	SHIFT right logical byte or word in mem/reg by [CL]	Operation same as SHR mem/reg, 1; however, shift count is specified in CL for shifts up to 255; zero and negative shifts are illegal
STC	CF ← 1	Set carry to one
STD	DF ← 1	Set direction flag to one
STI	IF ← 1	Set interrupt enable flag to one to enable maskable interrupts
STOS BYTE or STOSB STOS WORD or STOSW	FOR BYTE [[DI]] ← [AL] [DI] ← [DI] ± 1 FOR WORD [[DI]] ← [AX] [DI] ← [DI] ± 2	Store 8-bit data from AL or 16-bit data from AX into a memory location addressed by DI in segment ES; segment register ES cannot be overridden; if DF = 0, then DI is incremented by one for byte or incremented by two for word after the store
SUB mem/reg 1, mem/reg 2	[mem/reg 1] ← [mem/reg 1] - [mem/reg 2]	No memory-to-memory SUB permitted; all flags are affected; mem uses DS as the segment register
SUB mem, data	[mem] ← [mem] – data	Data can be 8- or 16-bit; mem uses DS as the segment register; all flags are affected
SUB reg, data	[reg] ← [reg] – data	Data can be 8- or 16-bit; all flags are affected
TEST mem/reg 1, mem/reg 2	[mem/reg 1]- [mem/reg 2], no result; flags are affected	No memory-to-memory TEST is allowed; no result is provided; all flags are affected with CF and OF cleared to zero; [mem], [reg 1] or [reg 2] can be 8-or 16-bit; no segment registers are allowed; mem uses DS as the segment register
TEST mem, data	[mem] - data, no result; flags are affected	Mem and data can be 8- or 16-bit; no result is provided; all flags are affected with CF and OF cleared to zero; mem uses DS as the segment register

Instructions	Interpretation	Comments
TEST reg, data	[reg]- data, no result; flags are affected	Reg and data can be 8- or 16-bit; no result is provided; all flags are affected with CF and OF cleared to zero; reg cannot be segment register;
WAIT	8086 enters wait state	Causes CPU to enter wait state if the 8086 TEST pin is high; while in wait state, the 8086 continues to check TEST pin for low; if TEST pin goes back to zero, the 8086 executes the next instruction; this feature can be used to synchronize the operation of 8086 to an event in external hardware
XCHG mem/reg, mem/reg	[mem] ↔ [reg]	reg and mem can be both 8- or 16-bit; mem uses DS as the segment register; reg cannot be segment register; no flags are affected; no mem to mem .
XCHG reg,reg	[reg] ↔ [reg]	reg can be 8-or 16-bit; reg cannot be segment register; no flags are affected
XLAT	[AL] ← [AL] + [BX]	This instruction is useful for translating characters from one code such as ASCII to another such as EBCDIC; this is a no-operand instruction and is called an instruction with implied addressing mode; the instruction loads AL with the contents of a 20-bit physical address computed from DS, BX, and AL; this instruction can be used to read the elements in a table where BX can be loaded with a 16-bit value to point to the starting address (offset from DS) and AL can be loaded with the element number (0 being the first element number); no flags are affected; the XLAT instruction is equivalent to MOV AL, [AL] [BX]
XOR mem/reg 1, mem/reg 2	[mem/reg 1] ← [mem/reg 1] ⊕ [mem/reg 2]	No memory-to-memory operation is allowed; [mem] or [reg 1] or [reg 2] can be 8- or 16-bit; all flags are affected with CF and OF cleared to zero; mem uses DS as the segment register
XOR mem, data	[reg] ← [mem] ⊕ data	Data and mem can be 8- or 16-bit; mem uses DS as the segment register; mem cannot be segment register; all flags are affected with CF and OF cleared to zero
XOR reg, data	[reg] ← [reg] ⊕ data	Same as XOR mem, data.

BIBLIOGRAPHY

Breeding, K., *Digital Design Fundamentals*, 2nd ed., Prentice-Hall, 1992.
Burns, J., "Within the 68020," *Electronics and Wireless World*, pp 209-212, February 1985; pp 103-106, March 1985.
Daconta, M., *Java for C/C++ Programmers*, John Wiley, 1996.
Dewey, A., *Analysis and Design of Digital Systems with VHDL, PWS Publishing, 1997.*
Feibus, M. and Slater, M., "Pentium Power," *PC Magazine*, April 27, 1993.
Hamacher, V.C., Vransic, Z.G., and Zaky, S. G., *Computer Organization*, McGraw-Hill, 1990, 1984.
Hamacher, V., Vrasenic, Z., and Zaky, S., *Computer Organization*, McGraw-Hill, 1978.
Hartman, B., "16-Bit 68000 Microprocessor Concepts on 32-Bit Frontier," MC 68000 Article Reprints, Motorola, pp 50-57, March 1981.
Hayes, J., *Introduction to Digital Logic Design*, Addison-Wesley Publishing Company, 1993.
Hayes, J., *Computer Architecture and Organization*, McGraw-Hill, 1978.
Hayes, J., *Digital System Design and Microprocessors*, McGraw-Hill, 1984.
Hwang, K. and Briggs, F.A., *Computer Architecture and Parallel Processing,* McGraw-Hill, 1984.
IEEE Computer Society " Computer," August 1996.
IEEE "Spectrum," January, 1998.
Intel, *Microprocessors and Peripheral Handbook, Vol.1, Microprocessors*, Intel Corporation, 1988.
Intel, *Microprocessors and Peripheral Handbook, Vol.2, Peripheral*, Intel Corporation, 1988.
Intel, *80386 Programmer's Reference Manual*, Intel Corporation, 1986.
Intel, *80386 Hardware Reference Manual*, Intel Corporation, 1986.
Intel, *80386 Advance Information*, Intel Corporation, 1985.
Intel, *80387 Programmer's Reference Manual*, 1987.
Intel, *Intel 486 Microprocessor Family Programmer's Reference Manual*, Intel Corporation, 1992.
Intel, *Intel 486 Microprocessor Hardware Reference Manual*, Intel Corporation, 1992.
Intel, *Pentium Processor User's Manual*, 1993.
Intel, *The 8086 Family User's Family*, Intel Corporation, 1979.
Intel, *Intel Component Data Catalog*, Intel Corporation, 1979.

Intel, *MCS-86 User's Manual*, Intel Corporation, 1982.

Intel, *Memory Components Handbook*, Intel Corporation, 1982.

Intel, "Marketing Communications," *The Semiconductor Memory Book*, John Wiley & Sons, 1978.

Johnson, "A Comparison of MC68000 Family Processors," *BYTE*, pp 205-218, September 1986.

Katz, R., *Contemporary Logic Design*, The Bejamin/Cummings Publishing Co. Inc., 1994.

Lee, S., *Design of Computers and other Complex Digital Devices,*Prentice-Hall, 2000.

MacGregor, Mothersole, Meyer, "The Motorola MC68020," IEEE MICRO, pp 101-116, August 1984.

MacGregor, "Diverse Applications Put Spotlight on 68020's Improvements," *Electronic Design*, pp 155-164, February 7, 1985.

MacGregor, "Hardware and Software Strategies for the MC68020," *EDN*, pp 163-168, June 20, 1985.

Mano, M., *Computer System Architecture*, Prentice-Hall, 1983.

Mano, M., *Digital Design*, 2nd ed., Prentice-Hall, 1991.

Mano, M., *Computer Engineering*, Prentice-Hall, 1988.

McCartney, Groepler, "The 32-Bit 68020's Power Flows Fully Through a Versatile Interface," *Electronic Design*, pp 335-343, January 10, 1985.

Miller, M., Raskin, R., and Rupley, S., "The Pentium That Stole Christmas," *PC Magazine*, February 27, 1995.

Motorola, *MC68000 User's Manual*, Motorola Corporation, 1979.

Motorola, *16-Bit Microprocessor - MC68000 User's Manual*, 4th ed., Prentice-Hall, 1984.

Motorola, *MC68000 16-Bit Microprocessor User's Manual*, Motorola Corporation, 1982.

Motorola, *MC68000 Supplement Material (Technical Training)*, Motorola Corporation, 1982.

Motorola, *Microprocessor Data Material*, Motorola Corporation, 1981.

Motorola, *MC68020 User's Manual*, Motorola Corporation, 1985.

Motorola, "MC68020 Course Notes,"MTTA20 REV 2, July 1987.

Motorola, "MC68020/68030 Audio Course Notes," 1988.

Motorola, *68020 User's Manual*, 2nd ed., MC68020 UM/AD Rev. 1, Prentice-Hall, 1984.

Motorola, *MC68040 User's Manual*, 1989.

Motorola, *Power PC 601, RISC Microprocessor User's Manual*, 1993.

Motorola, *Technical Summary, 32-bit Virtual Memory Microprocessor*, MC68020 BR243/D. Rev. 2, Motorola Corporation, 1987.

National Semiconductor, *Fast ® Advanced Schottky TTL Logic Data Book*, 1990.

National Semiconductor, *Programmable Logic Devices Data book and Design Guide*, 1989.

National Semiconductor, *CMOS Logic Data Book*, 1988.

National Semiconductor, *LS/S/TTL Logic Data Book*, 1989.

Osborne, A., *An Introduction to Microprocessors, Vol. 1, Basic Concepts*, rev. ed., Osborne/McGraw-Hill, 1980; 2nd ed., 1982.

Pellerin, D., and Holley, M., *Digital Design using ABEL*, Prentice-Hall, 1994.

Rafiquzzaman, M., *Microprocessors and Microcomputer Development Systems - Designing Microprocessor-Based Systems*, Harper and Row, 1984.

Rafiquzzaman, M., *Microprocessors and Microcomputer Development Systems*, John Wiley & Sons, 1984.

Rafiquzzaman, M., *Microcomputer Theory and Applications with the INTEL SDK-85*, 2nd ed., John Wiley & Sons, 1987.

Rafiquzzaman, M., *Microprocessors - Theory and Applications - Intel and Motorola*, Prentice-Hall, 1992.

Rafiquzzaman, M., and Chandra, R., *Modern Computer Architecture*, West, 1988.

Rafiquzzaman, M., *Microprocessors and Microcomputer-Based System Design*, 1st ed., CRC Press, 1990.

Rafiquzzaman, M., *Microprocessors and Microcomputer-Based System Design*, 2nd ed. CRC Press, 1995.

Smith, J. and Weiss, S., "Power PC 601 and Alpha 21064: A Tale of Two RISCs," *IEEE Computer*, June 1994.

Solomon, "Motorola's Muscular 68020," *Computers & Electronics*, pp 74-79, October 1984.

Tanenbaum, A.S., *Structured Computer Organization*, Prentice-Hall, 1984.

Texas Instruments, *The TTL Data Book*, Vol. 1, 1984.

Texas Instruments, *Linear Circuits Data Book*, 1990.

Texas Instruments, *The TTL Data Book for Design Engineers*, 2nd ed., 1976.

Tocci, R. J. and Widmer, N. S., *Digital Systems, 7th Edition, Prentice-Hall, 1998.*

Triebel, W., *The 80386 DX Microprocessor*, Prentice-Hall, 1992.

Wakerly, J., *Digital Design Principles and Practices*, Prentice-Hall, 1990.

White, R., *How Computers Work, Millennium Edition,* Que Corporation, 1999.

www.activewin.com, " Windows 2000", Internet Web site, Active Windows, 2000.

www.activewin.com, " DVD FAQs", Internet Web site, 2000.

Zorpette, G., "Microprocessors - The Beauty of 32-Bits," *IEEE Spectrum*, Vol. 22, No.9, pp 65-71, September 1994.

CREDITS

The following material is reprinted by permission of the sources listed below:

Intel Corporation: Figure 9.3, Figure 9.4, Figure 9.8, Figure 9.9, Figure 9.10, Figure 9.19, Figure 9.33, Figure 9.35, Appendix E, Appendix F. All mnemonics of Intel microprocessors are courtesy of Intel Corporation. The 80386 microprocessor referred to in the text as the i386™, the 80486 as the i486™, and the Pentium as the Pentium™, trademarks of Intel Corporation.

Motorola Corporation: Table 10-1, Table 10-2, Table 10-3, Table 10-13, Table 10-15, Table 10-16, Table 10-19, Table 10-20, Table 10-21, Table 10-23, Figure 10.1, Figure 10.2, Figure 10.3, Figure 10.6, Figure 10.7, Figure 10.7, Figure 10.13, Figure 10.17, Figure 10.19, Figure 10.28, Figure 10.28, Figure 10.32, Figure 10.33, Figure 10.34, Appendix C, Appendix D . All mnemonics of Motorola microprocessors are courtesy of Motorola Corporation.

Rafiquzzaman, M. and Chandra, Rajan, "Modern Computer Architecture," 1988, West/PWS Publishing/Rafi Systems Inc. used with permission; Figure 4.16, Figure 4.32, Figure 4.34, Figure 4.35, Figure 4.40, Figure 4.41, Figure 5.40, Figure 5.41, Figure 5.42, Figure 5.43, Figure 5.44, Figure 7.7, Figure 7.9, Figure 7.13, Figure 7.15, Figure 7.16, Figure 7.17, Figure 7.18, Figure 7.28, Figure 7.29, Figure 7.34, Figure 7.38, pages 169-174, pages 186-188, pages 306-326, pages 220-232, pages 242-267.

INDEX

B

C

E

F

J

K

L

M